D1162010

KIRK-OTHMER

ENCYCLOPEDIA OF CHEMICAL TECHNOLOGY

FOURTH EDITION

VOLUME **10**

EXPLOSIVES AND PROPELLANTS
TO
FLAME RETARDANTS FOR TEXTILES

KIRK-OTHMER

ENCYCLOPEDIA OF CHEMICAL TECHNOLOGY

FOURTH EDITION

VOLUME **10**

EXPLOSIVES AND PROPELLANTS
TO
FLAME RETARDANTS FOR TEXTILES

A Wiley-Interscience Publication
JOHN WILEY & SONS
New York • Chichester • Brisbane • Toronto • Singapore

This text is printed on acid-free paper.

Library of Congress Cataloging-in-Publication Data

Encyclopedia of chemical technology / executive editor, Jacqueline
I. Kroschwitz; editor, Mary Howe-Grant.—4th ed.
p. cm.
At head of title: Kirk-Othmer.
"A Wiley-Interscience publication."
Includes index.
Contents: v. 10, Explosives and propellants to flame retardants for textiles.
ISBN 0-471-52678-9 (v. 10)
1. Chemistry, Technical—Encyclopedias. I. Kirk, Raymond E.
(Raymond Eller), 1890–1957. II. Othmer, Donald F. (Donald
Frederick), 1904– . III. Kroschwitz, Jacqueline I., 1942– .
IV. Howe-Grant, Mary, 1943– . V. Title: Kirk-Othmer encyclopedia
of chemical technology.
TP9.E685 1992 91-16789
660′.03—dc20

Printed in the United States of America

10 9 8 7 6 5 4 3 2

CONTENTS

EDITORIAL STAFF FOR VOLUME 10

CONTRIBUTORS TO VOLUME 10

Darryl C. Aubrey, *Chem Systems, Inc., Tarrytown, New York,* Petrochemicals (under Feedstocks)

Malcolm H. I. Baird, *McMaster University, Hamilton, Ontario, Canada,* Extraction, liquid–liquid

John E. Boliek, *E. I. du Pont de Nemours & Co., Inc., Wilmington, Delaware,* Elastomeric (under Fibers)

Donald E. Brownlee, *University of Washington, Seattle,* Extraterrestrial materials

Timothy A. Calimari Jr., *U.S. Dept. of Agriculture, New Orleans, Louisiana,* Flame Retardants for Textiles

James Corbin, *University of Illinois, Urbana,* Pet foods (under Feeds and feed additives)

George C. Fahey, Jr., *University of Illinois, Urbana,* Ruminant feeds (under Feeds and feed additives)

James S. Falcone, Jr., *West Chester University, Pennsylvania,* Fillers

Richard G. Gann, *National Institute of Standards and Technology, Gaithersburg, Maryland,* Overview (under Flame retardants)

S. M. Hansen, *E. I. du Pont de Nemours & Co., Inc., Kinston, North Carolina,* Polyester (under Fibers)

Robert J. Harper, Jr., *U.S. Dept. of Agriculture, New Orleans, Louisiana,* Flame Retardants for Textiles

G. L. Hasenhuettl, *Kraft General Foods, Glenview, Illinois,* Fats and fatty oils

Jun-ichi Hikasa, *Kuraray Company, Ltd., Osaka, Japan,* Poly(vinyl alcohol) (under Fibers)

George Hoffmeister, *Consultant, Florence, Alabama,* Fertilizers

Norman C. Jamieson, *Mallinckrodt Specialty Chemicals Company, St. Louis, Missouri,* Standards (under Fine chemicals)

Sei-Joo Jang, *Pennsylvania State University, University Park,* Ferroelectrics

Arnold W. Jensen, *E. I. du Pont de Nemours & Co., Inc., Wilmington, Delaware,* Elastomeric (under Fibers)

Raymond S. Knorr, *Monsanto Company, Pensacola, Florida,* Acrylic (under Fibers)

F. X. N. M. Kools, *Philips Components, Evreux, France,* Ferrites

Sandra Kosinski, *AT&T Bell Laboratories, Murray Hill, New Jersey,* Fiber optics

L. M. Landoll, *Hercules Inc., Wilmington, Delaware,* Olefin (under Fibers)

Victor Lindner, *Armament Research, Development, and Engineering Agency, Dover, New Jersey,* Explosives; Propellants (both under Explosives and propellants)

Teh C. Lo, *T. C. Lo & Associates, Wayne, New Jersey,* Extraction, liquid–liquid

John B. MacChesney, *AT&T Bell Laboratories, Murray Hill, New Jersey,* Fiber optics

K. J. Mackenzie, *Consultant, Greenville, South Carolina,* Film and sheeting materials

Robert C. Monroe, *Hudson Products Corporation, Houston, Texas,* Fans and blowers

Roy E. Morse, *Consultant, Clemmons, North Carolina,* Fat replacers

Alex Pettigrew, *Ethyl Technical Center, Baton Rouge, Louisiana,* Halogenated Flame Retardants (under Flame retardants)

Peter Pollak, *Lonza Ltd., Basel, Switzerland,* Production (under Fine chemicals)

Ludwig Rebenfeld, *TRI/Princeton, Princeton, New Jersey,* Survey (under Fibers)

P. B. Sargeant, *E. I. du Pont de Nemours & Co., Inc., Kinston, North Carolina,* Polyester (under Fibers)

Surjit S. Sengha, *Sterling Winthrop, Inc., West Chester, Pennsylvania,* Fermentation

George A. Serad, *Hoechst-Celanese Corporation, Charlotte, North Carolina,* Cellulose esters (under Fibers)

Norman Singer, *Nutrasweet, Mount Prospect, Illinois,* Fat replacers

D. Stoppels, *Philips Components, Eindhoven, the Netherlands,* Ferrites

Gregory D. Sunvold, *University of Illinois, Urbana,* Ruminant feeds (under Feeds and feed additives)

Ladislav Svarovsky, *Consultant Engineers and Fine Particle Software, West Yorkshire, UK,* Filtration

Irving Touval, *Touval Associates, Sparta, New Jersey,* Antimony and other inorganic flame retardants (under Flame retardants)

Samuel M. Tuthill, *Mallinckrodt Specialty Chemicals Company, St. Louis, Missouri,* Standards (under Fine chemicals)

Lambertus van Zelst, *Conservation Analytical Laboratory, Smithsonian Institution, Washington, D.C.,* Fine art examination and conservation

Richard J. Wakeman, *University of Exeter, Devon, UK,* Extraction, liquid–solid

Park Waldroup, *University of Arkansas, Fayetteville,* Nonruminant feeds (under Feeds and feed additives)

Edward D. Weil, *Polytechnic University, Brooklyn, New York,* Phosphorus flame retardants (under Flame retardants)

Calvin R. Woodings, *Courtaulds, Coventry, UK,* Regenerated cellulosics (under Fibers)

Paul R. Worsham, *Eastman Chemical Company, Kingsport, Tennessee,* Coal chemicals (under Feedstocks)

C. J. Wust, Jr., *Hercules Inc., Wilmington, Delaware,* Olefin (under Fibers)

Raymond A. Young, *University of Wisconsin, Madison,* Vegetable (under Fibers)

NOTE ON CHEMICAL ABSTRACTS SERVICE REGISTRY NUMBERS AND NOMENCLATURE

Chemical Abstracts Service (CAS) Registry Numbers are unique numerical identifiers assigned to substances recorded in the CAS Registry System. They appear in brackets in the *Chemical Abstracts* (CA) substance and formula indexes following the names of compounds. A single compound may have synonyms in the chemical literature. A simple compound like phenethylamine can be named β-phenylethylamine or, as in *Chemical Abstracts*, benzeneethanamine. The usefulness of the *Encyclopedia* depends on accessibility through the most common correct name of a substance. Because of this diversity in nomenclature careful attention has been given to the problem in order to assist the reader as much as possible, especially in locating the systematic CA index name by means of the Registry Number. For this purpose, the reader may refer to the CAS Registry Handbook—Number Section which lists in numerical order the Registry Number with the *Chemical Abstracts* index name and the molecular formula; eg, **458-88-8**, Piperidine, 2-propyl-, (*S*)-, $C_8H_{17}N$; in the *Encyclopedia* this compound would be found under its common name, coniine [*458-88-8*]. Alternatively, this information can be retrieved electronically from CAS Online. In many cases molecular formulas have also been provided in the *Encyclopedia* text to facilitate electronic searching. The Registry Number is a valuable link for the reader in retrieving additional published information on substances and also as a point of access for on-line data bases.

In all cases, the CAS Registry Numbers have been given for title compounds in articles and for all compounds in the index. All specific substances indexed in *Chemical Abstracts* since 1965 are included in the CAS Registry System as are a large number of substances derived from a variety of reference works. The CAS Registry System identifies a substance on the basis of an unambiguous computer-language description of its molecular structure including stereochemical detail. The Registry Number is a machine-checkable number (like a Social Security number) assigned in sequential order to each substance as it enters the registry system. The value of the number lies in the fact that it is a concise and unique means of substance identification, which is independent of, and therefore bridges, many systems of chemical nomenclature. For polymers, one Registry Number may be used for the entire family; eg, polyoxyethylene (20) sorbitan monolaurate has the same number as all of its polyoxyethylene homologues.

Cross-references are inserted in the index for many common names and for some systematic names. Trademark names appear in the index. Names that are incorrect, misleading, or ambiguous are avoided. Formulas are given very frequently in the text to help in identifying compounds. The spelling and form used, even for industrial names, follow American chemical usage, but not always the usage of *Chemical Abstracts* (eg, *coniine* is used instead of *(S)-2-propylpiperidine, aniline* instead of *benzenamine*, and *acrylic acid* instead of *2-propenoic acid*).

There are variations in representation of rings in different disciplines. The dye industry does not designate aromaticity or double bonds in rings. All double bonds and aromaticity are shown in the *Encyclopedia* as a matter of course. For example, tetralin has an aromatic ring and a saturated ring and its structure

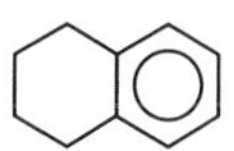

appears in the *Encyclopedia* with its common name, Registry Number enclosed in brackets, and parenthetical CA index name, ie, tetralin [*119-64-2*] (1,2,3,4-tetrahydronaphthalene). With names and structural formulas, and especially with CAS Registry Numbers, the aim is to help the reader have a concise means of substance identification.

CONVERSION FACTORS, ABBREVIATIONS, AND UNIT SYMBOLS

SI Units (Adopted 1960)

The International System of Units (abbreviated SI), is being implemented throughout the world. This measurement system is a modernized version of the MKSA (meter, kilogram, second, ampere) system, and its details are published and controlled by an international treaty organization (The International Bureau of Weights and Measures) (1).

SI units are divided into three classes:

BASE UNITS

length	meter† (m)
mass	kilogram (kg)
time	second (s)
electric current	ampere (A)
thermodynamic temperature‡	kelvin (K)
amount of substance	mole (mol)
luminous intensity	candela (cd)

SUPPLEMENTARY UNITS

plane angle	radian (rad)
solid angle	steradian (sr)

†The spellings "metre" and "litre" are preferred by ASTM; however, "-er" is used in the *Encyclopedia*.

‡Wide use is made of Celsius temperature (t) defined by

$$t = T - T_0$$

where T is the thermodynamic temperature, expressed in kelvin, and $T_0 = 273.15$ K by definition. A temperature interval may be expressed in degrees Celsius as well as in kelvin.

DERIVED UNITS AND OTHER ACCEPTABLE UNITS

These units are formed by combining base units, supplementary units, and other derived units (2–4). Those derived units having special names and symbols are marked with an asterisk in the list below.

Quantity	Unit	Symbol	Acceptable equivalent
*absorbed dose	gray	Gy	J/kg
acceleration	meter per second squared	m/s^2	
*activity (of a radionuclide)	becquerel	Bq	1/s
area	square kilometer	km^2	
	square hectometer	hm^2	ha (hectare)
	square meter	m^2	
concentration (of amount of substance)	mole per cubic meter	mol/m^3	
current density	ampere per square meter	$A//m^2$	
density, mass density	kilogram per cubic meter	kg/m^3	g/L; mg/cm^3
dipole moment (quantity)	coulomb meter	C·m	
*dose equivalent	sievert	Sv	J/kg
*electric capacitance	farad	F	C/V
*electric charge, quantity of electricity	coulomb	C	A·s
electric charge density	coulomb per cubic meter	C/m^3	
*electric conductance	siemens	S	A/V
electric field strength	volt per meter	V/m	
electric flux density	coulomb per square meter	C/m^2	
*electric potential, potential difference, electromotive force	volt	V	W/A
*electric resistance	ohm	Ω	V/A
*energy, work, quantity of heat	megajoule	MJ	
	kilojoule	kJ	
	joule	J	N·m
	electronvolt[†]	eV[†]	
	kilowatt-hour[†]	kW·h[†]	
energy density	joule per cubic meter	J/m^3	
*force	kilonewton	kN	
	newton	N	$kg{\cdot}m/s^2$

[†]This non-SI unit is recognized by the CIPM as having to be retained because of practical importance or use in specialized fields (1).

Quantity	Unit	Symbol	Acceptable equivalent
*frequency	megahertz	MHz	
	hertz	Hz	1/s
heat capacity, entropy	joule per kelvin	J/K	
heat capacity (specific), specific entropy	joule per kilogram kelvin	J/(kg·K)	
heat transfer coefficient	watt per square meter kelvin	$W/(m^2 \cdot K)$	
*illuminance	lux	lx	lm/m^2
*inductance	henry	H	Wb/A
linear density	kilogram per meter	kg/m	
luminance	candela per square meter	cd/m^2	
*luminous flux	lumen	lm	cd·sr
magnetic field strength	ampere per meter	A/m	
*magnetic flux	weber	Wb	V·s
*magnetic flux density	tesla	T	Wb/m^2
molar energy	joule per mole	J/mol	
molar entropy, molar heat capacity	joule per mole kelvin	J/(mol·K)	
moment of force, torque	newton meter	N·m	
momentum	kilogram meter per second	kg·m/s	
permeability	henry per meter	H/m	
permittivity	farad per meter	F/m	
*power, heat flow rate, radiant flux	kilowatt	kW	
	watt	W	J/s
power density, heat flux density, irradiance	watt per square meter	W/m^2	
*pressure, stress	megapascal	MPa	
	kilopascal	kPa	
	pascal	Pa	N/m^2
sound level	decibel	dB	
specific energy	joule per kilogram	J/kg	
specific volume	cubic meter per kilogram	m^3/kg	
surface tension	newton per meter	N/m	
thermal conductivity	watt per meter kelvin	W/(m·K)	
velocity	meter per second	m/s	
	kilometer per hour	km/h	
viscosity, dynamic	pascal second	Pa·s	
	millipascal second	mPa·s	
viscosity, kinematic	square meter per second	m^2/s	
	square millimeter per second	mm^2/s	

Quantity	Unit	Symbol	Acceptable equivalent
volume	cubic meter	m^3	
	cubic decimeter	dm^3	L (liter) (5)
	cubic centimeter	cm^3	mL
wave number	1 per meter	m^{-1}	
	1 per centimeter	cm^{-1}	

In addition, there are 16 prefixes used to indicate order of magnitude, as follows:

Multiplication factor	Prefix	Symbol	Note
10^{18}	exa	E	
10^{15}	peta	P	
10^{12}	tera	T	
10^{9}	giga	G	
10^{6}	mega	M	
10^{3}	kilo	k	
10^{2}	hecto	h[a]	[a]Although hecto, deka, deci, and centi are SI prefixes, their use should be avoided except for SI unit-multiples for area and volume and nontechnical use of centimeter, as for body and clothing measurement.
10	deka	da[a]	
10^{-1}	deci	d[a]	
10^{-2}	centi	c[a]	
10^{-3}	milli	m	
10^{-6}	micro	μ	
10^{-9}	nano	n	
10^{-12}	pico	p	
10^{-15}	femto	f	
10^{-18}	atto	a	

For a complete description of SI and its use the reader is referred to ASTM E 380 (4) and the article UNITS AND CONVERSION FACTORS which appears in Vol. 24.

A representative list of conversion factors from non-SI to SI units is presented herewith. Factors are given to four significant figures. Exact relationships are followed by a dagger. A more complete list is given in the latest editions of ASTM E 380 (4) and ANSI Z210.1 (6).

Conversion Factors to SI Units

To convert from	To	Multiply by
acre	square meter (m^2)	4.047×10^3
angstrom	meter (m)	$1.0 \times 10^{-10\dagger}$
are	square meter (m^2)	$1.0 \times 10^{2\dagger}$

†Exact.

To convert from	To	Multiply by
astronomical unit	meter (m)	1.496×10^{11}
atmosphere, standard	pascal (Pa)	1.013×10^{5}
bar	pascal (Pa)	$1.0 \times 10^{5\dagger}$
barn	square meter (m^2)	$1.0 \times 10^{-28\dagger}$
barrel (42 U.S. liquid gallons)	cubic meter (m^3)	0.1590
Bohr magneton (μ_B)	J/T	9.274×10^{-24}
Btu (International Table)	joule (J)	1.055×10^{3}
Btu (mean)	joule (J)	1.056×10^{3}
Btu (thermochemical)	joule (J)	1.054×10^{3}
bushel	cubic meter (m^3)	3.524×10^{-2}
calorie (International Table)	joule (J)	4.187
calorie (mean)	joule (J)	4.190
calorie (thermochemical)	joule (J)	$4.184^{\dagger}$
centipoise	pascal second (Pa·s)	$1.0 \times 10^{-3\dagger}$
centistokes	square millimeter per second (mm^2/s)	$1.0^{\dagger}$
cfm (cubic foot per minute)	cubic meter per second (m^3/s)	4.72×10^{-4}
cubic inch	cubic meter (m^3)	1.639×10^{-5}
cubic foot	cubic meter (m^3)	2.832×10^{-2}
cubic yard	cubic meter (m^3)	0.7646
curie	becquerel (Bq)	$3.70 \times 10^{10\dagger}$
debye	coulomb meter (C·m)	3.336×10^{-30}
degree (angle)	radian (rad)	1.745×10^{-2}
denier (international)	kilogram per meter (kg/m)	1.111×10^{-7}
	tex[‡]	0.1111
dram (apothecaries')	kilogram (kg)	3.888×10^{-3}
dram (avoirdupois)	kilogram (kg)	1.772×10^{-3}
dram (U.S. fluid)	cubic meter (m^3)	3.697×10^{-6}
dyne	newton (N)	$1.0 \times 10^{-5\dagger}$
dyne/cm	newton per meter (N/m)	$1.0 \times 10^{-3\dagger}$
electronvolt	joule (J)	1.602×10^{-19}
erg	joule (J)	$1.0 \times 10^{-7\dagger}$
fathom	meter (m)	1.829
fluid ounce (U.S.)	cubic meter (m^3)	2.957×10^{-5}
foot	meter (m)	$0.3048^{\dagger}$
footcandle	lux (lx)	10.76
furlong	meter (m)	2.012×10^{-2}
gal	meter per second squared (m/s^2)	$1.0 \times 10^{-2\dagger}$
gallon (U.S. dry)	cubic meter (m^3)	4.405×10^{-3}
gallon (U.S. liquid)	cubic meter (m^3)	3.785×10^{-3}
gallon per minute (gpm)	cubic meter per second (m^3/s)	6.309×10^{-5}
	cubic meter per hour (m^3/h)	0.2271

†Exact.

‡See footnote on p. xiii.

To convert from	To	Multiply by
gauss	tesla (T)	1.0×10^{-4}
gilbert	ampere (A)	0.7958
gill (U.S.)	cubic meter (m^3)	1.183×10^{-4}
grade	radian	1.571×10^{-2}
grain	kilogram (kg)	6.480×10^{-5}
gram force per denier	newton per tex (N/tex)	8.826×10^{-2}
hectare	square meter (m^2)	$1.0 \times 10^{4\dagger}$
horsepower (550 ft·lbf/s)	watt (W)	7.457×10^{2}
horespower (boiler)	watt (W)	9.810×10^{3}
horsepower (electric)	watt (W)	$7.46 \times 10^{2\dagger}$
hundredweight (long)	kilogram (kg)	50.80
hundredweight (short)	kilogram (kg)	45.36
inch	meter (m)	$2.54 \times 10^{-2\dagger}$
inch of mercury (32°F)	pascal (Pa)	3.386×10^{3}
inch of water (39.2°F)	pascal (Pa)	2.491×10^{2}
kilogram-force	newton (N)	9.807
kilowatt hour	megajoule (MJ)	$3.6^{\dagger}$
kip	newton(N)	4.448×10^{3}
knot (international)	meter per second (m/S)	0.5144
lambert	candela per square meter (cd/m^3)	3.183×10^{3}
league (British nautical)	meter (m)	5.559×10^{3}
league (statute)	meter (m)	4.828×10^{3}
light year	meter (m)	9.461×10^{15}
liter (for fluids only)	cubic meter (m^3)	$1.0 \times 10^{-3\dagger}$
maxwell	weber (Wb)	$1.0 \times 10^{-8\dagger}$
micron	meter (m)	$1.0 \times 10^{-6\dagger}$
mil	meter (m)	$2.54 \times 10^{-5\dagger}$
mile (statute)	meter (m)	1.609×10^{3}
mile (U.S. nautical)	meter (m)	$1.852 \times 10^{3\dagger}$
mile per hour	meter per second (m/s)	0.4470
millibar	pascal (Pa)	1.0×10^{2}
millimeter of mercury (0°C)	pascal (Pa)	$1.333 \times 10^{2\dagger}$
minute (angular)	radian	2.909×10^{-4}
myriagram	kilogram (kg)	10
myriameter	kilometer (km)	10
oersted	ampere per meter (A/m)	79.58
ounce (avoirdupois)	kilogram (kg)	2.835×10^{-2}
ounce (troy)	kilogram (kg)	3.110×10^{-2}
ounce (U.S. fluid)	cubic meter (m^3)	2.957×10^{-5}
ounce-force	newton (N)	0.2780
peck (U.S.)	cubic meter (m^3)	8.810×10^{-3}
pennyweight	kilogram (kg)	1.555×10^{-3}
pint (U.S. dry)	cubic meter (m^3)	5.506×10^{-4}

†Exact.

To convert from	To	Multiply by
pint (U.S. liquid)	cubic meter (m^3)	4.732×10^{-4}
poise (absolute viscosity)	pascal second (Pa·s)	$0.10^{\dagger}$
pound (avoirdupois)	kilogram (kg)	0.4536
pound (troy)	kilogram (kg)	0.3732
poundal	newton (N)	0.1383
pound-force	newton (N)	4.448
pound force per square inch (psi)	pascal (Pa)	6.895×10^{3}
quart (U.S. dry)	cubic meter (m^3)	1.101×10^{-3}
quart (U.S. liquid)	cubic meter (m^3)	9.464×10^{-4}
quintal	kilogram (kg)	$1.0 \times 10^{2\dagger}$
rad	gray (Gy)	$1.0 \times 10^{-2\dagger}$
rod	meter (m)	5.029
roentgen	coulomb per kilogram (C/kg)	2.58×10^{-4}
second (angle)	radian (rad)	$4.848 \times 10^{-6\dagger}$
section	square meter (m^2)	2.590×10^{6}
slug	kilogram (kg)	14.59
spherical candle power	lumen (lm)	12.57
square inch	square meter (m^2)	6.452×10^{-4}
square foot	square meter (m^2)	9.290×10^{-2}
square mile	square meter (m^2)	2.590×10^{6}
square yard	square meter (m^2)	0.8361
stere	cubic meter (m^3)	$1.0^{\dagger}$
stokes (kinematic viscosity)	square meter per second (m^2/s)	$1.0 \times 10^{-4\dagger}$
tex	kilogram per meter (kg/m)	$1.0 \times 10^{-6\dagger}$
ton (long, 2240 pounds)	kilogram (kg)	1.016×10^{3}
ton (metric) (tonne)	kilogram (kg)	$1.0 \times 10^{3\dagger}$
ton (short, 2000 pounds)	kilogram (kg)	9.072×10^{2}
torr	pascal (Pa)	1.333×10^{2}
unit pole	weber (Wb)	1.257×10^{-7}
yard	meter (m)	$0.9144^{\dagger}$

†Exact.

Abbreviations and Unit Symbols

Following is a list of common abbreviations and unit symbols used in the *Encyclopedia*. In general they agree with those listed in *American National Standard Abbreviations for Use on Drawings and in Text (ANSI Y1.1)* (6) and *American National Standard Letter Symbols for Units in Science and Technology (ANSI Y10)* (6). Also included is a list of acronyms for a number of private and government organizations as well as common industrial solvents, polymers, and other chemicals.

Rules for Writing Unit Symbols (4):

1. Unit symbols are printed in upright letters (roman) regardless of the type style used in the surrounding text.
2. Unit symbols are unaltered in the plural.
3. Unit symbols are not followed by a period except when used at the end of a sentence.
4. Letter unit symbols are generally printed lower-case (for example, cd for candela) unless the unit name has been derived from a proper name, in which case the first letter of the symbol is capitalized (W, Pa). Prefixes and unit symbols retain their prescribed form regardless of the surrounding typography.
5. In the complete expression for a quantity, a space should be left between the numerical value and the unit symbol. For example, write 2.37 lm, *not* 2.37lm, and 35 mm, *not* 35mm. When the quantity is used in an adjectival sense, a hyphen is often used, for example, 35-mm film. *Exception:* No space is left between the numerical value and the symbols for degree, minute, and second of plane angle, degree Celsius, and the percent sign.
6. No space is used between the prefix and unit symbol (for example, kg).
7. Symbols, not abbreviations, should be used for units. For example, use "A," not "amp," for ampere.
8. When multiplying unit symbols, use a raised dot:

   N·m for newton meter

   In the case of W·h, the dot may be omitted, thus:

   Wh

   An exception to this practice is made for computer printouts, automatic typewriter work, etc, where the raised dot is not possible, and a dot on the line may be used.
9. When dividing unit symbols, use one of the following forms:

   $$\mathrm{m/s} \quad or \quad \mathrm{m{\cdot}s^{-1}} \quad or \quad \frac{\mathrm{m}}{\mathrm{s}}$$

   In no case should more than one slash be used in the same expression unless parentheses are inserted to avoid ambiguity. For example, write:

   $$\mathrm{J/(mol{\cdot}K)} \quad or \quad \mathrm{J{\cdot}mol^{-1}\cdot K^{-1}} \quad or \quad \mathrm{(J/mol)/K}$$

   but *not*

   J/mol/K

10. Do not mix symbols and unit names in the same expression. Write:

joules per kilogram *or* J/kg *or* $J{\cdot}kg^{-1}$

but *not*

joules/kilogram *nor* joules/kg *nor* $joules{\cdot}kg^{-1}$

ABBREVIATIONS AND UNITS

A	ampere
A	anion (eg, HA)
A	mass number
a	atto (prefix for 10^{-18})
AATCC	American Association of Textile Chemists and Colorists
ABS	acrylonitrile–butadiene–styrene
abs	absolute
ac	alternating current, *n.*
a-c	alternating current, *adj.*
ac-	alicyclic
acac	acetylacetonate
ACGIH	American Conference of Governmental Industrial Hygienists
ACS	American Chemical Society
AGA	American Gas Association
Ah	ampere hour
AIChE	American Institute of Chemical Engineers
AIME	American Institute of Mining, Metallurgical, and Petroleum Engineers
AIP	American Institute of Physics
AISI	American Iron and Steel Institute
alc	alcohol(ic)
Alk	alkyl
alk	alkaline (not alkali)
amt	amount
amu	atomic mass unit
ANSI	American National Standards Institute
AO	atomic orbital
AOAC	Association of Official Analytical Chemists
AOCS	Americal Oil Chemists' Society
APHA	American Public Health Association
API	American Petroleum Institute
aq	aqueous
Ar	aryl
ar-	aromatic
as-	asymmetric(al)
ASHRAE	American Society of Heating, Refrigerating, and Air Conditioning Engineers
ASM	American Society for Metals
ASME	American Society of Mechanical Engineers
ASTM	American Society for Testing and Materials
at no.	atomic number
at wt	atomic weight
av(g)	average
AWS	American Welding Society
b	bonding orbital
bbl	barrel
bcc	body-centered cubic
BCT	body-centered tetragonal
Bé	Baumé
BET	Brunauer-Emmett-Teller (adsorption equation)
bid	twice daily
Boc	*t*-butyloxycarbonyl
BOD	biochemical (biological) oxygen demand
bp	boiling point
Bq	becquerel

C	coulomb
°C	degree Celsius
C-	denoting attachment to carbon
c	centi (prefix for 10^{-2})
c	critical
ca	circa (approximately)
cd	candela; current density; circular dichroism
CFR	Code of Federal Regulations
cgs	centimeter-gram-second
CI	Color Index
cis-	isomer in which substituted groups are on same side of double bond between C atoms
cl	carload
cm	centimeter
cmil	circular mil
cmpd	compound
CNS	central nervous system
CoA	coenzyme A
COD	chemical oxygen demand
coml	commercial(ly)
cp	chemically pure
cph	close-packed hexagonal
CPSC	Consumer Product Safety Commission
cryst	crystalline
cub	cubic
D	debye
D-	denoting configurational relationship
d	differential operator
d	day; deci (prefix for 10^{-1})
d-	*dextro*-, dextrorotatory
da	deka (prefix for 10^1)
dB	decibel
dc	direct current, *n.*
d-c	direct current, *adj.*
dec	decompose
detd	determined
detn	determination
Di	didymium, a mixture of all lanthanons
dia	diameter
dil	dilute
DIN	Deutsche Industrie Normen
dl-; DL-	racemic
DMA	dimethylacetamide
DMF	dimethylformamide
DMG	dimethyl glyoxime
DMSO	dimethyl sulfoxide
DOD	Department of Defense
DOE	Department of Energy
DOT	Department of Transportation
DP	degree of polymerization
dp	dew point
DPH	diamond pyramid hardness
dstl(d)	distill(ed)
dta	differential thermal analysis
(*E*)-	entgegen; opposed
ϵ	dielectric constant (unitless number)
e	electron
ECU	electrochemical unit
ed.	edited, edition, editor
ED	effective dose
EDTA	ethylenediaminetetra-acetic acid
emf	electromotive force
emu	electromagnetic unit
en	ethylene diamine
eng	engineering
EPA	Environmental Protection Agency
epr	electron paramagnetic resonance
eq.	equation
esca	electron spectroscopy for chemical analysis
esp	especially
esr	electron-spin resonance
est(d)	estimate(d)
estn	estimation
esu	electrostatic unit
exp	experiment, experimental
ext(d)	extract(ed)
F	farad (capacitance)
F	faraday (96,487 C)
f	femto (prefix for 10^{-15})

FAO	Food and Agriculture Organization (United Nations)
fcc	face-centered cubic
FDA	Food and Drug Administration
FEA	Federal Energy Administration
FHSA	Federal Hazardous Substances Act
fob	free on board
fp	freezing point
FPC	Federal Power Commission
FRB	Federal Reserve Board
frz	freezing
G	giga (prefix for 10^9)
G	gravitational constant = $6.67 \times 10^{11}\ N{\cdot}m^2/kg^2$
g	gram
(g)	gas, only as in $H_2O(g)$
g	gravitational acceleration
gc	gas chromatography
gem-	geminal
glc	gas–liquid chromatography
g-mol wt; gmw	gram-molecular weight
GNP	gross national product
gpc	gel-permeation chromatography
GRAS	Generally Recognized as Safe
grd	ground
Gy	gray
H	henry
h	hour; hecto (prefix for 10^2)
ha	hectare
HB	Brinell hardness number
Hb	hemoglobin
hcp	hexagonal close-packed
hex	hexagonal
HK	Knoop hardness number
hplc	high performance liquid chromatography
HRC	Rockwell hardness (C scale)
HV	Vickers hardness number
hyd	hydrated, hydrous
hyg	hygroscopic
Hz	hertz
^{i} (eg, Pr^i)	iso (eg, isopropyl)
i-	inactive (eg, *i*-methionine)
IACS	International Annealed Copper Standard
ibp	initial boiling point
IC	integrated circuit
ICC	Interstate Commerce Commission
ICT	International Critical Table
ID	inside diameter; infective dose
ip	intraperitoneal
IPS	iron pipe size
ir	infrared
IRLG	Interagency Regulatory Liaison Group
ISO	International Organization Standardization
ITS-90	International Temperature Scale (NIST)
IU	International Unit
IUPAC	International Union of Pure and Applied Chemistry
IV	iodine value
iv	intravenous
J	joule
K	kelvin
k	kilo (prefix for 10^3)
kg	kilogram
L	denoting configurational relationship
L	liter (for fluids only) (5)
l-	*levo-*, levorotatory
(l)	liquid, only as in $NH_3(l)$
LC_{50}	conc lethal to 50% of the animals tests
LCAO	linear combination of atomic orbitals
lc	liquid chromatography
LCD	liquid crystal display
lcl	less than carload lots
LD_{50}	dose lethal to 50% of the animals tested

LED	light-emitting diode
liq	liquid
lm	lumen
ln	logarithm (natural)
LNG	liquefied natural gas
log	logarithm (common)
LPG	liquefied petroleum gas
ltl	less than truckload lots
lx	lux
M	mega (prefix for 10^6); metal (as in MA)
M	molar; actual mass
$\overline{M}_w$	weight-average mol wt
$\overline{M}_n$	number-average mol wt
m	meter; milli (prefix for 10^{-3})
m	molal
m-	meta
max	maximum
MCA	Chemical Manufacturers' Association (was Manufacturing Chemists Association)
MEK	methyl ethyl ketone
meq	milliequivalent
mfd	manufactured
mfg	manufacturing
mfr	manufacturer
MIBC	methyl isobutyl carbinol
MIBK	methyl isobutyl ketone
MIC	minimum inhibiting concentration
min	minute; minimum
mL	milliliter
MLD	minimum lethal dose
MO	molecular orbital
mo	month
mol	mole
mol wt	molecular weight
mp	melting point
MR	molar refraction
ms	mass spectrometry
MSDS	material safety data sheet
mxt	mixture
μ	micro (prefix for 10^{-6})
N	newton (force)
N	normal (concentration); neutron number
N-	denoting attachment to nitrogen
n (as n_D^{20})	index of refraction (for 20°C and sodium light)
^{n} (as Bu^n), *n*-	normal (straight-chain structure)
n	neutron
n	nano (prefix for 10^9)
na	not available
NAS	National Academy of Sciences
NASA	National Aeronautics and Space Administration
nat	natural
ndt	nondestructive testing
neg	negative
NF	*National Formulary*
NIH	National Institutes of Health
NIOSH	National Institute of Occupational Safety and Health
NIST	National Institute of Standards and Technology (formerly National Bureau of Standards)
nmr	nuclear magnetic resonance
NND	New and Nonofficial Drugs (AMA)
no.	number
NOI-(BN)	not otherwise indexed (by name)
NOS	not otherwise specified
nqr	nuclear quadruple resonance
NRC	Nuclear Regulatory Commission; National Research Council
NRI	New Ring Index
NSF	National Science Foundation
NTA	nitrilotriacetic acid
NTP	normal temperature and pressure (25°C and 101.3 kPa or 1 atm)

NTSB	National Transportation Safety Board	qv	quod vide (which see)
O-	denoting attachment to oxygen	R	univalent hydrocarbon radical
o-	ortho	(*R*)-	rectus (clockwise configuration)
OD	outside diameter	*r*	precision of data
OPEC	Organization of Petroleum Exporting Countries	rad	radian; radius
o-phen	*o*-phenanthridine	RCRA	Resource Conservation and Recovery Act
OSHA	Occupational Safety and Health Administration	rds	rate-determining step
owf	on weight of fiber	ref.	reference
Ω	ohm	rf	radio frequency, *n.*
P	peta (prefix for 10^{15})	r-f	radio frequency, *adj.*
p	pico (prefix for 10^{-12})	rh	relative humidity
p-	para	RI	Ring Index
p	proton	rms	root-mean square
p.	page	rpm	rotations per minute
Pa	pascal (pressure)	rps	revolutions per second
PEL	personal exposure limit based on an 8-h exposure	RT	room temperature
pd	potential difference	RTECS	Registry of Toxic Effects of Chemical Substances
pH	negative logarithm of the effective hydrogen ion concentration	^{s} (eg, Bu^{s}); *sec*-	secondary (eg, secondary butyl)
phr	parts per hundred of resin (rubber)	S	siemens
p-i-n	positive-intrinsic-negative	(*S*)-	sinister (counterclockwise configuration)
pmr	proton magnetic resonance	*S*-	denoting attachment to sulfur
p-n	positive-negative	*s*-	symmetric(al)
po	per os (oral)	s	second
POP	polyoxypropylene	(s)	solid, only as in $H_2O(s)$
pos	positive	SAE	Society of Automotive Engineers
pp.	pages	SAN	styrene-acrylonitrile
ppb	parts per billion (10^9)	sat(d)	saturate(d)
ppm	parts per million (10^6)	satn	saturation
ppmv	parts per million by volume	SBS	styrene–butadiene–styrene
ppmwt	parts per million by weight	sc	subcutaneous
PPO	poly(phenyl oxide)	SCF	self-consistent field; standard cubic feet
ppt(d)	precipitate(d)	Sch	Schultz number
pptn	precipitation	sem	scanning electron microscope(y)
Pr (no.)	foreign prototype (number)	SFs	Saybolt Furol seconds
pt	point; part	sl sol	slightly soluble
PVC	poly(vinyl chloride)	sol	soluble
pwd	powder		
py	pyridine		

soln	solution
soly	solubility
sp	specific; species
sp gr	specific gravity
sr	steradian
std	standard
STP	standard temperature and pressure (0°C and 101.3 kPa)
sub	sublime(s)
SUs	Saybolt Universal seconds
syn	synthetic
^{t} (eg, Bu^t), *t*-, *tert*-	tertiary (eg, tertiary butyl)
T	tera (prefix for 10^{12}); tesla (magnetic flux density)
t	metric ton (tonne)
t	temperature
TAPPI	Technical Association of the Pulp and Paper Industry
TCC	Tagliabue closed cup
tex	tex (linear density)
T_g	glass-transition temperature
tga	thermogravimetric analysis
THF	tetrahydrofuran
tlc	thin layer chromatography
TLV	threshold limit value
trans-	isomer in which substituted groups are on opposite sides of double bond between C atoms
TSCA	Toxic Substances Control Act
TWA	time-weighted average
Twad	Twaddell
UL	Underwriters' Laboratory
USDA	United States Department of Agriculture
USP	*United States Pharmacopeia*
uv	ultraviolet
V	volt (emf)
var	variable
vic-	vicinal
vol	volume (not volatile)
vs	versus
v sol	very soluble
W	watt
Wb	weber
Wh	watt hour
WHO	World Health Organization (United Nations)
wk	week
yr	year
(*Z*)-	zusammen; together; atomic number

Non-SI (Unacceptable and Obsolete) Units		Use
Å	angstrom	nm
at	atmosphere, technical	Pa
atm	atmosphere, standard	Pa
b	barn	cm^2
bar[†]	bar	Pa
bbl	barrel	m^3
bhp	brake horsepower	W
Btu	British thermal unit	J
bu	bushel	m^3; L
cal	calorie	J
cfm	cubic foot per minute	m^3/s
Ci	curie	Bq
cSt	centistokes	mm^2/s
c/s	cycle per second	Hz

[†]Do not use bar (10^5 Pa) or millibar (10^2 Pa) because they are not SI units, and are accepted internationally only for a limited time in special fields because of existing usage.

Non-SI (Unacceptable and Obsolete) Units		Use
cu	cubic	exponential form
D	debye	C·m
den	denier	tex
dr	dram	kg
dyn	dyne	N
dyn/cm	dyne per centimeter	mN/m
erg	erg	J
eu	entropy unit	J/K
°F	degree Fahrenheit	°C; K
fc	footcandle	lx
fl	footlambert	lx
fl oz	fluid ounce	m^3; L
ft	foot	m
ft·lbf	foot pound-force	J
gf den	gram-force per denier	N/tex
G	gauss	T
Gal	gal	m/s^2
gal	gallon	m^3; L
Gb	gilbert	A
gpm	gallon per minute	(m^3/s); (m^3/h)
gr	grain	kg
hp	horsepower	W
ihp	indicated horsepower	W
in.	inch	m
in. Hg	inch of mercury	Pa
in. H_2O	inch of water	Pa
in.-lbf	inch pound-force	J
kcal	kilo-calorie	J
kgf	kilogram-force	N
kilo	for kilogram	kg
L	lambert	lx
lb	pound	kg
lbf	pound-force	N
mho	mho	S
mi	mile	m
MM	million	M
mm Hg	millimeter of mercury	Pa
mμ	millimicron	nm
mph	miles per hour	km/h
μ	micron	μm
Oe	oersted	A/m
oz	ounce	kg
ozf	ounce-force	N
η	poise	Pa·s
P	poise	Pa·s
ph	phot	lx
psi	pounds-force per square inch	Pa
psia	pounds-force per square inch absolute	Pa
psig	pounds-force per square inch gage	Pa
qt	quart	m^3; L
°R	degree Rankine	K
rd	rad	Gy
sb	stilb	lx
SCF	standard cubic foot	m^3
sq	square	exponential form
thm	therm	J
yd	yard	m

BIBLIOGRAPHY

1. The International Bureau of Weights and Measures, BIPM (Parc de Saint-Cloud, France) is described in Appendix X2 of Ref. 4. This bureau operates under the exclusive supervision of the International Committee for Weights and Measures (CIPM).
2. *Metric Editorial Guide (ANMC-78-1)*, latest ed., American National Metric Council, 5410 Grosvenor Lane, Bethesda, Md. 20814, 1981.
3. *SI Units and Recommendations for the Use of Their Multiples and of Certain Other Units (ISO 1000-1981)*, American National Standards Institute, 1430 Broadway, New York, N.Y. 10018, 1981.
4. Based on *ASTM E 380-89a (Standard Practice for Use of the International System of Units (SI))*, American Society for Testing and Materials, 1916 Race Street, Philadelphia, Pa. 19103, 1989.
5. *Fed. Regist.*, Dec. 10, 1976 (41 FR 36414).
6. For ANSI address, see Ref. 3.

R. P. LUKENS
ASTM Committee E-43 on SI Practice

Continued

EXPLOSIVES AND PROPELLANTS

EXPLOSIVES

Propellants and explosives are chemical compounds or mixtures that rapidly produce large volumes of hot gases when properly initiated. Propellants burn at relatively low rates measured in centimeters per second; explosives detonate at rates of kilometers per second. Pyrotechnic materials evolve large amounts of heat but much less gas than propellants and explosives (see PYROTECHNICS).

Many compounds explode when triggered by a suitable stimulus; however, most are either too sensitive or fail to meet cost and production-scale standards, requirements for safety in transportation, and storage stability. Propellants and explosives in large-scale use are based mostly on a relatively small number of well-proven ingredients. Propellants and explosives for military systems are manufactured in the United States primarily in government owned plants where they are also loaded into munitions. Composite propellants for large rockets are produced mainly by private industry, as are small arms propellants for sporting weapons.

A comparison of the characteristics associated with propellant burning, explosive detonation, and the performance of conventional fuels (see COAL; GAS, NATURAL; PETROLEUM) is shown in Table 1. The most notable difference is the rate at which energy is evolved. The energy liberated by explosives and propellants depends on the thermochemical properties of the reactants. As a rough rule

Table 1. Characteristics of Burning and Detonation[a]

	Burning		
Characteristics	Fuel	Propellant	Explosive detonation
material	coal in air	propellants	explosives
linear reaction rate, m/s	10^{-6}	10^{-2}	$2\text{–}9 \times 10^3$
reaction completion time, s	10^{-1}	10^{-3}	10^{-6}
factor controlling reaction rate	heat transfer	heat transfer	shock transfer
energy output, J/g[b]	10^4	4×10^3	4×10^3
power output, W/cm^2	10	10^3	10^9
most common initiation mode	heat	hot particles and gases	high temperature–high pressure shock waves
pressure developed, MPa[c]	0.07–0.7	$3\text{–}7 \times 10^2$	$7 \times 10^3\text{–}7 \times 10^4$
uses	source of heat and electricity	controlled gas pressure, guns, and rockets	blast, fragmenting munitions, civil engineering

[a]Refs. 1,2. All reactions are oxidation–reduction.
[b]To convert J to cal, divide by 4.184.
[c]To convert MPa to psi, multiply by 145.

of thumb, these materials yield about 1000 cm^3 of gas and 4.2 kJ (1000 cal) of heat per gram of material.

Explosives and propellants have relatively large amounts of available energy stored compactly and readily deliverable. The power output depends on the rate at which energy is liberated. Propellants are used wherever a readily controllable source of energy is required for periods of time ranging from milliseconds in guns to seconds in rockets. The gases evolved are employed as a working fluid for propelling projectiles and rockets, driving turbines, moving pistons, shearing bolts and wires, operating pumps, and starting engines. Explosives are used wherever very rapid rates of energy application and high pressures are essential. They are employed to produce high intensity shock waves in air, water, rock, and metal; for blasting, cratering, mining, and other civil engineering purposes; for metal welding and forming, cutting, and fragmentation; in shaped charges and many specialty devices requiring high rates of energy transmission; and for initiation of detonation phenomena. The terms burning and deflagration are often used synonymously to describe the gradual consumption of a propellant grain by a flame off the surface and to contrast it with the much more rapid, violent, and destructive phenomena associated with the detonation of explosives.

The development of explosives began in Europe with the formulation and use of black powder in about the middle of the thirteenth century, accelerated in the nineteenth century with the nitration of many compounds to produce high energy explosives, and greatly intensified during World War II. There has been enormous growth in the explosives field in the latter half of the twentieth century made possible in part by modern electronic instrumentation, high speed photography (qv), computers, military and space research, and the opportunities in worldwide mining and civil engineering programs (see COMPUTER TECHNOLOGY; METALLURGY; MINERAL RECOVERY AND PROCESSING). The most common explosive compounds are listed in Table 2 (3–7).

General Characteristics

Exothermic oxidation–reduction reactions provide the energy released in both propellant burning and explosive detonation. The reactions are either internal oxidation–reductions, as in the decomposition of nitroglycerin and pentaerythritol tetranitrate, or reactions between discrete oxidizers and fuels in heterogeneous mixtures.

An activation energy of 125–250 kJ/mol (30–60 kcal/mol) is usually required to initiate the reaction. Once initiated, the heat evolved is sufficient to cause the reaction to continue and become self-sustaining. Most explosives and propellants are organic compounds or mixtures of compounds that contain carbon, hydrogen, oxygen, and nitrogen. Metallic fuels such as aluminum may be added to increase the heat of reaction. Industrial dynamites have traditionally used nitroglycerin or nitroglycol, nitrocellulose, and inorganic salts as sources of oxygen, but these have been mostly replaced by formulations where ammonium nitrate is the primary source of oxygen. Composite propellants commonly use ammonium perchlorate to supply the oxygen required. The most common gaseous products of the

Table 2. Explosive Substances

Name	CAS Registry Number	Code	Formula	Use
		Inorganic salts		
ammonium nitrate	[*6484-52-2*]	AN	NH_4NO_3	solid oxidizer
ammonium perchlorate	[*7790-98-9*]		NH_4ClO_4	solid oxidizer
lead azide	[*13424-46-9*]		$Pb(N_3)_2$	primary explosive
ammonium picrate	[*131-74-8*]	AP	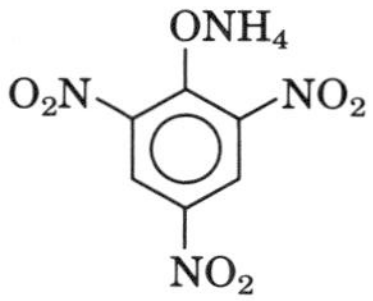 ONH_4, O_2N, NO_2, NO_2	secondary high explosive
2,4-diamino-1,3,5-trinitrobenzene	[*1630-08-6*]	DATB	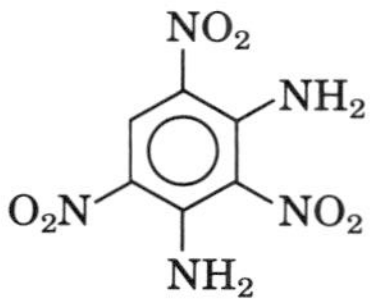 NO_2, NH_2, O_2N, NO_2, NH_2	secondary high explosive
diazodinitrophenol		DDNP	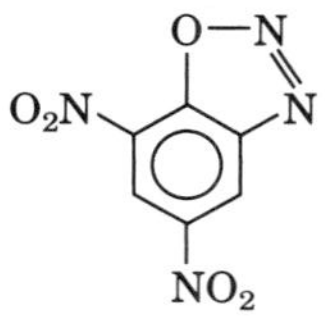 O—N, N, O_2N, NO_2	primary explosive
ethylene glycol dinitrate	[*628-96-6*]	EGDN	$O_2NOCH_2CH_2ONO_2$	liquid explosive

lead styphnate	[*15245-44-0*]		$[C_6H(NO_2)_3(O^-)_2]$ PbH_2O^{2+}	primary explosive
2,4,6-trinitrotoluene	[*118-96-7*]	TNT	$CH_3C_6H_2(NO_2)_3$	secondary high explosive
picric acid	[*88-89-1*]	PA	$HOC_6H_2(NO_2)_3$	secondary high explosive

Aliphatic nitrate esters

mannitol hexanitrate	[*15825-70-4*]	MN	H_2CONO_2–$HCONO_2$–$HCONO_2$–$HCONO^2$–$HCONO^2$–H^2CONO^2	primary explosive

Table 2. (*Continued*)

Name	CAS Registry Number	Code	Formula	Use
nitrocellulose	[*9004-70-0*], [*9046-47-3*]	NC	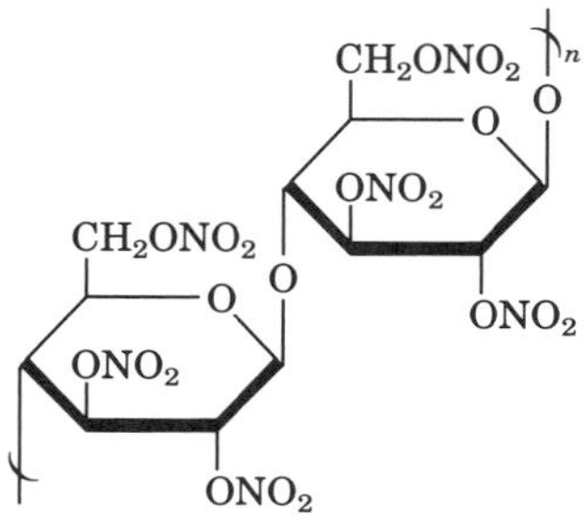 (cellulose nitrate repeat unit: two glucose rings, each with CH_2ONO_2, ONO_2, ONO_2, linked by O)$_n$	secondary explosive used in propellants
nitroglycerin	[*55-63-0*]	NG	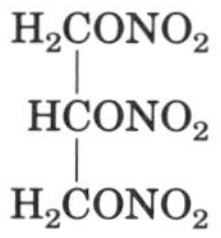 H_2CONO_2–$HCONO_2$–H_2CONO_2	liquid secondary explosive ingredient in commercial explosives and propellants
nitromethane	[*75-52-5*]	NM	CH_3NO_2	liquid secondary explosive
pentaerythritol tetranitrate	[*78-11-5*]	PETN	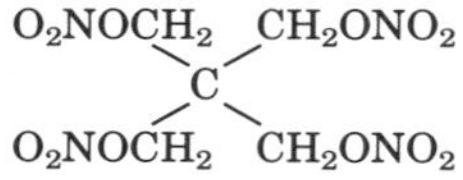 $C(CH_2ONO_2)_4$ (O_2NOCH_2, CH_2ONO_2, O_2NOCH_2, CH_2ONO_2 on central C)	secondary high explosive used as booster

Nitramines

cyclotrimethylenetrinitramine	[*121-82-4*]	RDX	NO_2, N, O_2N–N, N–NO_2	secondary high explosive
trinitrophenylmethylnitramine	[*479-45-8*]	tetryl	O_2NNCH_3, O_2N, NO_2, NO_2	secondary explosive used as booster
ethylenedinitramine	[*505-71-5*]	EDNA	$O_2NNHCH_2CH_2NHNO_2$	secondary high explosive
tetrazene	[*31330-63-9*]		N–N, N, N–NH, –N=NNHCNH$_2$·H$_2$O, NH	primary explosive
tetranitromethane	[*509-14-8*]	TNM	$C(NO_2)_4$	liquid explosive
cyclotetramethylenetetranitramine	[*2691-41-0*]	HMX	NO_2, NO_2, N, N, N, N, NO_2, NO_2	secondary high explosive

oxidation–reduction reactions are hydrogen, water, carbon monoxide, carbon dioxide, and nitrogen. Other products depend on the reactants involved.

Propellants and explosives evolve about the same amount of energy per unit weight. The specific energy is considerably less than that produced in the combustion of common fuels, because the latter take oxygen from the ambient air. A high performance rocket propellant produces 5–6.3 kJ/g (1.2–1.5 kcal/g); the detonation of nitroglycerin yields 6.3 kJ/g (1.5 kcal/g); burning a gram of carbon in air yields 33.5 kJ (8 kcal).

The specific stimulus that triggers an explosive detonation or propellant burning depends on the material involved and the system environment. In most cases, heat is the ultimate cause of the activation. Gun and rocket propellants are initiated by igniter compositions that produce hot gases and solids at relatively low pressures. Explosives are detonated by high pressure, high temperature shock waves that heat the explosive to rapid reaction temperatures by adiabatic compression. Mechanical impact, frictional forces, electric discharge, and other ultimate sources of heat may also act as initiating stimuli. The minimum quantity of energy required for initiation depends on the chemical characteristics of the material, its physical properties including mass, geometry, density, the degree of confinement, the rate at which the energy is delivered, the environment in which the energy release occurs, and the type of initiating stimulus provided. Explosives and propellants are often ranked in order of their sensitivity and response to a specific stimulus in a given environment (1,8).

Propellant Burning. Propellants generally operate at low pressures down to about 3–4 MPa (500 psi) in rockets and up to about 689 MPa (100,000 psi) in high performance guns (see EXPLOSIVES AND PROPELLANTS). The process is characterized by a reaction front that moves in a direction normal to the exposed surface of the grain, proceeding from the outside in laminar layers. The rate of burning depends on the intrinsic rate of decomposition of the propellant formulation and the rate of heat transfer from the hot gases above the propellant surface. One equation that defines the rate of propellant burning is

$$dx/dt = bP^n$$

where x = a coordinate normal to the grain surface, t = time, b = a constant that depends on the propellant temperature and composition, P = pressure at which burning occurs, and n = pressure exponent that depends on the propellant composition and the pressure. Values of n range from nearly 0 to 0.5 for low pressure rocket propellants, and 0.7 to 1.0 for high pressure gun propellant.

Explosive Detonation. Detonations proceed as a result of a reaction front moving in a direction normal to the surface of the explosive. However, detonaticon is a hydrodynamic phenomenon that differs in a fundamental sense from burning. Upon initiation, burning first occurs at an increasing rate for a period of time up to several microseconds. A high pressure shock wave is rapidly formed which passes through the explosive at high velocity. As it does so it causes exothermal decomposition of the explosive. The continued passage of the wave is supported by transfer of energy from the spent reacted explosive to the unreacted explosive by shock compression. Thus the rate of reaction depends on the rate of transmission of the shock wave rather than on the rate of heat transfer associated with propellant burning. The detonation rate, D, is stable and constant and is primarily governed by the physical and chemical properties of the explosive, the diameter of the charge, its degree of confinement, and particularly its density, ρ, with which it varies in a linear fashion for most explosives:

$$D_i = D_o + M(\rho_i - \rho_o)$$

where D_i = linear detonation rate, dx/dt, at density ρ_i, D_o = linear detonation rate at density ρ_o, and M = a constant characteristic of the explosive composition. Typical values of D_o at ρ_o = 1.0 g/cm are about 5000–6000 m/s. Values of M are about 3000/4000 m/s (1,9).

Deflagration to Detonation Transfer. The same compound or mixture may burn or detonate, depending on the type and intensity of initiation, the degree of confinement, and the physical and geometric characteristics of the material. Many explosives that ordinarily detonate may burn under carefully controlled conditions involving gentle ignition that avoids shock-wave formation. Burning characteristics are similar to those encountered in propellants of about the same energy level. Nonporous propellants containing colloidal nitrocellulose and nitroglycerin or ammonium perchlorate in a polymeric matrix normally burn quickly and controllably on initiation using conventional igniters. These are difficult to detonate under such conditions even when they contain relatively large quantities of highly energetic explosives. However, a detonation may occur if the propellant is initiated by a high intensity shock wave; if the physical conditions are appropriate, eg, the presence of a large degree of porosity and heavy confinement; or if the material is in the form of a large mass of finely divided material. Granular TNT initiated with black powder burns quietly if the TNT is spread in thin layers on the ground. It is apt to detonate if piled up in a large mound. The disastrous explosion in 1947 of a cargo of fertilizer-grade ammonium nitrate packed in the hold of a burning freighter illustrates how conditions may cause an ordinarily inert material to detonate (10,11).

Because unwanted detonations of propellants are likely to be catastrophic, the conditions at which deflagration to detonation transformations occur have been intensively studied. A simplified view of the process by which a deflagration is transformed to a detonation suggests that an initial burning of the type associated with propellant combustion occurs. This is followed by convective burning in which the hot gases penetrate the pores of the explosive. The combustion front may then build up into shock-wave pressures that produce a low velocity detonation which is rapidly converted to a full-fledged detonation with its characteristic shock pressure and temperature. Conditions that minimize energy losses and increase the likelihood of buildup of a shock wave tend to enhance the likelihood of a deflagration converting to a detonation. The scaling up of explosive processes increases the possibility of detonation because of the mass effect, as does heavy confinement and the use of large concentrations of crystalline oxidizers (12–15).

Pyrotechnic Compositions. Pyrotechnic compositions engage in oxidation–reduction reactions that resemble those of propellants and explosives, but generally produce little or no gas. These are heterogeneous mixtures of a finely powdered metal, metal alloy, or organic fuel and inorganic oxidizers. Such compositions are commonly used for flares, signals, tracers, incendiaries, delays, igniters, heating mixtures, and in devices where the formation of much gas is unacceptable either because the gas pressure causes unwanted changes in the reaction rate or the system is not designed to withstand the pressure without rupturing. The properties of typical pyrotechnic and explosive compositions are compared in Table 3.

Table 3. Comparison of Pyrotechnic Compositions with High Explosives[a]

Component	Composition, wt %	Heat of reaction, kJ/g[b]	Gas volume,[c] cm^3/g	Relative brisance, % TNT	Ignition temp,[d] °C	Impact test,[e] % TNT
	Pyrotechnic composition					
delay						
barium chromate	90					
boron	10	2.010	13	0	450	12
delay						
barium chromate	60					
zirconium–nickel alloy	26					
potassium perchlorate	14	2.081	12	0	485	23
flare						
sodium nitrate	38					
magnesium	50					
laminac	5	6.134	74	17	640	19
smoke						
zinc	69					
potassium perchlorate	19					
hexachlorobenzene	12	2.579	62	17	475	15
photoflash						
barium nitrate	30					
aluminum	40					
potassium perchlorate	30	8.989	15	15	700	26
		High explosive				
TNT		4.560	710	100	310	100
RDX		5.694	908	140	260	35

[a]Ref. 16.
[b]To convert J to cal, divide by 4.184.
[c]At STP.
[d]5-s value.
[e]Pyrotechnic compositions produce only a mild ignition on impact.

Although pyrotechnic compositions are composed of inert ingredients, accidental initiation during the manufacturing process may be accompanied by the same catastrophic consequences that attend explosive detonations. Mixtures of finely divided oxidizers and metals are sensitive to initiation by friction or by spark, particularly when in the powdered form. As little as 10 microJoules (2.4 μcal) can initiate some pyrotechnic dust mixtures. Safety measures include mixing in liquid media, using nonsparking tools, maintaining the moisture content of the plant atmosphere at 60% humidity, and using ionizing and grounding devices to reduce the possibility of electrostatic charge accumulation (16–21).

Safety Considerations

The catastrophic effects and the increasingly severe legal and economic implications of a disastrous explosion have caused a large effort to be devoted to the safety of explosive operations. Many governmental regulations control the classification, shipping, and handling of explosive materials. The publications in the field of safety have increased greatly, and numerous symposia have been held on the subject. There is an annual explosive safety seminar conducted by the Explosives Safety Board of the United States Department of Defense (22,23).

When a system containing an explosive detonates, the principal sources of hazard are the blast overpressures, high velocity metal fragments, and flying debris, which can include large pieces of structural material such as concrete, steel, and wood. Explosive materials are classified into five divisions by the Department of Defense Explosives Safety Board depending on detonation characteristics and likelihood of propagation of the event to other systems (24). These divisions are (*1*) systems that detonate and propagate into a mass detonation almost instantaneously; (*2*) systems that do not mass detonate when a single element is initiated; (*3*) systems that do not mass detonate but may burn vigorously and present a fire hazard and either a minor blast, fragment hazard, or both; (*4*) systems that do not present a blast hazard or significant fragmentation hazard on initiation. A fire must not cause instantaneous explosion of an entire package of such a system; and (*5*) very insensitive systems that are extremely unlikely to initiate, even in a fire.

The hazard posed can be limited by maintaining a zone free of people and property around a storage area of explosive material. The minimum radius of the zone depends on the type and quantity of explosive, the extent and type of barricading, and the magnitude of loss that would be encountered if an explosive incident occurred. The maximum distance to which hazardous explosive effects propagate depends on the blast overpressure created, which as a first approximation is a function of the cube root of the explosive weight, W. This is termed the quantity distance and is defined as

$$D = KW^{1/3}$$

where D is the allowable distance from the explosive site and K is a constant that depends on the classification of the explosive, the storage conditions, and the potential effect of a possible explosion on the people, structures, and material within the zone specified. Specific values for K are available (24,25).

Highly detailed and systematic techniques have been developed to identify possible failure modes, and to quantify the probability of occurrence and impact on operations, including hazards analysis and risk or reliability assessment. These are characterized by the step-by-step detailed examination of a process, often using fault-tree analysis to define and quantify the likelihood of undesirable events. Experiments are conducted simulating the actual application and environment of the explosives in the process to obtain valid quantitative data for calculation. The estimated probabilities of an accident occurring for each of the operations in a process are generally computerized to enable an assessment to be made of the overall hazard of the process to eliminate the dangerous elements (see HAZARD ANALYSIS AND RISK ASSESSMENT) (26–29).

The concept of TNT equivalency is used for evaluating the magnitude of an accidental explosion that may occur at some stage of an operation involving explosive materials. The TNT equivalency value, expressed as a percentage, is the relative weight of TNT, when tested as a hemispherical charge in the ground burst mode, which yields the same peak pressure or positive impulse as that produced at a given distance by explosive or explosive system tested. This information enables conversion of the data in barricade design manuals, which are usually presented in terms of surface bursts of TNT, to the explosive as it occurs in process operations. Typical TNT equivalency values are lead azide (35%), nitroglycerin (135%), Composition B (125%) and black powder (50%), and high energy gun propellant (125%).

A great deal of experimental work has also been done to identify and quantify the hazards of explosive operations (30–40). The vulnerability of structures and people to shock waves and fragment impact has been well established. This effort has also led to the design of protective structures superior to the conventional barricades which permit considerable reduction in allowable safety distances. In addition, a variety of techniques have been developed to mitigate catastrophic detonations of explosives exposed to fire.

Environmental Impact. Federal and State environmental legislation to limit, control, and remove pollutants entering the environment have resulted in numerous programs relating to propellant and explosive manufacture, storage, use, and disposal. The U.S. Federal Resources Consideration and Recovery Act (1976), the Water Pollution Control Act (1972), the Clean Water Quality Act (1977), the Clean Air Act (1967), The Resource Conservation Recovery Act (1976), The Federal Facilities Compliance Act (1992), and their amendments are examples of national legislation to establish controls and standards for numerous contaminants. The Superfund Act authorizes the federal government to clean up the worst hazardous waste sites. The National Environmental Policy Act (1969) requires that the environmental impact of federally sponsored programs be acceptable before implementation. These U.S. federal acts have been further reinforced by strong state and municipal limitations on permissible pollutants (see also AIR POLLUTION; AIR POLLUTION CONTROL METHODS).

Pollutant Reduction. Pollutants from explosives are primarily produced by waste from the explosives manufacture, such as the acids used in nitration (qv). Pollutants may also be produced during incorporation of the explosives in munitions, in the use of industrial explosives, and in clean-up and disposal operations. Table 4 lists the most common types of pollutants found in the manufacture of explosives, as well as effects and various procedures for reduction (41–54).

The wastewater effluent from a plant may have to be pH modified and otherwise treated to lower oxygen demand, dissolved and particulate solids, soluble nitrates and sulfates, and oil and grease. The overall principles for decreasing water contamination in explosive manufacture include (*1*) minimizing the quantity of water leaving the plant by recycling process and cooling waters; (*2*) segregating and treating highly polluted waters before dilution; (*3*) using settling reservoirs for water treatment and removal of suspended particles by sedimentation; (*4*) centrifuging to remove suspended solids; (*5*) employing ion-exchange (qv) resins to concentrate pollutants; and (*6*) biologically destroying organic compounds under aerobic conditions and biologically denitrifying nitrates under anaerobic conditions.

Table 4. Pollutants from Explosives Production

Pollutant	Effect	Reduction methods
	Water-soluble materials	
acids	toxic, corrosive	neutralization with limestone and landfill with solid products, biodenitrification of HNO_3, recycling and recovery
nitrates	toxic, increased solids content, eutrophication	biodenitrification, ion exchange
sulfates	increased solids content, odorific at anaerobic conditions	ion exchange, reverse osmosis, precipitation as the calcium or barium salts
phosphates	eutrophication	precipitation as calcium or rare-earth phosphates
acetates and organic esters	toxic, increases dissolved oxygen demand, increased acidity	biodegradation, neutralization, incineration
nitrobodies (pink water)	toxic, discolored water	carbon absorption, polymer resin absorption, biodenitrification, electrolytic oxidation, catalyzed ozonolysis
red water	toxic, discolored water	concentration and sale, incineration, reduction and carbonation
	Solid wastes	
propellants and explosives	dangerous, may be toxic, cannot be landfilled or open burned	fluidized-bed incineration, composting
inert contaminants	may be toxic, unsightly	dual-chamber or air curtain incineration, composting
process sludges	hazardous	incineration
contaminated activated carbon	pollutant if burned	thermal regeneration in indirectly heated rotary furnace, replacement with polymeric materials, solvent regeneration, use of powdered activated carbon followed by pyrolytic regeneration
	Air pollutants	
oxides of nitrogen	toxic and irritant corrosive reacts with ozone photochemically to produce smog	scrubbing with water or sulfuric and nitric acids, catalytic absorption with a molecular sieve, reduction to nitrogen
oxides of sulfur	toxic and irritant corrosive	catalytic reduction absorption in glycol
acid mist	toxic and irritant corrosive	mist eliminator
particulates	toxic and irritant unsightly	cyclone removal electrostatic precipitation, bag filters, water scrubbing

Investigations have also been conducted that show the feasibility of utilizing the energy content of explosives and propellants by mixing with fuel oils and burning the slurry. Preparation of the energetic material may involve grinding to reduce size, solution in organic solvents and subsequent solvent recovery, or slurrying in a water mix. The economic advantages depend on the plant requirements to assure safety and the comparative cost of available fuel oil.

Demilitarization and Disposal of Explosive Material. An important consequence of international agreements to greatly reduce the stockpiles of conventional and nuclear munitions is the intensification of a program to develop procedures to destroy, recycle, and/or reclaim explosives, propellants, and pyrotechnic material efficiently and without significant environmental impact.

The procedures commonly used to demilitarize conventional munitions include munitions disassembly, washout or steamout of explosives from projectiles and warheads, incineration of reclaimed explosives, and open burning or detonation. Open burning and detonation of large quantities of energetic compositions are no longer permitted in many areas of the United States. Increasing constraints are being placed even on the uncontrolled destruction of small quantities of these substances.

A wide variety of special-purpose incinerators (qv) with accompanying gas scrubbers and solid particle collectors have been developed and installed in various demilitarization facilities. These include flashing furnaces that remove all vestiges of explosive from metal parts to assure safety in handling; deactivation furnaces, to render safe small arms and nonlethal chemical munitions; fluidized-bed incinerators that burn slurries of ground up propellants or explosives in oil; and rotary kilns to destroy explosive and contaminated waste and bulk explosive.

Procedures investigated for removal of explosives and propellants from munitions include the use of steam (qv), low pressure hot water jets, high pressure hot or cold water jets, and high pressure water at supercritical temperature and autoclave melting. Organic solvents have been used for extraction and reclamation of soluble components from polymeric explosives and propellants. Similarly, supercritical fluid extraction uses high pressure to convert compounds that are gases at atmospheric pressure and ambient temperatures, eg, carbon dioxide and ammonia, to liquids. The fluids are then used at elevated temperatures and pressure to extract stabilizers, plasticizers (qv), and acetate esters from nitrocellulose propellants or ammonium perchlorate from composite propellant. After extraction, the pressure and temperature are reduced and the liquid reverts to its gaseous state. These procedures may be supplemented with mechanical methods such as saws, presses, and drills to facilitate the recovery process, particularly for large rocket motors (42,52,55–57).

Performance of Explosives

A wide variety of procedures have been developed to evaluate the performance of explosives. These include experimental methods as well as calculations based on available energy of the explosives and the reactions that take place on initiation. Both experimental and calculational procedures utilize electronic instrumentation and computer codes to provide estimates of performance in the laboratory and the field.

The experimental procedures depend to a large extent on the use to which the explosive is to be put. Comparison is often made to proven explosives of known performance. Many of the most commonly used tests in the various categories of concern are as follows:

Sensitivity

impact (drop)
detonation propagation
gap tests
friction
skid tests
pendulum tests
electrostatic discharge
susan tests
cap sensitivity
deflagration to detonation transfer
sensitivity to fragment initiation

Thermal characteristics

shelf-life test
stability tests
  Taliani
  vacuum stability
  potassium iodide
cook-off temperature
compatibility
unconfined burning
determination of Arrhenius constants
specific heat
differential scanning calorimetry
differential thermal and gravimetric analysis
thermal conductivity
accelerated rate calorimetric tests
self-heating tests
ignition temperature

Physical properties

coefficient of expansion
density
solubility
thermal/mechanical analysis
susceptibility to humidity

Performance

detonation velocity
detonation pressure
critical diameter
cylinder expansion test
shaped charge performance
fragmentation efficiency
plate dent output

Vulnerability

slow cook-off
fast cook-off
bullet impact
fragment sensitivity
shaped charge sensitivity
sympathetic detonation
effects of thermal cycling
effects of transportation, vibration, and drop

Environment and safety

toxicity
hazard classification
pollutant characteristics

Producibility

pressing/machining studies
castability studies
pilot lot production
capital investment costs
special safety requirements
waste product disposal
demil

Military requirements place a premium on maximum energy explosives having very long shelf lives, capable of being used in a wide variety of environment conditions, and often subject to extraordinary stresses of handling and use. By contrast, design of explosives for industrial use emphasizes minimum cost and moderate energy formulations that may be tailored for specific uses. These latter do not require the length of shelf life imposed on military material (58–69).

Computer codes are used for the calculational procedures which provide highly detailed data, eg, the Ruby code (70). Rapid, short-form methods yielding very good first approximations, such as the Kamlet equations, are also available (71–74). Both modeling approaches show good agreement with experimental data obtained in measures of performance. A comparison of calculated and experimental explosive detonation velocities is shown in Table 5.

Table 5. Calculated and Experimental Detonation Pressures and Rates of Explosives

Explosive	Density, g/mL	P_{Kamlet},[a] MPa[b]	P_{ruby},[c] MPa[b]	D_{calc},[a] mm/μs	D_{exp}, mm/μs
HMX	1.90	38,200	39,100	9.117	9.160
PETN	1.78	33,400	32,100	8.600	8.695
tetryl	1.70	24,700	25,200	7.680	7.560
TNT	1.65	20,600	21,400	6.959	6.950
DATB	1.79	26,100	28,200	7.685	7.520
RDX	1.80	34,300	34,400	8.780	8.754

[a]Kamlet equations used (73).
[b]To convert MPa to kilobars, multiply by 0.01.
[c]Ruby computer code used (70).

Primary Explosives

Explosives are commonly categorized as primary, secondary, or high explosives. Primary or initiator explosives are the most sensitive to heat, friction, impact, shock, and electrostatic energy. These have been studied in considerable detail because of the almost unique capability, even when present in small quantities, to rapidly transform a low energy stimulus into a high intensity shock wave.

Primary explosives are used to initiate the next element in an explosive chain which consists of explosive charges of increasing mass and decreasing sensitivity. They are arranged in sequence to amplify the input stimulus to an output level of sufficient intensity to maximize the probability of initiating the main charge. Overall energy intensification is about 10 million to one. Primary explosives are used in military detonators, commercial blasting caps, and in stab and percussion electrical primers. Primary explosives may be initiated electrically, by penetration of the element using a firing pin (stab detonator), or by the shock from an exploding wire. The explosives used in a detonator blasting cap are pressed into a metallic capsule, and are designed so that the transformation from burning to a detonation shock wave can occur within small (6 mm) diameters and lengths. A typical stab detonator consists of a primary charge that is readily ignited by friction or low velocity impact, an intermediate charge in which the transition from burning to detonation occurs, and a base charge that amplifies the energy output of the intermediate charge and assures initiation of the next element in the explosive series. The properties of commonly used primary explosives are shown in Table 6. The characteristics of other materials as initiators have also been studied (75–81).

Primary explosives have lower detonation rates and produce less energy than secondary explosives. They must be powders having good flow transfer and pressing characteristics to permit high speed automatic loading of detonators. Their high sensitivity to electrostatic energy requires special precautions in the detonator loading plants including atmospheric humidification, grounding of all equipment and even the operators at times, and antistatic clothing and materials. Because of this sensitivity, initiator materials are generally loaded into detonators in the plant where they are made. When transportation is necessary, initiator explosives are shipped wet with water or water and alcohol, suspended in rubberized cloth bags, and well cushioned with sawdust or other inert media to prevent frictional initiation. The manufacture of detonators is accompanied by all necessary precautions to prevent the accumulation of substantial masses of explosive, dusting, and accidental initiation by static electricity or friction. However, detonators are safe to use upon proper handling once the initiating explosives have been pressed into the metal capsule (82–85).

Some primary explosives are also used in mixtures with other ingredients in nondetonating stab and percussion primers to accomplish mechanical work or to initiate burning actions rather than detonation in pyrotechnic and propellant systems. The hot gases produced may be used to expand bellows, drive pins, and perform other types of controlled work or they may be used to initiate the igniter for a propellant charge. These primers are very small (eg, 4 mm dia by 2.5 mm long) elements that contain mixtures of an explosive ingredient and fuels and oxidizers to increase the amount of heat and gas evolved. Sometimes additional compounds and abrasives (qv) are incorporated to increase the sensitivity to mechanical action. A typical composition is NOL 130 which consists of 15 wt % antimony sulfide, 20 wt % lead azide, 40 wt % basic lead styphnate, 20 wt % barium nitrate [*10022-31-8*], $Ba(NO_3)_2$, and 5 wt % tetrazene.

Only relatively few compounds can act as primary explosives and still meet the restrictive military and industrial requirements for reliability, ease of manufacture, low cost, compatibility, and long-term storage stability under adverse environmental conditions. Most initiator explosives are dense, metalloorganic compounds. In the United States, the most commonly used explosives for detonators include lead azide, PETN, and HMX. 2,4,6-Triamino-1,3,5-trinitrobenzene (TATB) is also used in electric detonators specially designed for use where stability at elevated temperatures is essential.

Mercury Fulminate. Mercury fulminate [*628-86-4*], $Hg(CNO)_2$, slowly decomposes when stored at elevated temperature. Although nonhygroscopic, it reacts with metals in the presence of water, liberating mercury and forming metallic fulminates. After 30 months storage at 50°C, its purity is reduced from 99.75% to 90%, and its initiating properties are greatly impaired. At a density of 3.6 g/mL, it has a calculated detonation temperature of 6900 K, a detonation pressure of 22 GPa (3.2×10^6 psi) and the following detonation products in mol/kg: 0.4 CO, 3.3 CO_2, 3.5 N_2, 3.5 Hg, and 3.3 C (86,87).

Lead Azide. The azides belong to a class of very few useful explosive compounds that do not contain oxygen. Lead azide is the primary explosive used in military detonators in the United States, and has been intensively studied (see

Table 6. Properties of Primary Explosives

Property	Mercury fulminate	Lead azide	Silver azide	Normal lead styphnate	DDNP	Tetrazene
molecular weight	285	291	150	468	210	188
color	gray	white	white	tan	yellow	light yellow
crystal density, g/cm^3	4.43	4.38	5.1	3.10	1.63	1.7
crystal form	orthorhombic	orthorhombic monoclinic		cubic	tabular	
melting point, °C	160[a]		252	[a]	157	140–160[a]
heat of formation, kJ/g[b,c]	−0.925	−1.45	−2.07	17.9	4.00	1.13
heat of combustion, kJ/g[b,c]	3.93	2.64	4.34	5.24	13.58	
heat of detonation, kJ/g[b,c]	1.79	1.54	1.90	1.91	3.43	2.75
gas volume, cm^3/g at STP	316	308		368	876	
activation energy, kJ/mol[b]	29.8	172	146	230	230	
collision constant, log_{10}/s	10.8	14.0				
detonation rate, km/s	5.4	5.10	6.8	5.2	6.9	
density, g/cm^3	4.2	4.0	5.1	2.9	1.60	
specific heat, J/(g·K)[b]	0.50	0.46	0.50	0.67		
thermal conductivity, W/(m·K) $\times 10^{-4}$	1837	1256	837			
vacuum stability at 100°C, mL/gas per 40 h at STP	[a]	<1	<1	<1	<1	>5
weight loss at 100°C, %	<1	<1	<1	<1	<5	<5
explosion temperature at 5 s, °C	190–260	345	290	265–280	195	160
effect of prolonged storage	detonates at 80°C	stable	stable	stable	stable	stable
relative impact test value, % TNT	5	11	18	8	15	5
friction pendulum	reacts	reacts	reacts			
static discharge max energy for nonignition, J[b]	0.07	0.01	0.007	0.001	0.25	0.036
relative energy output, % TNT						
lead block	50	40	45	40	110	50
sand test	45	40		25	105	50

[a]Material explodes. [b]To convert J to cal, divide by 4.184. [c]Based on liquid H_2O formed.

also LEAD COMPOUNDS). However, lead azide is being phased out as an ignition compound in commercial detonators by substances such as diazodinitrophenol (DDNP) or PETN-based mixtures because of health concerns over the lead content in the fumes and the explosion risks and environmental impact of the manufacturing process.

Lead azide is stable at ambient and elevated temperatures, and has good flow characteristics. It has been stored for 25 months at 50°C without change in purity or performance, and is compatible with most explosives and priming mixture ingredients. Because lead azide is less sensitive to ignition than mercury fulminate, a more readily ignitable material such as lead styphnate or NOL 130 is often used as a cover charge to ensure initiation. The crystalline forms of the azide are sensitive to impact, and become even more sensitive if small amounts of foreign matter are present. Lead azide builds up to detonation velocity extremely rapidly on ignition. Even single crystals detonate when ignited. Although it has been reported that large crystals are supersensitive to shock, impact, and friction, little evidence supports this contention. Of the four polymorphic forms isolated (alpha to delta), the alpha is the most common. The detonation velocity of lead azide is relatively independent of its confinement in steel, aluminum, or brass, ranging from 4.52 km/s in steel to 4.85 km/s in brass sleeves at a density of 3.22 g/mL. Its calculated detonation temperature and pressure at a density of 4.0 g/mL are 5600 K and 25 GPa (3.6×10^6 psi). Its principal detonation products at the same density are 10.3 mol/kg N_2 and 3.4 mol/kg Pb.

Lead azide tends to hydrolyze at high humidities or in the presence of materials evolving moisture. The hydrazoic acid formed reacts with copper and its alloys to produce the sensitive cupric azide [*14215-30-6*], $Cu(N_3)_2$. Appropriate protection must be provided by hermetic sealing and the use of noncopper or coated-copper metal.

Lead azide is not readily dead-pressed, ie, pressed to a point where it can no longer be initiated. However, this condition is somewhat dependent on the output of the mixture used to ignite the lead azide and the degree of confinement of the system. Because lead azide is a nonconductor, it may be mixed with flaked graphite to form a conductive mix for use in low energy electric detonators. A number of different types of lead azide have been prepared to improve its handling characteristics and performance and to decrease sensitivity. In addition to the dextrinated lead azide commonly used in the United States, service lead azide, which contains a minimum of 97% lead azide and no protective colloid, is used in the United Kingdom. Other varieties include colloidal lead azide (3–4 μm), poly(vinyl alcohol)-coated lead azide, and British RD 1333 and RD 1343 lead azide which is precipitated in the presence of carboxymethyl cellulose (88–92).

Manufacture. Lead azide is typically made from sodium azide [*26628-22-8*] in small (eg, 5 kg) batches buffered by the reaction solutions of lead nitrate or lead acetate:

$$Pb(NO_3)_2 + 2\,NaN_3 \rightarrow 2\,NaNO_3 + Pb(N_3)_2$$

Sodium azide is insensitive but highly toxic. Contact must be avoided with acid, from which the dangerous hydrazoic acid forms, and copper, lead, cadmium, silver, mercury, or their alloys, from which sensitive azides may be formed. Nucleating agents, such as poly(vinyl alcohol) (PVA), sodium carboxymethylcellulose (CMC), or dextrin, may be added during precipitation to produce free-flowing crystals or rounded agglomerates required for the large-scale, automatic loading of detonators. The presence of hydrophilic polymeric substances also tends to eliminate the small possibility of spontaneous explosions occurring during the precipitation process. Wetting agents may also be added.

All phases of the manufacturing process are conducted by remote control in stainless steel vessels using either distilled or demineralized water and filtered solutions. The overall precipitation time is about 60 minutes. In the manufacture of dextrinated lead azide, lead nitrate stock solution is prepared by dissolving lead nitrate, dextrin, and sodium hydroxide in water at pH 4.6–4.8. The solution is cooled, filtered, pumped to a storage tank, and allowed to settle for eight hours or longer. A sodium azide stock solution is similarly prepared. The precipitation vessel is a precisely made, open-topped, round-bottom, double-walled, polished stainless steel tilting pot equipped with an agitator, feed tubes, and a water spray ring. The lead nitrate solution at 60°C is transferred to the precipitation vessel from a measuring tank, and the sodium azide solution is added at a rate of about 2 L/min while maintaining the 60°C temperature. Lead azide precipitates as free-flowing, fine white agglomerates. After settling, the mother liquor is decanted through a filter, collected, and neutralized using 30% sodium nitrate and then 30% nitric acid, or using ceric ammonium nitrate to decompose the azide ion. Excess acid is neutralized using soda ash. Any soluble lead is precipitated as the insoluble carbonate. The lead azide precipitate is washed repeatedly with water, vacuum filtered, and dried. Lead azide made without dextrin (RD 1333) usually contains more than 99% azide. When made with dextrin it contains about 92% lead azide, 4–5% lead hydroxide, 3% dextrin, and other impurities. Lead azide must be free of needle-shaped crystals longer than 0.1 mm. Dextrinated lead azide is less sensitive but is somewhat more hygroscopic, less dense, and less efficient as an initiator than the 99% product.

A low incidence of explosions has been reported when precipitation is effected without a nucleating agent and during the screening process. Precautions must be taken during detonator loading to prevent dusting and to maintain a scrupulously clean operation.

Silver Azide. Silver azide [*13863-88-2*], AgN_3, has received attention primarily in the United Kingdom as a potential replacement for lead azide because it may be used in smaller quantities as an initiator and, therefore, offers the possibility of miniaturization of fuse components. Silver azide requires somewhat less energy for initiation than lead azide and fires with a shorter time delay. It is less apt to hydrolyze and is more sensitive to heat. It is incompatible with sulfur compounds, with tetrazene, and with some metals, including copper. Silver azide is made in the same manner as lead azide, except that silver nitrate is used (91,93–96).

Lead Styphnate. Lead styphnate or lead 2,4,6-trinitroresorcinate [*15245-44-0*], $C_6H_3N_3O_8Pb$, is one of a number of compounds used in priming compositions to start the ignition-to-detonation process in the explosive sequence. Its sensitivity to stab, flame, heat, or impact ensures ignition of the true primary explosive. Lead styphnate is stable and noncorrosive. It is used as the top charge in stab primers, as a spot charge in electric detonators, or as a component of top charge mixtures. The addition of graphite enhances its electrical conductivity in systems designed for electrical initiation. Dry lead styphnate is the most sensitive of the primary explosives to electrostatic discharge. However, it is much less sensitive after being pressed into a detonator capsule.

Lead styphnate monohydrate is precipitated as the basic salt from a mixture of solutions of magnesium styphnate and lead acetate followed by conversion to the normal form by acidification using dilute nitric acid (97–99).

Diazodinitrophenol. DDNP is an orange-yellow compound made by the diazotization of picramic acid [*96-91-3*], $NH_2(NO_2)_2C_6H_2OH$, using sodium nitrite and hydrochloric acid, washing the product with ice water, and recrystallizing from hot acetone. It is almost nonhygroscopic, sensitive to impact and friction, and somewhat less stable than lead azide. It is about equivalent to TNT in brisance, and thus is more effective for some purposes than lead azide. DDNP is used as an initiator in commercial blasting caps to alleviate the lead produced in the atmosphere when large numbers of lead azide caps are detonated in large-scale blasting operations (100,101).

Tetrazene. Tetrazene is a pale yellow crystalline explosive made by adding sodium nitrite to a solution of 1-aminoguanidine hydrogen carbonate in dilute acetic acid at 30°C. The crystals are water-washed and dried at ambient. The compound is stable up to about 75°C, ignites readily, and has a relatively high explosion energy. Tetrazene is used as a component in priming compositions (102,103).

Secondary Explosives

Aliphatic Nitrate Esters. Aliphatic nitrate esters, such as glycerol trinitrate (nitroglycerin), ethylene glycol dinitrate (nitroglycol), cellulose nitrate (nitrocellulose), and pentaerythritol tetranitrate (PETN), are among the most powerful explosives available. These nitrate esters are generally less stable than aromatic nitro compounds or nitramines because the former tend to hydrolyze autocatalytically to form nitric and nitrous acids, which further accelerate decomposition. Liquid nitrates, ie, nitroglycerin, are usually less stable than crystalline compounds because of the higher energy state of the liquid phase. The properties of the more commonly used aliphatic nitrate esters are listed in Table 7.

Nitroglycerin. Nitroglycerin (NG), glyceryl trinitrate [*55-63-0*], $C_3H_5N_3O_9$, is primarily used as an explosive in dynamites and as a plasticizer for nitrocellulose in double- and multibase propellants. It is very sensitive to shock, impact, and friction, and is employed only when desensitized using other liquids or absorbent solids or when compounded with nitrocellulose. When desensitized with other liquids, such as triacetin [*102-76-1*] or dibutyl phthalate [*84-74-2*], it may be transported with care. NG is readily soluble in many organic solvents and acts as a solvent for many explosive ingredients. It is completely miscible with homologous nitrate esters such as ethylene and diethylene glycol dinitrates. Nitroglycerin is sufficiently soluble in water (0.18 g/100 g H_2O at 20°C) to pose a contamination problem in the disposal of process water produced during its manufacture or used in formulations.

Table 7. Properties of Explosive Aliphatic Nitrate Esters

Property	NC[a]	NG	EGDN	DEGN[b]	BTN[b]	TMETN[b]	PETN
molecular weight	$(286)_n$	227	152	196	241	255	316
color	white	yellow	colorless	colorless	yellow		white
density, g/cm^3	1.66	1.59	1.49	1.38	1.47	1.47	1.78
crystal form		rhombic					tetragonal
melting point, °C		13.2	−22.8	−11.3	−27	−3	143
hygroscopicity at 90% rh, 20°C, %	[c]	0.06	[c]		0.1	0.07	[c]
solubility in H_2O at 20°C, g/100 g	[c]	0.2	0.5	0.4	0.1	0.05	[c]
oxygen balance, % to CO_2	−29	3.5	0	−41	−17	−35	−10
viscosity at 20°C, mPa·s(=cP)		36	4.2	8	62	156	
heat of formation, kJ/g[d]	2.35	1.63	1.53	2.17	1.54	1.60	1.70
heat of combustion, kJ/g[d]	9.68	6.80	7.38	11.68	9.07	11.05	8.20
heat of detonation, kJ/g[d]	4.04	6.29	7.13	4.86	6.10	5.17	6.28
gas volume at STP, cm^3/g		715	740	880	840	855	784
activation energy, kJ/mol[d]	205	169	149	176			197
collision constant, $\log_{10}/s$	21.0	17.1	14.3				19.8
applicable temp range, °C	90–135	75–105	140–170				160–225
detonation velocity, km/s	7.3	7.60	7.8–8.0	6.75		7.05	8.31
at density, g/cm^3	1.20	1.59	1.49	1.38		1.47	1.77
detonation pressure, GPa[e]	25.30						32.00
at density, g/cm^3[f]	1.59						1.77
detonation temp, K[f]	3470						3400
vacuum stability at 100°C, mL/g gas per 40 h at STP		explode	explode	<1	1–3	1–3	<1
weight loss at 100°C, %	>1	3–5			1–3	1–3	1
explosion temp at 5 s, °C	170	220(dec)		240	230(dec)	235	225
relative impact test value, % TNT	10	15		100		45	15
friction pendulum	explode	explode		explode			explode
relative energy output, % TNT							
lead block	130	180	205	150		140	175
ballistic mortar	125	140		125		135	145
sand test	105	125	135	100	105	105	135

[a]Characteristics vary with nitrogen content; values given are for 13.4% nitrogen. [b]See Table 8. [c]Value is negligible.
[d]To convert J to cal, divide by 4.184. [e]To convert GPa to kilobars, multiply by 10. [f]Value is calculated.

The sensitivity of nitroglycerin decreases with decreasing temperature. Solid nitroglycerin is less sensitive to high pressure shock than the liquid, although some tests indicate it to be more sensitive to friction because of intercrystalline contact. It has a relatively high (13.2°C) freezing point possibly accounting for the reported increase in accident rate in loading pure NG dynamites and the significant decrease in mechanical properties of nitroglycerin–nitrocellulose propellants at very low temperatures, unless freezing point depressants are added. Ethylene glycol dinitrate (freezing point −22.8°C) is added to nitroglycerin-bearing dynamites for this purpose.

Pure nitroglycerin is a stable liquid at temperate conditions. It decomposes above 60°C to form nitric oxides which in turn catalyze further decomposition. Moisture increases the rate of decomposition under these conditions. Double- and multibase propellants containing nitroglycerin have substantially shorter stability lives at 65 and 80°C than do single-base propellants. The decomposition of nitroglycerin proceeds as

$$2\ C_3H_5(NO_3)_3 \rightarrow 3\ CO + 2\ CO_2 + 6\ NO + 4\ H_2O + H_2CO$$

A rate equation applicable to the liquid phase is $k = 10^{20.2}e^{-46,000/RT}$.

Unconfined nitroglycerin burns without exploding if present in thin layers and in small quantities, but detonates if confined. Nitroglycerin, like other liquid explosives when aerated or containing microbubbles, is especially sensitive to shock. Nitroglycerin exhibits a low velocity propagation regime and a detonation rate of about 2000 m/s compared to its high energy detonation rate of about 7800 m/s. The low velocity detonation is associated with the propagation of a precursor shock wave in the container wall, which causes cavitation bubbles to form in the liquid. Low velocity detonation is uncommon in high explosives, and is found primarily in nitroglycerin and its homologues, in some gelatin dynamites, and in PETN, TNT, RDX, and HMX in powder form. Absorption of nitroglycerin through the skin or by vapor inhalation expands blood vessels and may result in severe headaches. The accumulation of condensed pockets of nitroglycerin, as may occur in some propellant air-drying systems, should be prevented because of potential hazard (104–110).

Manufacture. Although nitroglycerin has traditionally been made by the batch process (106,107), the hazard of handling large quantities led to the development and wide use of continuous processes. A number of continuous and semicontinuous processes are in use including the Biazzi process, the Nitro Nobel injector process, and the Schmlid-Meissner process. In general, the methods used to make nitroglycerin are applicable to the manufacture of other liquid aliphatic polyhydroxy alcohol nitrates.

Nitroglycerin is made from pure glycerol [*56-81-5*] to ensure stability of the final product. Mixed acid (90% nitric acid and 25–30% oleum) is used in both the batch and continuous processes. The theoretical yield ratio of nitroglycerin from glycerol (qv) is 2.467:1. The actual yield using concentrated acid is about 2.36:1 and slightly higher in the continuous process than in the batch process. Affecting yields are solubility in spent acid and wash waters, acid concentration and composition, nitric acid–glycerol ratio, temperature stirring efficiency, and the efficiency of separating the nitroglycerin from spent acid and wash fluids.

In the Biazzi continuous process, the equipment consists of a nitrator and acid separator with an associated drowning tank, soda washers with a separator, and water washers each with an additional separator. Emulsification of the nitroglycerin increases safety because emulsions containing three or more parts of water to nitroglycerin are relatively insensitive to initiation. The stainless steel equipment is precision engineered and has highly polished inner surfaces. Operations are totally automatic, include numerous fail-safe features, controls, and signaling devices, and are monitored by a closed-television circuit from a remote control bunker. The nitrator is a small cylindrical vessel equipped with a multiple bank of spiral, closely packed cooling coils and a high speed turbostirrer. It has an outlet at the bottom for emergency drowning of the contents, an inlet for entry of displacement acid, and an emulsion overflow pipe attached ca 10 cm below the top. The nitrator cover has port holes for the impeller shaft, the glycerol nozzle, the mixed-acid fed pipe, and a fume pipe. A 250-L capacity nitrator produces ca 1700 kg/h of nitroglycerin.

The mixed acid and glycerol are metered into the nitrator, quickly submerged, emulsified, and forced up the nitrator past the cooling coils. Part of the nitroglycerin overflows into the separator, and part returns to the vortex in the nitrator fluid. The temperature is kept at 10–20°C. The emulsified mixture of nitroglycerin and spent acid enters the acid separator where a slight rotating action imparted to the upper layer of the liquid breaks the emulsion and prevents overheating. The spent acid flows continuously through an overflow from the bottom of the acid separator to its storage tank. The nitroglycerin may also be separated in a specially designed centrifuge. The remaining acid is neutralized, and the product washed with water until it passes the specification test for stability. The Biazzi process is also used for the manufacture of other nitrate esters such as triethylene glycol dinitrate, butanetriol trinitrate, and trimethylolethane trinitrate (metriol trinitrate).

In the Nitro Nobel injector process, an injector is used to mix the glycerol or other polyol with precooled nitration acid. The flow of the acid through the injector creates a vacuum so that a metered ratio of the polyhydroxy alcohol is sucked in through the throat, mixing the two quickly and thoroughly to form an emulsion. Increased safety is achieved because the quantity of alcohol being nitrated is automatically dependent on the quantity of acid entering the injector. Glycerol at about 48°C is sucked into the injector where it reacts almost instantaneously with the acid. The emulsion produced is rapidly cooled to 15°C, and flows by gravity to a centrifuge or gravity separator where the nitroglycerin is separated from the spent acid on a continuous basis. The spent acid can be recycled or denitrated. The acidic-containing nitroglycerin is emulsified immediately with a water jet to form a nonexplosive mixture, neutralized with sodium carbonate, and washed. The stabilized nitroglycerin is emulsified by passage through an injector to form a nonexplosive water emulsion for safe temporary storage (106,107,111–114).

Environmental and Safety Considerations. Soda ash and water washes are the primary sources of pollution from this process. These wastewaters may contain nitroglycerin, nitrates, and sulfates, and vary in pH from acid for the first wash after nitration to alkaline for the washes with soda ash. The process water is first sent through catch basins to remove nonsoluble nitroglycerin. The overflow may be discharged without further treatment or sent to percolation–evaporation ponds or earthen sumps. Acidic wastewater may be neutralized by passage through crushed lime beds. Caustic soda and sodium sulfide may also be used to decompose and dissolve nitroglycerin in the effluent, followed by use of an activated sludge system as the secondary treatment. Other methods investigated include biodegradation of the nitroglycerin, reverse osmosis, absorption by polymeric resins, treatment with lime and caustic, and oxidation with ozone and permanganate.

Safety has been greatly increased by use of the continuous nitration processes. The quantity of nitroglycerin in process at any one time is greatly reduced, and emulsification of nitroglycerin with water decreases the likelihood of detonation. Process sensors (qv) and automatic controls minimize the likelihood of runaway reactions. Detonation traps may be used to decrease the likelihood of propagation of an accidental initiation; eg, a tank of water into which the nitrated product flows and settles on the bottom.

Other Glycol Nitrates. Other liquid nitrates have been used as explosive plasticizers for nitrocellulose (Table 8). These may be made by mixed-acid nitration using procedures similar to those used for nitroglycerin.

The two most commonly used compounds are ethylene glycol dinitrate (nitroglycol) and diethylene glycol dinitrate. Their physical properties are listed in Table 7. Nitroglycol is an excellent solvent for low grade nitrocellulose, is comparable to nitroglycerin in explosive energy, and is less sensitive and more stable. It can detonate at high and low velocities (8000 and 1000–3000 m/s). The relatively high volatility of nitroglycol precludes its use in military propellants, although it is an excellent lower cost substitute for nitroglycerin in dynamite. It also appears to be more toxic than nitroglycerin, possibly because of its higher vapor pressure. The relative vapor pressures of various nitric esters at 15–55°C are listed in Table 9.

Nitroglycol may be made by nitration of ethylene glycol [*107-21-1*] with mixed acid with a yield of ca 93%. The demand for both NG and nitroglycol has been greatly decreased (115,116).

Table 8. Explosive Plasticizers

Compound	CAS Registry Number	Code	Use
ethylene glycol dinitrate	[*628-96-6*]	EGDN	as a freezing point depressant in low temperature dynamites
diethylene glycol dinitrate	[*693-21-0*]	DEGDN	propellant, coolant
triethylene glycol dinitrate	[*111-22-8*]	TEGDN	propellant, coolant
2-methyl-2-[(nitrooxy)methyl]-1,3-propanediol dinitrate ester	[*3032-55-1*]	TMETN	propellant, coolant
butanetriol trinitrate	[*41407-09-4*]	BTN	propellant, coolant
1,3-propanediol dinitrate	[*3457-90-7*]	PDN	propellant, coolant

Table 9. Vapor Pressures of Nitrate Esters, Pa[a]

Temperature, °C	Glycerol trinitrate	Ethylene glycol dinitrate	Diethylene glycol dinitrate	1,2-Propylene glycol dinitrate	1,3-Propanediol dinitrate
15	0.17	3.09	0.30	5.14	1.54
25	0.24	9.39	0.79	13.1	4.36
35	0.61	29.1	1.97	33.7	8.26
45	1.72	59.7	3.78	65.9	19.5
55	4.77	128	11.54	132.3	42.9

[a]To convert Pa to mm Hg, multiply by 7.50.

Diethylene glycol dinitrate (DEGDN) is the most widely used explosive plasticizer, other than nitroglycerin, in the formulation of military propellants for gun use. It is a better plasticizer for nitrocellulose than nitroglycerin. Because of its lower calorific value, it has been used to replace nitroglycerin to obtain cooler propellants, thereby decreasing erosion in gun tubes and increasing tube life. It also decreases muzzle flash and minimizes the need for flash reducers in propellant charges. It is employed in some double-base cast rocket propellants because of its superior physical properties in case-bonded motors. Although DEGDN is used in European propellants, it has not been generally used in U.S. gun propellants because its greater volatility poses problems during prolonged high temperature storage. DEGDN is made similarly to nitroglycerin, having a yield of about 85%. It is superior to nitroglycerin in terms of stability and handling safety, freezes at a lower temperature than nitroglycerin (−11.3 vs 13.2°C), and has a high (6760 m/s) and low (1800–2300 m/s) stable detonation rate. It does not appear to have harmful physiological effects (105,106,117).

Triethylene glycol dinitrate (TEGDN) is an explosive plasticizer of low sensitivity that has been used in some nitrocellulose-base propellant compositions, often in combination with metriol trinitrate. Butanetriol trinitrate has been used occasionally as an explosive plasticizer coolant in propellants. Its physical properties are listed in Table 7.

Trimethylolethane trinitrate (metriol trinitrate) is not satisfactory as a plasticizer for nitrocellulose, and must be used with other plasticizers such as metriol triacetate. Mixtures with nitroglycerin tend to improve the mechanical properties of double-base cast propellants at high and low temperatures. Metriol trinitrate has also been used in combination with triethylene glycol dinitrate as a plasticizer for nitrocellulose. Its physical properties are listed in Table 7 (118–122).

Nitrocellulose. The exceptional properties of nitrocellulose (NC), also called cellulose nitrate, are derived from the polymeric and fibrous structure of the cellulose from which it is made. Nitrocellulose, which provides mechanical strength as well as readily available energy to gun and rocket propellants, is manufactured by nitration of cellulose with mixed acid. The primary source of cellulose for this purpose is wood (qv) pulp (qv); short cotton (qv) fibers known as linters are also used. Cellulose molecules may contain up to 3500 glucose anhydride units, whereas cotton linters and wood pulp, which have been chemically treated and partially degraded, have from 900–1300 units. The polymerization of nitrocellulose may decrease to 400–700 units because of the additional molecular breakdown that occurs during the nitration process. The properties of nitrocellulose are given in Tables 7 and 10 (123–126).

Nitration of cellulose may result in the addition of from one to three nitrate groups per glucose anhydride unit. The following compounds and their nitrogen content have been postulated:

Compound	Formula	Nitrogen, %
cellulose trinitrate	$C_6H_7O_2(ONO_2)_3$	14.15
cellulose dinitrate	$C_6H_8O_3(ONO_2)_2$	11.11
cellulose mononitrate	$C_6H_9O_4(ONO_2)$	6.76

Neither the mononitrate nor the dinitrate have been isolated as distinct compounds. The trinitrate cannot be prepared with mixed acid, but is made from nitric anhydride or a mixture of phosphoric and nitric acids. The nitrogen content of nitrocellulose prepared using mixed acid does not exceed 13.4–13.5%. Conventionally made nitrocellulose retains the fibrous structure of cellulose, although x-ray diffraction studies show a crystalline structure for the higher nitrogen grades. It may also be made in pelletized form by precipitation from solution in an organic solvent, eg, ethyl acetate. The nitrogen content of nitrocellulose is most important in defining the significant properties of nitrocellulose and the propellants made

Table 10. Explosive and Thermochemical Characteristics of Nitrocellulose[a]

	Nitrogen, %		
Characteristic	12.60	13.15	14.0
brisance, sand test, g	45	48	52
heat of combustion, kJ/g[b]	10.1	9.81	9.36
heat of formation, kJ/g[b]	2.58	2.41	2.14
heat of explosion, kJ/g[b]			
H_2O liquid	3.91	4.25	4.77
H_2O vapor	3.58	3.91	4.43
gas vol, cm^3/g at STP			
H_2O liquid	744	722	687
H_2O gas	919	893	853
approximate gas composition, H_2O liquid, %			
CO_2	21	24	
CO	46	44	
H_2	19	17	
N_2	14	15	
estimated explosion temperature, K	3060	3292	
impetus, J/g[b]	998	1074	
ratio of specific heats	1.23	1.225	

[a]Ref. 126.

[b]To convert J to cal, divide by 4.184.

from it, including energy content and mechanical characteristics. Nitrocellulose used in propellants is most likely to have a nitrogen content of 12.6 or 13.15%, although lower grades (12.0 and 12.2% *N*) have also been used. Blasting gelatin uses 11–12% nitrogen nitrocellulose, whereas the nitrocellulose in commercial products may vary from 8 to 11.5% nitrogen. It is impossible to manufacture nitrocellulose to an exact nitrogen content. Required nitrocelluloses are produced by careful blending.

Nitrocellulose is among the least stable of common explosives. At 125°C it decomposes autocatalytically to CO, CO_2, H_2O, N_2, and NO, primarily as a result of hydrolysis of the ester and intermolecular oxidation of the anhydroglucose rings. At 50°C the rate of decomposition of purified nitrocellulose is about 4.5×10^{-6} %/h, increasing by a factor of about 3.5 for each 10°C rise in temperature. Many values have been reported for the activation energy, E, and Arrhenius frequency factor, Z, of nitrocellulose. Typical values for E and Z are 205 kJ/mol (49 kcal/mol) and 10.21 s^{-1}, respectively. The addition of small percentages of mildly alkaline compounds such as diphenylamine or diphenyldiethylurea greatly increases the stability by neutralizing oxides of nitrogen produced during decomposition. Compatibility with other substances must be well established before it is combined with new materials.

Dry nitrocellulose, which burns rapidly and furiously, may detonate if present in large quantities or if confined. Nitrocellulose is a dangerous material to handle in the dry state because of sensitivity to friction, static electricity, impact, and heat. Nitrocellulose is always shipped wet with water or alcohol. The higher the nitrogen content the more sensitive it tends to be. Even nitrocellulose having 40% water detonates if confined and sufficiently activated. All large-scale processes use nitric–sulfuric acid mixtures for nitration (127–132).

Manufacture. The batch nitration processes for nitrocellulose have included the pot process, the centrifugal process, the Thompson displacement process, and the mechanical dipper process. Semicontinuous nitration processes are also widely used for military and industrial grades. Cotton linters or wood pulp are nitrated using mixed acid followed by treatment with hot acidified water, pulping, neutralization, and washing. The finished product is blended for uniformity to a required nitrogen content. The controlling factors in the nitration process are the rates of diffusion of the acid into the fibers and of water out of the fibers, the composition of mixed acid, and the temperature (see CELLULOSE ESTERS, INORGANIC ESTERS).

In the mechanical dipper batch process, all raw materials are moved to the top floor of the plant by pump or conveyor, and then proceed by gravity through the processing operations. A battery of mechanically stirred nitrators (dipping pots) are located on a floor above a bottom discharge centrifuge into which the nitrocellulose is dropped after completion of nitration. The acid–cellulose ratio may vary from 20:1 to 50:1, depending on the type of cellulose used and the nitrogen content of the product. The composition of the mixed acid is closely controlled, and depends on the grade of nitrocellulose to be prepared and whether the starting material is cotton or wood pulp.

The linters or wood pulp or mixtures of the two are passed through picking rolls to form a fluffy mass of material and dried at 80–100°C to less than 1% moisture content to minimize dilution of acid and the possibility of fires in the nitrator. A measured volume of mixed acid corresponding to exactly one dipping charge is drawn into a measuring tank and into the stainless steel nitrator where it is stirred at high speed. The weighed cellulose is rapidly transferred by an operator to the pot where it is immediately drawn below the surface of the acid. The nitration of cellulose occurs rapidly at first, and then slows down as the maximum nitrogen content is approached in 15–20 minutes. The contents of the pot are then discharged by gravity into the centrifuge where the nitrocellulose is separated from the spent nitrating acid. The nitrators are staggered so that nitrations are carried out in sequence by a small team of workers moving from one to the other as nitrations are completed. Eight to twelve centrifuges may be used for a line of nitrators.

A filter cake from the wringer is washed to remove absorbed acid, transferred to a slurry tank of water, and quickly submerged, after which the nitrocellulose is pumped to the stabilization operation as a diluted water slurry. Exhaust systems are installed to protect personnel and equipment from acid fumes, and water sprays and cyclone separators are used for acid fume recovery before venting to the air.

The procedures for purification are essentially the same for both the batch and continuous processes. The nitrocellulose is completely submerged in water and the contents brought to about 98°C using steam; the acid boil is continued for 40–60 h for pyrocellulose (12.6 wt % *N*) to 60–96 h for guncotton (13.15% *N*). This is followed by several washings and neutral boils, pulping, and treatment with sodium carbonate. After poaching and further washings, the nitrocellulose slurry is screened and blended to the required uniform nitrogen content. The large quantities of water used in this process contain considerable amounts of acids and suspended solids (131–134).

The two procedures primarily used for continuous nitration are the semicontinuous method developed by Bofors-Nobel Chematur of Sweden and the continuous method of Hercules Powder Co. in the United States. The latter process, which uses a multiple cascade system for nitration and a continuous wringing operation, increases safety, reduces the personnel involved, provides a substantial reduction in pollutants, and increases the uniformity of the product. The cellulose is automatically and continuously fed into the first of a series of pots at a controlled rate. It falls into the slurry of acid and nitrocellulose and is submerged immediately by a turbine-type agitator. The acid is delivered to the pots from tanks at a rate controlled by appropriate instrumentation based on the desired acid to cellulose ratio. The slurry flows successively by gravity from the first to the last of the nitration vessels through under- and overflow weirs to ensure adequate retention time during nitration. The overflow from the last pot is fully nitrated cellulose.

The centrifuge is a horizontal basket designed to operate so that the cake formed on the screen is pushed as an increment from the loading end of the basket to the discharge end by a pusher plate operating on a timed cycle. On completion, the nitrocellulose cake is discharged into water in a slurry tub on a lower floor, and purified by conventional procedures.

Although the purification techniques used with the semicontinuous system are similar to that for the batch system, much less wastewater is produced because much less acid is retained on the nitrocellulose that is discharged from the centrifuge (135–137).

Pentaerythritol Tetranitrate. Specification-grade pentaerythritol tetranitrate (PETN) can be stored up to 18 months at 65°C without significant deterioration. However, many materials have been found to be incompatible with PETN and the presence of as little as 0.01% occluded acid or alkali greatly accelerates decomposition. The decomposition of PETN is autocatalytic with reported kinetic constants of $E = 196.6$ kJ/mol (47 kcal/mol) and $Z = 6.31 \times 10^{19}\ s^{-1}$. The decomposition products of PETN at 210°C in wt % are 47.7 NO, 21.0 CO, 11.8 NO_2, 9.5 N_2O, 6.3 CO_2, 2.0 H_2, and 1.6 N_2. The vapor pressure, P, of PETN in Pascals as a function of temperature is

$$\log_{10} P = 16.90 - 7380/K$$

The calorimetrically determined heat of detonation of the unconfined explosive is 6.1 kJ/g (1.47 kcal/g) and the calculated principal products of detonation (mol/mol PETN) are 3.89 CO_2, 4.00 H_2O, 2.00 N_2, 0.22 CO, and 0.90 C (solid). The detonation rate D in millimeters per microsecond as a function of density d in grams per cubic centimeter between 0.4–1.65 may be expressed as $D = 1.608 + 3.93\ d$. The thermal conductivity of PETN is 0.11 W/(m·K); coefficient of linear expansion, $8.3 \times 10^{-5}\ °C^{-1}$; compressive strength, about 10.3 MPa (1500 psi); modulus of elasticity, 6895 MPa (1×10^6 psi); and bulk modulus, 4826 MPa (7×10^5 psi). Additional physical properties are specific heat 1.13 J/(g°C) (0.27 cal/(g°C)); heat of fusion 318 J/g (76 cal/g); and heat of vaporization, 305 J/g (73 cal/g). Maximum electrostatic energy for noninitiation of PETN powder is about 0.2 J (0.05 cal). Other characteristics are listed in Table 7 (138–143).

Pentaerythritol may be nitrated by a batch process at 15.25°C using concentrated nitric acid in a stainless steel vessel equipped with an agitator and cooling coils to keep the reaction temperature at 15–25°C. The PETN is precipitated in a jacketed diluter by adding sufficient water to the solution to reduce the acid concentration to about 30%. The crystals are vacuum filtered and washed with water followed by washes with water containing a small amount of sodium carbonate and then cold water. The water-wet PETN is dissolved in acetone containing a small amount of sodium carbonate at 50°C and reprecipitated with water; the yield is about 95%. Impurities include pentaerythritol trinitrate, dipentaerythritol hexanitrate, and tripentaerythritol acetonitrate. Pentaerythritol tetranitrate is shipped wet in water–alcohol in packing similar to that used for primary explosives.

The Biazzi continuous process is also used. The reactants are continuously fed to a series of nitrators at 15–20°C followed by separation of the PETN, water washing, solution in acetone at 50°C, neutralization with gaseous ammonia, and precipitation by dilution with water. The overall yield is more than 95%. The acetone and the spent acid are readily recovered.

Pentaerythritol tetranitrate is a high energy explosive that is used as a pressed base charge in blasting caps and detonators, as the core explosive in commercial detonating cord, and as the main explosive ingredient in sheet explosives. It is also mixed in various proportions with TNT to form the less sensitive pentolites, eg, PETN 50/TNT 50. PETN is easily initiated, its responses are reproducible, and it is readily available (144–146).

Nitramines. The four most important nitramine explosives are cyclotrimethylenetrinitramine (RDX), cyclotetramethylenetetranitramine (HMX), nitroguanidine [*556-88-7*] (NQ), and 2,4,6-trinitrophenylmethylnitramine (tetryl). Tetryl has been increasingly replaced by HMX and RDX and is no longer used as a booster explosive in the United States. Both RDX and HMX are used as high energy explosives. They are also incorporated in high performance rocket propellants and propellants of reduced sensitivity to stimuli such as fragment impact. Nitroguanidine is employed almost exclusively in gun propellants. Properties of nitramines are listed in Table 11.

RDX and HMX. The properties of RDX and HMX are quite similar; HMX has a higher density and a higher detonation rate, yields more energy per unit volume, and has a higher melting point and higher explosion and cook-off temperatures. The vapor pressure, *P,* of RDX in Pascals at various temperatures may be calculated from $\log_{10} P = 16.26 - 6785/K$, heat of vaporization is 489 J/g (117 cal/g), and thermal conductivity is 0.29 W/(m·K).

Both RDX and HMX are stable, crystalline solids, somewhat less sensitive to impact than PETN. Both may be handled with no physiological effect if appropriate precautions are taken to assure cleanliness of operations. Both RDX and HMX detonate to form mostly gaseous, low molecular weight products and some intermediate formation of solid carbons. The calculated molar detonation products of RDX are 3.00 H_2O, 3.00 N_2, 1.49 CO_2, and 0.02 CO. RDX has been stored for as long as 10 months at 85°C without perceptible deterioration.

HMX, the highest density and highest energy solid explosive produced on a large scale, primarily for military use, exists in four polymorphic forms. The beta form is the least sensitive, most stable, and the type required for military use. The mole fraction products of detonation of HMX in a calorimetric bomb are 3.68 N_2, 3.18 H_2, 1.92 CO_2, 1.06 CO, 0.97 C, 0.395 NH_3, and 0.30 H_2.

Both RDX and HMX are substantially desensitized by mixing with TNT to form cyclotols (RDX) and octols (HMX) or by coating with waxes, synthetic polymers, and elastomeric binders. Most of the RDX made in the United States is converted to Composition B (60% RDX, 40% TNT, 1 part wax added). Composition A5 (RDX 98.5/stearic acid 1.5) and composition C4 (RDX91/nonexplosive plasticizer) account for the next largest uses. HMX is used as a propellant and in maximum-performance plastic bonded explosives such as PBX 9401 and PBX N5 and the octols (147–150).

Manufacture. The two most common processes for making RDX and HMX use hexamethylenetetramine (hexamine) as starting material. The Woolwich or direct nitrolysis process used in the United Kingdom proceeds according to:

$$\text{hexamine} + 6\,HNO_3 \rightarrow \text{RDX} + 3\,CO_2 + 2\,N_2 + 6\,H_2O$$

hexamine RDX

Table 11. Properties of Nitramine Explosives[a]

Property	NQ	EDNA	RDX	HMX	Tetryl
molecular weight	104	150	222	296	287
color	white	white	white	white	yellow
crystal density, g/cm^3	1.78	1.71	1.80	1.96 (beta)	1.73
crystal form	orthorhombic		orthorhombic	polymorphic	monoclinic
melting point, °C	245 (dec)	175 (dec)	204	286	129.5
hardness, Mohs'			2.5	2.3	<1
solubility in water at 20°C, g/100 g	0.4	negligible	0.005	negligible	0.75
oxygen balance, % to CO_2	−31	−32	−22	−22	−47
heat of formation, kJ/g[b]	0.950	0.561	0.277	0.253	0.1067
heat of combustion, kJ/g[b]	8.35	10.36	9.46	9.88	12.24
heat of detonation, kJ/g[b,c]	3.01	5.33	4.54	5.67	4.63
gas volume at STP, cm^3/g	1,077	908	780 (calc)	755 (calc)	760
activation energy, kJ/mol[b]	87.5		197	221	161
collision constant, $\log_{10}$/s	7.5	12.8	18.3	19.7	15.4
applicable temperature range, °C		>185	215–300	270–295	210–260
detonation velocity, km/s at	8.16	8.24	8.57	9.16	7.92
density, g/cm^3	1.72	1.66	1.80	1.90	1.73
detonation pressure, GPa[d] at	27.3	26.5	33.8	39.3	26.2
density, g/cm^3 (calc)	1.78	1.51	1.77	1.90	1.73
detonation temperature, K (calc)			2793	2365	2915
explosion temperature at 5 s, °C	275 (dec)	190 (dec)	260	335	260
effect of prolonged storage	negligible	negligible at 50°C	negligible	negligible	negligible at 65°C
relative impact test value, % TNT	200	70	35	35 (beta)	50
friction pendulum	no explosion	no explosion	explodes	explodes	explodes
relative gap test value, % TNT	390	80	40	40	50
performance, % TNT					
lead block	100	120	155	150	130
ballistic mortar	105	135	150	150	130
sand test	75	120	140	150	125
plate dent test	95	120	135	155	115
specific heat, J/(g·K)[b]	1.00		1.26	1.04	0.88

[a]All five compounds are nonhygroscopic at 90% rh at 20°C; vacuum stability at 100°C, mL/g gas/40 h, is less than 1%; weight loss at 100°C is less than 1%.
[b]To convert J to cal, divide by 4.184. [c]Per gram of H_2O. [d]To convert GPa to kilobars, multiply by 10.

The Bachmann process, used in the United States and in some European countries, is a simplification of a series of complex reactions. In this process, a solution of one part hexamine in 1.65 parts acetic acid, and a solution of 1.50 parts ammonium nitrate dissolved in 2.0 parts nitric acid and 5.20 parts acetic anhydride are used. The reaction may be summarized as:

$$C_6H_{12}N_4 + 4\ HNO_3 + 2\ NH_4NO_3 + 6\ (CH_3CO)_2O \rightarrow 2\ RDX + 12\ CH_3COOH$$

In the Bachmann process an 80–84% yield is obtained, ca 10% of which is cyclotetramethylenetetranitramine (HMX). The Woolwich process gives a 70–75% yield containing only a trace of HMX.

The RDX particle size distribution must be carefully controlled to produce castable slurries of RDX and TNT having acceptable viscosity. Several classes of RDX are produced to satisfy requirements for the various pressed and cast RDX-based compositions. A continuous process for medium-scale production of RDX has been developed by Biazzi based on the Woolwich process (79,151–154).

HMX is manufactured via a modification of the Bachmann process. The same starting materials are employed. The reaction temperature is lower, 44°C vs the 68°C for RDX, and the raw materials are mixed in a two-step process. The yield of HMX per mole of hexamine is about 55–60%, as compared to 80–85% for RDX (136,155–162).

Nitroguanidine. Nitroguanidine [*556-88-7*] (NQ), $CH_4N_4O_2$, has been used to some extent as an industrial explosive. It has not been used as a military explosive because of its relatively low energy content and difficulty of initiation. Although its detonation rate at comparable densities is almost as high as that of RDX, it has an anomalously low detonation energy and pressure, indicating a decomposition mechanism different from that of conventional explosives. NQ is approximately as effective as TNT. The high nitrogen content of the products of combustion and lower flame temperature per unit of energy accounts for the widespread use of NQ in multibase gun propellants to reduce barrel wear, muzzle blast, and flash. Very small particle size, spherical granules of nitroguanidine have also been experimentally tested as a component of low sensitivity explosives. Many guanidine derivatives have been prepared as possible substitutes for nitroguanidine in propellants including triaminoguanidine [*2203-24-9*], nitroaminoguanidine [*18264-75-0*], and trinitroethylguanidine [*8065-53-0*].

Nitroguanidine is stable and nonhygroscopic. It is produced for propellant incorporation in the alpha-crystalline form with a bulk density of about 0.2 g/cm^3. The crystals are needle-like and often hollow, and are about 5 μm in diameter and 15 μm long. Nitroguanidine appears to act as a stabilizer in nitrocellulose propellants by forming an addition compound with nitrogen oxides. NQ is less sensitive than TNT to impact, friction, and shock. Its brisance and blast characteristics are similar to that of TNT. Nitroguanidine has been studied as a source of explosive energy in systems where low detonation pressure and low detonation rates are required. The rate of detonation at infinite diameter, D_i in kilometers per second as a function of density d in g/cm^3 may be expressed by $D_i = 1.44 + 4.015\ d$. Nitroguanidine may be dead-pressed at densities greater than 1.63 g/cm^3. The estimated detonation products (wt %) are 38 N_2, 25 CO, 20 CO_2,

16 H_2O, and 1 H_2. Additional properties of nitroguanidine are presented in Table 11 (163–168).

Manufacture. Nitroguanidine may be made by several methods. In all the processes guanidine nitrate is the intermediate which is then dehydrated with sulfuric acid. When used in propellants, the average particle size of nitroguanidine has to be carefully controlled.

In the commonly used Welland process, calcium cyanamide, made from calcium carbonate, is converted to cyanamide by reaction with carbon dioxide and water. Dicyandiamide is fused with ammonium nitrate to form guanidine nitrate. Dehydration with 96% sulfuric acid gives nitroguanidine which is precipitated by dilution. In the aqueous fusion process, calcium cyanamide is fused with ammonium nitrate in the presence of some water. The calcium nitrate produced is removed by precipitation with ammonium carbonate or carbon dioxide. The filtrate contains the guanidine nitrate that is recovered by vacuum evaporation and converted to nitroguanidine. Both operations can be run on a continuous basis (see CYANAMIDES). In the Marquerol and Loriette process, nitroguanidine is obtained directly in about 90% yield from dicyandiamide by reaction with sulfuric acid to form guanidine sulfate followed by direct nitration with nitric acid (169–172).

Tetryl. 2,4,6-Trinitrophenylmethylnitramine (tetryl) was used in pressed form, mostly as a booster explosive and as a base charge in detonators and blasting caps because of its sensitivity to initiation by primary explosives and its relatively high energy content. Properties are presented in Table 11 (173). Batch and continuous processes for the production of tetryl have been developed. Tetryl is no longer used in the United States and has been replaced by RDX (174–178).

Nitroaromatics. The commonly used nitroaromatic explosives contain three NO_2 groups, generally in the 1, 3, and 5 positions. Aromatics are most often nitrated to the trinitro stage with mixed acid. Further nitration is difficult, and aromatics having four or more nitro groups attached to the ring tend to be relatively unstable. The most extensively used explosive is trinitrotoluene (TNT); however, hexanitrostilbene [*20062-22-0*] (HNS), $C_{14}H_6N_6O_{12}$; hexanitroazobenzene [*19159-68-3*] (HNAB), $C_{12}H_4N_8O_{12}$; and di- and triaminotrinitrobenzene (DATB and TATB) have also found important application because of low sensitivity to impact, shock, and friction, and their excellent stability at elevated temperatures. Ammonium picrate (AP) has been used in armor-piercing gun projectiles because of its insensitivity to impact and shock. The properties of the more commonly used nitroaromatics are in Table 12 (179).

Trinitrotoluene. α-2,4,6-Trinitrotoluene (TNT) is very stable and may be stored indefinitely at temperate conditions without deterioration. It does not show any evidence of significant decomposition even after having been cycled more than 50 times through the liquid–solid phases. The decomposition mechanism of TNT at elevated (200°C) temperatures is complex, producing at least 25 different compounds as well as large amounts of undefined polymeric material. Trinitrotoluene is nonhygroscopic and relatively insensitive to impact, friction, shock, and electrostatic energy. It has been fired in high acceleration gun projectiles with reported premature rates of less than one in one million. Its characteristics change only slightly at very low temperature. It is used in large quantities by itself as a sensitizer in industrial explosives and mixed with RDX as a cast military explosive.

Like most other nitroaromatics, TNT is toxic. Its solubility in water (0.013 g/100 g at 20°C) is high enough to destroy aquatic life. The solidification point of military-grade TNT is about 80.2°C, and that of pure TNT is 80.75°C. The viscosity of TNT is 8 mPa·s(=cP) at 99°C. The viscosity of TNT slurries with RDX and other explosives has been studied intensively because of its importance in the cast loading of projectiles. The enthalpy, ΔH, in J/mol from 25°C to the mp, and density, d, in g/mL from 83–120°C of TNT as functions of temperature, t, in °C, are represented by the following equations (180):

$$\Delta H = 0.045 - 0.246\,t + 4.205 \times 10^{-4}\,t^2$$

and

$$d = 1.5446 - 1.016 \times 10\,t$$

The following equations (181) may be used to calculate the maximum detonation rate, D, of pressed TNT as a function of density, d, from $d = 0.9 - 1.53$ g/mL

$$D = 1873 + 3187\,d$$

$$\text{for } d = 1.54 \text{ to } 1.64 \text{ g/mL}$$

$$D_\infty = 6762 + 3187\,(d - 1.534) - 2510\,(d - 1.534) + 115056\,(d - 1.542)^3$$

The principal detonation products per mole of heavily confined TNT as experimentally determined in a calorimetric bomb are 3.65 C, 1.98 CO, 1.60 H_2O, 1.32 N_2, 0.46 H_2, 0.16 NH_3, and 0.10 CH_4.

Bombs and projectiles are filled with steam-melted TNT by the casting process. Melted TNT also serves as the liquid carrier for RDX, HMX, aluminum, ammonium nitrate, and other solid ingredients to form a wide range of castable slurries. Because TNT expands about 10–12% on liquefaction and contracts on solidification, cracks and cavities tend to form in the cast TNT on cooling, unless special techniques are used to prevent this. These include slow-programmed cooling, hot probing, and temperature cycling following casting or compositional modifications to prevent crack and cavity formation. Undesirable exudation of isomers of TNT may occur during long-term storage at elevated temperature (182–188).

Manufacture. Trinitrotoluene is made by batch or continuous processes. Both processes, which are safe operations when conducted under proper conditions, are relatively low cost, in part because the products are only slightly soluble in the mixed nitration acids and are readily separable. The low solidification temperatures of the intermediate and final products permit pumping as fluids from one stage to the next at readily attainable temperatures. Toluene is batch nitrated in a three-stage operation by using increasing temperatures and mixed-acid concentrations to successively introduce nitro groups to form mononitrotoluene (MNT), dinitrotoluene (DNT), and trinitrotoluene. Numerous other compounds are also formed. Unless these impurities are removed, the TNT may be unstable

Table 12. Properties of Nitroaromatic Explosives

Properties	TNT	PA	AP	DATB	TATB	HNS	HNAB
molecular weight	227	229	246	243	258	450	452
color	beige	yellow	yellow	yellow	yellow	yellow	orange
density, g/cm^3	1.65	1.76	1.72	1.84	1.86	1.74	1.79
crystal form	orthorhombic	orthorhombic	rhombic	triclinicorthorhombic	triclinic	orthorhombic	
melting point, °C	80.75	122.5	271(dec)	286(dec)	448(dec)	415(dec)	215
hygroscopicity at 20°C, % at 90% rh	0.03	0.05	0.1	negligible	negligible	negligible	
solubility in H_2O at 20°C, g/100 g	0.01	1.2	1.1	negligible	negligible	negligible	
heat of formation, kJ/g[a,b]	0.293	0.938	1.56	0.41	0.54	−0.130	−0.536
heat of combustion, kJ/g[a]	15.02	11.17	12.09	12.25	12.23	0.17	
heat of detonation, kJ/g[a]	4.23	4.31	4.27	3.54	5.02	5.94	6.15
specific heat, C_p, J/(g·K)[a]	1.56 1.38	1.09		1.37	0.91	0.94	1.68
heat of sublimation, J/g[a]	447			577	652	397	
thermal conductivity, W/(m·K)	4.8	0.26		0.25	0.54	0.24	
coefficient of linear-expansion, % × 10^{-3}/°C	6.7			3.9		9.2	8.0
vapor pressure, kPa[c,d]	5.43_{373}			45(108°C)	220(177°C)	50(206°C)	
gas volume at STP, cm^3/g	710	740	680	625	651	590	
activation energy, kJ/mol[a]	143	245		193	250	126	121

collision constant, $\log_{10/s}$	11.4	22.5		15.1	19.5	9.2	6.8
				1.17×10^{15}			
detonation velocity, km/s	6.94	7.35	7.05	7.59	7.90	7.12	7.25
at density, g/cm^3	1.64	1.71	1.60	1.79	1.89	1.70	1.77
detonation pressure, GPa[e]	18.9	26.5		25.9	29.1	26.2	
(calc) at density, g/cm^3	1.64	1.76		1.79	1.89	1.70	
vacuum stability at 100°C, mL/g gas per 40 h at STP	<1	<1	<1	<1	<1	<1	<1
weight loss at 100°C in 48 h, %	<1	<1	<1	1	<1	<1	
explosion temperature, 5 s, °C	458 (dec)	320(dec)	320(dec)	260	315		
relative impact test value, % TNT	100	100	120	>200	150	>200	>200
relative gap test value, % TNT	100		80		160		
relative energy output, % TNT							
lead block	100	100	100				
ballistic mortar	100	100	100				
sand test	100	110	85				
plate dent test	100	105	85	120	120	120	

[a]To convert J to cal, divide by 4.184.
[b]Product is liquid H_2O.
[c]To convert kPa to mm Hg, multiply by 7.5.
[d]Subscript is temperature in °C at which measurement was obtained.
[e]To convert GPa to kilobars, multiply by 10.

at elevated temperatures, and may also form low melting eutectics that separate over a period of time. Following nitration, the TNT is treated with sodium sulfite (sellited) which reacts preferentially with the unsymmetrical meta isomers to form the water-soluble sodium salts of the corresponding sulfonic acids. The formation of isomeric impurities and reaction by-products together with the losses in the selliting operation reduce the theoretical yield of alpha-TNT to about 85–88%. TNT is now made predominantly by continuous process methods that operate at lower temperatures and offer significant advantages in safety and controllability than the batch process.

In the batch process which finds occasional use, the steps used in the successive nitrations are similar and include acid mixing, addition of the oil, digesting (cooking) the reaction to completion, cooling and settling the mix, and separating the oil from the acid. The nitrators are made of stainless steel plate and are equipped with concentric cooling water coils, stainless steel propellers, thermometer wells, and fume ducts. Drowning tubs are provided near and below each nitrator.

Trinitrotoluene is purified by crystallization and water washing, neutralization, selliting, and filtering in a process that takes about 25 minutes. The spent sellite solution is deep red (red water) and contains the sulfonated derivatives of the unsymmetrical TNT isomers, and sodium sulfite, sulfate, nitrite, and nitrate. It is never discharged into streams because it is highly toxic. The TNT cake on the filter is washed, reslurried, and pumped to the melter, where it is liquefied with live steam. After decanting the excess water, the melted TNT is dried by bubbling hot air at about 100°C through the charge. After drying, molten TNT is transferred to a cast-iron, water-cooled, brass-faced flaker. After flaking, the product is ready for packaging and shipping.

The continuous processes used in many countries are based on nitration of toluene where the organic phase flows countercurrent to the acid phase. This has the advantage of lower cost, increased safety, electronic controls, fewer pollutants, and operations at lower nitration temperature. Examples are the Bofors-Chematur process and the Canadian Industries Limited (CIL) process. The CIL process uses a series of eight nitrator–separator reactors to conduct the nitration in six stages. An agitator is fitted into a draft tube at the center of the nitrator to assure the required interphase mixing and circulation of the liquids over the cooling coils in the nitrator. The differential head created between the liquid on either side of the draft tube causes continuous fluid flow and internal recycling to occur between nitrator and separator. All the toluene is introduced in the first stage where mononitration and a small amount of dinitration occur. The nitrobody stream moves from reactor to reactor through the separators, becoming increasingly nitrated in the process. Oleum (100% sulfuric acid plus SO_3) is added to the last stage, and moves toward the first, decreasing in strength. Nitric acid is added to each stage as needed to maintain the required mixed acid composition. Mononitration occurs in stage 1, dinitration occurs primarily in stages 2 and 3, and trinitration in the remaining stages. The temperature and the acid concentration of the mix in the reactors is progressively increased. Nominal temperatures for the six nitration stages (eight reactors) are 55, 60, 70, 80, 85, 90, 95, and 100°C, respectively. The crude TNT is treated with countercurrent water washes, neutralized, and pelleted. It is then given a final wash and treated as in the batch process. A typical facility is schematically depicted in Figure 1.

Toxicity and Environmental Considerations. Prolonged exposure and skin contact with TNT in the workplace may lead to rashes, skin eruptions, and more serious consequences such as nose bleeds and hemorrhage of the skin, as well as mucose and blood disorders. Dust inhalation may result in nausea, vomiting, toxic hepatitis, and anemia. Occupational cleanliness is critically important in TNT manufacture. Wastewater from TNT contains mostly dissolved TNT and possible traces of dinitrotoluene and isomers of TNT. The water from loading plants generally contains TNT, HMX, RDX, and wax. The washings initially are colorless but turn pink if neutral or basic and exposed to sunlight. The dissolved products are removed by filtration through diatomaceous earth (see DIATOMITE) and activated carbon. The disposal of the explosive-contaminated carbon by open burning or as landfill in hazardous waste sites is increasingly unacceptable. An alternative process possible for future application consists of using ozone (qv) in the presence of uv light to decompose the organics in the pink water. Red water is produced in the selliting process, and has been either burned in rotary kiln separators or sold to the paper industry. These options are no longer viable, and alternative approaches are under study including process changes and modifications of current incineration technology (189–204).

Picric Acid and Ammonium Picrate. Picric acid (PA) (2,4,6-trinitrophenol) was the first modern high explosive to be used extensively as a burster in gun projectiles. It was first obtained by nitration of indigo, and used primarily as a fast dye for silk and wool. It offered many advantages: when compressed, it was used as a booster for other explosives, and when cast (melting point 122.5°C) served as a burster in shell; it was stable, insensitive, nonhygroscopic, relatively nontoxic, and of high density when cast, and could be made economically by simple nitration.

Picric acid has an energy content somewhat greater than that of TNT and a higher detonation rate. Its calculated molar detonation products at a density of 1.76 g/cm^3 are 3.16 C, 2.66 CO_2, 1.50 N_2, 1.50 H_2O, and 0.18 CO. A primary disadvantage is its tendency to form sensitive salts with calcium, lead, zinc, and other metals. Furthermore, its high melting point necessitates the use of superheated steam in shell-filling plants. For these reasons, picric acid is no longer used as a military explosive.

Picric acid may be made by gradually adding a mixture of phenol and sulfuric acid at 90–100°C to a nitration acid containing a small excess of nitric acid. The picric acid crystals are separated by centrifuging, washed, and dried. The wash water is reused to decrease losses owing to the water solubility of the picric acid. A yield of about 225% of the weight of phenol is commonly obtained.

Ammonium picrate (AP) is used only where a high explosive is required that is particularly insensitive to shock. It has been employed in pressed form primarily as a burster in naval projectiles for armor penetration. The compound is very stable and does not form sensitive metallic salts. Its explosive characteristics are comparable to those of TNT. Ammonium picrate is prepared by neutralizing a solution–suspension of picric acid in hot water with aqueous or gaseous ammonia. The salt, which crystallizes on cooling, is filtered, washed with cold water,

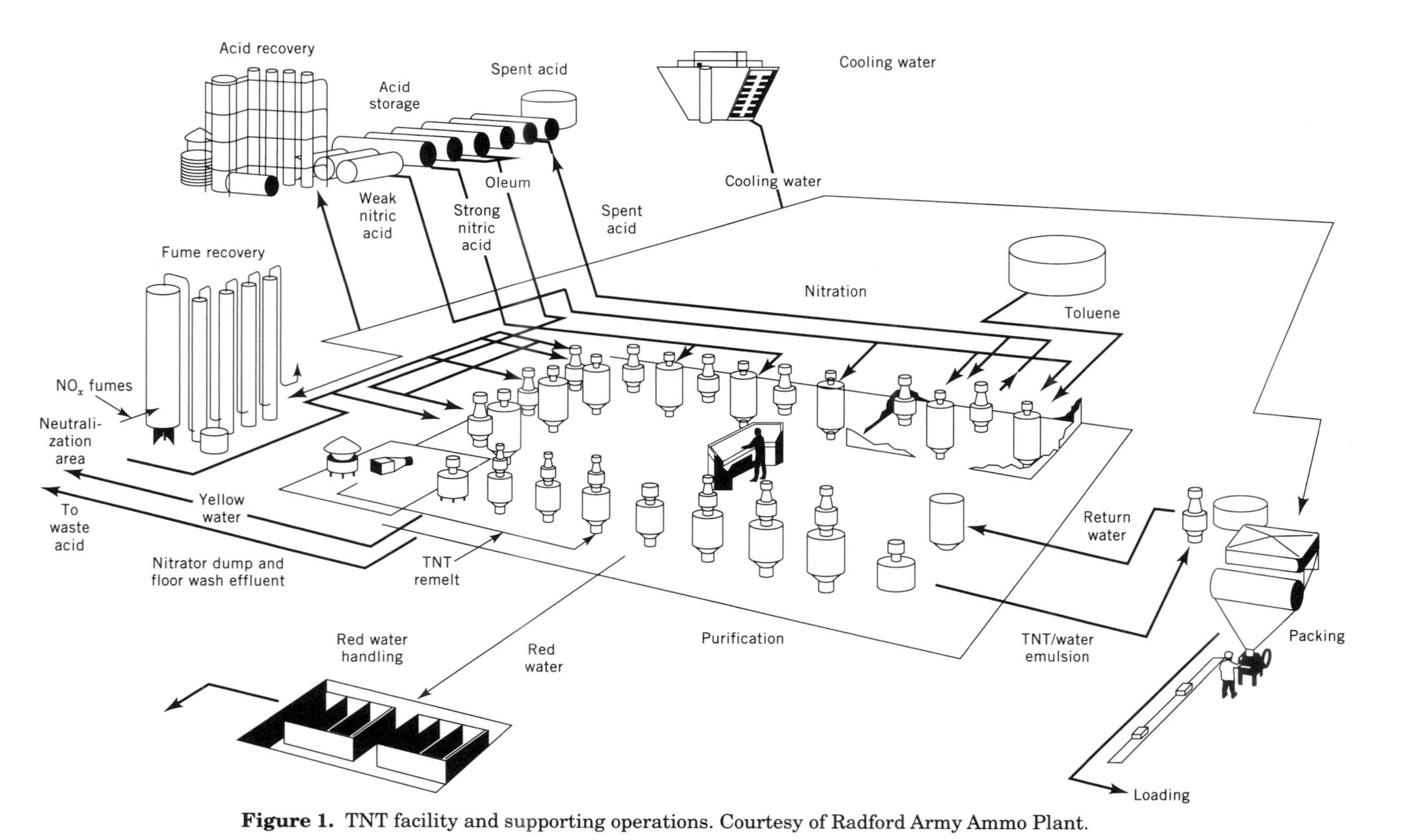

Figure 1. TNT facility and supporting operations. Courtesy of Radford Army Ammo Plant.

and dried. Additional properties of picric acid and ammonium picrate are included in Table 12 (204–206).

Explosives for Use at High Temperatures. Explosives that can withstand prolonged exposure to elevated temperatures without detonating find considerable use in such applications as detonators for deep oil wells, space travel, and componentry for specialized applications. The explosives developed for this purpose are primarily nitroaromatic compounds, the most important of which are hexanitrostilbene and triaminotrinitrobenzene. Although the energy level of these explosives lies between that of TNT and RDX, it is not sufficient for use in high performance of military systems that utilize HMX or RDX. The physical properties are listed in Table 12.

Hexanitrostilbene may be prepared by adding a solution of TNT in tetrahydrofuran and methanol at 5°C to aqueous sodium hypochlorite. To this mixture a 20% solution of trimethylamine hydrochloride is added at 5–15°C. Hexanitrostilbene precipitates, and is filtered and washed with methanol and acetone at about 50% yield (207–211).

1,3-Diamino-2,4,6-trinitrobenzene (DATB) may be obtained by nitration of *m*-phenylenediamine at 120–140°C. 1,3,5-Triamino-2,4,6-trinitrobenzene [*3058-38-6*] (TATB), $C_6H_6N_6O_6$, may be made by nitration of trichlorobenzene with mixed acid followed by treatment with ammonia in alcohol. It has about 70% of the energy of HMX and is among the safest of high explosives available for special uses. It is used in molding powders employing techniques similar to those for making polymer-coated, plastic-bonded explosives (PBX) (212–218).

Applications

Military. The single-component explosives most commonly used for military compositions are TNT, RDX or HMX, nitrocellulose, and nitroglycerin. The last two are used almost exclusively to make propellants. The production volume of TNT far exceeds that of any other explosive. It is used as manufactured, as a base of binary slurries with other high melting explosives, or in ternary systems generally containing a binary mix and aluminum.

Munitions are filled with TNT and TNT-based slurry mixes by the melt-cast process. The compositions are liquefied using plant steam to form a low viscosity fluid which is poured into munitions. Binary mixes are made by adding the high melting component to the liquified TNT. Composition B is the most widely used binary explosive, and many studies have been made of its casting characteristics and performance. It is made by adding water-wet RDX to TNT at 95–100°C, decanting excess water, and evaporating the remainder. Other compounds may be added to decrease sensitivity and exudation and increase the mechanical strength of the cast. Composition B or minor modifications are used in loading projectiles and warheads and as the starting material for making aluminized explosives. Other binary mixtures of explosives include the octols (HMX + TNT), cyclotols (RDX + TNT), pentolites (PETN + TNT), tetrytols (tetryl + TNT), amatols (ammonium nitrate + TNT), and picratols (ammonium picrate + TNT). The properties are listed in Table 13 (219–222).

The incorporation of aluminum increases the blast effect of explosives but decreases the rates of detonation, fragmentation effectiveness, and shaped charge

Table 13. Properties of TNT-Based Mixtures Without Aluminum

Property	Composition B	Cyclotol 75/25	Octol 75/25	Pentolite 50/50	Baratol 67/23	Amatol 80/20
density, g/cm^3	1.72	1.77	1.81	1.71	2.52	
heat of formation, kJ/g[a]	−0.042	−0.126	0.123	1.01	2.97	
heat of combustion, kJ/g[a]	11.67	10.98	11.20	6.48		4.19
heat of detonation, kJ/g[a] $H_2O(l)$	5.28	5.12	6.56	5.10	3.09	4.10
gas volume at STP, cm^3/g		865		815		860
activation energy, kJ/mol[a]	137					171
collision constant, log_{10}/s	13.3					15.5
detonation velocity, km/s	7.90	8.25	8.64	7.52	5.00	5.20
at density, g/cm^3	1.69	1.74	1.81	1.64	2.53	1.60
detonation pressure, GPa[b]	29.5	31.3	34.3	28.0	14.0	
at density, g/cm^3	1.72	1.74	1.81	1.68	2.53	1.60
detonation temperature,[c] K	2,750	2,710	2,580			
specific heat ratio	2.94	2.95	2.98			
vacuum stability at 100°C, cm^3 gas per g per 40 h	<1	<1		1–3	<1	<1
relative impact test value, % TNT	60	40	50	45	120	85
relative gap test value, % TNT	60	70	60	45	255	150
explosion temperature, 5 s, °C	250	270	350	220	385	280

compressive strength[d]						
stress at rupture, MPa[e]	12.1	11.6	10.3	14.5	34.5	11.0
compression at rupture, %	0.24	0.12	0.20			
modulus of elasticity, kPa[f]	8.1	20.9	9.3			
work to produce rupture, J/cm^3[a]	0.02	0.01	0.01			
tensile strength[d]						
stress at rupture, MPa[e]	0.65	1.55	1.00			
elongation at rupture, %	0.01	0.11	0.01			
modulus of elasticity, MPa $\times 10^{-3}$[e]	10.9	1.9	10.8			
shear strength[d]						
stress at rupture, MPa[e]	5.2	4.3	5.3			4.8
charpy impact strength, J[a]	1.5	1.6	1.8			
specific heat, J/(g·K)[a]	1.06[g]		1.13			0.8
relative energy output, % TNT						
lead block	130			120		125
ballistic mortar	135	140		135		130
sand test	120	120		125		80

[a]To convert J to cal, divide by 4.184.
[b]To convert GPa to atm, divide by 10^{-4}.
[c]Calculated.
[d]At load rate of 0.13 cm/min at 20°C.
[e]To convert MPa to psi, multiply by 145.
[f]To convert kPa to atm, divide by 101.3.
[g]The thermal conductivity of composition B is 0.5 W/(m·K).

performance. Mixes with aluminum are made by first screening finely divided aluminum, adding it to a melted RDX–TNT slurry, and stirring until the mix is uniform. A desensitizer and calcium chloride may be incorporated, and the mixture cooled to ca 85°C then poured. Typical TNT-based aluminized explosives are the tritonals (TNT + Al), ammonals (TNT, AN, Al), minols (TNT, AN, Al) torpexes pexes, and HBXs (TNT, RDX, Al) (Table 14) (223–226).

Casting is the most economical process for large-scale filling of many munitions. It requires little capital equipment and lends itself to automation. Gun projectiles, which in wartime have to be used in large quantity, may be filled automatically. Although precise charges may be cast if considerable care is taken, the casting process for TNT-based explosives has a number of limitations in plant-loading operations, including the possibility of component segregation and nonuniformity of cast, the introduction of porosity and cavitation, and the difficulty of filling components with small openings. Casting is not used for high melting explosives except as a slurry with TNT, nor is it used to form explosive configurations of maximum mechanical strength and geometric stability. However, the casting procedures used in the manufacture of large composite rocket propellant grains have been adopted to polymer-based RDX or HMX compositions (PBXs) which cure at elevated temperatures.

High speed automatic mechanical pressing is commonly used to volumetrically load small quantities of primary explosives into blasting caps and detonators and to make small explosive components. Primary explosives may be mixed with graphite to improve flow and antistatic properties, or may be desensitized with waxes, stearates, or polymeric compounds. Secondary explosives and explosive mixtures may be pressed to form booster pellets or to load components directly as in the case of armor-penetrating projectiles. Where the explosive is too sensitive in its pure crystalline state to permit press loading or lacks the required mechanical properties in its compressed state for subsequent use, it is coated with polymeric materials such as polystyrene and polybutadiene, to form molding powders, often referred to as plastic-bonded explosives. Desensitization is obtained when the explosive crystals are thoroughly and uniformly coated. A typical procedure for making PBX-type explosives involves making a lacquer of a solution of the organic polymer in a solvent, eg, ethylacetate, and adding it to a water slurry of the explosive. The solvent is distilled off under vacuum while the mix is agitated, precipitating the polymer on the explosive. The coated explosive forms small agglomerates as the solvent removal process continues. It is filtered, washed, and vacuum dried to form a free-flowing, dustless, high density powder. Bi- or trimodal size distributions of spherical shaped explosive particles are often used to improve the flow characteristics and packing density of the molding powder. Antistatic agents (qv) such as carbon black may be added to prevent dust explosions. In another coating technique, the required amount of low melting wax is added to a water slurry of the explosive at a temperature high enough to melt the wax. After agitation to distribute the wax on the crystals, the temperature is lowered, the water decanted, and the remaining mass filtered and dried.

The plastic-bonded explosive powders may be mechanically pressed, using special procedures including vacuum pressing and hot pressing. Hydrostatic pressing and isostatic pressing techniques are also used involving compression of the explosive by a high pressure fluid acting on the explosive through a flexible

Table 14. Properties of Explosive Mixtures Based on Ternary Systems Containing TNT and Aluminum[a]

	Composition				
Property	HBX 1	HBX 3	H 6	Tritonal 80/20	Minol 2
density, as cast, g/cm^3	1.75	1.86	1.75	1.77	1.70
heat of formation[b], kJ/g[c]	0.096	0.092	0.054	0.026	6.78
heat of combustion[b], kJ/g[c] H_2O[d]	14.99	18.80	16.61	18.74	13.22
heat of detonation[b], kJ/g[c] H_2O[d]	3.85	3.71	3.86		
gas volume[b], cm^3/g at STP	758	491	723		
detonation velocity, km/s	7.22	7.15	7.19	6.70	7.18
at density, g/cm^3	1.75	1.84	1.71	1.72	1.68
detonation pressure, GPa[e]	22.04		23.7	18.9	22.0
at density, g/cm^3	1.70		1.73	177	1.65
explosion temperature, 5 s, °C			275	470	260
relative impact test value, % TNT	75	70	75	75	95
relative gap test value, % TNT	80	90	75	80	120
rifle bullet, uneffected, % TNT	75	80	80	100	55
compressive strength[b]					
stress at rupture, MPa[f]		28.9	24.8	14.5	
compression at rupture, %		0.35	0.35	0.30	
modulus of elasticity, kPa[g]		14.5	11.0	7.9	
tensile strength[b]					
stress at rupture, MPa[f]		2.8	2.6	1.7	
elongation at rupture, %		0.02	0.20	0.03	
modulus of elasticity, kPa[g]		14.5	12.4	10.7	
work to produce rupture, J/cm^3		0.25	0.25	0.33	
shear strength[h]					
stress at rupture, MPa[f]		7.58	5.86	5.00	
coefficient of linear expansion, $\% \times 10^{-3}/°C$	4.4	5.4	5.7	8.7	
thermal conductivity, W/(m·K)	0.41	0.71	0.46	0.46	0.69
specific heat, J/(g·K)[c]	1.05	1.06	1.13	1.00	1.26
relative energy output, % TNT					
ballistic mortar	135	110	135	130	145
sand test	100		105	95	95

[a] For the compounds listed, vacuum stability at 100°C, cm^3 gas per g per 40 h, less than 1; friction pendulum test, no explosion.
[b] Molecular weight assumed to be 100 g.
[c] To convert J to cal, divide by 4.184.
[d] Product is liquid H_2O.
[e] To convert GPa to kilobars, multiply by 10.
[f] To convert MPa to psi, multiply by 145.
[g] To convert kPa to atm, divide by 101.3.
[h] At load rate of 0.13 cm/min at 20°C.

membrane. Polymer-based mixtures of RDX or HMX have also been formulated to permit extrusion from a press followed by subsequent curing.

The explosives may be first deaerated to a pressure below 133 Pa (1 mm Hg) and then consolidated at pressures to 200 MPa (2000 atm) and temperatures to 120°C. Densities of 97–98% of the maximum theoretical are produced. Final machining to shape is often required. High energy, polymer-bonded explosives prepared in this manner may be pressed into many configurations of high density, considerable mechanical strength, and good dimensional stability. They are generally prepared using 90–95% of RDX or HMX, and are much less sensitive to heat, friction, shock, and impact than pure RDX or HMX. Extrudable explosive mixtures consisting of a high energy explosive in a plastic matrix have also been prepared which are putty-like in consistency or may be cured to a rigid mass. The properties of a number of plastic-bonded explosives are presented in Table 15. Many inert formulations have also been developed for systems where it is desired to simulate the mechanical properties of live explosives (227–233).

Industrial. In the United States, private corporations operate most of the government-owned plants that make explosives for military use. Explosives and explosive components are also made for industrial use. Ammonium nitrate explosives typified by the water gels, slurries, emulsions, and the ammonium nitrate fuel oil blasting agents (ANFO) and unprocessed ammonium nitrate dominate the sales of industrial explosives and blasting agents in the United States. The U.S. sales of industrial explosives and blasting agents in 1988 and 1989 are shown in Table 16 (234).

Dynamites. The first dynamite, developed by Nobel, contained about 50% nitroglycerin absorbed into and desensitized by 50% of diatomaceous earth (guhr), nitroglycerin, and sodium nitrate. The inert guhr was later replaced by fuel and oxidizer components to increase the energy of the dynamites. The strength of the dynamites may be varied by modifying the percentage of nitroglycerin present and introducing other combinations of energetic oxidizers and fuels. Ethylene glycol dinitrate is mixed with or may entirely replace the nitroglycerin to decrease cost and the freezing point of the compositions. Other additives include carbonaceous fuel such as wood meal, antiacidic components, and sulfur. Straight dynamites commonly use sodium nitrate as the crystalline oxidizer; ammonia dynamites use ammonium nitrate. Straight gelatin and ammonia gelatin dynamites are similar to the straight and ammonia dynamites but contain sufficent nitroglycerin and nitrocellulose to act as a binder and impart plasticity and increased water resistance to the compositions.

The dynamites are formulated for specific applications and have a wide range of densities (0.8–1.7 g/mL) and detonation rates (2000–7000 m/s). Permissible dynamites are granular, semigel, or gelled mixtures of low to medium strength that are designed to be used in underground coal (qv) mines having inflammable coal dust and methane–air atmospheres. These generally contain about 10% sodium chloride to reduce the flame temperature of the decomposition products. Dynamites are ordinarily sufficiently sensitive to permit initiation in small diameter charges by no. 6 or no. 8 blasting caps. A list of typical substances used in dynamites follows (235–238).

Table 15. Properties of Plastic-Bonded Explosives

Property	Composition A 3	Composition C 4	HMX-KEL-F 95/5	LX-04-1	PBXN3	PBX 9404	PBX 9010
density, g/cm^3							
as pressed	1.64	1.64	1.87	1.87	1.80	1.83	1.78
heat of formation[a], kJ/g[b] H_2O[c]	−0.104	−0.138		0.899	−0.131	−0.003	0.330
heat of detonation[a], kJ/g[b] H_2O[c]	5.062	6.610[d]		5.481	6.49	5.481	6.150
detonation velocity, km/s	8.47	8.34	8.89	8.48	8.60	8.77	8.50
at density, g/cm^3	1.64	1.58	1.88	1.87	1.80	1.83	
detonation pressure, GPa[e]	27.7	25.7		35.4	38.0	36.8	
at density, g/cm^3	1.59	1.58		1.84		1.84	
explosion temperature, 5 s, °C	280	263		337	337	309	295
relative impact test value, % TNT	125	140	75	75	135	140	75
compressive strength[f]							
stress at rupture, MPa[g]	3.8		28.2	7.2	6.9	16.9	9.3
compression at rupture, %	0.25		1.5	0.9	1.3	1.5	0.8
modulus of elasticity, MPa[g] $\times 10^{-3}$	5.0		7.6	1.9		5.2	3.4
tensile strength[f]							
stress at rupture, MPa[g]	1.1		4.1	3.1	3.0	4.5	2.8
elongation at rupture, %	0.04		0.06	0.35	0.05	0.01	0.16
modulus of elasticity, MPa[g] $\times 10^{-3}$	5.0		7.6	2.6	13.8	6.6	6.6
shear strength[f]							
stress at rupture, MPa[g]	2.07		8.61	5.17		9.65	5.86
charpy impact strength, J[b]	0.15		0.18	0.37	0.17	0.18	0.18
coefficient of linear expansion, % $\times 10^{-3}$/°C	8.0	6.6	4.6	5.1	7.0	5.8	6.7
thermal conductivity, W/(m·K)		0.25		0.29	0.51	0.42	
specific heat, J/(g·K)[b]					1.2	1.2	
relative energy output, % TNT fragmentation velocity	125	125	140	130	125	140	130

[a]Molecular weight assumed to be 100 g. [b]To convert J to cal, divide by 4.184. [c]Product is liquid H_2O. [d]Calculated values.
[e]To convert GPa to atm, divide by 10^{-4}. [f]At load rate of 0.13 cm/min at 20°C. [g]To convert MPa to psi, multiply by 145.

Type	Primary purposes
oxidizers	energy modifiers
nitroglycerin	
nitrostarch	
ammonium nitrate	
sodium nitrate	
nitrocellulose	
fuels	energy modifiers and absorbents
sawdust	
wood metal	
flour	
wood pulp	
dextrin	
starch	
sulfur	
freezing point depressants	
ethylene glycol dinitrate	
sensitizers	increase initiability of small diameter charges (particularly ammonium nitrate explosives)
tetryl	
dinitrotoluene	
trinitrotoluene	
nitrostarch	
PETN	
smokeless powder	
pentolite	
waterproofing compounds	essential where ammonium nitrate or hygroscopic salts are present and in explosives for use in underwater demolition work
pregelatinized	
stearates	
silicon resins	
waxes	
swelling agents such as carboxymethylcellulose	
liquid film forming compounds	
coolants	reduce flame temperature to permit use in coal mines
ammonium nitrate	
sodium chloride	
ammonium chloride	
sodium bicarbonate	
antiacidic substances	increase stability
calcium carbonate	
magnesium carbonate	
zinc and magnesium oxides	

Ammonium Nitrate Explosives. Ammonium nitrate is the cheapest and safest source of readily deliverable oxygen for explosive applications (see AMMONIUM COMPOUNDS). The extensive use of ammonium nitrate in ammonium nitrate–fuel

Table 16. Sales of Industrial Explosives and Blasting Agents in the United States, t × 10^3

Class	1988	1989	1990	1991
permissible	12	10	9	6
other high explosives	65	64	60	49
water gels, slurries, and emulsions	298	291	295	276
ammonium nitrate–fuel oil blasting agents	393	347	311	264
unprocessed ammonium nitrate	1375	1465	1486	1258
Total	*2143*	*2177*	*2161*	*1853*

oil (ANFO) and water-based commercial explosives have largely displaced the nitroglycerin-based dynamites. Ammonium nitrate industrial explosives are low cost, safe, versatile in performance and application, and have better storage stability than dynamites. A large number of formulations are available for almost all purposes. Although hygroscopicity is the principal disadvantage of ammonium nitrate, coating techniques have been developed to reduce susceptibility to high humidity.

Pure ammonium nitrate decomposes in a complex manner in a series of progressive reactions having different thermochemical effects (Table 17). Oxygen is liberated from combination with combustibles only at temperatures above 300°C. When a combustible material such as fuel oil is present in stoichiometric proportions (ca 5.6%) the energy evolved increases almost threefold

$$3\ NH_4NO_3 + (CH_2)_n \rightarrow 3\ N_2 + 7\ H_2O + CO_2 + 4.288\ \text{kJ/g}\ (1.025\ \text{kcal/g})$$

Ammonium nitrate exists in five crystal forms that may transform from one to the other as the temperature changes with accompanying volume, crystal structure, and heat changes. It is very hygroscopic and deliquesces at above 60% relative humidities. Pure ammonium nitrate is stable and insensitive to impact and friction. It is impossible to initiate using conventional blasting caps unless

Table 17. Thermochemical Data for the Decomposition of Ammonium Nitrate

Reaction	°C	Heat liberated,[a] J/g[b]	Gas volume, mL/g at STP	Temperature of reaction, K
$NH_4NO_3 \rightarrow NH_3 + HNO_3$	180	−2144	560	endothermic
$NH_4NO_3 \rightarrow N_2O + 2\ H_2O$	250	525	840	770
$NH_4NO_3 \rightarrow N_2 + 2\ H_2O + \frac{1}{2}\ O_2$	300	1465	981	1560

[a]At 20°C, constant volume, water vapor.
[b]To convert J to cal, divide by 4.184.

strongly confined and in powder form. When mixed with organic materials such as hydrocarbons or cellulose, ammonium nitrate requires a powerful high explosive booster for initiation. Although the total energy evolved by ANFOs is comparable to that of low grade military explosives, the reaction rate is much slower and is much more dependent on charge diameter and confinement. Ammonium nitrate has been used in military explosives such as amatols, ammonals, minols, and amatexes as a partial replacement for TNT or RDX. It is made from anhydrous ammonia and nitric acid, and ranges from dense crystals to porous agglomerates (prills). Prills used in industrial explosives are made by spraying a 95% solution of the nitrate against a countercurrent stream of air in a prilling tower. The particles are dried in a series of rotary towers, and coated to improve flow characteristics and moisture resistance.

Ammonium nitrate-based explosives account for about 97% of total U.S. industrial explosive consumption. Coal mining in the United States formed about 65–68% of the demand for explosives in 1991. The remaining uses were quarrying and nonmetal mining, 15%; metal mining, 10%; construction, 7%; miscellaneous uses, 3–4%. The properties of ammonium nitrate are given in Table 18 (173,239–242).

ANFOs. The use of AN mixed with a fuel was proposed as a commercial explosive as early as 1867. It was only with the development of anticaking agents in the 1950s that ANFO became practically useful for rock blasting. Ammonium nitrate for oil compositions (ANFOs) consist of 94% ammonium nitrate prills coated with an anticaking agent and 6% absorbed fuel oil. In the United States they are classified as blasting agents, which are mixtures of nonexplosive ingredients used for blasting purposes that cannot be detonated by a no. 8 blasting cap. If they are sensitized by the addition of a high explosive component so that they detonate when initiated by a no. 8 blasting cap, they are defined as blasting explosives. The ANFOs are usually initiated with a high explosive booster such as 50/50 pentolite or composition B.

The sensitivity of ANFOs to initiation is significantly affected by their composition, physical characteristics, and environment. Decreasing the particle size and density of ammonium nitrate or increasing its porosity increases the sensitivity of the mix to initiation. Maximum sensitivity occurs at oil concentrations of ca 2–4%. The presence of water decreases the sensitivity. The detonation velocity increases as the oil content increases to a maximum at ca 6% oil. Maximum velocity is about 4300 m/s for large-diameter ANFO charges. The detonation velocity of ANFO charges increases with charge diameter up to ca 13 cm for compositions at conventional loading densities of 0.90–1.10 g/cm^3. Confinement also increases the detonation velocity. The addition of metallic fuels, such as aluminum or ferrosilicon, increases the energy content. Stabilizers and inhibitors may be added and the fuel oil may be dyed to identify specific compositions. The ANFOs may be mixed on site simply by adding oil to a bag of prills. More effectively, they can be prepared in on-site trucks equipped for the purpose and then augered into boreholes (243–250).

Water-Based and Oil–Water Emulsion Blasting Agents and Explosives. These explosives were developed to capitalize on the low cost of ammonium nitrate, to increase the available energy per unit volume beyond that obtainable by ANFOs, improve initiability in small diameter charges, and particularly to eliminate the

Table 18. Properties of Ammonium Nitrate

Property	Value
molecular weight	80.0
color	white
density at 20°C, g/cm^3	1.725
specific volume, cm^3/g	0.580
melting point, °C	169.6
hygroscopicity at 90% rh, 20°C, %	increase up to 156[a]
solubility in H_2O at 20°C, g/100 g	66
oxygen content, %	60
oxygen balance, to H_2O, %	20
hardness, Mohs'	1.1
heat of formation, kJ/g[b]	4.60
heat of combustion, kJ/g[b]	2.62
heat of detonation (H_2O), kJ/k[b]	2.63
heat of solution, kJ/g[b]	−0.33
gas volume (H_2O vapor), cm^3/g at STP	980
activation energy, kJ/mol[b]	163–167
estimated flame temperature, °C	1500
detonation velocity, km/s	2.70
at density, g/cm^3	0.98
detonation pressure, MPa[c]	1100
at density, g/cm^3	0.98
vacuum stability at 150°C, mL gas per g per 40 h at STP	<1
75°C international heat test, weight loss in 48 h	0
effect of long-term storage	very stable below melting point in absence of organic matter
rifle bullet test, initiations at density of 1.2 g/cm^3	0
impact test	no action
relative energy output, % TNT	
ballistic pendulum	80
lead block	75
specific heat from 0–31°C, J/mol[b]	1.72
thermal conductivity, W/(m·K)	0.25
coefficient of thermal expansion at 20°C, %/°C	9.82×10^{-4}
heat of fusion, kJ/g[b]	0.075
latent heat of sublimation, kJ/g[b]	2.18
maximum specific impulse in optimum formulations, N·s/kg	1965

[a] After 8 days exposure.
[b] To convert J to cal, divide by 4.184.
[c] To convert MPa to psi, multiply by 145.

problems associated with the solubility of ANFOs in wet drill holes. The use of water- or oil-based formulations increased overall safety, reduced cost, and improved fume characteristics. Water-based compositions are thickened suspensions of oxidizers, fuels, and a sensitizer dispersed in a saturated aqueous salt solution. Ammonium nitrate with or without other oxidizers, such as sodium nitrate or calcium nitrate, is dissolved and suspended in water. Explosive sensitizers, such as pentolite, monomethylamine nitrate, TNT, smokeless powder, nitrotoluene, and nitrostarch, may be incorporated. Inert sensitizers include finely divided aluminum, gas bubbles in suspension, gas enclosed in small glass spheres, and finely divided porous solids. Fuels, such as ethylene glycols, coal dust, urea, sulfur, and various types of hydrocarbons (qv), are added. Guar gums, gelatin-forming compounds, such as carboxymethylcellulose, resins and synthetic thickeners, such as the polyacrylamides, are incorporated to thicken the mix. Cross-linking agents, as sodium tetraborate and potassium dichromate, are used to control viscosity. Slurry explosives are made water resistant by the addition of hydrophilic colloids that bond the solid particles and prevent diffusion of water in and out of the system. Antifreezes, such as glycerol, methanol, and diethylene glycol, may be used. Typical slurry blasting agents or explosives may contain ammonium nitrate (30–70%), sodium nitrate (10–15%), calcium nitrate (15–20%), aliphatic amine nitrates (to 40%), aluminum (3–10%), TNT or other explosive sensitizer (5–25%), gellants (1–2%), stabilizers (0.1–2%), ethylene glycol (3–15%), and water (10–20%).

Many of the components used in slurry explosives and the functions are as follows:

Oxidizers	*Fuels*	*Explosives*
ammonium nitrate	aluminum	nitrostarch
sodium nitrate	silicon	TNT
sodium chlorate	coal	smokeless powder
calcium nitrate	sugar	dinitrotoluene
	urea sulfur	RDX

Chemical sensitizers		*Other*
hexamethylenetetramine mononitrate		ethylene glycol
monomethylamine nitrate		diesel oil
urea		lignosulfonates (bubble stabilizers)
isopropyl nitrate		anticaking agents
		surfactants
		waxes
		oils

Thickeners	*Cross-linking agents*	*Micropores (generally filled with inert gas)*
guar gum	potassium dichromate	hollow glass beads
carboxymethylcellulose	zinc chromate	urea–formaldehyde spheres
polyacrylamides	borates	
	sodium tetraborate	

pH control additives	*Gassing agents*	*Antifreeze*
acetic acid	$NaNO_2$	glycerol
fumaric acid	H_2O_2	methanol diethylineglycol

Water-in-oil emulsion explosives have been made as typified by a formulation containing 20% water, 12% oil, 2% microspheres, 1% emulsifier, and 65% ammonium nitrate. The micro droplets of an emulsion explosive offer the advantage of intimate contact between fuel and oxidizer, and tend to equal or outperform conventional water-based slurries.

The performance of several slurries in comparison with TNT and ANFO 94/6 is shown in Table 19. Aluminum-bearing slurries are among the highest energy producers of all industrial explosives, and are extensively used in wet conditions in open-pit blasting. The high loading densities and fluidity and a greater intrinsic energy output enable increased efficiencies to be attained in rock fragmentation. Slurry mixes have been studied and used for many military applications. They may be bulk mixed in a plant, transported to the site of operations, and pumped into boreholes after adding a thickening agent, or may be prepared on-site using a pump truck. Slurries may also be prepared hot in the plant and poured into lay-flat, sausage-shaped polyethylene bags for hand loading into bore holes.

A typical procedure for making an oil emulsion blasting agent begins with a concentrated solution of ammonium nitrate in water pumped from a tanker to a steam heated tank into which sodium nitrate is fed and dissolved in the heated ammonium nitrate solution. The hot (85–90°C) solution is transferred to an emulsifier to which is added oil, wax, and a surfactant. The mix is emulsified to 1–2 micrometer-sized droplets using shear energy for the purpose, microballoons are added to sensitize the formulation, and powdered aluminum to increase its energy. The formulation may be packaged into plastic sausage-shaped cylinders of various diameters depending on use.

Specialized Uses of Explosives. In addition to the conventional use of explosives for mining, civil engineering, and military purposes, an increasing variety of highly specialized applications as well as unusual forms of explosives have

Table 19. Cratering Characteristics of ANFO, Slurry Explosives, and TNT[a]

Explosive	Detonation pressure, GPa[b]	Bulk specific gravity	Detonation velocity, km/s	Contains high explosives	Heat of detonation, kJ/g[c]	Excavated vol relative to equal wt of TNT
ANFO	6.0	0.93	4.56	no	3.76	1.0–1.1
AN slurry	10.4	1.40	6.05	yes	3.05	1.0–1.2
2% Al	6.0	1.30	4.30	no	3.14	1.0–1.2
8% Al	6.6	1.33	4.50	no	4.64	1.2–1.4
20% Al	8.5	1.20	5.70	no	6.07	1.5–1.7
35% Al	8.1	1.50	5.00	no	8.16	1.6–1.8
TNT	18.7	1.64	6.93		4.61	1.00

[a]Ref. 248.
[b]To convert GPa to atm, divide by 10^{-4}.
[c]To convert J to cal, divide by 4.184.

been developed to meet unique requirements. Included among the unusual applications are metal forming and metal cladding (see METALLIC COATINGS, EXPLOSIVELY CLAD METALS) or metal welding to molecularly bond layers of different metals, metal cutting, and compaction of metal powders. Among the specialty forms of explosives used for these and many purposes are flexible sheet explosive generally consisting of a polymeric binder and a high performance explosive such as RDX or PETN; flexible or very rigid explosive foams formulated from nitromethane and a polymeric foam producer; fuel–air explosive systems in which a combustible hydrocarbon such as propane is dispersed in air and then ignited to form a high pressure blast wave; plastic bonded explosive (PBXs) containing RDX or HMX and a polymer designed to give very high mechanical strength; and low sensitivity liquid explosives such as hydroxyl ammonium nitrate used as a component in liquid propulsion guns (180,181,251–258).

More Powerful Explosives. The most energetic explosives commonly used are RDX and HMX. The search for higher energy, yet safe explosives, primarily for military purposes, is concentrated to a large extent on the synthesis of thermally stable molecules having mass densities $>$1.9 g/mL, the density of HMX, and at least 10% more energy than HMX. Hydrogen-free molecules containing only carbon, nitrogen, and oxygen have been synthesized. The atoms are often bound together in compact strained rings. These small-scale molecules provide additional latent strain energy. Research has also been conducted to find high energy polymeric binders for use with RDX, HMX, and higher energy explosives.

Among the molecules that have received greatest attention are 2,6-bis(picrylamino)-3,5-dinitropyridine (PYX); 3,6-dinitro-*s*-tetrazine; 2,4,6-trinito-*s*-trizene; octanitrocubane (ONC); 1,3,3-trinitroazetidine polynitroadamantane (TNAZ); and 1,4-dinitroglycouril (DINGU). The structures of these molecules are shown in Figure 2. Performance characteristics are listed in Table 20 (259–266).

Table 20. Calculated Crystal Densities and Releasable Energy Density for HEDCs and In-Service Explosives

Material	Density g/mL	Detonation performance index[a]
1,3,5,7-tetranitro-2,4,6,8-tetraazacubane[b]	2.21[c]	1.49
	2.20[c]	1.39
TNAZ	1.84	1.07
HMX	1.91	1.00
RDX	1.81	0.93
TNT	1.65	0.54
PYX	1.75	
DINGU	1.98	0.97
3,6-dinitro-5-tetrazine	1.98[c]	
2,4,6-trinitro-5-tetrazine	1.97[c]	

[a]Calculated relative energy of explosive released on detonation and subsequent expansion of detonation gases relative to HMX.
[b]Theoretical cage explosive molecules, not yet synthesized.
[c]Values are calculated.

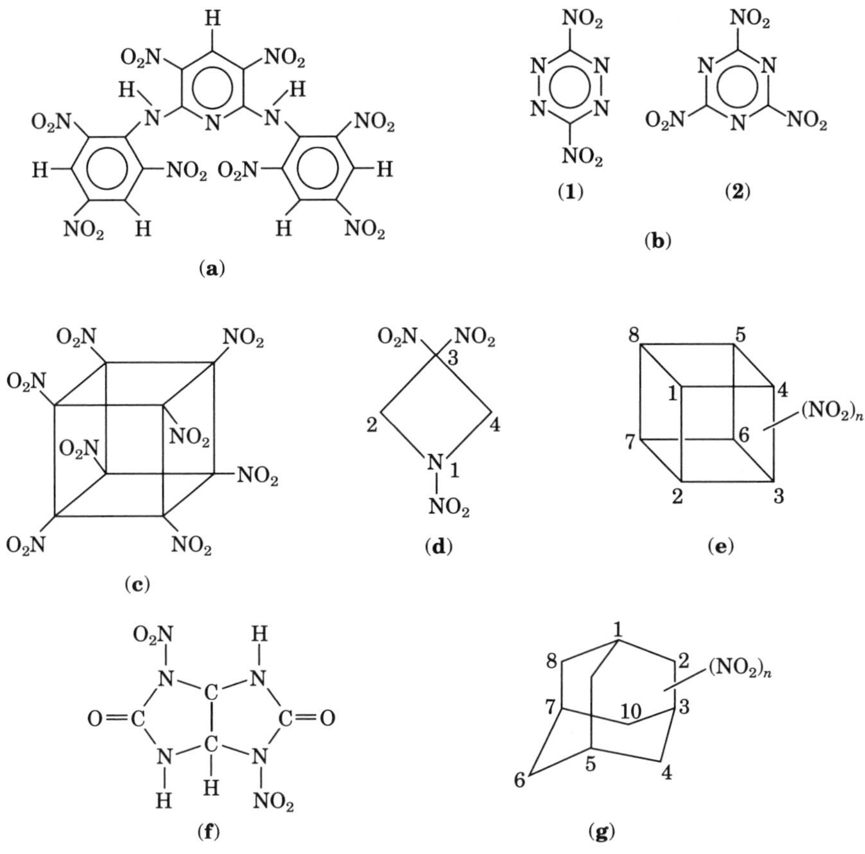

Fig. 2. High energy explosive molecules: (**a**) PYX; (**b**) nitroheterocycles (**1**) 3,6-dinitro-*s*-tetrazine and (**2**) 2,4,6-trinitro-*s*-triazine; (**c**) ONC; (**d**) TNAZ; (**e**) PNC; (**f**) DINGU; and (**g**) polynitroadamantane (PNA).

Insensitive Explosives. The catastrophic propagation of explosive detonations that have been accidentally initiated during peacetime or by military action during wartime, and the high costs of shipping and storing explosive systems have led to a search for low sensitivity explosives and explosive and propellant formulations. The support for these programs has come from military agencies to reduce quantity distances between storage sites, to decrease the battlefield vulnerability of tanks, self-propelled howitzers, and troop-carrying fighting vehicles that carry large numbers of high energy explosive and propellant loaded munitions, and to enhance warship survivability, particularly aircraft carriers, carrying large quantities of aircraft ordnance.

The criteria for insensitive explosives subjected to hazard tests permit no reaction more violent than burning in slow and fast cook-off tests and fragment and bullet tests, no propagation in sympathetic detonation tests, no detonation when struck by a shaped charge jet, no sustained burning when hit by a small fragment, and such special tests as may be required by the use of the explosive.

The general approach to this difficult problem of maximizing the energy of the explosive while minimizing sensitivity lies in the use of RDX or HMX (80–90%) thoroughly coated with a plasticized polymer similar to that used in some plastic bonded explosives. The HMX is usually in a bimodal crystal form; polymeric bonding components such as Hycar and hydroxy-terminated polybutadine (HTPB) have been used, as have energetic, insensitive plasticizers. Other components that have been used include laurylmethylacrylate and dioctyladipate.

Relatively insensitive explosives of medium energy levels have also been formulated using propellant-type formulations containing ammonium perchlorate, aluminum, and a polybutadiene or polyester binder (267–270).

BIBLIOGRAPHY

"Explosives" in *ECT* 1st ed., Vol. 6, pp. 1–91, by W. H. Rinkenbach, D. R. Cameron (Propellants), both of Picatinny Arsenal, U.S.A., Ordnance Dept., and W. O. Snelling (Blasting Explosives), Trojan Powder Co.; in *ECT* 2nd ed., Vol. 8, pp. 581–658, by Wm. H. Rinkenbach, Consulting Chemist; "Explosives and Propellants, Explosives" in *ECT* 3rd ed., Vol. 9, pp. 561–620, by V. Lindner, U.S. Army Armament Research and Development Command (ARRADCOM).

1. W. C. Davis, "Detonation Phenomena," in *12th Annual Symposium on Behaviour and Utilization of Explosives in Engineering Design,* University of New Mexico, Albuquerque, 1972.
2. M. A. Cook, *Science of High Explosives,* Reinhold Publishing Corp., New York, 1958.
3. J. R. Partington, *A History of Greek Fire and Gunpowder,* Heffer, Cambridge, UK, 1960.
4. R. Connor, in W. A. Noyes, Jr., ed., *Chemistry in World War II,* Little Brown and Co., Boston, 1948.
5. W. S. Dutton, *One Thousand Years of Explosives from Wildfire to the H Bomb,* Winston, Philadelphia, 1960.
6. T. Urbanski, *Chemistry and Technology of Explosives,* Vols. 1–4, Macmillan, New York, 1964–1984.
7. Y. Tran and L. J. Burke, *Acronyms and Terminology for Navy Explosives,* MP87-304, NSWC, White Oaks, Md., Jan. 1988.
8. J. Taylor, *Detonations of Condensed Explosives,* Clarendon Press, Oxford, UK, 1952.
9. A. W. Campbell and R. P. Engelke, "The Diameter Effect in High Density Explosives," in *6th Symposium (International) on Detonation ACR-221,* ONR, Dept. of Navy, Arlington, Va., 1977.
10. D. Price, J. F. Wehner, and G. E. Robertson, *Transition from Slow Burning to Detonation: Role of Confinement, Pressure Loading and Shock Sensitivity,* TR68-138, Naval Surface Weapons Center (NSWC), White Oaks, Md., 1968.
11. C. M. Kintz, G. W. Jones, and C. B. Carpenter, *Explosions of Ammonium Nitrate Fertilizer on Board the S. S. Grand Camp and S. S. High Flyer at Texas City, Tex.,* R.I. 4245, U.S. Bureau of Mines (BuM), Dept. of the Interior, Pittsburgh, Pa., 1948.
12. M. W. Beckstead and co-workers, "Convective Combustion Modelling Applied to Deflagation Detonation Transition," in *Proceedings of the 12th JANNAF Combustion Meeting,* Pub. No. 273, Chemical Propulsion Information Agency (CPIA), Johns Hopkins University, Laurel, Md., 1975.
13. W. A. Blaine and co-workers, "Detonation Characteristics of Gun Propellants," preprints of *9th Symposium (International) on Detonation,* Portland, Oreg., Aug. 1989, p. 233.

14. S. Wachtel, "Prediction of Detonation Hazards in Solid Propellants," in *Proceedings of the 145th National Meeting of the Division of Fuel Chemistry,* New York, 1963.
15. P. Benhaine and co-workers, "Investigation on Gun Propellant Breakup and its Effect on Interior Ballistics," in *Proceedings of 4th International Symposium on Ballistics,* Oct. 1989.
16. *Military Pyrotechnics,* AMCP 706-185 to 189, *Engineering Design Handbook Series,* U.S. Army Material Command (AMC), Alexandria, Va., 1974.
17. K. O. Brauer, *Handbook of Pyrotechnics,* Chemical Publishing Co., New York, 1974.
18. H. Ellern, *Military and Civilian Pyrotechnics,* Chemical Publishing Co., New York, 1968. See also *Proceedings of the International Pyrotechnics Seminars,* Co.
19. R. Lancaster and co-workers, *Fireworks: Principles and Practices,* Chemical Publishing Co., New York, 1972.
20. J. C. Cackett, *Monograph on Pyrotechnic Compositions,* RARDE, Fort Halstead, UK, 1965.
21. J. H. McLain, *Pyrotechnics from the Viewpoint of Solid State Chemistry,* Franklin Institute Press, Philadelphia, 1980.
22. C. S. Robinson, *Explosives, Their Anatomy of Destructiveness,* McGraw Hill Book Co., New York, 1944.
23. D. H. Chamberlain and R. H. Stresau, in Ref. 1.
24. *Explosives Hazard Classification Procedures,* TB700-2, U.S. Dept. of Defense, Explosives Safety Board, Alexandria, Va., Sept. 1982.
25. *Ammunition and Explosives Safety Standard,* U.S. Dept. of Defense Explosives Safety Board, no. 6055, DOD, Alexandria, Va., July 1984.
26. R. H. Richardson and co-workers, "Hazards Analysis Through Quantitative Interpretation of Sensitivity Testing," in *Proceedings of the International Conference on Sensitivity and Hazards of Explosives,* ERDE, 1963, p. 269
27. R. H. Richardson and J. M. Sutton, "Risk Analysis," in *Proceedings of the 14th Annual Explosives Safety Seminar,* U.S. Dept. of Defense, NTIS, Springfield, Va., 1972.
28. P. O. Chelsau, Reliability Computation Using Fault Tree Analysis, TR32-1542, NASA, Airport, Md., 1971.
29. R. Pape and co-workers, "Total System Hazard Analysis for the Western Area Demilitarization Facility," in *Proceedings of the 21st Explosive Safety Symposium,* Alexandria, Va., 1984.
30. M. M. Swisdak, Jr., "Maximum TNT Equivalence of Naval Propellants", V2, in *21st Explosives Safety Symposium,* U.S. Dept. of Defense Explosive Safety Board, Alexandria, Va., 1984.
31. M. Held, *Propellants, Explosives and Pyrotechnics* **8,** 158 (1983).
32. I. G. Bowen, E. R. Fletcher, and D. R. Richmond, *Estimate of Man's Tolerance to the Direct Effects of Air Blast,* DASA-2113, Defense Atomic Support Agency, Washington, D.C., 1968.
33. *Protective Construction Design Manual 1984–1989,* Applied Research Assoc., Albuquerque, N. Mex.
34. *Structures to Resist the Effect of Accidental Explosions,* Dept. of the Army Technical Manual TM5-1300, Washington, D.C., 1986.
35. B. W. Jezek, "Suppressive Shielding for Hazardous Munitions Production Operations" in *Symposium on Processing Propellants, Explosives, and Ingredients,* American Defense Preparedness Association (ADPA), Washington, D.C., 1977.
36. A. Copland, *Effective Techniques for the Modification of Violent Cook-off Reactions,* MR3452, Ballistics Research Laboratory (BRL), Aberdeen, Md., June 1985.
37. S. Glasstone, ed., *The Effects of Nuclear Weapons,* Supt. of Documents, U.S. Government Printing Office (USGPO), Washington, D.C., 1962.

38. *A Manual for the Prediction of Blast and Fragment Loadings on Structures,* DOE/TIC 112681, Southwest Research Inst., San Antonio, Tex.
39. N. Dobbs and co-workers, "Design of Steel, Masonry and Precast Concrete Structures to Resist the Effects of HE Explosions" in *17th Annual Dept. of Defense Explosive Safety Seminar,* NTIS, Springfield, Va., 1976.
40. J. A. E. Hannum, ed., *Hazards of Chemical Rockets and Propellants,* 3 vols., Illinois Institute of Technology, Chicago, June 1985.
41. *Symposia on Environmental Pollution and Energy Research,* ADPA Washington, D.C., ND-01P to ND-08P, 1970-1976; NC-02T, 3 Vols., 1975.
42. *Alternatives to Open Burning / Open Detonation of Propellants and Explosives,* CPIA Publication 540, CPIA, Laurel, Md., Mar. 1990.
43. N. A. Shapira, ed., *Wastewater Treatment in the Military Explosives and Propellant Production Industry,* Report NC-02T, 3 Vols., ADPA, Washington, D.C., 1975.
44. P. J. Unkefer, in Ref. 42, p. 307.
45. B. Kozlorowski and J. Kucharski, *Industrial Waste Disposal,* Pergamon Press, New York, 1972.
46. B. H. Carpenter and co-workers, *Specific Air Pollutants from Munitions Processing and Their Atmospheric Behavior,* 4 Vols., Research Triangle Institute, Research Triangle Park, N.C., 1978.
47. J. W. Patterson and R. A. Minear, *State of the Art for the Inorganic Chemicals Industry: Commercial Explosives,* 600/2-74-009b, Environmental Protection Agency, Washington, D.C., 1975.
48. J. A. Hathaway, "A Review of Reported Dose-Related Effects Providing Documentation for a Work Place Standard," in *Proceedings of the 17th Explosive Safety Seminar,* Dept. of Defense Explosives Safety Board, Washington, D.C., 1975, p. 693.
49. R. K. Andren and co-workers, "Removal of Explosives from Wastewater", in *Proceedings of the 30th Purdue Industrial Waste Conference,* Purdue University, Lafayette, Ind., 1975.
50. F. D. Lonadier and co-workers, "Biodegradability of Explosives," in *Minutes of the 15th Explosives Safety Seminar,* AD 775 660, NTIS, Springfield, Va., 1973, p. 797.
51. S. Kaplowitz and L. Sotsky, *Environmental Challenges in Energetics Manufacture,* Armament Research, Development and Engineering Center (ARDEC), Dover, N.J., Nov. 1990.
52. C. A. Myler and J. L. Mahannah, *Energy Recovery From Waste Explosives and Propellants Through Cofiring,* in CPIA Publication 556, CPIA, Laurel, Md., Oct. 1990, p. 61.
53. B. R. White, "Reclamation and Recycling of Propellants and Explosive," in *JANNAF Safety and Environmental Protection Subcommittee Meeting,* Monterey, Calif., May 1988.
54. B. C. Pol and M. B. Ryan, *Database Assessment of Pollution Control in the Military Explosives and Propellant Production Industry,* final report ORNL-22, Oak Ridge National Lab., Tenn., Feb. 1986.
55. S. Rosenberg and J. Carrayza, *Conventional Munition Demilitarization and Disposal,* private communication, ARDEC, Dover, N.J. (includes a list of activities at all U.S. government facilities).
56. M. A. McHugh and V. J. Krukonis, *Supercritical Fluid Extraction,* Butterworth Publishers, Stoneham, Mass., 1986.
57. D. Layton and co-workers, *Conventional Weapons Demilitarization, A Health and Environmental Effects Data Base Assessment: Explosives and Their Co-Contaminants,* UERL-21109, Livermore National Lab., University of California, Livermore, Dec 1987.

58. *Safety and Performance Tests for Qualification of Explosives,* NOL, White Oaks, Md., D44811, Jan. 1972.
59. *Qualification and Final (Type) Qualification Procedures for Navy Explosive Materials,* NAVSEA Inst. 8020.5A Sept. 1984, and 8020.5B May 1988.
60. J. M. Pakulak, Jr., and C. M. Anderson, "NWC Standard Methodology for Determining Thermal Properties of Propellants and Explosives," NWCTP 6118, Mar. 1980.
61. Mary R. Senn and co-workers, *Chemicals and Processing Assessment of Candidate Explosives for the Advanced Bomb Family,* IHTR 1370, NOS, Indian Head, Md., June 1990.
62. *Assessing the Thermal Stability of Chemicals by Methods of Differential Thermal Analysis,* American Society for Testing and Materials, Philadelphia.
63. G. R. Walker and co-workers, *Manual of Sensitivity Tests,* The Technical Cooperation Program, TICP Panel 0-2C Canadian Armament Research and Development Establishment, Quebec, Canada, 1966.
64. B. T. Federoff and O. E. Sheffield, *Encyclopedia of Explosives and Related Items,* Picatinny Arsenal, Dover, N.J., 1962.
65. T. R. Gibbs and A. Popolato, eds., *LASL Explosive Property Data,* University of California Press, Berkeley, 1980.
66. D. L. Ornellos, *Calorimetric Determination of the Heats and Products of Detonation for Explosives,* Lawrence Livermore Lab. (LLL), Livermore, Calif., Apr. 1982.
67. F. Volk, *Propellants and Explosives,* (Mar. 1978).
68. D. A. Cichra and R. M. Doherty, in *9th Symposium (International) on Detonation,* Vol. 3, Sept. 1989, p. 914.
69. I. B. Akst, in Ref. 68, p. 920.
70. D. Tuce and H. Hurivitz, *Ruby Code Calculations of Detonation Properties: ICHNO Systems,* TR63-216, NOL, White Oaks, Md., 1963.
71. W. Fickett and W. Davis, *Detonation,* University of California Press, Berkeley, 1979.
72. M. Suceska, *Propellants, Explosives, Pyrotechnics* **16** (1991).
73. M. J. Kamlet and co-workers, *J. Chem. Physics* **48** (1968).
74. V. K. Mohen and T. B. Tang, *Propellants, Explosives, Pyrotechnics* **9,** 30–36 (1984).
75. F. P. Bowden and A. D. Yoffe, *Initiation and Growth of Explosions in Liquids and Solids,* Cambridge University Press, New York, 1952.
76. T. J. Tucker, "Explosive Initiators," in *Behavior and Utilization of Explosives in Engineering Design, 12th Annual Symposium,* University of New Mexico, Albuquerque, 1972.
77. *Explosive Trains,* AMCP 706-179, AMC, Alexandria, Va., 1965.
78. *Proceedings of the International Conference on Research in Primary Explosives, Explosives Research, and Development Establishment,* Waltham Abbey, Essex, UK, 1975.
79. T. Urbanski, *Chemistry and Technology of Explosives,* Vol. 3, Pergamon Press, New York, 1967.
80. B. T. Federoff and O. Sheffield, *Encyclopedia of Explosives and Related Items,* Vol. 1, A545 TR 2700, Picatinny Arsenal (PTA), Dover, N.J., 1960.
81. W. G. Chace and H. K. Moore, eds., *Exploding Wires,* Vols. 1–3, Plenum Press, New York, 1959, 1962, 1964.
82. D. P. Donegan, "Trends in Electrostatic Precautions in Filling Factories," in *Minutes of the 17th Explosives Safety Seminar,* AD A036015, DDC, 1976.
83. A. F. Schlack, "Susceptibility of Electric Primers and Electrostatic Discharges" in *Minutes of the 15th Explosives Safety Seminar,* AD-775 580, NTIS, Springfield, Va., 1973.
84. E. Demberg, in Ref. 83.

85. W. B. Leslie, R. W. Dietzel, and J. A. Searcy, "A New Inherently Safe Explosive for Low Voltage Detonator Applications," in *Proceedings of the 6th (International) Symposium on Detonation,* Office of Naval Research (ONR), Washington, D.C., 1976, p. 144.
86. R. Thorpe and W. R. Fearheller, "Development of Processes for Reliable Detonator Grade Very Fine Explosive Powder," in *Proceedings of Joint International Symposium on Compatibility of Plastics and other Materials with Explosives, Propellants, Pyrotechnics,* ADPA, New Orleans, La., Apr. 1988.
87. B. T. Federoff and O. Sheffield, *Encyclopedia of Explosives and Related Items,* Vol. 6, F 217, TR 2700, PTA, 1960.
88. D. E. Seeger, and R. H. Stresau, "Lead Azide Precipitated with Polyvinyl Alcohol," in *2nd ONR Symposium on Detonation,* Office of Naval Research, Washington, D.C., 1955, p. 92.
89. I. Kabek and S. Urman, "Hazards of Copper Azide in Fuzes," in *Minutes of the 14th Annual Explosives Safety Seminar,* NTIS, 1972, p. 533.
90. R. L. Wagner, *Lead Azide, Its Properties and Use in Detonators,* TR 2662, PTA, Dover, N.J., 1960.
91. H. Fair and R. Walker, eds., *The Inorganic Azides,* Vols. 1–2, Plenum Press, New York, 1977.
92. Spec. MIL-L-14758, *Lead Azide, Special Purpose,* USGPO, Washington, D.C., 1968.
93. Brit. Pat. 887,141 (Jan. 17, 1962), E. Williams, S. Peyton, and R. C. Harris.
94. T. Costain, *A New Method for Making Silver Azide,* TR 4595, PTA, Dover, N.J., 1974.
95. C. A. Taylor and W. H. Rinkenbach, *Army Ordnance 5,824* (1925).
96. D. A. Young, *Recent Physico-Chemical Investigations on Silver Azide,* NAVORD 5746, Washington, D.C., 1958.
97. T. A. Bronner and H. J. Jackson, *Normal Lead Styphnate; Development of Standardized Preparatory Methods,* TR 3079, PTA, Dover, N.J., 1963.
98. B. F. Husten, *Basic Lead Styphnate, Fine Milling,* NAVORD OD 23996, Washington, D.C., 1963.
99. T. Urbanski, *Chemistry and Technology of Explosives,* Vol. 3, Pergamon Press, New York, 1967.
100. R. L. Grant and J. E. Tiffany, *J. Ind. Eng. Chem.* **37,** 661 (1945).
101. "Diazodinitrophenol," *Encyclopedia of Explosives and Related Items,* Vol. 5 D1160, PATR 2700, ARDEC, Dover, N.J., 1972.
102. S. M. Kaye, "Tetracene," *Encyclopedia of Explosives and Related Items,* Vol. 6, G169 PATR 2700, ARDEC, Dover, N.J.
103. W. H. Rinkenbach and O. Burton, *Army Ordnance 12,* 120 (1939).
104. R. Barefoot, "Compatibility of Nitrate and Nitrate Esters," in *Conference on Compatibility of Propellants, Explosives, and Pyrotechnics with Plastics and Additives,* Report NC-02P, ADPA, Washington, D.C., 1976, p. 1-E-22.
105. P. Naoum, *Nitroglycerin and Nitroglycerin Explosives,* Williams and Wilkins Co., Baltimore, Md., 1928.
106. T. Urbanski, *Chemistry and Technology of Explosives,* Vol. 2, Pergamon Press, New York, 1965.
107. S. Kaye, ed., *Encyclopedia of Explosives and Related Items,* Vol. 3, TR 2700, PTA, Dover, N.J., p. 501; Vol. 6, p. 699; Vol. 8, p. N56.
108. C. Boyars, "Sensitivity and Desensitization of Nitroglycerin," in *Proceedings of 2nd Symposium on Chemical Problems Connected with the Stability of Explosives,* Jonkoping, Sweden, 1970, p. 197.
109. F. P. Bowden and A. D. Yoffe, *Initiation and Growth of Explosion in Liquids and Solids,* Cambridge University Press, New York 1952.

110. O. E. Waring and G. Krastins, *The Kinetics and Mechanism of Thermal Decomposition of Nitroglycerin,* Report 5746, NOL, White Oaks, Md., 1958.
111. *Worldwide Technology Assessment of Nitroglycerin Manufacture,* 4 Vols., (Proprietary Information) Mason and Hanger, Silas Mason, 1985.
112. G. S. Biasutti, *Safe Manufacture and Handling of Liquid Nitric Esters, ACS Symposium Series No. 22,* ACS, Washington, D.C., 1975.
113. Personal communication, G. S. Biasutti, Vevey, Switzerland.
114. *Nitroglycerin Plants, S-71030,* Nitro Nobel A B ROC Division, Gyttorp, Sweden.
115. W. H. Rinkenbach, *J. Ind. Eng. Chem.* **18,** 1196 (1926).
116. B. T. Federoff and O. E. Sheffield, eds., "Ethylene Glycol Dinitrate," in *Encyclopedia of Explosives and Related Items,* Vol. 6, PATR 2700, Picatinny Arsenal, Dover, N.J., 1974, p. 259.
117. "Diethylene Glycol Dinitrate," Ref. 104, Vol. 5, D-1232.
118. "Triethylene Glycol Dinitrate," Ref. 104, Vol. 9, T56.
119. W. G. Clark, *Evaluation of 1,2,4-Butanetriol Trinitrate as the Liquid Explosive Plasticizer for Cast Double Base Propellant,* Report 4, PTA, Dover, N.J., 1960.
120. "Butanetriol Trinitrate," Ref. 104, Vol. 2, B371.
121. *Thermal Stability, Aging and Exudation of METN Propellants,* Publication LL 77-29, CPIA, Laurel, Md., 1977.
122. "Metriol Trinitrate," Ref. 104, Vol. 6, E152.
123. F. D. Miles, *Cellulose Nitrate,* Interscience Publishers, New York, 1945.
124. J. C. Arthur, Jr., ed., *Cellulose Chemistry and Technology, ACS Symposium Series No. 48,* ACS, Washington, D.C., 1977.
125. N. M. Bikales and L. Segal, *Cellulose and Cellulose Derivatives,* Parts 4–5, Wiley-Interscience, New York, 1971.
126. J. Jessup and E. J. Prosen, *J. Res. Nat. Bur. Stand.* **44,** 387 (1950).
127. J. M. Goldman and E. H. Zeigler, Jr., "The Shock Nitration Process for Spherical Propellant Manufacture," in *Symposium on Processing Propellants, Explosives, and Ingredients,* ADPA, Washington, D.C., 1977, p. 23.
128. E. K. Rideal and A. J. B. Robertson, "The Spontaneous Ignition of Nitrocellulose," in *Proceedings of 3rd Symposium on Combustion, Flame, and Explosion Phenomena,* U. S. BuM, Pittsburgh, Pa., p. 536, 1948.
129. R. Van Dolah and S. Newman, *Nitrocellulose: A Review of Some of Its Fundamental Properties,* NOTS and Alleghany Ballistics Laboratory, Cumberland, Md., 1953.
130. *Nitrocellulose Degradation,* Classified Publication no. LS 76-39, CPIA, Laurel, Md., 1977.
131. Ref. 106, p. 362.
132. B. T. Federoff and O. Sheffield, *Encyclopedia of Explosives and Related Items,* Vol. 2, p. C100, TR 2700 PTA, Dover, N.J., 1962.
133. *Industrial Nitrocellulose,* ICI Ltd., Nobel Division, Kynoch Press, Birmingham, Ala., 1961.
134. *Nitrocellulose,* U. S. Spec. MIL-N-244, USGPO, Washington, D.C., 1975.
135. E. Dodgen, "Continuous Nitration of Cellulose: SNIA Viscosa Process," in *Symposium on Processing Propellants, Explosives and Ingredients,* ADPA, Washington, D.C., 1977, p. 4.2-1.
136. N. Shapira, ed., *Wastewater Treatment in the Military Explosives and Propellants Production Industry,* Vol. 3, ADPA, Washington, D.C., 1975, p. 103.
137. M. R. Olsen and R. K. Major, *Comparative Study of Dehydrating Processes in the Manufacture of Nitrocellulose,* United Technology Chemical Systems Division, Sunnyvale, Calif., 1975.
138. E. Berlow, R. H. Barth, and J. E. Snow, *The Pentaerythritols,* Reinhold Publishing Co., New York, 1926.

139. Ref. 106, p. 175.
140. A. B. Coates, E. Friedman, and L. P. Kuhn, *Characteristics of Certain Military Explosives,* Rpt. 1507, BRL, Aberdeen, Md., 1970.
141. D. M. Coleman and R. N. Rogers, "Pentaerythritol Tetranitrate (PETN) Stability and Compatibility," in *Proceedings of Conference on Compatibility of Propellants, Explosives, and Pyrotechnics with Plastics and Additives,* ADPA, Washington, D.C. 1974, p. 11-B-1.
142. M. R. Kantz, *Pentaerythritol Tetranitrate: A Bibliography,* NTIS, Springfield, Va., 1965.
143. *Pentaerythritol Tetranitrate (PETN),* U. S. Spec. MIL-P-00387B, USGPO, Washington, D.C., 1967.
144. Private communication, G. S. Biasutti, Biazzi Co., Vevey, Switzerland.
145. J. W. Patterson and R. W. Minear, *State of the Art for the Inorganic Chemicals Industry: Commercial Explosives PB 240 960,* Illinois Institute of Technology, Chicago, 1975.
146. J. Roth, in *Encyclopedia of Explosives and Related Items,* TR 2700, Vol. 8, ARDEC, Dover, N.J., 1978, p. 86.
147. F. J. Hildebrandt and R. T. Schimmel, *Suitability of RDX Compositions for Replacing Tetryl in Booster Explosives,* TR 4637, PTA, Dover, N.J., 1973.
148. T. L. Boggs in K. K. Kus and M. Summerfield, eds., *Fundamentals of Solid-Propellant Combustion, Progress in Astronautics and Aeronautics,* Vol. 90, AIAA, New York, 1984, p. 121.
149. O. H. Johnson, *HMX as a Military Explosive,* Rpt. 4371, NAVORD, Washington, D.C., 1956.
150. J. T. Rogers, *Physical and Chemical Properties of RDX and HMX,* Control Rpt. 20-P-26A, Holston Defense Corp., Kingsport, Tex., 1962.
151. *A Photographic Tour of the Holston Army Ammunition Plant,* Holston Defense Corp., Kingsport, Tenn.
152. *Standard Operating Procedures for Production of RDX,* Holston Defense Corp., Kingsport, Tenn., Jan. 1978.
153. "RDX," *Encyclopedia of Explosives and Related Items,* 9 R120-146 PATR 2700, ARDEC, Dover, N.J., 1980.
154. W. Auyeung and co-workers, *An Evaluation of Technologies to Produce Fine Particle Size RDX,* ARLCD-TR-85024, ARDEC, Dover, N.J., Sept. 1985.
155. J. Solomon, *A Study of the Nitrolysis of Hexamine to Increase HMX Yields,* Illinois Institute of Technology, Chicago, 1973.
156. R. Robbins, *The Preparation, Properties, and Uses of HMX,* Rpt. RR-GC-149, Holston Defense Corp., Kingsport, Tenn., 1958.
157. C. P. Achuthan, *Propellants, Explosives and Pyrotechnics* **15,** 271–275 (1990).
158. V. I. Scale and co-workers, *Propellants, Explosives and Pyrotechnics* **6,** 67–73 (1981).
159. D. Burrows, *Literature Review of the Toxicity of RDX and HMX,* U. S. Army Medical and Bioengineering Research and Development Lab., Washington, D.C., 1973.
160. J. A. Hathaway and C. R. Buck, "Report of Absence of Health Hazards Associated with RDX Manufacture and Use in Shell Loading Plants," in *Minutes of 17th Explosives Safety Seminar,* Dept. of Defense Explosives Safety Board, Washington, D.C., 1976, p. 683.
161. L. Silberman and S. M. Adelman, "Improved Yields in the Manufacture of HMX and RDX by the Bachmann Process," in *Symposium on Processing Propellants, Explosives and Ingredients,* ADPA, Washington, D.C., 1977, p. 2.2-1.
162. *Safety, Pollution, and Conservation Energy Review (Spacer) for Munitions Plant Modernization,* ARLCD-SP-77001, ARDEC, Dover, N.J., 1977.

163. D. Price and A. R. Clairmont, Jr., "Explosive Behavior of Nitroguanidine," in *Proceedings of the 12th Symposium on Combustion,* Combustion Institute, Pittsburgh, Pa., 1969, p. 761.
164. S. Levmore, *Air Blast Parameters and Other Characteristics of Nitroguanidine and Guanidine Nitrate,* TR 4865, PTA, Dover, N.J., 1975.
165. J. L. Block, *Thermal Decomposition of Nitroguanidine,* NAVORD 2705, NOL, Washington, D.C., 1953.
166. J. Savitt, "Some Properties and Uses of Nitroguanidine," paper no. 38 in *Proceedings of Institute for Chemie der Traub-und-Explosivestaffe, ICT,* International Jahrsteig, Karlsuike, FRG, 1985.
167. A. J. Tulis, "On Intermediate Explosive Compositions," in *Proceedings of the 5th (International) Pyrotechnic Seminar,* Denver Research Institute, Colo., 1976, p. 522.
168. *Encyclopedia of Explosives and Related Items,* Vol. 6, PATR 2700, ARDEC, Dover, N.J., 1978, p. G154.
169. C. H. Nichols, *Evaluation of Technologies to Produce Nitroguanidine,* TR 4566, PTA, Dover, N.J., 1974.
170. Ref. 68, p. 22.
171. V. Milan and co-workers, *The Preparation of High Bulk Density Nitroguanidine,* Rpt. 3037, NAVORD, Washington, D.C., 1957.
172. C. E. Duerr and co-workers, *Nitroguanidine Process Optimization,* ARLCB-CR-85031, ARDEC, Dover, N.J., Sept. 1985.
173. J. G. Stites, M. D. Barnes, and R. F. McFarlin, "A Survey of the Physical and Chemical Characteristics of Fertilizer-Grade Ammonium Nitrate" in *Proceedings of 5th Annual Symposium on Mining Research,* University of Missouri, Rolla, 1960, p. 1.
174. C. J. Bain, *Army Ordnance 6,* 435 (1926).
175. Ref. 79, p. 40.
176. R. C. Elderfield, *Study of the British Continuous Tetryl Process,* Rpt. 661, Office of Scientific Research and Development (OSRD), Washington, D.C., 1942.
177. R. J. Spear and V. Nanut, *A Comparative Assessment of U.S. and U.K. Explosives Qualified as Replacements for Tetryl,* MRL-R-1094, Materials Research Lab., Australia, 1987.
178. S. M. Kaye, ed, in *Encyclopedia of Explosives and Related Items,* Vol. 9, T148, ARDEC, Dover, N.J., 1978.
179. *Ibid.,* Vols. 3 p. C501 and 8, p. N51, ARDEC, Dover, N.J., 1978.
180. *New Explosive Specialties Brochure on Line Wave Generators,* E. I. du Pont de Nemours & Co., Inc., Explosives Products Division, Wilmington, Del.
181. S. A. Moses, "Explosive Components for Aerospace Systems," in *Behaviour and Utilization of Explosives in Engineering Design,*
182. A. O. Long, *Viscosities of Some Castable High Explosives,* NAVORD 2910, Washington, D.C., 1953.
183. J. C. Dacons, H. G. Adolph, and M. J. Kamlet, *J. Phys. Chem.* **74,** 3035 (1970).
184. V. M. Titov and co-workers, "Investigation of Some Cast TNT Properties at Low Temperatures," in *6th Symposium (International) on Detonation,* ONR, Washington, D.C., 1976.
185. H. H. Cady and W. H. Rogers, *Enthalpy, Density, and Thermal Coefficient of Cubical Expansion of TNT,* LA-2696, LLL, Livermore, Calif., 1962.
186. M. J. Urizar, E. James, Jr., and L. C. Smith, "The Detonation Velocity of Pressed TNT," in *3rd (International) Symposium on Detonation,* ONR, Washington, D.C., 1960, p. 337.
187. D. J. Ornellas, *J. Phys. Chem.* **72,** 2390 (1968).
188. H. J. Reitsma, in J. Hansen ed. *Symposium on Chemical Problems Connected with Stability of Explosives,* Bastad, Sweden, 1979.

189. F. T. Kristoff and J. Digiovanni, *A Hazards Analysis Study of the Continuous TNT Manufacturing Plants,* U.S. Army Radford Ammunition Plant, Radford, Va., 1971.
190. J. Dunsten, ed., *Joint US/UK Seminar on TNT Chemistry and Manufacture,* Rpts. 26 (1971) and 106 (1972), ERDE, Waltham Abbey, Essex, UK.
191. W. T. Bolleter, "Recent Improvements in the Continuous TNT Manufacturing Process," in *Symposium on Processing Propellants, Explosives, and Ingredients,* ADPA, Washington, D.C., 1977, p. 2.4-1.
192. T. Urbanski, *Chemistry and Technology of Explosives,* Vols., 1–3, 1964, Vol. 4, 1984, Pergamon Press, New York.
193. Y. O. Dova, *The Chemistry and Technology of High Explosives,* Wright Patterson Air Force Base Translation, Dayton, Ohio, 1961.
194. A. B. Bofors, *TNT Manufacture by the Continuous Bofors-Norell Method,* brochure, Sweden, 1956.
195. R. L. Goldstein, "Recent Developments in the Optimization and Control of Nitration in the Continuous Manufacture of TNT," in *Symposium on Processing Propellants, Explosives, and Ingredients,* ADPA, Washington, D.C., 1977, p. 2.5-1.
196. A. E. Tatyrek, *Treatment of TNT Munitions Wastewaters, The Current State of the Art,* Rpt. 4909, PTA, Dover, N.J., 1976.
197. E. E. Gilbert, in S. M. Kaye, ed., *Encyclopedia of Explosives and Related Items,* Vol. 9, PATR 2700, ARDEC, Dover, N.J., 1980, p. 232.
198. E. E. Gilbert and co-workers, *Propellants, Explosives, Pyrotechnics* **7,** 150 (1982).
199. O. Sanders and co-workers, *Mechanism of Formation of Pink Water,* ARLCD-TR-70025, ARDEC, Dover, N.J., 1978.
200. Ref. 136, Vol. 1, 1972, p. 55.
201. L. L. Smith and co-workers, in Ref. 42.
202. *Pollution Abatement and Conservation of Energy Review for Munitions Plant Modernization,* TR 2210, PTA, Dover, N.J., 1976.
203. S. Zaheri and co-workers, *Occupational Health and Safety of 2,4,6 Trinitritoluene (TNT),* final report contract DAMD 17-77-C-7020, Franklin Institute Research Laboratory, Philadelphia, Pa., 1978.
204. *Properties of Explosives of Military Interest,* Engineering Design Handbook Series No. 707-177, AMC, Alexandria, Va., 1971.
205. J. M. Roth and S. M. Kaye, eds., "Picric Acid," in *Encyclopedia of Explosives and Related Items,* Vol. 8, PATR 2700, ARDEC, Dover, N.J., 1978, p. 285.
206. R. L. Beauregard, *History of Navy Uses of Composition A-3 and Explosive D in Projectiles,* TR 70-1, NAVORD, Washington, D.C., 1970.
207. H. Heller and A. L. Bertram, *HNS-Teflon: A New Heat Resistant Explosive,* TR 73-163, NOL, White Oaks, Md., 1973.
208. J. C. Dacons and E. E. Kilmer, "HNS Specifications", paper presented at *Annual Meeting of the Pyrotechnics and Explosives Application Section,* ADPA, Washington, D.C., 1976.
209. E. E. Kilmer, *Detonating Cords Loaded with HNS Recrystallized from Acid and Organic Solvents,* TR 75-142, NSWC, White Oaks, Md., 1975.
210. A. C. Schwartz, *Application of Hexanitrostilbene HNS in Explosive Components,* SC-RR-710673, Sandia Laboratories (SAN), Albuquerque, N. Mex., 1972.
211. F. Z. and B. Chen, *Propellants, Explosives, Pyrotechnics,* 16 (1991).
212. V. Evens, "Optimization of TATB Processing," in *Symposium on Processing Propellants, Explosives, and Ingredients,* ADPA, Washington, D.C., 1977.
213. R. K. Jackson and co-workers, "Initiation and Detonation Characteristics of TATB," in *Proceedings of Symposium on Compatibility of Plastics and Other Materials with Explosives, Propellants, and Pyrotechics,* ADPA, Washington, D.C., 1976.

214. R. H. Pritchard, "Compatibility of TATB-PBM with Weapon Material," in *Symposium on Compatibility of Plastics and Other Materials with Explosives, Propellants, and Pyrotechnics,* ADPA, Washington, D.C., 1976.
215. A. G. Osborn, "TATB Formulation Processing," in *Symposium on Processing Propellants, Explosives and Ingredients,* ADPA, Washington, D.C., 1977.
216. L. J. Bement, "Application of Temperature Resistant Explosives to NASA Mission," in *Symposium on Thermally Stable Explosives,* NOL, White Oaks, Md., June 1970.
217. C. D. Hutchinson and co-workers, "Initiation and Detonation Properties of the Insensitive High Explosive TATB/KEL-F800 95/5," Vol. 1, in *9th Symposium (International) on Detonation,* NAVORD, Washington, D.C., 1985, p. 28.
218. R. K. Jackson and co-workers, "Initiation and Detonation Characteristics of TATB," in *Proc. 6th Symposium (International) on Detonation,* San Diego, Calif., Aug. 1976.
219. B. T. Federoff and O. E. Sheffield, in *Encyclopedia of Explosives and Related Items,* Vol. 7, TR2700, PTA, Dover, N.J., 1975, p. 46.
220. J. E. Ablard, *Composition B: A Very Useful Explosive,* NAVSEA-03-TR-058, Naval Sea Systems Command, Washington, D.C., 1977.
221. R. Pellon and K. Russel, *Study of Melt Loading the 105mm M1 Projectile with Composition B Containing Grade B Wax,* TR 4854, PTA, Dover, N.J., 1975.
222. S. D. Stein, *The Problem of TNT Exudation,* TR2493, PTA, Dover, N.J., 1958.
223. J. E. Ablard, *HBX-1: Its History and Properties,* NAVSEA-03-TR-021, Washington, D.C., 1975.
224. *Explosive Compositions, HBX Type,* U.S. Military Spec. E-22267A, USGPO, Washington, D.C., 1963.
225. C-Y. Chen and J. Hwashiuan, *Propellants, Explosives, Pyrotechnics* **17,** 20–26 (1992).
226. B. T. Federoff and O. E. Sheffield, eds., "Aluminum Containing Explosives," in *Encyclopedia of Explosives and Related Items,* Vol. 1, PATR 2700, Dover, N.J., 1960, p. 146.
227. E. James, Development of Plastic Bonded Explosives, UCRL 12439-T, University of California Press, 1965.
228. E. Y. McGann, *A Safety, Quality, and Cost Effectiveness Study of Composition,* Press Loading Parameters, TR 76-1, Naval Ordnance Laboratory, White Oaks, Md., 1976.
229. S. M. Kaye, ed., *Encyclopedia of Explosives and Related Items,* Vol. 8, TR 2700, ARDEC, Dover, N.J., 1978, p. 60.
230. J. R. Polson, *Mechanical Pressing of Explosives,* Iowa Army Ammunition Plant, Burlington, Iowa, 1973.
231. D. Kite, Jr., A. K. Behlert, and E. Jerzcerewski, *Plastic Bonded Explosives for Use in Ammunition,* PATM 2-2-62, PTA, Dover, N.J., 1962.
232. J. McDevitt, "Processing Characteristics of Castable PBX Explosives," in *Symposium of Processing Propellants, Explosives, and Ingredients,* ADPA, Washington, D.C., 1977, p. 3.2-1.
233. J. R. Wanninger and E. Kleinschmidt, "Pressed Plastic Bonded Charges," in *Proceedings of Joint International Symposia on Compatibility of Plastics and Other Materials with Explosives, Propellants and Pyrotechnics,* ADPA, New Orleans, La., Apr. 1985.
234. *Apparent Consumption of Industrial Explosives and Blasting Agents in the United States,* Annual Mineral Industry Surveys. U.S. Dept. of Interior, Washington, D.C., 1989.
235. R. W. Watson, J. E. Hay, and R. W. Van Dolah, "Commercial Explosives in the United States: Generalities and Some Details," in *Symposium on Military Applications of Commercial Explosives,* DREV M-2241/72, Defense Research Establishment, Valcartier, Canada, 1972, p. 13.
236. C. E. Gregory, *Explosives for North American Engineers,* Trans Tech Publications, Clausthal, Germany, 1973.

237. T. Urbanski, *Chemistry and Technology of Explosives,* Vol. 3, Pergamon Press, New York, 1967, p. 395.
238. B. T. Federoff and O. Sheffield, *Encyclopedia of Explosives and Related Items,* Vol. 5, TR 2700 D 1584, PTA, Dover, N.J., 1972.
239. J. J. Yancik and G. B. Clark in Ref. 173, p. 67.
240. G. W. Brown and E. J. Styskala in Ref. 239, p. 126.
241. Ref. 238, p. A311.
242. T. Urbanski, *Chemistry and Technology of Explosives,* Vol. 2, Pergamon Press, New York, 1965, p. 450.
243. D. T. Baily and co-workers, in Ref. 173.
244. J. N. Johnston and co-workers, *Propellants, Explosives and Pyrotechnics* **8,** 8 (1983).
245. S. R. Brinkley and W. E. Gordon, "Explosive Properties of the Ammonium Nitrate-Fuel Oil System," in *Proceedings of 31st Inst. Congress of Industrial Chemistry,* Liege, Belg., 1958.
246. *ANFO Manual,* Monsanto Co., St. Louis, Mo., 1969.
247. W. E. Tournay and co-workers, in Ref. 173, p. 164.
248. J. Briggs, "A Safer Blast for the Modern Army," in *Proceedings of the 14th Annual Explosives Safety Seminar,* U.S. Department of Defense Explosive Safety Board, NTIS, Springfield, Va., 1972, p. 313.
249. *Proceedings of the Symposium on Military Applications of Commercial Explosives,* DREV M-2241/72 Defense Research Establishment, Valcartier, Can., 1972.
250. R. A. Dick, *Factors in Selecting and Applying Commercial Explosives and Blasting Agents,* I. C. Rpt. 9405, U.S. BuM, Pittsburgh, Pa., 1968.
251. B. Wells, *High Energy Flexible Explosive,* TR4846, Picatinny Arsenal, Dover, N.J., 1976.
252. M. E. Lackey, *Utilization of Energetic Materials in an Industrial Combustor,* AMXTHE-TE-R 85003, U.S. Army Toxic and Hazardous Materials Agency, Edgewood, Md., June 1985.
253. G. Cohen in Ref. 236, p. 189.
254. H. S. Yadav and co-workers, *Propellants, Explosives and Pyrotechnics,* **15,** 194 (1990).
255. S. C. Alford *Explosives Engineer,* 1985.
256. S. B. Marray and K. B. Gorrard, *Field Studies of Fuel-Air Explosive Line Charges,* Canadian Defense Research Est., Alberta, Canada, July 1990.
257. S. N. Shoukry and A. A. Hegazy, *Propellants, Explosives and Pyrotechnics* **13,** 144 (1988).
258. S. Schumann and co-workers, *Propellants, Explosives and Pyrotechnics* **11,** 133 (1986).
259. S. Iyer and co-workers, "New High Energy Density Materials for Propellant Applications," in *5th International Gun Propellant and Propulsion Symposium,* ARDEC, Picatinny Arsenal, N.J., Nov. 1991, pp. 18–21.
260. S. Iyer and co-workers, *Research Toward More Powerful Explosive,* technical report ARAED-TR-89010, ARDEC, Picatinny Arsenal, N.J., June 1989.
261. Milton Finger, *New Directions in Energetic Materials Research,* LLL, Livermore, Calif., Mar. 1981.
262. T. G. Archibald and co-workers, *J. Org. Chem.* **55,** 2920 (1990).
263. P. E. Eaton, in S. Iyer, ed., *Proceedings of the Ninth Annual Working Group Institute on the Synthesis of High Energy Density Materials,* U.S. Army, ARDEC, Picatinny Arsenal, N.J., June 90.
264. P. Dave and co-workers, in S. Iyer, ed., *Proceedings of the Tenth Annual Working Group Institute on the Synthesis of High Energy Density Materials,* U.S. Army ARDEC, Picatinny Arsenal, N.J., June 91.

265. *Toward New Explosive Molecules in Dynamic Testing,* Los Alamos National Laboratory, N. Mex., Mar. 1984.
266. M. M. Stine and L. A. Strety, "Sensitivity and Performance Characterization of DINGO," in *Eighth Symposium (International) on Detonation,* Albuquerque, N. Mex., 1983.
267. C. H. Dettling and co-workers, *Insensitive Munitions Characteristics of Air Launched In-Service Weapons: Summary Report of Fast Cook-off Times, Reactions and Hazards of Bombs, Rockets, Aircraft Guns, Air Launched Missiles, Mines and Torpedoes,* Naval Weapons Center, China Lake, Calif., Sept. 1989.
268. "Study of Insensitive High Explosives and Propellants," in *Proceedings of Conference USDR&E 79-653,* Mar. 1979.
269. *Hazard Assessment Tests for Non-Nuclear Ordnance,* Military Standard 2105A (Draft), 1991.
270. R. D. Lynch and co-workers, "Characterization of Insensitive High Explosives Developed with Propellant Technology," in *Proc. 1990 JANNAF Propulsion Meeting, VIII,* 3-5 CPIA Publication 550, CPIA, Laurel, Md., Oct. 1990, pp. 3–5.

General References

Annual Proceedings of the Safety Seminars, Dept. of Defense, Explosive Safety Board, Washington, D.C. International symposia on explosives and closely related subjects are excellent sources of information, ie, international symposia on detonation; symposia on combustion; symposia on chemical problems connected with the stability of explosives; international pyrotechnics seminars; symposia on compatibility of plastics and other materials with explosives, propellants, and pyrotechnics, and processing of explosives, propellants, and ingredients; and symposia on explosives and pyrotechnics

Mineral Industry Surveys, U.S. Bureau of Mines, Pittsburgh, Pa. Periodic publications dedicated primarily to explosive studies in: *Propellants and Explosives; Journal of Hazardous Materials;* and apparent consumption of industrial explosives and blasting agents in the United States.

B. T. Federoff, O. E. Sheffield, and S. Kaye, eds., *The Encyclopedia of Explosives and Related Items,* PATR 2700, Vols. 1–10, ARDEC, Dover, N.J. This provides a very wide variety of information

D. C. Ascani, "The Literature of Chemical Technology," in *Advances in Chemistry Series No. 76,* American Chemical Society, Washington, D.C., 1968. Bibliography on explosives.

Handbook of Foreign Explosives, FSTC 381-5042, AMC, Washington, D.C., 1965.

M. A. Cook, *The Science of High Explosives,* Reinhold Publishing Corp., New York, 1958.

M. A. Cook, *The Science of Industrial Explosives, Graphic Services and Supply,* IRECO Chemicals, Salt Lake City, Utah, 1974.

G. S. Biasutti, *History of Accidents in the Explosive Industry,* Vevey, Switzerland, 1980.

T. W. Urbanski, *Chemistry and Technology of Explosives,* Vols. 1–3, Pergamon Press, New York, 1967, Vol. 4, 1984.

S. Fordham, *High Explosives and Propellants,* 2nd ed., Pergamon Press, New York, 1980.

J. Taylor, *Detonation in Condensed Explosives,* Clarendon Press, Oxford, 1952.

R. H. Cole, *Underwater Explosives,* Princeton University Press, Princeton, N.J., 1948.

E. W. Baker, *Explosives in Air,* University of Texas Press, Austin, 1973.

K. O. Brauer, *Handbook of Pyrotechnics,* Chemical Publishing Co., New York, 1974.

J. H. McLain, *Pyrotechnics,* Franklin Institute Press, Philadelphia, 1980.

J. H. McLain, *Principles of Explosive Behavior,* Engineering Design Handbook, AMC 706-180, AMC, Alexandria, Va., 1972.

W. Fickett and W. C. Davis, *Detonation,* University of California Press, Berkeley, 1979.

Blasters Handbook: A Manual Describing Explosives and Practical Methods of Using Them, E. I. duPont de Nemours & Co., Inc., Wilmington, Del., 1977.

J. W. Patterson and R. W. Minear, *State of the Art for the Inorganic Chemicals Industry: Commercial Explosives,* PB240960, Illinois Institute of Technology, Chicago, 1979.
J. Glacken, *Elements of Explosives Production,* Paladin Press, 1976
R. J. Lewis, Sr., ed., *Registry of Toxic Effects of Chemical Substances,* Tracor Jitco, Inc., Rockville, Md., Jan. 1979.
C. S. Robinson, *Explosives, Their Anatomy of Destructiveness,* McGraw-Hill Book Co., Inc., New York, 1944.
R. Meyer, *Explosives,* 2nd ed., Verlag Cheme, Deerfield Beach, Fla., 1981.
T. N. Hall and J. R. Holden, *Navy Explosives Handbook, Explosion Effects and Properties, Part 3, Properties of Explosives and Explosive Compositions,* NSWC, White Oak, Md., MP-8116, Oct. 1988.
Military Explosives, Dept. of the Army technical manual, TM9-1300-214, Sept. 1984.
Sources of publications in the United States: Chemical Propulsion Information Agency, Applied Physics Laboratory, Johns Hopkins University, Laurel, Md.; and Defense Documentation Center.
Sources of information on the testing evaluation and properties of explosives include the following:
A Manual for the Prediction of Blast and Fragment Loadings on Structures, DOE/TIC 11268, U.S. Dept. of Energy, Amarillo, Tex., Nov. 1980.
J. M. Pakulak and C. M. Anderson, *Naval Weapons Center Standard Methods for Determining Thermal Properties of Propellants and Explosives,* NWC TP 6118, Naval Weapons Center, China Lake, Calif., Mar. 1980.
J. M. Pakulak and E. Kuletz, *Thermal Analyses Studies on Candidate Solid JPL Propellants for Heat Sterilizable Motors,* Naval Weapons Center, China Lake, Calif., TP 4258, July 1970.
Explosives Sampling, Inspection and Testing, U.S. Military Standard Specification 650, ARDEC, Dover, N.J., May 1973.
Navy Bank of Explosives Data (NAVBED), Naval Surface Weapons Center MP83-230, June 1983.
T. R. Gibbs and A. Popolato, eds., *LASL Explosive Property Data,* University of California Press, Berkeley, 1980.
B. M. Dobratyz, *LLNL Explosives Handbook, Properties of Chemical Explosives and Explosive Simulants,* UCRL 52997, LLNL, University of California, Livermore, Mar. 1981.
M. M. Swisdak, Jr., ed., *Explosive Effects and Properties,* TR75-116 NSWC, Oct. 1975, and *Explosion Effects in Water,* NSWC-TR-76-116 Feb. 1978, NSWC/NOL, White Oaks, Md.
U.S. Military Specifications. These are informative publications describing characteristics of materials, methods of testing, sampling, packaging and criteria for acceptance, U.S. Government Printing Office, Washington, D.C.

VICTOR LINDNER
U.S. Armament Research, Development and Engineering Agency

PROPELLANTS

Propellants are mixtures of chemical compounds that produce large volumes of high temperature gas at controlled, predetermined rates, and can sustain combustion without requiring atmospheric oxygen for the purpose. Principal applications are in launching projectiles from guns, rockets, and missile systems. Propellant-actuated devices are used to drive turbines, move pistons, operate rocket vanes, start aircraft engines, eject pilots, jettison stores from jet aircraft, pump fluids, shear bolts and wires, and act as sources of heat in special devices. Propellants are applicable wherever a well-controlled force must be generated for a relatively short period of time. Solid propellants are compact, have a long storage life, and may be handled and used without exceptional precautions.

General Characteristics

Gun Propellants. Solid gun propellants are employed in the form of dense cylindrical or spherical grains, elongated hollow or split sticks, or as sheets of plasticized nitrocellulose. Gun propellants are almost always based on nitrocellulose to provide mechanical strength. These also may contain inert or energetic liquid plasticizers (qv) or a combination of the two to improve physical and processing characteristics, high explosives to increase available energy, stabilizers to prolong storage life, and a small amount of inorganic additives to facilitate handling, improve ignitibility, and decrease muzzle flash. Single-based propellants, used exclusively in guns, derive energy primarily from nitrocellulose [*9004-70-0*]. Double-based nitrocellulose propellants contain liquid energetic plasticizers such as nitroglycerin [*55-63-0*], and are used in rockets as well as guns. Triple-based propellants contain crystalline additives, eg, nitroguanidine [*556-88-7*], as well as nitrocellulose and energetic additives. Both double- and triple-based propellants are used in guns and rockets. Low sensitivity propellants (LOVA) have also been developed for use in guns; these contain a high energy component, eg, cyclotrimethylene trinitramine (RDX) [*121-82-4*] in a polymeric binder. Gun propellants are made mostly by an extrusion process that produces small grains in large numbers. Nitrocellulose serves on the energetic binder. Typical components of nitrocellulose propellants and their functions are

nitrocellulose	energetic binder
polyglycol diols	nonenergetic binder
nitroglycerin, metriol trinitrate, diethylene glycol dinitrate, triethylene glycol dinitrate, dinitrotoluene	plasticizers, energetic
dimethyl, diethyl, or dibutyl phthalates, triacetin	plasticizers, fuels
diphenylamine, diethyl centralite, 2-nitrodiphenylamine, acardite, diethyl diphenylurea	stabilizers
lead salts, eg, lead stannate, lead stearate, lead salicylate	ballistic modifiers in rocket propellants
carbon black	opacifier
lead stearate, graphite, wax	lubricants

potassium sulfate, potassium nitrate, cryolite (potassium aluminum fluoride)	flash reducers in gun propellants
ammonium perchlorate, ammonium nitrate	oxidizers, inorganic
RDX, HMX, nitroguanidine, and other nitramines	oxidizers, organic
lead carbonate, tin	decoppering agents in gun propellants

Rocket Propellants. Solid rocket propellants are mostly based on chemically cross-linked polymeric elastomers to provide the mechanical properties required in launchings and the environmental conditions experienced in storage, shipment, and handling (see ELASTOMERS, SYNTHETIC). Double- and triple-based nitrocellulose propellants are also employed as rocket propellants.

Polymer-based rocket propellants are generally referred to as composite propellants, and often identified by the elastomer used, eg, urethane propellants or carboxy- (CTPB) or hydroxy- (HTPB) terminated polybutadiene propellants. The cross-linked polymers act as a viscoelastic matrix to provide mechanical strength, and as a fuel to react with the oxidizers present. Ammonium perchlorate and ammonium nitrate are the most common oxidizers used; nitramines such as HMX or RDX may be added to react with the fuels and increase the impulse produced. Many other substances may be added including metallic fuels, plasticizers, stabilizers, catalysts, ballistic modifiers, and bonding agents. Typical components are listed in Table 1.

Nitrocellulose-based rocket propellant grains contain energetic liquid plasticizers such as nitroglycerin, stabilizers, ballistic modifiers, nonenergetic plasticizers, inorganic oxidizer salts, organic explosives, and metallic fuels similar to those used in gun propellants. When these latter components are included, the composition is referred to as a composite-modified double-based propellant (CMDB). Nitrocellulose-based propellants have also been made using isocyanate-curable elastomers which permit a reduction in the amount of nitrocellulose used and an increase in the nitroglycerin contents. The composition of a typical elastomer-modified composite double-base composition (EMCDB), compared to characteristic compositions of straight double-base (DB), composite-modified double-base, and conventional composite rocket propellants, can be found in Table 2.

Rocket propellants are made mostly by a casting process as distinct from the extrusion process used to make the very much smaller and more numerous gun propellant grains (1,2).

Selection Criteria

Energy Considerations. The selection of gun and rocket propellants involves two principal considerations: the total amount of energy required and the mass rate at which the hot gases produced must be delivered to meet system performance requirements. The energy delivered per unit mass depends on the chemical energy of the propellant components, the characteristics of the products of combustion, the chemical equilibria which prevail among the reaction products, and the efficiency with which the system converts thermal to kinetic energy. The

Table 1. Typical Components of Composite Rocket Propellants

Component	Characteristics
	Binders
polysulfides	reactive group (mercaptyl, —SH), is cured by oxidation reactions; low solids loading capacity and relatively low performance; mostly replaced by other binders
polyurethanes, polyethers, polyesters	reactive group (hydroxyl, —OH), is cured with isocyanates; intermediate solids loading capacity and performance
polybutadienes copolymer of butadiene and acrylic acid	reactive group (carboxy, —COOH, or hydroxyl, —OH), is cured with difunctional epoxides or aziridines; intermediate solids loading capacity and better performance than polyurethanes; less than adequate cure stability and mechanical characteristics
terpolymers of butadiene, acrylic acid, and acrylonitrile	superior physical properties and storage stability
carboxy-terminated polybutadiene	cured with difunctional epoxides or aziridines; have good solids loading capacity, high performance, and good physical properties
hydroxy-terminated polybutadiene	cured with diisocyanates; have good solids loading and performance characteristics, and good physical properties and storage stability
	Oxidizers
ammonium perchlorate	most commonly used oxidizer; it has a high density, permits a range of burning rates, but produces smoke in cold or humid atmosphere
ammonium nitrate	used in special cases only, it is hygroscopic and undergoes phase changes, has a low burning rate, and forms smokeless combustion products
high energy explosives (RDX–HMX)	have high energy and density; produce smokeless products; have a limited range of low burning rates
	Fuels
aluminum	most commonly used; has a high density; produces an increase in specific impulse and smoky and erosive products of combustion
metal hydrides	provide high impulse, but generally inadequate stability; produce smoky products and have a low density
	Ballistic modifiers
metal oxides	iron oxide most commonly used
ferrocene derivatives	permit a significant increase in burning rate
other	coolants for low burning rate and various special types of ballistic modifiers
	Modifiers for physical characteristics
plasticizers	improve physical properties at low temperatures and processibility; may vaporize or migrate; can increase energy if nitrated
bonding agents	improve adhesion of binder to solids

Table 2. Composition of Rocket Propellants, wt %

	Propellant type[a]			
Constituent	DB	CMDB	EMCDB	Composite
nitrocellulose	53	25	15	
nitrate ester	40	25	30	
nitramines			43	
ammonium perchlorate		30		70
aluminum		13		18
stabilizers	2	2	2	
polymeric binder		5	5	12
ballistic additives	5	5	5	

[a] Terms are defined in text.

rate at which energy is produced depends on the intrinsic burning characteristics of the propellant, its burning surface area, and the operating pressure and temperature of the system. Control of the burning surface area is obtained by using appropriate grain geometries and the required number of grains. Uncontrolled burning can result in intolerably high pressure or, in the worst case, catastrophic detonation.

The thermochemical–thermodynamic factors affecting gun and rocket performance are essentially the same. Both guns and rockets convert thermal energy into kinetic energy through physical–chemical processes. The highest energy propellants produce the largest volume of gas per unit weight of propellant at the highest flame temperature. The selection of propellant compositions for maximum performance focuses on high density compositions that form highly exothermic low molecular weight combustion products that are stable with minimum dissociation at gun or rocket operating pressures. Many practical considerations limit the attainment of the theoretical maximum performance. High flame temperature propellants used in rockets may cause excessive nozzle erosion and dissociation of the gaseous products at the relatively low operating pressures in rocket chambers. Use in guns may cause excessive gun tube wear and muzzle flash. The incorporation of large percentages of nitramines such as nitroguanidine in triple-base propellants or RDX in LOVA gun propellants is intended to produce the maximum energy at the lowest possible flame temperature. The isochoric adiabatic flame temperatures of propellants in use ranges from ca 2000 to 3500 K. The impetus of gun propellants is ca 822 to 1196 J/g (275,000–400,000 ft·lb/lb). The specific impulse of high performance rocket propellants is ca 2455–2700 N·s/kg (250–275 lbf·s/lb). Factors influencing the selection of propellants for guns and rockets are as follows:

Manufacturing Characteristics

Availability and cost of raw materials and processing equipment

Simplicity and cost of manufacture and inspection

Manufacturing hazards

Propellant viscosity and flowability

Environmental considerations

Energy Delivery Requirements

Specific impulse or force

Loading density in terms of required burning characteristics

Metal parts requirements in terms of operating pressure over required temperature range

Temperature Dependence

Ignition, pressure, burning rate, and thrust characteristics over temperature range

Mechanical Characteristics Over Temperature Range

Effects of High–Low Temperature Cycling

Reliability of Performance

Lot-to-lot variations in burning rate and pressure

Effect of small variations in metal parts on performance

Effect of small variations in composition and dimensions on performance

Long-Term Storage Characteristics

Deformation changes

Performance changes

Moisture absorption

Exudation or migration of plasticizer

Effects of Mechanical Characteristics

Long-term storage

High–low temperature cycling

Acceleration forces

Rough handling

Case bonding

Compatibility

With process equipment

With personnel (toxicity)

With metal and plastic parts and other components

Of reaction products with personnel, metal parts, and electronic equipment

Erosive effects of reaction products

System Requirements

Smokeless exhaust

Combustion stability

Effect of exhaust plume on radar

Absence of ignition peaks or reinforcing pressure waves

Minimum gun smoke, flash, and blast pressure

Detonation-free in event of malfunction

Minimum sensitivity to fire, high velocity fragments, and other evolved stimuli

Performance Calculations. The energy evolved by a propellant may be estimated from the percentage composition, reaction products, the heats of formation of the reactants and the products, the propellant density, and the gases and solids produced. The composition and flame temperatures of the products are determined from the applicable enthalpy–temperature and chemical equilibrium functions of the various molecular species and the operating conditions in the combustion chamber. The most important thermodynamic–thermochemical characteristics of propellant combustion products, in addition to gas volume and flame temperature, are heat capacity, heat capacity ratio, and the covolume of the gases at high pressures. Rigorous calculations require the solution of numerous equations which describe the mass and enthalpy balance and the chemical equilibria of the reaction products at elevated temperatures and pressures. Many computer programs have been developed for predicting rocket or gun performance. Simplified first approximation techniques are also available (3–15).

Mechanical Characteristics. *Rocket Propellants.* Large rocket grains in particular must have adequate mechanical properties to enable them to withstand the stresses imposed during handling and firing. These must be capable of performing satisfactorily after undergoing the thermal stresses produced during long-term exposure and cycling at temperature extremes. The development of high energy rocket propellants emphasizes maximum toughness and low shock and impact sensitivity. The mechanical properties depend primarily on the characteristics of the binder, the percentage of solids present, and the particle size distribution of the solids.

Rocket propellants are subjected to a large number of tests and inspections to establish mechanical and physical characteristics. Well-established laboratory methods determine the tensile and compressive strengths, the modulus in tension and compression, elongation under tension, and deformation under compression. Empirical techniques correlate these data and field performance. High rate of load application of tensile forces simulate those generated during ignition. Low rate tests simulate the stresses produced by differential thermal expansion. Compression test data are related to the forces experienced by rocket grains supported at the aft end of the motor. Drop tests of loaded rocket motors and vibrational, centrifugal, and sled tests that impose acceleration forces comparable to those expected in use are among the techniques employed. The linear coefficient of thermal expansion is important in rocket systems in which the propellant is coated with an inhibitor or bonded to the motor. Typical values are $3.6–7.2 \times 10^{-5}$ m/(m·K). Thermal diffusivities generally range between 7.7 and 15.5×10^{8} m^2/s. Thermal conductivities are ca 0.22–0.33 W/(m·K) (0.13–0.19 Btu·ft/(h·ft^2·°F)). Specific heat values are ca 1.26×10^{3} J/(kg·K) (0.3 Btu/(lb·°F)). Low thermal conductivity may cause severe thermal stresses that sometimes lead to cracking in large rocket grains when abrupt changes in storage temperatures occur.

Rocket propellants must not contain sizable cracks, pores, or cavities. They are inspected using x-rays and ultrasonics (qv), and firings are conducted in strand burners, interrupted burners, and in reduced or full-scale rocket motors (see also NONDESTRUCTIVE TESTING) (16–20).

Gun Propellants. Although the stresses on individual gun propellant grains are less severe because of the small size, these propellants must withstand much higher weapon pressures and accelerations. Formulation options are usually more limited for gun propellants than for rocket propellants because the products of combustion must not foul or corrode a gun, should have a low flame temperature, and should exhibit minimum flash and smoke characteristics. Gun propellants are examined microscopically for porosity, are tested for mechanical characteristics, and fired in closed bombs to determine the burning characteristics.

Shelf Life Characteristics. The chemical safe life of all standard propellants is measured in years. Both gun and rocket propellants contain chemical stabilizers that combine with the products of decomposition to prevent autocatalytic breakdown of the propellant composition. The useful service life of gun propellants may be as long as 25 to 50 years. The useful service life of rocket propellants may be significantly less than the chemical safe life if gassing occurs, motor bonding deteriorates, or significant physical changes take place. Generally such effects are produced by high temperature storage or high–low temperature cycling, particularly if moisture is present. Relatively little degradation occurs at ambient temperature conditions. Gassing can produce internal pressures which may crack a large rocket grain or cause propellant–inhibitor or propellant–motor bond failure. The likelihood of performance failure in standard rocket systems as a result of gassing is low because of the use of chemical stabilizers and the selection of compatible inhibitors, cements, and insulation materials.

The procedures used for estimating the service life of solid rocket and gun propulsion systems include physical and chemical tests after storage at elevated temperatures under simulated field conditions, modeling and simulation of propellant strains and bond line characteristics, measurements of stabilizer content, periodic surveillance tests of systems received after storage in the field, and extrapolation of the service life from the detailed data obtained (21–33).

The Burning Process

The mass rate of a propellant grain burning at a given pressure and temperature depends on the amount of heat evolved during decomposition and the amount of heat transferred to the burning surfaces of the propellant from the hot gases above it. This rate is also influenced by the tangential velocity of the propellant gases and the radiation from the surroundings. Propellants burn in parallel layers so that the surface recedes in all directions normal to the original surface. The geometry of the grain on completion of burning is similar to its geometry at the start. Propellant burning at high gun pressures proceeds more smoothly and is less subject to erratic behavior than burning at very low pressures because the conditions are appropriate for maximum energy transfer in minimum time. The burning rate at gun pressures usually varies somewhat less than the first power of the pressure. It changes more slowly at rocket pressures of 3.45–10.34 MPa (500–1500 psi), often to less than the square root of the pressure.

The composition of the propellant determines the rate of exothermic molecular breakdown at a given temperature and pressure. As the reaction rate increases, the rates of heat production and transfer increase with associated increases in the linear burning rate of the propellant. The heat evolved per gram of propellant in an inert atmosphere is its heat of explosion, Q. It may be readily calculated or experimentally determined in a calorimetric bomb (see THERMAL,

GRAVIMETRIC, AND VOLUMETRIC ANALYSIS). Values range from ca 2.09 kJ/g (500 cal/g) for cool propellants to ca 6.27 kJ/g (1.5 kcal/g) for maximum energy propellants. The flame temperatures and the burning rates of uncatalyzed propellants of similar compositions are generally linearly related to the calorific values except at very low pressures. The presence of volatile solvents or water significantly reduces the propellant burning rate. The addition of crystalline oxidizers such as ammonium perchlorate or RDX also modifies the burning rate to a degree that depends on the physical and chemical characteristics of the compound and the percentage present. The burning rate of many propellants increases ca 0.1 to 0.4%/°C as the temperature of the propellant increases.

The operating pressure of the system is the dominant influence on the burning rate of propellants. Photographic evidence shows that increasing the pressure decreases the distance between the flame zone in the gas phase and the propellant surface. The rate of heat transfer to the propellant surface increases accordingly. The reaction rate among the gaseous components of the zones above the propellant also increases in accordance with established relationships between pressure and the rate of gas reactions in equilibrium. The velocity of the gases passing over the propellant at the dynamic conditions prevailing in a rocket motor or in a gun tube may further increase the burning rate (erosive burning). When a turbulent flow of gas occurs behind the reaction zone, part of the turbulence may penetrate the zone and increase heat transfer to the propellant surface (34–38).

Mechanism of Burning. *Nitrocellulose-Based Propellants.* Much of the information available on the burning process of nitrocellulose propellants is based on the decomposition of nitrate esters and the reaction of oxides of nitrogen with the products of decomposition. A one-dimensional physiocochemical description of a model of the burning of a double-base propellant at low (rocket) pressures is often used, and more complex models have been developed. The three reaction zones identified (Table 3) are (*1*) the foam zone where molecular bond breakage, primarily the O–N bond in cellulose nitrate–nitroglycerin-type propellants, occurs. Large volatile molecules are produced, such as aldehydes, alcohols, and low molecular weight oxygenated compounds. The rate of bond breakage depends on temperature in accordance with the applicable Arrhenius equation. (*2*) The fizz zone which is above the foam zone and results from partial reaction among the materials ejected from the foam surface. Aldehydes and alcohols are converted to smaller molecules; nitrogen, water, carbon monoxide, carbon dioxide, and nitric oxide are also formed. About half the total heat evolved by the propellant is liberated in the fizz zone, which is prominent at low pressures and disappears at

Table 3. One-Dimensional Model of Propellant Burning Process[a]

Parameter	Foam	Fizz	Flame	Final
zone description	solid-phase reaction	nonluminous gas-phase reaction	luminous flame reaction	final flame equilibrium
thickness[b], cm	10^{-2}	5×10^{-3}	10^{-3}	
temperature[b], K	300–600	600–1500	1500–3000	>3000

[a]Propellant is a solid at ambient temperature.
[b]Approximate value.

high pressure. And (*3*), the flame zone, where thermodynamic equilibrium is established. This zone defines the flame temperature of the propellant, which may range from ca 1500 K for cool propellant to 3500 K for very hot ones. Here nitric oxide reacts with the reaction products formed in the fizz zone to produce carbon monoxide, carbon dioxide, hydrogen, water, nitrogen, and a small percentage of other molecules.

Because the gaseous products are in thermodynamic equilibrium at the flame temperature, quite accurate calculations of gas composition, maximum temperature, and other thermodynamic properties may be readily derived from the propellant formulation and the thermochemical characteristics of its components. Typical values of these characteristics for many nitrocellulose gun propellants are given in Table 4.

Calculations of the burning rates are not as rewarding. A number of models have been developed that consider the gas-phase reactions and the rate of energy transfer from the gas phase to the propellant surface to be rate determining. Surface models and combination surface–gas-phase models have also been developed. Although the surface–gas-phase models have shown approximate agreement between calculated and experimental burning rates, the correlation is not good enough for design purposes.

Additives. Although the burning rate of nitrocellulose propellants at high gun pressures is not significantly affected by the presence of additives, the addition of 1 to 2% of some metallic salts such as lead acetyl salicylate, lead stearate [*1072-35-1*], and lead stannate [*1344-41-8*] to double-based propellants increases their burning rates at much lower rocket pressures (see LEAD COMPOUNDS). The effect decreases to that of the unleaded propellant as the pressure increases so that burning rate–pressure curves having very low pressure exponents are obtained only over limited pressure regions. It has been estimated that reducing the value of the pressure exponent, n, from 0.6 to 0.1 is equivalent to increasing the specific impulse by 15 seconds. The addition of solid oxidizers and metallic fuels tends to eliminate this catalytic effect. In plateau propellants, so named because of the shape of the log pressure–log burning rate curve, the catalytic effect disappears slowly, whereas in mesa propellants the catalytic effect disappears rapidly. The extent of the rate increase is affected markedly by the type and quantity of other components present (39–45).

Composite Propellants. A number of analytical models have been developed to quantitatively define the burning characteristics of composite propellants. The granular diffusion model postulates that the primary reaction zone of ammonium perchlorate propellants lies almost entirely in the gas phase. This zone is less than 0.01 cm thick at rocket pressures, and its thickness decreases as the pressure increases. The oxidant and fuel are decomposed and converted into gases by pyrolysis or sublimation as a result of energy transferred primarily by thermal conduction from the gas phase to the propellant surface. The gases evolved leave pockets on the surface. Pocket size is related to the particle size of the solid components. As burning occurs, the aluminum powder accumulates on the surface of the propellant and then agglomerates as clusters to form molten droplets up to 20 times the diameter of the individual particles in the propellant.

The Beckstead-Derr-Price model (Fig. 1) considers both the gas-phase and condensed-phase reactions. It assumes heat release from the condensed phase, an

Table 4. Thermochemical, Thermodynamic, and Performance Characteristics of Nitrocellulose Gun Propellants[a]

Characteristics	Designation												
	M1	M2	M5	M6	M8	M9	M10	M15	M17	M26	M30	M31	IMR
heat of explosion, J/g[b]	3140	4522	4354	3182	5192	5422	3936	3350	4019	4082	4082	3370	3601
heat of formation, $-\Delta H_f$, J/g[b]	2261	2366	2407	2261	1989	1989	2533	1256	1361	2114	1549	1465	2366
flame temperature, T_v, K	2435	3370	3290	2580	3760	3800	3040	2555	2975	3130	3000	2600	2835
impetus, J/g[b]	911	1121	1091	956	1181	1142	1031	980	1088	1082	1065	1000	1007
heat capacity, C_V, J/(g·K)[b]	1.46	1.51	1.46	1.46	1.42	1.51	1.42	1.51	1.51	1.46	1.51	1.52	1.46
mean heat capacity products, J/(mol·K)[b]	1.84	1.76	1.76	1.80	1.76	1.72	1.80	1.88	1.88	1.80	1.80	1.88	1.80
mean mol wt of products, g/mol	22.0	25.1	25.4	22.6	26.8	26.4	24.6	21.5	23.1	24.1	23.4	21.6	23.9
specific heat ratio of gases	1.26	1.22	1.22	1.25	1.21	1.21	1.23	1.25	1.24	1.24	1.24	1.25	1.24
gas volume, mol/g	0.045	0.040	0.040	0.044	0.038	0.038	0.041	0.046	0.043	0.042	0.042	0.044	0.042
burning rate at 20°C and 137.9 MPa,[c] cm/s	7.6	12.7	14.0	8.4	17.8	23.0	11.4	10.2	14.0	11.4	12.2	7.9	
pressure exponent	0.66	0.73		0.66	0.81	0.85	0.67	0.66	0.60	0.85	0.70	0.65	
combustion product composition, mol/g × 10^2													
CO	2.33	1.54	1.61	2.24	1.28	1.13	1.81	1.45	1.15	1.89			
CO_2	0.19	0.51	0.48	0.22	0.66	0.74	0.40	0.14	0.25	0.33			
H_2	0.88	0.31	0.34	0.78	0.19	0.15	0.44	0.92	0.57	0.52			
H_2O	0.64	1.10	1.08	0.72	0.11	0.09	0.99	0.83	1.07	0.95			
N_2	0.44	1.49	0.48	0.45	0.54	0.54	0.46	1.29	1.30	0.50			

[a]At loading density of 0.2 g/cm^3.
[b]To convert J to cal, divide by 4.184.
[c]To convert MPa to psi, multiply by 145.

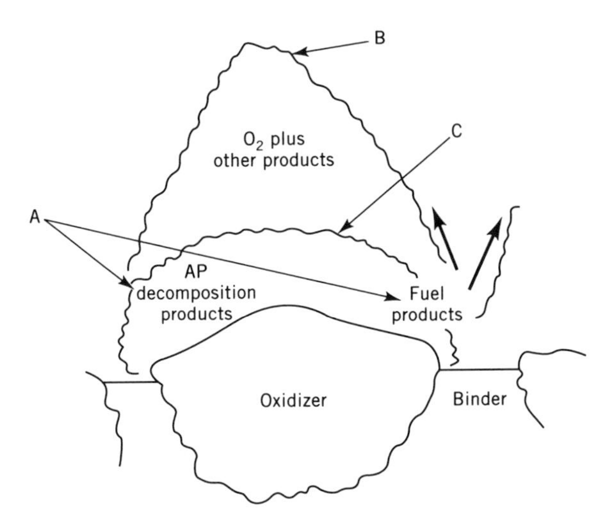

Fig. 1. The postulated flame structure for an AP composite propellant, showing A, the primary flame, where gases are from AP decomposition and fuel pyrolysis, the temperature is presumably the propellant flame temperature, and heat transfer is three-dimensional; followed by B, the final diffusion flame, where gases are O_2 from the AP flame reacting with products from fuel pyrolysis, the temperature is the propellant flame temperature, and heat transfer is three-dimensional; and C, the AP monopropellant flame where gases are products from the AP surface decomposition, the temperature is the adiabatic flame temperature for pure AP, and heat transfer is approximately one-dimensional. AP = ammonium perchlorate.

oxidizer flame, a primary diffusion flame between the fuel and oxidizer decomposition products, and a final diffusion flame between the fuel decomposition products and the products of the oxidizer flame. Examination of the physical phenomena reveals an irregular surface on top of the unheated bulk of the propellant that consists of the binder undergoing pyrolysis, decomposing oxidizer particles, and an agglomeration of metallic particles. The oxidizer and fuel decomposition products mix and react exothermically in the three-dimensional zone above the surface for a distance that depends on the propellant composition, its microstructure, and the ambient pressure and gas velocity. If aluminum is present, additional heat is subsequently produced at a comparatively large distance from the surface. Only small aluminum particles ignite and burn close enough to the surface to influence the propellant burn rate. The temperature of the surface is ca 500 to 1000°C compared to ca 300°C for double-base propellants.

The burning rates of composite propellants containing ammonium perchlorate can be significantly modified by changes in the particle size of the ammonium perchlorate. The burning rates of propellants containing ammonium nitrate or the nitramines, RDX or HMX, have high pressure exponents and are only slightly affected by particle size changes. Catalysts such as the oxides and chromates of iron, copper, and chromium also affect the burning rate of ammonium perchlorate propellants. However, these may have an adverse effect on the shelf life of the propellants (46–55).

Burning-Rate Equations. The design of propellants for gun or rocket performance requires a knowledge of the exact rate at which the products of combustion are produced under the prevailing conditions of pressure and temperature. Although burning rates may be estimated by various computational procedures, the required accuracy can only be obtained experimentally. Burning-rate equations have been developed to describe the performance of solid propellants based on the assumption that all the exposed propellant surfaces are ignited simultaneously and burn at the same linear rate. For example:

$$r = a + bP \tag{1}$$

$$r = cP^n \tag{2}$$

where r is the linear burning rate, P is the pressure, n is the pressure exponent, and a, b, and c are constants that vary with temperature. Equation 1 is often used for propellants burning at high gun pressures, whereas equation 2 is associated with low pressure rocket systems.

Experimental Determination of the Burning Rate. Experimental determinations of the burning rate are made with the closed bomb for gun propellants and the strand burner for rocket propellants. The closed bomb is essentially a heavy-walled cylinder capable of withstanding pressures to 689 MPa (100,000 psi). It is equipped with a piezoelectric pressure gauge and the associated apparatus required to measure the total chamber pressure, which is directly related to the force of the propellant. It also measures the rate of pressure rise as a function of pressure which can then be related to the linear burning rate of the propellant via its geometry. Other devices, such as the Dynagun and the Hi–Low bomb, have also been developed for the measurement of gun propellant performance.

The strand burner is a bomb pressurized with an inert gas-to-rocket pressure and equipped with auxiliary apparatus consisting primarily of electrical timers for determining the time to burn an accurately known distance on the propellant strand being tested. Tests are run on thin strands of propellant as extruded or machined from a grain. The data are directly converted to burning rates (56–63).

Burning Control. In order to produce propellant gas at a predetermined rate, a propellant composition is selected having the required burning rate for the operating pressures in the gun or rocket. The geometry of the propellant is then designed so that the necessary burning surface is available to provide the required mass rate of gas evolution. The individual propellant grains range from very small and numerous, eg, spherical grain dia = 0.01 cm, used for small arms, to very large, eg, dia = 3 to 5 m, and of complex geometry used for rocket boosters. Control of the total burning surface is achieved by establishing the number of grains to be used, the geometrical configuration, and, in the case of rocket propellants, the cementing of noncombustible inhibitors on grain surfaces to prevent burning, or by bonding the exterior of the propellant grain surface to the motor wall. The effect of grain shape on performance of gun and rocket propellants is shown in Figures 2 and 3, respectively. Propellant grains designed to provide a relatively uniform rate of gas evolution during the burning process have neutral geometries and undergo neutral burning. End burning grains are neutral burning, and

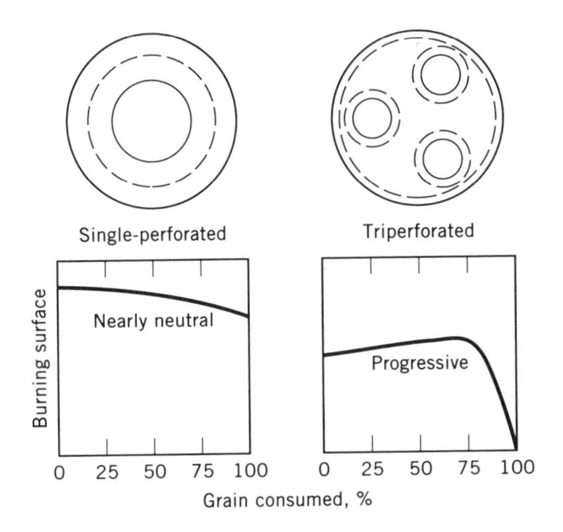

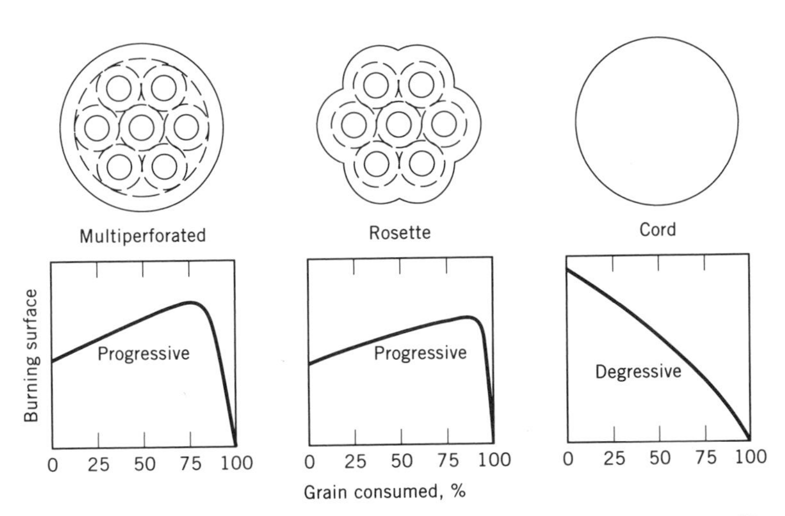

Fig. 2. Effect of grain shape on surface exposed during burning of gun propellants. Cross section of grains are shown.

single-perforated cylinders are almost neutral burning. Grains that increase in surface area during burning, eg, the seven or nineteen perforated grains used in large-caliber gun propellant charges, are said to burn progressively. Grains that decrease in surface area, eg, spherical grains, burn regressively. Regressive burning in a gun is often undesirable, thus liquid or solid deterrents such as dibutyl phthalate, dinitrotoluene, and diphenylamine may be applied to the surface of the spherical grains used in small-caliber weapons. The geometry of rocket propellant grains is tailored to meet the specific performance required. It varies considerably, and may be much more complex than that of gun propellants.

Uncontrolled Burning. Because propellants contain potentially explosive components, a controllable burning process may change under certain exceptional

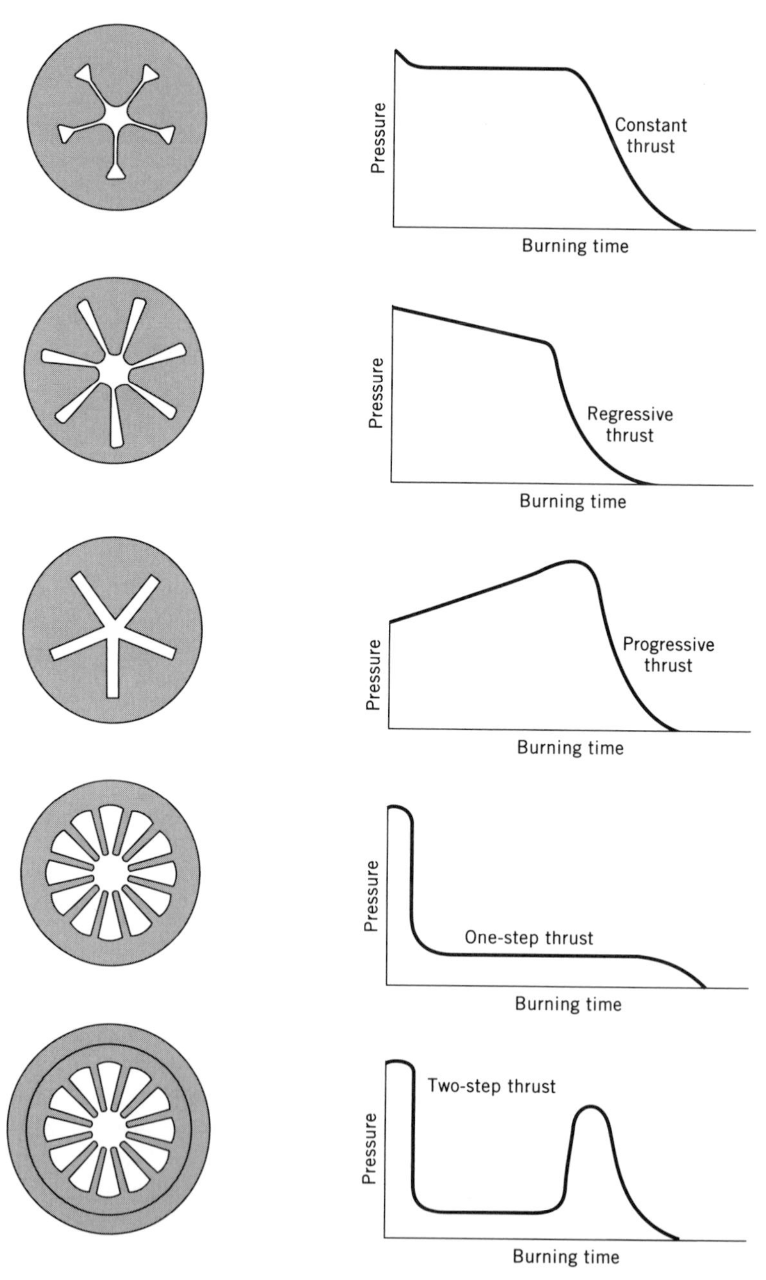

Fig. 3. Effect of grain shape on pressure–time traces of rocket propellants. Cross section of grains are shown.

conditions to uncontrollable burning with consequences comparable to a detonation. The transition from deflagration to detonation in explosives and propellants has been intensively studied, and the available evidence indicates that detonation is most likely to occur when the burning conditions can lead to the initiation and maintenance of a high pressure shock wave. If mechanical breakup of a rocket propellant occurs during the burning process, the large burning surface produced results in a high rate of gas evolution with correspondingly high pressures. An increasingly steep pressure front may evolve, accompanied by a pressure wave that transforms to a shock wave. Steady-state detonation can occur shortly thereafter. Detonation may also occur if fragments produced during grain breakup ricochet and rebound with sufficient kinetic energy to initiate an impact-sensitive propellant composition. Uncontrolled burning in rocket propellants is most likely to occur when using high energy propellants that have a large crystalline filler content of high energy explosives such as RDX or HMX and energetic plasticizer. It may occur in gun propellants if high loading densities are used to attain maximum velocities and ignition is not rapid, uniform, and nearly simultaneous in the charge; if the grains can be substantially compacted; or if the entire propellant charge is accelerated by localized ignition occurring at the charge base. A very high rate of pressure increase may then develop which produces a stress wave that is reflected from the base of the projectile back into the burning charge, reinforcing existing pressure waves and possibly leading to high pressure shock waves. The potential for transition from burning to detonation in a gun is minimized by designing for nearly simultaneous ignition; selecting propellants that have high mechanical strength at all temperatures; and using compact, well-supported charges. Firing of gun and rocket propellants at low temperatures, eg, $-50°C$, are more likely to produce failures than at ambient or higher temperatures (64–72).

Component Characteristics

Rocket Propellants. *Binders.* Composite propellants are broadly classified in terms of the binder used because it is the fuel that reacts with the oxidizer and has a fundamental effect on the stability properties of the propellant. The most commonly used binders are polymers that chemically cross-link during the curing process. These polymers generally show better mechanical properties at temperature extremes than plasticized binders. The mechanical properties depend on the number of cross-links and dangling chains. Thus the degree of cross-linking must be controlled to provide for polymer strength at elevated temperature, while allowing for the required elasticity at low temperature. The addition of trifunctional components to the composition and control of the ratio of trifunctional to bifunctional units establishes the number of branch points in the polymer and prevents excessive cross-linking.

A large number of polymeric compounds have been investigated, but most modern propellants utilize prepolymers that are hydroxy-functional polybutadienes (HTPB), carboxy-functional polybutadienes (CTPB), or a family of polyethylene oxides (PEGs) to form urethanes. Typical cure reactions are

$$\underset{\text{alcohol}}{RCH_2OH} + \underset{\text{isocyante}}{R'NCO} \rightarrow \underset{\text{urethane}}{RCH_2OOCNHR'}$$

$$\underset{\text{acid}}{RCH_2COOH} + \underset{\text{azeridine}}{\triangleright NCOR'} \rightarrow \underset{\text{amido ester}}{RCH_2COOCH_2CH_2NHCOR'}$$

$$\underset{\text{acid}}{RCH_2COOH} + \underset{\text{epoxide}}{CH_2CHR'\ (\text{O bridge})} \rightarrow \underset{\text{hydroxy ester}}{RCH_2COOCH_2CH(OH)R'}$$

Considerable work has also been conducted to try to find thermoplastic elastomers that can be used to simplify processing by enabling dry blending and melt casting instead of the conventional mixing and curing process (see ELASTOMERS).

Thermoplastic elastomers are copolymers in which a thermoplastic segment is linked to an elastomer to produce copolymers having a central elastomeric core and thermoplastic ends. Mechanical strength is obtained by physical molecular linkages rather than by the chemical bonds obtained by curing at elevated temperatures. Ionomers (qv) are also under investigation. These materials produce a nonchemically cured composite matrix by linking polymers having ionic end groups and metal ions in the polymer matrix to effect the linkage.

Binders must be fluid prepolymers even when filled with 85 to 90% granular material. They must not react with the crystalline filler or other components, and should polymerize or cross-link without the formation of gaseous reaction products. Binders must be chemically and physically stable over long periods of time under severe environmental and operational conditions. They must form a durable and tough coating around the oxidizer and metallic ingredients and be capable of bonding to the interior wall of the motor after it has been suitably prepared by coating with an insulating liner and a bonding polymer. The rheological characteristics of the binder–filler and its pot life are also critical (73–78).

Plasticizers. Plasticizers are added to the binders to improve processibility and flexibility at low temperatures. The plasticizer must have a very low melting point, dissolve in the polymer, and if possible provide oxygen in the combustion process to minimize any reduction in the specific impulse of the propellant. Many of the plasticizers used are esters of long-chain aliphatic alcohols and long-chain aliphatic acids. Typical compounds used with polybutadiene binders are isodecyl pelargonate [*109-32-0*] (mp −80°C) and diisooctyl adipate (mp −70°C). High energy plasticizers are designed to increase the energy level of the propellant. Typical nitrato- or nitroesters that have been investigated include nitroglycerin, butanetriol trinitrate [*41407-09-4*], trimethylethane trinitrate, bis(dinitropropylethyl) formal (FEFO), and a 1:1 mixture of bis(dinitropropyl) acetol and formal. Unfortunately many of the high energy plasticizers also tend to increase the propellant sensitivity.

Oxidizers. The characteristics of the oxidizer affect the ballistic and mechanical properties of a composite propellant as well as the processibility. Oxidizers are selected to provide the best combination of available oxygen, high density, low heat of formation, and maximum gas volume in reaction with binders. Increases in oxidizer content increase the density, the adiabatic flame temperature, and the specific impulse of a propellant up to a maximum. The most commonly used inorganic oxidizer in both composite and nitrocellulose-based rocket propellant is ammonium perchlorate. The primary combustion products of an ammonium perchlorate propellant and a polymeric binder containing C, H, and O are CO_2, H_2, O_2, and HCl. Ammonium nitrate has been used in slow burning propellants, and where a smokeless exhaust is required. Nitramines such as RDX and HMX have also been used where maximum energy is essential.

Characteristics of common inorganic oxidizers are listed in Table 5. In any homologous series, potassium perchlorate-containing propellants burn fastest; ammonium nitrate propellants burn slowest (79,80).

Ammonium perchlorate (AP) (see PERCHLORIC ACID AND PERCHLORATES) is hygroscopic between ca 75 to 95% relative humidity, and begins to deliquesce above 95%. AP starts to decompose at 439°C, and the decomposition may be catalyzed by metallic salts such as iron oxide and copper chromite at a lower temperature. Very finely divided ammonium perchlorate is more sensitive to impact and friction than the coarse material, and the presence of hydrocarbons greatly increases the likelihood of a detonable reaction. The burning rate of ammonium perchlorate propellants is also influenced by the particle size of the oxidizer, eg, the rate may increase by a factor of six by decreasing the average diameter from 400 to 1 μm. Bimodal and even trimodal distributions are used to load the binder with the maximum oxidizer content. Average particle size and particle size distribution affect the burning rate as well as the presence of other ingredients such as aluminum and catalysts. Particle size distribution of the perchlorate has a negligible effect on the pressure exponent and no effect on the specific impulse of the propellant (79–84).

Metallic Fuels. Aluminum is most commonly used to increase the impulse of both composite and nitrocellulose-base propellants because of its highly exothermic reaction with the oxidizer. Its heat of reaction with oxygen is 10.25 kJ/g (2.450 kcal/g). Materials such as aluminum hydride, beryllium, beryllium hydride, and boron offer theoretical advantages in increased impulse, but are not used because of increased cost, toxicity, or long-term instability, or because actual per-

Table 5. Properties of Inorganic Oxidizers

Oxidizer	Available oxygen	Melting point, °C	Density, g/cm³	Heat of formation, kJ/mol[a]	Heat capacity, J/(mol·K)[a]	Gas, moles per 100 g[b]
potassium perchlorate[c]	46.0		2.53	−433.4	112.5	0
ammonium perchlorate	34.0	dec	1.95	−290.3	128.0	2.55
ammonium nitrate[d]	20.0	169	1.72	−365.2	137.2	3.75
lithium perchlorate	60.6	236	2.43	−368.6	104.6	0

[a]To convert J to cal, divide by 4.184.
[b]Gas produced by oxidizer other than that formed by reaction of oxygen and fuel components.
[c]Propellants with potassium perchlorate have relatively high burning rates (1.75 cm/s at 6.9 MPa (1000 psi) and 21°C) and high burning rate exponents (0.6–0.7).
[d]Propellants with ammonium nitrate have very low burning rates (0.01 cm/s).

formance does not live up to calculated performance. Increasing the aluminum content of a propellant increases its density. The flame temperature and specific impulse sharply approach a maximum near the stoichiometric ratio of metal–oxidizer–binder. The aluminum increases the hydrogen content of the reaction products and, by minimizing the formation of water vapor, reduces the energy losses caused by dissociation of water at elevated temperatures. Incorporation of aluminum staples to replace a small part of the aluminum powder may quadruple the burning rate while maintaining the specific impulse. The presence of aluminum in rocket propellants also reduces or eliminates combustion instability caused by the formation of pressure waves in the motor chamber.

Aluminum-containing propellants deliver less than the calculated impulse because of two-phase flow losses in the nozzle caused by aluminum oxide particles. Combustion of the aluminum must occur in the residence time in the chamber to meet impulse expectations. As the residence time increases, the unburned metal decreases, and the specific impulse increases. The solid reaction products also show a velocity lag during nozzle expansion, and may fail to attain thermal equilibrium with the gas exhaust. An overall efficiency loss of 5 to 8% from theoretical may result from these phenomena. However, these losses are more than offset by the increase in energy produced by metal oxidation (85–87).

Liquid Propellants. *Rocket Propellants.* Liquid propellants have long been used to obtain maximum controllability of rocket performance and, where required, maximum impulse. Three classes of rocket monopropellants exist that differ in the chemical reactions that release energy: (*1*) those consisting of, eg, hydrogen peroxide, H_2O_2; ethylene oxide, C_2H_4O; and nitroethane, $CH_3CH_2NO_2$; that can undergo internal oxidation–reduction reactions; (*2*) those consisting of unstable molecules such as hydrazine, N_2H_2, and acetylene, C_2H_2; and (*3*) those consisting of stable mixtures of two or more compounds that are mutually compatible. These mixtures include methyl nitrate and methyl alcohol, CH_3NO_3/CH_3OH; hydrazine, hydrazine nitrate, and water, $N_2H_4/N_2NO_3/H_4/H_2O$. Hydrazine, which freezes at 275 K, is mixed with hydrazine nitrate and water (68:20:12) to meet low temperature requirements. Table 6 lists common and experimental liquid rocket bipropellants. Among the most commonly used are liquid oxygen and liquid hydrogen for maximum energy, nitrogen tetroxide and monomethyl hydrazine, and liquid oxygen and hydrocarbon fuels such as JP4.

Gun Propellants. Liquid propellants for guns have been investigated in two different types of systems: bulk and regenerative loading. Bulk loading involves insertion in the gun breech of the required quantity of propellant as a single unit, similar to a solid propellant charge, and subsequent ignition of the mass. The lack of reliable ignition and uniform combustion led to the abandonment of bulk loading.

Regenerative loading requires continuous injection and combustion of the liquid propellant. Typical components of a regenerative liquid propellant gun are shown in Figure 4. Regenerative loading has progressed to the stage in the United States where extensive firings have been successfully conducted in guns up to 155 mm in caliber. A monopropellant mixture is used that consists of hydroxyl ammonium nitrate, triethylene ammonium nitrate, and water (20:63:17). This mixture is difficult to ignite accidentally, very stable, and biodegradable. The products of combustion are approximately 71% H_2O, 17% N_2, and 12% CO_2. Its flame tem-

Table 6. Liquid Rocket Bipropellants

Oxidant	Fuel	Ratio oxidant/fuel	Specific impulse, s[a]
O_2	H_2	4.0	341
O_2	B_2H_6	2.0	344
O_2	N_2H_4	0.90	313
O_2	JP4	2.60	301
F_2	H_2	9.00	410
F_2	JP4	2.40	317
F_2	N_2H_4	2.30	363
IRFNA[b]	C_2H_5OH	2.50	219
IRFNA[b]	UDMH[c]	3.00	288
H_2O_2	C_2H_5OH	4.0	230
H_2O_2	JP4	6.5	233
H_2O_2	N_2H_4		245
N_2O_4	N_2H_4		249
ClF_3	H_2	11.50	318
ClF_3	N_2H_4		292

[a]Calculated values.
[b]IRFNA contains 20–40% lithium nitrate, 55–75% red fuming nitric acid (RFNA), and 4–5% SiO_2; mp = −54°C.
[c]UDMH = unsymmetrical dimethylhydrazine.

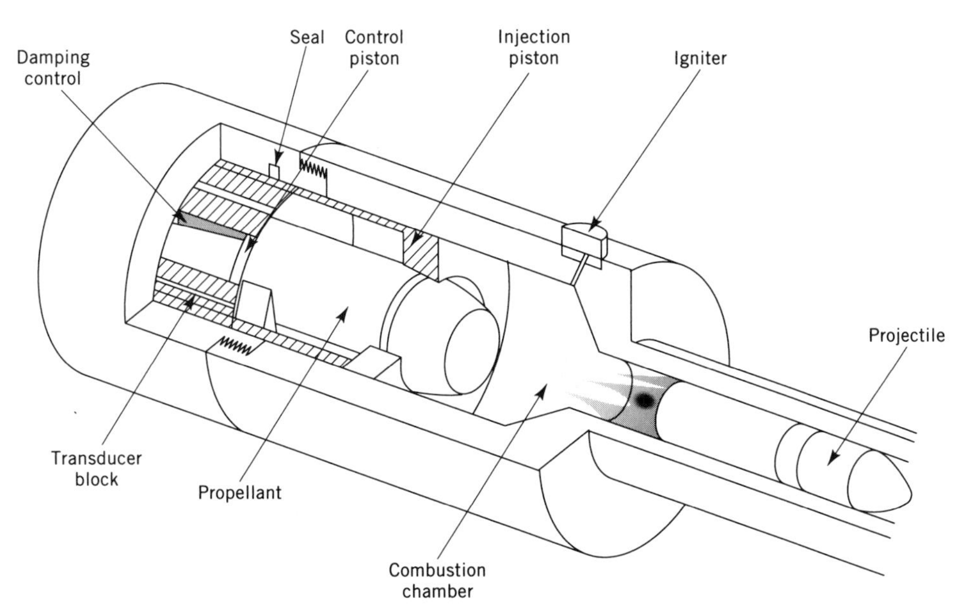

Fig. 4. 155-mm regenerative LP gun.

perature is 2590 K, freezing point −100°C, impetus 934 J/g (223 cal/g), specific heat ratio of the combustion gases, 1:22. It has low flammability and shock sensitivity, and does not detonate when subjected to the standard tests for insensitive munitions. The propellant is readily decomposed by transition metals and by nitric acid, and care must be taken to avoid contamination with ferrous materials (88–92).

Low Sensitivity Propellants. The initiation of gun and rocket propellants by fire, high velocity steel fragments and bullets, shaped charges, or electrostatic discharge, and the propagation of the resultant detonation shockwaves has resulted in catastrophic events on board military ships and on the battlefield, and has created problems in maintaining required safety distances in the storage of military materiel. As a result, gun and rocket propellants have been formulated to have minimum sensitivity to external stimuli and maximum energy content. Low sensitivity rocket propellants have also been designed to exhibit minimum smoke signatures.

Gun Propellants. Low sensitivity gun propellants, often referred to as LOVA (low vulnerability ammunition), use RDX or HMX as the principal energy components, and desensitizing binders such as cellulose acetate butyrate or thermoplastic elastomers (TPE) including polyacetal–polyurethane block copolymers, polystyrene–polyacrylate copolymers, and glycidyl azide polymers (GAP) to provide the required mechanical characteristics. Other high energy, low sensitivity plasticizers investigated include bis-dinitropropyl acetal formal, *n*-butyl-2 nitratoethylnitramine, and 1,3,3-trinitroazetidine. The weight percent composition and characteristics of a typical LOVA propellant, such as M43, for use with tank ammunition is 76% RDX, 12% cellulose acetate butyrate, 7.6% bis-2,2-dinitropropyl acetal/formal, 0.5% neoalkoxy tri(dioctylphosphato)titanate, 4.0% nitrocellulose, 12.6% N, and 0.4% ethyl centralite [*85-98-3*]; heat of explosion 196 J/g (820 cal/g); impetus 1070 J/g (256 cal/g); and flame temperature 3065 K.

Rocket Propellants. Ammonium nitrate is the most common low cost oxidizer used to reduce sensitivity in solid rocket propellants. It is used extensively in formulations for gas generators. The propellants are cool, clean burning, and insensitive but have relatively low impulse. When processed and maintained at very low relative humidities, the volume changes characteristic of ammonium nitrate do not occur. Introducing 8–10% of potassium nitrate into the crystal lattice also eliminates phase transitions within the normal operating temperature range. The specific impulse of ammonium nitrate compositions may be increased by the use of energetic polymeric binders, such as polyglycidylazide or the addition of low concentrations of nitramine compounds such as RDX or HMX, although at some sacrifice of sensitivity (93–101).

Minimum Signature Propellants. Rocket propellants may produce undesirable smoke-forming combustion products in the exhaust plumes, ie, these products become visible signatures of the location of the source of smoke, and can interfere with optical guidance systems. Smoke formation is caused primarily by particulate matter, such as aluminum oxide, from aluminum in the rocket propellant, or to a lesser extent by compounds of iron, lead, or copper. This is referred to as primary smoke. Ionizing gases such as HCl serve as nucleation centers for the condensation of water vapor. Water in the propellant combustion products produces the secondary smoke that forms the contrails associated with missile flight.

Primary smoke can be nearly eliminated by deletion of aluminum from the compositions or reduction in the amount added from ca 20 to 1.2%. Aluminum, added primarily to increase specific impulse, may be replaced by high energy nitramine compounds such as RDX or HMX and energetic polymeric binders such as nitrate ester plasticized polyesters. These compounds tend to increase the sensitivity of the propellants. Very dry ammonium nitrate or phase stabilized ammonium nitrate, neither of which undergoes volume changes in the useful temperature range, may be used where lower specific impulses can be accepted. Ammonium nitrate is clean burning, has low sensitivity, and low cost. The inclusion of glycidyl azide polymers as an energetic binder component has been proposed to offset the reduced specific impulse.

Secondary smoke is produced mostly by the condensation of water in humid or cold air. The presence of hydrogen chloride or hydrogen fluoride in the combustion products increases the extent and rate of condensation. Composition modifications to reduce primary smoke may reduce secondary smoke to some extent, but complete elimination is unlikely. The relatively small amount of smoke produced in gun firings by modern nitrocellulose propellants, although undesirable, is acceptable (102–109).

Nonconventional Methods of Gun Propulsion

Advanced gun propulsion programs are pursued primarily to obtain projectile velocities considerably greater than the approximately 1.5–2.0 km velocity obtainable using conventional propellants and guns that rely on traditional interior ballistics. Hypervelocities would offer the possibility of achieving extraterrestrial orbits using gun-type systems instead of missiles. The advanced propulsion programs are of three types: those using chemical propellants for accelerating projectiles by unusual methods; those using electrical sources of energy; and those combining these two procedures.

The Traveling Charge. Conventional chemical propellants are designed to produce hot, high pressure gases at the breech of a weapon to accelerate the projectile down the bore of the weapon tube. A significant limitation in attainable velocity occurs because of the energy lost in accelerating the propellant gases as well as the projectile. Utilizing the traveling charge enables a theoretical increase of 10–20% in propulsion efficiency to be attained at velocities exceeding two kilometers per second. This is accomplished by attaching to the projectile a propellant having a very high (ca 100–500 m/s) burning rate to produce hot gases at a sufficient rate to maintain a constant thrust and pressure on the projectile base. The decrease in the work to accelerate the combustion gases combined with the impulse from the burning propellant accounts for the increased efficiency obtained by the traveling charge. The lower breech chamber pressures and increased downbore pressures characteristic of the traveling charge also produce a flatter, more efficient pressure–time trace similar to that obtainable with a rocket propellant. The required very high burning rate propellants have been experimentally obtained by inducing high porosity in the propellant during its manufacture, or by brittle break up of the propellant or incorporation of compounds such as decaborane, $B_{10}H_{14}$, in the propellant composition. The necessary predictability and controllability of very high burning rate propellants has not been achieved as of this writing (110–116).

The Two-Stage Light Gas Gun. The light gas gun, designed to accelerate small projectiles to velocities up to 6.1 km/s, is an experimental tool used primarily to investigate hypervelocity penetrator/target interaction phenomena. The gun (Fig. 5) consists of a necked-down compression tube and a launch tube. The compression tube contains a firing chamber at the breech end filled with conventional propellant, a piston that closes the firing chamber, and a projectile that seals the mouth of the compression tube. Before firing, the front end of the compression tube is evacuated and filled with hydrogen or helium. The launch tube is also evacuated to minimize resistance to the projectile as it accelerates down the tube.

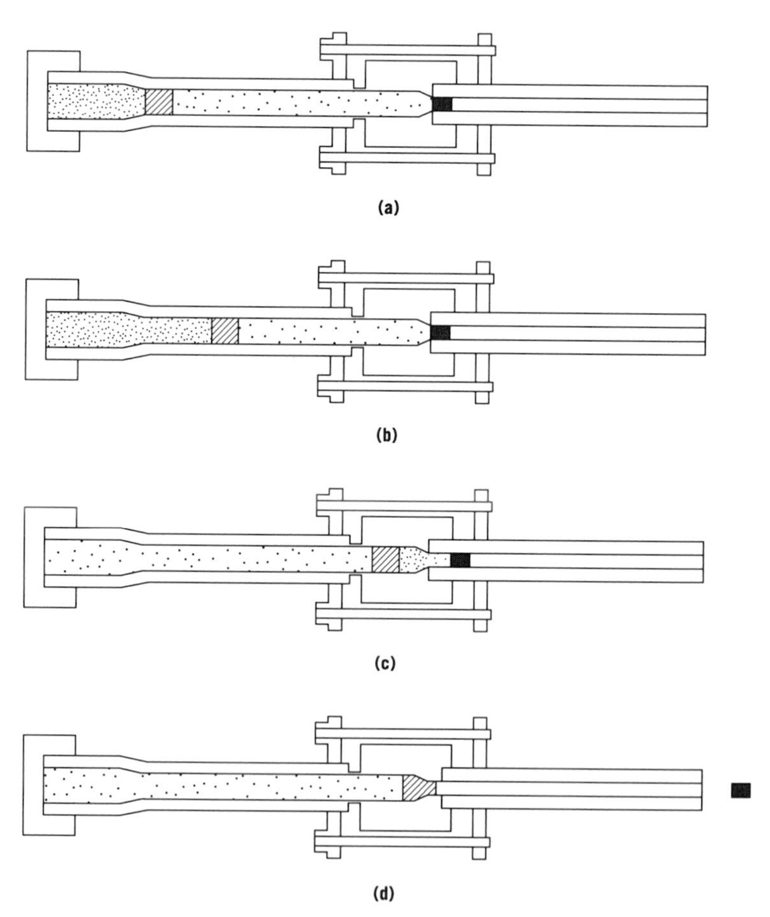

Fig. 5. Two-stage light gas gun showing the ▨ piston and ■ projectile where ⬚ = propellant charge and ⬚ = light gas: (**a**), before firing; (**b**), after firing propellar charge; (**c**), as piston nears necked-down mouth of the launch tube; (**d**), after completio of firing cycle. Piston is removed from neck of launch tube before refiring or launch tub is replaced.

On firing, the gases from the propellant accelerate the piston that compresses the light gas in front of it. At a preestablished pressure, the projectile is propelled down the launch tube accelerated by the low molecular weight gas which follows the projectile to the mouth of the tube. The target material is placed in front of the launch tube, and appropriate instrumentation used to establish the characteristics of the interface reaction between projectile and target (117–120).

The Ram Accelerator. High velocities are obtained by causing combustion to occur continuously in the RAM accelerator as the projectile travels down the tube. The RAM accelerator consists of a subcaliber projectile similar to the centerbody of a conventional ramjet. The subcaliber projectile is initially accelerated in a conventional tube to velocities of about 1 m/s by a conventional propellant, or for experimental purposes by a light gas gun. As the projectile moves forward in the tube, it is further accelerated by the combustion of a reactive, premixed fuel–oxidizer gas mixture introduced under pressure into a long, eg, 10–20 m, accelerator tube. The gas that flows around the projectile is thermally choked so that it is initiated by the projectile and provides additional propulsive energy to the projectile. Diaphragms are used to seal off the end of the accelerator tube and individual sections into which different gaseous mixtures may be introduced. Gas mixtures of oxygen and methane, ethylene, or hydrogen have been used as fuels with inert gases such as argon and helium as diluents. Acceleration of the projectile depends on the propellant energy evolved, and the ratio of the mass of propellant gas to projectile mass. Velocities up to 10 km/s may theoretically be obtained (121–123).

Electric Guns. Electric gun approaches that have been under considerable study include the electromagnetic (EM) gun and the electrothermal–chemical (ETC) gun. These use electrical pulse power to generate the energy required to achieve increased velocity and/or muzzle energy. The most commonly used energy storage devices are pulsed rotating machines or capacitors. The electromagnetic guns use intense magnetic fields, require special launchers, and can achieve maximum projectile velocities. The electrothermal guns use electrical energy to create a plasma, which adds energy to conventional propellant gun systems and produces greater control of the interior ballistic process.

EM guns, which offer the possibility of obtaining very high velocities with relatively short length accelerators, fall into two basic classes: railguns and coilguns. These differ in the geometry of achieving confined magnetic fields, and of coupling the resultant forces to achieve projectile acceleration, as shown in Figure 6. As a rule, railguns are conceptually and geometrically simpler, and have lower impedance, ie, require higher current and lower voltage for a specific propulsion task. They have received more developmental attention, despite the potential for greater energy efficiency from coilguns.

In the direct current electromagnetic railgun, a prime power source provides the energy to a pulsed power generator that produces electric pulses compatible with gun firing times. The current flows in one rail, across an armature on the projectile, and back through the second rail. The gun barrel and armature constitute a one-turn coil. The current generates a magnetic field inside the gun tube; the interaction between the field and the current generates a Lorentz force which accelerates the projectile.

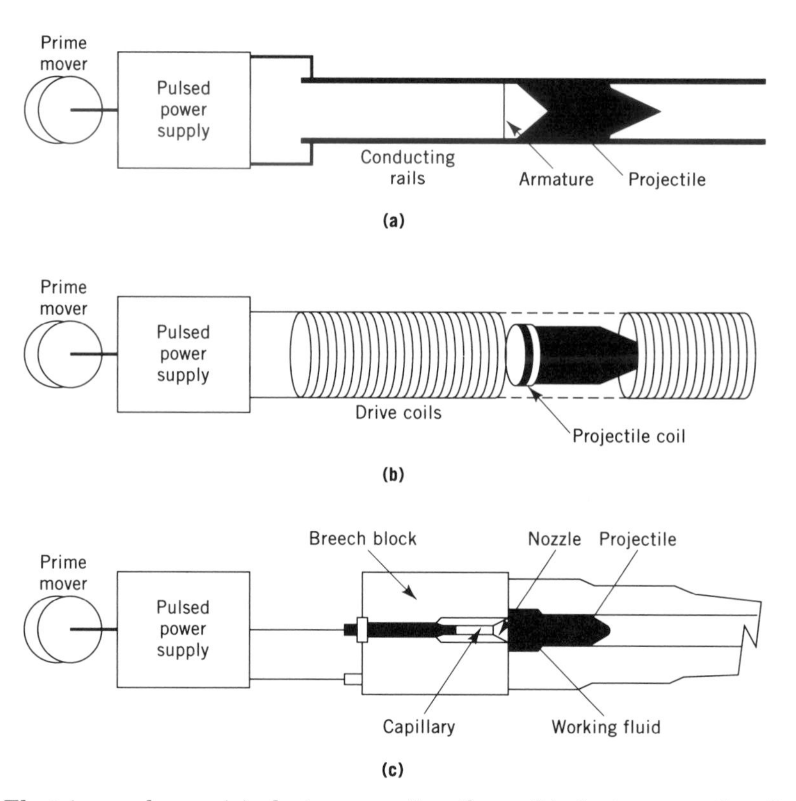

Fig. 6. Electric gun classes: (**a**), electromagnetic railgun; (**b**), electromagnetic coilgun; and (**c**), electrothermal gun.

The electromagnetic coilgun requires pulse power on the same scale as the railgun. The current alternating in the multicoil barrel generates a changing magnetic field at the site of the projectile coil. The current in the projectile coil is attracted and/or repelled by suitably activated barrel coils. A sequence of Lorentz forces is produced like a surfboard riding a magnetic field wave.

Electrothermal–chemical propulsion is based on control of the propellant combustion process to produce a constant breech pressure–time relationship until all propellant is consumed. This contrasts with a peak pressure–time relationship produced by conventional solid propellant. The overall procedure provides high energy electrical pulses produced by a pulsed power generator. The electrical pulses generate a thermal plasma in a cartridge that initiates and controls the propellant burning process that accelerates the projectile down the bore of the tube. The velocity of the projectile depends on the electrical energy produced by the cartridge and the chemical energy of the propellant. The plasma cartridge delivers 10–20% additional energy to the propulsion system. Typically, the plasma is produced by using a high voltage, high current source to explode a foil or wire which pyrolyzes a plastic liner to yield a high (10,000–15,000 K) temperature low mass output of ionized species and hydrogen. A current continues to flow as long as the ionized species exists. Electrothermal propellants under investigation include metal–water mixes, peroxide–hydrocarbon mixes, and metal–metal hydride and water formulations (124–132).

Manufacture of Solid Propellants

Gun Propellants. Large numbers of small perforated grains or long sticks are used in gun propellant charges to provide the high mass rate of burning required to accelerate projectiles to maximum velocity in the relatively short distances of travel in gun tubes. These geometries are produced in enormous quantities most often by plasticizing nitrocellulose in simple mixers and extruding the soft propellant mix, which can be readily cut to the specification grain lengths and dried to a hard, horn-like texture. Stick propellant is made in the same way as granular propellant, but cut to form long strands on extrusion. Uniformity of performance is obtained by control of the composition, the volatile material present, the grain dimensions, and by blending on a large scale. Some variation in burning characteristics is permissible, because gun propellant charges can be modified to a limited extent to meet ballistic requirements by the addition or removal of propellant grains. Gun propellants are best evaluated by composition analysis, measurements of the heat of explosion, the grain dimensions, the mechanical characteristics, and the closed-bomb characteristics of relative force and quickness followed by confirmatory weapon firings.

Batch processes, which have been widely used in the manufacture of gun propellants, offer advantages of flexibility of operations and low capital investment. However, highly automated continuous and semicontinuous processes are rapidly replacing the batch processes and have been developed for single-, double-, and triple-base gun propellants and ball powder. These latter procedures offer increased safety, reduced lot to lot variation, lower labor requirements, decreased overall costs, and fewer pollution problems. Continuous processes often incorporate a variety of in-process monitoring sensors (qv) and analytical devices, automatic sampling, nondestructive testing (qv), and a high degree of automation and feedback controls. Twin-screw extruders are widely used as part of the automation process. Special procedures for increasing safety and reducing cost of gun propellants involve conducting mixing, conveying, and cutting operations under water (133–142).

Extruded Nitrocellulose Propellants. Nitrocellulose propellants are made with or without incorporation of a solvent as plasticizer by five processes:

Process	Propellant use
solvent extrusion	cannon
	fast-burning rockets
	casting powder
	ignition powder
	rifles, small-caliber weapons, expulsion charges
solvent emulsion	rifles
	small-caliber weapons
solventless extrusion	small rockets
	cannon
solventless rolling	mortars
casting	small rockets
	large rockets

All five processes require plasticization of the nitrocellulose to eliminate its fibrous structure and cause it to burn predictably in parallel layers. Mechanical working of the ingredients contributes to plasticization and uniformity of composition. The compositions of representative nitrocellulose-based gun propellants are shown in Table 7.

Solvent Extrusion Batch Process. Almost all standard gun propellants and small-webbed rocket propellant grains are made by the solvent extrusion process. Grains having webs greater than ca 1.30 cm are produced by solventless or casting processes. The removal of solvents from large-web grains would require long periods of time and could introduce stresses that would lead to grain cracking. Triple-base propellants (M15, M17, M30, M31), having a high nitroguanidine content, are made similarly to double-base propellants. The manufacture of single-base propellants such as M1 and M6 compositions differs primarily in the mixing and drying operations.

In the typical process, purified, blended, and centrifuged nitrocellulose (NC) having the required nitrogen content and wet with ca 30% water is received from the nitrocellulose plant, transferred to a double-acting hydraulic dehydration press, and compressed at low pressure to remove some of the water. The remaining water is removed by pumping 95% ethyl alcohol through the nitrocellulose. The final blocks, containing ca 18% alcohol and ca 2.0% water, are broken up and screened to remove lumps or oversized particles. Mixing is conducted in a water-jacketed bladed mixer, and consists essentially of solid–solid and solid–liquid incorporation, the solution of stabilizers and possibly ballistic modifiers, and the absorption of solvents and liquid plasticizers by the nitrocellulose. This operation is governed by the heat generated, the heat-exchange characteristics of the operation, the method and sequence of incorporation and solvent addition, and the effects of specific equipment. The premixing operation for double- and triple-base propellants is designed to incorporate the nitroglycerin in the nitrocellulose and to begin to distribute the remaining ingredients in a slow and uniform manner. The final mixing blends the composition for an extended period of time until the ingredients are completely incorporated and plasticization occurs. The operation is generally conducted in a sigma-bladed, water-jacketed mixer to which several premix charges have been transferred. The temperature is maintained between 40 and 50°C, depending on the equipment and the colloid formation. The mix is cooled and discharged. All equipment must be grounded, and nonsparking tools used to avoid a solvent vapor–air explosion.

Single-base propellants are mixed in a similar fashion by adding the ingredients to the nitrocellulose in the mixer together with the required amounts of ether and alcohol. The mixing time is about one-half hour, and the temperature is kept below 25°C. The partly colloidal mixture looks like moist crude sugar. A maceration step may be included to increase homogeneity.

After mixing, the dough-like composition is transferred to a vertical block screening press where it is consolidated. The block is ejected from the press. These operations remove lumps and foreign particles from the mix, increase the uniformity of ingredient distribution, and improve the colloiding of the nitrocellulose and the density of the propellant. The strands from the screening press are again consolidated in a blocking press. The blocks of propellant, whether single-, double-, or triple-base, are transferred to a vertical or horizontal graining press

Table 7. Gun Propellant Composition[a], wt %

Component	M1[b]	M2	M5	M6	M8	M9	M10	M15	M17	M26	M30	M31[c]	IMR
nitrocellulose	85.0	77.5	82.0	87.0	52.2	57.8	98.0	20.0	22.0	67.5	28.0	20.0	100.0
(% *N*)	(13.15)	(13.25)	(13.25)	(13.15)	(13.25)	(13.25)	(13.15)	(13.15)	(13.15)	(13.15)	(12.6)	(12.6)	(13.15)
nitroglycerin		19.5	15.0		43.0	40.0		19.0	21.5	25.0	22.5	19.0	
nitroguanidine								54.7	54.7		47.7	54.7	
ethyl centralite		0.6	0.6		0.6	0.7		6.0	1.5	6.0	1.5		
diphenylamine	1.0[d]						1.0						0.7[d]
dinitrotoluene	10.0			10.0									8.0[e]
dibutylphthalate	5.0			3.0	3.0							4.5	
potassium nitrate		0.7	0.7		1.2	1.50				0.75			
barium nitrate		1.4	1.4							0.75			
potassium sulfate	1.0[f]			1.0[d]			1.0						1.0[d]
cryolite								0.3	0.3		0.3	0.3	
graphite		0.3	0.3			0.10[d]			0.15[d]				

[a]All compositions are solvent extruded as grains except M8 which is solventless-rolled as sheet.
[b]Also may contain 1.0 wt % basic lead carbonate.
[c]Also contains 1.5 wt % 2-nitrodiphenylamine.
[d]On added basis.
[e]Added as a coating.
[f]If required, on added basis.

and extruded at relatively low pressures of 10.3–17.2 MPa (1500 to 2500 psi) through dies designed to produce the required dimensions. To ensure safety during pressing, explosive mixtures of air and solvent have to be carefully excluded during the ramming operation.

The strands of propellant are fed to a mechanical cutter and sliced to specified lengths, either as small grains or long sticks. Grains of double-, triple-, and some single-base compositions are dried in trays with warm air. The drying process for single-base propellant is entirely different from that of double- and triple-base propellant. First the alcohol–ether wet propellant is air-dried in transfer carts or large tanks to recover the solvents and reduce the volatile solvent content to ca 6%. The initial drying process is carefully controlled to prevent skin hardening or grain cracking. Temperatures are gradually raised to 50–65°C over a period of days, depending on grain size. The vapors are condensed and the solvent recovered. The propellant is then immersed in circulating water at 50–60°C. The solvent diffuses into the water, which prevents case hardening of the propellant surface and permits the solvent to be removed more rapidly than exposure to air alone would allow. After a period of time, up to ca 30 days for large single-base cannon grains, the temperature is slowly increased to reduce the solvent to a controlled minimum, depending on grain size. Final air drying at ca 55°C removes surface water.

The propellant may be tumbled in drums with a small amount of graphite to improve its flow characteristics and bulk density and to decrease the likelihood of formation of an electrostatic charge as well as to perform a degree of blending. It is then screened to remove foreign matter and blended into large lots which may range from 20 to 225 metric tons. The blending operation is essential to provide as homogeneous a lot as possible for ballistic uniformity. Although propellant grains are relatively insensitive to static electricity, propellant dust may be as sensitive as dry nitrocellulose. Fires in drying buildings and blending towers have been attributed in some cases to the electrostatic ignition of dust.

The wastewater produced in this process consists mostly of water used in cleanup and propellant conveyance and sorting operations. Techniques such as the use of activated carbon and biological treatment are being investigated for the removal of solvents and dissolved organic compounds (143).

Continuous Solvent–Extrusion Process. A schematic for a typical continuous process, widely used for making solvent propellant for cannons, is shown in Figure 7. This continuous process produces ca 1100 metric tons of single-base propellant per month at the U.S. Army Ammunition Plant (Radford, Virginia). Continuous processes have also been developed for double- and triple-base propellants and for stick as well as granular geometries. A principal aspect of these processes has been the extensive use of single- and double-screw extruders instead of the presses used in the batch process.

The main features in which the Radford process differs from the batch operation are in thermal dehydration and compounding. Water-wet nitrocellulose on a continuous vacuum belt filter is vacuum-dried followed by hot air transfusion (80°C) to reduce the moisture to less than 2%. After cooling, alcohol is sprayed on the nitrocellulose to a concentration of 15–20%. The alcohol-wet nitrocellulose is then transferred from a surge feeder to a compounder by a continuous weigh-belt

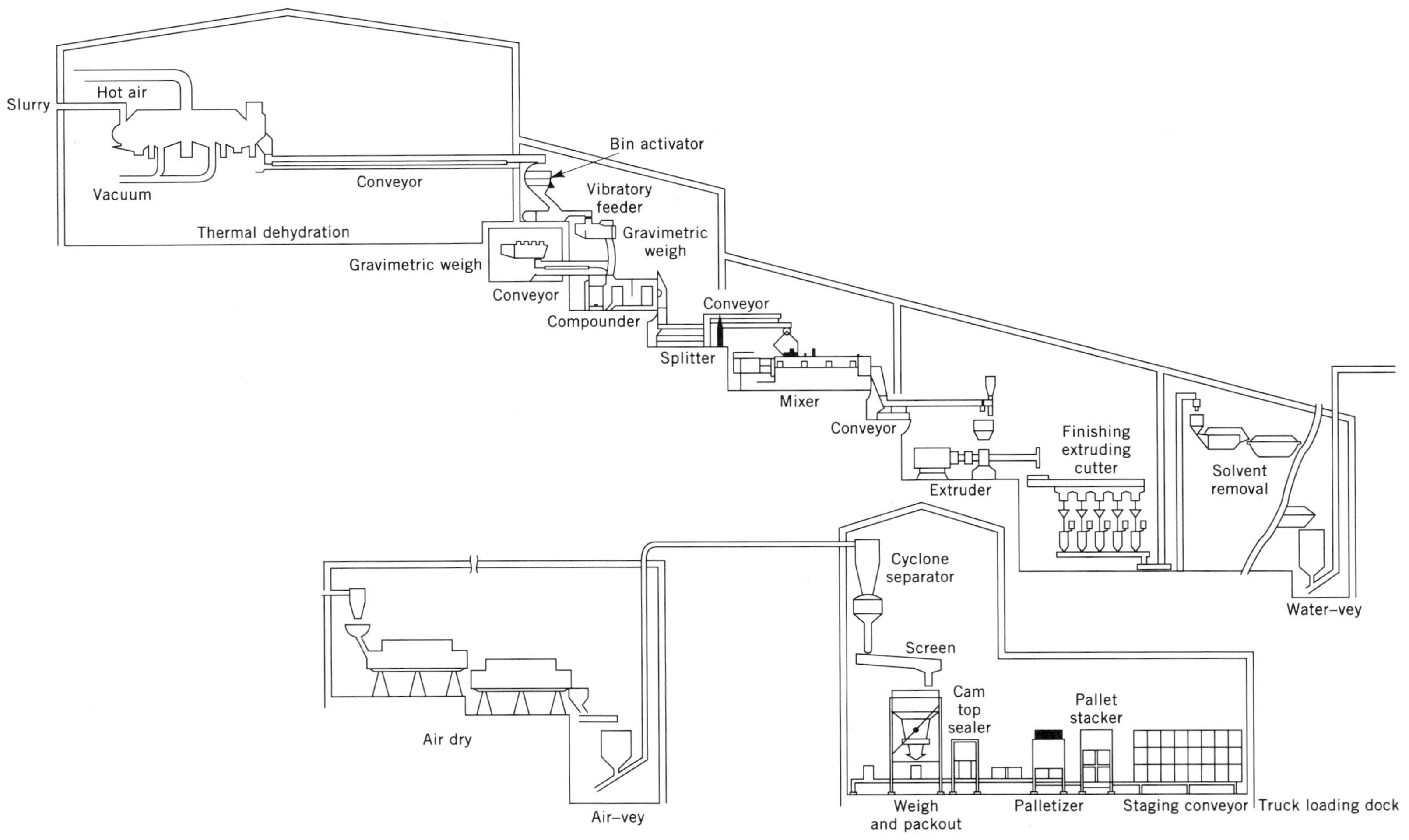

Figure 7. Continuous process for solvent-extruded single-base propellant (automated single-base line). Vey = conveyor. Courtesy of John Horvath, U.S.A. Radford Plant (Radford, Virginia).

along with the other ingredients of the composition, which are also weighed and added automatically.

The compounder is a water-jacketed horizontal rotary plow that blends the ingredients to produce a homogeneous premix paste. The mixed paste is fed by a conveyor to a heavy-duty reciprocating screw mixer that is temperature controlled and specially designed to thoroughly mix and work the paste by forcing it past pins in the mixer barrel and out through a die plate. As the paste is extruded, it is cut into small pellets that are fed continuously to water-jacketed screw extruders and forced through multiple dies which provide the final shape. The strands are cooled as they are extruded to facilitate cutting by an adjustable roll-type cutter. After the cut grains are screened to remove clusters and odd sizes, the solvent is removed in the solvent recovery-water dry system where the propellant is treated first with hot inert gas and then hot water. Finally, in a series of air dryer units, the moisture content is reduced from 12 to less than 0.8%.

The entire continuous automatic process is computer controlled so that continuous performance information is available. Pressure relief is permitted wherever possible to minimize the likelihood of a detonation. Continuous-screw extrusion processes may be employed for making nitrocellulose single-, double-, and triple-base gun propellants, for some composite propellants, and for some plastic bonded explosives (144).

Solventless Extrusion Process. The solventless process for making double-base propellants has been used in the United States primarily for the manufacture of rocket propellant grains having web thickness from ca 1.35 to 15 cm and for thin-sheet mortar (M8) propellant. The process offers such advantages as minimal dimensional changes after extrusion, the elimination of the drying process, and better long-term ballistic uniformity because there is no loss of volatile solvent. The composition and properties of typical double-base solvent extruded rocket and mortar propellant are listed in Table 8.

In the water-slurry process for making solventless propellants, developed in 1889, explosive and nonexplosive liquid plasticizers and water-insoluble constituents are incorporated into nitrocellulose suspended as a slurry in a large volume of hot (50°C) water. (The ratio of water to total propellant components is ca 10:1, ca 20:1 on a nitrocellulose base). After removing the excess water in a basket-type centrifugal wringer equipped with a wire-mesh screen, the resulting wet (ca 15% water) mass is partially dried at ambient temperature. During this aging process, the nitrocellulose absorbs and is partially gelatinized by the plasticizers. Water-soluble salts are then incorporated during a blending operation in a rotating drum.

The rolling operations that follow take place first on hot (95°C) differential-speed rolls which dry and colloid the paste and convert it into sheet form, and then on even-speed rolls which produce smoothly surfaced propellant sheets in which all ingredients have been uniformly incorporated. The roll gap in the differential rolls is adjustable to produce sheets of various thicknesses, and rolling is continued until the moisture is reduced to a predetermined level, usually less than 0.5%. The sheet is then cut off the roll. Differential rolling is potentially hazardous, and fires are not uncommon, although detonations are not apt to occur. Operations are conducted by remote control.

Table 8. Composition and Properties of Double-Base Solventless Propellant

Parameter	Extruded for	
	Rockets	Mortar sheets
Approximate composition, wt %		
nitrocellulose[a]	51.5	52.5
nitroglycerin	43.0	43.0
potassium nitrate		1.0
diethylphthalate	3.0	3.0
ethyl centralite	1.0	0.50
potassium sulfate	1.25	
carbon black	0.20	
wax	0.05	
Thermochemical properties		
flame temperature, K		
isochoric	3660	3695
isobaric	3010	
specific impulse, N·s/kg[b]	2317	
heat of explosion, J/g[c]	5108	5209
heat of combustion, J/g[c]	9295	
gas volume, mol/g	0.038	0.037
ratio of specific heats	1.22	1.21
burning rate at 5.89 MPa[d] and 21°C, cm/s	1.52	
pressure exponent	0.68	
products' composition, wt %		
hydrogen	6.5	
water	27.0	
carbon monoxide	33.0	
carbon dioxide	18.0	
nitrogen	14.0	
other	1.5	

[a]Nitrogen content is 13.25%.
[b]To convert N·s/kg to lbf·s/lb, divide by 9.82.
[c]To convert J to cal, divide by 4.184.
[d]To convert MPa to psi, multiply by 145.

Typical even-speed rolls, about the same dimensions as the differential rolls, are highly polished, heated to ca 60°C, and revolved at ca 10 rpm. If rocket propellant grains are being made, the sheet is slit into strips and rolled to form carpet rolls. A charge of large enough diameter is made to fit snugly into an extrusion press, which may be jacketed for temperature control and equipped with vacuum pumps for removal of air. The diameter of the press bore may be up to 60 cm, and the press may be horizontal or vertical with pressures up to 103 MPa (15,000 psi) used for extrusion. The press is loaded with the propellant, evacuated, and extrusion is begun. On extrusion, the strand is cut into the required grain lengths. The grains may be solid or have central perforations of various shapes, depending on the configuration of the die pin. The grains are visually inspected, annealed at elevated temperatures, and inspected by x-ray (143,145).

Ball Powder. Ball powder, typically used in small-caliber weapons such as 5.56-mm, 7.62-mm, and 20-mm projectiles, has also been proposed for large-caliber weapons. The product consists of spherically shaped or flattened ellipsoidal grains, ca 0.04–0.09 cm in diameter. The process permits the recovery and use of nitrocellulose from obsolete granular propellant. It eliminates the need for the conventional mixers, extruders, and cutters used to make granular and stick propellant, and is relatively inexpensive to operate. The process is safe because mixing and extrusion take place in the presence of water. The operations are flexible so that either single- or double-base ball propellants may be produced. The use of surface layer deterrents on ball powder reduces flame temperatures during the initial stages of burning and thereby reduces barrel erosion at the time when maximum gun pressures occur. Typical composition and characteristics are shown in Table 9. The product has desirable flow characteristics because of particle shape so that small arms ammunition can be rapidly loaded by high speed automatic equipment. LOVA compositions using ball powder have been investigated, as well as procedures for achieving higher loading density by compaction of the grains.

Batch Process. A flow chart for the ball powder batch operation is shown in Figure 8. Water-wet fresh or extracted nitrocellulose is transferred as a slurry to a graining still. Calcium carbonate is added to neutralize any free acid released by the dissolved nitrocellulose. Ethyl acetate is added as are other soluble components such as diphenylamine. The contents are heated to ca 70°C and agitated. When the proper viscosity of the lacquer is attained, a protective colloid such as animal glue is added to form an emulsion of nitrocellulose globules. Sodium sulfate is added ca one-half hour after the beginning of globule formation to extract water from the lacquer by establishing an osmotic pressure differential between the water-laden nitrocellulose globules and the concentrated salt solution in the still. Under these conditions, small spheres of dissolved nitrocellulose and the other soluble ingredients are formed. The ethyl acetate is distilled at 70–100°C, leaving spherical particles. This graining operation requires ca 1 to 1.5 h. Grain density and size are determined by the concentration of salt in solution, the temperature and time of the dehydration, agitation speed, and the rate of distillation of the ethyl acetate.

After graining, the slurry is water-washed and then wet-screened. Over- and undersize grains are returned to the graining operation for reworking. Nitroglycerin or an organic coating material are added following the transfer of the screened ball powder as a water slurry to a coating still. The water level is adjusted, the temperature increased to 60–65°C, and a solution of nitroglycerin in ethyl acetate added with slow agitation to form an emulsion. The ethyl acetate is distilled under vacuum at 70 to 85°C, leaving the nitroglycerin present in a gradient of decreasing concentration from the surface toward the center of the spheres. If a deterrent is used, such as dinitrotoluene in dibutylphthalate, it is then transferred to the still, the slurry is slowly agitated at ca 75°C, cooled, transferred to a wash tub, and water-washed. The coating cycle may take ca 24 to 30 h, depending on the propellant.

Table 9. Composition and Properties of Ball Powder Propellants

Parameter	Value
Composition, wt %	
graphite[a]	0.4
potassium nitrate	1.0–1.5
sodium sulfate[a]	0.5
calcium carbonate[a]	1.0
nitroglycerin	8.0–12.0
diphenylamine	0.75–1.50
dibutyl phthalate	3.5–7.5
total volatiles[a]	2.0
residual solvents[a]	1.2
dust and foreign matter[a]	0.1
dinitrotoluene	as required
nitrocellulose	remainder
nitrogen, %	13.0–13.2
Physical properties	
hygroscopicity	1.75
granulation	
through no. 20 (840 μm) sieve, %[b]	95
through no. 40 (420 μm) sieve, %[a]	50
through no. 45 (350 μm) sieve, %[a]	3
bulk density, g/cm^3	0.95–1.0
Thermochemical properties	
heat of explosion, J/g[c]	3350–3768
flame temperature, K	2700–3000
volume of gaseous products at STP mol/g	0.04
impetus, J/g[d]	1000

[a]Value given is maximum value.
[b]Value given is minimum value.
[c]To convert J to cal, divide by 4.184.
[d]To convert J/g to ft·lb/lb, multiply by 335.

Rolling may be used for shape and size modifications to increase burning surface area of the propellant. A thick water slurry of the spheres is passed through a set of appropriately spaced even-speed polishing rolls, rotating toward each other. Centrifuging the slurry reduces excess moisture to ca 6%. Surface ballistic modifiers such as dinitrotoluene, tin oxide, potassium nitrate, or small amounts of other water-soluble salts are added as an alcohol slurry. The alcohol is removed with hot air, and graphite is added to glaze the propellant. The moisture content is adjusted by the addition of water, if required, or drying.

The propellant is dry-screened to remove dust and foreign material, and excess graphite. Unacceptable propellant from the dry-screen operation may be

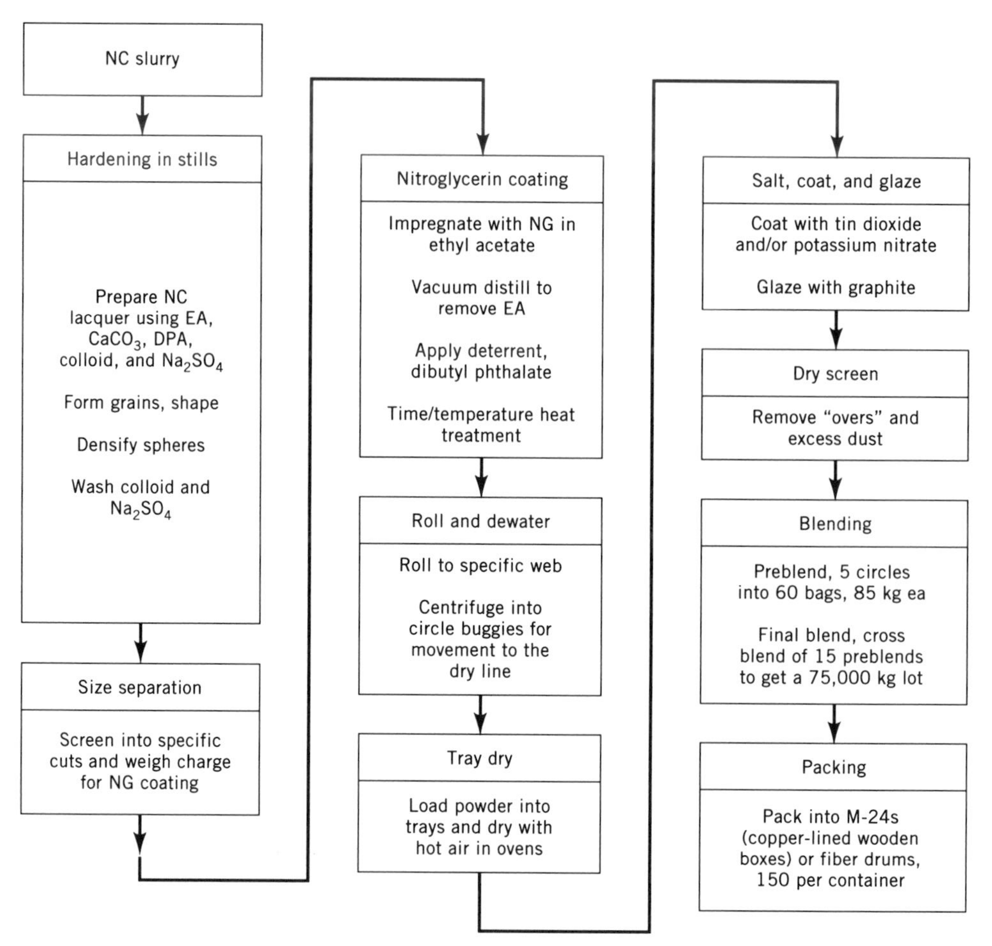

Fig. 8. Ball powder batch process. NC = nitrocellulose, EA = ethyl acetate, DPA = diphenylamine, NG = nitroglycerine.

returned to the graining stage for reworking. The product is blended in large rotating barrels for ballistic uniformity.

Continuous Process. In the continuous process, developed by Olin Corp. and now used more extensively than the batch process, the nitrocellulose and stabilizing additives are dissolved in the solvent in a continuous-screw mixer to form a dough-type lacquer, which is pumped continuously through filters to the graining operation. The lacquer is extruded as cylindrical strands that are cut with a rotary knife into grains having a length to diameter ratio of ca 1.5 to 1. The cut cylinders are flushed away from the exit side of the graining plate using a solution of water, colloid, and salt, and are transferred to the shaping and dehydration lines. Automatic controls and monitors are used throughout.

The shaping and dehydration line is a long bank of pipes connected by short-radius U-shaped bends that are hot-water jacketed to provide a gradually increasing temperature of ca 60 to 80°C. On passage through the pipe, the viscosity of the particles decreases, and they become spherical and are dehydrated by the dissolved salt present to ca 10% entrained water. The large percentage of entrained solvent is removed by passing the lacquer–water through a series of evaporators. The propellant slurry is then passed over a screen and vacuum filtered in series to recycle the salt and colloid solution and to wash residual colloid from the grains. After washing, the propellant may be vacuum-dried to remove excess water. It is size classified on a series of continuously rotating screens, and may then be impregnated with nitroglycerin; ethyl acetate is removed by vacuum distillation. The product is coated with deterrent, and rolled if necessary, by the same methods used in the batch process. Moisture is removed by a series of continuous vibrated semifluidized-bed dryers. Surface coatings are applied in a continuous drum or a batch barrel blender. Blending is carried out in a static internal tube-type blender or a large barrel blender. The propellant is packed in drums for shipment.

The wastewater in the ball powder processes arises primarily from the washing and wet-screening process. Wash and screen waters are passed through clarifiers to remove suspended solids. The overflow from the clarifiers may be accumulated in lagoons. The wash waters contain a considerable amount of protective colloid, organic solvent, and sodium sulfate, and have to be treated before they are discharged into local streams. The colloid foams in the effluent and increases the BOD by accumulating on the bottom of the collection ponds. Cooling water can be completely recycled, and it is possible to design the washing and wet screening operations to decrease the contaminants in the plant effluent (146).

Rocket Propellants. The manufacture of propellants for rocket systems poses problems that do not exist with gun propellants. Rocket grains are generally much larger and may have more complex shapes. Each rocket grain must be made free of flaws to avoid the possibility of internal burning and breakup. In-process variations of a minor and not readily identifiable nature may produce significant changes in performance. Once the rocket grain is produced, it cannot be readily changed if it does not meet requirements. Differences in lot performance cannot be blended out as with gun propellants. Many propellant compositions have been developed to meet specific needs (147).

The most common method for producing large rocket grains involves casting a prepared mix into a mold and causing it to solidify using a solvation process, as for nitrocellulose-based propellants, or a polymerization process, as for composite propellants. Extrusion procedures may be used for the smaller (dia <15–20 cm), rocket grains but are not feasible for large grains. The difficulty of controlling the curing operation during extrusion, and particularly the relatively limited numbers of grains required as compared to gun propellant requirements, limits the applicability of extrusion to large-volume tactical rocket applications (133).

Cast Propellants. *Nitrocellulose-Based.* Cast nitrocellulose propellant is made by a two-step process. In the first stage, casting powder is produced by procedures that are almost identical to those used for the manufacture of conventional solvent-extruded small-grain gun propellants. The second stage consolidates the casting powder by filling the interstices of the granules with a fluid plasticizer that diffuses into the powder and causes swelling and ultimate coalescence of the granules into a monolithic grain. The plasticizer generally consists of a mixture of an explosive energy-producing liquid such as nitroglycerin and an inert fluid such as triacetin. The process of consolidation is a physical one. No chemical reaction occurs, and there is virtually no shrinkage during curing.

A typical high performance composite-modified double-base cast rocket propellant starts with a single- or a double-base casting powder consisting of 30% nitrocellulose, 10% plasticizer, 30% solid oxidizer, 28% metallic fuel, and 2% stabilizer. The final propellant composition contains ca 22% nitrocellulose, 32% plasticizer, 24% solid oxidizer, 20% fuel, and ca 2% stabilizer. The type and percentage of nitrocellulose significantly affects the mechanical characteristics of the propellant. Tensile strength and the modulus of elasticity increase, and elongation decreases as the percentage of nitrocellulose increases from 12.0 to 13.15%. The mechanical properties improve at a 12.6% nitrogen content of the nitrocellulose which is most commonly used. Tougher propellants having favorable heats of formation and a satisfactory carbon–hydrogen–oxygen balance are obtained using low molecular weight nitrocellulose as a binder and the addition of compounds such as poly(ethylene glycol).

Because double-base propellants cannot be directly bonded to the walls of a rocket motor to maximize the propellant weight in the motor, an adhesive resin is sprayed into the interior while the motor is rotated. A small amount of casting powder may also be sprayed into the tacky resin. The liner is cured and becomes an integral part of the propellant charge after casting. The motor is fitted with the required casting attachments, placed in a casting pit if necessary, and the casting powder dispenser and associated equipment are installed. The casting powder flows from a hopper through a distributor screen and a screen plate to disperse the powder uniformly into the motor. A high velocity air stream may also be used to carry the powder into the motor.

The fluid plasticizer (solvent) consists of an energetic compound, eg, nitroglycerin, an inert carrier, and a stabilizer. The system is evacuated to remove volatiles, moisture, and air, and the plasticizer is then pressurized and passed slowly upward through the powder bed while the powder is held stationary by a pressure plate on the powder column. Casting solvent may also be added from the top of the mold.

The cast-loaded rocket motor is cured at 45 to 60°C for as long as two weeks, depending on grain size. The gelatinizing solvent and the casting grains mutually diffuse so that the final rocket grain is a tough, pore-free, sturdy structure. The compositions of several typical cast double-base and composite-modified double-base propellants are given in Table 10. Cast propellant may also be made similarly in plastic inhibitor cases or uninhibited for use in cartridge loaded applications (148–150).

Polymer-Based. The advantages of polymeric-based cast propellants are the extensive range of performance characteristics, excellent thermal and mechanical stability, and relatively low cost. Maximum performance is obtained by the use of maximum energy propellants, maximum loading density in the rocket motor, and lightweight graphite composite cases. The facilities used in composite propellant manufacture do not compete with those required for making nitrocellulose propellants. A number of programs have been developed to convert batch to continuous processes with varying degrees of success. However, batch processes are generally employed, although various operations may be automated and made semicontinuous. Polymer-based propellants have also been made for small and

Table 10. Composition and Properties of Nitrocellulose-Based Cast Propellants

Parameter	Type: Low energy A	High energy B	High energy C
Composition, wt %			
nitrocellulose, 12.6% *N*	59.0	20.0	22.0
nitroglycerin	24.0	30.0	30.0
triacetin	9.0	6.0	5.0
dioctylphthalate	3.0		
aluminum		20.0	21.0
HMX		11.0	
stabilizer	2.0	2.0	2.0
ammonium perchlorate		11.0	20.0
lead stearate	3.0		
Ballistic properties			
specific impulse, N·s/kg[a]	2062	2651	2602
burning rate at 6.9 MPa[b] and 20°C, cm/s	0.65	1.40	2.00
pressure exponent		0.45	0.40
pressure coefficient		0.025	0.04
Thermochemical–thermodynamic properties			
heat of explosion, J/g[c]	2931	7718	7432
heat of formation, $-\Delta H_f$, J/g[c]		1570	1842
flame temperature, K	1925	3850	3900
mean heat capacity, J/(g·K)[c]			
products	1.80	1.76	1.76
gases	1.80	1.26	1.21
mean molecular weight, g/mol			
products	21.8	27.9	28.9
gases	21.8	30.9	21.0
specific heat ratio, gas	1.27	1.18	1.17
Combustion products composition, mol / 100 g			
C	2.12		
CO_2	0.31	0.05	0.07
CO	2.12	1.30	1.15
H_2	1.06	0.75	0.66
H_2O	0.66	0.27	0.33
N_2	0.43	0.49	0.38
Pb	0.004		
Al_2O_3		0.35	0.37
H		0.20	0.23
OH		0.05	
other		0.5	0.10[d]

[a]To convert from N·s/kg to lbf·s/lb, divide by 9.82. [b]To convert MPa to psi, multiply by 145.
[c]To convert J to cal, divide by 4.184. [d]HCl.

medium size rocket motors using injection flow forming techniques and die extrusion processes (151).

The manufacturing operations for making different composite propellants are very similar, although a variety of polymeric binders may be used. Processing variations, even small ones, may have a significant effect on the mechanical properties of a propellant and the volumetric loading in the motor. Because the viscous propellant mix must flow uniformly and rapidly into all parts of the rocket motor assembly during the casting operation, the processibility of a formulation is fundamentally related to its rheological characteristics. These depend primarily on the cure characteristics of the polymeric binder, the volume of solids loaded into the binder, and the particle shape and size distribution of the solids. Relative humidity control at 40% or less is used in most of the process operations because degradation of the polymer, the liner insulation, and the bond between the liner and the propellant as well as shorter pot life and increased mix viscosity may occur in the presence of moisture. The perchlorate and the other components of the system may also be significantly affected.

A flow chart of a typical batch process is shown in Figure 9. The oxidizer most commonly used is ammonium perchlorate which is rigidly controlled for moisture, impurities, and particle size and shape, and may contain a flow additive such as tricalcium phosphate. Slow and high speed grinding are accomplished by hammer mills which may be coupled to an air classifier to provide the range of particle size distributions required. Fluid energy pulverizers are also used. Typical particle size ranges from 3–9 μm for microatomizers to 20–160 μm for micropulverizers. The oxidizer is blended, screened, and transferred to a storage hopper for subsequent use.

In the premix operation, a uniform slurry of all components, except the oxidizer, is prepared. The premixes may contain cross-linking, wetting, opacifying, and antifoaming agents, plasticizers, metallic fuels, catalysts, and curing compounds. Automated techniques ensure formulation uniformity and reproducibility. The polymer and other large-volume fluids required are pumped from the storage tanks to weigh tanks and then to the premix vessels. These may be up to 5000 L in capacity and equipped with turbine-driven agitators designed for the specific materials being handled. The secondary liquid components, including a portion of the curing agent, are weighed, added, and mixed at a controlled temperature after first purging the premix vessel with nitrogen. The necessary solids other than the oxidizer are screened and added, followed by further mixing under nitrogen, and finally under vacuum to remove entrapped gases. Batch mixers are temperature controlled and designed to deaerate the viscous mass while imparting a shear action to ensure thorough and rapid incorporation of solids. They include relatively conventional horizontal mixers such as the sigma-blade dough mixer used in making nitrocellulose propellants, mixers with heavy-duty bear claw blades, and ribbon mixers. Vertical change-can planetary mixers are commonly used to meet requirements for increased mix capacity. Mixing times and temperatures are tightly controlled to maximize mix uniformity and minimize viscosity changes without accelerating the cure reactions to the stage where the pot life is excessively reduced. The mix temperature increases as a result of work input on the viscous mass and the exothermicity of the initial cure reaction. The required quantity of the premix is transferred to the mixer bowl, which is moved

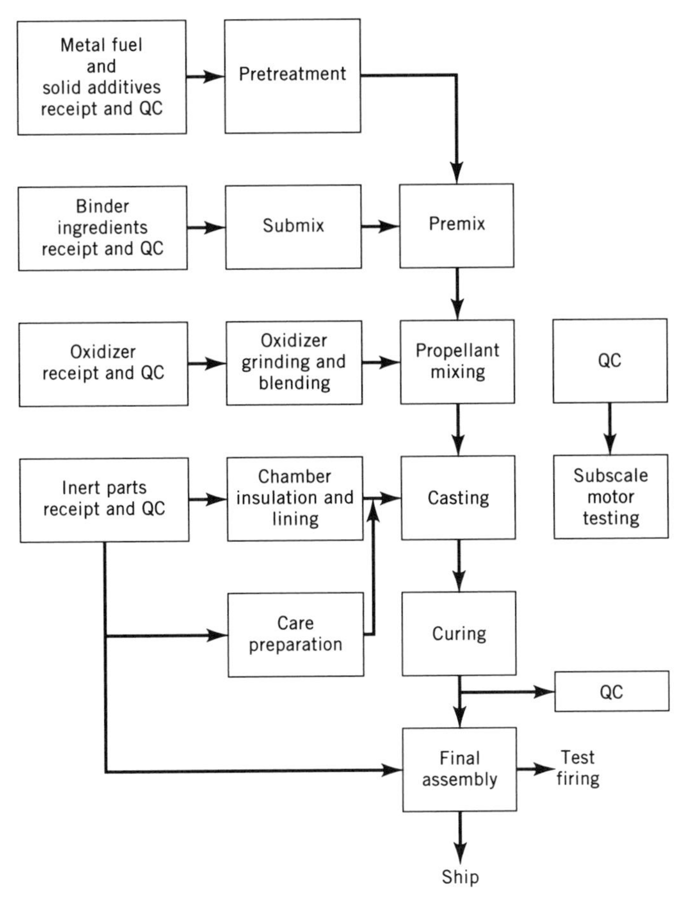

Fig. 9. Batch process for cast-composite polymer-based propellants. QC = quality control.

into position and assembled to the mixer. Mixing is begun after purging with nitrogen. When the process control conditions have been attained, the oxidizer is added followed by the curing agent. Mixing then proceeds under vacuum.

Propellants cast into rockets are commonly case-bonded to the motors to achieve maximum volumetric loading density. The interior of the motor is thoroughly cleaned, coated using an insulating material, and then lined with a composition to which the propellant binder adheres under the environmental stresses of the system. The insulation material is generally a rubber-type composition, filled with silica, titanium dioxide, or potassium titanate. Silica-filled nitrate rubber and vulcanizable ethylene–propylene rubber have been used. The liner generally consists of the same base polymer as is used in the propellant. It is usually applied in a thin layer, and may be partially or fully cured before the propellant is poured into the rocket.

In the cast loading of large booster rockets, the motor is fixed in a vertical position and surrounded with necessary handling gear to facilitate subsequent operations. Very large motors are inserted in huge cylindrical pits. A shroud or similar enclosure may be used to surround the motor so that dry, warm air can be passed into it to preheat the motor and control the temperature of the casting and curing operation. The central mandrel required for grain geometry is inserted into the motor with controls for rigid alignment to close tolerances. The exact casting technique used depends on the rheological characteristics of the propellant and the quantity being processed. Several methods are commonly used for large grains including bayonet, bottom, and vacuum casting.

Upon completion of the casting operation, the motor is maintained at a closely controlled temperature–time regime to cure the propellant. Composite propellants are usually cured between 40 to 60°C. After curing, the core is withdrawn from the motor, and the associated casting equipment removed. The final physical characteristics of the propellant are highly dependent on the cure conditions, which in turn depend on the characteristics of the composition and the thermal conductivities of the metal and motor lining. Completeness of the cure is best determined by measuring the mechanical properties of the propellant. The reaction is finished when no change occurs on additional curing. The formulations and characteristics of a number of composite propellants are shown in Table 11 (151–163).

Safety. The facilities for the manufacture of composite cast nitrocellulose-based propellants incorporate the latest techniques for hazard detection and prevention and for damage control. The processes are monitored and controlled from central stations. Closed-circuit television is used for direct observation and deluge sprinkler systems having frangible seals on the nozzles are installed which can rapidly respond to a fire. Battery power is available for emergencies. Pressure relief valves incorporating frangible disks, which are fragmented by small quantities of explosives, have been used in the lines to prevent pressure buildup and permit discharge if a fire occurs. Infrared detectors capable of sensing the light of burning propellant but indifferent to room light are mounted on the head of the batch mixers. Static pressure detector units to detect excess pressure may also be mounted in the mixer head. Both the light and pressure sensors can actuate a deluge system. In addition, numerous studies have been made of the factors that affect the sensitivity of the propellant and the rocket systems that utilize them (164,165).

Pollution Prevention. Procedures have been developed for recovery of composite ammonium perchlorate propellant from rocket motors, and the treatment of scrap and recovered propellant to reclaim ingredients. These include the use of high pressure water jets or compounds such as ammonia, which form fluids under pressure at elevated temperature, to remove the propellant from the motor, extraction of the ammonium perchlorate with solvents such as water or ammonia as a critical fluid, recrystallization of the perchlorate and reuse in composite propellant or in slurry explosives or conversion to perchloric acid (166,167).

Black Powder. Black powder is mainly used as an igniter for nitrocellulose gun propellant, and to some extent in safety blasting fuse, delay fuses, and in firecrackers. Potassium nitrate black powder (74 wt %, 15.6 wt % carbon, 10.4 wt % sulfur) is used for military applications. The slower-burning, less costly, and more hygroscopic sodium nitrate black powder (71.0 wt %, 16.5 wt % carbon, 12.5 wt % sulfur) is used industrially. The reaction products of black powder are complex (Table 12) and change with the conditions of initiation, confinement, and

Table 11. Composition and Properties of Polymer-Based Cast Composite Propellants

Parameter	Propellant type[a]							
	Poly-sulfide	Poly-urethane	CTPB	CTPB	HTPB	PBAN	PBAA	Buta-diene
			Composition, wt %					
ammonium perchlorate	63.0	70.0	73.0	63.0	70.0	69.0	68.0	80[b]
binder	36.0	21.0	12.0	10.0	12.0	11.0	15.0	14
aluminum		8.0	15.0	17.0	18.0	15.0	16.0	
other	1.2[c]	1.0		10.0[d]		5.0[e,f]	5.0[f]	6
			Ballistic properties					
specific impulse, N·s/kg[g]	2259	2406	2602	2602	2553	2602	2553	1866
burning rate at 6.9 MPa[h] and 20°C, cm/s	0.90	0.80	0.98	0.75	0.60	1.37	1.70	0.30
pressure exponent, n	0.45	0.20	0.30	0.30	0.30	0.33	0.20	0.50
pressure coefficient, c	0.012	0.07	0.06	0.03	0.03	0.06	0.14	0.004
			Thermochemical–thermodynamic properties[i]					
heat of explosion, J/g[j]		4543	6280		6448	5966		2721
heat of formation, $-\Delta H_f$, J/g[j]	2156	2470	1842	1549		1999	1842	3768
flame temperature, T_P, K	2375	2850	3500	3650	3450	3400	3300	1000
mean heat capacity, J/(g·K)[j]			1.80					
products	2.34			1.80	1.84			2.01
gases	2.34			1.97	1.97			2.01

Table 11. (*Continued*)

Parameter	Propellant type[a]							
	Poly-sulfide	Poly-urethane	CTPB	CTPB	HTPB	PBAN	PBAA	Buta-diene
mean molecular weight								
products	25.2			26.9				19.5
gases	25.2	24.8	28.1	20.5		28.1	27.2	19.3
specific heat ratio, gas	1.20			1.19	1.19			
Combustion products, composition, mol / 100 g[k]								
CO_2	0.28	0.15	0.08	0.06		0.07		0.37
CO	0.80	1.02	0.79	0.78	0.95	0.89	0.96	0.93
H_2	0.55	0.94	0.88	0.85	1.15	0.96	1.14	1.40
H_2O	1.17	1.00	0.73	0.57	0.41	0.64	0.38	1.37
N_2	0.23	0.31	0.30	0.40	0.30	0.30	0.28	1.00
Al_2O_3		0.14	0.26	0.29	0.30	0.26	0.25	
HCl	0.50	0.58	0.50	0.42	0.47	0.48	0.43	
H_2S	0.10							
H			0.12	0.17	0.13	0.10	0.09	
$AlCl + AlCl_2$			0.04		0.03	0.03		

[a] CTPB = carboxy-terminated polybutadiene; HTPB = hydroxy-terminated polybutadiene; PBAN = polybutadiene–acrylic acid–acrylonitrile; and PBAA = polybutadiene–acrylic acid.
[b] Ammonium nitrate.
[c] 1.0 wt % MgO, 0.2 wt % added sulfur.
[d] HMX.
[e] 4.0 wt % dioctyl adipate.
[f] 1.0 wt % iron catalyst.
[g] To convert N·s/kg to lbf·s/kg, divide by 9.82.
[h] To convert MPa to psi, multiply by 145.
[i] All gas volumes at standard temperature and pressure.
[j] To convert J to cal, divide by 4.184.
[k] Principal products only.

density. The reported thermochemical and performance characteristics vary greatly and depend on the source of material, its physical form, and the method of determination. Typical values are listed in Table 13.

The critical relative humidity of black powder is 60%. It gains ca 2% moisture in 48 h at 90% rh and 25°C. Ignitability decreases rapidly at ca 3 to 4% moisture

Table 12. Reaction Products of Black Powder

Component	Quantity, wt %
Gases	
carbon dioxide	49
carbon monoxide	12
nitrogen	33
hydrogen sulfide	2.5
methane	0.5
water	1
hydrogen	2
Total	*44*
Solids	
potassium carbonate	61
potassium sulfate	15
potassium sulfide	14.3
potassium thiocyanate	0.2
potassium nitrate	0.3
ammonium carbonate	0.1
sulfur	9
carbon	0.1
Total	*56*

Table 13. Characteristics of Black Powder

Characteristic	Value
flame temperature, K[a]	ca 2800
gas, mol/g	0.0128–0.0159
heat of explosion, J/g[b,c]	3015–3140
impetus, J/g[b]	239–284
burning rate at 6.9 MPa[d], cm/s	ca 1 to 1.5
temperature coefficient of pressure, %/°C	0.4
pressure exponent	0.25–0.5
ignition temperature, °C	450
activation energy, kJ/mol[b]	87.9

[a]Isochoric.
[b]To convert J to cal, divide by 4.184.
[c]Water as liquid.
[d]To convert MPa to psi, multiply by 145.

level. The structure of black powder granules also deteriorates during cycling through high humidity atmospheres. However, it can be stored satisfactorily for many years if dry. The hygroscopicity of black powder is caused by the carbon and impurities in the potassium nitrate.

The performance of black powder is critically dependent on the degree of intimacy of the components in the product. The manufacture of black powder is essentially a procedure for bringing the ingredients into maximum mutual contact. A detailed flow chart for the conventional process is presented in Figure 10.

Typically, dry potassium nitrate is pulverized in a ball mill. Sulfur is milled into cellular charcoal to form a uniform mix in a separate ball mill. The nitrate and the sulfur–charcoal mix are screened and then loosely mixed by hand or in a tumbling machine. Magnetic separators may be used to ensure the absence of ferrous metals. The preliminary mix is transferred to an edge-runner wheel mill with large, heavy cast iron wheels. A clearance between the pan and the wheels is required for safety purposes. The size of this gap also contributes to the density of the black powder granules obtained. Water is added to minimize dusting and improve incorporation of the nitrate into the charcoal. The milling operation requires ca 3 to 6 h.

The moist milled powder is transferred to a hydraulic press where it is consolidated in layers into cakes at pressures of ca 41.3 MPa (6000 psi) applied for ca 30 min. Each cake is ca 2.5 cm thick and 60 cm square. The density of the powder increases to 1.6 to 1.8 g/cm^3, depending on the pressure applied. The cakes are then transferred to a corning mill consisting of adjustable corrugated rollers that are cascaded so that a series of crushing actions occur. These are followed by automatic screening to form a product that approximates the granulation requirements. The dust and fines that have been screened are recycled to the press feed or used in fuse powder or fireworks. Coarse material is recycled. The grains are polished and dried; graphite is added followed by blending by tumbling in a large hardwood rotating drum. From 1360 to 2265 kg powder may be tumbled at 10–20 rpm for up to 8 h. Warm air may be forced through the barrel to assure drying and to decrease cycle time. The powder is screened before packing into airtight metal drums.

A number of techniques have been developed to eliminate the hazardous wheel milling process and reduce personnel exposure by increasing automation for the continuous transport of the product from operation to operation. The jet-mill air-attrition process, which has no moving parts, has replaced the wheel-mill operation of the conventional process. Potassium nitrate, sulfur, and charcoal are automatically weighed by transferring each ingredient to a weighing and mixing bin with a vibrating transporter. Air jets are applied to the bottom of the weigh-mix bin to blend the components. The air pressure is then increased to continuously transfer the mix pneumatically to a storage bin and then to the jet mill by air injection. A high velocity stream of air entering the mill forces the particles to collide and breaks them up by attrition. The product consists of a finely divided powder. The small particles exit through a cyclone separator where they are separated from the air, which is exhausted to the atmosphere. Coarse particles drop back to the attrition section where the milling action continues. The mill may be adjusted to produce powders of different granulations.

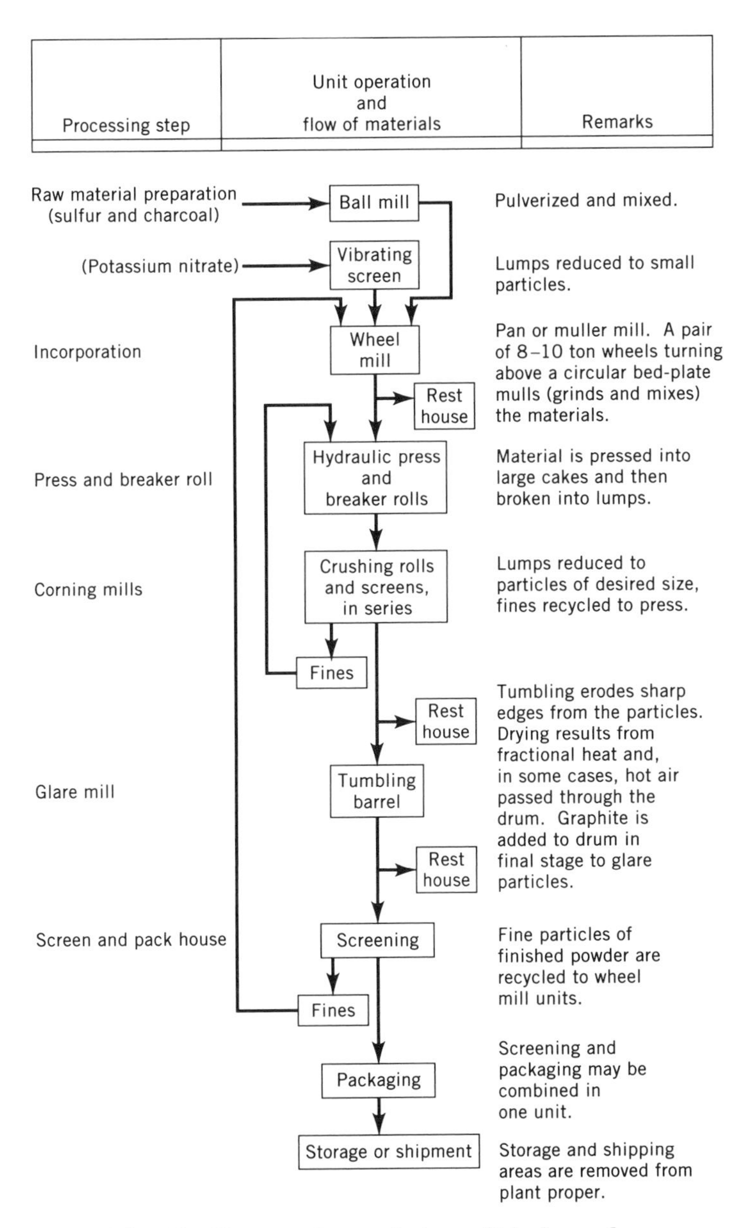

Fig. 10. Process of manufacture of black powder.

Pressing, corning, screening, and glazing are comparable to the conventional procedures except that automation is employed wherever possible. Deluge systems that are activated by uv light sensors and can respond in milliseconds are installed for additional safety. All operations are monitored and controlled from central process control areas. The presence of operators is restricted to the receipt of raw materials and packing of the final product. The pollution aspects of black

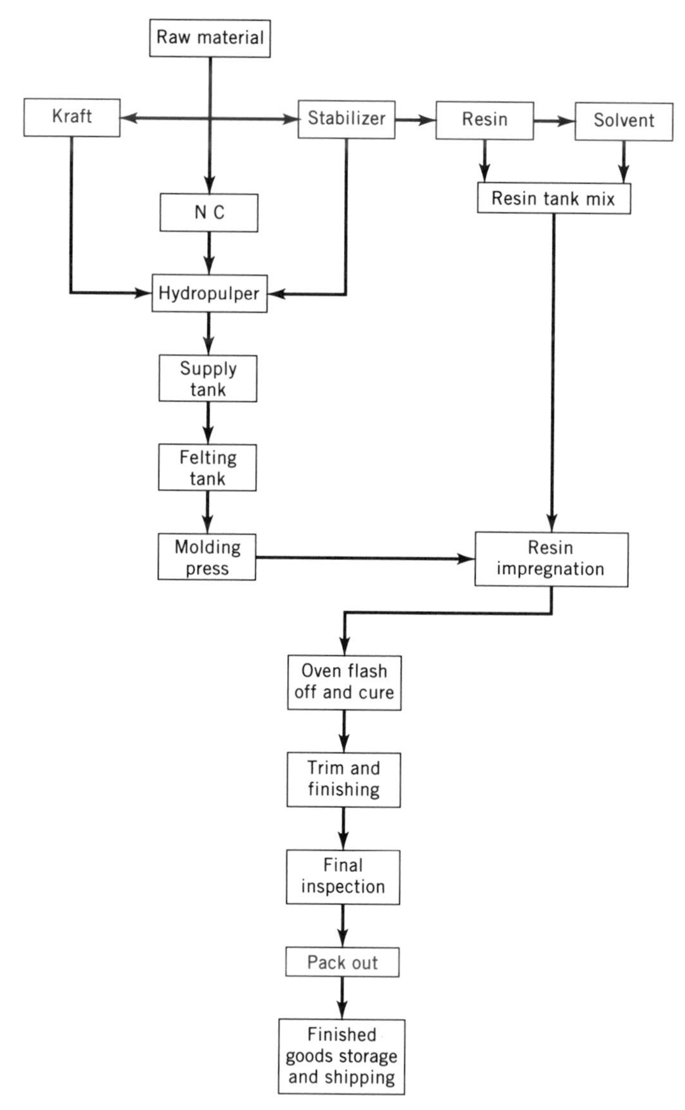

Fig. 11. Flow diagram for the beater additive process. Kraft represents the kraft process wood pulp and NC is nitrocellulose used as starting materials (182).

powder manufacture are relatively insignificant in view of the small quantities made. No wastewater or noxious fumes are produced (168–178).

Benite. Benite is an extrudable composition consisting of ca 60 parts of black powder in a matrix of ca 40 parts of plasticized nitrocellulose. It is used as a propellant igniter to reduce the residue formed compared to use of black powder alone. Benite can be extruded as strands, permitting a less obstructed flow of ignition gases and particles than granular black powder. Its approximate weight composition is nitrocellulose (13.15% N), 40%; potassium nitrate, 44%; sulfur, 6.5%; carbon, 9.5%; ethyl centralite added; 0.5%. It is made by the single-base process, followed by air-drying to remove volatile solvents (179).

Felted Nitrocellulose Compositions. A combustible case containing the propellant charge offers tactical, logistic, and performance advantages in certain types of munitions such as those used for tank weapons, mortars, and howitzers. This case is rigid and completely combustible, replacing metallic cases or flexible cases having low mechanical strength. Because nitrocellulose itself cannot be molded into a structure having the desired mechanical characteristics, inert fibers and a resin are added. A typical composition (wt %) consists of nitrocellulose (12.6% N), 55; kraft fiber, 9; acrylic fiber, 25; poly(vinyl acetate) resin, 10; and diphenylamine, 1.

The finished case has a density of ca 0.85 g/cm and a tensile strength of ca 24 MPa (3500 psi). Typical ballistic characteristics of combustible cases material are impetus, 578 J/g (138 cal/g); flame temperature, 1619 K; average molecular weight of the gaseous products, 23.3 g/mol; covolume, 1.17 mL/g; and specific heat ratio, 1.25.

Two processes may be used in the manufacture of combustible cases: the original post-impregnation process and the more recently and more widely employed beater additive process. The processes differ primarily in the point at which the required resin is added to the composition. A schematic of the beater additive process is shown in Figure 11.

The beater additive process starts with a very dilute aqueous slurry of fibrous nitrocellulose, kraft process woodpulp, and a stabilizer such as diphenylamine in a felting tank. A solution of resin such as poly(vinyl acetate) is added to the slurry of these components. The next step, felting, involves use of a fine metal screen in the shape of the inner dimensions of the final molded part. The screen is lowered into the slurry. A vacuum is applied which causes the fibrous materials to be deposited on the form. The form is pulled out after a required thickness of felt is deposited, and the wet, low density felt removed from the form. The felt is then molded in a matched metal mold by the application of heat and pressure which serves to remove moisture, set the resin, and press the fibers into near final shape (180–182).

BIBLIOGRAPHY

"Propellants," under "Explosives and Propellants," in *ECT* 3rd ed., Vol. 9, pp. 620–671, by V. Lindner, U.S. Army Armament Research and Development Command.

1. C. Boyars and K. Klager, eds., *Proceedings of Symposium on Propellants Manufacture, Hazards and Testing,* American Chemical Society, Washington, D.C., 1969.
2. A. E. Oberth, *Principles of Solid Propellant Development,* CPIA, Publication 469, Chemical Propulsion Information Agency (CPIA), Johns Hopkins University, Laurel, Md., Sept. 1987.
3. *Solid Propellant Manual Publication M2,* (classified), CPIA, Johns Hopkins University, Laurel, Md., 1969.
4. *Solid Propellant Selection and Characterization, Space Vehicle Design Criteria,* Monograph no. SP 8064, NASA, Airport, Md., 1971.
5. *Formulae and Calculated Thermochemical Values for Propellants,* Picatinny Arsenal (PTA), Dover, N.J., 1956.

6. T. C. Minor, *Interior Ballistic Calculations for the 155mm Cannon Launched Guided Projectile, XM712, Fired with the XM211 Propellant Charge,* interim memo report, Ballistics Research Laboratory (BRL), Aberdeen, Md., Aug. 1977, p. 569.
7. M. Shorr and A. J. Zachringer, eds., *Solid Propellant Technology,* John Wiley & Sons, Inc., New York, 1968.
8. D. R. Cruise, *Theoretical Computation of Equilibrium Composition, Thermal Dynamic Properties, and Performance Characteristics of Propellants Systems,* NWC TR6037, Naval Weapons Center, China Lake, Calif., Apr. 1979.
9. P. S. Gough, *The XNOVA KTC Code, BRL-CR-627,* BRL, APG, Aberdeen, Md., Feb. 1990.
10. L. S. Lussier and co-workers, "The Use of Thermodynamic Codes for Comparison of Propellant Performance," in *Fifth International Gun Propellant and Propulsion Symposium,* ARDEC, Dover, N.J., Nov. 1990.
11. S. Goldstein in H. B. Krier and M. Summerfield, eds., *Interior Ballistics for Guns,* Vol. 66, American Institute of Aeronautics and Astronautics Inc., 1979, pp. 67–86.
12. P. S. Gough, *Numerical Simulation of Current Artillery Charges Using the TDNOVA Code, BRL-TR-555,* BRL, Aberdeen, Md., June 1986.
13. G. R. Nickerson and co-workers, *A Computer Program for the Prediction of Solid Propellant Motor Performance,* Vols. 1, 2, and 3, ARFPL-TR-80-34, AFRPL, Dayton, Ohio, Apr. 1981.
14. M. G. Kurschner, *Propellants, Explosives and Pyrotechnics,* **1,** 81 (1976).
15. S. Gordon and B. J. McBride, *Computer Program for Calculation of Complex Equilibrium Compositions. Rocket Performance Incident and Reflected Shocks and Chapman-Jouquet Detonations,* NASA-Lewis Research Center, NASA, Airport, Md., Mar. 1976.
16. J. W. Cole, "Non-Destructive Testing of Large Rocket Motors: A State of the Art Survey," in *Bulletin of the Joint Meeting—JANNAF Panel of Physical Properties and Surveillance of Solid Propellants,* CPIA Publication PP-13/SPSP-8, CPIA, Johns Hopkins University, Laurel, Md., 1960, p. 345.
17. F. N. Kelley, in C. Boyars and K. Klager, eds., *Propellants Manufacture, Hazards and Testing, Advances in Chemistry Series 88,* American Chemical Society, Washington, D.C., 1969, p. 188.
18. *Handbook for the Engineering Structural Analysis of Solid Propellants,* CPIA Publication 214, CPIA, Johns Hopkins University, Laurel, Md., 1971.
19. G. J. Svob, in Ref. 2, Chapt. 10.
20. *JANNAF Solid Propellant Structural Integrity Handbook,* CPIA Publication 230, Johns Hopkins University, Laurel, Md., Sept. 1972.
21. *Solid Propellant Aging, Mechanical Behavior and Grain Structural Integrity,* CPIA Publication LS 77-27, CPIA, Johns Hopkins University, Laurel, Md., 1977.
22. R. Stenson, *Factors Governing the Storage Life of Solid Propellant Rocket Motors,* report no. TN047, Explosive Research and Development Establishment, Waltham Abbey, UK, 1972.
23. *Tools Required for a Meaningful Service Life Prediction,* CPIA Publication 259, Johns Hopkins University, Laurel, Md., 1974.
24. C. W. Fong, *Propellant, Explosives, Pyrotechnics* **10,** 91 (1985).
25. *JANNAF Structures and Mechanical Behavior Subcommittee Meeting,* CPIA Publication 566, NASA Marshall Space Flight Center, Ala., May 1991.
26. W. H. Stein and co-workers, in *Proceedings of the 4th International Gun Propellant Symposium,* Vol. 5, 1988, Dover, N.J., p. 363.
27. F. Volk and co-workers, *Propellants, Explosives, Pyrotechnics* **12,** 81 (1987).
28. H. Heloma and E. Kovene, "A Modern Method for Determining the Stability and Stabilizers in Solid Gun Propellants," in *Proceedings of the 6th Symposium on Chem-*

ical Problems Connected with the Stability of Explosives," Kungalo, Sweden, June, 1982.

29. H. J. Hoffman, *Moisture Effects on Mechanical Properties of Solid Propellants,* CPTR 84-29, CPIA Publications, Johns Hopkins University, Laurel, Md., Feb. 1984.
30. K. C. Schenk and Andrew Krause, *Lessons Learned: Deterioration of Munitions Stored Under Extreme Conditions (Bulk Ship Cargo, etc.),* ARDC-LL-86002, ARDEC, Dover, N.J., July 1986.
31. L. P. Piper, *HTPB Propellant Aging,* CPTR 82-12, CPIA Publications, Johns Hopkins University, Laurel, Md., June 1988.
32. F. Volk and JG. Wunsch, *Propellants, Explosives and Pyrotechnics* **10,** 181 (1985).
33. E. R. Bixon and D. Robertson, "Lifetime Predictions for Single Base Propellant Based on the Arrhenius Equation," in *Fifth International Gun Propellant and Propulsion Symposium,* ARDEC, Dover, N.J., Nov. 1991.
34. R. C. Strittmater, E. M. Wineholt, and M. E. Holmes, *The Sensitivity of Double Base Propellant Burning Rate to Initial Temperature,* MR-2593, BRL, Aberdeen, Md., 1976.
35. B. B. Grollman and C. W. Nelson, in *Bulletin of the 13th JANNAF Combustion Meeting,* CPIA Publication 281, Vol. 1, CPIA, Johns Hopkins University, Laurel, Md., 1976, p. 21.
36. D. W. Reifler, *Linear Burning Rates of Ball Propellants Based on Closed Bomb Firings,* CR 172, BRL, Aberdeen, Md., 1974.
37. B. T. Federoff and O. Sheffield, eds., "Burning Characteristics of Propellants for Rockets," in *Encyclopedia of Explosives and Related Items,* Vol. 2, TR 2700, PTA, Dover, N.J., 1962.
38. D. Guirie and D. H. Ensel, "Development of Spatial Laws of Burning for Single Base, Double Base and Heterogeneous Solid Propellant," in *Fifth International Gun Propellant and Propulsion Symposium,* ARDEC, Dover, N.J., Nov. 1991.
39. B. L. Crawford and co-workers, *J. Phys. and Colloid Chem.* **54**(6), 854 (1950).
40. O. K. Rice and R. Ginell, in Ref. 39, p. 885.
41. R. G. Parr and B. L. Crawford, in Ref. 39, p. 929.
42. G. K. Adams, in *Proceedings of the Fourth Symposium of Naval Structural Mechanics: Mechanics and Chemistry of Solid Propellants,* Pergamon Press, Inc., New York, 1967, p. 117.
43. G. A. Heath and R. Hirst, in *8th International Symposium on Combustion,* The Williams and Wilkins Co., Baltimore, Md., 1962, p. 711.
44. M. M. Ibercu and F. A. Williams, in *Proceedings of the 12th JANNAF Combustion Meeting,* CPIA Publication 273, Vol. 2, CPIA, Johns Hopkins University, Laurel, Md., 1975, p. 283.
45. H. Singh and K. Rao, *Propellants, Explosives and Pyrotechnics* **7,** 61 (1982).
46. C. K. Adams, B. H. Newman, and A. B. Robins, *Proceedings 8th Symposium (International) on Combustion,* The Williams and Wilkins Co., Baltimore, Md., 1960, p. 693.
47. M. Summerfield and co-workers, in M. Summerfield, ed., *Solid Propellant Research,* Academic Press, Inc., New York, 1960.
48. J. A. Stein, P. L. Stang, and M. Summerfield, *The Burning Mechanism of Ammonium Perchlorate-Based Composite Propellants,* Aerospace and Mechanical Sciences Report 830, Princeton University, N.J., 1969.
49. N. S. Cohen, R. L. Derr, and C. E. Price, in *Proceedings of the 9th JANNAF Combustion Meeting,* CPIA Publication 231, CPIA, Johns Hopkins University, Laurel, Md., 1972, p. 25.

50. M. W. Beckstead, in *Proceedings of the 14th JANNAF Combustion Meeting,* Vol. 1, CPIA Publication 292, CPIA, Johns Hopkins University, Laurel, Md., 1977, p. 281.
51. N. Kubota, "Survey of Rocket Propellants and Their Combustion Characteristics," in K. K. Kuo and M. Summerfield, eds., *Fundamentals of Solid Propellant Combustion,* Vol. 90, Progress in Astronautics and Aeronautics, AJAA, 1984.
52. M. W. Beckstead and K. P. Brooks, *A Model for Distributed Combustion in Solid Propellants,* Vol 2, CPIA Publication 557, CPIA, Johns Hopkins University, Laurel, Md., Nov. 1990, p. 227.
53. Ref. 2, Chapt. 9.
54. R. R. Miller and co-workers, *Ballistic Control of Solid Propellants,* Hercules, Inc., Rpt. AFRPL-81-058, Dayton, Ohio, 1981.
55. N. S. Cohen and D. A. Flanigan, *Effects of Propellants Formulation on Burn Rate—Temperature Sensitivity,* CPIA Publication 390, Vol. 3, CPIA, Johns Hopkins University, Laurel, Md., 1984.
56. A. O. Pallingston and M. Weinstein, *Method of Calculation of Interior Ballistic Properties of Propellants from Closed Bomb Data,* report 2005, PTA, Dover, N.J., 1959.
57. H. Jahnk, *Propellants, Explosives, Pyrotechnics* **1,** 47 (1976).
58. D. Vittal, *Propellants, Explosives, Pyrotechnics* **5,** 914 (1980).
59. K. Atlas, *A Method of Computing Web for Gun Propellant Grains from Closed Bomb Burning Rates,* memo report 73, Naval Powder Factory, U.S. Navy, Indian Head, Md., 1954.
60. J. Maillette and L-S. Lussier, in *10th International Symposium on Ballistics,* Vol. 2, San Diego, Calif., 1987, p. 52.
61. W. Langlotz, "Combustion Behavior of Classical Propellants," in *Fifth International Gun Propellant and Propulsion Symposium,* ARDEC, Dover, N.J., Nov. 1991.
62. E. B. Fisher and K. P. Tripper, in *Proceedings of the 12th JANNAF Combustion Meeting,* CPIA Publication 273, Vol. 1, CPIA, Johns Hopkins University, Laurel, Md., 1975, p. 323.
63. M. J. Adams and H. Crier, in Ref. 62, p. 303.
64. S. Wachtel, "Prediction of Detonation Hazards in Solid Propellants," paper presented at the *145th National Meeting of the Division of Fuel Chemistry,* New York, 1963.
65. J. Maillette and co-workers, "Pressure Wave Generation in Three Inch/50 Gun," in *Proceedings of the 10th International Symposium on Ballistics,* American Defense Preparedness Association, (ADPA), San Diego, Calif., 1987.
66. R. J. Lieb and J. J. Rocchio, "The Effects of Grain Fracture on the Interior Ballistics Performance of Gun Propellants," in *Proceedings of 8th International Symposium on Ballistics,* American Defense Preparedness Association, Washington, D.C., 1983.
67. A. W. Horst, I. W. May, and E. V. Clarke, *The Missing Link Between Pressure Waves and Breech Blows,* BRL Rpt. ARBRL-MR-02849, BRL, Aberdeen, Md., July 1978.
68. W. A. Blaine and co-workers, in *Preprints of the 9th Symposium (International) on Detonation,* Portland, Oreg., Aug. 28–Sept. 1, 1989, pp. 233–239.
69. A. W. Horst, T. C. Smith, and S. E. Mitchell, in *Bulletin of the 13th JANNAF Meeting,* CPIA Publication 281, Vol. 1, CPIA, Johns Hopkins University, Laurel, Md., 1976, p. 225.
70. J. F. Kincaid, *The Determination of the Propensity for Detonation of High Performance Propellants,* CPTR-6, CPIA Publication 334, CPIA Publications, Johns Hopkins University, Laurel, Md., Feb. 1981.
71. P. Benhaine, J. L. Paulin, and B. Zeller, "Investigation on Gun Propellant Breakup and its Effect in Interior Ballistics," *Proceedings of 10th International Symposium on Ballistics,* San Diego, Calif., Oct. 1989.
72. A. M. Miller and co-workers, *Prog. Energy and Combustion Sci.* **14**(3) (1988).

73. T. D. Wilson, *New Solid Rocket Propellant Polymer Binder Materiels,* CPTR 87-42, Aug. 1987.
74. E. J. Mastrolia and K. Klager, in Ref. 17.
75. A. E. Oberth, in Ref. 17.
76. F. Arendale, in Ref. 17, p. 67.
77. R. A. Henry and co-workers, *Polymeric Binders Which Reversibly Dissociate at Elevated Temperatures,* Tech Rpt. NWC-TP-5995, Naval Weapons Center, China Lake, Calif., May 1978.
78. B. D. Nahloosky and G. A. Zimmerman, *Thermoplastic Elastomers for Solid Propellant Binders,* AFRPL-TR-86-069, Aerojet Tactical Systems Co., Sacramento, Calif., Dec. 1986.
79. R. Muracarf and L. L. Taylor, *The Hygroscopicity of Lithium and Ammonium Nitrates and Perchlorates,* Progress Rpt. 20-347, Jet Propulsion Lab., Institute of Technology, Pasadena, Calif., 1958.
80. Ref. 2, Chapt. 5.
81. J. P. Renie, J. A. Condon, and J. R. Osborn, in Ref. 50, p. 325.
82. R. H. Waesche, "Workshop in the Relationship of Ammonium Perchlorate Decomposition to Deflagration," *Proceedings of the 7th JANNAF Combustion Meeting,* Vol. 1, CPIA Publication 204, CPIA, Johns Hopkins University, Laurel, Md., 1971, p. 15.
83. R. R. Miller, M. T. Donohue, and J. P. Peterson, in Ref. 62, p. 371.
84. J. Wenograd, in *Proceedings 6th ICRPG Combustion Conference,* Vol. 1, CPIA Publication 192, CPIA, Johns Hopkins University, Laurel, Md., 1969, p. 367.
85. P. Kuentzmann, *Specific Impulse Losses of Metallized Solid Propellants,* FTD-NC-23-2717-74, Air Force Foreign Technology Division, Wright Patterson Air Force Base, Dayton, Ohio, 1974.
86. R. L. Derr, in Ref. 69, p. 185.
87. R. L. Derr and co-workers, in Ref. 62.
88. T. E. Goddard, "The Future of Bulk Loaded Fluid Propellant Guns," in *3rd International Gun Propellant Symposium,* American Defense Preparedness Association, Washington, D.C., Oct. 1984.
89. W. F. Morrison and co-workers, "The Interior Ballistics of Regenerative Liquid Propellant Guns," *Proceedings of 8th International Symposium on Ballistics,* Apr. 1983.
90. S. Murad and P. Ravi, *Fifth Annual Conference on HAN-Based Propellants,* BRL, Aberdeen, Md., 1990.
91. P. S. Gough, *A Model of the Interior Ballistics of Hybrid Liquid Propellant Guns,* BRL-CR-566, BRL, Aberdeen, Md., Mar. 1987.
92. W. O. Seals and co-workers, in *Proceedings of Joint International Symposium on Compatibility of Plastics and Other Materials with Explosives, Propellants, and Pyrotechnics,* American Defense Preparedness Association, New Orleans, La., 1988, p. 261.
93. P. A. Kaste and co-workers, in Ref. 52, p. 109.
94. S. T. Peters, *LOVA Propellants—The State of the Art,* CPIA Publication 446, Vol. 2, CPIA, Johns Hopkins University, Laurel, Md., March 1986.
95. C. H. Dettling and co-workers, *Insensitive Munitions Characteristics of Air Launched In Service Weapons: Summary Report of Fast Cook Off Times, Reactions and Hazards of Bombs, Rockets, Aircraft Guns, Air Launched Missiles, Mines and Torpedoes,* Naval Weapons Center, NWC, China Lake, Calif., 1989.
96. B. D. Strauss and co-workers, "New Energetic Plasticizers for LOVA Propellants," in *3rd International Gun Propellant Symposium,* Dover, N.J., Oct. 1984.
97. *Polymeric Binder Materials for LOVA Propellants,* Battelle Columbus Labs, Columbus, Ohio, 1985.

98. F. Schledbauer, "LOVA Gun Propellants with GAP Binder," in *Fifth International Propellant and Propulsion Symposium,* ARDEC, Dover, N.J., Nov. 1991.
99. T. Boggs and A. Victor, *JANNAF Propulsion Meeting,* CPIA Publication 455, Vol. 3, CPIA, Johns Hopkins University, Laurel, Md., Aug. 1986, pp. 407–426.
100. T. L. Boggs and R. L. Derr, eds., *Hazard Studies for Solid Propellant Rocket Motors,* NATO Advisory Group for Aerospace Research and Development, Agardograph 316, Technical Press, London, Sept. 1990.
101. C. A. Dettling, in *1991 JANNAF Propulsion Systems Hazard Subcommittee Meeting,* CPIA Publication 562, Sandia National Labs, Albuquerque, N.Mex., Mar. 1991, pp. 18–22.
102. D. A. Ciaremitari and M. L. Chan, *High Performance Reduced-Smoke Propellant Formulations,* CPIA Publication 580, Vol. 2, CPIA, Johns Hopkins University, Laurel, Md., Jan. 1992.
103. Eugene Miller, in Ref. 51.
104. S. H. Hasty, *Ammonium Nitrate Rocket Propellants,* CPTR 91-48, CPIA Publications, Johns Hopkins University, Laurel, Md., Sept. 1991.
105. L. Asaoka, *Phase Stabilized Ammonium Nitrate Effects on Minimum Signature Propellant Properties,* Vol. 5, 3 CPIA Publication 550, Anaheim, Calif., 1990.
106. H. H. Weyland and co-workers, *Clean Propellants for Large Rocket Motors,* Final report AL-TR-90-016, Astronautics Laboratory, Edwards Air Force, Calif., Aug. 1990.
107. S. L. Newey and G. M. Clark, in Ref. 105.
108. D. C. Ferguson and co-workers, in Ref. 55, p. 541.
109. W. Klohn and S. Eisele, *Propellants, Explosives and Pyrotechnics* **12,** 71 (1987).
110. A. Arpad and co-workers, "Advanced Applications for Hypervelocity Gun Applications," in *Proceedings of the 6th International Symposium on Ballistics,* Batelle Columbus Lab, Columbus, Ohio, Oct. 1981.
111. I. W. May and co-workers, *The Travelling Charge Effect,* BRL memo report BRL-MR-03034, BRL, Aberdeen, Md., July 1980.
112. W. F. Eile and co-workers, "The Travelling Charge: An Experimental and Computational Investigation," in *Proceedings of Tenth International Symposium on Ballistics,* Vol. 1, ADPA, San Diego, Calif., 1987.
113. C. S. Leveritt, *Ultra-High Burning Rate Propellants for Travelling Charge Guns,* BRL-CR-00447, BRL, Aberdeen, Md., Feb. 1981.
114. P. Gough, *A Model of the Travelling Charge,* BRL TR-00432, BRL, Aberdeen, Md., July 1980.
115. A. A. Juhasz and I. W. May, "Advanced Propellants for Hypervelocity Gun Applications," in Ref. 110.
116. J. T. Barnes and E. B. Fisher, in Ref. 52, p. 1.
117. S. K. Combs and co-workers, *Repetitive Small Bore, Two Stage Light Gas Gun,* CONF-9110244-1, Oak Ridge National Lab, Tenn., 1991.
118. J. W. Hunter and R. A. Hyde, *Light Gas Gun System for Launching Building Material into Low Earth Orbit,* VCRL-99623, Lawrence Livermore Lab (LLL), Livermore, Calif., July 1989.
119. S. L. Milora and co-workers, *Quickgun: An Algorithm for Estimating the Performance of Two-Stage Light Gas Guns,* ORNL/TM-11561, Oak Ridge National Lab, Tenn., Sept. 1990.
120. F. K. Boutwell and co-workers, *GM DRL's 2¼ Inch Gun Project,* GM Defense Research Lab, Santa Barbara, Calif., Aug. 1966.
121. A. P. Bruckner and co-workers, in Ref. 52, p. 300.
122. A. Hertzberg and co-workers, *AIAA Journal* **26,** 195 (1988).
123. D. L. Kruczyski, in Ref. 52, p. 319.
124. W. J. Creighton, *IEE Transactions on Magnetics* **25**(1), 133 (Jan. 1989).

125. A. L. Brooks and R. S. Hawke, *IEE Transactions on Magnetics* **25**(1), 145 (Jan. 1989).
126. Y-C. Chio, in Ref. 110, p. 352.
127. A. L. Brooks and co-workers, *Design and Fabrication of Small and Large Bore Railguns,* UCRL-84876, Lawrence Livermore National Lab, Livermore, Calif., 1984.
128. M. Guillemot and co-workers, *IEE Transactions on Magnetics* **25,** 20 (Jan. 1989).
129. A. A. Juhasz, *Activities in Electrothermal Gun Propulsion,* CPIA Publication 528, Vol. 1, CPIA, Johns Hopkins University, Laurel, Md., 1989, p. 103.
130. A. A. Juhasz and co-workers, *Introduction to Electrothermal Gun Propulsion,* CPIA Publication , Vol. 4, CPIA, Johns Hopkins University, Laurel, Md., Oct. 1988, p. 223.
131. A. A. Juhasz and co-workers, in Ref. 52, p. 141.
132. W. Morelli and W. Oberle, "Electrothermal-Chemical Gun Propulsion in the United States," in *Fifth International Gun Propellant and Propulsion Symposium,* ARDEC, Dover, N.J., Nov. 1991.
133. T. D. Wilson, *New Solid Propellant Processing Techniques,* CPTR 191-47, CPIA Publications, Johns Hopkins University, Laurel, Md., Aug. 1991.
134. D. Mueller, *Procedures for the Production of Gun Propellant by Use of Different Extruders,* Frauerhofer Institute fuer Treb-und Explosivestaffe, Pfinztal-Berghausen, Germany, June 1982.
135. *A Continuous Automated Cannon Propellant Facility,* Radford Army Ammunition Plant, Radford, Va., 1970.
136. W. J. Nolan, in *Proceedings Symposium on Processing Propellants, Explosives and Ingredients,* ADPA, Washington, D.C., 1977, p. 1.3–I.
137. D. Mueller, "The Continuous Processing of Gun Propellants by the Twin Screw Extruder," in *Proceedings of the Third International Gun Propellant Symposium,* Dover, N.J., Oct. 1984.
138. M. Olsson, in Ref. 105.
139. J. Knobloch and co-workers, in Ref. 105.
140. D. M. Husband, "A Review of Technology Developments in Continuous Processing of Solid Propellants: (1960–1987)," in *Proceedings of 1987 JANNAF Propellant Characterization Subcommittee,* CPIA Publications, Johns Hopkins University, Laurel, Md., Oct. 1987.
141. F. S. Baker and R. E. Carter, *Propellants, Explosives and Pyrotechnics* **7,** 139 (1982).
142. J. J. Rutkowski, *Blending Stick Propellant,* ARLCD-TR 85018, ARDEC, Dover, N.J., Aug. 1985.
143. S. M. Kaye, ed., "Propellants, Solid," in *Encyclopedia of Explosives and Related Items,* Vol. 8, PATR 2700, Armament Research Development and Engineering Center, Dover, N.J., Mar. 1978, p. 402.
144. C. J. V. Helle, *Propellants, Explosives, Pyrotechnics* **13,** 55 (1988).
145. T. Urbanski, *Chemistry and Technology of Explosives,* Vol. 3, Pergamon Press, Inc., New York, 1967, Chapts. 7–8.
146. R. D. Anderson and R. T. Puhalla, "Parametric Study of Temperature Insensitivity of Ball Propellants," in *Proceedings of the 26th JANNAF Combustion Meeting,* Vol. 3, CPIA Publication 529, CPIA, Johns Hopkins University, Laurel, Md., Oct. 1989.
147. Ref. 2, Chapt. 10.
148. R. Steinberger, in S. S. Penner and J. DuCarme, eds., *The Chemistry of Propellants,* Pergamon Press, Inc., Oxford, 1960, p. 246.
149. R. Steinberger and P. D. Drechsel, in Ref. 17, p. 1.
150. R. Steinberger, in *Proceedings of the 7th International Symposium on Space Technology and Science,* AGNE Pub. Inc., Tokyo, 1967, p. 63.
151. R. A. McKay, *A Study of Selected Parameters in Solid Propellant Processing,* Jet Propulsion Lab, Pasadena, Calif., Aug. 1986; J. L. Brown and co-workers, *Manufacturing Technology for Solid Propellant Ingredients / Preparation Reclamation,* Morton Thio-

kol, Inc., Brigham City, Utah, Apr. 1985; W. P. Sampson, *Low Cost Continuous Processing of Solid Rocket Propellant,* A1-TR-90-008, Astronautics Laboratory/TSTR, Edwards AFB, Oct. 1990.
152. G. A. Fluke, in Ref. 17, p. 165.
153. E. M. G. Cooke, *Propellants, Explosives and Pyrotechnics* **15,** 235 (1990).
154. *Solid Propellant Selection and Characterization,* SP-8064, NASA, Cleveland, Ohio, 1971.
155. J. F. Tormey, "Processing and Manufacture of Composite Propellants," *Proceedings of the AGARD Colloquim, Advances in Tactical Rocket Propulsion,* CIRA Pub., Pelham, New York, 1968.
156. J. L. Brown and co-workers, *Manufacturing Technology for Solid Propellant Ingredients / Preparation Reclamation,* Morton Thiokol, Inc., Brigham City, Utah, Apr. 1985.
157. L. W. Collins, *The Feasibility of Utilizing a Fluid Energy Mill in the Grinding of Potassium Perchlorate,* MLW 2444, Monsanto Research Corp., Miamisburg, Ohio, 1977.
158. *Particle Size Analyses and Effects—Selected Papers,* CPIA Publication SP91-04, R2/92, CPIA, Johns Hopkins University, Laurel, Md., Aug. 1991.
159. *Solid Propellant Processing Factors in Rocket Motor Design,* Space Vehicle Design Criteria Monograph, SP-8075, NASA, Airport, Md., 1971.
160. *Solid Rocket Motor Insulation,* Space Vehicle Design Criteria Monograph, NASA, Airport, Md., 1971.
161. W. P. Killian, "Loading Composite Solid Propellant Rockets—Current Technology," *Proceedings Symposium on Selected Topics in Aerospace Chemistry, 64th National Meeting,* AICE, Orlando, Fla., 1968.
162. J. T. Carver, *Improved Specifications for Composite Propellant Binders for Army Weapon Systems,* TR-T-79-76, Army Missile Command, Huntsville, Ala., July 1979.
163. G. E. Herriott and J. M. Lilly, in *Proceedings of the 1990 JANNAF Propulsion Meeting,* Anaheim, Calif., 1990, p. 37.
164. D. S. Gaardner, ed., *Safety and Hazards of High Energy Propellants,* CPIA Publication 284, CPIA, Johns Hopkins University, Laurel, Md., 1977.
165. T. L. Boggs and co-workers, in *Proceedings of Eastern States Section,* Combustion Institute, Pittsburgh, Pa., Nov. 1985.
166. T. D. Wilson and O. T. Moskios, *Disposal of Solid Rocket Motor Propellants,* CPTR89-45, CPIA Publications, Johns Hopkins University, Laurel, Md., July 1989.
167. L. W. Poulter and co-workers, *Solid Propellant Ingredient Reclamation,* CPIA Publication 556, CPIA, Johns Hopkins University, Laurel, Md., Oct. 1990, p. 107.
168. *Black Powder,* Mil. Spec. 2123-B, ARDEC, Dover, N.J., 1973.
169. Ref. 37, p. 165.
170. T. Urbanski, *Chemistry and Technology of Explosives,* Vol. 3, Pergamon Press, Inc., New York, 1967.
171. J. Isaksson and L. Rittfeldt, "Characterization of Black Powders," *Proceedings of 3rd Symposium on Chemical Problems Connected with the Stability of Explosives,* Jonkoping, Sweden, 1952, p. 242.
172. F. A. Williams, in Ref. 62.
173. J. V. White and co-workers, in Ref. 69, p. 405.
174. K. Lvold, in Ref. 171, p. 266.
175. D. R. Mouta and co-workers, *Hazards Analysis of the Final Design of the Improved Black Powder Process,* Vols. 1–2, Rpt. J6329, Illinois Institute of Technology, Chicago, 1963.
176. S. I. Morrow and B. Haywood, *Propellants, Explosives, Pyrotechnics* **8,** 133–138 (1983).

177. H. Hahn, *PEP* **5**, 129–134 (1980).
178. R. A. Sasse, *7th International Pyrotechnics Seminar,* Vol. 2, 1980, CPIA Publications, Johns Hopkins University, Laurel, Md., p. 565.
179. E. Husselton and S. Kaplowitz, *Benite Manufacture,* DB-TR-5-60, PTA, Dover, N.J., 1961.
180. F. W. Robbins and J. W. Colburn, in Ref. 52, p. 61.
181. M. E. Sever and H. Hassmann, "155mm M203E2 Artillery Propeling Charge with Combustible Cartridge Case and Stick Propellant," in *Proceedings of the Third International Gun Propellant Symposium,* ARDEC, Dover, N.J., Oct. 1984.
182. H. A. Appleman, *Combustible Ordnance in the United States,* ARMTEC Defense Products Co., Cochella, Calif., 1988.

General References

The Annual Proceedings of the Joint Army-Navy-Air Force (JANNAF) Propulsion Meetings, the reports of the special committees, and the periodic literature surveys published by the Chemical Propulsion Information Agency including the annual Chemical Propulsion Abstracts are invaluable sources of information on all aspects of liquid and solid gun and rocket propellants. They may be classified.
R. T. Holzmann, *Advanced Propellant Chemistry, Advances in Chemistry Series 54,* ACS, Washington, D.C., 1966.
A. E. Oberth, *Principles of Solid Propellant Development,* CPIA Publication 469, CPIA, Johns Hopkins University, Laurel, Md., Sept. 1987.
B. Siegel and L. Schieler, *Energetics of Propellant Chemistry,* John Wiley & Sons, Inc., New York, 1964.
Proceedings of the International Gun Propellant Symposia.
Proceedings of the International Symposia on Combustion, Combustion Institute, Pittsburgh, Pa.
International Symposia on Ballistics, sponsored and published by the American Defense Preparedness Association (ADPA), Washington, D.C.
Proceedings of the Symposia on Compatibility of Plastics and Other Materials with Explosives, Propellants and Pyrotechnics and Processing of Propellants, Explosives, and Ingredients, ADPA, Washington, D.C.
Proceedings of the Annual Seminars on Explosive Safety, sponsored by the U.S. Explosives Safety Board, Washington, D.C.
Solid Propellant Manuals, issued by the Army Material Command, AMC 706-175, 1961, and AMC 706-176, 1961.
CPIA/M2 Solid Propellant Manual, CPIA, Johns Hopkins University, Laurel, Md., 1969.
F. X. Hartman, *Solid Propellants Safety Handbook,* Springfield, Va., 1965.
Military Specifications, U.S. Military Standards 286-B, 1971, and P-270-A, U.S. Government Printing Office, Washington, D.C., 1959.
Publications of the Chemical Propulsion Information Agency (CPIA) obtainable from Applied Physics Laboratory, Johns Hopkins University, Laurel, Md. (may be classified).
B. T. Federoff and O. E. Sheffield, *The Encyclopedia of Explosives and Related Items,* Vols. 1–10, TR 2700, ARDEC, Dover, N.J.
T. Urbanski, *Chemistry and Technology of Explosives,* Vol. 1–3, Pergamon Press, Inc., New York, 1967, and Vol. 4, 1984.
Thermochemical Data, Dow Chemical Co., Midland, Mich., 1967.
C. Huggett, C. E. Bartley, and M. M. Mills, *Solid Propellant Rockets,* Princeton University Press, N.J., 1960.
Proceedings of Symposium on Advanced Propellant Chemistry, American Chemical Society, Washington, D.C., 1965.

Rocket propellants

M. H. Smith, "The Literature of Rocket Propulsion," in *The Literature of Chemical Technology, Advances in Chemistry, Series No. 74,* American Chemical Society, Washington, D.C., 1968, p. 581.

Monographs on rockets and rocket propellants by the National Aeronautics and Space Administration (NASA), Lewis Research Center, Cleveland. These include the following: *Solid Propellant Selection and Characterization,* Report SP-8064, 1971; *Solid Rocket Motor Performance,* Report SP-8039, 1971; *Solid Rocket Motor Igniters,* Report SP-8051, 1971; *Solid Rocket Motor Metal Cases,* Report SP-8025, 1970, and *Captive Fire Testing of Solid Rocket Motors,* Report SP-8041, 1971.

G. P. Sutton, *Rocket Propulsion Elements,* 5th ed., John Wiley & Sons, Inc., New York, 1986.

M. Shorr and A. J. Zaehringer, eds., *Solid Propellant Technology,* John Wiley & Sons, Inc., New York, 1968.

K. K. Kuo and M. Summerfield, *Progress in Astronautics and Aeronautics,* Vol. 90, AIAA, New York, 1984.

F. Sarner, *Propellant Chemistry,* Reinhold Publishing Corp., New York, 1966.

G. P. Sutton, *Rocket Propulsion Elements,* 3rd ed., John Wiley & Sons, Inc., New York, 1963.

R. F. Gould, ed., *Propellant Manufacture, Hazards and Testing, Advances in Chemistry Series 88,* American Chemical Society, Washington, D.C., 1969.

Proceedings of the Symposium on Kinetics of Propellants, Division of Physical and Inorganic Chemistry, 112th Meeting of the American Chemical Society, New York, 1957, published in *J. of Phys. Colloid Chem.* **54** (1950).

F. A. Williams and co-workers, *Fundamental Aspects of Solid Rocket Propellants,* Agardograph No. 116, Technical Press, London, 1969.

S. S. Penner, *Chemical Rocket Propulsion and Combustion Research,* Gordon and Breach Science Publishers, New York, 1962.

F. A. Warren, "Solid Propellant Technology," in R. A. Gess, ed., *Selected Reprint Series,* AIAA, New York, 1970.

R. L. Wilkins, *Theoretical Evaluation of Chemical Propellants,* Prentice Hall, Inc., New York, 1963.

C. Huggett, C. F. Bartley, and M. Mills, *Solid Propellant Rockets,* Princeton University Press, N.J., 1960.

D. Altman and co-workers, *Liquid Rocket Propellants,* Princeton University Press, N.J., 1960.

C. Boyers and K. Klager, eds., *Proceedings of Symposium on Propellants Manufacture, Hazards and Testing,* American Chemical Society, Washington, D.C., 1969.

Gun propellants

J. Corner, *Theory of Interior Ballistics of Guns,* John Wiley & Sons, Inc., New York, 1950.

F. R. W. Hunt, ed., *Internal Ballistics,* Philosophical Library, New York, 1951.

Interior Ballistics of Guns, Engineering Design Handbook, AMCP 706-150, DARCOM, 1965.

J. H. Grese "Ballistic Calculations," in *Encyclopeia Brittanica,* Vol. 3, 1976.

P. Baer and J. Frankle, *Simulation of Interior Ballistic Performance of Guns by Digital Computer Program,* Rpt. 1183, BRL, Aberdeen, Md., 1962.

Journal of Ballistics, Memorials de Poudres, annual summaries of the Chemical Propulsion Information Agency (classified).

H. Krier and M. Summerfield, eds., *Interior Ballistics of Guns,* AIAA in Astronautics and Aeronautics, 1979, p. 66.

VICTOR LINDNER
Armament Research, Development and Engineering Agency

EXT. D&C DYES. See COLORANTS FOR FOOD, DRUGS, COSMETICS, AND MEDICAL DEVICES.

EXTRACTION

LIQUID–LIQUID

Liquid–liquid extraction, often loosely referred to as solvent extraction, was carried out as early as Roman times when silver and gold were extracted from molten copper (qv) using lead (qv) as a solvent. The first significant industrial application of solvent extraction was in the petrochemical industry (see PETROLEUM). This was followed by applications for the recovery of vegetable oils (qv) and the purification of penicillin (see ANTIBIOTICS, β-LACTAMS), and since 1945 in the nuclear industry in the refining of uranium, plutonium, and other radioisotopes (qv). Since 1960, solvent extraction has been applied on a large scale in the refining of other nonferrous metals, particularly copper (see METALLURGY, EXTRACTIVE). Most recently it has gained increasing importance as a separation technique in biotechnology (qv) (see also EXTRACTION, LIQUID–SOLID).

The physical process of liquid–liquid extraction separates a dissolved component from its solvent by transfer to a second solvent, immiscible with the first but having a higher affinity for the transferred component. The latter is sometimes called the consolute component. Liquid–liquid extraction can purify a consolute component with respect to dissolved components which are not soluble in the second solvent, and often the extract solution contains a higher concentration of the consolute component than the initial solution. In the process of fractional extraction, two or more consolute components can be extracted and also separated if these have different distribution ratios between the two solvents.

The principle of liquid–liquid extraction, and some of the special terminology, are illustrated in Figure 1 which shows a single contacting stage. If equilib-

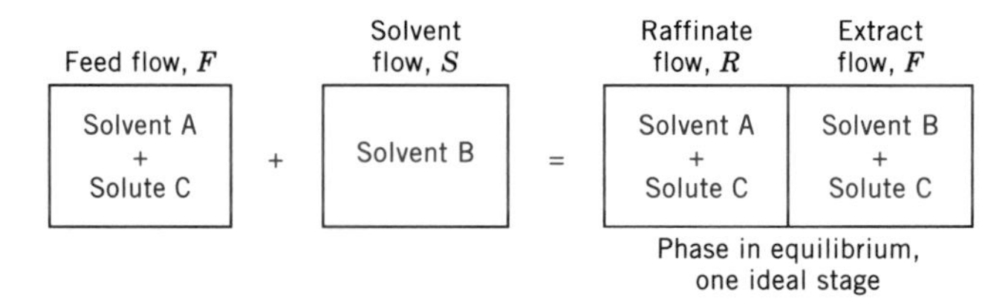

Fig. 1. Single contacting stage.

rium is fully established after contact, the stage is defined as an ideal or theoretical stage. The two resulting liquid phases are the raffinate from which most of solute C has been removed, and the extract, consisting mainly of solvent B and C.

In the simplest case, the feed solution consists of a solvent A containing a consolute component C, which is brought into contact with a second solvent B. For efficient contact there must be a large interfacial area across which component C can transfer until equilibrium is reached or closely approached. On the laboratory scale this can be achieved in a few minutes simply by hand agitation of the two liquid phases in a stoppered flask or separatory funnel. Under continuous flow conditions it is usually necessary to use mechanical agitation to promote coalescence of the phases. After sufficient time and agitation, the system approaches equilibrium which can be expressed in terms of the extraction factor ϵ for component C:

$$\epsilon = \frac{\text{quantity of C in B-rich phase}}{\text{quantity of C in A-rich phase}} = m\,\frac{\text{B}}{\text{A}} \tag{1}$$

where B and A refer to the quantities of the two solvents and m is the distribution coefficient.

The component C in the separated extract from the stage contact shown in Figure 1 may be separated from the solvent B by distillation (qv), evaporation (qv), or other means, allowing solvent B to be reused for further extraction. Alternatively, the extract can be subjected to back-extraction (stripping) with solvent A under different conditions, eg, a different temperature; again, the stripped solvent B can be reused for further extraction. Solvent recovery (qv) is an important factor in the economics of industrial extraction processes.

Whereas Figure 1 assumes a physical extraction based on different solubilities as expressed by the distribution coefficient, many extractions depend on chemical changes. In such cases the component C in the feed solvent may not itself have any solubility in the extracting solvent B, but can be made to react with an extractant to produce a compound or species which is soluble in B. Many metals can be extracted from aqueous solutions of their salts into organic carrier solvents by using organic extractants which can form organometallic compounds or complexes. Stripping of the metals from the organic to an aqueous phase can be effected by changing a chemical condition such as pH.

Extraction, a unit operation, is a complex and rapidly developing subject area (1,2). The chemistry of extraction and extractants has been comprehensively described (3,4). The main advantage of solvent extraction as an industrial process lies in its versatility because of the enormous potential choice of solvents and extractants. The industrial application of solvent extraction, including equipment design and operation, is a subject in itself (5). The fundamentals and technology of metal extraction processes have been described (6,7), as has the role of solvent extraction in relation to the overall development and feasibility of processes (8). The control of extraction columns has also been discussed (9).

The rapid development of the fundamentals and technology of solvent extraction is reflected in research papers found in chemistry, chemical engineering, and hydrometallurgical journals. Review articles (10–12) provide a critical summary of some of the ongoing research. Leading researchers and practitioners of solvent extraction meet regularly at the International Solvent Extraction Conferences (ISEC) whose proceedings (13–21) provide a useful indication of progress and trends. Papers published in the *ISEC Proceedings* numbered over 400 in 1990 (20), yet these make up less than 20% of all papers published in the solvent extraction area annually (12).

Principles

Physical Equilibria and Solvent Selection. In order for two separate liquid phases to exist in equilibrium, there must be a considerable degree of thermodynamically nonideal behavior. If the Gibbs free energy, G, of a mixture of two solutions exceeds the energies of the initial solutions, mixing does not occur and the system remains in two phases. For the binary system containing only components A and B, the condition (22) for the formation of two phases is

$$\frac{d^2G}{dx_A^2} > 0 \tag{2}$$

The stability criteria for ternary and more complex systems may be obtained from a detailed analysis involving chemical potentials (23). The activity of each component is the same in the two liquid phases at equilibrium, but in general the equilibrium mole fractions are greatly different because of the different activity coefficients. The distribution coefficient m', based on mole fractions, of a consolute component C between solvents B and A can thus be expressed in terms of activity coefficients.

$$m' = \frac{x_{CB}}{x_{CA}} = \frac{\gamma_{CA}}{\gamma_{CB}} \tag{3}$$

If the mutual solubilities of the solvents A and B are small, and the systems are dilute in C, the ratio m' can be estimated from the activity coefficients at infinite dilution. The infinite dilution activity coefficients of many organic systems have been correlated in terms of structural contributions (24), a method recommended by others (5). In the more general case of nondilute systems where there is significant mutual solubility between the two solvents, regular solution theory must be applied. Several methods of correlation and prediction have been reviewed (23). The universal quasichemical (UNIQUAC) equation has been recommended (25), which uses binary parameters to predict multicomponent equilibria (see ENGINEERING, CHEMICAL DATA CORRELATION).

In addition to thermodynamically based predictions of liquid–liquid equilibria, a great deal of experimental data is to be found in the research literature (26). A liquid–liquid equilibrium data bank is also available (27).

Because of the nonideal nature of liquid–liquid systems, it is common engineering practice to quote data as mass fractions rather than mole fractions. Ternary systems can be represented graphically on a triangular diagram. The triangle can be equilateral or right-angled; the latter has the advantage that ordinary graph paper can be used, and also the vertical and horizontal scales can be changed independently. Figure 2**a** shows a typical phase diagram for a system where solvents A and B are partially miscible, and the solute C is a liquid. The ordinate scale represents mass fraction x_C and the abscissa represents x_B. The composition x_A can be obtained as $(1 - x_B - x_C)$.

The triangular diagram shows the two-phase envelope which encloses regions of overall composition in which two phases exist in equilibrium. In Figure 2**a** there is only one two-phase region; other types of systems can have two or more such regions (5). The equilibrium compositions of two liquid phases are connected by tie-lines. Typical tie-lines are shown. It is often convenient for a triangular diagram to be accompanied by a tie-line location curve (Fig. 2**b**) which allows any tie-line to be drawn by simple geometric construction. The procedure is shown for a given composition x_C of 0.10 in the A-rich phase. A line is drawn across from the triangular diagram to Figure 2**b**, then reflected from the equilibrium curve to the diagonal line, to give a composition y_C representing the B-rich

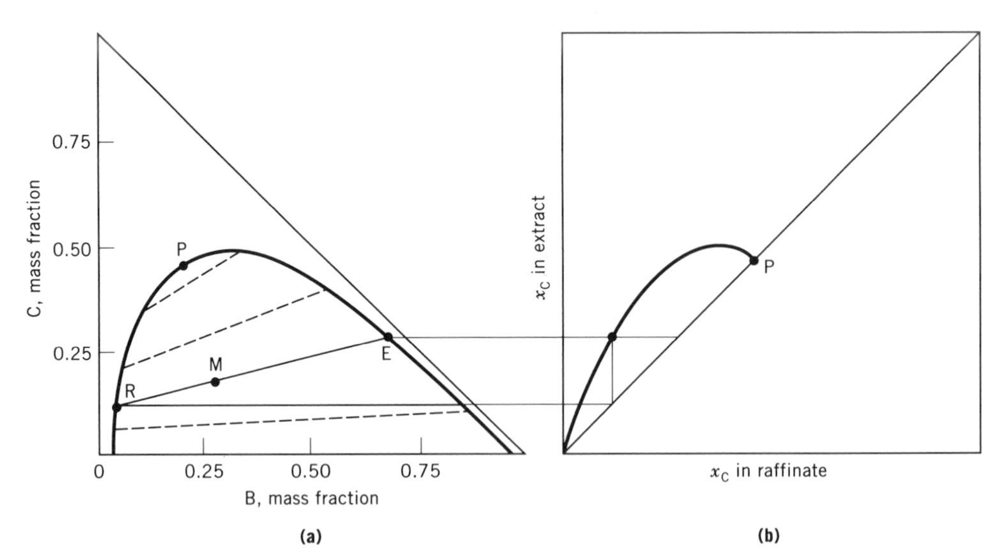

Fig. 2. (**a**) Triangular diagram, where the dashed lines represent tie-lines, and (**b**) tie-line location curve. Terms are defined in the text.

phase. A horizontal line is then drawn back to the appropriate point on the two-phase envelope in the triangular diagram, as shown.

It is seen that as the concentration of C is increased, the tie-lines become shorter because of the increased mutual miscibility of the two phases; at the plait point, P, the tie lines vanish. However, P does not necessarily represent the highest possible loading of C which can exist in the system under two-phase conditions. In Figure 2**b** the plait point lies on the diagonal because the compositions of the two phases approach each other at P.

An important use of the triangular equilibrium diagram is the graphical solution of material balance problems, such as the calculation of the relative amounts of equilibrium phases obtained from a given overall mixture composition. As an example, consider a mixture where the overall composition is represented by point M on Figure 2**a.** If the A-rich phase is denoted by point R (raffinate) and the B-rich phase is denoted by point E (extract), it can be shown that points R, M, and E are collinear, and also

$$\frac{\text{mass of extract}}{\text{mass of raffinate}} = \frac{\text{distance MR}}{\text{distance ME}} \tag{4}$$

This is a statement of the inverse lever rule, which is the basis of techniques for graphical multistage calculations in ternary systems. Although triangular diagrams are widely used, there is the disadvantage that at low concentrations of C the tie-lines become almost horizontal and are hard to draw accurately. This problem can be overcome by plotting the compositions on a solvent-free (B-free) basis using a rectangular diagram. Detailed descriptions of this method and its application are given elsewhere (5,28).

Liquid–liquid equilibria having more than three components cannot as a rule be represented on a two-dimensional diagram. Such systems are important in fractional extraction, for example, operations in which two consolute components C and D are separated by means of two solvents A and B. For the special case where A and B are immiscible, the linear distribution law can be applied to components C and D independently:

$$x_{\mathrm{CB}} = m_{\mathrm{C}} x_{\mathrm{CA}} \tag{5}$$

$$x_{\mathrm{DB}} = m_{\mathrm{D}} x_{\mathrm{DA}} \tag{6}$$

The selectivity or separation factor between the two solutes is defined as the ratio of the distribution ratios:

$$\beta_{\mathrm{CD}} = \frac{m_{\mathrm{C}}}{m_{\mathrm{D}}} = \frac{x_{\mathrm{CB}} x_{\mathrm{DA}}}{x_{\mathrm{CA}} x_{\mathrm{DB}}} \tag{7}$$

By convention, the components C and D are assigned so that the ratio β_{CD} exceeds unity. The greater the selectivity, the easier is the separation of C and D using solvents A and B. Selectivity can be defined in terms of mass ratio, mole ratio, or concentration.

The selection of solvents for a given separation depends largely on equilibrium considerations. Other important factors include cost, ease of solvent recovery by distillation (qv) or other means, safety and environmental impact, and physical properties which must permit easy phase dispersion and separation. Solvent selection is therefore a broad-based exercise which is hard to quantify (8). However a useful quantitative approach has been proposed (22) for comparing simplified equilibrium estimations on the basis of regular solution theory. The polar and hydrogen bonding parameters for solutes and possible solvents are plotted as points on a solvent selection diagram. The selectivities can be approximately related to the distances between the points, providing a quick comparison between a large number of potential solvents. Computer-aided solvent selection has been developed using a molecular graphics system (29).

Chemical Equilibria. In many cases, mass transfer between two liquid phases is accompanied by a chemical change. The transferring species can dissociate or polymerize depending on the nature of the solvent, or a reaction may occur between the transferring species and an extractant present in one phase. An example of the former case is the distribution of benzoic acid [*65-85-0*] between water and benzene. In the aqueous phase, the acid is partially dissociated:

$$C_6H_5COOH \rightleftharpoons C_6H_5COO^- + H^+ \tag{8}$$

In the organic phase the monomer is in equilibrium with the dimer:

$$2\ C_6H_5COOH \rightleftharpoons (C_6H_5COOH)_2 \tag{9}$$

Whereas the linear distribution law can be applied to the undissociated monomer, the interfacial distribution of total benzoic acid, as determined by analysis, is nonlinear.

Many industrial processes involve a chemical reaction between two liquid phases, for example nitration (qv), sulfonation (see SULFONATION AND SULFATION), alkylation (qv), and saponification. These processes are not always considered to be extractions because the main objective is a new chemical product, rather than separation (30). However these processes have many features in common with extraction, for example the need to maintain a high interfacial area with the aid of agitation and the importance of efficient phase separation after the reaction is completed.

In addition to the liquid–liquid reaction processes, there are many cases in both analytical and industrial chemistry where the main objective of separation is achieved by extraction using a chemical extractant. The technique of dissociation extraction is very valuable for separating mixtures of weakly acidic or basic organic compounds such as 2,4-dichlorophenol [*120-83-2*], $C_6H_4Cl_2O$, and 2,5-dichlorophenol [*583-78-8*], which are difficult to separate by other means (31). The technique involves the use of a controlled amount of a strong base or acid in the aqueous phase, so that the overall distribution of each species is affected by its dissociation constant as well as the distribution constant of the undissociated form (see eqs. 7 and 8). Dissociation extraction has been applied extensively to mixtures of closely similar phenols, amines, and organic acids (32).

In hydrometallurgical separations (7), a metal ion in aqueous solution can be selectively converted to an organometallic compound or complex which is soluble in an organic carrier solvent. The metal extractants may be classified (33) as: (*1*) acid and chelating acid extractants, (*2*) anion exchangers, and (*3*) solvating extractants. Acid extractants typically contain one or more long hydrocarbon chains, an example being di(2-ethyl-hexyl)phosphoric acid [*298-07-7*], $[C_4H_9CH(C_2H_5)CH_2]_2HPO_4$. The extractant is dissolved in an organic carrier solvent which may be a pure compound such as hexane, but in industry is commonly a kerosene. When the extractant in the carrier solvent is contacted using an aqueous solution of a metal cation, eg, M^{2+}, equilibria are set up which can be simplified as the overall equation:

$$M^{2+} + \overline{2\,LH} \rightleftharpoons 2\,H^+ + \overline{ML_2} \tag{10}$$

where the overbar denotes species in the organic phase. The control of pH is obviously an important factor in determining the degree of metal extraction and the selectivity of an extractant for two or more cations. Often the logarithm of the distribution coefficient can be plotted as a linear function of pH (34). After a metal has been selectively extracted to the organic phase it can be back-extracted or stripped to a strongly acidic aqueous phase, giving a purer and more concentrated solution than the initial aqueous feed.

Chelating extractants owe effectiveness to the attraction of adjacent groups on the molecule for the metal (see CHELATING AGENTS). Compounds containing long hydrocarbon chains and the hydroxyoxime group, or those based on 8-quinolinol [*148-24-3*], C_9H_6NOH, form the basis of various commercial formulations (8,33). These extractants often show an amphoteric behavior depending on pH, changing for example from L^- to LH to LH_2^+ as the pH is increased; these agents can be applied in basic as well as acidic conditions (35).

Anionic extractants are commonly based on high molecular weight amines. Metal anions such as MnO_4^- or ReO_4^- can be exchanged selectively with inorganic anions such as Cl^- or SO_4^{2-}. The equilibrium for a quaternary onium compound of organic radicals R for two anion species A^- and B^- might be:

$$\overline{R_4N^+A^-} + B^- \rightleftharpoons \overline{R_4N^+B^-} + A^- \tag{11}$$

Solvating extractants contain one or more electron donor atoms, usually oxygen, which can supplant or partially supplant the water which is attached to the metal ions. Perhaps the best known example of such an extractant is tri-(n-butyl) phosphate [*126-73-8*] (TBP), $C_{12}H_{27}O_4P$, which forms the basis of the PUREX process (36) for uranium extraction:

$$UO_2^{2+} + 2\,NO_3^- + \overline{2\,TBP} \rightleftharpoons \overline{UO_2(NO_3)_2{\cdot}2\,TBP} \tag{12}$$

TBP and nitric acid also tend to form a complex with each other, but at sufficiently high uranyl nitrate concentrations the nitric acid is mainly displaced into the aqueous phase.

Interfacial Mass-Transfer Coefficients. Whereas equilibrium relationships are important in determining the ultimate degree of extraction attainable, in practice the rate of extraction is of equal importance. Equilibrium is approached asymptotically with increasing contact time in a batch extraction. In continuous extractors the approach to equilibrium is determined primarily by the residence time, defined as the volume of the phase contact region divided by the volume flow rate of the phases.

The rate of mass transfer (qv) depends on the interfacial contact area and on the rate of mass transfer per unit interfacial area, ie, the mass flux. The mass flux very close to the liquid–liquid interface is determined by molecular diffusion in accordance with Fick's first law:

$$N = -D \frac{\partial c}{\partial z} \tag{13}$$

where N refers to the flux in the z direction, c is the concentration of the consolute component, and D is its molecular diffusivity in the solvent. It is accepted practice to use the concentration gradient as the driving force for mass transfer, although it has been suggested that the gradient of chemical potential is more appropriate (37,38). Molecular diffusivities D of solutes are commonly in the range 10^{-10} to 5×10^{-9} m^2/s at ambient temperatures. Data for many systems are available in the literature or can be measured experimentally or predicted from correlations (39,40). For dilute solutions of nonelectrolytes, the correlation of Wilke and Chang (41) is recommended (see also ENGINEERING, CHEMICAL DATA CORRELATION).

Although molecular diffusion itself is very slow, its effect is nearly always enhanced by turbulent eddies and convection currents. These provide almost perfect mixing in the bulk of each liquid phase, but the effect is damped out in the vicinity of the interface. Thus the concentration profiles at each side of a liquid–liquid interface have the appearance shown in Figure 3, with essentially uniform bulk composition, and a film on each side of the interface. The films are the regions in which there are significant concentration gradients. The equivalent film thicknesses depend on hydrodynamic conditions and are in the order of 100 μm. The film thicknesses and concentration profiles are therefore almost impossible to measure directly, and the mass fluxes are expressed by means of mass-transfer coefficients, k_A and k_B, and the concentration differences for each phase.

$$N = k_A (c_A - c_{Ai}) = k_B (c_{Bi} - c_B) \tag{14}$$

The mass-transfer coefficients are typically between 10 to 100 μm/s, depending on hydrodynamic conditions and the values of D.

The solute concentrations very close to the interface, c_{Ai} and c_{Bi}, are assumed to be in equilibrium, in the absence of any slow interfacial reaction. According to the linear distribution law, $c_{Bi} = mc_{Ai}$ and thus from equation 14 the mass-transfer flux can be expressed in terms of an overall mass-transfer coefficient K_A and an overall concentration driving force:

$$N = K_A (c_A - c_A^*) \tag{15}$$

where

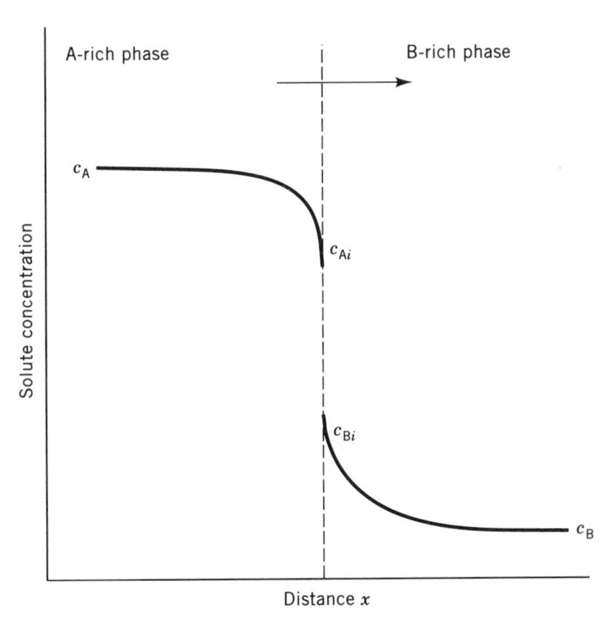

Fig. 3. Concentration profiles near an interface where the arrow represents the direction of mass transfer, c_A = concentration of C in A-rich phase, c_B = concentration of C in B-rich phase, and the subscript i denotes the interface.

$$c_A{}^* = c_B/m \tag{16}$$

and

$$1/K_A = 1/k_A + 1/(mk_B) \tag{17}$$

This is the important rule of additivity of resistances. In practice, k_A and k_B are often of the same order of magnitude, but the distribution coefficient m can vary considerably. For solutes which preferentially distribute toward solvent B, m is large and the controlling resistance lies in phase A. Conversely, if the distribution favors solvent A the controlling mass-transfer resistance lies in phase B.

Values of the mass-transfer coefficient k have been obtained for single drops rising (or falling) through a continuous immiscible liquid phase. Extensive literature data have been summarized (40,42). The mass-transfer coefficient is often expressed in dimensionless form as the Sherwood number:

$$Sh = kd/D \tag{18}$$

The values of k and hence Sh depend on whether the phase under consideration is the continuous phase, c, surrounding the drop, or the dispersed phase, d, comprising the drop. The notations k_c, k_d, Sh_c, and Sh_d are used for the respective mass-transfer coefficients and Sherwood numbers.

Mass-transfer coefficients are strongly affected by the degree of mobility of the drop surface. When the surface is free of adsorbed molecules it can move in response to surface shear stress, and the drops tend to circulate as they move through the continuous phase. However, even a trace of surface-active material can cause the drop surface to resist shear and the drop circulation is suppressed, resulting in greatly reduced mass transfer (43,44). Table 1 lists some of the proposed equations for mass transfer (expressed as Sh) under various conditions. Nearly a threefold variation in dispersed-phase mass transfer exists between circulating and noncirculating drops. For the continuous phase the effect of circulation is greatest at high Reynolds number. For a more detailed assessment the specialized reviews (11,40,42) are recommended.

Although the adsorption of surfactants tends to reduce mass-transfer coefficients by suppressing drop circulation, a sharp increase in mass transfer can occur if the transferring solute strongly reduces the interfacial tension. This effect, known as the Marangoni effect, results from interfacial turbulence induced by interfacial tension gradients (45). Extensive research in this area has been reviewed (46).

Mass-Transfer Coefficients with Chemical Reaction. Chemical reaction can occur in any of the five regions shown in Figure 3, ie, the bulk of each phase, the film in each phase adjacent to the interface, and at the interface itself. Irreversible homogeneous reaction between the consolute component C and a reactant D in phase B can be described as

$$\mathrm{C} + z\,\mathrm{D} \rightarrow \text{products} \tag{19}$$

The equations of combined diffusion and reaction, and their solutions, are analogous to those for gas absorption (qv) (47). It has been shown how the concentration profiles and rate-controlling steps change as the rate constant increases (48). When the reaction is very slow and the B-rich phase is essentially saturated with C, the mass-transfer rate is governed by the kinetics within the bulk of the B-rich phase. This is defined as regime 1. Concentration profiles are shown in Figure 4**a.** (48). For a slow reaction defined as regime 2, the consolute component C is almost entirely depleted in the bulk of the B-rich phase (Fig. 4**b**) and the mass transfer of C between the phases controls the rate of the reaction. For a very fast reaction the depletion of C affects the concentration profile in the diffusion film. The steepening of the concentration profile as shown in Figure 4**c** for regime 3 leads to an enhancement in the film mass-transfer coefficient in the B-rich phase. Finally, the case of an instantaneous reaction (regime 4) leads to the formation of a thin reaction zone to which components C and D diffuse in stoichiometric amounts. This is shown in Figure 4**d.**

Table 1. Equations for Liquid–Liquid Mass Transfer in Single Drops[a]

	Sherwood numbers	
Type of drop	Dispersed phase, Sh_d	Continuous phase, Sh_c
noncirculating	6.58	$2 + 0.6\,Re_c^{0.5}Sc_c^{0.333}$
circulating	17.9[b]	$1.13\,(Re_c \cdot Sc_c)^{0.5}$

[a]Ref. 40.
[b]For laminar conditions.

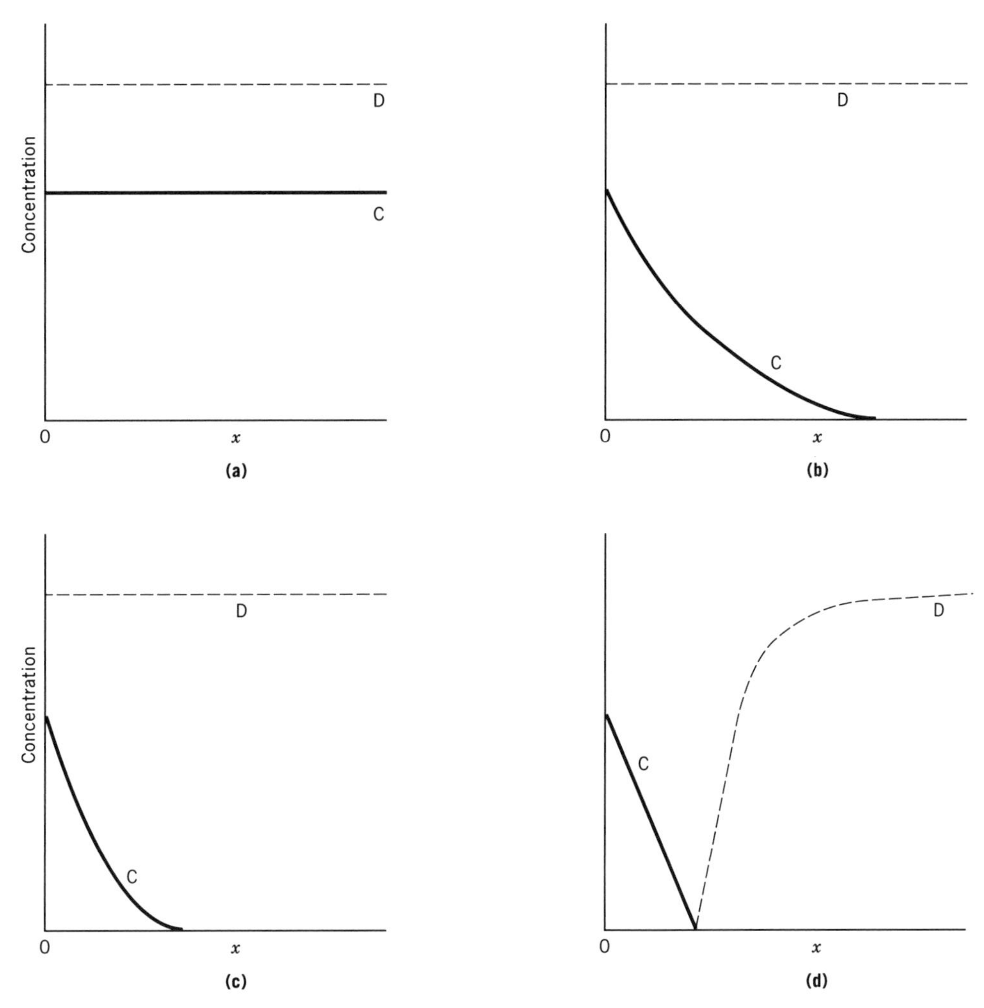

Fig. 4. Concentration profiles for the reaction of equation 19 (43) where (——) is the concentration of C; (–––) the concentration of D; and x is the distance from the interface. (**a**) Regime 1; (**b**) regime 2; (**c**) regime 3; and (**d**) regime 4, as described in the text.

The enhanced rate expressions for regimes 3 and 4 have been presented (48) and can be applied (49,50) when one phase consists of a pure reactant, for example in the saponification of an ester. However, it should be noted that in the more general case where component C in equation 19 is transferred from one inert solvent (A) to another (B), an enhancement of the mass-transfer coefficient in the B-rich phase has the effect of moving the controlling mass-transfer resistance to the A-rich phase, in accordance with equation 17. Resistance in both liquid phases is taken into account in a detailed model (51) which is applicable to the reversible reactions involved in metal extraction. This model, which can accommodate the case of interfacial reaction, has been successfully compared with rate data from the literature (51).

Interfacial Contact Area and Approach to Equilibrium. Experimental extraction cells such as the original Lewis stirred cell (52) are often operated with a flat liquid–liquid interface the area of which can easily be measured. In the single-drop apparatus, a regular sequence of drops of known diameter is released through the continuous phase (42). These units are useful for the direct calculation of the mass flux N and hence the mass-transfer coefficient for a given system.

In industrial equipment, however, it is usually necessary to create a dispersion of drops in order to achieve a large specific interfacial area, a, defined as the interfacial contact area per unit volume of two-phase dispersion. Thus the mass-transfer rate obtainable per unit volume is given as

$$(N{\cdot}a) = K_A a(c_A - c_A{}^*) \tag{20}$$

Drop dispersions are hardly ever uniform, and size distribution must be allowed for in calculating a. This can be done by means of the Sauter mean drop diameter, d_m, based on the average volume-to-area ratio for N drops.

$$d_m = \sum_{i=1}^{N} d_i^3 \Big/ \sum_{i=1}^{N} d_i^2 \tag{21}$$

The specific interfacial area based on unit volume of two-phase dispersion is given by

$$a = 6h/d_m \tag{22}$$

where h is the holdup of dispersed phase. The specific rate of interfacial mass transfer can be summarized from equations 20 and 22 as:

$$(N{\cdot}a) = K_A\,(6h/d_m)\,(c_A - c_A{}^*) \tag{23}$$

showing that the rate of attainment of equilibrium is proportional to the concentration difference (driving force) and to a physical rate constant ($K_A{\cdot}6h/d_m$) which has the units of s^{-1}. The extraction rate constant can be increased operationally by maintaining a high holdup and a small Sauter mean droplet diameter. Typically, in the absence of chemical effects, K_A would be on the order of 10 μm/s, h could be on the order of 0.2, and d_m could be 1 mm; hence ($K_A{\cdot}6h/d_m$) is approximately 0.012 s^{-1}. In a batch extraction under these conditions, the deviation from equilibrium would decay exponentially with a half-life of 0.693/0.012 s or just under one minute.

There are certain limits to how far h can be increased and d_m can be reduced; however, if the contact time in a well-mixed extractor can be maintained at several minutes, it can usually be assumed that equilibrium between the exit phases is attained, justifying the use of the equilibrium stage concept represented by Figure 1 and equation 1.

Calculation of Equilibrium Stages. Multistage contacting can be arranged in a cocurrent, crosscurrent, or countercurrent manner as shown in Figure 5. The sequence of stages is sometimes referred to as a cascade, referring to the early

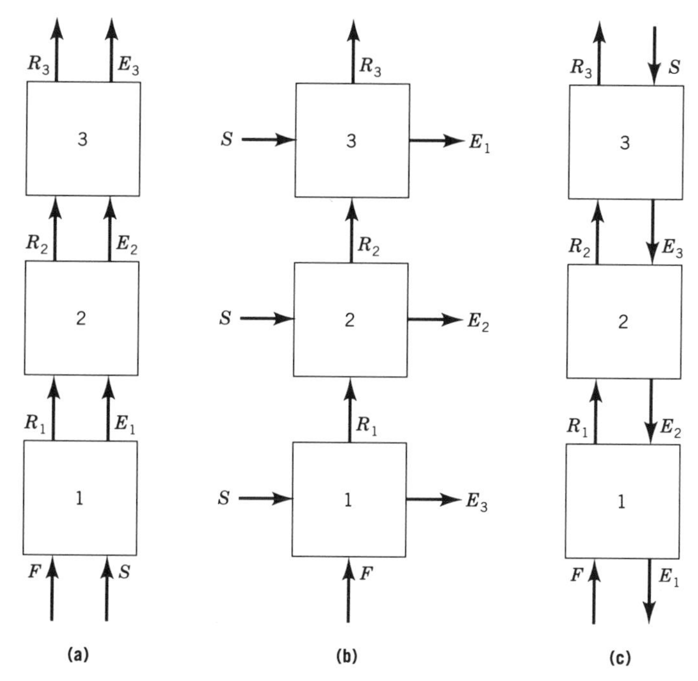

Fig. 5. Arrangement of multistage contactors where F = feed flow (A-rich), R = raffinate flow, S = solvent flow (B-rich), and E = extract flow. (**a**) Cocurrent; (**b**) crosscurrent; and (**c**) countercurrent.

use of gravity overflow from stage to stage. Cocurrent stagewise contact (Fig. 5**a**) is not usually necessary when the stages are ideal, because equilibrium is reached between the streams after stage 1. A crosscurrent cascade (Fig. 5**b**) in which fresh solvent is added at each stage gives an improvement over the separation obtainable in a single stage for a given ratio of solvent to feed (5,28).

The countercurrent arrangement (Fig. 5**c**) represents the best compromise between the objectives of high extract concentration and a high degree of extraction of the solute, for a given solvent-to-feed ratio. The feed entering stage 1 is brought into contact with a B-rich stream which has already passed through the other stages, while the raffinate leaving the last stage has been in contact with fresh solvent. Because of the economic advantages, continuous countercurrent extraction is normally preferred for commercial-scale operations. For the case of a partially miscible ternary system, the number of ideal stages in a countercurrent cascade can be estimated graphically on a triangular diagram, using the Hunter-Nash method (53). The feed and solvent compositions and the resulting mixture point M are first located on the diagram as in Figure 2**a.** If in addition one of the exit stream (extract or raffinate) compositions is given, a point representing the composition of the net flow in the countercurrent cascade can be located. This point, called the delta point, provides the basis for construction of material balance lines and tie-lines representing a sequence of ideal stages for the countercurrent extractor. The Hunter-Nash procedure is well known and useful (5,28). For dilute systems, it is often more convenient to use the delta point construction on a diagram with solvent-free coordinates (5,28). In this case a rectangular diagram is plotted in which the horizontal axis is the mass fraction of the solute C on a B-free basis, and the vertical axis is the mass ratio of B to A + C.

If the feed, solvent, and extract compositions are specified, and the ratio of solvent to feed is gradually reduced, the number of ideal stages required increases. In economic terms, the effect of reducing the solvent-to-feed ratio is to reduce the operating cost, but the capital cost is increased because of the increased number of stages required. At the minimum solvent-to-feed ratio, the number of ideal stages approaches infinity and the specified separation is impossible at any lower solvent-to-feed ratio. In practice the economically optimum solvent-to-feed ratio is usually 1.5 to 2 times the minimum value.

The design of countercurrent contactors is considerably simplified when the solvents A and B are not significantly miscible. The mass flows of A and B then remain constant from one stage to the next, and the material balance at any stage can be written

$$\mathrm{A}(X_0 - X) = \mathrm{B}\,(Y_1 - Y) \tag{24}$$

where A and B are the mass flows of A and B, X is the mass ratio of C to A in the feed, and Y is the ratio of C to B in the extract. The compositions X and Y, expressed as mass ratios, thus vary linearly and equation 24 can be plotted as the operating line on Figure 6. Also shown is the equilibrium curve.

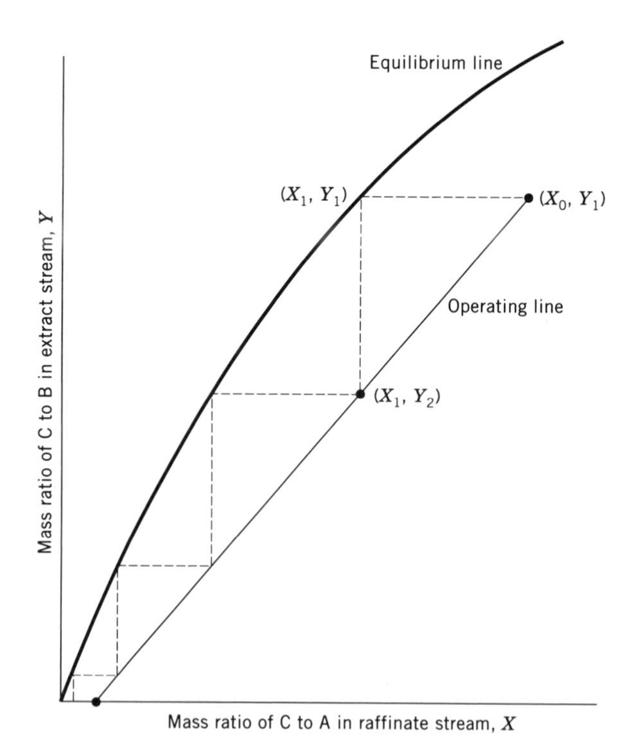

Fig. 6. Countercurrent extraction showing the equilibrium stages (horizontal dashed lines) where A and B are immiscible.

The number of ideal stages can readily be found from Figure 6 by the same type of stepwise construction as used for countercurrent distillation columns. Starting at point (X_0, Y_1), the first horizontal dashed line represents the establishment of equilibrium in stage 1 to give (X_1, Y_1). The second, vertical dashed line represents the solution of the material balance (eq. 24), giving a point on the operating line relating X_1 and Y_2, and so on. The example shown in Figure 6 indicates that between three and four ideal stages are required and in practice the designer would specify four ideal stages.

Although the triangular diagram is normally used when the solvents are partially miscible, it is also possible to construct an X–Y (mass ratio) or x–y (mass fraction) operating diagram (28) and use the stepwise procedure for calculating stages. When solvents A and B are partially miscible, the operating lines are curved. An important advantage of the stepwise procedure is that it can be adapted to the case where the rate of mass transfer (eq. 23) and contact time are not sufficient to ensure equilibrium between exit streams in each contacting stage. By analogy with distillation, a Murphree stage efficiency can be defined for either phase (28). If the efficiency is less than 1.0, more stages are needed for a given separation than for the ideal case.

When the solvents are substantially immiscible and the equilibrium curve is linear, $Y = mX$, the number of ideal stages can be calculated without the graphical constructions (54,55). When the extraction factor ϵ (eq. 1) is not equal to unity, it can be shown that

$$N_s = \frac{\log\left[\left(\dfrac{X_0 - Y_S/m}{X_N - Y_S/m}\right)\left(1 - \dfrac{1}{\epsilon}\right) + \dfrac{1}{\epsilon}\right]}{\log \epsilon} \tag{25}$$

and for $\epsilon = 1$, denoting parallel operating and equilibrium lines,

$$N = (X_0 - X_N)/(X_N - Y_S/m) \tag{26}$$

where X_0 is the ratio C/A in the feed, X_N is the ratio C/A in the raffinate after N ideal stages, and Y_S is the ratio C/B in the entering solvent. For very dilute systems, the mass ratios X and Y in the above equations can be replaced by mass fractions or concentrations.

Fractional Extraction. Fractional extraction is the separation of two or more consolute components by solvent extraction. In single solvent fractional extraction (Fig. 7**a**) the feed mixture of A and C is added at some point in a countercurrent cascade through which a solvent B is passed. The solvent B preferentially dissolves component C as it passes in the downward direction. The mixture of B and C leaving the cascade is then split into two phases, for example by cooling. Part of the C-rich layer is removed as product and part is returned to the cascade as reflux. The B-rich layer (solvent) is sent to the other end of the cascade where it strips component C from the A + C mixture, allowing an A-rich phase to leave. The two sections of the cascade are analogous to the stripping and enriching sections in a distillation column and the design procedure for estimating the number of stages is somewhat similar (28). Single-solvent fractional extraction has been

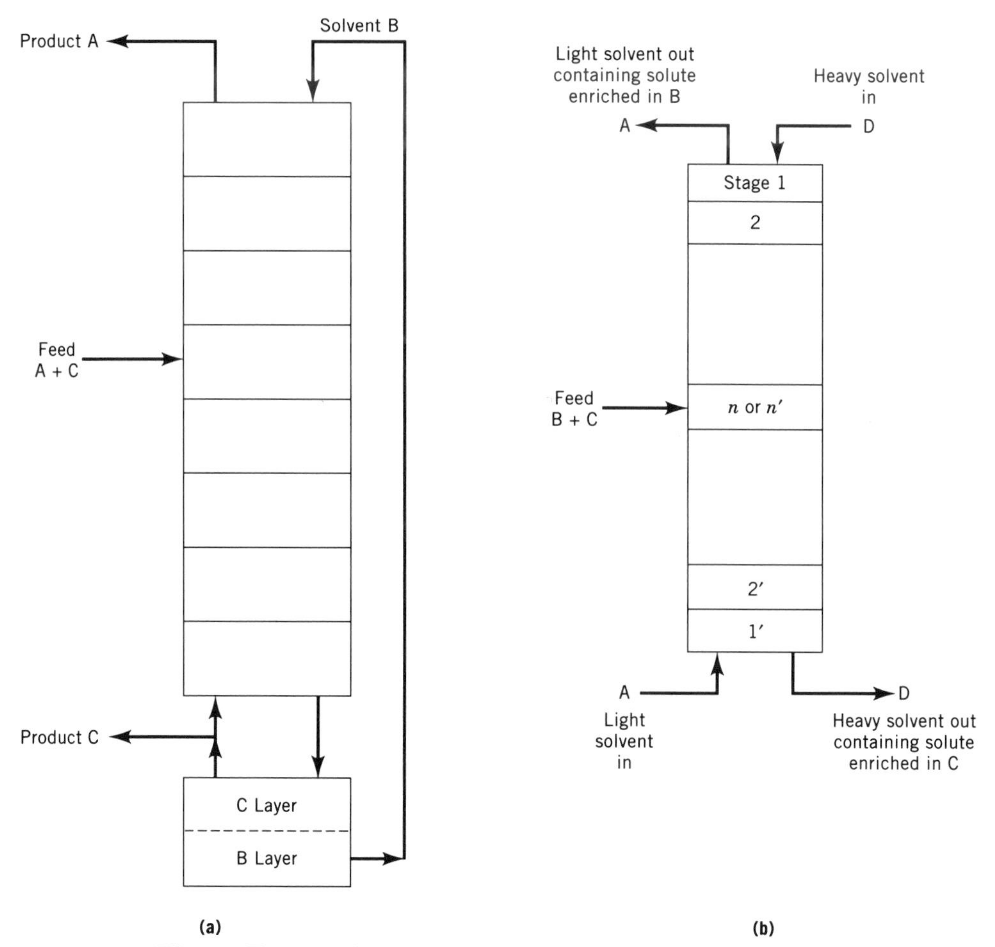

Fig. 7. Fractional extraction: (**a**) one solvent; (**b**) two solvents.

known for many years, but the range of solvents available is limited because of the requirement that the solvents must be sparingly miscible with each of the components A and C.

Dual solvent fractional extraction (Fig. 7**b**) makes use of the selectivity of two solvents (A and B) with respect to consolute components C and D, as defined in equation 7. The two solvents enter the extractor at opposite ends of the cascade and the two consolute components enter at some point within the cascade. Solvent recovery is usually an important feature of dual solvent fractional extraction and provision may also be made for reflux of part of the product streams containing C or D. Simplified graphical and analytical procedures for calculation of stages for dual solvent extraction are available (5) for the cases where β_{CD} is constant and the two solvents A and B are not significantly miscible. In general, the accurate calculation of stages is time-consuming (28) but a computer technique has been developed (56).

Differential Contacting. Although the equilibrium stage concept has proved extremely useful in describing the performance of mixer-settlers and plate columns having discrete stages, it is not appropriate for spray towers, packed columns, etc, in which no discrete stages can be identified. In such differential types of contactors, equilibrium between phases is never reached and therefore the mass-transfer rate is important in the design procedure.

A differential countercurrent contactor operating with a dilute solution of the consolute component C and immiscible components A and B is shown in Figure 8. Under these conditions, the superficial velocities of the A-rich and B-rich streams can be assumed not to vary significantly with position in the contactor, and are taken to be U_A and U_B, respectively. The concentration of C in the A-rich stream is c_A and that in the B-rich stream is c_B.

A steady-state material balance can be carried out on a small section of length dz and volume dz (on the basis of unit cross-sectional area) in the contactor:

$$U_B dc_B = U_A dc_A = K_A a(c_A - c_A)dz \tag{27}$$

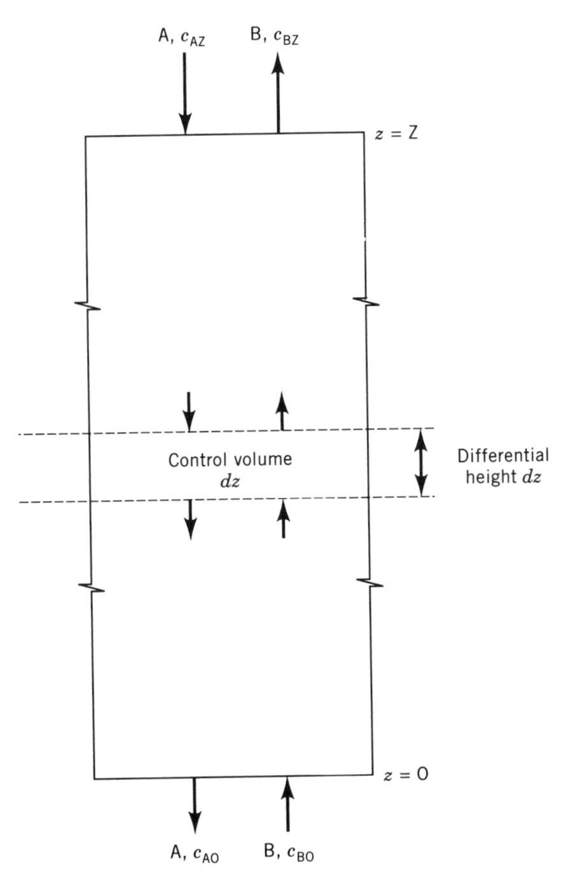

Fig. 8. Mass transfer in a differential contactor. Terms are defined in the text.

Rearrangement and integration give a relationship for the contactor height in terms of the concentration change:

$$dz = \frac{U_A}{K_A a} \cdot \frac{dc_A}{(c_A - c_A{}^*)} \tag{28}$$

$$Z = \left(\frac{U_A}{K_A a}\right) \int_{c_{A0}}^{c_{AZ}} \frac{dc_A}{(c_A - c_A{}^*)} \tag{29}$$

The integral can be found graphically if the equilibrium line is curved. An analytical expression for the integral is available for the case where both the equilibrium and operating lines are linear (5):

$$\int_{c_{A0}}^{c_{AZ}} \frac{dc_A}{(c_A - c_A^*)} = \frac{1}{1 - 1/\epsilon} \ln \left[\left(\frac{c_{A0} - c_{AZ}}{c_{AZ} - c_{AZ}^*} \right) (1 - 1/\epsilon) + 1/\epsilon \right] \tag{30}$$

where $\epsilon = mU_B/U_A$. The integral is unitless and is known as the number of transfer units (NTU) based on the overall A-rich-phase driving force. Obviously the NTU, and hence the contactor length Z required, increase as the difference between c_{AZ} and c_{A0} is increased.

The factor $(U_A/K_A a)$ in equation 29 is known as the height of a transfer unit (HTU). It is a characteristic of the hydrodynamic conditions such as the flow rate A and the specific interfacial area a, but is independent of changes in c_A. It is important that the HTU be specified correctly in regard to the phase and driving force considered; in this case it relates to the overall mass-transfer driving force in the A-rich phase. The HTU may vary with height because of changes in drop size, etc; an average value is usually taken, assuming no variation.

Mass-transfer theory (eq. 17) indicates that the overall mass-transfer resistance $1/K_A$ consists of contributions from each phase, so that the overall HTU is also the sum of two contributions:

$$\begin{aligned} (\mathrm{HTU})_{OA} &= \frac{U_A}{K_A a} + \frac{U_A}{mK_B a} \\ &= \frac{U_A}{K_A a} + \left(\frac{U_B}{K_B a}\right)\left(\frac{U_A}{mU_B}\right) \\ &= (\mathrm{HTU})_A + (\mathrm{HTU})_B/\epsilon \end{aligned} \tag{31}$$

The heights of a transfer unit in each phase thus contribute to the overall heights of a transfer unit. Data on values of HTU for various types of countercurrent equipment have been reviewed (1,10). In normal operating practice, the extraction factor is chosen to be not greatly different from unity, within the range of 0.5–2.

Although the stagewise model is not physically realistic for differential contactors, it is sometimes used. The number of equivalent theoretical stages N can be determined graphically using the stepwise construction illustrated in Figure 7. For the case where both the equilibrium and operating lines are linear, it can be shown that:

$$\frac{N}{(\mathrm{NTU})_{\mathrm{OA}}} = \frac{1 - 1/\epsilon}{\ln\epsilon} \tag{32}$$

If $\epsilon = 1$, the number of theoretical stages is equal to $(\mathrm{NTU})_{\mathrm{OA}}$.

Equations 27–32 are applicable only to dilute, immiscible systems. If the amount of mass transfer is significant in comparison to the total flow rates, more complicated treatments of differential contactors are required (5,28).

Axial Dispersion. The development following equation 27 has assumed that the two phases move countercurrently in plug flow, ie, all the fluid in each phase has the same residence time in the equipment. In practice this is rarely the case, because of axial mixing which arises from the action of turbulent eddies, circulation currents, or the effects of drop wakes (52). The effect is to flatten the axial concentration profiles within each phase. Axial mixing can lead to a reduction in the effective driving force for mass transfer which in turn reduces the NTU below that expected for the plug flow case (eq. 30). An important feature of the profile is the discontinuity or "jump" in concentration which occurs at entry to the contactor when the liquid in the feed line enters the mixed region of the column.

Two alternative approaches are used in axial mixing calculations. For differential contactors, the axial dispersion model is used, based on an equation analogous to equation 13:

$$N = -E\,\frac{\partial c}{\partial z} \tag{33}$$

Values of E, several orders of magnitude greater than the molecular diffusion coefficient D, are typically in the 10^{-4} to 10^{-3} m^2/s range for packed columns, and even larger for spray columns in which circulation currents are unimpeded. For contactors in which discrete well-mixed compartments can be identified, for example sieve-plate columns, axial mixing effects are incorporated into the stagewise model by means of the backflow ratio α which is defined as the fraction of the net interstage flow of one phase which is considered to flow in the reverse direction. For a contactor in which there are many compartments, the axial dispersion coefficient and the backflow ratio, α, are interrelated as follows:

$$E = \frac{UH}{\ln\left((1 + \alpha)/\alpha\right)} \tag{34}$$

where H is the height of one compartment and U is the superficial velocity. The detailed calculations of concentration profiles and mass-transfer rates with axial mixing require the solution of a fourth-order differential equation (dispersion model) or the equivalent difference equation (backflow model) along with appropriate boundary conditions. The methodology was developed in the 1950s and early 1960s and has been concisely reviewed (57). Pratt (58,59) has shown how the profile solutions can be rearranged to give a direct calculation of column height. In the case of the dispersion model, the relative effect of axial mixing is a function of the axial Peclet number, defined as

$$Pe = UZ/E \tag{35}$$

The Peclet numbers are different for each phase because U and E are different. Axial mixing effects are usually greater in the continuous phase than in the dispersed phase. Plug flow conditions can be assumed only if Pe exceeds about 50. Experimentally measured values of E or α are widely available for laboratory-scale columns with a diameter of up to 15 cm (1). Typically at low agitation rates, circulation effects (hydrodynamic nonuniformity) can lead to large values of E. The circulation effects are mainly a result of the motion of the dispersed-phase droplets, but unstable axial density gradients may also contribute to increased mixing (60,61). Circulation effects are reduced by mechanical agitation which promotes radial uniformity, but at high levels of agitation the increased turbulence leads to an overall increase in E, resulting in a minimum in the plotted curve of E versus agitation.

The effect of increasing column diameter is to increase the tendency for circulation, and hence to increase the axial mixing (62,63). However, extremely few measurements of axial mixing at the industrial scale are available, so large-scale contactor design must still rely quite heavily on empirical experience with the particular type of equipment.

Drop Diameter. In extraction equipment, drops are initially formed at distributor nozzles; in some types of plate column the drops are repeatedly formed at the perforations on each plate. Under such conditions, the diameter is determined primarily by the balance between interfacial forces and buoyancy forces at the orifice or perforation. For an ideal drop detaching as a hemisphere from a circular orifice of diameter d_0 and then becoming spherical:

$$d = (6\sigma d_0/g\Delta\rho)^{1/3} \quad (36)$$

Equation 36 must be corrected for changes in the drop shape and for the effects of the inertia of liquid flowing through the orifice, viscous drag, etc (64). As the orifice or aperture diameter is increased, d_0 has less effect on the drop diameter and the mean drop size then tends to become a function only of the system properties:

$$d_m \simeq (\sigma/g\Delta\rho)^{1/2} \quad (37)$$

This type of equation has been found useful in correlating drop diameters in packed columns where the packing size exceeds the drop diameter (65).

In many types of contactors, such as stirred tanks, rotary agitated columns, and pulsed columns, mechanical energy is applied externally in order to reduce the drop size far below the values estimated from equations 36 and 37 and thereby increase the rate of mass transfer. The theory of local isotropic turbulence can be applied to the breakup of a large drop into smaller ones (66), resulting in an expression of the form

$$d_m = K'\sigma^{0.6}\rho_m^{-0.6}\Psi^{-0.4} \quad (38)$$

In this equation, Ψ represents the rate of energy dissipation per unit mass of fluid. In pulsed and reciprocating plate columns the dimensionless proportionality constant K' in equation 38 is on the order of 0.3. In stirred tanks, the proportionality constant has been reported as 0.024(1 + 2.5 h) in the holdup range 0 to 0.35 (67). The increase of drop size with holdup is attributed to the increasing tendency for coalescence between drops as the concentration of drops increases. A detailed survey of drop size correlations is given by the literature (65).

The value of d_m is a mean value, based on a broad distribution of sizes. In a mass-transfer situation the smallest drops, because of the very high specific surface area, quickly come to equilibrium; conversely the largest drops, which typically have a diameter of about 2 d_m, are much slower to come to equilibrium with the continuous phase. The effects of drop size distribution on extractor performance are being studied (68–70), although the single parameter d_m is still widely in use for design work.

Holdup and Flooding. The volume fraction of the dispersed phase, commonly known as the holdup h, can be adjusted in a batch extractor by means of the relative volumes of each liquid phase added. In a continuously operated well-mixed tank, the holdup is also in proportion to the volume flow rates because the phases become intimately dispersed as soon as they enter the tank.

$$h = Q_d/(Q_c + Q_d) \tag{39}$$

However, in a countercurrent column contactor as sketched in Figure 8, the holdup of the dispersed phase is considerably less than this, because the dispersed drops travel quite fast through the continuous phase and therefore have a relatively short residence time in the equipment. The holdup is related to the superficial velocities U of each phase, defined as the flow rate per unit cross section of the contactor, and to a slip velocity U_s (71,72):

$$U_s = U_d/h + U_c/(1 - h) \tag{40}$$

In the case of a packed column, the terms on the right-hand side should each be divided by the voidage, ie, the volume fraction not occupied by the solid packing (71). In unpacked columns at low values of h, the slip velocity U_s approximates the terminal velocity of an isolated drop, but the slip velocity decreases with holdup and may also be affected by column internals such as agitators, baffle plates, etc. The slip velocity can generally be represented by (73):

$$U_s = U_k(1 - h)^{\beta} \tag{41}$$

where the characteristic velocity U_k is a function of drop size and system properties and the exponent β relates to system properties and the degree of flow uniformity in the contactor.

As the throughput in a contactor represented by the superficial velocities U_c and U_d is increased, the holdup h increases in a nonlinear fashion. A flooding point is reached at which the countercurrent flow of the two liquid phases cannot be maintained. The flow rates at which flooding occurs depend on system properties, in particular density difference and interfacial tension, and on the equipment design and the amount of agitation supplied (40,65).

The nonuniformity of drop dispersions can often be important in extraction. This nonuniformity can lead to axial variation of holdup in a column even though the flow rates and other conditions are held constant. For example, there is a tendency for the smallest drops to remain in a column longer than the larger ones, and thereby to accumulate and lead to a localized increase in holdup. This phenomenon has been studied in reciprocating-plate columns (74). In the process of drop breakup, extremely small secondary drops are often formed (64). These drops, which may be only a few micrometers in diameter, can become entrained in the continuous phase when leaving the contactor. Entrainment can occur well below the flooding point.

Coalescence and Phase Separation. Coalescence between adjacent drops and between drops and contactor internals is important for two reasons. It usually plays a part, in combination with breakup, in determining the equilibrium drop size in a dispersion, and it can therefore affect holdup and flooding in a countercurrent extraction column. Secondly, it is an essential step in the disengagement of the phases and the control of entrainment after extraction has been completed.

The role of coalescence within a contactor is not always obvious. Sometimes the effect of coalescence can be inferred when the holdup is a factor in determining the Sauter mean diameter (67). If mass transfer occurs from the dispersed (d) to the continuous (c) phase, the approach of two drops can lead to the formation of a local surface tension gradient which promotes the drainage of the intervening film of the continuous phase (75) and thereby enhances coalescence. It has been observed that d-to-c mass transfer can lead to the formation of much larger drops than for the reverse mass-transfer direction, c to d (76,77).

Phase disengagement occurs at a layer or wedge in which the holdup of the drop phase is very high, providing good opportunities for close contacts between drops (65). Coalescence between drops occurs by a mechanism of drainage of the intervening film of the continuous phase. This process is favored by low viscosity, high interfacial tension, and a relatively large difference in density between the liquid phases. For difficult systems which do not have these properties, various types of mesh-packing coalescence enhancers have been developed (78). These provide a large surface which is preferentially wetted by the drop phase. Another effective technique for enhancing coalescence and phase separation is the application of pulsed electrical fields (79).

Membrane Extraction. An extraction technique which uses a thin liquid membrane or film has been introduced (80,81). The principal advantages of liquid-membrane extraction are that the inventory of solvent and extractant is extremely small and the specific interfacial area can be increased without the problems which accompany fine drop dispersions (see MEMBRANE TECHNOLOGY).

Figure 9**a** shows an early form of liquid-membrane extraction in which the solute is transferred from the continuous phase to a thin spherical film of an immiscible phase; within this film there is a quantity of strip solution which preferentially removes the solute from the membrane. Thus the membrane is analogous to a selective filter for the diffusional transport of the solute (see FILTRATION). The spherical film can be stabilized by surfactants (qv), but a more convenient arrangement is the emulsion globule as shown in Figure 9**b**, in which the strip solution is dispersed as very small drops within a globule of solvent. This technique lends itself particularly to chemically driven extraction stripping, for example, hydrometallurgical extractions according to equation 10 (82,83,84). In this case the extractant acts as a carrier to transport the metal complex across the membrane.

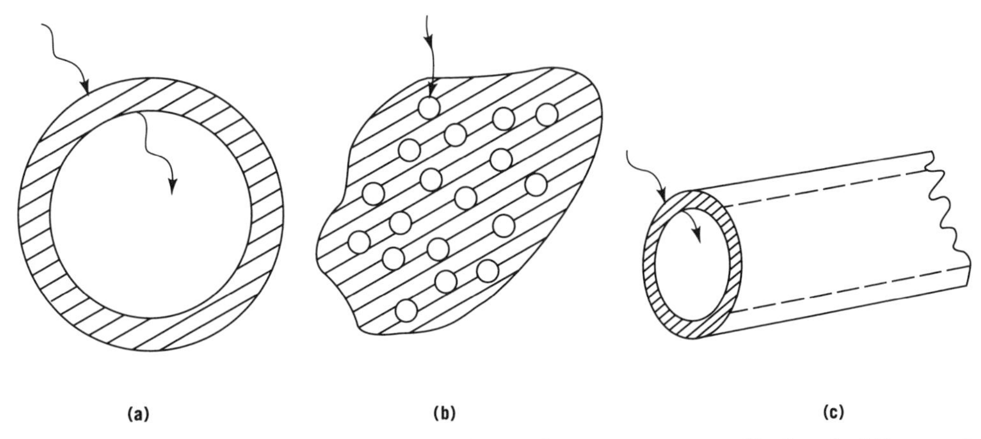

Fig. 9. Membrane extraction where the solvent phase is represented by hatched lines and the arrows show the direction of mass transfer. (**a**) Spherical film; (**b**) emulsion globule where the strip solution is represented by circles; and (**c**) hollow fiber support.

In order to maintain a definite contact area, solid supports for the solvent membrane can be introduced (85). Those typically consist of hydrophobic polymeric films having pore sizes between 0.02 and 1 μm. Figure 9**c** illustrates a hollow fiber membrane where the feed solution flows around the fiber, the solvent–extractant phase is supported on the fiber wall, and the strip solution flows within the fiber. Supported membranes can also be used in conventional extraction where the supported phase is continuously fed and removed. This technique is known as dispersion-free solvent extraction (86,87). The level of research interest in membrane extraction is reflected by the fact that the 1990 International Solvent Extraction Conference (20) featured over 50 papers on this area, mainly as applied to metals extraction. Pilot-scale studies of treatment of metal waste streams by liquid membrane extraction have been reported (88). The developments in membrane technology have been reviewed (89). Despite the research interest and potential, membranes have yet to be applied at an industrial production scale (90).

Supercritical Extraction. The use of a supercritical fluid such as carbon dioxide as extractant is growing in industrial importance, particularly in the food-related industries. The advantages of supercritical fluids (qv) as extractants include favorable solubility and transport properties, and the ability to complete an extraction rapidly at moderate temperature. Whereas most of the supercritical extraction processes are solid–liquid extractions, some liquid–liquid extractions are of commercial interest also. For example, the removal of ethanol from dilute aqueous solutions using liquid carbon dioxide (91) or a supercritical hydrocarbon solvent (92) is under active investigation and several potential applications in food technology have also been reported (92).

Two-Phase Aqueous Extraction. Liquid–liquid extraction usually involves an aqueous phase and an organic phase, but systems having two or more aqueous phases can also be formed from solutions of mutually incompatible polymers such as poly(ethylene glycol) (PEG) or dextran. A system having as many as 18 aqueous phases in equilibrium has been demonstrated (93). Two-phase aqueous extraction, particularly useful in purifying biological species such as proteins (qv) and enzymes, can also be carried out in combination with fermentation (qv) so that the fermentation product is extracted as it is formed (94).

Because of the growth in biotechnology, two-phase aqueous extraction is becoming more important industrially. Two-phase aqueous systems have low interfacial tension, low interphase density difference, and high viscosity in comparison with most aqueous–organic systems. Although interfacial contact is very efficient, the separation of the phases after contact can be slow, requiring centrifugation. The performance of a spray column for two-phase aqueous extraction has also been reported (95).

Equipment and Processing

The earliest large-scale continuous industrial extraction equipment consisted of mixer–settlers and open-spray columns. The vertical stacking of a series of mixer–settlers was a feature of a patented column in 1935 (96) in which countercurrent flow occurred because of density difference between the phases, avoiding the necessity for interstage pumping. This was a precursor of the agitated column contactors which have been developed and commercialized since the late 1940s. There are several texts (1,2,6,97–98) and reviews (99–100) available that describe the various types of extractors.

The unique ability of solvent extraction to achieve separation according to chemical type rather than physical characteristics, such as vapor pressure, enables a great variety of processes ranging from nuclear-fuel enrichment and reprocessing to fertilizer manufacture (see FERTILIZERS; NUCLEAR REACTORS), and from petroleum refining to biochemical and food processing (qv). Probably more types of contactors have been developed for solvent extraction than for any other chemical engineering unit operation. Contactors have been developed for specific processes with which they then tend to become associated. As a result, selection of extractors for a new process application is not necessarily simple, and the choice of a contactor remains both an art and a science, largely based on practical experience.

The following criteria should be considered when selecting a contactor for a particular application: (*1*) stability and residence time, (*2*) settling characteristics of the solvent system, (*3*) number of stages required, (*4*) capital cost and maintenance, (*5*) available space and building height, and (*6*) throughput. The preliminary choice of an extractor for a specific process is primarily based on consideration of the system properties and number of stages required for the extraction. A qualitative chart of the economic operating range of various classes of extractors is shown in Figure 10 (101). A useful selection chart is also available (102) (Table 2). The vendor's experience, pilot-testing procedures, scaling-up methods, costs for capital equipment and maintenance, and reliability of operation should be considered and evaluated at an early stage, before the pilot-plant tests are committed. Although cost ought to be a primary balancing consideration, in many cases previous experience and practice are the deciding factors.

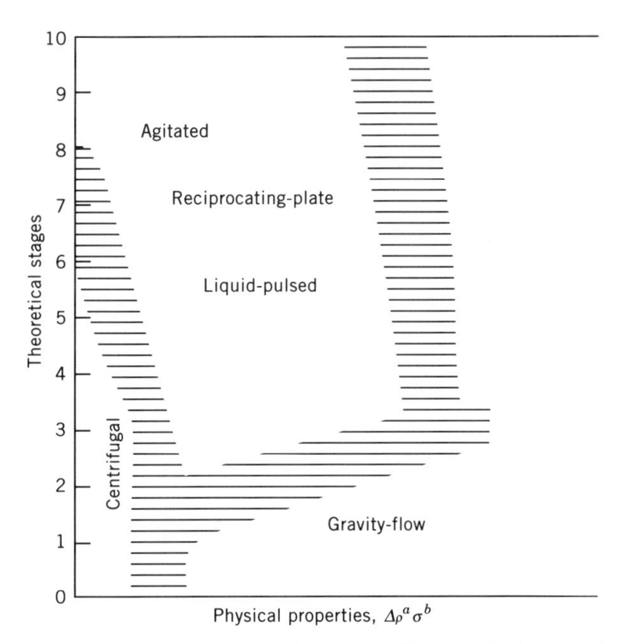

Fig. 10. Economic operating range of extractors. Superscripts a and b are constants. Courtesy of Luwa AG (101).

An extraction plant should operate at steady state in accordance with the flow-sheet design for the process. However, fluctuation in feed streams can cause changes in product quality unless a sophisticated system of feed-forward control is used (103). Upsets of operation caused by flooding in the column always force shutdowns. Therefore, interface control could be of utmost importance. The plant design should be based on (*1*) process control (qv) decisions made by trained technical personnel, (*2*) off-line analysis or limited on-line automatic analysis, and (*3*) control panels equipped with manual and automatic control for motor speed, flow, interface level, pressure, temperature, etc.

Laboratory Extractors, Pilot-Scale Testing, and Scale-Up. Several laboratory units are useful in analysis, process control, and process studies. The AKUFVE contactor (104,105) incorporates a separate mixer and centrifugal separator. It is an efficient instrument for rapid and accurate measurement of partition coefficients, as well as for obtaining reaction kinetic data. Miniature mixer–settler assemblies set up as continuous, bench-scale, multistage, countercurrent, liquid–liquid contactors (106) are particularly useful for the preliminary laboratory work associated with flow-sheet development and optimization because these give a known number of theoretical stages. Laboratory-scale columns are typified by the 2.5-cm diameter reciprocating-plate extraction column, in which a minimum height of an equivalent theoretical stage (HETS) of 7.1 cm and high volumetric efficiencies were achieved employing a methyl isobutyl ketone (MIBK)–acetic acid–water system (107).

Because the factors relating to mass transfer and fluid dynamics of the systems in an extractor are extremely complex, particularly for mixed solvents and

Table 2. Extractor Selection Chart[a,b]

	Gravity-separated extractors																Centrifugally separated			
	Continuous contact					Discontinuous contact							Mixer–settlers							
						Without interstage settling		With settling												
	Non-mechanical		Mechanical			Mechanical		Non-mechanical		Mechanical			Horizontal		Vertical		Continuous contact		Mixer–settler	
Design requirements	Spray column	Baffle-plate column	Packed column	Pulsed-packed column	Raining-bucket contractor	Rotary-agitated columns	Reciprocating-plate column	Pulsed-plate column	Perforated-plate column	Scheibel column	ARDC column	Rotary film contactor	Pump–settler	Agitated mixer–settler	Pump–settler	Agitated mixer–settler	Perforated plate	Film-flow type (de Laval)	Luwesta	Rotabel
total throughput, m^3/h																				
<0.25	3	3	3	3	3	3	3	3	3	3	3	1	0	1	0	1	3	1	0	0
0.25–2.5	3	3	3	3	3	3	3	3	3	3	3	3	1	3	1	3	3	3	1	1
2.5–25	3	3	3	3	3	3	3	3	3	3	3	3	3	3	3	3	3	3	3	3
25–250	3	1	3	3	1	3	1	1	3	1	1	1	3	3	3	1	0^c	0^c	0^c	0^c
>250	1	0	1	1	0	1	0	1	1	0	0	1	5	5	1	1	0^c	0^c	0^c	0^c
number of stages																				
<1.0	5^d	3	3	3	3	3	3	3	3	3	3	3	3	3	3	3	3	3	3	3
1–5	1,^{e}0	3	3	3	3	3	3	3	3	3	3	3	3	3	3	3	3	3	5^f	3
5–10	0	3	3	3	3	3	3	3	3	3	3	3	3	3	3	3	1^c	1	0^c	3
10–15	0	1	1	3	1	1	3	3	1	3	1	1	3	3	1	1	0^c	0^c	0^c	0^c
>15	0	1	1	1	1	1	1	1	1	1	1	0	3	3	1	1	0^c	0^c	0^c	0^c
physical properties[g]																				
$(\sigma/\rho g)^{1/2} > 0.60$	1	1	1	3	1	3	3	3	1	3	3	3	3	3	3	3	5	5	5	5
density difference, g/cm^3																				
$0.05 > \Delta\rho > 0.03$	3	3	3	0	3	0	0	0	1	0	1	3	1	1	1	1	5	5	5	5
viscosity,[e] mPa·s(=cP)																				
μ_c or $\mu_d > 20$	1	1	1	1	1	1	1	1	1	1	1	1	1	1	1	1	1	1	1	1

slow heterogeneous reaction																				
$k_t < 4 \times 10^{-5}$ m/s	0	1	1	3	1	3	3	3	1	3	3	0	0	3	0	3	3	3	3	3
slow homogeneous reaction																				
$t_{1/2}$ = 0.5–5 min	1	1	1	1	3	1	1	1	3	1	1	1	3	5	3	3	0	0	1	1
> 5 min	0	0	0	0	1	0	0	0	1	0	0	0	3	5	3	3	0	0	0	0
extreme phase ratio																				
$F_d/F_f < 0.2$ or >5	1	1	1	1	3	1	1	1	3	1	1	3	1	5[h]	1	5[h]	3	3	3	3
short residence time	0	0	0	1	0	1	1	1	0	0	0	0	0	0	0	0	5	5	3	3
ability to handle solids																				
trace (<0–0.1% in feed)	3	3	1	1	5	3	3	3	1	1	1	5	3	3	3	3	1	1	1	1
appreciable (0.1–1% in feed)	1	1	1	1	3	3	3	3	0	0	0	5	1	1	1	1	1	1	1	1
heavy (>1% in feed)	1	1	0	0	1	1	1	1	0	0	0	5	1[i]	1[i]	1[i]	1[i]	0	0	1	1
tendency to emulsify																				
slight	3	3	1	1	3	1	1	1	3	1	1	3	1	1	1	1	5	5	5	5
marked	1	1	1	0	1	0	0	0	1	0	0	1	0	0	0	0	3	3	3	3
limited space available																				
height	0	1	1	1	5	1	1	1	1	1	1	3	5	5	0	0	5	5	5	5
floor	5	5	5	5	0	5	5	5	5	5	5	0	0	0	5	5	5	5	5	5
special materials required																				
metals (stainless steel, Ti, etc)	5	3	5	5	3	3	3	3	3	3	3	3	3	3	3	3	5	5	5	5
nonmetals	5	3	5	1	1	0	0	1	1	0	0	1	1	5	1	1	0	0	0	0
radioactivity present																				
weak (mainly α, β)	5	5	5	3	1	1	1	3	3	1	1	1	3	5	3	1	1	1	1	1
strong γ	5	5	5	3	0	0	0	3	3	0	0	0	3	5	1	0	0	0	0	0
ease of cleaning	5	3	1	1	3	3	3	3	1	1	3	5	3	3	3	3	1	3	3	1
low maintenance	5	5	5	3	3	3	3	3	5	3	3	3	3	3	3	3	1	1	1	1

[a]Ref. 102. [b]Rating of 5, very strongly recommended; 3, satisfactory; 1, may be used; and 0, not suitable. [c]Multiple units in series or parallel can be used.
[d]For immeasurably fast homogeneous reaction. [e]For diameters <15 cm. [f]Two or three stages only in single machine.
[g]See text for effect of direction of transfer. [h]With recirculation of separated phases to mixer. [i]Requires provision for solids removal from settler.

feedstocks of commercial interest, pilot-scale testing remains an almost inevitable preliminary to a full-scale contactor design. These tests provide: (*1*) total throughput and agitation speed; (*2*) HETS or HTU; (*3*) stage efficiency; (*4*) hydrodynamic conditions, such as droplet dispersion, phase separation, flooding, emulsion layer formation, etc; (*5*) selection of direction of mass transfer; (*6*) solvent-to-feed ratio; (*7*) material of construction and its wetting characteristics; and (*8*) confirmation of the desired separation in cases where equilibrium data are not available.

For design of a large-scale commercial extractor, the pilot-scale extractor should be of the same type as that to be used on the large scale. Reliable scale-up for industrial-scale extractors still depends on correlations based on extensive performance data collected from both pilot-scale and large-scale extractors covering a wide range of liquid systems. Only limited data for a few types of large commercial extractors are available in the literature.

Commercial Extractors. Extractors can be classified according to the methods applied for interdispersing the phases and producing the countercurrent flow pattern. Figure 11 summarizes the classification of the principal types of commercial extractors; Table 3 summarizes the main characteristics.

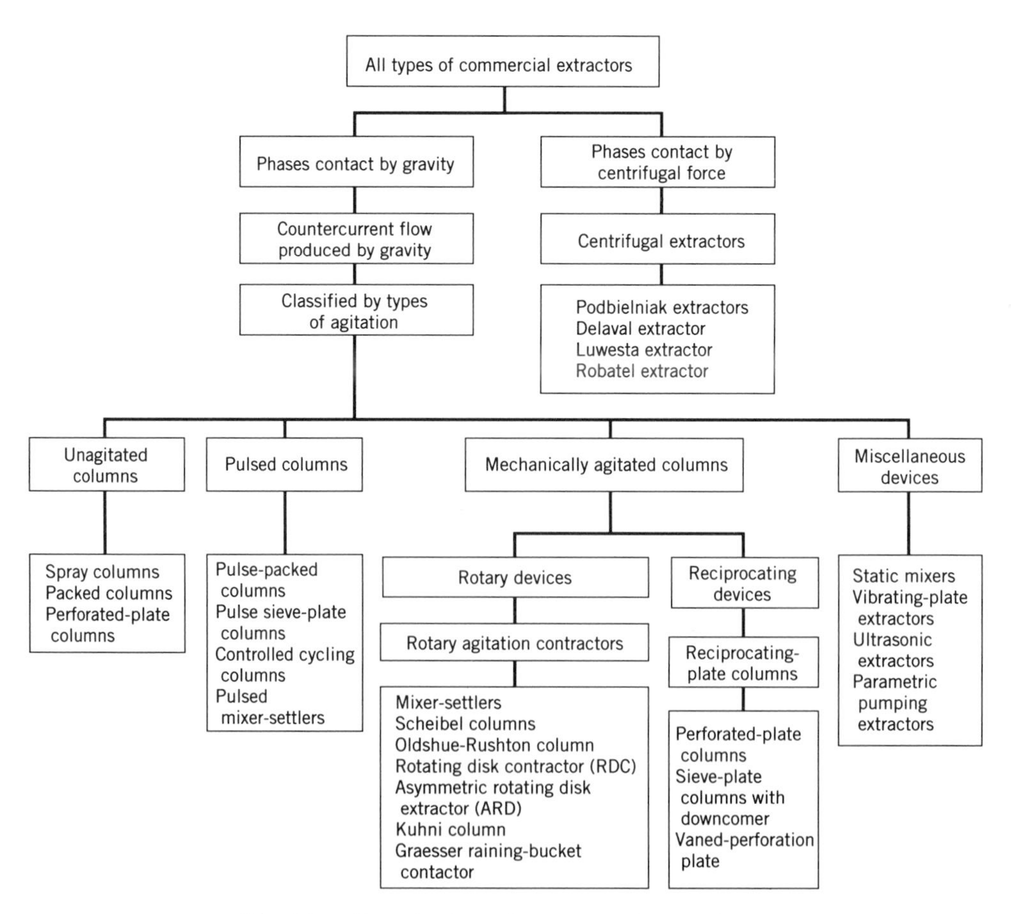

Fig. 11. Classification of commercial extractors.

Table 3. Summary of Commercial Extractors

Types of extractor	General features[a]	Fields of industrial application
unagitated columns	low capital cost, low operating and maintenance cost, simplicity in construction, handles corrosive material	petrochemical, chemical
mixer–settlers	high stage efficiency, handles wide solvent ratios, high capacity, good flexibility, reliable scale-up, handles liquids with high viscosity	petrochemical, nuclear, fertilizer, metallurgical
pulsed columns	low HETS, no internal moving parts, many stages possible	nuclear, petrochemical, metallurgical
rotary agitation columns	reasonable capacity, reasonable HETS, many stages possible, reasonable construction cost, low operating and maintenance cost	petrochemical, metallurgical, pharmaceutical, fertilizer
reciprocating-plate columns	high throughput, low HETS, great versatility and flexibility, simplicity in construction, handles liquids containing suspended solids, handles mixtures with emulsifying tendencies	pharmaceutical, petrochemical, metallurgical, chemical
centrifugal extractors	short contacting time for unstable material, limited space required, handles easily emulsified material, handles systems with little liquid density difference	pharmaceutical, nuclear, petrochemical

[a]HETS = height of an equivalent theoretical stage.

Unagitated Columns. Because of the simplicity and low cost, unagitated columns are widely used in industry despite low efficiency, particularly for processes requiring few theoretical stages and for corrosive systems where absence of mechanical moving parts is advantageous (Table 3). Three types of unagitated column extractors are shown in Figure 12. Spray columns (Fig. 12**a**) are the simplest in construction mechanically but have very low efficiency because of poor phase contacting and excessive backmixing in the continuous phase. These generally provide one or, at the most, two equilibrium stages. For example, a baffled spray tower, 2.7 m in diameter and 24 m in height for propane deasphalting of residue was reported to have only 3 to 3.5 theoretical stages (108). Because of the simple construction, however, spray columns are used for industrial operations requiring only a few stages.

Packed columns (Fig. 12**b**) have better efficiency because of improved contacting and reduced backmixing; it is important that the packing material should be wetted by the continuous phase to avoid coalescence of the dispersed phase.

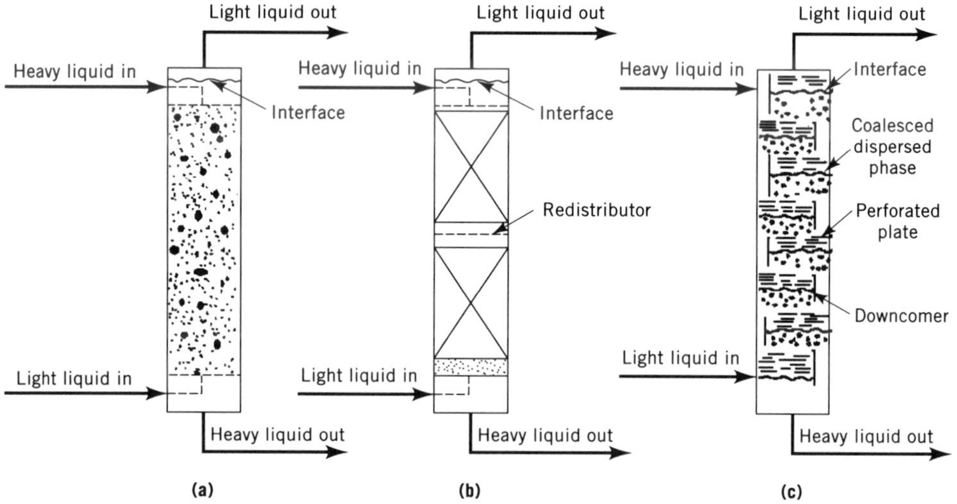

Light liquid out
Heavy liquid in
Interface
Light liquid in
Heavy liquid out
Redistributor
Coalesced dispersed phase
Perforated plate
Downcomer
(a) (b) (c)

Fig. 12. Unagitated column extractors: (**a**) spray column; (**b**) packed column; and (**c**) perforated-plate column.

To reduce the effects of channeling, redistribution of the liquids at fixed intervals is normally required in the taller columns. Packed columns should not be used if the ratio of the phase-flow rates is beyond the range 0.5 to 2.0 because of probable flooding when suitable holdup and interfacial area are provided (109). Normally, a packed column is preferred over a spray column because the reduced flow capacity is less important than the improved mass transfer. Sulzer static mixers have been reported as a packing in liquid–liquid extraction. The overall values of height of a transfer unit range from 0.6 to 1.6 m depending on the system and direction of mass transfer (110). Sulzer static mixers have also been used in a column for multistage supercritical fluid extraction. A description of high efficiency packing for liquid–liquid extraction has recently been reported (110).

Perforated-plate columns (Fig. 12**c**) are operated semistagewise and are reasonably flexible and efficient. If the light phase is dispersed, the light liquid flows through the perforations of each plate and is dispersed into drops which rise through the continuous phase. The continuous phase flows horizontally across each plate and passes to the plate beneath through a downcomer. If the heavy phase is dispersed, the column is reversed and upcomers are used for the continuous phase. A perforated-plate tower 2.13 m in diameter and 24.38 m in height used for extraction of aromatics was reported to have the equivalent of 10 theoretical stages (111). Mass-transfer data obtained in various types of perforated-plate columns up to 225 mm in diameter using different extraction systems have been summarized (112). The data are generally correlated in terms of overall heights of transfer units vs flow velocities of the phases for a specific column and system. There are many variations of basic designs for perforated-plate (sieve-plate) columns and detailed information is given in the literature (108).

Mixer–Settlers. Mixer–settlers are widely used in the chemical process industry because of reliability, flexibility, and high capacity. These extractors are particularly economical for operations that require high throughput and few stages. Mixer–settlers having capacity up to 22.7 m^3/min (6000 gal/min) have been used in the mining industry. The main disadvantages of mixer–settlers are size and the inventory of material held up in the equipment. Considerable development work has been done to improve mixer–settler design, and many newer devices have been reported. Figure 13 shows some of these extractors.

The simple box-type mixer–settler (113) has been used extensively in the UK for the separation and purification of uranium and plutonium (114). In this type of extractor, interstage flow is handled through a partitioned box construction. Interstage pumping is not needed because the driving force is provided by the density difference between solutions in successive stages (see PLUTONIUM AND PLUTONIUM COMPOUNDS; URANIUM AND URANIUM COMPOUNDS).

A widely used type of pump–mixer–settler, developed by Israeli Mining Industries (IMI) (115), is shown in Figure 13**a**. A unit having capacity 8.3 m^3/min (2000 gal/min) has been used in phosphoric acid plants (116). The unique feature of this design is that the pumping device is not required to act as the mixer, and the two phases are dispersed by a separate impeller mounted on a shaft running coaxially with the drive to the pump.

The General Mills mixer–settler (117), shown in Figure 13**b,** is a pump–mix unit designed for hydrometallurgical extraction. It has a baffled cylindrical mixer fitted in the base and a turbine that mixes and pumps the incoming liquids. The

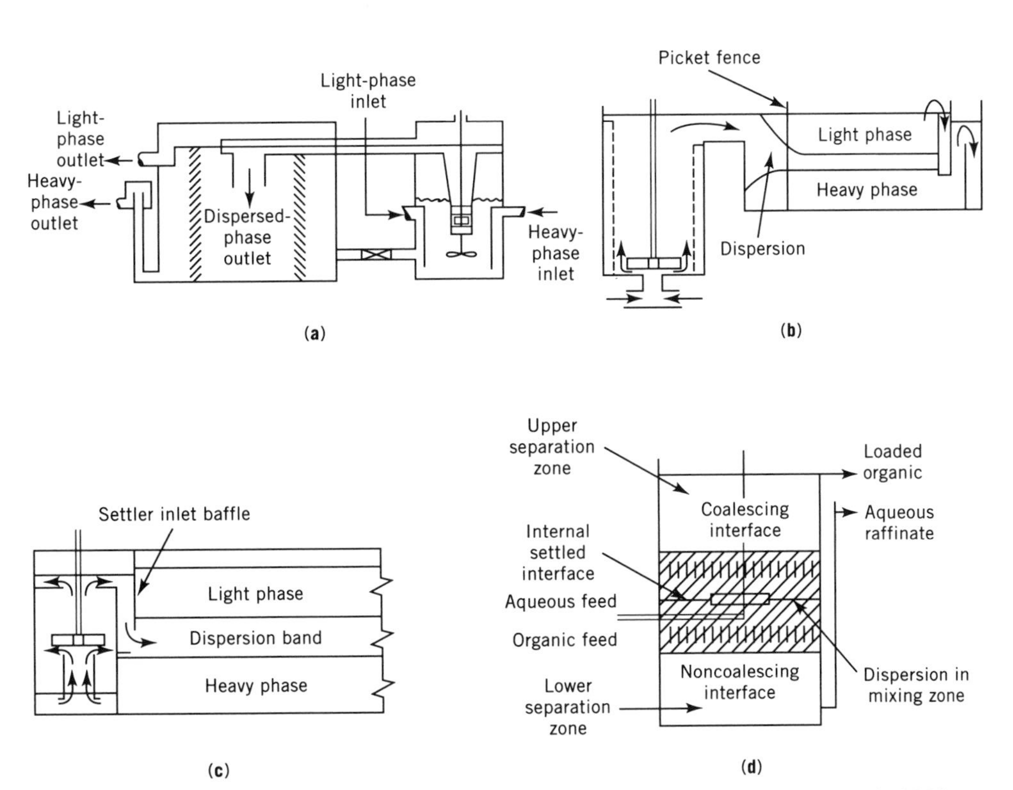

Fig. 13. Mixer–settlers: (**a**) IMI; (**b**) General Mills; (**c**) Davy-McKee; and (**d**) CMS.

dispersion leaves from the top of the mixer and flows into a shallow rectangular settler designed for minimum holdup.

In the Davy-Powergas unit (118–120), shown in Figure 13**c**, the liquids run through a draft tube and are pumped by an impeller running directly above the draft tube. The dispersion flows out from the top of the mixer and down through a channel into a rectangular settler. Large units of this type are used for copper extraction (7).

The development of the novel Davy-McKee combined mixer–settler (CMS) has been described (121). It consists of a single vessel (Fig. 13**d**) in which three zones coexist under operating conditions. A detailed description of units used for uranium recovery has been reported (122), and the units have also been studied at the laboratory scale (123). Application of the Davy combined mixer electrostatically assisted settler (CMAS) to copper stripping from an organic solvent extraction solution has been reported (124).

The Lurgi contactor (125), developed in Germany, consists of stacked mixer–settler units. Mixing and interphase transfer take place in pumps attached to the side of the settling column. It has a capacity of 1600 t/h and columns up to 3 m in diameter have been used for aromatic extractions. The Holmes and Narver mixer–settler (126) incorporates a multicompartment mixer and has many other special features. The Kemira mixer–settler (127) developed in Finland also uses the pump–mix concept, in which the phase to be mixed is drawn from a point in line with mixing impeller. Only the heavy phase is pumped into the mixer, and the light phase is allowed to flow freely from the settler. A large auxiliary space is provided between the mixer and the settler. The unit has been successfully used in extraction of the rare earths (see LANTHANIDES) and nitrophosphate fertilizer processes (128) and found to be particularly flexible when there are great variations in flow rate from stage to stage. A new type of mixer–settler (EC-D) having a delta-type pump–mixing impeller has been developed in China (129). The delta impeller is reported to have the advantages of developing high flow velocities in both the axial and radial directions (130), resulting in high efficiency and relatively low energy consumption. Applications in a large rare-earth extraction plant have been reported (131).

Motionless inline mixers obtain energy for mixing and dispersion from the pressure drops developed as the phases flow at high velocity through an array of baffles or packing in a tube. Performance data on the Kenics (132) and Sulzer (133) types of motionless mixer have been reported.

The scale-up and design of mixer–settlers is relatively reliable because they are practically free of interstage backmixing and stage efficiencies are high, typically 80 to 90%. Various studies (134–136) have shown that (*1*) the rate of extraction is a function of power input, and (*2*) mixers can be reliably scaled up by geometric similitude at constant power input per unit mixer volume, up to a 200-fold factor of throughput (137,138). The processes taking place in the settler are complex. In large industrial mixer–settlers, the settlers usually represent at least 75% of the total volume of the units. The flow capacity of a settler depends on the behavior of a band of dispersion at the interface. The thickness of the band is a measure of the approach to flooding (97). The thickness increases exponentially with increasing flow per unit interfacial area, and settlers can be scaled up by factors of up to 1000 on this basis. A practical means to increase the throughput per unit settler area is needed so that the size of the settler can be reduced and the inventory of solvent lowered. The efficiency of the settler can be enhanced by minimizing turbulence and the formation of small drops, and maintaining low values of the linear velocity along the settler to avoid entrainment of small drops from the dispersion band.

Pulsed Columns. The efficiency of sieve-plate or packed columns is increased by the application of sinusoidal pulsation to the contents of the column. The well-distributed turbulence promotes dispersion and mass transfer while tending to reduce axial dispersion in comparison with the unpulsed column. This leads to a substantial reduction in HETS or HTU values.

The pulsed-plate column is typically fitted with horizontal perforated plates or sieve plates which occupy the entire cross section of the column. The total free area of the plate is about 20–25%. The columns are generally operated at frequencies of 1.5 to 4 Hz with amplitudes 0.63 to 2.5 cm. The energy dissipated by the pulsations increases both the turbulence and the interfacial areas and greatly improves the mass-transfer efficiency compared to that of an unpulsed column. Pulsed-plate columns in diameters of up to 1.0 m or more are widely used in the nuclear industry (139,140).

Figure 14 shows that several regions of operation in the pulsed-plate column can be distinguished, depending on the flow rate and intensity of pulsation (141). At low pulsation velocities (expressed as amplitude × frequency), a discrete layer of liquid appears between plates during each reversal of the pulse cycle. At higher velocities there is little or no layer formation and the column then behaves as a differential contactor. Extensive studies have been made on flooding, mass transfer, and the development of empirical correlations for the column design (142–144), and on the hydrodynamics and performance of columns of various sizes in uranium extraction (139).

Pulsed-packed columns consist of vertical cylindrical vessels filled with packing. The light and dense liquids passing countercurrently through the packing

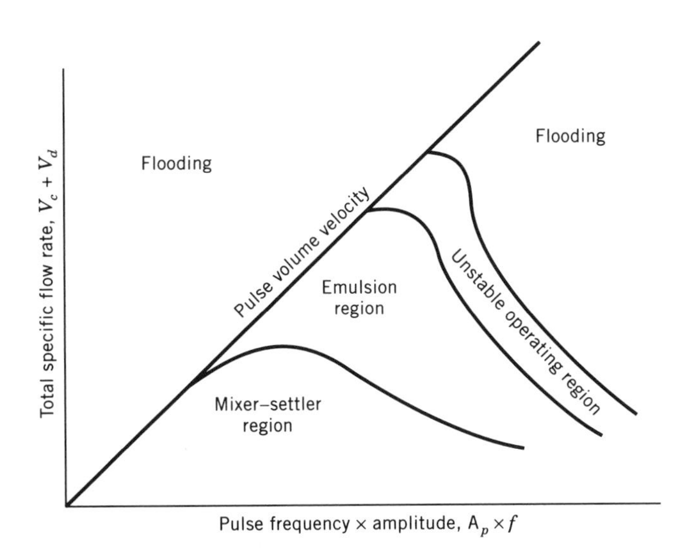

Fig. 14. Regions of operation of a pulsed, perforated-plate column (141).

are acted on by pulsations transmitted hydraulically to form a dispersion of drops. The pulsation device is connected to the side of the column, usually at the base, through a pulse leg. Mechanical difficulties with the generation of the pulse formerly limited pulsed columns to comparatively small diameters, but the installation of pulsed-packed columns up to 2.74 m in diameter has been reported (145,146). The generation of pulses by compressed air has received increasing attention (147). A detailed model of pulsed-packed column behavior has been developed (147). The controlled cycling column (148) has a high throughput, but no large-scale application has been reported.

Mechanically Agitated Columns. *Rotary Agitated Columns.* Because of the mechanical advantages of rotary agitation, most modern differential contactors employ this method. The best known of the commercial rotary agitated contactors are shown in Figure 15. Features and applications of these columns are given in Table 3.

In the Scheibel column, developed in 1948 (149), every alternate compartment is agitated by an impeller, and the unagitated compartments are packed with open woven wire mesh. Capacity and mass-transfer data are given in the literature (149–151). A newer type of Scheibel column (Fig. 15**a**) using horizontal baffles with or without wire mesh packing was developed in 1956 (152). Performance data for a 30.5-cm column, with or without wire mesh packing, have shown that the HETS varies as the square root of diameter. A third design (153) is basically similar, but a pumping impeller instead of a turbine is used in the mixing stage.

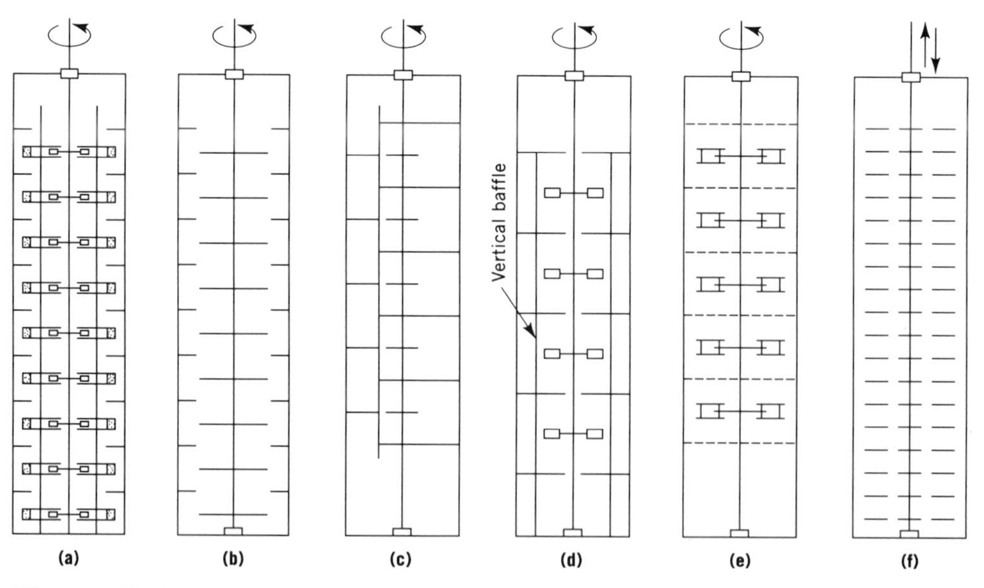

Fig. 15. Mechanically agitated columns: (**a**) Scheibel column; (**b**) rotating-disk contactor (RDC); (**c**) asymmetric rotating-disk (ARD) contactor; (**d**) Oldshue-Rushton multiple-mixer column; (**e**) Kuhni column; and (**f**) reciprocating-plate column.

Scale-up and performance of a 1.47-m Scheibel column have been reported (98,154,155), as have detailed description and design criteria for the Scheibel column (156) and scale-up procedures (157). The same stage efficiency can be maintained on scale-up, and total throughput can be increased by three and one-half times at the expense of higher HETS. As of this writing, Scheibel columns up to 2.75 m in diameter are in service.

The rotating-disk contactor (RDC), developed in the Netherlands (158) in 1951, uses the shearing action of a rapidly rotating disk to interdisperse the phases (Fig. 15**b**). These contactors have been used widely throughout the world, particularly in the petrochemical industry for furfural [*98-01-1*] and SO_2 extraction, propane deasphalting, sulfolane [*126-33-0*] extraction for separation of aromatics, and caprolactam (qv) [*105-60-2*] purification. Columns up to 4.27 m in diameter are in service. An extensive study (159) has provided an excellent theoretical framework for scale-up. A design manual has also been compiled (160). Detailed descriptions and design criteria for the RDC may also be found (161).

The Oldshue-Rushton column (Fig. 15**d**) was developed (162) in the early 1950s and has been widely used in the chemical industry. It consists essentially of a number of compartments separated by horizontal stator-ring baffles, each fitted with vertical baffles and a turbine-type impeller mounted on a central shaft. Columns up to 2.74 m in diameter have been reported in service (162–167). Scale-up is reported to be reliably predictable (168) although only limited performance data are available (169). A detailed description and review of design criteria are available (170).

The asymmetric rotating disk (ARD) contactor (Fig. 15**c**) was developed in Czechoslovakia (160,171–174) and has been increasingly used in western Europe. Its design aims at retaining the efficient shearing and dispersing action of the RDC while reducing backmixing by means of the coalescence–redispersion cycle produced in the separate transfer and settling zones. The ARD extractor is used for extraction of petrochemicals, pharmaceuticals (qv), and caprolactam, as well as for propane deasphalting, phenol removal from wastewater, furfural refining of oils, etc. Columns up to 2.4 m in diameter are in service and a detailed description and review of design procedures are given in the literature (175).

Kuhni contacters (Fig. 15**e**) have gained considerable commercial application. The principal features are the use of a shrouded impeller to promote radial discharge within the compartments, and a variable hole arrangement to allow flexibility of design for different process applications. Columns up to 5 m in diameter have been constructed (176). Description and design criteria for Kuhni extraction columns have been reported (177,178).

The RTL contactor, formerly known as the Graesser raining bucket contactor (179), is a horizontal design having the phases interdispersed by slowly rotating waterwheel-type impellers. This unit, which has the feature of dispersing each phase into the other, was developed for handling the difficult settling systems found in the coal-tar industry. It is also suitable for solid–liquid systems (180) and data on mass transfer and axial mixing have been reported (181). Units have been built from 100 mm (4 in.) to 1.8 m (6 ft) in diameter.

There are many other types of rotary agitated contactors (182) which have been less widely used.

Reciprocating-Plate Columns. Phase dispersion can also be achieved by reciprocating or vibrating of plates in a column. Improvement of extraction efficiency in a perforated-plate column by pulsing the liquid contents or by reciprocating the plates was proposed in 1935 (183). A reciprocating-plate column (RPC) was later developed (184) and scale-up was shown to be effectively accomplished by adding fixed baffles to the column (185). Many different types of column employing reciprocating plates or packing have been described (186–191) (Fig. 15**f**).

Reciprocation of plates requires less energy than pulsing the entire volume of liquid in a column, and has the same effect in terms of mixing patterns and uniform dispersion. This is a considerable advantage in large-scale commercial extractors (98). The main difference between the different types of RPC that have been built for industrial use lies in the plate design, as shown in Figure 16. Table 2 outlines the general features and industrial applications of the different types of RPC. Also available are reviews (191,192) and a more detailed description of the design criteria (193).

The open-type (Karr) RPC plate (183) has relatively large (12 to 15 mm) diameter perforations and free area of about 58% (Fig. 16**a**). It operates only in the emulsion regime. A minimum HETS of 50.8 cm has been measured in a 0.91-m diameter column using the relatively difficult (high interfacial tension) system *o*-xylene–acetic acid–water. Empirical correlations for scale-up have been

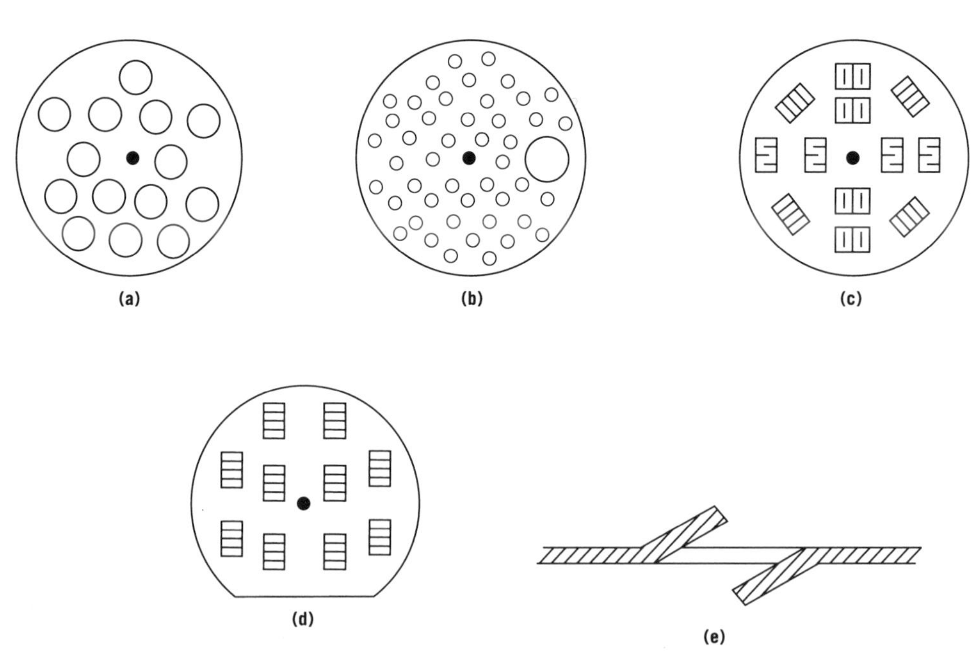

Fig. 16. Types of RPC plate in industrial use: (**a**) Karr RPC plate, $\phi = 0.5–0.6$, $d_o = 10–16$ mm; (**b**) Prochazka RPC plate, $\phi = 0.04–0.3$ (excluding (○) downcomer), $d_o = 2–5$ mm; (**c**) KRIMZ RPC plate, $\phi \approx 0.45$, which has vaned rectangular perforations; (**d**) GIAP II RPC plate, $\phi \approx 0.05–0.15$; and (**e**) sectional view of rectangular perforations for (**d**).

proposed (97,107,185,194–196). Hydrodynamics and axial mixing have also been studied (197,198). As well as being operated countercurrently, the Karr RPC can be operated in cocurrent flow. This type of column has gained increasing industrial application in the pharmaceutical, petrochemical, and hydrometallurgical industries, and in wastewater treatment (97,192,194); the Karr RPCs are in service in North America and Europe in diameters up to 1.7 m (199).

RPCs having perforated plates and downcomers (Fig. 16**b**) have been developed industrially in the former Czechoslovakia (200,201) under the trademark VPE (vibrating plate extractor). For large columns a segmental downcomer is used instead of a tubular downcomer. The downcomers permit a much higher throughput than would be possible using perforations alone. The largest units for phenol [*108-95-2*] extraction have a diameter of 1.2 m and plate stack height 9.1 m. These have a capacity of 80 m^3/h for phenolic wastewater (97,194).

The KRIMZ and GIAP types of RPC (190,191) (Figs. 16**c** and 16**d**, respectively) were developed in the former USSR and the plate designs feature rectangular punched perforations where the displaced metal strips remain attached as inclined vanes. The purpose of the vanes is to deflect the liquid and give it radial motion, which can be beneficial in reducing axial mixing in larger diameter columns. The modeling, design, and scale-up (202) have been based on theoretical principles (203). Industrial applications of KRIMZ and GIAP plates in RPCs up to 1.5 m in diameter have been reported (192,204).

Other types of RPC have been proposed but are not in industrial use as of this writing. These include a reciprocated wire-mesh packing (188), a reciprocating screen-plate (205), and the multistage vibrating disk column (MVDC) developed in Japan (189,206,207). These types of RPC may be useful for gas–liquid contact as well as liquid–liquid contact.

Centrifugal Extractors. In centrifugal extractors, contact time between the phases is reduced and phase separation is accelerated by the application of centrifugal forces which greatly exceed gravitational forces. The units are compact and a relatively high throughput per unit volume can be achieved. Centrifugal extractors are particularly useful for systems which are chemically unstable, eg, extraction of antibiotics (qv), or for systems in which the phases are slow to settle. General features and fields of application are given in Table 3 and a detailed review including design criteria is available (208).

The first differential centrifugal extractor to be used in industry was the Podbielniak extractor which was introduced in the 1950s (209,210) and can be regarded as a perforated-plate column wrapped around a rotor shaft. Rotation creates a centrifugal force which results in a great reduction in the equivalent height and contact time that would be needed in a conventional perforated-plate column.

The behavior of drops in the centrifugal field has been studied (211) and the residence times and mass-transfer rates have been measured (212). Podbielniak extractors have been widely used in the pharmaceutical industry, eg, for the extraction of penicillin, and are increasingly used in other fields as well. Commercial units having throughputs of up to 98 m^3/h (26,000 gal/h) have been reported.

The Alfa-Laval extractor (213) can give up to 20 theoretical stages in one unit. Depending on the system being handled, the capacity of the standard unit ranges between 5.7 and 21.2 m^3/h (1500–5600 gal/h). Antibiotic extractions and petrochemical processing are typical applications.

A new countercurrent continuous centrifugal extractor developed in the former USSR (214) has the feature that mechanical seals are replaced by liquid seals with the result that operation and maintenance are simplified; the mechanical seals are an operating weak point in most centrifugal extractors. The operating units range between 400 and 1200 mm in diameter, and a capacity of 70 m^3/h has been reported in service. The extractors have been applied in coke-oven refining (see COAL CONVERSION PROCESSES), erythromycin production, lube oil refining, etc.

The class of discrete stage centrifugal extractors includes the Westfalia centrifugal extractor (215,216) which rotates about a vertical axis and is available with up to three contact stages. Its advantage is that the light phase does not have to be introduced under pressure. The capacity of the largest model is reported as ranging from 7.6 m^3/h (2000 gal/h) for three stages, to 49.2 m^3/h (13,000 gal/h) for a single stage. Another important member of this class is the Robatel extractor (220) which consists of a series of mixer–settlers stacked on their sides with the mixing in each stage being provided by a stationary disk attached to the shaft while the chamber rotates. Typical units provide three to eight stages, and throughputs up to 6.2 m^3/h (1600 gal/h) have been reported. Robatel extractors have found general application in the chemical, pharmaceutical, and petrochemical industries, and particularly the nuclear industry. Technical and economic comparisons of the Robatel extractor with mixer–settlers and pulsed columns have been made (217). Research and development on other nondispersive forms of contactors, eg, Hi-Gee solvent extractors that give a high efficiency per unit volume, and contactors effective with very short residence times, eg, improvement on the centrifugal extractor, has been reported (218).

Economics of Extraction. Economic considerations for solvent extraction include both capital and operating costs. Capital cost is made up of the installed cost of equipment and the cost of the inventory of material (including solvent and extractant) held within the plant. Operating costs include the cost of extractor operation, solvent recovery, and solvent losses. Solvent recovery is often the dominant factor because of the high energy consumption involved. Process economy can often be improved by increasing the number of stages, which reduces the solvent recovery despite increasing the capital cost.

Organic Processes

Petroleum and Petrochemical Processes. The first large-scale application of extraction was the removal of aromatics from kerosene [*8008-20-6*] to improve its burning properties. Jet fuel kerosene and lubricating oil, which require a low aromatics content (see AVIATION AND OTHER GAS TURBINE FUELS), are both in demand. Solvent extraction is also extensively used to meet the growing demand for the high purity aromatics such as benzene, toluene, and xylene (BTX) as feedstocks for the petrochemical industry (see BTX PROCESSING; FEEDSTOCKS, PETROCHEMICALS). Additionally, the separation of aromatics from aliphatics is one of the largest applications of solvent extraction (see PETROLEUM, REFINERY PROCESSES SURVEY).

Lubricating Oil Extraction. Aromatics are removed from lubricating oils to improve viscosity and chemical stability (see LUBRICATION AND LUBRICANTS). The solvents used are furfural, phenol, and liquid sulfur dioxide. The latter two solvents are undesirable owing to concerns over toxicity and the environment and most newer plants are adopting furfural processes (see FURAN DERIVATIVES). A useful comparison of the various processes is available (219).

Separation of Aromatic and Aliphatic Hydrocarbons. Aromatics extraction for aromatics production, treatment of jet fuel kerosene, and enrichment of gasoline fractions is one of the most important applications of solvent extraction. The various commercial processes are summarized in Table 4.

The Udex process (220) was popular in the United States in the 1970s. The original process produced high purity gasoline by removing aromatics using diethylene glycol as a solvent. The process has also been used for the manufacture of BTX; aqueous tetraethylene glycol [*112-60-7*] appears to be the best solvent (221). The sulfolane process (222–224), introduced by the Shell Co. in 1962 (Fig. 17), is used in many large units all over the world. Sulfolane [*126-33-0*] [(tetrahydrothiophene)-1,1-dioxide], is a strongly polar compound that is highly selective for aromatic hydrocarbons and has much greater solvent capacity for hydrocarbons than glycol systems. Additional features in its favor are high density, heat capacity, and chemical stability. The sulfolane process uses the rotating-disk contactor (RDC) (225). The Lurgi Arosolvan process (226) has been used in over a dozen commercial installations. Two process arrangements are available, using as solvent either a mixture of *N*-methyl-2-pyrrolidinone [*872-50-4*] (NMP) and water or a mixture of NMP and ethylene glycol [*107-21-1*]. The polar mixing component (water or ethylene glycol) increases the selectivity of the solvent for aromatics. The Lurgi multistage mixer–settler is used with towers up to 6 m in diameter and 35 m high. The NMP (Arosolvan) process for BTX separation has been described (227). A dimethyl sulfoxide [*67-68-5*] (DMSO) process which employs two separate extraction steps has been developed (228). The selectivity and low viscosity of the solvent (DMSO plus a few percent water) allow the extraction to take place entirely at ambient temperatures. In addition, DMSO is nontoxic and relatively inexpensive. The process uses the Kuhni column (Fig. 15**e**) in diameters up to 2.7 m. In the Union Carbide process (229), the solvent (tetraethylene glycol) is recovered by a second extraction step rather than from the extract by distillation. However, it is necessary to distill the raffinate from the first extractor in order to recover the dissolved process solvents. A useful description of the Union Carbide TETRA process is available (230). The Formex process (231) which employs *N*-formylmorpholine [*4394-85-8*] and a few percent water as solvent, has the flexibility to handle different feedstocks and product ranges. Either distillation or secondary extraction may be used to regenerate the solvent, depending on the range of aromatics which is to be produced. The Redox process (232) (recycle extract dual extraction) improves the octane number of diesel fuels by extracting an aromatic concentrate (see GASOLINE AND OTHER MOTOR FUELS). The solvents include furfural–furfuryl alcohol–water mixtures, aqueous tetrahydrofurfuryl alcohol, and aqueous dimethylformamide.

Table 4. Extractive Processes for the Separation of Benzene–Toluene–Xylene Mixture from Light Feedstocks[a]

Process	Solvent	Solvent additives and reflux conditions	Operating temperature, °C	Contacting equipment	Comments
Shell process, Universal Oil Products	sulfolane	sulfolane selectivity and capacity insensitive to water content caused by steam-stripping during solvent recovery; heavy paraffinic countersolvent used	120	rotating-disk contactor, up to 4 m in diameter	the high selectivity and capacity of sulfolane leads to low solvent–feed ratios, and thus smaller equipment
Udex process, Universal Oil Products	glycol–water mixture	solvent can be diethylene glycol and water, or mixture of diethylene and dipropylene glycols and water, or tetraethylene glycol and water; light hydrocarbon reflux	150 for diethylene glycol and water	sieve-tray extractor	tetraethylene glycol and water mixtures are claimed to increase capacity by a factor of 4 and also require no antifoaming agent; the extract requires a two-step distillation to recover BTX
Union Carbide Corp.	tetraethylene glycol (TETRA)	the solvent is free of water; a dodecane reflux is used which is later recovered by distillation	100	reciprocating-plate extractor	the extract leaving the primary extractor is essentially free of feed aliphatics, and no further purification is necessary; two-stage extraction uses dodecane as a displacement solvent in the second stage
Institut Français du Pétrole	dimethyl sulfoxide (DMSO)	solvent contains up to 2% water to improve selectivity; reflux consist of aromatics and paraffins	ambient	rotating-blade extractor, typically 10–12 stages	low corrosion allows use of carbon steel equipment; solvent has a low freezing point and is nontoxic; two-stage extraction has displacement solvent in the second stage
Arosolvan process, Lurgi	*N*-methyl-2-pyrrolidinone (NMP)	a polar mixing component, either water (12–20) or monoethylene glycol (40–50 wt %) must be added to the NMP to increase the selectivity and to decrease the boiling point of the solvent; the NMP–water processes use pentane countersolvent	NMP–glycol, 60; NMP–water, 35	vertical multistage mixer-settler, 24–30 stages, up to 8 m in diameter	the quantity of mixing component required depends on the aromatics content of the feed
Formex process, Snamprogetti	*N*-formylmorpholine (FM)	water is added to the FM to increase its selectivity and also to avoid high reboiler temperatures during solvent recovery by distillation	40	perforated-tray extractor, FM density at 1.15 aids phase separation	low corrosion allows use of carbon steel equipment

[a]Ref. 176.

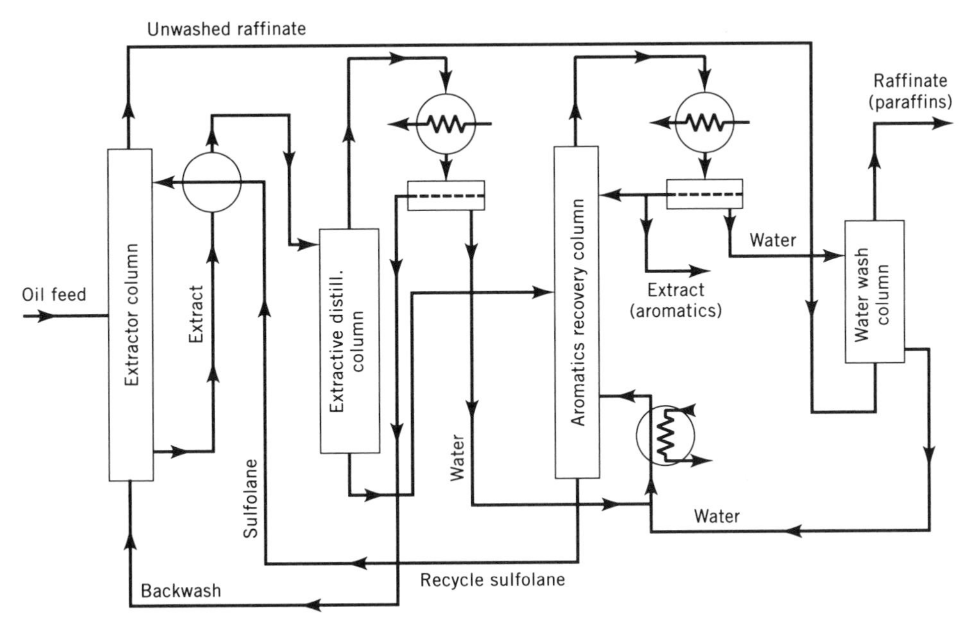

Fig. 17. Aromatic separation, sulfolane process (222–224).

Desulfurization. The sulfur compounds in petroleum oil include hydrogen sulfide, mercaptans, thiophenols, and thioethers in amounts ranging from a few tenths to several percent. Sulfur compounds have objectionable odors and adversely affect the stability of light distillate and the antiknock and oxidation characteristics of gasoline. Sulfurs are generally removed by multistage countercurrent extraction using a relatively large volume of dilute alkali solution (see SULFUR REMOVAL AND RECOVERY).

Butadiene Separation. Solvent extraction is used in the separation of butadiene (qv) [*106-99-0*] from other C-4 hydrocarbons in the manufacture of synthetic rubber. The butadiene is produced by catalytic dehydrogenation of butylene and the liquid product is then extracted using an aqueous cuprammonium acetate solution with which the butadiene reacts to form a complex. Butadiene is then recovered by stripping from the extract. Distillation is a competing process.

Caprolactam Extraction. A high degree of purification is necessary for fiber-grade caprolactam, the monomer for nylon-6 (see POLYAMIDES). Crude aqueous caprolactam is purified by solvent extractions using aromatic hydrocarbons such as toluene as the solvent (233). Many of the well-known types of column contactors have been used; a detailed description of the process is available (234).

Extraction of C-8 Aromatics. The Japan Gas Chemical Co. developed an extraction process for the separation of *p*-xylene [*106-42-3*] from its isomers using $HF{-}BF_3$ as an extraction solvent and isomerization catalyst (235). The highly reactive solvent imposes its own restrictions but this approach is claimed to be economically superior to more conventional separation processes (see XYLENES AND ETHYLBENZENE).

Anhydrous Acetic Acid. In the manufacture of acetic acid by direct oxidation of a petroleum-based feedstock, solvent extraction has been used to separate acetic acid [*64-19-7*] from the aqueous reaction liquor containing significant quantities of formic and propionic acids. Isoamyl acetate [*123-92-2*] is used as solvent to extract nearly all the acetic acid, and some water, from the aqueous feed (236). The extract is then dehydrated by azeotropic distillation using isoamyl acetate as water entrainer (see DISTILLATION, AZEOTROPIC AND EXTRACTIVE). It is claimed that the extraction step in this process affords substantial savings in plant capital investment and operating cost (see ACETIC ACID AND DERIVATIVES). A detailed description of various extraction processes is available (237).

Synthetic Fuel. Solvent extraction has many applications in synthetic fuel technology such as the extraction of the Athabasca tar sands (qv) and Irish peat using *n*-pentane [*109-66-0*] (238) and a process for treating coal (qv) using a solvent under hydrogen (qv) (239). In the latter case, coal reacts with a minimum amount of hydrogen so that the solvent extracts valuable feedstock components before the solid residue is burned. Solvent extraction is used in coal liquefaction processes (240) and synthetic fuel refining (see COAL CONVERSION PROCESSES; FUELS, SYNTHETIC).

Pharmaceutical Processes. The pharmaceutical industry is a principal user of extraction because many pharmaceutical intermediates and products are heat-sensitive and cannot be processed by methods such as distillation. A useful broad review can be found in the literature (241).

Antibiotics. Solvent extraction is an important step in the recovery of many antibiotics (qv) such as penicillin [*1406-05-9*], streptomycin [*57-92-1*], novobiocin [*303-81-1*], bacitracin [*1405-87-4*], erythromycin, and the cephalosporins. A good example is in the manufacture of penicillin (242) by a batchwise fermentation. Amyl acetate [*628-63-7*] or *n*-butyl acetate [*123-86-4*] is used as the extraction solvent for the filtered fermentation broth. The penicillin is first extracted into the solvent from the broth at pH 2.0 to 2.5 and the extract treated with a buffer solution (pH 6) to obtain a penicillin-rich solution. Then the pH is again lowered and the penicillin is re-extracted into the solvent to yield a pure concentrated solution. Because penicillin degrades rapidly at low pH, it is necessary to perform the initial extraction as rapidly as possible; for this reason centrifugal extractors are generally used.

Fractional extraction has been used in many processes for the purification and isolation of antibiotics from antibiotic complexes or isomers. A 2-propanol–chloroform mixture and an aqueous disodium phosphate buffer solution are the solvents (243). A reciprocating-plate column is employed for the extraction process (154).

Vitamins. The preparation of heat-sensitive natural and synthetic vitamins (qv) involves solvent extraction. Natural vitamins A and D are extracted from fish liver oils and vitamin E from vegetable oils (qv); liquid propane [*74-98-6*] is the solvent. In the synthetic processes for vitamins A, B, C, and E, solvent extraction is generally used either in the separation steps for intermediates or in the final purification.

Miscellaneous Pharmaceutical Processes. Solvent extraction is used for the preparation of many products that are either isolated from naturally occurring materials or purified during synthesis. Among these are sulfa drugs, methaqualone [*72-44-6*], phenobarbital [*50-06-6*], antihistamines, cortisone [*53-06-5*], estrogens and other hormones (qv), and reserpine [*50-55-5*] and alkaloids (qv). Common solvents for these applications are chloroform, isoamyl alcohol, diethyl ether, and methylene chloride. Distribution coefficient data for drug species are important for the design of solvent extraction procedures. These can be determined with a laboratory continuous extraction system (AKUFVE) (244).

Food Processing. Food processing (qv) makes use of solvent extraction in several ways. Industrial refining of fats and oils using propane is known as the Solexol process (245). Vegetable oils are refined by extraction using furfural as solvent (246). Solvent extraction is used in many protein refining processes, for example the extraction of fish protein from ground fish using *i*-propyl alcohol (247). Recovery of lactic acid by an extractive fermentation has recently been reported (248). The applications of extraction in the food industry have been reviewed (249).

Other Organic Processes. Solvent extraction has found application in the coal-tar industry for many years, as for example in the recovery of phenols from coal-tar distillates by washing with caustic soda solution. Solvent extraction of fatty and resimic acid from tall oil has been reported (250). Dissociation extraction is used to separate *m*-cresol from *p*-cresol (251) and 2,4-xylenol from 2,5-xylenol (252). Solvent extraction can play a role in the direct manufacture of chemicals from coal (253) (see FEEDSTOCKS, COAL CHEMICALS).

Treatment of Industrial Effluents. Solvent extraction appears to have great potential in the field of effluent treatment, both for the economic recovery of valuable materials and for the removal of toxic materials to comply with environmental requirements.

The Phenox process (254) removes phenol (qv) from the effluent from catalytic cracking in the petroleum industry. Extraction of phenols from ammoniacal coke-oven liquor may show a small profit. Acetic acid can be recovered by extraction from dilute waste streams (255). Oils are recovered by extraction from oily wastewater from petroleum and petrochemical operations. Solvent extraction is employed commercially for the removal of valuable by-products from wool industry effluents (256) and is applied in the same way in the pharmaceutical industry. A successful extraction process to recover *p*-nitrophenol [*100-02-7*] from a waste solution containing 8000 ppm has been developed (257). A combination of solvent extraction and wet air oxidation is used in the treatment of toxic pharmaceutical effluent prior to discharge for biological treatment. Several schemes for organic industrial wastewater treatment have been reported (258). Amphiphilic polymer solutions have high capacity for trace organics and can be used with hollow fiber membrane extractors to treat contaminated aqueous streams for environmental applications (258) (see WASTE REDUCTION; WASTES, INDUSTRIAL).

Biopolymer Extraction. Research interests involving new techniques for separation of biochemicals from fermentation broth and cell culture media have increased as biotechnology has grown. Most separation methods are limited to small-scale applications but recently solvent extraction has been studied as a potential technique for continuous and large-scale production and the use of two-phase aqueous systems has received increasing attention (259). A range of enzymes have favorable partition properties in a system based on a PEG–dextran–salt solution (97):

Enzyme	Industrial application
α-amylase	glues–food ingredients
glucoamylase	cornstarch–glucose conversion; starch–glucose conversion
α-glucosidase	maltose–glucose conversion
glucose-6-phosphate dehydrogenase	medicinal indicator
formate dehydrogenase	oxalate–formate determination
formaldehyde dehydrogenase	aldehyde–alcohol conversion
catalase	cold milk sterilization
pullunanase	starch–maltose conversion
glucose isomerase	glucose–fructose conversion
β-glucosidase	food processing
interferon	pharmaceutical applications

In many cases rapid and effective removal of contaminants and undesirable products such as nucleic acids (qv) and polysaccharides is achieved.

Difficult Separations. Difficult separations, characterized by separation factors in the range 0.95 to 1.05, are frequently expensive because these involve high operating costs. Such processes can be made economically feasible by reducing the solvent recovery load (260); this approach is effective, for example, in the separation of *m*- and *p*-cresol, linoleic and abietic components of tall oil (qv), and the production of heavy water (see DEUTERIUM AND TRITIUM, DEUTERIUM).

Inorganic Processes

The first significant application of liquid–liquid extraction in inorganic chemical technology was the separation of uranium and plutonium from nuclear reactor fission products in the late 1940s (261). A few years later, extraction was successfully applied at the front end of the nuclear fuel cycle in separating uranium from ore leach liquors as an alternative to ion exchange (qv). Since then, many other hydrometallurgical applications of liquid–liquid extraction have been developed (1,2,7,262) as well as a number of applications involving nonmetallic inorganic products (263) (see METALLURGY, EXTRACTIVE).

Most inorganic compounds are insoluble or sparingly soluble in organic solvents, whereas metal ions in aqueous solution are stable because water has a high dielectric constant and because of the solvation of ions by water molecules. Aqueous affinity must be overcome usually by an extractant which can react with the metal ion to displace the solvated water and form an uncharged species having significant solubility in the organic solvent, as illustrated in equations 10–12. The organic carrier solvent, or diluent, is usually regarded as being chemically inert although the relative aliphatic–aromatic content of the diluent can affect the extraction rates and equilibria (264). Physical properties of the carrier solvent should include a low viscosity, low flash point, and low vapor pressure to minimize evaporative losses (8). The interfacial tension between the extractant–diluent phase and the aqueous phase should preferably be high in order to provide good phase separation and minimize entrainment losses. For reasons of cost, the carrier solvent is usually a cut from petroleum distillation having flash points in the 40–80°C range.

As metal extraction into a diluent–extractant solution proceeds, there is sometimes a tendency for formation of two organic phases in equilibrium with the aqueous phase. A third phase is highly undesirable and its formation can be prevented by adding to the organic phase a few percent of a modifier which is typically a higher alcohol or tri-*n*-butyl phosphate (TBP) (7).

Nuclear Fuel Reprocessing. Spent fuel from a nuclear reactor contains ^{238}U, ^{235}U, ^{239}Pu, ^{232}Th, and many other radioactive isotopes (fission products). Reprocessing involves the treatment of the spent fuel to separate plutonium and unconsumed uranium from other isotopes so that these can be recycled or safely stored (261,264,265) (see NUCLEAR REACTORS, NUCLEAR FUEL).

The spent fuel is dissolved in nitric acid and the solution is extracted by an appropriate solvent. The Purex process (264–266) uses a 30% solution of tri-*n*-butyl phosphate (TBP) as extractant, in an aliphatic diluent such as a kerosene. The uranium and plutonium are present in the aqueous phase in the hexavalent state as $UO_2(NO_3)_2$ and $PuO_2(NO_3)_2$ and are selectively extracted (see eq. 12). Fission products remain in the aqueous raffinate. The organic extract is treated with an aqueous strip solution containing nitric acid and a cationic reducing agent which converts Pu to its trivalent state, which is preferentially stripped to the aqueous phase. Finally the uranium is stripped from the organic phase by contact with a dilute aqueous solution of nitric acid. The aqueous raffinate from the Purex process contains actinides and rare-earth fission products which are long-lived and radiotoxic. Bifunctional extractants of the carbamoylmethylphosphoryl family have been developed with the capability to remove many of these substances and thus improve the economics of safe disposal of the bulk raffinate (265,267). This objective can also be achieved by *n,n*-dialkyl amides (268). The various alternative approaches to fuel reprocessing have been critically reviewed (268).

Special safety constraints apply to equipment selection, design, and operation in nuclear reprocessing (269). Equipment should be reliable and capable of remote control and operation for long periods with minimal maintenance. Pulsed columns and remotely operated mixer–settlers are commonly used (270). The control of criticality and extensive monitoring of contamination levels must be included in the process design.

Uranium Extraction from Ore Leach Liquors. Liquid–liquid extraction is used as an alternative or as a sequel to ion exchange in the selective removal of uranium [*7440-61-1*] from ore leach liquors (7,265,271). These liquors differ from reprocessing feeds in that they are relatively dilute in uranium and only slightly radioactive, and contain sulfuric acid rather than nitric acid.

In the Amex process, the feed typically containing 5 g/L uranium and 100 g/L sulfuric acid is first filtered and treated to remove interfering anions such as molybdates and vanadates. The extractant is a commercially available formulation containing tertiary amines having C-8 to C-12 alkyl chains. A kerosene diluent is used, and the extractant is at a concentration of about 5%, plus about 2% of a higher alcohol such as decanol as a modifier. The extracted species is an amine uranyl sulfate. Mixer–settler extractors are commonly used; stripping of the uranium can be carried out under acidic, basic, or neutral conditions. The alternative

Dapex process employs alkyl phosphoric acids such as di-(2-ethylhexyl) phosphoric acid [*298-07-7*] (D2EHPA) as extractants. Although the feed pretreatment requirements are less rigorous than for the Amex process, the extractant is somewhat less selective for uranium (271).

Uranium is present in small (50–200 ppm) amounts in phosphate rock and it can be economically feasible to separate the uranium as a by-product from the crude black acid (30% phosphoric acid) obtained from the leaching of phosphate for fertilizers (qv). The development and design of processes to produce 500 t U_3O_8 per year at Freeport, Louisiana have been detailed (272).

Copper. The recovery of copper [*7440-50-8*], Cu, from ore leach liquors as a stage in the hydrometallurgical route to the pure metal is one of the largest applications of liquid–liquid extraction (7,198,209). It has been estimated (262) that in 1984 the total copper production capacity of solvent extraction plants was in excess of 700 t per day.

The most common type of copper feed entering a liquid–liquid extraction plant is produced by dilute sulfuric acid leaching and contains between 1 and 10 g/L Cu. This concentration is too low for electrowinning and the purpose of solvent extraction is to raise the concentration as well as to purify the copper solution. A typical extraction circuit for acid leach liquors is shown in Figure 18 (273). Extraction is carried out at a pH of 2 to 4 using an aliphatic kerosene diluent containing chelating extractants based on hydroxyoximes or quinolinol derivatives. These extractants effectively exchange cupric ions and hydrogen ions in the aqueous phase as in the case of acid extractants (eq. 10), so the equilibrium is pH-dependent. Stripping is effected by a strongly acid solution having zero or negative pH, as shown in Figure 18. Only a few stages are needed for extraction and stripping, so mixer–settlers rather than columns are used. Rapid and efficient phase separation in the settlers is an important element of the plant design (7,274).

Nickel and Cobalt. Often present with copper in sulfuric acid leach liquors are nickel [*7440-02-0*] and cobalt [*7440-48-4*]. Extraction using an organophosphoric acid such as D2EHPA at a moderate (3 to 4) pH can readily take out the nickel and cobalt together, leaving the copper in the aqueous phase, but the cobalt–nickel separation is more difficult (274). In the case of chloride leach liquors, separation of cobalt from nickel is inherently simpler because cobalt, unlike

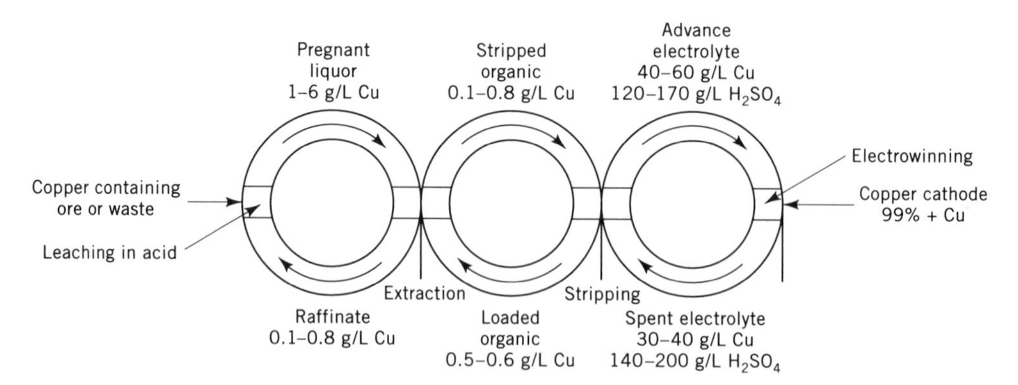

Fig. 18. Diagrammatic representation of copper extraction using solvent extraction (273).

nickel, has a strong tendency to form anionic chloro-complexes. Thus cobalt can be separated by amine extractants, provided the chloride content of the aqueous phase is carefully controlled. A successful example of this approach is the Falconbridge process developed in Norway (274).

Other Metals. Because of the large number of chemical extractants available, virtually any metal can be extracted from its aqueous solution. In many cases extraction has been developed to form part of a viable process (275). A review of more recent developments in metal extraction including those for precious metals and rare earths is also available (262). In China a complex extraction process employing a cascade of 600 mixer–settlers has been developed to treat leach liquor containing a mixture of rare earths (131).

The depressed prices of most metals in world markets in the 1980s and early 1990s have slowed the development of new metal extraction processes, although the search for improved extractants continues. There is a growing interest in the use of extraction for recovery of metals from effluent streams, for example the wastes from pickling plants and electroplating (qv) plants (276). Recovery of metals from liquid effluent has been reviewed (277), and an AM-MAR concept for metal waste recovery has recently been reported (278). Possible applications exist in this area for liquid membrane extraction (88) as well as conventional extraction. Other schemes proposed for effluent treatment are a wetted fiber extraction process (279) and the use of two-phase aqueous extraction (280).

Extraction of Nonmetallic Inorganic Compounds. Phosphoric acid is usually formed from phosphate rock by treatment with sulfuric acid, which forms sparingly soluble calcium sulfate from which the phosphoric acid is readily separated. However, in special circumstances it may be necessary to use hydrochloric acid:

$$Ca_3(PO_4)_2 + 6\ HCl \rightleftharpoons 3\ CaCl_2 + 2\ H_3PO_4 \tag{42}$$

A process developed in Israel (263) uses solvent extraction using a higher alcohol or other solvating solvent. This removes phosphoric acid and some hydrochloric acid from the system driving the equilibrium of equation 42 to the right. The same principle can be applied in other salt–acid reactions of the form

$$MX + HY \rightleftharpoons MY + HX \tag{43}$$

where M is a metal cation and X and Y are anions. An organic solvent is chosen to remove HX, again driving the equilibrium to the right. Examples of this type of reaction are (*1*) the production of potassium nitrate from potassium chloride and nitric acid and (*2*) the production of alkali metal phosphates from the alkali chloride and phosphoric acid (263).

NOMENCLATURE

a	= specific interfacial area	m^{-1} or cm^{-1}
A	= quantity or flow of component A	kg or kg/s
B	= quantity or flow of component B	kg or kg/s

c	= concentration	kg/m^3 or g/cm^3
D	= molecular diffusivity	m^2/s or cm^2/s
d	= drop diameter	m or cm
d_m	= Sauter mean drop diameter	m or cm
d_o	= orifice diameter	m or cm
E	= axial dispersion coefficient	m^2/s or cm^2/s
	Figs. 1 and 5 extract flow	kg/s or g/s
F	= feed flow	kg/s or g/s
g	= acceleration owing to gravity	m/s^2 or cm/s^2
G	= molar Gibbs free energy	J/mol
H	= height of a compartment	m or cm
HTU	= height of a transfer unit	m or cm
h	= holdup of dispersed phase	
k	= mass-transfer coefficient	m/s or cm/s
K	= overall mass-transfer coefficient	m/s or cm/s
K'	= dimensionless constant in equation 38	
m	= distribution ratio based on mass fraction	
m'	= distribution ratio based on mole fraction	
N	= flux of solute (eq. 21) number of drops	kg/(m^2·s) or g/(cm^2·s)
N_s	= number of stages	
NTU	= number of transfer units	
Q	= volume flow rate	m^3/s or cm^3/s
R	= raffinate flow	kg/s or g/s
Re	= Reynolds number	
S	= solvent flow	kg/s or g/s
Sc	= Schmidt number	
Sh	= Sherwood number	
U	= superficial velocity	m/s or cm/s
U_s	= slip velocity	m/s or cm/s
U_K	= characteristic velocity	m/s or cm/s
X	= mass ratio of C to A	
x	= mole fraction	
Y	= mass ratio of C to B	
Z	= contactor height	m or cm
z	= distance (eq. 18) stoichiometric factor	m or cm
α	= backflow ratio	
β	= selectivity or (eq. 41) exponent	
$\Delta\rho$	= density difference	kg/m^3 or g/cm^3
γ	= activity coefficient	
ϵ	= extraction factor	
σ	= interfacial tension	N/m
Φ	= fractional open area, perforated plate	
Ψ	= specific energy dissipation rate	W/kg

SUBSCRIPTS

A	= component A
B	= component B
C	= component C
CA	= component C in solvent A

CB = component C in solvent B
c = continuous phase
D = component D
DA = component D in solvent A
DB = component D in solvent B
d = dispersed phase
i = at the interface or (eq. 21) identity of drop
N = exit from stage N
O = overall
0 = feed stream
1 = exit from stage 1
2 = exit from stage 2, etc

BIBLIOGRAPHY

"Liquid–Liquid Extraction" under "Extraction" in *ECT* 1st ed., Vol. 6, pp. 122–140, by E. G. Scheibel and A. J. Frey, Hoffmann-La Roche Inc.; "Extraction, Liquid–Liquid" in *ECT* 1st ed., Suppl. 1, pp. 330–365, by Marcel J. P. Bogart, the Lummus Co.; "Liquid–Liquid Extraction" under "Extraction" in *ECT* 2nd ed., Vol. 8, pp. 719–761, by E. G. Scheibel, Cooper Union School of Engineering and Science; "Extraction, Liquid–Liquid" in *ECT* 3rd ed., Vol. 9, pp. 672–721, by T. C. Lo and M. H. I. Baird.

1. T. C. Lo, M. H. I. Baird and C. Hanson, eds., *Handbook of Solvent Extraction,* Wiley-Interscience, New York, 1983.
2. J. D. Thornton, ed., *The Science and Practice of Liquid–Liquid Extraction,* Oxford University Press, Oxford, 1992.
3. T. Sekine and Y. Hasegawa, *Solvent Extraction Chemistry; Fundamentals and Applications,* Marcel Dekker, New York, 1977.
4. S. Alegret, ed., *Developments in Solvent Extraction,* Ellis Horwood, Chichester, UK, 1988.
5. R. E. Treybal, *Liquid Extraction,* 2nd ed., McGraw-Hill, New York, 1963.
6. G. M. Ritcey and A. W. Ashbrook, *Solvent Extraction: Principles and Applications to Process Metallurgy,* Part I, Elsevier, Amsterdam, the Netherlands, 1984.
7. G. M. Ritcey and A. W. Ashbrook, *Solvent Extraction: Principles and Applications to Process Metallurgy,* Part II, Elsevier, Amsterdam, the Netherlands, 1979.
8. R. Blumberg, *Liquid–Liquid Extraction,* Academic Press, London, 1988.
9. K. Najim, *Control of Liquid–Liquid Extraction Columns,* Gordon and Breach, New York, 1988.
10. J. L. Humphrey, J. A. Rocha, and J. R. Fair, *Chem. Eng.,* 76 (Sept. 17, 1984).
11. E. Blass, G. Goldmann, K. Hirschmann, P. Mihailowitsch, and W. Pietzch, *Ger. Chem. Eng.* **9,** 222 (1986).
12. M. H. I. Baird, *Can. J. Chem. Eng.* **69,** 1287 (1991).
13. J. G. Gregory, B. Evans, and P. C. Weston, eds., *Proceedings of the International Solvent Extraction Conference 1971,* Vols. 1 and 2, Society of Chemical Industry, London, 1971.
14. G. V. Jeffreys, ed., *Proceedings of the International Solvent Extraction Conference, 1974,* Vols. 1–3, Society of Chemical Industry, London, 1974.
15. B. H. Lucas, ed., *Proceedings of the International Solvent Extraction Conference, 1977,* Vols. 1–2, Canadian Institute of Mining and Metallurgy, Montreal, 1979.
16. *Proceedings of the International Solvent Extraction Conference 1980,* Vols. 1–3, Association des Ingenieurs Sortis de l'Université de Liege, Liege, Belgium, 1980.

17. *Proceedings of the International Solvent Extraction Conference 1983,* AIChE, New York, 1983.
18. *Preprints of the International Solvent Extraction Conference 1986,* Vols. 1–3, Dechema, Frankfurt, Germany, 1986.
19. *Proceedings of the International Solvent Extraction Conference 1988,* USSR Academy of Sciences, Moscow, Russia, 1990.
20. T. Sekine, ed., *Proceedings of the International Solvent Extraction Conference 1990,* Elsevier, Amsterdam, the Netherlands, 1992.
21. D. H. Logsdail and M. J. Slater, eds., *Proceedings of the International Solvent Extraction Conference, 1993,* Vols. 1–3, Elsevier, Amsterdam, the Netherlands, 1993.
22. J. W. Gibbs, *Collected Works,* Yale University Press, New Haven, Conn., 1928.
23. N. F. Ashton, C. McDermott, and A. Brench, in Ref. 1, Chapt. 1.
24. G. J. Pierotti, C. H. Deal, and E. L. Derr, *Ind. Eng. Chem.* **51,** 95 (1959).
25. D. S. Abrams and J. M. Prausnitz, *AIChE J.* **21,** 116 (1975).
26. J. Wisniak and A. Tamir, *Liquid–Liquid Equilibrium and Extraction,* Physical Science Data Series 7, Elsevier, Amsterdam, the Netherlands, 1981; J. Wisniak and A. Tamir, *Liquid–Liquid Equilibrium and Extraction,* Physical Science Data Series 23, Elsevier, Amsterdam, the Netherlands, 1985.
27. J. M. Sorenson and W. Arlt, *Liquid–Liquid Equilibrium Data Collection,* Dechema Chemistry Data Series, Frankfurt, Germany, Part 1, 1980.
28. H. R. C. Pratt in Ref. 1, Chapt. 5.
29. A. H. Meniai and D. M. T. Newsham, *Chem. Eng. Res. Design* **70,** 78 (1992).
30. C. Hanson in Ref. 1, Chapt. 22.
31. M. M. Milnes in Ref. 11, Vol. 1, p. 983.
32. V. V. Wadekar and M. M. Sharma, *J. Separ. Proc. Tech.* **2,** 1 (1981).
33. M. Cox and D. S. Flett in Ref. 1, Chapt. 2.2.
34. M. Aguilar in Ref. 4, Chapt. 5.
35. A. Leveque and J. Helgorsky in Ref. 12, Vol. 2, p. 439.
36. P. Danesi in Ref. 4, Chapt. 12.
37. G. S. Hartley, *Phil. Mag.* **12,** 473 (1931).
38. Y. Marcus in Ref. 4, Chapt. 2.
39. R. C. Reid and T. K. Sherwood, *The Properties of Gases and Liquids,* McGraw-Hill Book Co., New York, 1958.
40. G. S. Laddha and T. E. Degaleesan, *Transport Phenomena in Liquid Extraction,* Tata-McGraw-Hill, New Delhi, India, 1976.
41. C. R. Wilke and P. C. Chang, *AIChE J.* **1,** 264 (1955).
42. H. R. C. Pratt in Ref. 1, Chapt. 3.
43. K. P. Lindland and S. G. Terjesen, *Chem. Eng. Sci.* **5,** 1 (1956).
44. G. Thorsen and S. G. Terjesen, *Chem. Eng. Sci.* **17,** 137 (1962).
45. J. T. Davies, *Turbulence Phenomena,* Academic Press, Inc., New York, 1972.
46. E. S. Perez de Ortiz in Ref. 2, Chapt. 3.
47. P. V. Danckwerts, *Gas–Liquid Reactions,* McGraw-Hill Book Co., Inc., New York, 1970.
48. M. M. Sharma in Ref. 1, Chapt. 2.1.
49. S. Sarkar, C. J. Mumford, and C. R. Philips, *Ind. Eng. Chem. Proc. Des. Dev.* **10,** 665 (1980).
50. *Ibid.,* p. 672.
51. M. A. Hughes and V. Rod, *Faraday Discuss. Chem. Soc.* **77,** paper 7 (1984).
52. J. B. Lewis, *Chem. Eng. Sci.* **3,** 248, 260 (1954).
53. T. G. Hunter and A. W. Nash, *J. Soc. Chem. Ind. (London)* **53,** 95T (1932).
54. A. Kremser, *Natl. Pet. News* **22**(21), 42 (1930).
55. M. Souders and G. G. Brown, *Ind. Eng. Chem.* **24,** 519 (1932).
56. J. Prochazka and V. Jiricny, *Chem. Eng. Sci.* **31,** 179 (1976).

57. H. R. C. Pratt and M. H. I. Baird in Ref. 1, Chapt. 6.
58. H. R. C. Pratt, *Ind. Eng. Chem. Process Des. Dev.* **14,** 74 (1975).
59. H. R. C. Pratt, *Ind. Eng. Chem. Process Des. Dev.* **15,** 544 (1976).
60. T. L. Holmes, A. E. Karr, and M. H. I. Baird, *AIChE J.* **37,** 360 (1991).
61. M. H. I. Baird and N. V. Rama Rao, *AIChE J.* **37,** 1019 (1991).
62. A. M. Rosen and V. S. Krylov, *Chem. Eng. J.* **7,** 85 (1974).
63. A. E. Karr, S. Ramanujan, T. C. Lo, and M. H. I. Baird, *AIChE J.* **65,** 373 (1987).
64. R. Clift, J. R. Grace, and M. E. Weber, *Bubbles, Drops and Particles,* Academic Press, Inc., New York, 1978.
65. G. S. Laddha and T. E. Degaleesan in Ref. 1, Chapt. 4.
66. R. Shinnar and J. M. Church, *Ind. Eng. Chem.* **52,** 253 (1960).
67. J. W. van Heuven and W. J. Beek in Ref. 13, p. 70.
68. P. M. Bapat, L. L. Tavlarides, and G. W. Smith, *Chem. Eng. Sci.* **38,** 2003 (1983).
69. J. J. C. Cruz-Pinto and W. J. Korchinski, *Chem. Eng. Sci.* **36,** 687 (1981).
70. J. F. Milot, J. Duhamet, C. Gourdon, and G. Cassamatta, *Chem. Eng. J.* **45,** 111 (1991).
71. R. Gayler, N. W. Roberts, and H. R. C. Pratt, *Trans. Inst. Chem. Eng.* **31,** 57 (1953).
72. L. Lapidus and J. C. Elgin, *AIChE J.* **3,** 63 (1957).
73. J. C. Godfrey and M. J. Slater, *Chem. Eng. Res. Design* **69,** 130 (1990).
74. V. Jiriczny and J. Prochazka, *Chem. Eng. Sci.* **35,** 2237 (1980).
75. H. Groothuis and F. J. Zuiderweg, *Chem. Eng. Sci.* **12,** 288 (1960).
76. G. V. Jeffreys and G. B. Lawson, *Trans. Inst. Chem. Engrs.* **43,** 294 (1965).
77. Z. J. Shen, M. H. I. Baird, and N. V. Rama Rao, *Can. J. Chem. Eng.* **63,** 29 (1985).
78. G. A. Davies, G. V. Jeffreys, and M. Azfal, *Br. Chem. Eng.* **17,** 709 (1972).
79. P. J. Bailes and S. K. L. Larkai, *Trans. Inst. Chem. Engrs.* **60,** 115 (1982).
80. N. N. Li, *AIChE J.* **17,** 459 (1971).
81. N. N. Li, *Ind. Eng. Chem. Proc. Des. Dev.* **10,** 215 (1971).
82. N. N. Li, R. P. Cahn, D. Naden, and R. W. M. Lai, *Hydrometallurgy* **9,** 277 (1983).
83. K. Osseo-Asare, *Sepn. Sci. Technol.* **23,** 1269 (1988).
84. J. Draxler, W. Furst, and R. Marr in Ref. 18, Vol. 1, p. 553.
85. P. Danesi in Ref. 4., Chapt. 9.
86. R. Prasad and K. K. Sirkar, *J. Mem. Sci.* **47,** 235 (1989).
87. R. Prasad and K. K. Sirkar, *J. Mem. Sci.* **50,** 153 (1990).
88. J. Draxler and R. Marr in Ref. 20, p. 37.
89. W. S. Ho, *Membrane Handbook,* Van Nostrand Reinhold, New York, 1992.
90. D. S. Flett, Ref. 21, Vol. 1, p. vi.
91. M. A. McHugh and V. Krukonis, *Supercritical Fluid Extraction. Principles and Practice,* Butterworths, Stoneham, Mass., 1986; T. Suzuki, N. Tsuge, and K. Nagahama in Ref. 20 (Area 11).
92. H. Horizoe, T. Tanimoto, I. Yamamoto, K. Ogawa, M. Maki, and Y. Kano in Ref. 20 (Area 11); B. Simandi and co-workers, in Ref. 21, Vol. 2, p. 676; Y. Shibuya, in Ref. 21, Vol. 2, p. 684.
93. P. A. Albertsson, *Partition of Cell Particles and Macromolecules,* 2nd ed., Wiley–Interscience, New York, 1971.
94. I. Kuhn, *Biotech. Bioeng.* **22,** 2393 (1980); C. Weilnhammer and E. Blass, in Ref. 21, Vol. 2, p. 1072.
95. S. B. Sawant, S. K. Sikdar, and J. B. Joshi, *Biotech. and Bioeng.* **36,** 109 (1990).
96. U.S. Pat. 2,000,606 (May 7, 1935), D. F. Othmer.
97. T. C. Lo and M. H. I. Baird, in R. A. Meyers, ed., *Encyclopedia of Physical Science and Technology,* Academic Press, Inc., San Diego, Calif., 1987.
98. T. C. Lo, in P. Schweitzer, ed., *Handbook of Separation Techniques for Chemical Engineers,* 2nd ed., Sec. 1.10, McGraw-Hill Book Co., Inc., New York, 1988.

99. R. W. Cusack and P. Fremeaux, *Chem. Eng.*, 132 (Mar. 1991).
100. J. L. Humphrey, J. A. Rocha, and J. R. Fair, *Chem. Eng.*, 76 (Sept. 17, 1984).
101. J. Marek, technical report, Luwa AG, Zurich, Switzerland, Mar. 1970.
102. H. R. C. Pratt and C. Hanson, in Ref. 1, pp. 476–477, Chapt. 16.
103. S. Ochia, *Automatica* **13,** 435 (1977).
104. J. Rydberg, H. Reinhardt, and J. O. Liljenzin, in A. Marinsky and Y. Marcus, eds., *Ion Exchange and Solvent Extraction,* Vol. 3, Marcel Dekker, Inc., New York, 1973, p. 111.
105. H. Reinhardt and J. Rydberg, *Chem. Ind. (London)* **11,** 488 (1970).
106. M. M. Anwar, C. Hanson, and M. W. T. Pratt, *Chem. Ind. (London)* **9,** 1090 (1969).
107. T. C. Lo and A. E. Karr, *Ind. Eng. Chem. Process Des. Dev.* **11**(4), 495 (1972).
108. S. D. Cavers in Ref. 1, Chapt. 10.
109. R. E. Treybal, *Mass Transfer Operations,* 2nd ed., McGraw-Hill Book Co., Inc., New York, 1968.
110. R. Akell and C. J. King, eds., *New Developments in Liquid–Liquid Extractors: Selected papers from ISEC '83,* Vol. 80, no. 238, AIChE Symposium Series, New York, 1984; W. Y. Fei and co-workers, in Ref. 21, Vol. 1, p. 49.
111. G. H. Reman, *Chem. Eng. Prog.* **62**(9), 56 (1966).
112. R. H. Perry and D. Green, *Chemical Engineers Handbook,* 6th ed., McGraw-Hill Book Co., Inc., New York, 1984, pp. 21–55.
113. L. Lowes and M. J. Larkin, *IChemE Symposium Series No. 26,* Institute of Chemical Engineers, London, 1967, p. 111.
114. B. F. Warner, *Proc. 3rd U.N. Conf. Peaceful Uses Atom. Ener.* **10,** 224 (1964).
115. J. Mizrahi, E. Barnea and D. Meyer, in Ref. 14, Vol. 1, p. 141.
116. IMI staff in Ref. 13, Vol. 2, pp. 1, 386.
117. D. W. Ager and E. R. Dement, *Proceedings of the International Symposium on Solvent Extraction in Metallurgical Processes,* Technologisch Instituut K. VIV, Antwerp, Belgium, 1972, p. 27.
118. G. C. I. Warwick, J. B. Scuffham, and J. D. Lott in Ref. 13, Vol. 2, p. 1373.
119. G. C. I. Warwick and J. B. Scuffham in Ref. 117, p. 36.
120. I. D. Jackson and co-workers, *IChemE Symposium Series No. 26,* Institute of Chemical Engineers, London, 1967, p. 111.
121. J. B. Scuffham, *Chem. Eng.*, 328, July 1981.
122. G. C. I. Warwick and J. B. Scuffham in Ref. 1, Chapt. 9.3.
123. N. V. R. Rao and M. H. I. Baird, *Can. J. Chem. Eng.* **62,** 498 (1984).
124. M. Dilley, M. T. Errington, and D. Nadden, in Ref. 21, Vol. 1, p. 140.
125. W. Mehner, G. Hochfeld, and E. Mueller in Ref. 117, Vol. 2, p. 1265.
126. Exhibition during *International Solvent Extraction Conference ISEC 1974,* Lyon, France, 1974.
127. T. K. Mattile in Ref. 14, Vol. 1, p. 169.
128. L. Niinimaki and J. R. Orjans, *Chem. Eng. Symp. Series* **78**(1), 63 (1971); *Kemira Liquid–Liquid Extraction,* Bulletin, Kemira Oy, Helsinki, Finland.
129. Z. J. Shen, Q. Y. Zhang, B. Y. Sun, and Y. F. Sun in Ref. 17, pp. 24–25.
130. Z. J. Shen and M. H. I. Baird, *Chem. Eng. Res. Design* **69,** 143 (1991).
131. Z. J. Shen, J. Li, and K. G. Song, *Proceedings of the International Conference on Separation Science and Technology,* Vol. 1, Canadian Society for Chemical Engineering, Hamilton, Canada, 1989, p. 244.
132. *Kenics Static Mixers,* Bulletin KTEK-5, Kenics Corp., Danvers, Mass., 1972.
133. *Chem. Eng. (N.Y.)* **80**(7), 111 (1973).
134. R. E. Treybal, *Chem. Eng. Progr.* **62**(9), 67 (1966).
135. S. A. Miller and C. A. Mann, *Trans. Am. Inst. Chem. Eng.* **40,** 709 (1944).
136. A. W. Flynn and R. E. Treybal, *AIChE J.* **1,** 324 (1955).

137. A. D. Ryon, F. L. Daley, and R. S. Lowry, *Chem. Eng. Prog.* **55**(10), 71 (1959).
138. B. F. Warner, joint symposium, *The Scaling-Up of Chemical Plant and Processes,* London, 1957, p. 44.
139. H. Rouyer and co-workers in Ref. 14, Vol. 3, p. 2339.
140. *Liquid–Liquid Extraction in C.E.A. Establishments,* Bulletins 25/74, Commissariat à l'Energie Atomique, Genas, France, 1974.
141. G. Sege and F. W. Woodfield, *Chem. Eng. Prog.* **50**(8), 396 (1954).
142. J. D. Thornton, *Br. Chem. Eng.* **3,** 247 (1958).
143. J. D. Thornton, *Trans. Inst. Chem. Eng.* **35,** 316 (1957).
144. D. H. Logsdail and J. D. Thornton, *Trans. Inst. Chem. Eng.* **35,** 331 (1957).
145. *The Bronswerk Technical Bulletin on Pulsed Packed Column,* Bronswerk, PC ES, Amersfoort, the Netherlands.
146. N. U. Spaay, A. J. F. Simons, and G. P. ten Brink in Ref. 24, Vol. 1, p. 281.
147. M. H. I. Baird and G. M. Ritcey, in Ref. 14, Vol. 2, p. 1571.
148. M. E. Weech and B. E. Knight, *Ind. Eng. Chem. Process Des. Dev.* **6,** 480 (1967); **7,** 157 (1968).
149. U.S. Pat. 2,493,265 (Jan. 3, 1950), E. G. Scheibel (to Hoffmann-La Roche Inc.); *Chem. Eng. Prog.* **44**(9), 681 (1948).
150. A. E. Karr and E. G. Scheibel, *Chem. Eng. Prog. Symp. Ser.* **50**(10), 73 (1954).
151. E. G. Scheibel and A. E. Karr, *Ind. Eng. Chem.* **42**(6), 1048 (1950).
152. U.S. Pat. 2,856,362 (Sept. 2, 1958), E. G. Scheibel (to Hoffmann-La Roche Inc.).
153. E. G. Scheibel, *AIChE J.* **2,** 74 (1956).
154. T. C. Lo in *Engineering Foundation Conference on Mixing Research, Rindge, N.H., 1975,* The Engineering Foundation, New York, 1975.
155. U.S. Pat. 3,389,970 (June 25, 1968), E. G. Scheibel.
156. E. G. Scheibel in Ref. 1, Chapt. 13.3.
157. *Scale-Up Procedures for a Scheibel Extraction Column,* NTIS Report No. DE3-013576, National Technical Information Service, U.S. Department of Commerce, Washington, D.C., 1983.
158. G. H. Reman, *Proceedings of the 3rd World Petroleum Congress,* the Hague, the Netherlands, 1951, Sect. III, P. 121.
159. C. P. Strand, R. Olney, and G. H. Ackerman, *AIChE J.* **8,** 252 (1962).
160. T. Misek, *Rotating Disc Extractors and Their Calculation,* State Publishing House, technical literature, Prague, Czechoslovakia, 1964.
161. W. C. G. Kosters in Ref. 1, Chapt. 13.1.
162. J. Y. Oldshue and J. H. Rushton, *Chem. Eng. Prog.* **48**(6), 297 (1952).
163. J. Y. Oldshue, *Biotech. Bioeng.* **8**(1), 3 (1966).
164. R. Bibaud and R. Treybal, *AIChE J.* **12,** 472 (1966).
165. H. F. Haug, *AIChE J.* **17,** 585 (1971).
166. J. Ingham, *Trans. Inst. Chem. Eng.* **50,** 372 (1972).
167. T. Miyauchi, H. Mitsutake, and I. Harase, *AIChE J.* **12,** 508 (1966).
168. J. Y. Oldshue, private communication, Mixing Equipment Co., Inc., Rochester, N.Y., 1970.
169. J. Y. Oldshue, F. Hodgkinson, and J. C. Pharamond in Ref. 14, Vol. 2, p. 1651.
170. J. Y. Oldshue in Ref. 1, Chapt. 13.4.
171. J. Oldshue in Y. Marcus, ed., *Solvent Extraction Reviews,* Vol. 1, Marcel Dekker, Inc., New York, 1976.
172. T. Misek and J. J. Marek, *Br. Chem. Eng.* **15,** 202 (1970).
173. J. Marek and co-workers, paper presented at the *Society of Chemical Industry Symposium,* Bradford, UK, 1967.
174. B. Seidlova and T. Misek in Ref. 14, Vol. 3, p. 2365.
175. T. Misek and J. Marek in Ref. 1, Chapt. 13.2.

176. P. J. Bailes, C. Hanson, and M. A. Hughes, *Chem. Eng. (N.Y.)* **83**(2), 86 (1976).
177. A. Mogli and U. Buhlman, in Ref. 1, Chapt. 13.5.
178. U. Buhlman, in Ref. 21, Vol. 1, p. 17.
179. UK Pats. 860,880 (Feb. 15, 1961); 972,035 (Oct. 7, 1964); 1,037,573 (July 27, 1966), J. Coleby.
180. J. Hu, Z. J. Shen, and Y. F. Su, paper presented in *International Meeting on Chemical Engineering and Biotechnology,* ACHEMA'88, Frankfurt, Germany, 1988.
181. A. R. Sheikh, C. Hanson, and J. Ingham, *Trans. Inst. Chem. Eng.* **50,** 199 (1972); Z. J. Shen, J. Hu, Y. F. Su, and M. H. I. Baird, *Proceedings of the Second International Conference on Separation Science Technology,* Vol. 1, Canadian Society for Chemical Engineering, Hamilton, Canada, 1989, p. 282.
182. M. H. I. Baird in Ref. 1, Chapt. 14.
183. U.S. Pat. 2,011,186 (Aug. 13, 1935), W. J. D. Van Dijek.
184. A. E. Karr, *AIChE J.* **5,** 446 (1959).
185. A. E. Karr and T. C. Lo in Ref. 15, Vol. 1, p. 229.
186. A. Guyer, A. Guyer, Jr., and K. Mauli, *Helv. Chim. Acta* **38,** 790, 995 (1955); N. Issac and R. L. DeWitte, *AIChE J.* **4,** 498 (1958), *Dechema Monogr.* **32,** 218 (1959); J. Prochazka and co-workers, *Br. Chem. Eng.* **16,** 42 (1971).
187. D. Elenkov and co-workers, *Khim. Inst. Sof.* **4,** 181 (1966).
188. R. Wellek and co-workers, *Ind. Eng. Chem. Process Des. Dev.* **8,** 515 (1969).
189. K. Tojo, T. Miyanami, and T. Yano, *J. Chem. Eng. Jpn.* **7,** 123 (1974).
190. USSR Pat. 175,489 (1965), S. M. Karpacheva, E. I. Zakharov, L. S. Raginski, V. M. Muratov, and A. V. Romanov.
191. I. Y. Gorodetski, A. A. Vasin, V. M. Olevski, and P. A. LuPanov, *Vibratoionnye Massoobmennye Apparaty,* Khimia, Moscow, 1980.
192. T. C. Lo, M. H. I. Baird, and N. V. R. Rao, *Chem. Eng. Comm.* **116,** 67–88 (1992).
193. M. H. I. Baird, N. V. R. Rao, J. Prochazka, and H. Sovova, in M. J. Slater and J. Godfrey, eds. *Solvent Extraction Equipment,* John Wiley & Sons, Inc., New York, 1994 (in press).
194. T. C. Lo and J. Prochazka, in Ref. 1, Chapt. 12.
195. A. E. Karr and T. C. Lo in Ref. 15, paper 8a.
196. A. E. Karr and T. C. Lo, *Chem. Eng. Prog.* **72**(11), 68 (1976).
197. M. H. I. Baird, R. G. McGinnis, and G. C. Tan in Ref. 13, Vol. 1, p. 251.
198. S. D. Kim and M. H. I. Baird, *Can. J. Chem.* **54,** 81 (1976).
199. A. E. Karr and S. Ramanujam, in Ref. 18, Vol. 3, p. 493.
200. J. Prochazka, *Dechema Monogr.* **65,** 325 (1970).
201. U.S. Pat. 3,583,856 (1971), J. Landau, J. Prochazka, and F. Souhrada.
202. I. J. Gorodetskii, A. A. V. M. Olevskii, A. E. Konstanyan, and co-workers, *Khim. Prom.* **8,** 480 (1984).
203. A. M. Rosen, I. G. Martyushin, and co-workers, *Scale Transition in Chemical Industry,* Khimia, Moscow, 1980.
204. I. J. Gorodetskii, and co-workers in Ref. 19, Vol. 2, p. 225.
205. N. S. Yang, B. H. Chen, and A. F. McMillan, *Can. J. Chem. Eng.* **64,** 387 (1986).
206. K. Takeba, preprint of *The 10th General Symposium of the Society of Chemical Engineers,* Japan, 1971, p. 124.
207. K. Miyanami, K. Tojo, and T. Yano, *J. Chem. Eng. (Japan)* **6,** 518 (1973).
208. M. M. Hafez in Ref. 1, Chapt. 15.
209. W. J. Podbielniak, *Chem. Eng. Prog.* **49**(5), 252 (1953).
210. D. B. Todd and G. R. Davis in Ref. 14, Vol. 3, p. 2379.
211. D. B. Todd, G. R. Davis, and H. A. Lange in Ref. 18, Vol. 3, p. 345.
212. F. Otillinger and E. Blass in Ref. 18, Vol. 3, p. 445; in Ref. 19, Vol. 2, p. 235.

213. E. Broadwell, paper presented at *The Society of Chemical Industry Symposium,* Bradford, UK, 1967.
214. Yu. A. Dulatav and I. I. Poniharov in Ref. 18, Vol. 2, p. 259.
215. H. Einsenlohr, *Dechema Monogr.* **19,** 222 (1951).
216. Paper presented at *The Society of Chemical Industry Symposium,* Bradford, UK, 1967.
217. C. Bernard, P. Michel, and M. Amero in Ref. 13, Vol. 2, p. 1282.
218. C. R. Howarth, J. G. M. Lee, and C. Ramshaw, in Ref. 21, Vol. 1, p. 25; C. Judson King, in Ref. 21, Vol. 1, p. 3.
219. B. M. Sankey and D. A. Gudolis in Ref. 1, Chapt. 18.3; W. S. Nogueria and M. F. Moraes, in Ref. 21, Vol. 2, p. 1081.
220. D. Read, paper presented at *American Petroleum Institute Meeting,* San Francisco, Calif., May 1952.
221. T. S. Hoover, *Hydrocarbon Process.* **12,** 69 (1969).
222. H. Voetter and W. C. G. Kosters, *Proceedings of the World Petroleum Congress,* Vol. III, 1963, p. 131.
223. F. S. Beadmore and W. C. G. Kosters, *J. Inst. Pet.* **49,** 469 (1963).
224. W. C. G. Kosters, paper presented at *The Institute for Chemical Engineering Symposium Liquid Extraction,* Newcastle-upon-Tyne, UK, 1967.
225. W. C. G. Kosters in Ref. 1, Chapt. 18.2.3.
226. E. Muller and G. Hoehfeld, *Proceedings of the 7th World Petroleum Congress,* Vol. IV, 1967, p. 13.
227. E. Muller in Ref. 1, Chapt. 18.2.1.
228. *Hydrocarbon Process. Pet. Refiner,* 185 (Sept. 1972).
229. G. S. Somekh in Ref. 1, Vol. 13, p. 323.
230. J. A. Vidueira in Ref. 1, Chapt. 18.2.2.
231. E. Cinelli, S. Noe, and G. Paret, *Hydrocarbon Process. Pet. Refiner,* 141 (Apr. 1972).
232. A. L. Benham and co-workers, *Hydrocarbon Process. Pet. Refiner* **46**(9), 134 (1967).
233. J. Coleby in C. Hanson, ed., *Recent Advances in Liquid–Liquid Extraction,* Pergamon Press, Oxford, UK, 1971.
234. A. J. F. Simon and N. F. Hassen, in Ref. 1, Chapt. 18.4.
235. T. Ueno, in Ref. 1, Chapt. 18.6.
236. E. Lloyd-Jones, *Chem. Ind. (London),* 1590 (1967).
237. C. J. King, in Ref. 1, Chapt. 18.5.
238. F. Panzner, S. R. M. Ellis, and T. R. Bott, in Ref. 15, paper 15g, p. 685.
239. G. H. Beyer, *Proceedings of the International Solvent Extraction Conference ISEC 1977,* The Canadian Institute of Mining and Metallurgy, Ottawa, Canada, 1977, p. 715.
240. J. M. Fox, *Hydrocarbon Process. Pet. Refiner,* 2 (Sept. 1963).
241. K. Ridgeway and E. E. Thorpe, in Ref. 1, Chapt. 19.
242. A. L. Edler, ed., *Chem. Eng. Prog. Symp. Ser.* **66** (1970).
243. U.S. Pat. 3,572,750 (Sept. 8, 1970), A. E. Karr (to Hoffmann-La Roche Inc.); A. E. Karr and co-workers, Hoffmann-La Roche Inc., Nutley, N.J., unpublished report, 1970.
244. S. S. Davis and co-workers, *Chem. Ind. (London),* 677 (Aug. 1976).
245. H. J. Passino, *Ind. Eng. Chem.* **41,** 280 (1949).
246. S. W. Glover, *Ind. Eng. Chem.* **40,** 228 (1948).
247. *Chem. Eng. Prog.* **67**(5), 131 (1971).
248. T. Hano, M. Matsumoto, and T. Ohtake, in Ref. 21, Vol. 2, p. 1025.
249. W. Hamm, in Ref. 1, Chapt. 20.
250. J. M. Nogueria and J. C. Pereira, in Ref. 21, Vol. 2, p. 1088.
251. M. W. T. Pratt and J. Spokes in Ref. 15, paper 31C.

252. J. Coleby, paper presented at *Symposium on Solvent Extraction,* Institute of Chemical Engineers, Newcastle-upon-Tyne, UK, 1967.
253. L. Crainger and W. S. Wise, *Chem. Br.* **4,** 12 (1968).
254. W. L. Lewis and W. L. Martin, *Hydrocarbon Process.* **46**(2), 131 (1967); Manual on *Disposal of Refinery Wastes,* American Petroleum Institute, Washington, D.C., 1969, Chapt. 10.
255. *Chem. Eng. (N.Y.),* 58 (Mar. 15, 1976).
256. P. Ramsden, paper presented at *Institute of Chemical Engineering Research Meeting on Solvent Extraction,* Bradford, UK, 1965.
257. A. E. Karr and S. Ramanujan, *International Solvent Extraction Conference 1986,* Vol. 2, Dechema, Frankfurt, Germany, 1986.
258. D. Mackay and M. Medir in Ref. 1, Chapt. 23; P. N. Hurter and co-workers, in Ref. 21, Vol. 3, p. 1663.
259. M. R. Kula in Ref. 18, Vol. 3, p. 567; T. A. Hatton, in Ref. 19, Part A., p. 23.
260. E. G. Scheibel, *Chem. Eng. Prog.* **62**(9), 66 (1966).
261. J. T. Long, *Engineering for Nuclear Fuel Reprocessing,* Gordon and Breach, Inc., New York, 1967.
262. M. Cox in Ref. 4, Chapt. 11.
263. R. Blumberg in Ref. 1, Chapt. 26.
264. G. R. Choppin and J. Rydberg, *Nuclear Chemistry,* Pergamon Press, New York, 1980.
265. P. Danesi in Ref. 4, Chapt. 12.
266. D. A. Orth in Ref. 18, Vol. 1, p. 75.
267. E. P. Horwitz, H. Diamond, D. Kalina, L. Kaplan, and G. W. Mason in Ref. 17, p. 451.
268. C. Musikas in Ref. 20, p. 297; W. L. Wilkinson, in Ref. 21, Vol. 3, p. 1455.
269. J. A. Williams and W. J. Bowers in Ref. 1, Chapt. 31.
270. A. Naylor and P. D. Wilson in Ref. 1, Chapt. 25.12.
271. P. J. D. Lloyd in Ref. 1, Chapt. 25.11.
272. P. D. Mollere in Ref. 18, Vol. 2, p. 49.
273. J. F. C. Fisher and C. W. Notebaart in Ref. 1, Chapt. 25.1.
274. G. M. Ritcey in Ref. 1, Chapt. 25.2.
275. Ref. 1, Chapt. 25.3–25.14.
276. D. S. Flett in Ref. 20, p. 1.
277. S. O. S. Anderson and H. Reinhardt, in Ref. 1, Chapt. 25.10.
278. H. Reinhardt, in Ref. 21, Vol. 3, p. 1625.
279. G. Angelov and N. Panchev, in Ref. 21, Vol. 3, p. 1649.
280. R. D. Rogers, A. H. Bond, and C. B. Bauer, in Ref. 21, Vol. 3, p. 1641.

TEH C. LO
T. C. Lo & Associates

MALCOLM H. I. BAIRD
McMaster University

LIQUID–SOLID

Liquid–solid extraction or leaching is a unit operation that predates large-scale industrial operations, with a history of known uses that goes back to Roman times. The early leaching process was known as lixiviation, a term used to describe the extraction of alkaline salts from wood (qv) and other plants. The term leaching, in use in the eighteenth century, was used to describe the process of percolating a liquid through a solid material. It is presumed that a soluble component was removed from the solid phase during the percolation so that the operation was distinguishable from sand filters which were quite widely used by that date, but which served a quite different function. Extraction is used in a wide variety of process industries and traditional terms have evolved in different fields. These include leaching, washing, percolation, digestion, steeping, lixiviation, and infusion, among others.

Many substances used in modern processing industries occur in a mixture of components dispersed through a solid material. To separate the desired solute constituent or to remove an unwanted component from the solid phase, the solid is contacted with a liquid phase in the process called liquid–solid extraction, or simply leaching. In leaching, when an undesirable component is removed from a solid with water, the process is called washing.

In the biological and food processing industries many products are extracted from their original structure by liquid–solid extraction (see BIOTECHNOLOGY; FOOD PROCESSING). Sugar (qv) is extracted from sugar beets using hot water; instant coffee (qv) is leached from ground roasted coffee using water; soluble tea (qv) is leached from tea leaves; pharmaceutical components, flavors, and essences are leached from plant roots, leaves, and stems (see FLAVORS AND SPICES; PHARMACEUTICALS); and oil is extracted from peanuts, soybeans, sunflower and cotton (qv) seeds, and halibut livers by solvents such as hexane, acetone, or ether (see SOLVENTS, INDUSTRIAL; SOYBEANS AND OTHER OILSEEDS). These are all examples of liquid–solid extraction.

Large-scale leaching also occurs in the metal processing industries, where useful metals frequently occur mixed with large quantities of unwanted matter, and leaching is used to remove the metals as soluble salts. For example, gold is leached from its ore using aqueous sodium cyanide solutions (see CYANIDES; GOLD AND GOLD COMPOUNDS); cobalt and nickel by sulfuric acid–ammonia–oxygen mixtures (see COBALT AND COBALT ALLOYS; NICKEL AND NICKEL ALLOYS); and copper salts by sulfuric acid or ammoniacal solutions (see METALLURGY, EXTRACTIVE; MINERAL RECOVERY AND PROCESSING).

Mechanisms of Extraction

If the solute is uniformly distributed through the solid phase the material near the surface dissolves first to leave a porous structure in the solid residue. In order to reach further solute the solvent has to penetrate this outer porous region; the process becomes progressively more difficult and the rate of extraction decreases. If the solute forms a large proportion of the volume of the original particle, its removal can destroy the structure of the particle which may crumble away, and further solute may be easily accessed by solvent. In such cases the extraction rate does not fall as rapidly.

In general, the following steps can occur in an overall liquid–solid extraction process: solvent transfer from the bulk of the solution to the surface of the solid; penetration or diffusion of the solvent into the pores of the solid; dissolution of the solvent into the solute; solute diffusion to the surface of the particle; and solute transfer to the bulk of the solution. The various fundamental mechanisms and processes involved in these steps make it impracticable or impossible to describe leaching by any rigorous theory.

Any one of the five basic processes may be responsible for limiting the extraction rate. The rate of transfer of solvent from the bulk solution to the solid surface and the rate into the solid are usually rapid and are not rate-limiting steps, and the dissolution is usually so rapid that it has only a small effect on the overall rate. However, knowledge of dissolution rates is sparse and the mechanism may be different in each solid (1).

The overall extraction process is sometimes subdivided into two general categories according to the main mechanisms responsible for the dissolution stage: (*1*) those operations that occur because of the solubility of the solute in or its miscibility with the solvent, eg, oilseed extraction, and (*2*) extractions where the solvent must react with a constituent of the solid material in order to produce a compound soluble in the solvent, eg, the extraction of metals from metalliferous ores. In the former case the rate of extraction is most likely to be controlled by diffusion phenomena, but in the latter the kinetics of the reaction producing the solute may play a dominant role.

Diffusion and Mass Transfer During Leaching. Rates of extraction from individual particles are difficult to assess because it is impossible to define the shapes of the pores or channels through which mass transfer (qv) has to take place. However, the nature of the diffusional process in a porous solid could be illustrated by considering the diffusion of solute through a pore. This is described mathematically by the diffusion equation, the solutions of which indicate that the concentration in the pore would be expected to decrease according to an exponential decay function.

To obtain an indication of the rate of solute transfer from the particle surface to the bulk of the liquid, the concept of a thin film providing the resistance to transfer can be used (2) and the equation for mass transfer written as:

$$\frac{dM}{dt} = \frac{D^* A(c_s - c)}{\delta} \tag{1}$$

where A = the area of the solid–liquid interface, c = the concentration of the solute in the bulk of the liquid at time t, c_s = the concentration of the saturated solution in contact with the particles, D^* = a diffusion coefficient (approximated by the liquid-phase diffusivity), M = the mass of solute transferred in time t, and δ = the effective thickness of the liquid film surrounding the particles. For a batch process where the total volume V of solution is assumed to remain constant, $dM = V\,dc$ and

$$\frac{dc}{dt} = \frac{D^* A(c_s - c)}{\delta V} \tag{2}$$

The time t taken for the concentration of the solution to rise from its initial value c_0 to a value c is obtained by integration of this equation, assuming that both A and δ remain constant, to give:

$$c = c_s - (c_s - c_0) \exp\left(\frac{-D^* At}{\delta V}\right) \tag{3}$$

This simple analysis also shows that the bulk solution approaches a saturated condition exponentially. That A and δ are constant are both significant assumptions rarely met in extraction, although the change with time may be slow. The interfacial area tends to increase as extraction proceeds, and is of course dependent on the extent to which the solid material has been ground prior to extraction. D^* should be treated as an effective mass-transfer coefficient which would be sensitive not only to the composition and properties of the solution surrounding the particle, but also to the hydrodynamic conditions in the bulk of the solution. For larger particles, which are usually present in leaching, equations are available (3,4) to predict the mass-transfer coefficient in agitated vessels.

Process Design

In most leaching operations the maintenance of constant fluid flows, pressures, and temperatures are important. These, together with the need to provide a sufficient contact time between the solvent and the solids, usually indicate a need for continuous, multistage, countercurrent processes in which fresh solvent is fed to the final stage while the solids are fed to the first stage. The objective is to be able to operate at steady conditions, and to be able to avoid extraction of undesirable material while preventing loss of solvent for both economic and safety reasons. This is usually achieved through the use of the usual control equipment, and recording instruments provide a useful means of studying plant performance. There are other factors which must be taken into account in the early stages of a design such as the particle size of the solid and the solvent employed.

Particle Size. The smaller the particle size, the greater is the relative interfacial area between the solid and the liquid and the shorter are diffusional path lengths, and therefore the higher is the rate of transfer of solute. However, smaller particle sizes tend to lead to lower drainage rates from the solid residue and can create problems in the solids flow through countercurrent extraction equipment. A compromise has to be made to select a particle size which offers an acceptable extraction rate but yet does not unduly impede flow of solvent through a percolation process or of solids through a countercurrent process. If the extractable material is a minor proportion of the starting material the disposal of the solids residue may present a problem. If the residues have some commercial value as a product after further processing, excessive grinding of the feed solids may render the residue unsuitable for the product and hence have an adverse effect on the economics of the extraction process.

Solvent. Solvent choice is determined by the chemical structure of the material to be extracted, and the rule that like dissolves like provides useful guidance. Thus vegetable oils (qv) consisting of triglycerides of fatty acids are normally extracted with hexane [*110-54-3*], whereas for free fatty acids, which are more polar than the triglycerides, more polar alcohols are used. Halogenated hydrocarbons and hexane are both widely used as solvents, and liquid carbon dioxide [*124-38-9*] (qv) appears to be suitable for extracting flavor components from plants (5) (see SUPERCRITICAL FLUIDS). Where a choice of solvent other than water exists on the grounds of comparable solubility of the solute, the following criteria are likely to be considered.

Selectivity. Solvent selectivity is intimately linked to the purity of the recovered extract, and obtaining a purer extract can reduce the number and cost of subsequent separation and purification operations. In aqueous extractions pH gives only limited control over selectivity; greater control can be exercised using organic solvents. Use of mixed solvents, for example short-chain alcohols admixed with water to give a wide range of compositions, can be beneficial in this respect (6).

Physical Properties. Low surface tension facilitates wetting of the solids in the first extraction stage, and low viscosity assists diffusion rates in the solvent phase. A low solvent density is desirable to reduce the mass of solvent held up in the solid being extracted, but solvent choice is usually dictated by other factors. A high boiling solvent with a high latent heat of evaporation requires recovery conditions that may be adverse for thermally sensitive extracts and increases the cost of solvent recovery. In chemical leaching the thermodynamics of the leaching reaction must be considered in terms of the redox potential–pH diagram (Pourbaix diagram) which can be constructed from standard free-energy data (7). This provides the basis for choosing equilibrium leaching conditions (acidic vs basic, oxidizing vs reducing), leaving the need for a kinetic study to provide data on leaching rates.

Thermal Stability. At processing temperatures in both the extraction and recovery plants the solvent should be completely stable to avoid expensive solvent losses; contamination of the solvent by any solvent breakdown products must be avoided.

Hazards. The solvent should be nontoxic and nonhazardous; adequate design must take into account flammability and explosivity characteristics of the solvent.

Cost. The cost of fresh solvent is reflected in the operating costs in the form of solvent make-up charges. Avoidance of solvent losses, and hence a reduction of operating costs, may be obtainable through better plant design which is usually associated with increased capital costs.

Temperature. Both the solubility of the material being extracted and its diffusivity usually increase with temperature, and higher extraction rates are obtained. In some cases the upper limit for the operating temperature is determined by factors such as the need to avoid undesirable side reactions.

Agitation of the Fluid. Agitation of the solvent increases local turbulence and the rate of transfer of material from the surface of the particles to the bulk of the solution. Agitation should prevent settling of the solids, to enable most effective use of the interfacial area.

Equilibrium Relationships and Mass Balances

The solid can be contacted with the solvent in a number of different ways but traditionally that part of the solvent retained by the solid is referred to as the underflow or holdup, whereas the solid-free solute-laden solvent separated from the solid after extraction is called the overflow. The holdup of bound liquor plays a vital role in the estimation of separation performance. In practice both static and dynamic holdup are measured in a process study, other parameters of importance being the relationship of holdup to drainage time and percolation rate. The results of such studies permit conclusions to be drawn about the feasibility of extraction by percolation, the holdup of different bed heights of material prepared for extraction, and the relationship between solute content of the liquor and holdup. If the percolation rate is very low (in the case of oilseeds a minimum percolation rate of 3×10^{-3} m/s is normally required), extraction by immersion may be more effective. Percolation rate measurements and the methods of utilizing the data have been reported (8,9); these indicate that the effect of solute concentration on holdup plays an important part in determining the solute concentration in the liquor leaving the extractor.

Single-Stage Leaching. A single-stage leaching process is shown in Figure 1. The solution overflow rate is V kg/h; the mass fraction of solute in the overflow solution is x_A; and the liquid in the slurry is flowing at L kg/h, and has a composition y_A. The mass flow of dry inert solids in the slurry is B kg/h.

The material balance equations are, for the total solution:

$$L_0 + V_2 = L_1 + V_1 = M \tag{4}$$

where M is the total input flow rate of solution to the unit; for the solute component A:

$$L_0 y_{A0} + V_2 x_{A2} = L_1 y_{A1} + V_1 x_{A1} = M x_{Am} \tag{5}$$

where $x_1 = y_1 = x_{Am}$; and for the solids:

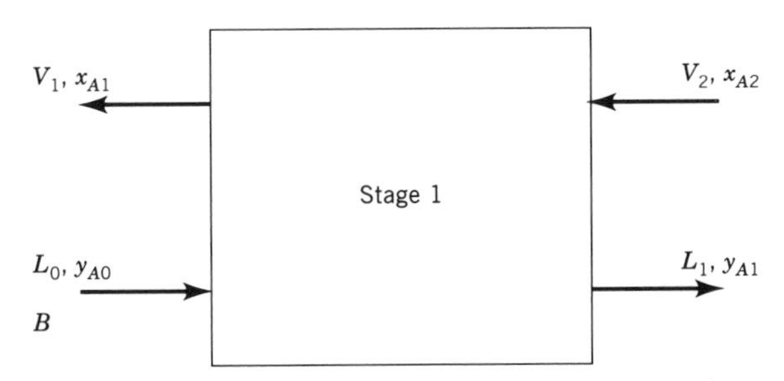

Fig. 1. Flow diagram for single-stage leaching.

$$B = L_0 N_0 = L_1 N_1 \tag{6}$$

where N_i is the mass concentration of inert solids in the ith stream, ie, kg of inert solid per kg solution. From these balances the concentration of the discharged solution can be estimated.

Countercurrent Multistage Leaching. Countercurrent extraction offers the most economical use of solvent, permitting high concentrations in the final extract and high recovery from the initial solid but utilizing the least amount of solvent. In a multistage operation fresh solid enters the first stage and fresh solvent enters the final stage; the latter is gradually enriched in solute until it leaves the extraction battery as overflow from the first stage. The operation is usually discontinuous in that the solvent is pumped from one vessel to the next intermittently and allowed to remain until equilibrium extraction is approached. When the amount of solvent removed with the insoluble solid in the underflow is constant, it is convenient to define the ratio

$$R = \frac{\textit{amount of solvent in overflow}}{\textit{amount of solvent in underflow}} = \frac{V_n}{L_n} \tag{7}$$

If perfect mixing occurs in each stage and the solute is not adsorbed preferentially at the surface of the solid, then the concentration of the solution in the underflow is the same as that in the overflow and

$$R = \frac{\textit{amount of solute in overflow}}{\textit{amount of solute in underflow}} = \frac{V_n x_{An}}{L_n y_{An}} \tag{8}$$

Referring to Figure 2, by considering solute mass balances over n, $(n - 1)$, . . . 2, 1 units in turn and eliminating intermediate solute mass fractions and flow rates, the amount of solute associated with the leached solid may be calculated in terms of the composition of the solid and solvent streams fed to the system. The resulting equation is (2)

$$L_0 y_{A0} = \frac{R^{n+1} - 1}{R - 1} L_n y_{An} - \frac{R^n - 1}{R - 1} V_{n+1} x_{An+1} \tag{9}$$

In many cases the amount of solute associated with the leached solid must not exceed a certain value, and it is possible to compute directly the minimum number of units needed by putting $n = N$.

Alternative approaches are to be found in the literature. Derivations of the above equations are given in numerous texts (2,10–12), which also describe graphical or analytical solutions to the problem. Many of these have direct analogues

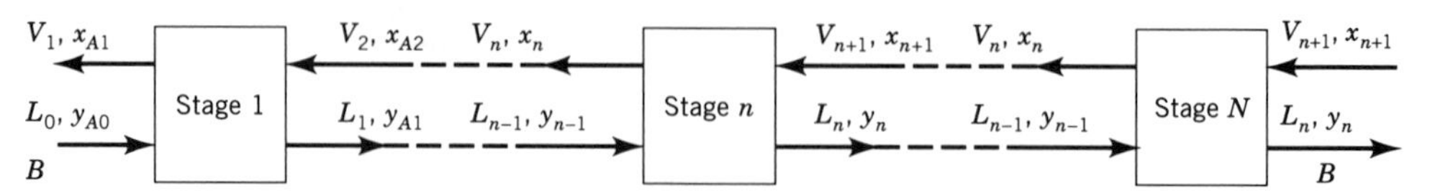

Fig. 2. Flow diagram for countercurrent multistage leaching.

in other separation processes such as distillation (qv) and liquid–liquid extraction, and use plots such as the McCabe-Thiele diagram or Ponchon-Savarit diagram.

Countercurrent Leaching With Variable Underflow. In practice most cases of interest exhibit a variable underflow rate, which is normally greatest at that point in the process where the solute concentration in the solvent is highest. In some leaching operations the viscosity and density of the solution changes appreciably as the solute concentration. Consequently, the operating line derived from mass balance equations for the McCabe-Thiele diagram has a slope which varies from stage to stage, and equation 9 is no longer valid. In the lower numbered stages the solute concentrations are higher and the underflows may retain more solution than the underflows from the higher numbered stages. Clearly, if the underflow rate L_n varies then so does the overflow rate. As in the previous case, each unit is assumed to be well mixed and the solute mass fractions in the overflow and underflow are related by:

$$x_{An} = y_{An} \tag{10}$$

The solute mass fraction in the overflow solution from the first unit ($n = 1$) is

$$x_{A1} = \frac{V_{n+1}x_{An+1} + L_0 y_{A0} - L_n y_{An}}{V_{n+1} + L_0 - L_n} \tag{11}$$

and the solute mass fraction in the solution fed to unit n is

$$x_{An+1} = \frac{V_{n+1}x_{An+1} - L_n y_{An} + L_n y_{An}}{V_{n+1} - L_n + L_n} \tag{12}$$

It is often the case that all quantities in equation 11 except x_{A1} are known. If, instead of V_{n+1}, the concentration of the solution leaving the system is specified, then equation 11 can be used to calculate V_{n+1}.

Extractors

Calculations serve as a guide to the analysis of an extraction plant and, as for the analysis of equipment performance in any other sphere of process engineering, these may be supplemented by empirical correlations or process models pertinent to the particular equipment under consideration. Another process which usually deserves special attention is that needed for solvent regeneration and for solute recovery. Solvent recovery (qv) is often energy intensive and a full process energy analysis is recommended to reduce costs. Recovery of organic solvents from the exhausted solids is also important and can be more troublesome than recovery from a liquid, and consideration should be given to the use of superheated solvent vapor for this purpose (13) (see PROCESS ENERGY CONSERVATION).

Extractors rely on either percolation or agitation to ensure intimate contact between the solids and solvent. Percolation and extraction rate data provide guidance on whether extraction should be by percolation or immersion and what extraction time is needed to give an acceptable approach to equilibrium. For a percolation system to be viable the extraction rate needs to be high, as the solvent residence time is often relatively short, although where percolation rates are so high (residence times very short) that extraction becomes inefficient, upward flow through the bed of solids can sometimes be advantageous (14). A percolation process can be carried out either stagewise or in a differential contactor; for an immersion process stagewise contact is often more practicable, particularly where a low extraction rate requires an extended residence time or multiple contact with the solvent. Under these circumstances, or when extraction is accompanied by chemical reaction, a countercurrent multistage operation is often beneficial. When percolation and extraction rates are being measured for equipment specification and sizing purposes, the conditions in the test extractor should match as closely as possible those anticipated in the full-scale extractor in order to provide the most reliable data.

Extractors often contribute substantially to the capital and operating costs of a plant, which provides the impetus to seek ways to reduce the extraction load in order to increase extractor capacity and reduce specific solvent requirements. When the feed material is of plant origin and the solute is contained in cells that can be ruptured by heat or pressure, pre-treatment frequently involves removing part of the solute by pressing. The variety of extractors used in liquid–solid extraction is diverse, ranging from batchwise dump or heap leaching for the extraction of low grade ores to continuous countercurrent extractors to extract materials such as oilseeds and sugar beets where problems of solids transport have dominated equipment development.

Batch Extractors. Coarse solids are leached by percolation in fixed or moving-bed equipment. Both open and closed tanks (qv) having false bottoms are used, into which the solids are dumped to a uniform depth and then treated with the solvent by percolation, immersion, or intermittent drainage methods.

The pot extractor is a batch extraction plant in which extraction and solvent recovery from the exhausted solids can be carried out in a single vessel. These extractors are normally agitated vessels having capacities in the range of 2 to 10 m^3, beyond which the battery system becomes a preferred technical alternative.

The diffuser is a closed percolation vessel which is used when the pressure drop is too high for gravity flow of solvent, when evaporative losses of solvent would be too high, or when it is necessary to use elevated temperatures. The solvent is circulated through the tank by pumping, or leaching may be achieved without solvent circulation. A diffuser battery is a semibatch extraction system operating on a cyclical basis. The system comprises a battery of diffusers or vessels, each of which is charged with the solids to be extracted. Fresh solvent is fed to the first one or two vessels, sometimes with the solvent being heated before being fed to the unit. The underflow from one unit is fed to the next, again with the option of interunit heating available. The underflow from the final unit is a solute-rich solution. The actual number of diffusers in the battery depends on extraction and equilibrium conditions, but an additional diffuser above the estimated number is required to permit cyclical operation. When a battery consists of more than four units a close approximation to countercurrent flow is achieved, and owing to the cyclic nature of the operation each unit changes its position in the extraction sequence at each cycle changeover. In the cyclic operation, the most exhausted diffuser is bypassed and emptied, and an empty one is charged with fresh solids. This rather cumbersome plant layout is used, for example, for the extraction of coffee solubles using hot (150–180°C) water, but can be largely replaced by fully continuous devices.

Continuous Extractors. Continuous extractors are available in a variety of forms. The main difference between them is the way by which the solids are transported through the equipment. For convenience the method of solids transport is used as a means of equipment classification.

Moving-bed percolation systems are used for extraction from many types of cellular particles such as seeds, beans, and peanuts (see NUTS). In most of these cases organic solvents are used to extract the oils from the particles. Pre-treatment of the seed or nut is usually necessary to increase the number of cells exposed to the solvent by increasing the specific surface by flaking or rolling. The oil-rich solvent (or miscella) solution often contains a small proportion of fine particles which must be removed, as well as the oil separated from the solvent after leaching.

The Bollman extractor (15) (Fig. 3) is a moving-bed, perforated-basket type of extractor. The solids are loaded into baskets fixed to a chain conveyor in a closed vessel. Solid is fed to the top basket on the downward side of the conveyor and is discharged from the top basket on the upward side. Fresh solvent is sprayed on the solid about to be discharged, leaving some time for drainage from the basket before discharge is effected, and passes downward through the baskets to effect a countercurrent flow. The partially rich solvent (half-miscella) from the bottom of the upward side is pumped to the basket at the top of the downward side, from which solvent flows from basket to basket in cocurrent fashion. The final solvent solution, miscella, is collected from the bottom of the downward side. Control of flake size during pre-treatment is desirable, as is control of the thickness and bulk density of the bed. A typical extractor moves at about 0.3 m/s, each basket contains some 350 kg of seeds, about equal masses of seeds and solvent are used, and the miscella contains about 25% oil by mass (2). Advantages of this design of extractor are that a solids-free miscella can be obtained, the residue is well drained when the equipment is properly controlled, and large quantities of solids can be extracted continuously.

The Rotocel extractor (16) achieves countercurrent extraction through a sequence of discrete liquid–solid contacts. The solids to be extracted are fed continuously as a dry material or as a slurry to sector-shaped cells arranged around a horizontal rotor. Each cell has a perforated base which allows easy drainage of solvent into a basin at the base of the cell from which the solvent is pumped into the next cell in the countercurrent direction. Fresh solvent is supplied to the last cell, which also occupies a larger sector than the other cells to allow for drainage of the extracted solids prior to discharge. The miscella is filtered by the bed of solids in each cell, and miscella from such rotary-type extractors can be expected to contain less than 5 ppm suspended solids, sometimes effecting a saving on the cost of subsequent solid–liquid separation equipment.

Tipping pan and horizontal filters are also used for leaching: the modus operandi of the Rotocel extractor resembles that of a tipping pan filter, although the details of its design differ slightly.

An alternative tower design, the Bonotto extractor (15) (Fig. 4), is a series of slowly rotating horizontal trays equispaced vertically in a tall cylindrical vessel.

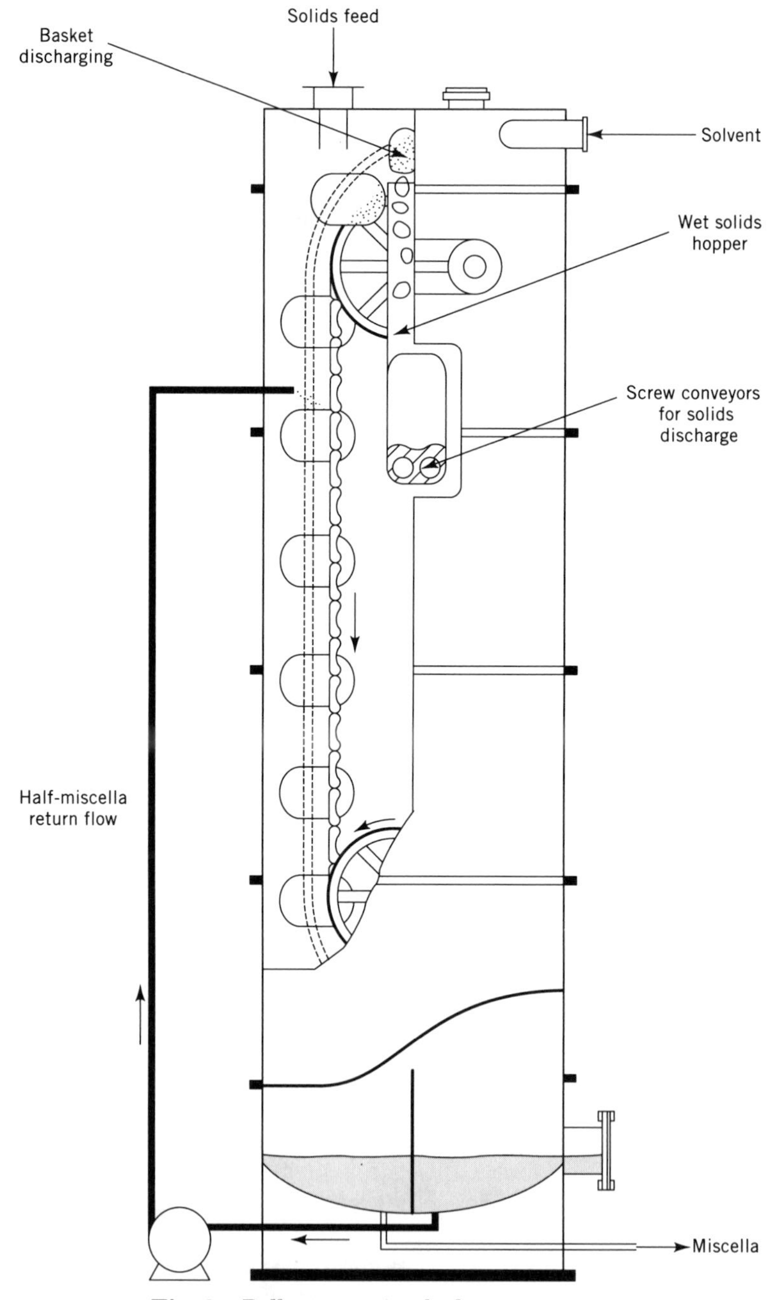

Fig. 3. Bollman moving-bed-type extractor.

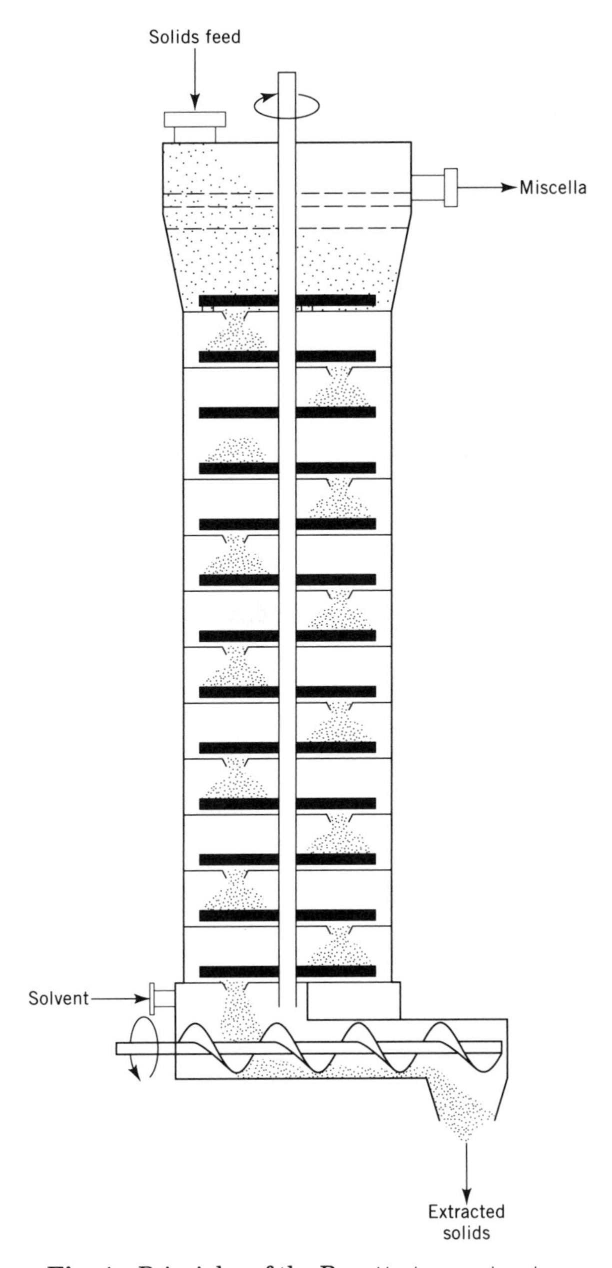

Fig. 4. Principles of the Bonotto-type extractor.

The solid is fed continuously close to the outside edge on the top tray and a stationary scraper attached to the vessel causes the solid to cross the tray. The solid then falls through an opening onto the tray beneath, where another scraper moves the solid across the tray in the opposite direction toward a similar opening near the periphery of this tray. This sequence of moving the solid across each plate in opposite directions on alternate plates is continued until the solid reaches the bottom of the tower. It is then transported from the tower by a screw conveyor, although alternative types of solids conveyor could be used. The solvent is fed to the bottom of the vessel and flows upward to give a flow countercurrent to the solids flow direction. Clearly the upward velocity of the solvent should be lower than the fall velocity of the solids to prevent entrainment of the solids, and the density of the solvent may change markedly up the column as the concentration of solute increases.

Endless belt percolation extractors (Fig. 5) such as the uncompartmented de Smet belt extractor and the compartmented Lurgi frame belt extractor are similar in principle and closely resemble a belt filter, and are probably the simplest type of percolation extractor from a mechanical point of view. These are fitted with a slow-moving perforated belt. The belt is made from steel mesh cloths when the solids are fine, or coarser screens when the solids are larger, and is attached to chains which pass over sprockets at each end of the extractor. The solid is fed from a hopper at one end of the extractor to the moving belt, and the bed height is controlled by an adjustable damper at the outlet of the feed hopper. The two side walls of the extractor provide support for the bed on the moving belt. Fresh solvent is fed by spraying it onto the bed close to the discharge end of the belt, but leaving sufficient distance for adequate drainage of the bed prior to discharge. Miscella draining from the bed is collected in a pan below the belt and circulated back to be sprayed onto the bed at a point closer to the solids-feed end of the belt; this process is repeated to achieve extraction operating with a countercurrent flow. The top of the bed is scraped by a hinged rake which has two functions: (*1*) it prevents a layer of fine solids from accumulating at the top of the bed thereby reducing permeability, and (*2*) it form a solids pile which helps to prevent intermingling of miscella from different feed points at the surface of the bed. The belt is effectively washed twice: once by fresh solvent just after the solids discharge point, and then at the other end of the belt return by miscella. The extraction time and percolation rate determine the belt speed and the amount of drainage area, and hence linear length of belt, required. These parameters control the plant capacity as the bed height is fixed by the mechanical design of the extractor.

Immersion extraction systems are useful in handling finely ground material or when the percolation rate through the material to be extracted is too rapid to

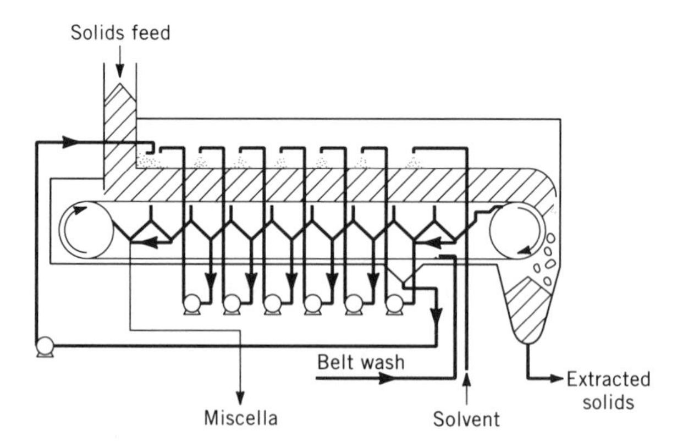

Fig. 5. Principles of the belt-type extractor.

allow effective diffusion from the solids. These systems are applied extensively in the sugar industry, in extraction from oilseeds having a high oil content, and from plant materials. The stepwise extraction by immersion, or continuous countercurrent leaching systems, can be carried out by a number of techniques analogous to the mixer–settler widely used in liquid–liquid extraction. The method is only viable if the solids settle more or less completely so that a clear supernatant liquor remains for decantation and when the slurry formed is pumpable, but it is nonetheless a most important method of leaching. If this is not so, other solid–liquid separation stages, such as filtration (qv) or centrifugation (see SEPARATION, CENTRIFUGAL), have to be considered. However, such mixer–settler methods are continuous only by virtue of repeating a sequence of similar stages to achieve a given degree of extraction. More fully continuous methods of extraction were designed as tower systems and later as screw conveyor systems as effective methods of solids transport became reliable.

Continuous countercurrent decantation systems are not uncommon, employing a cascade of thickeners to wash solute from fine particles or to wash the solids formed by chemical reactions. The capacity of each thickener in the cascade is designed so that the residence time of the particles is long enough to allow the reaction to go to completion or to allow an acceptable degree of leaching to be achieved. Interthickener filtration permits use of much smaller volumes of solvent, sometimes also allowing greater removal of solute. The drawback of interstage filtering is that filters tend to be more expensive than equivalent thickeners, but this also has to be set against the smaller amount of space required for the whole installation.

The BMA diffusion tower (17), in common with some other tower systems, employs a central shaft fitted with a series of inclined plates or wings that direct movement of the solids. The tower shell is fitted with staggered guide plates for the same purpose. The solids are fed to the bottom of the tower and transported upward. Such units are found most widely in sugar beet refining. Tower heights of 10 to 15 m are used, with the diameter being dependent on the solids throughput capacity. For a capacity of 3000 metric tons of beet per day (17), a tower diameter of 5.5 m is required and the power consumption is of the order of 40 kW. In general, immersion extractors take up less space than the percolation types and have a lower power consumption.

Immersion-type extractors have been made continuous through the inclusion of screw conveyors to transport the solids. The Hildebrandt immersion extractor (18) employs a sequence of separate screw conveyors to move solids through three parts of a U-shaped extraction vessel. The helix surface is perforated so that solvent can pass through the unit in the direction countercurrent to the flow of solids. The screw conveyors rotate at different speeds so that the solids are compacted as they travel toward the discharge end of the unit. Alternative designs using fewer screws are also available.

The De Danske Sukkerfabriker (DDS) diffuser extractor (Fig. 6) is a relatively simple version of this family of machines, employing a double screw rotating in a vessel mounted at about 10° to the horizontal. The double screw is used to transport the solids up the gradient of the shell, while solvent flows down the gradient. Equipment using a single screw in a horizontal shell for countercurrent extraction of solids under pressure has been described (19).

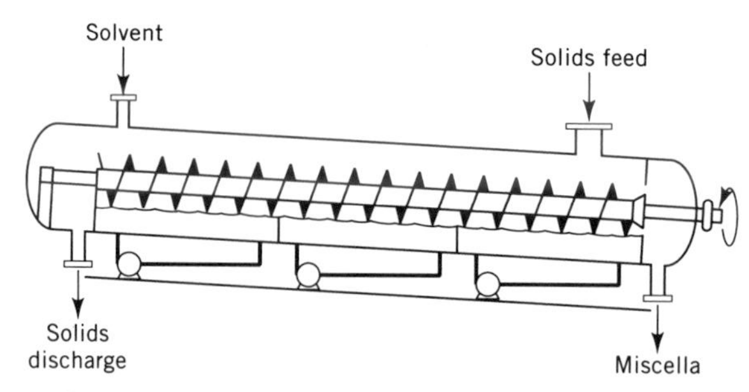

Fig. 6. Principles of the De Danske Sukkerfabriker (DDS) diffuser extractor.

Safety and Environmental Considerations

Solvent flammability, the solvent, and dust loading in the atmosphere of the working environment and of the products in the case of edible materials are the main factors that constitute health and safety hazards in extraction plants (20). General safety and environmental standards must therefore be applied (see PLANT SAFETY) and due recognition taken of the most recently published national regulations relating to acceptable threshold limit values (TLVs) for solvents and dusts. The permissible levels of solvent residues and emissions have repeatedly been lowered in recent years, making it important to ensure that the most up-to-date regulations are available to and acted upon by plant designers and engineers.

Disposal of exhausted solids can be easily overlooked at the plant design stage, particularly when these have no intrinsic value; alternative disposal methods might include landfill of inert material or incineration, hydrolysis, or pyrolysis of organic materials. Liquid, solid, and gaseous emissions are all subject to the usual environmental considerations.

BIBLIOGRAPHY

"Liquid–Solid Extraction" under "Extraction" in *ECT* 1st ed., Vol. 6, pp. 91–122, by F. Lerman, The Vulcan Copper & Supply Co.; in *ECT* 2nd ed., Vol. 8, pp. 761–775, by E. G. Scheibel, The Cooper Union for the Advancement of Science and Art; "Extraction, Liquid–Solid" in *ECT* 3rd ed., Vol. 9, pp. 721–739, by W. Hamm, Unilever Ltd.

1. G. Karnofsky, *J. Am. Oil Chemists Soc.* **26**, 564 (1949).
2. J. M. Coulson, J. F. Richardson, J. R. Backhurst, and J. H. Harker, *Chemical Engineering*, Vol. 2, 4th ed., Pergamon Press, Oxford, UK, 1991.
3. A. W. Hixson and S. J. Baum, *Ind. Eng. Chem.* **33**, 478 (1941).
4. N. Blakeborough, *Biochemical and Biological Engineering Science*, Vol. 1, Academic Press, Inc., New York, 1968.
5. D. R. J. Laws, *J. Inst. Brew. London* **83**(1), 39 (1977).
6. J. H. Hildebrand, J. M. Prausnitz, and R. L. Scott, *Regular and Related Solutions*, Van Nostrand, New York, 1970.
7. A. R. Burkin, *The Chemistry of Hydrometallurgical Processes*, E. & F. N. Spon, London, 1966.
8. J. D. Keane and C. T. Smith, *J. Am. Oil Chem. Soc.* **35**, 199 (1958).

9. H. Tomschke, M. Meiners, and E. Frohnert, *Tech. Mitt. Krupp Werksber.* **35**(1), 9 (1977).
10. W. L. McCabe and J. C. Smith, *Unit Operations in Chemical Engineering*, 3rd ed., McGraw-Hill Book Co., Inc., London, 1976.
11. R. H. Perry and D. Green, eds., *Perry's Chemical Engineers Handbook*, 50th ed., McGraw-Hill Book Co., Inc., 1984.
12. C. J. Geankoplis, *Transport Processes and Unit Operations*, Allyn and Bacon, Boston, 1978.
13. K. Weber, *Fette Seifen Anstrichmittel* **76**, 495 (1974).
14. *Food Process.* **36**(3), 71 (1975).
15. W. H. Goss, *J. Am. Oil Chem. Soc.* **23**, 348 (1946).
16. K. W. Becker, *AIChE. Symposium Ser.* **64**(86), 60 (1968).
17. F. Schneider, ed., *Technologie des Zuckers*, 2nd ed., M. & M. Schaper, Hannover, Germany, 1968.
18. R. N. Rickles, *Chem. Eng.*, 157 (Mar. 15, 1965).
19. F. A. Cantazini, *Braz. Pedido*, PI BR 80 03788 (Nov. 1981).
20. K. N. Palmer, *Dust Explosions and Fires*, Chapman and Hall, London, 1973.

RICHARD J. WAKEMAN
University of Exeter

EXTRACTIVE METALLURGY. See METALLURGY, EXTRACTIVE.

EXTRATERRESTRIAL MATERIALS

Extraterrestrial materials are samples from other bodies in the solar system that can be studied in Earth-bound laboratories. Sensitive and ever-improving analytical techniques are used to provide information at levels of detail and sophistication that cannot be matched by telescopic or spacecraft investigations (see ANALYTICAL METHODS). Much of the knowledge of early solar system bodies, processes, environments, and chronology has come from the study of these samples. Extraterrestrial materials that are available for laboratory study include meteoritic materials that fall naturally to the Earth, some meteoritic material that has been captured in space, and lunar samples that were recovered by the Apollo and Luna sample-return missions flown to the Moon during the years 1969 to 1972 (1). Missions to return samples from Mars, asteroids, and comets have been studied but have never been successfully implemented. The meteoritic materials in existing collections include samples from asteroids, comets, the Moon, and probably Mars. The comet and asteroid samples are the best preserved solids from the early solar system and are the oldest and most cosmochemically primitive samples available for direct study. Because of the primitive and unfractionated nature, these samples provide the best determination of the composition of the Sun and the solar system as a whole. It has been shown that many meteorites contain preserved interstellar grains, particles older than the Sun that formed around other stars and served as the initial building blocks of the solar system.

Meteorites

Meteorites by definition are extraterrestrial materials that fall from the sky and actually hit the surface of the Earth. In space they are considered to be meteoroids and during their luminous entry into the atmosphere they are called meteors. Meteorite strictly applies only after impacting the Earth. Conventional meteorites are rocks ranging in size from a centimeter to a few meters. The largest known meteorite is the 70-ton Hoba that resides at its discovery site in South Africa. Larger meteoroids do not sufficiently decelerate from cosmic velocity in the atmosphere and are destroyed upon impact, forming an explosion crater. Meteorites fall randomly to Earth but are not found randomly distributed on the Earth's surface. The highest general concentrations of meteorites occur in Antarctica where long exposure time and the combined effects of ice movement and sublimation concentrate meteorites on top of blue ice fields. Because of the scarcity of country rocks, it has been possible to collect over 10,000 meteorites from Antarctica since the early 1970s. In Antarctica and elsewhere, meteorites are often found in clusters created by the breakup of a larger body during hypervelocity entry into the atmosphere. When a meteor breaks up at high altitude, the resulting fragments impact over an elliptical region several kilometers across the ground, forming a strewn field where sometimes thousands of individual specimens are found. Because of atmospheric breakup, the number of individual meteorite specimens that are collected is much larger than the actual number of meteoroids that produced them. Meteoroids are themselves fragments of bodies that broke up in space, and the actual lineage of meteoritic samples may trace back to a relatively limited number of initial parent bodies.

All meteorites enter the atmosphere at velocities in excess of 11.2 km/s, the velocity of escape from the Earth. For the kilogram bodies that produce most meteorites, the high initial kinetic energy of $>10^8$ J/kg (2×10^7 cal/kg) is lost in only a few seconds time scale by collision with air molecules in the 100–30 km altitude range (2). During the period of deceleration to a terminal velocity on the order of 100 m/s, a meteoroid undergoes fragmentation resulting from mechanical and thermal stress. It also loses a significant fraction of its original mass by ablation. The ram pressure of atmospheric air is high and only strong materials survive without fragmentation into dust. The resulting selection process prevents weak meteoroid types from becoming meteorites. This process prevents the fragile matter observed in the annual cometary meteor showers from producing conventional meteorites. In many cases, the thermal effects of atmospheric entry results in ablation of more than half of the initial meteoroid mass. Because of the short duration of atmospheric heating, the heat pulse penetrates only a short distance into the surviving meteorite mass. All freshly fallen stone meteorites have a glassy fusion crust a few 100 μm thick. In general, discernible thermal effects rarely penetrate more than a few millimeters.

Types. Most meteorites can be classified into definite groups distinguished by elemental, mineralogical, petrographic, and isotopic composition (3). The general groups are the chondrites, achondrites, irons, and stony irons (Table 1). Although fragments of one meteorite class are often found inside another as a result of collisional mixing, in general the bulk properties of meteorites fall into quantified groups without a continuum of compositions between established groups. It is likely that some groups are samples of single asteroids, the apparent source of most meteorites. The majority of asteroids are located in the asteroid belt between Jupiter and Mars, and are believed to be relic solar nebula planetismals that escaped incorporation into planets. It is not known if asteroid compositions are quantized or if there is actually a continuum of compositions and the existing meteorites are just an incomplete sampling of a broader population. Comparison of spectral reflectance data from laboratory meteorites with that from asteroids indicates that meteorites are not a representative sampling of the asteroid belt (4).

Chondrites. Over 90% of meteorites that are observed to fall out of the sky are classified as chondrites, samples that are distinguished from terrestrial rocks in many ways (3). One of the most fundamental is age. Like most meteorites, chondrites have formation ages close to 4.55 Gyr. Elemental composition is also a property that distinguishes chondrites from all other terrestrial and extraterrestrial samples. Chondrites basically have undifferentiated elemental composi-

Table 1. Meteorite Classification

Meteorite type	Class	Frequency of occurrence,[a] %
chondrites	CI	0.7
	CM	2.0
	CV	1.1
	CO	0.8
	CR	<0.5
	H	32
	L	39
	LL	7.2
	E	1.5
achondrites	aubrites	1.1
	urelites	0.4
	howardites	1.1
	diogenites	2.4
	eucrites	2.7
	lunar	<0.5
irons	all types	4.7
stony irons	pallasites	0.3
	mesosiderites	0.8

[a]Percentage of meteorites seen to fall.

tions for most nonvolatile elements and match solar abundances except for moderately volatile elements. The most compositionally primitive chondrites are members of the type 1 carbonaceous (CI) class. The analyses of the small number of existing samples of this rare class most closely match estimates of solar compositions (5) and in fact are primary source solar or cosmic abundances data for the elements that cannot be accurately determined by analysis of lines in the solar spectrum (Table 2).

Another unique property of chondrites is the presence of chondrules, objects found in nearly all chondrites except those of the CI class. Chondrules (Fig. 1) are millimeter-sized, spheroidal bodies composed predominantly of olivine [*1317-71-1*], $(Mg,Fe)_2SiO_4$, pyroxene, and a glass of approximate feldspathic composition. The textures of chondrules range from very fine-grained cryptocrystalline to well-formed crystals contained in glass. Chondrule textures of silicate minerals and interstitial glass indicate that these were once molten spherical bodies that cooled rapidly (time scales from minutes to hours) (7). The shape resulted from surface tension acting on small molten droplets freely suspended in the solar nebula. Rapid cooling times indicate local transient heating events in the solar nebula, but details of the origin of chondrules, the source of heating, and the nature of their precursors are very poorly known. It is believed that these were objects individually orbiting the sun formed by rapid heating and cooling of millimeter-sized precursors (7). The processes must have been highly efficient, as chondrules comprise $> 75\%$ of the mass of many meteorites. It is possible that chondrules were the primary building blocks of the Earth and the terrestrial planets. The material between chondrules is a fine-grained matrix that has a complementary relationship with chondrules in the sense that it contains an excess abundance of elements such as volatiles and iron in which chondrules are depleted. Although the ratio of chondrules to matrix varies among different chondrites the combined bulk composition is approximately constant. In some chondrites, the chondrules have well-defined rims of fine-grained matrix, suggesting that at least some matrix material directly accreted onto chondrules before the chondrules accreted to form the meteorite parent bodies (8).

Chondrites are divided into eight subclasses distinguished by elemental, isotopic, and mineralogical composition. Seventy-eight percent of falls, meteorites actually seen to fall (eliminating discovery bias), are the H, L, and LL groups that together are termed the ordinary chondrites. The second most abundant group is the carbonaceous chondrites, where the prefix begins with C. The second letter in each carbonaceous chondrite group designates the name of the town nearest to the fall location of the prototypical specimen. For example, the CI and CM classes are named after the CI chondrite Ivuna and CM chondrite Murray, respectively. The rarest of the C chondrites are the CIs. Many of the distinguishing properties of the chondrite classes are the result of fractionation processes that occurred in the solar nebula during or before the time when the meteorites accreted from small grains.

A characteristic distinguishing different chondrite groups is the abundance and oxidation state of iron, Fe. As shown in Figure 2, the Fe:Si ratio of these groups varies by a factor of up to 2, and the oxidation state of Fe varies from totally reduced in the case of E chondrites, to the totally oxidized CI chondrites. The relative abundances of many of the siderophile (iron-loving) elements such

Table 2. Solar System Abundances of the Elements[a]

Element	Solar system[b]	Mean CI chondrite, ppb	Orgueil, ppb	Element	Solar system	Mean CI chondrite, ppb	Orgueil, ppb
H	2.79×10^{10}		2.02	Ru	1.86	712	714
He	2.72×10^{9}		56×10^{3}	Rh	0.344	134	134
Li	57.1	1.50×10^{3}	1.49×10^{3}	Pd	1.39	560	556
Be	0.73	24.9	24.9	Ag	0.486	199	197
B	21.2	870	870	Cd	1.61	686	680
C	1.01×10^{7}		3.45[c]	In	0.184	80	77.8
N	3.13×10^{6}		3.18×10^{6}	Sn	3.82	1720	1680
O	2.38×10^{7}		46.4[c]	Sb	0.309	142	133
F	843	60.7×10^{3}	58.2×10^{3}	Te	4.81	2320	2270
Ne	3.44×10^{6}		203[d]	I	0.90	433	433
Na	5.74×10^{4}	5×10^{6}	4.9×10^{6}	Xe	4.7		8.6
Mg	1.074×10^{6}	9.89[c]	9.53[c]	Cs	0.372	187	186
Al	8.49×10^{4}	8.680×10^{6}	8.69×10^{6}	Ba	4.49	2340	2340
Si	1.00×10^{6}	10.64[c]	10.67[c]	La	0.4460	234.7	236
P	1.04×10^{4}	1.220×10^{6}	1.18×10^{6}	Ce	1.136	603.2	619
S	5.15×10^{5}	6.25[c]	5.25[c]	Pr	0.1669	89.1	90
Cl	5240	70.4×10^{4}	6.98×10^{5}	Nd	0.8279	452.4	463
Ar	1.01×10^{5}		751[d]	Sm	0.2582	147.1	144
K	3770	5.58×10^{5}	5.66×10^{5}	Eu	0.0973	56.0	54.7
Ca	6.11×10^{4}	9.28×10^{6}	90.2×10^{4}	Gd	0.3300	196.6	199
Sc	34.2	5.82×10^{3}	5.83×10^{3}	Tb	0.0603	36.3	35.3

Table 2. *(Continued)*

Element	Solar system[b]	Mean CI chondrite, ppb	Orgueil, ppb	Element	Solar system	Mean CI chondrite, ppb	Orgueil, ppb
Ti	2400	4.36×10^5	4.36×10^5	Dy	0.3942	242.7	246
V	293	56.5×10^3	56.2×10^3	Ho	0.0889	55.6	55.2
Cr	1.35×10^4	2.66×10^6	2.66×10^6	Er	0.2508	158.9	162
Mn	9550	1.99×10^6	1.98×10^6	Tm	0.0378	24.2	22
Fe	9.00×10^5	19.04[c]	18.51[c]	Yb	0.2479	162.5	166
Co	2250	50.2×10^4	50.7×10^4	Lu	0.0367	24.3	24.5
Ni	4.93×10^4	1.10[c]	1.10[c]	Hf	0.154	104	108
Cu	522	1.26×10^5	11.9×10^4	Ta	0.0207	14.2	14.0
Zn	1260	3.12×10^5	31.1×10^4	W	0.133	92.6	92.3
Ga	37.8	10.0×10^3	10.1×10^3	Re	0.0517	36.5	37.1
Ge	119	32.7×10^3	32.6×10^3	Os	0.675	486	483
As	6.56	1.86×10^3	1.85×10^3	Ir	0.661	481	474
Se	62.1	18.6×10^3	18.2×10^3	Pt	1.34	990	973
Br	11.8	3.57×10^3	3.56×10^3	Au	0.187	140	145
Kr	45		8.7[d]	Hg	0.34	258	258
Rb	7.09	2.30×10^3	2.30×10^3	Tl	0.184	142	143
Sr	23.5	7.80×10^3	7.80×10^3	Pb	3.15	2470	2430
Y	4.64	1.56×10^3	1.53×10^3	Bi	0.144	114	111
Zr	11.4	3.94×10^3	3.95×10^3	Th	0.0335	29.4	28.6
Nb	0.698	246	246	U	0.0090	8.1	8.1
Mo	2.55	928	928				

[a]Ref. 6.
[b]Atoms per 1.00×10^6 Si atoms.
[c]Value given is percentage.
[d]Value given is in pL/g.

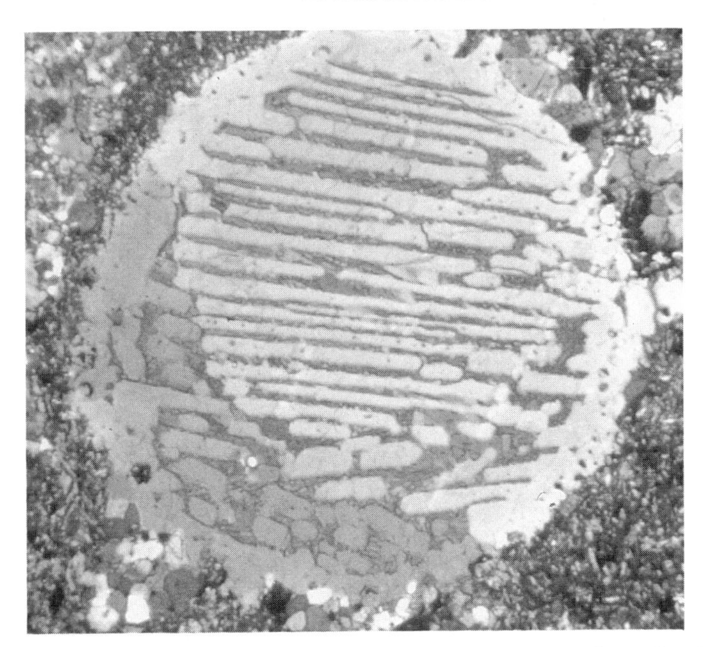

Fig. 1. A "barred olivine" chondrule from the Allende-type CV chondrite that fell in Mexico in 1979. The transmitted polarized light image of the 0.5 mm-diameter chondrule was taken from a polished thin section. The bars are composed of olivine, $(Mg,Fe)_2SiO_4$. The interstitial material is glass quenched by rapid cooling.

as Ni, Co, Ir, and the other elements in Groups 8–10 (VIII) closely follow the iron content. The Fe–silicate fractionation in chondrites is most probably the result of differences in the efficiency in formation and accretion of the silicate and metal grains from which chondrites formed. The variations of Fe oxidation state are related to the temperature at which the solar nebula gas and grains last equilibrated and the regional H_2:H_2O ratio in nebular gas. This ratio strongly influences the condensation of many compounds and determines the oxidation state of Fe in condensed solids. Under equilibrium conditions, Fe condenses above 1000 K as metal from nebular gas having solar H_2:H_2O ratios. It forms oxides only at lower temperatures or in environments where the H_2:H_2O ratio is low (9). The H_2:H_2O ratio in the nebula can vary widely owing to concentration of oxygen-rich solids such as ice and silicates followed by subsequent vaporization and enrichment of the surrounding nebular gas.

Chondrite classes are also distinguished by their abundances of both volatile and refractory elements (3). For volatile elements the variation among groups results from incomplete condensation of these elements into solid grains that accrete to form meteorite parent bodies. Volatile elements such as C, S, N, Zn, Cd, Bi, In, and Pb show large and systematic variations between the most volatile rich meteorites, ie, the CI and CM chondrites, and the ordinary chondrites. Refractory elements such as Ca, Al, Ti, Zr, and Sc show remarkably tight clustering in their Si normalized ratios (Fig. 3), with distinct differences between the chondrite classes. If the CI chondrites represent the actual solar-system abundances of these elements, then most of the other chondrites are depleted in refractory

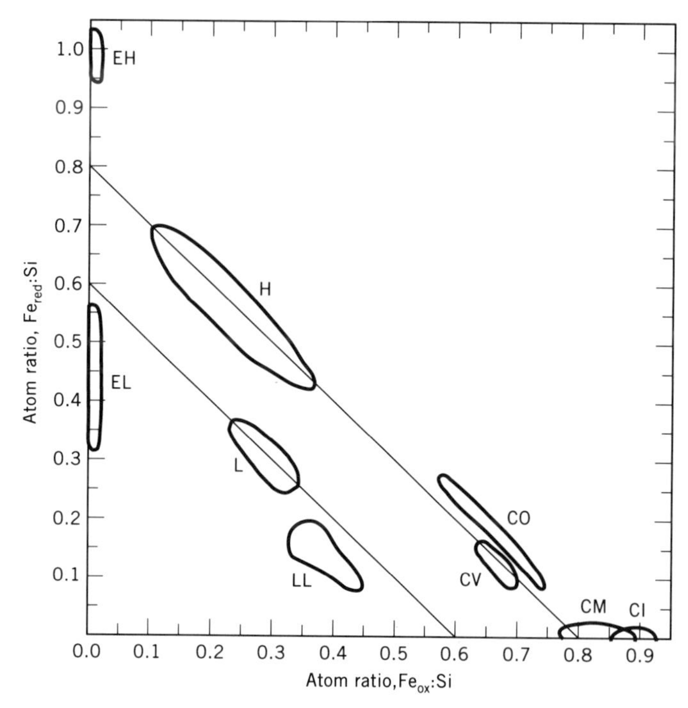

Fig. 2. The plot of total reduced iron, Fe_{red}, and oxidized iron, Fe_{ox}, normalized to Si abundance shows how the chondrite classes fall into groups distinguished by oxidation state and total Fe:Si ratio. The solid diagonal lines delineate compositions having constant total Fe:Si ratios of 0.6 and 0.8. The fractionation of total Fe:Si is likely the result of the relative efficiencies of accumulation of metal and silicate materials into the meteorite parent bodies. The variation in oxidation state is the result of conditions in the solar nebula when the solids last reacted with gas. Terms are defined in Table 1 (3).

elements whereas the CV class is enriched. These elements condense as a group before the condensation of Mg silicates and Fe metal that dominate the mass of stony objects. The compositional variation among different classes is presumed to be related to differing processes of accretion. The earliest solids to condense in the solar nebula were composed of refractory elements. If these solids preferentially accreted into certain bodies or were otherwise separated from the remaining nebular materials, then later-forming objects would accumulate from reservoirs depleted in refractory elements. Examples of the first generation of refractory condensates (or evaporative residues) are found in carbonaceous chondrites as calcium–aluminum-rich inclusions (CAIs). These inclusions range in size from tens of micrometers to several centimeters in diameter and are dominated by refractory silicates and oxides such as melilite [*12173-94-3*], perovskite [*9003-99-0*], spinel [*1302-67-6*], diopside [*14483-19-3*] and hibonite. CAIs have very low initial ^{87}Sr:^{86}Sr ratios, implying formation in the earliest history of the solar system (10), and contain isotopic anomalies owing to the presence of presolar components and the decay of extinct isotopes such as ^{26}Al (11). CAIs are most abundant in the CV chondrites and are responsible for the unusually high refractory element abundances in this class (Fig. 3).

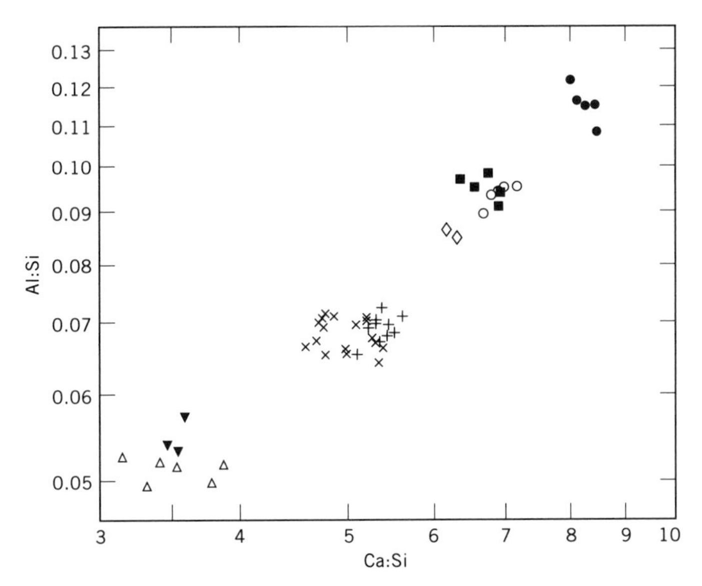

Fig. 3. The bulk Al:Si and Ca:Si ratios of chondrites fall into discrete groups where ● represent CV; ■, CM; ○, CO; ◇, CI; +, H; ×, L; ▼, EL; and △, EH. The fractionation of these refractory elements is believed to be the result of relative efficiencies of incorporation of condensed solids rich in early high temperature phases into the meteorite parent bodies at different times and locations in the solar nebula. The data are taken from Reference 3.

Most meteorite classes can also be distinguished by the oxygen isotope compositions (12). This remarkable distinction illustrated in the three-isotope plot shown in Figure 4, correlates with compositional and mineralogical classification. Before the solar system formed, oxygen existed in the precursor materials in both solid and gaseous species each potentially having its own oxygen isotopic composition traceable to variable inputs from materials of different nucleosynthetic sources. Oxygen arising from at least three isotopically distinct reservoirs, plus mass-dependent fractionation in the solar system, resulted in clear separation of the isotopic compositions. All natural terrestrial materials lie on the terrestrial fractionation line and a truly unique property of extraterrestrial samples is that, with the exception of the lunar samples, the extraterrestrials do not fall on this line. The highest deviations from terrestrial values are found in the CAIs where materials have been found that have as much as 5% excess ^{16}O.

Within each chondrite class there are petrographic grades that relate to alteration processes that occurred within the meteorite parent body. The grades range from 1 to 6 (13), although no class has examples in more than four grades. Grades 3 to 6 represent the effects of thermal metamorphism where the higher number is the more strongly altered. Grades 1 and 2 occur only for the CI and CM chondrites, respectively. CI and CM chondrites have been extensively altered by aqueous alteration in their parent bodies probably as a result of the melting of ice followed by reactions of preexisting phases. The sulfates, magnetite, carbonates, and hydrated silicates in CI chondrites are largely secondary minerals resulting from aqueous processes that apparently occurred at near freezing tem-

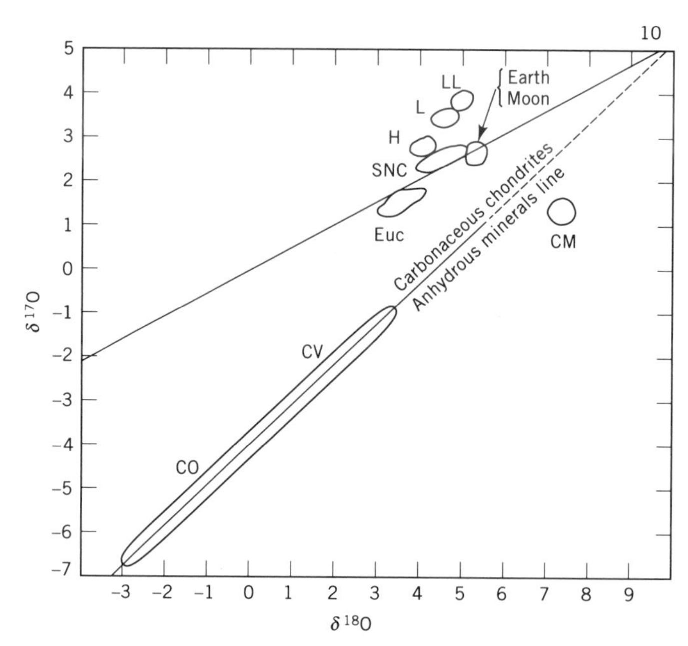

Fig. 4. The bulk oxygen isotopic composition of different meteorite classes where (—) is the terrestial fractionation line. The δ notation refers to the normalized difference between $^{17}O{:}^{16}O$ or $^{18}O{:}^{16}O$ ratios to those in standard mean ocean water (SMOW) in relative units of parts per thousand. The meteorites formed from materials that were enriched or depleted in ^{16}O and their bulk compositions plot off of the terrestrial line which has a slope of 1/2 owing to mass dependent fractionation. Some of the anhydrous minerals from carbonaceous chondrites fall on a line having a slope of 1. These anomalous compositions may be produced by mixing with an ^{16}O-rich component that has a different nucleosynthetic history than mean solar system material. The data are taken from Reference 10.

peratures. Veins of water-soluble sulfates deposited from solution exist in CI and some CM chondrites. Primitive $^{87}Sr{:}^{86}Sr$ ratios imply that vein formation occurred more than 4.4 Gyr ago (14).

The ordinary chondrites show a range of thermal alteration that is evidence of parent-body heating to temperatures above 800°C for the petrographic grade 6 chondrites. These effects show up in many ways. The least-heated ordinary chondrites are petrographic grade 3 meteorites, also known as the unequilibrated ordinary chondrites (UOC). These are characterized by the high degree of disequilibrium among different phases. For example, the Fe:Mg ratios of olivine are highly variable in H3 chondrites, whereas all olivines in the heated H6 have the same Fe:Mg ratio because they were equilibrated owing to thermal diffusion at elevated temperature. There are many effects that occur in the metamorphic sequence ranging from 3 to 6. Glass, which is common in grade 3, is completely devitrified above grade 5; Ni, which exists in sulfides at several percent in lower grades, is less than 0.5% in grades above 4 because it has diffused into metal; and secondary feldspar is absent in low petrographic grades but occurs as clear interstitial grains in grade 6. Metamorphic heating causes extensive alteration of the initial phases. A primitive grade 3 chondrite is composed of well-defined chondrules surrounded by dark, fine-grained matrix; the metamorphosed chondrites have poorly defined chondrules, and the matrix is composed of coarse grains that have grown at the expense of the original fine-grained matrix. For grades 3 to 6 a precise determination of petrographic grade can be determined by measurement of thermoluminescence, a property indicative of the amount of feldspar produced from glass as a result of thermal metamorphism (15).

Achondrites. The achondrites are differentiated stony meteorites that are apparently derived from parent bodies that were heated to at least partial melting temperatures. Achondrites do not contain chondrules and do not have elemental compositions that match solar abundances for condensable elements. These materials are old but their properties were more determined by planetary processes such as melting and differentiation than by primary nebular processes, such as condensation and accretion. Many of the achondrite subclasses can be combined into three basic groups: HED, SNC, and lunar groups. The HED group comprises the howardite, eucrite, and diogenite subclasses, which together are responsible for more than 6% of all meteorite falls. Eucrites are basalts that formed by rapid cooling of a basaltic composition magma and are largely composed of plagioclase and pryoxene. The diogenites are coarser grained, dominated by orthopyroxene, and have a mantle-like composition from which the eucrite basalts could have been derived. The howardites are breccias composed of mechanically mixed fragments of eucrite and diogenite components. The mineralogical composition and identical oxygen isotopic compositions of the HED meteorite clan is consistent with possible derivation from a single differentiated parent body. The spectral reflectivity of the HED meteorites is a strong and essentially unique match with the 500-km asteroid Vesta and its family of fragments. This suggests that this single asteroid is the source of the HED meteorite group (16).

The SNC achondrites are composed of the Shergotty, Nakhla, and Chassigny subgroups. Although the different specimens in this group have separate lithologies, they are all highly differentiated igneous rocks having properties consistent with derivation from a geochemically evolved body of planetary size. Their crystallization ages of only a Gyr clearly set them apart from nearly all other meteorites. Because it can be shown that SNC achondrites are not lunar material and are unlike the largest known asteroids, by default the most likely parent body of adequate size is Mars (17). Geochemical evidence and the similarity of trapped gas to actual atmospheric measurements on Mars have provided strong evidence that the SNC meteorites are samples of Mars. The only known ejection mechanism capable of launching rocks from a planetary surface at speeds in excess of the escape velocity involves significant impacts of asteroids and comets. At one time it was thought that this mechanism was implausible because the shock energy would melt or vaporize any fragments, but the discovery of impact ejected lunar meteorites on the Earth has demonstrated that this is a viable mechanism.

Among the rarest of all meteorites are the lunar meteorites. Isotopic, mineralogical, and compositional properties of these samples provide positive identification as lunar samples because of the unique properties of lunar materials that have been discovered by extensive analyses of lunar materials returned by the manned Apollo and unstaffed Luna missions. All but one of the lunar meteorites that have been found to date have been recovered from Antarctica.

Irons. Approximately 4% of meteorite falls are irons. Because they are distinctive rocks and weather relatively slowly, most meteorites that were not seen to fall, but were found accidentally, are irons. Iron meteorites represent metallic iron and siderophile elements that fractionated from molten parent bodies. They may have been cores of asteroids or they may have only been localized metal accumulations. The measured cooling rates of these objects at the time of formation imply that they are derived from bodies smaller than planets. Irons are predominantly composed of metallic iron having nickel, Ni, contents ranging from 6% to as much as 18%. The dominant minerals are kamacite and taenite, although other phases such as troilite [*1317-96-0*], graphite, schreibersite [*12424-46-3*], diamond, and even silicates occur as inclusions. Many of the irons have a distinctive crystal structure called the Widmanstatten pattern (Fig. 5), produced by slow cooling. Iron meteorites fall into more than a dozen subgroups distinguished by trace-element composition. These groups are most clearly isolated on the basis of the gallium, germanium, and iridium abundances normalized to nickel.

Stony Irons. The stony iron meteorites are composed of substantial iron and silicate components. The pallasites contain cm-sized olivine crystals embedded in a solid FeNi metal matrix and have properties consistent with formation at the core mantle boundary of differentiated asteroids. The mesosiderites are composed of metal and silicates that were fractured and remixed, presumably in the near-surface regions of their parent bodies.

Origin. Typical meteorites have formation ages of 4.55 Gyr and exposure ages of only 10^7 years, during which time they existed as meter-sized bodies unshielded to the effects of cosmic rays. With the exception of the SNC (Martian) and lunar meteorites it is widely believed that most conventional meteorites are asteroid fragments liberated relatively recently by collisions and transferred to

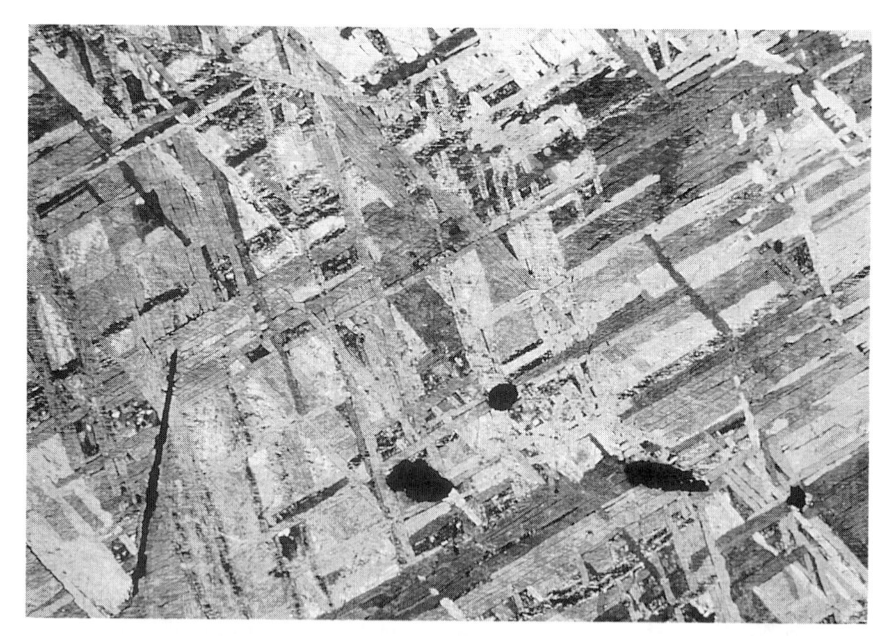

Fig. 5. The Widmanstatten pattern in this polished and etched section of the Gibbeon iron meteorite is composed of intergrown crystals of kamacite and taenite, NiFe phases that differ in crystal structure and Ni content. Ni concentration gradients at crystal boundaries in this 3-cm-wide sample can be used to estimate the initial cooling rates and corresponding size of the asteroid from which the meteorite was derived.

the Earth by gravitational perturbations. The principal source location is thought to be the zone 2.5 AU (one AU is equal to the mean distance between the Earth and the Sun) from the Sun where the orbital frequency about the Sun is exactly three times that of Jupiter. Objects in this zone undergo chaotic perturbations that change relative circular orbits to Earth-crossing orbits on a 10^6-yr time scale. The three meteorites for which accurate entry paths into the atmosphere have been determined can be traced to asteroidal origins.

The general scenario for the history of meteorites is that they are fragments of asteroids that have been stored in the asteroid belt 2.2 to 3.3 AU from the Sun. Early in the history of the solar system many if not all of the asteroids were heated to temperatures ranging from 0°C to as high as 1300°C. The source of this heating may have been resistive heating from electrical currents by magnetic fields in the solar nebula or the decay of ^{26}Al and ^{60}Fe, short-lived radionuclides that have been shown to have been present inside meteorites at the time of their formation. Many of the chondrite parent bodies were not heated severely and still retain many of the properties of the nebular materials from which they formed. These properties include chondrules formed in the nebula by yet unknown processes, reworked nebular materials, primary nebular condensates, and interstellar grains that actually predate the solar system. One of the most remarkable aspects of chondrites is that they contain preserved interstellar grains retaining isotopic and mineralogical properties that have survived since they formed in circumstellar regions around other stars. These phases include diamonds; silicon carbide [*409-21-2*], SiC; titanium nitride [*25583-20-4*], TiN; and graphite (18).

Interplanetary Dust

Interplanetary dust particles (IDPs) are the submillimeter-size regime of the solar system meteoroid inventory ranging in size from tens of nanometers to 1000 km in diameter. These particles are short lived in the interplanetary medium because of the effects of self-collisions and orbital decay caused by the drag component of sunlight (the Poynting-Robertson effect). Most of the IDPs that now reach the Earth were liberated from comets and asteroids within the last 10^6 years. Over 40,000 tons of IDPs impinge on the Earth annually and cumulatively they are the dominant meteoritic mass input on time scales shorter than 10^7 years. Interplanetary dust in the size range of a few micrometers to a millimeter can be collected in and below the atmosphere, and recovered particles provide an important sampling of asteroids and comets. These samples complement conventional meteorites because they include specimens of objects that for a variety of reasons do not either reach the Earth or survive atmospheric entry in greater than cm-sized pieces to become conventional meteorites (19). Dust provides a broader and less biased sampling of interplanetary materials because small samples of fragile materials can survive atmospheric entry without crushing. Additionally, sunlight pressure effects cause all dust beyond the Earth's orbit to evolve toward the Sun where collisions with Earth are possible.

Small meteoroids are not subjected to high ram pressures during entry because they decelerate at altitudes near 90 km, where the air density and corresponding ram pressure from collisions with air molecules is small. Conventional meteorites are larger and penetrate deeper into the atmosphere with high velocity, where they are subjected to crushing ram pressure. Unlike conventional meteorites which reach the Earth only by rare gravitational perturbation, dust from all bodies beyond the Earth's orbit, if not immediately ejected from the solar system by radiation pressure, experiences the effects of radiation drag causing it to inexorably spiral toward the Sun and past the Earth's orbit.

Collection. IDPs can be collected in space although the high relative velocity makes nondestructive capture difficult. Below 80 km altitude, IDPs have decelerated from cosmic velocity and collection is not a problem; however, particles that are large or enter a very high velocity are modified by heating. Typical 5-μm IDPs are heated to 400°C during atmospheric entry whereas most particles larger than 100 μm are heated above 1300°C, when they melt to form cosmic spherules (Fig. 6).

The flux of 10-μm particles is $1/(m^2 \cdot d)$, a value high enough that these IDPs can be collected directly using high altitude aircraft (20). The spatial density of 10-μm IDPs at 20-km altitude is $10^{-3}/m^3$. Particles > 100 μm fall at a rate of only $1/(m^2 \cdot yr)$ and can only effectively be collected after they have fallen to the ground and concentrated in a surface deposit. The small particles are collected primarily using stratospheric aircraft although they have also been recovered by melting pristine Antarctic ice. The larger IDPs have been collected from deep sediments, Greenland ice, and Antarctic ice, and a few other selected terrestrial environments that allow the extraterrestrial to be efficiently isolated and distin-

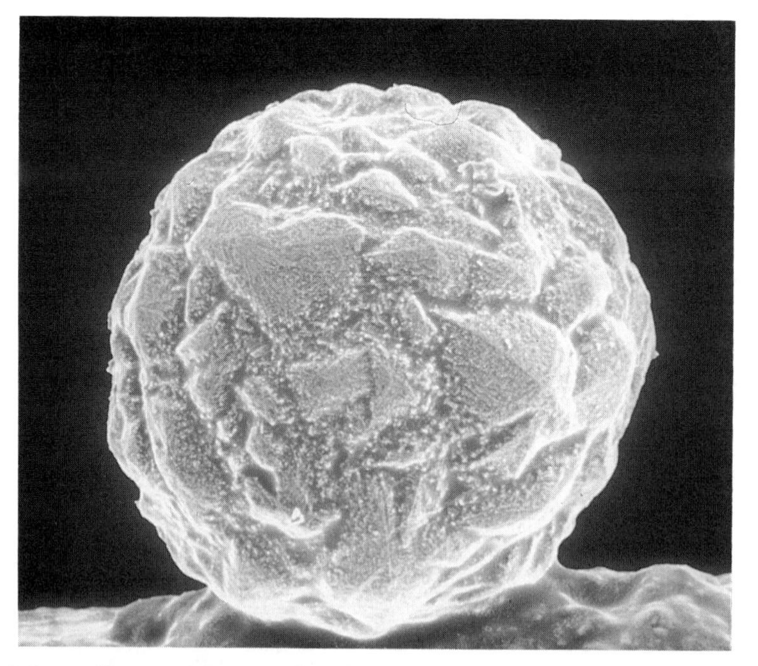

Fig. 6. A 0.3-mm-diameter cosmic spherule collected from the ocean floor. The particle is composed of olivine, glass, and magnetite and has a primary element composition similar to chondritic meteorites for nonvolatile elements. The shape is the result of melting and rapid recrystallization during atmospheric entry.

guished from terrestrial particles. Many of the > 100-μm cosmic particles are spherules and their shape assists in making a distinction from other materials.

Extraterrestrial dust particles can be proven to be nonterrestrial by a variety of methods, depending on the particle size. Unmelted particles have high helium, He, contents resulting from solar wind implantation. In 10-μm particles the concentration approaches $1/(cm^3 \cdot g)$ at STP and the $^3He:^4He$ ratio is close to the solar value. Unmelted particles also often contain preserved tracks of solar cosmic rays that are seen in the electron microscope as randomly oriented linear dislocations in crystals. For larger particles other cosmic ray irradiation products such as ^{53}Mn, ^{26}Al, and ^{10}Be can be detected. Most IDPs can be confidently distinguished from terrestrial materials by composition. Typical particles have elemental compositions that match solar abundances for most elements. Typically these have chondritic compositions, and in descending order of abundance are composed of O, Mg, Si, Fe, C, S, Al, Ca, Ni, Na, Cr, Mn, and Ti.

The most common IDPs are black objects having approximately solar elemental composition except for very volatile elements such as the noble gases. There are particles that deviate strongly from this pattern but they are rare and are usually dominated by a single mineral such as FeS, olivine, or FeNi metal. Most of the particles can be grouped into two classes: one contains hydrated minerals such as serpentine [*12168-92-2*] and smectite [*12199-37-0*]; the other, ones that are anhydrous. The hydrated particles are often nonporous (Fig. 7) and have some mineralogical properties similar to the CM and CI chondrites. Many of the anhydrous IDPs are porous materials having an aggregate structure (Fig. 8). The anhydrous particles are glass- and carbon-rich materials that contain FeNi, FeS, olivine, and pyroxene but have no close analogue among conventional meteorites.

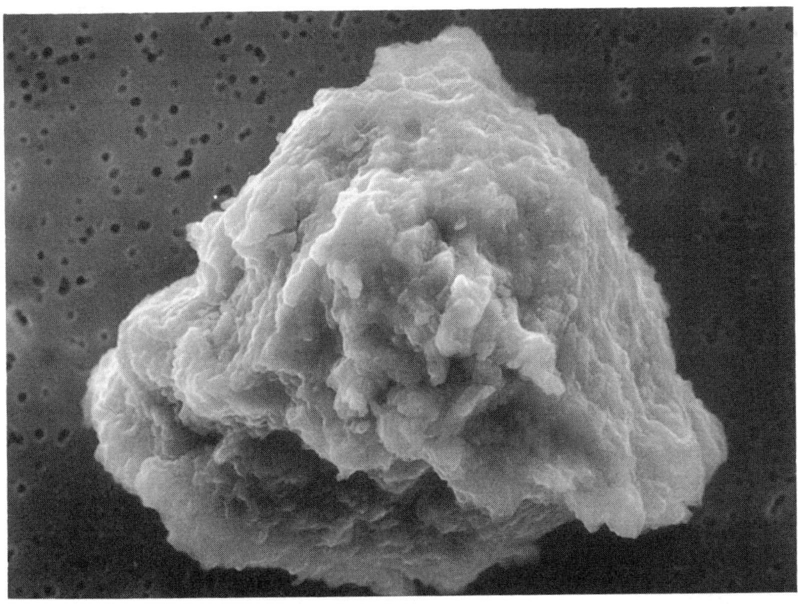

Fig. 7. A 10-μm interplanetary dust particle that is not porous and contains hydrated silicates. The particle's elemental composition is a good match to solar composition except for very volatile elements like the noble gases.

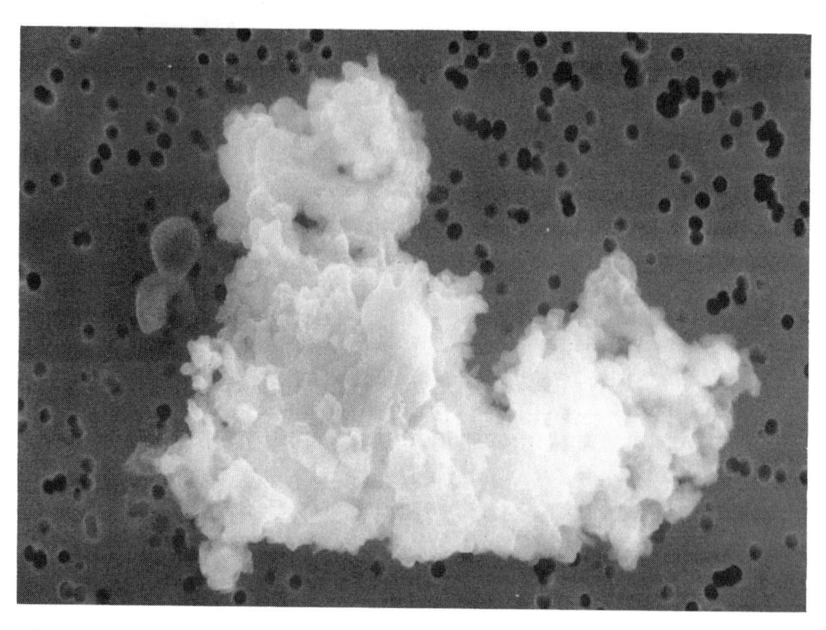

Fig. 8. A porous interplanetary dust particle collected in the stratosphere. The particle is 10 μm across and is composed of anhydrous minerals.

Origin. Individual IDPs are short-lived and the long-term presence of dust in the solar system implies that there must be sources capable of generating the approximately 10^7 kg/s of new dust required to balance losses by collisions, ejection by radiation pressure, and spiraling into the Sun owing to the Poynting-Robertson effect. For sizes > 10 μm, it has long been known that comets and asteroids are the principal source. Particles are liberated from asteroids by collisions of both asteroids and asteroid debris. Whereas collision velocities are high, most of the liberated dust is not strongly modified by the collision process and is ejected with a space velocity close to that of the more massive of the two colliding bodies. Dust in the asteroid belt has been detected by its thermal emission in the infrared at wavelengths > 10 μm. The orbiting IRAS infrared telescope discovered dust bands in the asteroid belt that correlate with the principal collisionally-produced asteroid families (21) (see INFRARED AND RAMAN SPECTROSCOPY). Dust generated in the asteroid belt undergoes orbital decay owing to light pressure drag and spirals toward the Sun reaching the Earth on relatively low velocity orbits. Dust from comets is released when solar heating sublimes the ice matrix in comets. Dust is ejected, and because of the effects of light-pressure drag and ejection velocity, it forms the dust tail than can extend to lengths of over 10^8 km. Most of the comet dust that is collectable on Earth is believed to have been derived from the Kuiper belt comets that reside in a flattened distribution extending from the region of the outer planets to distances of a few 100 AU from the Sun. At the time of dust release these bodies have evolved to short-period comets whose orbits pass close to the Sun. Dust from these comets reaches the Earth on relatively high velocity eccentric orbits. The dust from both comets and asteroids is believed to be samples of early solar system materials that has been relatively well preserved over the age of the solar system inside moderately small bodies.

Methods used to determine the temperature and hence entry velocity of 10-μm particles entering the atmosphere have provided direct information of the relative abundances of asteroid and comet dust reaching the Earth. Helium content, volatile content, and mineralogical indicators can be used to determine the maximum temperature reached during atmospheric entry. Comet dust enters faster than asteroid dust and accordingly is more strongly heated. These studies indicate that ~20% of the 10-μm IDPs in the Earth's atmosphere are of cometary origin and that > 50% are derived from asteroids. Taking into account the gravitational focusing enhancement at Earth, the abundance of comet dust in space is thought to be higher than 20%.

BIBLIOGRAPHY

"Space Chemistry" in *ECT* 3rd ed., Vol. 21, pp. 442–465, by D. E. Brownlee, University of Washington.

1. S. R. Taylor, *Planetary Science: A Lunar Perspective*, Lunar and Planetary Institute, Houston, Tex., 1982.
2. E. Opik, *Physics of Meteor Flight in the Atmosphere*, Interscience Publishers, New York, 1958.
3. J. T. Wasson, *Meteorites, Their Early Record of Early Solar-System History*, W. H. Freeman and Co., New York, 1985.
4. D. T. Britt, D. J. Tholen, J. F. Bell, and C. M. Pieters, *Icarus* **99**, 153 (1992).
5. E. Anders and N. Grevesse, *Geochim. Cosmochim. Acta* **53**, 197 (1989).
6. *Ibid.*, 198.
7. P. M. Radomsky and R. H. Hewins, *Geochim. Cosmochim. Acta* **54**, 3475 (1990).
8. K. Metzler, A. Bischoff, and D. Stoffler, *Geochim. Cosmochim. Acta* **56**, 2873 (1992).
9. J. A. Wood, *Ann. Revs. Earth Planet. Sci.* **16**, 53 (1988).
10. S. R. Taylor, *Solar System Evolution*, Cambridge University Press, Cambridge, UK, 1992.
11. T. Lee, D. A. Papanastassiou, and G. J. Wasserburg, *Geophys. Res. Lett.* **3**, 109 (1976).
12. R. N. Clayton, T. K. Mayeda, J. N. Goswami, and E. J. Olsen, *Geochim. Cosmochim. Acta* **55**, 2317 (1991).
13. W. R. Van Schmus and J. A. Wood, *Geochim. Cosmochim. Acta* **31**, 747 (1967).
14. J. D. Macdougall, G. W. Lugmair, and J. F. Kerridge, *Nature* **307**, 249 (1984).
15. D. W. Sears and co-workers, *Nature* **287**, 791 (1980).
16. R. P. Binzel and X. Shui, *Science* **260**, 186 (1993)
17. H. Y. McSween, *Rev. Geophys.* **23**, 391 (1985).
18. R. S. Lewis, S. Amari, and E. Anders, *Nature* **348**, 293 (1990).
19. J. P. Bradley, S. A. Sandford, and R. M. Walker, in J. F. Kerridge and M. S. Matthews, eds., *Meteroites and the Early Solar System*, University of Arizona Press, Tucson, Ariz., 1988, p. 861.
20. D. E. Browlee, in *Ann. Revs. Earth Planet. Sci.* **13**, 147 (1985).
21. M. V. Sykes, R. Greenberg, S. F. Dermott, P. D. Nicholson, and J. A. Burns, in R. P. Binzel, T. Gehrels, and M. S. Matthews, eds., *Asteroids II,* University of Arizona Press, Tucson, Ariz., 1989, p. 336.

DONALD E. BROWNLEE
University of Washington

FACE POWDER. See COSMETICS.

FACILITATED TRANSPORT. See MEMBRANE TECHNOLOGY.

FANS AND BLOWERS

Fan is the generic term for low pressure air- and gas-moving devices using rotary motion. Fans are subdivided into centrifugal and axial-flow types, depending on the direction of air flow through the impeller. In centrifugal fans, the air is introduced into the center of a revolving wheel or rotor with peripheral blades. Air is drawn through the blades and forced out in centrifugal flow into a scroll or volute housing where a portion of the kinetic energy is converted to pressure or static head. In axial-flow fans the air continues to move directly forward through the fan along the axis of the shaft. Kinetic energy is imparted to the air by the shape and arrangement of the blades. After discharge through the blades, although the general flow direction is still forward, a spiral component of velocity generally has been added to the air. A propeller-type fan is the most common axial-flow fan but more complicated types are in use where the blades resemble vanes in a turbine.

Blower is a term applied to a centrifugal fan generally when it is used to force air through a system under positive pressure. It generally implies a fan developing a reasonably high static pressure of at least 500 Pa (several inches of water). High speed centrifugal blowers (≥3600 rpm) are also available in one or more stages to compress air to pressures of 108–150 kPa (1–7 psig). The term blower is also applied to relatively low pressure positive displacement compressors of the rotary lobe, screw, or sliding vane types where the discharge pressure is usually less than 205 kPa (15 psig). Positive displacement blowers are outside the scope of this article.

When a fan is placed at the end of a system so that most of the system pressure drop is on the suction side of the fan, it is commonly called an exhaust fan or an exhauster. This term may also be applied to a ventilating fan where the primary function is to exhaust air from a room or an open hood.

Centrifugal compressors or turbocompressors are high volume centrifugal devices capable of gas compression varying from 105 to >1500 kPa (0.5 to several hundred psig). These generally consist of a number of stages of alternating rotating and stationary turbine blades and turn at very high speeds (see HIGH PRESSURE TECHNOLOGY).

The total pressure produced by a fan can be measured with an impact probe pointed directly upstream. The pressure so measured is a combination of both the static pressure and the kinetic energy pressure equivalent. Static pressure can be measured using a properly designed static wall tap or using the static pressure parts of a pitot tube. The latter represents the true pressure head exclusive of velocity effects. The difference between the total (impact) pressure and the static pressure is the velocity pressure or velocity head. Pressure readings are normally expressed in millimeters or inches of water (1 mm water = 9.807 Pa; 1 in. water = 248.8 Pa) and are referred to atmospheric pressure (101.3 kPa) as the reference base. Thus barometric pressure must be added to obtain absolute pressure. The total pressure rise produced by a fan is the difference in total pressure between the fan outlet and inlet. The fan static pressure is the total pressure rise for the fan reduced by the discharge velocity pressure. Inlet velocity head is assumed to be zero for fan rating purposes.

Centrifugal Fans

Figure 1 shows parts and names commonly associated with centrifugal fan components. The rotation of the wheel causes air between the blades to be rotated. The resulting centrifugal force causes this air to be compressed and ejected radially from the wheel. The compression results in an increase in static pressure in the fan scroll. The static pressure produced at the blade tips depends on the ratio of the velocity of air leaving the tips to the velocity of air entering at the heel of the blades. Thus the longer the blades, the greater the static pressure developed by the fan at a constant speed.

As the air leaves the blade tips, it contains kinetic energy by virtue of its velocity. The directional component of this velocity is both rotative and radial. When the fan blades are inclined forward, these components are cumulative. With backward-inclined blades, the components are in opposition. The purpose of the fan volute or scroll-shaped casing is to convert a portion of the kinetic energy of the air leaving the blades into static pressure.

Design operating efficiencies of fans under test conditions are in the range of 40–80%. Actual efficiency can be affected appreciably by the arrangement of inlet and outlet duct connections.

The air power in watts of a fan is given by equation 1:

$$\text{air power} = Q\Delta p \tag{1}$$

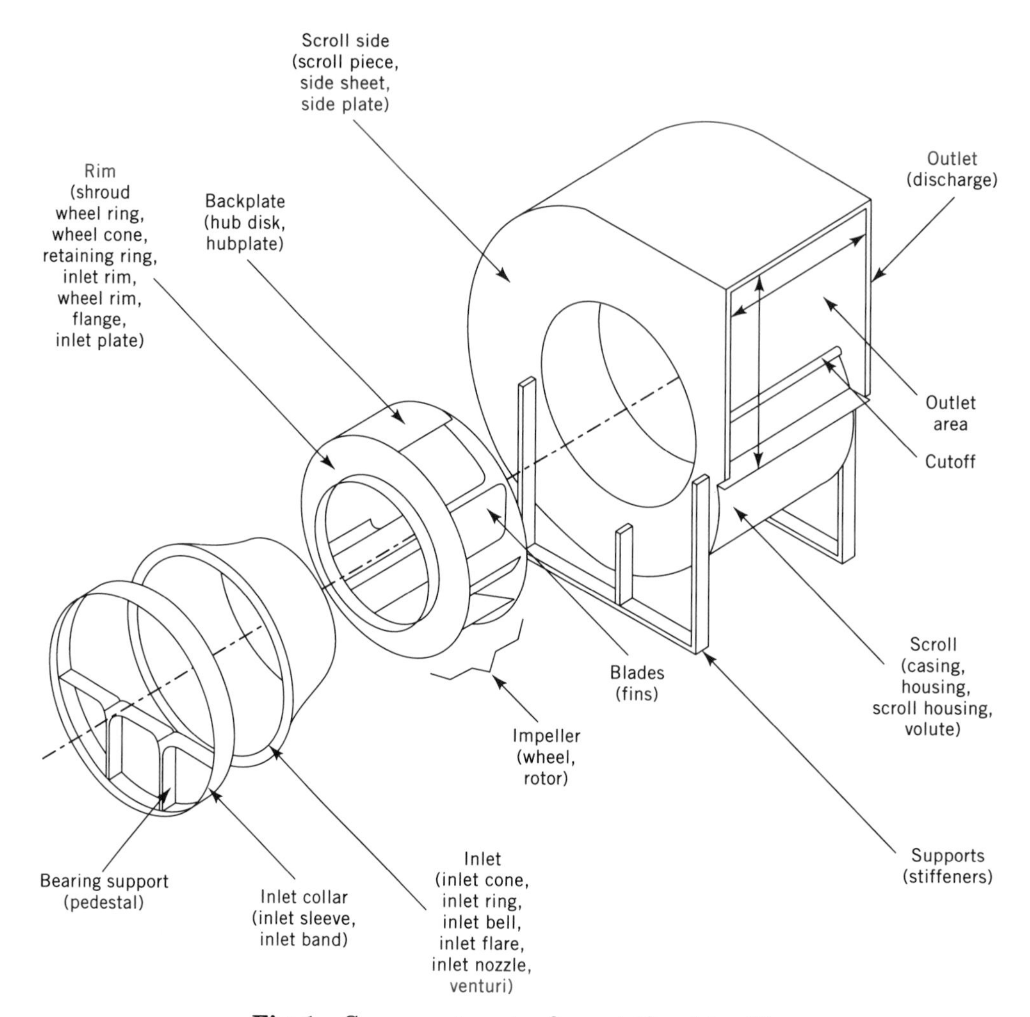

Fig. 1. Component parts of a centrifugal fan (1).

where Q is the volume of gas handled in m^3/s, and Δp is the pressure rise across the fan, Pa. (In units of hp, the air power $= 144\ Q'\Delta p'/33{,}000$; Q' is in ft^3/min; and $\Delta p'$ in psi.) In many fan installations, the velocity head of the fan discharge is wasted. In such cases, the fan static pressure may be used in equation 1 instead of the total pressure. Fan efficiency is expressed by equation 2:

$$\text{efficiency} = \frac{\text{air power}}{\text{shaft power}} \tag{2}$$

Performance Testing. Although fan performance characteristics can be roughly estimated during the early stages of design, fan efficiency losses and slip cannot be estimated accurately from theory alone. Therefore, the exact performance characteristics of a new fan design must be determined by testing. Test conditions must be carefully controlled, such as provision for steady and uniform flow of air approaching the fan inlet, because any inlet disturbances can affect performance. For this reason, fan field tests are seldom reliable, and most testing is performed in a laboratory on a test block following procedures set forth in standards for performance (2) including sound (3).

Figure 2 illustrates one of the several available test methods (2) and a typical performance curve. Fans designed for a duct as illustrated have a section of straight discharge duct attached. Straightening vanes are provided to eliminate swirl, reduce turbulence, and aid flow equalization across the duct. Air flow is determined using a pitot traverse while the fan is operated at a constant speed. The measured pressures are corrected for duct losses back to fan outlet conditions. A fan performs in accordance with the performance curve only if there is an equivalent duct present to convert velocity head efficiently to static head. At the end of the test, the duct is blanked to measure discharge pressure and shaft power at a shutoff (no flow) condition. The opposite extreme of the curve, free delivery (equivalent to duct removal), is extrapolated from nearly wide-open conditions. Intermediate points at sufficiently close intervals to define the curve would be measured by replacing the blank at the end of the duct with restricting orifices of varying cross section.

Types and Characteristics. The four basic fan wheel and blade designs and the corresponding performance curves are illustrated in Figure 3.

Forward-Curved Blades. In the forward-curved design (Fig. 3**a**), both the heel and tip of the blade are curved forward in the direction of rotation. Air leaves the tip of the wheel at a velocity greater than the wheel-tip speed. Blades are generally quite shallow and spaced much closer together than in other blade designs; 24–64 blades are typical. For a given fan duty, the wheel would have the smallest diameter and operate at the lowest speed of the various blade types. Such fans are commonly used for low pressure, high volume ventilating applications. As such, the wheel is often constructed of lightweight, low cost materials. Its mechanical efficiency is generally somewhat lower than that of the backward-curved blade fan. The pressure curve has a dip to the left of the peak which can

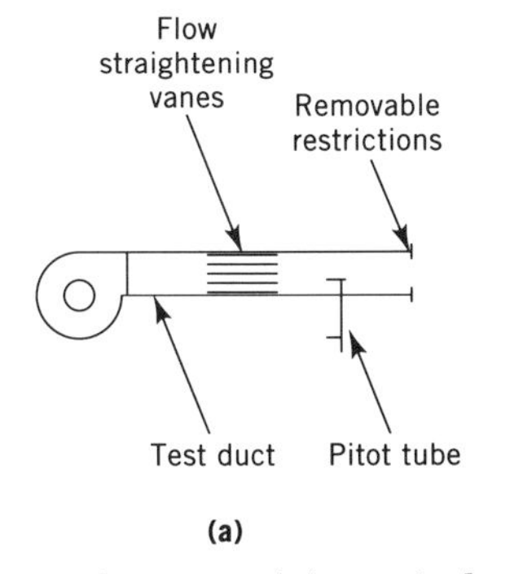

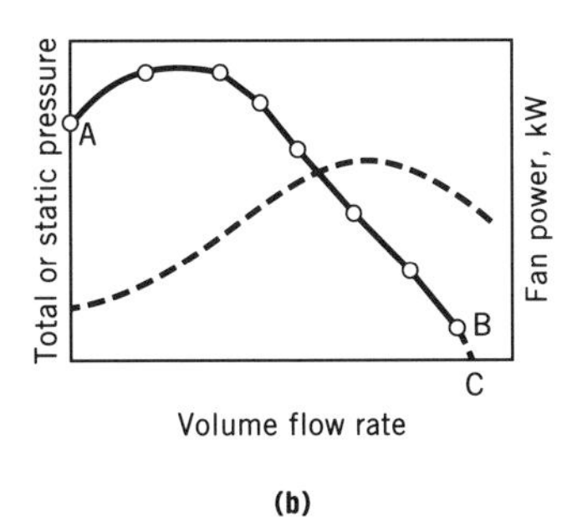

Fig. 2. Fan performance: (**a**) a typical test arrangement; (**b**) performance curve where the solid line showing data points is the pressure; the dashed line is the power. At point A the duct is blanked off, and at point B the flow is wide open; the data points in between represent progressively less restricted flow. Point C represents free delivery.

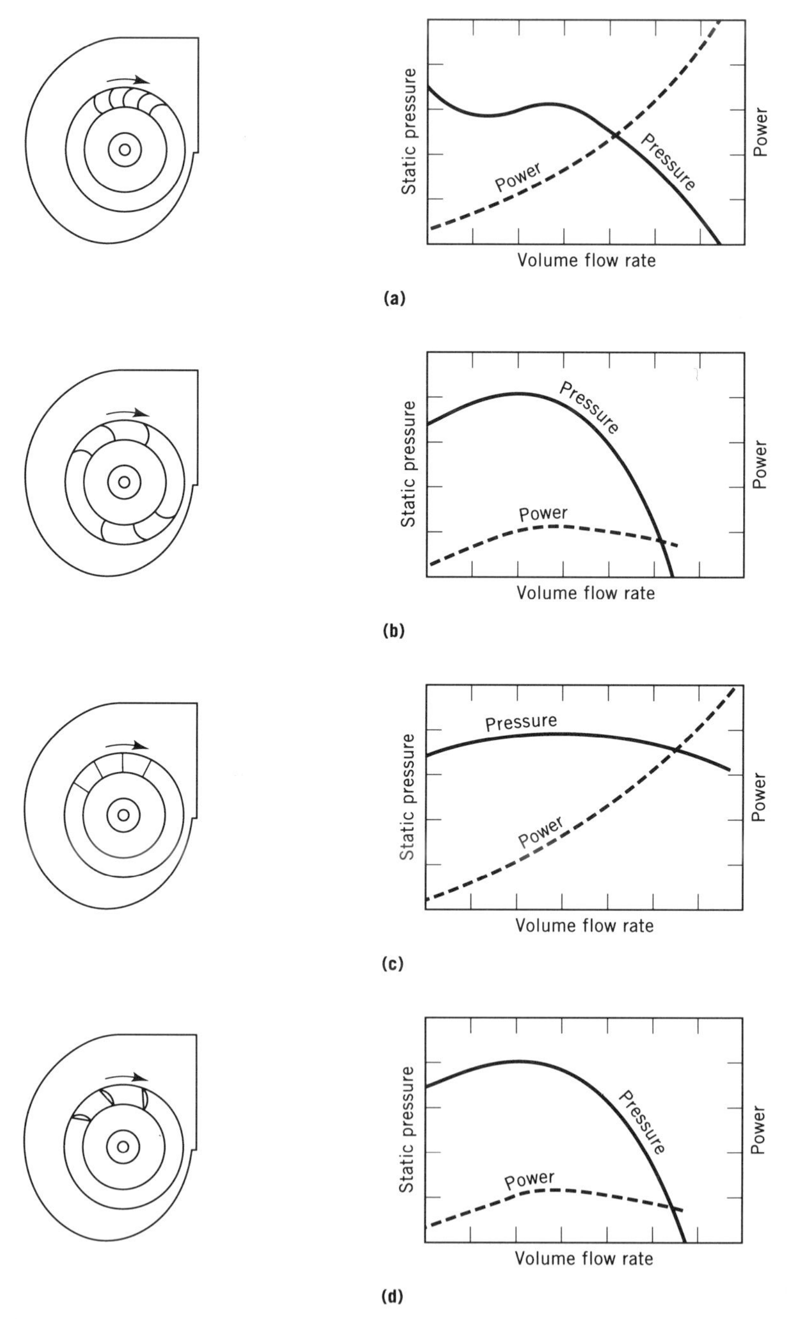

Fig. 3. Shape of fan blades and typical performance curves: (**a**) forward-curved blades; (**b**) backward-inclined blades, at left, straight backward blades at top, curbed backward blades at bottom; (**c**) straight radial blades; (**d**) airfoil blades.

cause operating problems (fan instability). Flow control in this region is difficult. Highest efficiency is reached to the right of the pressure peak, usually at 40–50% of wide-open flow. The fan is usually operated and rated to the right of the pressure peak. Power rises continually toward free delivery, which must be considered in motor selection.

Backward-Curved Blades. In the backward-curved design (Fig. 3**b**), the blades incline backward (opposite to rotation direction) from the point of heel attachment on the wheel. The single-thickness blades may be either straight or curved, usually 12–16 blades to a wheel. Air leaves the blade at a velocity less than wheel-tip speed because the increasing flow passage through the blade provides for expansion of the air. This feature improves the mechanical efficiency over that of the forward-curved blade. The deep blades lend themselves to developing a high static pressure. Wheel diameter and speed are generally higher for a given performance than the forward-curved blade. Close clearance and alignment of the wheel with the inlet bell are important aspects of this design to obtain maximum efficiency, especially at high static pressures. Inaccurate clearances allow leakage of compressed air back to the suction side of the wheel. The pressure curve rises somewhat from shutoff with increase in flow until a maximum pressure is reached. The maximum efficiency is reached at 50–65% of wide-open volume. The power curve reaches a maximum near the point of peak efficiency and then tends to drop off slightly with increased flow resulting in a nonoverloading design from the standpoint of motor sizing.

Straight Radial Blades. The straight radial blade design is the simplest of all centrifugal fans and also the least efficient (Fig 3**c**). However, the wheel can be designed with great mechanical strength and is easily repaired. It is useful for two different applications. For a given speed, it tends to develop a higher static pressure than other wheel designs and thus is attractive for high speed, high pressure fans compressing air to 108–120 kPa (1–3 psig). Such designs often have pressure performance curves that are fairly flat to 70% of the wide-open flow. This is desirable for applications where a constant output pressure is needed, such as for primary combustion air. Another use is in material handling. The blades can be coated with abrasion-resistant coatings or equipped with replaceable liners (see COATINGS). When large objects must occasionally be handled, such as a loose bag from a bag-filter house, the rims of the wheel may be omitted entirely and the blades supported with stiff struts from the hub. The shape of the power curve leads to overloading characteristics.

Airfoil Design. The airfoil design is similar to the backward-curved blade (Fig. 3**d**), except that it is designed for maximum mechanical efficiency. Each blade is composed of two pieces, with the upper surface contoured to reduce air friction and provide for most efficient compression of the air. For a given performance, this fan has the highest rotational speed of any of the wheel designs. The scroll is usually designed for the most efficient conversion of velocity head to static pressure. Performance characteristics are generally similar to a backward-curved blade but power requirements are somewhat less. Such fans are generally more expensive to construct and are used only in larger sizes with higher pressures or large flow volumes where reduced operating cost justifies the increased initial expense.

Fan Laws and Their Applications

Manufacturers' performance ratings are generally based on atmospheric pressure at sea level, 20°C, and 50% rh. Changes in temperature, gas density, and fan speed affect the performance. Fan laws predict these effects, are invaluable in predicting fan performance at various operating points, and are accurate over an extremely wide range of speeds. Some authorities (4) list as many as 10 fan laws relating variables such as size, speed, capacity, gas density, discharge pressure, power, efficiency, and sound level. For most fan users, the following four laws are adequate.

When fan speed is changed: (*1*) the capacity or flow rate varies directly with the speed ratio; (*2*) discharge pressure varies directly with the square of the speed; (*3*) power varies directly with the cube of the speed (at constant inlet density with no change in temperature, absolute pressure, or composition); and (*4*) discharge pressure and power requirements at a constant capacity and fan speed vary directly with gas density, p. These laws apply to varying the operating conditions of the same fan, and can be used to predict the performance of a given fan if sped up or slowed down. Other laws (4) describe the effects of varying the diameter of a fan or the solidity ratio. For example, when considering performance of different diameter fans, airflow capability is a function of diameter squared, although not necessarily at the same power requirement.

Fan Selection. A fan is selected according to its location in the air-flow system, system performance and control characteristics, cost, efficiency, control stability, flexibility, and noise level (see INSULATION, ACOUSTIC; NOISE POLLUTION AND ABATEMENT METHODS). Location in the flow system is very important. A fan operating on the highest density inlet gas available is smaller and less expensive, has lower operating costs, and requires less maintenance. This is partly evidenced by the fourth fan law, which states that the fan pressure varies directly with the density. Because a system has a required pressure drop at a given flow, a fan operating on a low density gas has to be operated at a higher speed than if it were operating on a more dense gas. At this higher speed the low density fan requires more power, and bearing life is shorter. Frequently, a fixed mass of air must be moved through the system rather than a fixed volume. Under this condition, the capacity of a low density fan has to be greater than that of a high density fan. Thus the benefits of using a forced-draft fan located near the air introduction point of the system become evident. Inlet density is reasonably high and most of the system pressure drop occurs on the discharge side of the fan. Placement of an exhaust fan at the end of the system with most of the system pressure drop on the suction side ensures that the fan handles a lower density gas. Similarly, if throttling flow control is used, or if the air is to be heated, it is desirable to place the throttling damper or the air heater on the fan discharge.

System flow resistance as a function of flow rate is needed to select the proper fan size. For calculation of system pressure drop see References 5–8. The resistance pressure curve for a typical system (Fig. **4a**) shows that the pressure required to force air through the system increases with the flow rate. The pressure–volume curve of a proposed centrifugal fan has a different shape. This fan curve must be drawn for the anticipated fan inlet density expected at its location in the system. The point of intersection of these two curves locates the flow rate and pressure rise at which the fan and system operate. This intersection represents

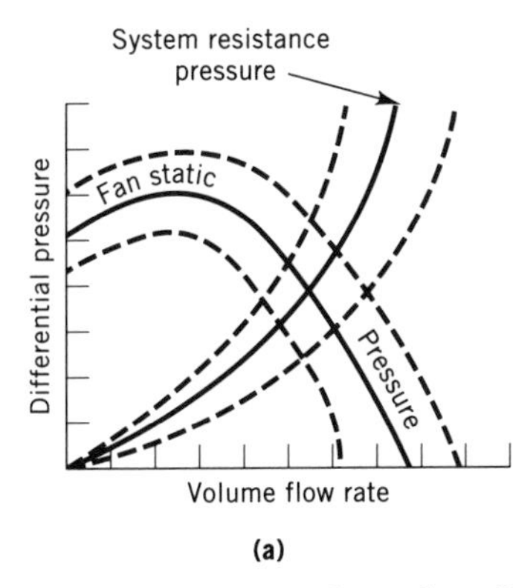

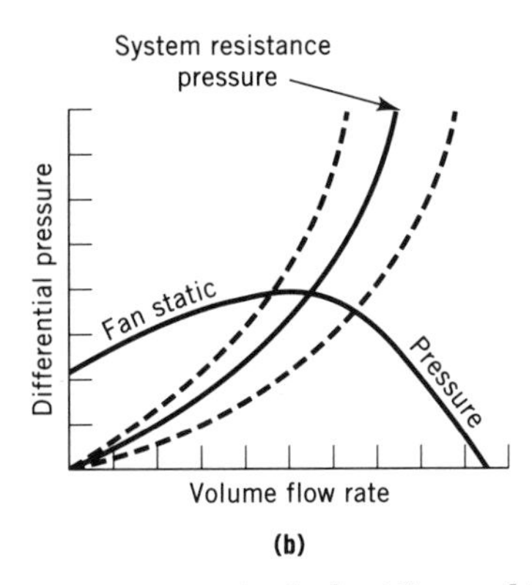

Fig. 4. Selection of fan size where the solid line represents a typical setting and the dashed lines the operating extremes. (**a**) Desirable sizing. The system resistance curve intersects the fan curve near its maximum efficiency. Changes in system resistance from a flow-control element also intersect the fan curve at desirable points for good flow control. The dashed curves also intersect system resistance curves at desirable locations. (**b**) A fan essentially too large for the system. The intersection of the system curve near the peak of the fan curve results in poor system flow control and perhaps surging.

a desirable operating combination for fan and system. The system curve intersects the fan curve in the middle of its maximum efficiency range and also at a point where the fan pressure produced varies smoothly but distinctly in a constant trend with flow rate which is desirable for flow control.

For air-flow control, the system may contain a control valve or damper that automatically or manually modulates system pressure drop. The dotted curves in Figure 4**a** on each side of the system resistance curve might represent operating extremes of the system resistance as the control valve is varied from maximum to minimum opening. These curves also intersect the fan curve at desirable operating portions of its range both for efficiency and flow control.

If a much larger fan as in Figure 4**b** had been considered so that the system resistance curve intercepted the fan curve close to its pressure peak, flow control would be much poorer. Fan pressure rise changes very little over the anticipated flow control range so that larger changes in flow volume accompany small changes in system pressure drop. In addition, fan pressure decreases on both a flow rise and decrease. This is a situation likely to cause surging and out-of-phase hunting between the fan and an automatic control system. Higher flow rates may be required for future expansion. Lower flow rates may also be desirable seasonally. These flow changes might best be achieved through changes in speed. The dashed lines in Figure 4**a** on each side of the fan curve represent higher and lower wheel speeds for which this fan is suitable.

The wisest fan choice is frequently not the cheapest fan. A small fan operates well on its curve but may not have adequate capacity for maximum flow control, future needs, or process upset conditions. It may be so lightly constructed that it is operating near its peak speed with no provision for speed increases in the future, if needed. As fan size is increased, efficiency generally improves and wheel speed is lower. These factors decrease operating cost and provide reserve capacity for the future. However, it is also possible to oversize a fan and impair its performance.

Noise level has to be considered in fan selection. Most manufacturers provide tables of operating ranges of quietest operation. There is no set fan discharge velocity that is applicable to all fans to ensure quiet operation. Fans do not operate as quietly when throttled back as when allowed to handle substantial quantities of air. Figure 5 illustrates the range of quiet operation of a specific airfoil fan as a function of outlet velocity and discharge pressure. Outlet velocity and hence fan capacity must be allowed to increase with static pressure to stay in the quiet region. Table 1 lists typical fan outlet velocities for quiet operation. Industrial process fans having backward-inclined blades should usually be selected with discharge velocities somewhat higher than those for quiet operation to achieve best all around performance and to provide pressure reserve.

Duct Connections. Performance curves are measured under ideal laboratory conditions. However, to obtain the same performance curve from a fan in a field

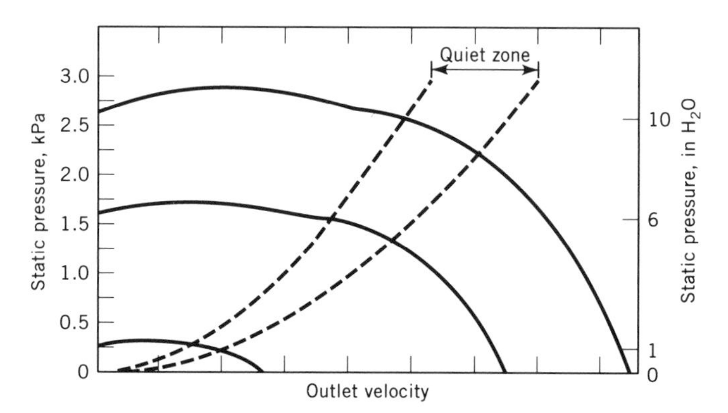

Fig. 5. Static pressure vs outlet velocity for a specific airfoil fan, where the dashed lines define the quiet operating range of an airfoil fan.

Table 1. Fan Outlet Velocities for Quiet Operation[a]

Static pressure		Forward-curved fan		Flow-nozzle airfoil fans[b]	
kPa	in. of water	m/s	ft/min	m/s	ft/min
0.25	1	8.1–10.4	1600–2050	4.3–7.4	850–1450
0.50	2	11.2–14.4	2200–2840	6.4–10.2	1250–2000
0.75	3			7.6–12.7	1500–2500
1.0	4			8.6–14.5	1700–2850
1.2	5			9.4–16.3	1850–3200
1.5	6			10.7–17.8	2100–3500
1.7	7			11.7–19.3	2300–3800
2.0	8			12.7–20.3	2500–4000
2.2	9			13.5–21.8	2650–4300
2.5	10			14.2–22.9	2800–4500
2.7	11			14.7–24.4	2900–4800
3.0	12			15.2–25.4	3000–5000

[a]Recomputed from data of New York Blower Co.
[b]Somewhat higher outlet velocities should normally be used for industrial processes using backward-inclined blade fans.

installation, the system must approach the characteristics of the test conditions at least in that part of the system close to the fan. Both inlet and outlet duct connections can influence fan performance significantly. These connections can actually change the shape of the fan curve. Therefore, no single correction factor can account for the performance change over its entire range. Although poor outlet connections affect performance, improper inlet connections generally hurt performance more, reducing it the most near free-delivery conditions and the least at peak pressure.

Poor performance can result from fan inlet eccentric or spinning flow, and discharge ductwork that does not permit development of full fan pressure. Sometimes inlet restrictions starve a fan and limit performance. To obtain rated performance, the air must enter the fan uniformly over the inlet area without rotation or unusual turbulence. This allows all portions of the fan wheel to do equal work. If more air is distributed to one side of the wheel, such as with an elbow on the inlet, the work performed by the lightly loaded portions of the wheel is reduced and capacity is decreased by 5–10%. The use of an inlet box duct on a fan can reduce capacity by as much as 25% unless there are turning vanes in the duct. Use of the vanes reduces the capacity loss to around 5%.

Spinning or vortex flow of air entering a fan can have as much effect on performance characteristics as the installation of inlet vanes to provide for reduced flow. If the air spins in the direction of wheel rotation, the bite of the blades on the air is reduced and both air flow and pressure are reduced. If the spin opposes wheel rotation, the wheel must overcome the momentum of the air: power requirements increase and efficiency is reduced. Spiral flow in a duct can be set up by a series of bends and elbows forming a corkscrew path, cyclones, and tangential inlets. A full diameter inlet duct that is straight for 10 diameters is desirable. Where such inlet connections are not possible, corrective devices should be provided in the ductwork. Spiral flow can be eliminated with the use of eggcrate straightening vanes. Turning vanes in the ductwork can largely eliminate problems of eccentric flow.

The velocity of air discharging from a fan is not uniform across the discharge outlet but tends to be higher toward the outside of the scroll as shown in Figure 6. The discharge duct evens the velocity distribution into the standard turbulent-

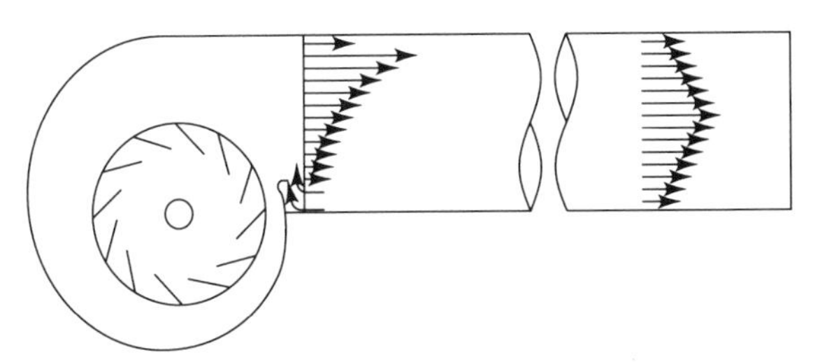

Fig. 6. Illustration of variation of velocity of air at the outlet of a centrifugal fan and the function filled by several diameters of straight discharge duct in converting velocity head to static head and establishing normal turbulent flow distribution. Bends or obstructions at the discharge outlet cause turbulence and prevent conversion of velocity head to static pressure.

flow distribution some distance downstream and converts part of the discharge velocity to static pressure. If a fan is operated without an outlet duct (discharging into a large plenum or the atmosphere), it loses 1–1.5 velocity heads. Such a loss must be added to the calculated system resistance. The addition of a straight discharge duct for several diameters and of the same size as the fan outlet can obviate this loss. An expanding outlet can increase static pressure beyond the curve performance if an efficient expander (included angle no more than 17°) is used. An elbow placed directly on a fan discharge destroys most of the velocity pressure leaving the fan. Suggestions for improving fan duct connections and effect factors to be applied to fan performance curves are given in various publications (7–12).

Flow Control. In many applications, it is desirable to be able to change the quantity of air being handled through the system. The need to change the flow may be frequent, such as every few minutes or every hour, or less frequent such as daily, weekly, or even seasonally. The choice of control method can be influenced by the frequency with which the flow must be changed. In order to control flow, either the system characteristics or the fan characteristics must be changed. Generally, flow control affects the energy input to the fan. Low cost control devices often result in reduced fan efficiency and increased power consumption. Thus if flow reduction is to occur for a long time with powerful fans, more energy-efficient control devices should be considered.

The simplest and cheapest control device is a damper, butterfly valve, or an orifice placed in the duct to throttle the flow and change the system resistance characteristics. As the flow is throttled more, system resistance is increased as illustrated in Figure 4**a**. To produce a higher discharge pressure, flow through the fan has to decrease. The power input is reduced but the energy expended in pressure drop across the throttling element is wasted. Placement of a throttling device on the discharge side of a fan is often preferable because the density of the air entering the fan is not reduced.

Changing centrifugal fan characteristics usually results in greater energy savings than changing system characteristics. If fan pressure can be reduced together with flow, the most desirable method of energy conservation is to change the speed, because that leaves the efficiency unchanged. If fan capacity is to be changed only infrequently, speeds of belt-driven fans can be adjusted easily with sheave changes. Where frequent speed changes are required, variable-speed motors and drives (electric or hydraulic) are the best but the most expensive. Multispeed motors and motors having step speed control can be used when infinitely variable control is not needed. The effects of speed control on a fan can be predicted from the fan laws. An alternative to speed change for axial-flow fans is blade-pitch control.

Inlet-vane control can be used to change the shape of the fan performance curve through imparting spin to the air entering the fan. As more spin is imparted, less energy can be transferred to the air from the blades and static pressure output is reduced. Figure 7 illustrates how the performance is reduced as more and more spin is imparted to the inlet air. Each setting of the inlet vanes has a separate power curve. The intersection of the system curve with the various fan pressure curves is shown, as is the equivalent power. The power required using inlet-vane

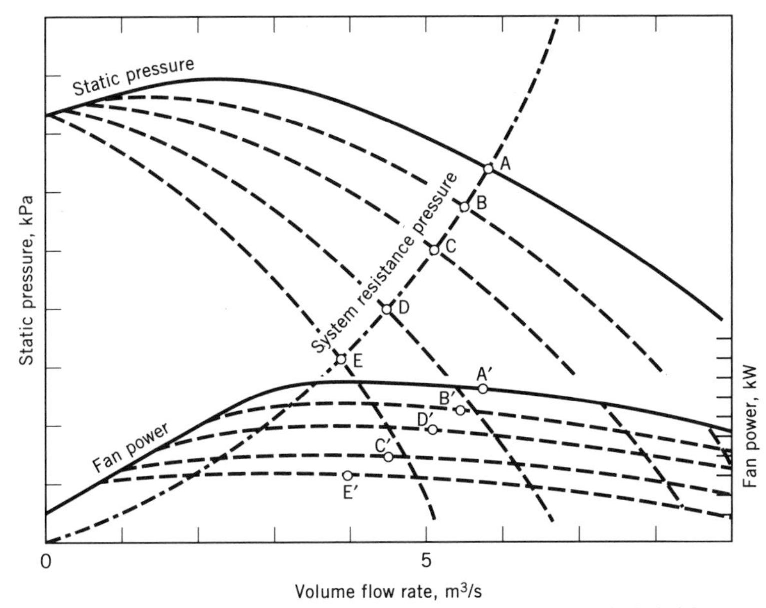

Fig. 7. Control of fan performance with inlet vane control. Solid lines marked A and A′ show normal performance without vanes (vanes wide open). As vanes are progressively closed, static and power curves are modified as indicated by dashed lines. Intersection (—‐ ‐—) of the system resistance curve with these reduced pressure curves at points B, C, D, and E shows how imparting more spin to the inlet air reduces flow. Projecting points A to E vertically downward to the corresponding power curve locates fan power points A′ through E′. Power savings achieved over throttling control can be estimated by projecting points B through E vertically downward to the A′ power curve and comparing the value with that from the proper reduced power curve. To convert kPa to in. H_2O, divide by 0.249; to convert m^3/s to ft^3/min, multiply by 2119.

control is usually intermediate between that required using throttling control and speed control.

Motor and Drive. The preferred prime mover for a fan is usually an electric motor. For fans of low to moderate power, V-belt drives are frequently employed. This permits selection of fans that can be operated over a wide range of speeds rather than being limited to motor synchronous speeds. Furthermore, change of speed is less expensive with V-belt drives. However, fans requiring powerful motors, 37–75 kW (50–100 hp) and higher, are generally directly connected to the motor and driven at synchronous speed.

When selecting the motor, power requirements, effect of temperature changes on load, and motor starting current and torque have to be considered. Calculation of system air-flow resistance is subject to some error and cannot always be predicted precisely. Therefore, the fan power predicted by the intersection of the fan and system curves may not be precise. If the system resistance is higher, it may be necessary to speed up the fan, which makes it draw more power. If the resistance is less than anticipated, the flow increases (unless dampered) also resulting in higher power consumption. A general rule is to size the motor for the power required for a system pressure drop both 25% greater and less than that predicted. Air temperature can also affect power requirements. A fan normally operated on a hot gas may have to be started when the system is cold. Under such conditions, the inlet gas density is much higher. The fan develops more head and a greater mass of air is delivered. Unless the system flow can be throttled back until normal operating temperatures are reached, the motor has to be sized for the cold-starting conditions based on density ratios, often two or three times normal running power.

In starting a fan, the air power increases gradually with speed which is a desirable starting load. However, in large heavy fans considerable torque is required to overcome the fan wheel inertia (referred to as WR^2, where W is the mass of the wheel and shaft and R is the radius of gyration). Figure 8 illustrates typical fan wheel and motor torques as a function of system speed during the starting process. Fan torque is that required for overcoming wheel inertia and for running power for the speed attained. Motor torque at every point on the starting curve must be greater than fan torque. The vertical difference shown in Figure 8 is the torque available for acceleration. If the motor is started across the line and the length of time required to reach full load is too long (usually 10 s is desirable), the motor may become overheated and overload controls shut off the power. Thus the fan cannot be started. On long power lines, the inrush of starting current may also drop line voltage sufficiently so that fan and motor are too slow in coming up to speed. Alternatively, reduced-voltage starters can be used, which permit extended starting periods without motor overheating, or special motors with winding that can be bypassed during starting can be used. In calculating the starting time of a fan (13), in addition to the WR^2 of the fan wheel, the flywheel effect of large drive sheaves and the motor rotor itself must also be included.

Other Selection Problems. Additional considerations can arise when fans must handle solids or gases of low density, or must be operated in parallel or series. A complicated flow system involving several fans in parallel, all of which are in series with a common exhaust fan, can lead to surging and vibration unless selected carefully. Maximum tip speed, bearing types, single- and double-inlet fans, and wheel and shaft natural frequency and rigidity must also be considered.

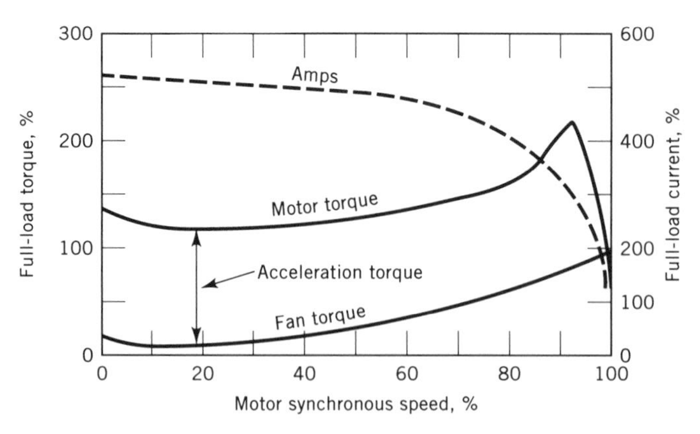

Fig. 8. Plot of fan and motor starting-torque curves.

Low Density Gases. A fan may have to operate on low density gas because of temperature, altitude, gas composition (high water vapor content of the gas can be a cause of low density), reduced process pressure, or a combination of such causes. To develop a required pressure, the fan has to operate at a considerably higher speed than it would at atmospheric pressure, and hence it must operate much closer to top wheel speed. Bearing life is shorter, and the fan tends to vibrate more or can be overstressed more easily by a slight wheel unbalance. Abrasion of the blades from dust particles is more severe. Therefore, a sturdier fan is needed for low density gas service.

Near top speed, a fan may operate at a speed that is near or above the natural frequency of the wheel and shaft. Under such conditions, the fan can vibrate badly even when the wheel is clean and properly balanced. Whereas manufacturers often do not check the natural frequency of the wheel and shaft in standard designs, many have suitable computer programs for such calculations. Frequency calculations should be made on large high speed fans. The first critical wheel and shaft speed of a fan that is subject to wheel deposits or out-of-balance wear should be about 25–50% above the normal operating speed.

Mechanical Considerations. The mechanical design of a fan and the various forces that fan parts must withstand are discussed in Reference 14. The forces result from a combination of fluid, inertial, and vibrational effects.

Tangential forces from air compression act on the blades and are transmitted through the fan hub to the shaft in the form of a resisting torque. Axial thrust may be developed on the fan wheel and shaft because of pressure differences about the wheel and the directional change in momentum of the air at the wheel inlet. The net unbalanced axial thrust must be taken by a thrust bearing that transmits it through the bearing supports to the fan foundation. At maximum fan efficiency, the radial fluid forces acting on the wheel are nearly balanced, but the volute can be correctly designed for only one rating condition. Therefore, as fan operation departs from maximum efficiency, unbalanced radial thrust increases which must be carried by the bearings to the foundation. Centrifugal forces also act on the wheel. If the center of the wheel and shaft rotation does not coincide exactly with the center of the rotating mass, a flexural force produces bending of the shaft, apparent as vibration. In a rotating elastic system, dangerous vibrations are likely to occur at critical speeds. The application of repeated external forces such as flow surging or wheel unbalance excites the elastic structure and causes it to vibrate. If the excitation frequency is close to the natural frequency, resonance can occur with large amplitude vibrations. All of these forces must be carried by the bearings. Thus it is common to use heavier components as fans are called on to operate at higher speeds or higher pressure differentials. Many fan designs are available in different construction strengths designated Class I, II, III, and IV. The higher numbers denote a fan capable of operation at a higher speed and higher pressure rise. The required class of the fan needed must be considered in its selection.

Bearings used on fans may be either sleeve or antifriction type and must be designed to withstand loads resulting from dead weight, unbalance, and rotor thrust and be able to operate at the intended maximum speed without excessive heating (see BEARING MATERIALS). When natural convection from the bearings is inadequate, some other cooling method must be provided. Lubricating oil may be circulated through an external cooler, or the pillow blocks may be cored with passages for forced circulation of air or water. Fans operated at high temperatures increase the bearing cooling problem caused by heat conduction along the shaft. A small external fan wheel on the shaft, called a heat slinger, is frequently provided, or forced-circulation water cooling is used. In addition to the bearings of fans operating on hot, low density gas at high pressure rise, special attention is needed to ensure high rigidity of the wheel and shaft. Fan wheels should be balanced both statically and dynamically, eg, in the field with chalk and weights (15). Elaborate electronic test instruments are also available. An unbalanced condition causes a vibrational displacement of the bearings which is frequently checked. Table 2 lists typical displacements of fans operating at various speeds and various degrees of unbalance.

Table 2. Fan-Bearing Vibrational Displacement[a]

	Bearing displacement,[b] μm			
Wheel speed, rpm	Smooth	Fair	Rough	Very rough
600	50	100	200	380–500
900	38	70	150	200–250
1200	25	50	115	150–200
1800	19	38	90	125–180
3600	10	18	65	100–125

[a]Ref. 15.
[b]For qualitative degrees of unbalance.

Small volume fans, usually designed with an air inlet on only one side of the wheel and casing, are known as single-inlet fans. With an enclosed wheel, the fan hub is fastened to a solid backplate that supports the blades. Larger capacity fans can be either single- or double-inlet fans. A double-inlet fan has an inlet on both sides of the wheel and casing. The hub is usually fastened to a common backplate midway between the two inlets. A double-inlet fan is generally more efficient or runs at a slower speed than a large single-inlet fan for the same capacity because the air is better distributed over the width of the wheel. Finally, air volumes are reached with large fans such that only double-inlet designs are feasible.

Vibration in a fan may be caused by mechanical problems or by the flowing air, eg, surging, poor fan-curve operating position, poor design of fan-duct connections resulting in poor air distribution, etc. A double-inlet fan is expected to have little axial unbalance because the symmetrical design of the air flow between the two halves of the wheel tends to result in a balancing of opposing forces. Such fans are frequently supplied with bearings suitable for only small thrust loads. Poor inlet ductwork arrangements can result in excessive thrust if unequal air flows are provided to opposite sides of the wheel. An unsteady air-flow unbalance that alternates between inlets can set up an alternating thrust pattern which can be very damaging to bearings designed for low thrust load. Mechanical vibration and elastic deformation problems and diagnostic techniques for structural inadequacies in fan design are discussed in Reference 16.

Axial Flow Fans

Axial flow fans, in which the air flow is parallel to the fan axis, are the workhorse fans in many petrochemical and utility industry applications. These are the first choice of air mover whenever large volumes of air at low (most commonly up to 500 Pa (2.0 in. H_2O)) pressures are needed. Axial flow fans range in size from 25 mm diameter (cooling computers) up to 12.3 m in diameter (cooling condenser water in power plant cooling towers). These fans are used in air-cooled heat exchangers for process cooling in many chemical plants in sizes of 1.8–4.3 m (see HEAT-EXCHANGE TECHNOLOGY). Axial fans from 0.6 to 9 m diameter are used in heating, ventilation, and air conditioning (HVAC) applications in homes and office buildings around the world. Most commonly, axial flow fans are used in short ducts called fan rings or cylinders, discharging into the atmosphere. Most large fans in cooling towers have velocity recovery stacks that capture the wasted velocity pressure energy and convert it back into useful work.

Axial fans are classified as propeller, tube-axial, and vane-axial (Fig. 9). The choice of fan required is determined by the resistance (static pressure) the fan must work against as well as the volume flow required. Propeller fans usually discharge into a plenum or directly into the atmosphere. Tube-axial fans are usually mounted in ducts as in an air conditioning system. Vane-axial fans are also mounted in ducts but feature a stationary guide vane on the discharge side that

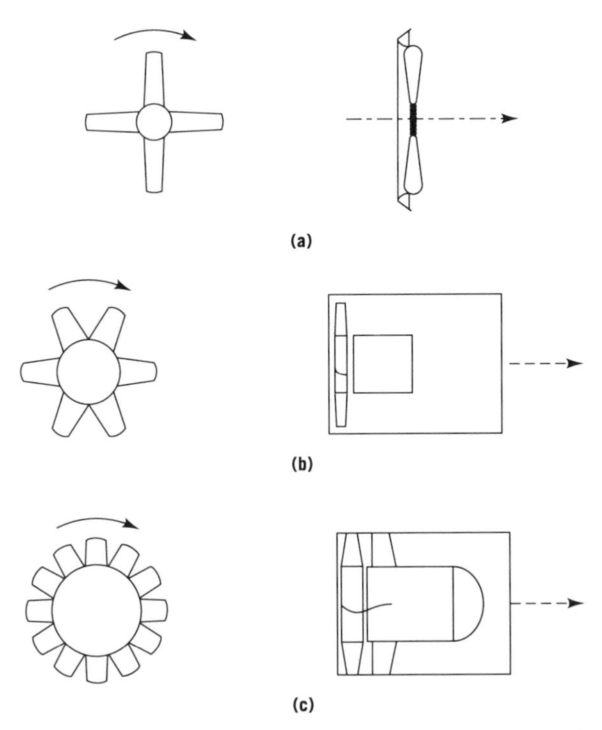

Fig. 9. Types of axial-flow fans where the dashed arrows denote the direction of air flow: (**a**) propeller fans; (**b**) tube-axial fans; (**c**) vane-axial fans (17).

straightens the air flow to improve efficiency. Tube-axial fans can work at static pressures up to 623 Pa (2.5 in. H_2O); vane-axial fans can work up to 2000 Pa (8 in. H_2O).

Design Elements. Ideal conditions are obtained in the design of an axial-flow fan when energy transfer from the blade to the gas is uniform along the length of the blade, resulting in uniform pressure generation, minimum losses, and maximum efficiency and stability. Because the blade linear velocity varies with position from tip to hub, attainment of a uniform pressure rise along the blade at different radii requires variation of the blade angle from hub to tip. The choice of blade section is dictated by the required aerodynamic characteristics and varies in practice from cast or molded precise airfoil profiles to formed materials to single-thickness plate materials. Hub size is increased for higher pressure designs where it is impractical to generate equal pressures nearer the center of the wheel. Low pressure designs have hubs ranging from 1/3 to 1/2 wheel diameter whereas hubs in higher pressure designs may occupy 75–85% of the tip diameter. The number of blades must also be increased as pressure rise is increased; 3 to 5 may be used with lower pressure designs, as many as 24 with higher pressures. Close clearance between blade tips and fan housing is a stringent requirement to prevent backflow losses at the housing wall. High pressure designs require clearances of less than 0.79 mm. The cylindrical housing of a vane-axial fan may be cast or rolled. To attain the close clearance at the blade tips, either very careful forming or machining is required. Inlet and outlet connections are carefully designed to minimize turbulence and connecting inlet and outlet ducts should be straight for at least 2–3 diameters to avoid undue effect on fan performance. Performance curves are shown in Figure 10.

Propeller Fans. Propeller fans may have from 2 to 6 blades mounted on a central shaft and revolving within a narrow mounting ring, either driven by belt drive or directly connected. The form of the blade in commercial units varies from a basic airfoil to simple flat or curved plates of many shapes. The wheel hub is

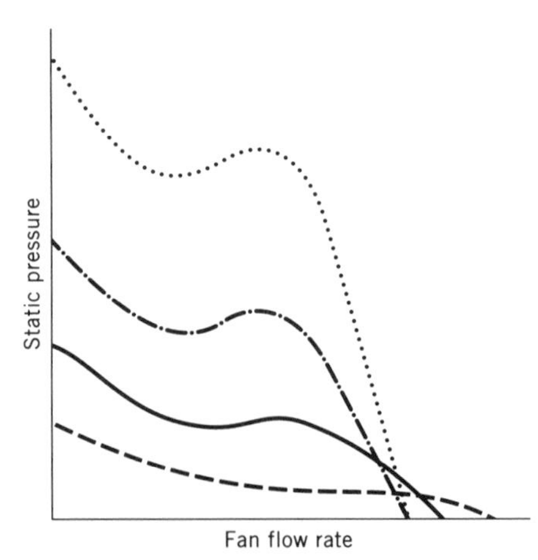

Fig. 10. Performance characteristics for axial flow fans: (– – –), propeller fan; (——), tube-axial fan; and (— · —), single-stage and (.....), two-stage vane-axial fans.

small in diameter compared to the wheel. The blades may even be mounted to a spider frame or tube without any hub. The housing surrounding the blades can range from a simple plate or flat ring to a streamlined or curved bell–mouth orifice.

Some type of close-clearance shroud at the blade tips is desirable to prevent air recirculation from the discharge side of the blades back to the suction side. A propeller fan with no shroud has fairly low efficiency because of air recirculation. A curved orifice-like ring greatly reduces air recirculation and improves efficiency. An angle-shaped ring essentially eliminates recirculation, and optimum efficiency is achieved with an angle-like ring with streamlined edges. Power requirements for most propeller fans increase as flow decreases and static pressure increases.

Tube-Axial Fans. The tube-axial fan is a refinement of the propeller fan in both wheel design and mechanical strength, having improved capacity, pressure level, and efficiency. Designs are often capable of operating over a greater range of speeds. The cheapest fans may have an open-type propeller wheel with the motor enclosed in a tube if directly connected. Belt-drive models are also available. In more refined types, the blades are shorter and of airfoil cross section mounted on a large diameter hub which may approach 50% of the wheel diameter. The hub and motor tube are normally of the same diameter and reduce the back flow of higher pressure air, which might recycle through less effective central portions of the wheel if a smaller hub were utilized. The performance curve (Fig. 10) may have a dip to the left of the pressure peak which would constitute an unstable region for fan operation and which should be avoided. Commercial models are available having static pressures up to 750 Pa (3 in. of H_2O). The general range of application is for pressures of 125–375 Pa permitting use of appreciable ductwork. Maximum efficiencies are in the range of 65–75%. Principal applications are in industrial processes and ventilation requiring moderate static pressures and the need for simplicity of fan installation in a straight duct.

Vane-Axial Fans. The vane-axial fan resulted from the development of the propeller fan using refined aerodynamic principles and precise manufacturing procedures and control. Where such principles and techniques are applied, excellent capacity, pressure, efficiency, and sound emission levels are attained. Some units have mechanical efficiencies above 90%. High efficiency vane-axial fans are more efficient than comparable centrifugal fans, and have been used for energy conservation in Europe for some time. U.S. industry has also shown interest in large vane-axial fans for energy conservation in applications such as electric power boiler service (18,19).

The vane-axial fan wheel has short, stubby airfoil blades mounted on a hub which may be as large as 75% of the wheel diameter. The air leaving the axial-flow wheel has an appreciable rotational component which can be converted to static pressure in a suitably designed set of stationary straightening vanes. The straightening vanes are shaped to pick up the air leaving the wheel blades without shock. Although straightening vanes of airfoil cross section are theoretically desirable, vanes formed of pressed heavy sheet metal are less expensive. The motor is enclosed in a housing having the same diameter as the hub and has either a rounded cap or a bullet-shaped tail to reduce eddy losses. The straightening vanes surround the motor housing and can serve as structural supports for the housing. Generally, the number of guide vanes exceeds the number of propeller vanes by one, with the numbers selected so that there is no common divisor for the number of hub vanes and guide vanes. This minimizes flow pulsation and noise. Single-stage fans can develop pressures to 1.5 kPa (6 in. of water) with some designs going as high as 2.25 kPa (9 in. of water). Standard designs are available either belt-driven or directly connected to motors with speeds as high as 3450 rpm. In addition, two-stage units have been developed that produce considerably higher pressures but have received little industrial use. Performance curves show a dip to the left of the pressure peak (Fig. 10). Whereas vane-axial-flow fans can be designed that do not have such dips, those that do have dips should be operated to the right of the pressure peak. The principal advantage of the vane-axial fan is compactness and convenience of use in inline ducts, plus its better efficiency when carefully designed. The higher manufacturing precision required generally eliminates any cost savings that might result from its smaller size.

Capacity Control. *Variable Air Flow Fans.* Variable air flow fans are needed in the process industry for steam or vapor condensing or other temperature critical duties. These also produce significant power savings. Variable air flow is accomplished by (*1*) variable speed motors (most commonly variable frequency drives (VFDs); (*2*) variable pitch fan hubs; (*3*) two-speed motors; (*4*) selectively turning off fans in multiple fan installations; or (*5*) variable exit louvers or dampers. Of these methods, VFDs and variable pitch fans are the most efficient. Variable louvers, which throttle the airflow, are the least efficient. The various means of controlling air flow are summarized in Table 3.

Variable frequency drives are based on the principle that motor speed is a direct function of the frequency of the alternating current. In other words, a frequency of 60 Hz produces 100% speed; 30 Hz frequency produces 50% speed. The development of these drives is receiving much attention and both costs and the size of the controllers are steadily decreasing. Many options for control are programmable via keyboards mounted on the control boxes. One advantage of VFDs is that often several types of soft start options, ie, variable ramp times, are available as well as digital readout of many functions, etc. The main advantage of the VFD is that once the operating point of the fan is selected, that efficiency is carried at all speeds and flows. Capacity is directly related to motor frequency and thus to speed. Another important benefit is that as the fan speed decreases, fan noise and vibration also decrease significantly.

Table 3. Variable Air Flow Devises

Device	Air flow control	Cost		Noise
		Initial	Operating	
variable speed	continuously variable	high	lowest	decreases with speed
variable pitch	continuously variable	medium	low	constant
two-speed motor	full/half	medium–low	medium–low	decreases at half-speed
outlet louvers	variable	medium	highest	constant

Variable pitch fans or controllable pitch fans operate by means of specifically designed hubs which permit changing blade pitch while the fan is in operation. These are available in fans of 1.5 m to 6 m diameter for use in the petrochemical industry consuming from 3.7 to 56 kW. Large fans in the utility industry can consume up to 373 kW and work at up to 12.5 kPa (50 in. H_2O). These fans typically are pneumatically operated by 20–103 kPa (3–15 psi) or 4–20 mA signals, and are designed to fail to maximum air flow in the event of signal pressure loss. When fan speed is constant, fan noise remains almost constant, even if air flow is essentially zero. Typically the range of air flow is an essentially linear decrease from maximum to zero flow, the pitch being regulated by the controller. Typically, a 20 kPa (3 psig) signal relates to 100% (design duty) whereas a 103 kPa (15 psig) signal relates to essentially zero air flow.

A variable pitch fan has the unique capability to produce significant amounts of negative air flow when adjusted to do so. Negative flow, in the opposite direction from normal flow, is used in some refinery services to keep from over cooling some types of fluids in the tubes which have low pour points, and in cooling tower services to periodically deice the tower inlet louvers. The variable pitch fan can produce approximately 60% of its upward flow in the negative direction at maximum power. Another advantage of variable pitch fans is that the initial cost is generally 50% or less compared to VFD-type drives.

Two-speed motors are typically used on noncondensing services where the process is not sensitive to temperature but mostly seasonal or variable throughput of fluids in the air cooler requires some degree of air flow control. This is a simple, rather inexpensive means to control air flow when volume air flow is not critical. Typical motor ratings are 1800/900 rpm, although 1800/1200 rpm types are available.

Air control louvers or dampers, popular in the past for air flow control, are used primarily for only very low scale air flow control. Louvers are used in many winterized heat exchangers in extremely low ambient temperature locations to retain and recirculate warm air in completely enclosed heat exchangers, sometimes in very complicated control schemes. The use of louvers on the discharge side of a fan to control air flow is inefficient and creates mechanical problems in the louvers because of the turbulence. A fan is a constant volume device, thus use of louvers to control flow is equivalent to throttling flow through a valve. The fan keeps moving the original amount of air at the original power even when the net flow is reduced by the partially closed louver. This is very wasteful of energy. Louvers are controlled by pneumatic or electric actuators utilizing the 20–103 kPa (3–15 psig) or 4–20 mA signal. Large banks of louvers are linked together by mechanical linkages.

Fan Rating. Axial fans have the capability to do work, ie, static pressure capability, based on their diameter, tip speed, number of blades, and width of blades. A typical fan used in the petrochemical industry has four blades, operates near 61 m/s tip speed, and can operate against 248.8 Pa (1 in. H_2O). A typical performance curve is shown in Figure 11 where both total pressure and velocity pressure are shown, but not static pressure. However, total pressure minus velocity pressure equals static pressure. Velocity pressure is the work done just to collect the air in front of the fan inlet and propel it into the fan throat. No useful work is done but work is expended. This is called a parasitic loss and must be

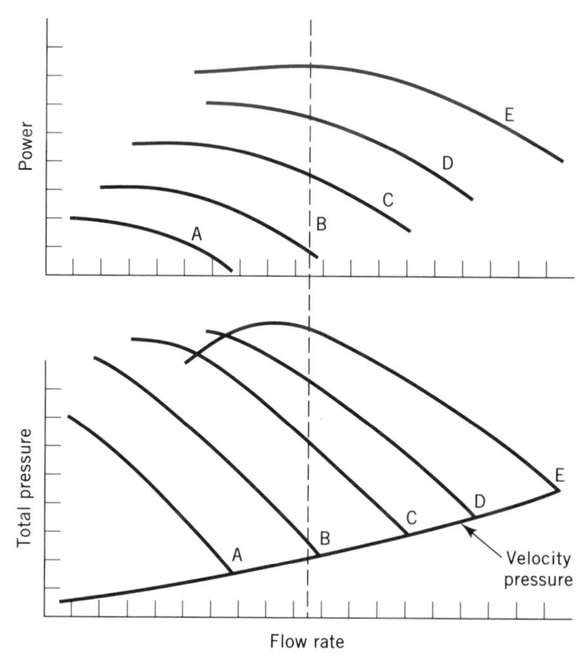

Fig. 11. Performance curve for an axial fan where A, B, C, D, and E represent fan pitch angles of 6°, 10°, 14°, 18°, and 22°, respectively.

accounted for when determining power requirements. Some manufacturers' fan curves only show pressure capability in terms of static pressure vs flow rate, ignoring the velocity pressure requirement. This can lead to grossly underestimating power requirements.

Efficiency. Fan efficiency describes a fan's ability to do work and is calculated as total efficiency (Eff) using total pressure (TP) or static efficiency (Eff) using static pressure (SP). Total efficiencies for axial fans range from 55 to 80%; static efficiencies range from about 40 to 65%. When pressure is in pascals and flow rate in m^3/s,

$$\text{total Eff} = \frac{\text{TP} \times \text{flow rate} \times 10^{-3}}{\text{kW}}$$

$$\text{static Eff} = \frac{\text{SP} \times \text{flow rate} \times 10^{-3}}{\text{kW}}$$

For flow rate in ft^3/min and power in brake horsepower, a correction factor of 0.157 must be applied to each equation.

An axial fan is a constant volume device. That is, a fan at a certain pitch moves a constant volume of air or gas at a constant speed and resistance (static pressure). If the density changes, the static pressure and wattage change, but the volume remains constant, ie, if the density (temperature) decreases, the static pressure and kW go up, but air flow remains the same.

Performance Curves. Fan manufacturers furnish fan performance curves for each type fan available. These are typically based on 61 m/s (12,000 ft/min) tip speed and 1.20 kg/m^3 (0.075 lb/ft^3) density. To select a fan for a specific duty requires knowledge of the flow, static pressure resistance, and density of the actual operating conditions. Usually the fan diameter is known as well as some idea of operating speed; a 61 m/s tip speed can often be assumed.

Selection. The fan selection process consists of determining the exact operating point that coincides with the design static pressure, air flow, and density required by the system resistance line. A typical procedure would be (*1*) select a fan curve of the appropriate diameter, assuming four blades for small fans, eight blades for larger fans; (*2*) calculate the required operating point which relates tip speed and pressures to the tip speed and density of the fan curve; (*3*) select a fan pitch which relates the flow rate and total pressure required; (*4*) read curve power requirement and pitch angle; and (*5*) relate power curve to actual rpm and density by using the fan laws. Most fan manufacturers gladly assist in fan selections. A fan operating point, the point where fan output exactly satisfies the system requirements, is shown in Figure 12.

Application Criteria. The design and construction of axial fans is dictated by size and function. Small fans are usually molded plastic having fixed blades. Most fans larger than one meter in diameter feature hollow fiber glass or extruded or cast aluminum blades that can be adjusted to the proper pitch when the fan is at rest, to provide the required air flow at the design speed. To perform properly, the output of the fan in terms of air flow and static pressure capability must match the system resistance at the design air density. This requirement dictates the fan diameter, rpm, number and types of blades, and blade pitch settings. If a fan has fixed pitch blades, the fan speed must be adjusted. Another increasingly important requirement is fan noise. To meet the maximum allowable noise, fan speed is normally limited. Often fan diameters are determined by space limitations as well as by volume flow requirements.

In petrochemical plants, fans are most commonly used in air-cooled heat exchangers that can be described as overgrown automobile radiators (see HEAT-

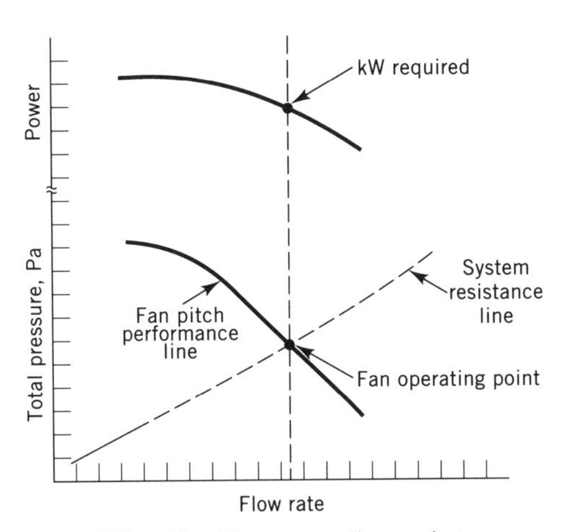

Fig. 12. Fan operating point.

EXCHANGE TECHNOLOGY). Process fluid in the finned tubes is cooled usually by two fans, either forced draft (fans below the bundle) or induced draft (fans above the bundles). Normally, one fan is a fixed pitch and one is variable pitch to control the process outlet temperature within a closely controlled set point. A temperature indicating controller (TIC) measures the outlet fluid temperature and controls the variable pitch fan to maintain the set point temperature to within a few degrees.

The utility industry utilizes fans typically from 6.7–10 m diameter in banks of 8 to 12 fans in wet cooling towers. These towers cool the water used to condense the steam from the turbines. Many towers may be needed in large plants requiring as many as 50 to 60 fans 12 m in diameter. These fans typically utilize velocity recovery stacks to recoup some of the velocity pressure losses and convert it to useful static pressure work.

Noise of Fans

Fan noise is demanding and receiving much attention because of environmental laws. The basic control document is the federal OSHA limitation of 90 dB(A) at an operator's work place for 8-h exposure. There are other limitations on entire plant noise at the boundary of new plants from local ordinances which are typically more severe than the OSHA limitation (see NOISE POLLUTION AND ABATEMENT METHODS).

Two terms are important: the sound pressure level (SPL) and sound power level (PWL). SPL or Lp is expressed as decibels (dB). PWL or Lw is a number representing the energy level of the noisemaking device, also in dB. The terms dB(A) and octave bands refer to the sound spectrum as a whole, where the dB(A) is a weighted sum of a spectrum represented by a single number. The A weighting system represents the response of the human ear, ie, low frequency sounds are not distinguished as well as high frequencies. For a more demanding description of noise, the octave band center frequencies are used. The most frequently used octave bands are 32, 63, 125, 250, 500, 1k, 2k, 4k, and 8k Hz.

To utilize a noise limitation specification, several items must be known: (*1*) the specification of SPL in decibels expressed either in dB(A) or octave band levels; (*2*) the distance from the measurement point to the geometric center of the noisemaking device or array; and (*3*) the quantity of noisemaking devices (fans) in the array. Additionally, noise attenuates with distance by the relation $10 \log_{10}(1/12.57\, R^2)$ when R is line of sight distance in meters from source to measurement point (the term is $-20 \log R$ when R is in feet). Noise from multiple sources increases by the relation $10 \log N$ where N is the number of identical sources in an array. For example, two identical fans produce 3 dB more noise than one fan ($10 \log 2 = 3$). The terms SPL and PWL are related by $\text{SPL} = \text{PWL} + 10 \log_{10}(1/12.57\, R^2) + 10 \log N$. For example, if a unit having two fans must not exceed a SPL of 64 dB(A), and it is 10.7 m (35 ft) from the measurement point to the center of the noise source, to determine the fan operating conditions the PWL must be used. Solving for PWL and using R in m, $\text{PWL} = \text{SPL} - 10 \log_{10}(1/12.57\, R^2) - 10 \log N = 91.9$ dB(A). Therefore the fan designer must select a fan that meets the design airflow needs at the design static pressure but produces no more than the PWL of 91.9 dB(A).

The required PWL is used to determine the tip speed of the fan. The velocity of the blade tips equals the rpm $\times$ dia $\times$ π. The American Petroleum Institute (API) determines PWL through the equation

$$\text{PWL dB(A)} = 36 + 30 \log U\text{tip} + 10 \log \text{kW}$$

where Utip is tip speed in m/s and the power absorbed by the fan is in kW. For tip speed in ft/min and power in brake horsepower (HP), the equation becomes

$$\text{PWL dB(A)} = 56 + 30 \log \frac{\text{tip speed}}{1000} + 10 \log \text{HP}$$

Those values also determine the rpm to operate the fan so that the PWL can be met.

A typical axial fan produces most noise at low frequencies, eg, 63, 125, and 250 Hz. A frequency spectrum for a 4.3 m diameter fan operating at 61 m/s is shown in Figure 13.

Effect of Vibration. All objects have a natural frequency of vibration when struck sharply and fan rings, blades, structure, etc are no exception. Vibration is usually sinusoidal and its frequency measured in Hertz. The travel or displacement of the vibration is measured in mils (1/1000 of an inch) in the United States but in micrometers elsewhere. Another measurement is velocity (mm/s) of movement.

A fan blade is continuously vibrating millions of cycles up and down in operation over a short period of time. Each time a blade tip moves past an obstruction it is loaded and then unloaded. If forced by virtue of tip speed and number of blades to vibrate at its natural frequency, the amplitude is greatly increased and internal stresses result. It is very important when selecting or rating a fan to avoid operation near the natural frequency. The most common method of checking for a resonance problem is by using the relation:

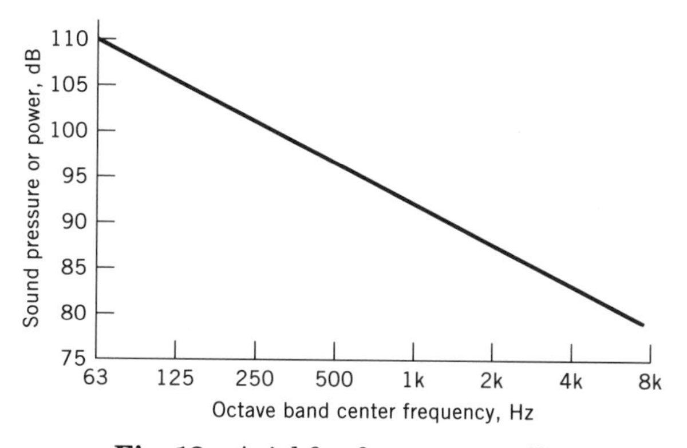

Fig. 13. Axial fan frequency profile.

$$\mathrm{BPF} = \frac{\text{no. blades} \times \text{rpm}}{60}\ \mathrm{Hz}$$

when BPF is the blade passing frequency. This is compared to the blades first and second mode natural frequency to make sure there is sufficient safety margin. Conventional margins are 5–20% difference between resonant frequencies and BPF. Another term used is that of critical speed which is a way of checking that the first or second mode resonant frequency is not selected for fan operation.

$$\mathrm{rpmc1} = \frac{\mathrm{Fn1} \times 60}{\text{no. blades}}$$

where rpmc1 = critical speed and Fn1 = first mode critical frequency. Similarly, rpmc2 can be calculated to check for second mode resonance. Important points regarding vibration and fans are fan rotor unbalance occurs only at 1 times running speed; vibration at blade pass frequency is usually caused by aerodynamic problems or structure resonance; and vibration at motor speed usually indicates misalignment or unbalance of a drive shaft. Resolving fan vibration problems usually requires knowledge not only of amplitude but also frequency of vibration. Fan rotor unbalance can be corrected by a dynamic fan balance operation in-place. Vibration at blade pass frequency is harder to resolve because it often relates to a resonance in the structure or some component, and the only means to resolve this type of problem is to change fan speed which is often expensive, or to change the number of blades in the fan.

A design guide for air-cooled heat exchangers in refinery services specifies 0.15 mm (6 mils) maximum vibration on motors and principal structural components (20). Generally, large fans are slow-moving compared to large, high speed turbines so amplitudes can generally be larger without damage.

Fatigue from Vibrations. To avoid the possibility of fatigue caused by a resonance, more extensive use of fiber glass blades is being made rather than metallic blades such as aluminum. Fiber glass composite blades, which are often hollow, are not notch sensitive, ie, a small scratch or crack does not spell the disaster it would in a metal blade. Secondly, the lighter mass of a fiber glass blade means less kinetic energy to dissipate in the event of an accident, because the destructive energy in a fan wreck is directly proportional to the mass weight of the blades.

Uses

Fans and blowers are the most widely used mechanical devices for moving air and gases in both large and small volumes (21). Uses include ventilation, mechanical draft for combustion (including forced- and induced-draft fans and primary- and secondary-air fans), local exhaust for fume and dust containment at hoods and equipment enclosures, forced- and induced-draft cooling for spray towers, cooling towers and ponds, and air-cooled heat exchangers, and conveying of solids (see also HEAT-EXCHANGE TECHNOLOGY). Other applications include air or gas movement in dryers, gas-recirculation fans, air supply for air curtains and air-blast operations, and a great many miscellaneous process industry uses often involving hot and corrosive gases. The range of performance required by fans for these various applications is enormous. Most ventilating applications require pressures ranging from 25 to 1500 Pa (0.1–6 in. of water). Induced-draft fans must often handle gases of 150–425°C containing various levels of suspended erosive particles. Such fans are frequently equipped with replaceable wear pads of abrasion-resistant materials or are coated with wear-resistant surfaces.

Forced-draft fans generally operate on clean air and at pressures from a few hundred Pa to as high as 20 kPa (80 in. of water) for pressurized furnaces. Backward-inclined blading is used almost exclusively for high efficiency. Blades with airfoil contours give improved structural strength, higher efficiency, and lower sound levels in large fans. Conveying systems in which the solids pass through the fan almost always use low speed wheels of the radial-blade paddle-wheel-type of construction. In the area of hot and corrosive process gas handling, fan designs are adapted to the specific need of the process. Frequently stainless steel or other alloy construction is required. Where dilution of the gas with atmospheric air is objectionable, the fan shaft is equipped with a stuffing box, a rubber labyrinth seal, or even a purged rotary seal depending on the degree of contamination control required. Occasionally, fans must handle gases having sticky or tarry particulates where solids buildup can occur. Continuous or intermittent flushing of the fan with a liquid spray in the fan inlet is helpful. In addition to deliberate flushing, fans may be called on to handle gases containing mist or entrained liquid droplets. A large percentage of such mist may be collected and agglomerated in the fan, particularly if operated at a high tip speed. Liquid-handling fans must be equipped with oversize motors as the acceleration of the liquid within the fan can utilize considerable power. Particular attention must be paid to the corrosiveness of the wet–dry environment within the fan. The presence of chloride ions and high wheel stress can lead to stress corrosion cracking in stainless-steel wheels. Although elimination of the chlorides is the best solution, the use of much lower wheel-tip speeds and wheels that can be stress-relieved to remove residual fabrication stresses is often helpful. Fans with rubber and polymeric coatings are often useful in moist environments, but special considerations in fan design are necessary to assure thorough bonding of such coatings to the wheel. Buildup of liquid within the fan casing can also be a problem with liquid-handling fans. The use of bottom–horizontal discharge designs with a large discharge duct drain is generally more satisfactory than a small fan-housing drain.

The choice between axial-flow and centrifugal fans in certain applications is by no means clear-cut. Axial fans have an advantage when compactness is important and when straight through flow benefits the installation and frequently, they also have higher efficiency which is important in energy conservation. Advantages of the centrifugal fan are the better ability to cope with fluctuating operating conditions, conditions that could result in unstable fan operation; greater ability to vary fan performance through speed changes; better access to the fan motor and greater facility to provide sturdy structural support for the fan and motor; generally lower noise level; and natural adaptability to design situations where a 90° turn in gas direction is desirable.

Centrifugal blowers may be obtained from American Fan Co., Fairfield, Ohio; TLT-Babcock, Inc., Akron, Ohio; Buffalo Forge Co., Buffalo, New York; Zurn

Industries, Clarage Fan Division, Birmingham, Alabama; Lamson Corp., Centrifugal Air Systems Division, Syracuse, New York; and Hartzell Fan, Inc., Piqua, Ohio. Axial fans are available from Hudson Products Corp., Houston, Texas; Buffalo Forge Co., Buffalo, New York; Hartzell Fan, Inc., Piqua, Ohio; Howden Sirocco Inc., Hyde Park, Massachusetts; Moore Co., Marceline, Missouri; and Aerovent, Inc., Piqua, Ohio.

BIBLIOGRAPHY

"Fans and Blowers" in *ECT* 3rd ed., Vol. 9, pp. 768–794, by B. B. Crocker, Monsanto Co.

1. *ASHRAE 1992 Handbook - HVAC Systems and Equipment*, American Society of Heating, Refrigerating and Air-Conditioning Engineers, Inc., Atlanta, Ga., 1992.
2. *Laboratory Methods of Testing Fans for Rating AMCA Standard 210-85*, Air Moving and Conditioning Association, Arlington Heights, Ill., 1990.
3. *Methods for Calculating Fan Sound Ratings from Laboratory Test Data, Standard 301*, Air Moving and Conditioning Association, Arlington Heights, Ill., 1990.
4. R. Jorgensen, *Fan Engineering*, 8th ed., Buffalo Forge Co., Buffalo, N.Y., 1983.
5. J. L. Alden and J. M. Kane, *Design of Industrial Exhaust Systems for Dust and Fume Removal*, 4th ed., Industrial Press, New York, 1970.
6. *AMCA Fan Application Manual, AMCA Publication B200-3*, Air Moving and Conditioning Association, Arlington Heights, Ill., 1990.
7. J. W. Market, *Heat., Piping, Air Cond.* **42**, 100 (Oct. 1970).
8. D. G. Traver, in *Fan Application - Testing and Selection Symposium Papers, San Francisco, Calif., Jan. 1970*, ASHRAE, Atlanta, 1972.
9. *AMCA Fan Application Manual, AMCA Publication 201*, Air Moving and Conditioning Association, Arlington Heights, Ill., 1973, Sect. 1.
10. D. H. Cristie, *ASHRAE Trans.* **77**, 84 (1971).
11. L. S. Marks and E. A. Winzenburger, *Trans. ASME* **54**(21), 213 (1932).
12. H. F. Farquhar, "Outlet Ducts–Effect on Fan Performance" in Ref. 8, pp. 7–10.
13. Ref. 4, pp. 295–313.
14. Ref. 4, pp. 271–283.
15. Ref. 4, pp. 285–288.
16. J. W. Martz and R. R. Pfahler, *Hydrocarbon Process.* **54**, 57 (June 1975).
17. Ref. 1, p. 3.2.
18. C. E. Wagner, *Combustion* **47**, 20 (Mar. 1976).
19. C. C. Curley and P. Olesen, *Combustion* **48**, 23 (Sept. 1976).
20. *Air-Cooled Heat Exchangers for General Refinery Service*, API 661, American Petroleum Institute, Washington, D.C., Apr. 1992.
21. Ref. 4, Chapt. 14–22.

General References

T. Baumeister Jr., *Fans*, McGraw-Hill Book Co., Inc., New York, 1935.

W. C. Osborne, *Fans*, 2nd ed., Pergamon Press, New York, 1977.

J. K. Salisbury, ed., *Kent's Mechanical Engineers Handbook*, Power Vol., 12th ed., John Wiley & Sons, Inc., New York, 1950.

R. Pollak, *Chem. Eng.* **80** (Jan. 1973).

J. B. Graham, in *Fan Application - Testing and Selection Symposium Papers, San Francisco, Calif., Jan. 1970*, ASHRAE, New York, 1972.

J. Thompson, *Plant Eng.* **31** (May 1977).

R. C. Monroe, *Hydrocarbon Proc.*, (Dec. 1980).

R. C. Monroe, *Chem. Eng.*, (May and June 1985).
American Petroleum Institute STD 661, Washington, D.C., Apr. 1992.

Noise Control

L. L. Beranek, *Noise and Vibration Control,* Institute of Noise Control Engineering, Cambridge, Mass., 1988.
A. Thumann and R. Miller, *Fundamentals of Noise Control Engineering,* 2nd ed., The Fairmont Press, Inc., Lilburn, Ga., 1990.
W. Neise, *Sound Vibration* (Apr. 1976).
J. B. Graham, *Can. Min. J.*, (Oct. 1975).
T. W. Rimmer and R. J. Anderson *In-Duct Measurement of Centrifugal Fan Sound Power,* ASHRAE, Sept. 1976.

ROBERT C. MONROE
Hudson Products Corporation

FAST COLOR SALTS. See AZO DYES.

FAT REPLACERS

Early workers in the nutrition field analyzed foods for measureable constituents including fat, protein, carbohydrate, water, minerals, and caloric content. A few centers of nutritional research excellence arose in the late 1910s and early 1920s. Early work on lipids in the human diet and health was done at the University of Minnesota (1). Following this early work, increased interest and emphasis on the quantity and nature of edible fats led to accumulating knowledge that fats, under some circumstances, cause health problems for at least a segment of the population.

Massive studies of the health problems associated with fats have included the Anti-Coronary Club study in New York City (2), the Framingham study (3) in Massachusetts, and the Chicago study (4). These studies were financed by the United States federal government and supported by the Surgeon General's office, which issued a statement recommending that the fat component of the diet be reduced from an estimated 40% of calories to 30% (5).

The food producing industry has responded to consumer demand for foods with lower fat content (Table 1). Foods with low or no cholesterol claims leaped 78% from 1980 to 1990, in spite of the fact that many of the principal food producers reduced the amount of new product introductions during 1989 and 1990 (7). Table 2 indicates the change in the market for various food industry segments, especially those suspected as fat problem generators, including dairy and meat

Table 1. Food Products With Health Claims[a]

	Products per year			
Product type	1980	1989	1990	Change, %
low calorie[b]	475	962	1165	+21.0
reduced fat	275	626	1024	+64.0
low/no cholesterol	126	390	694	+78.0

[a]Ref. 6.
[b]According to the Food Labeling and Education Act (1992), low calorie must contain less than 40 calories per serving and 100 grams of food; reduced calorie must be reduced by at least one-third and by at least 40 calories; and calorie-free must contain fewer than 4 calories per serving.

Table 2. Retail Sales, Low Fat, Low Calorie[a] **Foods, 10^9 $**[b]

Product	1989	1990	1991	1992	1993	Growth, %
dairy	16.3	17.5	18.5	19.6	20.5	+5.9
frozen desserts	1.6	1.9	2.2	2.5	2.8	+15.0
frozen dinners	1.0	1.2	1.5	1.8	2.2	+21.8
salted snacks	0.8	1.0	1.2	1.4	1.6	+18.9
cookies and crackers	0.5	0.6	0.7	0.8	1.0	+18.9
baked goods	0.3	0.4	0.5	0.6	0.7	+23.6
mayonnaise and margarine	0.7	0.7	1.1	1.4	1.5	+21.6

[a]1 food calorie = 1 kcal = 4.184 kJ.
[b]Ref. 8.

foods. Many low fat and low cholesterol foods were created by adding claims to food that have always been low in fat and/or cholesterol.

Some nutritionists have voiced concern at the rapid transition to reduced fat consumption, eg, a diet tending to fat omission might lead to an even less desirable replacement (9). One study has shown that replacement of dietary fat with other macronutrients causes no significant difference in total macronutrient intake (10); therefore, fat-free diets help reduce fat intake but at no loss of energy intake (11). In another study where subjects were fed diets at different energy levels, when fat was reduced, the subjects made up for the energy loss by increasing intake of other foods to yield a similar energy intake level in both low fat and usual fat level groups (12). One explanation is the postulation of a dietary intake set point which seems to govern obese as well as normal intake subjects (13).

A USDA report indicates that between 1967 and 1988, butter consumption remained stable at 2 kg per capita, margarine dropped from 5.1 to 4.7 kg, and measured total fat intake per day dropped from 84.6 to 73.3 g (14). This study also projects that the reduced consumption of tropical oils is only temporary and will return to former use levels, possibly even higher. One reason for this projected rise in tropical oil consumption is the knowledge of the beneficial effects of medium-chain length acids high in lauric oils. There is a keen interest in omega-3 fatty acids, as well as linoleic acid, contained in fish oils.

Chemistry

Several texts provide information on fat chemistry, ie, industrial edible fats and oils (15); shortening and margarine (16,17); fish oils, high in omega-3 fatty acids (18); and soy oil widely used in food processing (19). There are general information texts on general properties of food fats and uses in the food processing industry (20–22), and a brief but thorough coverage of the role of fats in human nutrition (23).

The simplest nutritional role of fats in the diet is that of energy supply. There are differences between members of the same class of food materials, but the accepted convention attributes a value of 9 kcal (37.7 kJ) of energy per gram of fat, and 4 kcal (16.7 kJ) of energy per gram of all carbohydrates and proteins. This is a serious consideration in generating weight loss diets.

The basic structure of most edible fats is that of a triglyceride, ie, a glycerol backbone with up to three fatty acid residues attached via an ester linkage. There are mono-, di-, and triglycerides depending on the number of fatty acids attached to the glycerol (see FATS AND FATTY OILS). Fatty acids of physiological interest may have from 4 to 30 carbon atoms in the chain, but most are under 20. They may be fully saturated, ie, have only single bonds connecting the C atoms, or unsaturated fats, ie, having double bonds between one or more of the carbon atoms (see CARBOXYLIC ACIDS). Unsaturated fats are of extreme interest to those working in public health and fat ingestion.

The state of knowledge in the early 1990s of the effects of fat on health lacks clarity and general agreement. There is great support for the thesis that fully saturated fats are associated with problems of atherosclerosis and arterial fatty deposit, but there is evidence that stearates, which are saturates, are only poorly utilized in human digestion. Another body of work has established a connection between unsaturated fatty acids and a better state of arterial health and lowered fat body attachment to the arterial wall (23); contrary evidence exists that highly unsaturated fats polymerize more readily and thus contribute to arterial plaque formation.

The unsaturated fatty acids, linoleic [*60-33-3*] and linolenic [*463-40-1*], contain two and three double bonds and are considered beneficial components of the diet. The double bond is an essential ingredient for human nutrition when it is in the correct position on the fat molecule. Humans are unable to insert the double bond at the omega-3 and -6 position. Therefore, fatty acids containing double bonds at these positions are essential in the diet, including linoleic and linolenic acids. They are accordingly described as essential fatty acids (EFA) (23).

The saturated fatty acids, stearic [*57-11-4*] and palmitic [*57-10-3*], are found in animal fats and dairy products. Extensive studies point to the deleterious effect of these acids on arterial walls; as a result it is recommended that saturated fatty acid intake be carefully controlled and intake limited (23).

There are physical–chemical differences between fats of the same fatty acid composition, depending on the placement of the fatty acids. For example, cocoa butter and mutton tallow share the same fatty acid composition, but fatty acid placement on the glycerin backbone yields products of very different physical properties.

The metabolism of triglycerides has been viewed traditionally as unaffected by fatty acid composition, but work since the 1970s has demonstrated otherwise.

Most fats, including soy, corn, and safflower oils, are composed of long-chain fatty acids. A group of oils, including coconut, babassu, and palm kernel oil, are composed of fatty acids of C-14 and less and are known as the lauric fats because they contain as high as 45–50% lauric acid [*143-07-7*]. The difference between the long-chain and lauric fats is the route of absorption in the human body. Long-chain fatty acid oils are absorbed after emulsification by the lymphatic system. Lauric fats are transported directly to the liver by the portal system, where they are oxidized and metabolized as rapidly as carbohydrates. These medium-chain glycerides lower blood and tissue cholesterol in animals and humans. They have a very low tendency to deposit as depot fat, and thus are of interest in the control of obesity (24). This leads to the postulation that one could structure fats by adding medium-chain length fatty acids to the glycerol backbone and produce fats that are to be used for energy. The medium-chain length fats produce only 4 or 5 kcal of energy as opposed to the 9 kcal of long-chain fatty acid oils (25). A biotechnology company has been able to insert the laurate gene into rapeseed and produce high levels of lauric acid in the oil fraction. Commercial production of high lauric canola oil from rapeseed is expected in 1993.

Fatty acids are susceptible to oxidative attack and cleavage of the fatty acid chain. As oxidation proceeds, the shorter-chain fatty acids break off and produce progressively higher levels of malodorous material. This condition is known as rancidity. Another source of rancidity in fatty foods is the enzymatic hydrolysis of the fatty acid from the glycerol. The effect of this reaction on nutritional aspects of foods is poorly understood and little research has been done in the area.

Fat in Foods

Gums and starches were used in early attempts to replace the viscosity and lubricity of oils in foods. These were not well received by consumers because they assumed fats merely supplied mouthfeel and a bit of flavor. On closer examination, it became evident that fats in food and in the diet performed many roles, some simple, some extremely complex, some understood, and some not understood.

A good compilation of the functions of fats in various food products is available (26). Some functions are quite subtle, eg, fats lend sheen, color, color development, and crystallinity. One of the principal roles is that of texture modification which includes viscosity, tenderness (shortening), control of ice crystals, elasticity, and flakiness, as in puff pastry. Fats also contribute to moisture retention, flavor in cultured dairy products, and heat transfer in deep fried foods. For the new technology of microwave cooking, fats assist in the distribution of the heating patterns of microwave cooking.

The deleterious effect of some fat substitutes has been demonstrated in cake frosting (27); the result is an unacceptable frosting, filled with air bubbles. In another example, some low fat cheeses are quite acceptable when cold, but when heated result in a product texture that changes to a sticky, gummy mass. Attempts to replace fat must be viewed as a total systems approach (28,29). It is likely that no one material will replace fats in food; rather, replacement will consist of mixtures with each ingredient addressing one or more of the roles played by fats in food.

When fat is present as the disperse phase of an emulsion, it contributes characteristic rheological and visual properties to foods thought of as creamy, eg, cream, mayonnaise, and pourable salad dressings. In standardized ice creams, fat is a significant factor in ice cream overrun, ie, the ability of the chilled mix to entrap air during freezing agitation, thus producing a lighter texture. When a separating salad dressing is compared to a creamy dressing, the dispersed oil droplets of the creamy dressing provide the rheological properties of increased viscosity, thixotropy, and viscoelasticity as well as visual whiteness and opacity.

In chocolate, cocoa butter is the continuous phase. The characteristic meltability of cocoa butter constitutes a puzzle in chemical structure and poses difficulty in replacement; cocoa butter has a sharp melting point at body temperature.

Fats of high melting point contribute significantly to the textural stability of many sausages and meat mixtures, eg, the smearability of pâté de foie gras in contrast to the solidity of summer sausage.

One disadvantage of fats contained within foodstuffs is the deterioration of the fat through oxidative rancidity. Many consumers find the aroma and flavor of deteriorated fats in foods repulsive, while others are fond of country ham and butter which owe their aroma and flavor to fat rancidity and other breakdown products. The use of antioxidants (qv) makes such products commercially viable.

Fats can be an important source of lubrication in the preparation and consumption of foods (30). Marble slabs on which hot candy is poured are lubricated with fat to prevent sticking. Also, bread and cake pans are treated with heat-stable edible oil.

Fats also contribute gloss to foodstuffs. Of special interest are chocolates and chocolate frostings as well as other frostings and glazes. Many foods are sprayed with a light coating of oil after fabrication to achieve gloss on the finished product and to retard desiccation. Gloss is particularly useful when applied to some crackers and cookies. The residual oil on oil-roasted nuts yields gloss, resulting in an attractive appearance.

Lubricity appears to be easier to measure and to replace in lower fat foods. Carbohydrates and particulated proteins can be formulated so that they help in this important aspect of fats.

Another feature contributed to foods by fats and oils is mouthfeel. Mouthfeel is a difficult attribute to emulate since it appears to be a combination of several factors including viscosity, body, lubricity, and mouth coating. There are effects on the cheeks, tongue, and back of the throat. Other mouthfeel properties include resistance to chewing or change in viscosity during mastication, and other factors yet to be identified.

Some pioneering work has been done on the effect of particle size on mouthfeel and texture perception (31). When particles of food materials are smaller than 0.1 μm they impart no sense of substance and the consumer calls the product watery. Particles of 0.1–3.0 μm are sensed as a smooth rich fluid, but when the particles exceed 3 μm the food is perceived as chalky or powdery. By controlling particle size, desirable creaminess can be obtained (32).

Fats contribute to the rheological properties in flowable and pastry foods. By combining with starches to form a clathrate, a product different from the native starch is formed, eg, shortening in baked goods. The highly developed shortness of pies baked in earlier times resulted from the use of high levels of lard. The use of less fat in pie crusts is evident, ie, the crusts are harder and readily become soggy.

The most difficult property of fat to replace is flavor. Great expenditure of effort has gone into producing a true butter flavor as flavor boosters in nondairy fat products and in dairy products including milk, cream, butter, and ice cream. Results have led to a successful duplication of buttery flavors which closely match the intended target.

Fat Altered Foods. The Food Labeling and Education Act, effective November 1992 (33,34), defines low fat, low cholesterol, and fat-free foods. Low fat must contain less than 3 grams of fat per serving and per 100 grams of food, reduced fat must be 50% of the usual fat content and reduction must exceed 3 grams per serving, fat-free must have less that 0.5 grams of fat per serving, and percent fat-free must describe foods that fit into the definition of low fat. Foods labeled "lite" must be one-third lower in calories. If more than 50% of the calories are from fat, it must have the fat content reduced by 50%. If lite means other than calories, it must be specified.

With the extensive knowledge available in oil chemistry, development of designer fats and oils is possible (34). This is of special interest to nutritionists who see the possibilities for structurally designed fats to meet developing knowledge in clinical nutrition and food product development. There is more activity in dairy products than anywhere else in the food industry. Ice milk and frozen yogurt, early leaders in the field, rose rapidly in sales then plummeted. Fat-free ice cream has been marketed, but final results are not yet available. Sales of these products have not cannibalized traditional ice cream (35). Standards for traditional ice cream call for a minimum of 10% butterfat. One fat-free ice cream product is prepared from nonfat milk (skim) and cellulose gum. Fat-free ice creams have encountered strong resistance in some segments of the retail trade. Retailers in Maine and New York, states with important dairy producing industries, refuse to sell such products (36).

One frozen dessert is made with Simplesse, a protein-based fat mimetic that contains no fat (37). Other dairy product developments include a fat flavor, produced by encapsulating milk fatty acids in maltodextrins (38); fat-free cottage cheeses; and 2% fat milk, prepared by steam stripping cream with partial fat addback, with a cholesterol level about 60% lower than the starting material (39).

Activity in the cereal field includes the introduction of a full line of fat-free, cholesterol-free loaf cakes (23), crunch cakes, and cookies (40); a light frosting mix and a light pancake mix are also included (41).

Another significant fat source in everyday diets is salad dressing. Salads are difficult to masticate without some lubrication, which is usually furnished by dressing with its oil component. The fabrication of salad dressing without a fat ingredient is a difficult technical problem. Nevertheless there is available a fat- and cholesterol-free dressing, and a fat-reduced, but not fat-free, dressing (42). There are Federal Standards of Identity for dressings; one labeling requirement is a minimum fat content. The standards and labeling for lowered fat constitute a problem for the regulatory agencies (43). In 1990 there were 201 new salad dressings introduced, about one-fourth of which were low or reduced fat.

Another food product which faces standards problems is processed meat (44). A great deal of interest has been focused on hamburgers, especially by the fast food restaurants. Carrageenan and hydrolyzed vegetable protein has been used to produce a hamburger with less than 10% fat (45). Another approach uses frozen soy protein isolate to admix with ground beef to produce hamburger, with an accompanying drop in fat content from 42% down to 24% (46). Oatrim, a hydrolyzed oat flour, has also been employed as a mixer to produce a low fat hamburger, although at a somewhat elevated price (47).

The complexity of total replacement has slowed the rate of introduction of new materials, but most ingredient producers introduce a product which replaces one or two aspects of fat functionality and has already been cleared for use in foods by the FDA.

Fat Extenders

Carbohydrates. The materials offered for fat replacement are either carbohydrate or protein and protein-like compounds (29). Table 3 lists carbohydrate fat-sparing agents. Another listing is available (34) which includes materials offered on the European market. New compounds appear at such a rate that it is

Table 3. Carbohydrate Fat-Sparing Agents[a]

Trade name	Chemical composition	Producer
Amalean	high amylose corn starch	American Maize
Sta-Slim	modified potato and tapioca starches	Staley Mfg.
Stellar	hydrolyzed corn starch	Staley Mfg.
Paselli SA2	low DE hydrolyzed potato starch[b]	Avebe America
Oatrim	oat maltodextrin	Rhone-Poulenc
Maltrin	corn maltodextrins	Grain Processing
Sta-Dri	waxy maize maltodextrins, DE series	Staley Mfg.
Rice*Complete	hydrolyzed rice solids	Zumbro
Rice*Trin	rice maltodextrins	Zumbro
Rice*Pro	rice protein and maltodextrin	Zumbro
N-Lite B	corn maltodextrin	Nat'l Starch
N-Lite D	modified starch	Nat'l Starch
N-Lite F	mix of starch, milk, and guar	Nat'l Starch
N-Lite L	modified starch	Nat'l Starch
N-Lite LP	pre-gelled starch	Nat'l Starch
N-Oil	tapioca dextrin	Nat'l Starch
Slenderlean	modified starch	Nat'l Starch
Lycasin	hydrogenated corn starch hydrolysate	Roquette
Trim Choice	oat maltodextrin	Con Agra
Superbase	mix of rice maltodextrin, starch xanthan, and whey protein	Excel
Litesse	polydextrose	Pfizer

[a]Many of these products have multiple suppliers.
[b]DE = dextrose equivalent.

difficult to keep a current compilation; proceedings from an annual meeting are published (48).

Table 4 demonstrates that starches for fat replacement originate from corn, potato, tapioca, oat, and rice. Starches are comprised chiefly of straight (amylose) and branched (amylopectin) chains of glucose. The ratio of branched to straight chains has an effect on the nature of the resultant starch; this ratio varies with the starch source. Hydrolysis of these starches yields dextrins and maltodextrins, which are also useful in replacing food fats. Hydrolysis can be an enzymatic or acid reaction, and can be stopped at different stages of breakdown to yield a host of products from a single starch, eg, Rice*Complete, 10% protein and 90% maltodextrins; Rice*Pro, 25% protein and 75% maltodextrins; and Rice*Trin, 100% maltodextrins (Zumbro). Starting with whole rice, hydrolyzates of the complete cereal are also available. Hydrolysis is controlled to allow a predetermined degree of chain shortening, identified by the term dextrose equivalent (DE). The higher the DE value, the greater the chain-length reduction, eg, DE 3 is very high in

Table 4. Plant Gums Used as Fat Replacers[a]

Plant gum	Composition	Trade name	Producer
xanthan	D-glucosyl, D-mannosyl, D-glucosyluronic acid residues		Rhone-Poulenc
guar	galactomannan	Edicol	Indian Gum
locust bean	galactomannan		Gumix Int'l
arabic (acacia)	protein with branched polysaccharides		Colloides Naturels, Inc.
karaya	acetylated polysaccharide		TIC
tragacanth	bassorin and tragacanthin		TIC
konjac	glucomannan	Nutricol	FMC Corp.
ghatti	Ca–Mg salt of L-arabinose, D-galactose, D-mannose, D-xylose		TIC
agar	galactose sulfate polymer		Gumix Int'l
alginate	sodium polymannuronate		Kelco Merck
carrageenan	galactose and anhydro-galactose polymer		Kelco Merck
pectin	methyl polygalacturonate	Slendid	Hercules
dried plums	dried fruit	Prune Paste	Mariani
colloidal cellulose	cellulose	Avicel	FMC Corp.
methylcellulose	repeating methylcellulose	Methocel	Dow Chemical
hydroxylpropyl methylcellulose	oxypropylated cellulose methyl ether	HPMC	Dow Chemical
carboxymethyl-cellulose	sodium cellulose glycolate	CMC	FMC Corp./ Hercules
cellulose	cellulose	Solka-Floc	Fiber Sales
particulate protein	whey protein	Simplesse	Nutrasweet
particulate protein	zein and maltodextrin	LITA	Enzytech
particulate protein	whey protein	Trailblazer	Kraft

[a]Many of these products have multiple suppliers.

maltodextrins and is starch-like because hydrolysis has not been carried very far toward the simple sugars. Other rice-derived products offer DEs of 10, 18, and 25. Also available is a series of products prepared using only rice starch as a starting material, offered in DEs of 10, 18, and 25.

Somewhat analogous to these rice products is Oatrim, a material based on oat flour. It was developed at the Northern Laboratory of the USDA and is offered commercially by several firms. Oatrim contains 5% protein, 5% β-glucan, 2% pentosans, and 83% maltodextrins. Unique properties are claimed based on the β-glucan component, and preparation of Oatrim is disclosed in USDA publications and patents. Briefly, oat flour is broken down by α-amylase, then the water-soluble component is dried and is the product of commerce, aimed at ground meat product usage.

Similar materials are available based on potato starch, eg, Paselli SA2 which claims DE below 3 and has unique properties based on its amylose–amylopectin ratio peculiar to potato starch. The product contains only 0.1% protein and 0.06% fat which helps stabilize dried food mixes compounded with it. Another carbohydrate raw material is waxy-maize starch. Maltodextrins of different DE values of 6, 10, and 15, using waxy-maize starch, are available (Staley Co.). This product, called Stellar, is offered in several physical forms such as agglomerates and hollow spheres, and is prepared by acid modification (49). Maltodextrins based on corn starch are offered with DEs of 5, 10, 15, and 18 as powders or agglomerates (Grain Processing Corp.).

Each member of the N-Lite Series (National Starch) is aimed at a specific role in fat replacement, ie, N-Lite L for soups and sandwich spreads; N-Lite LP for use in dressings and dips. N-Lite LP is an instant modified food starch which yields a very oily texture and is stable to heat, acid, and shear; N-Oil is a tapioca dextrin suggested for fat-like mouthfeel; and the N-Flate and N-Oil Series are propriety mixtures of starches and other fat replacers.

Celluloses. Complex carbohydrates including gums, cellulose, methylcellulose, and carboxymethylcellulose also have found application in fat replacement. A good summary of the application of these materials is available (21).

Microcrystalline cellulose is a nonfibrous form of cellulose obtained by breaking fibrous plant cell walls into fragments, sized from 25 μm to a few tenths of a μm. In commercial preparations 60% of the fragments are below 0.2 μm. The fragments are of the same chemical composition as native cellulose and exhibit the same x-ray diffraction pattern. In use, microcrystalline cellulose is combined with (sodium) carboxymethylcellulose (CMC) [*9004-32-4*] to produce a colloidal gel. In this combination, CMC serves as a protective colloid, and permits ready dispersion of the powdered mix. When dispersed in a food, the mixture sets up a three-dimensional network held together by weak hydrogen bonds. This structure gives the mix its functional properties, ie, emulsion and foam stabilization, high temperature stability, thickening, suspension, and ice crystal control in frozen desserts.

Methylcellulose [*9004-67-5*] and hydroxypropylmethylcellulose [*9004-65-3*] are water-soluble gums derived from cellulose. They possess the ability to gel upon heating, a property unique among gums. Both products have polymeric backbones of cellulose, ie, a repeating structure of anhydroglucose units. They differ in the substitution units, methoxyl and hydroxypropoxyl, and both may vary depending on the degree of substitution. Degree of substitution influences solubility and thermal gel point. The properties of cellulose derivatives can be altered by admixture with other gums. One available mixture has different levels of guar gum, plus added maltodextrins and xanthan.

Plant Gums. There are a large number of plant gums that find application as fat replacers (Table 4) (see GUMS, INDUSTRIAL). Many are dried plant exudates. A good review of plant gums exists, giving sources, uses, and an extensive list of references (50). There is also a well-known text available (51).

Polydextrose (Pfizer) is prepared by high temperature polymerization of glucose in the presence of a catalyst. It is a water-soluble, amorphous solid used primarily as a bulking agent (52). Dried fruit, including prunes, and dried plum, date, and grape juice is used for similar applications (53).

A wide variety of plant exudates have been used in foods and medicines for centuries, including acacia, karaya, and ghatti. Plant gums derived from seeds include arabic, guar, locust bean, tamarind, and tara. All play a role in fat replacement either singly or in mixtures.

Gum arabic (acacia) is derived from the acacia tree grown primarily in Sudan and Senegal. It is a complex of calcium, magnesium, and potassium salts of arabic acid, with a molecular weight of about 250,000. Guar is derived from the endosperm of the guar plant. The guar plant is cultivated commercially in India and Pakistan; some is grown in the southwestern United States. On average, guar gum contains 80% galactomannan, 12% water, and 5% protein.

Locust bean (carob) is derived from the endosperm portion of seeds of a tree widely cultivated in the Mediterranean area. It is a polysaccharide built of mannose units with short branches of single galactose units, with an average molecular weight of 310,000.

Gum tragacanth is obtained from the large tap root and branches of a small perennial shrub found in the Middle East, especially Iran. Chemically, it is a mixture of water-insoluble polysaccharides. It is stable to heat, acidity, and aging, and is used extensively in pourable low calorie salad dressings.

Xanthan, although a gum, is derived from the pure culture fermentation of an organism, *Xanthomonas campestris*. The organism is filtered from the growth medium and the gum recovered by alcoholic precipitation, followed by drying. It is composed primarily of D-glucose and D-mannose units.

Konjac flour, derived from the konjac plant tuber, has a long history of use in the Far East, but is a newcomer to the United States. It reacts with many starches to enhance the viscosity of both, and is used in gels that are stable in boiling water.

Pectin, which occurs in most plants as the glue which binds the cells together, is extracted commercially from citrus peel and has been extracted from apple pomace. It is suggested for many no-fat products including sauces, desserts, and dressings.

Marine Gums. There are several related gums of marine origin. Carrageenan, from red seaweed, is probably the best known and yields three basic types of gum when extracted, ie, kappa, iota, and lambda; the difference in properties results from sulfate linkages on the repeating galactose units. There are red seaweed farms in the Southern Philippines. Carrageenan is used in low fat pourable salad dressing and low fat meats. It is the ingredient in a patented process used in a low fat hamburger, and is used in low fat frankfurters with a fat reduction of 30%. Agar, a gum known for its gel properties, is derived from red algae gathered from inshore sources, mainly in the Far East. Its unique property is firm gel formation that does not liquefy below 85°C. It is used for starch replacement in cereals (54).

Microparticulate Protein Fat Mimetics

There are only a few fat replacement products based on protein. LITA is a corn protein–polysaccharide compound; the role of the polysaccharide is to stabilize the protein (zein). The final product is 87% protein and 5% polysaccharide. The mixture, spray dried after processing, claims to look like cream on rehydration. It is low in viscosity, flavor, and lubricity, and is stable to mild heating. The protein particle size is 0.3–3 μm (55).

Simplesse, the best known entry in the protein field, is the subject of several patents (32). A good scientific description of the product and the process is available (56). The original product was prepared by treating acidified whey concentrate to simultaneous pasteurization and homogenization. The resultant product is spherical and is in the 1-μm particle size range. The product owes its effectiveness to this particle size, which is below the threshold of particle sensing and serves as a surrogate disperse phase. The effect is that of creaminess in the absence of fat. A later product is based on combined egg white and skim milk (33,57), and a further development employs unacidified whey protein concentrate. The original product has found application, on a commercial scale, in a frozen dessert. It has also been employed in a low fat cheesecake, low fat cheese spread, and several types of natural and processed cheeses. A low fat mozzarella has been employed in the manufacture of a low fat pizza. Salad dressings, table spreads, and sour cream have been commercialized. Trailblazer (Kraft-General Foods), a microparticle protein suspension produced from cheese whey (57), is similar to Simplesse. The long-standing practice of admixing isolated soy protein with ground meat to lower fat content should not be overlooked. Soy protein has found its best application in hamburger patties, though there are applications in other ground meat products. The previous limitation of beany flavor has been sufficiently reduced to allow these new applications commercial viability.

Synthetic Fats

In contrast to the above approaches is the strategy of creating new molecules with fat and oil-like physical properties, but a molecular configuration not recognized by the human digestive system; hence they are noncaloric. Olestra, the first such product, achieves its effect by attaching fatty acids to a sucrose, rather than a glycerol backbone. It has most of the qualities for fat replacement including utility in deep fat frying. It has been delayed since the 1970s because of lack of FDA clearance. There is a claim that the potential volume of use is so huge that it constitutes a new development in the approval process (58). A thorough history of this product and its origins is available (59,60).

The size of the fat replacement market has stimulated many attempts to synthesize fat replacer molecules. These efforts, described as attempts to produce "acaloric compounds with fat-like properties, but whose ester bonds have been modified" (60), would include glycerol ethers, pseudofats, and carbohydrate fatty acid esters. Research into this group of compounds indicates that as the number of ester groups increases, there is decreased digestion; ie, at eight ester groups there is virtually no digestive lipolysis and the compound is passed through undigested (61). The octaester has been shown to lower cholesterol, especially low density cholesterol. Mitsubishi and Unilever are both working on partially esterified sucrose.

Another group of synthetic fat compounds is centered around polycarboxylic esters. Trialkoxycarballylate (TATCA) and trialkoxycitrate (TAC) (62) (Best Foods) have been tried in mayonnaise and margarine. Similar materials have been developed based on malonic acid esters (60). Ether–triglyceride linkages also have been investigated, but preparation was found to be too expensive to sustain commercial viability (63). Work on propoxylated glycerols (Atlantic Refining Co. (ARCO)) is still at the bench level. Silicone oils have been investigated as fat replacers including phenylmethylpolysiloxane and organosilanes, but the controversy over breast implants and the possible side effects of silicones may cast a cloud over this product area.

Caprenin, suggested as a cocoa butter substitute, has useful functional properties with the added value of fewer calories than cocoa butter. It is a triglyceride comprised of two short-chain fatty acids and behenic acid (docosanoic acid) [*112-85-6*], a fully saturated 22-carbon fatty acid. Caprenin is found in high concentration in canola (rapeseed) oil, and can be synthesized from erucic acid, which is also high in earlier strains of canola. The obvious application of caprenin is in the candy and chocolate industry which has been hurt by its image as a high calorie food. Caprenin has had its first sales (64), probably for use in chocolate coatings. The Chocolate Marketing Association claims there is little demand for low fat chocolate (64). Candy, now consumed at 19 pounds per capita, is predicted to go to 25 pounds by 1995. The postulation is that the consumer focus is now on fat and cholesterol and this has led to a deemphasis of sugar rejection.

BIBLIOGRAPHY

1. A. Keys, *J. Chron. Dis.*, 364–380 (1956).
2. G. Christakis and co-workers, *Am. J. Pub. Health*, 299–315 (Feb. 1966).
3. T. Gordon, W. Castelli, M. C. Hjortland, W. B. Kammel, and T. R. Dawber, *Am. J. Med.* **67**, 707–714 (1977).
4. J. Stamler, in B. Rifkind, B. Dennis, and N. Ernst, eds., *Nutrition of Lipids and Coronary Heart Diseases*, Raven Books, New York, 1979, pp. 32–50.
5. C. E. Koop, *The Surgeons Report on Nutrition and Health*, Prima Publishing Co., Rocklin, Calif., 1988, 726 pp.
6. *Food Products Magazine, New Products Annual*, Gorman Publishing Co., Chicago, 1991.
7. *Prepared Foods, New Products Annual*, Gorman Publishing Co., Chicago, 1991, p. 60.
8. *Fiber Facts*, Williamson Fiber Co., Louisville, Ky., 1991.
9. R. Johnson, *Wall Street Journal*, New York, Feb. 3, 1991, p. 12.
10. A. Gillis, *JAOCS* **65**(11), 1708–1711 (1988).
11. B. J. Lyle, K. E. McMahon, and P. A. Kreutler, *J. Nutr.* **122** 211–216 (1992).

12. R. W. Foltin, N. W. Fishman, T. H. Moran, B. J. Rolls, and T. H. Kelly, *Am. J. Clin. Nutr.* **52**, 969–980 (1990).
13. A. J. Stunkard and T. A. Wadden, *Nutr. Rev.* **48**(2), 4 (1990).
14. *Reports, Food Consumption, Prices and Expenditures, Years 1967–88*, U.S.D.A., Washington, D.C., 1991.
15. M. W. Forma, E. Jungermann, F. A. Noris, and N. O. V. Sonntag, in D. Swern, ed., *Bailey's Industrial Oil and Fat Products*, Interscience Publishers, John Wiley & Sons, Inc., New York, 1979, 1022 pp.
16. M. T. Gillies, *Shortenings, Margarines and Food Oils*, Noyes Data Corp., Park Ridge, N.J., 1974, 330 pp.
17. W. H. Meyer, *Food Fats and Oils*, Institute of Shortening and Edible Oils, Washington, D.C., 1982, 22 pp.
18. M. E. Stansby, *Fish Oils*, Avi Publishing Co., Westport, Conn., 1967, 407 pp.
19. D. R. Erickson, *Handbook of Soy Oil Processing and Utilization*, American Oil Chemists Society, Champaign, Ill., 1980, 598 pp.
20. N. N. Potter, *Food Science*, 4th ed., Avi Publishing Co., Westport, Conn., 1986, 620 pp.
21. *Fat Replacers*, Calorie Control Council, Atlanta, Ga., 1991, 12 pp.
22. J. D. Dziezak, *Food Technol.*, 66–74 (July 1989).
23. R. L. Wysong, *Lipid Nutrition*, Inquiry Press, Midland, Mich., 1990, 170 pp.
24. V. K. Babayan, *J. Am. Oil Chem. Soc.* **45**(23), 23–25 (1968).
25. T. A. Voelker and co-workers, *Science* **257**, 72–74 (1992).
26. S. Latta, *Inform* **1**(4), 258–259 (1989).
27. D. Best, *Prepared Foods*, Gorman Publishing Co., Chicago, Mar. 1991, pp. 47–48.
28. D. Best, in Ref. 27, May 1991, pp. 72–77.
29. J. Zallie, *Proceedings of VII Eastern Institute of Food Technologists*, Baltimore, Md., 1991, pp. 42.
30. D. Chapman, in Ref. 29, pp. 41.
31. N. Singer and N. Desai, personal communication, Simplesse Co., Deerfield, Ill., 1992, 15 pp.
32. U.S. Pat. 4,734,287 (Mar. 29, 1988), N. S. Singer, S. Yamamato, and J. Latella.
33. A. M. Thayer, *C & E News*, 9–12, (June 3, 1991).
34. S. M. Lee, *Int. Food Ingred.* **1**, 28–39 (1992).
35. M. Friedman, in Ref. 7, p. 80.
36. D. Crothers, *Dairy Field*, Chicago, Mar. 1991, pp. 16–18.
37. E. Dexhelmer, *Dairy Foods*, 23–24 (1991).
38. *Prepared Foods*, Gorman Publications, Chicago, Mar. 1991, pp. 37.
39. *Butter Buds, A Dried Cream Extract*, Cumberland Packing Co., Racine, Wis., 1991, p. 1.
40. Ref. 38, May 1991, pp. 39.
41. L. Dornblaser, in Ref. 7, p. 131.
42. D. Crawford, in Ref. 7, p. 91.
43. Ref. 38, p. 73.
44. J. L. Marsden, personal communication, American Meat Institute, Washington, D.C., 1991.
45. W. R. Egbert, D. L. Huffman, C. Chen, and D. P. Dylewski, *Food Technol.* **45**, 64–73 (1991).
46. C. M. Amundson, *Protein Technology International*, Ralston Co., St. Louis, Mo., 1991.
47. T. Simon, *Shape*, 40–41 (Feb. 1989).
48. C. Andrews, personal communication, IBC, USA Conferences, South Natick, Mass., 1991.

49. A. M. Thayer, *Chem. Eng. News*, 26–44 (June 15, 1992).
50. M. Glicksman, *Food Technol.*, 94–103 (Oct. 1991).
51. M. Glicksman, *Food Hydrocolloids*, CRC Press, Boca Raton, Fla., 1982, 412 pp.
52. *Pfizer Polydextrose for the Market*. Pfizer Co., Groton, Conn., 1985, p. 4.
53. *Dried Fruit Based Fat Replacers*, Mariani Packing Co., San Jose, Calif., 1992, 5 pp.
54. *Agar Agar*, Gumix International, Fort Lee, N.J., 1992, 4 pp.
55. R. Cook, E. T. Finocchiaro, N. Shulman, and F. Mallee, *Enzytech Food Ingredients*, MIT, Cambridge, Mass., 1991, 4 pp.
56. N. S. Singer and J. M. Dunn, *J. Am. Coll. Nutr.* **9**(4), 388–397 (1990).
57. R. Gibson, *Wall Street Journal*, New York, Sept. 13, 1992, p. 11.
58. K. A. Harrigan and W. M. Breene, *Cereal Foods World* **34**(3), 261–263 (1989).
59. B. Summerkamp and M. Hesser, *Bulking Agents and Fat Substitutes: Analysis of a Dynamic Industry*, HRA, Inc., Prairie Village, Kans., 1989.
60. R. G. LaBarge, *Food Technol.*, 84–90 (Jan. 1988).
61. F. H. Mattson and R. A. Volpenhein, *J. Lipid Res.* **13**, 325–328 (1972).
62. D. J. Hamm, *J. Food Sci.* **49**, 419–423 (1984).
63. U.S. Pat. 4,582,927 (1986), J. Fulcher.
64. "P&G Introduces Fat Replacement to Candy Industry," *Wall Street Journal*, New York, Jan. 23, 1991, p. 4.

ROY E. MORSE
Consultant

NORMAN SINGER
Nutrasweet

FATS AND FATTY OILS

Fats and oils are one of the oldest classes of chemical compounds used by humans. Animal fats were prized for edibility, candles, lamp oils, and conversion to soap. Fats and oils are composed primarily of triglycerides (**1**), esters of glycerol and fatty acids. However, some oils such as sperm whale (1), jojoba (2), and orange roughy (3) are largely composed of wax esters (**2**). Waxes (qv) are esters of fatty acids with long-chain aliphatic alcohols, sterols, tocopherols, or similar materials.

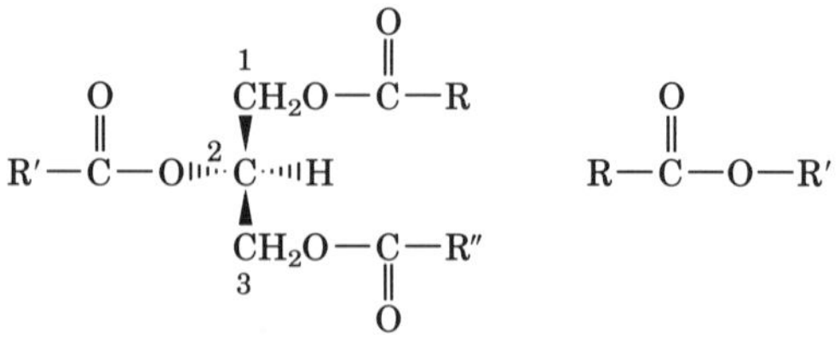

$$\begin{array}{c} \overset{1}{C}H_2O-\overset{O}{\overset{\|}{C}}-R \\ R'-\overset{O}{\overset{\|}{C}}-O\cdots\overset{2}{C}\cdots H \\ \underset{3}{C}H_2O-\underset{O}{\underset{\|}{C}}-R'' \end{array} \qquad\qquad R-\overset{O}{\overset{\|}{C}}-O-R'$$

(1) **(2)**

Fatty acids derived from animal and vegetable sources generally contain an even number of carbon atoms since they are biochemically derived by condensation of two carbon units through acetyl or malonyl coenzyme *A*. However, odd-numbered and branched fatty acid chains are observed in small concentrations in natural triglycerides, particularly ruminant animal fats through propionyl and methylmalonyl coenzyme *A*, respectively. The glycerol backbone is derived by biospecific reduction of dihydroxyacetone.

Structure (**1**) shows the stereochemistry of the triglyceride molecule. Positions are numbered by the stereochemical numbering (sn) system. In chemical processes the 1 and 3 positions are not distinguishable. However, for biological systems, the enantiomeric (*R* or *S*) form is important. Simple triglycerides contain only one type of fatty acid, eg, tristearin [*555-43-1*], and since the fatty acid residues in the 1 and 3 position are identical, do not exhibit enantiomeric forms. If more than one fatty acid is present, mixed triglycerides are distinguished. Naming mixed triglycerides without regard to stereochemistry involves two conventions: (*1*) the fatty acid with the shortest carbon chain is named first, eg, palmitodistearin, and (*2*) for fatty acids with an equal number of carbon atoms, the acid with the lesser number of double bonds is named first, eg, stearodiolein. When stereochemistry is taken into account, the acids are numbered as they occur, eg, *sn*-1-oleo-2-palmito-3-stearin.

Fatty acids may be saturated, monounsaturated, or polyunsaturated according to the number of double bonds in the alkyl chain. Naturally occurring double bonds are almost exclusively *cis* (*Z*) in configuration. Table 1 lists the fatty acids found in representative triglycerides. Fatty acids are often referred to by their common names (see CARBOXYLIC ACIDS). For example, 9-*cis*-octadecenoic acid has long been known as oleic acid. A convenient shorthand notation for fatty acids identifies the chain length followed by a colon and the number of double bonds in the chain, eg, oleic acid is 18:1. The most common fatty acids in animal and vegetable fats and oils are dodecanoic (lauric, 12:0), hexadecanoic (palmitic, 16:0), octadecanoic (stearic, 18:0), 9-*cis*-octadecenoic (oleic, 18:1), 9-*cis*,12-*cis*-octadecadienoic acid (linoleic, 18:2), and 9-*cis*,12-*cis*,15-*cis*-octadecatrienoic acid (linolenic, 18:3).

Fats and oils are distinguished by their physical state; fats are solid at ambient temperature, whereas oils are liquid. Some edible triglycerides, such as butter, lard, vegetable oils, shortenings, and margarines, have substantial quantities of both liquid and solid components at ambient temperature. Commercial products may be derived from animal carcasses by rendering, or vegetable sources by pressing or solvent extraction (4).

Composition

Natural fats and oils are composed principally of triglycerides, but other components may be present in minor quantities. These components may have important effects on the nature and quality of the oil or fat.

Free Fatty Acids and Partial Glycerides. After harvest, many crude oil crops contain lipase enzymes that cleave triglycerides into fatty acids and partial glycerides. For example, free fatty acid content of rice bran oil and palm oil can

Table 1. Fatty Acids Found in Naturally Occurring Triglycerides

Fatty acid	CAS Registry Number	Common name	Designation[a]	Principal sources
butanoic	[*107-92-6*]	butyric	4:0	butter
hexanoic	[*142-62-1*]	caproic	6:0	butter
octanoic	[*124-07-2*]	caprylic	8:0	coconut
decanoic	[*334-48-5*]	capric	10:0	coconut
dodecanoic	[*143-07-7*]	lauric	12:0	coconut, palm kernel, butter
tetradecanoic	[*544-63-8*]	myristic	14:0	coconut, palm kernel, butter
hexadecanoic	[*57-10-3*]	palmitic[b]	16:0	palm, cottonseed, butter, animal fat, marine fats
cis-9-hexadecenoic	[*373-49-9*]	palmitoleic	16:1 (9*c*)	butter, animal fats
octadecanoic	[*57-11-4*]	stearic[b]	18:0	butter, animal fats
cis-9-octadecenoic	[*112-80-1*]	oleic[b]	18:1 (9*c*)	olive, tall oil, peanut, canbra[c], animal fats, butter, marine fats
cis,cis-9,12-octadecadienoic	[*60-33-3*]	linoleic[b]	18:2 (9*c*, 12*c*)	safflower, sunflower, corn, soy, cottonseed
cis,cis,cis-9,12,15-octadecatrienoic	[*463-40-1*]	linolenic	18:3 (9*c*, 12*c*, 15*c*)	linseed
cis,cis,cis,cis-6,9,12,15-octadecatetraenoic	[*20290-75-9*]		18:4 (6*c*, 9*c*, 12*c*, 15*c*)	marine fat
cis,trans,trans-9,11,13-octadecatrienoic	[*506-23-0*]	α-eleostearic	18:3 (9*c*, 11*t*, 13*t*)	tung
12-hydroxy-*cis*-9-octadecenoic	[*141-22-0*]	ricinoleic	18:1 (9*c*) 12-OH	castor
cis-9-eicosenoic	[*29204-02-2*]	gadoleic	20:1 (9*c*)	marine fat
cis-11-eicosenoic	[*5561-99-9*]		20:1 (11*c*)	rapeseed
all *cis*-5,8,11,14-eicosatetraenoic	[*506-32-1*]	arachidonic	20:4 (5*c*, 8*c*, 11*c*, 14*c*)	animal, marine fats
all *cis*-8,11,14,17-eicosatetraenoic	[*24880-40-8*]		20:4 (8*c*, 11*c*, 14*c*, 17*c*)	marine fats
all *cis*-5,8,11,14,17-eicosapentaenoic	[*10417-94-4*]		20:5 (5*c*, 8*c*, 11*c*, 14*c*, 17*c*)	marine fats
docosanoic	[*112-85-6*]	behenic	22:0	
cis-11-docosenoic	[*506-36-5*]	cetoleic	22:1 (11*c*)	marine fats
cis-13-docosenoic	[*112-86-7*]	erucic	22:1 (13*c*)	rapeseed
all *cis*-7,10,13,16,19-docosapentaenoic	[*24880-45-3*]		22:5 (7*c*, 10*c*, 13*c*, 16*c*, 19*c*)	marine fats
all *cis*-4,7,10,13,16,19-docosahexaenoic	[*6217-54-5*]		22:6 (4*c*, 7*c*, 10*c*, 13*c*, 16*c*, 19*c*)	marine fats

[a]Number of carbon atoms: number of double bonds (geometric (cis, trans) isomerism).
[b]Constituent of most fats.
[c]Low erucic rapeseed.

reach high concentrations if the enzymes are not denatured promptly by heat treatment. Acids, bases, and heat abuse in the presence of moisture can lead to high free fatty acid concentrations as well as oxidative degradation. Elevated free fatty acid concentrations are undesirable because they cause high losses during further processing of the oil.

Diglycerides and monoglycerides are formed by hydrolysis. Diglycerides formed by lipolysis may initially have a 1,2-configuration. However, a 1,2 acyl shift often occurs to form the less sterically crowded 1,3 isomer. This rearrangement has even been shown to occur in the solid state (5). 2-Monoglycerides undergo a 1,2 acyl shift to form the more stable 1-monoglycerides (6). Monoglycerides are interfacially active and may form emulsions in the presence of water.

Free fatty acids are removed by refining, physical refining, or deodorization. Mono- and diglycerides are not removed by alkali refining or bleaching and may have an adverse effect on the quality of the oil. However, during deodorization (vacuum steam distillation), partial glycerides may disproportionate to triglycerides and glycerol [*56-81-5*]. The glycerol (qv) is removed with the deodorization distillate. A method has been reported where monoglycerides and other surface-active materials may be removed by treatment with Florisil (7).

Phospholipids. Glycerides esterified by fatty acids at the 1,2 positions and a phosphoric acid residue at the 3 position constitute the class called phospholipids (**3**). In older literature and in commercial practice, these materials are described as phosphatides.

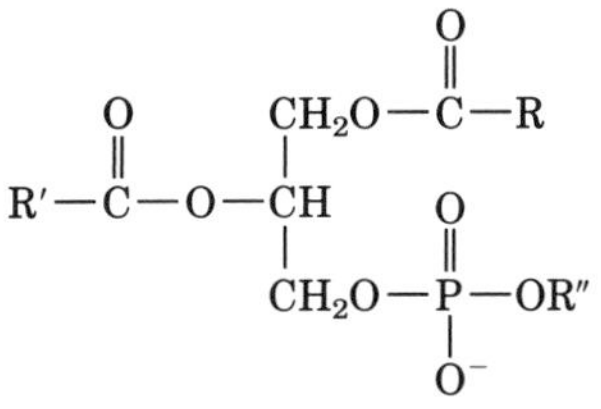

$$
\begin{array}{l}
\phantom{R'-\overset{O}{\overset{\|}{C}}-O-}CH_2O-\overset{O}{\overset{\|}{C}}-R \\
\phantom{R'-\overset{O}{\overset{\|}{C}}-O-}| \\
R'-\overset{O}{\overset{\|}{C}}-O-CH \\
\phantom{R'-\overset{O}{\overset{\|}{C}}-O-}| \\
\phantom{R'-\overset{O}{\overset{\|}{C}}-O-}CH_2O-\overset{O}{\overset{\|}{P}}-OR'' \\
\phantom{R'-\overset{O}{\overset{\|}{C}}-O-CH_2O-}\underset{O^-}{|}
\end{array}
$$

(**3**)

where R″ = $-CH_2CH_2\overset{+}{N}(CH_3)_2$ (**3a**), $-CH_2CH_2NH_2$ (**3b**), or [inositol ring with five OH groups] (**3c**)

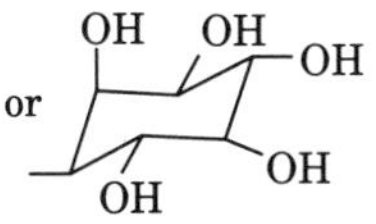

The identity of the moiety (other than glycerol) esterified to the phosphoric group determines the specific phospholipid compound. The three most common phospholipids in commercial oils are phosphatidylcholine or lecithin [*8002-43-5*] (**3a**), phosphatidylethanolamine or cephalin [*4537-76-2*] (**3b**), and phosphatidylinositol [*28154-49-7*] (**3c**). These materials are important constituents of plant and animal membranes. The phospholipid content of oils varies widely. Lauric oils, such as coconut and palm kernel, contain a few hundredths of a percent. Most oils contain 0.1 to 0.5%. Corn and cottonseed oils contain almost 1% whereas soybean oil can vary from 1 to 3% phospholipid. Some phospholipids, such as dipalmitoylphosphatidylcholine (R = R′ = palmitic; R″ = choline), form bilayer structures known as vesicles or liposomes. The bilayer structure can microencapsulate solutes and transport them through systems where they would normally be degraded. This property allows their use in drug delivery systems (qv) (8).

Sterols. Sterols (**4**) are tetracyclic compounds derived biologically from terpenes. They are fat-soluble and therefore are found in small quantities in fats and oils. Cholesterol [*57-88-5*] (**4a**) is a common constituent in animal fats such as lard, tallow, and butterfat. The hydroxyl group can be free or esterified with a fatty acid.

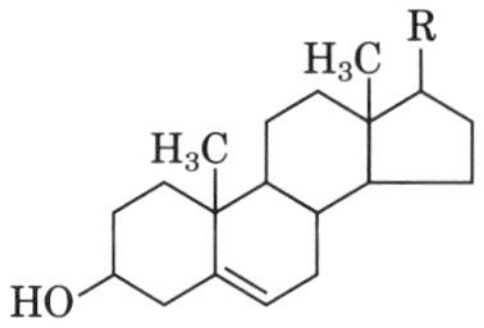

(**4**)

where R = $-CH(CH_3)(CH_2)_3CH(CH_3)_2$ (**4a**), $-CH(CH_3)CH_2CH_2CH(CH_2CH_3)CH(CH_3)_2$ (**4b**), or

$-CH(CH_3)CH{=}CHCH(CH_2CH_3)CH(CH_3)_2$ (**4c**).

In milk fat, cholesterol is associated with lipoproteins in the milk fat globule. It is also a component of animal membranes and controls rigidity and permeability of the membranes. Cholesterol has interesting surface properties and can occur in liquid crystalline forms. Plants contain sterols such as β-sitosterol [*83-46-5*] (**4b**) or stigmasterol [*83-48-7*] (**4c**). Their functions in plant metabolism are not yet well understood. Analysis of sterols has proven useful for detection of adulteration of edible fats (9).

Tocopherols and Tocotrienols. Algae and plants used as sources of edible oils contain tocopherols, phenolic materials that function as antioxidants (qv). Mammals do not synthesize these compounds and residues present in their bodies are present because of ingestion. Concentrations of tocopherols in commercial vegetable oils vary widely from 0.01% or less in coconut and olive oils; up to 0.05% in palm and peanut oils; approximately 0.1% in corn, cottonseed, and linseed oils; up to 0.2% in soybean oil; and to as high as 0.5% in wheat germ oil. Tocopherols are designated as being vitamin E active (10). Other structural features of the tocopherols are the chromane heterocyclic ring and the extensively branched side chain that arises biologically from condensation of isoprene units. The α-, β-, γ-, and δ-tocopherols are represented by (**5**), where $R_1 = R_2 = R_3 = CH_3$ is α-tocopherol [*59-02-9*]; $R_1 = R_3 = CH_3$, $R_2 = H$ is β-tocopherol [*148-03-8*]; $R_2 = R_3 = CH_3$, $R_1 = H$ is γ-tocopherol [*7616-22-0*]; and $R_1 = R_2 = H$, $R_3 = CH_3$ is δ-tocopherol [*119-13-1*].

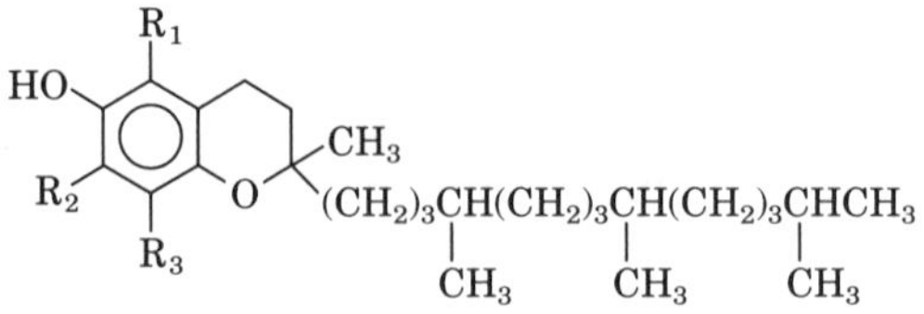

(**5**)

Tocotrienols differ from tocopherols by the presence of three isolated double bonds in the branched alkyl side chain. Oxidation of tocopherol leads to ring opening and the formation of tocoquinones that show an intense red color. This species is a significant contributor to color quality problems in oils that have been abused. Tocopherols function as natural antioxidants (qv). An important factor in their activity is their slow reaction rate with oxygen relative to combination with other free radicals (11).

Several other naturally occurring antioxidants have been identified in oils. Sesamol [*533-31-3*] (**6**) occurs as sesamoline [*526-07-8*], a glycoside, in sesame seed oil. Ferulic acid [*1135-24-6*] (**7**) is found esterified to cycloartenol [*469-38-5*] in rice bran oil and to β-sitosterol in corn oil. Although it does not occur in oils, rosemary extract has also been found to contain powerful phenolic antioxidants (12).

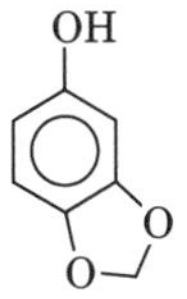

(**6**) 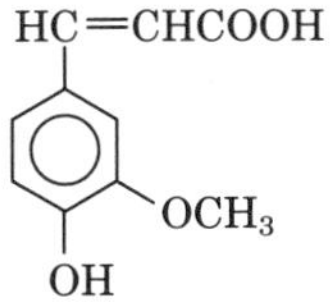 (**7**)

Carotenoids and Other Pigments. Carotenoids contain conjugated double bonds, a strong chromophore which produces red and yellow coloration in vegetable oils. Carotenoids are tetraterpene hydrocarbons formed by the condensation of eight isoprene units. Another class of compounds, the xanthophylls, is produced by hydroxylation of the carotenoid skeleton. β-Carotene [*7235-40-7*] is the best known component of the carotenoids because it is the precursor for vitamin A. Carotenoid pigments are unstable toward heat, light, and oxidation. Oxidative bleaching is occasionally used but great care must be taken to remove excess peroxides. Reaction of peroxide with tocopherols may produce an even deeper red coloration. In the process of deodorization, heat destroys the red carotenoid pigments (heat bleaching).

Green coloration, present in many vegetable oils, poses a particular problem in oil extracted from immature or damaged soybeans. Chlorophyll is the compound responsible for this defect. Structurally, chlorophyll is composed of a porphyrin ring system, in which magnesium is the central metal atom, and a phytol side chain which imparts a hydrophobic character to the structure. Conventional bleaching clays are not as effective for removal of chlorophylls as for red pigments, and specialized acid-activated adsorbents or carbon are required.

Processing of Fats and Oils

Fats and oils are derived from animals, plants, or fish by rendering (animal tissues), pressing, or solvent extraction. Animal fat may be obtained from carcasses by trimming the fat from the carcass, treating the trimmings with live steam to melt the fat, and filtering to remove proteinaceous material and debris, such as plastic wrapping, that may be carried along with the trimmed fat.

Vegetable seeds having high oil contents may be extruded through a mechanical press to separate oil from the other components. Peanut oil and cocoa butter are often obtained using this process. With the most modern mechanical screw press, the pressing method leaves about 3–5% residual oil in the meal. Older presses may leave as much as 6% oil in the press cake. Solvent extraction provides higher yields because it leaves less ($\leq$1%) oil residue in the meal. Soybeans, for example, are mechanically pressed into flakes and then extracted using a solvent such as hexane. After filtration, the solvent is recovered for reuse from the oil and meal. In a few specialized cases, the solvent is not removed but refining is performed on the solution. This process is known as miscella refining.

Crude oils from these processes are often of insufficient quality to be used directly, particularly for edible products. Impurities such as pigments, phosphatides, volatile odorous compounds, and certain metals must be removed by further processing.

Degumming and Dewaxing. Some oils, such as soybean, contain appreciable amounts of phospholipids. These materials are often referred to as gums because of high viscosity and sticky texture. Hydratable phosphatides are removed by treating the crude oil with hot water or dilute acid. In older plants, hydration is carried out as a batch process. The hydrated oil is allowed to stand in a tank and then the precipitated phospholipids are skimmed from the top. In modern plants, the oil is hydrated continuously and the phosphatides are removed in a centrifuge. The vacuum-dried sludge from soybean processing is known as lecithin. This material may be used crude or may be further purified. Lecithin (qv) is a surfactant with several applications in foods or specialty chemical products.

Many seed oils, especially sunflower and linseed, contain waxes which serve as a protective coating for the seed. These waxes solidify at colder temperatures and impart turbidity to the oil and interfere with subsequent processing. They are commonly removed from the crude oil by refrigeration followed by filtration, a process commonly known as winterization.

Refining. Stored crops undergo gradual hydrolysis because of the presence of lipolytic enzymes and moisture. The resulting presence of free fatty acids is undesirable for an edible oil and is referred to as hydrolytic rancidity. Free fatty acids in cooking oils cause the oils to smoke when they are heated. Removal of the free fatty acid may be accomplished by vacuum steam distillation (physical or steam refining) or extraction with aqueous alkaline solution (alkali refining) (13). Physical refining has the advantages of selectively removing free fatty acids with minimal oil loss and deodorizing the oil simultaneously. In addition, the environmental problem of disposing of acidified caustic solution is eliminated. However, physical refining only works well for oils that are low in phosphatides and are stable to heat in an unrefined form. Palm, coconut, and palm kernel oils are well suited to physical refining. Other oils may be subjected to extremely efficient degumming followed by physical refining (14).

In alkali refining, a solution of caustic soda (sodium hydroxide) is mixed with the oil to form soaps. These soaps are dispersed in the aqueous phase together with phospholipids, some pigments, and other compounds. Aqueous and oil phases are separated by centrifugation. The oil is subsequently extracted with water to remove the bulk of the residual soap. Refining losses may occur during the process because of saponification of the oil or emulsification of some neutral oil in the aqueous phase, particularly if the free fatty acid content of the incoming oil is high resulting in a high soap concentration. Phosphatides act as surfactants and tend to increase refining losses if they are not removed by degumming. Conversely, the alkali process effectively removes residual phosphatides.

Strength of the caustic solution, temperature, contact time, and amount used are the critical parameters used to set optimum refining conditions. The type of oil and its free fatty acid and phosphatide content greatly influence refining behavior. An oil with a high phosphatide content requires a greater excess of sodium hydroxide than an oil with low phosphatide concentration. Soapstock is a by-product of alkali refining. It is commonly acidified and the fatty acids are utilized for animal feed. Alternatively, the acidified soapstock may be treated to split fatty acids from residual glyceride and the fatty acids distilled and sold as commercial products. A modification of alkali refining has minimized or eliminated the water stream associated with alkali refining (15). Silica is utilized with a lower concentration of alkali to refine the oil. Soap is absorbed on the silica and subsequent bleaching clay rather than being washed out and neutralized. Soap is therefore disposed of as solid waste along with the bleaching clay and silica rather than acidified and concentrated to crude fatty acid.

Bleaching. Light color and a clear bright appearance have long been associated with oil quality. Removal of pigments is accomplished by absorption on solid materials (absorption bleaching) or by treatment with oxidizing agents such as hydrogen peroxide or hypochlorite solutions. Bleaching absorbents are derived from bentonite or montmorillonite clays (qv). The activity of these clays may be increased by treatment with acid. In addition to removing pigment, an important function of bleaching clays is the reduction of concentrations of metals such as copper, iron, manganese, and cobalt which catalyze oxidation of oils and reduce the shelf life of edible products. Activated carbon is an effective bleaching agent but also absorbs a high percentage of its own weight in oil resulting in increased oil loss. Carbon may be used effectively in small amounts in combination with activated bleaching earths.

Absorbents may be selected according to the specific pigments to be removed from an oil. Carotenoid pigments, which produce red colors, are effectively removed by activated clays. Carbon is quite effective for removing green chlorophyll pigments from soybean oil obtained from damaged beans. Brown pigments, which may be extremely difficult to remove, result from oxidation of carotenoids and are encountered in abused oils. Type of absorbent and concentration are selected considering activity, absorbent cost, oil loss, and metal removal.

The efficiency of a particular absorbent or blend of absorbents is determined using the Freundlich equation:

$$\log \frac{x}{m} = K + n \log c$$

where x is the amount of pigment absorbed, m is the amount of absorbent, c is the amount of residual pigment, and K and n are constants. Absorption isotherms are generated by a plot of x/m vs c on a log–log scale. Concentrations for the initial $(x + c)$ concentration of pigment and the residual (c) may be measured by spectrophotometry or comparison to standard reference colors.

The principal disadvantage of absorption bleaching is the problem of disposal of spent bleaching clay. Oil absorbed on the clay is exposed to air and is generally too oxidized to recover. Furthermore, spontaneous combustion of the oil-laden clay is a possibility in a landfill. Incineration of the spent clay along with solid municipal waste to recover otherwise wasted energy is an attractive possibility.

Chemical bleaching is never used on oils intended for edible use because it oxidizes unsaturated fatty acids to cause off-flavors. However, it does find wide usage for specialty linseed oil, for the paint industry, and fatty chemicals such as sorbitan esters of fatty acids and sodium stearoyl lactylate. Residual peroxide is destroyed by heating above its decomposition temperature.

Hydrogenation. Solid and semisolid fats have been used in food products such as margarines and shortenings and in nonedible applications such as candles (16). Historically, such solid fatty products were derived from animal fats or solid fractions of these fats. The discovery of catalytic hydrogenation in the early 1900s (17) allowed liquid oils to be converted into solid and semisolid fats with similar consistency. Solid supported nickel catalysts are most commonly used for addition of hydrogen to unsaturated and polyunsaturated fatty acid side chains (16). Addition of hydrogen to the double bond is the primary reaction, but isomerization of the double bond can also occur. As shown in Figure 1, the semihydrogenated transition state may (*1*) add another hydrogen to yield the saturated fatty acid, (*2*) lose another hydrogen atom to reform the double bond at a different position, or (*3*) rotate 180° around a sigma bond and lose a hydrogen atom to form a trans fatty acid. Double bonds may migrate along the entire length of the chain. Trans isomers increase until the cis monoenes are hydrogenated (18). Because naturally occurring oils contain a mixture of unsaturated triglycerides, hydrogenation results in a complex mixture of products.

Multiply unsaturated linolenic and linoleic acid residues make triglycerides more vulnerable to oxidative degradation than oleic acid which is relatively stable. It is therefore desirable to hydrogenate the most unsaturated residues selectively without production of large quantities of stearic (fully saturated) acid. The stepwise reduction of an unsaturated oil may be visualized as:

$$\text{linolenic} \xrightarrow{k_1} \text{linoleic} \xrightarrow{k_2} \text{oleic} \xrightarrow{k_3} \text{stearic}$$

If first-order kinetics are assumed, k_2/k_1 is the linoleic selectivity ratio and k_2/k_3

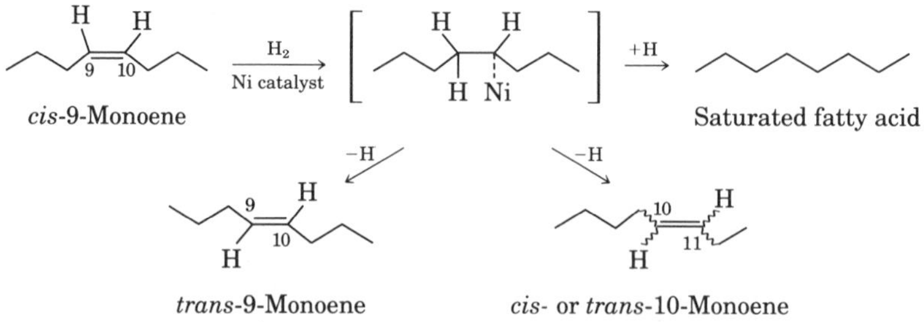

H H
9 10
H_2
Ni catalyst
H H
H Ni
+H
cis-9-Monoene
Saturated fatty acid
−H
−H
9 H
10
H
10 H
11
H
trans-9-Monoene
cis- or *trans*-10-Monoene

Fig. 1. Alternative pathways for hydrogenation in oils.

is the selectivity ratio for reduction of linoleic acid to stearic acid. Figure 2 shows a typical course of hydrogenation for soybean oil; the rate constants are k_1 = 0.367, k_2 = 0.159, and k_3 = 0.013. With a selective nickel catalyst, linolenic acid may be reduced without an appreciable buildup of undesirable saturates. Platinum catalyst, on the other hand, hydrogenates straight through to saturates, ie, shows no selectivity.

Conditions of hydrogenation also determine the composition of the product. The rate of reaction is increased by increases in temperature, pressure, agitation, and catalyst concentration. Selectivity is increased by increasing temperature and negatively affected by increases in pressure, agitation, and catalyst. Double-bond isomerization is enhanced by a temperature increase but decreased with increasing pressure, agitation, and catalyst. Trans isomers may also be favored by use of reused (deactivated) catalyst or sulfur-poisoned catalyst.

Commercially, hydrogenation is still carried out using a batch process. A typical "dead end" reactor is shown in Figure 3. The heated oil is stirred with the heterogeneous catalyst while hydrogen is bubbled through the bottom. Efficiency of mixing gas, solid, and liquid phases is critical. Agitation is reported to be increased by ultrasonic energy (19). Continuous hydrogenation processes have been developed but have not yet been widely applied. Development of rapid process

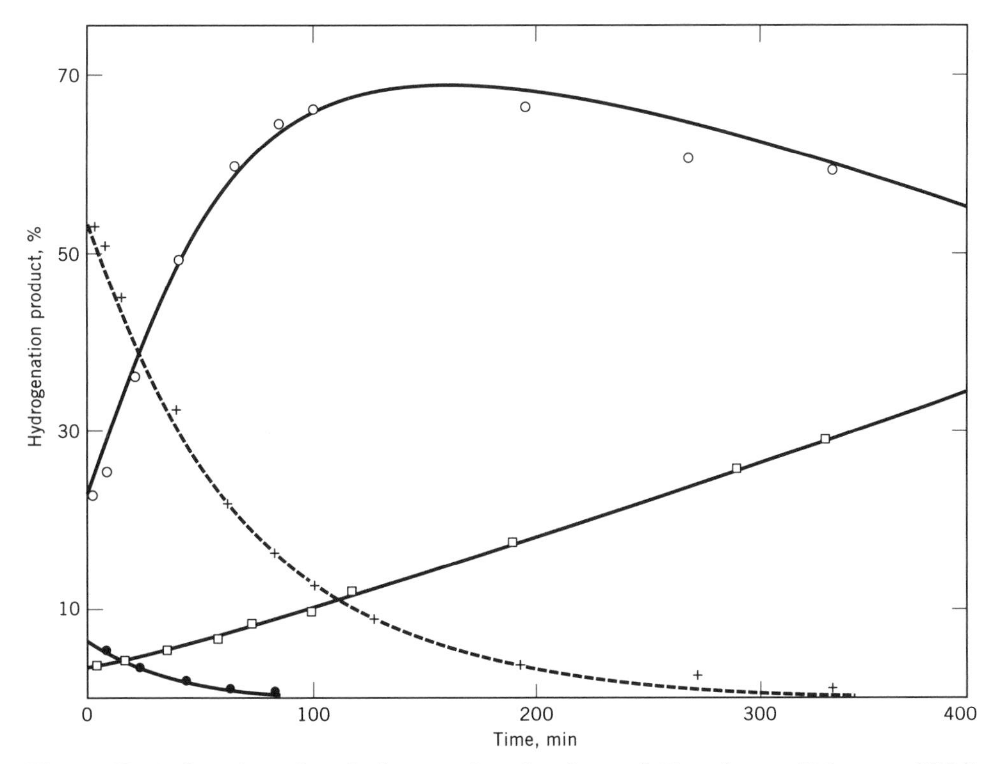

Fig. 2. Typical products from hydrogenation of soybean oil. Reaction conditions are 175°C, 0.02% Ni, 113 kPa (15 psig), and 600 rpm agitation. ○ is oleic; +, linoleic; □, stearic; and ●, linolenic (16).

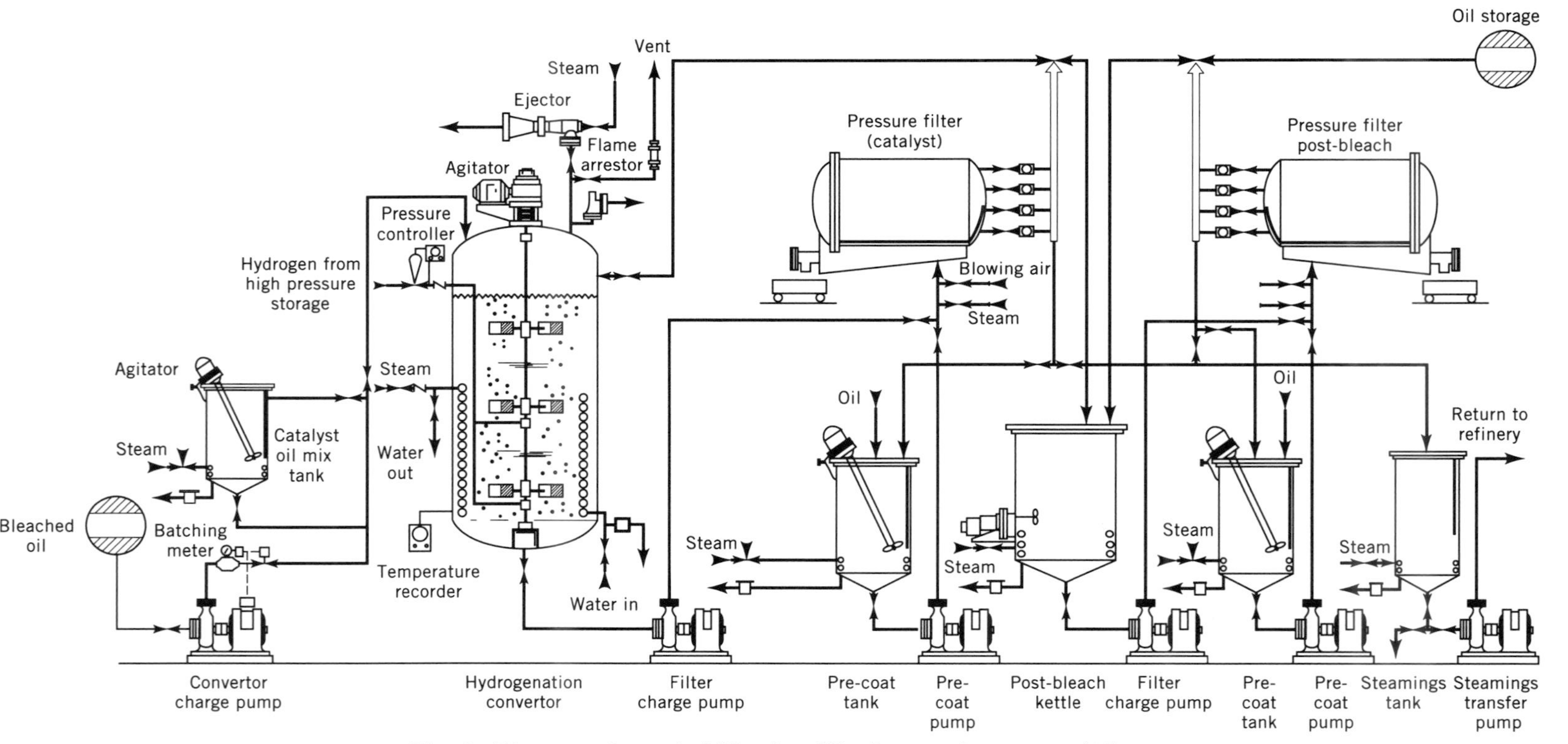

Fig. 3. Diagram of a typical "dead end" hydrogenation reactor (16).

control systems and consolidation of the industry to larger plants with dedicated process lines may forecast an increase in this approach.

Hydrogenation is an exothermic reaction. The heat generated is a function of the number of hydrogenated double bonds in a glyceride and the molecular weight. Initially, the oil is heated to an intermediate temperature. When hydrogen is introduced, heat is evolved and the desired reaction temperature is reached. Additional energy savings may be achieved if the finished product is passed through a heat exchanger to heat incoming oil for the next batch.

Disposal of spent hydrogenation catalyst requires a special chemical waste landfill because of its nickel content and the fact that oil-soaked catalysts tend to be pyrophoric. Compared to disposal costs, reprocessing to recover the nickel may become economically viable.

Randomization/Interesterification. Transesterification occurs when a carboxylic acid (acidolysis) or alcohol (alcoholysis) reacts with an ester to produce a different ester (20). Ester–ester interchange is also a form of transesterification. If completely unsaturated triglyceride oil (UUU) reacts with a totally saturated fat (SSS) in the presence of an active catalyst such as sodium, potassium, or sodium alkoxide, triglycerides of intermediate composition may be formed.

$$\text{SSS} + \text{UUU} \xrightleftharpoons[\text{vacuum or } N_2]{NaOCH_3} \text{SSU} + \text{SUS} + \text{SUU} + \text{USU}$$

SSS	UUU	SSU	SUS	SUU	USU
solid	liquid	semisolid	semisolid	semiliquid	semiliquid

Many seed oils have specific triglyceride structures with unsaturated fatty acids in the 2 position. Methoxide rearrangement of these oils randomizes the distribution of fatty acids on the glyceride. The most apparent result is a change in melting behavior (21). For animal fats, randomization has little effect on the melting point. However, the polymorphic crystal pattern and functional properties may be changed. Lard, for example, has been rearranged to improve its functionality in baking applications (22).

Recently, the use of lipase enzymes to interesterify oils has been described (23). In principle, if a 1,3-specific lipase is used, the fatty acid in the 2 position should remain unchanged and the randomization occur at the terminal positions. However, higher temperatures, needed to melt solid fats, may cause a 1,2-acyl shift and fatty acids are scrambled over all positions.

The reaction is generally carried out in a batch reactor equipped with agitation and vacuum. Heated oil is first dried under vacuum to remove traces of water. Water has a low molecular weight and deactivates sodium methoxide catalyst by reaction to form sodium hydroxide, inactive as a catalyst under these conditions, and methanol. A freshly refined oil must be used to minimize catalyst deactivation by free fatty acid. A small amount (~0.1–0.3%) of sodium methoxide catalyst is added and the mixture is allowed to react for a limited time. Darkening of the oil indicates onset of reaction, probably resulting from formation of enolate ions. Following the reaction, catalyst may be removed by water washing or acidification of the oil.

Filtration through an acidic filter aid or silica removes the last traces of soap from the oil. The finished oil is heated under vacuum to remove small amounts of fatty acid methyl esters.

Physical Fractionation. Specific fractions of glyceride oils may be separated according to their melting points by fractional crystallization and filtration (winterization) or solvent fractionation. Melting points of triglycerides are determined by the number and position of unsaturated fatty acids and fatty acid chain length. Short-chain fatty acids are liquid whereas longer chains are solid. For fatty acids of the same chain length, melting points of triglycerides in decreasing order are SSS >> SUS,SSU > SUU,USU > UUU where S = saturated and U = unsaturated fatty acid.

Fats high in trisaturates (SSS) are used for encapsulation and as crystal nuclei and stiffening agents for applications such as shortenings and margarine oils. Oils with high levels of disaturates (SSU,SUS) are utilized for confectionery fats because they melt sharply near body temperature. Monosaturates (SUU,USU) are semisolid triglycerides and function as consistency modifying agents in margarine and frying oils. Oils with a high content of triunsaturates are utilized as salad oils or base oils for salad dressings and mayonnaise.

Fractional crystallization may be accomplished on a batch, continuous, or semicontinuous basis. Oil is chilled continuously while passing through the unit and is then passed over a continuous belt filter which separates solid fat from the liquid oil. The process gives poorer separation compared to solvent fractionation because oils are viscous at crystallization temperatures and are entrained to a significant extent in the solid fraction. The liquid fraction, however, is relatively free of saturated material.

Winterization is a specialized application of fractional crystallization that is utilized to remove saturates or waxes from liquid oils. Salad oils, which do not cloud at refrigerator temperature, have been produced by winterizing lightly hydrogenated soybean oil. However, many producers now use refined, bleached, deodorized oils for this purpose (24).

Solvent fractionation requires greater capital cost than fractional crystallization but produces more efficient separation of saturated triglycerides. Because of dilution and lower viscosity of the filtrate, little liquid oil is entrained in the solid fraction. Acetone (25), hexane (26), and 2-nitropropane (27), have been used as solvents for fractionation of oils. Partially hydrogenated soybean oil, palm oil, and palm kernel oil are common feedstocks that yield products for extension or replacement of cocoa butter in confectionery coatings. The historical high cost of cocoa butter has been the economic driving force for the capital-intensive construction of solvent fractionation plants.

Deodorization. Removal of volatile odorous material and residual fatty acids is the final step in oil processing prior to packaging or filling for bulk shipment (28). The oil is heated to 230–260°C under vacuum. Steam is passed through the oil to assist in carrying over the volatile material. Aldehydes, enals, dienals, ketones, and hydrocarbons, which are responsible for disagreeable odors, generally boil at lower temperatures than fatty acids. Analysis showing a free fatty acid concentration of less than 0.05% is an indication that deodorization is sufficiently complete. Some of the dienals have very low odor thresholds and sensory evaluation of the finished oil is a judicious quality assurance step.

Deodorization can be carried out in batch, continuous, or semicontinuous systems. Figure 4 shows a typical design for a semicontinuous deodorizer. The heated oil is passed through a series of trays under vacuum. Steam is passed through the oil through a steam sparge in the bottom of the tray. Volatiles are carried through the headspace and condensed. In addition to fatty acids and compounds responsible for odor, some tocopherols and sterols are also distilled into the condensate. The amount of tocopherols distilled depends on deodorization temperature and vacuum.

Sources of Fats and Oils

Fats and oils may be synthesized in enantiomerically pure forms in the laboratory (30) or derived from vegetable sources (mainly from nuts, beans, and seeds), animal depot fats, fish, or marine mammals. Oils obtained from other sources differ markedly in their fatty acid distribution. Table 2 shows compositions for a wide variety of oils. One variation in composition is the chain length of the fatty acid. Butterfat, for example, has a fairly high concentration of short- and medium-chain saturated fatty acids. Oils derived from cuphea are also a rich source of capric acid which is considered to be medium in chain length (32). Palm kernel and coconut oils are known as lauric oils because of their high content of C-12 saturated fatty acid (lauric acid). Rapeseed oil, on the other hand, has a fairly high concentration of long-chain (C-20 and C-22) fatty acids.

Another variation of the fatty acid is the degree of unsaturation. Cocoa butter and butterfat have a high concentration of saturated fatty acids which accounts for their solid physical state. Oils such as sunflower, soybean, linseed, and safflower have a high proportion of unsaturated and polyunsaturated fatty acids. Olive oil has a high level of monounsaturated (oleic) fatty acid. Even within a species, there may be some variation in unsaturation depending on the climate in which it is grown. In cold climates, polyunsaturates occur at higher levels. This is apparently important for the plant to maintain fluidity of storage fats and membranes at colder temperatures. The most striking seasonal variation is seen with sunflower oil (33). Fish oils display even higher unsaturation as well as elongated fatty acid chains. They are sources for the biologically important eicosapentaenoic acid [*90417-94-4*] (EPA, 20:5) and docosahexaenoic acid [*6217-54-5*] (DHA, 22:6). Double bonds in polyunsaturated oils usually occur as methylene-interrupted dienes. Linoleic acid, for example, is 9,12-octadecadienoic acid. One exception to this generalization is tung oil. Approximately 85% of its fatty acid composition consists of 9,11,13-octadecatrienoic acid. The conjugated triene system allows rapid thermal polymerization of tung oil making it useful as a protective coating material.

Castor oil (qv) contains a predominance of ricinoleic acid which has an unusual structure inasmuch as a double bond is present in the 9 position while a hydroxyl group occurs in the 12 position. The biochemical origin of ricinoleic acid [*141-22-0*] in the castor seed arises from enzymatic hydroxylation of oleoyl-CoA in the presence of molecular oxygen. The unusual structure of ricinoleic acid affects the solubility and physical properties of castor oil.

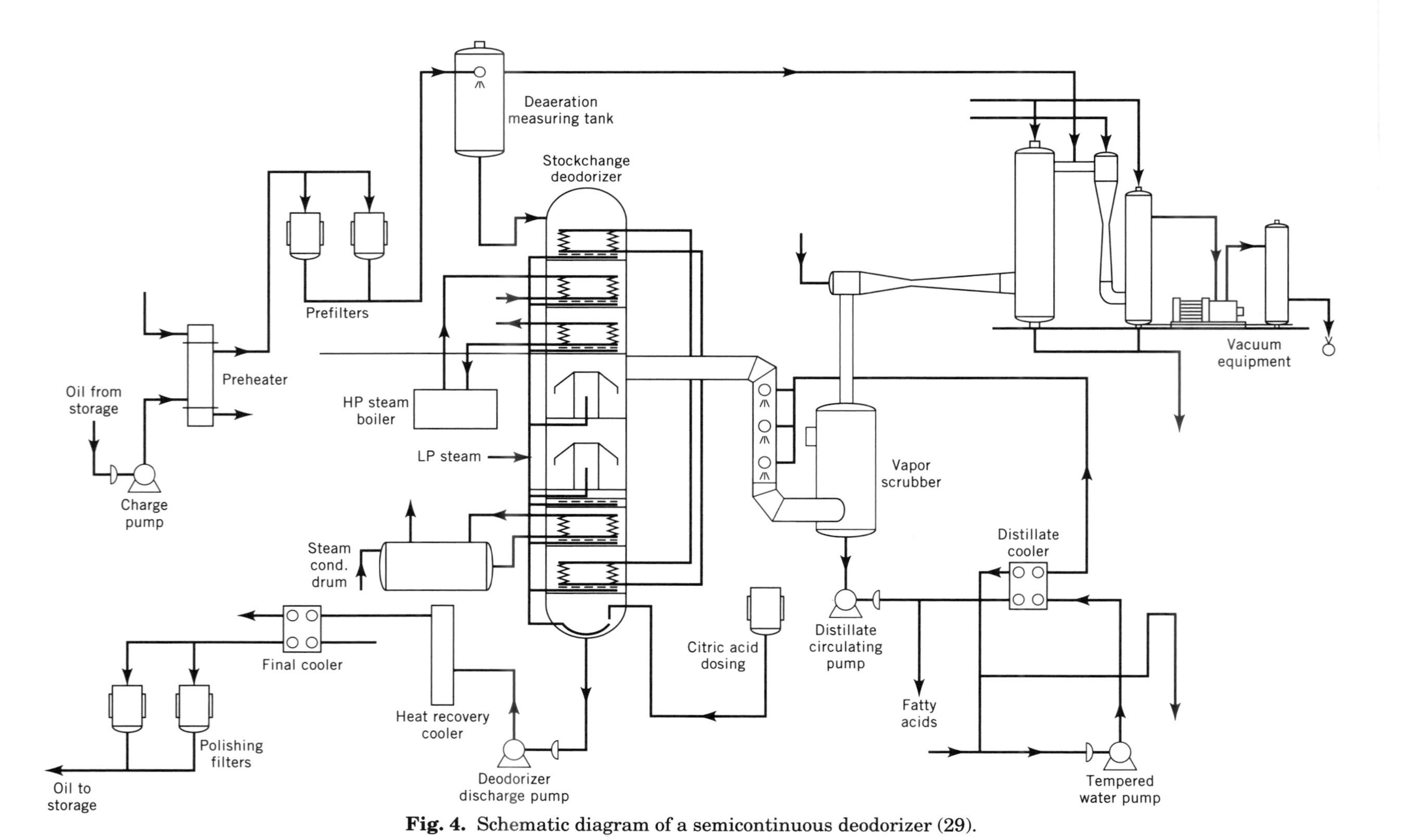

Fig. 4. Schematic diagram of a semicontinuous deodorizer (29).

Table 2. Fatty Acid Compositions of Naturally Occurring Fats and Oils[a]

Fat or oil	CAS Registry Number	Fatty acid,[b] %													
		6:0	8:0	10:0	12:0	14:0	16:0	16:1	18:0	18:1	18:2	18:3	20:0	22:0	22:1
butterfat[c]		2.3	1.1	2.0	3.1	11.7	28.6	1.9	12.5	28.2	1.7		0.1		
canola oil							3.9	0.2	1.9	64.1	18.7	9.2	0.6	0.2	
castor oil[d]	*[8001-79-4]*						0.9	0.2	1.2	3.3	3.7	0.2			
chicken fat					0.2	1.3	23.2	6.5	6.4	41.6	18.9	1.3			
citrus seed oil					0.1	0.5	28.4	0.2	3.5	23.0	37.8	5.7	0.8		
cocoa butter						0.1	25.8	0.3	34.5	35.3	2.9		1.1		
coconut oil	*[8001-31-8]*	0.5	8.0	6.4	48.5	17.6	8.4		2.5	6.5	1.5		0.1		
cod liver oil[e]	*[8001-69-2]*					2.8	10.7	6.9	3.7	23.9	1.5	0.9			5.3
corn oil	*[8001-30-7]*						12.2	0.1	2.2	27.5	57.0	0.9	0.1		
cottonseed oil	*[8001-29-4]*					0.9	24.7	0.7	2.3	17.6	53.3	0.3	0.1		
cuphea oil				84.6	2.3	1.7	2.6			4.0	3.1				
lard					0.1	1.5	24.8	3.1	12.3	45.1	9.9	0.1	0.2		
linseed oil	*[8001-25-0]*						7.0		4.0	39.0	15.0	35.0			
menhaden oil[f]	*[8002-50-4]*					6.6	15.7	8.7	2.7	14.3	1.8	1.5	0.2		0.8
oat oil						0.2	17.1	0.5	1.4	33.4	44.8		0.2		
olive oil	*[8001-25-0]*						13.7	1.2	2.5	71.1	10.0	0.6	0.9		
palm oil	*[8002-75-3]*				0.3	1.1	45.1	0.1	4.7	38.5	9.4	0.3	0.2		
palm kernel oil	*[8023-79-8]*	0.3	3.9	4.0	49.6	16.0	8.0		2.4	13.7	2.0		0.1		
peanut oil	*[8002-03-7]*					0.1	11.6	0.2	3.1	46.5	31.4		1.5	3.0	
rapeseed oil	*[8002-15-9]*					0.1	2.8	0.2	1.3	21.8	14.6	7.3	0.7	0.4	34.8
rice bran oil			0.1	0.1	0.4	0.5	16.4	0.3	2.1	43.8	34.0	1.1	0.5	0.2	
safflower oil	*[8001-23-8]*					0.1	6.5		2.4	13.1	77.7		0.2		
safflower oil (high oleic)						0.1	5.5	0.1	2.2	73.7	12.0	0.2	0.2		
sesame oil	*[8008-74-0]*						9.9	0.3	5.2	41.2	43.2	0.2			
soybean oil	*[8001-22-7]*						0.1	11.0	4.0	23.4	53.2	7.8	0.3	0.1	
sunflower oil	*[8001-21-6]*				0.5	0.2	6.8	0.1	4.7	18.6	68.2	0.5	0.4		
sunflower oil (high oleic)							3.3		4.8	81.9	7.6	0.4	0.5	1.5	
beef tallow				0.1	0.1	3.3	25.5	3.4	21.6	38.7	2.2	0.6	0.1		
tung oil	*[8001-20-5]*						3.1		2.1	11.2	14.6	69.0[g]			

[a]Ref. 31. [b]Designations as explained in Table 1. [c]Also, 3.9% 4:0.
[d]Castor oil is 89% C-18 acid with a 9,10 double bond and an OH substituent on the chain at C-12. [e]Also 14.3% 22:6 and 6.0% 20:5.
[f]Also 12.1% 22:6 and 15.6% 20:5. [g]The 18:3 in tung oil is *cis*-9,*trans*-11,*trans*-13-octadecatrienoic acid *[506-23-0]*.

Solid fats may show drastically different melting behavior. Animal fats such as tallow have fatty acids distributed almost randomly over all positions on the glycerol chain. These fats melt over a fairly broad temperature range. Conversely, cocoa has unsaturated fatty acids predominantly in the 2 position and saturated acids in the 1 and 3 positions. Cocoa butter is a brittle solid at ambient temperature but melts rapidly just below body temperature.

Historically, many attempts have been made to systematize the arrangement of fatty acids in the glyceride molecule. The even (34), random (35), restricted random (36), and 1,3-random (37) hypotheses were developed to explain the methods nature utilized to arrange fatty acids in fats. Invariably, exceptions to these theories were encountered. Plants and animals were found to biosynthesize fats and oils very differently. This realization has led to closer examination of biosynthetic pathways, such as chain elongation and desaturation, in individual genera and species.

Biotechnology is slowly revolutionizing the edible oils industry. Crop breeding (38), genetic engineering (39), tissue culture (40), and mutation selection (41) are avenues being pursued to deliver desirable fatty acid compositions into agronomically favored plants. Oils from microbial sources may offer unique fatty acid compositions (42). Canola, high oleic sunflower, and high oleic canola oils are recent successes in harnessing the biosynthetic factories.

Separation of a fat or oil from its source material can be accomplished by several different methods. Selection of an extraction process is based on: (*1*) obtaining oil substantially undamaged and relatively free of undesirable impurities, (*2*) achieving the highest practical yield, and (*3*) obtaining the maximum economic return on the oil and coproducts.

Rendering. Separation of animal fats from water and connective tissue in crude trimmings from carcasses is accomplished by treatment with heat. The process may also be applied to whole small fish such as sardines or herring. Nonedible fats are generally obtained by dry rendering where moisture is removed and the melted fat is drained and pressed away from connective tissue. Wet rendering can be carried out by treating fat-bearing material with boiling water (low temperature) or with high temperature steam. Edible materials with good color, flavor, and stability can be obtained from wet rendering.

Mechanical Pressing. Historically, the first large commercial production of oils from seeds and nuts was carried out using labor-intensive hydraulic presses. These were gradually replaced by more efficient mechanical and screw presses. Solvent extraction was developed for extraction of seeds having low oil content. For seeds and nuts having higher oil content, a combination of a screw press followed by solvent extraction is a common commercial practice (prepress–solvent extraction).

Prior to pressing, the material is heat treated to inactivate enzymes which are detrimental to oil or meal quality and to allow oil droplets to coalesce into larger drops. Proteins in the structure are denatured and set into a structure that is more porous to the flow of the oil. Residual oil is commonly left in the meal after pressing. This can range from a few percent as in cocoa butter to a much higher percentage as seen in peanuts. The pressing operation can be a batch process, such as a hydraulic box press, or continuous by use of a mechanical screw press. The latter is more efficient for removal of oil from materials such as cottonseed. Some higher value oils, such as olive, may be extracted by prepress–solvent extraction. The method is also practiced for lower value seeds such as flax, rape, sunflower, and cottonseed.

Solvent Extraction. Treatment of oil-bearing materials with solvent can effect virtually complete removal of oil from meal. However, the capital cost of equipment is higher and subsequent processing to remove traces of solvent from both oil and meal is necessary. Hexane obtained as a light petroleum distillate is used almost exclusively as the solvent. Other solvents such as 2-propanol and methylene chloride have been used alone or as blends with hexane. The more polar solvents promote the extraction of more polar components such as phospholipids. Gossypol, a toxic substance to domestic animals, may be removed from cottonseed meal with 2-propanol and then removed from the oil by alkali refining. Soybeans are almost exclusively solvent-extracted in the United States and other developed countries. The soybeans are heat treated and then mechanically pressed into flakes in order to render the structure more porous to solvent. The use of expanders to further open the structure has become widely used (29). Another exciting development in extraction technology is the use of supercritical fluids, such as carbon dioxide, to extract oil from its source (43,44). The greatest advantage is that flammable potentially toxic solvents are avoided. The supercritical fluid is a gas at ambient conditions and is easily removed. The principal barrier to commercialization is the high capital cost of continuous equipment designed for use at high pressures.

Oil Production. World vegetable and marine oil production in 1991 was forecast at 60.1 million t with all significant oils except sunflower and coconut showing increasing trends (45). Figure 5 shows world production of commercial vegetable oils by oil and by country. Some specialization for specific countries and regions are evident: palm oil is produced mostly in Malaysia, coconut in the Philippines, and rapeseed and sunflower in Europe. Commodity prices can range from \$0.37/kg for palm kernel oil to \$2.68/kg for cocoa butter (47). These price differentials have stimulated a great deal of research effort to develop functional equivalents to cocoa butter using less expensive fats.

Physical Properties of Fats and Oils

The physical properties of fats and oils have been reviewed (48).

Crystallization and Melting Behavior. Pure compounds usually display sharp melting points and impure compounds show broad melting behavior. However, even pure triglycerides show complex melting behavior because of their tendency to pack in several different crystal lattice forms (polymorphism). Complete crystal structures have been obtained for tricaprin (49), trilaurin [*538-24-9*] (50), and the triglyceride of 11-bromoundecanoic acid (51). Structures of other triglycerides relate back to these determinations by analogy. Triglycerides having three identical fatty acids pack into three distinct polymorphs: (*1*) β, the most stable form shows a triclinic subcell, (*2*) β′, a less stable crystal which suggests orthorhombic packing, and (*3*) α, a loosely packed triglyceride which packs hexagonally. Rapid cooling of the triglyceride leads initially to the α form followed by slow reorganization to β′ and β forms. Mixed glycerides, with more than one type of saturated fatty acid, pack with defects in the structure and chains appear to tilt to correct for these defects. Glycerides with unsaturated fatty acids must pack to

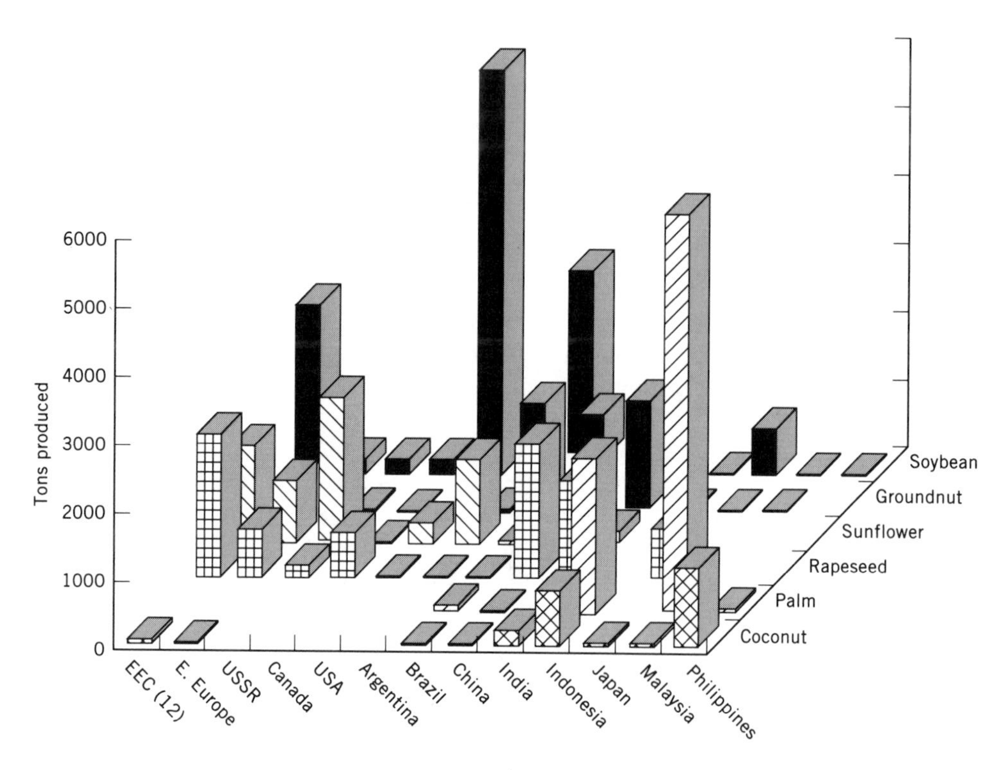

Fig. 5. Production of specific oils by country (1989–1990) (46).

accommodate the bend in the alkyl chain caused by the cis double bond. Perhaps the most widely studied fat, cocoa butter, may show as many as seven distinct polymorphic forms.

The crystal structure of glycerides may be unambiguously determined by x-ray diffraction of powdered samples. However, the dynamic crystallization may also be readily studied by differential scanning calorimetry (dsc). Crystallization, remelting, and recrystallization to a more stable form may be observed when liquid fat is solidified at a carefully controlled rate in the instrument. Enthalpy values and melting points for the various crystal forms are shown in Table 3 (52).

Specific heats of fats and oils may be calculated with some precision as a function of temperature, t, in °C (53):

Liquid oils

$$C_p = 1.93 + 0.0025\,t\ (t = 15\text{–}60°\text{C})$$

Tallow, palm oil, and partially hydrogenated fats

$$C_p = 1.99 + 0.0023\,t\ (t = 40\text{–}70°\text{C})$$

Fully hydrogenated fats

$$C_p = 1.92 + 0.003\,t \ (t = 60\text{–}80°\text{C})$$

Some general trends in specific heats have been suggested: (*1*) for solid fats, there is little variation in specific heat for saturated fats and their fatty acids as chain length varies. Specific heat varies directly with the degree of unsaturation. Specific heat of a solid is less than that of the liquid at the same temperature; (*2*) specific heat of liquid fatty acids and glycerides increases with increasing chain length but decreases with increasing unsaturation. For both liquids and solids, specific heat increases with increasing temperature; and (*3*) mixed-acid glycerides have lower specific heats than their corresponding simple glycerides.

Practical consequences of polymorphism may be seen in some food products. In the preparation of chocolate products it is difficult to blend other fats with cocoa butter. Cocoa butter is required to be in the β form for optimal melting and mouthfeel. If another fat with differing crystallization tendencies is incorporated, the

Table 3. Melting Points and Enthalpy Values for Selected Triglycerides[a]

	Mp, °C			ΔH, kJ/mol[c]	
C:Δ[b]	α	β′	β	α	β
	Simple saturated triacylglycerols				
6:0			−25		
8:0		−21	8.2		67.4
10:0	−15	18.0	31.5		93.3
11:0	1.0	26.5	30.5		
12:0	15	35.0	46.5		123
13:0	25	41.0	44.0		
14:0	33	46.5	57.0	105	148
15:0	40	51.5	54.0		
16:0	44.7	56.5	66.4	126	172
17:0	50	61.0	64.0		
18:0	54.9	64.0	73.1	145	197
20:0	61.8	69.0	78.0		223
22:0	68.2	74.0	82.5		262
	Simple unsaturated triacylglycerols				
cis-18:1-Δ9	−32	−12	5.5		95.3
trans-18:1-Δ9	15.5	37	42		146
cis-18:1-Δ6			26.2		
18:2-Δ9,12	−43		−13.1		83.3
18:3-Δ9,12,15	−44.6		−24.2		
cis-22:1-Δ13	6	17	30		
trans-22:1-Δ13	43	50	59		

[a]Ref. 52.
[b]Δ designates double bond.
[c]To convert kJ to kcal, divide by 4.184.

foreign fat crystallizes separately and a whitish coating referred to as "bloom" appears on the surface of the product. In margarine manufacture, a β' crystal is desired because it forms small crystals that feel smooth and melt easily in the mouth. If a fat forms a β crystal, the margarine appears "grainy" or "sandy" and is unacceptable to consumers; x-ray diffraction clearly shows the two forms. Examples of mesomorphic forms, or liquid crystals, may be of importance to other industrial technologies, such as manufacture of waxes.

Partial glycerides tend to melt higher than their triglyceride counterparts, as shown in Figure 6. This observation is consistent with the presence of free hydroxyl groups which can participate in increased intermolecular hydrogen bonding.

Viscosity. Fats, oils, fatty acids, and other fatty acid derivatives show relatively high viscosities compared to other liquids because of the intermolecular interaction of long alkyl chains (Table 4). They are similar to paraffinic hydrocarbons in this respect. Some general trends are that longer chain lengths and lower unsaturation produce higher viscosities. Fatty acids are more viscous than esters because of the greater tendency to form hydrogen bonds. Castor oil is in a class by itself because of its side chain hydroxyl group which can form hydrogen bonds. Derivatives of castor oil are consequently useful as specialty lubricants. Fats and oils behave as Newtonian liquids except at very high shear rates where degradation may begin to occur.

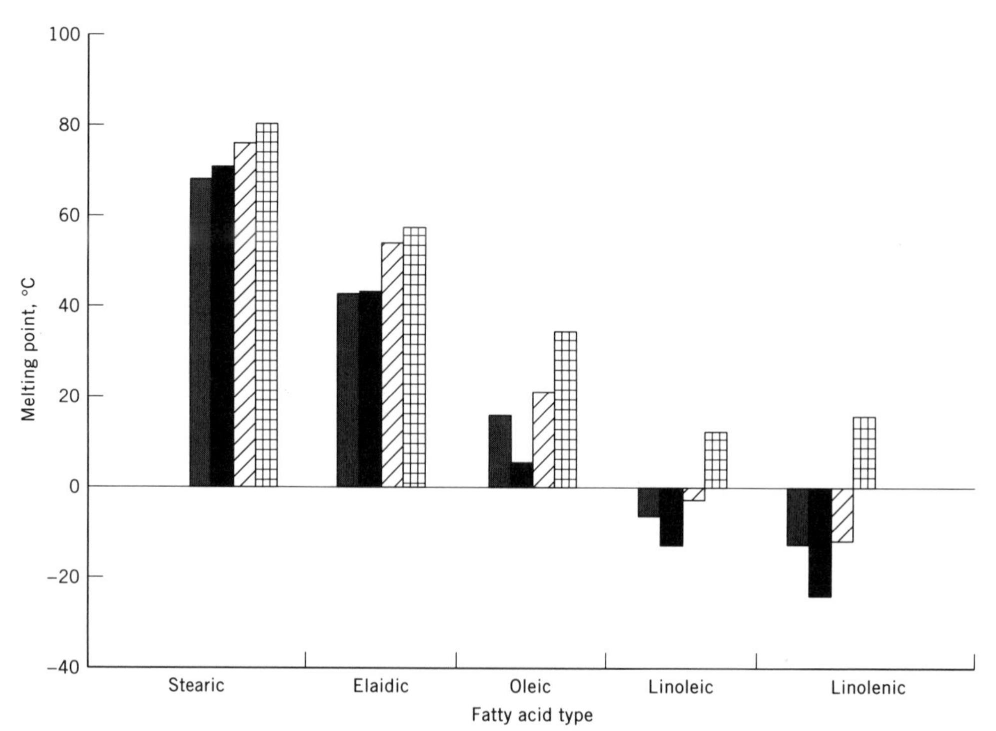

Fig. 6. Melting points of mono-, di-, and triglycerides, where ■ is fatty acid; ■, triglyceride; ▨, 1,3-diglyceride; and ⊞, 1-monoglyceride.

Table 4. Viscosity of Deodorized Oils and Fats at Different Temperatures[a]

	Viscosity,[b] mPa·s(=cP)				Constants[c]	
Oil or fat	IV[d]	20°C	40°C	60°C	a	b
soybean oil	134	60	28	15	−0.073	46.6
medium-chain triglycerides	0		21	11	−0.306	50.1
sunflower seed oil	132	63	29	16	−0.038	44.8
corn oil	122	70	30	16	−0.142	49.9
coconut oil	9		27	14	−0.242	51.0
hydrogenated soybean oil, mp 28°C	101		33	18	−0.148	51.1
butterfat	38		34	17	−0.151	51.2
groundnut oil	89	81	36	19	−0.080	50.5
olive oil	83	82	35	17	−0.102	50.1
hydrogenated cottonseed oil, mp 32°C	76		45	23	−0.166	55.9
rapeseed oil	104	93	41	21	−0.023	50.1
lard olein	73		36	18	−0.151	51.9
palm olein	64		37	19	−0.145	52.2
palm oil	51		37	19	−0.192	53.8
lard	63		36	19	−0.068	48.2
hydrogenated rapeseed oil, mp 32°C	81		49	24	−0.140	56.0

[a]Ref. 54.
[b]Standard deviation of replicates 1%.
[c]Constants in the equation $\log \eta = a + 10^6 b t^{-3}$ (t = °C).
[d]IV = iodine value, a measure of unsaturation.

Surface and Interfacial Tension. Data for fats, fatty acids, and derivatives are relatively sparse. Commercial oils tend to have lower surface and interfacial tensions because of the presence of polar surface-active components such as monoglycerides, phospholipids, and soaps. Purification of oils on a Florisil column can be used to obtain higher and more consistent values (55). Monoglycerides and phospholipids can reduce the interfacial tension between an oil and water. Emulsions (qv) may be formed that are relatively stable. Food products such as mayonnaise, margarine, and nonseparating salad dressings are commercial examples of stable emulsions.

Density. The density of liquid oils at 15°C does not vary markedly with changes in composition. Values generally range from 0.912 to 0.964 g/mL. Density increases with decreasing molecular weight and increasing unsaturation. An approximate equation (56) for determination of specific gravity from saponification number (SN) (function of chain length) and iodine value (IV) (function of unsaturation) at 15°C is sp gr = 0.8475 + 0.00030 (SN) + 0.00014 (IV). Density decreases approximately linearly with increasing temperature. Densities of fats in the solid state are much higher, approximating the values for water (1.0 g/mL).

Smoke, Flash, and Fire Points. These thermal properties may be determined under standard test conditions (57). The smoke point is defined as the temperature at which smoke begins to evolve continuously from the sample. Flash point is the temperature at which a flash is observed when a test flame is applied. The fire point is defined as the temperature at which the fire continues to burn. These values are profoundly affected by minor constituents in the oil, such as fatty acids, mono- and diglycerides, and residual solvents. These factors are of commercial importance where fats or oils are used at high temperatures such as in lubricants or edible frying fats.

Refractive Index. Refractive index of a fat or oil increases with molecular weight and unsaturation. For this reason, the method is used to monitor the progress of processing operations such as hydrogenation. In order to obtain values for both liquids and oils, most control laboratories operate the instrument at increased temperature (48–60°C), rather than the standard 25°C used for other liquids. Refractive indexes of simple glycerides are significantly higher than those of fatty acids. Values for partial glycerides are higher than their triglyceride counterparts. The presence of conjugated double bonds, as in tung oil or dehydrated castor oil, leads to a marked elevation of the refractive index.

Absorption Spectra. Double bonds in naturally occurring fats and oils are interrupted by methylene groups and, except for a few unusual oils, they are not conjugated. For this reason they are transparent to most of the ultraviolet spectrum. However, conjugated systems can be produced by autoxidation, chemical reactions, or processing and the conjugated isomers can be conveniently detected by uv spectroscopy. Bands for dienes (233 nm), trienes (270 nm), and tetraenes (305 nm) are diagnostic for conjugated systems. Caution must be exercised in interpretation because minor constituents in fats or oils may also contain chromophores absorbing in the same regions. Absorption of light in the visible range results in a colored appearance to oils. Color is an important specification in trading of oils.

Infrared spectra of fats and oils are similar regardless of their composition. The principal absorption seen is the carbonyl stretching peak which is virtually identical for all triglyceride oils. The most common application of infrared spectroscopy is the determination of trans fatty acids occurring in a partially hydrogenated fat (58,59). Absorption at 965–975 cm^{-1} is unique to the trans functionality. Near infrared spectroscopy has been utilized for simultaneous quantitation of fat, protein, and moisture in grain samples (60). The technique has also been reported to be useful for instrumental determination of iodine value (61).

Absorption of x-rays by a powdered sample of solid fat has been a useful method for determination of polymorphic character as discussed earlier. The α, β', and β forms may be distinguished; however, interpretation is made more difficult because subsets of the β' and β forms have often been encountered. Also, a fat may contain mixtures of polymorphic forms and properties may therefore be difficult to relate to the spectra.

Proton chemical shift data from nuclear magnetic resonance has historically not been very informative because the methylene groups in the hydrocarbon chain are not easily differentiated. However, this can be turned to advantage if a polar group is present on the side chain causing the shift of adjacent hydrogens downfield. High resolution ^{13}C-nmr has been able to determine position and stereochemistry of double bonds in the fatty acid chain (62). Broad band nmr has also been shown useful for determination of solid fat content.

Solubility Properties. Fats and oils are characterized by virtually complete lack of miscibility with water. However, they are miscible in all proportions with many nonpolar organic solvents. True solubility depends on the thermal properties of the solute and solvent and the relative attractive forces between like and unlike molecules. Ideal solubilities can be calculated from thermal properties. Most real solutions of fats and oils in organic solvents show positive deviation from ideality, particularly at higher concentrations. Determination of solubilities of components of fat and oil mixtures is critical when designing separations of mixtures by fractional crystallization.

Chemical Properties

Most triglyceride fats and oils have only two reactive functional groups: the ester linkage joining the fatty acid to the glycerol backbone and double bonds in the alkyl side chain. There is a free hydroxyl group in the side chain of ricinoleic acid found in castor oil and a carbonyl group in the licanic acid side chain of oiticica oil. The double bond influences the reactivity of the adjacent allylic carbon atom, particularly when multiple double bonds are present. Figures 7 and 8 show some important reactions at the primary reactive sites of triglycerides and fatty acids. Minor constituents, such as sterols, may have other functional groups in their structures presenting an opportunity for separation and enrichment of those components.

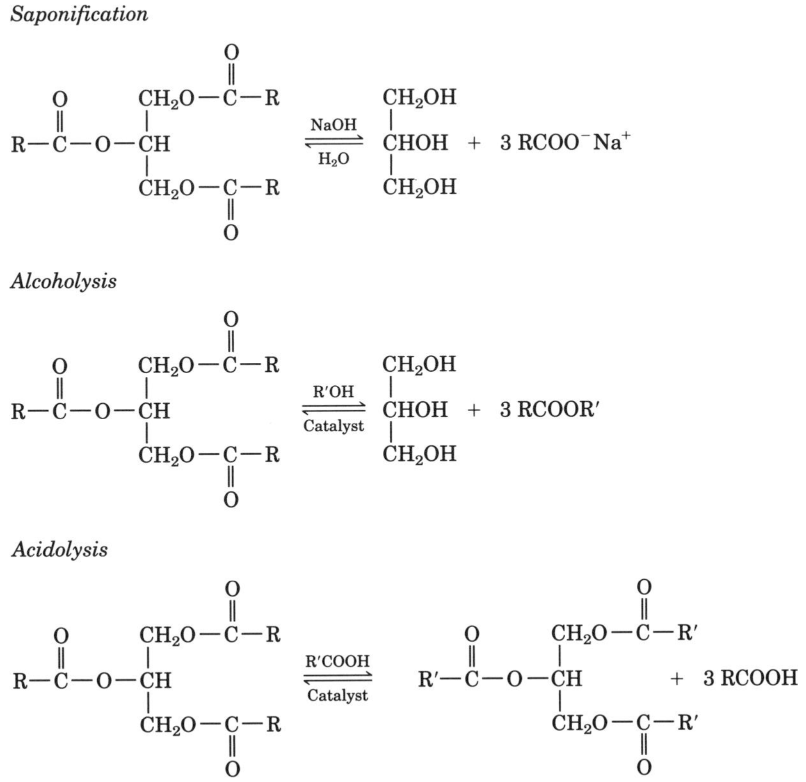

Saponification

$$\begin{array}{l} \mathrm{CH_2O{-}C({=}O){-}R} \\ \mathrm{R{-}C({=}O){-}O{-}CH} \\ \mathrm{CH_2O{-}C({=}O){-}R} \end{array} \underset{\mathrm{H_2O}}{\overset{\mathrm{NaOH}}{\rightleftharpoons}} \begin{array}{l} \mathrm{CH_2OH} \\ \mathrm{CHOH} \\ \mathrm{CH_2OH} \end{array} + 3\ \mathrm{RCOO^-Na^+}$$

Alcoholysis

$$\begin{array}{l} \mathrm{CH_2O{-}C({=}O){-}R} \\ \mathrm{R{-}C({=}O){-}O{-}CH} \\ \mathrm{CH_2O{-}C({=}O){-}R} \end{array} \underset{\text{Catalyst}}{\overset{\mathrm{R'OH}}{\rightleftharpoons}} \begin{array}{l} \mathrm{CH_2OH} \\ \mathrm{CHOH} \\ \mathrm{CH_2OH} \end{array} + 3\ \mathrm{RCOOR'}$$

Acidolysis

$$\begin{array}{l} \mathrm{CH_2O{-}C({=}O){-}R} \\ \mathrm{R{-}C({=}O){-}O{-}CH} \\ \mathrm{CH_2O{-}C({=}O){-}R} \end{array} \underset{\text{Catalyst}}{\overset{\mathrm{R'COOH}}{\rightleftharpoons}} \begin{array}{l} \mathrm{CH_2O{-}C({=}O){-}R'} \\ \mathrm{R'{-}C({=}O){-}O{-}CH} \\ \mathrm{CH_2O{-}C({=}O){-}R'} \end{array} + 3\ \mathrm{RCOOH}$$

Fig. 7. Reactions of triglycerides at the carbonyl ester linkage. R represents the alkyl side chain.

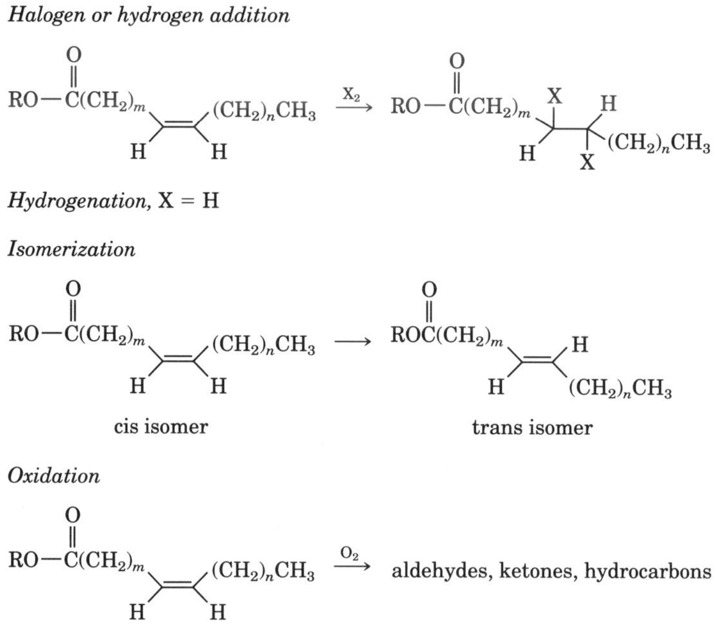

Halogen or hydrogen addition

$$RO{-}\overset{O}{\overset{\|}{C}}(CH_2)_m{-}CH{=}CH{-}(CH_2)_nCH_3 \xrightarrow{X_2} RO{-}\overset{O}{\overset{\|}{C}}(CH_2)_m{-}CHX{-}CHX{-}(CH_2)_nCH_3$$

Hydrogenation, X = H

Isomerization

$$RO{-}\overset{O}{\overset{\|}{C}}(CH_2)_m{-}CH{=}CH{-}(CH_2)_nCH_3 \longrightarrow RO\overset{O}{\overset{\|}{C}}(CH_2)_m{-}CH{=}CH{-}(CH_2)_nCH_3$$

cis isomer trans isomer

Oxidation

$$RO{-}\overset{O}{\overset{\|}{C}}(CH_2)_m{-}CH{=}CH{-}(CH_2)_nCH_3 \xrightarrow{O_2} \text{aldehydes, ketones, hydrocarbons}$$

Fig. 8. Reactions of triglycerides at double bonds in the alkyl chain. R represents the glycerol backbone.

Chemical reactions can cause serious quality problems for oils. Hydrolytic and oxidative rancidity can cause oils to become unacceptable to consumers for edible or other uses. Hydrolysis can occur if crops are not dried after harvesting or moisture is in contact with the oil over a large surface area. Hydrolysis may be catalyzed by acids, bases, or lipase enzymes. Oxidation may occur in any oil that contains double bonds. The highly unsaturated oils containing 18:2, 18:3, 20:5, and 22:6 fatty acids are particularly vulnerable. Oxidation is catalyzed by light, as in oxidation of milk lipids, by lipoxygenase enzymes, or by transition metals. After an initiating event, the chain reaction is self-catalyzed (autoxidation). The mechanism of lipoxygenase-catalyzed oxidation (63) and autoxidation (64) have been reviewed. In addition to producing secondary volatile oxidation products having rancid odors and flavors, lipid peroxides may cause oxidation of proteins and thereby cause loss of their nutritional values (65).

Analytical Methods

Throughout the history of the development of fats and oils, many wet chemical methods have been developed to assess the quality of the raw materials and products. As sophisticated instrumentation develops, many of the wet methods are being replaced. Particular attention is being given to methods that eliminate the use of solvents which cause an environmental disposal problem. Many in-line sensors are also being developed to allow corrections of critical parameters to be made more quickly in the process.

Specifications. The quality of individual crude oils is specified by trading rules established by organizations such as the National Soybean Processors Association, National Renderers Association, or National Institute of Oilseed Processors. Standardized tests are defined by the American Oil Chemists Society (AOCS), the Association of Official Analytical Chemists (AOAC), and the American Society for Testing Materials (ASTM). Crude oils must contain minimal amounts of foreign material, protein, volatile or toxic solvents, pesticides, heat-transfer media, moisture, and foreign adulterating fats. They must also not be abused or mishandled which causes them to become oxidized. Crude oils must not show excessive loss on refining which adds to costs and waste disposal problems. Oil processors must also meet specifications for their customers which include measures of oil quality, such as free fatty acid level, color, and peroxide value. Other specifications may relate to functionality in the customer's product, such as melting range or fatty acid composition.

Free Fatty Acid and Saponification Value. High concentrations of free fatty acid are undesirable in crude triglyceride oils because they result in large losses of neutral oil during refining. They are also undesirable in finished oils because they can cause off-flavors and shorten the shelf life of the oil. Free fatty acid is determined by titration with base to a phenolphthalein end point (66). Saponification value is determined by splitting triglycerides into free fatty acids in alcoholic base and back-titrating with acid (67). The number obtained is an inverse function of molecular weight and therefore alkyl chain length.

Fatty Acid and Triglyceride Composition. Triglycerides may be alcoholyzed using alkaline methanol or derivatized as methyl esters using diazomethane. The resulting fatty acid methyl esters may be separated and quantitated by gas–liquid chromatography (glc) (68). The development of capillary columns has greatly improved the resolution and sensitivity of this method (69). Determination of fatty acid composition is the first step in identification of an unknown oil. Triglycerides may be broadly grouped into trisaturates, monounsaturates, diunsaturates, and triunsaturates. These broad classes may be determined by thin-layer chromatography (tlc) on a silver-impregnated support (70). A semiautomated tlc unit is commercially available for tlc of lipids. To separate and quantitate mono-, di-, and triglycerides, tlc and glc may also be used.

The arrangement of fatty acids on the glycerol backbone is also an important consideration, particularly where polymorphic properties are critical. Triglycerides may be separated by high temperature glc according to their molecular weight and therefore the number of carbon atoms in the fatty acid chain (71). High performance liquid chromatography (hplc) on reversed-phase columns separates glycerides by the number of double bonds and the number of carbon atoms in the side chain (72). Components elute in order of increasing chain length. The presence of a double bond reduces the retention time the same amount as a decrease of two carbon atoms in the chain. Combination of hplc and high temperature glc can yield a complete triglyceride profile. An enzymatic technique is used to determine the distribution of fatty acids at the 2 position. Pancreatic lipase hydrolyzes fatty acids at the primary position leaving 2-monoglycerides which are derivatized and injected into a glc (73).

Measurement of Unsaturation. The presence of double bonds in a fatty acid side chain can be detected chemically or through use of instrumentation. Iodine value (IV) (74) is a measure of extent of the reaction of iodine with double bonds; the higher the IV, the more unsaturated the oil. IV may also be calculated from fatty acid composition. The cis–trans configuration of double bonds may be determined by infrared (59) or nmr spectroscopy. Naturally occurring oils have methylene-interrupted double bonds that do not absorb in the uv; however, conjugated dienes may be determined in an appropriate solvent at 233 nm.

Measurement of Solid Fat. Many commercially used fats and oils are mixtures of liquids and solids and the ratio is largely dependent on composition and temperature. Solid Fat Index (SFI) (75) is a dilatometric method to determine solid fat. Samples are conventionally equilibrated (tempered) and read at several different temperatures and a profile is reported. The profile gives a visualization of how sharp or broad the melting range is for the solid fat. Some applications such as margarines or shortenings benefit from a broad melting range, whereas others such as confectionery products require a sharp melting fat. The method for determination of SFI is laborious and time consuming. Many efforts are currently in progress to develop a rapid and automated technique to acquire the information. Differential scanning calorimetry has also been used to show the dynamic melting properties of fat (76). More recently another profile, Solid Fat Content (SFC), has been developed. Two low resolution nuclear magnetic resonance techniques, wide-line or continuous nmr (77) and pulsed nmr (78), are utilized to determine the SFC index. Both methods discriminate differences in protons in a liquid state from those in a solid state. Wide-line nmr measures the spin–lattice relaxation time whereas pulsed nmr measures the spin–spin relaxation time. Efforts to correlate nmr methods with dilatometric SFI have met with mixed results. A practical applications-oriented measure of solid fats is the cold test in which oils are chilled to a low temperature and held for a specified period.

Melting Properties. As previously discussed, because fats are mixtures of triglycerides and exhibit polymorphic forms they have broad melting points. Several methodologies have been devised to measure melting points. Capillary melting point (79) and the Wiley melting point were subject to operator variability and have largely been replaced by the Mettler dropping point (80). This technique involves heating a disk of fat at a programmed rate until the disk melts sufficiently to fall through its holder. The motion breaks an electric eye circuit and the temperature is recorded.

Color. The most common colored materials in oil are naturally occurring pigments such as red from carotenoids and green from chlorophyll. These colors are generally removed substantially in the bleaching and deodorization processes. Measurement of color has been carried out by comparison to reference standards. The three most common scales are the Lovibond (81), the FAC (82), and the Gardner (83). Lovibond colors are generally used to specify vegetable oils for edible purposes whereas the Gardner scale is used for other industries such as paints, cosmetics, and plastics. The FAC scale is useful for animal fats which are mostly yellow with various greenish tones. Spectrophotometric methods in the region of

400–700 nm have been developed but have been somewhat slow in gaining acceptance. Their potential for automation and on-line monitoring will drive their use in the near future.

Measurement of Oxidation. The most common cause of rancidity in fat-based products is oxidation. In the initial stages a free radical is produced that captures oxygen and then a hydrogen atom to produce hydroperoxide. The peroxide then decomposes into lower molecular weight secondary oxidation products that have disagreeable odors and flavors. An extraordinary number of tests have been developed to determine the extent of oxidation that has occurred in the oil (84). Peroxide value (85) is an indication of the concentration of hydroperoxides (primary oxidation products). Another method, the thiobarbituric acid test (TBA), presumably measures the concentration of malonaldehyde, a secondary oxidation product (86). During the course of oxidation, conjugated dienes are formed which are measurable by uv absorption. Peroxide value is generally accepted in the edible oil industry whereas the latter two methods are popular in the measurement of biomedical materials. A comparison of methods for biological samples has been published (87).

Stability of an oil to oxidation has been measured by several techniques. In the Schaal oven test (88) an oil is heated in an oven and the peroxide value is periodically measured. In the Active Oxygen method (AOM) a calibrated stream of air is bubbled through a sample and the peroxide value is measured as a function of time. Alternatively, the oil may be oxidized in a pressurized vessel (bomb). The value is reported in hours to reach a peroxide value of 100 (89). A more automated procedure has been reported where air is bubbled through a heated oil and then through a conductivity cell (90). The induction time is measured by an inflection point where the conductivity begins to increase. Sampling of a head-space above an oil can be used to determine the concentrations of volatile oxidation products. Off-flavor descriptors may be rated on a scale by sensory panels to determine the acceptability of the oil for edible purposes. Many attempts have been made to correlate the sensory experience scores with volatile components by glc (91). Measurement of hexanal may also be used for evaluation of oxidation in biological systems (92). Polar and polymeric oxidation products may be determined by hplc (93).

Analysis of Trace or Minor Components. Minor or trace components may have a significant impact on quality of fats and oils (94). Metals, for example, can catalyze the oxidative degradation of unsaturated oils which results in off-flavors, odors, and polymerization. A large number of techniques such as wet chemical analysis, atomic absorption, atomic emission, and polarography are available for analysis of metals. Heavy metals, iron, copper, nickel, and chromium are elements that have received the most attention. Phosphorus may also be detectable and is a measure of phospholipids and phosphorus-containing acids or salts.

Antioxidants (qv) have a positive effect on oils when present in the proper concentration. Sterols and tocopherols, which are natural antioxidants, may be analyzed by gas–liquid chromatography (glc), high performance liquid chromatography (hplc), or thin-layer chromatography (tlc). Synthetic antioxidants may be added by processors to improve the performance or shelf life of products. These compounds include butylated hydroxyanisole (BHA), butylated hydroxytoluene

(BHT), *tert*-butylhydroquinone (TBHQ), and propyl gallate. These materials may likewise be analyzed by glc, hplc, or tlc. Citric acid (qv), which functions as a metal chelator, may also be determined by glc.

Uses of Fats and Oils

Food Components and Cooking Oils. Fats and oils have a relatively large number of carbon–carbon and carbon–hydrogen bonds. When burned as fuel in the body they provide approximately 37.7 kJ/g (9 kcal/g) in energy. By comparison, proteins and carbohydrates contribute only about 16.7 kJ/g (4 kcal/g) (one food calorie = 1 kcal). Fat is therefore the most energy dense food available. In addition, fats contribute essential fatty acids. Linoleic acid is the most effective component for prevention of fatty acid deficiency. Fortunately, the daily requirement is very low and extremely severe restrictions are required to induce deficiencies. Fats produce a feeling of satiety and attempts to severely curtail fat may result in a "fat hunger." Vitamins A, D, E, and K are fat-soluble and are generally associated with fats in foods. Fats contribute desirable flavor and texture to food products. In addition to native or modified flavor, fats and oils serve as solvents and modulators for other flavor components present in foods.

Unfortunately, excess consumption of fatty foods has been correlated with serious human disease conditions. Effects on cardiovascular disease (95), cancer (96), and function of the immune system (97) have been shown. Numerous studies have been conducted to determine the effects of saturated, monounsaturated, and polyunsaturated fatty acids on serum cholesterol and more recently high density lipoprotein (HDL) and low density lipoprotein (LDL) present in serum. The effects of saturates, polyunsaturates, and dietary cholesterol on total serum cholesterol have been quantitated (98,99) using the following equations:

$$\Delta SC = 1.35(2\ \Delta S - \Delta P) + 1.4\ \Delta C$$

$$\Delta SC = 2.16\ \Delta S - 1.65\ \Delta P + 0.0677\ \Delta C - 0.53$$

where ΔSC = total serum cholesterol, ΔS = saturated fat, ΔP = polyunsaturated fat, and ΔC = dietary cholesterol. Significant effects were found for saturated fatty acids having 12–16 carbon chain lengths.

Relative amounts of saturated, monounsaturated, and polyunsaturated fatty acids for several oils are shown in Figure 9. Saturated fatty acids have been associated in some studies with increases in total cholesterol and LDL. Trans unsaturates, produced as by-products of hydrogenation, allegedly raise LDL and lower HDL, but this point is currently under some debate in the nutritional literature (101). Linoleic acid has been found to be a promoter for tumor growth where malignancy has been initiated with chemical carcinogens; however, no evidence has yet been presented to show an effect on noninduced tumors. Fat and calorie consumption have been correlated with some types of cancer. However, some fat constituents, such as conjugated linoleic acid (CLA) (102), and omega-3 (n-3) oils, such as fish oil, have been shown to retard tumor growth. Dietary recommendations (103) have advised the consumption of less than 30% of calories

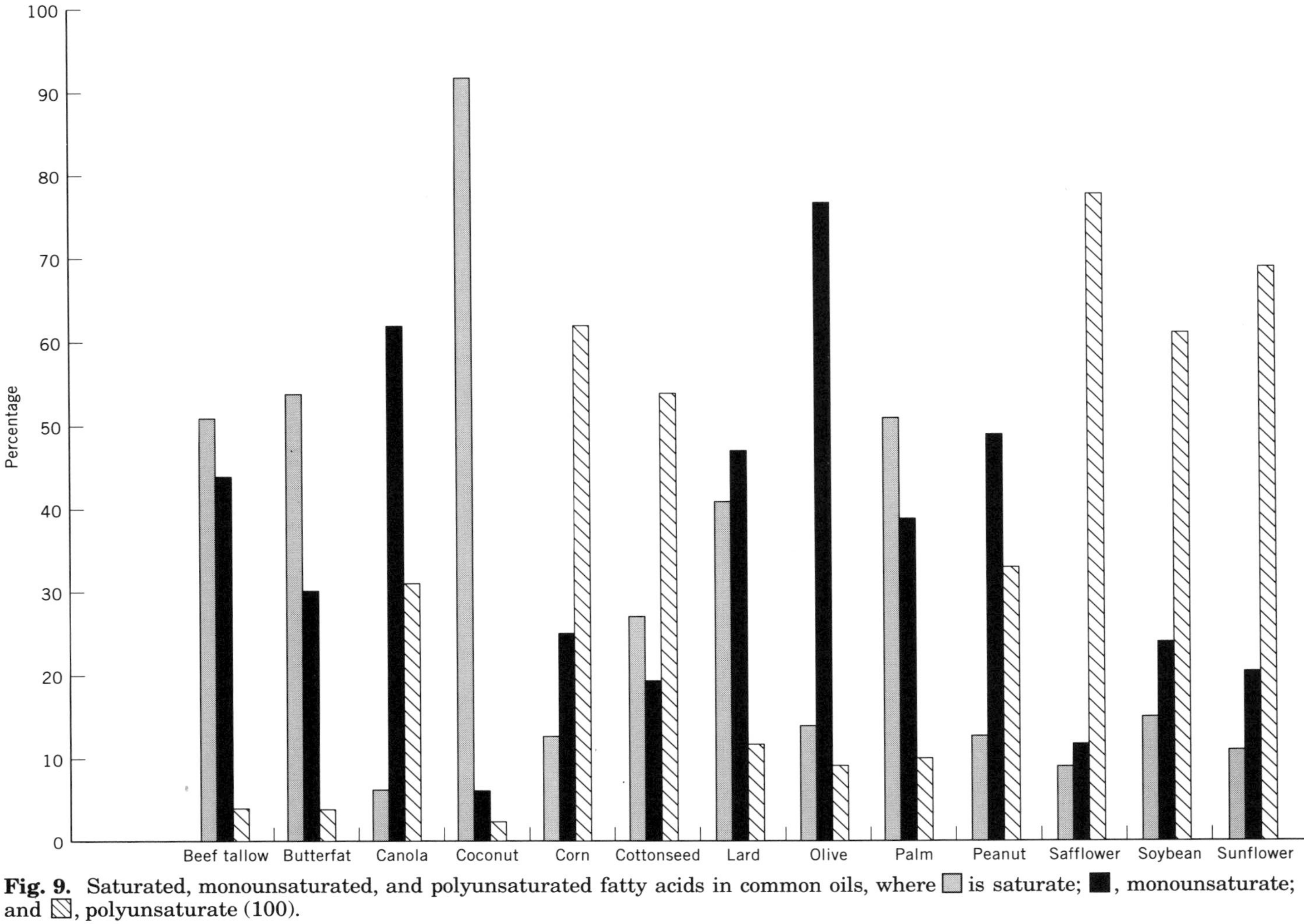

Fig. 9. Saturated, monounsaturated, and polyunsaturated fatty acids in common oils, where ▢ is saturate; ■, monounsaturate; and ▨, polyunsaturate (100).

from fat and equal contributions should be derived from saturates, monounsaturates, and polyunsaturates.

Although most of the fat consumed is indigenous to foods such as meat, poultry, fish, and dairy products, processed fats and oils make a significant contribution in the diet. Cooking oils provide a heat-transfer medium and contribute to flavor development in foods. Partially hydrogenated fats are added to bakery products, icings, and whipped toppings to improve aeration, texture, and mouthfeel. Oil contributes flavor and texture effects to margarine, mayonnaise, salad dressing, and confectionery products. Although many fat replacers (qv) have been developed to allow reduction or elimination of fats in food, technology has not yet advanced to a point where these products are perceived as equally preferred to their full fat counterparts.

Derivatives of fats and oils may be used as emulsifiers, stabilizers, and antioxidants in foods. Examples are mono- and diglycerides, propylene glycol esters, lactylated monoglycerides, sodium and calcium stearoyl lactylates, monoglyceride citrates, succinylated monoglycerides, diacetyltartaric esters of monoglycerides (DATEM), sucrose esters, sorbitan esters, polysorbates, and polyglycerol esters.

Soaps and Detergents. Soap is one of the earliest chemical substances known (104), dating from before 2500 BC. Its original derivation was by boiling ashes rich in alkali with fats to achieve saponification of the fat to soap and glycerol. Modern soaps may be derived from batch or continuous saponification or by neutralization of fatty acids which are split from fats. Coconut oil is a very popular starting oil since its soap has high surface activity and residues on the skin are less greasy than soaps with longer alkyl chains.

Detergents may be produced by the chemical reaction of fats and fatty acids with polar materials such as sulfuric or phosphoric acid or ethylene oxide. Detergents emulsify oil and grease because of their ability to reduce the surface tension and contact angle of water as well as the interfacial tension between water and oil. Recent trends in detergents have been to lower phosphate content to prevent eutrification of lakes when detergents are disposed of in municipal waste.

Drying Oils. Liquid oils have a relatively long history of use in paints, lacquers, and varnishes. Flax was grown by the ancient Egyptians as a source of drying oil for their decorative paints. Linseed, fish, tung, soybean, tall oil, and dehydrated castor oil are the most widely used in oil-based coatings. Drying oils (qv) are highly unsaturated and because of their structure are readily polymerized to form cross-linked protective films. Several mechanisms may be operating in the polymerization process. Rearrangement of double bonds into conjugation allows Diels-Alder reactions to occur to produce cross-linking. Oxidation produces free radicals which induce and promote chain-reaction polymerization. Drying may be accelerated by heat, light, or the presence of transition metals (driers) (see DRIERS AND METALLIC SOAPS). Drying oils react with a variety of chemicals to yield resins that produce harder and more durable films. Phenolic resins, alkyd resins (polyesters of polyfunctional alcohols and polybasic acids), and urethanes are the most widely used.

Triglyceride oils have declined since the 1980s and have been replaced by petroleum-derived products. However, as fossil fuels deplete the supply of petrochemicals, triglyceride-based oils are available as a renewable resource.

Manufacture of Fatty Acids and Derivatives. Splitting of fats to produce fatty acids and glycerol (a valuable coproduct) has been practiced since before the 1890s. In early processes, concentrated alkali reacted with fats to produce soaps followed by acidulation to produce the fatty acids. Acid-catalyzed hydrolysis, mostly with sulfuric and sulfonic acids, was also praticed. Pressurized equipment was introduced to accelerate the rate of the process, and finally continuous processes were developed to maximize completeness of the reaction (105). Lipolytic enzymes may be utilized to split fatty acids from oils which would be adversely affected by high temperatures and strongly acidic conditions. The high cost of enzymes and relatively slow rate of reaction have prevented wide-scale commercial use of this process. Regioselective enzymes have been used for analysis of fatty acid residues in the 2 position of triglycerides. Transesterification (interesterification) of fats or oils with monohydric alcohols produces alkyl esters of fatty acids and glycerol. Methyl esters from this process are widely used industrial fatty acid derivatives. Figure 7 showed saponification and alcoholysis which corresponds to interesterification.

Miscellaneous Uses of Fats, Oils, and their Derivatives. Fatty-derived materials have been important in a wide variety of industrial applications for many years, although they have been supplanted in some areas with less expensive petroleum-based compounds. Fats, oils, fatty acids, and their derivatives remain in use in industrial areas where their unique structure produces a functional effect. Drying oils for protective coatings depend on oxidation of multiple double bonds. Aldehydes derived from oil oxidation react with collagen in the process of leather tanning. Fatty acids, present in low concentration in oils, allow the formation of thin films on metal parts, providing useful lubricating oils. Wax esters are useful plasticizers because they have a critical balance of alkyl chains and polar functionality. As fossil fuels become more critically managed in the future, renewable fats and oils are likely to claim an additional share in many industrial applications.

BIBLIOGRAPHY

"Fats and Fatty Oils" in *ECT* 1st ed., Vol. 6, pp. 140–172, by A. E. Bailey, The Humko Co.; in *ECT* 2nd ed., Vol. 8, pp. 776–811, by F. A. Norris, Swift & Co.; in *ECT* 3rd ed., Vol. 9, pp. 795–831, by T. H. Applewhite, Kraft, Inc.

1. R. T. Hamilton, M. Long, and M. Y. Raie, *J. Am. Oil Chem. Soc.* **49**, 307–310 (1972).
2. J. A. Clark and D. M. Yermanos, *J. Am. Oil Chem. Soc.* **58**, 176–178 (1980); G. F. Spencer and R. D. Plattner, *J. Am. Oil Chem. Soc.* **61**, 90–94 (1984); T. K. Mina, *J. Am. Oil Chem. Soc.* **61**, 407–411 (1984); M. L. Tonnet, R. L. Dumstome, and A. Shani, *J. Am. Oil Chem. Soc.* **61**, 1061–1063 (1984).
3. M. Mori and T. Saito, *Bull. Jpn. Soc. Sci. Fish.* **32**, 730–736 (1966); D. H. Buisson, D. R. Body, G. J. Dougherty, L. E. Eyres, and P. Vlieg, *J. Am. Oil Chem. Soc.* **59**, 390–395 (1982).
4. F. A. Norris, in D. Swern, ed., *Bailey's Industrial Oil and Fat Products*, Vol. 2, John Wiley & Sons, Inc., New York, 1982, pp. 178–245.
5. W. Th. M. DeGroot, *Lipids* **7**, 626 (1972).

6. G. Y. Brokaw, E. S. Perry, and W. C. Lyman, *J. Am. Oil Chem. Soc.* **32**, 194–197 (1955).
7. A. Gaonkar, *J. Am. Oil Chem. Soc.* **66**, 1090–1092 (1989).
8. G. Gregoriadis, ed., *Liposome Technology*, Vol. 3, *Targeted Drug Delivery and Biological Interactions*, CRC Press, Boca Raton, Fla., 1985.
9. J. Eisner and D. Firestone, *J. Assoc. Off. Agric. Chem.* **46**, 542 (1963).
10. J. C. Bauernfeind, *CRC Crit. Rev. Food Sci. Nutr.* **8**, 337 (1977).
11. T. Doba, G. W. Burton, K. U. Ingold, and M. Matsuo, *J. Chem. Soc., Chem. Commun.*, 461–462 (1984).
12. S. S. Chang and co-workers, *J. Food Sci.* **4**, 1102–1106 (1977); J. W. Wu, M. H. Lee, C. T. Ho, and S. S. Chang, *J. Am. Oil Chem. Soc.* **59**, 339–346 (1982); R. Inatani, N. Nakatani, and H. Fuwa, *Agric. Biol. Chem.* **47**, 521–528 (1983); C. M. Houlihan, C. T. Ho, and S. S. Chang, *J. Am. Oil Chem. Soc.* **61**, 1036–1039 (1984).
13. F. A. Norris, in Ref. 4, pp. 253–292. For an overview of refining and bleaching.
14. J. C. Segers, *J. Am. Oil Chem. Soc.* **60**, 262–264 (1983).
15. K. F. Carlson and J. D. Scott, *INFORM* **2**, 1048–1049 (1991).
16. R. R. Allen, in Ref. 4, pp. 1–96. For a review of hydrogenation.
17. P. Sabatier, *Catalysis In Organic Chemistry*, Van Nostrand Reinhold, New York, 1922; Brit. Pat. 1,515 (1903), W. Norman.
18. R. R. Allen and A. A. Kiess, *J. Am. Oil Chem. Soc.* **32**, 400–405 (1955).
19. K. J. Moulton, S. Koritala, and E. N. Frankel, *J. Am. Oil Chem. Soc.* **60**, 1257–1258 (1983).
20. N. O. V. Sonntag, in Ref. 4, pp. 127–174.
21. G. Juriens and A. C. J. Kroesen, *J. Am. Oil Chem. Soc.* **42**, 9–14 (1965).
22. Ref. 20, p. 151.
23. F. X. Malcata, H. R. Reyes, H. S. Garcia, C. G. Hill, and C. H. Amundson, *J. Am. Oil Chem. Soc.* **67**, 890–910 (1990).
24. A. E. Thomas, in T. H. Applewhite, ed., *Bailey's Industrial Oil and Fat Products*, Vol. 3, John Wiley & Sons, Inc., New York, 1985, pp. 1–40.
25. G. D. Revankar, D. P. Sen, J. Hemavathy, and G. Matthew, *J. Oil Technol. Assoc. India* **7**, 85–87 (1975); E. Tacchino, *Riv. Ital. Sostanze Grasse* **58**, 331–333 (1981); S. Adhikari, *Fette Seifen Anstrichm.* **84**, 185–188 (1982); B. P. Baliga and A. D. Shitole, *J. Am. Oil Chem. Soc.* **58**, 110–114 (1981).
26. N. W. Lam, L. Hg, and S. C. Oh, *Malays. J. Sci.* **6**, 145–157 (1980); E. Tacchino, *Riv. Ital. Sostanze Grasse* **58**, 394–398 (1981); J. Catalano, *Ind. Aliment. Agr.* **15**, 119–122 (1976).
27. M. Bernardini and E. Bernardini, *Rev. Fr. Corps Gras.* **18**, 439–443 (1971); U.S. Pat. 3,345,389 (1967), K. T. Zilch; U.S. Pat. 3,541,123 (1970), K. Kaiwada.
28. F. A. Norris, in Ref. 24, pp. 127–166.
29. Ref. 15, pp. 1034–1060.
30. R. G. Jensen and R. E. Pitas, *Adv. Lipid Res.* **14**, 213 (1976); D. Buchnea, in A. Kuksis, ed., *Handbook of Lipid Research*, Vol. 1, Plenum Press, New York, 1978, pp. 233–288; C. M. Lok, *Chem. Phys. Lipids* **22**, 323–337; A. Singh, *J. Lipid Res.* **31**, 1522–1525 (1990).
31. Technical data, "Typical Composition and Chemical Constants of Common Edible Fats and Oils", Durkee Industrial Foods, 1992.
32. A. E. Thompson and R. Kleiman, *J. Am. Oil Chem. Soc.* **65**, 139–146 (1988); A. E. Thompson, D. A. Dierig, S. J. Knapp, and R. Kleiman, *J. Am. Oil Chem. Soc.* **67**, 611–617 (1990).
33. J. A. Robertson, *J. Am. Oil Chem. Soc.* **49**, 239–244 (1972).
34. T. P. Hilditch, *J. Am. Oil Chem. Soc.* **26**, 41–45 (1949).

35. H. E. Longnecker, *Chem. Rev.* **29**, 201–224 (1941); K. F. Matil and F. A. Norris, *Science* **105**, 257–259 (1947); *J. Am. Oil Chem. Soc.* **24**, 274–275 (1947).
36. A. R. S. Kartha, *J. Am. Oil Chem. Soc.* **29**, 109–110 (1952); **30**, 280–282 (1953); **31**, 85–88 (1954).
37. R. J. Vander Wal, *J. Am. Oil Chem. Soc.* **37**, 18–20 (1960); **40**, 242–247 (1963).
38. M. Ratner, *Bio/Technol.* **7**, 337–341. (1989); W. Friedt, *Fat Sci. Tech.* **90**, 51–55 (1988).
39. M. B. Oliveira and M. A. Ferreira, *Rev. Port. Farmacia* **38**, 1–17 (1988); H. M. Davies, U. C. Knauf, U. C. Kridl, and G. A. Thompson, *INFORM* **2**, 368 (1991); C. Somerville and co-workers, *J. Cell Biochem., Suppl.* **14E**, 267 (1990); D. J. Murphy, *Trends Biotechnol.* **10**, 84–87 (1992).
40. K. P. Pauls, *J. Am. Oil Chem. Soc.* **66**, 455–456 (1989).
41. A. G. Green, N. B. Hulse, and M. L. Tonnet, *INFORM* **2**, 316 (1991); D. W. James and H. K. Dooner, *Theor. Appl. Genet.* **82**, 409–412 (1991).
42. D. Kyle, S. Bingham, and R. Radmer, *Hortscience* **25**, 1523–1526 (1990).
43. J. M. Snyder, J. P. Friedrich, and D. D. Christianson, *J. Am. Oil Chem. Soc.* **61**, 1851–1856 (1984).
44. G. R. List and J. P. Friedrich, *J. Am. Oil Chem. Soc.* **66**, 98–101 (1989).
45. *USDA-ARS Oil Crop Situation and Outlook Report*, OCS-31, U.S. Department of Agriculture, Washington, D.C., Oct. 1991.
46. *Oil World*, 45–50 (Aug. 25, 1989).
47. *Chem. Mark. Rep.* (Dec. 9, 1991).
48. M. N. Formo, in D. Swern, ed., *Bailey's Industrial Oil and Fat Products*, Vol. 1, John Wiley & Sons, Inc., New York, 1978, pp. 177–232; D. M. Small, *The Physical Chemistry of Lipids, Handbook of Lipid Research*, Vol. 4, Plenum Press, New York, 1986.
49. L. H. Jensen and A. J. Mabis, *Acta Crystallogr.* **B27**, 977 (1966).
50. M. S. Gray, N. V. Lovegren, and D. Mitchum, *J. Am. Oil Chem. Soc.* **53**, 196 (1976).
51. K. Larsson, *Proc. Chem. Soc.* **87**, 87 (1963).
52. *The Physical Chemistry of Lipids, Handbook of Lipid Research*, Vol. 4, Plenum Press, New York, 1986, p. 623.
53. J. Baites, *Gewinning und Werzarbeitung Von Nahrungsfetten*, P. Parey, Berlin, 1975.
54. A. J. Haighton, K. Van Putte, and L. F. Vermas, *J. Am. Oil Chem. Soc.* **49**, 153 (1972).
55. Ref. 7, Fig. 1.
56. D. Swern, ed., *Bailey's Industrial Oil and Fat Products*, Vol. 1, John Wiley & Sons, Inc., New York, 1978, p. 187.
57. D. Firestone, ed., *Official Methods and Recommended Practices of the American Oil Chemists' Society*, Champaign, Ill., 1991, Method Cc9a-48.
58. *Ibid.*, Method Cd 14-61.
59. B. L. Madison, R. A. Depalma, and R. P. D'Alonzo, *J. Am. Oil Chem. Soc.* **59**, 178–181 (1982).
60. J. A. Robertson and F. E. Barton, *J. Am. Oil Chem. Soc.* **61**, 543–547 (1984); R. A. Hartwig and C. R. Horburgh, *J. Am. Oil Chem. Soc.* **67**, 435–437 (1990).
61. R. T. Sleeter, *J. Am. Oil Chem. Soc.* **60**, 343–349 (1983).
62. F. D. Gunstone, *Chem. Ind.*, 802–803 (1991).
63. C. Kemal, P. Louis-Flamberg, R. Krupinski-Olsen, and A. L. Shorter, *Biochemistry* **26**, 7064–7072 (1987).
64. E. N. Frankel, *Prog. Lipid Res.* **23**, 127 (1985).
65. K. M. Schaich, *CRC Crit. Rev. Food Sci. Nutr.*, 189–224 (Nov. 1980).
66. Ref. 57, Method Ca 5a-40.
67. Ref. 57, Method Cd 3-25.

68. F. WQ. Karasek, F. I. Onuska, F. J. Yang, and R. E. Clement, *Anal. Chem.* **56**, 174R–199R (1984).
69. W. W. Christie, *Gas Chromatography and Lipids*, The Oily Press, Ltd., Ayr, Scotland, 1989, pp. 85–125.
70. F. D. Gunstone and F. B. Padley, *J. Oil Chem. Soc.* **42**, 957–961 (1965); J. N. Roehm and O. S. Privett, *Lipids* **5**, 353–358 (1970); N. R. Bottino, *J. Lipid Res.* **12**, 24–30 (1971).
71. W. R. Eckert, *Fette Seifen Anstrichm.* **79**, 360–362 (1977); Ref. 69, pp. 186–204.
72. W. N. Christie, *HPLC and Lipids: A Practical Guide*, Pergamon Press, New York, 1987, pp. 169–190.
73. *Global Engineering Documents*, International Organization for Standardization, Method ISO-6800, Irvine, Calif., 1985.
74. Ref. 57, Method Cd 1-25.
75. Ref. 57, Method Cd 10-57.
76. W. J. Miller, W. H. Koester, and F. E. Freeberg, *J. Am. Oil Chem. Soc.* **46**, 341–343 (1969).
77. K. Van Putte and J. Van Den Enden, *J. Am. Oil Chem. Soc.* **51**, 316–320 (1974); E. Sambuc and M. Naudet, *Rev. Fr. Corps Gras.* **21**, 309–312 (1974); K. Van Putte, L. Vermaas, J. Van Den Enden, and C. den Hollander, *J. Am. Oil Chem. Soc.* **52**, 179–181 (1975).
78. J. C. Van Den Enden, J. B. Rossel, L. F. Vermaas, and D. Waddington, *J. Am. Oil Chem. Soc.* **59**, 433–438 (1982); V. K. S. Shukla, *Fette Seifen Anstrichm.* **85**, 467–471 (1983).
79. Ref. 57, Method Cc 1-25.
80. Ref. 57, Method Cc 18-80; J. M. De Man, L. De Man, and B. Blackman, *J. Am. Oil Chem. Soc.* **60**, 91–94 (1983).
81. Ref. 57, Method Cc 13b-45.
82. Ref. 57, Method Cc 13a-43.
83. Ref. 57, Method Td 1a-64.
84. R. T. Sleeter, in Ref. 24, pp. 183–199. For a review of methods used in quality assurance.
85. Ref. 57, Method Cd 8-53.
86. R. Marcuse and L. Johanssen, *J. Am. Oil Chem. Soc.* **50**, 387–391 (1973).
87. R. S. Kim and F. S. LaBella, *J. Lipid Res.* **28**, 1110–1117 (1987).
88. N. T. Joyner and J. E. McIntyre, *Oil Soap* **15**, 184–186 (1938); H. S. Olcott and E. Einset, *J. Am. Oil Chem. Soc.* **35**, 161–162 (1958).
89. A. E. King, H. L. Roschen, and W. H. Irwin, *Oil Soap* **10**, 204–207 (1933).
90. J. Frank, J. Geil, and R. Freaso, *Food Technol.* **36**, 71 (1982); W. J. Woestengerg and J. Zaalberg, *Fette Seifen Anstrichm.* **88**, 53 (1986); M. N. Laeuble and P. A. Bruttel, *J. Am. Oil Chem. Soc.* **63**, 792 (1986); M. W. Laeuble, P. Bruttel, and E. Schaich, *Fette Wiss. Technol.* **90**, 56 (1988).
91. P. K. Jarvi, G. D. Lee, D. R. Erickson, and E. A. Butkus, *J. Am. Oil Chem. Soc.* **48**, 121–124 (1971); H. P. DuPuy, E. T. Rayner, J. I. Wadsworth, and M. G. Legendre, *J. Am. Oil Chem. Soc.* **54**, 445–449 (1977); J. L. Williams and T. H. Applewhite, *J. Am. Oil Chem. Soc.* **54**, 461–463 (1977); A. E. Walking and H. Zmachinski, *J. Am. Oil Chem. Soc.* **54**, 454–457 (1977); H. W. Jackson and D. J. Giarcherio, *J. Am. Oil Chem. Soc.* **54**, 458–460 (1977); W. H. Morrison, B. G. Lyon, and J. A. Robertson, *J. Am. Oil Chem. Soc.* **58**, 23–27 (1981); D. B. Min, *J. Food Sci.* **46**, 1453–1456 (1981); D. B. Min, *J. Am. Oil Chem. Soc.* **60**, 544–545 (1983); J. L. Gensic, B. F. Szuhaj, and J. G. Endres, *J. Am. Oil Chem. Soc.* **61**, 1246–1249 (1984).
92. E. N. Frankel, M. L. Hu, and A. L. Tappel, *Lipids* **24**, 976–981 (1989); E. N. Frankel and A. L. Tappel, *Lipids* **26**, 479–484 (1991).

93. C. W. Fritsch, D. C. Egberg, and J. S. Magnuson, *J. Am. Oil Chem. Soc.* **56**, 746–752 (1979).
94. R. T. Sleeter, in Ref. 24, pp. 215–225.
95. G. D. Nelson, ed., *Health Effects of Dietary Fatty Acids*, American Oil Chemists' Society, Champaign, Ill., 1991, pp. 50–135; A. K. Sen Gupta, *Dtsch. Lebensrm. Rundsch.* **87**, 282–287 (1991).
96. G. D. Nelson, ed., *Health Effects of Dietary Fatty Acids*, American Oil Chemists' Society, Champaign, Ill., 1991, pp. 136–167.
97. Ref. 96, pp. 167–208.
98. A. Keys, J. T. Anderson, and F. Grande, *Metabolism* **14**, 747–787 (1965).
99. D. M. Hegsted, R. B. McGandy, M. L. Myers, and F. J. Stare, *Am. J. Clin. Nutr.* **17**, 281–295 (1965); D. M. Hegsted, in R. J. Nicolosi, ed., *Proceedings from the Scientific Conference on the Effects of Dietary Fatty Acids on Serum Lipoproteins and Hemostasis*, American Heart Association, 1989, pp. 103–114.
100. J. B. Reeves and J. L. Weinrouch, *Composition of Foods, Agriculture Handbook*, No. 8-4, U.S. Department of Agriculture, Washington, D.C., 1979.
101. R. P. Mensink and M. B. Katan, *N. Engl. J. Med.* **323**, 439–445 (1990); E. A. Emken, Ref. 96, pp. 245–263; P. L. Zock and M. B. Katan, *J. Lipid Res.* **33**, 399–410 (1992).
102. Y. L. Ha, J. Storkson, and M. W. Pariza, *Cancer Res.* **50**, 1097–1101 (1990); S. G. Chin, J. A. Scimeca, and M. W. Pariza, *Cancer Res.* **51**, 6118–6124 (1991).
103. National Research Council, *Diet and Health: Implications for Reducing Chronic Disease Risk*, National Academy Press, Washington, D.C., 1989.
104. E. Jungermann, in Ref. 56, pp. 511–586.
105. N. O. V. Sonntag, in Ref. 4, pp. 97–173.

G. L. HASENHUETTL
Kraft General Foods

FATTY ACIDS.

See CARBOXYLIC ACIDS.

FATTY ACIDS FROM TALL OIL.

See CARBOXYLIC ACIDS; TALL OIL.

FEEDS AND FEED ADDITIVES

NONRUMINANT FEEDS

Meat and eggs produced by nonruminants, ie, swine and poultry, in the United States contribute significantly toward meeting the nutritional requirements of the human population. In 1990, the average per capita food consumption in the United States included 22.6 kg of pork, 8.2 kg of turkey, 31.8 kg of broilers, and 235 eggs. This was supplied by 85 million hogs, 283 million turkeys, 5.3 billion broilers, and 274 million laying hens (1). The feed required to produce such a quantity of meat and eggs is considerable.

Most poultry production, and a growing percentage of swine production, takes place in intensive, confinement operations. Much of the poultry production is carried out under a system of vertical integration in which a producer hatches the chicks, grows them in the producer's facilities or in contract facilities, provides the feed, processes the animals, and markets the product. This system of vertically integrated production is not as common in the swine industry.

Feed Ingredients

Both swine and poultry diets are comprised primarily of grains such as corn and grain sorghum, with occasional use of wheat, barley, and other small grains (see WHEAT AND OTHER CEREAL GRAINS). Soybean meal is the primary source of protein in these diets, but animal by-products such as meat and bone meal, poultry by-product meal, feather meal, and fish meal contribute significant amounts of protein and provide some of the minerals required for growth, maintenance, reproduction, and lactation (see SOYBEANS AND OTHER OILSEEDS). Many human and industrial by-products are also used in swine and poultry feeds, eg, dried bakery products, produced from leftover bread and other bakery waste; inedible fats from the processing of vegetable oils (qv) for human consumption; large amounts of fats and oils from the restaurant and fast food trade that are produced as by-products of cooking food; and numerous other products that would otherwise go unused. An excellent review of the characteristics of many common feed ingredients is available (2), as are nutrient composition tables for ingredients most commonly used in animal feeds (3). An estimation of the principal feedstuffs used in animal feeds is given in Table 1.

Because of the simplicity of swine and poultry feeds, most feed manufacturers add vitamins (qv) and trace minerals to ensure an adequate supply of essential nutrients. Amino acids (qv) such as methionine [*7005-18-7*], lysine [*56-87-1*], threonine [*36676-50-3*], and tryptophan [*6912-86-3*], produced by chemical synthesis

Table 1. Feed Ingredient Usage by the Livestock and Poultry Industry,[a] 1989–1991[b,c] 10^3 t

Ingredient	1989	1990	1991
	Grain		
corn	113,200	119,600	121,900
sorghum	13,100	10,200	9,900
oats	3,800	4,300	3,600
barley	4,100	4,000	3,800
wheat and rye	4,000	13,500	9,700
	Oilseed meal		
soybean	20,464	20,626	21,115
cottonseed	1,239	1,476	1,609
linseed	126	115	125
peanut	112	99	135
sunflower	296	287	415
canola	316	361	375
	Animal protein		
tankage and meat meal	2,320	2,250	2,350
fish meal and solubles	321	260	265
milk products	400	390	380
	Grain protein feed		
gluten feed and meal	218	164	200
brewer's dried grains	125	na	na
distiller's dried grains	850	na	na
	Other		
wheat millfeed	5,716	5,965	5,995
rice millfeed	556	555	555
dried and molasses beet pulp	758	1,051	925
alfalfa meal	300	351	400
fats and oils	972	1,010	1,050
molasses, inedible	1,988	2,160	2,100
miscellaneous by-product feed	1,342	1,562	1,403

[a]Includes ruminants and nonruminants.
[b]Values given are estimates.
[c]Ref. 4.

or by fermentation (qv), are used to fortify swine and poultry diets. The use of these supplements to provide the essential amino acids permits diets with lower total crude protein content.

Virtually all broiler and turkey diets, and much of the swine feeds, are pelleted prior to feeding. Pelleted feeds are consumed in greater quantity than are feeds in a mash or meal form, and generally result in more rapid weight gain and better feed conversion. Proper pelleting of feed also aids in reducing the potential of salmonellas or other bacterial contamination of feeds. Pelleting is accomplished by forcing the feed through a die having many small holes. Pelleting is improved by steaming the feed to gelatinize the starch provided by the grains, by adding low (<2%) levels of fat to the feed, or by addition of various types of pellet binders. There are several types of pellet binders, including bentonite clays, lignosulfonates, and grain starch products, that result in firmer, more durable pellets able to withstand the rigors of mechanical feed handling systems.

Nutrient Requirements of Swine and Poultry

Numerous researchers at state and governmental research institutes have defined the requirements of swine and poultry for virtually all known nutrients. In addition, the nutrient composition of common ingredients has been determined. Through the use of a mathematical technique known as linear programming, aided by the use of high speed computers, poultry and swine nutritionists are able to formulate nutritionally balanced diets for all species of animals. As ingredient prices change as a result of supply and demand, diet composition can be changed almost instantly.

There is no best feed composition because animals thrive on diets composed of many different types of ingredients. Swine and poultry generally adapt readily and rapidly to changes in ingredient composition, as long as the diets provide adequate levels of essential nutrients. Tables 2 through 6 list information on the nutrient requirements of various types of swine and poultry.

Reference Diets for Chickens. Poultry can be grown on many diverse types of diets. Because of the high percentage of chickens grown under an integrated system, it is sometimes difficult to purchase small quantities of high quality feeds. Persons who sometimes utilize chickens for laboratory or research animals may require information regarding formulas that can be mixed from readily available ingredients. Three such formulas are given in Table 7. The first of these is a practical diet, composed primarily of corn and soybean meal and appropriate vitamin and mineral supplements. The second is a purified diet, based principally on isolated soybean protein and glucose. The third is a chemically defined diet, composed primarily of amino acids, glucose, and vitamin and mineral supplements. These diets are designed to allow adequate, but not necessarily optimal, growth of young chicks.

Feed Additives for Nonruminants

Feed additives are common to swine and poultry feeds for a number of purposes. Antibiotics (qv), used to promote growth and improve feed utilization, are poorly digestible and function primarily by controlling the bacterial flora of the intestinal tract. Antibiotics fed to animals for this purpose generally are not used for human antibiotic therapy. Other antibiotics, used for disease therapy, are generally injected or are absorbed into the body tissues where they are effective against the disease-causing organisms. Other feed additives include antioxidants (qv) to protect feeds against oxidative rancidity, mold inhibitors to prevent development of

Table 2. Nutrient Requirements of Broiler Chickens at Specific Ages,[a] Weeks

Nutrient	0 to 3	3 to 6	6 to 8
metabolizable energy, kJ/g[b]	13.39	13.39	13.39
protein, %	23.0	20.0	18.0
arginine, %	1.44	1.20	1.00
glycine and serine, %	1.50	1.00	0.70
histidine, %	0.35	0.30	0.26
isoleucine, %	0.80	0.70	0.60
leucine, %	1.35	1.18	1.00
lysine, %	1.20	1.00	0.85
methionine and cystine, %	0.93	0.72	0.60
methionine, %	0.50	0.38	0.32
phenylalanine and tyrosine, %	1.34	1.17	1.00
phenylalanine, %	0.72	0.63	0.54
threonine, %	0.80	0.74	0.68
tryptophan, %	0.23	0.18	0.17
valine, %	0.82	0.72	0.62
linoleic acid, %	1.00	1.00	1.00
calcium, %	1.00	0.90	0.80
phosphorus, available, %	0.45	0.40	0.35
potassium, %	0.40	0.35	0.30
sodium, %	0.15	0.15	0.15
chlorine, %	0.15	0.15	0.15
magnesium, mg/kg	600	600	600
manganese, mg/kg	60	60	60
zinc, mg/kg	40	40	40
iron, mg/kg	80	80	80
copper, mg/kg	8	8	8
iodine, mg/kg	0.35	0.35	0.35
selenium, mg/kg	0.15	0.15	0.15
vitamin A, IU/kg	1500	1500	1500
vitamin D, ICU/kg[c]	200	200	200
vitamin E, IU/kg	10	10	10
vitamin K, mg/kg	0.50	0.50	0.50
riboflavin, mg/kg	3.60	3.60	3.60
pantothenic acid, mg/kg	10	10	10
niacin, mg/kg	27	27	11
vitamin B_{12}, μg/kg	9	9	9
choline, mg/kg	1300	850	500
biotin, mg/kg	0.15	0.15	0.10
folacin, mg/kg	0.55	0.55	0.25
thiamin, mg/kg	1.80	1.80	1.80
pyridoxine, mg/kg	3	3	2.5

[a]Ref. 5.
[b]To convert kJ to kcal, divide by 4.184.
[c]Requirements of poultry for vitamin D are expressed in international chick units (ICU) which are based on the activity of vitamin D_3 in chick bioassays.

Table 3. Nutrient Requirements of Leghorn-Type Chickens at Specific Ages,[a] Weeks

Nutrient	0 to 6	6 to 14	14 to 20	Laying hens[b]	Breeder hens
metabolizable energy, kJ/g[c]	12.14	12.14	12.14	12.14	12.14
protein, %	18.0	15.0	12.0	14.5	14.5
arginine, %	1.00	0.83	0.67	0.68	0.68
glycine and serine, %	0.70	0.58	0.47	0.50	0.50
histidine, %	0.26	0.22	0.17	0.16	0.16
isoleucine, %	0.60	0.50	0.40	0.50	0.50
leucine, %	1.00	0.83	0.67	0.73	0.73
lysine, %	0.85	0.60	0.45	0.64	0.55
methionine and cystine, %	0.60	0.50	0.40	0.55	0.55
methionine, %	0.32	0.25	0.20	0.32	0.32
phenylalanine and tyrosine, %	1.00	0.83	0.67	0.80	0.80
phenylalanine, %	0.54	0.45	0.36	0.40	0.40
threonine, %	0.68	0.57	0.37	0.45	0.45
tryptophan, %	0.17	0.14	0.11	0.14	0.14
valine, %	0.62	0.52	0.41	0.55	0.55
linoleic acid, %	1.00	1.00	1.00	1.00	1.00
calcium, %	0.80	0.70	0.60	3.40	3.40
phosphorus, available, %	0.40	0.35	0.30	0.32	0.32
potassium, %	0.40	0.30	0.25	0.15	0.15
sodium, %	0.15	0.15	0.15	0.15	0.15
chlorine, %	0.15	0.12	0.12	0.15	0.15
magnesium, mg/kg	600	500	400	500	500
manganese, mg/kg	60	30	30	30	60
zinc, mg/kg	40	35	35	50	65
iron, mg/kg	80	60	60	50	60
copper, mg/kg	8	6	6	6	8
iodine, mg/kg	0.35	0.35	0.35	0.30	0.30
selenium, mg/kg	0.15	0.10	0.10	0.10	0.10
vitamin A, IU/kg	1500	1500	1500	4000	4000
vitamin D, ICU/kg[d]	200	200	200	500	500
vitamin E, IU/kg	10	5	5	5	10
vitamin K, mg/kg	0.50	0.50	0.50	0.50	0.50
riboflavin, mg/kg	3.60	1.80	1.80	2.20	3.80
pantothenic acid, mg/kg	10	10	10	2.2	10
niacin, mg/kg	27	11	11	10	10
vitamin B_{12}, μg/kg	9	3	3	4	4
choline, mg/kg	1300	900	500		
biotin, mg/kg	0.15	0.10	0.10	0.10	0.15
folacin, mg/kg	0.55	0.25	0.25	0.25	0.35
thiamin, mg/kg	1.8	1.3	1.3	0.80	0.80
pyridoxine, mg/kg	3.0	3.0	3.0	3.0	4.5

[a]Ref. 5.
[b]Assumes an average daily intake of 110 g of feed per hen.
[c]To convert kJ to kcal, divide by 4.184.
[d]ICU = international chick units.

Table 4. Nutrient Requirements of Growing Turkeys at Specific Ages,[a] Weeks

Nutrient	0 to 4	4 to 8	8 to 12	12 to 16	16 to 20	20 to 24
metabolizable energy, kJ/g[b]	11.72	12.14	12.56	12.98	13.39	13.81
protein, %	28	26	22	19	16.5	14
arginine, %	1.60	1.50	1.25	1.10	0.95	0.80
glycine and serine, %	1.00	0.90	0.80	0.70	0.60	0.50
histidine, %	0.58	0.54	0.46	0.39	0.35	0.29
isoleucine, %	1.10	1.00	0.85	0.75	0.65	0.55
leucine, %	1.9	1.75	1.5	1.3	1.1	0.95
lysine, %	1.6	1.5	1.3	1.0	0.8	0.65
methionine and cystine, %	1.05	0.90	0.75	0.65	0.55	0.45
methionine, %	0.53	0.45	0.38	0.33	0.28	0.23
phenylalanine and tyrosine, %	1.8	1.65	1.4	1.2	1.05	0.9
phenylalanine, %	1.0	0.9	0.8	0.7	0.6	0.5
threonine, %	1.0	0.93	0.79	0.68	0.59	0.5
tryptophan, %	0.26	0.24	0.2	0.18	0.15	0.13
valine, %	1.2	1.1	0.94	0.8	0.7	0.6
linoleic acid, %	1.0	1.0	0.8	0.8	0.8	0.8
calcium, %	1.2	1.0	0.85	0.75	0.65	0.55
phosphorus, available, %	0.6	0.5	0.42	0.38	0.32	0.28
potassium, %	0.7	0.6	0.5	0.5	0.4	0.4
sodium, %	0.17	0.15	0.12	0.12	0.12	0.12
chlorine, %	0.15	0.14	0.12	0.12	0.12	0.12
magnesium, mg/kg	600	600	600	600	600	600
manganese, mg/kg	60	60	60	60	60	60
zinc, mg/kg	75	65	50	40	40	40
iron, mg/kg	80	80	80	80	80	80
copper, mg/kg	8	8	8	8	8	8
iodine, mg/kg	0.4	0.4	0.4	0.4	0.4	0.4
selenium, mg/kg	0.2	0.2	0.2	0.2	0.2	0.2
vitamin A, IU/kg	4000	4000	4000	4000	4000	4000
vitamin D, ICU/kg[c]	900	900	900	900	900	900
vitamin E, IU/kg	12	12	10	10	10	10
vitamin K, mg/kg	1	1	0.8	0.8	0.8	0.8
riboflavin, mg/kg	3.6	3.6	3.0	3.0	2.5	2.5
pantothenic acid, mg/kg	11	11	9	9	9	9
niacin, mg/kg	70	70	50	50	40	40
vitamin B_{12}, μg/kg	3	3	3	3	3	3
choline, mg/kg	1900	1600	1300	1100	950	800
biotin, mg/kg	0.2	0.2	0.15	0.125	0.10	0.10
folacin, mg/kg	1.0	1.0	0.8	0.8	0.7	0.7
thiamin, mg/kg	2	2	2	2	2	2
pyridoxine, mg/kg	4.5	4.5	3.5	3.5	3.0	3.0

[a]Ref. 5.
[b]To convert kJ to kcal, divide by 4.184.
[c]ICU = international chick units.

Table 5. Nutrient Requirements of Growing Swine Allowed Feed *Ad Libitum*[a,b]

Intake levels	Swine liveweight, kg				
	1–5	5–10	10–20	20–50	50–110
expected weight gain, g/day	200	250	450	700	820
expected feed intake, g/day	250	460	950	1900	3110
	Nutrients				
metabolizable energy, kJ/g[c]	13.47	13.56	13.60	13.64	13.70
protein, %	24	20	18	15	13
arginine, %	0.60	0.50	0.40	0.25	0.10
histidine, %	0.36	0.31	0.25	0.22	0.18
isoleucine, %	0.76	0.65	0.53	0.46	0.38
leucine, %	1.00	0.85	0.70	0.60	0.50
lysine, %	1.40	1.15	0.95	0.75	0.60
methionine and cystine, %	0.68	0.58	0.48	0.41	0.34
phenylalanine and tyrosine, %	1.10	0.94	0.77	0.66	0.55
threonine, %	0.80	0.68	0.56	0.48	0.40
tryptophan, %	0.20	0.17	0.14	0.12	0.10
valine, %	0.80	0.68	0.56	0.48	0.40
linoleic acid, %	0.1	0.1	0.1	0.1	0.1
calcium, %	0.90	0.80	0.70	0.60	0.50
phosphorus, %					
total	0.70	0.65	0.60	0.50	0.40
available	0.55	0.40	0.32	0.23	0.15
sodium, %	0.10	0.10	0.10	0.10	0.10
chlorine, %	0.08	0.08	0.08	0.08	0.08
magnesium, %	0.04	0.04	0.04	0.04	0.04
potassium, %	0.30	0.28	0.26	0.23	0.17
copper, mg/kg	6.0	6.0	5.0	4.0	3.0
iodine, mg/kg	0.14	0.14	0.14	0.14	0.14
iron, mg/kg	100	100	80	60	40
manganese, mg/kg	4	4	3	2	2
selenium, mg/kg	0.3	0.3	0.25	0.15	0.1
zinc, mg/kg	100	100	80	60	50
vitamin A, IU/kg	2200	2200	1750	1300	1300
vitamin D, IU/kg	220	220	200	150	150
vitamin E, IU/kg	16	16	11	11	11
vitamin K, mg/kg	0.5	0.5	0.5	0.5	0.5
biotin, mg/kg	0.08	0.05	0.05	0.05	0.05
choline, g/kg	0.6	0.5	0.4	0.3	0.3
folacin, mg/kg	0.3	0.3	0.3	0.3	0.3
niacin, available, mg/kg	20	15	12.5	10	7
pantothenic acid, mg/kg	12	10	9	8	7
riboflavin, mg/kg	4	3.5	3	2.5	2
thiamin, mg/kg	1.5	1	1	1	1
pyridoxine, mg/kg	2	1.5	1.5	1	1
vitamin B_{12}, μg/kg	20	17.5	15	10	5

[a]Ref. 6.
[b]Ninety percent dry matter basis. Requirements based on the following diets: 1–5 kg pig diet includes 25–75% milk products; 5–10 kg pigs, a corn–soybean meal diet that includes 5–25% milk products; 10–110 kg pigs, a corn–soybean meal diet. In corn–soybean meal diets the corn and soybean meal contain 8.5 and 44% crude protein, respectively.
[c]To convert kJ to kcal, divide by 4.184.

Table 6. Nutrient Requirements of Breeding and Lactating Swine[a,b]

Nutrient	Breeding[c]	Lactating[d]
metabolizable energy, kJ/g[e]	13.43	13.43
protein, %	12	13
arginine, %	0.00	0.40
histidine, %	0.15	0.25
isoleucine, %	0.30	0.48
leucine, %	0.30	0.48
lysine, %	0.43	0.60
methionine and cystine, %	0.23	0.36
phenylalanine and tyrosine, %	0.45	0.70
threonine, %	0.30	0.43
tryptophan, %	0.09	0.12
valine, %	0.32	0.60
linoleic acid, %	0.1	0.1
calcium, %	0.75	0.75
phosphorus, %		
total	0.60	0.60
available	0.35	0.35
sodium, %	0.15	0.20
chlorine, %	0.12	0.16
magnesium, %	0.04	0.04
potassium, %	0.20	0.20
copper, mg/kg	5	5
iodine, mg/kg	0.14	0.14
iron, mg/kg	80	80
manganese, mg/kg	10	10
selenium, mg/kg	0.15	0.15
zinc, mg/kg	50	50
vitamin A, IU/kg	4000	2000
vitamin D, IU/kg	200	200
vitamin E, IU/kg	22	22
vitamin K, mg/kg	0.5	0.50
biotin, mg/kg	0.20	0.20
choline, g/kg	1.25	1.00
folacin, mg/kg	0.30	0.30
niacin, available, mg/kg	10	10
pantothenic acid, mg/kg	12	12
riboflavin, mg/kg	3.75	3.75
thiamin, mg/kg	1	1
pyridoxine, mg/kg	1	1
vitamin B_{12}, μg/kg	15	15

[a]Ref. 6.

[b]Requirements based on corn–soybean meal diets with typical feed intakes and performance levels. In corn–soybean meal diets the corn and soybean meal contain 8.5 and 44% crude protein, respectively.

[c]Gilts, sows, and adult boars.

[d]Gilts and sows.

[e]To convert kJ to kcal, divide by 4.184.

Table 7. Reference Diets for Growing Chickens[a]

Ingredient	Practical	Purified[b]	Chemically defined[c,d]
ground yellow corn, g/kg	580		
soybean meal, 48% protein, g/kg	350		
isolated soybean protein, g/kg		250	
DL-methionine, g/kg	2.5	6	
glycine, g/kg		4	
corn oil, g/kg	30	40	50–150
cellulose, g/kg		30	30
choline chloride, 50%, g/kg	1.5	2.0	2.0
thiamin HCl, mg/kg	1.8	15.0	100.0
riboflavin, mg/kg	3.6	15.0	16.0
calcium pantothenate, mg/kg	10.0	20.0	20.0
niacin, mg/kg	25.0	50.0	100.0
pyridoxine HCl, mg/kg	3.0	6.0	6.0
folacin, mg/kg	0.55	6.0	4.0
biotin, mg/kg	0.15	0.6	0.6
vitamin B_{12}, μg/kg	10	20	20
p-aminobenzoic acid, mg/kg			2.0
ascorbic acid, mg/kg			250.0
vitamin A, IU/kg	1500	4500	10,000
vitamin D_3, ICU/kg[e]	400	4500	600
vitamin E, IU/kg	10	50	20
vitamin K, mg/kg	0.55	1.5	5.0
ethoxyquin, mg/kg	125	100	125
iodized salt, g/kg	5		
NaCl, g/kg		6	8.8
$CaCO_3$, g/kg	10	14.8	3
$CaHPO_4 \cdot 2H_2O$, g/kg	20	20.7	
$MgSO_4 \cdot H_2O$, g/kg		10	9
K_2HPO_4, g/kg		10	9
$NaHCO_3$, g/kg			15
$MnSO_4 \cdot 5H_2O$, mg/kg	170	350	650
$ZnSO_4 \cdot H_2O$, mg/kg	110		
$ZnCO_3$, mg/kg		150	100
$Fe_2(SO_4)_3 \cdot 7H_2O$, mg/kg	500		
ferric citrate pentahydrate, mg/kg	500		500
$CuSO_4 \cdot 5H_2O$, mg/kg	16	30	20
Na_2SeO_3, mg/kg	0.2	0.2	0.2
glucose or starch,[f] kg	1	1	1

[a]Ref. 5.

[b]Also contains 1 g/kg KCl, 2 mg/kg KIO_2, and 1.7 mg/kg $CoCl_2$.

[c]Based on an amino acid mixture: 11.5 g L-arginine HCl, 4.5 g L-histidine $HCl \cdot H_2O$, 11.4 g L-lysine HCl, 4.5 g L-tyrosine, 1.5 g L-tryptophan, 5 g L-phenylalanine, 3.5 g DL-methionine, 3.5 g L-cystine, 6.5 g L-threonine, 10 g L-leucine, 6 g L-isoleucine, 6.9 g L-valine, 6 g glycine, 4 g L-proline, and 120 g L-glutamic acid.

[d]Also contains 28 g/kg $Ca_3(PO_4)_2$; 40 mg/kg KI; 1 mg/kg $CoSO_4 \cdot 7H_2O$; 9 mg/kg H_3BO_3; and 9 mg/kg $Na_2MoO_4 \cdot 2H_2O$.

[e]ICU = international chick units.

[f]Value given is maximum.

potentially toxic mold products, anticoccidial compounds that protect chickens against this severe parasitic disease, anthelmintics, and other types of growth promoters (see GROWTH REGULATORS).

Feed additives fall into two general categories, ie, medicated and generally recognized as safe (GRAS). Medicated feed additives consist of products where usage level and purpose are strictly regulated by the United States Food and Drug Administration (FDA). Feed manufacturers who utilize medicated feed additives must first register as a drug establishment with the FDA. Individuals mixing feed for their own animals are subject to the same rules as commercial feed mills. Annual registration is required and once the mill is registered it must obtain approval from the FDA for each of the medicated feed additives used. The feed manufacturer must determine whether each animal drug used has been approved by the FDA for its intended use and what further approvals, if any, are needed.

The various FDA approved (ca 1992) feed additives for chicken, turkey, and swine, grouped according to their usage, are as follows:

Nutritional Feed Additives

Food efficiency/growth promotion in chicken, turkey, swine
  arsanilic acid
  bacitracin methylene disalicylate
  bacitracin zinc
  bambermycins
  chlortetracycline
  oxytetracycline
  penicillin
  roxarsone
  virginiamycin
Food efficiency/growth promotion in chicken and swine
  tylosin
  lincomycin
Food efficiency/growth promotion in swine
  carbadox
  tiamulin
  tylosin/sulfamethazine
Pigmentation in chicken and turkey
  arsanilic acid
  roxarsone
Egg hatchability/production in chicken and turkey
  oxytetracycline
Egg hatchability/production in chicken
  chlortetracycline
Egg hatchability in chicken
  bacitracin zinc
Egg production in chicken
  bacitracin methylene disalicylate
Eggshell texture, and quality in chicken
  oxytetracycline

Medicinal Feed Additives for Chicken and Turkey

Blackhead
  nitarsone
Blue comb (nonspecific enteritis)
  chlortetracycline
  oxytetracycline
  penicillin
Breast blisters
  novobiocin
Cholera, fowl
  novobiocin
  sulfadimethoxine and ormetoprim
Chronic respiratory disease
  erythromycin
  penicillin
Coccidiosis
  amprolium
  halofuginone hydrobromide
  monensin
  sulfadimethoxine and ormetoprim
  zoalene

Mycosis, crop
  nystatin
Mycotic diarrhea
  nystatin
Stress
  chlortetracycline
  oxytetracycline
Synovitis
  chlortetracycline
  novobiocin
  oxytetracycline

Medicinal Feed Additives for Turkey

Airsacculites
  oxytetracycline
Hexamatiasis
  chlortetracycline
  oxytetracycline
Leucocytozoonosis
  clopidol
Paratyphoid
  chlortetracycline
Sinusitis, infectious
  chlortetracycline
  oxytetracycline
Transmissable enteritis
  bacitracin methylene disalicylate

Medicinal Feed Additives for Chicken

Cholera, fowl
  oxytetracycline
Chronic respiratory disease
  chlortetracycline
  oxytetracycline
  tylosin
Coccidiosis
  chlortetracycline
  clopidol
  decoquinate
  lasalocid
  maduramycin ammonia
  narasin
  narasin/nicarbazin
  nicarbazin
  oxytetracycline
  robenidine hydrochloride
  salinomycin
Colibacilosis
  sulfadimethoxine and ormetoprim
Coryza, infectious
  erythromycin
  sulfadimethoxine and ormetoprim
Fly control
  larvadex
Hepatitis, infectious
  oxytetracycline
Necrotic enteritis
  bacitracin methylene disalicylate
  lincomycin
  virginiamycin
Worms, capillary and cecal (Heterakis)
  coumaphos
  hygromycin B
Worms, common roundworms
  coumaphos
Worms, large roundworms (ascaris)
  hygromycin B

Medicinal Feed Additives for Swine

Atrophic rhinitis
  chlortetracycline
  oxytetracycline
  tylosin
  tylosin/sulfamethazine
Bacterial swine enteritis (scours)
  apramycin
  carbadox
  chlortetracycline
  oxytetracycline
Cervical abcesses
  chlortetracycline
Colibacillosis
  apramycin
  colimix
Dysentery
  arsanilic acid
  bacitracin methylene disalicylate
  carbadox
  lincomycin
  roxarsone
  tiamulin
  virginiamycin
Dysentery, vibrionic
  carbadox

oxytetracycline
tylosin
tylosin/sulfamethazine
Fly control
rabon
Leptospirosis
chlortetracycline
oxytetracycline
Mycoplasma pneumonia
lincomycin
Necrotic enteritis
carbadox
oxytetracycline
Stress
chlortetracycline
Worms, kidney
fenbendazole
levamisole hydrochloride
Worms, large roundworms
dichlorvos
fenbendazole
hygromycin B
levamisole hydrochloride
pyrantel tartrate
thiabendazole
Worms, lungworms
fenbendazole
levamisole hydrochloride
Worms, nodular
dichlorvos
fenbendazole
hygromycin B
levamisole hydrochloride
pyrantel tartrate
Worms, small stomach
fenbendazole
Worms, thick stomach
fenbendazole
Worms, threadworms
levamisole hydrochloride
Worms, whipworms
dichlorvos
fenbendazole
hygromycin B

Official information concerning FDA approval of antibiotics and other drugs is available in the *Code of Federal Regulations* (7). This document is revised at least once per year and updated in individual issues of the *Federal Register*. These two publications are utilized together to determine the latest status of any given product. The *Code of Federal Regulations* is published in six parts: animal feeds, drugs, and related products are included (7).

An effective and less expensive way to maintain information regarding the status of approved feed additives for animal feeds is to subscribe to the *Feed Additive Compendium,* published yearly and updated monthly (8). This publication gives detailed information on specific antimicrobial agents, levels of usage, sources of product, and legal requirements for use in the United States. Although effective as a source of information on approved feed additives, it cannot be considered as a legal authority.

There are a large number of feed additive products classified as generally recognized as safe (GRAS). These are products that have been considered by a group of qualified experts to be safe for the intended use in animal feeds; no permission or registration is required for use at recommended levels based on scientific procedures or experience in common use in feed. There must be reasonable evidence to support the safety of such products. Some restrictions are placed on quantity of some products, such as selenium and ethoxyquin. GRAS products include a wide range of materials, ranging from ammoniated cottonseed meal to xanthan gum. A list of all GRAS substances for animal feeds is available (7).

The GRAS listing does not include widely used, historical products such as grains, sugar, salt, etc. In general, feed ingredients listed in the *American Asso-*

ciation of Feed Control Officials (AAFCO) official publication are considered in the GRAS category (9).

A number of products designated GRAS are being scrutinized by the FDA because of advertisements and claims made by producers or manufacturers of these products. Statements that indicate that feeding such products improve animal performance may require substantive data to support such claims in the future.

BIBLIOGRAPHY

"Feeds, Animal," in *ECT* 1st ed., Vol. 6, pp. 299–312, by H. M. Briggs, Oklahoma Agricultural and Mechanical College; in *ECT* 2nd ed., Vol. 8, pp. 857–870, by H. M. Briggs, South Dakota State University; "Pet and Livestock Feeds," in *ECT* 3rd ed., Vol. 17, pp. 90–109, J. Corbin, University of Illinois.

1. *Feedstuffs Reference Issue,* Miller Publishing Co., Minnetonka, Minn., 1991.
2. M. S. Ash, *Animal Feeds Compendium, Agricultural Economic Report No. 656,* U.S. Dept. of Agriculture, Economic Research Service, Washington, D.C., 1992.
3. National Academy of Science, *Atlas of Nutritional Data on United States and Canadian Feeds,* National Academy Press, Washington, D.C., 1971.
4. *Feed Situation and Outlook Report,* U.S. Dept. of Agriculture, Washington, D.C., 1991.
5. National Research Council, *Nutrient Requirements of Poultry,* 8th ed., National Academy Press, Washington, D.C., 1984.
6. National Research Council, *Nutrient Requirements of Swine,* 9th ed., National Academy Press, Washington, D.C., 1988.
7. *Code of Federal Regulations,* Title 21, part 500–599, and *Federal Register,* Superintendent for Documents, U.S. Government Printing Office, Washington, D.C.
8. *Feed Additive Compendium,* Miller Publishing Co., Minneapolis, Minn.
9. *American Association of Feed Control Officials (AAFCO) Directory,* AAFCO, College Station, Tex.

PARK W. WALDROUP
University of Arkansas

PET FOODS

All pet foods sold in the United States are subject to scrutiny by both competitors and feed control officials, including the Food and Drug Administration (FDA), Association of American Feed Control Officials (AAFCO), U.S. Department of Agriculture (USDA), Federal Trade Commission (FTC), American Animal Hospital Association (AAHA), American Veterinary Medical Association (AVMA), and Pet Food Institute (PFI). A European group organized to assure fair trade and free circulation of products through Europe (FEDIAF) also monitors every aspect of U.S. pet foods and follows American trends. More is known about the nutrition of dogs and cats than is known about the nutrition of humans.

Pet foods are different from other animal feeds. Most pet foods are processed in highly sophisticated plants using equipment, sanitation, and quality control exceeding standards observed in many plants producing human-grade foods. Pet foods may be stored for up to a year following manufacture before being consumed.

This possible delay in the consumption of pet foods requires more careful ingredient selection, preservation of freshness with antioxidants (qv), packaging and processing to avoid insect infestations and rancidity, and careful storage. Pet foods may contain expensive ingredients to provide desirable promotional and marketing copy.

Pets are fed a wide range of commercial foods, which vary on a dry basis from 15 to 60% protein and 5 to 50% fat. Some pet foods are expensive, nutrient-rich foods that contain twice as much nutrition density as needed. Although small quantities of a high calorie food may be consumed, approximately equivalent nutrition may be obtained by pets consuming larger quantities of a less concentrated food. For example, dogs may consume foods containing up to twice as much protein as they require; the excess digested protein is metabolized, the protein used as energy, and the nitrogen excreted, causing no problems in normal dogs. The protein content of cat foods is much higher than that required by dogs and more realistically conforms to the amounts that cats actually require for growth, gestation, lactation, and maintenance. Most pet foods are carefully balanced to meet the pet's needs without nutrient deficiencies or significant excesses.

The first pet food, a baked mixture of meat, vegetables, and wheat flour, was produced in the late 1800s. Early canned dog foods were composed mostly of meat from horses or dead stock. In the 1950s, high quality, nutritionally balanced, oven-baked, and pelleted dog foods became popular with dog owners and provided the most economical and satisfactory sources of dog nutrition. The extrusion process for pet foods was developed in 1954 and by 1957 extruded dog food had become the nation's leading dry pet food.

Types of Commercial Pet Foods

Pet foods are produced in canned, semimoist, and dry forms. Canned pet foods contain approximately 78 to 82% water and have a strong appeal to both pets and owners. Semimoist foods have moisture contents of 25 to 50%. Dry-type foods contain 10 to 12% moisture and supply about 90% of the nutrition consumed by dogs and 72% of the nutrition eaten by cats.

Therapeutic foods have been developed to meet the needs of pets that have nephritic failure, allergies, thyroid problems, geriatric difficulties, and obesity. Most of these therapeutic diets are dispensed by veterinarians, though some are available in pet food outlets and human-food stores stocking pet foods. Treats are usually snacks that may be nutritionally complete or may provide a tasty morsel as a reward. The number of treat products has escalated rapidly.

Canned and Semimoist Foods. Canned and dry foods are nutritionally comparable on a moisture-free basis. Some canned foods are basically dry foods to which gravy, moisture, and flavor enhancers have been added. Almost all animals tend to prefer moist foods to dry, and canned foods are desirable for geriatric dogs and cats, particularly those having gum and dental deterioration. Canned foods can be gulped by dogs and consumed quickly by cats.

Semimoist foods usually contain about 30% moisture and provide pets with a meat-like food. This meat-like appearance has been the primary basis for the popularity of semimoist pet foods.

Dry Foods. Dry foods are concentrated sources of nutrition and provide the most economical nutritional value because water in canned foods is expensive. Dry foods tend to scrape the teeth as pets eat, minimizing tartar deposition. When dry food is moistened prior to being consumed, tartar accumulates in a manner comparable to deposits observed with canned foods. Approximately 95 to 98% of dry-type cat and dog foods are made by the extrusion process; the remainder is made by pelleting or baking.

Pelleted Pet Food. Pelleted pet foods for dogs are processed and sold in relatively small quantities. These are seldom used for cats. In the pelleting process, slightly moistened ingredients are placed between rollers and huge dies containing thousands of holes. As the die rotates, rollers with tremendous pressures push the food mixture through the die holes and scrapers or cutters on the external surface of the die then cut or shear the pelleted particles into desirable lengths. The temperature, moisture, and time during pelleting are too low to gelatinize carbohydrates (qv) and carbohydrate ingredients are precooked in an extruder or baked prior to pelleting. Outdated bread, cookies, and other pastries are often recycled as the ingredients for pelleted foods. Moistened pelleted foods become sticky or soggy and disintegrate rapidly. Pellets are much more dense than extruded pet foods, and pelleted and extruded bags of equal weight are considerably different in size. Pelleted foods can be made in square, round, or oval pellets depending on the die holes and in different lengths depending on the position of cutters. Extruded foods, by contrast, can be made in many more shapes and sizes.

Extruded Food. Extrusion is an extremely versatile method of producing pet foods. The concept of extrusion for pet foods is based on expeller screw pressing of oil seeds. An adaptation to remove the oil-escape orifices of the screw press resulted in an extruder that produced elevated temperatures and high pressure through internal friction. An experimental extruder was used first in an attempt to produce human cereals. In 1954, the machine successfully produced a formed dog food with the carbohydrates dextrinized. The food was highly palatable, did not disintegrate into a mush in the dog's mouth, and had an attractive appearance. The first extruder produced quantities on the order of kilogram per hour whereas modern extruders produce many tons per hour.

During the extrusion process, ingredients, primarily dry, are blended into mixtures that normally do not include labile nutrients. The blend is moistened to approximately 30% moisture and then heated in a conditioning chamber prior to extrusion. Fresh meat and moist ingredients including animal by-products also may be added in the conditioning chamber. The heated and moistened mixture then enters the commercial extruder, ie, an immense enclosed screw with flights arranged in progressively restricted capacities, thus producing an enormous amount of heat by friction, up to 150°C, and pressures of nearly 6.89 MPa (1000 psi). After a few seconds moving through the extruder, the food is forced through forming dies and out into the room environment. The sudden release of pressure lets the entrapped moisture produce an expansion similar to exploding popcorn. Orifices within the die can produce almost any form, including stars, fish, chunks, and pellets of many sizes. Following extrusion, the food is moved by conveyor or negative air pressure to large ovens and spread in thin layers for drying. The moisture is reduced to about 10%, and the food is cooled. Additional fats, vitamins, and flavor enhancers are then added and the food packaged.

Extrusion processing is highly automated. Some extruders may process over 9 t/h, and in one Ralston Purina plant (Davenport, Iowa) 30 extruders were operating in a single location. With computer assistance, one person can operate many different extruders, and several different foods can be produced simultaneously. These may be different formulations or different colors and shapes to be packaged singly or combined into one variety pack. The differences in variety may be attributable only to added colors or different shapes.

The heat and pressure of extrusion cooks (gelatinizes) carbohydrates and helps prevent diarrhea and flatulence. Extruding also inactivates several types of trypsin inhibitors found in legumes; most antivitamins, such as avidin; and some mineral-binding factors. Extrusion destroys the 13 different species of proteinase inhibitors found in the raw potato (1). It destroys thiaminase found in fish spleen, liver, and intestines; and inactivates lipoxidase present in raw soybeans which can oxidize carotene. Canned foods are processed in retorts using elevated temperatures and pressure which also gelatinizes carbohydrates.

Pet Food Formulation

Weights of adult cats in normal physical condition vary from 2 to 6 kg, which is contrasted with the 1 to 100 kg encountered in adult dogs of different breeds. Dogs have proportionately longer digestive tracts and can digest foods more efficiently than cats. This difference in digestibility helps account for the requirement by cats for higher protein diets.

Animal food ingredients are selected to provide desirable contributions of nutrient availability, digestibility, droppings condition, palatability, processing characteristics, ethical desirability, and economics. Modern commercial pet foods contain about 50 nutrient and nonnutrient additives. Each nutrient is supplied at a near-optimum bioavailable level. Some nutrients have a symbiotic relationship with other nutrients, whereas some combinations are antagonistic, eg, an excess of zinc may be antagonistic to the availability of dietary copper. Copper is a part of many biological functions and closely linked with iron metabolism. A copper deficiency may decrease iron transport, produce anemia and bone disorders, and impair hemoglobin synthesis. Copper levels recommended for and tolerated by most strains of dogs may be toxic to some lineage of Bedlington terriers (2), with inborn errors of metabolism resulting in excess levels of copper accumulated in the liver and often death. High levels of copper have been used in some dog foods, but generally low levels are added to dog and cat foods to minimize copper accumulation in susceptible strains of dogs.

Ingredients used in pet foods are usually high in nutritional quality but generally not desirable as human foods primarily because they do not conform to human taste or processing expectations. By-products such as rendered proteins and fat converted into pet foods may have a derivation unappealing to humans, yet after processing may actually be more free of microorganisms and toxins than foods consumed by humans.

Nutritive Ingredients. Nutrients include amino acids (qv), fats, carbohydrates, fibers, minerals, and vitamins (qv). Some ingredients, such as niacin, supply only niacin, whereas salt provides both essential sodium and chlorine. Meat and bone meal may contain all of the nutrients, but not the correct quantities and ratios needed by dogs and cats, and the minimum required level of some nutrients for some species may be toxic to others. Nutrient concentrations in ingredients may be given in absolute amounts, as determined by chemical or physical laboratory procedures which have little direct nutritional application, because there may not be a relationship to bioavailability, ie, the amount of a nutrient absorbed from the digestive tract and available for the animal's use. An allowance must be made for the bioavailability of specific nutrients, and absorption alone is not proof that nutrients are bioavailable. Some peptides are absorbed and excreted in the urine without being utilized (3). Utilization of some nutrients depends on the dietary concentration. Calcium from calcium carbonate at low dietary levels is utilized efficiently, but at high dietary calcium concentrations increased quantities of calcium are excreted in the feces. Thus the bioavailability of calcium tends to be correlated with the quantity of the chemical component, dietary concentration, and grind or particle size. The iron in minerals such as ferric oxide and ferrous carbonate is not readily available; that from ferrous sulfate is highly available. Phosphorus in phosphoric acid, animal muscle, or organ tissue is highly available; plant phosphorus is poorly available to dogs and cats. Acidulants including phosphoric acid are used in cat foods to produce an acidic urine and help prevent urinary calculi, primarily struvite (magnesium ammonium phosphate hexahydrate) deposition. Too much phosphoric acid may produce a urinary pH of less than 6.5. This excess urinary acidity produces hypokalemia unless dietary potassium is increased to provide ample dietary potassium quantities.

Proteins. Proteins (qv) supply amino acids (qv), palatability enhancement, and, when present in more than required amounts, energy as the proteins are degraded and nitrogen compounds excreted. Dogs and cats can consume and meet amino acid requirements in the form of pure amino acids with complete success. However, animal tissue cannot differentiate between pure, plant, or animal sources of those amino acids, and those amino acids can be obtained much more economically from either plant or animal proteins.

Huge amounts of concentrated proteins, available as oilseed plant by-products from the brewing, distilling, starch, and oil industries, provide excellent sources of amino acids for pets. The world production of oilseed meals is estimated to be 109×10^6 t. Horses, sheep, cattle, swine, and poultry also use oilseeds efficiently and provide intense competition for the use of these plant proteins. Plant proteins are heated during processing to inactivate enzymes that could otherwise be detrimental. Some plant proteins, such as soybean meals, contain enough relatively indigestible oligosaccharides, including stachyose [*470-55-3*] and raffinose [*512-69-6*], to limit usage in pet foods. Microorganisms in the large intestine associating with stachyose and raffinose produce undesirable skatoles, indoles, and flatulence. Accompanying fiber concentrations in most plant proteins act like thousands of tiny sponges in the digestive tract, absorbing large amounts of water. This helps prevent constipation, but decreases fecal dry matter and increases fecal volume. The extra fecal volume is undesirable in kennel and pet management.

Soybean products that have been processed to remove a portion or all of the carbohydrates and minerals are used to make textured vegetable proteins which can be formed into various shapes and textures (see SOYBEAN AND OTHER OILSEEDS). Many canned dog foods utilize the textured vegetable protein chunks with added juices, flavor enhancers, vitamins, and minerals to produce canned dog foods that have the appearance of meat chunks. Similarly, those proteins can be combined with uncolored ingredients to imitate marbling and form pet foods with chunk-meat appearance. This processing is commonly used in semimoist pet foods.

Plant proteins from single sources, such as soybean meal, may be abundant in specific amino acids that are deficient in some cereal grains. Thus a combination of soybean meal and corn with their amino acid symbiosis may provide an excellent amino acid profile for dogs. Plant protein mixtures alone do not meet the amino acid needs for cats, because taurine [*107-35-7*] is not generally present in plant proteins.

Plant proteins are less expensive than animal proteins and are used in formula quantities at the greatest extent possible while still retaining the maximum desirable food characteristics. Plant proteins are extremely important in the nutrition of pets.

In the United States, more than 16.3×10^9 kg of human-inedible raw materials are available each year, and the rendering industry is a valuable asset in diverting these into valuable ingredients for use primarily in animal foods (4). The three largest meat packers are responsible for nearly four-fifths of all red meat production (5) and enormous amounts of rendered meat meal and animal fat. Three broiler producers account for about 40% of the total broiler production. American Proteins, Inc. (Roswell, Georgia), the world's largest processor of poultry by-products, produces more than 450,000 t of poultry meal, feather meal, and poultry fat each year. It also produces more than 100,000 t of fish meal, fish oil, and fish products each year. Fish meal production worldwide in 1986 was estimated at 6.23×10^6 t, which with the 125×10^6 t of meat and bone meal plus 6.67×10^6 t of feather meal and poultry by-product meal (6) is the primary source of animal proteins used by the pet food industry.

New Zealand rendering industries have low temperature rendering systems with cooking (rendering) temperatures ranging from 75 to 100°C for two to ten minutes with separation of liquid and solid phases. This produces high quality animal protein by-products with the least amount of nutrient degradation, yet permits destruction of most microorganisms and fat separation from the protein solids (7). This method of processing retains higher than normal levels of arginine, lysine, and other amino acids with less nonnitrogen protein. This processing helps account for the popularity of New Zealand sheep by-products, primarily lamb meal, gaining popularity in U.S. pet foods.

Milk and egg products are highly desired in pet foods since they supply the highest quality amino acid profiles with nearly 100% digestibility. Most milk protein concentrates are used for human foods, but some are available to pets (see MILK AND MILK PRODUCTS). An enormous quantity of whole eggs (qv) derived from egg graders, egg breakers, and hatchery operations are handled as dehydrated, liquid, or frozen ingredients.

Meat derived from crippled, old, discarded, injured animals, and those that have recently died (designated as 4-D beef), as well as USDA rejected meats, are used in canned pet foods. Fresh meats of human-grade also are used, including wing-tips, gizzards, livers, necks, backs, and meat still attached to bones. Edible meat removed by deboning machines from backs, necks, and bones in USDA inspected plants are used primarily for soup and meat-filled human foods; the excess is used in pet foods. Those parts that are not deboned are sold fresh and frozen to the pet food industry. These meat products (qv) are shipped in sealed-containers marked USDA inedible or denatured with charcoal or dye to prevent any use as nonanimal food.

Fats and Oils. Fats and oils from rendering animal and fish offal and vegetable oilseeds provide nutritional by-products used as a source of energy, unsaturated fatty acids, and palatability enhancement. Fats influence the texture in finished pet foods. The use and price of the various melting point fats is determined by the type and appearance of the desired finished food appearance.

Large quantities of fat are used from the fast food industry; these fats may have dissolved plastics from restaurant wrappers which can restrict spray nozzle orifices as the fats cool during spraying on pet foods (see FATS AND FATTY OILS).

Vegetable oils (qv), which have become increasingly desirable in human foods because of the high levels of polyunsaturated fatty acids, are generally too expensive to be used in pet foods. An excess of vegetable oils in the nutrition of show dogs tends to produce less-firm fat deposition, which may be objectionable in conformation competition. Canned dog foods may have extremely high levels of fat, especially those containing 4-D meat sources; they may exceed 50% fat on a dry matter basis.

Carbohydrates and Plant Products. The world supply of excess grains provides desirable sources of carbohydrates (qv) and fibers (qv) for animals, including pets. Most grains are relatively low in proteins and, unless processed for starch or alcohol, are generally ground whole and used in animal feeds. Thus the contribution of the accompanying protein, vitamins, minerals, and fibers can be accounted for advantageously during pet food formulation.

Corn, wheat, and rice are the most desirable common grains and are used extensively in pet foods. Oats and barley often tend to have excess fiber, which can be objectionable. However, barley is a preferred grain for moisture absorption and form in canned foods because the turgid white form is desired in some canned dog foods. Milo has enormous variations in tannin content which can influence digestibility and acceptability, thus limiting its use in pet foods (see WHEAT AND OTHER CEREAL GRAINS).

Fibers and Fiber Sources. Fibers are present in varying amounts in food ingredients and are also added separately (see DIETARY FIBER). Some fibers, including beet pulp, apple pomace, citrus pulp, wheat bran, corn bran, and celluloses are added to improve droppings (feces) form by providing a matrix that absorbs water. Some calorie-controlled foods include fibers, such as peanut hulls, to provide gastrointestinal bulk and reduce food intake. Peanut hulls normally have a high level of aflatoxins. They must be assayed for aflatoxin and levels restricted to prevent food rejection and undesirable effects of mycotoxins.

Normally, fecal moisture increases and dry matter digestibility decreases with added dietary fibers. When 500 grams of common dog foods sold in food stores are consumed, about 300 grams of droppings containing 30% dry matter are produced. For the inclusion of 12% dietary beet pulp, 500 grams of food can produce up to 800 grams of feces having 19 to 20% dry matter content.

Nonnutrient Additives. Nonnutritional dietary additives provide antioxidants to preserve freshness, flavor enhancers to stimulate food selection, color to meet the owner's expectations, pellet binders to minimize fine particles, mycostats to minimize mold growth, and ingredient-flow enhancers. Pet foods do not include coccidiostats, antibiotics, added hormonal materials, and fly-larval insecticides used in other animal feeds.

Antioxidants. Naturally occurring and synthesized antioxidants are added to help protect and spare vitamin E, selenium, vitamin C, taurine, xanthophylls, and other nutrients with antioxidant properties that are needed to protect cell membranes against peroxidation and the destructive effects of free radicals. Vitamin E's most potent biological form, α-tocopherol [*59-02-9*], is unavailable in adequate quantities to meet demands for human foods. There is speculation that vitamin E may impede the formation of low density lipoproteins (LDL) in humans and ameliorate atherosclerotic effects (8,9). Controversy has developed relating to a synthesized antioxidant, ethoxyquin [*91-53-2*]. Owners of dogs that perhaps have genetic or management problems have pointed to ethoxyquin as the cause of the undesirable problems observed with their dogs. That unsubstantiated association has stimulated activism for the removal of ethoxyquin, which spares other dietary antioxidants. As of 1993, some pet food producers have complied.

Early cat foods without antioxidants produced steatitis, ie, yellow fat disease, when the food contained quantities of fish, particularly tuna, and high levels of polyunsaturated oil. Early (1950s) shipments of Peruvian fish meal often spontaneously combusted, and ships' fish-meal cargo sometimes ignited. Ethoxyquin, considered the most highly effective antioxidant, combined with fish meal immediately after rendering, eliminates the combustion hazard. Because pet foods have such a long shelf life, ethoxyquin has been the choice of antioxidants because of efficacy, durability, and economics. Although ethoxyquin has been used in dog foods for almost 40 years with excellent success, additional extensive long-term testing of ethoxyquin in beagles is under way. Natural tocopherols are expensive and, because of the increasing demand in human diets, may become almost unavailable for pet nutrition.

Flavor Enhancers. Competition for the dog owner's money places enormous emphasis on instant acceptance of foods by pets. This has created an industry to supply flavor enhancers for dog and cat foods having aromas that appeal to owners. Different types of spices, essential oils (qv), amino acids, natural extractives, and synthetic flavoring substances have been used. Dogs and cats tend to prefer fats and fatty acids, amino acids, onion and garlic, hydrolysates, and most meat and fish flavors. Many flavor chemists have researched flavor enhancers (10), eg, they have identified and elucidated a series of factors contributing to high quality beef flavors, and have identified the distinctive beef flavor as a peptide chain of eight amino acids that develops as beef ages (see FLAVORS AND SPICES).

Newer flavor enhancers include hydrolyzed animal and plant proteins. Hydrolyzed proteins are used in dry-type dog and cat foods to provide enhanced acceptability. These are highly effective either sprayed on as a liquid or dehydrated and dusted on the outside of the pet food after extrusion and drying. Inclusion of flavor enhancers at the pre-extruder conditioner and heating just prior to extrusion significantly decreases flavor enhancement. That flavor loss apparently is a result of changes associated with high pressures and temperatures during extrusion, plus flavor loss during flashing as the extrudate is released from the high pressure into the atmospheric conditions.

Color Additives. Color additives, for the benefit of dog and cat owners, help simulate food richness, which is evaluated in many different ways. The addition of color helps minimize variations in appearance associated with batch difference in food ingredients and fineness of grind. Cats and dogs are practically color blind; colors have little influence on them.

Other Nonnutrient Additives. Mycostats are included in most dry-type and semimoist pet foods to prevent mold development. When pet food packages are stacked against cold surfaces, internal moisture within the bag migrates toward those cold surfaces. That concentration of water along the periphery is conducive to growth of any viable fungi, thus producing mold. Small quantities of fungistats help prevent mold growths.

Additional nonnutrient additives include sequestrants to provide ingredient separation and stabilizers such as gums. Spices, essential oils, oleoresins, synthetic flavorings, and adjuvants are also used in pet foods.

Cat Specific Additives. Cats are more sensitive to some nutritional deviations than are dogs. A dietary deficiency of arginine [*7004-12-8*] is more severe in cats than in dogs. This difference is associated with the higher dietary protein in cat foods. An arginine-free diet, intentionally produced with purified ingredients, was designed to evaluate arginine requirements of cats. Graded levels of arginine were added to the basal diet to determine requirements. It was observed (11) that an arginine deficiency in the cat produces ammonia toxicity. Accumulation of ammonia in the blood appears within three hours following consumption of a single arginine-free meal and is accompanied by emesis, hyperactivity, ataxia, tetanic spasms, and other abnormalities. Although severe deficiencies of arginine are unlikely to be encountered, arginine and taurine are most likely to be the limiting indispensable amino acids.

Taurine. Taurine is a sulfonic amino acid derived from methionine and cystine and functions in many biological systems. Although taurine is plentiful in most mammalian tissues as a free acid, the cat's synthesis of taurine is insufficient to meet its biological needs (12). The cat's synthesis of taurine by the cysteine sulfinic acid pathway is limited by the low activity and concentration of cysteine sulfinic acid decarboxylase. Taurine scavenges strong oxidants, including free radicals. Taurine deficiency has caused abnormal reproduction by queens and abnormal development of kittens, eg, abortion and resorption of fetuses, stillbirths, low neonatal weights, and abnormal brain and neurological development have been observed with taurine-deficient queens (13). Ocular lesions, associated with feline central retinal degeneration (FCRD), terminating in blindness develop in growing kittens and are the initial observable signs of deficiencies associated with taurine (14–16). Cats have been used in long-term studies to determine that no FCRD abnormalities occur with purified diets containing 375 mg of taurine kg/diet; 400 mg taurine/kg diet is recommended for growing kittens (17). Taurine also is associated with dilated cardiomyopathy (18), a relaxation of the heart wall muscles that decreases cardiac efficiency; prolonged taurine deprivation can cause death.

The taurine status of cats is easily measured using assays for blood or serum taurine levels. Serum taurine levels of 50 nmol/mL are generally considered adequate. Because heat processing during canning inactivates considerably more taurine or forms an inhibition against taurine uptake by the feline, less taurine is required in dry foods than in canned foods. The Feline Nutrition Expert Subcommittee of the Association of American Feed Control Officials (AAFCO), in the nutrient profiles for complete and balanced cat foods (19), suggests 0.1% taurine in extruded food and 0.2% in canned foods as a result of the extra loss of taurine during the canning processing.

Feather meal, first hydrolyzed and then oxidized, produces cysteic acid [*13100-82-8*] an excellent precursor for taurine in cats (20). Hydrolyzed feather meal may supplement the taurine provided by other dietary animal proteins and help replace part or all of the synthetic taurine in cat food formulations with considerable cost savings.

Phosphoric Acid. Cats generally consume about two-thirds as much water per unit of food dry matter intake as do dogs (21). This water conservation is often associated with feline urine obstruction, primarily formed by the deposition of struvite (magnesium ammonium hexahydrate) in the urinary tract. Dietary magnesium in amounts required for growth is near the concentration which also precipitates in the urinary tract and may cause obstruction. To provide safe levels of dietary magnesium and also prevent feline urinary syndrome (FUS), ingredients such as phosphoric acid [*7664-38-2*], which acidulates the urine, are added at carefully controlled levels to produce an acidic urine of approximately pH 6.5. This keeps practically all struvite in suspension in the urine. Relatively high levels of dietary potassium are required with low urine acidity to prevent hypokalemia.

Other Additives. Cats cannot convert tryptophan to niacin (22), or carotene to vitamin A in sufficient amounts to meet their needs (23). These deviations, as compared with other animals, need not produce problems because added dietary sources of niacin and vitamin A provide the needs of cats.

AAFCO Nutrient Profiles. Pet food products provide package claims of "complete and balanced for specific physiological states" to provide the pet owner with confidence and to assure that pets receive nutritionally desirable foods. Before the promulgation and acceptance of the Association of American Feed Control Official (AAFCO) Nutrient Profiles (Table 1), a number of references were used for complete and balanced recommendations (17,24,25). The Canine Nutrition Expert (CNE) subcommittee was formed to establish new profiles for complete and balanced dog foods. The AAFCO–CNE nutrient profiles are considered the AAFCO-recognized authority on canine nutrition (26). Dog foods bearing the label claim of nutritional adequacy by reference to the AAFCO dog food nutrient profile must meet all of the minimum and maximum levels for nutrients as established by the CNE subcommittee or must meet successful feeding test criteria based on published feeding protocols (27,28). The Feline Nutrition Expert (FNE) subcommittee was appointed following the development of AAFCO–CNE recommendations to compile profiles for complete and balanced cat foods. These AAFCO–FNE

Table 1. AAFCO Nutrient Profiles

	Dog foods[a]			Cat foods[b]		
Nutrient	Growth and reproduction, min	Adult maintenance, min	Maximum suggested level	Growth and reproduction, min	Adult maintenance, min	Maximum suggested level
proteins, %	22.0	18.00		30.0	26.0	
arginine	0.62	0.51		1.25	1.04	
histidine	0.22	0.18		0.31	0.31	
isoleucine	0.45	0.37		0.52	0.52	
leucine	0.72	0.59		1.25	1.25	
lysine	0.77	0.63		1.20	0.83	
methionine–cystine	0.53	0.43		1.10	1.10	
methionine				0.62	0.62	1.50
phenylalanine–tyrosine	0.89	0.73		0.88	0.88	
phenylalanine				0.42	0.42	
taurine						
extruded				0.10	0.10	
canned				0.20	0.20	
threonine	0.58	0.48		0.73	0.73	
tryptophan	0.20	0.16		0.25	0.16	
valine	0.48	0.39		0.62	0.62	
fat, %	8.0	5.0		9.0	9.0[c]	
linoleic acid	1.0	1.0		0.5	0.5	
arachidonic acid				0.02	0.02	
minerals, %						
calcium	1.0	0.6	2.5	1.0	0.6	
phosphorus	0.8	0.5	1.6	0.8	0.5	
Ca:P ratio	1:1	1:1	2:1			
potassium	0.6	0.6		0.6	0.6	
sodium	0.3	0.06		0.2	0.2	
chloride	0.45	0.09		0.3	0.3	
magnesium	0.04	0.04	0.3	0.08	0.04[d]	
iron,[e] mg/kg	80	80	3,000	80	80	

copper, mg/kg	7.3	7.3	250	5	5	
manganese, mg/kg	5.0	5.0		7.5	7.5	
zinc, mg/kg	120	120	1,000	75	75	2,000
iodine, mg/kg	1.5	1.5	50	0.35	0.35	
selenium, mg/kg	0.11	0.11	2	0.1	0.1	
vitamins						
vitamin A, IU/kg	5,000	5,000	50,000	9,000	5,000	750,000
vitamin D, IU/kg	500	500	5,000	750	500	10,000
vitamin E,[f] IU/kg	50	50	1,000	30	30	
vitamin K,[g] mg/kg				0.1	0.1	
thiamine,[h] mg/kg	1.0	1.0		5.0	5.0	
riboflavin, mg/kg	2.2	2.2		4.0	4.0	
pantothenic acid, mg/kg	10	10		5.0	5.0	
niacin, mg/kg	11.4	11.4		60	60	
pyridoxine, mg/kg	1.0	1.0		4.0	4.0	
folic acid, mg/kg	0.18	0.18		0.8	0.8	
vitamin B_{12}, mg/kg	0.022	0.022		0.02	0.02	
biotin,[i] mg/kg				0.07	0.07	
choline,[j] mg/kg	1,200	1,200		2,400	2,400	

[a]Presumes a metabolizable energy density of 14.64 kJ/g of dry matter. Ratings greater than 16.74 kJ/g should be corrected for energy density.

[b]Presumes a metabolizable energy density of 16.74 kJ/g of dry matter based on the modified Atwater values of 14.64, 35.56, and 14.64 kJ/g for protein, fat, and carbohydrate (nitrogen-free extract, NFE), respectively. Rations greater than 18.83 kJ/g should be corrected for energy density; rations less than 16.74 kJ/g should not be corrected for energy. To convert kJ to kcal, divide by 4.184.

[c]Although a true requirement for fat *per se* has not been established, the minimum level was based on recognition of fat as a source of essential fatty acids, as a carrier of fat-soluble vitamins, to enhance the palatability, and to supply an adequate caloric density.

[d]If the mean urine pH of cats fed *ad libitum* is not below 6.4, the risk of struvite urolithiasis increases as the magnesium content of the diet increases.

[e]Because of very poor bioavailability, iron from carbonate or oxide sources that are added to the diet should not be considered as components in meeting the minimum nutrient level in cats.

[f]For cats, add 10 IU vitamin E above minimum level per gram of fish oil per kilogram of diet.

[g]Vitamin K needs to be added to cat food only when diet contains greater than 25% fish on a dry matter basis.

[h]Because processing may destroy up to 90% of the thiamin in the diet, allowances in formulation should be made to ensure the minimum nutrient level is met after processing.

[i]Biotin needs to be added to cat food when diet contains antimicrobial or antivitamin compounds.

[j]In cat food, methionine may substitute for choline as methyl donor at a rate of 3.75 parts for 1 part choline by weight when methionine exceeds 0.62%.

nutrient protocols for cats were published, and include protocols for adequate testing of pet food products, which are monitored by AAFCO (28). The AAFCO nutrient profiles for both dogs and cats offer maximum suggested nutrient levels; excess nutrient levels can be as harmful as deficiencies. Also, added nutrient levels for the stages of growth and reproduction are additional features. The AAFCO nutrient profiles for dogs and cats are given in Table 1.

Economic Aspects

The annual production of pet foods is approximately 6.35×10^6 t, valued at $7.6 billion (Table 2). It has been estimated that there are as many as 15,000 different brand labels and package sizes of pet foods, marketed by 3000 manufacturers

Table 2. U.S. Dog and Cat Food Sales[a,b]

Type	Total sold, t	Dry matter, %	Total dry matter basis, t	Sales, $ × 10^6
		Dog food		
food store				
dry	1,977,633	90	1,779,896	1,986.0
canned	762,489	22	167,747	964.0
semimoist	78,471	70	54,930	138.3
snacks/treats	125,645	90	113,080	440.2
total	*2,944,268*		*2,115,653*	*3,528.5*
feed mills	1,982,471			1,246.0[c]
total, dog	*4,926,739*			*4,774.5*
		Cat food		
food store				
dry	518,910	90	467,019	883.0
canned	615,253	22	135,352	1,365.0
semimoist	57,606	70	40,324	180.0
snacks/treats	3,810	90	3,447	
total	*1,195,579*		*646,142*	*2,428.0*
feed mills	31,479			410.0
total, cat	*1,227,058*			*2,838.0*
		Dog and cat total food		
food stores	4,139,847			5,956.5
feed mills	2,013,950[d]			1,656.0
Total	*6,153,797*			*7,612.5*

[a]For year ending June 1990.
[b]Ref 29.
[c]Estimated to be equal to 27% of value of sales through food outlets.
[d]Feed mills include sales through feed stores and pet stores; sales direct to some large retailers, veterinarians, larger kennels, and research facilities; and export. It is estimated to account for 40.3% of total dog food production plus 9.3% of canned and dry cat food production.

(30). Conservative estimates are closer to 3000 brands (31) and sizes, with 1800–1900 registered in the state of Texas.

Pet food purchases are based on the satisfaction of the owner, and pet food proliferation is enormous with accompanying advertising descriptors including natural, lite, low calorie, high calorie, low protein, and high protein. New therapeutic series, sizes, densities, colors, and attractive packaging have also added to the proliferation.

Commercial flavor enhancers for pet foods have become big business. Flavor enhancers, primarily so-called digests, provide high acceptance of pet foods and enable the pet to select one food over another. Commercial companies compete with flavors based on the types that pets like. However, owner objections minimize the use of some acceptability enhancers such as some fish products, onions, and garlic.

Digest is the most widely used flavor enhancer. Digests are mostly hydrolyzed proteins and fats, primarily of animal origin. When meat or animal byproducts are subjected to acids or enzymes and disintegrate into components, some have potent attraction to both dogs and cats. Digests are sprayed or dusted on the outside of many dog and cat foods, and are much more effective as palatability enhancers than are those same digests added with the other ingredients and processed through pelleting or extrusion machines.

Cost per kilogram of dog food ranges from ca $0.30 to $30/kg of dry matter; demands exist for pet foods in each range. Some of the highest quality pet foods are manufactured by companies conducting an enormous amount of applied research and producing only foods that are sold relatively inexpensively through private label channels and not with their own labels.

BIBLIOGRAPHY

"Feeds, Animal," in *ECT* 1st ed., Vol. 6, pp. 299–312, by H. M. Briggs, Oklahoma Agricultural and Mechanical College; in *ECT* 2nd ed., Vol. 8, pp. 857–870, by H. M. Briggs, South Dakota State University; "Pet and Livestock Feeds," in *ECT* 3rd ed., Vol. 17, pp. 90–109, J. Corbin, University of Illinois.

1. R. Bernard, *Cornell Vet.* **32**, 29 (1942).
2. D. C. Tinedt, I. Sternlieb, and S. A. Gilberson. *J. Am. Vet. Med. Assoc.* **175**, 269 (1979).
3. J. E. Ford and C. Shorrock, *Br. J. Nutr.* **26**, 311–322 (1971).
4. P. Brown and co-workers, *J. Infectious Diseases* **161**, 467 (1990).
5. F. Burton, *Render* **19**(4), 7 (1990).
6. F. Burton, *Render* **15**(3), 7, 25 (1986).
7. F. Burnham, *Render* **15**(3), 10 (1986).
8. M. Stampfer, *The Champaign-Urbana* (Illinois) *News-Gazette*, 4A (Nov. 18, 1992).
9. D. Steinberg and co-workers, *New Engl. J. Med.* **310**, 915–924 (1989).
10. R. W. Hitzman, *Identification and Elucidation of Factors Contributing to High Quality Beef Flavors*, Ph.D. dissertation, University of Illinois, Urbana, 1986.
11. J. G. Morris and Q. R. Rogers, *J. Nutr.* **106** (abstr. 63) (1976).
12. I. Burger and K. Earle, *Waltham International Focus* **2**(2), 9–13 (1992).
13. J. A. Sturman and co-workers, *J. Nutr.* **116**, 655–667 (1986).
14. K. C. Hayes, A. R. Rabin, and E. L. Berson, *Am. J. Pathol.* **78**, 505–515 (1975).

15. P. A. Anderson, *Significance of Indispensable Amino Acids, Choline, and Taurine in Feline Nutrition*, Ph.D. dissertation, University of Illinois at Urbana-Champaign, 1979.
16. Q. R. Rogers and J. G. Morris, *3rd Annual Pet Food Inst. Technical Symposium*, Sept. 24, 1982, Kansas City, Mo., 1982.
17. National Research Council, *Nutrient Requirements of Cats*, National Academy Press, Washington, D.C., 1986.
18. P. D. Pion and co-workers, *Science* **237**, 764–768 (1987).
19. D. A. Dzanis, *FDA Veterinarian* **7**(6), 1–2 (Nov./Dec. 1992).
20. Q. R. Rogers, personal communication, 1992.
21. H. N. Waterhouse and D. S. Carver. *Proc. Animal Care Panel* **12**, 271 (1962).
22. B. Ahmad, *Biochem. J.* **25**, 1195 (1931).
23. M. Ikeda and co-workers, *J. Biol. Chem.* **240**, 1395 (1965).
24. National Research Council, *Nutrient Requirements of Dogs*, National Academy Press, Washington, D.C., 1974.
25. Ref. 17, 1978 ed.
26. D. A. Dzanis, *FDA Veterinarian* **6**(5), 1,2,4 (Sept./Oct. 1991).
27. Association of American Feed Control Officials, official publication, AAFCO, Inc., Atlanta, Ga., 1992.
28. Association of American Feed Control Officials, official publication, AAFCO, Inc., Atlanta, Ga., 1993.
29. ARBITON/SAMI report of 52 weeks ending June 13, 1990, Warehouse Withdrawals, modified with *Maxwell Report Food Industry,* Jan./Feb. 1991, p. 9, plus Feed Mill sales data.
30. R. H. Brown, *Feedstuffs* **64**(52), 5 (1992).
31. R. Sellers, personal communication, 1992.

General Reference

Official Publication of Association of American Feed Control Officials, 1993, Association of American Feed Control Officials, Georgia Department of Agriculture, Capitol Square, Atlanta, Ga.

JAMES CORBIN
University of Illinois at Urbana-Champaign

RUMINANT FEEDS

Many species of ruminants exist worldwide (1). The feeds and feed additives common to U.S. agriculture for the nutrition and management of domesticated ruminant animals, ie, cattle, sheep, and goats, are discussed herein.

Ruminants, which consume plant material grown on land that may be unsuitable for crop farming, need not compete with humans and nonruminant livestock for feed resources. At least one-third of the world's land area is more suitable for grazing than for cultivation (1,2). The high fiber content forage produced on this land is largely undigested by monogastrics. Ruminant animals, whose ruminal microflora ferment and digest cellulose (qv), the predominant component of fiber and the earth's most abundant carbohydrate (3), utilize a large amount of the plant energy produced on this land (3). Anatomical differences between monogastric and ruminant animals allow the ruminant to be more efficient in digesting cellulose, but generally less efficient in gaining weight and converting feed to gain, because of energetic losses resulting from the fermentation (qv) process.

The ruminant has a four-compartment stomach, as opposed to the single-compartment stomach of monogastrics. The esophagus delivers food after oral ingestion to the reticulum, the first compartment. The reticulum is attached directly to the rumen, the principal fermentation compartment. The rumino-reticulum compartment makes up at least 60% of the total stomach compartment in cattle and sheep (4). A sphincter muscle between the rumen and omasum (the third stomach compartment) regulates feed outflow from the rumen. Following the omasum is the abomasum. The omasum filters contents flowing between the rumen and the abomasum to maintain proper particle size in the abomasum (5). The abomasum empties into the small intestine and is considered the gastric stomach of the ruminant, equivalent to the monogastric stomach. Gastric juices, such as hydrochloric acid [*7647-01-0*], as well as proteolytic enzymes are the primary secretions of the abomasum (5). The reticulum and rumen serve as a large fermentation sac where anaerobic bacteria, protozoa, and fungi reside. When particle size is reduced sufficiently, the solid material passes into the omasum from the rumen. A portion of the solid food is regurgitated (ruminated) between meals so that it may be remasticated, reswallowed, and fermented further.

Compartmentalization of the stomach is the principal trait allowing ruminants to utilize fibrous feeds (5). Certain species of microorganisms present in the rumen secrete cellulase [*9025-56-3*] which degrades cellulose by fermentation, primarily to acetate, propionate, and butyrate. The ruminal microbial population also degrades and resynthesizes nitrogen compounds in the ingested feeds into microbial protein, a relatively high quality amino acid mixture that passes to the small intestine. The lower pH of the abomasum lyses the microorganisms passing from the rumen. The released microbial protein is more similar to the animal's amino acid requirements than that contained in plant protein sources.

Feeds

Forages/Roughages. Approximately 75–80% of the feed fed to ruminants during their lifetime production cycle is forage/roughage material (6). Roughages are made up predominantly of the stem or stalk portion of plants and usually include the seeds and leaves of these plants. These feeds are typically high in fiber, >50% neutral detergent fiber; low in starch, <4%; and moderately low in crude protein, <20%. Roughages are not only a source of nutrition to the ruminant but also help to maintain normal rumen function. Roughages are important to the ruminant animal for a variety of reasons including maintenance-level feeding of herd or flock animals; weaning of young ruminants from milk onto solid feed; preventing metabolic diseases, eg, bovine ketosis and ovine pregnancy toxemia; and maintaining proper fat level, approximately 3.5%, in milk produced by dairy cows. Under certain circumstances, eg, maintenance of nonpregnant females, roughages are the preferred feed choice because they generally cost less than other feedstuffs such as grain concentrates. Newly weaned ruminants require forages for purposes of establishing a gastrointestinal tract microbial population. Proper milk fat levels are maintained by including an adequate amount of forage in the lactating dairy cow diet.

Ruminants consume forages either by grazing or being fed harvested material. Grazing reduces input costs to the producer but does not allow any control over the amount and quality of forage consumed by the animal. Approximately one-third of the world's land area is classified as grassland unsuitable for cultivation (2), and grazing by ruminants is essential in optimizing the world's natural resources.

The moisture content at which a plant is harvested usually determines the storage method. Low (15–25%) moisture forages are often stored in some type of bale form in the presence of air. Stave silos, oxygen-limiting silos, concrete bunker silos, and large plastic bags are all methods of storing high (40–75%) moisture forages (7). High moisture forages are sealed from air exposure in order to allow anaerobic bacteria to ferment the available soluble carbohydrates (qv). The acidic by-products produced by the bacteria, ie, lactic [*598-82-3*], acetic [*64-19-7*], propionic [*79-09-4*], and butyric acid [*107-92-6*], decrease the pH of the forage to a point where no other bacteria can grow, thus preserving forage quality. Common examples of high moisture forages are silages and haylages. Silages can be made from harvesting and chopping whole corn plants, sorghum plants, or oats. Haylages generally contain less moisture than silages, ie, 40–60% vs 60–75% moisture. Alfalfa haylage is the most commonly fed haylage in the United States.

Several different sources of low moisture forage, eg, prairie hay, alfalfa, bromegrass, orchard grass, and blends of hay grown specifically for the purpose of harvesting; and roughage, eg, crop residues such as corn stalks, soybean stubble, or small grains straw, are available. Because the source of these forages/roughages is highly variable, the relative feed quality is also highly variable. Alfalfa is a relatively high quality forage source and is harvested in both the high moisture (haylage) and low moisture (hay) forms. It contains a relatively high (20%) amount of crude protein and a moderate (40–50%) amount of neutral detergent fiber. More information on alfalfa is available (8).

High Energy Feeds. Concentrated sources of energy are fed to ruminants to allow young ruminant animals to grow more quickly and efficiently. Feedstuffs of this nature are generally high in readily fermentable carbohydrates, ie, they are high starch-containing feedstuffs. Feedstuffs containing high amounts of starch (qv) are often from the seeds of plants such as corn, grain sorghum, oats, and barley (9). Wheat is also a highly digestible feedstuff although its demand as a human food makes its use as a feed cost-prohibitive. Millets are of minor importance except in areas of Asia, Africa, and the Commonwealth of Independent States (formerly Soviet Union) where millet can be grown successfully in drought-stricken areas (5). Rye grain is sometimes fed to ruminants. However, several compounds in rye have been identified (10) that decrease its usefulness as an energy source. Crossing wheat and rye produces a hybrid grain called triticale (5). Animals consuming triticale have not grown as efficiently as expected, perhaps because of the presence of trypsin and chymotrypsin inhibitors (11), alkyl resorcinols (12), water-soluble pentosans (13), and ergot (5), as well as low acceptability by the animal (14) (see WHEAT AND OTHER CEREAL GRAINS).

By-products of agricultural commodities are used as readily fermentable energy-containing feedstuffs. Molasses is a commonly used by-product of the sugar-refining industry. It contains at least 46% sugar, a low protein content, and 15–25% water (5). The addition of molasses in feed increases feed acceptability, reduces dustiness of the feed, and improves feed pelleting. Molasses from citrus and wood (qv) processing also is available (5). Other useful, energy-containing by-products include wheat bran, wheat middlings and shorts, dried citrus pulp, dried beet pulp, dried bakery products, hominy, oat groats, potato meal, whey, corn gluten feed, and rice bran. Since over 450 million metric tons of agricultural by-products and residues exist (15), a great quantity of ruminant feedstuffs is available that is not used by humans and nonruminant livestock. These by-products, often used as feed because of low cost and availability, may present other problems such as lower energy content than corn and low acceptability (9).

Besides changing the source of energy, feed processing methods influence the amount of energy available to the ruminant, ie, by influencing the sites of digestion and absorption in the ruminant animal. Fermentation products produced in the rumen and absorbed through the ruminal wall do not contain as much energy as the carbohydrates absorbed through the small intestine (6). Methods of processing include grinding, rolling, cracking, extruding, steam flaking, heating, wetting, and gelatinizing. More information on feed processing methods is available (5,9).

Various sources of lipid have been incorporated into ruminant diets to increase the energy density and provide the large amount of energy needed for slaughter animals to achieve market weight or for dairy cows to produce milk (see MILK AND MILK PRODUCTS). Fats also reduce the dustiness of feeds, increase the feedstuffs' ability to pellet, and improve feed acceptability.

The predominant feed source of lipid is from animal origin (5). Animal fat is typically higher in saturation and is often referred to as grease. Various vegetable sources of lipid also are available. Oils, containing a higher amount of unsaturation than animal fat, are extracted from corn, cotton seeds, soybeans, olives, safflowers, sunflowers, rapeseeds, and peanuts (5). Whole oilseeds increasingly are being fed especially to dairy cows during lactation (16) (see NUTS; SOYBEANS AND OTHER OILSEEDS).

Lipids present in the diet may become rancid. When fed at high (> 4–6%) levels, lipids may decrease diet acceptability, increase handling problems, result in poor pellet quality, cause diarrhea, reduce feed intake, and decrease fiber digestion in the rumen (5). To alleviate the fiber digestion problem, calcium soaps or prilled free fatty acids have been developed to escape ruminal fermentation. These fatty acids then are available for absorption from the small intestine (5). Feeding whole oilseeds also has alleviated some of the problems caused by feeding lipids. A detailed discussion of lipid metabolism by ruminants can be found (16).

Supplements

Protein. Although most feedstuffs contain protein, supplemental protein or nitrogen often is needed to meet animal physiological requirements (see PROTEINS). Practical situations in which supplemental protein is required include the feeding of growing/immature animals, lactating females, females in the last trimester of pregnancy, and ruminants grazing rangelands. The ruminal microflora not only require an energy source but also a nitrogen source. The nitrogen source does not need to be entirely in the form of protein but can be in the form of nonprotein nitrogen (NPN), eg, urea, biuret, ammonia, or dried poultry waste. Substitution of protein with an NPN source does not always result in the same level of animal performance. This may be because of the additional energy and minerals provided by the protein source (17). Increased performance from protein feeding may occur because some ruminal bacteria require branched-chain volatile fatty acids (VFA), derived from branched-chain amino acids (qv) (18) for growth. Nonprotein nitrogen should be limited to no more than 50% of the nitrogen fed in the diet.

Soybean meal is the most frequently used source of supplemental protein in the United States (5). Cottonseed meal is another important protein supplement. Both meals are by-products from oil extraction of the seeds. Canola meal is derived from rapeseed low in erucic acid [*112-86-7*] and glucosinolates. Linseed (derived from flax seed), peanut, sunflower, safflower, sesame, coconut, and palm kernel meals are other sources of supplemental protein that are by-products of oil extraction (4).

Raw soybeans also may be used as a supplemental protein source. Dry beans, ie, beans normally harvested in the green/immature state, fava beans, lupins, field peas, lentils, and other grain legumes are potential supplemental protein sources; however, several of these may have deleterious effects, predominantly enzyme inhibition, on the animal. The supply of each is limited (5).

Various milling, distilling, and brewing by-products are available as supplemental protein sources. Corn gluten meal, 60% crude protein (CP), and corn gluten feed, 20–25% CP, are derived from wet milling of corn. Wheat middlings, 15–20% CP, are the offal from milling of wheat into flour. Distilleries produce distillers' dried solubles and grains, 26–35% CP, as the by-products of liquor and wine (qv) production. Brewers grains, 26–29% CP, are by-products of beer (qv) produced from barley fermentation (see BEVERAGE SPIRITS, DISTILLED).

Legume forages, such as alfalfa or clover, are considered high quality, readily available protein sources. Animal sources of supplemental protein include meat and bone meal; blood meal, 80% CP; fish meal; other marine products; and hydrolyzed feathermeal, 85–90% CP. Additionally, synthetic amino acids are available commercially. Several sources (3,9,19) provide information about the protein or amino acid composition of feedstuffs.

Protein nutrition of ruminants also considers the rate of ruminal degradability and the amount of protein escaping the rumen. Ruminal escape (bypass) protein is a term used to describe the amount of protein that escapes ruminal fermentation and is passed to the gastric portion of the gut where it may be subsequently digested and absorbed. This aspect of protein utilization is of great interest to those developing ruminant diets. Protein sources are analyzed routinely for the amount of ruminal degradation that occurs. A discussion of the rate and extent of ruminal protein degradation for a variety of feedstuffs (20), and estimates of the amount of ruminal escape for various protein sources (3,5,17) are available.

Supplemental energy may be needed when the majority of a ruminant's diet is from bulky feedstuffs such as poor quality roughages. In the case of cows grazing rangeland, protein intake from the forage may be adequate, but the microflora in the rumen of the animal may lack a readily fermentable source of carbohydrate. In this case, a highly fermentable starch such as corn or sorghum grain is provided. In a similar situation, feeding a limited amount, ca 0.6% of body weight, of a readily fermentable source of carbohydrate to steers raised on vegetative, immature pastures may enhance their efficiency.

Minerals. Supplementation of macrominerals to ruminants is sometimes necessary. Calcium and phosphorus are the minerals most often supplemented in ruminant diets. One or both may be deficient, and the level of one affects the utilization of the other. Limestone, 36% calcium, is commonly used as a source of supplemental calcium. Dolomite, 22% calcium; oyster shells, 35% calcium; and gypsum, 29% calcium, are sources of calcium. Bone meal, 29% calcium, 14% phosphorus; dicalcium phosphate, 25–28% calcium, 18–21% phosphorus; and defluorinated rock phosphate, 32% calcium, 18% phosphorus, are sources of both calcium and phosphorus. Diammonium phosphate, 25% phosphorus; phosphoric acid, 32% phosphorus; sodium phosphate, 22% phosphorus; and sodium tripolyphosphate, 31% phosphorus, are additional sources of phosphorus (5).

Magnesium is deposited largely in the bones of the body. Magnesium oxide and magnesium sulfate are supplemental sources of magnesium. Sodium chloride is relatively inexpensive and is provided either free or incorporated directly into animal feed to prevent sodium and chloride deficiencies. Potassium is usually not deficient because most forages have adequate quantities. Therefore, it should be supplemented only when animals consume poor quality roughages or a high concentrate diet, or when they are under stress, dehydrated, or suffering from diarrhea (5). Potassium deficiency usually is alleviated by changing the diet or by supplementing with potassium sulfate.

Sulfur deficiency usually is not a problem for ruminants because the ruminal microflora can utilize sulfur-containing amino acids. A deficiency can occur, however, when an NPN source is fed. L-Methionine [*63-68-3*] is the most biologically available source of sulfur (21). Various sulfates are intermediate in sulfur availability, and elemental sulfur is the least available source of sulfur.

Cobalt, copper, molybdenum, iodine, iron, manganese, nickel, selenium, and zinc are sometimes provided to ruminants. Mineral deficiency or toxicity in sheep, especially copper and selenium, is a common example of dietary mineral imbalance (21). Other elements may be required for optimal ruminant performance (22). Excellent reviews of trace elements are available (5,22).

Vitamins. The B-vitamins and vitamin K [*84-80-0*], $C_{31}H_{46}O_2$, are synthesized by ruminal microorganisms and their supplementation is usually unnecessary (see VITAMINS). However, supplementation with B-vitamins is beneficial in certain situations. Polioencephalomalacia (PEM), a nervous disorder, is alleviated by intravenous injections of thiamine hydrochloride [*67-03-8*], $C_{12}H_{18}Cl_2N_4OS$. More information about PEM is available (23). Niacin supplementation has been shown to partially alleviate subclinical ketosis, increase milk production, and increase average daily weight gain under some conditions (23). Certain studies have shown higher milk fat percentages and fat-corrected milk yields after supplementing with choline (qv) (23). Dicumarol [*66-76-2*], $C_{19}H_{12}O_6$, a metabolic inhibitor of vitamin K, is present in sweet clover and its negative effects are overcome by supplementing with vitamin K.

Vitamins A, D, and E are required by ruminants and, therefore, their supplementation is sometimes necessary. Vitamin A [*68-26-8*] is important in maintaining proper vision, maintenance and growth of squamous epithelial cells, and bone growth (23). Vitamin D [*1406-16-2*] is most important for maintaining proper calcium absorption from the small intestine. It also aids in mobilizing calcium from bones and in optimizing absorption of phosphorus from the small intestine (23). Supplementation of vitamins A and D at their minimum daily requirement is recommended because feedstuffs are highly variable in their content of these vitamins.

Vitamin E acetate [*58-95-7*], $C_{31}H_{52}O_3$, serves primarily as an antioxidant and is closely associated with selenium. Vitamin E usually is present in normal feedstuffs at levels high enough to meet requirements except if the feedstuff has undergone excessive heating or prolonged storage. To ensure that adequate levels of vitamin E are present, α-tocopherol [*59-02-9*] is sometimes added to the diet.

Additives

Several feed additives for ruminants are available. All additives increase animal growth or efficiency of weight gain, and many provide additional benefits (see GROWTH REGULATORS). Additives can be classified into groups based on function.

Ionophores, widely used feed additives that alter ruminal fermentation, interact with the transport of ions across ruminal bacterial cell membranes. They also change the relative proportions of fermentative products produced by bacteria, ie, acetate production decreases and propionate production increases. Ionophores do this by altering the proportion of various ruminal bacteria present. The two FDA-approved ionophores for nonlactating ruminants are monensin [*17090-79-8*], $C_{36}H_{62}O_{11}$, and lasalocid [*25999-31-9*], $C_{34}H_{54}O_8$. These usually improve feed efficiency and may improve average daily gain of immature (< 16 months old) cattle consuming a high energy diet. Other potential benefits of feeding ionophores include decreased incidence of lactic acidosis, control of coccidiosis, control of feedlot bloat, and reduction in the number of face and horn fly larvae in feces (24).

Zeolites (qv), ion-exchange compounds, have been researched to some extent and have been proposed to improve NPN utilization (25). However, no improvement in NPN utilization was found with lambs fed zeolites (26).

Direct-fed microbials are feed additives composed of microbes and/or ingredients to stimulate microbial growth (27) which allegedly result in a more efficient microbial population (28). This may result from changing the gut microflora and reducing *E. coli*, producing antibiotics (qv), synthesizing lactic acid, colonizing the intestinal mucosa, or preventing toxic amine synthesis in the gut (29). Yeasts (qv), especially *Saccharomyces cerevisiae* and *Aspergillus oryzae*, have been researched as direct-fed antibiotics for cattle (5). Several experiments have shown beneficial changes in ruminal fermentation and/or increases in fiber digestion after adding direct-fed antibiotics to diets (5). However, other reports (30) cite conflicting results as to the benefit of adding direct-fed microbials to ruminant diets.

A problem common to animals consuming a high energy diet or lush, immature legume vegetation is increased susceptibility to bloat. Bloat is a condition where ruminal fermentation occurs too rapidly, leading to excess gas buildup in the rumen. This gas is not expelled owing to heavy foam production in the rumen. Antifoaming agents available to prevent bloat include silicones, detergents, vegetable oils, animal fats, animal mucins, and liquid paraffins (31). Poloxalene [*9003-11-6*] is an example of a commonly used surfactant developed primarily to prevent bloat on pasture.

Buffers are used to stabilize ruminal pH at 6.0–6.8. Available buffers include bicarbonate, calcium carbonate [*471-34-1*] (limestone), $CaCO_3$, and bentonite [*1302-78-9*], $Al_2O_3{\cdot}SiO_2{\cdot}H_2O$ (5). Sodium bicarbonate is the most effective buffer at ruminal pH and, consequently, the buffer most widely used. High concentrate diets are rapidly fermented and, therefore, result in a decrease in pH as a result of production of a large amount of fermentation products. The decreased pH is deleterious to the animal not only because it slows fermentation, but also because it may physically harm the animal, eg, cause ruminal lesions. Therefore, feedlot diets with readily fermentable energy sources often include sodium bicarbonate, and dairy diets high in readily fermentable energy sometimes include buffers (32). High fiber diets benefit the least from adding buffers, because of less fermentative activity or a higher buffering capacity of these forages. The addition of limestone to the diet may increase the pH of the small intestine because it increases the buffering capacity of intestinal contents (5). This, in turn, could increase the activity of pancreatic α-amylase which is most active at pH 6.9 (21).

Many ruminal bacteria require one or more branched-chain volatile fatty acids (VFA) for proper growth. The branched-chain VFA, ie, *n*-valeric [*109-52-4*], $C_5H_{10}O_2$; isobutyric [*79-31-2*], $C_4H_8O_2$; 2-methylbutyric [*623-42-7*], $C_5H_{10}O_2$; and isovaleric acids [*503-74-2*], $C_5H_{10}O_2$, are normally derived from branched-chain amino acids (qv) (18). Ruminal microbial metabolism may be limited if low amounts of branched-chain VFA are present (5). Therefore, supplementation of these VFA has been practiced; however, results have been variable (5,30,33). Because branched-chain VFA are derived from deaminated amino acids, diets high in NPN or low in ruminally degradable protein would perhaps be unable to provide adequate ruminal levels (5). Although branched-chain VFA are no longer produced commercially, feeding protein sources that are ruminally degradable can provide these necessary nutrients.

Defaunation is a term used to describe the elimination of protozoa from the rumen. Ruminal protozoa allegedly have both positive and negative effects on animal performance (5,34). Defaunation may increase ruminal microbial efficiency because less methane is produced and less proteolysis occurs (34). However, under certain conditions, protozoa may help stabilize the ruminal environment

(5). This may occur in the case of those animals consuming large amounts of starch. The ruminal protozoa engulf a portion of the starch and prevent it from undergoing rapid bacterial fermentation. Delaying starch fermentation by bacteria may lessen the incidence of lactic acidosis. Defaunation can be accomplished using copper sulfate and nonionic and anionic detergents (5). Defaunation is not yet (ca 1993) practical to implement in a production system.

Increased hormonal levels have resulted in better ruminant animal performance. Estrogen-containing compounds, eg, zearalenone [*17924-92-4*], $C_{18}H_{22}O_5$ (5), are available to improve growth and feed efficiency. These estrogens are given as ear implants. Melengestrol acetate [*2919-66-6*] (MGA), $C_{25}H_{32}O_4$, an additive that is a synthetic progesterone, suppresses estrus in heifers and results in increased weight gain and feed efficiency (5).

Antibiotics (qv) have been fed at subtherapeutic levels to promote ruminant animal growth. Possible reasons for the observed growth include decreased activity of microbes having a pathogenic effect on the animal, decreased production of microbial toxins, decreased microbial destruction of essential nutrients, increased vitamin synthesis or synthesis of other growth factors, and increased nutrient absorption because of a thinner intestinal wall (35). Antibiotics fed at subtherapeutic levels may help alleviate the effects of stress on an animal (5). Tylosin [*1401-69-0*], $C_{46}H_{77}NO_{17}$, and avoparcin [*37332-99-3*], $C_{89}H_{101}ClN_9O_{36}$, are examples of antibiotics used as feed additives in ruminant diets (18).

Propionic acid, $C_3H_6O_2$, and ammonia [*7664-41-7*], NH_3, are additives used to prevent molding of feed (5). Bentonite, hemicellulose extracts, and lignin sulfonate are used to hold feed pellets together.

Young Animal Feeds

When a ruminant is born, it consumes colostrum within a few hours. Colostrum contains antibodies from the mother's milk that serve to immunize the neonate against disease (37). These antibodies can be absorbed by the neonate only within the first few days of its life; there is no placental transfer of antibiotics in ruminants (37).

The rumen is not functional at birth and milk is shunted to the abomasum. One to two weeks after birth, the neonate consumes solid food if offered. A calf or lamb that is nursing tends to nibble the mother's feed. An alternative method of raising the neonate is to remove it from its mother at a very young age, < 1 week. A common example of an early weaning situation is the dairy calf that is removed from the cow soon after birth so that the cow's milk supply might be devoted entirely to production. In this instance, the neonate requires complete dietary supplementation with milk replacer. Sources of milk replacer protein have traditionally included milk protein but may also include soybean proteins, fish protein concentrates, field bean proteins, pea protein concentrates, and yeast protein (4). Information on the digestibility of some of these protein sources is available (4).

By approximately 8 weeks after birth, the ruminant has developed a fully functional rumen capable of extensive fermentation of feed nutrients (4). The rate of development of the ruminal environment depends on the amount of milk consumed by the neonate in relation to its growth requirements, the availability and consumption of readily digestible feedstuffs, and the physical form of the feedstuffs (4). The rumen develops much faster with hay than with milk (36). Concentrates, ie, high cereal grain diets, increase the absorptive surface of the rumen but ruminal size and musculature develops much more slowly with a concentrate diet than with a forage diet (4).

Several sources of energy are available for the neonate. Lipids are highly digestible, ca 90%, by neonatal calves and continue to increase in digestibility as the calf matures (4). Lipid sources include milk fat, tallow, and corn oil (4). Carbohydrates are another source of energy for the young ruminant. Lactose [*63-42-3*], glucose [*50-99-7*], and galactose [*26566-61-0*] are efficiently digested whereas starch, maltose [*69-79-4*], sucrose [*57-50-1*], and fructose [*57-48-7*] are poorly utilized by the young ruminant (4). Hydrolyzed starch has been used successfully to replace a portion of the energy in diets fed to young ruminants (4). Protein sources given to calves just starting to consume solid food should contain protein from natural plant sources or from natural plant sources with milk by-products (37).

Little research is available regarding the amounts of vitamins and minerals needed by young ruminants. However, it is common to supply calcium, phosphorus, trace-mineralized salt, and vitamins A, D, and E (4). In the absence of a functional rumen, B-vitamins and vitamin K should be supplemented.

Creep feeding often is used in the production of beef cattle. Nursing calves are given access to feedstuffs physically separated from their mothers. One means of creep feeding is to allow the calves access to concentrate feeds, eg, corn, oats, or barley. Several commonly used concentrate-based creep feeds are reported (5). Another means of creep feeding is to allow calves access to an ungrazed forage stand. This form of creep feeding allows calves to consume a much higher quality forage than otherwise available (38).

BIBLIOGRAPHY

"Feeds, Animal," in *ECT* 1st ed., Vol. 6, pp. 299–312, by H. M. Briggs, Oklahoma Agricultural and Mechanical College; in *ECT* 2nd ed., Vol. 8, pp. 857–870, by H. M. Briggs, South Dakota State University; "Pet and Livestock Feeds," in *ECT* 3rd ed., Vol. 17, pp. 90–109, J. Corbin, University of Illinois.

1. D. C. Church, in D. C. Church, ed., *The Ruminant Animal*, Prentice Hall, Inc., Englewood Cliffs, N. J., 1988, p. 1.
2. M. E. Heath and C. J. Kaiser, in M. E. Heath, R. F. Barnes, and D. S. Metcalfe, eds., *Forages: The Science of Grassland Agriculture*, 4th ed., The Iowa State University Press, Ames, 1985, p. 3.
3. P. J. Van Soest, *Nutritional Ecology of the Ruminant: Ruminant Metabolism, Nutritional Strategies, the Cellulolytic Fermentation and the Chemistry of Forages and Plant Fibers*, Cornell University Press, Ithaca, N.Y., 1987.
4. S. J. Lyford, in Ref. 1, p. 44.
5. P. R. Cheeke, *Applied Animal Nutrition: Feeds and Feeding*, MacMillan Publishing Co., New York, 1991.
6. G. C. Fehey, Jr. and L. L. Berger, in Ref. 1, p. 269.
7. W. E. Larsen and A. R. Rider, in Ref. 2, p. 452.
8. D. K. Barnes and C. C. Sheaffer, in Ref. 2, p. 89.

9. A. C. Cullison, *Feeds and Feeding*, 2nd ed., Reston Publishing Co., Reston, Va., 1979.
10. D. C. Honeyfield, J. A. Froseth, and J. McGinnis, *Nutr. Rep. Int.* **28**, 1253 (1983).
11. J. P. Erickson and co-workers, *J. Anim. Sci.* **48**, 547 (1979).
12. B. C. Radcliffe, C. J. Driscoll, and A. R. Egan, *Aust. J. Exp. Agric. Anim. Husb.* **21**, 71 (1981).
13. M. Rundgren, *Anim. Feed Sci. Tech.* **19**, 359 (1988).
14. A. Shimada, T. R. Cline, and J. C. Rogler, *J. Anim. Sci.* **38**, 935 (1974).
15. H. G. Walker and G. O. Kohler, *Agric. Environ.* **6**, 229 (1981).
16. F. M. Byers and G. T. Schelling, in Ref. 1, p. 298.
17. F. N. Owens and R. Zinn, in Ref. 1, p. 227.
18. M. T. Yokoyama and K. A. Johnson, in Ref. 1, p. 125.
19. D. C. Church, in D. C. Church, ed., *Livestock Feeds and Feeding*, O & B Books, Corvallis, Oreg., 1977, p. 97.
20. J. E. Nocek and J. B. Russell, *J. Dairy Sci.* **71**, 2070 (1988).
21. R. Kincaid, in Ref. 1, p. 326.
22. J. K. Miller, N. Ramsey, and F. C. Madsen, in Ref. 1, p. 342.
23. J. Huber, in Ref. 1, p. 313.
24. R. D. Goodrich and co-workers, *J. Anim. Sci.* **58**, 1484 (1984).
25. F. A. Mumpton and P. H. Fishman, *J. Anim. Sci.* **45**, 1188 (1977).
26. W. G. Pond, *J. Anim. Sci.* **59**, 1320 (1984).
27. M. E. Ensminger and C. G. Olentine, Jr., *Feeds & Nutrition*, Ensminger Publishing Co., Clovis, Calif., 1978.
28. R. Fuller, *J. Appl. Bacteriol.* **66**, 365 (1989).
29. D. S. Pollmann, D. M. Danielson, and E. R. Peo, Jr., *J. Anim. Sci.* **51**, 577 (1980).
30. R. A. Britton, in M. L. Pinkston, ed., *Proceedings of the 39th Annual Pfizer Research Conference*, New York, 1991, p. 124.
31. H. W. Essig, in Ref. 1, p. 468.
32. R. A. Erdman, *J. Dairy Sci.* **71**, 3246 (1988).
33. J. I. Andries and co-workers, *Anim. Feed Sci. Tech.* **18**, 169 (1987).
34. F. N. Owens and A. L. Goetsch, in Ref. 1, p. 145.
35. W. J. Visek, *J. Anim. Sci.* **46**, 1447 (1978).
36. R. G. Warner, W. P. Flatt, and J. K. Loosli, *J. Agr. Food Chem.* **4**, 788 (1956).
37. A. J. Kutches, in Ref. 1, p. 191.
38. L. L. Wilson and V. H. Watson, in Ref. 2, p. 560.

GREGORY D. SUNVOLD
GEORGE C. FAHEY, JR.
University of Illinois, Urbana

FEEDSTOCKS

COAL CHEMICALS

Coal is used in industry both as a fuel and in much lower volume as a source of chemicals. In this respect it is like petroleum and natural gas whose consumption also is heavily dominated by fuel use. Coal was once the principal feedstock for chemical production, but in the 1950s it became more economical to obtain most industrial chemicals from petroleum and gas. Nevertheless, certain chemicals continue to be obtained from coal by traditional routes, and an interest in coal-based chemicals has been maintained in academic and industrial research laboratories. Much of the recent activity in coal conversion has been focused on production of synthetic fuels, but significant progress also has been made on use of coal as a chemical feedstock (see COAL CONVERSION PROCESSES).

The term feedstock in this article refers not only to coal, but also to products and coproducts of coal conversion processes used to meet the raw material needs of the chemical industry. This definition distinguishes between use of coal-derived products for fuels and for chemicals, but this distinction is somewhat arbitrary because the products involved in fuel and chemical applications are often identical or related by simple transformations. For example, methanol has been widely promoted and used as a component of motor fuel, but it is also used heavily in the chemical industry. Frequently, some or all of the chemical products of a coal conversion process are not isolated but used as process fuel. This practice is common in the many coke plants that are now burning coal tar and naphtha in the ovens.

Because of the overlapping roles of coal in industry, many of the technologies covered here have been developed for synthetic fuel applications, but they also have been used or have demonstrated potential for production of significant quantities of chemicals. The scope of an article on coal as a chemical source would not be complete without coverage of synfuel processes, but the focus will be on the chemical production potential of the processes, looking toward a future when coal again may become the principal feedstock for chemical production.

Coal Chemical Origins

The discovery that useful chemicals could be made from coal tar provided the foundation upon which the modern chemical industry is built. Industrial chemistry expanded rapidly in the late nineteenth century in German laboratories and factories where coal-tar chemicals were refined and used in synthesis of dyes and pharmaceuticals. But coal-tar production has an earlier origin, dating back to the discovery by William Murdock in 1792 that heating coal in the absence of air generated a gas suitable for lighting. Murdock commercialized this technology, and by 1812 the streets of London were illuminated with coal gas (1).

Coal tar is produced as a coproduct during coal gasification. Up until the middle of the nineteenth century coal tar was regarded as a waste product, although John Kidd had isolated naphthalene from coal tar as early as 1819. The German chemist August Wilhelm von Hofmann earned a Ph.D. in 1841 under Liebig with a dissertation on the properties of aniline from coal tar. Hofmann later taught in England at the Royal College of Chemistry in London, where in 1855 he employed seventeen-year-old William Perkin as his assistant. Hofmann's suggestion that the drug quinine might be made from coal-tar chemicals prompted Perkin to undertake experiments to attempt this synthesis. In 1856 Perkin mixed aniline and potassium dichromate, but instead of quinine he isolated a purple compound which proved to be an excellent dye for silk and later became known as the color mauve (1). Perkin obtained a patent on the process at the age of eighteen and left school to start a successful dye business. Hofmann returned to Germany in 1864, and with Baeyer, who synthesized indigo, and other chemists advanced Germany to world leadership in synthetic organic chemistry and established a strong chemical industry (see DYES AND DYE INTERMEDIATES).

Coal feedstocks dominated the chemical industry up until the 1950s. However, the volume of chemicals available from this source was not sufficient to meet wartime demands or keep up with a rapidly expanding chemical industry after the war. The emergence of the petrochemical industry provided the required quantities, and petroleum rapidly became a more economical feedstock for chemical production (see FEEDSTOCKS, PETROCHEMICAL). In Germany, petroleum did not surpass coal as a chemical feedstock until after 1961 (2). Even in the 1990s some aromatic and heterocyclic chemicals are most easily obtained from coal tar. Since the oil supply disruption problems of the 1970s, much work has been done to develop conversion processes that can transform coal economically to liquid and gaseous fuels and chemical products. Methods to accomplish this goal include gasification, direct coal liquefaction, and indirect liquefaction.

Although significant progress has been made in reducing costs of coal conversion processes, the economics of a specific project depend heavily on process complexity, the cost of competing raw materials, and project or site-specific criteria such as desired end product and proximity to raw material supplies. The relatively stable prices of petroleum and natural gas since the 1970s have decreased the urgency of transition to coal-based fuels and chemicals, but competing coal conversion processes stand ready, and will again become favorable if oil and gas prices climb out of proportion to coal feedstocks. Even now, under certain conditions, the advent of modern coal conversion technology has allowed coal to become a significant raw material for chemical manufacture.

Coal Carbonization

The thermal degradation of coal in the absence of air is known as pyrolysis or carbonization. This reaction also is referred to as coking because of its large volume use to prepare coke for blast furnaces. The type and quantity of products produced by coal pyrolysis depend on type of coal, rate of heating, and final temperature. In general, rapid heating affords higher liquid yields than slow pyrolysis. Low temperature carbonization or mild gasification is performed at

450–700°C and is used in some areas to prepare smokeless fuels. High temperature carbonization is performed in the range of 900–1100°C and is generally the process used to prepare blast furnace coke. Low temperature carbonization gives fewer gaseous and more liquid products than carbonization at higher temperature (1).

Most coal-tar chemicals are recovered from coproduct coke ovens. Since the primary product of the ovens is metallurgical coke, production of coal chemicals from this source is highly dependent on the level of activity in the steel industry. In past years most large coke producers operated their own coproduct recovery processes. Because of the decline in the domestic steel industry, the recent trend is for independent refiners to collect crude coal tars and light oils from several producers and then separate the marketable products.

When coal is coked at a temperature of approximately 1000°C, about 70–75% of the product is coke. Nearly 20% of the product is a light gas, mostly methane and hydrogen, that typically is used as fuel to heat the ovens. Coal tars amount to about 4% of the product and light oil or naphtha is about 1%. Ammonia is recovered in an amount equal to about 0.3% of the feed coal. The ammonia is usually converted to ammonium sulfate and sold as a fertilizer. Little or no ammonia [*7664-41-7*] is produced in low temperature carbonization (3).

Many valuable chemicals can be recovered from the volatile fractions produced in coke ovens. For many years coal tar was the primary source for chemicals such as naphthalene [*91-20-3*], anthracene [*120-12-7*], and other aromatic and heterocyclic hydrocarbons. The routes to production of important coal-tar derivatives are shown in Figure 1. Much of the production of these chemicals, especially tar bases such as the pyridines and picolines, is based on synthesis from petroleum feedstocks. Nevertheless, a number of important materials continue to be derived from coal tar.

Benzene [*71-43-2*], toluene [*108-88-3*], xylene [*1330-20-7*], and solvent naphtha are separated from the light oil. Benzene (qv), toluene (qv), and xylene are useful as solvents and chemical intermediates (see XYLENES AND ETHYLBENZENE). The crude light oil is approximately 60–70% benzene, 12–16% toluene, 4–8% xylenes, 9–16% other hydrocarbons, and about 1% sulfur compounds (5) (see BTX PROCESSING).

Naphthalene, anthracene, carbazole [*86-74-8*], phenol [*108-95-2*], and cresylic acids are found in the tar. Phenol and cresylic acids are useful as chemical and resin intermediates. The aromatic chemicals are useful in the manufacture of pharmaceuticals, dyes, fragrances, and pesticides. Various grades of pitch are made from residues of tar refining. Coal-tar pitch is used for roofing and road tar, and as a binder mixed with petroleum coke to produce anodes for the aluminum industry.

Approximately 50–55% of the product from a coal-tar refinery is pitch and another 30% is creosote. The remaining 15–20% is the chemical oil, about half of which is naphthalene. Creosote is used as a feedstock for production of carbon black and as a wood preservative. Because of modifications to modern coking processes, tar acids such as phenol and cresylic acids are contained in coal tar in lower quantity than in the past. To achieve economies of scale, these tar acids are removed from crude coal tar with a caustic wash and sent to a central processing plant where materials from a number of refiners are combined for recovery.

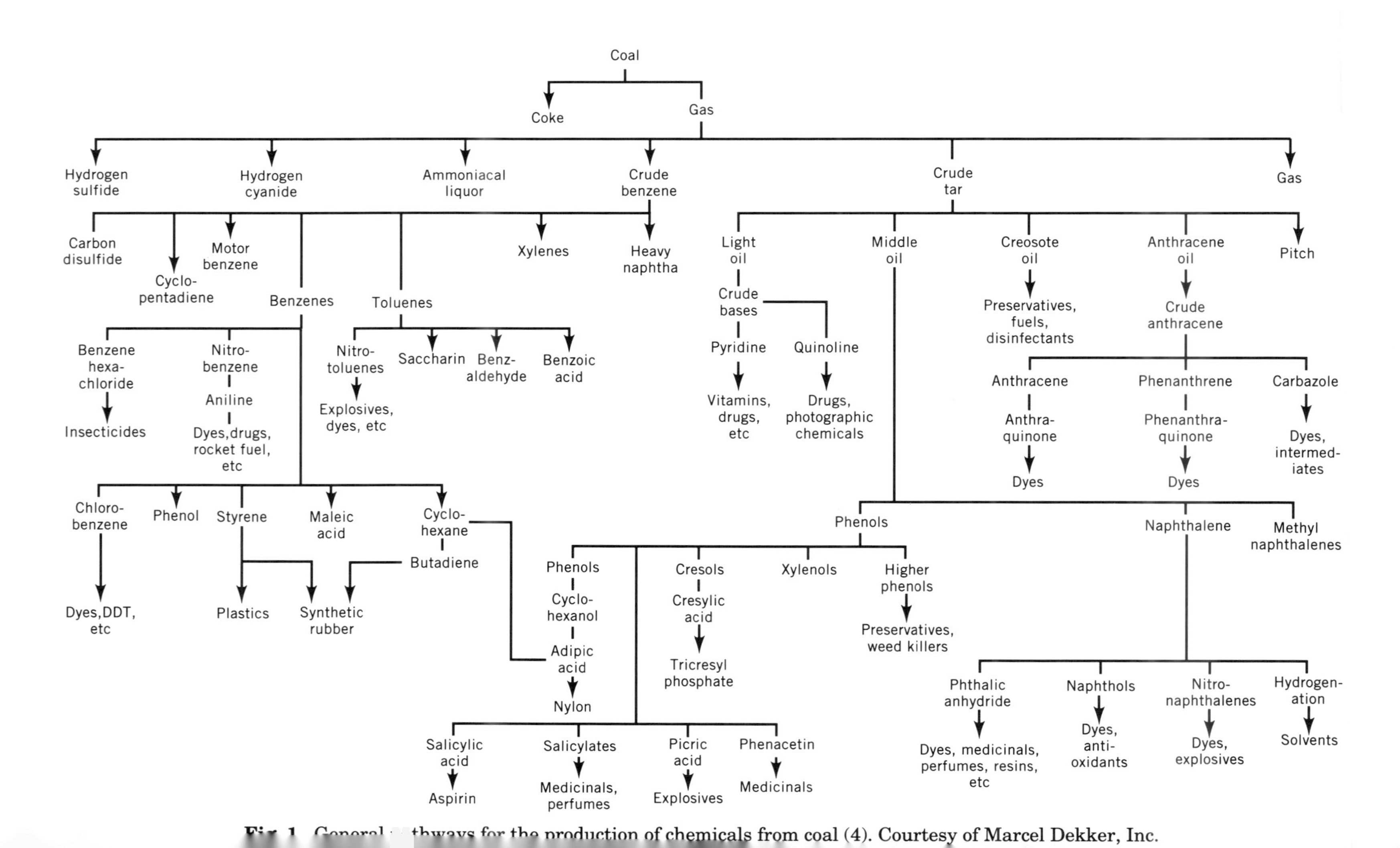

Fig. 1. General pathways for the production of chemicals from coal (4). Courtesy of Marcel Dekker, Inc.

In 1990, U.S. coke plants consumed 3.61×10^7 t of coal, or 4.4% of the total U.S. consumption of 8.12×10^8 t (6). Worldwide, roughly 400 coke oven batteries were in operation in 1988, consuming about 4.5×10^7 t of coal and producing 3.5 $\times 10^7$ t metallurgical coke. Coke production is in a period of decline because of reduced demand for steel and increasing use of technology for direct injection of coal into blast furnaces (7). The decline in coke production and trend away from recovery of coproducts is reflected in a 70–80% decline in volume of coal-tar chemicals since the 1970s.

In 1990, U.S. production of crude coal tar was 597,000 m^3 (700,000 t) and production of crude light oil was 255,000 m^3 (200,000 t). Crude coal tar and light oil were refined to produce 110,000 m^3 of crude naphthalene (freezing point 76–79°C), 8700 m^3 of crude tar acid oils (tar acid content 5–24%), and 297,000 m^3 of creosote oils. Coal-tar pitch production in 1990 amounted to 600,000 t (8).

In 1980, the last year for which a breakdown has been published, the amount of benzene derived from coal in the United States was 168,000 t or 2.5% of domestic benzene production. Coal-derived toluene was 0.8% of production, and xylenes from coal were only 0.1% of total chemical production (9). The amounts and proportions of BTX components derived from coal in the United States are expected to be nearly the same today as in 1980. Based on information submitted to the International Trade Commission, approximately 25 companies participated in the coal-tar industry in the United States in 1990.

World production of coal tar and light oil in 1987 was estimated to be 15–17 million metric tons and 5×10^6 t, respectively. Approximately 7.5×10^6 t of coal tar were processed by distillation in 1989, affording 950,000 t naphthalene, 20,000 t anthracene, and 10,000 t other two–four ring aromatics (7). Much of this coal-tar processing occurs in Eastern Europe, India, and Japan. In spite of Germany's past activity, only one company (Ruetgers) is known to be engaged in recovery of chemicals from coal tar. The principal producers of coal-tar chemicals in the United States include Allied-Signal, Aristech, Koppers Industries, and Cooper Creek Chemical.

As the economic value of coproducts has decreased, it has become more difficult to provide capital for environmental controls on air emissions and wastewater streams such as toxic phenolic effluents from chemical recovery operations. Some former coke and manufactured gas sites may require remediation to clean up contaminated soil and groundwater. These difficulties will force the shutdown of some operations and discourage recovery of coproducts in future installations.

One challenging problem is control of the approximately 300,000 t SO_2/yr that are emitted from the approximately 30 coke oven plants in the United States. Innovative technology to reduce these emissions is under development by Bethlehem Steel in conjunction with the U.S. Department of Energy's (DOE) Clean Coal Technology Program. The planned demonstration will clean 2.1×10^6 m^3/day (7.4×10^7 SCF/day) of coke oven gas using water produced in the coke oven batteries to absorb ammonia and hydrogen sulfide from the gas. The ammonia will be destroyed in a catalytic converter, and the hydrogen sulfide will be converted to sulfur in a Claus plant. Widespread adoption of this technology would substantially reduce the ammonium sulfate coproduct available from coke oven operations (10).

Coal Hydrogenation

Although small amounts of liquids are produced during coal pyrolysis, significant amounts of coal-derived liquids did not become available until after the discovery of coal hydrogenation. This discovery is attributed to Berthelot about 1869, but the first practical process was the Bergius hydrogen donor process, developed in 1913 in Germany. Work was continued at BASF. By the late 1920s a solvent extraction process also was developed (11). In 1927 the first pilot plant was erected at Leuna. At the height of production in Germany, 12 hydrogenation plants were producing 4×10^6 t/yr of fuel (12). Since 1945 only pilot-scale coal hydrogenation facilities have been built and operated. Perhaps the best known is the 200 t/day demonstration plant operated from 1981–1987 by Ruhrkohle AG at Bottrop. This plant, based on the Bergius-Pier process, had typical operating temperature of 450°C and pressure of 31 MPa (306 atm). An inexpensive iron catalyst was slurried with Ruhr coal and solvent in a single-stage process that yielded 66 wt % liquid products (13). Other direct liquefaction processes that reached an advanced stage of development in the 1970s and 1980s include the Exxon Donor Solvent (EDS), H-Coal, and Solvent Refined Coal (SRC I and SRC II) processes (14).

A new generation of coal liquefaction processes has been under study since the 1980s, differing from the previous versions in that a two-stage approach is employed. A two-stage process has the advantage of allowing optimum conditions to be established in separate reactors for the coal dissolution and upgrading reactions, resulting in higher distillate yield, lower gas yield, and higher quality products. All recent direct liquefaction work in the United States has been devoted to study of catalytic two-stage liquefaction. The Wilsonville, Alabama process development unit has been one of the most active facilities for study of coal liquefaction (13). In general, oils produced from direct coal liquefaction are lower boiling, have lower hydrogen content, and higher oxygen and nitrogen content than typical petroleum crudes. Also, they differ from petroleum in that they contain mostly condensed cyclic compounds, few paraffins, and no residuum. Virtually all of the work on coal liquid upgrading has concentrated on study of refining conditions for production of gasoline, diesel, jet fuel, and heating oil. Hydrotreated and hydrocracked naphthas from coal liquids are easy to reform and can be converted to benzene, toluene, and xylenes with yields as high as 70–80% (13). A typical aromatic product distribution from reforming of hydrotreated H-coal process-derived naphtha at 520°C over platinum catalyst is 13% benzene, 26% toluene, 20% xylene, and 15% C-9 aromatics for a total aromatics yield of 74% (5).

Synthesis Gas Generation

Selection of Feedstock. The variety of chemicals available by conversion of synthesis gas is illustrated in Figure 2. Synthesis gas can be produced by steam reforming of natural gas and naphtha, partial oxidation of heavy oil and petroleum coke, and coal gasification. In 1977 the relative ratio of synthesis gas produced by these methods was 87:10:3, respectively (16). Virtually any hydrocarbon source can be used to generate synthesis gas, and the choice of feedstock for a

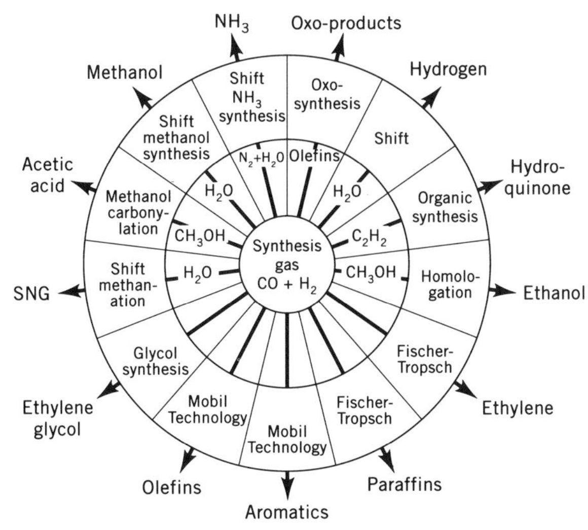

Fig. 2. The use of synthesis gas as a chemical feedstock (15). SNG is substitute natural gas.

given project can be made only after careful evaluation of all the alternatives. Most economic studies have shown that the cost of producing liquids from synthesis gas is dominated by the cost of producing the gas. This cost can be as high as 60–75% of the overall cost of the final product (16). Because of the high cost of producing synthesis gas from coal under economic conditions prevailing in the early 1990s, coal will be the favored feedstock for chemicals only under special circumstances.

Some of the factors that influence the choice of feedstock and potentially favor coal are feedstock cost, availability, and C/H ratio. The choice is most heavily influenced by the projected balance of raw material and capital costs. The size of the differential between coal and alternative feedstocks determines whether feedstock savings compensate for the increased capital required to install coal-based plants. Concern about the reliability of feedstock supply also enters into the decision, and in many areas of the United States this factor favors abundant, locally available coal. Transportation charges are a significant component of feedstock cost. For a given site, minimization of these charges requires procurement of a feedstock from the closest source. Clearly, this factor favors coal only for users located in coal producing areas. Another important consideration is the relative amount of hydrogen and carbon monoxide produced from different feedstocks. Project economics benefit by direct production of the required hydrogen to CO ratio. The ratio available by gasification of coal closely matches the ratio required for many desirable chemical products.

Gasification Chemistry. Gasification of coal involves many reactions. Some of the prominent reactions in a partial oxidation gasifier are shown in Table 1. For many gasifiers, particularly those operating at higher temperatures, these reactions approach equilibrium, and the product mixture is strongly dependent on the feed composition and gasifier temperature and pressure.

Table 1. Calculated Heats of Reaction at 298 K for Selected Gasification Reactions[a]

Reaction	Heat of reaction, kJ/mol[b]
$C + 1/2\ O_2 \rightarrow CO$	−110.5
$C + O_2 \rightarrow CO_2$	−393.5
$CO + H_2O \rightarrow CO_2 + H_2$	−41.2
$C + 2\ H_2 \rightarrow CH_4$	−74.8
$C + CO_2 \rightarrow 2\ CO$	172.5
$C + H_2O \rightarrow CO + H_2$	131.3

[a]Ref. 14.
[b]To convert kJ/mol to kcal/mol, divide by 4.184.

Coal Gasifier Designs. After many years of development three general gasifier designs have achieved commercial status. These types are fixed-bed, fluidized-bed, and entrained-bed gasifiers. Each design has advantages and disadvantages and may be more suitable than other designs for a given type of coal or process application (see COAL CONVERSION PROCESSES, GASIFICATION).

The fixed-bed gasifier is fed lump coal through the top of the gasifier. Oxidant and steam are introduced in the bottom of the gasifier and react with the coal generating a hot gas that passes up through the coal bed promoting further gasification reactions and coal pyrolysis. In traditional designs a dry ash removal system is provided at the bottom of the gasifier. A more recent version provides for removal of the ash as a molten slag. The most widely used fixed-bed design is the Lurgi dry ash gasifier.

As a consequence of the countercurrent feed mode, the average temperature gas in a fixed-bed reactor is lower than in alternative gasifier designs. The lower temperature requires longer residence times to achieve satisfactory conversion, but results in a high carbon inventory in the gasifier which provides some safety benefits. Also, the lower gas exit temperature reduces the need for expensive cooling and heat recovery equipment. The product from this type of gasifier has a high methane content and contains some hydrocarbon liquids and coal tars. These undesirable products may be recycled to the gasifier or reformed in a separate process to produce additional synthesis gas.

A fluidized-bed gasifier uses crushed or pulverized coal. The coal and oxidant feeds are injected into a fluidized bed of sand or ash particles maintained at a temperature below the melting point. A high flow of recirculating gas is used to keep the solids in the gasifier fluidized, which contributes to the very even temperature distribution through the reactor. The characteristics of the fluidized bed require that the particle size of the coal feed be tightly controlled, and that entrained char be recycled to obtain high coal conversion. The temperature achieved in fluidized-bed reactors allows most of the tars and liquids to be converted to gas. The High Temperature Winkler (HTW) process is one of the more advanced versions of a fluidized gasifier design.

In an entrained-bed reactor, the coal and oxidant are fed simultaneously at one end of the reactor and pass through the reaction zone together. A short residence time of less than a minute is typical for this type of reactor. Because of the short residence time the coal feed must be ground to a powder to ensure rapid reaction with the oxidant. Coal can be fed as a dry powder or slurried with water and sprayed in the gasifier. These reactors operate at very high temperature, which eliminates all tars and liquids from the product gas. Also, the coal ash is converted to a molten form that forms a glassy, granular slag when removed from the gasifier. Product gas cooling is accomplished either with a combination of radiant and convection coolers or by direct quench with water. In the quench mode, elaborate heat recovery equipment is required for efficient operation. Several entrained-bed processes including Texaco, Dow, Shell, Koppers-Totzek, and Prenflo have been commercialized or demonstrated on a commercial scale.

Synthesis Gas Chemicals

Fischer-Tropsch Process. The literature on the hydrogenation of carbon monoxide dates back to 1902 when the synthesis of methane from synthesis gas over a nickel catalyst was reported (17). In 1923, F. Fischer and H. Tropsch reported the formation of a mixture of organic compounds they called synthol by reaction of synthesis gas over alkalized iron turnings at 10–15 MPa (99–150 atm) and 400–450°C (18). This mixture contained mostly oxygenated compounds, but also contained a small amount of alkanes and alkenes. Further study of the reaction at 0.7 MPa (6.9 atm) revealed that low pressure favored olefinic and paraffinic hydrocarbons and minimized oxygenates, but at this pressure the reaction rate was very low. Because of their pioneering work on catalytic hydrocarbon synthesis, this class of reactions became known as the Fischer-Tropsch (FT) synthesis.

A systematic evaluation of catalysts and reaction conditions by Fischer and others through the 1920s and 1930s revealed that promoted cobalt and nickel catalysts also were active for higher hydrocarbon synthesis. This work led to construction in 1934 of a 1000 t/yr pilot plant by Ruhrchemie AG at Oberhausen-Holten. This pilot plant used a manganese oxide-promoted nickel catalyst. Subsequently, more active cobalt catalysts were developed, and by 1936 four commercial plants with a combined output of 200,000 t/yr were put in operation. Later, a catalyst with the composition Co–ThO_2–MgO–kieselguhr became the standard for nine commercial plants operating between 1938 and 1944 (19) (see COAL CONVERSION PROCESSES, LIQUEFACTION).

The FT process offers an effective alternative to direct liquefaction of coal for liquid hydrocarbon production and, along with methanol synthesis, represents a coal conversion strategy referred to as indirect coal liquefaction. Although the FT process was developed for fuel production, the value of aliphatic FT products as feedstocks for chemical production was recognized very early. By 1944 almost 40% of the primary FT product from German plants underwent further chemical processing. A process was developed for aromatization of the C-6–C-8 fraction over alkaline aluminum and chromium oxides to produce BTX compounds. The C-10–C-18 paraffin fraction was used for manufacture of lubricants by chlorination and condensation with aromatics, and the olefins were polymerized with aluminum chloride. Detergent feedstocks were produced by sulfochlorination of FT products (2). Although oxo alcohol chemistry was not practiced during the 1940s, the unbranched terminal olefins are particularly suitable feeds for this process.

The Fischer-Tropsch process can be considered as a one-carbon polymerization reaction of a monomer derived from CO. The polymerization affords a distribution of polymer molecular weights that follows the Anderson-Shulz-Flory model. The distribution is described by a linear relationship between the logarithm of product yield vs carbon number. The objective of much of the development work on the FT synthesis has been to circumvent the theoretical distribution so as to increase the yields of gasoline range hydrocarbons.

An enormous amount of research effort has been devoted to improving FT catalysts. Although many metals have been investigated, only iron and cobalt have sufficient activity and selectivity to be of interest. Iron is the preferred catalyst because of its low cost and high selectivity. Also, the water gas shift activity of iron catalysts allows the use of synthesis gas with a low H_2/CO ratio. Slurry-phase FT process and catalyst development work again is beginning to attract attention, and an active effort is underway within the U.S. Department of Energy's Indirect Liquefaction Program (20).

The FT process was used in Germany during World War II to supply motor fuels, and production reached a maximum of about 600,000 t/yr in early 1944 before Allied bombing destroyed much of the capacity. In the 1950s, developments included the Arge process, jointly developed by Ruhrchemie and Lurgi, and the Rheinpreussen-Koppers liquid-phase process. In 1953 a demonstration plant with a hydrocarbon production capacity of 11.5 t/day went into operation using the liquid-phase process. Both advanced processes employed improved iron catalysts that were more flexible and efficient than the previous cobalt catalysts (2). The Sasol facilities in South Africa are the only modern large-scale Fischer-Tropsch plants in operation.

Sasol produces synthetic fuels and chemicals from coal-derived synthesis gas. Two significant variations of this technology have been commercialized, and new process variations are continually under development. Sasol One used both the fixed-bed (Arge) process, operated at about 240°C, as well as a circulating fluidized-bed (Synthol) system operating at 340°C. Each FT reactor type has a characteristic product distribution that includes coproducts isolated for use in the chemical industry. Paraffin wax is one of the principal coproducts of the low temperature Arge process. Alcohols, ketones, and lower paraffins are among the valuable coproducts obtained from the Synthol process.

Recent advances in Fischer-Tropsch technology at Sasol include the demonstration of the slurry-bed Fischer-Tropsch process and the new generation Sasol Advanced Synthol (SAS) Reactor, which is a classical fluidized-bed reactor design. The slurry-bed reactor is considered a superior alternative to the Arge tubular fixed-bed reactor. Commercial implementation of a slurry-bed design requires development of efficient catalyst separation techniques. Sasol has developed proprietary technology that provides satisfactory separation of wax and solid catalyst, and a commercial-scale reactor is being commissioned in the first half of 1993.

The principal advantage of the SAS reactor is that it can be built for approximately half the cost of the previous Synthol reactor design. The SAS reactor concept was proven in commercial operation starting in 1987 when the first full-scale version was commissioned in Sasolburg at Sasol One. This reactor represented a 30-fold scale-up of the demonstration plant. The improved efficiency and operating characteristics of the commercial SAS reactor compared to the conventional circulating reactor design have been confirmed during operation of this plant. Other advantages of the new design are low pressure drop, increased flexibility in choosing flow rates, efficient operation at lower recycle ratio, and better temperature control. Finally, the new reactor is easier to operate and maintain than the conventional version.

Ammonia and Hydrogen Production. The earliest route for manufacture of ammonia from nitrogen was the cyanamide process commercialized in Italy in 1906. In this process calcium carbide manufactured from coal was treated with nitrogen at 1000°C to form calcium cyanamide, $CaCN_2$. The cyanamide was hydrolyzed with water affording ammonia and calcium carbonate. Production reached 140,000 t/yr in Germany in 1915, but this process was energy intensive and soon was displaced by the more efficient Bosch-Haber process. This process was developed by BASF and commercialized in 1913 and involves the high pressure reaction of nitrogen and hydrogen over an iron catalyst. Most of the world's hydrogen production is used in ammonia synthesis by the Bosch-Haber process. The hydrogen for ammonia synthesis generally is obtained from synthesis gas produced by steam reforming of natural gas or naphtha. Carbon monoxide in the synthesis gas is shifted to produce additional hydrogen, and the resulting carbon dioxide is removed. Several ammonia plants based on synthesis gas from coal are in operation.

Methanol Production. Like hydrogen production, steam reforming of natural gas is by far the largest source of synthesis gas for methanol [*67-56-1*]. But alternative feedstocks in use include heavy oil and coal. In 1927, the first methanol plant in the United States to use synthesis gas generated from coal was built by Du Pont in Belle, West Virginia. At least four other methanol plants are operating or have operated on coal in recent times. These facilities include the 165,000 t/yr plant at Eastman in Kingsport and a 250,000 t/yr plant at Leuna in Germany. The Leuna plant reportedly was closed in June 1990, but another 100,000 t/yr plant is operated by RWE-DEA at Wesseling. Finally, a methanol plant is operated in conjunction with ammonia production at AECI in South Africa near Johannesburg. Recently, plans were announced for a new 100,000 t/yr coal-based methanol plant that will be built in China and will use the Texaco gasification process.

Methanol (qv) is one of the 10 largest volume organic chemicals produced in the world, with over 18×10^6 t of production in 1990. The reactions for the synthesis of methanol from CO, CO_2, and H_2 are shown below. The water gas shift reaction also is important in methanol synthesis.

$$CO + 2\,H_2 \rightarrow CH_3OH \qquad -90.6\ \text{kJ/mol}\ (-21.7\ \text{kcal/mol})$$

$$CO_2 + 3\,H_2 \rightarrow CH_3OH + H_2O \qquad -49.5\ \text{kJ/mol}\ (-11.8\ \text{kcal/mol})$$

All commercial methanol plants use gas-phase heterogeneous catalytic reactors. Two main processes account for the majority of methanol produced; the Lurgi process, which uses a water-cooled, tubular, fixed-bed reactor, and the ICI process, which employs larger fixed-bed reactors with interstage cooling. Copper–zinc oxide alumina catalysts are now the standard for methanol plants, but they are subject to sulfur poisoning if sulfur species in the feed gas are not removed to less than 1 ppm. Until the advent of modern sulfur removal technology, this constraint prevented the use of copper catalysts with coal-derived synthesis gas.

Except in special situations, alternative feedstocks such as coal are not expected to be competitive with gas during the 1990s. As discussed previously, the principal reason for the higher cost of coal-based methanol is the large capital investment required, which has been estimated to be 2.8 to 3.1 times larger than the capital required for an equally sized gas-based plant. Construction of a gas-based plant in a remote location is expected to increase capital costs by about 60% (21). A related analysis by Lurgi suggests that, at somewhat higher gas prices, these infrastructure capital costs combined with transportation costs of $25–30/t raise the cost of remote methanol to a level potentially competitive with coal-based plants in developed plant locations (22).

The price differential at which coal becomes competitive with gas depends on plant size and the cost of capital, but based on estimates by the International Energy Agency (21) the required price ratio for gas to coal in North America falls into the range of 3.1 to 3.7 on an equivalent energy basis ($/MJ). Current prices give a gas/coal cost ratio nearer 1.5 to 2.0. As a result, all projected new methanol capacity is based on natural gas or heavy oil except for the proposed coal-based plant in China.

Methanol/Higher Alcohol Mixtures. Conversion of synthesis gas directly to higher alcohols or a mixture of methanol and higher alcohols is receiving increasing attention. These alcohols or derivatives are attractive for use as a constituent in reformulated gasoline, but have potential applications in the chemical industry. Processes for higher alcohol synthesis have been developed based on modified high pressure methanol synthesis with alkalized zinc oxide–chromia catalysts (Snamprogetti, Enichem, Haldor-Topsoe, and SEHT), a combination of Fischer-Tropsch and methanol-type chemistry using copper–cobalt oxide catalysts (Institut Francais du Petrol, IFP), modified low pressure methanol synthesis on alkalized copper oxide (Lurgi OCTAMIX), and the Dow/Union Carbide process over alkalized molybdenum sulfide catalysts. Also, an extensive investigation of cesium doped copper–zinc oxide catalysts has been performed at Lehigh University, and a number of rhodium-based catalysts have shown activity for higher alcohol synthesis, particularly ethanol (23).

The alkalized zinc oxide–chromia process developed by SEHT was tested on a commercial scale between 1982 and 1987 in a renovated high pressure methanol synthesis plant in Italy. This plant produced 15,000 t/yr of methanol containing approximately 30% higher alcohols. A demonstration plant for the IFP copper–cobalt oxide process was built in China with a capacity of 670 t/yr, but other higher alcohol synthesis processes have been tested only at bench or pilot-plant scale (23).

Isobutyl alcohol [*78-83-1*] forms a substantial fraction of the butanols produced by higher alcohol synthesis over modified copper–zinc oxide-based catalysts. Conceivably, separation of this alcohol and dehydration affords an alternative route to isobutylene [*115-11-7*] for methyl *t*-butyl ether [*1624-04-4*] (MTBE) production. MTBE is a rapidly growing constituent of reformulated gasoline, but its growth is likely to be limited by available supplies of isobutylene. Thus higher alcohol synthesis provides a process capable of supplying all of the raw materials required for manufacture of this key fuel oxygenate (24) (see ETHERS).

Liquid-Phase Methanol Process. Because of the highly exothermic nature of the methanol synthesis reaction, it would be advantageous to run the process in a liquid-phase slurry-bed reactor. Such a process would provide greater heat removal capability and higher production rates. One slurry-bed process, designated the Liquid-Phase Methanol Process (LPMEOH), has been under development for several years in a joint project sponsored by Air Products, Chem Systems, and the U.S. Department of Energy (DOE). Recent development work on the LPMEOH process has been directed to preparing the process for integration with gasification combined cycle power plants (25). Pilot-plant work on the process has been performed in a 4.5 t/day process development unit at LaPorte, Texas operated by Air Products under a DOE contract.

In this configuration the process employs an entrained slurry reactor in a once-through mode of operation. Synthesis gas from a source such as coal gasification is fed to the reactor without making adjustments to the hydrogen/carbon oxides ratio. The methanol reaction consumes a portion of the hydrogen and carbon monoxide in the feed producing fuel-grade methanol and a CO-rich fuel gas. About 32% of the heating value contained in the synthesis gas feed is contained in the methanol produced by this scheme.

The attractive feature of methanol synthesis for power plant operators is the ability to convert some of the energy produced in a coal gasification plant to a form that can be stored for use during peak generation hours. Alternatively, the methanol produced in the process could be sold to the chemical industry. However, it has been claimed that the LPMEOH process generates a high percentage of byproducts, making it difficult to refine the crude methanol to chemical grade (22).

Dimethyl Ether. Synthesis gas conversion to methanol is limited by equilibrium. One way to increase conversion of synthesis gas is to remove product methanol from the equilibrium as it is formed. Air Products and others have developed a process that accomplishes this objective by dehydration of methanol to dimethyl ether [*115-10-6*]. Testing by Air Products at the pilot facility in LaPorte has demonstrated a 40% improvement in conversion. The reaction is similar to the liquid-phase methanol process except that a solid acid dehydration catalyst is added to the copper-based methanol catalyst slurried in an inert hydrocarbon liquid (26).

By selection of appropriate operating conditions, the proportion of coproduced methanol and dimethyl ether can be varied over a wide range. The process is attractive as a method to enhance production of liquid fuel from CO-rich synthesis gas. Dimethyl ether potentially can be used as a starting material for oxygenated hydrocarbons such as methyl acetate and higher ethers suitable for use in reformulated gasoline. Also, dimethyl ether is an intermediate in the Mobil MTG process for production of gasoline from methanol.

Mobil MTG and MTO Process. Methanol from any source can be converted to gasoline range hydrocarbons using the Mobil MTG process. This process takes advantage of the shape selective activity of ZSM-5 zeolite catalyst to limit the size of hydrocarbons in the product. The pore size and cavity dimensions favor the production of C-5–C-10 hydrocarbons. The first step in the conversion is the acid-catalyzed dehydration of methanol to form dimethyl ether. The ether subsequently is converted to light olefins, then heavier olefins, paraffins, and aromatics. In practice the ether formation and hydrocarbon formation reactions may be performed in separate stages to facilitate heat removal.

In the early 1980s, the process was commercialized in New Zealand to convert offshore natural gas to 2200 m^3/day (14,000 barrels/day) gasoline. Since then some of the methanol has been diverted from fuel production to chemical-grade methanol production by adding additional methanol refining capacity.

The MTG process accomplishes the same end as the Fischer-Tropsch process, but each method has some advantages. The Fischer-Tropsch process benefits from a cheap, throwaway iron catalyst. The desirable features of the MTG process for high octane liquid fuel production include high selectivity, low methane yield, and high aromatics yield. However, environmental concerns about aromatics in gasoline may offset the octane benefits. It is interesting to note that Fischer-Tropsch technology was selected for the recently completed Mossgas project in South Africa, which also produces liquids from natural gas.

Oxo Synthesis. All of the synthesis gas reactions discussed to this point are heterogeneous catalytic reactions. The oxo process (qv) is an example of an industrially important class of reactions catalyzed by homogeneous metal complexes. In the oxo reaction, carbon monoxide and hydrogen add to an olefin to produce an aldehyde with one more carbon atom than the original olefin, eg, for propylene:

$$CH_3CH{=}CH_2 + CO + H_2 \rightarrow CH_3CH_2CH_2CHO + CH_3\underset{}{\overset{\overset{\displaystyle CHO}{|}}{C}}HCH_3$$

Often the aldehyde is hydrogenated to the corresponding alcohol. In general, addition of carbon monoxide to a substrate is referred to as carbonylation, but when the substrate is an olefin it is also known as hydroformylation. The early work on the oxo synthesis was done with cobalt hydrocarbonyl complexes, but in 1976 a low pressure rhodium-catalyzed process was commercialized that gave greater selectivity to linear aldehydes and fewer coproducts.

The only known application of coal-derived synthesis gas for oxo chemicals was the Synthesegasanlage Ruhr (SAR) plant in Germany. Synthesis gas was provided by a 750 t/day Texaco gasification plant. The design was based on the successful 150 t/day demonstration plant operated by Ruhrkohle and Ruhrchemie in Oberhausen-Holten between 1978 and 1985. The plant began production on coal in 1986, but because of the escalating costs of German coal, the plant was converted in 1989 to use heavy oil as feedstock. During over 10,000 hours of operation, the plant gasified 313,000 t of coal.

Acetic Acid and Anhydride. Synthesis of acetic acid by carbonylation of methanol is another important homogeneous catalytic reaction. The Monsanto acetic acid process developed in the late 1960s is the best known variant of the process.

$$CH_3OH + CO \rightarrow CH_3COOH \qquad -120.5\ \text{kJ/mol}\ (-28.8\ \text{kcal/mol})$$

This reaction is rapidly replacing the former ethylene-based acetaldehyde oxidation route to acetic acid. The Monsanto process employs rhodium and methyl iodide, but soluble cobalt and iridium catalysts also have been found to be effective in the presence of iodide promoters.

The Eastman acetic anhydride [*108-24-7*] process provides an extension of carbonylation chemistry to carboxylic acid esters. The process is based on technology developed independently in the 1970s by Eastman and Halcon SD. The Eastman acetic anhydride process involves carbonylation of methyl acetate [*79-20-9*] produced from coal-derived methanol and acetic acid [*64-19-7*].

$$CH_3COOCH_3 + CO \rightarrow CH_3COOCOCH_3 \qquad -50.6 \text{ kJ/mol } (12.1 \text{ kcal/mol})$$

In analogy to the acetic acid process, the acetic anhydride process is a homogeneous rhodium-catalyzed reaction that requires methyl iodide. Two key requirements for the commercial acetic anhydride process are the presence of a reducing agent and an iodide salt. But there are important differences that distinguish the anhydrous acetic anhydride system from the aqueous acetic acid process. A difference with significant process implications is the lower heat of reaction of the methyl acetate carbonylation reaction. The reduced thermodynamic driving force imposes equilibrium limitations on conversion in the acetic anhydride reactors (27).

Emerging Processes. There is extensive literature on production of chemicals from synthesis gas. In addition to the chemicals described in the foregoing, reports or patents have been published on ethylene glycol, vinyl acetate, aliphatic amines, acrylic acid, and homologation of carboxylic acids (28,29). The amidocarbonylation reaction is a new procedure for utilization of synthesis gas to introduce amido and carboxylate groups into olefins or aldehydes (30). The reaction is a potentially useful technique for industrial production of amino acids and derivatives. Also, biological conversion offers an alternative route for utilization of synthesis gas to make chemicals. Production of ethanol and acetic acid by bioconversion of synthesis gas are areas under active investigation (31).

Carbide Chemicals

The development of technology for production of calcium carbide and acetylene in 1892 provided a means for conversion of coke to chemical products. Further progress in Germany between 1920 and 1930 led to installation of electric arc processes for production of acetylene from coke oven gases. The commercialization of acetylenic chemicals is based on the work of the German chemist Julius Reppe. Both the electric arc process and calcium carbide process were used extensively in Germany during World War II for production of acetylene and derivatives vital to the war effort. Such industrial chemicals as acetaldehyde, acetic acid, ethylene, and butadiene have been made from coal using the acetylene route (32–34) (see ACETYLENE-DERIVED CHEMICALS).

Calcium carbide has been used in steel production to lower sulfur emissions when coke with high sulfur content is used. The principal use of carbide remains hydrolysis for acetylene (C_2H_2) production. Acetylene is widely used as a welding gas, and is also a versatile intermediate for the synthesis of many organic chemicals. Approximately 450,000 t of acetylene were used annually in the early 1960s for the production of such chemicals as acrylonitrile, acrylates, chlorinated solvents, chloroprene, vinyl acetate, and vinyl chloride. Since then, petroleum-derived olefins have replaced acetylene in these uses. The principal chemicals based on Reppe chemistry today are propargyl alcohol and butyn-1,4-diol and derivatives such as 2-pyrrolidinone, *N*-vinyl-2-pyrrolidinone, polyvinylpyrrolidinone, vinyl ethers, and tetrahydrofuran (33).

The first process for manufacture of calcium carbide [*75-20-7*] and acetylene [*74-86-2*] involved the reaction of coke and lime. The carbide process operates at a temperature of about 2000°C according to the following reaction:

$$CaO + 3\,C \rightarrow CaC_2 + CO \qquad -464\ kJ/mol\ (-111\ kcal/mol)$$

A typical large carbide furnace consumes 950 kg of lime, 550 kg of coke, and 15 kg of electrode carbon, and requires about 3200 kW·h of electrical energy per metric ton of carbide produced. This grade of carbide produces about 300 L C_2H_2/kg, and has an analysis of 80.5% CaC_2, 12.9% CaO, 1.3% Si, 1.1% Al, 0.2% Fe, 0.5% S, and 0.3% C (2).

Newer coal-based methods of acetylene manufacture under development include the AVCO process, based on the reaction of coal in a hydrogen plasma. Finely divided coal is passed through a hydrogen plasma arc generating temperature gradients of up to 15,000 K. About 67% of the coal is consumed, yielding char and acetylene in concentrations up to 16%. An energy requirement of 9.5 kW·h/kg acetylene has been reported (33).

Commercial Facilities Using Coal as a Chemical Feedstock

Eastman Coal Chemicals. In 1983 Eastman Chemical Co. became the first chemical producer in the United States to return to coal as a raw material for large-scale manufacture of industrial chemicals (35). In that year, Eastman started manufacturing acetic anhydride from coal. Acetic anhydride is a key intermediate for production of coatings, cellulosic plastics, and cellulose acetate fibers. Acetic anhydride from other sources also is used in the manufacture of pharmaceuticals, starches and sweeteners, and flavors and fragrances.

The Eastman Chemicals from Coal facility is a series of nine complex interrelated plants. These plants include air separation, slurry preparation, gasification, acid gas removal, sulfur recovery, CO/H_2 separation, methanol, methyl acetate, and acetic anhydride. A block flow diagram of the process is shown in Figure 3. The facility covers an area of 2.2×10^5 m^2 (55 acres) at Eastman's main plant site in Kingsport, Tennessee. The air separation plant is owned and operated by Air Products, and provides up to 965 t/day of oxygen for the gasification plant. The plant contains three complete cryogenic air separation trains to provide reliability for continuous operation.

The slurry preparation plant includes two rod mills to grind run-of-mine coal with sufficient water and additives to produce a fluid slurry with a particle size distribution and viscosity suitable for pumping to the gasifier. The plant produces a slurry that contains approximately 65–70% coal and 30–35% water. The slurry

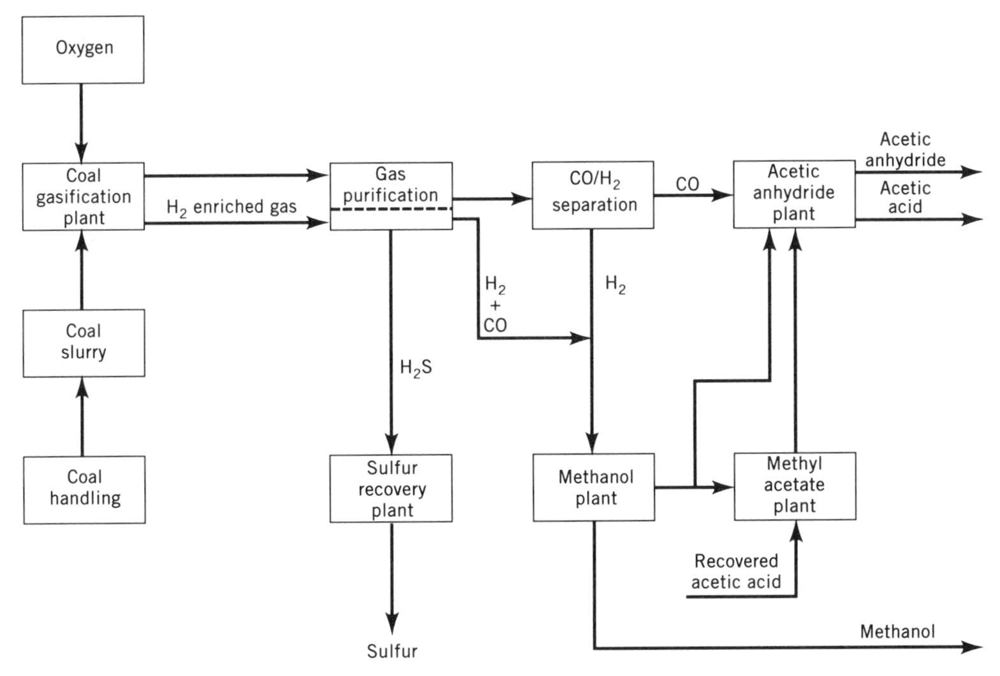

Fig. 3. Overall block flow diagram for Eastman's coal gasification–acetic anhydride complex (35).

is pumped from holding tanks to the gasifier with high pressure diaphragm pumps designed to handle slurries.

The gasification plant is equipped with two Texaco gasifiers, each capable of producing all of the synthesis gas required for operation of the complex. Eastman chose an entrained-bed gasification process for the Chemicals from Coal project because of three attractive features. The product gas composition using locally available coal is particularly suitable for production of the desired chemicals. Also, the process has excellent environmental performance and generates no liquids or tars. Finally, the process can be operated at the elevated pressure required for the downstream chemical plants.

The gasifiers convert up to 1000 t/day of coal at a temperature of approximately 1260–1370°C and pressure of 6.2–6.9 MPa (900–1000 psi). Normally, one gasifier is in operation while the other is on standby. The gas contains mainly CO, H_2, and CO_2. The sulfur in the coal is converted to H_2S and to a small amount of COS, which are removed and converted to elemental sulfur in a Claus plant. The ash becomes molten in the gasifier and is cooled and solidified in the quench section. The resulting granular slag is removed with a lock hopper and deposited in a landfill.

The crude product from the gasifier contains CO_2 and H_2S, which must be removed before the gas can be used to produce chemicals. The Rectisol process is used to remove these contaminants from the gas. This is accomplished by scrubbing the product with cold methanol which dissolves the CO_2 and H_2S and lets the H_2 and CO pass through the scrubber. The H_2S is sent to a Claus sulfur plant where over 99.7% of the sulfur in the coal feed is recovered in the form of elemental sulfur. A portion of the clean H_2 and CO are separated in a cryogenic distillation process. The main product from the cryogenic distillation is a purified CO stream for use in the acetic anhydride process. The remaining CO and hydrogen are used in the methanol plant.

The chemical complex includes the methanol plant, methyl acetate plant, and acetic anhydride plant. The methanol plant uses the Lurgi process for hydrogenation of CO over a copper-based catalyst. The plant is capable of producing 165,000 t/yr of methanol. The methyl acetate plant converts this methanol, purchased methanol, and recovered acetic acid from other Eastman processes into approximately 440,000 t/yr of methyl acetate.

The acetic anhydride process employs a homogeneous rhodium catalyst system for reaction of carbon monoxide with methyl acetate (36). The plant has capacity to coproduce approximately 545,000 t/yr of acetic anhydride, and 150,000 t/yr of acetic acid. One of the many challenges faced in operation of this plant is recovery of the expensive rhodium metal catalyst. Without a high recovery of the catalyst metal, the process would be uneconomical to operate.

Production of Eastman's entire acetic anhydride requirement from coal allows a reduction of 190,000 m^3/yr (1.2 million barrels/yr) in the amount of petroleum used for production of Eastman chemicals. Now virtually all of Eastman's acetyl products are made in part from coal-based feedstocks. Before the technology was introduced, these chemicals had been made from petroleum-based acetaldehyde. Reduced dependence on petroleum, much of which must be obtained from foreign sources, is important to maintain a strong domestic chemical industry.

Sasol Chemical Production. The best known and largest coal conversion facilities in the world are those of Sasol in South Africa. Sasol was formed in 1950 as the South African Coal, Oil, and Gas Corp., and began production of liquid transportation fuels from coal in 1955 at Sasolburg. Sasol's entry into the chemical industry began in earnest in 1964 when Sasol began production of styrene and butadiene for synthetic rubber, and ammonia for fertilizers. In 1965 a naphtha cracker for production of ethylene was put in operation. Beginning in the early 1980s, the facilities and product mix were significantly expanded, and at the present time Sasol is a principal producer of chemicals as well as fuel products.

Coal consumption at Sasol One is approximately 7×10^6 t/yr. Originally nine dry ash, oxygen blown Lurgi gasifiers with a diameter of approximately 3.7 m were installed at Sasol One. In 1981 an additional prototype Lurgi gasifier with a diameter of 4.7 m was installed (37). Gas production is approximately 10^7 m^3/day (3.5×10^8 SCF/day) from which liquid fuel production of 850 t/day is obtained. Also, about 1.7×10^6 m^3/day (6×10^7 SCF/d) of 19 MJ/m^3 fuel gas is produced (14).

In 1991, the relatively old and small synthetic fuel production facilities at Sasol One began a transformation to a higher value chemical production facility (38). This move came as a result of declining economics for synthetic fuel production from synthesis gas at this location. The new facilities installed in this conversion will expand production of high value Arge waxes and paraffins to 123,000 t/yr in 1993. Also, a new facility for production of 240,000 t/yr of ammonia will be added. The complex will continue to produce ethylene and process feedstock from other Sasol plants to produce alcohols and higher phenols.

Sasol Two and Sasol Three are essentially identical second generation FT plants located adjacent to one another in Secunda. Sasol Two started up in 1980 and Sasol Three in 1982. Each plant operates 40 Lurgi gasifiers and eight Synthol reactors. The combined coal requirement for gasification and steam production is over 40×10^6 t/yr, and output is approximately 1.5×10^4 m^3/day (100,000 barrels/day) of motor fuels. An approximate distribution of products from one of the second generation Sasol plants is shown in Table 2, and a block flow diagram for chemical and fuel production at Sasol Two is shown in Figure 4.

Secunda discharges no process water effluents. All water streams produced are cleaned and reused in the plant. The methane and light hydrocarbons in the product are reformed with steam to generate synthesis gas for recycle (14). Even at this large scale, the cost of producing fuels and chemicals by the Fischer-Tropsch process is dominated by the cost of synthesis gas production. Sasol has estimated that gas production accounts for 58% of total production costs (39).

Ammonia from coal gasification has been used for fertilizer production at Sasol since the beginning of operations in 1955. In 1964 a dedicated coal-based ammonia synthesis plant was brought on stream. This plant has now been deactivated, and is being replaced with a new facility with three times the production capacity. Nitric acid is produced by oxidation and is converted with additional ammonia into ammonium nitrate fertilizers. The products are marketed either as a liquid or in a solid form known as Limestone Ammonium Nitrate. Also, two types of explosives are produced from ammonium nitrate. The first is a mixture of fuel oil and porous ammonium nitrate granules. The second type is produced by emulsifying small droplets of ammonium nitrate solution in oil.

The tar and pitch coproducts from gasification of coal at Sasol are upgraded to a number of higher value products such as wood preservatives and pitch for cellulose fiber pipes. A plant to upgrade pitch to produce electrode coke for the aluminum and iron industries is being commissioned in 1993. The crude tar acid processing facilities have been upgraded to produce higher quality phenol and to add *o*-cresol as well as mixed cresylics to the product spectrum.

Table 2. Approximate Product Distribution from Sasol Two[a]

Product	Output, t/yr
motor fuels	1,500,000
ethylene	185,500
chemicals	85,500
tar products	185,500
ammonia (as N)	100,000
sulfur	90,000
Total	*2,146,500*

[a]Ref. 23.

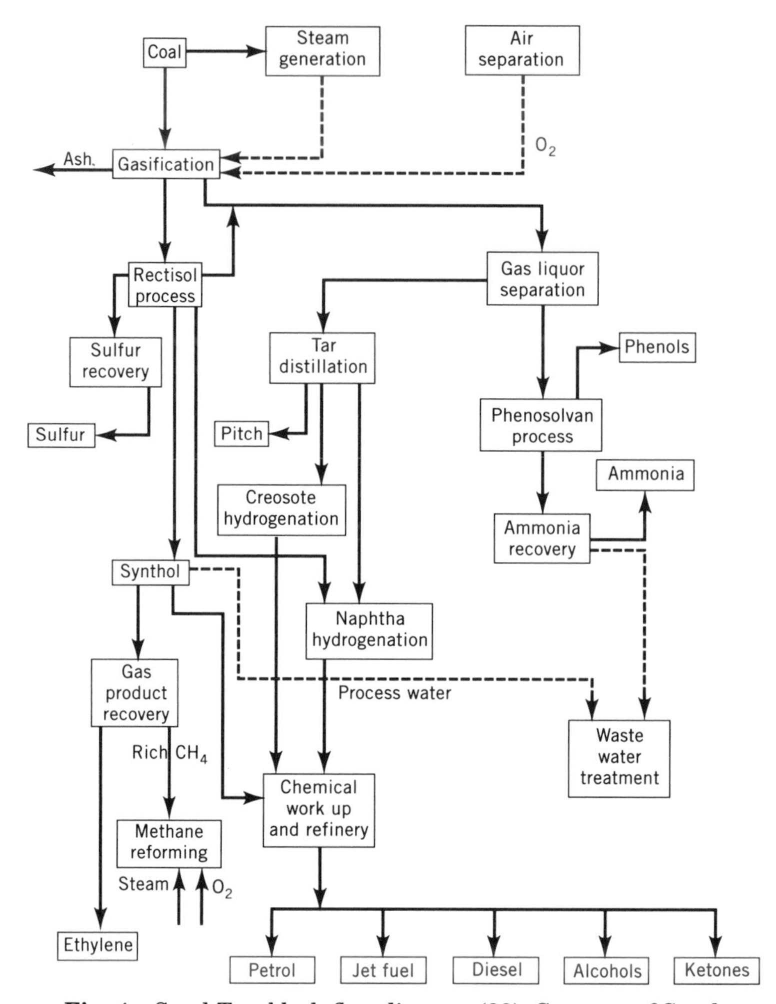

Fig. 4. Sasol Two block flow diagram (38). Courtesy of Sasol.

A number of chemical products are derived from Sasol's synthetic fuel operations based on the Fischer-Tropsch synthesis including paraffin waxes from the Arge process and several polar and nonpolar hydrocarbon mixtures from the Synthol process. Products suitable for use as hot melt adhesives, PVC lubricants, corrugated cardboard coating emulsions, and polishes have been developed from Arge waxes. Wax blends containing medium and hard wax fractions are useful for making candles, and over 20,000 t/yr of wax are sold for this application.

Synthol coproducts include alcohols, ketones, and lower paraffins. They are used mainly as solvents in the paint and printing industries, although some alcohols are blended into fuels. In 1992 Sasol began producing 17,500 t/yr 1-butanol [*71-36-3*] from acetaldehyde [*75-07-0*] and planned to start a plant to produce high purity ethanol [*64-17-5*] in 1993. Acetone [*67-64-1*] and methyl ethyl ketone [*78-93-3*] are two ketone coproducts sold as solvents.

The latest of three ethylene recovery plants was started in 1991. Sasol sold almost 300,000 t of ethylene in 1992. Sasol also produces polypropylene at Secunda from propylene produced at Sasol Two. In 1992 Sasol started construction of a linear alpha olefin plant at Secunda to be completed in 1994 (40). Initial production is expected to be 100,000 t/yr of pentene and hexene. Sasol also has a project under construction to extract and purify krypton and xenon from the air separation plants at Sasol Two. Other potential new products under consideration at Sasol are acrylonitrile, acetic acid, acetates, and alkylamines.

Chemical Production at Great Plains. The largest coal conversion facility in the United States is the Dakota Gasification Co. Great Plains plant at Beulah, North Dakota. This facility was constructed in the early 1980s with financing from the Federal Government's now defunct Synthetic Fuels Corp. The facility started operation in 1984, and currently gasifies 15,500 t/day North Dakota lignite in 14 Lurgi dry ash coal gasifiers, producing 4.5×10^6 m^3/day (1.6×10^8 SCF/day) of synthetic natural gas. The plant also produces 64 t/day of ammonia and 45 t/day of sulfur (41). Up until recently all coal liquid coproducts from gasification were burned to generate process steam. Now the company has begun isolating and marketing some coproducts. These products include phenol, cresylic acids, hydrotreated naphtha, anhydrous ammonia, and sulfur. Dakota Gasification's chemical sales exceeded $5 million in 1991. In 1992 Dakota Gasification entered into a contract to supply Merichem with crude cresylic acids. The volume of cresylic acids available from Great Plains coproduct streams has been estimated to correspond to 70% of U.S. and 15% of world demand (42). Potential coproduct yields from Great Plains are shown in Table 3.

Coal-Based Ammonia Production. Synthesis gas for the commercial production of ammonia has been made from coal in many locations around the world. The atmospheric pressure Koppers-Totzek coal gasification process was used almost exclusively for synthesis gas generation up until development of pressurized gasification processes, which are more economical because of the savings in synthesis gas compression costs. Between 1951 and 1981 contracts were obtained for

Table 3. Coproduct Yields From the Great Plains Plant[a]

Product	Yield, t/day
argon	104
krypton/xenon	0.04
carbon dioxide	11,800
phenol	60
o-cresol	15
m,*p*-cresol	34
mixed xylenols	29
creosote	320
aromatic naphtha	120
benzene	43
toluene	21
mixed xylenes	9

[a]Ref. 41.

approximately 20 Koppers-Totzek based ammonia plants. The largest of these plants is the African Explosives and Chemicals Industry (AECI) plant at Modderfontein in South Africa which started production in 1974. This plant operates six gasifiers with the capacity to produce 2.15×10^6 m^3/day (7.6×10^7 SCF/day) of synthesis gas, 1000 t/day of ammonia, and 100 t/day of methanol. In 1973, Krupp Koppers began development of a pressurized, entrained-flow process (Prenflo) based on Koppers-Totzek gasification.

In 1984, the Ube Ammonia Industry Co. began operating the largest Texaco coal gasification complex to date. This facility is located in Ube City, Japan, and has a rated gasification capacity of 1500 t/day of coal, and production capacity of 1000 t/day of ammonia. The plant has successfully gasified coals from Canada, Australia, South Africa, and China. At the present time the plant uses a mixture of petroleum coke and coal (43).

Ube has a long history of ammonia production from coal, and began operation in 1934 using the Koppers low temperature carbonization process to produce ammonium sulfate for fertilizers from locally mined subbituminous coal. Over the years, Ube's feedstock flexibility has been increased by adding oil gasification and naphtha steam reforming capacity. The oil supply disruptions of the 1970s and 1980s encouraged Ube to return to coal as a principal feedstock for ammonia production.

The Ube plant consists of four complete trains of Texaco quench-type gasifiers. During normal operation, three gasifiers are on line and one is on standby. Each gasifier consumes 500 t/day of coal to generate syngas for 350 t/day of ammonia. Up to the middle of 1990 the Ube plant gasified 2.2 million t of coal and petroleum coke.

Planned Coal Chemical Projects

Texaco Gasification Projects in China. Because of large domestic coal reserves, China is becoming very active in commercialization of coal conversion projects. Two coal gasification projects using the Texaco process are currently under construction, and two additional Texaco gasification projects were announced in early 1992. The Lunan Fertilizer plant in Shandong province is expected to start operation in 1993, and the Shougang plant in Beijing will begin producing fuel gas for industrial use in 1994. The Lunan plant will produce ammonia from 360 t/day of coal. The Shougang plant will have the capacity to gasify 1000 t/day of coal. Newly announced projects are the Shanghai Coking and Chemical Plant town gas production facility and the Weihe Chemical Fertilizer Plant in Shaanxi province.

The Shanghai plant will gasify 1100 t/day of coal to produce 1.7 m^3/day (60 ft^3/day) of town gas, and 200,000 t/yr of methanol. Plans call for the first phase of the new gasification facility to begin operation in 1994. Increased town gas production from this plant will eliminate the need for residents to burn coal for home heating, and reduce air pollution from this source. Also, coal slurry for the gasifier will be prepared using wastewater from existing processes and remove these discharges from the Huangpu River.

When completed in 1996, the Weihe plant will gasify 1500 t/day of coal to produce 300,000 t/yr of ammonia, which will be used to manufacture 520,000 t/yr of urea fertilizer. This project is the eighth Texaco oil or coal gasification plant licensed by Chinese industry.

TVA Urea Project. In Round 4 of DOE's Clean Coal Technology program, the Tennessee Valley Authority (TVA) proposed construction of a 250 MW IGCC power plant that would coproduce urea and electric power from 730,000 metric tons of coal a year. The project was not selected in Round 4, but is being submitted again for consideration in Round 5, which is expected to give chemical projects favorable treatment. Recently, TVA announced that the Shell coal gasification process has been selected for this plant. TVA has experience in coal gasification, having successfully operated a Texaco gasifier based demonstration plant for ammonia synthesis from coal in the late 1970s and early 1980s.

BIBLIOGRAPHY

"Feedstocks" in *ECT* 3rd ed., Vol. 9, pp. 831–845, by F. A. M. Buck, King, Buck, & Associates, and M. G. Marbach, Shell Chemical Co.; "Coal Chemical and Feedstocks," in *ECT* Suppl. Vol., pp. 191–194, by G. Kölling, Bergbau-Forschung GmbH.

1. R. N. Shreve, *Chemical Process Industries*, 3rd ed., McGraw-Hill Book Co., Inc., New York, 1967, Chapt. 5.
2. J. Falbe, ed., *Chemical Feedstocks from Coal*, John Wiley & Sons, Inc., New York, 1982.
3. P. J. Wilson and J. H. Wells, *Coal, Coke, and Coal Chemicals*, McGraw-Hill Book Co., Inc., New York, 1950, p. 289.
4. R. K. Hessley and R. H. Schlosberg, in J. G. Speight, ed., *Fuel Science and Technology Handbook*, Marcel Dekker, Inc., New York, 1990, p. 768.
5. G. O. Davies, *Coal Chem 2000*, Institute of Chemical Engineers Symposium Series No. 62, Rugby, Warks, UK, 1980, pp. D6–D10.
6. *U.S. Industrial Outlook 1992*, U.S. Department of Commerce, International Trade Administration, Washington, D.C., 1992.
7. C. Song and H. H. Schobert, *Proceedings of the 203rd American Chemical Society National Meeting, San Francisco, Apr. 5–10, 1992,* Vol. 37, Division of Fuel Chemistry Preprints, American Chemical Society, Washington, D.C., pp. 524–532.
8. *Synthetic Organic Chemicals, U.S. Production and Sales*, publication 2470 U.S. International Trade Commission, Washington, D.C., 1990.
9. Ref. 8, 1980.
10. *U.S. Department of Energy Clean Coal Technology Program Update 1991,* publication DOE/FE-0247P, U.S. Dept. of Energy, Washington, D.C.
11. M. G. Thomas, in B. R. Cooper and W. A. Ellingson, eds., *The Science and Technology of Coal and Coal Utilization*, Plenum Press, New York, 1984, p. 231.
12. J. Langhoff and co-workers, in Ref. 4, p. P2.
13. H. D. Schindler, ed., *Coal Liquefaction—A Research Needs Assessment Technical Background*, publication DOE/ER-0400, Vol. 2, U.S. Dept. of Energy, Washington, D.C., 1989, Chapts. 4 and 9.
14. R. F. Probstein and R. E. Hicks, *Synthetic Fuels*, pH Press, Cambridge, Mass., 1990.
15. *Surface Coal Gasification Program Fiscal Year 1991, Summary Program Plan*, publication DOE/FE-0235P, U.S. Dept. of Energy, Washington, D.C., 1991.
16. L. K. Rath and J. R. Longanbach, *Energy Sources* **13**, 443–459 (1991).
17. P. Sabatier and J. B. Senderens, *Compt. Rend.* **134**, 514–516 (1902).
18. F. Fischer and H. Tropsch, *Brennstoff-Chem.* **4**, 276–285 (1923).

19. H. H. Storch, N. Golumbic, and R. B. Anderson, *The Fischer-Tropsch and Related Syntheses*, John Wiley & Sons, Inc., New York, 1951, p. 337.
20. R. D. Srivastava and co-workers, *Hydrocarbon Process.*, 59–68 (Feb. 1990).
21. A. Stratton, D. F. Hemming, and M. Teper, *Methanol Production from Natural Gas or Coal*, report no. E4/82, International Energy Agency Coal Research, London, 1982.
22. E. Supp, *How to Produce Methanol from Coal*, Springer-Verlag, Berlin, 1990.
23. I. Wender and K. Klier, in Ref. 13, Chapt. 5.
24. G. A. Mills, *Proceedings of the 199th American Chemical Society National Meeting, Boston, Apr. 22–27, 1990,* Vol. 35, Division of Fuel Chemistry Preprints, American Chemical Society, Washington, D.C., pp. 241–257.
25. R. L. Mednick, *Proceedings of the Eleventh Annual EPRI Contractors' Conference on Clean Liquid and Solid Fuels*, EPRI AP-5043-SR, Electric Power Research Institute, Palo Alto, Calif., 1987.
26. D. M. Brown and co-workers, *Catal. Today* **8**, 279–304 (1991).
27. J. R. Zoeller and co-workers, *Catal. Today* **13**, 73–91 (1992).
28. R. L. Pruett, *Science* **211**, 11–16 (1981).
29. P. D. Sunavala and B. Raghunath, *J. Sci. Ind. Res.* **45**, 327–335 (1986).
30. J. J. Lin and J. F. Knifton, *Chemtech* **22**, 248–252 (1992).
31. K. T. Klasson and co-workers, *Proceedings of the 204th American Chemical Society National Meeting, Washington, D.C., Aug. 23–28, 1992,* Vol. 37, Division of Fuel Chemistry Preprints, American Chemical Society, Washington, D.C., pp. 1977–1982.
32. P. H. Spitz, *Chemtech* **19**, 92–100 (1989).
33. R. J. Tedeschi, *Acetylene Based Chemicals from Coal and Other Natural Resources*, Marcel Dekker, Inc., New York, 1982.
34. F.-W. Kampmann and W. Portz, in K. R. Payne, ed., *Chemicals from Coal: New Processes*, John Wiley & Sons, Inc., New York, 1987, Chapt. 2.
35. V. H. Agreda, D. M. Pond, and J. R. Zoeller, *Chemtech* **22**, 172–181 (1992).
36. S. L. Cook, in V. H. Agreda and J. R. Zoeller, eds., *Acetic Acid and Its Derivatives*, Marcel Dekker, Inc., New York, 1993, Chapt. 9.
37. A. Geertsema, *Proceedings of the 6th Annual Pittsburgh Coal Conference*, Sept. 25–29, 1989, University of Pittsburgh, pp. 582–587.
38. A. Geertsema, *Proceedings of the 7th Annual Pittsburgh Coal Conference*, Sept. 10–14, 1990, University of Pittsburgh, pp. 571–578.
39. M. E. Dry, *Proceedings of the American Chemical Society National Meeting, New York, Apr. 13–18, 1986,* Vol. 31, Division of Fuel Chemistry Preprints, American Chemical Society, Washington, D.C., pp. 288–292.
40. A. Geertsema, *Proceedings of the 9th Annual Pittsburgh Coal Conference*, Oct. 12–16, 1992, University of Pittsburgh, pp. 395–400.
41. G. Baker and co-workers, in Ref. 38, pp. 561–570.
42. M. C. Bromel, A. K. Kuhn, and D. C. Pollock, *Proceedings of the 4th Annual Pittsburgh Coal Conference*, Sept. 28–Oct. 2, 1987, University of Pittsburgh, pp. 524–531.
43. E. Matsunaga, in Ref. 42, pp. 482–489.

General References

R. F. Probstein and R. E. Hicks, *Synthetic Fuels*, pH Press, Cambridge, 1990.

H. D. Schindler, ed., *Coal Liquefaction—A Research Needs Assessment Technical Background*, publication DOE/ER-0400, Vol. 2, U.S. Dept. of Energy, Washington, D.C., 1989.

K. R. Payne, ed., *Chemicals from Coal: New Processes*, John Wiley & Sons, Inc., New York, 1987.

J. Falbe, ed., *Chemical Feedstocks from Coal*, John Wiley & Sons, Inc., New York, 1982.

R. J. Tedeschi, *Acetylene Based Chemicals from Coal and Other Natural Resources*, Marcel Dekker, Inc., New York, 1982.

Coal Chem 2000, Institution of Chemical Engineers Symposium Series No. 62, Rugby, Warks, UK, 1980.
H. H. Lowry, *Chemistry of Coal Utilization*, Suppl. Vol., John Wiley & Sons, Inc., New York, 1963.
P. J. Wilson and J. H. Wells, *Coal, Coke, and Coal Chemicals*, McGraw-Hill Book Co., Inc., New York, 1950.

PAUL R. WORSHAM
Eastman Chemical Company

PETROCHEMICALS

By definition petrochemicals are either isolated from or derived from natural gas or petroleum. Other sources of feedstocks such as coal and agricultural products remain very small sources of petrochemicals. The choice of feedstock for a given operator is an economic decision, but may be constrained by hardware limitations on the part of the operator or the availability of indigenous hydrocarbon supplies in a given geographical region. Because the primary uses of natural gas and petroleum are as sources of energy, the costs of petrochemical feedstocks are closely related to the alternative energy values of the various feedstocks. The petrochemical producer must buy feedstocks out of the alternative energy markets. Hence the structure and economics of the energy industry in a given area affect the choice of petrochemical feedstocks. Seasonal effects in the prices of energy products introduce variability in petrochemical feedstock prices and affect the choice of preferred feedstock for a given operator at a point in time. International events such as the OPEC oil embargo of 1973–1974, the Iranian revolution of 1979, or the Middle East war of 1990–1991 have dramatically affected the costs and choice of petrochemical feedstocks for a period of time. Government regulations, such as the Clean Air Act Amendments of 1990, also impact petrochemical feedstock availability, relative cost, and choice of alternative feedstocks.

It is convenient to divide the petrochemical industry into two general sectors: (*1*) olefins and (*2*) aromatics and their respective derivatives. Olefins are straight- or branched-chain unsaturated hydrocarbons, the most important being ethylene (qv) [*74-85-1*], propylene (qv) [*115-07-1*], and butadiene (qv) [*106-99-0*]. Aromatics are cyclic unsaturated hydrocarbons, the most important being benzene (qv) [*71-43-2*], toluene (qv) [*108-88-3*], *p*-xylene [*106-42-3*], and *o*-xylene [*95-47-5*] (see XYLENES AND ETHYLBENZENE) There are two other large-volume petrochemicals that do not fall easily into either of these two categories: ammonia (qv) [*7664-41-7*] and methanol (qv) [*67-56-1*]. These two products are derived primarily from methane [*74-82-8*] (natural gas) (see HYDROCARBONS, C_1–C_6).

Olefins are produced primarily by thermal cracking of a hydrocarbon feedstock which takes place at low residence time in the presence of steam in the tubes of a furnace. In the United States, natural gas liquids derived from natural gas processing, primarily ethane [*74-84-0*] and propane [*74-98-6*], have been the dominant feedstock for olefins plants, accounting for about 50 to 70% of ethylene production. Most of the remainder has been based on cracking naphtha or gas oil hydrocarbon streams which are derived from crude oil. Naphtha is a hydrocarbon fraction boiling between 40 and 170°C, whereas the gas oil fraction boils between about 310 and 490°C. These feedstocks, which have been used primarily by producers with refinery affiliations, account for most of the remainder of olefins production. In addition a substantial amount of propylene and a small amount of ethylene are recovered from waste gases produced in petroleum refineries.

In most of the rest of the world the olefins industry was originally based on naphtha feedstocks. Naphtha is the dominant olefins feedstock in Europe and Asia. In the middle 1980s several large olefins complexes were built outside of the United States based on gas liquids feedstocks, most notable in western Canada, Saudi Arabia, and Scotland. In each case the driving force was the production of natural gas, perhaps associated with crude oil production, which was in excess of energy demands.

Since the early 1980s olefin plants in the United States were designed to have substantial flexibility to consume a wide range of feedstocks. Most of the flexibility to use various feedstocks is found in plants with associated refineries, where integrated olefins plants can optimize feedstocks using either gas liquids or heavier refinery streams. Companies whose primary business is the production of ethylene derivatives, such as thermoplastics, tend to use ethane and propane feedstocks which minimize by-product streams and maximize ethylene production for their derivative plants.

Flexibility allows the operator to pick and choose the most attractive feedstock available at a given point in time. The steam-cracking process produces not only ethylene, but other products as well, such as propylene, butadiene, butylenes (a mixture of monounsaturated C-4 hydrocarbons), aromatics, etc. With ethane feedstock, only minimal quantities of other products are produced. As the feedstocks become heavier (ie, as measured by higher molecular weights and boiling points), increasing quantities of other products are produced. The values of these other coproduced products affect the economic attractiveness and hence the choice of feedstock.

Aromatics are produced primarily from two sources. The most important in the United States is refinery catalytic reformer operations. Catalytic reformers are high temperature catalytic dehydrogenation processes which convert naphthenes into aromatics for the purpose of increasing the octane level and hence the gasoline blending quality of the stream being processed. Reformers produce large quantities of the primary aromatic chemicals. Benzene, toluene, and a mixed xylene stream are subsequently recovered by extractive distillation using a solvent (see BTX PROCESSING). Recovery of *p*-xylene from a mixed xylene stream requires a further processing step of crystallization and filtration or adsorption followed by desorption on beds of molecular sieves.

The second source of aromatics is pyrolysis gasoline from olefins plants. Pyrolysis gasoline is one of the by-product streams produced in olefin plants when cracking heavier feedstocks such as naphtha or gas oil. Pyrolysis gasoline contains hydrocarbon fractions from C-5 olefins through C-9 aromatics. Recovery of aromatics from this stream is by extraction and fractionation, just as in recovery from reformate. Pyrolysis gasoline is a more important aromatics feedstock in Europe and Asia than in the United States. This is because naphtha is the dominant feedstock in these regions of the world, and less gasoline is produced in the refineries of these regions; hence there is less reforming capacity. Otherwise there are only minimal differences in aromatics production in various regions of the world.

Olefin Feedstocks

Olefins are produced primarily by steam cracking of hydrocarbon feedstock. Steam cracking is a thermal cracking process in the presence of steam which takes place at low residence times in the tubes of a furnace. The thermal cracking process was originated in the early 1920s and has been the dominant route to olefins since that time (1). Olefin plants are complex and costly, with both complexity and cost increasing as the design feedstock increases in heaviness (molecular weight) from the lightest feed (ethane) to the heaviest feed (gas oil). Heavier feeds produce a wider array and larger quantities of products which require more complex separation and recovery processes. A new world-scale ethylene plant (680×10^6 kg per year including off-site facilities) based on naphtha feedstock will cost approximately $750 million (2). Originally olefins plants were designed to consume a specific feedstock, such as ethane or naphtha, or a narrow range of feedstocks, most commonly ethane–propane mixtures. Starting around 1980 producers in the United States which were based on heavier feedstocks such as naphtha or gas oil began to introduce feedstock flexibility to consume a wide range of feedstocks in their olefin plants. This was done to allow the operator to pick and choose the most attractive feedstock at a given point in time in order to maximize profitability. Olefin plants in the United States have incorporated more feed flexibility than those in Europe, and are therefore able to take advantage of a wider array of potential feedstocks.

Most of the flexibility to use various feedstocks is found in plants with associated refineries, where integrated olefins plants can optimize feedstocks using either gas liquids or heavier refinery streams. Figure 1 illustrates the many linkages between an olefin plant and a refinery which is typical of integrated complexes in the United States.

Companies whose primary business is the production of ethylene derivatives, such as thermoplastics, tend to use ethane and propane feedstocks which minimize by-product streams and maximize ethylene production for their derivative plants. Table 1 provides a summary of the 1990 production quantity and value of primary olefins petrochemicals and olefin feedstocks in the United States.

The cracking process produces not only ethylene, but other products as well. With ethane feedstock, only minimal quantities of other products are produced. As the feedstocks become heavier, increasing quantities of other products are produced. Most operators characterize these other products as either coproducts or by-products. In this article coproducts are defined as other primary olefins (propylene and butadiene), and primary aromatics (benzene, toluene, and xylenes). By-products are everything else ranging from C-4 monoolefins to tars. The principal distinction between coproducts and by-products is that coproducts are the more desired and valuable products where operators often adjust feedstocks and operating conditions to maximize or minimize production, depending on the economics at the point in time. The values of these coproducts and by-products affect the economic attractiveness and hence the choice of feedstock. Table 2 pre-

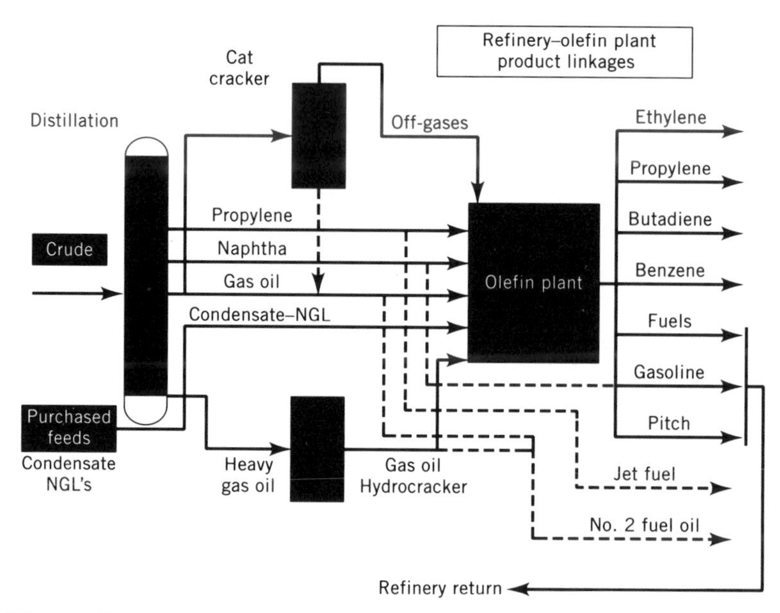

Fig. 1. Typical refinery–olefin plant complex. Courtesy of Shell Oil Co.

Table 1. Olefin Feedstocks and Olefins Production, 1990[a]

Material	Volume, 10^6 kg	Cost, 10^6 $
	Feedstocks	
ethane	9,490	1,410
propane	7,940	1,240
butane	2,030	370
naphtha	9,390	1,730
gas oils	5,000	770
	Primary olefin petrochemicals	
ethylene	16,998	9,410
propylene	10,085	3,760
butadiene	1,431	850

[a]Ref. 3. Courtesy of Chem Systems Group.

sents typical yields for a new plant employing the latest technology from various feedstocks as well as a characterization of the feedstock density and boiling range. Yields from older existing plants differ somewhat from these yields, generally being lower in ethylene production (see ETHYLENE).

It can be seen that in the case of ethane, the yields of ethylene are the greatest and only minimal quantities of other products are produced. In the case

Table 2. Yields[a] From Typical Ethylene Feedstocks,[b] wt %

Feedstock	Ethane	Propane	Naphtha[c]	Naphtha[d]	Gas oil
density, g/L			683.0	706.3	890.5
ASTM ibp,[e] °C			37	37	310
ASTM ebp,[f] °C			141	167	488
		Products			
hydrogen	3.64	1.41	1.03	0.92	0.58
methane	3.13	23.64	17.50	15.80	9.40
acetylene	0.33	0.49	1.05	0.82	0.30
ethylene	50.04	37.15	35.40	31.70	20.50
ethane	40.00	4.23	3.85	3.80	2.90
propadiene	0.04	0.48	1.11	0.90	0.51
propylene	0.75	13.91	14.90	14.40	13.80
propane	0.16	8.00	0.35	0.30	0.27
butadiene	0.85	3.64	4.60	4.60	4.45
butylenes	0.19	0.98	3.90	3.90	4.10
butane	0.23	0.11	0.23	0.20	0.10
C-5–C-8 naphtha	0.39	1.83	3.40	4.90	4.75
benzene	0.20	2.26	5.50	5.80	6.50
toluene	0.05	0.52	1.60	2.80	2.25
C-8 aromatics	0.00	0.08	0.20	1.40	1.70
styrene	0.00	0.41	0.70	0.75	0.70
C-9-205 FBP naphtha	0.00	0.75	1.08	2.05	3.00
fuel oil	0.00	0.11	3.60	4.96	24.19
Total	*100.00*	*100.00*	*100.00*	*100.00*	*100.00*

[a]Single pass.
[b]Courtesy of ABB Lummus Crest Inc.
[c]Light naphtha.
[d]Full-range naphtha.
[e]Initial bp.
[f]End bp.

of gas oil, the yields of ethylene are the lowest and large quantities of heavier coproducts and by-products are produced. When cracking heavier feedstocks, for example naphtha, the severity of the cracking can be varied by adjusting the cracking temperature. Yields of ethylene can be maximized at the expense of propylene and butylene by increasing severity, or ethylene yields can be minimized and thus propylene and butylene yields maximized by lowering severity. Table 3 illustrates the effect of severity on yields for full-range naphtha cracking at high and low severity.

Olefin Feedstock Selection. The selection of feedstock and severity of the cracking process are economic choices, given that the specific plant has flexibility to accommodate alternative feedstocks. The feedstock prices are driven primarily by energy markets and secondarily by supply and demand conditions in the olefins feedstock markets. The prices of individual feedstocks vary widely from time to time as shown in Figure 2, which presents quarterly prices of the various feed-

Table 3. Yields at High and Low Severity For Full-Range Naphtha Feedstock[a,b]

Product	High severity yield, wt %	Low severity yield, wt %
hydrogen	0.92	0.74
methane	15.80	12.60
acetylene	0.82	0.35
ethylene	31.70	26.00
ethane	3.80	4.40
propadiene	0.90	0.60
propylene	14.40	17.50
propane	0.30	0.50
butadiene	4.60	4.40
butylenes	3.90	6.80
butane	0.20	0.85
C-5–C-8 naphtha	4.90	11.70
benzene	5.80	3.10
toluene	2.80	2.75
C-8 aromatics	1.40	2.50
styrene	0.75	0.57
C-9-205 FBP naphtha	2.05	2.30
fuel oil	4.96	2.34
Total	*100.00*	*100.00*

[a]Courtesy of ABB Lummus Crest Inc.
[b]Density = 706.3 g/L; ASTM ibp = 37°C; ASTM ebp = 167°C.

stocks in the United States from 1978 through 1991 in dollars per metric ton (1000 kg) (4).

It can be seen that the feedstock prices are not only volatile but also vary with respect to each other. In the 1979 to 1981 period, natural gas price controls tended to restrain the prices of natural gas liquids such that gas liquids feeds were much more attractive than heavier naphtha feeds. This situation encouraged olefin operators, whose plants were designed for heavy feeds, to modify their plants to increase their choice of feedstocks. On the other hand, at the end of 1991, temporary supply shortages of gas liquids caused ethane and propane prices to rise relative to heavier feeds. The relative cost of various feedstocks can be more easily understood by examining the ratios of one to another. In Figure 3 the ratios of the various alternative feeds to ethane are shown. The high ratio of naphtha to ethane in the 1979–1982 period clearly shows the relative disadvantage of naphtha feedstock. In late 1991, by contrast, all of the feed prices converged.

Simply looking at the feedstock prices or price ratios is insufficient to accurately identify the most attractive feedstock because the values of all of the coproducts and by-products must also be taken into account. This is usually accomplished by calculating the cost to produce ethylene with all other coproduct and by-product yields credited against the cost of ethylene. An example of the cost of ethylene is presented in Table 4. The cash costs of ethylene from various feedstocks are compared for the months of July and November of 1991. Cash costs

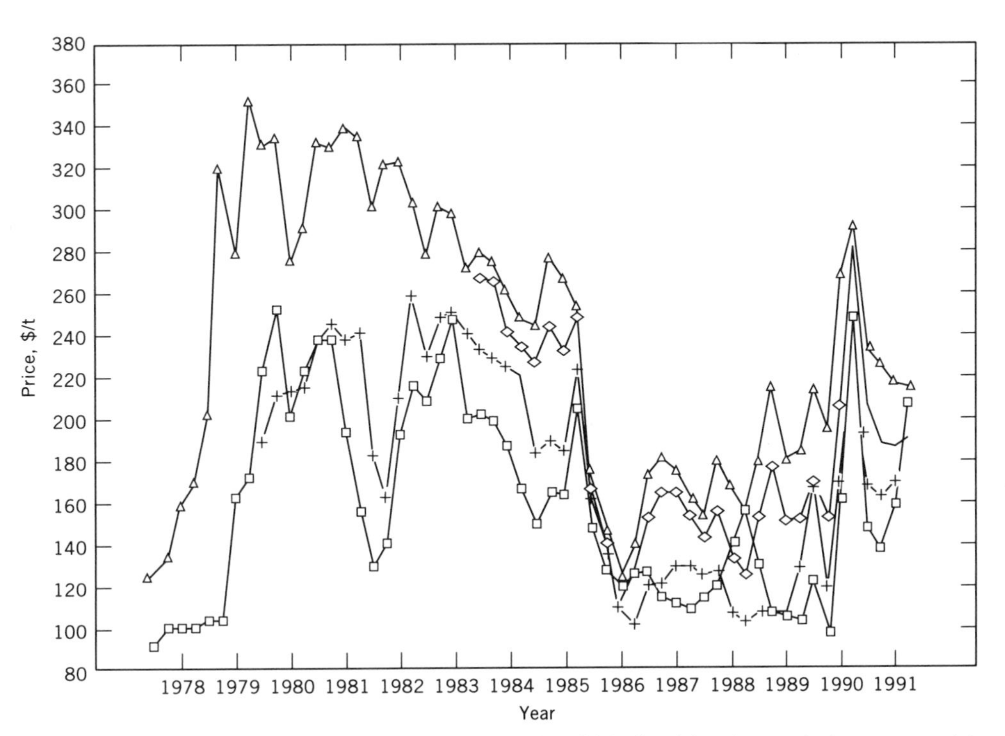

Fig. 2. Quarterly olefin feedstock prices, 1978–1991, for (□) ethane, (+) propane, (◇) light naphtha, and (△) naphtha. Courtesy of DeWitt & Co., Inc.

reflect all plant manufacturing costs except depreciation and are a measure of the out-of-pocket cash costs generated by the operation.

The July data are generally typical of ethylene costs in the summer season. Natural gas prices are weaker in the summer, hence the cost of gas liquids is seasonally lower. Conversely, gasoline demand and prices are seasonally higher due to the peak driving season, and the cost of naphtha reflects this strength. In July the cash costs of producing ethylene from naphtha are about five and one half to six cents per kg of ethylene higher than the cash costs from ethane or propane; hence gas liquids are the most economical source of feedstock. Table 4 also presents the cash costs of ethylene from the same feedstocks in November 1991; ethane remained the preferred feedstock in November, but now naphtha appears favorable to propane by more than four cents per kg. Seasonally higher fuel prices in the winter increased the costs of gas liquids while weaker gasoline markets lowered the relative price of naphtha. Thus an operator with feedstock flexibility would vary the feedslate over time in order to achieve the lowest possible costs. The historical feedslate for olefins production in the United States is presented in Table 5.

The most recent new plants have been based primarily on ethane and propane feeds because the companies constructing these facilities are predominantly thermoplastic producers requiring ethylene, but most industry observers expect that olefin feedslates in the United States will become progressively heavier in

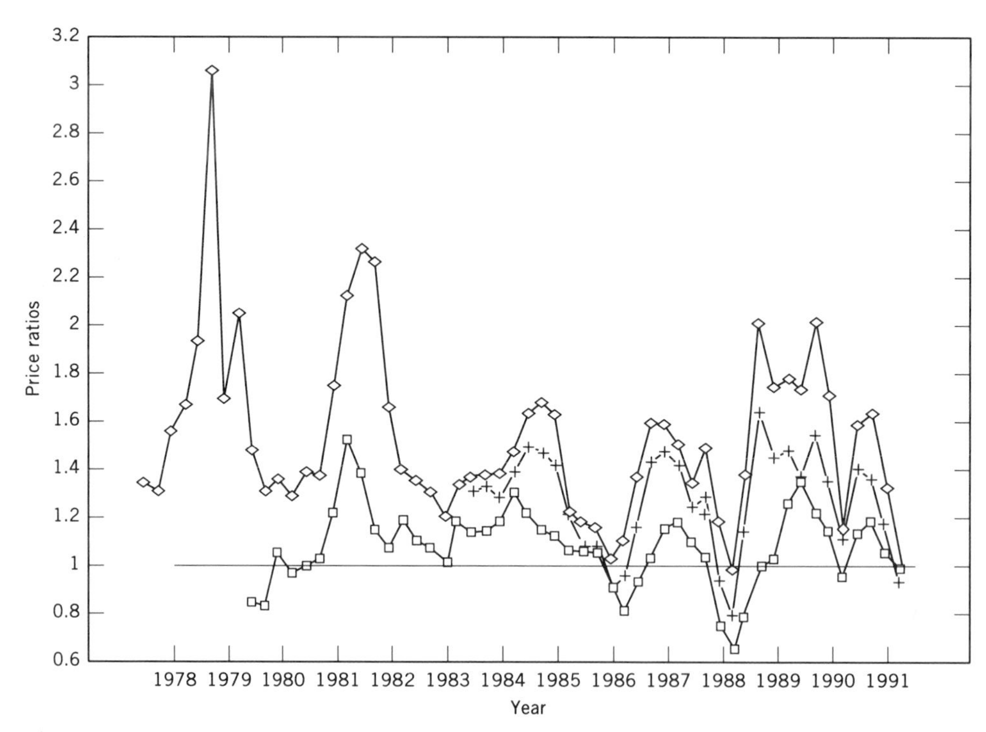

Fig. 3. Price ratio of various feeds to ethane, 1978–1991; (□) propane–ethane, (+) light naphtha–ethane, and (◇) full-range naphtha–ethane. Courtesy of DeWitt & Co., Inc.

Table 4. Cash Costs[a] of Ethylene From Various Feeds, ¢/kg

Cost	Ethane		Propane		Naphtha	
	July	Nov.	July	Nov.	July	Nov.
feedstock	14.32	26.49	30.65	49.17	48.03	56.96
by-product credits	(4.84)	(5.37)	(20.79)	(21.25)	(36.41)	(38.10)
catalyst and chemicals	0.18	0.18	0.18	0.20	0.31	0.33
utilities	1.72	1.94	2.11	2.35	3.74	4.20
operating	2.60	2.64	2.46	2.51	3.34	3.43
plant overhead	3.01	3.10	2.90	2.97	4.00	4.11
Total cash costs	*16.99*	*28.98*	*17.57*	*35.95*	*23.01*	*30.93*

[a]1991.

Table 5. Historical Ethylene Feedslate,[a] 10^6 kg

Product	1983	1984	1985	1986	1987	1988	1989	1990
ethane	19	20	21	19	19	20	19	21
propane	11	13	15	14	19	18	18	18
n-butane	0	0	0	1	1	1	4	4
naphtha	14	18	11	20	19	25	20	20
gas oil	13	12	9	13	12	11	10	11
Total	*57*	*63*	*57*	*68*	*70*	*76*	*71*	*74*
Percent light feedstock (ethane and propane)								
	53	52	64	50	54	50	51	52

[a]Courtesy of Chem Systems Group.

the future. Additional supplies of ethane will be limited, barring significant new discoveries of natural gas. Some additional propane will probably be imported from foreign sources. Lower volatility standards for gasoline will free up additional normal butane [*106-97-8*] now blended into gasoline (5) increasing the use of butane as feedstock. Adequate supplies of naphtha and gas oil are available to meet the remainder of growth demands for ethylene. On a worldwide basis, gas liquids accounted for about 40% of ethylene production in 1990 and are expected to continue at that level in the future because of additional recovery of gas liquids associated with crude oil production outside of the United States (6). As the United States olefins feedslate becomes somewhat heavier, the feedslate in the rest of the world will become somewhat lighter.

Refinery off-gases are a significant source of propylene and a minor source of ethylene in the United States. About one-half of the propylene produced in the United States is recovered directly from refinery gases, which are largely produced in fluid catalytic cracking (FCC) units (CATALYSTS, REGENERATION–FCC UNITS). FCC units are designed to convert heavy gas oil fractions into gasoline blend stocks and are used more extensively in the United States than in Europe and Asia. In the normal course of events this propylene stream would be converted into heavier gasoline blend stocks by alkylation or dimerization processes or blended into liquefied petroleum gas (LPG). Refinery-grade propylene is usually in the 50 to 70% purity range, with the principal impurity being propane. Propylene prices are normally such that it is more attractive for refiners to sell or transfer the refinery-grade propylene to operators for the purpose of purification to chemical-grade quality (minimum 92% propylene) or polymer-grade quality (about 99+% propylene). Purification consists of fractionation and removal of trace impurities. There are a few chemical processes that can utilize refinery-grade propylene without purification; cumene [*98-82-8*], 2-propanol [*67-63-0*], and higher olefins such as nonene (C-9 monoolefins) and tetramer (C-12 monoolefins).

Aromatics Feedstocks

Aromatics are produced primarily from two sources, refinery catalytic reformer operations and recovery from the pyrolysis gasoline fraction from olefin plants.

The most important source in the United States is refinery catalytic reformer operations. Catalytic reformers are high temperature catalytic dehydrogenation processes which convert naphthenes contained in virgin naphthas into aromatics for the purpose of increasing the octane value and hence the gasoline blend value of the stream being processed. Reformers produce large quantities of the primary aromatics, benzene, toluene, and isomers of xylene (see BTX PROCESSING).

The preferred feedstocks for aromatics production are naphthas containing high concentrations of the naphthene precursors to benzene, toluene, and xylenes. Naphthas are characterized according to their naphthene + aromatic (N + A) values and boiling range. Naphthas with high N + A values in the range of about 40 to 45 or above are preferred reformer feedstocks and command premium prices. Naphthas with lower N + A values are poor reformer feedstocks and have lower values, but happily these low N + A naphthas are preferred for olefins plant feedstocks. The production of aromatic chemicals in the reforming process can be maximized by adjusting the cut points on the naphtha feedstock to include the C-6 to C-8 cycloparaffins and aromatics. The resulting reformate contains from 4–8% benzene, with the higher level preferred. A typical composition (vol %) of the C-6 to C-8 aromatics fraction from a reformer operating at 96 research octane number (RON) severity is benzene, 12%; toluene, 43%; and xylenes fraction, 45%. The xylene fraction is 25% *o*-xylene, 20% *p*-xylene, 40% *m*-xylene, and 15% ethylbenzene.

Benzene, toluene, and a mixed xylene stream are subsequently recovered by extractive distillation using a solvent. Recovery of *p*-xylene from a mixed xylene stream requires a further process step of either crystallization and filtration or adsorption on molecular sieves. *o*-Xylene can be recovered from the raffinate by fractionation. In *p*-xylene production it is common to isomerize the *m*-xylene in order to maximize the production of *p*-xylene and *o*-xylene. Additional benzene is commonly produced by the hydrodealkylation of toluene to benzene to balance supply and demand. Less common is the hydrodealkylation of xylenes to produce benzene and the disproportionation of toluene to produce xylenes and benzene.

The second source of aromatics feedstock is pyrolysis gasoline from olefins plants. Pyrolysis gasoline is one of the by-product streams produced in olefin plants when cracking heavier feedstocks such as naphtha or gas oil. Pyrolysis gasoline contains hydrocarbon fractions from C-5 olefins through C-9 aromatics. Recovery of aromatics from this stream is by extraction and fractionation, just as in the recovery from reformate, following treatment to saturate diolefins. Pyrolysis gasoline is a more important aromatics feedstock in Europe and Asia than in the United States. This is because naphtha feedstocks are used more in these regions of the world and less gasoline is produced in these regions; hence there is less reforming in refineries. A typical composition of C-6 to C-8 aromatics stream from pyrolysis gasoline is benzene, 54%; toluene, 31%; and xylenes fraction, 15%. The xylene fraction is 17% *o*-xylene, 23% *p*-xylene, 37% *m*-xylene, and 23% ethylbenzene.

Pyrolysis gasoline is richer in benzene than reformate, 54% vs 12%. Essentially all of the pyrolysis gasoline in the United States is processed to recover aromatic chemicals because of the higher benzene content and problems associated with blending pyrolysis gasoline into motor gasoline. Pyrolysis gasoline contains higher molecular unsaturated compounds which tend to form gum materials in gasoline unless hydrogen treated.

The Clean Air Act Amendments of 1990 limit the amount of benzene in gasoline in the United States to 1% (7). Initially there was some concern that this would disrupt the benzene supply and demand balance in the chemical industry because at that time gasoline contained benzene above 1%. If refiners had to extract all of the benzene above 1%, substantial additional benzene would be produced. However, only modest increases in the quantity of benzene produced from reformer sources is expected as most refiners can adjust the composition of reformer feed and reformer severity to produce less benzene.

Other Feedstocks

The only other petrochemical feedstock of significant commercial use is methane (natural gas) which is used primarily to produce ammonia and methanol. Consumption factors are about 28 GJ and 31 GJ per metric ton, respectively (58,300 and 64,700 BTU/lb) (8). Approximately 460×10^6 GJ (436×10^{12} BTU) of methane was processed into $15,200 \times 10^6$ kg of ammonia (9) and 924×10^6 kg of methanol (3) in the United States in 1990. Natural gas prices generally follow crude oil prices in the United States because they compete in energy markets, but natural gas prices exhibit less volatility and have been lower in cost on a fuel basis. Historical natural gas and crude oil prices are shown in Figure 4.

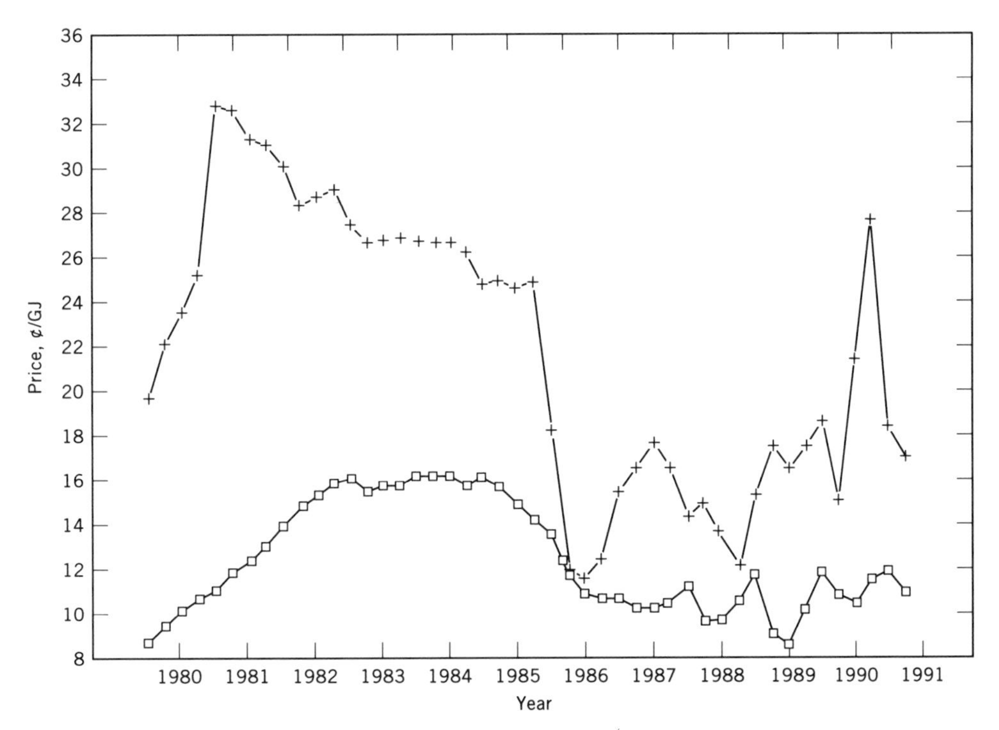

Fig. 4. Prices of natural gas (□) and crude oil (+), 1980–1991. To convert ¢/GJ to ¢/BTU, multiply by 1.055×10^{-6}. Courtesy of DeWitt & Co., Inc.

Alternative Feedstocks

Alternative feedstocks for petrochemicals have been the subject of much research and study over the past several decades, but have not yet become economically attractive. Chemical producers are expected to continue to use fossil fuels for energy and feedstock needs for the next 75 years. The most promising sources which have received the most attention include coal, tar sands, oil shale, and biomass. Near-term advances in coal-gasification technology offer the greatest potential to replace oil- and gas-based feedstocks in selected applications (10) (see FEEDSTOCKS, COAL CHEMICALS).

Because oil and gas are not renewable resources, at some point in time alternative feedstocks will become attractive; however, this point appears to be far in the future. Of the alternatives, only biomass is a renewable resource (see FUELS FROM BIOMASS). The only chemical produced from biomass in commercial quantities at the present time is ethanol by fermentation. The cost of ethanol from biomass is not yet competitive with synthetically produced ethanol from ethylene. Ethanol (qv) can be converted into a number of petrochemical derivatives and could become a significant source.

BIBLIOGRAPHY

"Petroleum Products" in *ECT* 1st ed., Vol. 10, pp. 161–177, by V. L. Shipp, Socony-Vacuum Oil Co., Inc.; "Petroleum Products" under "Petroleum" in *ECT* 2nd ed., Vol. 15, pp. 77–92, by G. B. Gibbs and H. L. Hoffman, *Hydrocarbon Processing*; "Petroleum (Products)" in *ECT* 3rd ed., Vol. 17, pp. 257–271, by H. L. Hoffman, Gulf Publishing Co.; "Feedstocks" in *ECT* 3rd ed., Vol. 9, pp. 831–845, by F. A. M. Buck, King, Buck, & Associates, and M. G. Marbach, Shell Chemical Co.

1. P. Spitz, *Petrochemicals, The Rise of an Industry*, Wiley-Interscience, 1988, pp. 77–78.
2. S. Field, *Hydrocarbon Process.* **69**, 3 (Mar. 1990).
3. *Synthetic Organic Chemicals Report*, U.S. International Trade Commission, Washington, D.C., 1990.
4. *Historical Pricing—Petrochemicals*, DeWitt & Co., 1991.
5. P. Savage, *Chem. Week* **141**(7), 6–10 (1987).
6. R. DiCintio, M. Picclotti, V. Kaiser, and C. A. Pocini, *Hydrocarbon Process.* **70**(7), 83 (1991).
7. *Chem. Week* **149**(16), 35 (1991).
8. *Hydrocarbon Process.* **70**(3), 134, 164 (1991).
9. *Fertilizer Materials*, Current Industrial Reports, U.S. Department of Commerce, Bureau of the Census, Washington, D.C.
10. A. Wood, *Chem. Week* **145**(13), 78–88 (1989).

DARRYL C. AUBREY
Sacred Heart University
Chem Systems, Inc.

FELTS. See PAPERMAKING MATERIALS AND ADDITIVES.

FEMTOCHEMISTRY. See KINETICS MEASUREMENTS; LASERS.

FERMENTATION

Fermentation is derived from the Latin verb *fervere* (to boil) and describes the anaerobic evolution of carbon dioxide (qv) from the action of yeast on fruit or malted grain (see YEASTS). The term fermentation has been retained even though anaerobic processes are now in the minority and much modern microbial technology essentially embraces aerobic processes. The origins of fermentation as an industry are deeply rooted in the alcoholic beverage industry; fermentation was the first industrial process for the production of a microbial metabolite. Despite numerous advances, existing fermentation technology still utilizes much of the equipment, process design, and terminology originally used for brewing beer (qv) and grain spirit liquors (see BEVERAGE SPIRITS, DISTILLED). Some of the traditional fermentation products, such as beer, wine (qv), vinegar (qv), and pickles, rely on the anaerobic metabolism of microbes such as yeast and some bacteria, and thus are fermentations in the original sense of the word. The processes for newer products such as antibiotics (qv), enzymes, and organic acids tend to rely on the aerobic metabolism of bacteria, molds, and fungi.

Although a tremendous number of fermentation processes have been researched and developed to various extents, only a couple of hundred are used commercially. Fermentation industries have continued to expand in terms of the number of new products on the market, the total volume (capacity), and the total sales value of the products. The early 1990s U.S. market for fermentation products was estimated to be in the $\$9–10 \times 10^9$ range. The total world market is probably three times that figure, and antibiotics continue to comprise a primary share of the industry. Other principal product categories are enzymes, several organic acids, baker's yeast, ethanol (qv), vitamins (qv), and steroid hormones (qv).

The different types of commercially important products can be classified into four groups: (*1*) the production of biomass, primarily yeast and some animal feed additives (see FEEDS AND FEED ADDITIVES); (*2*) the production of metabolic products or primary metabolites such as antibiotics, amino acids (qv), vitamins, organic acids, etc; (*3*) the conversion of a substance or groups of substances into another substance, often referred to as biotransformation or bioconversion (see MICROBIAL TRANSFORMATION), eg, steroid conversions; and (*4*) the production of microbial enzymes, eg, amylases, lipases, and chymosin (see ENZYME APPLICATIONS). The developments leading from genetically engineered microbes may constitute a newer fifth group composed of therapeutic proteins (qv), polypeptides, hormones, and other products not normally synthesized by the host microbe, eg, insulin and some interleukins (see GENETIC ENGINEERING; INSULIN AND OTHER ANTIDIABETIC DRUGS).

There are several advantages offered by fermentation processes. Often, it is the only practical way of synthesizing a complex compound. Examples are penicillin (see ANTIBIOTICS, β-LACTAMS) and vitamin B_{12} [*68-19-9*] (see VITAMINS, VITAMIN B_{12}). Microbes often accomplish in a single step an economically feasible molecular change that otherwise can only be achieved by a long chemical synthesis. Fermentation may also permit the use of cheaper raw materials. Additionally, fermentation often permits the manufacture of compounds having greater specificity and in a purer form. This is especially true when a specific chiral enantiomer is preferred. Moreover, use of microbial enzymes may replace the adverse condi-

Table 1. Chronological Development of the Fermentation Industry[a]

Main products	Equipment	Process control	Strain selection
Stage 1, pre-1900			
alcohol	wooden vats; copper used in later breweries	minimal; limited to use of thermometer, hydrometer, and heat exchanger	pure yeast used by the Carlsberg Brewery from 1896; vinegar inocula derived directly from previous fermentation
vinegar	barrels, shallow trays, and trickle filters		
yogurt	clay pots		
Stage 2, 1900–1945			
baker's yeast biomass, glycerol, citric acid, lactic acid, acetone–butanol, gluconic acid, amylase, invertase, riboflavin	steel vessels for acetone butanol; air sparging introduced for yeast manufacture; mechanical stirrers used in smaller vessels	pH electrodes with off-line control; temperature control	other pure cultures introduced and used
Stage 3, 1945–1960			
pencillin G, vitamin B_{12}, bacitracin, streptomycin, 7-chlortetracycline, neomycin, other antibiotics, semisynthetic antibiotics, amino acids, steroids (eg, cortisone), gibberellins, cellulase, dextran, pectinase, itaconic acid	mechanically aerated vessels; aseptic operation; continuous sterlization of feeds	steam-sterlizable electrodes for pH and DO_2; process control loops introduced	mutation and selection programs essential, especially for antibiotic production

Stage 4, 1960–1970			
fusidic acid, gentamicins, rifamycins, other antibiotics, lipases, glucose isomerase, glucoamylase, xanthan gum, bioinsecticides, 5′-nucleosides, single-cell proteins, valine	pressure cycle fermentors to minimize heat and O_2 transfer limitations; immobilized whole cells and enzymes; continuous flow centrifuges for harvesting	computer monitoring and control of many functions; supervisory control systems; mathematical modeling for control of yeast fermentations	genetic engineering techniques applied to production strains; deregulation of metabolic controls
Stage 5, 1970–1993			
antibiotics, malic acid, aspartic acid, xylitol, rennet, dextranase, lactase, proteases, phenylalanine, methionine, insulin, interleukins	fluidized-bed and air-lift systems; biological containment systems; pressure cycle and pressure jet fermentors; Vogel-Busch deep-jet[b]	direct digital controllers (DDC) introduced; polarographic DO electrodes[c]; limited success with artificial intelligence (AI) applications	automated screening of antibiotic producers; genetic engineering applied to make nontraditional therapeutics

[a]Ref. 1. Courtesy of the American Chemical Society.
[b]Ref. 2.
[c]DO electrodes monitor dissolved oxygen.

tions such as high temperatures, pressures, and drastic pHs often necessary in chemical synthesis (see ENZYMES IN ORGANIC SYNTHESIS).

The chronological development of the fermentation industry can be represented (Table 1) in five stages. Although beer was brewed by the ancient Egyptians, true large-scale breweries date from the early 1700s (3). Also, yogurt, soy sauce, and other fermented foods have been popular with many cultures for several centuries. Prior to 1900, alcoholic beverages and vinegar really constituted the total world fermentation industry. In the period between 1900 and 1940, chemicals such as lactic acid, glycerol, and some organic solvents were made by fermentation. Moreover, some advances in fermentation technology were made. Fed-batch cultures in yeast production were developed to overcome oxygen depletion. Sparge tubes were introduced to improve aeration. Additionally, the first truly aseptic fermentation, the acetone–butanol fermentation (Weizmann process) was developed. During this stage, the concepts of sterile media, pure culture technique, and the need to understand the immense role played by microbial physiology and biochemistry in fermentations, arose.

The third stage, from 1945–1960, in some ways resulted from the wartime need to produce large amounts of penicillin in submerged culture under aseptic conditions. The penicillin process required many developments, most of which emerged from the need to properly aerate viscous mycelial cultures and to isolate a compound present in very low concentrations. Sparging with sterilized air and mutation–selection programs aimed at creating enhanced or super-producing strains were pivotal to the success of most development programs. Pilot plants that mimicked larger fermentation and isolation equipment led to stepwise oriented process development plants and programs. Many other processes were developed during this period for commercial amounts of enzymes; amino acids, primarily glutamic acid [*56-86-0*], and lysine [*56-87-1*]; bioconverted steroids for use as antiinflammatory agents, eg, cortisone [*53-06-5*], hydrocortisone [*50-23-7*], and prednisolone [*50-24-8*] (see ANALGESICS, ANTIPYRETICS, AND ANTIINFLAMMATORY AGENTS); steroids as a base for the birth control pill (see CONTRACEPTIVES); other antibiotics, eg, erythromycin [*114-07-8*] and the tetracyclines; vitamins; and gibberellins. Overall, this period probably accounted for the most significant developments in fermentation technology.

From 1960–1970 the number of different products introduced grew rapidly. About 25 antibiotics, several important enzymes, polysaccharides, microbial insecticides, ie, biopesticides, and flavor-enhancing purine nucleosides were introduced. Of these products, glucose isomerase for the production of high fructose corn syrup (HFCS) and microbial rennet as a replacement and supplement for the calf-stomach product are two excellent commercial examples (see DAIRY SUBSTITUTES). This era also saw the emergence of products such as flavors (see FLAVORS AND SPICES) and fragrances (see PERFUMES) which are made in relatively small amounts but which continue to command relatively high prices. In the early 1960s there was also an abundance of relatively low priced potential and real fermentation carbon feedstocks (qv) such as methanol (qv) and agricultural by-products. Thus a number of large international companies began investigating the production of microbial biomass from feedstocks as a source of protein feedstock for poultry, cows, and pigs. This field was led by ICI in the development of an extremely large (ca 3000 m^3) continuous fermentor based on a pressure cycle and located in the UK (4). The large volume was needed to produce significant quantities of material at a price low enough to compete with the traditional sources such as fish meal. The sheer size of the equipment and the scale of the project meant that the system was plagued with contamination and other problems. As of this writing, the fermentor has only been operated continuously for relatively short periods of time to produce test material. This venture has not been a profitable one.

The stage was set for the fifth period of the industry by genetic engineering (qv) discoveries (5–7), ie, the *in vitro* manipulation of microorganisms. As a result of the ability to precisely alter genetic material and then readily transfer it between unrelated microorganisms, a revolution has occurred in the fermentation industries (see also CELL CULTURE TECHNOLOGY). Microbial cells can be endowed with the ability to synthesize products that are normally associated with higher cells. Both bovine and porcine somatotrophin [*9002-72-6*], insulin [*9004-10-8*], interleukins, and some monoclonal antibody fragments used in oncological imaging and therapy are made by fermentations. Use is often made of a technique commonly referred to as high cell density (HCD) fermentations where microbes are grown and later induced to produce the product of interest (8). Apart from enhancing the spectrum of microbial products, the techniques of genetic engineering have made it possible to increase the productivity of traditional microbial processes. The key is the construction of a microbial variant harboring multiple copies of a gene for an enzyme that might have been responsible for a bottleneck in the desired pathway. This is often termed pathway engineering, and as a technique has been applied with considerable success in the biosynthesis of phenylalanine [*63-91-2*] and other primary metabolites (9). Other notable developments were immobilized enzyme and whole cell systems (10) for producing amino acids such as aspartic acid [*56-84-8*] and fine chemicals (qv).

Fermentation Companies and Products

There are thousands of breweries worldwide. However, the number of companies using fermentation to produce therapeutic substances and/or fine chemicals number well over 150, and those that grow microorganisms for food and feed number nearly 100. Lists of representative fermentation products produced commercially and the corresponding companies are available (1). Numerous other companies practice fermentation in some small capacity because it is often the only route to synthesize biochemical intermediates, enzymes, and many fine chemicals used in minor quantities. The large volume of L-phenylalanine is mainly used in the manufacture of the artificial dipeptide sweetener known as aspartame [*22389-47-0*]. Prior to the early 1980s there was little demand for L-phenylalanine, most of which was obtained by extraction from human hair and other nonmicrobiological sources.

The modern fermentation industries developed from the early era of antibiotics. Over 4000 antibiotics have been discovered since the 1950s. However, only about 100 are produced on a commercial scale and over 40 of these are prepared by a combination of microbial synthesis and chemical modifications. Antibiotics produced by fermentation and used as starting materials in chemical syntheses are given in Table 2.

Table 2. Antibiotics Used as Starting Materials for Chemical Syntheses[a]

Antibiotic[b]	CAS Registry Number	Product
cephalosporin C	[*61-24-5*]	7-aminocephalosporanic acid[c]
demeclocycline	[*127-33-3*]	minocycline
kanamycin	[*8063-07-8*]	amikacin
oxytetracycline	[*79-57-2*]	doxycycline
pencillin G, penicillin V	[*87-08-1*]	6-aminopenicillanic acid[d]
rifamycin B	[*13929-35-6*]	rifampin
sisomicin	[*32385-11-8*]	netilmicin

[a]Ref. 1.
[b]Produced by fermentation.
[c]Used for the synthesis of many semisynthetic cephalosporins.
[d]Used for the synthesis of semisynthetic penicillins.

Yeast fermentations are typically carried out on a large (250 m^3) scale. Countries where yeast is manufactured for food and feed include: for bakers' yeast, Argentina, Australia, Austria, Belgium, Brazil, Canada, Chile, Colombia, Denmark, Egypt, Finland, France, Germany, Guatemala, Holland, Iran, Japan, Mexico, Peru, Philippines, Republic of Korea, South Africa, Spain, Sweden, Tunisia, Turkey, United Kingdom, United States, Uruguay, Zaire, and Zambia; for food and feed yeast, Belgium, Finland, France, and the United States. Pfizer Inc. has always ranked in the top few fermentation companies worldwide both in terms of volume and dollar value. However, the 1990 sale of its citric acid [*77-92-7*] business to the Archer Daniels Midland Co. (ADM) meant a significant decrease in overall fermentation capacity. Miles Inc. acquired the citric acid fermentation capacity of J. & E. Sturge in Yorkshire, UK, and now ranks with ADM as a premier producer. Most of the larger and well-established fermentation companies, eg, Pfizer Inc., Eli Lilly and Co., Merck & Co., Novo Labs, Ajinomoto, Miles Inc., Takeda, ICI plc (Zeneca subsidiary), and Gist-Brocades, have built plants around the world or are involved in joint ventures. Extensive mergers and acquisitions activity involving fermentation technologies and billions of dollars in the late 1980s resulted in Bristol-Myers-Squibb and SmithKline Beecham and the acquisition of Wyeth-Ayerst Labs by American Home Products. A small cross section of companies formed since the late 1970s is listed in Table 3. Many of these latter entities owe their existence to the use of genetic engineering to tailor suitable microbial strains.

Microbiological Aspects

The microbial strain chosen for development plays a significant role in determining the process design and engineering of the fermentation plant and downstream processes. Therefore, the isolation, preservation, and improvement or breeding of

Table 3. Biotechnology Companies Utilizing Fermentation Technology in 1993

Company	Country	Emphasis
Allelix	Canada	therapeutics, biochemicals
Amgen	United States	therapeutics
Bideco	Switzerland	therapeutics
Biocon	Ireland	biochemicals
Biogen	United States	therapeutics, diagnostics
Biotransplant	United States	post-surgery therapeutics
Green Cross	Korea	fine chemicals, biochemicals
British Biotechnology	United Kingdom	therapeutics, diagnostics
Calgene	Australia	agriculture
Cangene	Canada	therapeutics
Cetus (Chiron)	United States	strains, biochemicals, therapeutics
DNAP	United States	artificial snow/agriculture
Dowelanco	United States	crop protection
Envirogen	United States	environmental decontamination
Enzymatix	United Kingdom	biochemicals, enzymes
Genencore	United States	enzymes, biochemicals
Genentech (Roche)	United States	therapeutics
Genzyme	United States	biochemicals, therapeutics
Lucky Biotech	Korea	biochemicals
Mogen	Holland	agriculture
Mycogen	United States	crop protection
Regeneron	United States	therapeutics
Serono	Switzerland	therapeutics
Zeagen (Coors)	United States	vitamins and pigments

appropriate strains is of fundamental importance. Traditionally, microorganisms have been selectively cultured, purified (subcultured), and exposed to a battery of techniques that increase their ability to over-produce the product of interest (11,12). Genetic engineering is playing an ever increasing role in producing better strains. Historically, the streptomycetes and other fungi were not good candidates for engineered genetic changes. However, that is changing through some pioneering work (13) providing the ability to make rather precise genotypic and therefore phenotypic changes. Because the microorganism is the most valuable part of a successful process, the strain must be preserved in some manner, ie, its supply and integrity must remain intact. Many techniques have been developed and evaluated and the usefulness of each depends on the microorganism of interest. Nearly all methods rely on the suppression of multiplication while maintaining viability. One of the older techniques, lyophilization, is used throughout the world in more than 3000 laboratories. Although it is only suitable for a few microorganisms, storage on agar slants in a refrigerator (4°C) or in commercial freezers (−20°C) is practiced. Much higher survival rates of cells are obtained, using liquid nitrogen (−196°C) and a cryoprotective agent such as 25% glycerol. Also, newer compressor technology has delivered mechanical freezers capable of maintaining temperatures as low as −185°C, thus eliminating the need to regularly stock up systems using liquid nitrogen. Storage in the vapor phase of liquid nitrogen (−140°C) is a good lower cost alternative for large culture collections.

Strain development programs can run into millions of dollars. This combined with the status of the strain in the overall process makes the strains extremely valuable and manufacturers rarely deposit useful cultures in one of the commercial collections (14). Nevertheless, commercial collections exist and are a very valuable, often used source of microorganisms that have shown some promise of producing small or minute quantities of products having real or possible commercial applications. These strains are often useful in augmenting in-house activities or as starters for a development program.

Production microorganisms are generally kept as trade secrets as a result of the arduous task of obtaining and defending a patent for the microorganism. Most companies do not attempt to patent microorganisms because competitors would then be alerted to areas of interest. It has been reported (15) that Merck and Co. found one useful microorganism in 10,000; Eli Lilly and Co. (16) produced three antibiotics in 10 years which involved screening 400,000 microorganisms. The very high cost and long-term commitment necessary to obtain potent high titer strains has opened up a business of supplying and developing microorganisms. Panlabs Inc. specializes in penicillin and cephalosporin-producing microorganisms and Cetus Corp. (Chiron) is a supplier and developer of antibiotic-producing microorganisms.

The complexity of the biochemical pathways, certainly for antibiotic synthesis, means that the careful and rigorous selection of the microorganism is the key to obtaining higher titers. However, most microorganisms only show their true potential when cultured under optimal fermentation conditions. Of these, the medium formulation and physical conditions, primarily pH, temperature, and oxygenation, are key. Typically, a strain development program goes in tandem with medium and process conditions development. Examples of successful fermentation and strain development programs are commonly attained penicillin G titers well in excess of 75 g/L (a 15,000-fold increase over wild-type levels) and titers of 150 g/L of citric acid. Traditional mutation and selection techniques cannot be used alone to construct or genetically engineer microorganisms that produce the whole plethora of biotech products. Microbial rennin [*9001-98-3*] is a prime example of a product that could not be produced without the advent of genetic engineering.

Components of Fermentation Processes

With few exceptions, such as biotransformations, most existing fermentation processes can be broken down into several distinct operations. First, there is the formulation of the culture medium used to grow the microorganism during inoculum propagation and production fermentation. This is followed by the preparation of sterile medium, fermentors, and related equipment. Then there is propagation of a pure and active culture to seed the production fermentor. Fourth is the culturing of the microorganism in the production fermentor under optimum conditions for product formation. Then the product is separated from the fermentation broth and subsequently purified. Lastly, there is the treatment and disposal of the cellular and effluent by-products of the process.

All microorganisms require water, sources of carbon, energy, nitrogen, and minerals. Certain vitamins and growth factors are often required as well. Additionally, oxygen is necessary for all aerobic fermentations. Chemically defined or semidefined media are often used up to the seed stage of a fermentation to ensure rapid and reproducible growth. However, on a large scale, economics play a key role in determining the components. The material cost and availability, location of fermentation plant and commodity pricing are key variables. In the case of the ICI pressure cycle fermentor for single-cell proteins the whole fermentation process (4) was essentially designed around the availability of a relatively inexpensive and abundant raw material, ie, methanol derived from methane gas. Generally, the nutritional requirements for most conventional fermentations can be met by formulating a medium having, eg, molasses, cereal grains, starches, glucose, sucrose, and lactose as carbon sources. Corn steep liquor, soybean meal, cottonseed flour, slaughterhouse waste, fish meals, ammonia, ammonium salts, and fermentation residues often serve as the nitrogen base. Soya bean oil, lard, and cotton seed oil serve to control foam in aerated fermentations as well as providing carbon for nutrition. Often, other more traditional defoamers such as silicone oils and polypropylene glycol are necessary as well. Brewer's yeast is frequently used as a combination nitrogen, vitamin, and growth factor source. The overall contribution of the cost of the medium may be very high in some commodity-type fermentations; thus the producer microorganism is often selected or engineered to use components it traditionally would not, but which are available at a reasonable price.

Certain factors and product precursors are occasionally added to various fermentation media to increase product formation rates, the amount of product formed, or the type of product formed. Examples include the addition of cobalt salts in the vitamin B_{12} fermentation, and phenylacetic acid and phenoxyacetic acid for the penicillin G (benzylpenicillin) and penicillin V (phenoxymethylpenicillin) fermentations, respectively. Biotin is often added to the citric acid fermentation to enhance productivity and the addition of β-ionone vastly increases beta-carotene fermentation yields. Also, inducers play an important role in some enzyme production fermentations, and specific metabolic inhibitors often block certain enzymatic steps that result in product accumulation.

With few exceptions, the nutrient medium is charged or batched into the fermentor in a sterile form or is sterilized in the fermentor. Sterilization inactivates the indigenous fauna present including water that would foul up the fermentation of interest. Sterilization of the batch medium is usually accomplished by heating to 121–125°C under pressure of 103–124 kPa (15–18 psig) for at least 15 minutes. Heat is supplied by directly injecting live clean steam into the fermentor or into the fermentor jacket. Continuous sterilization of various feed components and often the initial batch charge itself is accomplished by pumping the medium through heat exchangers using an appropriate holdup time. Smaller amounts of various components may be autoclaved and heat-sensitive components are filter-sterilized under aseptic conditions prior to addition. The bioconversion of D-sorbitol [*50-70-4*] to L-sorbose [*87-79-6*] by *Gluconobacter suboxydans* can be accomplished using partial sterilization because the process only takes 24 h and the high concentration of D-sorbitol (20% w/w) inhibits the growth of most contaminating microorganisms. In the formation of lactic acid by *Lactobacillus delbruekii*, the fermentation occurs at 50°C, and this combined with a high substrate concentration guards against contamination. Apart from these few exceptions, any material must be sterilized prior to entering a fermentor. Air is routinely sterilized by passage through various filters such as packed columns of glass wool, stainless-steel mesh, or 0.2 micrometer hydrophobic (so-called absolute) bacteriological filters. A few companies prefer to heat sterilize air or heat sterilize the air in addition to a downstream filter, thus minimizing any chance of contamination. Following sterilization, the fermentor medium is adjusted in terms of pH and temperature prior to being seeded or inoculated with an actively growing culture of the producer microorganism. Inocula are typically developed in several stages and the final volume is usually 1–10% of the fermentation volume. Actively growing cultures are preferred because those that have entered stationary phase may lead to undesirable process changes and variations as a result of metabolic and genetic variations. It is particularly important to use actively growing cultures of sporulating bacteria. Sporulation leads to an excessively long lag phase. On the other hand, fermentations employing fungi and streptomycetes often have an inoculum development protocol that employs spores.

Once the culture is actively growing, the environment in the fermentor changes in part because of the depletion of medium components and in part because of the production of metabolites by the fermenting mass. In order to maximize productivity, a number of parameters must be maintained within close predetermined limits. The temperature is usually maintained at ± 0.5°C from the set point. Thermocouples (TC) or platinum resistance thermometer devices (RTD) are typically used to measure the temperature. The set point for the larger portion of the fermentation is often a few degrees lower than that used to grow the culture, and such changes or profiles are typically controlled by a computer or microprocessor based on time or some off-line measurement such as culture density, oxygen uptake rate, or carbon dioxide evolution rate. Experience has shown that in very large (250 m^3) fermentors when the culture is actively growing, such fine control is not normally possible and transient temperatures of a few degrees above the set point can be attained because the fermentation broth produces heat at a greater rate than can be removed by the cooling process employed. Cooling is achieved by pumping chilled water through coils submerged inside the vessel or by the half-pipe design using external coils. Cooling jackets are rarely employed on large-scale production fermentors.

In part because of the enormous cost of cooling large-scale production fermentors, much attention has been paid to the design and operation of both internal and external cooling coils (17). Also, because cooling water is not sterile, it has a tremendous potential to contaminate the fermentation if hairline fractures ever occur in internally located coils. A novel method of cooling is the use of a heat exchanger located external to the fermentor. This technique is employed in the Vogel-Busch deep-jet system for smaller (<10 m^3) volumes of fast growing *E. coli* fermentations producing genetically engineered products at high cell densities.

Other than temperature, pH is the second most important variable that must be controlled within defined limits. The actual pH used is, of course, a function of the microorganism employed. Control is usually achieved to within ± 0.20 pH units in large-scale operations through the use of steam sterilizable electrodes. Changes in pH are normally attributable to the excretion of organic acids by the culture. The automatic addition of base or a basic nutrient feed maintains the desired set point. Although fermentor pH meters and controllers may appear to be functioning normally, pH electrode membranes are prone to fouling that can result in electrode drift and inaccuracy over a period of time. Extended fermentations and those employing rich media containing relatively large proportions of proteinaceous material usually pose the most severe electrode drift problems. Two or more electrodes are often employed in large-scale fermentors to minimize this situation. The use of modern gel-filled electrodes is another step in overcoming electrode fouling. The pH set point is rarely altered during the course of a fermentation, but notable exceptions do exist in which a change in pH is permitted or even initiated to bring about the bioconversion of an accumulated precursor to its product. The *in situ* conversion of phenylpyruvic acid [*156-06-9*] to L-phenylalanine can be achieved by gradually lowering the pH in the latter part of the fermentation.

Aerobic fermentations must be supplied with an adequate supply of sterile air. Because of the low solubility of oxygen in water (10 mg/L) it must be constantly replenished. The amount of air supplied is typically 0.5–1.0 fermentor broth volumes per minute. For an industrial-scale fermentor (350 m^3) the enormous amount of sterile air needed lends itself to a significant operating expense. The air is typically introduced into the fermentor through a perforated pipe or sparger located at the bottom of a stainless steel draft tube. To maximize transfer of oxygen into the liquid phase, many design changes to the traditional ring assembly sparger have been proposed and used. The open pipe or ring spargers dominate design features as of this writing (see AERATION, BIOTECHNOLOGY).

Aeration by sparging alone can be insufficient to meet the demands of an active culture and oxygen starvation is often the limiting factor in both culture growth and productivity. The use of a stirrer or agitator (ca 40–70 rpm) having multiple impellers in combination with internal baffles and tank-over pressure helps overcome this problem to a large extent. The design and testing of impellers to improve both oxygen transfer and mixing of viscous fermentations broths is seemingly paying off with the use of hydrofoil mixers in conjunction with traditional Rushton turbines. However, in many fermentations, despite the use of stirrer motors of several hundred horsepower, oxygen transfer is still insufficient to provide an optimally aerated culture.

On a laboratory scale and in smaller (<1.5 m^3) fermentors the air can be enriched with oxygen and stirrer speeds (up to 600 rpm) can largely offset oxygen deprivation. Oxygen enrichment and high speed agitation ($>$ 100 rpm) is prohibitive from both safety and cost perspectives on a large scale. Because of mechanical simplicity and the ability to provide reasonable aeration and good mixing, air-lift fermentors and their draft-tubeless counterparts, ie, the bubble-columns, are increasingly being used at the largest end of the scale. Steam sterilizable probes provide real time information with respect to the oxygen tension in the fermentor. The more expensive, but more reliable, polarographic-type probe is preferred over the galvanic type. The probe output is usually referred to as the dissolved oxygen or simply DO. Although the DO reading is an important indicator of both culture growth and productivity, care is needed to make sure that the reading is an indication of what is really happening in the bulk phase of the fermentation broth rather than just a localized phenomenon. More so than pH probes, DO probes are prone to fouling and can give erroneous readings in various media that contain oil or silicone-based antifoams. Better control capabilities and understanding of the overall process have led to many sophisticated computerized process control schemes linking carbon and energy feeds to fermentation process variables. These schemes link feed rates to changes in DO, off-gas composition, heat evolution, and pH. In general, computerized process control (qv) is another area of fermentation processes that is receiving much attention (18,19) because of the cost savings that can be obtained from both a decrease in labor and improvements gained from finer control over the process. Of all the facets of fermentation design, engineering, and process control, the provision of sufficient oxygen to a culture remains a rich area for improvement.

The heart of any fermentation process is the fermentor. Figure 1 shows a schematic of a typical vessel. Whereas most fermentation processes differ one from an other from an operational viewpoint because of different microorganisms, scale, feed substrates, etc, these processes all share some common features. Typical features are depicted in Figure 2. A starter culture (usually 1–5 mL) is prepared from a working stock of the appropriate microorganism. Spore suspensions or cultures stored in liquid nitrogen or low temperature freezers are typical sources. The starter culture is grown in Erlenmeyer flasks, eg, 500 mL, containing 100 mL of suitable medium, for a period of 1–4 days at an appropriate temperature (25–40°C) in a rotary shaker (200–400 rpm). The culture is transferred to 4 L baffled flasks containing 2 L of medium and incubated for 12–48 h under similar conditions. The contents are transferred to a 1 m^3 stirred and air-sparged pre-seed vessel containing 700 L of medium for another 18–24 h. Temperature, pH, and foaming are automatically controlled in this vessel to provide an optimal growth environment.

The pre-seed vessel is used to inoculate a suitably equipped seed vessel, 10 m^3 or larger containing 7.5 m^3 of medium, where the culture is again grown for another 18–36 h under similar conditions before transfer to a production vessel. Alternatively, the pre-seed vessel is sometimes designated as the seed vessel and the pre-seed stage is bypassed. The production vessel (250 m^3 total volume containing 150 m^3 of medium) is equipped for automatic control of temperature and pH via addition of an acid and a base chosen from NaOH, NH_4OH, $NH_3(g)$, HCl, and H_2SO_4. Foaming is typically controlled by foam sensing devices (usually conductance probes) that meter in organic matter such as animal and vegetable oils or silicone-based antifoams. The former are often more desirable because they often act as slow-release forms of carbon for product synthesis as well.

Most large-scale fermentations are of the fed-batch variety; thus necessary provision is made for additions of sterile carbon feeds syrups, eg, glucose, sucrose, maltodextrins, and lactose. Frequently, oils, eg, lard and soybean, are used to extend fermentations by virtue of being metabolized more slowly than carbohydrates. For many fermentations employing genetically engineered microbes for the production of enzymes or therapeutic substances, it is often necessary for good product yields to provide a separate nitrogen-rich feed stream containing protein extracts, eg, yeast extracts. For penicillin synthesis, it is also necessary to meter in sterile synthetic side-chain derivatives such as phenylacetic acid to elicit pen-

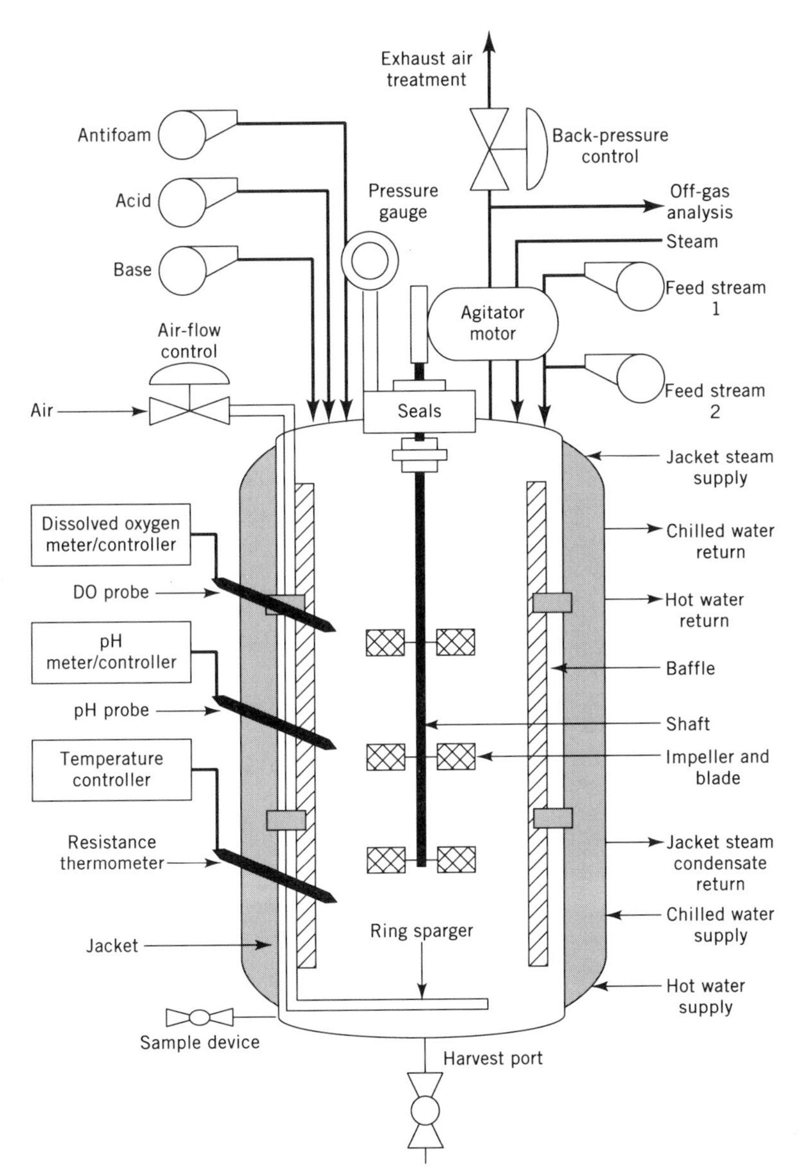

Fig. 1. Representation of a jacketed fermentor.

icillin G production. Generally, fermentation in the production vessel continues for 2–7 days depending on the nature of the product, the process, and the microorganism employed. Many products such as fine chemicals and some enzymes are manufactured on small scales necessitating production vessels of only a few thousand or even a few hundred liters. For these smaller systems, the associated seed-train vessels are also smaller.

From an operational viewpoint, there are many differences between some of the traditional fermentations and those employing cloned or genetically engi-

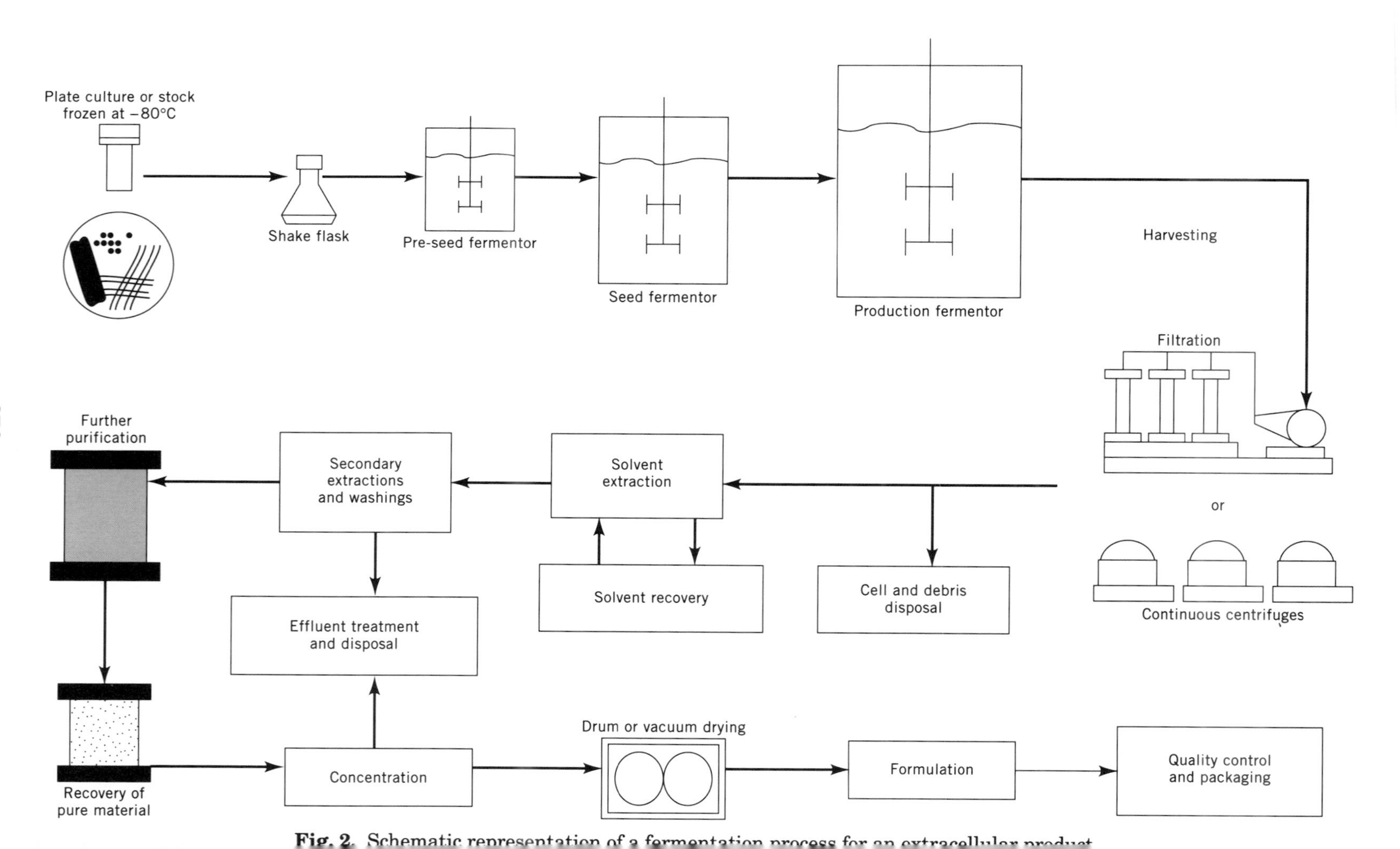

Fig. 2. Schematic representation of a fermentation process for an extracellular product.

neered microbes, referred to as recombinant fermentations. By virtue of employing strains harboring multiple genes copies, recombinant fermentations often have much higher specific productivities and can therefore utilize much smaller fermentors to produce the same amount of product. Recombinant fermentations are often used to produce materials having high market value which are normally required in much smaller quantities. Recombinant fermentations nearly always employ an induction phase for product synthesis and the type of inducer is entirely dependent on the genetic construct of the recombinant microbe. Chemicals, eg, lactose for the *lac* / *tac* promoters and tryptophan for the *trp* promoter; changes in inorganic components, eg, PO_4^{3-} in the *phoA* promoter; along with temperature or pH changes are the routinely used induction systems (20). The choice of promoter and hence induction system used depends, among other things, on the product, the host microbe, the scale of the fermentation, availability of suitable cloning vehicles, and the cost of the inducer. Recombinant fermentations are usually much shorter (ca 14–20 h) and therefore allow shorter turnaround times between runs. The problems of strain degeneration are often more severe in recombinant fermentations because of the loss of plasmids bearing the cloned gene. This latter problem has been overcome in some cases by promoting the incorporation of the foreign genetic material into the microbial chromosomes which provides a much more stable genotype.

Once a fermentation is complete, the product must be separated from the waste. The majority of the time the fermentation broth supernatant is the only material of interest, but there are many exceptions to this rule, such as baker's yeast and other whole-cells products formulations, eg, bacitracin and vitamin B_{12}, where the biomass is the desired portion. In addition, the biomass is retained in fermentations that produce intracellular products such as many enzymes and genetically engineered products. Separation of cells from the liquor is often achieved by filtration (qv), centrifugation, or flocculation. Although not yet universally accepted as a true manufacturing technology, separation using hollow fibers in ultrafiltration (qv) or microfiltration is potentially very powerful (see HOLLOW-FIBER MEMBRANES). Prior to any processing or separation work, the microorganisms often need to be inactived or killed by adjustments in pH, temperature elevations, or through the addition of chemical agents.

One of the key challenges is the separation of secreted soluble fermentation products from the broth. A few products are present at concentrations as low as 0.1%. Even the more typical concentrations of values of 5–15% entail formidable efforts. Then the products often need further purification prior to any form of final inspection or packaging. Numerous physical and chemical methods are employed in a variety of combinations with most fermentations for the isolation and purification steps (21–23). As an example, for the penicillin fermentation, cells are separated from the product containing supernatant by filtration. The filtrate is passed through a series of Podbelniak extractors and the penicillin moves into an organic phase (amyl or butyl acetate). Extraction using phosphate buffer separates the potassium penicillin into the aqueous phase. Crystallization from an *n*-butanol–water mixture renders a pure product that can be further polished or converted into other semisynthetic penicillins. For intracellular products, where whole-cell preparations are inappropriate, cells have to be broken open, eg, by high pressure homogenization or bead milling, and the cell debris has to be removed, usually by centrifugation, prior to additional purification steps. The challenge is even tougher for recombinant proteins and peptides. Isolation must be in an intact form appropriate for further purification and refolding into an active form. The elimination of the key contaminants is accomplished using traditional techniques along with some methods, eg, NH_4SO_4 fractionation and urea treatment, that are specifically used when working with proteins (24). Overall for most fermentations the cost of product recovery and purification often far exceeds that of the fermentation itself.

Economic Aspects

Worldwide, fermentation products span the spectrum from high value, low volume products, eg, vitamin B_{12}, gene-splicing enzymes, and insulin, that command in excess of $1000/kg, to low value, high volume products, eg, citric acid, monosodium L-glutamate [*142-47-2*] (MSG), and aspartic acid, that sell for a few dollars per kilogram. Process improvements, scale of production, and mounting competition have even enabled antibiotics such as penicillins, tetracycline, and erythromycin, and other products, eg, amino acids, almost to become commodity priced items. In 1992, largely fueled by the soft drinks industry (see CARBONATED BEVERAGES), worldwide sales of phenylalanine exceeded 12,000 metric tons.

Although there are hundreds of different products and as many companies producing them, relatively few products rank in annual worldwide sales above $50,000,000. Included in this small group are the antibiotics, bacitracin, cephalosporins, erythromycin, gentamicins, kanamycin, neomycins, penicillins, tetracycline, tylosin, and some newer additions; the enzymes amylases, diagnostic enzymes, glucose isomerase, rennet and proteases; citric and gluconic acids; ethanol; vitamin B_{12}; amino acids, MSG, L-lysine, L-tryptophan, L-phenylalanine, L-aspartic acid; and miscellaneous substances, eg, ergot alkaloids, L-sorbose for vitamin C synthesis, xanthan gum, and insulin. Setting aside the tremendous volume of potable ethanol produced by fermentation, the nearly 470,000 t of MSG and approximately 325,000 t of citric acid produced per year represent the top end of the volume scale (Table 4). Fermentation product sales in the United States are presented in Table 5. The 1992 U.S. market share by sales volume for the top 10 fermentation companies is as follows:

Company	Market share, %
Eli Lilly & Co.	18.5
Archer Daniels Midland Co.	13.9
Merck & Co.	12.3
Pfizer, Inc.	11.7
Abbott Labs	6.0
Lederle Labs (division of American Cyanamid)	5.2
Bristol-Myers-Squibb Co.	4.7
SmithKline-Beecham Co.	2.3
Miles Inc. (division of Bayer AG)	1.9
Ajinomoto Co.	1.6
others	21.9

Table 4. Worldwide 1992 Production and Sale Price for Fermentation Products[a]

Product	Production,[b] t/yr	Sale price,[b] $/kg
bacitracin	10,000	7.80[c,d]
citric acid	325,000	0.8–1.25
erythromycin	3,000	52.00
gluconic acid	44,000	5.50[e]
L-lysine	44,000	2.50[f]
DL-leucine		75.00
monosodium glutamate	470,000	1.90
neomycin sulfate		90.00[g]
L-phenylalanine	12,000	20.00[h]
penicillin[i]	44,000	50.00
riboflavin (vitamin B_2)		58.00[j]
streptomycin sulfate		150.00[g]
tetracycline HCl	15,000	55.00
L-threonine	10,000	13.08[j]
L-tryptophan	4,000	68.00
vitamin B_{12}	18	8,000[d]
xanthan		6.00
pure brewer's yeast		2.40

[a]China, the world's largest producer of antibiotics, had a volume that exceeded 10,000 t in 1991 (25).
[b]Values are estimates.
[c]Price per 10^6 units.
[d]USP grade, nonsterile.
[e]USP grade, calcium salt.
[f]MonoHCl, feed grade.
[g]USP grade.
[h]Estimated production cost. Majority is used to make aspartame.
[i]Potassium salt of benzyl and phenoxymethylpenicillin.
[j]Feed grade.

Generally, for most fermentation processes to yield a good quality product at a competitive price, at least six key criteria must be met. (*1*) Fermentation is a capital intensive business and investment must be minimized. (*2*) The raw materials should be as cheap as possible. (*3*) Only the highest yielding strains should be used. (*4*) Recovery and purification should be as rapid and as simple as possible. (*5*) Automation should be employed to minimize labor usage. (*6*) The process must be designed to minimize waste production and efficiently use all utilities (26,27).

Future Developments

Developments in gene cloning are expected to have the most effect on the fermentation industry. Whereas many industrial biotechnological applications are geared toward obtaining products from mammalian cell cultures, the restriction

Table 5. 1992 U.S. Fermentation Products Market

Product	Sales, $\$ \times 10^6$
ethanol	1900
antibiotics	
bacitracin	65
cephalosporins	2340
penicillins	810
tetracyclines	590
others	1135
total antibiotics	*4940*
amino acids	
L-aspartic acid	16
L-lysine	215
MSG	60
L-phenylalanine	95
others	65
total amino acids	*451*
enzymes	
amylases	58
dextrinases	13
glucose isomerase	84
pectinases	12
rennin	11
others	25
total enzymes	*203*
organic acids	
citric	365
gluconic	43
lactic	22
others	19
total organic acids	*449*
miscellaneous	
fine chemicals	10
flavors and fragrances	20
steroids	15
polysaccharides	465
vitamins	7
Total	*8454*

enzymes used to construct mammalian cloning vectors are produced by traditional fermentation. There are many examples of the success of biotechnology as related to production by fermentation. Insulin, blood factors, and immune system modulators are among those that have been commercialized. Moreover, the Human Genome Project (HGP) must rely on phage and bacteria as vectors in most sequencing and identification protocols. Top-tier biotech companies (Table 3), such as Amgen, Biogen, Genyzme, and Chiron, are now fully fledged pharmaceutical concerns that have combined annual sales well in excess of a billion dollars.

The fermentation industry is expected to benefit from the developments made in biotechnology through the ability to engineer more productive strains through pathway engineering (9); the production of other substances normally foreign to microorganisms, eg, human growth hormone, somatotropin, somatostatin, and human antibody fragments (8,28,29); and the ability to direct biosynthesis in multinucleated microorganisms, eg, streptomycetes, in a manner much more reliable than existing genetic techniques (13).

Overall, the distinction between traditional fermentation applications and spin-offs from biotechnology continues to blur. Microbiologists, cell biologists, and biochemical engineers use a potpourri of techniques leading to commercial quantities of substances manufactured in bioreactors that best approximate ideal conditions. Examples include antibiotics and bulk products such as amino acids and acidulants. With respect to citric acid, the world soft drinks market is growing at double-digit figures.

Commercialized products in the 1990s, such as an enzyme to prevent bread staling from Novo; a fungus-based protein source for human consumption from ICI; and Kelcotrol, a new water-based gum for food use from Merck & Co., should buoy the demand for fermentation capacity. Wider regulatory acceptance for microbial insecticides (*Bacillus thuringiensis* and spores of *Bacillus popilliae*) is also expected (see INSECT CONTROL TECHNOLOGY). Products such as microbially produced pigments, flavors, fragrances, therapeutic substances or components, and optically active enantiomers are at the threshold of commercialization. In addition, the application of microbes to handle solid wastes, eg, land farming with microbial cells, and to reclaim land contaminated with organics, eg, phenol, benzene, and derivatives, is rapidly becoming more important. The increasing demand for poultry and pork in preference to beef is expected to push the market for bacitracin, a feed additive, to above $75 million by the mid-1990s. *E. coli* and some yeasts (8,28,29) are proving to be good hosts for the production of antibody fragments that are expected to be useful in cancer treatment (by direct therapy, imaging, or as a drug delivery system) and for other therapeutic proteins.

BIBLIOGRAPHY

"Fermentation" in *ECT* 1st ed., Vol. 6, pp. 317–375, by G. I. de Becze, Schenley Distillers, Inc.; in *ECT* 2nd ed., Vol. 8, pp. 871–880, by R. F. Anderson, International Minerals & Chemical Corp., Bioferm Division; in *ECT* 3rd ed., Vol. 9, pp. 861–880, by D. Perlman, University of Wisconsin.

1. D. Perlman, *CHEMTECH* **7**, 434 (1977).
2. K. Schreier, *Chemiker Zeitung* **99**, 328–331 (1975).
3. H. S. Corran, *A History of Brewing*, David and Charles, Newton Abbott Press, UK, 1975.
4. S. R. L. Smith, in H. Dalton, ed., *Microbial Growth on C_1 Compounds*, Heyden Press, London, 1981, pp. 342–348.
5. S. N. Cohen and co-workers, *Proc. Nat. Acad. Sci. USA* **70**(11), 3240–3244 (1973).
6. U.S. Pat. 4,237,224 (Dec. 2, 1980), S. N. Cohen and H. W. Boyer (to Stanford University).
7. R. W. Old and S. B. Primrose, *Principles of Genetic Manipulation: An Introduction to Genetic Engineering*, University of California Press, Berkeley, 1985.
8. P. Carter and co-workers, *Bio/Technology* **10**, 163–167 (1992).

9. M. H. W. Huseman and E. T. Papoutsakis, *Appl. Microbiol. Biotechnol.* **31**, 435–444 (1989).
10. I. Chibata, T. Tosa, and T. Sato, *Adv. Biotechnol. Process.* **10**, 203–222 (1983).
11. R. L. Hamill, in J. D. Bu'Lock and co-eds., *Bioactive Microbial Products: Search and Discovery*, Academic Press, Inc., London, 1982, pp. 71–105.
12. A. A. Fantini, *Methods Enzymol.* **43**, 24–41 (1975).
13. D. A. Hopwood and co-workers, in *Genetic Manipulation of Streptomycetes: A Laboratory Manual*, John Innes Foundation, Norwich, UK, 1985.
14. S. M. Martin and B. D. Skerman, *World Directory of Micro-organisms*, Wiley-Interscience, New York, 1972.
15. H. B. Woodruff and L. E. MacDaniel, *8th Symposium of the Society for General Microbiology*, Society for General Microbiology, London, 1958, pp. 29–32.
16. T. C. Nelson, *Mutation and Plant Breeding*, publication no. 891, National Academy of Science, National Research Council, New York, 1961, pp. 331–349.
17. W. H. Bartholomew and H. B. Reisman, in H. J. Peppler and D. Perlman, eds., *Microbial Technology*, Vol. 2, 2nd ed., Academic Press, Ltd., London, 1979, pp. 463–496.
18. J. S. Alford, in D. R. Omstead, ed., *Computer Control of Fermentation Processes*, CRC Press Inc., Boca Raton, Fla., 1990, pp. 221–235.
19. J. A. Petersen, *Pharm. Tech.* **12**(7), 42–50 (1988).
20. G. J. Tolentino and co-workers, *Biotech. Lett.* **14**(3), 157–162 (1992).
21. G. J. Calton and co-workers in A. L. Demain and N. A. Solomon, eds., *Manual of Industrial Microbiology and Biotechnology*, American Society for Microbiology, Washington, D.C., 1986, pp. 436–447.
22. M. J. Weinstein and G. H. Wagman, *Antibiotics, Isolation, Separation and Purification*, Elsevier, Amsterdam, the Netherlands, 1978.
23. B. Atkinson and P. Sainter, *J. Chem. Tech. Biotechnol.* **32**, 100–108 (1982).
24. J. A. Asenjo and I. Patrick, in E. L. V. Harris and S. Angal, eds., *Protein Purification Applications*, IRL Press, Oxford, UK, 1990, pp. 1–28.
25. R. Yuan and M. Hsu, *Genet. Eng. News* **12**(3), 12–13 (1992).
26. J. D. Stowell and J. B. Bateson, in L. J. Nisbet and D. J. Winstanley, eds., *Bioactive Microbial Products*, Vol. 2, *Developments and Production*, Academic Press, Ltd., London, 1984, pp. 117–139.
27. R. R. Swartz, in G. T. Tsao, ed., *Ann. Rep. Ferm. Process.* **3**, 75–110 (1979).
28. B. E. Power and co-workers, *Gene* **113**, 95–99 (1992).
29. R. G. Buckholz and M. A. G. Gleeson, *Bio/Technology* **9**, 1067–1072 (1991).

General References

B. Atkinson and F. Mavituna, *Biochemical Engineering and Biotechnology Handbook*, 2nd ed., Macmillan Publishers Ltd., Basingstoke, UK, 1991. An exceptional collection of information on all aspects of fermentation with an exhaustive and up to date bibliography.

J. E. Bailey and D. F. Ollis, *Biochemical Engineering Fundamentals*, 2nd ed. McGraw-Hill Book Co., Inc., New York, 1986. A very good treatise describing the application of basic engineering principles to fermentation technology.

A. L. Demain and N. A. Solomon, eds., *Manual of Industrial Microbiology and Biotechnology*, American Society for Microbiology, Washington, D.C., 1986. An excellent source of practical applications technology.

W. M. Fogarty, *Microbial Enzymes and Biotechnology*, Applied Science Publishers Ltd., Barking, UK, 1983.

R. Gerals, ed., *Prescott and Dunns Industrial Microbiology*, 4th ed., The Avi Publishing Co. Inc., Westport, Conn., 1982.

H. C. Vogel, *Fermentation and Biochemical Engineering Handbook*, Noyes Publishing, Inc., Park Ridge, N.J., 1983.

H. J. Peppler and D. Perlman, eds., *Microbial Technology*, Vols. I & II, 2nd ed., Academic Press, Inc., New York, 1979.
A. H. Rose, ed., *Economic Microbiology*, Vols. 1–5, Academic Press, Ltd., London, 1977–1981. Provides good biochemical and physiological insights.
G. Tsao, ed., *Ann. Rep. Ferm. Process.*, Vols. 1–8, Academic Press, Inc., New York, 1978–1985. A collection of papers largely detailing and reviewing individual industrial processes.
D. I. C. Wang and co-workers, *Fermentation and Enzyme Technology*, John Wiley & Sons, Inc., New York, 1979.

SURJIT S. SENGHA
Alexin Pharmaceutical Corporation

FERMIUM. See ACTINIDES AND TRANSACTINIDES.

FERRICYANIDES. See IRON COMPOUNDS.

FERRITES

The term ferrite is commonly used generically to describe a class of magnetic oxide compounds which contain iron oxide as a principal component. In metallurgy (qv), however, the term ferrite is often used as a metallographic indication of the α-iron crystalline phase.

Some representatives of the ferrite family have long been known as magnetic minerals. For example, magnetite [*1317-61-9*], Fe_3O_4, or loadstone (1) was known in ancient times, ca 400 BC, and magnetoplumbite [*12173-91-0*], $Pb(MnFe)_{12}O_{19}$ (2), has been known at least since the beginning of the twentieth century. The real breakthrough, however, of ferrites as important materials for electrotechnical applications took place in the early 1950s (3,4). This breakthrough built upon pioneering work in the mid-1930s (5,6) followed by the theoretical framework in 1948 (7).

Ferrites can be classified according to crystal structure, ie, cubic vs hexagonal, or magnetic behavior, ie, soft vs hard ferrites. A systematic classification as well as some applications are given in Table 1 (see also MAGNETIC MATERIALS, BULK; MAGNETIC MATERIALS, THIN FILM).

The cubic ferrites, perovskite, garnet, and spinel, have quite different crystal structures and correspondingly different properties and applications. Perovskite ferrites, isostructural with the mineral perovskite [*9003-99-0*], $CaTiO_3$, contain iron in unusually high valency states, eg, +4, +5, +6 (8), giving rise to high and

Table 1. Systematic Classification of Ferrites

		Magnetic nature			Appearance			
Crystal chemistry[a]	Formula[b]	Soft	Intermediate	Hard	Poly-crystalline	Single crystalline	Bonded powder	Application
				Cubic				
spinel	$MeFe_2O_4$	X			X	X		recording heads
		X			X			core material for various inductors, transformers, and TV deflection units
	$CoFe_2O_4$, γ-Fe_2O_3		X				X	recording tape
garnet	RFe_5O_{12}		X			X		microwave, magnetooptics, bubble-information storage
perovskite	$R'FeO_3$					X		electroceramic devices
ortho-ferrite[c]	$RFeO_3$		X			X		bubble-information storage

(Continued)

Table 1. (*Continued*)

Crystal chemistry[a]	Formula[b]	Magnetic nature: Soft	Intermediate	Hard	Appearance: Poly-crystalline	Single crystalline	Bonded powder	Application
				Hexagonal				
magneto-plumbite	$R'Fe_{12}O_{19}$			X[d]	X		X	permanent magnets
	$R'Fe_{12-x}Me'O_{19}$			X[e]			X	tape for perpendicular recording
W (=MS)	$R'Me_2Fe_{16}O_{27}$			X[d]	X		X	permanent magnets[f]
	$R'Co_2Fe_{16}O_{27}$		X[e]			X		microwaves, shielding[f]
X (=M_2S)	$R'MeFe_{28}O_{46}$			X[d]	X		X	permanent magnets[f]
	$R'CoFe_{28}O_{46}$		X[e]			X		microwaves, shielding[f]
Y (=ST)	$R'_2Me_2Fe_{12}O_{22}$		X[e]		X	X		microwaves, shielding
Z (=MST)	$R'_3Me_2Fe_{24}O_{41}$			X[d]				permanent magnets[f]
	$R'_3Co_2Fe_{24}O_{41}$		X[e]		X			microwaves, shielding

[a] M = $BaFe_{12}O_{19}$; S = $Me_2Fe_4O_8$; T = $Ba_4Fe_8O_7$. See Figure 1.
[b] Me = Fe^{2+}, Ni^{2+}, Mn^{2+}, Co^{2+}, Zn^{2+}, etc; Me′ = Mn^{3+}, (Ti^{4+} + Co^{2+}), etc; R = Y, Nb, etc; R′ = Ba, Sr, Pb, Ca.
[c] These materials are orthorhombic.
[d] Preferred axis (uniaxial anisotropy).
[e] Preferred plane (planar anisotropy).
[f] Potential application.

almost metallic electrical conductivity. Representatives are $(CaSr)FeO_3$ (8) and $(PbCa)Fe_2O_5$ (9). Applications are mainly in the field of electroceramic devices, eg, electrical transducers (see CERAMICS AS ELECTRICAL MATERIALS). Closely related to the perovskite ferrites are orthoferrites where the Ca is replaced by a rare-earth element, resulting in a distorted perovskite structure, which is essentially orthorhombic. Orthoferrites, studied extensively in the early 1970s as potential data storage materials based on magnetic bubble domains (10), have been largely replaced by the garnet materials (see INFORMATION STORAGE MATERIALS).

Garnet ferrites, isostructural with the mineral pyrope, $Mg_3Al_2Si_3O_{12}$, have a special feature. These materials can only accommodate trivalent iron together with trivalent rare earths. The absence of other valencies gives rise to extremely high electrical resistivity and very low losses in the microwave region (11). As single crystals, these are transparent and exhibit interesting magnetooptical properties (12). In addition garnet ferrites are the preferred bubble-domain data storage materials (13).

Spinel ferrites, isostructural with the mineral spinel [*1302-67-6*], $MgAl_2O_4$, combine interesting soft magnetic properties with a relatively high electrical resistivity. The latter permits low eddy current losses in a-c applications, and based on this feature spinel ferrites have largely replaced the iron-based core materials in the r-f range. The main representatives are MnZn-ferrites (frequencies up to about 1 MHz) and NiZn-ferrites (frequencies $>>$ 1 MHz).

The soft magnetic spinel ferrites, by far the most important cubic ferrites, were first introduced by Philips under the trade name Ferroxcube (14) and are now widely commercially available under various trade names. The world market for soft magnetic ferrites amounts to about one billion dollars (1991), about 350 million dollars of which is in the United States.

Hexagonal ferrites cover a number of strongly related structures (Fig. 1) (15). The magnetization may be strongly bound to the hexagonal axis, the material exhibiting hard magnetic properties, or to the plane normal to this axis, such that the properties are mixed, ie, there is high permeability in the plane and low permeability in other directions. The group having mixed properties is called Ferroxplana (16). The hexagonal ferrites with uniaxial anisotropy are in principle interesting as permanent magnet material. However, W-, X-, and Z-type compounds (see Fig. 1) are not interesting economically because of relatively difficult processing.

Ferroxplana offer unique high frequency properties, being applicable where other high frequency materials such as NiZn ferrites fail (0.1–1 GHz). Some representatives are Co_2Z (17) and Zn_2Y (18). In spite of numerous investigations, application has always remained limited because of the difficult processing. Small-scale applications are in the field of microwaves (see MICROWAVE TECHNOLOGY) (19).

M-type ferrites, isostructural with the mineral magnetoplumbite, are by far the most important hexagonal ferrites. The M-type ferrites have a dominant position in the permanent magnet market, and are good candidates for another big volume market, ie, as base material for perpendicular recording tape. In addition there are some small-scale applications, eg, as microwave material at frequencies

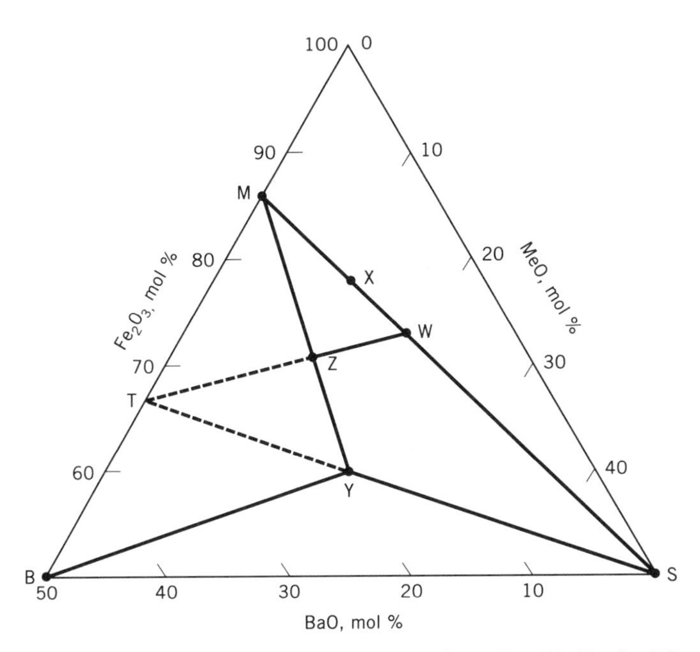

Fig. 1. Composition diagram for hexagonal ferrites where B = $BaFe_2O_4$, M = $BaFe_{12}O_{19}$, S = $MeFe_2O_4$, T = $BaFe_4O_7$, W = $BaMe_2Fe_{16}O_{27}$, X = $BaMeFe_{14}O_{23}$, Y = $BaMeFe_6O_{11}$, Z = $Ba_3Me_2Fe_{24}O_{41}$, and Me is as defined in Table 1.

> 20 GHz (19). Interest in barium-ferrite as a base material for particulate recording media has increased considerably since the invention of perpendicular recording in 1977. The Ba-ferrite medium is particularly suitable because it has a platelet crystallite shape with the preferred axis normal to the wide surface, thus allowing perpendicular alignment during the coating on the substrate film. Relatively low coercivity ($\simeq$ 100 kA/m) and small ($\simeq$ 50 nm) particle size are needed. This is realized by the partial substitution of Fe, eg, by using Co and Ti, and special powder preparation techniques. Despite promising characteristics (20,21), Ba-ferrite media have not yet found their way to the mass production market.

M-type ferrites are mainly used as permanent magnet material. They have largely replaced the alnicos as preferred permanent magnet material, as a result of the lower material and processing costs. These ferrites were first introduced under the trade name Ferroxdure, the isotropic form in 1952 (22) and the anisotropic (crystal oriented) form in 1954 (23), and are widely available commercially under various trade names such as Oxid and Koerox. They cover about 55% of the world market of permanent magnet materials, corresponding to 1100 million U.S. dollars (1991), as well as 55% of the U.S. market, at $300 million.

More information on noncommercially available ferrites can be found in the literature (4,24). Extended reviews on soft (4,24–29) and hard (4,29–32) ferrites are also available. Explanations of magnetic concepts used in this article, such as permeability and coercivity, may be found in magnetism textbooks.

Common Properties of Spinel Ferrites and M-Type Ferrites

The commercial sintered spinel and M-type ferrites have a porosity of 2–15 vol % and a grain size in the range of 1–10 μm. In addition, these materials usually contain up to about 1 wt % of a second phase, eg, $CaO + SiO_2$ on grain boundaries, originating from impurities or sinter aids.

Ferrites are oxides and thus rather inert with respect to water, bases, and organic solvents. However, they may be attacked by acids having sufficiently high strength (pH < 2), for instance

$$MeFe_2O_4(s) + 8\ HCl \rightarrow Me^{2+} + 2\ Fe^{3+} + 8\ Cl^- + 4\ H_2O \qquad (1)$$

where Me is defined as in Table 1. The reaction rate is rather limited because of the low specific surface. However, the second phase on the grain boundaries, eg, CaO, may be more sensitive to acids and this may induce a more serious attack, the more so when there is open porosity, ie, when porosity > 10 vol %. Being ceramic materials, ferrites are also resistant to high temperatures, at least up to the sintering temperature (1200–1400°C) (see CERAMICS). However, noticeable reduction may take place at temperatures > 1100°C and an oxygen partial pressure, P_{O_2}, below the equilibrium oxygen partial pressure, as described by the relation

$$\log P_{O2} = A/T + B \qquad (2)$$

where T is in Kelvin and A and B are specific for each ferrite (26). M-ferrites can hardly change in oxidation degree and the P_{O2}–T relation in fact represents the decomposition conditions. Spinel ferrites, however, allow considerable changes in the Fe^{2+} concentration without decomposition, but only accompanied by changes in the defect chemistry.

Ceramic ferrites cannot explode or release poisonous gases, and generally do not contain toxic elements. However, permanent magnets based on Sr-ferrite contain strontium [*7440-24-6*], Sr, which is in principle toxic. In dense (porosity < 10%) materials the Sr is firmly bound; however, in porous (porosity > 10%) materials the second phase may dissolve partially in water or acids giving rise to release of Sr. Even in the latter case the effect is limited. Such magnets are used in the stomach of cows (pH < 3) in order to collect iron-based particles eaten by the cow. Elements which can have different valencies, such as chromium, Cr; manganese, Mn; titanium, Ti; and vanadium, V, may be carcinogenic, in particular in the high valency state, eg, Cr^{6+}. Whereas these elements may be present in minor quantities in both hard and soft ferrites, generally they are firmly bound in the lattice and present in lower valency states. Nevertheless, upon chemical attack by an acid or during the preparation of those ferrites, some precautions are advisable.

Crystal Chemistry and Physical Properties

The magnetic properties of ferrites result from the electronic configuration and mutual interactions of the ions present. Thus investigation of the crystal structure is fundamental to the understanding of these materials. Although the specific structures of spinel ferrites and M-type ferrites differ, both classes can be considered to be composed of two sublattices: an anionic lattice having relatively large anions and a cationic lattice containing the smaller cations, which fill interstitial sites.

Spinel Ferrites. In spinel ferrites having the composition AB_2O_4, where A and B are metals, cubic close-packed oxygen ions leave two kinds of interstitial sites for the cations: tetrahedral or A-sites, surrounded by four oxygen ions; and octahedral or B-sites, surrounded by six oxygen ions. Figure 2 gives a schematic impression. The smallest crystallographic unit cell having the required symmetry contains eight formula units of AB_2O_4. Each unit cell has 64 tetrahedral sites, eight of which are occupied, and 32 octahedral sites, 16 of which are occupied. A wide variety of transition-metal cations can be fit into these interstitial sites. The most important family of spinels is $Me^{2+}Fe_2^{3+}O_4$, where Me = Mg, Mn, Fe, Co, Ni, Zn, Cu, etc, either singly or in combination. But similar ferrites having less than two Fe-ions per formula unit are also of industrial significance because of the high electrical resistivity.

The site preference of several transition-metal ions is discussed in References 4 and 24. The occupation of the sites is usually denoted by placing the cations on B-sites in structure formulas between brackets. There are three types of spinels: normal spinels where the A-sites have all divalent cations and the B-sites all trivalent cations, eg, Zn-ferrite, $Zn^{2+}[Fe_2^{3+}]O_4$; inverse spinels where all the divalent cations are in B-sites and trivalent ions are distributed over A- and B-sites, eg, Ni-ferrite, $Fe^{3+}[Ni^{2+}Fe^{3+}]O_4$; and mixed spinels where both divalent and trivalent cations are distributed over both types of sites, eg, Mn-ferrite $Mn_{0.8}^{2+}Fe_{0.2}^{3+}[Mn_{0.2}^{2+}Fe_{1.8}^{3+}]O_4$.

In the cases of existing unpaired d-electrons, transition-metal ions possess a net magnetic moment. In a spinel these magnetic moments interact through the anions (super exchange), resulting in a situation where the moments of both A-site and B-site ions are aligned, ie, A–A and B–B parallel, but A–B antiparallel, a ferrimagnetic ordering. The net magnetic moment per unit formula can be calculated from the distribution of the cations over these sites. For example, the inverse spinel $Fe^{3+}[Ni^{2+}Fe^{3+}]O_4$, where Fe^{3+} and Ni^{2+} have five and two un-

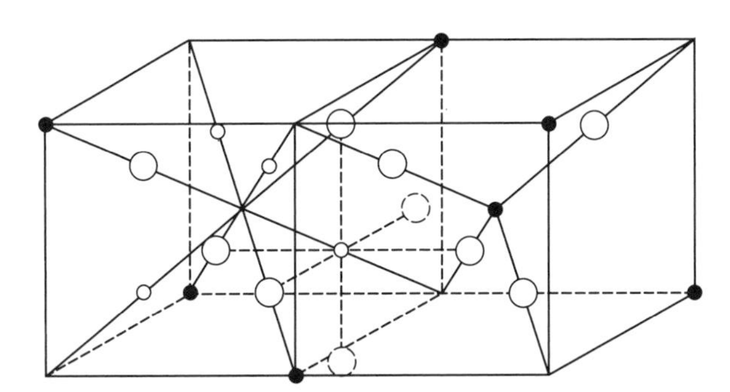

Fig. 2. Spinel structure where ○ is oxygen, ●, A-sites, ○, B-sites. A unit cell consists of eight fcc subcells with different cation occupants. Adapted from Reference 4.

paired electrons, respectively, has a magnetic moment of 5 Bohr magnetons, μ_B, on the A-site sublattice and a magnetic moment of $5 + 2 = 7\ \mu_B$ on the B-site sublattice. The opposite alignment of the two sublattices yields a net magnetic moment of 2 μ_B per formula unit. Because $1\ \mu_B = 1.1653 \times 10^{-29}$ Wb·m, there are eight formula-units per cell, and the lattice constant is 8.34×10^{-10} m, it can easily be calculated that this corresponds to a value for the saturation induction, B_s, of 0.32 tesla at 0 K. This is reasonably near the measured value of 0.35 tesla (Fig. 3). In comparing these data, it is assumed that all magnetic domains, in each of which the ferrimagnetic ordering holds often over many thousands of unit cells, are perfectly aligned by a strong applied magnetic field.

It is possible to systematically alter the net magnetic moment of ferrites by chemical substitutions. A very important industrial application is the increase of the magnetic moment in mixed MnZn-ferrites and NiZn-ferrites. When Zn ions are introduced in Mn-ferrite or Ni-ferrite, these ions prefer to occupy A-sites. Because Zn^{2+} is nonmagnetic, the A-sublattice magnetization is reduced and consequently the total net magnetic moment is increased. Figure 3**a** shows that for smaller Zn contents, magnetic data follow the expected relationships. However, high Zn content causes the net moments to decrease because some of the Fe^{3+} ions no longer have magnetic neighbors and interactions break up. Another decrease of the degree of magnetic moment alignment occurs when temperature is increased. Figure 3**b** shows the influence of thermal agitation. At a certain temperature, the Curie temperature T_c, the magnetic ordering completely vanishes.

The direction of the alignment of magnetic moments within a magnetic domain is related to the axes of the crystal lattice by crystalline electric fields and spin-orbit interaction of transition-metal d-ions (24). The dependency is given by the magnetocrystalline anisotropy energy E_a expression for a cubic lattice (33):

$$E_a = K_1(\alpha_1^2\alpha_2^2 + \alpha_1^2\alpha_3^2 + \alpha_2^2\alpha_3^2) + K_2\alpha_1^2\alpha_2^2\alpha_3^2 + \cdots, \quad (3)$$

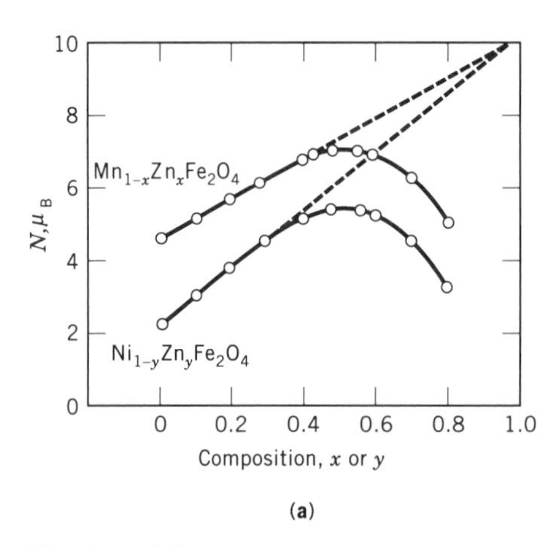

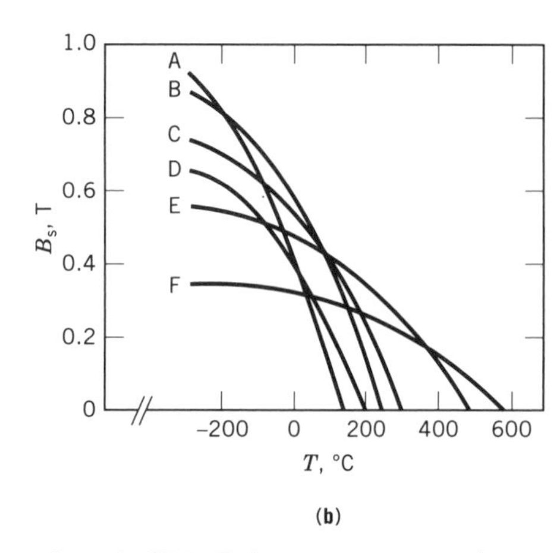

Fig. 3. (**a**) Saturation magnetic moment per formula unit, N, in Bohr magnetons vs chemical composition at 0 K and (**b**) saturation magnetic induction, B_s, in teslas vs temperature for mixed ferrites $Mn_{1-x}Zn_xFe_2O_4$ and $Ni_{1-y}Zn_yFe_2O_4$ where A, B, and C represent $x = 0.5$, 0.15, and 0, respectively, and D, E, and F represent $y = 0.65$, 0.2, and 0, respectively. To convert μ_B to J/T, multiply by 9.274×10^{-24}. Adapted from Reference 4.

where the α_i are direction cosines of the magnetization with respect to the three principal crystal axes and K_i are phenomenological anisotropy constants. The lowest order terms are the most relevant ones; terms higher than α_i^6 can be neglected. The constants K_1 and K_2 can be determined by measuring the torque exerted on a monocrystalline sample sphere by a rotated saturating magnetic field (33).

Figure 4 shows a typical result for K_1 as a function of temperature for a Fe^{2+}-containing MnZn-ferrite. K_1 changes sign at a certain temperature T_0 be-

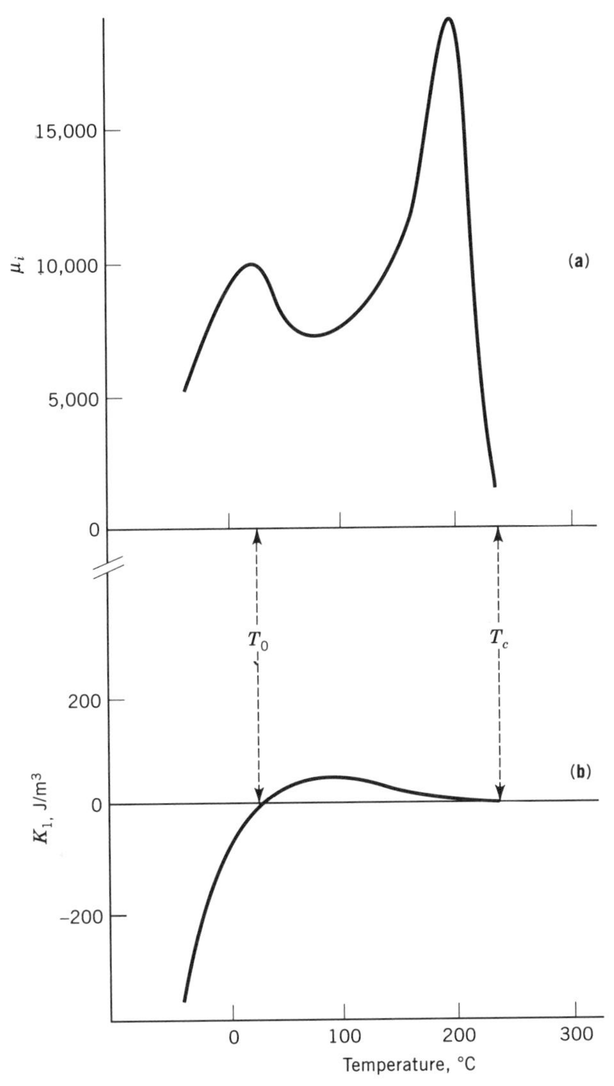

Fig. 4. The effect of temperature for $Mn_{0.6}Zn_{0.3}Fe^{2+}_{0.1}Fe^{3+}_2O_4$ on (**a**) initial magnetic permeability, μ_i, measured on a polycrystalline toroid applied as a core for a coil driven by a low ($B < 0.1$ mT) amplitude, low (10 kHz) frequency sinusoidal signal; and (**b**) magnetocrystalline anisotropy constant, K_1, measured on a monocrystalline sphere showing the anisotropy/compensation temperature T_0 and the Curie temperature, T_c. To convert joules to calories, divide by 4.184.

cause the positive contribution from the Fe^{2+} ions compensates the negative contributions from the other ions (24,34,35). From the expression for E_a it can be derived that the preferred direction of the magnetization changes from [111] to [100] at the temperature where $K_1 + (1/9)K_2 = 0$. This is not far from T_0 (35). Near T_0 the direction of the magnetization can easily be changed by applying just a small magnetic field H. In other words, the initial magnetic permeability $\mu_i = B/(\mu_0 \cdot H)$, where B is the resulting net macroscopic magnetic induction and $\mu_0 = 4\pi \times 10^{-7}$ Wb/(A·m), is high in that region (Fig. **4a**). The vanishing anisotropy near the Curie temperature leads to the so-called primary maximum in $\mu_i(T)$; the maximum at T_0 is known as the secondary maximum.

The net macroscopic B and the resulting μ_i result from two types of magnetization processes. First there is a contribution from the rotation of the magnetization inside each individual magnetic domain, from the preferred direction toward the direction of the applied magnetic field until the sum of the magnetostatic energy (minimal if B lies along H) and the anisotropy energy has reached its minimum value. Secondly, domain walls move. Domains having favorable magnetization directions with respect to the applied magnetic field grow at the expense of others, thus further minimizing the total magnetostatic energy. In view of the small magnitude of the applied field H, irreversible jumps of domain walls from pinning points such as grain boundaries, internal pores, or inclusions do not take place, but small reversible motions may readily occur. The extent to which each of the two mechanisms contribute to μ_i depends on quite a number of parameters: chemical composition, temperature, ceramic microstructure, stresses, frequency of H-field, etc. As for the microstructure, grain boundaries often play an essential role because of pinning of domain walls. The effective initial permeability depends more or less linearly on grain size (36,37). In small grains the rotational contribution dominates, whereas at increasing grain sizes the wall contribution becomes more and more important. A review of experimental data and recent theoretical models can be found in Reference 38.

Stresses, which can for example be introduced during cooling after ceramic sintering, during machining of sintered products, or simply when product parts are clamped together before use, lead to anisotropy by the magnetostriction effect (24,33,39). The influence of stress anisotropy can be large in cases of a small magnetocrystalline anisotropy, ie, in the regions of the $\mu_i(T)$ maxima, resulting in a considerable suppression of the maxima and sometimes even of the whole curve.

Magnetocrystalline anisotropy, magnetostriction, and magnetic permeability depend markedly on chemical composition. These dependencies have been extensively investigated within the ternary diagrams $MnFe_2O_4$–$ZnFe_2O_4$–Fe_3O_4 and $NiFe_2O_4$–$ZnFe_2O_4$–Fe_3O_4 (33,34,39,40). The saturation induction B_s and Curie temperature T_c are also composition dependent (40–42) as is the electrical resistivity, which varies markedly. Spinel ferrites are semiconductors (qv) (24). In case of an excess of iron appearing as Fe^{2+}, the conduction is of the n-type where electrons hop between different Fe-ions. The resistivity then is usually much smaller than for Fe-deficient ferrites. This relatively high (0.1 Ω·m to 10^8 Ω·m at 25°C) d-c resistivity is very important.

Properties can also be manipulated by adding specific dopants: Co^{2+} ions are, for example, introduced for extra anisotropy compensation (24,43–45); and

Ti^{+4} ions are substituted to form pairs with Fe^{2+} ions and thus to reduce electron hopping (45,46). Extensive investigations (47–50) have been carried out involving the addition of dopants such as CaO (typically 0.1 mol %) and SiO_2 (0.01 mol %) in order to provide ceramic grains having electrically insulating grain boundaries, thus markedly increasing the effective resistivity of the ferrite product.

At high frequencies ferrites exhibit energy losses resulting from various physical mechanisms at different frequencies and appearing as heat dissipation. Hysteresis losses arise from irreversible domain wall jumps. During each cycle of the H- and B-fields one hysteresis loop is completed and the loss per cycle is proportional to the area of the loop. A way to reduce hysteresis losses, ie, prevent domain wall jumps, is to reduce the number of inhomogeneities able to pin domain walls, eg, pores and impurities, and to reduce magnetocrystalline and stress anisotropy (51,52). Another method is to deliberately pin the walls, for instance by addition of Co^{2+} and Ti^{4+} ions or by using ceramic microstructures having small grains (53,54). A second important loss contribution comes from eddy currents, induced by alternating magnetic fluxes. This contribution can be limited by providing a high electrical resistivity (55,56). At high frequencies this is not easy at all, because insulating grain boundaries tend to become short circuited as a result of the permittivity of the ferrite (28). A third main loss contribution is from magnetic resonances. Above about 1 MHz this is usually the dominant contribution. When the driving frequency is in resonance with the natural frequency at which the magnetization rotates, there is a large peak in power absorption. This effect can be seen when the magnetic permeability is considered as the complex parameter $\mu = \mu' + j\mu''$. The ratio μ''/μ' is usually expressed as $\tan \delta$, where δ is the so-called loss angle, the phase-lag of the magnetic induction with respect to the applied magnetic field. For inductors in electrical circuits, δ expresses the phase difference between voltage and current. Figure 5 shows μ' and μ'' as functions of

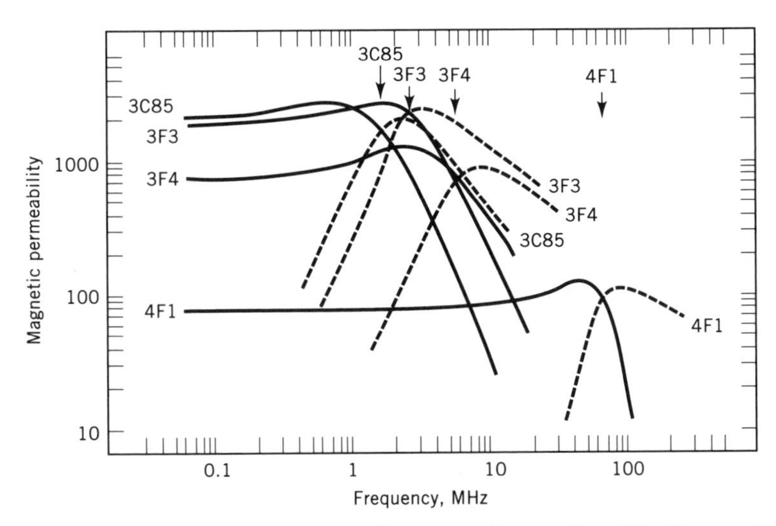

Fig. 5. Complex magnetic permeability vs frequency for a series of ferrites used for power transformers and inductors at 25°C, $B < 0.1$ mT: (——) represents real parts μ', (- - -) show the imaginary parts μ''. The arrows indicate the frequencies where $\tan \delta = \mu''/\mu' = 1$ (57).

frequency for some commercial ferrites. These ferrites show large differences concerning the frequencies where μ'' starts to come up and accordingly, in the resonance frequencies, where $\tan \delta = 1$. But also the low frequency permeabilities differ. The resonance frequency is inversely proportional to the low frequency permeability (58). Applying this, shifting resonance to high frequencies can be realized by small grain sizes. The ferrites of Figure 5 are optimized for a range of frequencies and induction levels (57).

M-Type Ferrites. The M-structure borrows its name from the lead-containing magnetic mineral magnetoplumbite on which this structure was based in 1938 (59). Once the technical importance of M-ferrites had become clear, the structure of $BaFe_{12}O_{19}$ was determined in more detail, together with the related structures of other hexagonal ferrites (60). Figure 6 shows the unit cell of the M-structure, which corresponds to two formula units. Its symmetry is characterized by the space group $P6_3/mmc$. The structure shows a closest packing of O- and Ba-ions; Fe-ions are in interstitial sites. The structure is built up from smaller units: a cubic block, S, having the spinel structure and a hexagonal block, R, containing the Ba-ions. The Fe-ions are located at five different crystallographic positions; cell dimensions and theoretical densities are given in Table 2 (59,61,62). Various substitutions are possible in $BaFe_{12}O_{19}$ (27). Some important ones are Sr and Pb for Ba; and Mn, Al, and Cr for Fe. If the substituting ion does not have the same charge, the charge must be compensated, eg, $BaFe_{12-x}Fe^{2+}_xO_{19-x}F^-_x$ and $BaFe_{12-2x}Ti^{4+}_xCo^{2+}_xO_{19}$.

The magnetism of $BaFe_{12}O_{19}$ comes from the Fe^{3+}-ions, each carrying a magnetic moment of 5 μ_B. These are aligned by either parallel or antiparallel ferromagnetic interaction. Ions of the same crystallographic position are aligned parallel, constituting a magnetic sublattice. The interaction between neighboring ions of different sublattices is a result of superexchange by oxygen. The theory predicts that the atomic moments are parallel when the Fe–O–Fe angle is about 180° and antiparallel when this angle is about 90° (4). The most probable spin configuration is that represented schematically in Figure 6 (66). Experimental verification was first based on the measurement of the saturation magnetization. Additional evidence for the existence of the five sublattices and their mutual alignment is obtained from electron diffraction analysis (67) and Mössbauer spectroscopy (68). It is the magnetic structure in terms of sublattices and their mutual orientation that governs magnetic behavior, which in turn is described in terms of intrinsic and material properties.

The intrinsic magnetic properties may be subdivided into primary and secondary. The primary properties (Table 2), such as the saturation magnetization J_s and the magnetocrystalline anisotropy constant K_1, are directly related to the magnetic structure. The secondary properties, such as the anisotropy field strength H_A and the specific domain wall energy (γ_w), are derived from the primary ones. These latter govern the actual magnetic behavior. The temperature dependence of the primary magnetic properties is shown in Figure 7.

The saturation magnetization, J_s, is the (maximum) magnetic moment per unit of volume. It is easily derived from the spin configuration of the sublattices: eight ionic moments and, hence, 40 μ_B per unit cell, which corresponds to J_s = 668 mT at 0 K. This was the first experimental evidence for the Gorter model (66). The temperature dependence of J_s (Fig. 7) is remarkable: the J_s–T curve is much

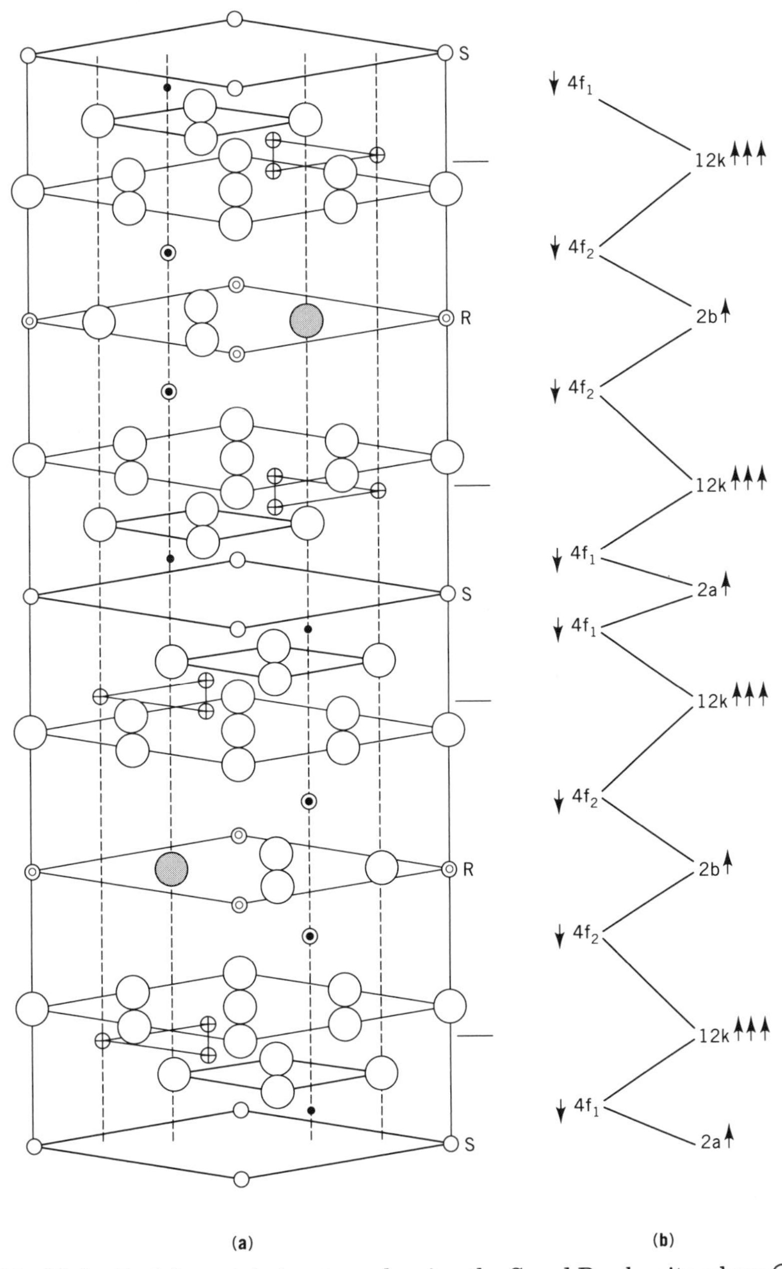

Fig. 6. The M-ferrite (**a**) crystal structure showing the S and R subunits where ○ is O^{2-}; ◍, Ba^{2+}; and ⊕, ⊙, ○, •, and ◎ are all Fe^{3+}at 12k, $4f_2$, $4f_1$, 2a, and 2b positions, respectively; (**b**) magnetic structure where the arrows represent size and spin direction of unpaired electrons at the various crystallographic positions.

Table 2. Properties of M-Ferrites[a]

Parameter	Ferrite[b]		
	BaM	SrM	PbM[c]
Crystallographic properties			
lattice constants, nm			
a	0.5893	0.588	0.588
c	2.3194	2.307	2.302
molecular weight	1112	1062	1181
density, g/cm^3	5.28	5.11	5.68
Reference	61	59	62
Primary magnetic properties			
saturation magnetization, J_s, mT	478	478	220
σ_s, μV·s·m/kg	90.4	93.4	70.8
anisotropy constant,[d] K_1, kJ/m^3	330	360	250
Curie temperature, T_c, K	740	750	725
Reference	63	63	64

[a]Measurements are at room temperature.
[b]M = $Fe_{12}O_{19}$.
[c]The exchange energy coefficient, A, is 5.1×10^{-12} J/m (1.2×10^{-12} cal/m) (69).
[d]To convert joules to calories, divide by 4.184.

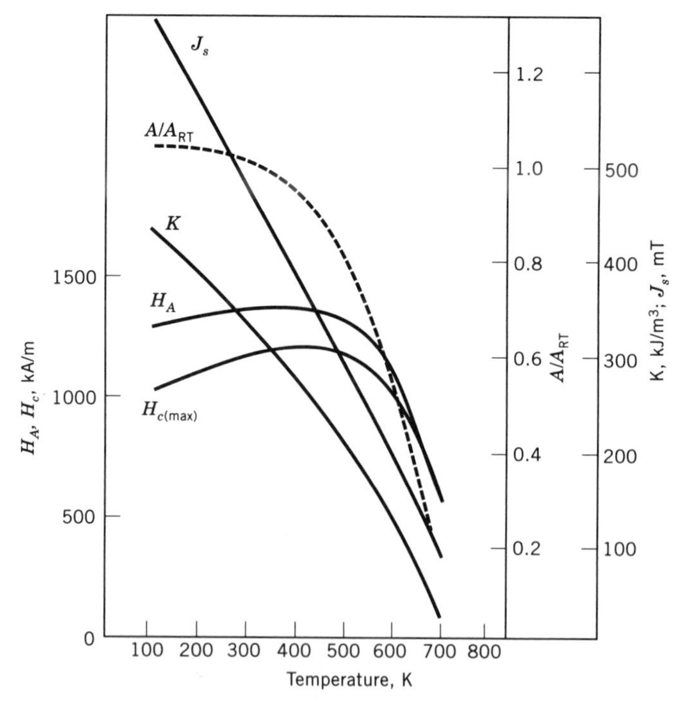

Fig. 7. Temperature dependence of J_s, K_1, H_A, and $H_{c(max)}$ for (——) BaM (4) and of A for (---) PbM (69). See text. To convert joules to calories, divide by 4.184.

less rounded than the usual Brillouin function (4). This results in a relatively low J_s value at RT (Table 2) and a relatively high (-0.2 %/°C) temperature coefficient of J_s. By means of Mössbauer spectroscopy, the temperature dependence of the separate sublattice contributions has been determined (68). It appears that the 12k sublattice is responsible for the unusual temperature dependence of the overall J_s.

The exchange energy coefficient A characterizes the energy associated with the (anti)parallel coupling of the ionic moments. It is directly proportional to the Curie temperature T_c (70). Experimental values have been derived from domain-width observations (69). Also the temperature dependence has been determined. It appears that A is rather stable up to about 300°C. Because the Curie temperatures and the unit cell dimensions are rather similar, about the same values for A may be expected for BaM and SrM.

The magnetization is strongly bound to the hexagonal c-axis, owing to spin-orbit coupling of the Fe-ions, in particular on the 2b sites (65). The energy associated with this phenomenon is characterized by the anisotropy constant K_1. Higher order constants (K_2, K_3) are negligibly small. Whereas there is some difference between the RT values (Table 2), the temperature behavior (Fig. 7) is quite normal.

The secondary magnetic properties characterize the actual magnetic state. The latter minimizes the three energies involved: the exchange energy, E_e, the anisotropy energy, E_a, and the magnetostatic energy, E_m, which are characterized by the values of A, K, and J_s, respectively. Some main secondary properties are given in Table 3. The specific wall energy, γ_w, represents a combination of both E_e and E_a, because walls are present to reduce E_m, but their internal structure is not favorable for E_e and E_a. The critical diameter for single-domain behavior, D_c, is the diameter below which magnetic domains are unfavorable in an isolated spherical particle. Although M-ferrite particles are not spherical and magnetostatic interactions between the particles also play a role, D_c remains an important indicator for the grain size needed in high quality magnets. In the absence of domains magnetization reversal proceeds by rotation. The ratio E_a/E_m determines the rotation mechanism. For M-ferrites, where $E_a/E_m > 0.36$, rotation is completely coherent (70). The anisotropy fieldstrength, H_A, is the maximum internal fieldstrength needed for magnetization reversal by coherent rotation.

The maximum coercivity $H_{c(\max)}$ corresponds to H_A, but refers to the external field. It explicitly takes into account the self-demagnetizing field of the crystal

Table 3. Secondary Magnetic Properties for SrM[a,b]

Parameter	Definition	Value
specific wall energy,[c] γ_w, J/m^2	$4(AK)^{1/2}$	54.2×10^{-4}
critical diameter single-domain behavior, D_c, μm	$18\ \mu_0 \gamma_w J_s^{-2}$	0.54
anisotropy energy/magnetostatic energy, E_a/E_m	$2\ \mu_0 K J_s^{-2}$	4.0
anisotropy field, H_A, kA/m	$2\ K J_s^{-1}$	1,506
maximum coercivity, $H_{c(\max)}$, kA/m	$H_A - N J_s\, \mu_0^{-1}$	1,240

[a] $A = 5.1 \times 10^{-12}$ J/m, $K = 360$ kJ/m^3, $J_s = 478$ mT (Table 2).
[b] Self-demagnetizing factor of crystals $N = 0.7$; $\mu_o = 4\ \pi \times 10^{-7}$ Wb//(A·m).
[c] To convert joules to calories, divide by 4.184.

(NJ_s/μ_0) as governed by the self-demagnetizing factor N. The latter ranges from 0 (for needles) to 1 (for thin plates). For platelet-shaped M-ferrite crystals N ranges from 0.6 to 0.9. $H_{c(\max)}$ represents an upper limit for the coercivity of an aligned assembly of noninteracting crystals and 0.48 $H_{c(\max)}$ the same for an isotropic assembly (71). Real coercivity values are much smaller resulting from the formation of transient domains and magnetostatic interactions.

The temperature dependence of H_A and $H_{c(\max)}$ is also shown in Figure 7. H_A appears to increase slightly and $H_{c(\max)}$ to increase clearly up to 300°C, implying a positive temperature coefficient for the coercivity. This is indeed observed, although $H_{c(\max)}$ describes only an idealized case.

The intrinsic properties may be modified by substitution (31). Ba can be fully replaced by Sr or Pb and partly by Ca (< 40 mol %). CaM, stabilized with 0.03 mol % La_2O_3, is also possible. The intrinsic properties of these M-ferrites vary somewhat and other factors such as sintering behavior and price of raw materials often dictate the commercial viability. Large-scale production is concentrated on BaM and SrM. High quality magnets are generally based on SrM, and somewhat lower priced magnets are based on BaM.

Substitution for Fe^{3+} has a drastic effect on intrinsic magnetic properties. Partial substitution by Al^{3+} or Cr^{3+} decreases J_s without affecting K_1 seriously, resulting in larger H_A and H_c values. Substitution by Ti^{4+} and Co^{2+} causes a considerable decrease in K_1; the uniaxial anisotropy ($K_1 > 0$) may even change into planar anisotropy ($K_1 < 0$). Intermediate magnetic structures are also possible. For example, preferred directions on a conical surface around the c-axis are observed for In^{3+} substitution (72). For a few substitutions the K_1 value is increased whereas the J_s value is hardly affected, eg, substitution of Fe^{3+} by Ru^{3+} (73) or by Fe^{2+} compensated by La^{3+} at Ba-sites (65).

The magnetic material properties of a permanent ferrite magnet are essentially characterized by the J–H loop and in particular by the demagnetization curve. The latter, in turn, is characterized by the two end points, remanence B_r and coercivity H_{cJ}, as well as by one overall performance figure. The material properties are governed by the intrinsic properties, but also by a number of microstructural factors such as grain size and shape, volume fraction of ferrite phase, and alignment.

For medium and high grade materials where domains are absent in the remanent state, the remanence is given by:

$$B_r = f(d/d_x)s\cdot J_s \tag{4}$$

where f is the degree of alignment: (d/d_x) the relative density, hence, the volume fraction of the solid; and s the fraction of pure ferrite in the solid. Typical values are $f = 0.5$ (isotropic) and $f = 0.9$ (anisotropic); $(d/d_x) = 0.9$ (sintered) and 0.6 (plastic bonded); $s = 0.96$. For high B_r material such as that used for loudspeakers, high density and high alignment are crucial.

Typical anisotropic sintered materials have a grain size of 1 μm, ie, somewhat larger than D_c, but sufficiently small to avoid domains down to considerable counter fields. Magnetization reversal proceeds by nucleation and growth of (transient) domains (74). On a macroscale the reversal process is nonuniform, being governed by the initiation and growth of multicrystal reversed regions (75). On this basis, the next expression for the coercivity has been proposed (75):

$$H_{cJ} = aH_A - bJ_s/\mu_0 = H_n - N(B_r + J_s)/\mu_0 \tag{5}$$

where the first term represents the average internal field needed to nucleate a reversed domain H_n. The coefficient a ranges from 0 to 1 and depends mainly on the grain size. The second term represents the effect of internal demagnetization fields. The coefficient b ranges from 0 to 2 and depends on the remanence and the crystal demagnetizing factor N, determined by the crystal shape. Although this coercivity expression has been derived for anisotropic sintered materials, it appears to apply also to isotropic ones (76).

For plastic-bonded materials, no clear-cut expression for the coercivity is known. It may be expected that it is rather similar, but with a smaller influence of B_r. For loosely packed powders, the B_r influence has become zero and H_n should be multiplied by 0.48 to account for the isotropy (71). In all cases high coercivity is obtained by using small grains with limited plate-like shape, ie, the value of N is not too high.

The performance of a magnet is characterized by a combination of B_r and H_c. For static applications at high $B/\mu_o H$, the $(BH)_{max}$ value can be used. For dynamic applications, however, $(BH)_{max}$ is less indicative, because it is hardly sensitive or even insensitive to improved H_c values. In that case the product of B_r and H_{cJ} is more suited (77). The parameter K (mT) $= B_r$ (mT) $+ 0.4\, H_{cJ}$ (kA/m) is a measure for the processing quality (costs). It is in principle independent of the Al^{3+} or Cr^{3+} content with which the B_r/H_{cJ} ratio can be varied, while the processing remains equivalent.

Physical properties other than magnetic ones are summarized in Table 4, including the influences of porosity and of the measuring direction, ie, parallel (‖) or perpendicular (⊥) to the preferred axis. Nearly all properties are clearly anisotropic, the ratio between both values being about 2. For application, the mechanical tensile strength is most important. It is very sensitive to microstructural factors and to the measuring method. For that reason there is always a large scatter in measuring results. Mostly, the flexural strength is measured instead of the tensile strength because the scatter in results is less. The intrinsic material strength depends on overall factors, such as porosity, second phase, and internal stress level. The latter is considerable when, upon cooling from sintering, the

Table 4. Physical Properties of BaM and SrM

Property	Value ⊥	Ratio ‖/⊥	M-ferrite Ba	M-ferrite Sr	Remarks	References
Vickers hardness, kN/mm^2	5.6	1.5		X	depends on porosity	79
strength, N/mm^2						
tensile	73	0.5	X		depends on porosity;	78
flexural	178–255	≃ 0.65		X	sensitive to	80
compressive	785				imperfections and measuring method	81
Young's modulus, kN/mm^2	211–317	1.2–2.0	X		depends on porosity	78,82
thermal expansion coefficient, %/K	10^{-5}	1.4			depends slightly on temperature until T_c	78

anisotropic thermal shrinkage is not free in both directions, eg, in the case of radially oriented rings (78). The actual strength of a product depends also on local factors such as inhomogeneities in density, alignment, grain size, and the presence of initial cracks. For complicated products, such as motor segments (arcs), the local factors are often predominant.

Processing

Commercial ferrites are produced by a ceramic process involving powder preparation, shaping, firing, and finishing (see CERAMICS). The powder preparation is usually the classical one involving the mixing of powder raw materials, prefiring, milling, and granulating. The raw materials are oxides or carbonates, the main component always iron oxide. The purity of the raw materials is an important factor with respect to the processing and final quality. Mixing can be done in different ways, depending on the nature and quality of raw materials and of the final product. During prefiring the different compounds react in the solid state to form the final compound or intermediate compounds, losing the volatile substances such as CO_2. Mostly this process is accompanied by homogenization on a local scale. In addition, there is some densification. To limit the effect of densification and to facilitate the handling of the material, it is often granulated before prefiring. To enable shaping and sintering, the prefired material has to be milled down to micrometer-sized particles. The last milling step is generally wet milling to prevent agglomeration effects. During or after milling, binders and lubricants are usually added to facilitate granulating and pressing.

In some cases it may be advantageous to deviate from the classical technology. For example, in wet-chemical preparation better chemical and morphological control may be achieved by starting from salt solutions.

Shaping is often done by dry pressing, which in fact is a simple and effective method to make the variety of shapes needed for electronic applications. Dry pressing requires a drying and granulating step after wet milling. For high grade, M-type ferrites, usually wet pressing is applied for reasons related to the field aligning of the powder particles. Special shaping techniques such as injection molding, extrusion, and isostatic pressing may be applied to realize special shapes or high product qualities.

During firing, formation of the proper compound is completed and densification occurs from about 50 to 90% solid by volume, implying a linear shrinkage of 10–25%. This shrinkage has to be anticipated by the pressing dimensions, constituting the shrinkage allowance. During the sintering process grain growth also occurs. Because the grain size and the grain boundary state are key factors for establishing the final electrical properties, control of grain size and grain boundaries is crucial. Generally a main factor is control of the second phase by ensuring the purity of raw materials and milling additives.

Pressing and sintering together determine the size and shape of the final product, but these are not well controlled with respect to the critical parts which determine the airgap in the magnetic circuit. For that reason a finishing touch by machining is necessary.

Spinel Ferrites. Prefiring is usually carried out in an air atmosphere in a continuous rotary kiln. In such a kiln the material is transported through a heat zone typically of 900–1100°C, in a rotating tube inclined at a small angle, which transports the powder downward along its length by a tumbling action. The angle is predesigned for a proper heating time and an economical throughput. When the mixture of raw materials enters the heat zone, carbonates and higher oxides decompose and a sequence of solid-state reactions occurs, starting with the formation of Zn-ferrite and ending with the formation of the desired MnZn-ferrite or NiZn-ferrite (83). Usually the aim is not a 100% spinel structure after prefiring. A 50–80% one usually suffices because the remaining conversions take place during the final sintering process after the forming step. Too high prefiring temperatures would result in considerable shrinkage in this stage, which makes the ferrite hard and thus difficult to mill. The prefired powder is characterized by x-ray diffraction, by the BET specific surface or the Fisher number, and sometimes by the inductance of a coil wound on a toroid pressed from the prefired powder. The prefired and subsequently milled powder has to be such that it results in a predictable and very constant shrinkage of pressed products during final sintering, in order to satisfy tight demands normally imposed on final product dimensions or to be able to realize these dimensions by grinding.

Dry pressing requires free-flowing spherical granules. These granules are usually made by spray drying slurries of 50–80 wt % ferrite in water, to which 1–4 wt % organic binder, eg, an acrylate or poly(vinyl alcohol) and a dispersant, eg, polyethylene glycol, are added (see DISPERSANTS). In this process the slurry is sprayed and atomized into droplets, which are subsequently dried in whirling hot air and collected as dry granules. To enhance the flowability, a lubricant such as zinc stearate or ammonium stearate is often added before spray drying. The granules must be solid enough to prevent the formation of dust during transportation, but must easily deform during pressing to facilitate good compaction. Granule sizes are usually in the 50–500 μm range depending on the size and the geometry of the products to be pressed. Applied pressures are 50–150 MPa (7,250–21,750 psi) typically, resulting in pressed densities of the order of 2–3 g/cm^3.

If products having large aspect ratios, eg, relatively long products with small cross sections, have to be pressed, powder–wall friction tends to result in locally low pressed densities. Abrupt changes of cross sections can also give rise to considerable density gradients. The result can be serious product deformation during sintering–shrinkage. These gradients can be reduced by compressing granulates from two sides by moving the die in the direction of the punch. In addition, powder–wall friction can be reduced by die wall coating, applying a thin, very smooth, wear-resistant layer.

The choice of the granulate binder system is also determined by the desirability of obtaining a high green strength of the pressed products in order to facilitate handling without damage and by the necessity of avoiding residues after binder burnout.

Binder burnout is usually the first part of the final firing cycle. In this cycle spinel structure formation is completed, shrinkage and formation of microstructure take place, and multivalent ions are given the desired valencies. Thus the final mechanical, electrical, and magnetic properties of the ferrite are determined. This firing is performed in large, very precisely computer-controlled kilns, which may be either continuous tunnel kilns through which the ferrite products travel or stationary batch kilns. NiZn-ferrites can usually be sintered in a simple constant atmosphere like air or nitrogen. MnZn-ferrites, however, especially those designed to contain Fe^{2+} ions, require sophisticated computer control of atmosphere during firing. The Fe^{2+} content is of paramount importance to the magnetic anisotropy, the permeability, electrical resistivity, energy losses, etc. The Fe^{2+}–Fe^{3+} balance depends on the concentration of oxygen ions in the spinel structure, and because this depends on the oxygen equilibrium between the ferrite and the surrounding atmosphere during firing, atmosphere is determining for the Fe^{2+}–Fe^{3+} balance. Thus a well-controlled oxygen partial pressure P_{O_2}, eg, in an N_2-atmosphere, is required. In order to obtain a desired Fe^{2+} content during cooling, P_{O_2} has to be adapted in such a way that log P_{O_2} varies linearly with $1/T$ (84,85). This may lead to P_{O_2} values as low as 0.01–0.001% at 1000–900°C, requiring very strict technical precautions. Below these temperatures, diffusion rates in the ferrites become sufficiently low to make the atmosphere less critical.

Actual temperature and atmosphere curves depend markedly on the material properties to be realized. High permeability ferrites, for example, require a microstructure with large (eg, 10–40 μm) grains without internal pores that could act as pinning points for magnetic domain walls and thus reduce permeability. This microstructure can be realized by using pure raw materials, applying reactive milling after prefiring, establishing relatively slow grain growth to ensure that the nonsoluble constituents present are transported to grain boundaries, and applying a high (up to 1400°C) top temperature for several hours. The drawback of a high sintering temperature, however, is that considerable Zn-evaporation readily occurs. A way to handle this is to provide a high P_{O_2} at the top temperature, then decreasing the P_{O_2} during cooling to correct the oxygen content of the ferrite.

High frequency power ferrites require small (a few μm) grains having electrically insulating grain boundaries. Small grains can be realized by less reactive milling after prefiring or by applying a sintering curve that has a relatively low top temperature during not too long a time. The cooling part of the firing curve has to be slow enough to facilitate segregation of additives to grain boundaries in order to provide for electrical insulation.

M-Type Ferrites. There are a variety of processing routes (30,32) for the four main classes of M-ferrite magnets: (*1*) sintered, (*2*) plastic bonded, (*3*) anisotropic, and (*4*) isotropic. Discussion herein is limited to the manufacture of high grade anisotropic SrM (30). As raw materials, dry powders of strontium carbonate [*1633-05-2*], $SrCO_3$, Fe_2O_3, and additives such as silica [*7631-86-9*], SiO_2, and boric trioxide [*1303-86-2*], B_2O_3, are employed. The desired quantities, eg, mole ratio Fe_2O_3–SrO $\simeq$ 5.5, are weighed and dry-mixed in a Müller-type mixer. The mixture is granulated in a disk agglomerator to granules of about 5 mm and subsequently prefired at about 1250°C in air, by using a rotary kiln. During prefiring the raw materials react to form the desired compound, $SrFe_{12}O_{19}$. The hard prefired granules are wet-milled in steel-ball mills to a fine powder. The resulting thick suspension or slurry is processed to an aligned pressed product by wet-pressing or pressure filtration in a magnetic field. During this complicated operation aligning of the crystallites, removal of water, and shaping to the desired product is performed (86). After drying, the compacts are sintered at about 1250°C in air by using electrical or gas-fired furnaces. During sintering anisotropic shrinkage occurs, $\simeq$ 15% perpendicular and $\simeq$ 30% parallel to the preferred direction. Accurate dimensional control is not possible. For that reason grinding, at least of the pole face, is necessary.

The most important microstructural demands are high aligning degree, high density, and small grain size. These impose demands on the sintering process and the preceding operations. During sintering considerable densification must be realized without allowing significant grain growth. These more or less contradictory demands are fairly well realized by using sinter additives such as SiO_2 (87) or B_2O_3 and SiO_2 (88). The effect of SiO_2 has been investigated extensively. When added in the right quantity, SiO_2 and the excess SrO form a temporary liquid phase which promotes shrinkage while grain growth is suppressed. The latter is related to the dissolving of SiO_2 in the M lattice and the formation of an Si-enriched region near the grain boundaries (segregation) (87). Important demands for the pressed products are high and homogeneous pressed density and high degree of aligning. The powder particles in the milling slurry have to be single-crystalline and free-movable in view of the aligning process, and sufficiently small in view of the sintering process. To enable a good milling performance, the grain size in the prefired granule must not be too large. Control of grain growth during prefiring can be attained by using SrO in excess in combination with additives such as B_2O_3 and SiO_2 (89). In addition, the conversion to SrM must be sufficient, otherwise a high B_r is no longer attainable.

Since the 1960s, the quality of M-ferrite manufactured has improved continuously, while the price has decreased considerably. Decisive progress in quality was obtained by the application of sinter additives (90), the introduction of pressing in a magnetic field (23), and the use of Sr instead of Ba, in combination with a sophisticated application of SiO_2 (91,92). Important contributions to price reduction came from the development of fast multiple-die pressing and the introduction of cheap raw materials (89). In spite of the latter, the performance, in particular the H_{cJ}, could be improved by a better control of prefiring, milling, and sintering. More recently, developments in material quality have been focused on realizing extremely high H_{cJ} values without losing too much in B_r.

Trends in the field of economics are the centralization of the powder fabrication to enable production on a large scale and the manufacture of low quality anisotropic materials by a much less expensive technology. An example of the latter is the introduction of alignment during pressing of the raw material mixture in the fabrication route of isotropic materials.

Uses

Spinel Ferrites. The number of applications of spinel ferrites is very large and growing. Table 5 gives a schematic impression of the main application areas and functions (93). In radio, television, and measuring equipment ferrite cores are extensively applied as inductors in LC-filters. In telecommunication ferrites serve the same purpose, but have clearly higher demands on quality factor Q and temperature-plus-time stability of the inductor. High Q-values require a ferrite core having low energy losses, $\tan\delta / \mu_i$, at small signals in the relevant frequency ranges, and the temperature factor $\alpha_F = (\Delta\mu_i/\Delta T) / \mu_i^2$, derived from the slope of the $\mu_i(T)$ curve, has to be within very narrow limits. Because the inductance L is proportional to μ_i, this also determines the temperature dependence of L, which has to be such that it compensates for the temperature drift of the capacitor in the filter. The inductance of a coil in a circuit can be fine-tuned by bridging the

Table 5. Applications and Functions of Spinel Ferrites[a]

Function	Measurement and control	Car electronics	Telecommunication	Electronic data processing	Consumer electronics	Power conversion	Lighting	Household appliances	Electric tools	EMC[b] equipment and services
coils										
tuning	X				X					
filter			X							
deflection				X	X					
proximity switches	X									
delay lines	X			X						
EMI[c] filters	X	X	X	X	X	X	X	X	X	X
absorbing surfaces										X
transformers										
wide-band			X							
power		X	X	X	X	X	X	X		
line-output				X	X					
current	X	X	X	X	X	X	X	X		
drive	X	X	X	X	X	X	X	X	X	
rotating					X					
output chokes			X	X	X	X	X	X		
magnetic regulators			X	X	X	X				
magnetic heads			X	X	X					

[a]Ref. 93.
[b]EMC = electromagnetic compatibility.
[c]EMI = electromagnetic interference.

gap between two ferrite core halves with an inserted ferrite adjuster. If a ferrite experiences some kind of magnetic, thermal, or mechanical disturbance, the initial permeability is instantaneously increased to an unstable value as a result of a changed configuration of magnetic domain walls. From that point it returns to its original level by relatively slow diffusion processes, often associated with preferred distribution of Fe^{2+} ions or cation vacancies in the spinel structure (94,95). Filter applications require high stability, expressed by a small disaccommodation factor $D_F = \Delta\mu_i / (\mu_i \log_{10} t_2/t_1)$, describing the relative change of μ_i during a time interval (t_1, t_2) after a disturbance. Table 6 presents some specifications for typical filter ferrite grades. Stability can, among other things, be promoted by substituting stabilizers, eg, Ti^{4+} or Sn^{4+}, into the ferrites (96).

As electronic equipment is increasingly used electromagnetic compatibility (EMC) has become an issue of fast-growing importance (see ELECTRONIC MATERIALS). Emission of unwanted signals has to be limited, as does the sensitivity of equipment to incoming interferences. These signals are subjected to international and national regulations. For both purposes ferrites are being increasingly

Table 6. Properties of MnZn and NiZn Ferrite Grades[a]

Parameter	Filter inductors				EMI-Suppression				Wide-band transformers				Power transformers and inductors			
ferrite																
grade	4C6	3D3	3H3	3H1	4C65	4A15	3S1	4S2	3E1	3E4	3E5	3E6	3C85	3F3	3F4	4F1
type	NiZn	MnZn	MnZn	MnZn	NiZn	NiZn	MnZn	NiZn	MnZn	MnZn	MnZn	MnZn	MnZn	MnZn	MnZn	NiZn
μ_i (±20%) at <10 kHz, <0.1 mT, 25°C	100	750	2,000	2,300	125	1,200	4,000	700	3,800	4,700	10,000	15,000	2,000	1,800	750	80
$\tan \delta/\mu_i \times 10^6$ at <0.1 mT, 25°C																
10 kHz				<1.5								<10				
30 kHz			<1.6								<25	<30				
100 kHz			<2.5	<5					<20	<20	<75					
300 kHz		<10							<150	<150						
1,000 kHz		<30				<250										
3,000 kHz	<60				<80	<1500										
10,000 kHz	<100				<130											
$\alpha_F \times 10^6$ at <10 kHz, <0.1 mT, K^{-1}																
5–25°C	3 ± 3		0.7 ± 0.3	1 ± 0.5												
25–55°C	3 ± 3		0.7 ± 0.3	1 ± 0.5												
25–70°C		1.5 ± 1	0.7 ± 0.3	1 ± 0.5												
$D_F \times 10^6$ at <0.1 mT, 25°C	[b]															
10 kHz		<12	<3	<4.5					<5	<5						
B at 10 kHz, mT																
250 A/m, 100°C	250	260	250	210	250	180	180	180	200	210	210	210	>330	>330	>300	>100
3,000 A/m, 25°C	380	400	450	400	380	350	400	350	400	400	400	400	500	500	450	320
H_c at 10 kHz, 25°C, A/m	300	70	15	15	300	20	10	30	12	10	5	3	15	15	60	170
B_r at 10 kHz, 25°C, mT	280	150	60	80	270	150	100	150	120	120	80	60	150	140	150	200
power losses at 100°C, kW/m³																
25 kHz, 200 mT													<140	<90		
100 kHz, 100 mT													<165	<80		
400 kHz, 50 mT														<150		
1,000 kHz, 30 mT															<300	
3,000 kHz, 10 mT															<300	<200
10,000 kHz, 5 mT																<200
T_c, °C	>350	>200	>160	>130	>350	>125	>125	>125	>125	>125	>125	>130	>200	>200	>220	>260
d-c resistivity at 25°C, Ωm	10^5	2	2	1	10^5	10^5	1	10^5	1	1	0.5	>0.01	2	2	10	10^5
density, kg/m³	4,500	4,700	4,700	4,800	4,500	5,100	4,900	5,000	4,800	4,800	4,900	4,900	4,800	4,800	4,700	4,600

[a] Ref. 93. [b] At 100 kHz, the value for the 4C6 grade of NiZn is <10.

used because ferrites can supply electromagnetic interference (EMI) suppression as inductor cores in low pass LC-filters, as well as serve as selective impedance inductors in series with circuit load impedances, without the use of capacitors.

The objective is to block unwanted signals having frequencies that differ sufficiently from those of wanted signals, which should pass with virtually no attenuation. Effective blocking is obtained in case of high impedances $Z = 2\pi fL$. Because L is proportional to $(\mu'^2 + \mu''^2)^{1/2}$ and because this expression as a function of f shows a resonant maximum in a frequency region depending on the ferrite's chemical composition and microstructure, the choice of the ferrite grade is determined by the frequency to be blocked. In filters for frequencies below about 500 kHz, high permeability 3E-materials, which according to Table 6 are primarily intended for wide-band transformer applications, are often used; between 500 kHz and a few MHz ferrites the 3C85 and 3F3 grades, meant in the first instance as power materials, offer effective solutions. Above a few MHz the grades 4A15 and 4C65 at $>$ 3 MHz and $>$ 30 MHz, respectively, become important as high frequency filter materials, and grades 3S1 and 4S2, up to 30 MHz and from 10 to 1000 MHz, respectively, are used as suppressors without using separate capacitors.

In communication systems and modern digital networks, signal and pulse transformers known as wide-band transformers are used. The functions are to transform signal amplitudes and to provide impedance matching and d-c isolation, usually at low signal power levels. Low distortion of analogue signals or digital pulses requires good wide-band characteristics, optimum coupling between the transformer windings, and high inductances in the case of pulse transformers. Well-known high grade materials other than those given in Table 6 are the Siemens-Matsushita T30 series (97) and the TDK H5 series (98). All these high permeability ferrites are used in closed cores, such as ring cores, or composed cores having carefully polished contact surfaces.

Another area having a wide range of applications for modern ferrites is power conversion such as the introduction of switched mode power supplies (SMPS) (99). SMPS devices are used for a-c–d-c conversion and for a-c–d-c transformation. An incoming a-c voltage is first rectified and filtered and then chopped by a high frequency switch. The chopped signals are transformed to the desired voltage levels, rectified, and filtered to provide the required d-c output. The result is measured by a control unit, which, in case of deviations, supplies corrective signals to the switching circuit. For the heart of the operation, the voltage conversion, several circuit designs are in use, making extensive use of ferrites as transformer core materials. Besides this, ferrites are used in SMPS as input filters and output chokes. Transformer ferrites must show low energy losses at high induction levels at higher and higher frequencies (51–54,100,101). The latter facilitates reduction of volume and weight of the SMPS devices. For this purpose highly efficient core shapes and winding configurations have also been developed (101,102). The decrease in weight of power supplies is illustrated in Figure 8.

Ferrites allowing for operation at frequencies well above 1 MHz have also become available, eg, 3F4 and 4F1 (Table 6). Other newer industrial power ferrites are the Siemens-Matsushita N-series (28,97) the TDK PC-series (28,100), and the Thomson B-series (28,103). While moving to higher frequencies, the ferrites have been optimized for different loss contributions, eg, hysteresis losses, eddy current

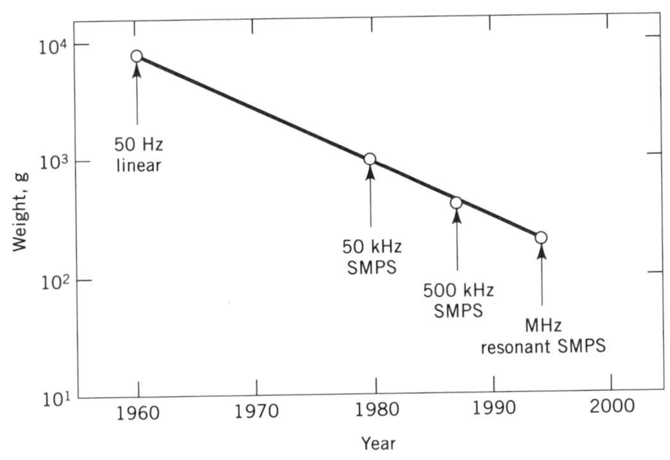

Fig. 8. SMPS = switch mode power supply. Weight reduction of 100 W power supplies from the 1960s through the early 1990s.

losses, and resonance losses. Loss levels are specified at 100°C because ambient temperature in power applications is about 60°C plus an increase caused by internal heat dissipation of about 40°C.

M-Type Ferrites. Since their introduction on the market in the early 1950s, M-ferrites have gradually acquired a dominant position in the permanent magnet market (Table 7), mainly by replacing the established alnico magnets in many applications. The most important feature behind the great economic success of

Table 7. Permanent Magnet Materials

Parameter	Ferrite	NdFeB	SmCo	Al–Ni–Co
B_r, mT	370	1100	890	1200, 700[a]
H_{cJ}, kA/m	255	> 1000	1200	50, 150[a]
$(BH)_{max}$,[b] kJ/m^3	25.5	215	150	40
B/$\mu_0 H$ at $(BH)_{max}$	1.0	1.0	1.0	20, 5[a]
density, kg/m^3	4650	7400	8300	7300
ρ, Ω·m	10^4	10^{-6}	5×10^{-7}	5×10^{-7}
T_c, K	750	585	995	860
$\alpha(B_r)$, % K^{-1}	−0.2	−0.13	−0.05	−0.02
$\alpha(H_c)$, % K^{-1}	+0.4	−0.6	−0.3	−0.03
world production 1990,				
t $\times$ 10^3	240	1.8	1.1	6.5
U.S. \$ $\times$ 10^6	1100	310	310	230
price ratio per unit[c]				
weight[d]	1	37	75	7.5
magnetic energy[d]	1	7.0	23	7.5

[a]Two extreme materials.
[b]To convert joules to calories, divide by 4.184.
[c]Price per unit magnetic energy = price per unit weight $\times$ density/$(BH)_{max}$.
[d]Ferrite is taken as reference.

ferrite magnets is the low price per unit of available magnetic energy, attributable to the relatively low cost and wide availability of the raw materials. As of this writing, however, dominance of ferrites in terms of market value has begun to decrease because of the growing importance of rare-earth-based metallic magnet material (see LANTHANIDES). This material, $Nd_2Fe_{14}B$, has largely extended the separate market of low volume, high price, high performance magnets for miniaturized systems, which had been represented only by the very expensive SmCo-based magnets.

A large variety of M-ferrite magnets and applications is available. Table 8 gives a survey of the grades F_1 through F_9. Bonding of the microscopic crystallites to solid bodies is performed either by sintering or by plastic bonding. The latter produces a plasto-ferrite. In both cases the crystallites may be either randomly oriented (isotropic) or aligned with the *c*-axis in one direction (anisotropic). Depending on the binder used, plasto-ferrites may be flexible or rigid. Special advantages of sintered materials are high magnetic quality, close dimensional control when machined, and relatively high mechanical strength. When these properties are not of prime importance, plasto-ferrites may be preferred because of lower price or special mechanical properties and shaping possibilities. Anisotropic materials are applied when high magnetic quality is essential. The less expensive isotropic material is preferred when low magnetic quality is acceptable or when the properties have to be isotropic.

The anisotropic sintered form is by far the most important one. It is produced in various grades which can be divided into four main groups. Group F_2 is used for loudspeakers, where a high B_r is essential, whereas a relatively low H_{cJ} is acceptable. Groups F_3–F_5 are used for different types of motors, requiring quite distinct H_{cJ} levels. In these cases a high B_r is also desired, but in fact it is limited by the H_{cJ} level and the performance factor K.

Figure 9 shows typical demagnetization curves for the materials of Table 8 (104,105). A material group may include more than one grade. In that case the values for the two extreme K-values are given.

Prices for the different material groups (F_1–F_9) are largely influenced by product size and shape and the quantity of production. Apart from product shape, the main price-determining factor is the performance factor K. Prices for the powder, a sintered anisotropic ring, and a sintered anisotropic segment are roughly on the order of \$1, \$2, and \$5, respectively.

A variety of shapes and sizes exists for the commercial M-magnets. Most M-ferrite products are flat owing to the relatively high H_c and moderate B_r. Basic shape groups are rings, disks, plates, rods, cylinders, blocks, strips, and segments. Each shape-group in turn contains a number of products. The more complicated forms are preferably made from rigid plasto-ferrite because of versatility in shaping without the necessity of finishing.

Products, whether or not anisotropic, may have different magnetization modes. That is, magnetization may be from one side to the opposite one, or along one side (lateral magnetization). The resulting polar surfaces may contain regions having opposite polarity (poles), separated by a neutral zone. Different pole numbers and configurations are possible. Combinations of these possibilities give rise to a variety of magnetization modes. All products may be anisotropic and products of the same appearance may have a different aligning mode. The most important

Table 8. Properties of Commercial M-Ferrites[a]

Type	Grade	M[b]	B_r, mT	$_BH_c$, kA/m	H_{cJ}, kA/m	K^c, mT	$(BH)_{max}$,[d] kJ/m^3	μ_{rec}[e]	H_s, kA/m	$\alpha(B_r)$,[f] %/K	$\alpha(H_c)$,[f] %/K	ρ, Ω·m	Density, kg/m^3	Typical application	Market share, %
Sintered															
isotropic	F_1	Ba	215	145	>240	>311	7.4	1.2		−0.2	0.4	10^4		rotors for cycle dynamos	~10
anisotropic low, H_c	F_2	Ba	400	160	165	466	29.5	1.1	560	−0.2	0.48	10^4	4900	flat rings for loudspeakers	~70
medium, H_{cJ}	F_3	Sr	370	245	255	472	25.5	1.1	875	−0.2	0.37	10^4	4650	segments for windshield wiper motors	
			425	250	260	529	33.6	1.1	955	−0.2	0.37	10^4	4900		
high, H_{cJ}	F_4	Sr	380	280	320	508	26.8	1.1	1115	−0.2	0.30	10^4	4700	segments for blower motors	
			400	295	330	532	30.5	1.1	1100	−0.2	0.29	10^4	4850		
very high, H_{cJ}	F_5		385	285	360	529	27.6	1.1	1200	−0.2	0.26	10^4	4850	segments for starter motors	
Plastic-bonded															
flexible isotropic	F_6	Ba	125	88	190	201	2.8	1.15	800	−0.2	0.4	10^7	3100	doorcatches (refrigerators), small d-c motors (fuel pumps)	~10
			145	96	190	221	3.6	1.15	800	−0.2	0.4	10^6	3700		
anisotropic	F_7		200	140	200	280	8.0	1.05		−0.2	0.4				
			250	176	240	346	12.0	1.05		−0.2	0.4				
rigid isotropic	F_8	Ba	80	58	190	156	0.9	1.15	800	−0.2	0.4	10^8	2500	correction magnets for TV	~10
			155	104	190	231	4.4	1.15	800	−0.2	0.4	10^4	3900		
anisotropic	F_9	Sr	245	100	260	349	12.0	1.05	800	−0.2	0.4	10^5	3500	small d-c motors	
			270	196	260	374	14.0	1.05	800			10^5	3900		

[a] Refs. 104, 105. [b] Normally applied *M*-ferrite. [c] $K(\text{mT}) = B_r(\text{mT}) + 0.4_JH_c(\text{kA/m})$, factor independent of the aluminum content. [d] To convert joules to calories, divide by 4.184. [e] Recoil permeability. [f] α = Temperature coefficient (0.8 kA/m·K for BaM and 0.95 kA/m·K for SrM).

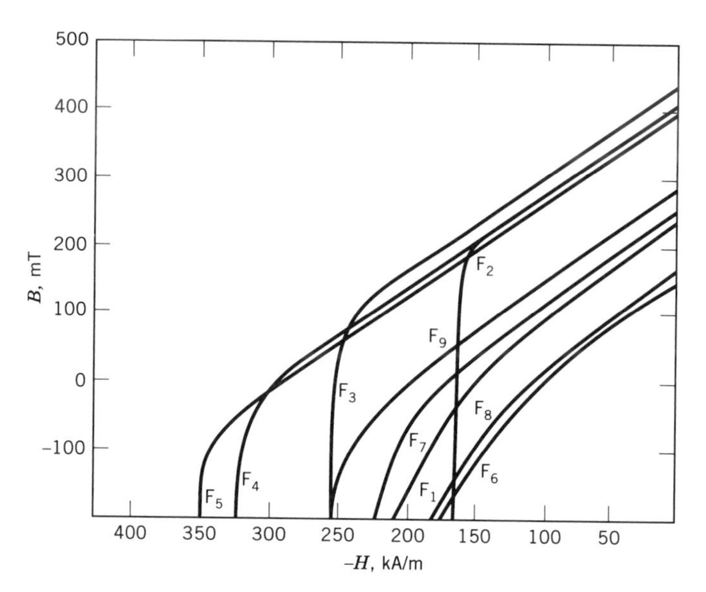

Fig. 9. Demagnetization curves for M-ferrites at high K values (see Table 8).

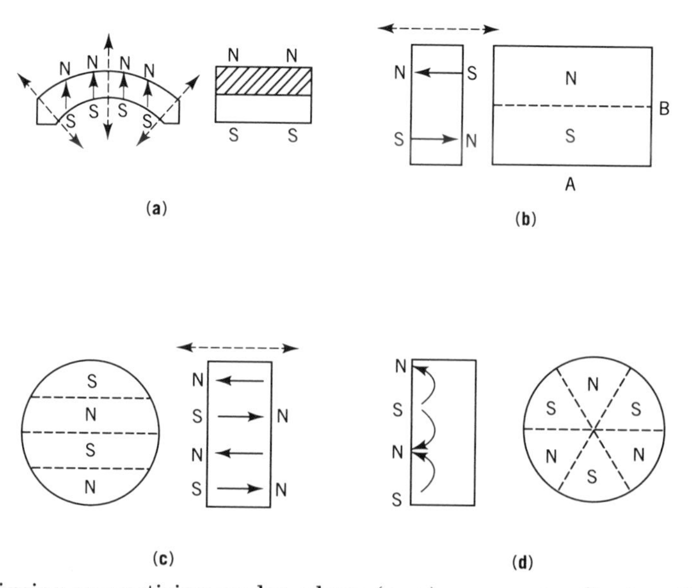

Fig. 10. Aligning–magnetizing modes where (←-→) represents aligning direction and (----) the neutral zones. (**a**) Segment, radially aligned, diametrically magnetized; (**b**) block, perpendicularly aligned and magnetized, two poles, neutral zone parallel to A; (**c**) disk, axially aligned and magnetized, four poles, neutral zone diametrical; and (**d**) disk, isotropic, laterally magnetized, six poles, where the neutral zone is radial.

aligning modes are shown in Figure 10. When a complex magnetization mode is desired, isotropic materials are preferred. Lateral magnetization, always in multipole, is only applied to isotropic materials.

As compared to the classical alnico-magnets, M-ferrite magnets have some distinct advantages. These are high H_c, high resistivity, low price, low density, high chemical resistance, and the suitability of being applied as (flexible) plastoferrite. The high H_c and the low price have especially contributed to economic success.

The most important disadvantages are moderate B_r and $(BH)_{max}$, relatively high temperature coefficients αB_r and αH_c, and poor mechanical properties (low strength, brittleness). The moderate B_r and $(BH)_{max}$ are perhaps the less serious, as a larger cross-sectional area produces the required flux. The other two disadvantages exclude certain applications where the magnet is exposed to strong mechanical stresses or impacts, or where the surrounding temperature temporarily drops far below the normal operating temperature. In the latter case the H_c may temporarily decrease so far that the knee in the *BH* curve is surpassed, resulting in a partial demagnetization of the magnet. A temporary increase in temperature does not cause an irreversible demagnetization, only a temporary decrease of B_r. In some cases, eg, with high precision instruments, that is also a problem, but in most cases it is acceptable.

Owing to low price, M-ferrites have replaced other magnet materials in existing systems, whether or not modification has been made to the system. This is

Table 9. Applications of M-Ferrites

Magnet type	Market, U.S. $ × 10^6 World	Market, U.S. $ × 10^6 U.S.A.	Market sector	%	Examples
motor-segments (arcs)	600	140	automotive	70	d-c motors in cars: windshield wiper, starter
			domestic	15	synchronous motors, kitchen appliances, recharg. tools
			industrial	15	servo motors, machine tools
loudspeaker rings	140	30	audio—car	40	car radio
			hi-fi	30	music centers, speaker systems
			portable	10	portable radio
			video—television	20	
miscellaneous (blocks, etc)	130	30	industrial	30	separators, couplings, industrial motors
			domestic	40	sticking devices, cycle dynamos
			others	30	small motors for audio/video, toys
Total (sintered)	*870*	*200*			

particularly so in static applications where small demagnetizing fields are involved. A typical example is the application of flat *M*-ferrite rings used instead of the high metallic center-core magnets in loudspeaker systems. The high H_c has stimulated the development of new systems. This applies especially to dynamic applications where periodically high demagnetizing fields are involved. A typical example is the electromotor, which has strong demagnetizing fields. Electromotors are being developed requiring very high H_{cJ} values which lie far outside the scope of the alnico materials (106), eg, the starter motor, requiring $H_{cJ} > 320$ kA/m.

Table 9 gives a survey of the most important applications of sintered *M*-ferrite magnets. The main products which together form about 50% of the total production are anisotropic segments for dc-motors in cars (F_3–F_5) and anisotropic rings for loudspeakers (F_2). Large-scale production is concentrated on systems requiring relatively large magnets. Thus the greatest advantage of these ferrites is the low material price.

BIBLIOGRAPHY

"Ferrites" in *ECT* 2nd ed., Vol. 8, pp. 881–901, by G. Economos, Allen-Bradley Co.; in *ECT* 3rd ed., Vol. 9, pp. 881–902, by T. G. Reynolds III, Ferroxcube Corp.

1. W. L. Bragg, *Phil. Mag.* **30,** 305 (1915).
2. G. Aminoff, *Geol. Fören. Förhandl.* **47,** 283 (1925).
3. J. L. Snoek, *New Developments in Ferromagnetic Materials,* Elsevier, Amsterdam, the Netherlands, 1947.
4. J. Smit and H. P. J. Wijn, *Ferrites,* Philips' Technical Library, Eindhoven, the Netherlands, 1959.
5. J. L. Snoek, *Physica* **3,** 463 (1936).
6. T. Takei, *J. Electrochem. Japan* **5,** 411 (1937).
7. L. Néel, *Ann. de Phys.* **3,** 137 (1948).
8. M. Takano, N. Nakanishi, Y. Takeda, and T. Shinjo, in H. Watanabe, S. Iida, and M. Sugimoto, eds., *Ferrites,* Proceedings of the 3rd International Conference on Ferrites (ICF-3), Center for Academic Publishing Japan, Tokyo, 1981, p. 389.
9. M. M. Abou-Sekkina, *Advan. Ceramics* **15,** 553 (1985).
10. A. H. Bobeck and E. Della Torre, *Magnetic Bubbles,* North Holland Publishing Co., Amsterdam, the Netherlands, 1979.
11. M. A. Gilleo, in E. P. Wohlfarth, ed., *Ferromagnetic Materials,* Vol. 2, North Holland Publishing Co., Amsterdam, the Netherlands, 1982, Chapt. 1.
12. P. Hansen and J. P. Krumme, *Thin Solid Films* **114,** 69 (1984).
13. S. L. Blank and co-workers, *J. Appl. Phys.* **50** 2155 (1979).
14. J. J. Went and E. W. Gorter, *Philips Techn. Rev.* **13,** 181 (1951–1952).
15. M. Sugimoto, in Ref. 11, Vol. 3, Chapter 6.
16. G. H. Jonker, H. P. J. Wijn, and P. B. Braun, *Philips Techn. Rev.* **18,** 145 (1956/57).
17. Y. H. Chang, C. C. Wang, T. S. Chin, and F. S. Yen, *J. Magn. Magn. Mats.* **72,** 343 (1988).
18. R. L. Harvey, I. Gordon, and R. A. Braden, *RCA Rev.* **22,** 648 (1961).
19. G. Winkler and H. Dösch, *Proceedings of the 9th European Microwave Conference,* Microwave Exhibitions and Publications, Sevenoaks, UK, 1979, p. 13.
20. M. P. Sharrock, *IEEE Trans. Magn.* **MAG25,** 4374 (1989).
21. M. Noda, Y. Okazaki, K. Hara, and K. Ogisu, *IEEE Trans. Magn.* **MAG26,** 81 (1990).
22. J. J. Went, G. W. Rathenau, E. W. Gorter, and G. W. van Oosterhout, *Philips Techn. Rev.* **13**(7), 194 (1952).

23. A. L. Stuyts, G. W. Rathenau, and G. H. Weber, *Philips Techn. Rev.* **16**(5/6), 141 (1954).
24. A. Broese van Groenou, P. F. Bongers, and A. L. Stuyts, *Mater. Sci. Eng.* **3,** 317–392 (1968–1969).
25. J. Nicolas, in Ref. 11, Vol. 2, Chapt. 2.
26. P. I. Slick, in Ref. 11, Vol. 2, Chapt. 3.
27. S. Krupicka and P. Novak, in Ref. 11, Vol. 3, Chapt. 4.
28. E. C. Snelling, *Soft Ferrites,* 2nd ed., Butterworth & Co. Publishers Ltd., Kent, UK, 1988.
29. A. Goldman, *Modern Ferrite Technology,* Van Nostrand Reinhold Co., Inc., New York, 1990.
30. C. A. M. van den Broek, and A. L. Stuyts, *Philips Techn. Rev.* **37**(7), 157 (1977).
31. H. Kojima, in Ref. 11, Vol. 3, Chapt. 5.
32. H. Stäblein, in Ref. 11, Vol. 3, Chapt. 7.
33. S. Chikazumi, *Physics of Magnetism,* John Wiley & Sons, Inc., New York, 1978.
34. K. Ohta, *J. Phys. Soc. Japan* **18,** 685 (1963).
35. D. Stoppels, *J. Appl. Phys.* **51,** 2789 (1980).
36. E. Röss, I. Hanke, and E. Moser, *Z. angew. Phys.* **17,** 504 (1964).
37. D. J. Perduijn and H. P. Peloschek, *Proc. Brit. Cer. Soc.* **10,** 263 (1968).
38. E. G. Visser, M. T. Johnson and P. J. van der Zaag, *Proceedings of the 6th International Conference on Ferrites (ICF-6),* The Japanese Society of Powder and Powder Metallurgy, Tokyo, 1992.
39. K. Ohta and N. Kobayashi, *Jap. J. Appl. Phys.* **3,** 576 (1964).
40. E. Röss and E. Moser, *Z. angew. Phys.* **13,** 247 (1961).
41. T. Iimura, T. Shinohara, and M. Kudo, in Ref. 8, 1981, p. 726.
42. D. Stoppels, P. G. T. Boonen, J. P. M. Damen, L. A. H. van Hoof, and K. Prijs, *J. Magn. Magn. Mats.* **37,** 123 (1983).
43. J. C. Slonczewski, *J. Appl. Phys.* **32,** 253 S (1962).
44. J. G. M. de Lau, *Philips Res. Repts.* **Suppl. 6,** 45 (1975).
45. M. T. Johnson, *Proceedings of the 5th International Conference on Ferrites (ICF-5),* C. M. Srivastava and M. J. Patni, eds. Oxford and IBH Publishing Co. PVT., Ltd., New Delhi, 1989, p. 605.
46. T. G. W. Stijntjes, J. Klerk, and A. Broese van Groenou, *Philips Res. Repts.* **25,** 95 (1970).
47. T. Akashi, *Trans. Jpn. Inst. Met.* **2,** 171 (1961).
48. T. Akashi, *NEC Res. Dev.* **8,** 89 (1966).
49. U. Wagner, *J. Magn. Magn. Mats.* **4,** 116 (1977).
50. A. D. Giles and F. F. Westendorp, *J. Phys. Colloq. Suppl.* **38,** C1-317 (1977).
51. T. G. W. Stijntjes and J. J. Roelofsma in F. F. Y. Wang, ed., *Proceedings of the 4th International Conference on Ferrites (ICF-4), Pt. II,* San Francisco, 1985, p. 493.
52. T. G. W. Stijntjes in Ref. 45, 1989, p. 587.
53. E. G. Visser, J. J. Roelofsma, and G. J. M. Aaftink in Ref. 45, 1989, p. 605.
54. T. Sano, A. Morita, and A. Matsukawa, *Power Electronics PCIM* **19** (July 1988) and Ref. 45, 1989, p. 595.
55. K. Ishino and Y. Narumiya, *Am. Cer. Bull.* **66,** 1469 (1987).
56. M. H. Berger, J. Y. Laval, F. Kools, and J. Roelofsma in Ref. 45, 1989, p. 619.
57. J. W. Waanders, *Data Handbook of Soft Ferrites,* Philips Components, Eindhoven, the Netherlands, 1993.
58. J. L. Snoek, *Physica* **14,** 207 (1948).
59. V. Aldelsköld, Arkiv Kemi, *Min. Geol.* **12A**(29), 1 (1938).
60. P. Braun, *Philips Res. Repts.* **12,** 491 (1957).
61. W. D. Townes, J. H. Frang, and A. J. Perrotta, *Z. Kristallogr.* **125,** 437 (1967).

62. A. J. Mountvala and S. F. Ravitz, *J. Am. Cer. Soc.* **45**(6), 285 (1962).
63. B. T. Shirk and W. R. Buessem, *J. Appl. Phys.* **40,** 1294 (1969).
64. R. Pauthenet and G. Rimet, *Compt. Rend.* **249,** 656 (1959).
65. F. K. Lotgering, P. R. Locher, and R. P. van Stapele, *J. Phys. Chem. Solids* **41,** 481 (1980).
66. E. W. Gorter, *Proc. IEEE* **104B,** 255 S (1957).
67. E. F. Bertaut, A. Deschamps, R. Pauthenet, and S. Pickart, *J. de Phys. Rad.* **20,** 404 (1959).
68. J. S. van Wieringen, *Philips Techn. Rev.* **28,** 33 (1967).
69. R. Gemperle, E. V. Shtolts and M. Zeleny, *Phys. Stat. Sol.* **3,** 2015 (1963).
70. H. Zijlstra in Ref. 11, Vol 3, Chapt. 5.
71. E. C. Stoner and E. P. Wohlfarth, *Phil. Trans. Roy. Soc.* **240A,** 599 (1948).
72. G. Albanese and A. Deriu, *Ceramurgia Int.* **5** (1), 3 (1979).
73. H. R. Zai, J. Z. Liv and M. Lu, *J. Appl. Phys.* **52,** 2323 (1981).
74. R. Schippan and K. A. Hempel in Ref. 51, Pt II, p. 579.
75. F. Kools, *Proc. Magn. Mats. & Appl. (MMA), J. Phys. (Paris)* **46,** C6-349, Grenoble (1985).
76. F. Kools in Ref. 45, p. 417.
77. H. J. H. van Heffen, *Electr. Compon. Appl.* **3**(1), 22 (1980).
78. F. Kools, *Sci. Ceramics Soc. Française Cér.* **7,** 27 (1973).
79. J. D. B. Veldkamp and R. J. Klein Wassink, *Philips Res. Rpts.* **31,** 153 (1976).
80. A. B. D. van der Meer, *Sci. of Ceramics Ned. Ker. Ver.* **9,** 535 (1979).
81. G. de With and N. Hattu, *Proc. Brit. Cer. Soc.* **32,** 191 (1982).
82. M. Inasa, E. C. Liang, R. C. Bradt, and Y. Nakamura, *J. Am. Cer. Soc.* **64,** 390 (1981).
83. O. Kimura and A. Chiba in Ref. 51, Pt. I, p. 115.
84. J. M. Blank, *J. Appl. Phys.* Suppl. **32,** 378S (1961).
85. R. Morineau and M. Paulus, *IEEE Trans. Magn.,* **MAG11,** 1312 (1975).
86. F. Kools and O. Fiquet, in G. de With, R. A. Terpstra and R. Metselaar, eds., *Euroceramics,* Vol. 1, Elsevier Applied Science, London, 1989, p. 258.
87. F. Kools, *Sci. Sintering* **17** (1), 49 (1985).
88. H. Harada, in Ref. 8, p. 354.
89. C. A. M. van den Broek. *Proceedings of the 3rd European Conference on Hard Magnetic Materials,* Bond Materialen Kennis, Amsterdam, 1973, p. 53.
90. Fr. Pat. 1,085,491 (1954) (to N. V. Philips Gloeilampen Fabrieken).
91. A. Cochardt, *J. Appl. Phys.* **34,** 123 (1963).
92. G. S. Krijtenburg, *Proceedings of the 1st European Conference on Hard Magnetic Materials,* Vienna (unpublished), 1966.
93. J. W. Waanders, *Soft Ferrite Selection Guide,* Philips Components, Eindhoven, The Netherlands, 1991.
94. A. Braginski, *Phys. Stat. Sol.* **11,** 603 (1965).
95. A. Fox, *J. Phys.* **D4,** 1239 (1971).
96. J. E. Knowles, *Philips Res. Rpts.* **29,** 93 (1974).
97. J. Hess, *Siemens Components* **26,** 142 (1991).
98. H. Tsunekawa, A. Nakata, T. Kamijo, K. Okutani, R. K. Mishra, and G. Thomas, *IEEE Trans. Magn.* **MAG15,** 1855 (1979).
99. G. Cryssis, *High Frequency Switching Power Supplies,* Mc. Graw-Hill Book Co., Inc., New York, 1984.
100. T. Mitsui and G. Van Schaick, *Powertechn. Mag.,* 15 *(Feb. 1991).*
101. W. Waanders and A. Shpilman, *Powertechn. Mag.,* 20 (May 1991).
102. S. Mulder, *Application Note on the Design of Low Profile High Frequency Transformers—A new Tool in SMPS,* Philips Components, Eindhoven, the Netherlands, 1990.

103. P. Gaudry and J. J. Putigny, *Electronique de Puissance* **32,** 54 (1989).
104. *Data Handbook on Permanent Magnets,* Philips Components, Eindhoven, the Netherlands, 1991.
105. M. McCraig and A. E. Clegg, *Permanent Magnets in Theory and Practice,* Pentech, London, 1987.
106. A. Mohr and J. Koch in Ref. 51, Pt. II, p. 515.

F. X. N. M. Kools
D. Stoppels
Philips Components

FERROCYANIDES.

See Iron compounds.

FERROELECTRICS

Polarization which can be induced in nonconducting materials by means of an externally applied electric field $\overline{E}$ is one of the most important parameters in the theory of insulators, which are called dielectrics when their polarizability is under consideration (1). Experimental investigations have shown that these materials can be divided into linear and nonlinear dielectrics in accordance with their behavior in a realizable range of the electric field. The electric polarization $\overline{P}$ of linear dielectrics depends linearly on the electric field $\overline{E}$, whereas that of nonlinear dielectrics is a nonlinear function of the electric field (2). The polarization values which can be measured in linear (normal) dielectrics upon application of experimentally attainable electric fields are usually small. However, a certain group of nonlinear dielectrics exhibit polarization values which are several orders of magnitude larger than those observed in normal dielectrics (3). Consequently, a number of useful physical properties related to the polarization of the materials, such as elastic, thermal, optical, electromechanical, etc, are observed in these groups of nonlinear dielectrics (4).

The most important materials among nonlinear dielectrics are ferroelectrics which can exhibit a spontaneous polarization $\overline{P}_s$ in the absence of an external electric field and which can split into spontaneously polarized regions known as domains (5). It is evident that in the ferroelectric the domain states differ in orientation of spontaneous electric polarization, which are in equilibrium thermodynamically, and that the ferroelectric character is established when one domain state can be transformed to another by a suitably directed external electric field (6). It is the reorientability of the domain state polarizations that distinguishes ferroelectrics as a subgroup of materials from the 10-polar-point symmetry group of pyroelectric crystals (7–9).

Ferroelectric crystals exhibit spontaneous electric polarization and hysteresis effects in the relation between polarization and electric field, as shown in Figure 1. This behavior is usually observed in a limited temperature range, ie, usually below a transition temperature (10).

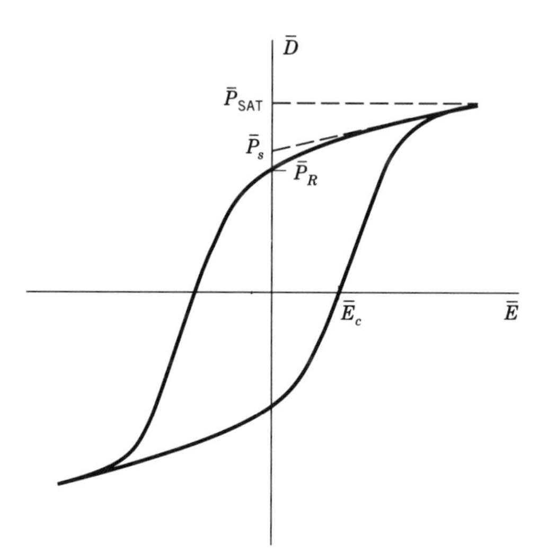

Fig. 1. Hysteresis loop of dielectric displacement, $\overline{D}$, versus applied electric field $\overline{E}$ where $\overline{E}_c$ is coercive field and $\overline{P}_{SAT}$, $\overline{P}_s$, and $\overline{P}_R$ are the saturated, spontaneous, and remanent polarization, respectively.

Properties

In considering the energy stored in a polarizable and deformable elastodielectric medium by separating the pure dielectric, pure elastic, and the cross-coupling terms, a phenomenological equation for the elastic Gibbs free energy may be written in the simple form (4,11)

$$\begin{aligned}\Delta G_1 &= \chi_{mn}P_mP_n + \text{higher order dielectric terms}\\ &\tfrac{1}{2}\, s_{ijkl}T_{ij}T_{kl} + \text{higher order elastic terms}\\ &-b_{mij}P_mT_{ij} - Q_{mnij}P_mP_nT_{ij} + \text{higher order cross terms}\end{aligned}$$

where χ_{mn} = dielectric stiffness at constant elastic stress, P_m, P_n = components of the dielectric polarization, s_{ijkl} = elastic compliance at constant polarization, T_{ij}, T_{kl} = components of the elastic stress, b_{mij} = piezoelectric coefficient tensor (in polarization notation), and Q_{mnij} = electrostrictive coefficient tensor. Taking the partial derivatives of the energy function with respect to the different variables results in relations involving the quantities and properties of the material in terms of the coefficients of the energy function.

The dielectric stiffness χ_{ij} can be expressed as a linear temperature dependence based on the Curie-Weiss law at above the Curie point T_c.

$$\chi_{ij} = \frac{1}{\epsilon_{ij}} = \chi_0(T - T_c) = \frac{1}{C}(T - T_c)$$

or, solving for ϵ_{ij},

$$\epsilon_{ij} = \frac{C}{T - T_c}$$

where ϵ_{ij} = dielectric susceptibility, C = Curie constant, and T = temperature, which is similar to the susceptibility law for ferromagnets. However, for the dielectric case the Curie constant C is several orders of magnitude larger than that of the ferromagnet. In practice this means that very high useful dielectric permittivities persist for a wide range of temperature above T_c, ie, in the paraelectric phase. In fact, it is this high intrinsic dielectric susceptibility response that is the phenomenon most used in the practical application of polycrystalline ceramic ferroelectrics. Ferroelectric ceramics having relative permittivities $\epsilon_{ij}/\epsilon_0 = K_{ij}$ ranging up to 10,000, where ϵ_0 is the dielectric permittivity of vacuum, are widely used in many types of capacitors including the multilayer variety (see ADVANCED CERAMICS; CERAMICS; CERAMICS AS ELECTRICAL MATERIALS) (12).

The strain tensor S_{ij} can be written for noncentrosymmetry point group crystals as:

$$S_{ij} = s_{ijkl}\,T_{kl} + b_{ijm}\,P_m + Q_{mnij}\,P_m\,P_n$$

or for $T_{kl} = 0$,

$$S_{ij} = b_{ijm}\,P_m + Q_{mnij}\,P_m\,P_n$$

or

$$S_{ij} = d_{ijm}\,E_m + M_{nmij}\,E_m\,E_n$$

where d_{ijm} = the piezoelectric voltage coefficient tensor, M_{mnij} = the electrostrictive voltage coefficient tensor, and E_m and E_n = the electric field vectors. The piezoelectric and electrostrictive voltage coefficients, d_{ijm} and M_{mnij}, of the ferroelectrics are very large because of the large polarizability (13). Thus a second principal application of the ferroelectrics uses this high electromechanical coupling for efficient transduction between electrical and mechanical signals in sonic and ultrasonic transducers and filter applications (see ULTRASONICS).

Both the spontaneous polarization $\overline{P}_s$ and the remanent polarization $\overline{P}_R$ are strong functions of temperature, particularly near the transition temperature T_c in ferroelectrics (7):

$$\Delta\overline{P}_R = \pi\Delta T$$

where π is the pyroelectric coefficient. Many ferroelectrics have large pyroelectric coefficients and can be used in thermometry and in bolometry sensing devices of infrared radiation (see INFRARED AND RAMAN SPECTROSCOPY; SENSORS).

Many ferroelectrics are high band gap insulating crystals and have good transparency in both the visible and near-ir spectral regions. Qualitatively, it may be expected that the large dielectric polarizability leads to an ability to modify the refractive index ellipsoid (indicatrix) under electric fields, thus resulting in high linear and quadratic electrooptic coefficients. Much research has been devoted to seeking to develop effective broad-band modulation and switching techniques for both bulk and thin-film materials by using ferroelectric electrooptic structures. Polycrystalline ceramic ferroelectrics have been processed to very high densities and good optical transparency (14). These materials possess new parameter combinations for modulation and imaging devices. Large photorefractive effects can also be generated in transparent ferroelectrics (15) (see CERAMICS, NONLINEAR OPTICAL AND ELECTROOPTIC CERAMICS).

In the single-domain state, many ferroelectric crystals also exhibit high optical nonlinearity and this, coupled with the large standing optical anisotropies (birefringences) that are often available, makes the ferroelectrics interesting candidates for phase-matched optical second harmonic generation (SHG).

One area of application utilizes the interaction between the dielectric polarization and the electrical transport processes in ferroelectrics. In single dielectric crystals the effects of the domain polarizations on the drift and retrapping of photogenerated carriers give most interesting photoferroelectric effects. Of more immediate applicability, however, are the large effects of the dielectric changes at the ferroelectric phase transition on the potential barriers at grain boundaries in suitably prepared semiconducting ceramic ferroelectrics. These barium titanate–based compositions show strong positive temperature coefficients of resistivity (PTC effects) and are finding widespread use in temperature and current control for domestic, industrial, and automotive applications.

Materials

Oxygen Octahedra. An important group of ferroelectrics is that known as the perovskites. The perfect perovskite structure is a simple cubic one as shown in Figure 2, having the general formula ABO_3, where A is a monovalent or diva-

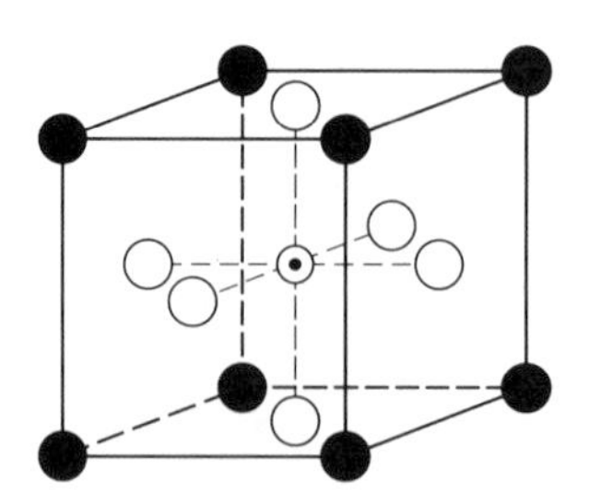

Fig. 2. Cubic (m3m) prototype structure of perovskite-type ABO_3 compounds where ●, A; ⊙, B; and ○, O.

lent metal such as Na, K, Rb, Ca, Sr, Ba, or Pb, and B is a tetra- or pentavalent cation such as Ti, Sn, Zr, Nb, Ta, or W. The first perovskite ferroelectric to be discovered was barium titanate [*12047-27-7*], and it is the most thoroughly investigated ferroelectric material (10).

Simple ABO_3 compounds in addition to $BaTiO_3$ are cadmium titanate [*12014-14-1*], $CdTiO_3$; lead titanate [*12060-00-3*], $PbTiO_3$; potassium niobate [*12030-85-2*], $KNbO_3$; sodium niobate [*12034-09-2*], $NaNbO_3$; silver niobate [*12309-96-5*], $AgNbO_3$; potassium iodate [*7758-05-6*], KIO_3; bismuth ferrate [*12010-42-3*], $BiFeO_3$; sodium tantalate, $NaTaO_3$; and lead zirconate [*12060-01-4*], $PbZrO_3$. The perovskite structure is also tolerant of a very wide range of multiple cation substitution on both A and B sites. Thus many more complex compounds have been found (16,17), eg, $(K_{1/2}B_{1/2})TiO_3$, $(Na_{1/2}Bi_{1/2})TiO_3$, $Pb(Sc_{1/2}Nb_{1/2})O_3$, $Pb(Fe_{1/2}Nb_{1/2})O_3$, and $Pb(Fe_{1/2}Ta_{1/2})O_3$.

The characteristic feature of the $BaTiO_3$ unit cell (Fig. 2) is the TiO_6-octahedron, which, because of its high polarizability, essentially determines the dielectric properties. The high polarizability results from the small Ti^{4+} ion having a relatively large space within the oxygen octahedron. The idealized unit-cell of Figure 2 where the Ti^{4+} ion is in the center of the oxygen octahedron is, however, stable only above the Curie point T_c of about 135°C. Below T_c the Ti^{4+} ions occupy off-center positions. This transition to the off-center position at T_c results in a series of important physical consequences (Fig. 3). The crystal structure changes from cubic ($T > 135°C$) via tetragonal ($5°C < T < 135°C$; $c/a = 1.01$) and orthorhombic ($-90°C < T < 5°C$) to rhombohedral ($T < -90°C$). At the same time a spontaneous polarization $\overline{P}_s$ (26 μC/cm² at room temperature) appears, the direction of which in the tetragonal phase is along one of the six edges, in the orthorhombic phase along one of the 12 surface diagonals, and in the rhombohedral

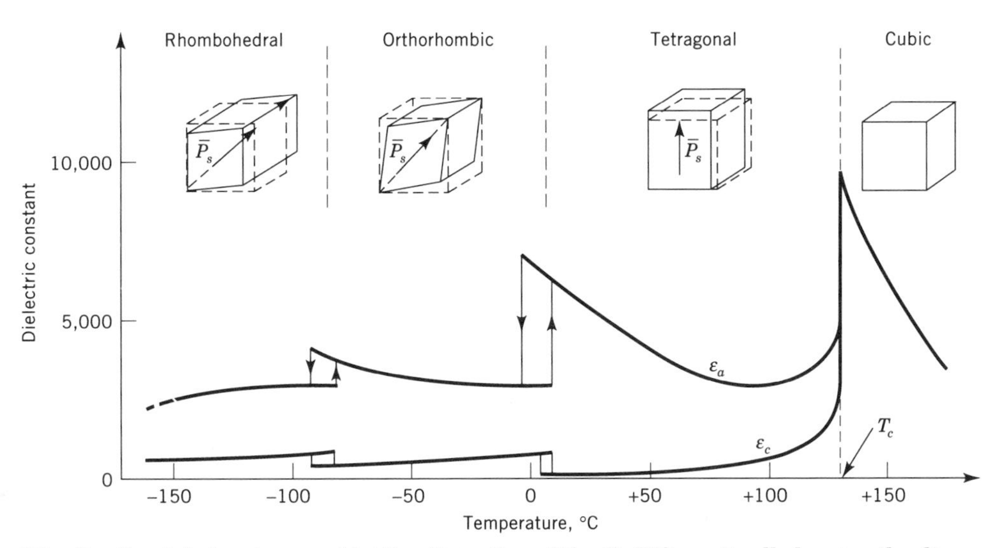

Fig. 3. Crystal structure and lattice distortion of the $BaTiO_3$ unit cell showing the direction of spontaneous polarization, and resultant dielectric constant ε vs temperature. The subscripts *a* and *c* relate to orientations parallel and perpendicular to the tetragonal axis, respectively. The Curie point, T_c, is also shown.

phase along one of the eight body diagonals of the ideal cubic unit cell. The direction of $\overline{P}_s$ can be switched by a high (1–2 kV/cm) electrical field between the different crystallographically allowed and thermodynamically equilibrium positions which are characteristic in each ferroelectric phase (10,16).

At the temperatures of the phase transitions, maxima of the dielectric constant up to 10,000 are found. Moreover, in the ferroelectric state below T_c the material becomes pyroelectric and shows high piezoelectric activity (see PIEZOELECTRICS).

Perovskite-type compounds, especially $BaTiO_3$, have the ability to form extensive solid solutions. By this means a wide variety of materials having continuously changing electrical properties can be produced in the polycrystalline ceramic state. By substituting Pb^{2+} ions for Ba^{2+} ions, T_c can be increased linearly up to 490°C for a 100% Pb^{2+} substitution. In the same manner, T_c can be continuously decreased by the substitution of Sr^{2+} for Ba^{2+} or of Zr^{4+} or Sn^{4+} for Ti^{4+} (Fig. 4). Simultaneous with the change of T_c by formation of solid solutions, the low temperature phase transitions between tetragonal–orthorhombic and orthorhombic–rhombohedral phases shift in a rather complex manner.

$PbZrO_3$–$PbTiO_3$-Based Materials. Since the middle of the 1950s, solid solutions of $PbZrO_3$–$PbTiO_3$ (PZT) ceramics having the perovskite structure have gained rising interest because of the superior piezoelectric properties (10,17,18). The phase diagram of the $Pb(Zr_xTi_{1-x})O_3$ system is shown in Figure 5. At high temperatures, ie, above T_c, the ideal cubic paraelectric structure is stable and no ferroelectric phenomena such as spontaneous polarization appear. At room temperature the materials are ferroelectric and for Ti-rich compositions ($0 \leq x \leq 0.52$) show a tetragonal distortion of the unit cell. Compositions having lower ($0.52 \leq x \leq 0.94$) Ti content have rhombohedrally distorted unit cells. Both phases are separated by a morphotropic phase boundary at $x = 0.48$. Compositions near the Zr side of the system ($0.94 \leq x \leq 1$) are antiferroelectric and have orthorhombic structure. The direction of the spontaneous polarization $\overline{P}_s$ is along one of the edges of the unit cell for tetragonal distorted compositions and along one of the space diagonals for rhombohedral distorted compositions.

A third or even a fourth and fifth phase of a complex perovskite, for example, $Pb(Mg_{1/3}Nb_{2/3})O_3$, may be added in addition to $PbTiO_3$ and $PbZrO_3$ when forming the solid solution (10,19–21). In the same manner, 20 different elements having similar ionic radii can be substituted in place of Mg^{2+} or Nb^{4+}, leading to a huge number of possible combinations and a multitude of compositions which have, however, generally comparable properties. At low concentrations of a complex perovskite addition, the phase relationships of the quasibinary composition are maintained. Increasing amounts of a complex perovskite reduces T_c, and pseudocubic phases begin to appear.

There is often a wide range of crystalline solid solubility between end-member compositions. Additionally the ferroelectric and antiferroelectric Curie temperatures and consequent properties appear to mutate continuously with fractional cation substitution. Thus the perovskite system has a variety of extremely useful properties. Other oxygen octahedra structure ferroelectrics such as lithium niobate [*12031-63-9*], $LiNbO_3$, lithium tantalate [*12031-66-2*], $LiTaO_3$, the tungsten bronze structures, bismuth oxide layer structures, pyrochlore structures, and order–disorder-type ferroelectrics are well discussed elsewhere (4,12,22,23).

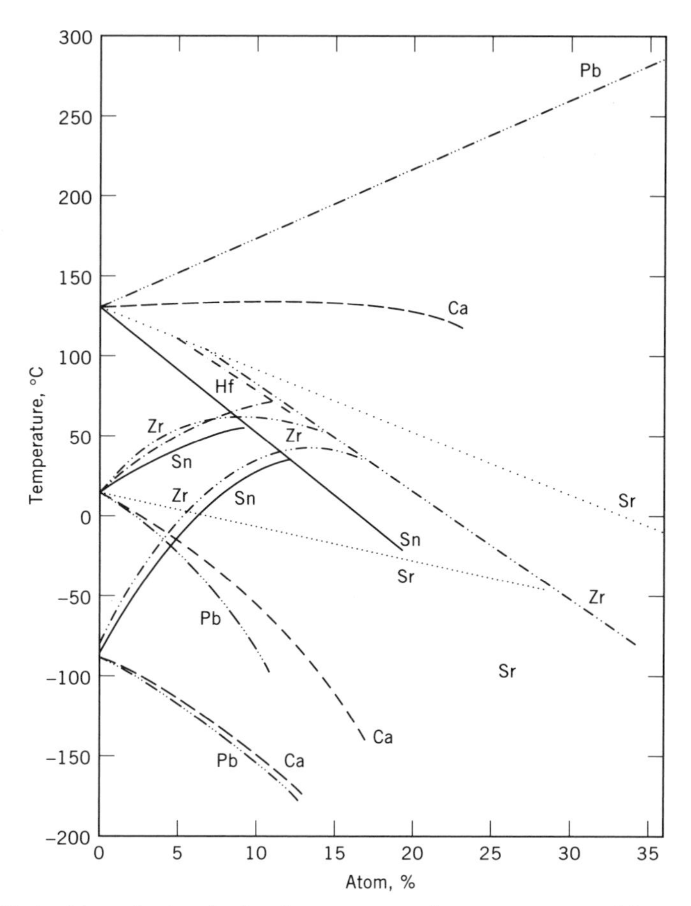

Fig. 4. Effect of isovalent substitutions on crystal structure transition temperatures of ceramic $BaTiO_3$ where (—···—) represents Pb^{2+}; (——), Ca^{2+}; and (····) Sr^{2+} substitution for Ba^{2+}; and (—), Sn^{4+}; (—··—), Zr^{4+}; and (—·—) Hf^{4+} substitution for Ti^{4+}. Transition temperatures for pure $BaTiO_3$ are 135, 15, and −90°C (see Fig. 3).

Preparation of Ferroelectric Materials

Ceramics. The properties of ferroelectrics, basically determined by composition, are also affected by the microstructure of the densified body which depends on the fabrication method and condition. The ferroelectric ceramic process is comprised of the following steps (10,24,25): (*1*) selection of raw oxide materials, (*2*) preparation of a powder composition, (*3*) shaping, (*4*) densification, and (*5*) finishing.

Raw Materials. Most of the raw materials are oxides (PbO, TiO_2, ZrO_2) or carbonates ($BaCO_3$, $SrCO_3$, $CaCO_3$). The levels of certain impurities and particle size are specified by the chemical supplier. However, particle size and degree of aggregation are more difficult to specify. Because reactivity depends on particle size and the perfection of the crystals comprising the particles, the more detailed the specification, the more expensive the material. Thus raw materials are usually selected to meet application-dependent requirements.

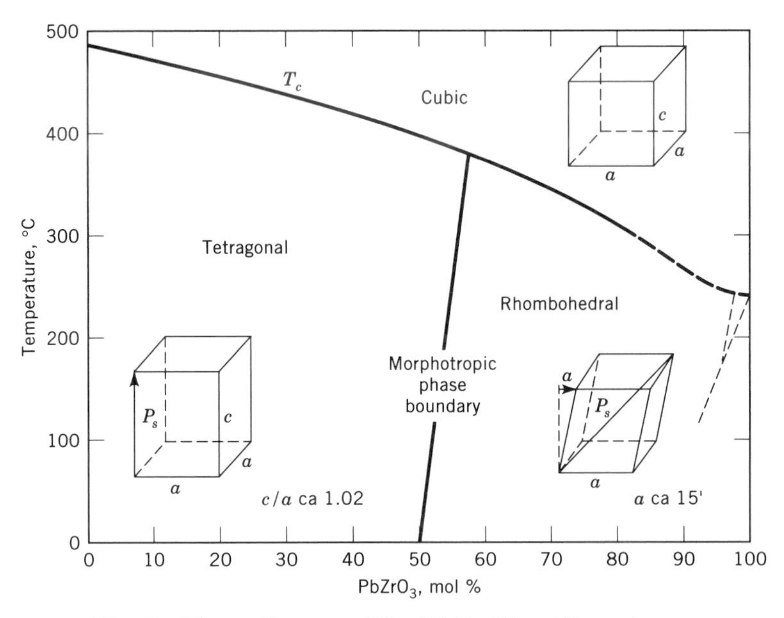

Fig. 5. Phase diagram of the $Pb(Zr_x,Ti_{1-x})O_3$ system.

Powder Preparation. Mixing. The most widely used mixing method is wet ball milling, which is a slow process, but it can be left unattended for the whole procedure. A ball mill is a barrel that rotates on its axis and is partially filled with a grinding medium (usually of ceramic material) in the form of spheres, cylinders, or rods. It mixes the raw oxides, eliminates aggregates, and can reduce the particle size.

Calcination. Calcination involves a low (<1000°C) temperature solid-state chemical reaction of the raw materials to form the desired final composition and structure such as perovskite for $BaTiO_3$ and PZT. It can be carried out by placing the mixed powders in crucibles in a batch or continuous kiln. A rotary kiln also can be used for this purpose to process continuously. A sufficiently uniform temperature has to be provided for the mixed oxides, because the thermal conductivity of powdered materials is always low.

Shaping. The calcined powders must be milled and a binder (usually organic materials) added if necessary for the forming procedure. Table 1 summarizes this procedure (25).

Densification. Sintering, hot-pressing, or hot-isostatic-pressing methods may be used to densify the shaped green ferroelectric ceramics to ~ 95–100% of theoretical value. Sintering takes place at high (>1000°C) temperatures for several hours. A green ferroelectric ceramic is converted into a denser structure of crystallite. The crystallites are joined to one another by grain boundaries the thickness of which vary from about 100 pm to over 1 μm. These usually consist of the second phase and have a great deal of influence on the electrical and mechanical properties of ferroelectric ceramic devices.

Grain growth which also affects the final properties occurs during the sintering. Using hot-isostatic-pressing (HIP) and hot-uniaxial-pressing (HUP), the

Table 1. Ceramic Shaping Method[a]

Shaping method	Type of feed material	Type of shape
dry pressing	free-flowing granules	small, simple shapes
isostatic pressing	fragile granules	larger, more intricate shapes
calendering	plastic mass based on an elastic polymer	thin plates
extrusion	plastic mass using a viscous polymer solution	elongated shapes of constant cross section
jiggering	stiff mud containing clay	large, simple shapes
injection molding	organic binder giving fluidity when hot	complex shapes
slip-casting	free-flowing cream	mainly hollow shapes
band-casting	free-flowing cream	thin plates and sheets
silk-screening	printing ink consistency	thin layers on substrates

[a]Ref. 25.

ceramic density close to the theoretical value can be obtained. However, these latter processes are usually a little more expensive.

Finishing. The densified ferroelectric ceramic bodies usually require machining and metallizing for dimension and surface roughness control and electrical contact. Ceramic preparations are discussed in detail in the literature (10,25,26).

Thin-Film Ferroelectrics. The trends in integrated circuits (qv) and packaging technologies toward miniaturization have stimulated the development of ferroelectric thin films (27) (see PACKAGING, SEMICONDUCTORS AND ELECTRONIC MATERIALS; THIN FILMS). Advances in thin-film growth processes offer the opportunity to utilize the material properties of ferroelectrics such as pyroelectricity, piezoelectricity, and electrooptic activity for useful device applications (28). The primary impetus of the activity in ferroelectric thin-film research is the large demand for the development of nonvolatile memory devices, also called FERRAMS (ferroelectric random access memories) (29). FERRAMS promise fast read and write cycles, low (3–5 V) switching voltages, nonvolatility, long (10^{12} cycles) endurance, and radiation hardness compatible with semiconductors (qv) such as gallium arsenide, GaAs (30).

Several techniques have been investigated for the preparation of ferroelectric thin films. The thin-film growth processes involving low energy bombardment include magnetron sputtering (31–34), ion beam sputtering (35,36), excimer laser ablation (37–39), electron cyclotron resonance (ECR) plasma-assisted growth (40,41), and plasma-enhanced chemical vapor deposition (PECVD) (42) (see PLASMA TECHNOLOGY). Other methods are sol-gel (43–45), metal organic decomposition (MOD) (46,47), thermal and *e*-beam evaporation (48,49), flash evaporation (50,51), chemical vapor deposition (CVD) (52), metal organic chemical vapor deposition (MOCVD) (53,54), and molecular beam epitaxy (MBE) (55).

The requirements of thin-film ferroelectrics are stoichiometry, phase formation, crystalization, and microstructural development for the various device applications. As of this writing multimagnetron sputtering (MMS) (56), multiion beam-reactive sputter (MIBERS) deposition (57), uv-excimer laser ablation (58), and electron cyclotron resonance (ECR) plasma-assisted growth (59) are the latest ferroelectric thin-film growth processes to satisfy the requirements.

Ferroelectric Ceramic–Polymer Composites. The motivation for the development of composite ferroelectric materials arose from the need for a combination of desirable properties that often cannot be obtained in single-phase materials. For example, in an electromechanical transducer, the piezoelectric sensitivity might be maximized and the density minimized to obtain a good acoustic matching with water, and the transducer made mechanically flexible to conform to a curved surface (see COMPOSITE MATERIALS, CERAMIC-MATRIX).

The development of active ceramic–polymer composites was undertaken for underwater hydrophones having hydrostatic piezoelectric coefficients larger than those of the commonly used lead zirconate titanate (PZT) ceramics (60–70). It has been demonstrated that certain composite hydrophone materials are two to three orders of magnitude more sensitive than PZT ceramics while satisfying such other requirements as pressure dependency of sensitivity. The idea of composite ferroelectrics has been extended to other applications such as ultrasonic transducers for acoustic imaging, thermistors having both negative and positive temperature coefficients of resistance, and active sound absorbers.

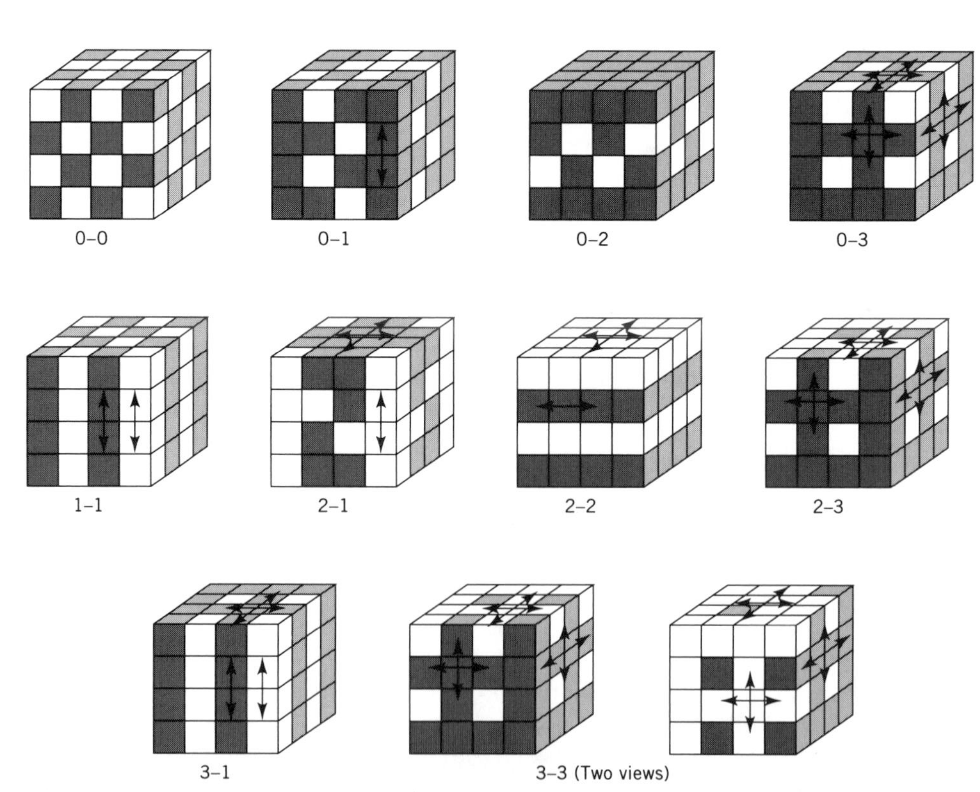

Fig. 6. Connectivity patterns for a diphasic solid showing zero-, one-, two-, or three-dimensional connectivity of each phase to itself. In the 3–1 composite, for instance, the shaded phase is three-dimensionally connected. Arrows are used to indicate the connected directions.

To optimize a ferroelectric ceramic and polymer composite device for a certain application, it is important to define a figure of merit which includes the most sensitive parameters. Maximizing a figure of merit requires not only choosing correct component phases having the right properties, but also designing the proper connectivity of the composite structures (60). Connectivity is a key feature in property development in multiphase solids because physical properties can change by many orders of magnitude depending on the manner in which connections are made. Each phase in a composite may be self-connected in zero, one, two, or three dimensions. It is natural to confine attention to orthogonal systems. The 10 connectivity patterns for a diphase solid are shown in Figure 6. Extrusion, dicing, tape-casting, injection-molding, and hot-rolling methods are examples of processing techniques used for making ferroelectric composites having different connectivities. The 3–1 connectivity pattern is ideally suited to extrusion processing. A ceramic slip is extruded through a die giving a three-dimensionally connected pattern with one-dimensional holes, which can later be filled with a second phase.

Applications

Multilayer Capacitors. Multilayer capacitors (MLC), at greater than 30 billion units per year, outnumber any other ferroelectric device in production. Multilayer capacitors consist of alternating layers of dielectric material and metal electrodes, as shown in Figure 7. The reason for this configuration is miniaturization of the capacitor. Capacitance is given by

$$C = \frac{\epsilon_0 KA}{t}$$

where C is in Farads, ϵ_0 is permittivity in a vacuum, A is the area, K is the relative dielectric permittivity, and t is the thickness. Therefore, capacitance increases with increasing area and decreasing thickness. A multilayer capacitor usually contains up to 100 thin layers typically 10 to 35 μm thick of dielectric materials (26).

The dielectric materials used in multilayer capacitors must satisfy several electrical property requirements. A high dielectric permittivity with a minimal temperature dependence is desired for a wide temperature-range application of the MLC. $BaTiO_3$-based ceramics show high dielectric permittivities; however, the dielectric permittivity of $BaTiO_3$ ceramics has a strong temperature dependence and a maximum at the Curie point (Fig. 3). Although the Curie point can be shifted to room temperature by a partial substitution of Ba or Ti (Fig. 4), dielectric permittivity of $BaTiO_3$ ceramics also depends on the grain size of materials. When the powder is prevented from growing grains larger than 1.5 μm, the 90° domain walls are not formed. The material then remains stressed and the relative dielectric permittivity goes to ~2500–3500. Figure 8 compares the dielectric constants vs temperatures of small- and large-grain undoped barium titanate (71).

Piezoelectric and Electrostrictive Device Applications. Devices made from ferroelectric materials utilizing their piezoelectric or electrostrictive properties range from gas igniters to ultrasonic cleaners (or welders) (72).

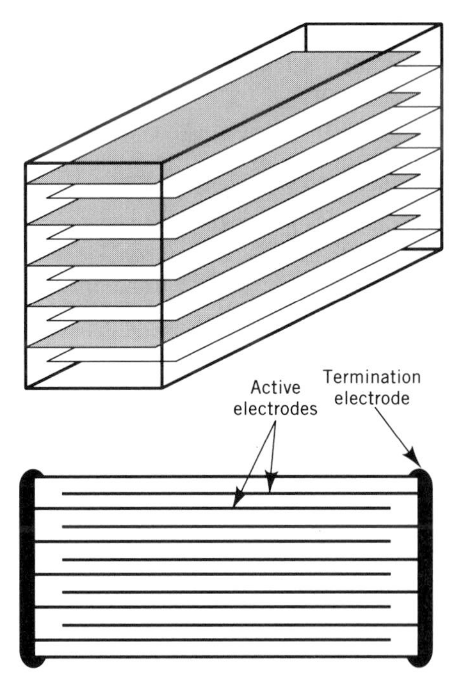

Fig. 7. Schematic of a conventional multilayer capacitor. The orientations of the internal and the termination electrodes are shown.

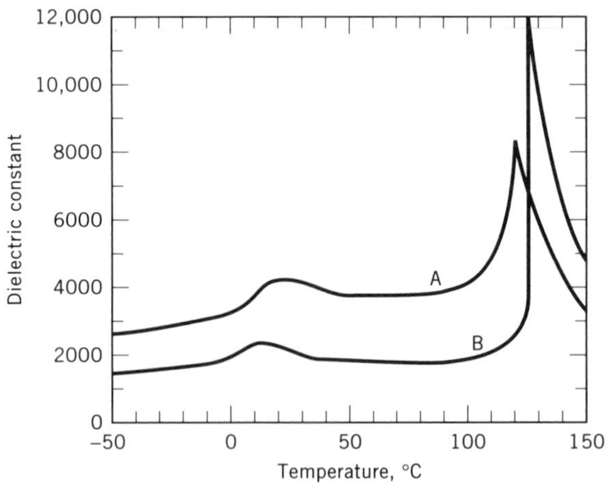

Fig. 8. Dielectric constant (1 kHz) vs temperature for $BaTiO_3$ ceramics of A, 1-μm grain size, and B, 50-μm grain size.

Applications of piezoelectric and electrostrictive ceramics include:

Classification	Applications
high voltage generators	gas appliances, cigarette lighters, fuses (igniters) for explosives, flash bulbs
high power ultrasonic generators	ultrasonic cleaners, sonar, echo sounding, ultrasonic machining, atomization, pulverization
transducers for sound and ultrasound in air	microphones (eg, for telephones), burglar alarm systems, remote control, loudspeakers (eg, tweeters), buzzers, medical ultrasonics equipment
sensors	phonograph pick-ups, accelerometers, hydrophones, detection systems in machinery, musical instruments
resonators and filters	radios and televisions, remote control, electronic instrumentation
delay lines	color televisions, computers, electronic instrumentation
keyboards	computers, printers, desk calculators, vending machines, telephones
actuators	micropositioners, ink-jet printers
smart materials	active noise control for automobiles, aircraft, and trains; active shock absorber
electrooptic devices	optical shutter
composites	hydrophones, medical imaging transducers
miscellaneous	voltage transformers, flow meters

Multilayer-type piezoelectric or electrostrictive actuators are used for several applications including the composite smart structure shown in Figure 9 (73).

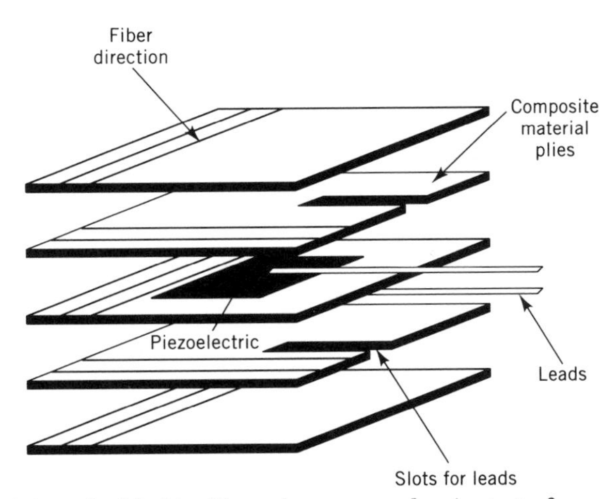

Fig. 9. Piezoelectric embedded inside a glass–epoxy laminate to form a composite smart structure.

Small ultrasonic motors such as the rotary actuator shown in Figure 10 have also been made and can be used for automobile windows, seats, and windshield wipers. Many small industrial motors could make use of the high torque, low speed, and precise stepping character of these actuators.

Composite Devices. Composites made of active-phase PZT and polymer-matrix phase are used for the hydrophone and medical imaging devices (see COMPOSITE MATERIALS, POLYMER-MATRIX; IMAGING TECHNOLOGY). A useful figure of merit for hydrophone materials is the product of hydrostatic strain coefficient d_h and hydrostatic voltage coefficient g_h where g_h is related to the d_h coefficient by (74)

$$g_h = d_h/\epsilon_0 \mathrm{K}$$

Dielectric and piezoelectric properties of 3–3 type (60,63,75), 1–3 type (62,76–78), diced (79,80), 3–0 type (81,82), 3–1 and 3–2 type (83,84), and 0–3 type composites (69,85–88) have been investigated. For medical ultrasonic transducers, the 1–3 PZT rod and polymer composites are used. Electromechanical coupling properties of these materials are available (89–96).

Relaxor Ferroelectrics. The general characteristics distinguishing relaxor ferroelectrics, eg, the $PbMg_{1/3}Nb_{2/3}O_3$ family, from normal ferroelectrics such as $BaTiO_3$, are summarized in Table 2 (97). The dielectric response in the paraelectric–ferroelectric transition region is significantly more diffuse for the former. Maximum relative dielectric permittivities, referred to as K_{max}, are greater than 20,000. The temperature dependence of the dielectric properties is shown in Figure 11. The dielectric permittivity and loss exhibit dispersion near the transition, and K_{max} decreases and shifts to higher temperatures with increasing measuring frequency. The temperature of the maximum dielectric loss does not coincide with T_{max} in relaxors; rather, the loss increases with measurement frequency. The polarization–electric field behavior exhibits typical ferroelectric hysteresis at temperatures well below the T_{max} as shown in Figure 12. The pronounced hysteresis slowly decays to slim-loop behavior at temperatures above T_{max}. Structural

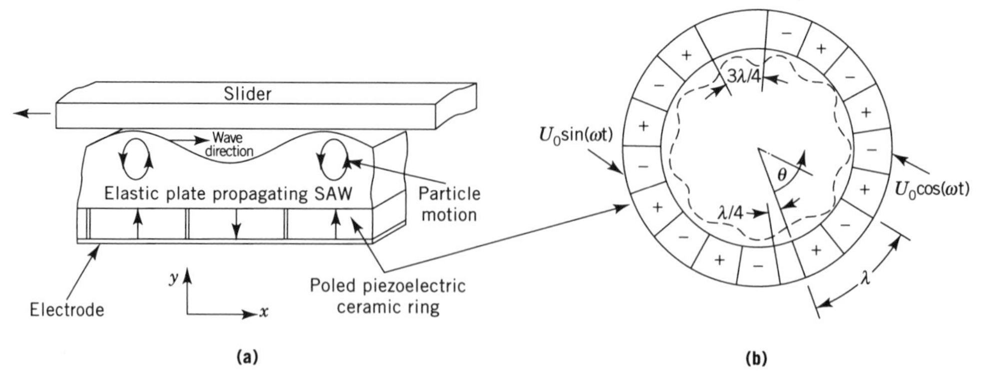

Fig. 10. The rotary actuator: (**a**) side view where SAW = surface acoustic wave; and (**b**) view of the poled piezoelectric ceramic ring showing poled segments and how temporal and spatial phase differences are established. Courtesy of Shinsei Kogyo Co.

Table 2. Properties of Relaxor and Normal Ferroelectrics[a]

Property	Normal ferroelectrics	Relaxor ferroelectrics
permittivity temperature dependence $\epsilon = \epsilon(T)$	sharp first- or second-order transition above Curie temperature	broad–diffuse phase transition about Curie maxima
permittivity temperature and frequency dependence $\epsilon = \epsilon(T,\omega)$	weak frequency dependence	strong frequency dependence
remanent polarization	strong remanent polarization	weak remanent polarization
scattering of light	strong anisotropy (birefringent)	very weak anisotropy (pseudocubic)
diffraction of x-rays	line splitting owing to spontaneous deformation from paraelectric to ferroelectric phase	no x-ray splitting giving a pseudocubic structure

[a]Ref. 97.

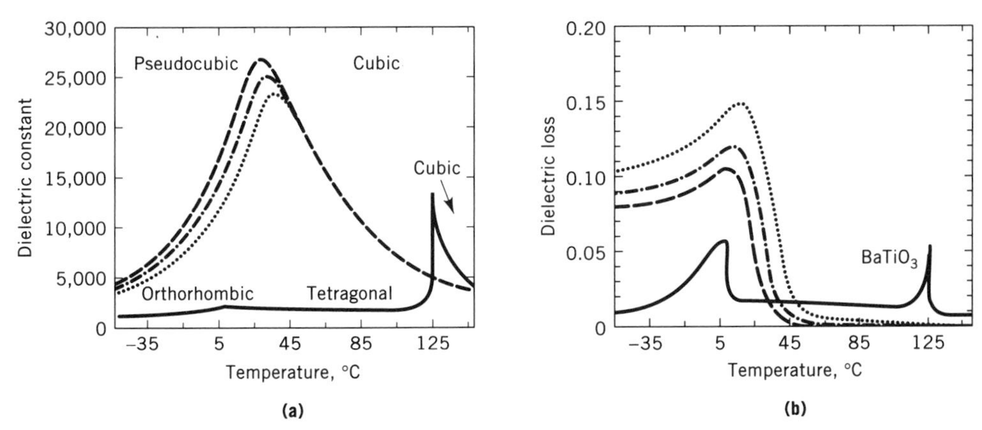

Fig. 11. Fundamental characteristics of relaxor materials compared to $BaTiO_3$. Temperature dependence for the relaxor ferroelectric 0.93 $PbMg_{1/3}Nb_{2/3}O_3$–0.07 $PbTiO_3$ at (———) 1, (—·—) 10, and (···) 100 kHz and for $BaTiO_3$ at (——) 1 kHz of (**a**) dielectric constant and (**b**) dielectric loss (98).

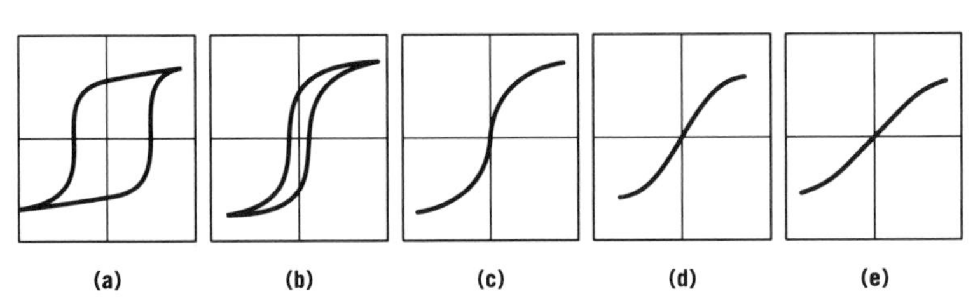

Fig. 12. Dielectric hysteresis in $Pb(Mg_{1/3}Nb_{2/3})O_3$ at (**a**) −150°C; (**b**) −80°C; (**c**) −20°C; (**d**) −10°C; and (**e**) 25°C (98).

ordering as evidenced by optical anisotropy or x-ray line splitting from optical birefringence and x-ray diffraction studies, respectively, has not been observed in relaxors as in normal ferroelectrics (97).

Relaxor ferroelectrics have been extensively investigated since the late 1970s because of the ability to generate large electrically induced strains, minimal hysteresis of the strain–electric field response, and minimal thermal strain (13,99,100). These materials also show promise in capacitor applications because of large dielectric permittivities (101,102). In addition to actuator and capacitor applications, a large piezoelectric effect can be induced by the application of an external electric field in relaxor ferroelectric materials possessing large polarizations (13). Field-induced piezoelectric transducers exhibit a nonlinear strain–electric field response and are not piezoelectrically active when the external bias is removed. These materials allow for the adjustment (tuning) of the piezoelectric coefficients between on and off states for specific operating conditions. The field-induced piezoelectric and elastic properties of relaxor ferroelectrics have also been investigated for transducer applications, including three-dimensional medical ultrasonic imaging devices.

Polymer Ferroelectrics. In 1969, it was found that strong piezoelectric effects could be induced in the polymer poly(vinylidene fluoride) (known as PVD_2 or PVDF) by application of an electric field (103). Pyroelectricity, with pyroelectric figures of merit comparable to crystalline pyroelectric detectors (104,105) of PVF_2 films polarized this way, was discovered two year later (106.)

These discoveries were important breakthroughs for the field of electromechanical and pyroelectric devices. Fabrications of an inexpensive, completely flexible, rugged, wide area piezoelectric transducer and pyroelectric detector were realized immediately. Within a few years, a wide variety of prototype devices had been fabricated. These include audio frequency transducers, such as microphones, headphones, and loudspeaker tweeters having excellent frequency response and low distortion because of the low density lightweight transducer film; ultrasonic transducers for underwater applications, such as hydrophones, and for medical imaging applications; electromechanical transducers for computer and telephone keypads and a variety of other contactless switching applications; and pyroelectric detectors for infrared imaging and intruder detection. At least 1–2 MV/cm of poling field is required to obtain 65 mC/m^2 polarization level of PVDF thin film which is consistent with 50% crystallinity and perfect alignment (107).

Economic Aspects

Ferroelectric–polymer composite devices have been developed for large-area transducers, active noise control, and medical imaging applications (see NOISE POLLUTION AND ABATEMENT METHODS). North American Philips, Hewlett-Packard, and Toshiba make composite medical imaging probes for in-house use. Krautkramer Branson Co. produces the same purpose composite transducer for the open market. NTK Technical Ceramics and Mitsubishi Petrochemical market ferroelectric–polymer composite materials (108) for various device applications, such as a towed array hydrophone and robotic use. Whereas the composite market is growing with the invention of new devices, total unit volume and dollar amounts are small compared to the ferroelectric capacitor and ferroelectric–piezoelectric ceramic markets (see MEDICAL IMAGING TECHNOLOGY).

Ferroelectric thin films have not, as of this writing, been commercialized. Demand for PTC ferroelectrics has been decreasing rapidly. Wide usage of the fuel injector in automobiles and other types of composite PTC devices is the main reason.

Approximately 40% of the U.S. electronic ceramics industry is represented by ferroelectrics. Table 3 shows U.S. consumption of ceramic capacitors and piezoelectric materials (109).

Table 3. U.S. Ceramic Capacitor and Ceramic Piezoelectric Material Consumption[a]

Product	Value, $ × 10^6 1987	Value, $ × 10^6 1992	Five-year growth, %
ceramic capacitors[b]	609	740	4
piezoelectric materials	70	115	10

[a]Ref. 109.
[b]Includes chip capacitors.

Japanese suppliers generally dominate the electronic ceramic business. Japanese production of ferroelectric devices in the first nine months of 1990 was valued at $711 $\times$ 10^6 for ceramic capacitors and $353 $\times$ 10^6 for piezoelectric devices, representing growths of 7 and 10.8%, respectively, over the previous year.

Principal producers of ferroelectric capacitors are (110) Murata Manufacturing Co., Ltd., Kyoto, Japan; Kyocera Corp., Kyoto, Japan; Philips Electronics, N.V., Eindhoven, the Netherlands; and NEC Corp., Tokyo, Japan. Principal piezoelectric ceramic component producers are (110) Murata Manufacturing Co., Ltd., Kyoto, Japan; Motorola, Inc., Schaumburg, Ill.; EDO Corp., College Point, N.Y.; Morgan Matroc, Inc., Bedford, Ohio; Kyocera Corp., Kyoto, Japan; and Philips Electronics, N.V., Eindhoven, the Netherlands.

BIBLIOGRAPHY

"Ferroelectrics" in *ECT* 2nd ed., Vol. 9, pp. 1–25, by E. C. Henry, General Electric Co.; in *ECT* 3rd ed., Vol. 10, pp. 1–30, by L. E. Cross, The Pennsylvania State University, and K. H. Härdtl, Philips Forschungslaboratorium Aachen GmbH.

1. M. Faraday, *Experimental Researches in Electricity*, Taylor and Francis, London, 1838; Dover, New York, 1965, Vol. I, § 1168.
2. J. Reitz and F. Milford, *Foundations of Electromagnetic Theory*, Addison-Wesley Publishing Co., Inc., Reading, Mass., 1969.
3. J. M. Herbert, *Ceramic Dielectric and Capacitors*, Gordon & Breach, New York, 1985.
4. M. E. Lines and A. M. Glass, *Principles and Applications of Ferroelectrics and Related Materials*, Clarendon, Oxford, UK, 1977.
5. F. Jona and G. Shirane, *Ferroelectric Crystals*, Pergamon Press, Inc., Elmsford, N.Y., 1962.

6. J. Burfoot, *Ferroelectrics, An Introduction to the Physical Principles*, D. Van Nostrand, Princeton, N.J., 1967; R. E. Cohen, *Nature* **358**, 136 (1992).
7. J. F. Nye, *Physical Properties of Crystals*, Clarendon, Oxford, UK, 1957.
8. K. Aizu, *J. Phys. Soc. Jpn.* **20**, 959 (1965); *Phys. Rev.* **146**, 423 (1966).
9. L. A. Shuvalov, *J. Phys. Soc. Jpn.* **285**, 38 (1970).
10. B. Jaffe, W. R. Cooke, Jr., and H. Jaffe, *Piezoelectric Ceramics*, Academic Press, New York, 1971.
11. A. F. Devonshire, *Phil. Mag.* **40**, 1040 (1949) and *Phil. Mag.* **42**, 1065 (1951).
12. Landolt-Börnstein, *Ferroelectric and Anti-ferroelectric Substances*, Vol. 3, New Series Group III, Springer-Verlag, Berlin, Germany, 1969.
13. S. J. Jang, *Electrostrictive Ceramics for Transducer Applications*, Ph.D. dissertation, The Pennsylvania State University, University Park, 1979.
14. G. H. Haertling and C. E. Land, *J. Am. Ceram. Soc.* **54**, 1 (1971).
15. P. Günter and J. P. Huignard, *Photorefractive Materials and their Applications I, II*, Springer-Verlag, New York, 1988.
16. I. S. Zheludev, *Physics of Crystalline Dielectrics*, Plenum Press, New York, 1971.
17. B. Jaffe, R. S. Roth, and S. Marzullo, *J. Res. Nat. Bur. Stand.* **55**, 239 (1955).
18. E. Sowaguchi, *J. Phys. Soc. Jpn.* **8**, 615 (1953).
19. H. Ouchi, K. Nagano, and S. Hayakawa, *J. Am. Ceram. Soc.* **48**, 630 (1965).
20. H. Ouchi, M. Nishida, and S. Hayakawa, *J. Am. Ceram. Soc.* **49**, 577 (1966).
21. H. Ouchi, *J. Am. Ceram. Soc.* **51**, 169 (1968).
22. Landolt-Börnstein, *Ferroelectric and Anti-ferroelectric Substances*, Vol. 9, New Series Group III, Springer-Verlag, Berlin, Germany, 1977.
23. Landolt-Börnstein, *Ferroelectric and Related Substances*, Vol. 28, New Series Group III, Springer-Verlag, Berlin, Germany, 1990.
24. J. M. Herbert, *Ferroelectric Transducers and Sensors*, Gordon & Breach, New York, 1982.
25. A. J. Moulson and J. M. Herbert, *Electroceramics*, Chapman and Hall, London, 1990.
26. J. H. Adair, D. A. Anderson, G. O. Dayton, and T. R. Shrout, *J. Mater. Educ.* **9**, 71–118 (1987).
27. S. B. Krupanidhi, *J. Vac. Sci. Tech.* **A10**, 1509 (1992).
28. M. H. Francombe and S. V. Krishnaswamy, *J. Vac. Sci. Tech.* **A8**, 1382 (1990).
29. J. F. Scott and C. A. Araujo, *Science* **246**, 14800 (1989).
30. L. E. Sanchez, S. Y. Wu, and I. K. Naik, *Appl. Phys. Lett.* **56**, 2399 (1990).
31. S. B. Krupanidhi, N. Maffei, M. Sayer, and K. El-Assal, *J. Appl. Phys.* **54**, 6601 (1983).
32. M. Okuyama and Y. Hamakawa, *Ferroelectrics* **63**, 243 (1985).
33. S. B. Krupanidhi and M. Sayer, *J. Vac. Sci. Technol.* **A2**, 203 (1984).
34. K. Sreenivas and M. Sayer, *J. Appl. Phys.* **64**, 1484 (1988).
35. R. N. Castellano and L. G. Feinstein, *J. Appl. Phys.* **50**, 4406 (1979); S. B. Krupanidhi, H. Hu, and V. Kumai, *J. Appl. Phys.* **71**, 376 (1992).
36. D. Xiao, Z. Xiao, J. Zhu, Y. Li, and H. Guo, *Ferroelectrics* **108**, 59 (1990).
37. R. Ramesh and co-workers, *Appl. Phys. Lett.* **57**, 1505 (1990).
38. S. B. Krupanidhi and co-workers, *J. Vac. Sci. Tech.*, **A10**, 1815 (1992).
39. K. L. Saenger, R. A. Roy, K. F. Etzold, and J. J. Cuomo, *Mater. Res. Soc. Symp. Proc.* **200**, 115 (1990).
40. M. Okuyama, Y. Togani, and Y. Hamakawa, *Appl. Surf. Sci.* **33/34**, 625 (1988).
41. Y. Masuda, A. Baba, H. Masomoto, T. Goto, M. Minikata, and T. Kirai, in Ref. 9, p. 337.
42. W. T. Petusky and S. K. Dey in *Proceedings of the 3rd International Symposium on Integrated Ferroelectrics, Colorado Springs, Colo., Apr. 1991,* Gordon and Breach Science Publishers, New York (in press).
43. K. D. Budd, S. K. Dey, and D. A. Payne, *Br. Ceram. Proc.* **36**, 107 (1985).

44. S. K. Dey and R. Zuleeg, *Ferroelectrics* **108**, 37 (1990).
45. G. A. C. M. Spierings, M. J. E. Ulenaers, G. L. Kampschoer, H. A. M. van Hal, and P. K. Larsen, *J. Appl. Phys.* **70**, 2290 (1991).
46. R. W. West, *Ferroelectrics* **102**, 53 (1990).
47. G. H. Haertling, *J. Vac. Sci. Technol.* **A9**, 414 (1976).
48. M. Iukawa and K. Toda, *Appl. Phys. Lett.* **29**, 491 (1976).
49. A. Mansingh and S. B. Krupanidhi, *J. Appl. Phys.* **51**, 5408 (1980).
50. J. R. Slack and J. C. Burfoot, *Thin Solid Films* **6**, 233 (1970).
51. A. Mansingh and S. B. Krupanidhi, *Thin Solid Films* **80**, 359 (1981).
52. J. Kojima, M. Okuyama, T. Nakagawa, and Y. Hamakawa, *Jpn. J. Appl. Phys.* **22**, Suppl. 2, 14 (1983).
53. M. Okada, S. Takai, M. Amemiya, and K. Tominaga, *Jpn. J. Appl. Phys.* **28**, 1030 (1989).
54. P. C. Van Buskirk, R. Gardiner, P. S. Kirlin, and S. B. Krupanidhi, *J. Vac. Sci. Technol.* (in press).
55. S. Sinharoy, H. Buhay, M. H. Francombe, W. J. Takei, N. J. Doyle, J. R. Rieger, D. R. Lampe, and E. Stepke, *J. Vac. Sci. Technol.* (in press).
56. H. Adachi, T. Mitsuyu, O. Yamazaki, and K. Wasa, *J. Appl. Phys.* **60**, 736 (1986).
57. J. J. Cuomo, S. M. Rossnagel and H. R. Kaufman, *Handbook of Ion Beam Processing Technology*, Noyes, Park Ridge, N.J., 1989.
58. R. K. Singh and J. Narayou, *Phys. Rev.* **B42**, 8843 (1990).
59. D. B. Beach, *IBM J. Res. Dev.* **34**, 795 (1990).
60. R. E. Newnham, D. P. Skinner, and L. E. Cross, *Mater. Res. Bull.* **13**, 525 and 599 (1978).
61. K. A. Klicker, J. V. Biggers, and R. E. Newnham, *J. Am. Ceram. Soc.* **64**, 5 (1981).
62. M. J. Haun, P. Moses, T. R. Gururaja, and W. A. Schulze, *Ferroelectrics*, **49**, 259 (1983).
63. K. Rittenmyer, T. Shrout, W. A. Schulze, and R. E. Newnham, *Ferroelectrics*, **41**, 189 (1982).
64. A. Safari, R. E. Newnham, L. E. Cross, and W. A. Schulze, *Ferroelectrics* **41**, 197 (1982).
65. R. E. Newnham, L. J. Bowen, K. A. Klicker, and L. E. Cross, *Mater. Eng.* **112**, 93 (1980).
66. J. Runt and E. C. Galgoci, *J. Appl. Polym. Sci.* **39**, 611 (1984).
67. T. R. Gururaja, W. A. Schulze, L. E. Cross, R. E. Newnham, B. A. Auld, and J. Wang, *IEEE Trans. Sonic Ultrasonics* **SU-32**, 481 and 499 (1985).
68. J. R. Giniewicz, R. E. Newnham, A. Safari, and D. Moffatt, *Ferroelectrics*, **73**, 405–417 (1987).
69. S. Sa-Gong, A. Safari, S. J. Jang, and R. E. Newnham, *Ferroelectric Lett.* **5**, 131 (1986).
70. K. A. Hu, J. Runt, A. Safari, and R. E. Newnham, *Ferroelectrics*, **68**, 115 (1986).
71. W. R. Buessem, L. E. Cross, and A. K. Goswami, *J. Am. Ceram. Soc.* **49**, 33 (1966).
72. J. Van Kanderaat and R. E. Setterington, *Piezoelectric Ceramics, Application Book*, Ferroxcube Corp., New York, 1974.
73. N. W. Hagood, E. F. Crawley, J. deLuis, and E. H. Anderson, in C. A. Rogers, ed., *Smart Materials, Structures, and Mathematical Issues*, Technomic Publishing Co., Lancaster, Pa., 1988, p. 80.
74. L. M. Levinson, *Electronic Ceramics*, Marcel Dekker, New York, 1988.
75. K. Hikita, K. M. Nichioka, and M. Ono, *Ferroelectrics*, **49**, 265 (1983).
76. R. Y. Ting, A. Halliyal, and A. S. Bhalla, *Jpn. J. Appl. Phys.* **24** (Suppl. 24-2), 982 (1985).
77. A. Halliyal, A. Safari, A. S. Bhalla, R. E. Newnham, and L. E. Cross, *J. Am. Ceram. Soc.* **67**, 331 (1984).

78. K. A. Klicker, W. A. Schulze, and J. V. Biggers, *J. Am. Ceram. Soc.* **65**, C208 (1982).
79. H. P. Savakus, K. A. Klicker, and R. E. Newnham, *Mater. Res. Bull,* **16**, 677 (1981).
80. N. M. Shorrocks, M. E. Brown, R. W. Whatmore, and F. W. Ainger, *Ferroelectrics* **54**, 215 (1984).
81. M. Kahn, R. W. Rice, and D. Shadwell, *Advan. Ceram. Mater.* **1**, 55 (1986).
82. S. Pilgrim and R. E. Newnham, "A New Type of 3-0 Composites," presented at the *American Ceramics Society, Electronics Division 86 Meeting*, Chicago, 1986.
83. T. R. Shrout, L. J. Bowen, and W. A. Schulze, *Mater. Res. Bull.* **15**, 1371 (1980).
84. A. Safari, A. Halliyal, R. E. Newnham, and I. M. Lachman, *Mater. Res. Bull.* **17**, 301 (1982).
85. L. A. Pauer, *IEEE Intl. Conv. Rec.* 1 (1973).
86. H. Banno and S. Saito, *Jpn. J. Appl. Phys.* **22** (Suppl. 22-2), 67 (1983).
87. R. Y. Ting, "Evaluation of New Piezoelectric Composite Materials for Hydrophone Applications," presented at the *Bernard Jaffe Memorial Colloquium*, American Ceramics Society, 86 Meeting, Pittsburgh, 1984.
88. D. L. Monroe, J. B. Blum, and A. Safari, *Ferroelectrics Lett.* **5**, 39 (1986).
89. T. R. Gururaja and co-workers, *Ferroelectrics* **39**, 1245 (1981).
90. W. A. Smith, A. A. Shaulov, and B. M. Singer in *Proceedings of the 1984 IEEE Ultrasonics Symposium*, Dallas, Tex., 1984, p. 539.
91. A. A. Shaulov, W. A. Smith, and B. M. Singer, in Ref. 90, p. 545.
92. W. A. Smith, A. Shaulov, and B. A. Auld in *Proceedings of the 1985 IEEE Ultrasonics Symposium,* San Francisco, Calif., 1985, p. 642.
93. A. A. Shaulov and W. A. Smith, in Ref. 92, p. 648.
94. H. Takeuchi, C. Nakaya, and K. Katakura in Ref. 90, p. 504.
95. H. Takeuchi and C. Nakaya, *Ferroelectrics* **68**, 53 (1986).
96. A. Fukumoto, *Ferroelectrics* **40**, 217 (1982).
97. L. E. Cross, *Ferroelectrics* **76**, 241–267 (1987).
98. T. R. Shrout and J. P. Dougherty, *Ceram. Trans.* **8**, 3–19 (1990).
99. J. Kuwata, K. Uchino, and S. Nomura, *Jpn. J. Appl. Phys.* **19**, 2099 (1980).
100. S. Nomura and K. Uchino, *Ferroelectrics* **41**, 117 (1982); **50**, 197 (1983).
101. S. L. Swartz and co-workers, *J. Am. Ceram. Soc.* **67**, 311 (1984).
102. T. R. Shrout and A. Halliyal, *Am. Ceram. Bull.* **66**, 704 (1987).
103. H. Kawai, *Jpn. J. Appl. Phys.* **8**, 975 (1969).
104. J. G. Bergman, J. H. McFee, and G. R. Crane, *Appl. Phys. Lett.* **18**, 203 (1971).
105. K. Nakamura and Y. Wada, *J. Polym. Sci.* **A-29**, 161 (1971).
106. A. M. Glass, J. H. McFee, and J. G. Bergman, *J. Appl. Phys.* **42**, 5219 (1971).
107. T. T. Wang, J. M. Herbert, and A. M. Glass, *The Applications of Ferroelectric Polymers,* Blackie & Sons, Ltd., London, 1988.
108. H. Banno and K. Ogura, *J. Ceram. Soc. Jpn.* **100**, 551 (1992).
109. T. J. Dwyer and R. B. McPhillips, *Ceram. Bull.* **67**, 1894 (1988).
110. *Ceram. Ind.*, Jul. 1988, June 1991, Aug. 1992.

SEI-JOO JANG
The Pennsylvania State University

FERROFLUIDS. See MAGNETIC MATERIALS, THIN FILM.

FERTILIZERS

Nutrient Requirements of Plants

Plants, in contrast to animals, have the ability to convert carbon dioxide from the atmosphere and inorganic components of the earth directly into high energy carbohydrates (qv) (see VEGETABLE OILS), fats, and proteins (qv). These plant materials are absolutely essential to human nutrition as well as to the nutrition of other animal species. Thus consumption of plant matter, either directly or through a food chain, is essential to animal life and humans are totally dependent on agricultural endeavors, ie, the culture and harvesting of plant matter.

Plant life has existed and thrived on the earth for millions of years. Until relatively recent times, this existence was without human interference. The elements essential to plant growth are, under normal conditions, sufficiently present in nature. Early agriculture utilized principally those elements provided by nature: air, water, soil, and seed. Irrigation, to supply crops with needed water, was probably the earliest example of an agricultural practice that, in the broadest sense of the word, would qualify as a fertilization practice. Numerous early civilizations developed irrigation practices. Some were quite sophisticated. Another fertilization practice widely adopted by early agriculturists was the addition of organic matter to the soil. The benefits of supplementing the soil with animal and human excrements (manures) or composts of decayed vegetation or animal remains were recognized and utilized. Isolated instances are reported also of soil application of other materials, for example wood ashes, gypsum, and saltpeter.

Until about the beginning of the nineteenth century, world agriculture depended totally on empirically developed agricultural practices. About that time, however, the growth in understanding of plant physiology, and thus of plant growth needs and fertilization possibilities, began to parallel the rapid advance of general chemical knowledge. Scientific study of plant needs became possible. As of this writing, the scientific understanding of plant physiology, nutrition, and fertilization is well advanced but by no means complete. Even photosynthesis, the most basic of the plant growth processes, still presents challenges to complete understanding. New techniques, for example radioactive tracing, mass spectrometry (qv), genetic alteration, etc, are being applied to increase knowledge of plant processes (see GROWTH REGULATORS, PLANTS).

For satisfactory growth, most plants must have access to at least 22 different chemical elements. Additionally, deficiency of any of these essential elements, regardless of how much is required by the plant, can become the limiting factor in retarding healthy plant growth. This important principle, known as the law of the minimum, was one of a number of important nineteenth century contributions of Justus von Liebig. Elements now recognized as absolutely essential to healthy plant growth are given in Table 1.

To agronomists concerned with ensuring an adequate supply of these elements to crops, important considerations are (*1*) the route by which the plant ingests each of the elements, (*2*) the amount of each element required, and (*3*) the chemical forms of the element that are accepted by the plant. Hydrogen and oxygen enter the plant in the form of water, which is absorbed almost entirely through the roots. In healthy plant tissue, up to 95% of the tissue weight consists of water obtained by this route. In the laboratory, low temperature drying of plant tissue can be used to expel this water. Analyses of the remaining dry matter show some variation of elemental content, depending on species, growth environment, and other factors. A typical analysis for the dry matter from a healthy plant is shown in Table 2.

Of the dry matter, 44% consists of carbon all of which enters the plant as gaseous carbon dioxide absorbed from the atmosphere through the leaves. This carbon is transformed to the myriad of organic plant constituents through photosynthesis. In the photosynthesis process, gaseous oxygen is expelled from the leaves into the atmosphere. The green pigment chlorophyll [*1406-65-1*], present in most leaves, is a participant in the photosynthesis process. All of the remaining 21 essential elements present in plants, under usual growing conditions, enter

Table 1. Chemical Elements Essential to Healthy Growth of Plants[a]

Essential macronutrients	Essential micronutrients	Beneficial	Essentiality not demonstrated
	Metals		
potassium	iron	aluminum	chromium
calcium	copper	strontium	tin
magnesium	manganese	rubidium	nickel
	zinc		
	molybdenum		
	cobalt		
	vanadium		
	sodium		
	gallium		
	Nonmetals		
carbon	boron	selenium	fluorine
hydrogen	silicon		bromine
oxygen	chlorine		
phosphorus	iodine		
nitrogen			
sulfur			

[a]Ref. 1.

Table 2. Partial Chemical Analysis of Healthy Plant Tissue[a]

Name	Symbol	CAS Registry Number	Amount in dried tissue, %
		Structural elements	
oxygen	O	[*17778-80-2*]	45
carbon	C	[*7440-44-0*]	44
hydrogen	H	[*12385-13-6*]	6
		Primary nutrients	
nitrogen	N	[*17778-88-0*]	2
phosphorus	P	[*7723-14-0*]	0.5
potassium	K	[*7440-09-7*]	1.0
		Secondary nutrients	
calcium	Ca	[*7440-70-2*]	0.6
magnesium	Mg	[*7439-95-4*]	0.3
sulfur	S	[*7704-34-9*]	0.4
		Micronutrients	
boron	B	[*7440-42-8*]	0.005
chlorine	Cl	[*22537-15-1*]	0.015
copper	Cu	[*7440-50-8*]	0.001
iron	Fe	[*7439-89-6*]	0.020
manganese	Mn	[*7439-96-5*]	0.050
molybdenum	Mo	[*7439-98-7*]	0.0001
zinc	Zn	[*7440-66-6*	0.0100
Total			99.9011

[a]Ref. 2.

the plant only through root absorption. These are the elements that can be provided by, or supplemented by, the application of fertilizers to the soil. Foliar absorption of at least some of these elements is also possible by spraying solutions onto leaves, but this is a highly specialized practice.

To determine the feasibility of, or need for, fertilization requires knowing (*1*) which of the required elements, if any, are deficient in the soil; (*2*) what chemical forms of the deficient elements are assimilable by the plants and thus suitable as fertilizers; (*3*) what quantity of fertilizer material is required to meet the needs of the crop; and (*4*) whether the crop yield increase resulting from fertilizer application would warrant the cost of the fertilizer production and application.

Plant nutrients usually are catagorized (Table 2) as being structural elements, primary or secondary nutrients, or micronutrients. Most micronutrient elements are sufficiently present in native soils or are impurities in nonmicronutrient fertilizers applied to soils. Thus fertilization to provide these micronutrient elements specifically can often be omitted. Important exceptions arise in cases of localized soil deficiencies, in special requirements of some crops, or soil depletion resulting from repeated cropping. In these special cases, significant benefits are obtained by the application of small amounts of micronutrient sources and one segment of the fertilizer industry is devoted to providing micronutrient fertilizers. On a volume basis this is only a small segment of the world fertilizer industry.

The secondary nutrients, calcium, magnesium, and sulfur, are in no way secondary in regard to plant need. They are, however, secondary regarding the need to be furnished through fertilizer application, because these elements are abundant components of soil minerals at most locations. These elements also are incidental components of many fertilizers. Moreover, the widespread agricultural practice of liming by application of pulverized limestone [*1317-65-3*] or dolomite [*17069-72-6*] to the soil, intended chiefly for control of soil pH, incidentally adds calcium and magnesium (see LIME AND LIMESTONE). Many soils contain sulfur owing to atmospheric emissions from the industrial burning of coal (qv) and from volcanic eruptions. Thus intentional fertilization to furnish the secondary elements can, in the majority of cases, be omitted. However, there are exceptions. For example, there are large areas in Australia and the southeastern United States that are naturally deficient in sulfur and thus can be made more fertile by application of sulfur-containing fertilizers.

The elements nitrogen, phosphorus, and potassium are primary nutrients not only because healthy plant growth requires them in relative abundance, but also because these are the primary elements that most often must be furnished by fertilizers. None of these elements is a principal component of the usual soil minerals. Elemental nitrogen is, of course, the primary component of the atmosphere (79% by volume). To most plants, however, this form of nitrogen is totally inaccessible. Agricultural crop growth causes relatively rapid depletion of the primary nutrients in the soil. Thus replenishment by fertilization becomes essential. The principal task of the chemical fertilizer industry worldwide is to furnish agriculture with chemical forms of nitrogen (N), phosphorus (P), and potassium (K) that, when applied to the soil, can be readily assimilated by crop plants. Ability to include secondary nutrients and micronutrients in fertilizers when needed is also a responsibility of the industry, but, volumewise, it is the production of N, P, and K fertilizers that defines the industry.

Nature of Chemical Fertilizers

Chemical Content. Numerous chemical compounds have been shown to be suitable as sources of primary nutrients in fertilizers. A partial listing is given in Table 3. Because the route of these nutrients into the plant is through root absorption from the soil solution, solubility of the compounds in the soil solution is of prime importance. All of the compounds in Table 3 are highly soluble in water, with the exception of dicalcium phosphate. Dicalcium phosphate, nevertheless, is suitably soluble in most soil solutions and is recognized as a highly acceptable fertilizer component. Agronomic response to phosphorus in fertilizers can be suitably predicated by laboratory measurement of solubility in certain neutral or alkaline citrate solution reagents. Solubility in pure water therefore is not a necessity, although a certain degree of water solubility is recommended by most agronomists.

Table 3. Compounds Agronomically Effective in Fertilizers

Compound	CAS Registry Number	Formula	Primary nutrient content, %[a] N	P_2O_5	K_2O
	Nitrogen sources				
ammonia	*[7664-41-7]*	NH_3	82.2		
ammonium sulfate	*[7783-20-2]*	$(NH_4)_2SO_4$	21.2		
ammonium nitrate	*[6484-52-2]*	NH_4NO_3	35.0		
sodium nitrate	*[7631-99-4]*	$NaNO_3$	16.5		
calcium nitrate	*[10124-37-5]*	$Ca(NO_3)_2$	17.0		
urea	*[57-13-6]*	$CO(NH_2)_2$	46.6		
calcium cyanamide	*[156-62-7]*	$CaCN_2$	34.9		
monoammonium phosphate	*[7722-76-1]*	$(NH_4)H_2PO_4$	12.1	61.7	
diammonium phosphate	*[7783-28-0]*	$(NH_4)_2HPO_4$	21.2	53.7	
potassium nitrate	*[7757-79-1]*	KNO_3	13.8		46.6
	Phosphorus sources				
monocalcium phosphate	*[10031-30-8]*	$Ca(H_2PO_4)_2$		60.6	
dicalcium phosphate	*[7789-77-7]*	$CaHPO_4$		52.1	
monoammonium phosphate	*[2722-76-1]*	$(NH_4)H_2PO_4$	12.1	61.7	
diammonium phosphate	*[7783-28-0]*	$(NH_4)_2HPO_4$	21.2	53.7	
potassium phosphate	*[7778-53-2]*	K_3PO_4	0.0	33.4	66.5
	Potassium sources				
potassium chloride	*[7447-40-7]*	KCL			63.1
potassium sulfate	*[7778-80-5]*	K_2SO_4			54.0
potassium magnesium sulfate	*[13826-56-7]*	$K_2SO_4 \cdot 2MgSO_4$			22.7
potassium nitrate	*[7757-79-1]*	KNO_3		13.8	46.6
potassium phosphate	*[7778-53-2]*	K_3PO_4		33.4	66.5

[a]Pure salt basis.

The legal basis for the sale of fertilizers throughout the world is laboratory evaluation of content as available nitrogen, phosphorus, and potassium. By convention, numerical expression of the available nutrient content of a fertilizer is by three successive numbers that represent the percent available of N, P_2O_5, and K_2O, respectively. Thus, for example, a 20–10–5 fertilizer contains available nitrogen in the amount of 20% by weight of N, available phosphorus in amount equivalent to 10% of P_2O_5, and available potassium in amount equivalent to 5% K_2O. The numerical expression of these three numbers is commonly referred to as the analysis or grade of the fertilizer. Accepted procedures for laboratory analysis are fixed by laws that vary somewhat from country to country.

In the United States the analytical methods approved by most states are ones developed under the auspices of the Association of Official Analytical Chemists (AOAC) (3). Penalties for analytical deviation from guaranteed analyses vary, even from state to state within the United States (4). The legally accepted analytical procedures, in general, detect the solubility of nitrogen and potassium in water and the solubility of phosphorus in a specified citrate solution. Some very slowly soluble nutrient sources, particularly of nitrogen, are included in some specialty fertilizers such as turf fertilizers. The slow solubility extends the period of effectiveness and reduces leaching losses. In these cases, the proportion and nature of the specialty source must be detailed on the labeling.

As is evident from the listing in Table 3, the fertilizer manufacturer has a wide array of compounds from which to choose. Final choices of products and processes therefore rest heavily on such other factors as availability and cost of raw materials, economy of processing, safety of product, economy of handling and shipping, acceptability of physical form and physical behavior of the product, and farmer acceptance.

Chemical fertilizers need not be pure compounds. In fact, impurities in products may beneficially furnish secondary or micronutrients, or the presence of impurities may improve the physical condition of the fertilizer. The presence of impurities in large proportion, however, is disadvantageous because of dilution, which lowers the plant food content (grade) of the product. Commercial chemical fertilizer is a large-volume product that receives considerable shipment and handling from the manufacturer to the farmer's field. High plant-food content (high analysis) minimizes these costs, and has been a primary goal of the industry. As a result, there has been a trend toward higher analyses, and the incidental benefits of impurities have been lost in some products. This has increased the need for intentional application of secondary nutrients and micronutrients.

Commercial fertilizers are produced as (*1*) single-nutrient materials, which contain only one of the primary nutrient elements; (*2*) binutrient materials, which contain two of the primary nutrients; or (*3*) multinutrient products, which contain all three primary nutrients. The two- and three-component products are frequently referred to as mixed fertilizers. The proportioning of nutrients in mixed fertilizers is variable, depending entirely on farmer preference or on soil analysis to determine relative needs of the soil and the intended crop. In 1990, mixed fertilizers of 174 different grades were sold in the United States in amounts exceeding 10,000 metric tons each. Thousands of other grades were sold in smaller amounts (5). Some mixed fertilizers are produced directly in the manufacturing process with fixed nutrient ratios; other mixed products are made by simple blending of single- or binutrient products. The latter method, which in its present popular form is referred to as bulk blending, offers great flexibility in grade, economies in manufacture, and other advantages. Yearly U.S. consumption of chemical fertilizers from 1965 to 1991 is shown in Figure 1.

Physical Properties. The physical form and stability of a fertilizer product is of an importance almost equal to that of its chemical content. Commercial fertilizers of importance include not only solids, but also fluids, both solutions and suspensions, and even a gas (anhydrous ammonia).

Factors of importance in regard to the physical properties of solid fertilizers include particle size, particle strength, caking tendency, chemical stability, and hygroscopicity (6). Most solid fertilizers are produced in granular form, ie, having a particle diameter range of 1 to 4 mm (Fig. 2). In comparison to earlier, pulverized products, the granular fertilizers are less dusty, have less tendency to cake in storage, and flow more freely in handling and application equipment. Granularity is of importance also in the bulk blending system of producing mixed fertilizers

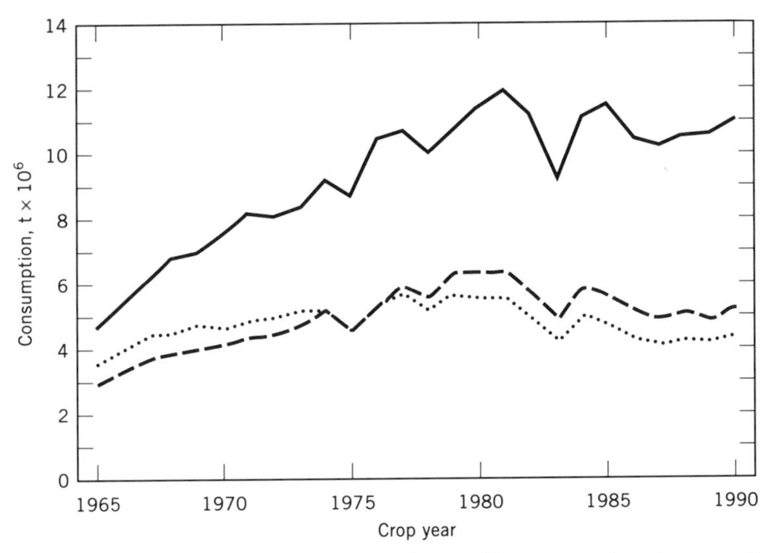

Fig. 1. U.S. consumption of plant nutrients in fertilizers: (——), nitrogen, N; (— — —), potassium, as K_2O; and (......), phosphorus, as P_2O_5. Courtesy of Tennessee Valley Authority (TVA).

by dry blending. Matching the granule sizes of the blend ingredients facilitates mixing and minimizes segregation during handling. Hygroscopicity, one of the other factors of importance in considering solid fertilizers, is the tendency of the product to become wet and unmanageable by absorption of moisture from the atmosphere during handling or field application. Fertilizer salts and mixtures vary considerably in regard to the critical humidity at which they begin to absorb moisture from the atmosphere and in their ability to tolerate absorbed moisture (6). In fertilizer products, hygroscopicity is controlled by choice of ingredients and sometimes by special treatments such as coating of particles with a moisture barrier.

Physical requirements of fluid fertilizers include freedom from sediments, suitably low viscosity, low vapor pressure, and noncorrosivity with regard to available handling equipment. Using anhydrous ammonia, the chief physical concerns, are in the safety of handling under pressure and the minimizing of vapor loss during injection into the soil.

Nitrogen Fertilizers

In the year ending June 30, 1991, 77.0 million metric tons of fertilizer N was used worldwide, and 10.1 million tons of this utilization was in the United States. Consumption of nitrogen, in general, is more than double that of either P_2O_5 or K_2O (Fig. 1) reflecting both the greater need for nitrogen exhibited by most food crops and the greater economic benefits obtainable from nitrogen fertilizer appli-

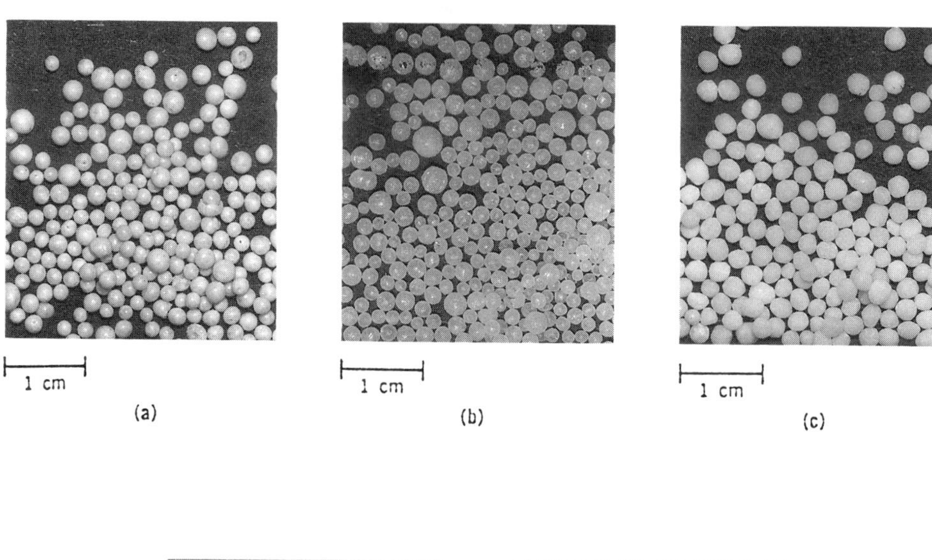

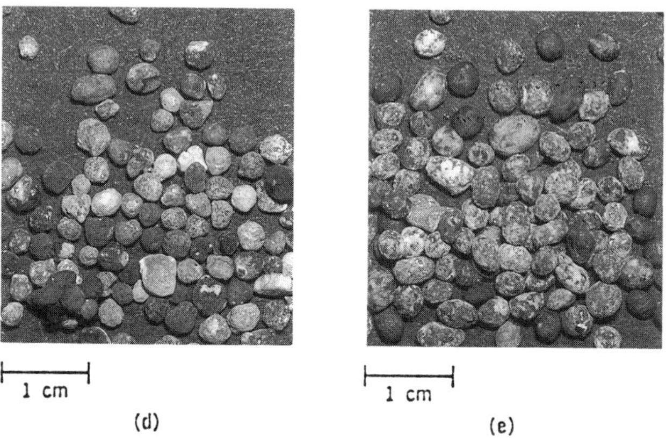

Fig. 2. Fertilizers in the prilled or granular form: (**a**), prilled ammonium nitrate; (**b**), prilled urea; (**c**), granular urea; (**d**), granular diammonium phosphate; and (**e**), a granular NPK mixed fertilizer. Courtesy Herbert O. Hester.

cation. Additionally, there is a tendency for applied nitrogen to be lost by leaching and volatilization, and a consequent requirement for more frequent replenishment than is usual for P_2O_5 and K_2O.

Nitrogen utilization by the first crop following application of fertilizer ranges from a low of about 30% for vegetables to a high of about 60% for grain crops. On average, no more than one-half of the applied fertilizer nitrogen is used by crops. Most nitrogen fertilizer is water soluble and, in time, much of it can be leached from most soils, especially during nongrowing seasons. Moreover, because of reactions in the soil, oxides of nitrogen and elemental nitrogen are volatilized. Usually, little of the fertilizer nitrogen applied in one season remains in the soil for use by later crops.

HISTORIC NITROGEN SOURCES

Nearly all commercial nitrogen fertilizer is derived from synthetic ammonia. However, prior to the introduction of ammonia synthesis processes in the early 1900s dependence was entirely on other sources. These sources are still utilized, but their relative importance has diminished.

Crop Rotation. Prior to about 1945, considerable dependence was placed on crop rotation as a means of supplying nitrogen. Certain plant varieties, particularly legumes such as peas and clovers, through a symbiotic relationship with certain soil bacteria, have the ability to utilize atmospheric nitrogen for nutrition. Soil bacteria infect the roots of these plants, causing the development of characteristic nodules. At the sites of these nodules, by mechanisms that are quite complicated, elemental atmospheric nitrogen becomes fixed and is absorbed into the root system of the plant (see NITROGEN FIXATION). In the farming system known as crop rotation, advantage is taken of this biological fixation by growing first a nitrogen-fixing legume, plowing it into the ground, and following with growth of a nonlegume farm crop. This is an effective system, said to have been recognized even by the early Greeks. However, for modern high yield agriculture, it is inefficient. Alternate growing seasons are, to a large extent, lost by growing the legumes. Also, modern hybrid varieties of grains are voracious consumers of nitrogen, having requirements beyond the supply capability of a plowed-under legume crop.

Natural Organics. Organic materials traditionally used as nitrogen fertilizers include manures (animal and human excrements), guano (deposits of accumulated bird droppings), fish meal (dried, pulverized fish and fish scrap), and packing-house wastes including bone meal and dried blood. The nitrogen content of these materials is very low compared to those of chemical fertilizers, but the organic content usually gives a supplementary benefit in physical conditioning of the soil. Use of these materials in the 1990s persists to some extent, but the overall impact is small. It is estimated that of the total fertilizer nitrogen now used worldwide, less than 1% is derived from these sources. Normally, these materials are not chemically treated. Processing consists mainly of drying and pulverizing. Because of pressure toward environmental awareness and utilization of wastes, there is considerable research activity directed toward producing fertilizer from waste materials. Such activity is likely to contribute to waste reduction but is not likely to have much impact on the fertilizer market.

Mineral Nitrogen. Nitrogen is rarely found in minerals because of the solubility of nitrates. Deposits of alkali metal nitrates are found in some dry desert areas of the world, but only in Chile has such a deposit been commercially exploited for fertilizer use. The Chilean ore is found in large rock-like deposits of impure sodium nitrate known as coliche. Mining and processing, chiefly for fertilizer use, began in the early part of the nineteenth century. Processing consists essentially of blasting, sizing, leaching, and crystallization. The principal product is sodium nitrate of 16% nitrogen content. In the fertilizer trade, this product is known as Chilean nitrate or Chilean saltpeter. Farmers often refer to it simply as soda. The first shipment of this material for fertilizer use in the United States was in 1830. Although the product was a primary nitrogen fertilizer into the early twentieth century, its importance in the United States is overshadowed by nitrogen products based on synthetic ammonia. In the year ended June 30, 1991, total fertilizer use of sodum nitrate in the United States was equivalent to only 5900 t of N, which was less than 0.1% of total fertilizer nitrogen usage.

By-Product Ammonia. Prior to the introduction of ammonia synthesis, the principal industrial source of ammonia was as a by-product from the coking of coal (qv). Raw coals normally contain about 1% of nitrogen. In the coking process, which involves heating in a nonoxidizing atmosphere, about half of this nitrogen is evolved as ammonia (see COAL CONVERSION PROCESSES, CARBONATION). To recover this ammonia, the coke oven effluent gases are scrubbed with sulfuric acid in saturator–crystallizers. The ammonia reacts with the sulfuric acid to form ammonium sulfate, and the heat of the gases concentrates the solution; the result is crystal formation. By-product ammonium sulfate made in this manner was a mainstay of the fertilizer industry until the advent of synthetic ammonia. Coke oven ammonia is still recovered in this manner and utilized in fertilizers, but it represents only a small fraction of total fertilizer nitrogen usage. Ammonium sulfate is an excellent fertilizer (Table 3) having a nominal nitrogen content of about 20%. In addition, it contains 24% of the secondary nutrient, sulfur. Ammonium sulfate for fertilizer use is not only made from coke oven gas but is also obtained from other sources.

NITROGEN FERTILIZERS FROM SYNTHETIC AMMONIA

Over 95% of all commercial nitrogen fertilizer in the 1990s is derived from synthetic ammonia (qv). Worldwide, the yearly production of synthetic ammonia is about 120 million t, of which about 85% finds use in fertilizers.

Essentially all the processes employed for ammonia systhesis are variations of the Haber-Bosch process developed in Germany from 1904–1913. One of the all-time breakthroughs of chemical technology, the synthesis process involves the catalytic reaction of a purified hydrogen–nitrogen mixture under high (14 to 70 MPa (2,030 to 10,150 psi)) pressure and temperature (400 to 600°C). The preferred catalysts consist of specially activated iron. The ammonia that forms is condensed by cooling with liquefied ammonia; the unreacted gases are recycled to the synthesis loop. In over 80% of the ammonia plants of the 1990s, the hydrogen–nitrogen feed mixture is prepared by a series of reactions known as steam reforming, for which the raw materials are steam, natural gas (methane), and air. Commercial plants have also been based on use of naphtha or coal (coke) as feedstock. All facets of ammonia production are highly sophisticated engineering processes requiring both a high level of technical know-how and a large capital investment.

The principal routes by which synthetic ammonia is processed into finished fertilizers are shown in Figure 3. Also included are U.S. consumption data on each of these products for the crop year ended June 30, 1990.

Anhydrous Ammonia. One of the striking features of nitrogen fertilizer usage in the United States is the large amount of anhydrous ammonia that is applied directly to the soil, without further processing. This practice was proven practical in the early 1940s, and has found heavy use ever since, particularly in the wheat and corn belts of the north central United States (see WHEAT AND OTHER CEREAL GRAINS). As indicated in Figure 3, direct application of anhydrous

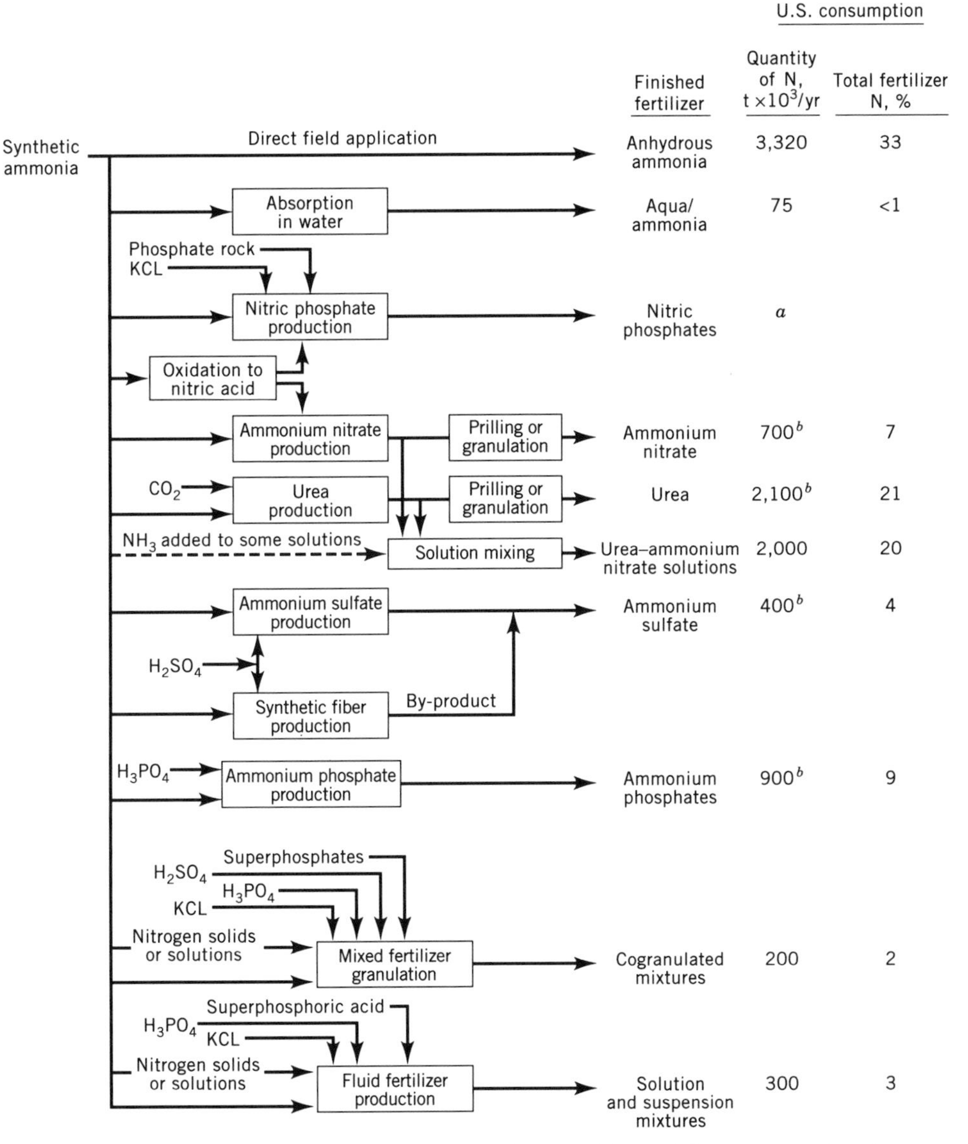

Fig. 3. Routes for making synthetic ammonia into fertilizers. The consumption data are for the year ended June 30, 1990 (5). [a]Significant quantities are made and used in foreign countries; [b]includes quantities applied in dry blends. Courtesy of TVA.

ammonia accounts for about one-third of the total nitrogen applied as fertilizer in the United States. Anhydrous ammonia is also a principal nitrogen fertilizer in Canada, Denmark, and Mexico.

The reason for the popularity of anhydrous ammonia is its economy. No further processing is needed and it has a very high (82.2%) nitrogen content. Additionally if held under pressure or refrigerated, ammonia is a liquid. Being a liquid, pipeline transport is practical and economical. A network of overland pipelines (Fig. 4) is in operation in the United States to move anhydrous ammonia economically from points of production near natural gas sources to points of utilization in farming areas (see PIPELINES).

Anhydrous ammonia is a gas at ambient temperatures and pressure, which dictates special procedures for storage, handling, and field application. Both pressure storage and refrigerated storage are utilized to maintain ammonia in liquid form. For large storage depots, the preference usually is for refrigerated storage in large, well-insulated vessels at or near atmospheric pressure. The vapor pressure of anhydrous ammonia is such that refrigeration to about −33°C is required to avoid boiling at atmospheric pressure. In the insulated storage vessels, this refrigeration is accomplished simply by allowing the ammonia to boil in amounts equivalent to the in-leakage of heat, and providing compressor and condenser capacity for reliquification of the vapor. Smaller storage depots sometimes use

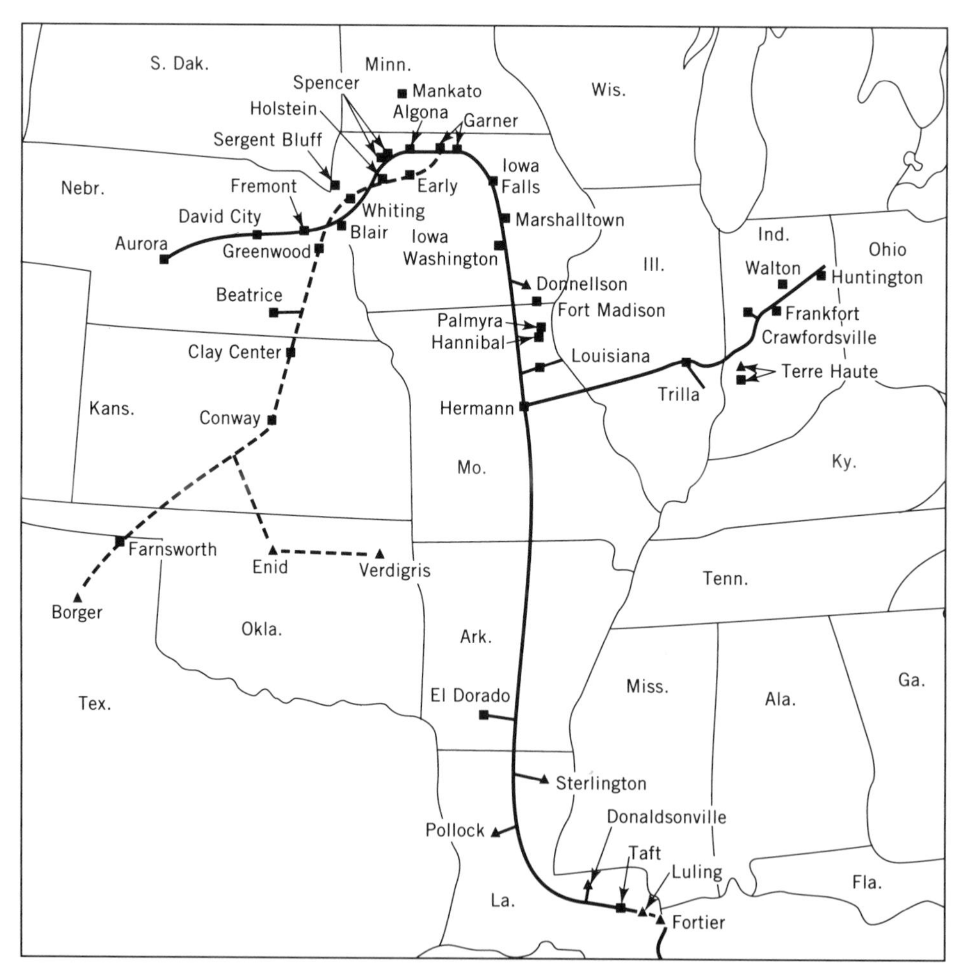

Fig. 4. Pipeline systems of transport for anhydrous ammonia within the United States (7), where ▲ represents an ammonia plant location; ■, storage terminals; (——), Gulf Central pipeline; and (– – –), Mapco pipeline.

semirefrigerated storage, ie, storage at somewhat elevated pressure but with compressor capacity for some recondensation. The advantage over storage at atmospheric pressure is that a higher storage temperature can be used, with resultant lowered refrigeration requirement (see REFRIGERATION AND REFRIGERANTS).

The third type of storage for anhydrous ammonia is pressurized storage without refrigeration. This is used at small terminals, in some transport vehicles, and in field application equipment. For safety reasons, maximum allowable storage pressure in vessels at retail locations and in the field has been set at 1.83 MPa (265 psig) (8). This pressure is sufficient to prevent boiling at temperatures up to 46°C. Safety pop-off valves are provided to vent ammonia should higher temperatures be encountered.

Anhydrous ammonia is moved worldwide in specially designed vessels, barges, rail cars, and trucks. All modes are covered by rigid safety restrictions (8). Ships and barges usually employ refrigeration or semirefrigeration to maintain the liquid, while rail cars and trucks normally depend only on pressure (see TRANSPORTATION).

Field application of anhydrous ammonia requires knifing of the gas into the soil considerably below the surface to prevent escape to the atmosphere as shown in Figure 5. The ammonia is metered and piped through tubing down the trailing edge of the blade to a single opening, usually 15 to 25 centimeters below the soil surface. Field applicators vary in size from small five-row tractor drawn units (five knives) to specially designed crawlers that cover 30 or more rows in a swath 20 meters wide.

Aqua Ammonia. Aqua ammonia, a simple solution of ammonia in water, is used to a limited extent as a nitrogen fertilizer. The solution, which is usually of 20% nitrogen content (23% NH_3), is prepared from anhydrous ammonia and water in simple mixing–cooling units known as converters. The chief advantage of aqua ammonia over anhydrous is the lower vapor pressure, which allows storage and handling without pressurization or cooling. Field application is usually by knifing below the soil surface, but a relatively shallow (8 to 12 cm) depth is sufficient. The main disadvantage of aqua ammonia is its relatively low nitrogen content as compared to anhydrous ammonia (20% vs 82.2%). This low concentra-

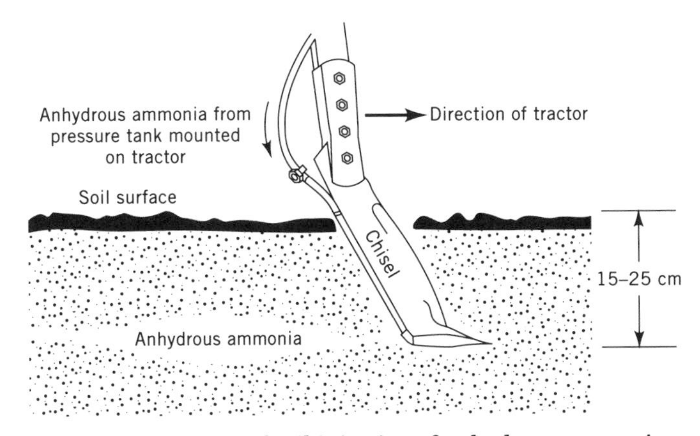

Fig. 5. Schematic of soil injection of anhydrous ammonia.

tion increases storage and handling requirements more than fourfold; thus the product is normally used only very close to the point of preparation. In the United States, aqua ammonia furnishes less than 1% of the total fertilizer nitrogen (Fig. 3).

Ammonium Nitrate. Broadly defined, fertilizer ammonium nitrates include straight ammonium nitrate (AN), containing 33–34% N, ammonium sulfate nitrate (ASN) 26% N, and calcium ammonium nitrate [*39368-85-9*] (CAN) 20–26% N. Worldwide, the estimated total production of these materials is about one-half straight AN. Most of the balance is CAN. The ASN and CAN grades are used primarily because of greater safety and better storage and handling qualities. Ammonium nitrate fertilizers were produced and used to some extent in Europe prior to 1942, but became of importance only after World War II when plants that had been producing ammonium nitrate for munitions were converted to fertilizer production. Ammonium nitrate rapidly became the leading solid nitrogen fertilizer both in the United States and worldwide, and held that position until about 1975, at which time its use was equalled and then surpassed by that of synthetic urea. Figure 6 gives world consumption of N fertilizers from 1955 through 1990. For the year ended June 30, 1990, world consumption of ammonium nitrate in solid form was 26.4×10^6 t (8.86×10^6 t N) or about 11% of total world consumption of fertilizer nitrogen (9). United States consumption data for the same period (Fig. 3) show that only 7% of U.S. nitrogen consumption was furnished as solid ammonium nitrate, although an additional 20% was furnished as urea–ammonium nitrate solutions.

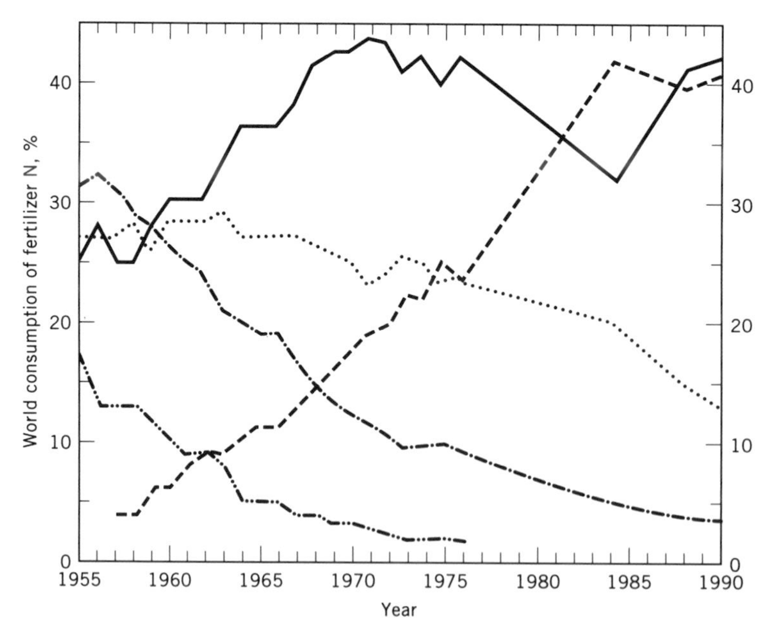

Fig. 6. World trends in types of nitrogenous fertilizers consumed, where (——) represents anhydrous ammonia, ammonium phosphates, cogranulated mixtures, fluid mixtures; (– – –), urea, including that in nitrogen solutions; (····), ammonium nitrate, including that in nitrogen solutions; (–·–), ammonium sulfate; and (–··–) others.

In the United States, only high grade (33–34% N) ammonium nitrate is produced, and, because of the oxidizing nature and explosive potential of this material, strict handling precautions are required (10) (see EXPLOSIVES AND PROPELLANTS, EXPLOSIVES). Several other countries, France, Russia, Romania, and the UK also allow production of the high grade product, whereas others, for safety reasons, allow only the CAN grades (20–26% N). Ammonium nitrate fertilizers are highly regarded because of the rapid agronomic response, especially true in the cooler latitudes. For assimilation by the roots of most plants (rice is an exception), all fertilizer nitrogen must first be converted in the soil to nitrate form (11, 12). For non-nitrate fertilizers in the cooler areas, this process may be rather slow. Ammonium nitrate is produced from ammonia and nitric acid (qv). The acid also is produced from anydrous ammonia (see AMMONIUM COMPOUNDS).

Prilling and Granulation. Ammonium nitrate is normally produced initially in the form of a 75 to 85% solution saturated at 40 to 77°C. Conversion to solid prills or granules suitable for fertilizer use requires high temperature concentration of the solution to 96–99+% ammonium nitrate content followed by solidification from that solution by cooling and drying in either a prilling tower or a granulation system.

The usual prilling operation and the design of prilling towers have been described in detail (13). Basically, a highly concentrated solution of ammonium nitrate is sprayed as small droplets into the top of a tall tower countercurrent to an updraft of cool air. Solidification occurs in the form of small spheres (prills), usually of 1 to 3 mm particle diameter. A relatively short tower, typically about 21 to 30 m in height, and a feed of 99.5–99.8% solution is used to produce high density prills. The prill hardness and particle size are about optimum under those conditions. If there is proper control of solution concentration and air flow, no further drying is required. When lower (96 to 99%) solution concentrations are used, a taller tower is required, but the prills are of lower density, and passage through a heated dryer often is required.

Treatment of prilling tower effluent air to avoid atmospheric pollution is a particularly difficult problem. Partly for this reason, granulation has, to some extent, been introduced as an alternative to prilling. Another advantage of granulation is greater particle-size flexibility, which assumes importance when producing closely sized product for use in bulk blending. The Spherodizer granulation process developed by Cominco Ltd. (Canada) and C & I/Girdler, Inc. (14,15) is one of the more advanced granulation processes. Very concentrated ammonium nitrate solution is sprayed onto a rolling bed of solid particles in a rotating drum, resulting in the formation of granules by layering and solidification. The granules contain only 0.1–0.5% moisture and require no further drying. From a screening operation, oversize granules are crushed and recycled to the granulator along with undersize particles to serve as nuclei for forming more product-size granules. The product size is controlled by choice of screen size. Air pollution (qv) is virtually eliminated by wet scrubbing in a manner that is simple and economical compared to that required for a standard prilling operation. The granules are harder than prills.

Pan granulation of ammonium nitrate also has been developed commercially (16,17). Essentially water-free melt (<0.5% moisture) is sprayed onto a cascading bed of fines (usually crushed recycle) in a tilted, rotating pan. Production of 18–20 t/h is common. Product made in this manner is said to require no conditioning. The pan is an excellent size classifier, making it easy to produce granules of any desired size from 1 to 11 mm diameter. Advantages of pan granulation are ease of granule size control, hardness of granule, and simplicity of pollution control.

Fire and Explosion Hazard. Ammonium nitrate is a strong oxidizing agent, and mixing it with certain materials, including fertilizer potassium chloride, can induce self-heating and combustion. In addition, ammonium nitrate has the potential for explosive decomposition under certain conditions of high temperature and pressure. Mixture of ammonium nitrate with small amounts (even $<1\%$) of carbonaceous material or certain other materials sensitizes ammonium nitrate to detonation. Serious fires and explosions have occurred as a result of mishandling ammonium nitrate. The worst of the explosions was that at Texas City, Texas in 1947 where two shiploads of ammonium nitrate fertilizer conditioned with petrolatum and paraffin wax exploded, killing 500 people and injuring 3000. Extensive studies of ammonium nitrate self-heating and detonation have since been carried out (18–20), and conditioning with organic materials no longer is allowed. As a result of extensive studies it has been concluded that ammonium nitrate can be safely produced, stored, and handled if simple and reasonable precautions are taken. The thermal stability of fertilizers containing ammonium nitrate has been evaluated and discussed at length (21), and a review of ammonium nitrate plant safety measures has been published (22).

Conditioning. The production of low moisture (0–0.3%) granules or prills greatly reduces the likelihood of any caking problems. However, in the absence of preventive measures, a volume change that accompanies a crystal transition at 32.3°C causes degradation of crystals, prills, and granules, which increases susceptibility to moisture absorption and caking. Additives (internal conditioners) that have been used commercially to shift the transition temperature away from usual storage temperatures include potassium nitrate, magnesium nitrate, and nucleating agents such as clay, talc, and other silicates. Coating dusts such as kieselguhr (see DIATOMITE) and clays (qv) frequently are applied to ammonium nitrate prills as parting agents to reduce caking tendency. A coating dosage of 1.5 to 2.0% is required for effectiveness, diluting product grades from 35 to about 33.5% N. Materials such as petrolatum, paraffin waxes, and oils no longer are used as conditioners for ammonium nitrate, because of the effect on increasing sensitivity to burning and explosion.

Packaging and Storage. Bags used for packaging ammonium nitrate must be impenetrable to water vapor because of the hygroscopicity of the product which has a critical relative humidity of 59% at 30°C. Simple paper, jute, or woven plastic bags are unsatisfactory because their porosity allows rapid absorption of moisture, with resultant wetting and caking of the ammonium nitrate. Suitable bags are of multiwall paper with a bitumen or polyethylene moisture-proofing layer. Monofilm bags of heavy (0.15–0.2 mm thickness) polyethylene also are satisfactory. Jute or woven polypropylene bags with monofilm plastic liners are used in some countries. In the United States, bags must carry precautionary labels indicating an oxidizing material. Any spilled material may become contaminated and should not be rebagged. Bagged materials should not be stored within 76 cm of building walls or partitions to allow for circulation of cooling air. Because pressure enhances caking, bagged material should not be stacked too high. Ammonium nitrate should not be placed in storage while its temperature is above 54°C. Higher temperatures accelerate caking.

Ammonium nitrate is extensively stored in bulk. It should be placed on a clean floor to avoid contamination. Ammonium nitrate in bulk storage must be kept separate from other materials and not close to sources of heat. It is preferable to store the bulk material in dehumidified areas but good quality granules and prills are often stored without dehumidification. Covering the floor and pile with plastic sheet reduces moisture absorption and essentially eliminates caking. Material caked in storage should never be dynamited or blasted. No smoking signs should be posted and observed, and all open flames should be kept away from the materials.

Transportation. Ammonium nitrate is safely transported by rail, road, and water. However, its transportation on U.S. navigable waterways is restricted. Good ventilation must be provided and precautions taken against leakage and contamination. The material must be completely isolated from other cargo and must be kept free of extraneous combustible materials.

Synthetic Urea. Urea (qv), $CO(NH_2)_2$, was first synthesized by Friedrich Wohler in 1828 by the reaction of ammonia and cyanuric acid, both of which are inorganic starting materials (see CYANURIC AND ISOCYANURIC ACIDS). Prior to that synthesis, it was generally believed that only living organisms could produce organic materials. In the 1990s, synthetic urea, albeit produced by a different synthesis route, is the leading nitrogen fertilizer worldwide and the leading solid nitrogen fertilizer in the United States. For the year ended June 30, 1991, world consumption of urea fertilizer in solid form (9) was 64.0×10^6 t (29.4×10^6 t N). An additional 4.5×10^6 t of urea (2.1×10^6 t N) was used in nitrogen solutions. In the United States, for the same period, consumption of solid urea fertilizer was 4.6×10^6 t (2.1×10^6 t N) and consumption of urea in nitrogen solutions was 2.2×10^6 t (1.0×10^6 t N). Urea in solid and solution form provided about 41% of total world fertilizer nitrogen and about 31% of total U.S. nitrogen.

Properties and Production. Several characteristics of urea make it particularly attractive as a fertilizer material. Its nitrogen content of 46.6% is considerably higher than that of ammonium nitrate (35.0%), ammonium sulfate (21.2%), and any other solid nitrogen fertilizer. This is a great advantage in storage, handling and shipping. Also, urea presents no burning or explosion hazard, which is a significant advantage over ammonium nitrate. Urea also is less hygroscopic than ammonium nitrate, having a critical relative humidity of 73% at 30°C as compared to 59% for ammonium nitrate. This lower hygroscopicity translates to better storage and handling properties, especially in bulk. Urea is cheaper to produce than ammonium nitrate, per unit of nitrogen, largely because the nitric acid intermediate is not required.

Agronomically, both advantages and disadvantages have been ascribed to urea. A principal agronomic advantage over ammonium nitrate is in the fertilization of flooded rice. In the anaerobic conditions of a flooded paddy the nitrate radical of ammonium nitrate is partially reduced to gaseous N_2O or N_2, which results in considerable volatilization loss of nitrogen. Urea, on the other hand, supplies only ammonia nitrogen, which is easily retained in the flooded paddy. The roots of rice plants are uniquely able to utilize the ammonia directly, whereas absorption by the roots of most crops requires that ammonia first be converted to nitrate through a nitrification process in the soil (11,12). This advantage of urea in rice fertilization resulted in its early preference in the Far Eastern markets. In other areas of the world, there has been some reluctance to accept urea because of certain perceived problems. For most crops, other than rice, urea in the soil must first undergo hydrolysis to ammonia and then nitrification to nitrate before it can be absorbed by plant roots. One problem is that in relatively cool climates these processes are slow; thus plants may be slow to respond to urea fertilization. Another problem, more likely in warmer climates, is that ammonia formed in the soil hydrolysis step may be lost as vapor. This problem is particularly likely when surface application is used, but can be avoided by incorporation of the urea under the soil surface. Another problem that has been encountered with urea is phytotoxicity, the poisoning of seed by contact with the ammonia released during urea hydrolysis in the soil. Placement of urea away from the seed is a solution to this problem. In view of the growing popularity of urea, it appears that its favorable characteristics outweigh the extra care required in its use.

Commercial urea generally contains a small amount of the compound biuret [*108-15-0*], $NH_2CONHCONH_2$. For fertilizer use other than foliar application, biuret contents of about 1.5% or less are not harmful, and under usual production conditions it is not difficult to avoid higher biuret levels. For most foliar application, however, the level should not exceed 0.1%, and special production modifications are usually required to ensure such a low level.

The technology of urea production is highly advanced. The raw materials required are ammonia and carbon dioxide. Invariably, urea plants are located adjacent to ammonia production facilities which conveniently furnish not only the ammonia but also the carbon dioxide, because carbon dioxide is a by-product of synthesis gas production and purification. The ammonia and carbon dioxide are fed to a high pressure (up to 30 MPa (300 atm)) reactor at temperatures of about 200°C where ammonium carbamate [*111-78-0*], $CH_6N_2O_2$, urea, and water are formed.

$$2\ NH_3 + CO_2 \rightleftarrows NH_2COONH_4 + \text{heat}$$

$$NH_2COONH_4 + \text{heat} \rightleftarrows NH_2CONH_2 + H_2O$$

The fundamental chemistry and engineering requirements of the process have been studied and are described elsewhere (23,24).

Because an excess of ammonia is fed to the reactor, and because the reactions are reversible, ammonia and carbon dioxide exit the reactor along with the carbamate and urea. Several process variations have been developed to deal with the efficiency of the conversion and with serious corrosion problems. The three main types of ammonia handling are once through, partial recycle, and total recycle. Urea plants having capacity up to 1800 t/d are available. Most advances have dealt with reduction of energy requirements in the total recycle process. The economics of urea production are most strongly influenced by the cost of the raw material ammonia. When the ammonia cost is representative of production cost in a new plant it can amount to more than 50% of urea cost.

Prilling and Granulation. Worldwide, prilling is the most widely used method of solidifying urea, but the use of granulation is increasing rapidly. In prilling, molten urea that is almost anhydrous is forced through spray heads or spinner buckets at the top of a tower to produce droplets that fall through a countercurrent stream of air in which they solidify to form prills. Urea prilling requires taller towers than ammonium nitrate prilling in order to achieve comparable particle size. The height of towers ranges from 21 to 52 m. The temperature and rate of flow of air also affect the size and quality of prills. Urea prilling is an economical method for finishing, but the prills have low strength and are generally too small for use in blending with granular materials such as diammonium phosphate. Also, the prilling operation is a serious polluter, the abatement of which is costly because of the large volume of dust-laden air that must be treated. For these reasons there is a strong trend toward granulation of urea.

Granulation is now the leading method of finishing urea in the United States. The granulation process that is used in the United States almost exclusively is the Spherodizer process (14,15). In this process, granules are formed by successive spraying and drying (layering) of concentrated urea solution on recycled granules in a rotating drum. Special design of the drum and sprays, together with control of air flow, results in hard granules of particle size favorable for blending and other use (Fig. **2c**). A somewhat similar process, known as curtain granulation, was developed by the Tennessee Valley Authority (TVA) (25) and was operated on an experimental basis. The internal design of the TVA drum includes a sloping baffle that provides a falling curtain of urea, onto which the concentrated urea solution is sprayed. In Europe, some urea is granulated by pan-granulation methods pioneered by Norsk Hydro (16). The method is particularly suited for producing extra large granules 4 mm or greater in diameter, known as forestry-grade urea, which are preferred for aerial application to forests.

Conditioning. Notwithstanding the more favorable critical humidity of urea, as compared with that of ammonium nitrate, unconditioned urea still exhibits a tendency to cake in storage. Causes for such caking include plasticity or softness of particles, pressure in storage, temperature changes, and absorption of moisture. Coating of prills or granules with 1 or 2% of a finely ground absorbent clay is an effective anticaking treatment and was used on prilled urea almost exclusively until about the mid-1960s. This procedure, however, undesirably diluted the nitrogen content of the product from 46 to 45%. Internal conditioning, that is inclusion of an anticaking agent in the urea melt before prilling or granulation, has supplanted coating. The agent used almost exclusively is formaldehyde [*50-00-0*], CH_2O, which is effective in dosage of only 0.2 to 0.5%. This amount of additive does not lower nitrogen content below 46%. In addition to providing anticaking of the final product, formaldehyde improves the granulation process by reducing dust formation. Action of the additive apparently is that of a crystal modifier and hardner. During a time period when possible health hazards of formaldehyde were being questioned, an alternative internal conditioner, sodium lignosulfonate [*8061-51-6*], was developed. This additive also is effective at very low dosage level and is considered to be a competitive conditioner (26).

Nitrogen Solutions. In the year ended June 30, 1990, 20% of the fertilizer nitrogen used in the United States was applied in the form of urea–ammonium nitrate solutions (Fig. 3). This amount equals the proportion furnished by solid urea (prills and granules). The solutions used are urea–ammonium nitrate–water mixtures designed to take advantage of the unusually high joint solubility of urea and ammonium nitrate. Properties of typical solutions are given in Table 4.

Table 4. Properties of Urea Ammonium Nitrate Nonpressure Solutions

Parameter	Solution		
grade, % N	28	30	32
composition, wt %			
ammonium nitrate	40.1	42.2	43.3
urea	30.0	32.7	35.4
water	29.9	25.1	20.3
specific gravity at 15.6°C	1.283	1.303	1.32
salt-out temperature, °C	−18	−10	−2

Solutions contain from 28 to 32% nitrogen, and have salt-out temperatures of −18 to −2°C. For either ammonium nitrate or urea alone in solution, nitrogen content higher than about 18% results in too high a salt-out temperature for practical use; thus the favorable effect of the eutectic solubility is obvious. Nonpressurized storage and handling of these solutions is entirely satisfactory, and, because of the fairly high nitrogen content, transportation costs from points of production are reasonable. Production costs per unit of nitrogen are usually lower than for nitrogen solids because of the elimination of concentration and prilling or granulation steps. Farmer preference for these solutions is usually based on the favorable price, ease of handling and application, and/or ease and effectiveness of pesticide incorporation in the solution. Many preemergent herbicides are compatible with the nitrogen solutions; thus it is economical and effective to combine early season fertilization with preemergent herbicide treatment. These nitrogen solutions are used both for direct field application and as a nitrogen ingredient in fluid mixed fertilizers.

The usual manufacturing process for nitrogen solutions involves blending urea and ammonium nitrate solutions in a mixing tank in either a batch or continuous manner. The feed solutions contain about 83% ammonium nitrate and 72–85% urea. The plants usually are part of a nitrogen fertilizer complex that makes both of the required ingredients. The ammonium nitrate solution is withdrawn from the production line and combined in the desired ratio with the urea solution also taken directly from the production line. Mixing, transfer, and storage constitute the process steps. Nitrogen solutions can, however, cause serious corrosion. Ammonium thiocyanate, sodium arsenite, sulfonate OA5, or a trace of free ammonia are often incorporated in the solution as corrosion inhibitors. Aluminum is one of the more resistant equipment materials.

Ammonium Sulfate. Historically ammonium sulfate was important as a fertilizer. However, since the introduction of ammonium nitrate and urea, the relative importance of ammonium sulfate worldwide has steadily decreased. In the year ended June 30, 1990, ammonium sulfate furnished only about 4% of the fertilizer nitrogen used in the United States (Fig. 3) and worldwide (Fig. 6).

At most locations, it is no longer economically attractive to synthesize ammonium sulfate from virgin acid and ammonia. The low (21.2% N) nitrogen content of ammonium sulfate translates to high shipping costs, and in most locations, farmers are unwilling to attribute any monetary value to the sulfur content (24.3% S) of ammonium sulfate, even though the sulfur is an effective secondary plant nutrient. There are exceptions, however. For example, in an extensive sulfur-deficient area of Australia ammonium sulfate is synthesized in granular form and is highly regarded for its sulfur content.

Essentially all the ammonium sulfate fertilizer used in the United States is by-product material. By-product from the acid scrubbing of coke oven gas is one source. A larger source is as by-product ammonium sulfate solution from the production of caprolactam (qv) and acrylonitrile, (qv) which are synthetic fiber intermediates. A third but lesser source is from the ammoniation of spent sulfuric acid from other processes. In the recovery of by-product crystals from each of these sources, the crystallization usually is carried out in steam-heated saturator–crystallizers. Characteristically, crystallizer product is of a particle size about 90% finer than 16 mesh (ca 1 mm dia), which is too small for satisfactory dry blending with granular fertilizer materials. Crystals of this size are suitable, however, as a feed material to mixed fertilizer granulation plants, and this is the main fertilizer outlet for by-product ammonium sulfate.

Some success has been achieved in bringing the size of by-product ammonium sulfate up to ca 1–3.4 mm (−6 +16 mesh), which makes it suitable for dry blending with granular fertilizer materials. Such size increase would be desirable to widen the possibilities for marketing ammonium sulfate. One approach involves increasing the retention time in the saturator–crystallizers. In this way, larger crystals are made, but crystallizer throughout is considerably reduced. Another approach involves agglomeration (granulation) of the fine crystals in a rotary drum granulator to which some ammonia and acid are also fed to provide binding (27). The drawback to this method is the expense of the equipment and the additional raw materials. A third method for agglomeration of the fine crystals is known as compaction. In this method (28), which is used to some extent commercially, the crystals are fed to a roll compactor that exerts extremely high pressure to form a dense sheet of ammonium sulfate. This sheet then is fragmented and screened to give desirably sized product. A serious problem has been excessive wear of the rolls and bearings under the high pressures required.

Ammonium sulfate is very resistant to moisture absorption, ie, it is nonhygroscopic, and has a critical relative humidity of 79% at 30°C. It has excellent storage properties in bags and bulk, requiring no anticaking conditioners. Agronomically, ammonium sulfate is highly regarded and requires no special precautions in its use.

Nitrogen in Multinutrient Fertilizers. Single-nutrient nitrogen materials supplied over 85% of the fertilizer nitrogen used in the United States during the year ended June 30, 1990. The remaining 15% was supplied as multinutrient materials (Fig. 3). This included 9% as ammonium phosphate, 2% as cogranulated mixtures, and 3% as fluid mixtures.

Phosphate Fertilizers

Worldwide, consumption of phosphate fertilizer in 1991 (9) was equivalent to 35.9 $\times\ 10^6$ t of P_2O_5. In the United States, consumption was equivalent to 3.8 $\times\ 10^6$ t of P_2O_5.

HISTORIC SOURCES

The phosphorus content of most virgin soils is very low, for example, only about 0.0 to 0.3% P in the United States (29). Furthermore, the phosphorus content of soils often is in forms unavailable to plants. Also, there are practically no natural routes of phosphorus replenishment operating, other than the natural recycling of plant material and animal wastes. The growth of legumes in a crop rotation pattern, although effective for supplying nitrogen, does nothing toward supplying phosphorus. Organic manures do furnish some phosphorus, but are much less effective than in supplying nitrogen. For these reasons, repeated cropping of land without fertilization can soon deplete the supply of phosphorus and the land becomes barren.

One fertilization practice, initiated in the early nineteenth century, that was effective in furnishing phosphorus, was the application of ground bones to soil. Raw bones normally contain 8 to 10% phosphorus (20 to 25% P_2O_5), and are thus rich phosphorus sources. Decomposition of raw bones in the soil is quite slow, however, limiting effectiveness. Also, the supply of bones is insignificant in regard to world need for phosphate fertilizer. In about 1830, in England, sulfuric acid pretreatment of bones was begun as an effective method of increasing the availability of the phosphorus to growing plants. About 10 years later, similar sulfuric acid treatment of mineral phosphate ore was begun to produce the effective fertilizer product now known as ordinary (or normal) superphosphate. A small commercial superphosphate operation, begun in England in 1842, is now recognized as the beginning of the chemical fertilizer industry.

PHOSPHATE FERTILIZERS FROM MINERAL PHOSPHATES

Essentially all fertilizer phosphate is derived from mineral phosphates. Deposits of mineral phosphate are abundant and widely dispersed throughout the world. Nearly all of the mineable deposits are minerals of the apatite group represented by the general formula $Ca_5(F,Cl,OH,0.5CO_3)(PO_4)_3$. As mined, essentially all the ores require beneficiation to reduce the content of clay, silica, or other extraneous material. Beneficiation methods commonly used are washing (gravity separation) and/or flotation (qv) using various flotation agents. Beneficiated ores, as supplied to fertilizer producers, range in P_2O_5 content from 28 to 38%. World average P_2O_5 content has been about 32%, but is decreasing as high grade ores become exhausted. The phosphorus content of ores and concentrates is, in the trade, usually expressed as bone phosphate of lime (% BPL), which is tricalcium phosphate, $Ca_3(PO_4)_2$. Concentrate of 32% P_2O_5 content has a grade of 69.9% BPL.

The routes by which mineral phosphates are processed into finished fertilizers are outlined in Figure 7. World and U.S. trends in the types of products produced are shown in Figures 8 and 9, respectively. Most notable in both instances is the large, steady increase in the importance of monoammonium and diammonium phosphates as finished phosphate fertilizers at the expense of ordinary superphosphate, and to some extent at the expense of triple superphosphate. In the United States, about 65% of the total phosphate applied is now in the form of granular ammonium phosphates, and additional amounts of ammonium phosphates are applied as integral parts of granulated mixtures and fluid fertilizers.

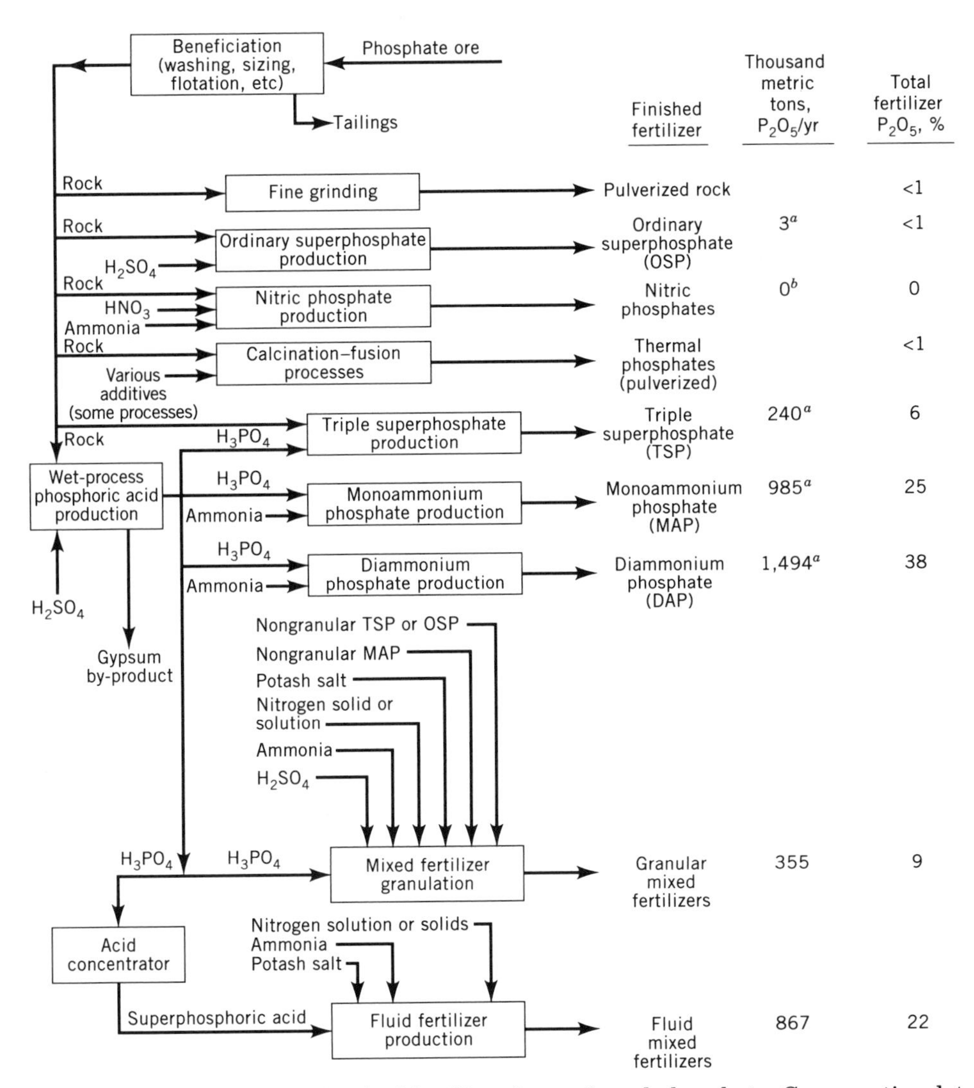

Fig. 7. Routes for making finished fertilizer from mineral phosphate. Consumption data are for year ending June 30, 1990 (5). [a]Includes quantities applied in dry blends; [b]significant quantities are made and used in some foreign countries.

Direct Application Rock. Finely ground phosphate rock has had limited use as a direct-application fertilizer for many years. There have been widely varying results. Direct application of phosphate rock worldwide amounts to about 8% of total fertilizer phosphate used, primarily in the former Soviet Union, France, Brazil, Sri Lanka, Malaysia, and Indonesia. The agronomic effectiveness of an apatitic rock depends not only on the fineness of the grind but also strongly on the innate reactivity of the rock and the acidity of the soil; performance is better on more acid soils. Probably more than half of the potentially productive tropical soils are acidic, some with pH as low as 3.5–4.5. Certain phosphate rocks may thus become increasingly important as fertilizer in those areas. The International

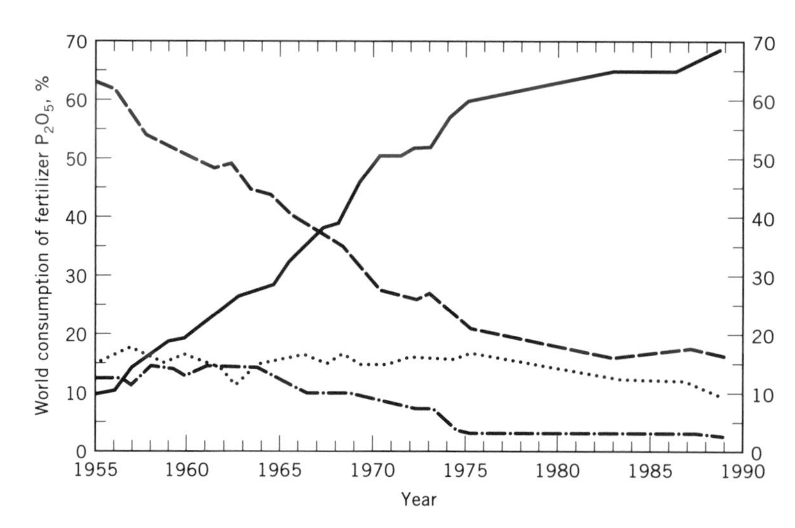

Fig. 8. World trends in types of phosphate fertilizers consumed, where (——) represents ammonium phosphates and multinutrient compounds; (– – –), normal superphosphate; (· · · ·), triple superphosphate; and (—·—), basic slag and raw rock.

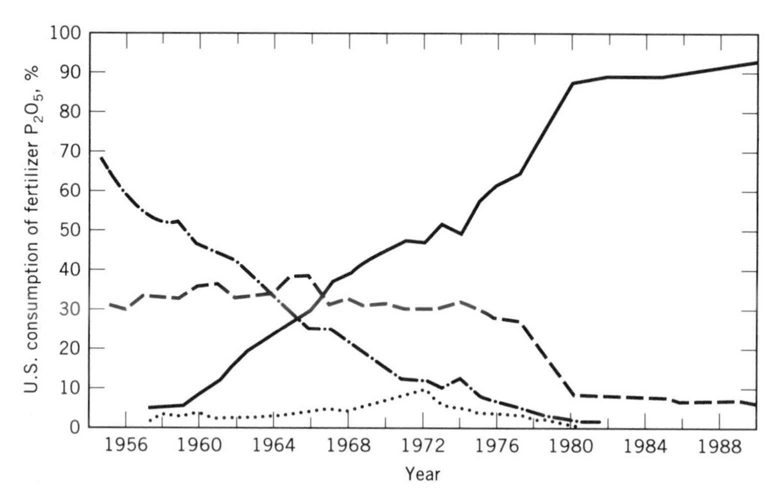

Fig. 9. U.S. trends in types of phosphate fertilizers consumed, where (——) represents ammonium phosphates, cogranulated mixtures, and fluids; (– – –) triple superphosphate (direct or bulk blend); (—·—), normal superphosphate (direct or bulk blend); and (· · · ·), slag and raw rock.

Fertilizer Development Center at Muscle Shoals, Alabama is active in researching this field (30).

The principles underlying the variable reactivity of phosphate rocks have been elucidated (31,32) and there are reliable laboratory methods for evaluating this reactivity. Broadly speaking, the reactivity increases with increase in lattice substitution by carbonate and fluoride for phosphate, and of sodium, magnesium, aluminum, rare earths, and other metals for calcium. These substitutions correlate well with changes in the unit cell α-axis dimension. Using this guideline, together with chemical methods, an absolute citrate solubility (ACS) value was defined, and a method for its determination was developed (33). An excellent correlation between ACS and plant response of several crops, including rice, was demonstrated in greenhouse and field trials. Thus a method for selecting reactive rocks is available. Among the more reactive are certain rocks from Algeria, Israel, the former Soviet Union, and North and South Carolina in the United States. In the United States, application of raw rock was once popular in the Midwest as a long-term phosphate treatment. However, that practice now has essentially disappeared, and much less than 1% of total phosphate application in the United States is as raw rock.

Normal Superphosphate. From its beginning as the first commercial phosphate fertilizer, normal superphosphate (NSP), also called ordinary or single superphosphate, has continued among the top fertilizers of the world (Fig. 8). Use of normal superphosphate decreased steadily on a percentage basis because of growing production of more concentrated materials, but grew on a P_2O_5 tonnage basis to a maximum of 6.7×10^6 t worldwide in 1967. Its production dropped to 6.1×10^6 t P_2O_5 in 1975 and remained at about that level in 1991. In the United States, use of NSP in 1991 was equivalent to less than 1% of total P_2O_5 utilization.

The sustained world popularity of NSP results from simplicity of production and high agronomic quality as a carrier of available P_2O_5, calcium, sulfur, and usually some incidental micronutrients. In terms of agronomic value for large numbers of crops, no phosphate fertilizer has been shown to be superior to NSP. It is likely to remain in strong demand in parts of the world where simplicity of production or sulfur fertilization has high priority and where transportation costs are not prohibitive.

NSP is produced by the reaction of phosphate rock and sulfuric acid. This reaction quickly yields a solid mass containing monocalcium phosphate monohydrate and gypsum, $CaSO_4 \cdot 2H_2O$, according to the simplified equation

$$Ca_{10}F_2(PO_4)_6 + 7\ H_2SO_4 + 5\ H_2O \rightarrow 3\ Ca(H_2PO_4)_2 \cdot H_2O + 7\ CaSO_4 \cdot 2\ H_2O + 2\ HF$$

Normally, a slight excess of sulfuric acid is used to bring the reaction to completion. There are, of course, many side reactions involving silica and other impurity minerals in the rock. Fluorine–silica reactions are especially important as these affect the nature of the calcium sulfate by-product and of fluorine recovery methods. Thermodynamic and kinetic details of the chemistry have been described (34).

Properties. Important properties of NSP are free acid content, as H_2SO_4, 1–2%; moisture content, 5–8%; P_2O_5 soluble, in neutral citrate solution, 20–21%; hygroscopicity at 30°C, 94% relative humidity; bulk density, nongranular, 800 kg/m^3. The porosity of nongranular NSP is important to ensure its capacity to react with ammonia, a requirement for much of its use in mixtures.

Production Technology. A moderately high (33.5% P_2O_5, 73% BPL) grade of phosphate rock is required for the production of a product that contains 20% available P_2O_5. Significant process variables in the manufacture of NSP are listed in Table 5.

Table 5. Process Variables in the Manufacture of Normal Superphosphate

Variable	Range	Average
H_2SO_4 concentration, %	68–75	71
acid temperature, °C	28–95	54
particle size of rock, %		
through 149 μm (100 mesh)	80–95	90
through 74 μm (200 mesh)	50–95	70
acid:rock ratio, kg 100% H_2SO_4 per 100 kg	55–65	60
rock curing time	several weeks	

A wide variety of mixing and curing equipment has been used over the long history of NSP production (34). Early plants utilized batch-type mixers and batch curing dens. Modern plants employ continuous-type technology, as illustrated in Figure 10. In this type plant, the acid and ground rock are combined by swirling in a cone mixer developed by TVA. The mixer discharges into a continuous den in which a slat conveyor moves the solid to a disintegrator. The disintegrator product moves by conveyor to the curing pile. Retention time between the mixer and the disintegrator is adjusted to provide time for the acid–rock reactions and for the product to become sufficiently dry and friable to permit disintegration. In

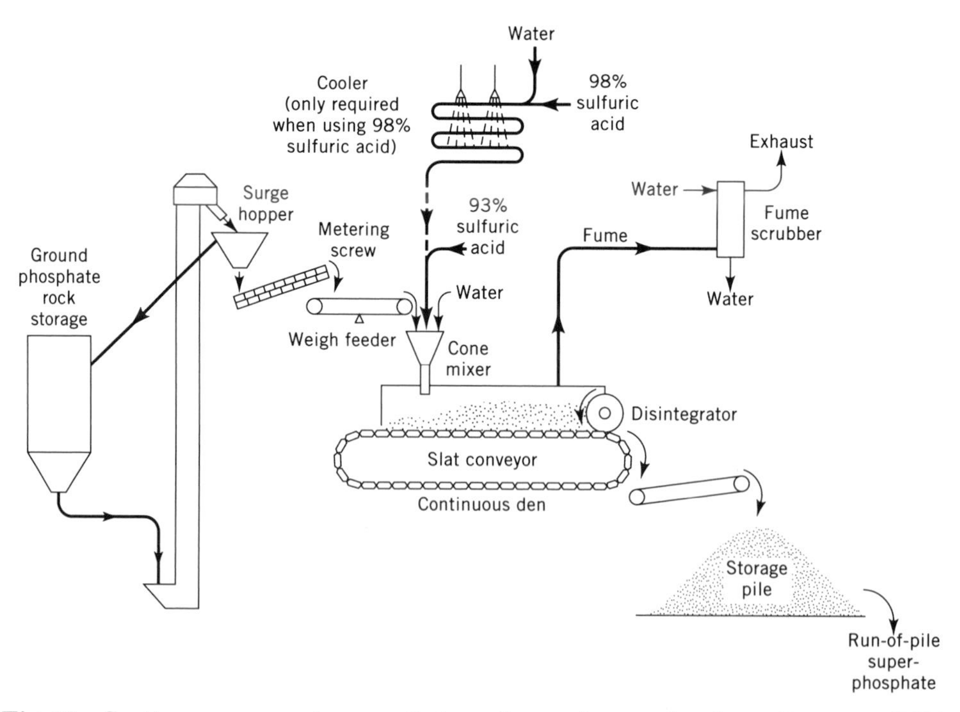

Fig. 10. Continuous process for manufacture of normal superphosphate. Courtesy of TVA.

a continuous den, 40 to 50 minutes retention is usually sufficient; batch dens require up to 2 hours. Curing time of 4 to 6 weeks in the curing pile promotes near completion of chemical reactions, increase in availability of the P_2O_5, reduced acidity, and better handling properties. The final product is known as run-of-pile normal superphosphate. Formerly, chief use of the run-of-pile product was as an ingredient in nongranular mixed fertilizers. As of 1993, at least in the United States, the more likely use is as feed to a mixed fertilizer granulation plant.

A modification of the NSP production process involves use of a mixture of sulfuric and phosphoric acids. The resultant product is referred to as enriched superphosphate and can contain up to 40% P_2O_5. The usual P_2O_5 content of enriched superphosphate is about 27%.

Cost Factors. The delivered costs of the phosphate rock and sulfuric acid raw materials often account for more than 90% of the cost of producing NSP, thus the production cost varies considerably with plant location. Because the rock is richer in P_2O_5 than is the low analysis NSP product, NSP need not be produced near the phosphate mine. However, delivery of sulfuric acid and shipment of product to market are important cost factors. Most United States NSP plants have been located east of the Mississippi river, with concentration in the southeastern and extreme southern parts of the country where the largest use of the product has occurred. Production and use of the product also has been high in California.

Raw materials requirements per ton of NSP are 0.610 t 75% BPL rock and 0.450 t H_2SO_4 (sp gr 1.70 or 60° Bé).

Wet-Process Phosphoric Acid. As indicated in Figure 7, over 95% of the phosphate fertilizer used in the United States is made by processes that require an initial conversion of all or part of the phosphate ore to phosphoric acid. On a worldwide basis also, the proportion of phosphate fertilizer made with phosphoric acid is very high. Thus processes for production of phosphoric acid are of great importance to the fertilizer industry (see PHOSPHORIC ACID AND THE PHOSPHATES).

There are two principal routes by which phosphoric acid is made from phosphate ore, a thermal route and a wet-process route. The thermal route involves electric-furnace smelting of the ore using coke and silica to produce elemental phosphorus, which then is converted to phosphoric acid by first burning (oxidizing) the phosphorus to P_2O_5 and then absorbing the P_2O_5 in water. This process results in a food-grade acid of high purity that has proven to be too expensive for general fertilizer use. Essentially all the acid used in fertilizer production is made by wet processes.

There are numerous variations of the wet process, but all involve an initial step in which the ore is solubilized in sulfuric acid, or, in a few special instances, in some other acid. Because of this requirement for sulfuric acid, it is obvious that sulfur is a raw material of considerable importance to the fertilizer industry. The acid–rock reaction results in formation of phosphoric acid and the precipitation of calcium sulfate. The second principal step in the wet processes is filtration to separate the phosphoric acid from the precipitated calcium sulfate. Wet-process phosphoric acid (WPA) is much less pure than electric furnace acid, but for most fertilizer production the impurities, such as iron, aluminum, and magnesium, are not objectionable and actually contribute to improved physical condition of the finished fertilizer (35). Impurities also furnish some micronutrient fertilizer elements.

World production of wet-process phosphoric acid in 1991 was equivalent to 26.4×10^6 t of P_2O_5. About one-third of this (10.3×10^6 t P_2O_5) was produced in North America. Over 90% of production is used for fertilizer manufacture.

Chemistry and Properties. The chemistry of phosphoric acid manufacture and purification is highly complex, largely because of the presence of impurities in the rock. The main chemical reaction in the acidulation of phosphate rock using sulfuric acid to produce phosphoric acid is

$$Ca_{10}F_2(PO_4)_6 + 10\ H_2SO_4 + 10x\ H_2O \rightarrow 6\ H_3PO_4 + 10\ CaSO_4{\cdot}xH_2O + 2\ HF$$

where $x = 0, 0.5–0.7$, or 2.0. This acidulation process is strongly exothermic, but using temperature control it can be operated to produce either the calcium sulfate dihydrate gypsum; the hemihydrate, $CaSO_4{\cdot}0.5H_2O$; or anhydrite, $CaSO_4$. In all cases, the calcium sulfate is filtered off to yield the wet-process filter-grade acid. Part of the fluorine content of the rock is volatized in the reaction and must be absorbed by scrubbing to abate air pollution. Some of the fluorine reacts with silica and other impurities in the rock and contaminates the phosphoric acid with fluosilicic acid. Twelve different fluorine compounds have been identified as possible precipitates in the production of wet-process acid (36). Up to 77% of the fluorine can be immobilized as insoluble compounds, yielding possible improvement in acid quality and reduction in fluorine emissions.

In addition to the main acidulation reaction, other reactions also occur. Free calcium carbonate in the rock reacts with the acid to produce additional byproduct calcium compounds and CO_2 gas which causes foaming. Other mineral impurities, eg, Fe, Al, Mg, U, and organic matter, dissolve, the result being that the wet-process acid is highly impure.

Israel Mining Industries developed a process in which hydrochloric acid, instead of sulfuric acid, was used as the acidulant (37). The acidulate contained dissolved calcium chloride which then was separated from the phosphoric acid by use of solvent extraction using a recyclable organic solvent. The process was operated commercially for a limited time, but the generation of HCl fumes was destructive to production equipment.

Nitric acid acidulation of phosphate rock produces phosphoric acid, together with dissolved calcium nitrate. Separation of the phosphoric acid for use as an intermediate in other fertilizer processes has not been developed commercially. Solvent extraction is less effective in the phosphoric–nitric system than in the phosphoric–hydrochloric system. Instead, the nitric acid acidulate is processed to produce nitrophosphate fertilizers.

Filter acid from a dihydrate process contains 28–32% P_2O_5. It is usually concentrated to 40–45% P_2O_5 when used at the site for fertilizer production. For shipping, the acid is concentrated to 52–54% P_2O_5. At all of these concentrations the product is orthophosphoric acid [*7664-38-2*], H_3PO_4.

On aging, many complex compounds precipitate from the concentrated acid to yield sludge. Examples of these compounds are (38) $(Fe,Al)_3KH_{14}(PO_4)_8{\cdot}4H_2O$; $FeNaH_5(PO_4){\cdot}H_2O$; $(Fe,Al)_3(K,H_3O)H_8(PO_4)_6{\cdot}6H_2O$; $(Na,K)SiF_6$; $Na_xMg_xAl_{2-x}$-

$(F,OH)_6 \cdot H_2O$; $Fe(H_2PO_4)_2 \cdot 2H_2O$; $CaSO_4 \cdot H_2O$; $CaSO_4 \cdot 5H_2O$; and $MgSiF_6 \cdot 6H_2O$. The largest portion of the sludge is removed by settling prior to shipment of the acid. The partial compositions of several desludged shipping-grade phosphoric acids produced in the United States are shown in Table 6. As a result of low rock quality, the impurity content of acid has increased over time.

Production Technology. Processes for extraction of P_2O_5 from phosphate rock by sulfuric acid vary widely, but all produce a phosphoric acid–calcium sulfate slurry that requires solids–liquid separation (usually by filtration (qv)), countercurrent washing of the solids to improve P_2O_5 recovery, and concentration of the acid. Volatilized fluorine compounds are scrubbed and calcium sulfate is disposed of in a variety of ways.

Phosphoric acid processes are dominated by the chemistry and crystallography of calcium sulfate because of the large amount produced as by-product and because its crystal form controls filterability and often plant capacity, as well as the P_2O_5 recovery efficiency of the process. Some P_2O_5 always remains incorporated in the crystal lattice of the calcium sulfate as it forms in the phosphoric acid solution, regardless of the crystal form precipitated. The most important process variations are those that affect crystal growth and form. The different types of processes are named for the crystal form that predominates: dihydrate process, $CaSO_4 \cdot 2H_2O$; hemihydrate process, $CaSO_4 \cdot 0.5H_2O$; anhydrite process, $CaSO_4$; dihydrate–hemihydrate process, in which the dihydrate is formed first, then dissolved, and the calcium sulfate is recrystallized as hemihydrate; and hemihydrate–dihydrate process in which the hemihydrate is formed first, dissolved, and the calcium sulfate is recrystallized as dihydrate. In the latter two processes, the purpose of the crystal transformation processes is to free the lattice P_2O_5, and to recrystallize the calcium sulfate under conditions of low P_2O_5 and relatively high sulfate concentration so that P_2O_5 is not a serious contaminant in the new crystals. The main process variables are temperature of the extraction and recrystallization steps.

A broad comparison of the main types of processes, the strength and quality of phosphoric acid, and the form and quality of by-product calcium sulfate are summarized in Table 7. Because the dihydrate process is the most widely used, the quality of its acid and calcium sulfate and its P_2O_5 recovery are taken as reference for performance comparisons. Illustrative flow diagrams of the principal

Table 6. Compositions of U.S. Desludged Wet Process Acids

	Shipping grades, H_3PO_4		
Constituent, wt %	Acid A	Acid B	Acid C
P_2O_5, total	53.2	51.8	54.4
Fe_2O_3	1.3	1.3	1.4
Al_2O_3	1.2	1.8	1.8
MgO	0.58	0.4	0.77
SO_4	3.1	4.8	3.1
F	0.7	1.0	1.0
organic material	0.5	0.9	0.6
water-insoluble solids	1.5	2.8	1.7

Table 7. Wet-Acid Processes and Products

Type of process	Operating temperature, °C		Acid concentration, % P_2O_5	Acid impurity level vs dihydrate acid	P_2O_5 recovery, %
	Extraction	Crystal conversion			
dihydrate	71–85	none	28–32		95
hemihydrate	91–99	none	45–50	ca same	91–94
anhydrite	102–238	none	40–50	lower	91
dihydrate–hemihydrate[a]	62–68	93–99	33–38[b]	higher	97
hemihydrate–dihydrate[a]	91–99	60–80	40–50[c]	can be lower	96–98

[a]Calcium sulfate is suitable for wallboard or cement.
[b]Value is 30–35% without filtration in crystal transition.
[c]With filtration in crystal transition.

variations in process types have been published (39). Numerous other variations in process details are also used (40–42). The majority of plants use a dihydrate process and some of these have production capacity up to 2100 t of P_2O_5 per day.

The single-stage hemihydrate process is energy efficient, producing 50% P_2O_5 acid directly, and therefore is attractive for some situations. However, P_2O_5 recovery is low, and there is a serious disposal problem for impure calcium sulfate. Processes that yield high quality calcium sulfate and give high P_2O_5 recovery are those that carry out a crystal-phase transition to release lattice phosphate. Because of local demand for domestic sources of high quality gypsum for wallboard and other uses, certain Japanese developers offer three variations of hemihydrate–dihydrate processes. All of these processes give high P_2O_5 recovery and high quality calcium sulfate for commercial use, but produce only dilute (30% P_2O_5) acid. A Fisons hemihydrate–dihydrate process produces high concentration (50% P_2O_5) acid having high (98%) P_2O_5 efficiency as well as high quality, market-grade gypsum.

The selection of a process can be complex, requiring careful evaluation of the many variables for each application. The hemihydrate process is energy efficient, but this may not be an overriding consideration when energy is readily available from an on-site sulfuric acid plant. The energy balance in the total on-site complex may be the determining factor.

Dry grinding, handling, and feeding of phosphate rock was at one time the most common practice but the trend in new plants has been to wet grinding and feeding, eliminating the energy required for drying the rock and also reducing dust problems. The H_2SO_4 concentration fed to dihydrate reactors must be increased when wet grinding is used in order to maintain a water balance in the plant. The reactor of a phosphoric acid plant normally consists of a series of vessels connected by underflow and overflow pipes or passages. A typical Prayon dihydrate process reactor is a lined concrete shell that contains up to nine compartments.

Many different filter types are used, including horizontal rotary table, belt, in-line pan, and horizontal rotary tilting pan. The dihydrate acid containing 28–32% P_2O_5 is concentrated further for most uses, usually in single-effect vacuum evaporators. A variety of types are used, including forced circulation, natural circulation, and falling film. Direct-heated spray towers and submerged combustion have also been used but these are no longer preferred. Because corrosion is a serious problem in wet-process acid plants, carbon brick linings and rubber linings are used extensively. Filters, pumps, and agitators are of stainless steel, and piping is made of rubber-lined steel and a variety of plastics.

The abatement of fluorine emissions and disposal of by-product calcium sulfate from phosphoric acid plants are environmental concerns.

Purification. The impurity content of wet-process acid, which contains sludge, dissolved metals, and F, and dissolved and colloidal organic compounds, is well known. The sludge is troublesome in storage, shipping, and handling of the acid. Most shipping-grade acid has been at least partially clarified by settling or other means for removal of most of the sludge. If made from uncalcined rock, the clarified acid retains the black color imparted by organic matter. A greater degree of purification of acid for fertilizer use is needed only to produce clear solution fertilizers that can be stored for extended periods without forming precipitates. However, most liquid fertilizers now are made from superphosphoric acid because of its capacity for sequestering (solubilizing) metallic impurities.

Numerous purification processes have been developed for application to wet-process acid (43–49) but these are not applied to most acid used in fertilizer production.

Storage, Handling, and Shipping. Properties that most affect the storage and handling of wet-process acid are sludge formation, corrosiveness, and viscosity. Often, 54% P_2O_5 acid is aged for about a month before shipping to permit formation and settling of sludge. One month unassisted settling of 100 t/d P_2O_5 as 54% P_2O_5 acid requires a minimum of 2830 m^3 (7.5×10^5 gal) of storage capacity. Acid made from calcined rock requires only about one-fourth as long for post-precipitation and settling and thus a correspondingly smaller storage capacity. Settling can be accelerated by mild movement, by flocculents, and by addition of a few percent of H_2SO_4. Sludge densities up to 40% solids can be produced in cyclones, and up to 70% in a conveyor-type centrifuge. Sludge frequently is utilized on-site in production of solid fertilizers such as superphosphate and monoammonium phosphate. The viscosity of shipping grade 50–54% P_2O_5 acid ranges from 300 to 30 mPa·s(=cP) over 4 to 38°C.

The container material most frequently used for storing and shipping phosphoric acid is rubber-lined mild steel (50). Both butyl and natural rubber are used. Type 316 stainless steel also is satisfactory for storage tanks and transfer lines. PVC plastics are sometimes used as tank liners but temperature-strength limitations must be considered. Below ground level pits lined with plastic sheet also are used (51). Fiber glass tanks, because of attack by flourine, have not proven fully satisfactory. Heating and insulation are not required for orthophosphoric acid tanks. Wet-process acid is shipped in railroad tank cars, trucks, river barges, and ocean-going tankers. Phosphoric acid is not usually classified as a hazardous chemical, but in contact with any part of the body it may cause burns. In appropriate situations, the usual chemical plant safety equipment should be used, ie, safety goggles, plastic face shields, breathing apparatus and masks, and rubber gloves.

Triple (Concentrated) Superphosphate. The first important use of phosphoric acid in fertilizer processing was in the production of triple superphosphate (TSP), sometimes called concentrated superphosphate. Basically, the production process for this material is the same as that for normal superphosphate, except that the reactants are phosphate rock and phosphoric acid instead of phosphate rock and sulfuric acid. The phosphoric acid, like sulfuric acid, solubilizes the rock and, in addition, contributes its own content of soluble phosphorus. The result is triple superphosphate of 45–47% P_2O_5 content as compared to 16–20% P_2O_5 in normal superphosphate. Although triple superphosphate has been known almost as long as normal superphosphate, it did not reach commercial importance until the late 1940s, when commercial supply of acid became available.

Worldwide, triple superphosphate, over the period 1955 to 1980, maintained about a 15% share of the phosphate fertilizer market (Fig. 8). World consumption for the year ended June 30, 1991 (9) was equivalent to 3.6×10^6 t of P_2O_5, which was about 10% of world fertilizer P_2O_5 consumption. In the United States, consumption for the year ended June 30, 1990 (Fig. 7) was equivalent to about 240×10^3 t of P_2O_5, which represented only 6% of U.S. fertilizer P_2O_5 consumption.

Simplicity of production, high analysis, and excellent agronomic quality are reasons for the sustained high production and consumption of TSP. A contributing factor is that manufacture of the triple superphosphate has been an outlet for so-called sludge acid, the highly impure phosphoric acid obtained as a by-product of normal acid purification.

Chemistry and Properties. TSP is essentially impure monocalcium phosphate monohydrate, $Ca(H_2PO_4)_2{\cdot}H_2O$, made by acidulating phosphate rock with phosphoric acid according to

$$Ca_{10}F_2(PO_4)_6 + 14\ H_3PO_4 + 10\ H_2O \rightarrow 10\ Ca(H_2PO_4)_2{\cdot}H_2O + 2\ HF$$

The complete chemistry of TSP production has been studied and reported in great detail (34). As in the production of NSP there are also reactions with impurity minerals. In fact, the increasing amounts of such impurities in U.S. commercial phosphate rocks, especially those from Florida, are now reflected in somewhat lowered amounts of citrate-soluble P_2O_5 in the product. The range of constituents in commercial TSP from wet-process acid and phosphate rocks are typically (34): $Ca(H_2PO_4)_2{\cdot}H_2O$, 63–73%; $CaSO_4$, 3–6%; $CaHPO_4$ and Fe and Al phosphates, 13–18%; silica, fluorosilicates, unreacted rock, and organic matter, 5–10%; and free moisture, 1–4%. The average citrate solubility of the P_2O_5 in best quality TSP is 98–99%, but products with citrate solubility values a few percentage points lower are not uncommon. The P_2O_5 citrate solubility of TSP made by a quick-cure process is ca 96% (34). Other common properties of TSP are bulk density, nongranular, 879 kg/m^3; granular, 1040–1200 kg/m^3; and critical relative humidity at 30°C, 94%.

Production Technology. Phosphate rock and wet-process phosphoric acid are the only raw materials required for manufacturing TSP. The grade of rock can be a little lower than that needed for NSP production. Over the years, a large number of process modifications, both batch and continuous, have been used. For the production of nongranular TSP, 52–54% P_2O_5 acid is used without dilution or heating. The P_2O_5:CaO mole ratio, including the P_2O_5 in the rock, is 0.92–0.95. The rock is ground to 70% <74 μm (200 mesh). Pile curing for a few weeks is typical, as for NSP.

Granular TSP (−6 + 16 mesh (1.19 to 3.35 mm dia)) is preferred for direct application and is used in bulk blend fertilizers. A widely used slurry granulation process, the Dorr-Oliver process, is illustrated in Figure 11. The ground rock is mixed with 38–49% P_2O_5 acid in a series of reaction vessels. Slurry from the reaction train then is mixed with a large proportion of recycled undersize granules and crushed oversize in either a pugmill or drum granulator. The granules are dried and screened. The product-size material is about 1 part in 13, thus the recycle ratio is 12:1. Other processes involve granulation of previously prepared nongranular TSP. A one-step process for simultaneous acidulation and granulation was studied by TVA (52), but the process has not been commercialized.

Economics. In contrast to NSP, the high nutrient content of TSP makes shipment of the finished product preferable to shipping of the raw materials. Plants, therefore, are located at or near the rock source. The phosphoric acid used, and the sulfuric acid required for its manufacture, usually are produced at the site of the TSP plant. As in the case of NSP, the cost of raw materials accounts for more than 90% of the total cost. Most of this is the cost of acid.

Since about 1968, triple superphosphate has been far outdistanced by diammonium phosphate as the principal phosphate fertilizer, both in the United States and worldwide. However, production of triple superphosphate is expected to persist at a moderate level for two reasons: (*1*) at the location of a phosphoric acid–diammonium phosphate complex, production of triple superphosphate is a convenient way of using sludge acid that is too impure for diammonium phosphate

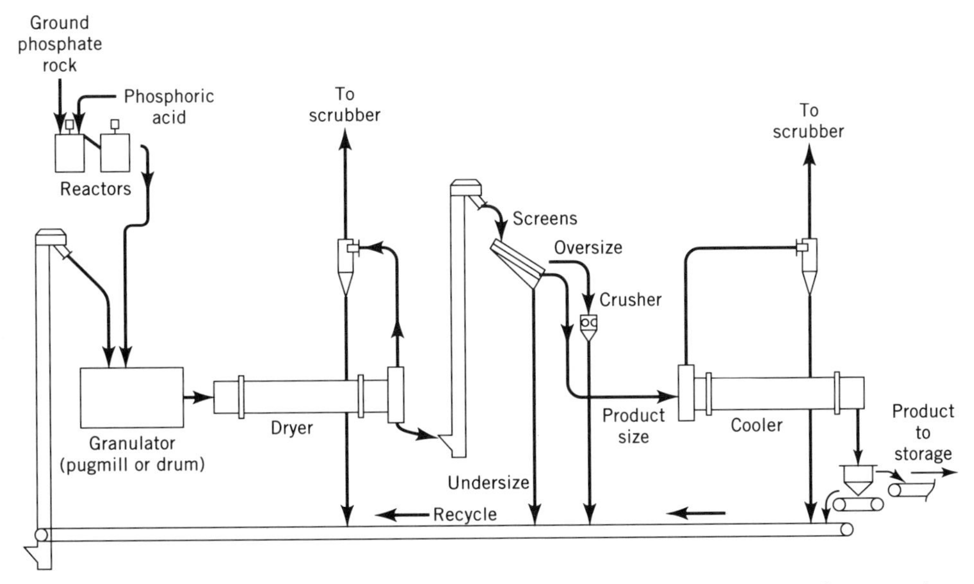

Fig. 11. Dorr-Oliver type slurry process for manufacture of granular triple superphosphate. Courtesy of TVA.

production; and (*2*) the absence of nitrogen in triple superphosphate makes it the preferred source of phosphorus for the no-nitrogen bulk-blend fertilizers that frequently are prescribed for leguminous crops such as soy beans, alfalfa, and clover.

Diammonium Phosphate. Diammonium phosphate (DAP), $(NH_4)_2HPO_4$, was introduced as a commercially viable fertilizer in the early 1950s. Since that time, its acceptance has been phenomenal. For the year ended June 30, 1990, about 3.2×10^6 t of diammonium phosphate (1.5×10^6 t P_2O_5) was produced in the United States and used by U.S. farmers (5). This consumption represented about 38% of total U.S. fertilizer P_2O_5 usage (Fig. 7). Additional quantities were produced for export. Worldwide, ammonium phosphate fertilizer consumption (diammonium and mono-) for the same period (9) amounted to 11.1×10^6 t of P_2O_5, which was about 31% of total fertilizer P_2O_5 utilization. Although the ammonium phosphates are, strictly speaking, mixed fertilizers, containing both nitrogen and phosphorus, it is appropriate to consider these materials as primarily phosphate suppliers.

Production Processes. The earliest commercial ammonium phosphate fertilizers were either monoammonium phosphates or mixtures of mono- and diammonium phosphates. Ammonium sulfate was usually a component of these products by virtue of the use of mixed sulfuric–phosphoric acids. Granulation equipment and procedures had already been developed for producing granular mixed fertilizers, and these methods were applied to production of the ammonium phosphate fertilizer products. Grades such as 11–48–0, 13–52–0, and 16–48–0 were made on a limited scale from wet-process acid and ammonia. Use of the higher degree of ammoniation that is required to produce chiefly diammonium phosphate (18–46–0 grade) was at first considered impractical because of high loss of ammonia in the production process and presumed chemical instability of diammonium phosphate as a fertilizer product. The feasibility of producing and using diammonium phosphate as a fertilizer was, however, convincingly demonstrated in work by TVA. Initially, the work concentrated on the use of pure, electric-furnace phosphoric acid to produce essentially pure, crystalline diammonium phosphate of 21–53–0 grade. Both saturator–crystallizer (53) and vacuum crystallizer (54) processes were explored and proved technically feasible. Later work concentrated on the use of wet-process acid and resulted in development (55) and patenting (56) of a practical process that, in various modifications, is still the principal process used for production of granular diammonium phosphate from wet-process acid.

A flow sheet of the basic TVA process for granular diammonium phosphate is given in Figure 12. The raw materials are wet-process phosphoric acid and anhydrous ammonia. Feed acid concentration of at least 40% P_2O_5 is required to give a satisfactory water balance. This average concentration usually is provided by two separate feed streams, one of 54% P_2O_5 concentration and one of about 30% P_2O_5. In the arrangement shown, the 54% acid is fed to a large, agitator-equipped vessel referred to as the preneutralizer. Detail of a typical preneutralizer vessel is shown in Figure 13. Anhydrous ammonia in gaseous form is sparged into the bottom of this vessel to provide partial ammoniation of the acid. The heat of reaction in the preneutralizer is sufficient to cause boiling and thus to effect desirable loss of water and concentration to slurry form. The degree of partial ammoniation carried out in the preneutralizer is critical and is a primary speci-

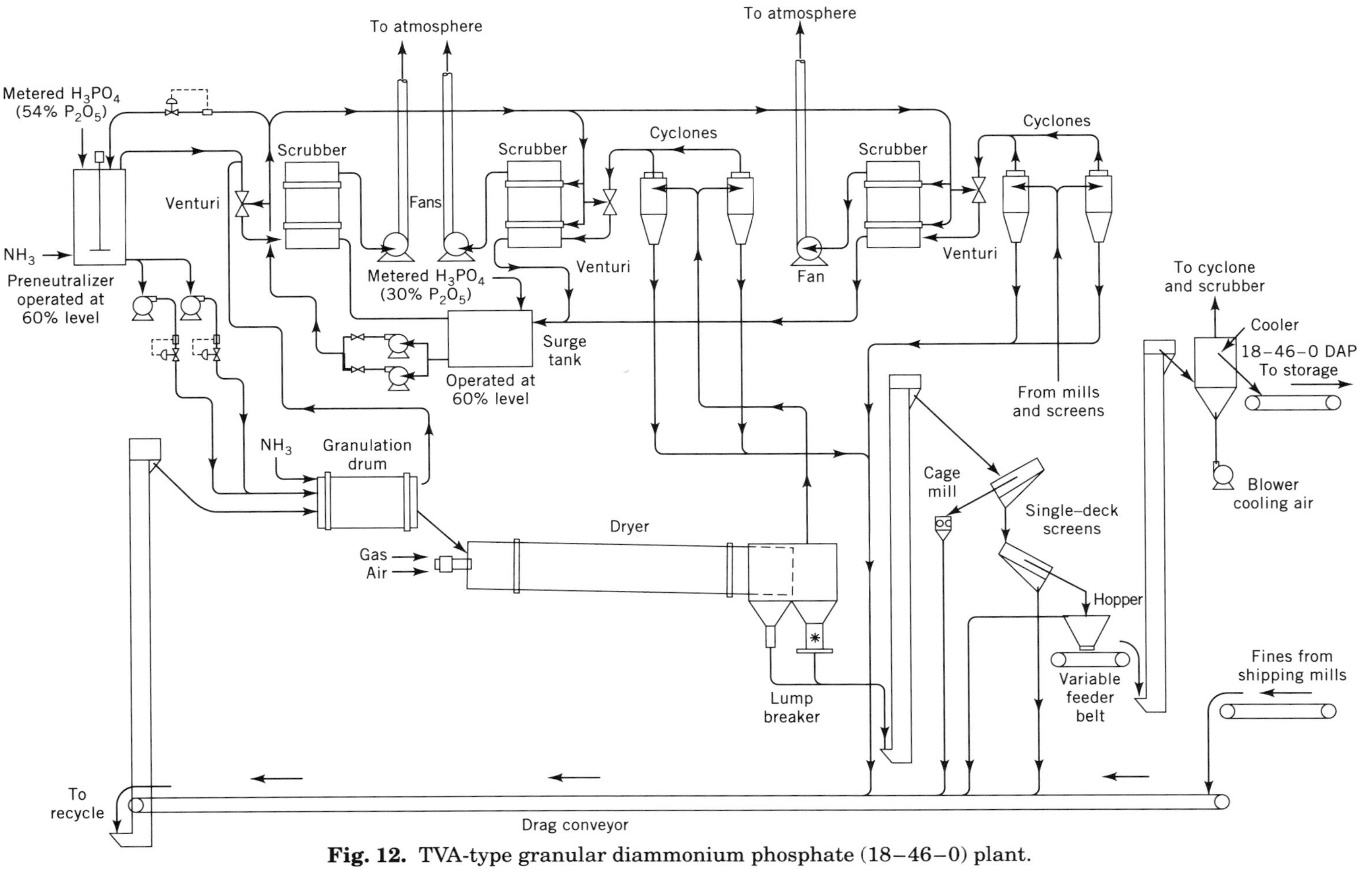

Fig. 12. TVA-type granular diammonium phosphate (18–46–0) plant.

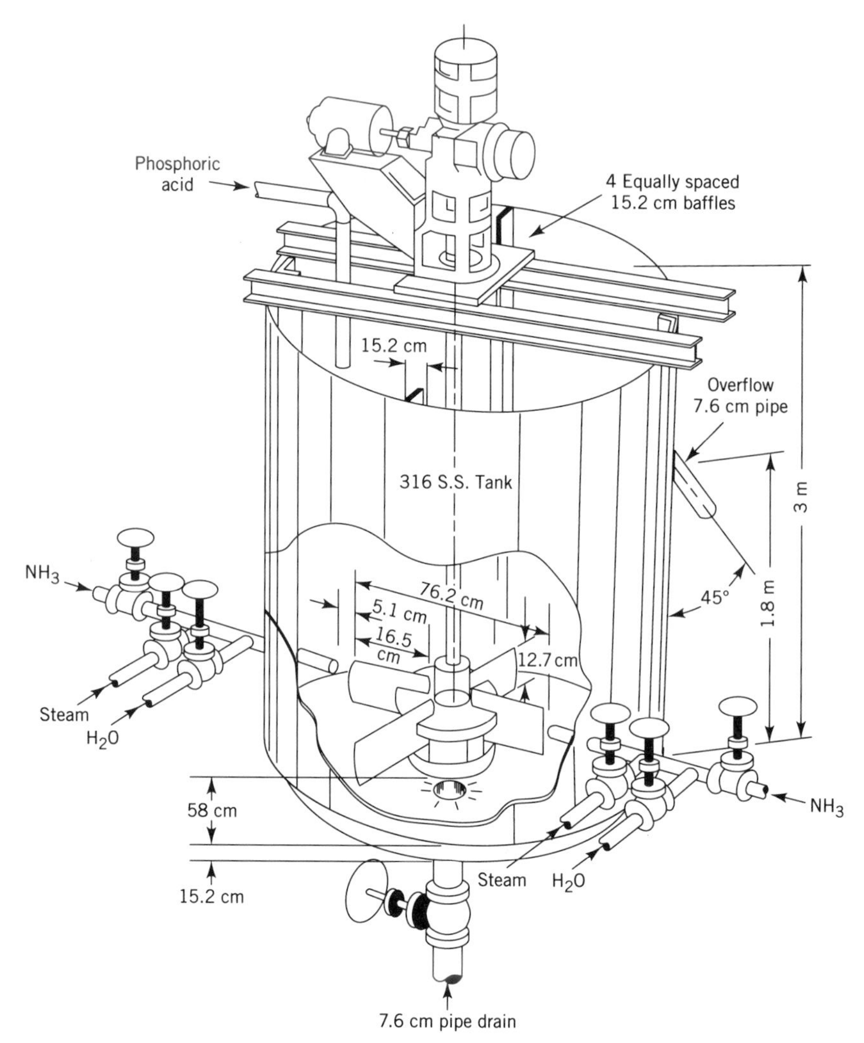

Fig. 13. Preneutralizer vessel of the type used in production of granular diammonium phosphate and mixed fertilizers.

fication of the process patent (56). The preferred NH_3:PO_4 mole ratio to be maintained is about 1.4. As shown in Figure 14, this corresponds to a maximum solubility peak in the system NH_3–H_3PO_4–H_2O. Thus, by maintaining this composition in the preneutralizer, it is possible to volatilize the maximum amount of water while still maintaining fluidity of the preneutralizer slurry. Typically, slurry discharged from the preneutralizer is at about 115–125°C and contains only 16 to 20% water. This hot slurry is pumped directly to a rotary, TVA-type, ammoniator–granulator drum where it is distributed over the surface of a rolling bed of ammonium phosphate solids. In this drum, additional anhydrous ammonia is sparged under the bed of rolling solids to complete the ammoniation of the

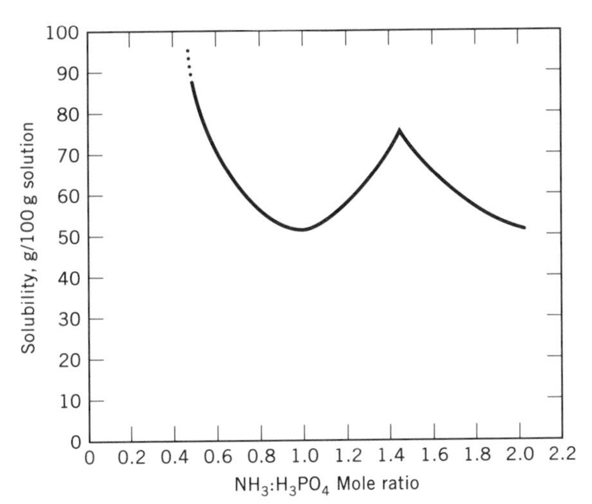

Fig. 14. Effect of NH_3:H_3PO_4 mole ratio on solubility in the system NH_3–H_3PO_4–H_2O.

slurry to diammonium phosphate; that is, to bring the overall NH_3:PO_4 mole ratio to approximately 2.0.

A sketch of a typical TVA-type ammoniator–granulator as used in diammonium phosphate production is shown in Figure 15. This general type of granulator was originally developed for the continuous ammoniation–granulation of superphosphates and mixed fertilizers. In the diammonium phosphate granulator, the further ammoniation promotes solidification of the feed slurry by increasing NH_3:PO_4 mole ratio to 2.0, which is a relatively low solubility point in the NH_3–H_3PO_4–H_2O system (Fig. 14), and by releasing further heat of reaction, which volatilizes additional moisture. Another essential feature of the process is the feeding of recycled, dry, undersize solids to the granulator. Such recycle provides the substrate for a layering process in which feed slurry from the preneutralizer coats the recycled solid particles, is ammoniated, solidifies, and thus contributes to granule formation and growth. As indicated in the flow sheet, this recycle is obtained from the final sizing (screening) of dried product.

The recycle consists usually of a mixture of three entities: screen undersize, crushed screen oversize, and some crushed product-size material. The usual rate of recycle required for satisfactory granulation is 3 to 7 kilograms per kilogram of final product. Variation of this recycle rate is the principal control over granulation efficiency in the drum, and such variation usually is accomplished by varying the amount of crushed product-size material that is included in the recycle. The total solids discharge from the granulator contains 2 to 4% moisture and must be dried prior to the screening operation.

Drying (qv) is accomplished in a rotary drum-type dryer, usually gas fired. Diammonium phosphate becomes unstable at relatively low temperatures; care must be taken to avoid overheating. For this reason, cocurrent drying is the preferred mode and product discharge temperature is not allowed to exceed about 88°C. Normally, the screening operation is carried out hot, and only the product-size fraction is cooled. Cooling is carried out by contact with air in either a rotary

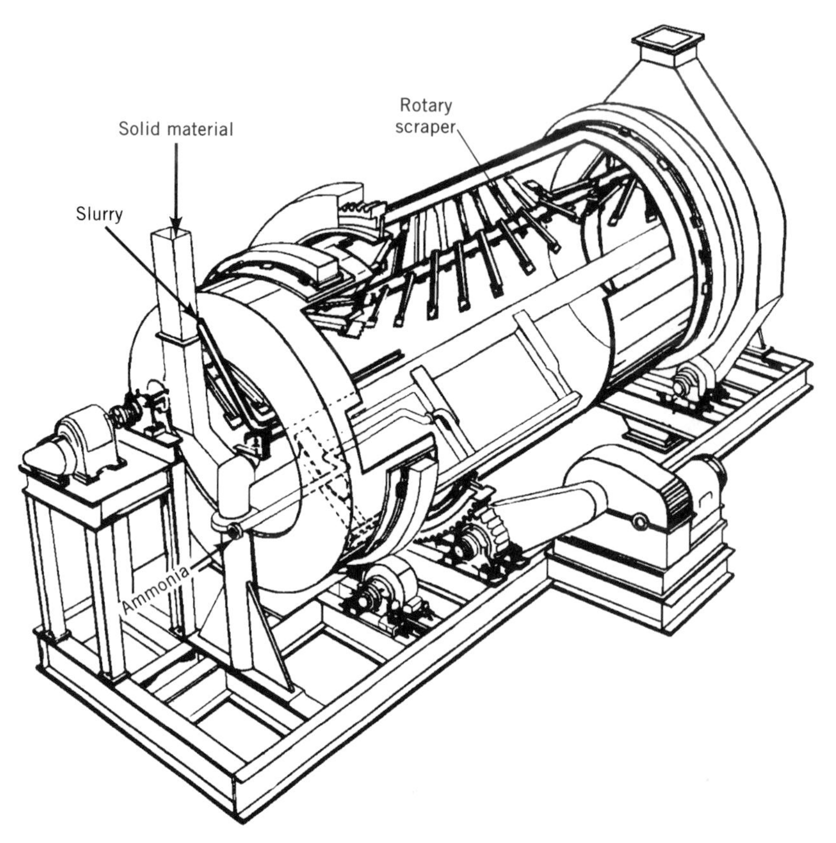

Fig. 15. Cutaway view of a TVA-type ammoniator–granulator for production of granular ammonium phosphates or NPK cogranulated mixtures.

drum or a fluidized bed. Cooling to about 50°C is considered desirable to avoid caking in bulk storage. Particle size of the granular product, as produced in the United States, normally is within the range of 1 to 3.5 mm particle diameter. An average diameter is about 2.3 mm. Moisture content is 1 to 2%. Normally, no granule coating or other anticaking treatment is necessary. Iron, aluminum, and other impurities derived from the wet-process feed acid exert sufficient hardening and anticaking effect (35).

In both the preneutralizer and the rotary ammoniator–granulator, the absorption of feed ammonia is incomplete; thus acid scrubbing of the effluent vapors from these units is an important feature of the process. A typical flow sheet, as in Figure 12, employs at least two scrubbers, one serving the relatively low vapor volumes from the preneutralizer and granulator and a second serving the high volume gaseous effluents from the dryer and cooler. Scrubber liquors are recirculated through an interlinked system. Fresh acid, usually of 30 to 40% P_2O_5 concentration, is fed to the scrubber system while a corresponding amount of circulating liquor is bled off to the preneutralizer. The overall effect is that of a closed system having very high recovery of ammonia and relatively clean stack effluent. Specifications of a typical granular diammonium phosphate, as produced in the United States, are shown in Table 8.

A relatively recent process innovation that has been introduced in connection with production of diammonium phosphate is the use of a pipe reactor to replace the preneutralizer. In this modification (57–60), the acid and a large proportion of the ammonia are mixed in a section of pipe that discharges a resultant melt directly into the granulator. Ammoniation is completed in the granulator. Utilization of the heat of reaction is said to be highly efficient, to the extent that the drying requirement is reduced or even eliminated. At present, application of the pipe reactor to production of DAP is still largely experimental, whereas its application to granulation of monoammonium phosphate and mixed fertilizers is well established.

Monoammonium Phosphate. Monoammonium phosphate [*7722-76-1*] (MAP), $NH_4H_2PO_4$, has become second only to diammonium phosphate as a phosphate fertilizer material of trade. During the year ended June 30, 1990, monoammonium phosphate used in the United States furnished 985 thousand t of P_2O_5 as compared to 1.5 million t furnished by diammonium phosphate and 240 thousand t by triple superphosphate (Fig. 7). Monoammonium phosphate furnished 25% of total P_2O_5 consumption.

Pure monoammonium phosphate has a grade 12.1–61.7–0, but is never made for fertilizer use. Made from wet-process acid, the grade for fertilizer use is typically 10–54–0 to 11–52–0, but may be as low as 10–50–0, depending mainly on the impurity level of the starting acid. The physical properties of monoam-

Table 8. Specifications of Fertilizer-Grade Granular Diammonium Phosphate[a,b]

Parameter	Value	
	Range	Typical
available nitrogen,[c] %	18.0–18.2	18.0
P_2O_5, %		
total available[d]	46.1–46.7	46.3
	46.0–46.5	46.2
water-soluble[e]	88–92	90
moisture, %	0.5–2.0	1.0
pH	7.5–7.8	7.7
sieve/size opening		
+6 mesh (3.36 mm)	0–12	6
+8 mesh (2.38 mm)	35–65	48
+10 mesh (1.68 mm)	75–95	88
+14 mesh (1.19 mm)	96–100	98
+20 mesh (0.84 mm)	98–100	99

[a]Courtesy of International Minerals and Chemical Corp.
[b]Typical bulk density, loose, 881–929 kg/m^3; angle of repose, 28–32 degrees.
[c]A minimum value of 18.0 is guaranteed.
[d]A minimum value of 46.0 is guaranteed.
[e]As percent of available P_2O_5.

monium phosphate made from wet-process acid are excellent for a fertilizer material. It is chemically and physically very stable and relatively nonhygroscopic (critical relative humidity 70 to 80%) and is noncaking if properly dried. Drying is relatively easy because monoammonium phosphate remains stable at relatively high drying temperatures. Monoammonium phosphate is made in both granular and nongranular (powder) forms. The granular form is used both for direct field application and in bulk blends. Agronomically, because of its higher acidity, MAP is preferred over diammonium phosphate for use on alkaline soils such as those in western Canada and parts of the western United States.

The nongranular forms of monoammonium phosphate find use as feed material to ammoniation–granulation plants that produce mixed fertilizers. In this application, the acidity of monoammonium phosphate, like that of superphosphate, provides an ammonia absorption capacity. Significant quantities of the nongranular monoammonium phosphate also are used as a phosphorus source in suspension-type fertilizers. The relatively high P_2O_5 content of monoammonium phosphate presents a distinct shipping cost advantage over other products in cases where a primary objective is the shipment of phosphate. On the other hand there are several reasons why granular monoammonium phosphate has not fared better as a blend material in the important bulk blending (dry blending) segment of the industry. One reason is that grade specifications for monoammonium phosphate, unlike those for diammonium phosphate, have not been standardized industry-wide. This presents blenders with difficulties in both storage and formulation. The practice of using MAP production as a route for utilization of sludge acid complicates standardization. Another drawback is that use of MAP instead of DAP in bulk blends requires that for many popular grades of blend a larger proportion of supplemental nitrogen material must be used, which often is not as economical as using DAP. Monoammonium phosphate, in spite of its high P_2O_5 content, does contain nitrogen and thus, like DAP, cannot serve as a substitute for triple superphosphate in production of no-nitrogen blends. For these reasons, the choice of many bulk blenders is diammonium phosphate for N–P and N–P–K blends and triple superphosphate for P–K blends.

Production Processes. An excellent review of most methods for MAP production has been published (61). Granular MAP can be produced in a plant similar to that used for diammonium phosphate (Fig. 12) with little process modification. For MAP production (62), the preneutralizer may be operated at the same high solubility $NH_3:PO_4$ mole ratio of 1.4 that is used for diammonium phosphate, in which case additional phosphoric acid is sparged into the granulation drum to bring the final product ratio to that of MAP ($NH_3:PO_4$ = 1.0). Another mode of operation that has been shown to be satisfactory is operating the preneutralizer at a $NH_3:PO_4$ mole ratio of 0.6, which also is a high solubility point (Fig. 14), followed by final ammoniation to 1.0 mole ratio in the granulation drum. A number of granular MAP plants worldwide now use a pipe reactor in MAP production. Use of the pipe reactor eliminates need for the preneutralizer and often also eliminates the need for a dryer. Ammoniation of the acid is carried out either totally in the pipe or in the pipe and granulator drum. Heat of reaction is sufficient to provide all the drying needed. If all of the ammoniation is carried out in the pipe it is necessary to limit feed acid concentration to 50% P_2O_5 or lower to avoid formation of polyphosphates.

The first to produce nongranular MAP commercially was Scottish Agricultural Industries, Ltd. under the trade name Phos SAI. In the SAI process, gaseous ammonia and wet-process acid are allowed to react in a tank to produce a slurry having an N:P mole ratio 1.3:1. Acid is added to the slurry in a pin mixer to adjust the mole ratio to 1.0 and to disengage water vapor (63). The powder product contains 6% moisture which is a claimed advantage over smaller amounts because of less dustiness and higher ammonia reactivity. A gel impurity phase covers the surface of the crystals and is claimed to have beneficial effect on the granulation quality of the material. In a process developed by Fisons Ltd., ammonia vapor and 48–52% P_2O_5 wet-process acid are mixed in a reactor at 308 kPa (30 psig) in an N:P mole ratio of 1. The hot reaction product containing 9–10% water is sprayed through a nozzle into a tower where water flashes off and powder MAP collects. The product contains about 6% moisture. A number of such plants have been built. The characteristic feature of another process developed by Swift, Inc., is a two-fluid nozzle that gives instantaneous mixing of liquid ammonia and 49–51% P_2O_5 wet-process acid. A reactor pipe discharges the slurry at about 127°C into the top of a tower, in which water is flashed and powder MAP collects. A typical product grade is 10.5–53.0–0 and moisture content is 4%. The heart of another powder MAP process, developed by Gardinier, is a novel pipe reactor. Wet-process acid and ammonia are introduced tangentially into the reactor through diametrically opposed inlet lines. The reactants flow in a spiral pattern to the nozzle end of the pipe where they are ejected into a tower. The product, 11–53–0 grade, contains 2–4% moisture. Phosphoric acid containing up to 30% solids can be used in this process, which makes it adaptable to the use of sludge acid. A single reactor can produce up to 500 t/d of product.

Ammonium Polyphosphate. By increasing the concentration of feed acid to a MAP pipe reactor to 52–54% P_2O_5, and/or preheating the acid, a high temperature can be developed in the pipe and a product can be produced that is a mixture of monoammonium phosphate and diammonium dihydrogen pyrophosphate [*13957-86-9*], $(NH_4)_2H_2PO_7$. Melt granulation of this type of low polyphosphate MAP, also called ammonium polyphosphate [*120124-31-9*] (APP), was carried out on a demonstration scale by TVA (7). Acid of 52 to 54% P_2O_5 content was ammoniated in a pipe reactor at about 218°C to produce a melt in which 15–25% of the P_2O_5 was nonorthopolyphosphate. The melt could be granulated in a pug mill, rotary granulator, or by prilling. The usual product grade was 11–55–0. The properties of this material were total N, 11–12%; ammonium N, 100% of total; total phosphorus, 54–58% P_2O_5; P_2O_5 availability, 100% of total; water-soluble phosphorus, 99–100% of total P_2O_5; moisture content, 2%; critical relative humidity at 30°C, 70–75%. When 20% of the phosphorus was polyphosphate, the compounds in the product were in the ratio of 3.5 mole MAP per mole of the pyrophosphate. The principal use of the material was in the production of suspension fertilizers. In this application the polyphosphate content imparted improved storage properties to the suspensions. The granular solid APP, however, also had excellent storage properties and was a good material for use in bulk blends and for direct application.

Nitric Phosphate. About 15% of worldwide phosphate fertilizer production is by processes that are based on solubilization of phosphate rock with nitric acid instead of sulfuric or phosphoric acids (64). These processes, known collectively as nitric phosphate or nitrophosphate processes are important, mainly because of the independence from sulfur as a raw material and because of the freedom from the environmental problem of gypsum disposal that accompanies phosphoric acid-based processes. These two characteristics are expected to promote eventual increase in the use of nitric phosphate processes, as sulfur resources diminish and/or environmental restrictions are tightened.

Production of nitric phosphates is not expected to expand rapidly in the near future because the primary phosphate exporters, especially in North Africa and the United States, have moved to ship upgraded materials, wet-process acid, and ammonium phosphates, in preference to phosphate rock. The abundant supply of these materials should keep suppliers in a strong competitive position for at least the short-range future. Moreover, the developing countries, where nitric phosphates would seem to be appealing for most crops except rice, have already strongly committed to production of urea, a material that blends compatibly with sulfur-based phosphates but not with nitrates.

Nonetheless, production and use of nitric phosphates in Europe are continuing to grow. In general, nitric phosphate processes are somewhat more complicated than sulfur-based processes and require higher investment. In the past, several attempts have been made to establish commercial acceptance of this type process in the United States, but plant operations have been relatively short lived because of low sulfur prices and resultant competition from sulfur-based processes.

Nitrophosphates are made by acidulating phosphate rock with nitric acid followed by ammoniation, addition of potash as desired, and granulation or prilling of the slurry. The acidulate, prior to ammoniation, contains calcium nitrate and phosphoric acid or monocalcium phosphate according to the following equations:

$$Ca_{10}F_2(PO_4)_6 + 20\ HNO_3 \rightarrow 10\ Ca(NO_3)_2 + 6\ H_3PO_4 + 2\ HF$$

$$Ca_{10}F_2(PO_4)_6 + 14\ HNO_3 \rightarrow 3\ Ca(H_2PO_4)_2 + 7\ Ca(NO_3)_2 + 2\ HF$$

Process variations include the use of phosphoric or sulfuric acid in mixtures with nitric acid, removal of part of the calcium nitrate by cooling, crystallization, and centrifuging, or addition of carbon dioxide to precipitate calcium carbonate. The purpose of each of those variations is to reduce the amount of calcium available for combining with phosphate, thus reducing or preventing the formation of high calcium phosphates that are less soluble than dicalcium phosphate. These modifications also eliminate calcium nitrate as a component of the ammoniated product and thus prevent excessive hygroscopicity. To achieve these benefits, at least 40% of the calcium must be removed as crystallized calcium nitrate or combined with anions such as phosphate, sulfate, or carbonate. To the extent that water-soluble phosphate is required (dicalcium phosphate is citrate soluble but water insoluble), the ratio of available CaO to P_2O_5 is adjusted to less than 2. When this adjustment is made by use of phosphoric or sulfuric acid, the process loses its independence of sulfur. Process variations thus provide a wide range of possibilities that affect the phosphate solubility and, to some extent, uses that can be made of the product.

There is extensive literature on nitrophosphates (65,66). A description of the Norsk Hydo nitrophosphate process, ie, using calcium adjustment by crystallization of calcium nitrate, emphasizing the environmental advantages is also available (64).

Potash Fertilizers

World consumption of potassium salts presently exceeds 28 million t of K_2O equivalent per year. About 93% of that is for fertilizer use (see POTASSIUM COMPOUNDS). The potash [*17353-70-7*] industry is essentially a mining and beneficiation industry. The two main fertilizer materials, KCl and K_2SO_4 are produced by beneficiating ores at the mine sites. The upgraded salts then are shipped to distributors and manufacturers of mixed goods.

Some 90% of all potash fertilizer is KCl; the proportion is slightly less in the United States. Most of the remainder of fertilizer potash is sulfate, either as K_2SO_4 or the double salt langbeinite [*14977-37-8*], $K_2SO_4{\cdot}2MgSO_4$. A small but increasing amount of KNO_3 is used, mostly as a specialty fertilizer. This latter salt does not occur in significant quantities in nature. It is generally produced by the reaction of KCl with HNO_3. Test production and marketing of potassium phosphate salts as fertilizer in the United States was begun by Pennzoil Inc. in 1976 but has been discontinued. Properties of the three principal potassium fertilizer salts are shown in Table 9.

For most of the decade prior to 1965, the United States was the world's largest single producer of potash. More than 90% of this production came from mines in New Mexico. As the grades of these deposits lowered and production costs rose, expanding production in Saskatchewan, Canada, increasingly met the growing United States market. In the 1990s, the United States is by far the largest single importer of potash, almost entirely from Canada. Potash mines have long been worked also in Germany and in the former Soviet Union.

Essentially all of the workable potassium salt deposits were formed in ancient times by the evaporation of seas or brine lakes. These salt deposits occur in

Table 9. Properties of Potassium Fertilizer Salts[a]

Property	KCl	K_2SO_4	KNO_3
crystal form	cubic	rhombic	rhombic
K content, %	52.44	44.89	38.7
K_2O content, %	63.17	54.08	46.6
specific gravity, 20°C	1.988	2.662	2.11
refractive index	1.4904	1.494	1.5056
melting point, °C	790.0	588.0	334.0
hygroscopic point, critical rh at 30°C, %	84.0	96.3	90.5
solubility in water, per 100 parts water			
at 0°C	27.6	7.35	13.3
at 100°C	56.7	24.1	246.0

[a]On the basis of 100% purity.

horizontal layers at depths ranging from a few hundred to several thousand meters. Thickness of the deposits ranges from a few centimeters to a few meters, but deposits less than 1 m thick are usually not considered workable. Both hard-rock shaft mining methods (as for coal) and solution methods (as for Frasch sulfur) are used, depending on depth and thickness of the deposit (see MINERAL RECOVERY AND PROCESSING). Chemically, the deposits vary with location but in general are composed of water-soluble double salts and salt mixtures. The mineral salts commonly encountered are listed in Table 10. These salts are almost invariably encountered in association with sodium chloride. Small but commercial quantities of potassium salts are also recovered from natural brines such as those of the Great Salt Lake in Utah, Searles Lake in California, and the Dead Sea on the Israel–Jordan border. In these operations, solar evaporation in shallow basins is the first step (see CHEMICALS FROM BRINE).

Potassium Chloride. The principal ore encountered in the U.S. and Canadian mines is sylvinite [*12174-64-0*], a mechanical mixture of KCl and NaCl. Three beneficiation methods used for producing fertilizer grades of KCl are thermal dissolution, heavy media separation, and flotation (qv). The choice of method depends on factors such as grade and type of ore, local energy sources, amount of clay present, and local fuel and water availability and costs.

The thermal method is based on the much higher solubility of KCl in hot water as compared to the solubility of NaCl. The KCl is recovered in vacuum crystallizers, filtered or centrifuged, dried, and sometimes granulated by compaction. Product from the thermal beneficiation method usually is of relatively high purity and is particularly suitable for use in formulating solution-type fertilizers. Guaranteed K_2O content of this product is usually 62% K_2O.

The heavy media and flotation processes depend on most potash deposits being mechanical mixtures of salts that can be separated after crushing. The flotation process uses collectors such as long-chain fatty acid amines or acetates and frothing agents such as pine oil in a cell containing saturated liquor through which fine air bubbles are forced. KCl floats to the top of the bath where it is recovered from the foam. Products from heavy media and flotation methods are guaranteed at only 60% K_2O content. Potash producers usually market products

Table 10. Common Potash Minerals[a]

Mineral	Formula	K content,[b] %	K_2O content,[b] %
sylvite	KCl	52.44	63.17
carnalite	$KCl \cdot MgCl_2 \cdot 6H_2O$	14.07	16.95
kainite	$KCl \cdot MgSO_4 \cdot 3H_2O$	15.71	18.92
langbeinite	$K_2SO_4 \cdot 2MgSO_4$	18.85	22.70
leonite	$K_2SO_4 \cdot MgSO_4 \cdot 4H_2O$	21.33	25.69
schoenite	$K_2SO_4 \cdot MgSO_4 \cdot 6H_2O$	19.42	23.39
polyhalite	$K_2SO_4 \cdot MgSO_4 \cdot 2CaSO_4 \cdot 2H_2O$	12.97	15.62

[a]Ref. 67.
[b]Pure basis.

of three particle sizes: regular grade, −12 mesh Tyler (<1.68 mm dia), is usable as feed to granulation plants that produce mixed-grade fertilizers; coarse grade, −8 to +16 mesh Tyler (from 1.19 to 2.38 mm dia), is particularly suitable in the granulation of high potash mixed fertilizers; and granular, −6 +12 mesh Tyler (from 1.68 to 3.36 mm dia), is specifically designed for use in bulk blending (dry mixing) with other granular fertilizer materials to make mixed blends.

Potassium Sulfate. Potassium sulfate is a preferred form of potash for crops that have a low tolerance for chloride. Tobacco and potatoes are two such crops. K_2SO_4 is produced most often from langbeinite by metathetical reaction in aqueous solution:

$$K_2SO_4{\cdot}2\ MgSO_4 + 4\ KCl \rightleftarrows 3\ K_2SO_4 + 2\ MgCl_2$$

Langbeinite, also commercially called Sul-Po-Mag and K-Mag, is also used directly as a fertilizer. Beneficiated langbeinite contains 22% K_2O and 18% MgO. Only about 5% of total fertilizer potash is furnished as potassium sulfate and other nonchloride forms including potassium nitrate.

Potassium Nitrate. Potassium nitrate, known but little used as a fertilizer for many years, may be reclaimed as a by-product of the production of sodium nitrate from natural deposits of caliche in Chile. KNO_3 also has been produced by the double decomposition reaction between sodium nitrate and potassium chloride:

$$NaNO_3 + KCl \rightleftarrows KNO_3 + NaCl$$

The product salts were separated by fractional crystallization. However, the decline of natural saltpeter mining has virtually eliminated these processes as significant sources of KNO_3.

The U.S. domestic commercial potassium nitrate of the 1990s contains 13.9% N, 44.1% K_2O, 0–1.8% Cl, 0.1% acid insoluble, and 0.08% moisture. The material is manufactured by Vicksburg Chemical Co. using a process developed by Southwest Potash Division of AMAX Corp. This process uses highly concentrated nitric acid to catalyze the oxidation of by-product nitrosyl chloride and hydrogen chloride to the more valuable chlorine (68). The much simplified overall reaction is

$$2\ KCl + 4\ HNO_3 \rightleftarrows 2\ KNO_3 + Cl_2 + 2\ NO_2 + 2\ H_2O$$

The NO_2 is absorbed in nitric acid to produce the highly concentrated acid required for the overall process.

Israel Mining Industries produces potassium nitrate by a process in which KCl is converted to KNO_3 in one step at an ambient temperature:

$$KCl + HNO_3 \rightarrow KNO_3 + HCl$$

An organic liquid extractant is used to remove the HCl and to shift the reaction in the desired direction. After concentration, the by-product hydrochloric acid is suitable for commercial use.

Potassium nitrate is being used increasingly on intensive crops such as tomatoes, potatoes, tobacco, leafy vegetables, citrus, and peaches. The properties that make it particularly desirable for these crops are low salt index, nitrate nitrogen, favorable $N:K_2O$ ratio, negligible Cl^- content, and alkaline residual reaction in the soil. The low hygroscopicity of KNO_3 (Table 9) leads to its use in direct application and in mixtures. It is an excellent fertilizer but the high cost of production limits its use to specialty fertilizers.

Potassium Phosphates. Because of the very high analysis of potassium phosphates, eg, KH_2PO_4 is 0–50–33; KPO_3, 0–60–39, and the freedom from chloride and high solubility in liquid fertilizers, methods for economical production of these materials have been sought for many years. These methods have generally been based on the reaction of phosphoric acid or P_2O_5 and KCl, where a variety of potassium phosphates are formed. The inevitable by-product is hydrochloric acid which is often contaminated with fluorine and other impurities. The value and the means for disposal of the by-product HCl have been critical components in the overall economic evaluation of most proposed processes. The savings in shipment costs of high analysis materials, the stability of high potash liquid fertilizers, and the possibility of improved fertilizing quality (Cl-free, low salt index) do, of course, justify some margin for premium cost. The only commercial venture into production in the United States was that by Pennzoil in 1976 (69). That operation has been discontinued. As of this writing no proposed process is in large-scale operation.

Mixed Fertilizers

For most crop–soil combinations, optimum fertilization calls for application of more than one of the three primary nutrients and sometimes for application of secondary or micronutrients. Exact requirements often are determined by soil analysis. Mixed fertilizers are provided to supply single fertilizers that contain these nutrients in the required proportions. Thus the farmer needs only to make a single application. There are, however, both economic and agronomic considerations, that sometimes favor multiple applications of single-nutrient fertilizers over application of a single, multinutrient product. One such example is the use of low cost anhydrous ammonia as the nitrogen application along with supplementary application of a primarily phosphate–potash fertilizer. A similar situation, but agronomically dictated, is in the growing of corn. Side dressing using only nitrogen fertilizer at the proper stage of growth often is profitable to corn growers. Phosphate and potash are usually supplied to the corn crop as a separate, initial application of a mixture.

For the year ended June 30, 1990, about 39% of the total primary nutrient used in the United States was applied in mixtures, whereas the remaining 61% was applied by direct application. Breakdown by plant nutrient is shown in Table 11. High usage of anhydrous ammonia and nitrogen solutions (Fig. 3) accounts for the large proportion of nitrogen applied by direct application. Because plant nutrient requirements are quite variable, depending on soils, crop varieties, locality, and previous fertilization history, producers of mixed fertilizers are faced with producing many grades (nutrient ratios) of product. In the year ending June

Table 11. Direct Application of Single-Nutrient Fertilizers vs Application in Mixtures

	Method, % applied	
Nutrient	Direct	Mixtures[a]
nitrogen	80	20
phosphate	8	92[b]
potash	65	35
Total	*61*	*39*

[a]Includes bulk blends.
[b]Includes ammonium phosphates applied alone.

30, 1990, a total of 21,358 different grades of mixed fertilizers were registered for sale in the United States (5).

Production of mixed fertilizers involves, in most cases, a bringing together of some combination of the basic nitrogen, phosphate, and potash fertilizers previously described. This production of mixtures, therefore, results in additional expense, over and above that incurred in producing the feed materials. The added convenience must be worth this added expense. Because of the need for a variety of locally tailored grades, mixed fertilizer production plants tend to be regionally located and to serve only local or regional markets. The ultimate in this respect are the bulk-blending plants and fluid fertilizer mixing plants which often serve farms only in a radius of 40 to 50 km and routinely prepare mixtures on a prescription basis.

Nongranular Mixtures. Prior to about 1945, all commercial mixed fertilizers were produced in powdered, ie, nongranular, form. The phosphate source almost always was nongranular normal superphosphate. The principal nitrogen source was crystalline ammonium sulfate, usually from coke-oven ammonia. The primary potash source was small-crystal potassium chloride, potassium sulfate, or potassium–magnesium sulfate. Production plants were local or regional in nature and the mixed fertilizers were bagged for distribution. The superphosphate was produced locally from sulfuric acid that was also often produced locally, usually by the now obsolete chamber process. Plants were located on railroad sidings to facilitate import of phosphate rock, elemental sulfur (for acid production), ammonium sulfate, and potash, and to facilitate shipment of the bagged mixtures.

Production facilities included the acid plant and nongranular superphosphate production and curing facilities. The mixing (blending) operation was usually carried out batchwise in a horizontal-axis, rotary, drum-type mixer of 1- to 5-ton capacity. A common practice was to meter aqua ammonia or ammonia–ammonium nitrate solution into the mixer to react with the superphosphate during the mixing operation, and to thus provide a relatively low cost nitrogen addition. Such ammoniation also lowered acidity of the product, improved physical condition, and reduced deterioration of bags. Other ingredients often were included during the mixing operation. Fish meal, packing house waste, and dried blood were materials sometimes included for their nitrogen content. Materials such as tobacco stems, cotton seed hulls, ground corn cobbs, vermiculite, and diatomaceous earths sometimes were added to improve storage and handling properties. Undesirable characteristics of these nongranular mixed fertilizers were the dustiness and caking tendency, which interfered with handling and application, and low analysis, which resulted in high handling costs. In the United States, production of such nongranular mixtures declined and essentially ceased during the late 1940s and early 1950s, in favor of the production of granulated mixtures. In a few parts of the world, however, production of nongranular mixtures by this method persists.

Granulation of Mixed Fertilizers. A granular fertilizer, in contrast to a pulverized or nongranular one, is a fertilizer in which the individual particles are relatively large and are restricted to closely specified upper and lower size limits. Generally accepted limits in the United States and Canada are 6 Tyler mesh (3.35 mm dia) as the upper limit and 16 Tyler mesh (1.19 mm dia) as the lower limit. A well-sized granular product has at least 90% of its particles within this range. In European countries, the preferred upper limit is somewhat higher, usually 4 mesh (4.75 mm dia). The prilled ammonium nitrate and urea shown in Figure 2, although produced by prilling, usually fall within the size specifications for granular fertilizers. Essentially all solid fertilizers intended for field application in the United States or Europe are produced in the granular or prilled form. Commercially, the switch from the earlier, pulverized, nongranular form began in the late 1940s and was essentially complete by 1960. Two important factors that fostered the development and adoption of granulation processes are (7) (*1*) the formulations of the pulverized mixed fertilizers were being modified to increase plant-food concentration and thus to minimize costs of transportation, bagging, storage, and handling, but the resultant, more concentrated nongranular products were presenting increased caking and handling problems; and (*2*) the nongranular products behaved poorly in mechanical applicators, which came into widespread use as a result of increased fertilizer use and high labor costs. Granulation promised to alleviate both of these problems.

Steam Granulation. The development of granulation methods for mixed fertilizers took somewhat divergent paths in the United States and Europe. U.S. granulation methods usually involve considerable chemical reaction among feed materials in the granulator as a method of developing heat and providing the plasticity required for granulation. With superphosphate as one of the feed materials, reactions during granulation typically result in some loss of P_2O_5 water solubility, but not of citrate solubility which is the basis for fertilizer sale in the United States.

In Europe, where fertilizer sale often is based on water solubility of P_2O_5, granulation methods were developed that largely avoid chemical reaction in the granulator. In such procedures, the feed materials are finely ground fertilizer materials such as ammonium sulfate, ammonium nitrate, urea, superphosphates, ammonium phosphates, and potash salts. Plasticity and agglomeration are promoted by the addition of water and steam while the mass is being subjected to vigorous rolling action in one of several types of granulation equipment. Granulator types used successfully include inclined rotating pans, rotary drums, and pug mills or pin mills. Because steam is essential to provide both heat and moisture, this general type of granulation is commonly referred to as steam granulation. Detailed discussion of the steam granulation process is available (70).

Plasticity, and hence granulation efficiency, varies considerably with the nature and proportion of feed materials. Pure salts, such as potassium chloride and ammonium sulfate, lend little or no plasticity and thus are difficult to granulate. Superphosphates provide good plasticity. The plasticity of ammonium phosphates depends chiefly on the impurity content of iron and aluminum. The higher the impurity the greater the plasticity. In some cases, binders such as clay are added to provide plasticity.

A flow sheet of a typical steam granulation plant is shown in Figure 16. Initial steps are crushing, screening, proportioning, and blending of the dry fertilizer ingredients. The dry mixture then is fed continuously to a rotating, cylindrical, nonflighted drum granulator. Water is metered into the drum, usually through a spray. Steam, to provide heating, is metered in through a sparger submerged beneath the rolling bed of material. Discharge from the granulator drum consists of soft granules of varying size together with some ungranulated feed material. The next step is drying by contact with a cocurrent flow of heated air in a rotary drum dryer. The dryer discharge then is sized by screening to recover the desired product size material, which then is sent to a rotary drum cooler. The flow sheet also shows a small drum in which the cooled granules are coated with clay, chalk, or diatomaceous earth to reduce caking tendency. Under the best of conditions, up to 80% of the dryer discharge is of the desired product size. Oversize material is crushed and, together with undersize, is recycled to the granulation drum.

Steam granulation is practiced in Europe, Australia, and elsewhere, chiefly in small plants in which superphosphate, either ordinary or triple, is a primary ingredient. However, for many of the larger operations, superphosphates have been replaced by ammonium phosphates as the principal P_2O_5 source, and granulation procedures involving chemical reactions are employed in Europe as well as in the United States.

Slurry Granulation. Slurry granulation is granulation in which at least one of the primary feed ingredients to the granulator is a slurry. As developed and employed in the United States, slurry granulation of mixed fertilizer involves preparation first of an ammonium phosphate slurry of closely controlled composition by the reaction of wet-process phosphoric acid with ammonia in a vessel known as a preneutralizer. This slurry then is fed to a granulator along with other solid, liquid, or gaseous, ie, NH_3, fertilizer materials.

Development of slurry granulation in the United States was preceded by nonslurry granulation procedures in which the chemical reactions were carried out entirely within the granulator. The pioneer development in this regard was the introduction in 1953 of the TVA continuous ammoniator–granulator (71,72). A sketch of the pilot-plant ammoniator–granulator is given in Figure 17. Solid feed ingredients, usually including ordinary or triple superphosphate, were fed into one end of the rotating drum to form a rolling bed of considerable depth. A submerged, perforated sparger was used to introduce ammonia or ammoniating solution which reacted with the superphosphate. This reaction developed heat and plasticity, and thus promoted granulation. It was soon found that introduction of sulfuric and/or phosphoric acids through additional submerged spargers, as shown, enhanced granulation and was an economically beneficial method of providing nitrogen and phosphorus. With the inclusion of phosphoric acid in this manner, as a partial replacement for superphosphate, the product grade was significantly increased, providing additional benefit. This type of granulation was widely accepted in the United States, and by 1962 there were about 250 granu-

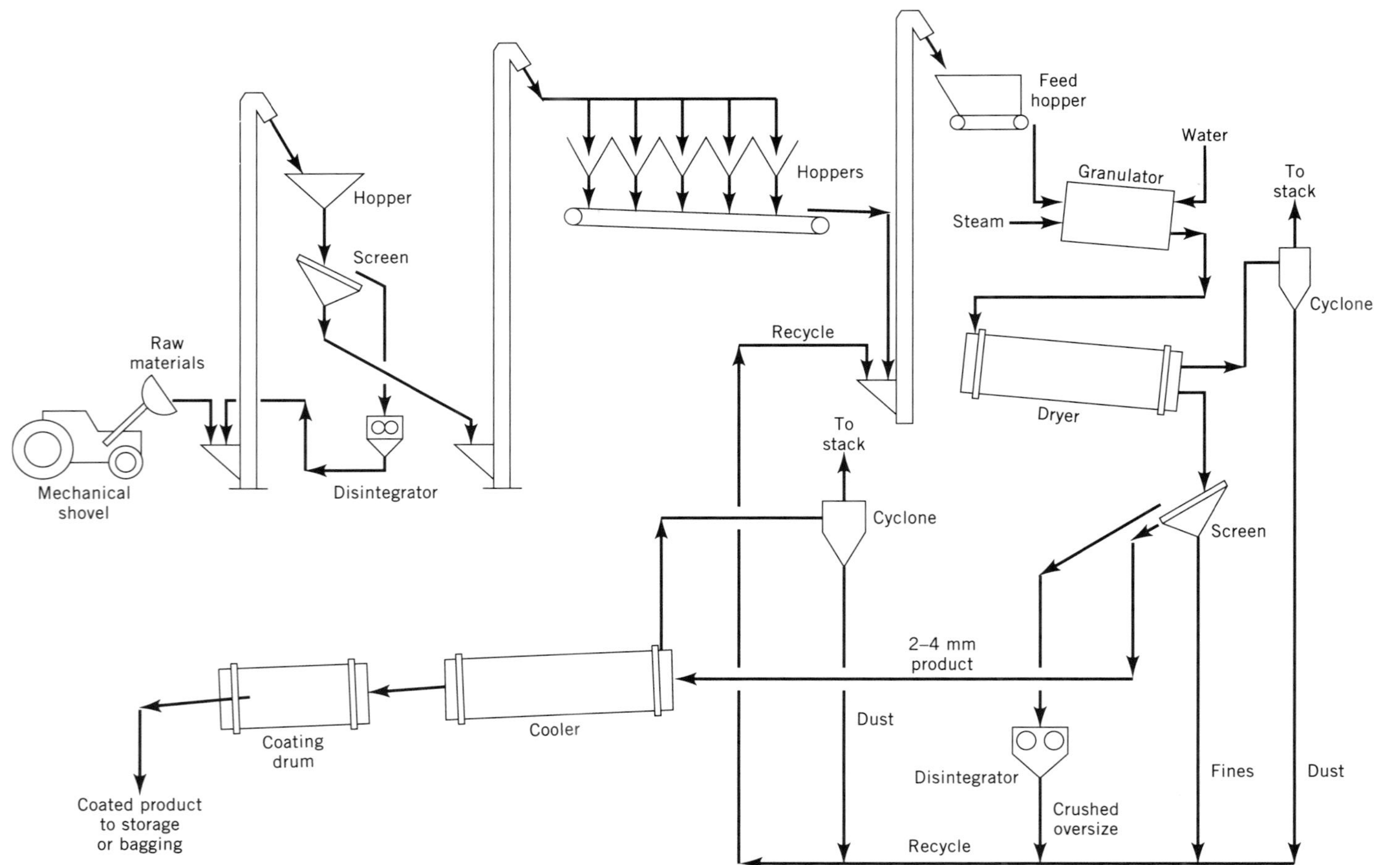

Fig. 16. Steam granulation process for production of granular mixed fertilizers from dry, pulverized feed materials (7). A granulator producing 12 t/h would have a dia = 2 m, be 6 m long, and would rotate at 9–12 rpm.

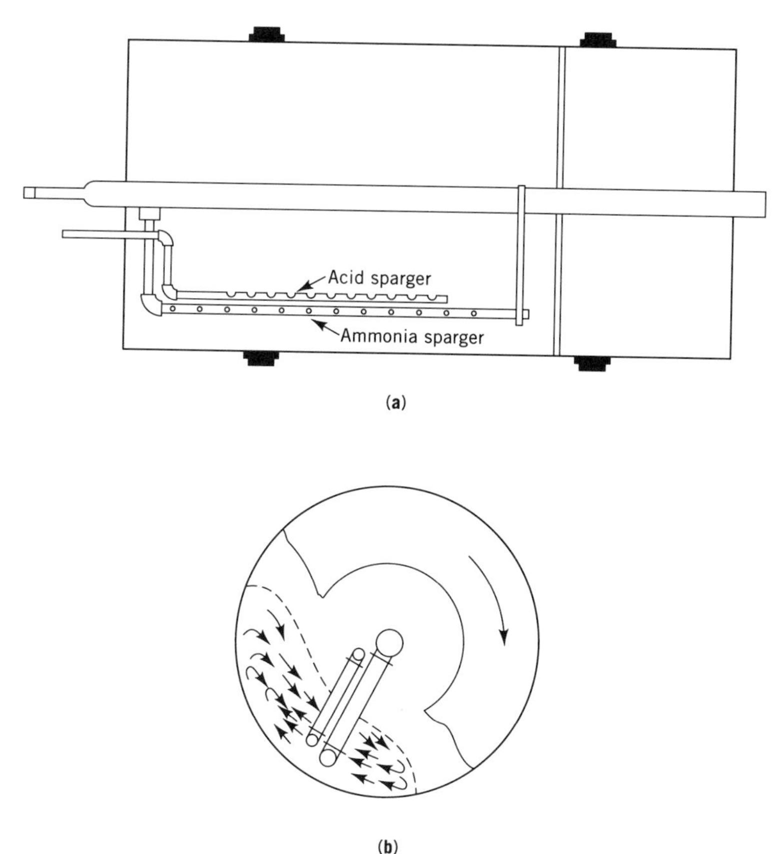

Fig. 17. Pilot-plant ammoniator–granulator. (**a**) Side view, showing placement of acid and ammonia spargers, and (**b**) feed end view (71,72).

lation plants operating in the country, each with an estimated production capacity of 100,000 to 300,000 tons of product per year.

Modification of the TVA granulation process by the addition of a preneutralizer was first demonstrated in 1962 (73). With this modification, all or part of the phosphoric acid feed is fed to a tank-type reaction vessel (preneutralizer) (Fig. 13) into which ammonia is sparged to effect a closely controlled partial neutralization. A flow sheet of a typical slurry granulation plant employing a preneutralizer is shown in Figure 18. The ammonium phosphate slurry is sparged into the ammoniator–granulator drum, where neutralization with ammonia is completed. The great benefit of the preneutralizer addition was the ability to feed a higher proportion of phosphoric acid to the process than could be tolerated with acid feed directly to the granulator; the higher proportion of acid increases the grade of products significantly. This tolerance of higher acid feed rates is a result of the heat-dissipation and water-evaporation capabilities of the preneutralizer. Such use of a preneutralizer in the granulation of mixed fertilizer was a precursor to later development of the diammonium phosphate granulation process described earlier. It was in connection with mixed fertilizer granulation that the benefit of

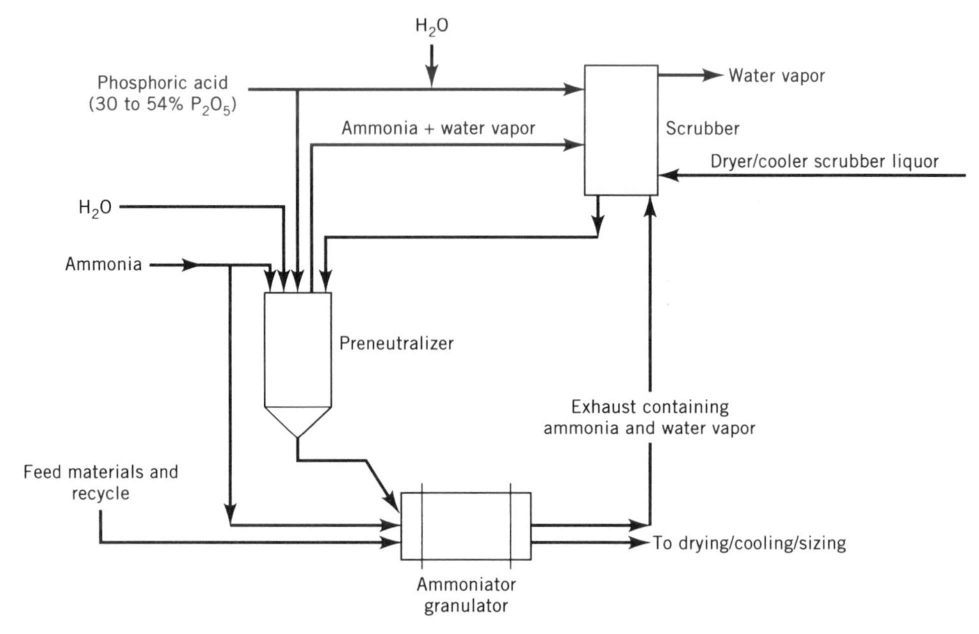

Fig. 18. TVA-type cogranulation process with preneutralizer, as used for production of granular mixed fertilizers. Feed materials such as ammonium sulfate, ammonium nitrate, urea, superphosphates, sulfuric acid, and potash are used.

maintaining the preneutralizer composition at the high solubility point of 1.4 moles of ammonia per mole of phosphoric acid was discovered. Additional ammoniation of the slurry in the granulator then decreases solubility and thus promotes solidification and granulation. By 1965 preneutralization had been adopted by essentially all the then existing U.S. granulation plants.

As of the end of 1988, there were only 50 mixed fertilizer granulation plants operating in the United States. Some of these employed the slurry granulation process described, although there has been considerable conversion to melt granulation using a pipe reactor. The decline in the number of U.S. regional granulation plants from the high of 250 in 1962 is a direct result of competition from the bulk blending method of mixed fertilizer production and from fluid mixed fertilizers.

Melt Granulation. Melt granulation is similar to slurry granulation except that feed of slurry to the granulator is replaced by feed of a hot, concentrated, almost anhydrous melt. The hot melt provides the plasticity required for granulation of the solid feeds, and hardening of the granules occurs as the melt cools in the granulator and in a subsequent cooler. An advantage of melt granulation is elimination of the need for a fuel-fired product dryer, resulting in savings in both investment and fuel costs. In most cases, contacting the hot product with ambient air in a rotary-drum or fluid-bed cooler is sufficient to both cool the product and remove all but tolerable remnants of moisture. In application of melt granulation to production of mixed fertilizers, the melt used is usually ammonium phosphate or ammonium phosphate sulfate.

The melt is prepared in a simple pipe reactor or pipe cross reactor as shown in Figure 19 (74–78). A typical pipe reactor for production of 16 tons of melt per hour is 15 cm in diameter by 3 m long. A number of variables dictate optimum pipe size. The open discharge end of the pipe is positioned in the granulator over the rolling bed of solids, usually with a downturned elbow as shown. Phosphoric acid and anhydrous ammonia are fed to the closed, external end of the pipe. In the case of the pipe cross reactor, sulfuric acid also is fed at that point. Conservation of the heat of the acid–ammonia reaction in this confined volume is such that high temperatures are developed and most of the water is evaporated. A frothy mixture of melt and water vapor discharges into the granulator, where a

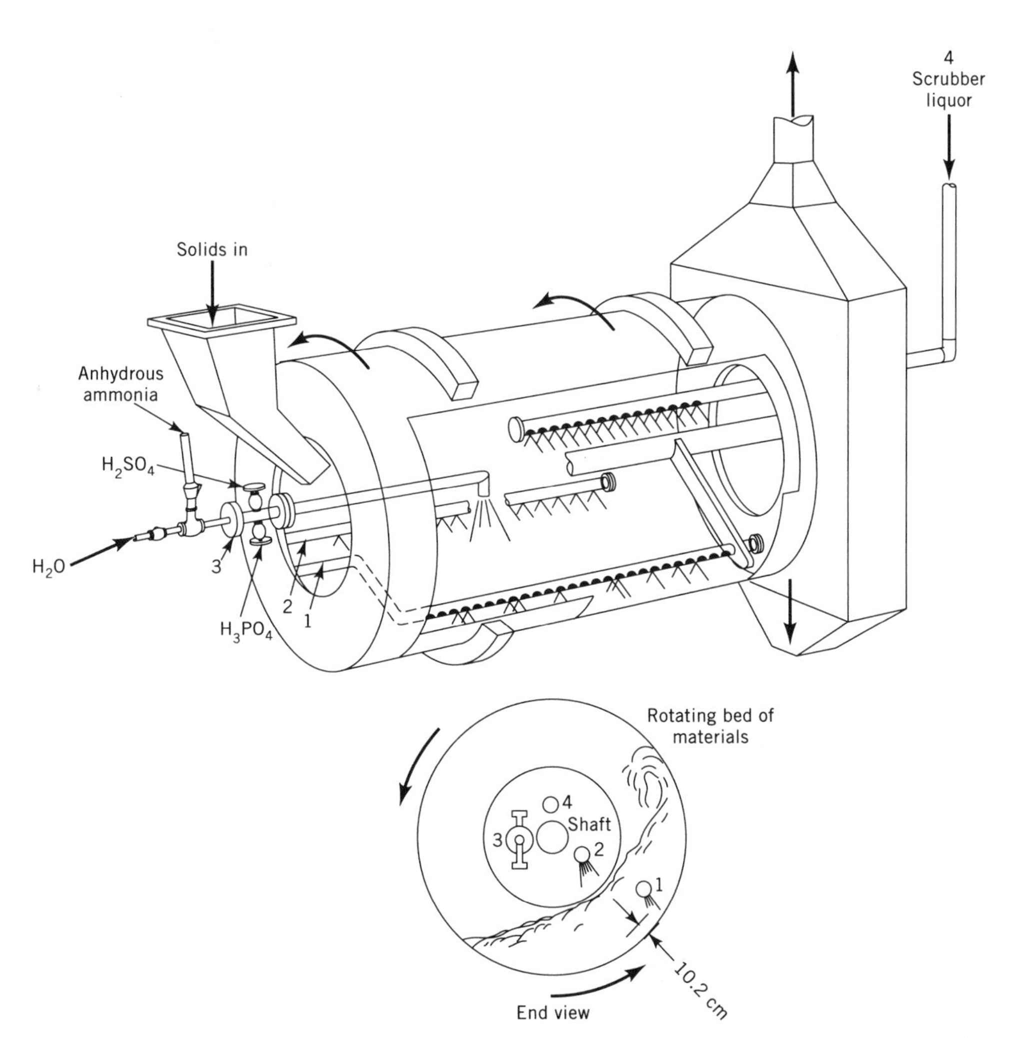

Fig. 19. TVA-type ammoniator–granulator incorporating a pipe cross reactor. 1, ammonia sparger, located at the 4 o'clock position 11.4 cm from granulation shell with holes facing the rotating stream; 2, phosphoric acid sparger, located to discharge phosphoric acid onto the top and near the center of the rotating bed of materials; 3, pipe cross reactor; 4, scrubber liquor distributor, located above the bed in granulator to dribble scrubber liquor onto bed.

flow of air is used to sweep away the water vapor. The temperature and water content of the melt are dependent on concentration of feed acid, temperature of feed acid, temperature of ammonia, and degree of ammoniation. Also, the proportion of sulfuric acid used has a large effect because of the relatively high heat release by the sulfuric acid–ammonia reaction.

Various modifications of the pipe and pipe cross-granulation system are coming into use in the United States and throughout the world. A variety of mixed grades are being made. In some modifications the pipe is operated at sufficiently high temperature to convert part of the phosphate to polyphosphate form (79). The polyphosphates are, by nature, notably viscous, which enhances granulation and also results in denser, harder granules having excellent storage properties (74). Corrosion of the pipe reactors is a critical factor. Both stainless steel and Hastelloy pipes are used, but selection of proper material is dependent on projected operating conditions.

In a melt granulation process employing a granulator of the type shown in Figure 19, product sizing (screening) and recycling of oversize and undersize are the same as for other granulation processes (Fig. 12). However, the flow sheet for a melt granulation process is simplified by omission of a preneutralizer and a dryer. Of the 50 granulation plants remaining in the United States at least 15 have at present completed adoption of pipe or pipe cross reactors.

Compaction Granulation. Compaction granulation is based on the principle that many solid materials are semiplastic and deform under high pressure. In compaction granulation of mixed fertilizers a powdered feed mixture containing at least one such plastic material is subjected to high pressure between counter-rotating steel rolls to form a dense, stable sheet of mixture. This sheet then is crushed and sized to yield granule-size chips. These chips, although irregular in shape, are generally acceptable as granular fertilizer because, in comparison to pulverized materials, they exhibit the favorable characteristics of improved flowability, nondustiness, and low caking tendency. Compaction is widely used by the basic potash producers to produce granular size potassium chloride from finer material. It is satisfactory for this purpose because crystalline potassium chloride is relatively plastic and thus requires only moderately high (6 t force per cm of active roll width) roll pressure for compaction. To a limited extent, compaction has been used also to produce granular ammonium sulfate from fine-crystal by-product material. This product is satisfactory, but, because of limited crystal plasticity of ammonium sulfate, very high (13 t force per centimeter of active roll width) roll pressures have been required, and roll wear has been a serious problem. Some applications of compaction to granulation of mixed fertilizers have been successful, mostly at locations outside the United States (80–83). In such applications, selection and proportioning of ingredients is important to provide sufficient plasticity. Materials, in addition to potassium chloride, that exhibit significant plasticity include ammonium phosphates and superphosphates. The method apparently is most applicable to relatively small volume production of specialty mixtures and is likely to find its future use mainly for such applications.

Bulk Blending. Over half of all the mixed fertilizer made and distributed in the United States is in the form of bulk blends. Bulk blends are simple dry mixtures of two or more previously prepared granular size fertilizer products. Because the starting materials are finished fertilizers of granular size, the resultant bulk blends have all the favorable handling properties characteristic of granular products. Preparation is a simple process of dry blending and can be carried out in relatively small, inexpensive plants that serve only local customers. The principal requirements are facilities for receiving and storing granular fertilizers from basic producers; equipment for proportioning and mixing; and means for moving the blends (dry mixes) to local farmers. Because of the local nature of bulk blending plants, the blends are almost always custom prepared in relatively small batches to meet individual farmer requests, and are moved to the farm and applied to the field without any need for bagging. Custom field application is a service usually provided, for a charge, by the bulk blender using the blender's own application equipment. Bulk blending was a natural outcome after nitrogen, phosphate, and potash fertilizers became available in granular size in the early 1950s.

Advantages of the bulk blending system are (*1*) the basic granular fertilizer materials (blend ingredients) can be made in large efficient plants at economically favorable locations. These products can be shipped to blenders in bulk without the expense of bagging; (*2*) local blend plants can custom blend to meet the specific nutrient ratio requirements of local farmers; (*3*) micronutrients, herbicides, and insecticides can be added as needed; and (*4*) bulk handling and field application eliminates the expense of bagging and bag handling.

Since its introduction in the 1950s, bulk blending has shown continued growth in the United States at the expense of the cogranulation (ammoniation–granulation) methods of mixed fertilizer production as shown in Figure 20. From 1960 to 1990 bulk blending's share of the mixed fertilizer market increased from 20 to about 50%, whereas that of cogranulated mixes decreased from 80 to less than 30%.

The bulk blending system has provided competition to the mixed fertilizer cogranulation plants because the complexity and high equipment and labor costs of the latter make only relatively large, high investment plants feasible. Annual output of about 100,000 ton per plant is about the minimum economically feasible. Marketing of the large output from such a plant involves shipment over relatively large distances and often handling (and profit taking) by an intermediate near

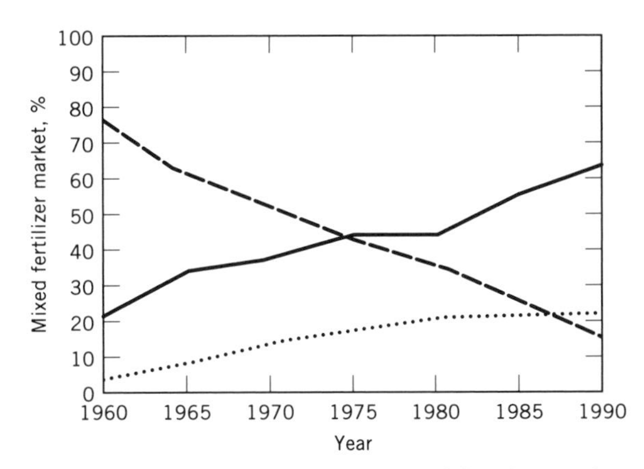

Fig. 20. U.S. market share for several types of mixed fertilizer, where (——) represents bulk blends; (— — —) ammoniation–granulation; and (····) fluids.

the point of use. Such handling often is feasible only in bags, which adds another expense. Finally, cogranulation is quite inflexible on a cost basis in regard to custom formulating to meet varying farmer requirements. Changing formulation in a cogranulation plant is a rather difficult operation and can be done economically only for relatively large production runs. Such a formula change also introduces complications in storage, labeling, and marketing of multiple grades. Figure 20 shows also that growth in the production of fluid mixed fertilizers has about paralleled that of bulk blends, although at a lower level. Fluid fertilizer plants also are usually small local units having many of the same advantages as bulk blend plants.

Plant Design and Operation. There are over 5000 bulk blending plants in the United States, each located in a farming area and serving farmers within a radius that seldom exceeds 40 km. The average annual production per plant is only 5000 tons, and more than half the plants produce only 1000 to 4000 tons (84). Designs of the plants vary considerably in regard to storage facilities and materials handling, proportioning, and mixing equipment. A popular design, and one that illustrates the operations common to all plants, is shown in Figure 21.

Almost invariably, bulk blend plants are located on a railroad siding to facilitate bulk receipt of granular fertilizer materials. A separate storage bay is provided for each material. Relative use of various materials is shown in Table 12 (84). Urea, ammonium nitrate, and ammonium sulfate are the most popular nitrogen sources, in that order. Granular diammonium phosphate is a mainstay phosphorus source used by 95% of all blenders in at least some blends. Granular

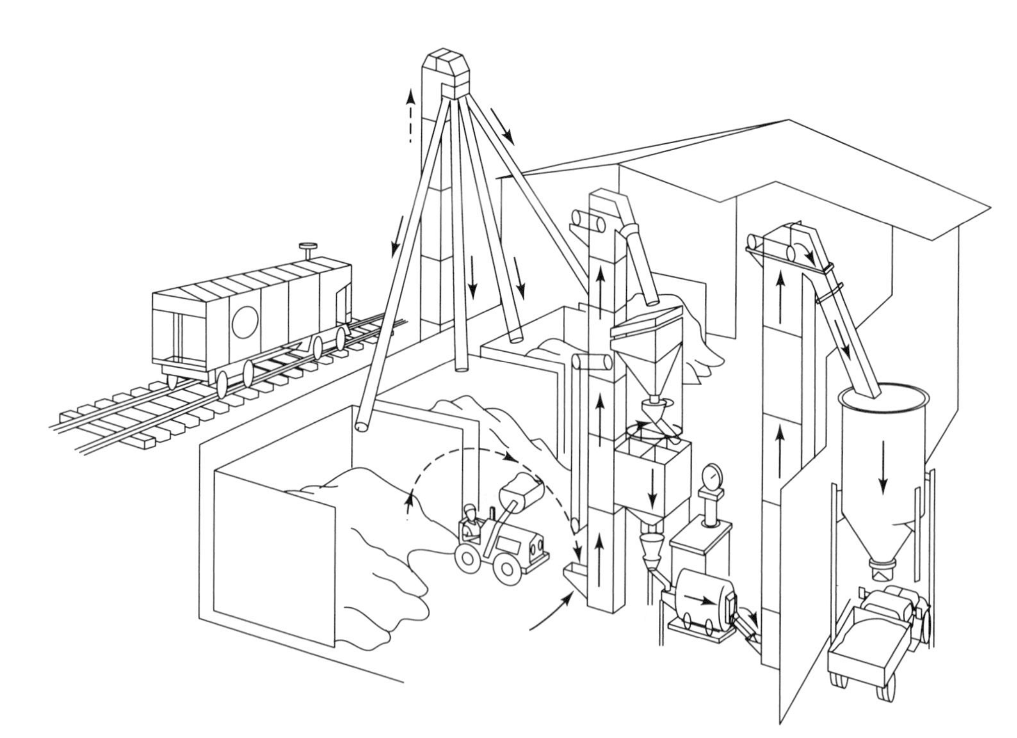

Fig. 21. Layout of a small bulk blending plant.

Table 12. Fertilizer Materials Used in Bulk Blends[a]

Compound	Typical grade, N–P_2O_5–K_2O	Plants using material, %
ammonium nitrate	34–0–0	41
urea	46–0–0	66
ammonium sulfate	21–0–0	22
diammonium phosphate (DAP)	18–46–0	95
monoammonium phosphate (MAP)	11–52–0	11
triple superphosphate (TSP)	0–46–0	78
normal superphosphate	0–20–0	4
potassium chloride	0–0–60	94

[a]Ref. 84.

triple superphosphate, being free of nitrogen, is an important source of phosphate for production of no-nitrogen blends. The chief source of potash in blends is granular potassium chloride. Granular potassium sulfate and potassium–magnesium sulfate are available for use on crops, such as tobacco, that cannot tolerate chloride. Typical formulations for some bulk blends are shown in Table 13.

Movement of blend ingredients from the storage bays to the proportioning–mixing equipment is usually by means of a single front-end loader. Various proportioning (weighing) systems are used. Proportioning is either into a holding bin or directly into the mixer. Mixers are almost always of the batch-type and batch capacities are in the range of 2 to 8 tons. The most popular types are the horizontal-drum, flighted type and the inclined, tapered-drum (cement mixer) type. Intensive mixing is not desirable because it tends to abrade granules and to promote segregation. Batch mixing times of only one to two minutes usually are optimum for good blending. Blend discharged from the mixer is elevated either directly into a spreader truck for immediate dispatch to the farm or into a holding bin for later discharge into a truck (Fig. 21).

The flexibility of the bulk blending system and the close relationship with the farmer allow the bulk blender to provide a number of valuable supplementary

Table 13. Formulations for Bulk Blends

Blend ingredient	Typical grade	Ingredient for indicated grade of blend, kg/t				
		17–17–17	19–19–19	9–23–30	8–32–16	0–26–26
ammonium nitrate	34–0–0	310				
urea	46–0–0		256			
DAP	18–46–0	376	421	500	462[a]	
TSP	0–46–0				261[a]	566
potassium chloride	0–0–60	288	323	500	277	434
filler (granular limestone)	0–0–0	26				

[a]Chemical reaction of DAP with TSP is possible if stored. See text.

services, such as adding herbicides, insecticides, micronutrients, or seeds to the blends; bagging blends; liming; and sampling soil. Consultation services and custom application can also be provided as can sale of anhydrous ammonia or nitrogen solution.

Problems in Bulk Blending. Technical problems are relatively few in the bulk blending process. Some precautions must be exerted to protect hygroscopic starting materials, such as ammonium nitrate and urea, from excessive humid exposure. Covering bins with plastic sheet is sometimes used if storage time is to be very long. There are, however, several material combinations that must be avoided in blends because of chemical incompatibility. In particular, urea must not be mixed with ammonium nitrate because the combination is extremely hygroscopic and spontaneously becomes wet. Another combination that usually is avoided is urea and superphosphate, because the mixture tends to react and release water of hydration, which causes wetting (85). Another problem that can become quite serious is particle size incompatibility of some blend ingredients. Bulk blends, like all fertilizers, are subject in the United States to check analyses by state regulators. In this connection, bulk blenders generally have experienced higher rates of analysis deviation, and hence more frequent penalties, than producers of, for example, cogranulated products. Studies have shown that the problem with such blends almost always is segregation of blend ingredients during handling, resulting from particle-size mismatch of the ingredients. Rather close matching of particle size is required to avoid segregation susceptability, whereas rather wide deviations in particle density and shape are not troublesome (86,87). Providing blend ingredients of a predetermined particle size for blending becomes the responsibility of the basic suppliers of these materials. Although there are no regulations in the United States specifying acceptable particle size, it is generally recognized among large producers that a material for blending should be sized within the limits of 6- to 16-mesh Tyler screens (3.3 to 1.0 mm particle diameter) and that 25 to 45% of the particles should be larger than 8 mesh (2.4 mm particle dia) (88). The government of Canada, to encourage the production of size-compatible blend materials, has established a relatively simple size matching index (Size Guide Number, SGN) for use by basic producers and blenders but, as of this writing, has not established it as a legal specification (89).

Bulk Blending Outside the United States. Outside the United States, bulk blending has found varying degrees of acceptance, depending on local conditions. In such places as Canada, Brazil, and Australia, which have vast agricultural areas similar to those in the United States, bulk blending is very successful. In Europe, where the farming areas are smaller and agronomic requirements are less varied, cogranulation in relatively large, centrally located plants is quite competitive. For some small countries and developing countries that are dependent on imported fertilizer materials, bulk blending at the port of entry plus bagging for distribution can be an effective system. Overall, it seems that bulk blending is likely to at least maintain its present large market share in the United States and probably increase its share elsewhere.

Fluid Mixtures. About one-third of all the nitrogen fertilizer used in the United States is applied as the fluid anhydrous ammonia, and an additional 20% is applied in the form of urea–ammonium nitrate solutions. Since the late 1950s, there has also been, as indicated in Figure 20, a steady increase in the amount of mixed fertilizer applied in fluid form. At present, about 20% of all mixed fertilizer applied in the United States is in fluid form, as compared to about 60% as bulk blends and about 20% as cogranulated solids. Of the fluid mixed fertilizers applied, about 60% are true solutions, whereas the remainder are suspensions. Fluid fertilizer plants, like bulk blending ones, are usually small, local enterprises serving local farmers. This is an understandable necessity because of the high water content of fluid fertilizers and the economic infeasibility of transporting products of high water content over long distances. These local plants enjoy some of the same beneficial characteristics as bulk blend plants, in that the products are tailored to local needs and can be delivered and applied in bulk. The estimated number of such plants in the United States as of 1993 is 3000.

Solution-Type Mixtures. The commercial production of fluid mixed fertilizers for farm use began in the early 1950s. The mixtures first made and marketed were solutions in which all the plant nutrient content is in solution. At that time, a somewhat erratic supply of pure electric-furnace orthophosphoric acid was commercially available, and that acid was used as the phosphate source. The acid was ammoniated to a grade of 8–24–0, and this ammonium phosphate base solution was shipped to mixing plants in farm areas. In the chemical system NH_3–H_3PO_4–H_2O, the 8–24–0 grade represents the solution of highest plant food content ($N + P_2O_5$) that does not salt out at −8°C. At the mixing plants, this base solution usually was mixed with urea–ammonium nitrate solution of 28 to 32% nitrogen content, and crystalline potassium chloride was dissolved in the mixtures to provide K_2O content. Some maximum mix grades feasible by this method were 13–13–0, 7–7–7, and 7–14–7. By 1960, an estimated 335 mixing plants were distributing such solution mixtures, largely in the mid and western United States. Whereas ease of application and accuracy of placement were particularly appreciated, there were several serious problems with the production system. One problem was the relatively low concentration (grade) of both the 8–24–0 base solution and the final fluid mixes. Another was the high cost and limited availability of electric-furnace phosphoric acid. Use of the cheaper wet-process orthophosphoric acid was out of the question because, on ammoniation of such acid, iron, aluminum, magnesium, and other impurities form voluminous, gelatinous, and crystalline precipitates that cannot be handled.

A breakthrough came in 1957 with the introduction by TVA of superphosphoric acid and of base solutions made by ammoniation of that acid (90,91). This superphosphoric acid, which at first was made by the electric-furnace process, contained 76% P_2O_5 as compared to 54% P_2O_5 in commercial orthophosphoric acid. At this higher P_2O_5 concentration, equilibrium in the system H_2O–P_2O_5 dictates the presence of about 50% of the P_2O_5 as nonorthopolyphosphoric acids such as pyrophosphoric acids, tripolyphoric, and tetrapolyphosphoric (92). Because of higher solubility of the ammonium salts of these polyphosphoric acids, ammoniation of superphosphoric acid allowed production of higher (10–34–0 vs 8–24–0) grade base solution and higher grade mixed fertilizer solutions. Figure 22 shows the effects of several polyphosphate levels and $N{:}P_2O_5$ ratio on the total plant food ($N + P_2O_5$) solubility in base solutions.

Although the introduction of superphosphoric acid was an effective solution to the problem of low analysis liquids, there was still the problem of the high cost of the electric-furnace acid process. In this connection, intensive studies were

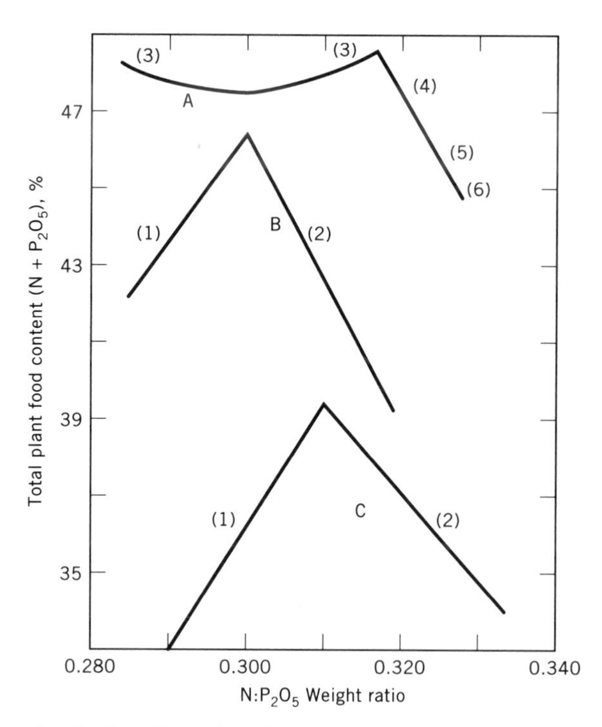

Fig. 22. Effects of polyphosphate level and $N:P_2O_5$ weight ratio on solubility of ammoniated phosphoric acids at 0°C, where A represents 70% of total P_2O_5 as polyphosphate; B, 45%; and C, 0%, and the various crystallizing phases are (1), $(NH_4)H_2PO_4$; (2), $(NH_4)_2HPO_4$; (3), $(NH_4)_3HP_2O_7{\cdot}H_2O$; (4), $(NH_4)_5P_3O_{10}{\cdot}2H_2O$; (5), $(NH_4)_2HPO_4{\cdot}2H_2O$; and (6), $(NH_4)_4P_2O_7{\cdot}H_2O$.

undertaken to produce usable superphosphoric acid by concentration of the cheaper merchant-grade (54% P_2O_5) wet-process acid. It was known that the polyphosphates are effective in sequestering, ie, preventing precipitation, of many metallic ions, thus it was hoped that concentration of wet-process acid not only would provide high concentration but also would prevent objectionable precipitation of the impurities that are characteristic of wet-process acid. Results of the concentration studies were successful, and usable acid of about 72% P_2O_5 content (polyphosphate level of 45 to 50%) was produced. Results were best with acids made from phosphate ores that had been calcined to remove organic content. Difficulties with the wet-process superacid of this concentration were its high viscosity and the poor storage properties of 10–34–0 base solution made from it. However, in 1973 the introduction of the now universally used pipe reactor process for production of high polyphosphate (70 + % polyphosphate level) 10–34–0 grade base solution from wet-process superphosphoric acid effectively solved both of these problems (93,94). By this process, the high polyphosphate 10–34–0 can be made from low conversion superphosphoric acid of only 68 to 70% P_2O_5 content with a polyphosphate level of about 30%. This acid has lower viscosity and better handling properties than the previously used, more concentrated acid. Storage properties of the 10–34–0 base solution made by the pipe reactor process were much improved because of its higher polyphosphate content. Typical grades of solution-type mixtures made from this base are 21–7–0, 7–21–7, and 8–8–8.

Figure 23 is a sketch of the typical equipment used in a pipe reactor plant for production of high polyphosphate 10–34–0 grade base solution from low conversion wet-process superphosphoric acid. The development of the high polyphosphate level in this system results from the conservation of the heat of reaction in the pipe, which results in a reaction temperature of 345°C. Rapid cooling of the product 10–34–0 also is essential to arrest hydrolysis of polyphosphates and thus to preserve the high polyphosphate level. There were about 150 such plants located throughout the United States in 1993. These were supplied with low conversion acid chiefly from one producer in North Carolina and one in Idaho. All solution-type mixed fertilizers for farm use are made using high poly 10–34–0,

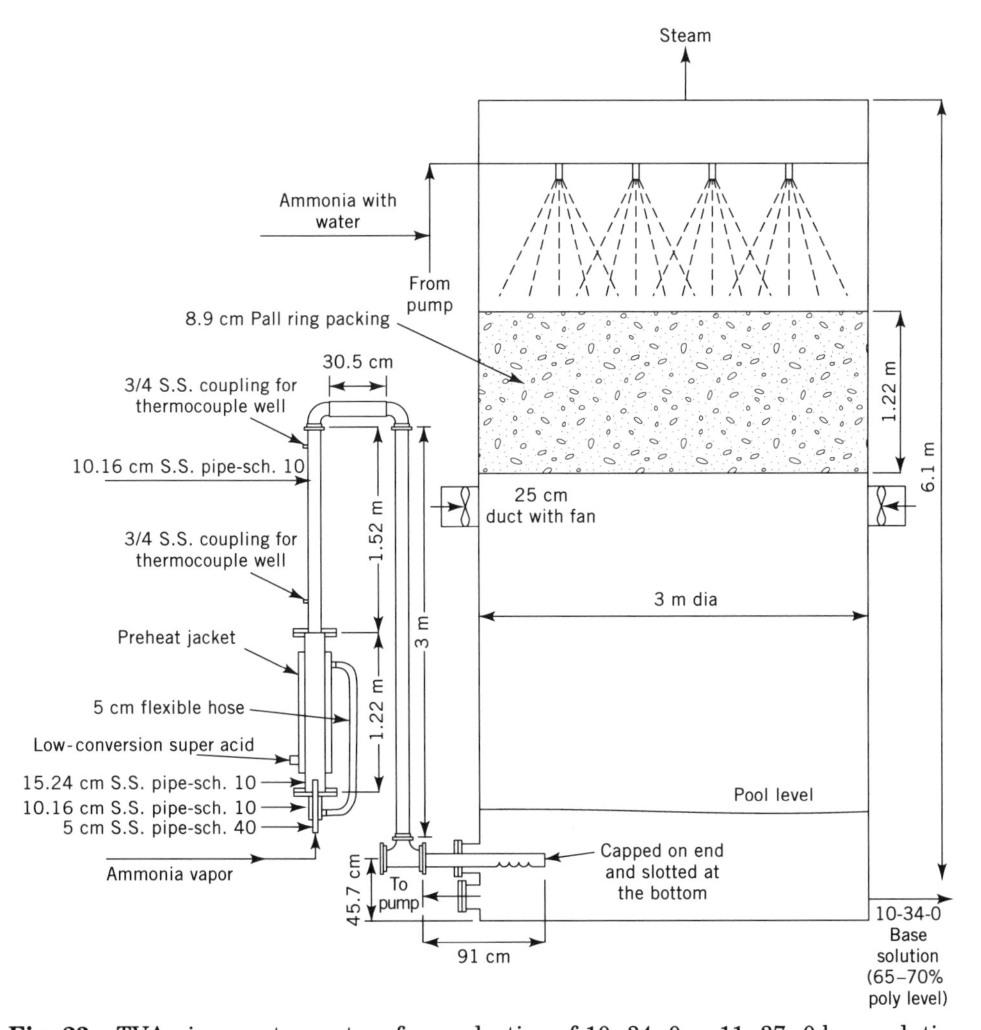

Fig. 23. TVA pipe reactor system for production of 10–34–0 or 11–37–0 base solution from low conversion superphosphoric acid.

or a similar product of 11–37–0 grade, as the source of phosphate. There is also a considerable amount of these two solutions applied directly to the soil, particularly in the U.S. wheat belt and other areas where potash fertilization often can be omitted. Also, these high polyphosphate solutions frequently are added to suspension-type fertilizers to improve handling and storage properties. A schematic summary of the methods by which solution- and suspension-type fertilizers are made is shown in Figure 24.

Suspension-Type Mixtures. Fluid mixed fertilizers of the solution type have two inherent shortcomings. First, the grade, ie, the plant food content, is limited by the solubility of the ingredients. This is a particular problem in regard to potash, because potassium chloride and other commonly available potash salts are not highly soluble. Second, some important micronutrient sources have very limited solubility in the liquid fertilizer environment. The problem regarding micronutrients can be circumvented to some extent by using specially prepared, highly soluble metalloorganic compounds known as chelates, but these are relatively expensive. The production of suspension-type fertilizers solves these problems.

In suspension fertilizers, the plant food content exceeds the solubility of the system, and the excess is present in the form of finely divided, suspended particles. In almost all suspension fertilizers, a suspending agent is included to prevent settling of the solids. The suspending agent that is most cost effective, and hence most widely used, is attapulgite, a swelling-type clay composed mainly of microscopic, needle-like crystals of hydrated magnesium aluminum silicate. Attapulgite, which takes its name from the town of Attapulgus, Georgia, U.S., is found extensively in that area and in northern Florida. Other deposits are found in France and in a few other locations. Related swelling-type minerals such as bentonite and sepiolite have also been used but are not as effective as attapulgite in the high electrolyte environment of fluid fertilizers. Attapulgite, as mined, consists of minute tightly bound bundles of the needle-like silicate crystals. To realize the suspending properties of this material, it is essential that these crystal bundles be thoroughly disrupted mechanically and that the resultant individual crystals be dispersed in an electrolyte solution. When this is done, the individual needles orient themselves in the arrangement of a weak gel having effective suspending properties. In the preparation of suspension fertilizers, therefore, provision is always made for gelling the attapulgite by providing either a high shear agitator, a high shear recirculation pump, or both. The usual proportion of attapulgite is 1 to 3% by weight. A procedure called pregelling often is used in which the attapulgite is subjected to high shear agitation using only a small amount of water or fertilizer solution prior to addition to the main batch. Some pregelled attapulgite also is marketed in the form of a weak electrolyte slurry known as liquid clay. The use of attapulgite in suspension fertilizers has been described in detail (95).

Because of the absence of solubility restraints, suspension fertilizers are much more flexible than solution fertilizers in regard to types and quantities of starting materials that can be used and the plant food contents of the products that can be made. The fertilizer grades producible as suspensions are almost as high as those of the solids mixtures normally made by bulk blending. Typical grades are 14–14–14, 20–10–10, 7–21–21, and 3–10–30. Other grades are also made (Fig. 24). There is less restraint on the addition of micronutrients to sus-

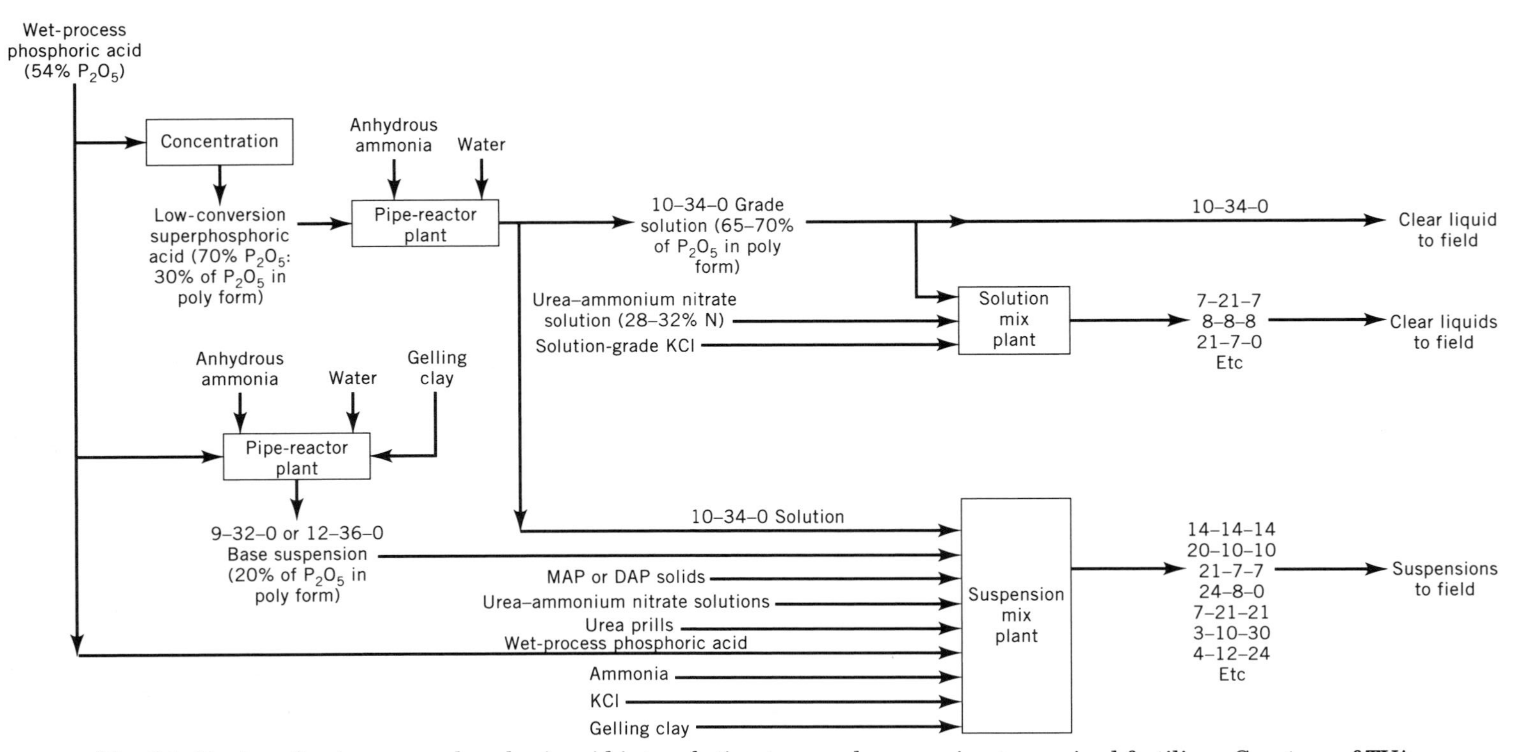

Fig. 24. Routes of wet-process phosphoric acid into solution-type and suspension-type mixed fertilizer. Courtesy of TVA.

pensions. Figure 24 gives an indication of the variety of raw materials that may be used in suspensions. Formulations are developed to fit individual needs. A common practice, and one that is not feasible in the preparation of solution-type mixtures, is the use of solid, wet-process monoammonium phosphate or diammonium phosphate as the sole or partial source of phosphate in the suspension. With such a suspension, inclusion of some 10–34–0 ammonium polyphosphate base solution improves storage and handling properties by virtue of the impurity-sequestering power of the polyphosphate. Other solids frequently included in suspension mixtures are urea, ammonium nitrate, and potassium chloride. Wet-process phosphoric acid also is incorporated sometimes along with ammonia to develop heat, improve disintegration and dissolution of solids, and to provide a cheap source of P_2O_5.

Suspension preparation is carried out batchwise, usually in relatively small local plants. All three primary plant nutrients may be supplied in solid form or the addition of wet-process acid and ammonia can be an option. Another option is the addition of 10–34–0 ammonium polyphosphate base solution. Inclusion of the polyphosphate improves storage properties if the suspension is not to be used immediately. Addition of attapulgite gelling clay is usually in a proportion of 1 to 3%. In a small plant, it is essential that the mix tank be equipped with a high intensity mixer in order to disintegrate and disperse the solid ingredients and to gell the suspending clay. A number of commercial mixing units are specifically designed and marketed for this purpose. Use of a high shear recirculation pump fed from the bottom of the mix tank and discharged into the top also improves dispersion and clay gelling.

A method for making a polyphosphate base suspension directly from 54% P_2O_5 wet-process orthophosphoric acid has been developed and demonstrated on a production scale (96,97). Such a base suspension can be field applied directly or used as a base for mixing. A flow sheet of the process is shown in Figure 25. Both the wet-process acid and the ammonia vapor feed materials are preheated before being brought together in a well-insulated pipe reactor. By this conservation of heat, pipe reactor temperatures of 230 to 240°C are reached, which is sufficient to develop a polyphosphate level of about 20% in melt discharged from the pipe. The hot melt from the pipe reactor discharges with considerable turbulence into a dissolution tank into which water is added to bring the fluid to a 9–32–0 or 12–36–0 grade. A pump is used to recirculate this hot (93°C) fluid through the heat exchangers that are used to preheat the acid and ammonia reactants. A product stream of fluid is bled off of the dissolution tank and is immediately cooled to arrest hydrolysis of the polyphosphate. Cooling is by countercurrent air contact in an evaporative-type cooler. The final processing of the fluid is the addition of 2% of attapulgite clay in a gelling tank equipped with a high shear recirculation pump. The purpose of the clay is to suspend any undissolved impurities, to suspend any crystals that may precipitate during storage, and to provide sufficient gelling clay for most mixed-grade suspensions that might subsequently be made from the base suspension. This process has created considerable interest in the industry. As of this writing, it is in commercial use at one location in the United States and five locations overseas.

Field Application of Fluids. A detailed discussion of methods and equipment for the field application of solution and suspension-type fertilizers has been given

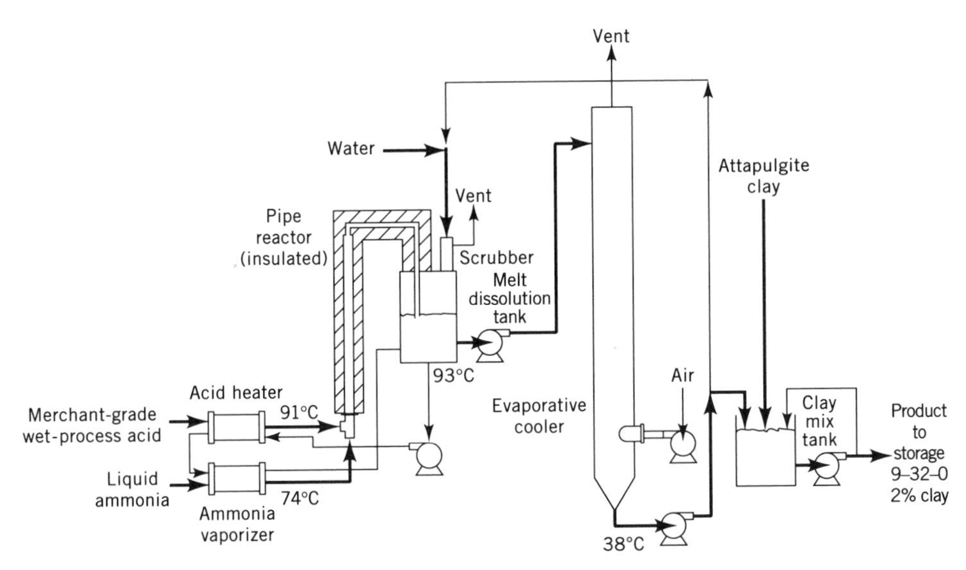

Fig. 25. TVA pipe reactor process for production of 9–32–0 and 12–36–0 grade base suspensions directly from wet-process orthophosphoric acid.

(98). In general, four different types of application are used: (*1*) broadcasting, (*2*) surface band application, (*3*) subsurface band application, and (*4*) inclusion in irrigation water. The type of application used is usually based on the particular soil–crop combination. Suspension fertilizers, which often are cheaper than solutions, work well for broadcast applications at relatively high application rates where spray nozzles having relatively large, nonplugging openings can be used. The spray booms on such an applicator often cover a swath width of 15 m or more. For the other three types of fluid applications, the choice often is to use solution-type fertilizers to avoid plugging problems. For band applications, the flow rates per fluid outlet usually are smaller and the nozzle openings are proportionately smaller and more subject to plugging. Some special equipment and techniques have been used successfully for band application with suspensions, however (98).

For addition of fertilizer to irrigation spray systems, complete water solubility, and hence use of the solution-type fertilizers, is essential. An additional requirement is that the fertilizer be of a composition that does not react with the normal mineral content ie, the hardness, of the irrigation water to form objectional scaling of equipment.

Secondary and Micronutrients in Fertilizers

The great majority of farm fertilizers are produced, marketed, and applied with regard only to the primary plant nutrient content. The natural supply of secondary and micronutrients in the majority of soils is usually sufficient for optimum growth of most principal crops. There are, however, many identified geographical areas and crop–soil combinations for which soil application of secondary and/or micronutrient sources is beneficial or even essential. The fertilizer industry accepts the responsibility for providing these secondary and micronutrients, most often as an additive or adjunct to primary nutrient fertilizers. However, the source chemicals used to provide the secondary and micronutrient elements are usually procured from outside the fertilizer industry, for example from mineral processors. The responsibilities of the fertilizer producer include procurement of an acceptable source material and incorporation in a manner that does not decrease the chemical or physical acceptability of the fertilizer product and provides uniform application of the added elements on the field.

Secondary Nutrients. *Calcium.* Soil minerals are a main source of calcium for plants, thus nutrient deficiency of this element in plants is rare. Calcium, in the form of pulverized limestone [*1317-65-3*] or dolomite [*17069-72-6*], frequently is applied to acidic soils to counteract the acidity and thus improve crop growth. Such liming incidentally ensures an adequate supply of available calcium for plant nutrition. Although pH correction is important for agriculture, and liming agents often are sold by fertilizer distributors, this function is not one of fertilizer manufacture.

Some commonly used primary nutrient fertilizers are incidentally also rich sources of calcium. Ordinary superphosphate contains monocalcium phosphate and gypsum in amounts equivalent to all of the calcium originally present in the phosphate rock. Triple superphosphate contains soluble monocalcium phosphate equivalent to essentially all the P_2O_5 in the product. Other fertilizers rich in calcium are calcium nitrate [*10124-37-5*], calcium ammonium nitrate [*39368-85-9*], and calcium cyanamide [*156-62-7*]. The popular ammonium phosphate-based fertilizers are essentially devoid of calcium, but, in view of the natural calcium content of soils, this does not appear to be a problem.

Magnesium. Like calcium, magnesium is abundantly present in soil minerals. Magnesium minerals, however, are more soluble than those of calcium and magnesium level in soils is lower than that of calcium. Pulverized dolomite, calcium–magnesium carbonate, the oft used liming agent for pH control, provides nutrient magnesium. Soils are occasionally deficient in magnesium. An example of the problems caused by magnesium deficiency in a crop results when a lush, spring growth of grass used for feed is low in magnesium content and produces the sickness grass tetany in cattle. Magnesium deficiency is also sometimes seen in citrus crops grown on calcareous and high sodium soils. Tomatoes grown in soil having magnesium deficiency exhibit characteristic black spots. It is therefore not unusual to add magnesium compounds to fertilizers.

Most of the supplemental magnesium is supplied as dolomitic limestone but other important sources are magnesium sulfate [*7487-88-9*] (Epsom salts), and potassium–magnesium sulfate (langbeinite), as well as organic chelate compounds. Magnesium can undergo reactions with phosphorus compounds, sometimes lowering phosphate solubility, but the addition of magnesium to fertilizers, and the correction of magnesium deficiency generally presents no insurmountable problems. Sometimes low magnesium content of pasture crops makes it necessary to feed a magnesium supplement in the diet for animals rather than to incorporate it in the fertilizer.

Sulfur. Sulfur occurs in plants at levels as high as that of phosphorus. It is absorbed by plants as the sulfate anion, SO_4^{2-}, which is a constituent of many fertilizers. Normal superphosphate contains much calcium sulfate [*7778-18-9*]. Ammonium sulfate is a good source. Wet-process phosphoric acid often contains a few percent of sulfur as sulfuric acid and calcium sulfate, and this sulfur appears in fertilizers, such as ammonium phosphates, made from the acid. Other sources of sulfur in soils are decreasing. In industrialized, humid regions sulfur additions to the soil from rain range from 28 to 168 kg/hm^2 yearly. Soils in such regions do not often show deficiency in sulfur, but as environmental goals of eliminating sulfur oxide emissions from industrial plants are met (see EXHAUST CONTROL, INDUSTRIAL), sulfur deficiency in soils can be expected to increase. All indications are that fertilizers need to contain increasing amounts of sulfur. In fact, symptoms of sulfur deficiency in soils of the United States have been increasing for some years. Sulfur deficiency had been identified in 13 states in 1960; the number grew to 31 states by 1973. The average sulfur concentration of fertilizers in the United States has been decreasing since the 1950s as the average NPK grade of fertilizers has increased.

Some of the principal forms in which sulfur is intentionally incorporated in fertilizers are as sulfates of calcium, ammonium, potassium, magnesium, and as elemental sulfur. Ammonium sulfate [*7783-20-2*], normal superphosphate, and sulfuric acid frequently are incorporated in ammoniation granulation processes. Ammonium phosphate–sulfate is an excellent sulfur-containing fertilizer, and its production seems likely to grow. Some common grades of this product are 12–48–0–5S, 12–12–12S, and 8–32–8–6.5S.

Granular sulfur-containing fertilizers used in bulk blends include ammonium sulfate, ammonium phosphate–sulfate, potassium sulfate, and potassium–magnesium sulfate. Sulfur-coated granular fertilizer can be used, but this expensive form of sulfur is generally used to achieve other advantages such as controlled release of the coated nutrient. A granular gypsum also is available. Finely divided elemental sulfur and sulfur compounds can be added to suspension fertilizer; ammonium thiosulfate is a common sulfur supplement in clear liquid fertilizer (see CONTROLLED RELEASE TECHNOLOGY, AGRICULTURAL).

Micronutrients. Attention to meeting the micronutrient needs of crops has greatly increased as evidenced in an analysis undertaken by TVA and the Soil Science Society in 1972 (99). The micronutrient elements most often found wanting in soil–crop situations are boron, copper, iron, manganese, molybdenum, and zinc. Some of these essential micronutrients can be harmful to plants when used in excess.

The problem in micronutrient fertilization is that of adequately identifying the need, reducing this to a prescription, compounding the fertilizer, and distributing it evenly in the soil. Prescription compounding is being used increasingly as deficiencies are better identified, but a shotgun approach is also used. In the shotgun method, a range of micronutrients is added to fertilizers at levels believed to be low enough to avoid harmful effects but adequate to prevent deficiencies. Fertilizers of this kind usually are produced in ammoniation–granulation plants. When these are sold as premium fertilizers because of the added micronutrients, the minimum guaranteed amounts of micronutrients allowed by law in most U.S. states are 0.02 wt % B, 0.05 wt % Cu, 0.10 wt % Fe, 0.05 wt % Mn, 0.0005 wt % Mo, and 0.05 wt % Zn.

Generally, soluble materials are more effective as micronutrient sources than are insoluble ones. For this reason, many soil minerals that contain the micronutrient elements are ineffective sources for plants. Some principal micronutrient sources and uses are summarized below. In this discussion the term frits refers to a fused, pulverized siliceous material manufactured and marketed commercially for incorporation in fertilizers. Chelates refers to metalloorganic complexes specially prepared and marketed as especially soluble, highly assimilable sources of micronutrient elements (see CHELATING AGENTS).

Boron. The principal materials used are borax [*1303-96-4*], sodium pentaborate, sodium tetraborate, partially dehydrated borates, boric acid [*10043-35-3*], and boron frits. Soil application rates of boron for vegetable crops and alfalfa are usually in the range of 0.5–3 kg/hm^2. Lower rates are used for more sensitive crops. Both soil and foliar application are practiced but soil applications remain effective longer. Boron toxicity is not often observed in field applications (see BORON COMPOUNDS).

Copper. Some 15 copper compounds (qv) have been used as micronutrient fertilizers. These include copper sulfates, oxides, chlorides, and cupric ammonium phosphate [*15928-74-2*], and several copper complexes and chelates. Recommended rates of Cu application range from a low of 0.2 to as much as 14 kg/hm^2. Both soil and foliar applications are used.

Iron. As with copper, some dozen or more materials are used as fertilizer iron sources. These include ferrous and ferric oxides and sulfides and ferrous ammonium phosphate [*10101-60-7*], ferrous ammonium sulfate [*10045-89-3*], frits, and chelates. In many instances, organic chelates are more effective than inorganic materials. Recommended application rates range widely according to both type of micronutrient used and crop. Quantities of Fe range from as low as 0.5 kg/hm^2 as chelates for vegetables to as much as a few hundred kg/hm^2 as ferrous sulfate for some grains.

Manganese. Commonly used manganese fertilizer materials are manganous and manganic sulfates, chlorides, carbonates, oxides, frits, and chelates. Soil application rates range from about 2 to 150 kg/hm^2 of Mn.

Molybdenum. The commonly used molybdenum materials are sodium molybdate [*7631-95-0*], ammonium molybdate [*12027-67-7*], molybdenum trioxide [*1313-27-5*], molybdenum sulfate [*51016-80-9*], and frits. Molybdenum is used in smaller amounts than any of the other micronutrients, ranging from a few grams to 3 kg/hm^2 of Mo.

Zinc. Zinc, one of the most widely needed and used micronutrients, is applied as sulfates (both basic and normal hydrates), carbonate, sulfide, phosphate, oxide, chelates, and other organic materials. Rates of Zn application range from 0.2 to 22 kg/hm^2.

Micronutrients in Granular Fertilizers. In the production of granular fertilizers, it is relatively simple and effective to incorporate micronutrient materials as feeds in the granulation process. A problem with this method, however, is that granulation processes are most efficient and economical when operated continuously to produce large tonnages of the same or similar composition. Frequent changes in product composition or storage of a wide range of grades is simply uneconomical. These factors seriously limit the practice of prescription micronutrient formulation in granular fertilizer production processes. Granulation processes more often use the shotgun approach. As a result, some unneeded elements are provided with no benefit.

Micronutrients in Bulk Blends. Prescription formulation of fertilizers containing micronutrients is more practical in the production of bulk blends and fluid fertilizers, because these processes are more adaptable to frequent changes in grades, small-batch preparation, and thus to custom formulation. The problems with formulation and distribution in bulk blends are in achieving even distribution and in avoiding segregation of the micronutrients in storage, transportation, and application. The fundamental concepts for handling these problems are the same as those for avoiding segregation of primary nutrients in bulk blends.

One way to achieve good distribution is to blend properly sized granular premium fertilizer with the other granular components of the blend. Micronutrient materials of granular size can also be blended, but a potential problem with this method is that, at recommended application rates, the micronutrient-containing granules in a blend may be too few to yield acceptable distribution of micronutrient in the field. Spaces between micronutrient-affected volumes of soil may be too large.

Coating granular fertilizers or bulk blends with powdered micronutrients overcomes the disadvantages of using granular micronutrient additives (100). When the coating is done well, segregation is not a problem because the micronutrient essentially becomes an integral part of each granule. The coating process generally consists of (*1*) dry mixing the granular blend with finely powdered micronutrient materials; (*2*) adding a small amount of liquid binder such as a fertilizer solution or an oil, during mixing; and (*3*) continuing the mixing for a short period. The total cycle requires only about 3 min in a rotary mixer.

Micronutrients in Fluid Fertilizers. In terms of homogeneity and even distribution, fluid fertilizers are probably the best micronutrient carriers. Fluid carriers of micronutrients usually are nitrogen solutions, clear liquid mixtures, or suspensions. Some micronutrients, however, are applied as simple water solutions or suspensions. Foliar micronutrient sprays often contain only the micronutrient material in water solution.

Several zinc and copper micronutrient compounds are soluble in a variety of nitrogen solutions. Ammonia–ammonium nitrate solutions containing 2.5% Zn and 1% Cu can be prepared (100). Micronutrients are not very soluble in urea–ammonium nitrate solution unless the pH is raised to 7 or 8 by adding ammonia, whereupon zinc and copper become much more soluble.

Orthophosphate liquid mixtures are ineffective as micronutrient carriers because of the formation of metal ammonium phosphates such as $ZnNH_4PO_4$. However, micronutrients are much more soluble in ammonium phosphate solutions in which a substantial proportion of the phosphorus is polyphosphate. The greater solubility results from the sequestering action of the polyphosphate. The amounts of Zn, Mn, Cu, and Fe soluble in base solution with 70% of its P as polyphosphate are 10 to 60 times their solubilities in ammonium orthophosphate solution. When a mixture of several micronutrients is added to the same solution, the solubility of the individual metals is lowered significantly. In such mixtures the total micronutrient content should not exceed 3% and the storage time before precipitates appear may be much shorter than when only one micronutrient is present.

Suspension fertilizers offer a simple and effective way of preparing homogeneous products containing high concentrations of micronutrients because the solubility of the micronutrient is immaterial, except as it affects agronomic response. Incorporation of the micronutrient may be carried out by adding the desired amount to the suspension either while it is being mixed or after. As for clear liquids, it is important to provide good agitation to disperse the powder.

Synthetic organic chelates and natural organic complexes are sometimes more effective agronomically per unit of micronutrient than inorganic forms, but the organic materials are more expensive. The chelates can be used with both orthophosphate and polyphosphate liquids and suspensions.

Micronutrient Reactions in Mixtures. The chemical systems that can be created in mixtures of six micronutrients in NPK mixed fertilizers are many and highly complex (101). Many chemical reactions occur, and the solubility of the reaction products determines the micronutrient solubility. The six principal micronutrients fall into two groups on the basis of the tendency to undergo adverse reactions in carrier fertilizers. Boron and molybdenum undergo few reactions that lower their solubility. Copper, iron, manganese, and zinc react to form more insoluble compounds. Of these adverse reactions those of copper are the least detrimental.

Because use of micronutrient fertilizers is expected to become increasingly important, their effective use requires cooperation between soil chemists and agronomists to identify and quantify needs, and fertilizer production technologists for effective incorporation.

Raw Material Resources for Fertilizers

Resources for Nitrogen Fertilizers. The production of more than 95% of all nitrogen fertilizer begins with the synthesis of ammonia, thus it is the raw materials for ammonia synthesis that are of prime interest. Required feed to the synthesis process (synthesis gas) consists of an approximately 3:1 mixture (by volume) of hydrogen and nitrogen.

Hydrogen. Technologies for extracting hydrogen (qv) from both hydrocarbons (qv) and water (qv) are available. The extent of reserves of the preferred raw material, which is natural gas, and the cost of extracting hydrogen are increasingly serious problems for the fertilizer industry. Raw materials used to produce hydrogen in the order of decreasing preference are natural gas, straight-run naphthas, fuel oil, coal (qv), and water. The availability of hydrogen for ammonia synthesis, thus, is an inextricable part of the world energy supply–demand situation (see FUEL RESOURCES; GAS, NATURAL).

Natural gas is by far the preferred source of hydrogen. It has been cheap, and its use is more energy efficient than that of other hydrocarbons. The reforming process that is used to produce hydrogen from natural gas is highly developed, environmental controls are simple, and the capital investment is lower than that for any other method. Comparisons of the total energy consumption (fuel and synthesis gas), based on advanced technologies, have been discussed elsewhere (102).

Since 1960, about 95% of the synthetic ammonia made in the United States has been made from natural gas; worldwide the proportion is about 85%. Most of the balance is made from naphtha and other petroleum liquids. Relatively small amounts of ammonia are made from hydrogen recovered from coke oven and refinery gases, from electrolysis of salt solutions, eg, caustic chlorine production, and by electrolysis of water. In addition there are about 20 ammonia plants worldwide that use coal as a hydrogen source.

Considering petroleum price trends and the fact that economically extractable natural gas reserves in the United States are being depleted at a rate above that for discovery of new reserves, a potentially serious problem looms for the fertilizer industry and agriculture in the United States and in other countries dependent on imported petroleum feedstocks. Rising prices for natural gas have led worldwide toward increased recovery of natural gas, and reduced flaring of gas in petroleum-rich regions. Estimates of potentially extractable hydrocarbons in the United States, including Alaska and the continental shelves, have been made (103). The global life expectancy of crude oil and natural gas and of alternative fossil energy resources has been reviewed (104).

Coal is expected to be the best domestic feedstock alternative to natural gas. Although coal-based ammonia plants have been built elsewhere, there is no such plant in the United States. Pilot-scale projects have demonstrated effective ammonia-from-coal technology (102). The cost of ammonia production can be anticipated to increase, leading to increases in the cost of producing nitrogen fertilizers.

Nitrogen. Nitrogen (qv) comprises about 79% of air, by volume, and is an unlimited reserve. The technologies for recovering nitrogen from air by liquefaction–distillation or by partial combustion to fix the oxygen are highly developed (see CRYOGENICS). Nitrogen, however, is a very stable element, and combining it with hydrogen to form ammonia is an energy-intensive operation. Future increases in energy costs therefore are expected to be reflected in higher ammonia costs.

Resources for Phosphate Fertilizers. Natural mineral deposits are the source of phosphorus for essentially all manufactured phosphate fertilizer (see PHOSPHORUS AND THE PHOSPHIDES; PHOSPHORUS COMPOUNDS). Phosphate deposits are numerous throughout the world but size and quality vary widely. Those minerals that contain enough phosphorus to be potential sources for industrial use are called phosphate rock and, sometimes, phosphate ore. Phosphorus makes up about 1.22% of the earth's crust. About 150 known minerals contain at least 1% P_2O_5. Nearly all mineable deposits of phosphate are of the apatite group represented by the general formula $Ca_5(F,Cl,OH,0.5CO_3)(PO_4)_3$. A small percentage comes from secondary aluminum deposits derived from apatite by weathering.

The following definitions of terms are used widely by the phosphate rock industry (105). Phosphate rock designates either (*1*) an apatite-bearing rock containing enough P_2O_5 to be processed into fertilizer or phosphorus, or (*2*) a beneficiated apatite concentrate. Phosphorite is a sedimentary rock in which a phosphate mineral is a principal constituent. This category includes phosphatic sandstones and limestones. Ore, or matrix, is material that can be mined at a profit. Grade of ore is expressed in terms of phosphate content, either as $\%P_2O_5$ or percent bone phosphate of lime $Ca_3(PO_4)_2$ (%BPL). Reserve is rock known to be mineable at a profit. Resource is rock which, for reasons such as low grade or inaccessability, is not mineable at a profit as of this writing.

A detailed and well-defined estimate of world phosphate resources and reserves was prepared in 1979 (105). A summary is presented in Table 14. Inacces-

Table 14. World Reserves and Resources of Phosphate Rock[a]

Location	Reserves,[b] 10^6 t		Resources,[c] 10^6 t	
	Igneous apatite	Marine phosphorite	Igneous apatite	Marine phosphorite
North Africa				
Algeria		500		600
Morocco		5,000		35,000
Tunisia		500		800
West Africa				
Angola		20		100
Senegal		190		3,000
Spanish Sahara		1,600		15,000
Togo		100		200
South Africa	100		1,300	
Uganda	40		160	
Rhodesia	10		10	
Middle East				
ARE (Egypt)		800		2,000
Iran		30		100
Iraq		60		600
Israel		100		200
Jordan		100		200
Saudi Arabia				1,000
Syria		400		400
Turkey				300
Europe				
CIS	400	1,450	400	2,950
Finland	50		100	
others		15		30
Asia				
People's Republic of China		100		1,000
India		70		200
Mongolia		250		700
Vietnam		100		400
Pakistan				150
Democratic People's Republic of Korea	5		30	
North America				
Mexico		1,140		0.07
United States				
eastern		1,600		6,000
western		6,000		7,000
South America				
Brazil	237	200	775	520
Colombia				600
Peru (Sechura)				6,100
Venezuela		20		0.1
Australia		500		1,500
others	10		30	
Total	*852*	*19,705*	*2,805*	*87,810*

[a]Ref. 105.

[b]Rock containing at least 30% P_2O_5, known mineable at a profit; variable with economic factors.

[c]Based on solid information, but cannot be mined and processed at a profit.

sibility of geographic location, as in the Sechura desert of Peru and also in northern Saudi Arabia, can cause otherwise high grade deposits to be considered as resource rather than reserve. In the case of the Savannah River deposit in Georgia, environmental considerations have caused downgrading of the deposit. This deposit contains at least 1.8×10^9 t of recoverable phosphate rock containing about 30% P_2O_5 (105), yet it is classified as a resource because its mining is not permitted.

At the present rate of world phosphate rock consumption (150×10^6 t/yr), the total world reserve (Table 14) is sufficient for about 200 years, and the resource would be sufficient for nearly 900 years. At expected increased rates of consumption, the reserves and resources are adequate for at least 150 years and 700 years, respectively. At projected rates of consumption, the high grade reserves in Florida probably will be exhausted by the year 2000. Rock production from the Florida reserve presently constitutes about 80% of all United States production and about one-third of world production (106). This rate of depletion is causing increased interest in western United States reserves which represent nearly 80% of present U.S. total reserves.

In addition to the indicated total world reserves of about 21×10^9 t having P_2O_5 content of at least 30%, and the 9.1×10^{10} t classified as resources, there are about 250×10^6 t of reserves and resources derived from guano which are not included in Table 14.

Other U.S. phosphate deposits having potential usage but not included in Table 14, are phosphorite formations in Idaho, Wyoming, Montana, and Utah which may contain up to 14.5×10^9 tons of phosphorus. Most of this material is deeply buried, is not weathered, and would require new mining and processing techniques. A formation in Florida known as the Hawthorn formation contains hundreds of billions of metric tons of phosphate pellets containing 10×10^9 tons of phosphorus, but new mining and processing methods would be necessary for that deposit also.

Billions of metric tons of phosphate rock also are present offshore in the oceans, eg, best estimates are that a billion tons of pellets that may contain about 30% P_2O_5 are present in a Baja California–Mexico deposit alone. Other areas in the world that contain large, unevaluated amounts of phosphate include Australia, Alaska, Africa, the Near East, Peru, Colombia, Brazil, the People's Republic of China, Mongolia, and the former Soviet Union.

Resources for Potash Fertilizers. Potassium is the seventh most abundant element in the earth's crust. The raw materials from which potash fertilizer is derived are principally bedded marine evaporite deposits, but other sources include surface and subsurface brines. Both underground and solution mining are used to recover evaporite deposits, and fractional crystallization (qv) is used for the brines. The potassium salts of marine evaporite deposits occur in beds in intervals of halite [*14762-51-7*], NaCl, which also contains bedded anhydrite [7778-18-9], $CaSO_4$, and clay or shale. The K_2O content of such deposits varies widely (see POTASSIUM COMPOUNDS).

The common ores of potassium include (*1*) sylvinite, a mixture of sylvite, KCl, and halite; (*2*) hartsalz, composed of sylvite, halite, and kieserite [*14567-64-7*], $MgSO_4 \cdot H_2O$, or anhydrite; (*3*) carnallitite, carnallite [*1318-27-0*], KCl·

$MgCl_2{\cdot}6H_2O$, and halite; (*4*) langbeinite ore, langbeinite, $K_2SO{\cdot}2MgSO_4$, and halite; and (*5*) kainite ore, kainite [*1318-72-5*], $4(KCl{\cdot}MgSO_4){\cdot}11H_2O$, and halite. More than 90% of the estimated potassium reserves occur principally as sylvinite and carnallitite (107). Sylvite, the richest of the minerals at 63% K_2O, is the principal economically exploitable reserve. In addition, there are four principal insolúble potassium silicate minerals, glauconite, leucite, nepheline, and orthoclase–sanidine which range in K_2O content from 7 to 22%. These minerals are plentiful but for economic reasons may never be exploited for potassium.

An assessment of world potash resources (108) is shown in Table 15. Of the 67×10^9 t of total estimated reserves and resources in Canada, nearly 5×10^9 t is recoverable by conventional mining methods and the remainder by solution mining. As of 1974, Canada had about half of the known world reserves and about 90% of known world resources of potassium.

Other potential sources of potassium include insoluble minerals and ores, and the oceans, which contain 3.9×10^5 t/(km)^3 of seawater (see OCEAN RAW MATERIALS). The known recoverable potash reserves are sufficient for more than 1000 years at any foreseeable rate of consumption.

Table 15. Assessment of World Potash Reserves and Resources,[a] t × 10^6 of K_2O

Location	Reserves	Resources	Total
North America			
Canada	4,537	62,613	67,150
United States	182	182	364
South America			
Chile	9	9	18
others	0	15	15
Europe			
France	91	91	182
Germany			
former East	2,450	2,450	4,900
former West	1,633	1,633	3,266
Italy	9	14	23
Spain	73	73	146
CIS	726	907	1,633
United Kingdom	23	23	46
others	0	64	64
Asia			
Israel and Jordan	218	907	1,125
People's Republic of China	5	5	10
Africa			
People's Republic of the Congo (Brazzaville)	18	18	36
others	0	68	68
Oceania and Australia	9	9	18
World total	*9,983*	*69,089*	*79,064*

[a]Ref. 108.

Resources of Sulfur. In most of the technologies employed to convert phosphate rock to phosphate fertilizer, sulfur, in the form of sulfuric acid, is vital. Treatment of rock with sulfuric acid is the procedure for producing ordinary superphosphate fertilizer, and treatment of rock using a higher proportion of sulfuric acid is the first step in the production of phosphoric acid, a production intermediate for most other phosphate fertilizers. Over 1.8 tons of sulfur is consumed by the world fertilizer industry for each ton of fertilizer phosphorus produced, ie, 0.8 t of sulfur for each ton of P_2O_5. Of the total 13.7×10^6 t of sulfur consumed in the United States for all purposes in 1991, 60% was for the production of phosphate fertilizers (109). Worldwide the percentage was probably even higher.

Total 1991 world production of sulfur in all forms was 55.6×10^6 t. The largest proportion of this production (41.7%) was obtained by removal of sulfur compounds from petroleum and natural gas (see SULFUR REMOVAL AND RECOVERY). Deep mining of elemental sulfur deposits by the Frasch hot water process accounted for 16.9% of world production; mining of elemental deposits by other methods accounted for 5.0%. Sulfur was also produced by roasting iron pyrites (17.6%) and as a by-product of the smelting of nonferrous ores (14.0%). The remaining 4.8% was produced from unspecified sources.

World resources of sulfur have been summarized (110,111). Sources, ie, elemental deposits, natural gas, petroleum, pyrites, and nonferrous sulfides are expected to last only to the end of the twenty-first century at the world consumption rate of 55.6×10^6 t/yr of the 1990s. However, vast additional resources of sulfur, in the form of gypsum, could provide much further extension but would require high energy consumption for processing.

Resources of Secondary and Micronutrients. *Calcium.* Calcium is the fifth most abundant element in the earth's crust. There is no foreseeable lack of this resource as it is virtually unlimited. Primary sources of calcium are lime materials and gypsum, generally classified as soil amendments (see CALCIUM COMPOUNDS). Among the more important calcium amendments are blast furnace slag, calcitic limestone, gypsum, hydrated lime, and precipitated lime. Fertilizers that carry calcium are calcium cyanamide, calcium nitrate, phosphate rock, and superphosphates. In addition, there are several organic carriers of calcium. Calcium is widely distributed in nature as calcium carbonate, chalk, marble, gypsum, fluorspar, phosphate rock, and other rocks and minerals.

Magnesium. The magnesium content of soils usually is less than that of calcium because magnesium is more soluble. Important sources of magnesium are dolomitic limestone, Epsom salts, calcined kieserite, magnesia, potassium–magnesium sulfate, and a few organic sources. Magnesium is the eighth most plentiful element in the earth, and in its many forms makes up about 2.07% of the earth's crust (108). Dolomite, seawater, and well and lake brines, the sources of most of the world's magnesium and magnesium compounds, are available in unlimited quantities. Seawater, having a magnesium content of 0.13 wt %, is an inexhaustible resource.

Boron. Virtually all United States boron production and about three-fifths of the world production comes from bedded deposits and lake brines in California. U.S. reserves are adequate to support high production levels. Turkey is the only other boron-producing country of significance. Only about 5% of boron production is used in agriculture.

Copper. World copper reserves are estimated at 408×10^6 t copper; one-fifth being in the United States. Requirements for fertilizer are very small.

Iron. World reserves are placed at 236×10^9 t of ore containing 90×10^9 t of iron; world resources are estimated at 180×10^9 t of iron. Only a small fraction of world production is required for fertilizer use.

Manganese. U.S. resources of manganese are estimated to be on the order of 67×10^6 t Mn. World reserves are about 1.8×10^9 t Mn, and resources 1.45×10^9 t Mn (108). Only small amounts are required for fertilizer use.

Molybdenum. U.S. reserves and resources are about 3 and 13×10^6 t, respectively. World reserves and resources are about 6 and 23×10^6 t, respectively (108). The requirement for fertilizer is very small.

Zinc. Zinc deposits in the United States extend from Maine through the Appalachian Mountains, and west through the Mississippi Valley into the Rocky Mountain states. U.S. reserves are estimated to be 27×10^6 t Zn (108). World reserves and resources are 135 and 110×10^6 t, respectively. The requirements for fertilizer are relatively small.

Environmental Aspects

Fertilizer Use. The worldwide use of fertilizers has an important, positive effect on the environment. Conservative estimates (112) indicate that about 30% of world food production is directly attributable to fertilizer use. Without fertilizer, therefore, at least 30% more virgin land would have to be devoted to agriculture, and 30% more labor and other resources would have to be expended. Even more serious would be the effects of land tillage and cropping without nutrient replenishment. Past experience has shown that, under such a condition, crop yields progressively decrease, the land eventually becomes barren, and forces of wind and water erosion prevail.

Possible negative environmental effects of fertilizer use are the subject of intensive evaluation and much discussion. The following negative effects of fertilizer usage have been variously suggested (113): a deterioration of food quality; the destruction of natural soil fertility; the promotion of gastrointestinal cancer; the pollution of ground and surface water; and contributions toward the destruction of the ozone layer in the stratosphere.

In regard to food quality, the bulk of the evidence appears to indicate no lowering by fertilizer use, but rather a frequent improvement in quality. Soil fertility, likewise, is generally improved by fertilizer use because of the higher return of crop residues to the soil. The suggestion of fertilizer use as a cancer cause is largely disproven by statistics that show a relatively low cancer rate in some areas, such as western Europe, where fertilizer use has been particularly intensive for a long time. Pollution of ground and surface water by fertilizer runoff does occur and is now the subject of much investigation (see GROUNDWATER MONITORING; SOIL STABILIZATION). The problem apparently is serious only under certain conditions of excessive fertilizer application, abnormal soil porosity, or improper choice of fertilizer type. Prudent use of fertilizer, in most applications, does not contribute to this problem. Finally, the possible effects of fertilizer emissions of nitrogen oxides on the ozone layer apparently were initially exaggerated. Some studies now place nitrogen oxides emission from fertilizers at only 1% of that from other sources. Studies are ongoing.

The suggestion frequently is made that substitution of organic fertilizers, namely manures and composts, for chemical fertilizers would be of ecological benefit. The reality is, however, that the supply and logistics of such materials could never be adequate for the present-day level of agriculture. Furthermore, intensive application of such materials to the soil would itself present ecological problems, such as run-off pollution and steady buildup of toxic heavy metals.

The fertilizer industry, through its various trade organizations, supports continued intensive study of the effects of fertilizer use on the environment. Support is also given to educational programs that promote prudent use of fertilizer.

Fertilizer Production. The fertilizer industry, like the chemical industry as a whole, is fully subject to stringent national, state, and local antipollution regulations. For the most part, abatement of pollution in the fertilizer industry is handled by employment of standard procedures such as the scrubbing of gaseous effluents and purification treatment for liquid effluents. Problems that are somewhat unique to the industry are nitrous oxide emissions from nitric acid plants and granular fertilizer plants, particulate emissions from ammonium nitrate and urea prilling towers, strip mining of phosphate ore, gypsum disposal from wet-process acid plants, and fluorine emissions from phosphate processing and from gypsum disposal ponds.

Economic Aspects

The 1993 prices and values of fertilizer products consumed in the United States are summarized in Table 16. The prices given are fob production sites or principal

Table 16. Prices and Value of Important Fertilizer Products Consumed in the United States

Product	Grade	Annual U.S. consumption, 10^3 t	Price,[a] $/t		Value of U.S. consumption, 10^6 $/yr
			Material	N basis	
anhydrous ammonia	82.2–0–0	4040	97	118	392
nitrogen solution	32–0–0	6250	118	369	738
urea	46–0–0	4565	140	304	639
ammonium nitrate	35–0–0	2000	160	457	320
diammonium phosphate	18–46–0	3250	120	304[b,c]	390
triple superphosphate	0–46–0	520	104[d]		54
potash	0–0–60	2830	110[e]		311
Total					*2844*

[a]Sept. 1993, fob production or terminal sites.
[b]Nitrogen content assigned value prevailing for urea.
[c]Price on the basis of P_2O_5 is $141/t.
[d]Price on the basis of P_2O_5 is $226/t.
[e]Price on the basis of K_2O is $183/t.

terminals; thus costs to farmers are greater by virtue of shipping and handling costs and local dealer profits.

The economic magnitude of the U.S. fertilizer industry is indicated by the yearly value of products which approaches $3 billion. U.S. fertilizer consumption represents only about 13% of the total world consumption. Thus the annual value of worldwide consumption is at least $22 billion. The world investment for production facilities to produce these quantities of fertilizer is also very high (114).

The tabulation of plant nutrient costs, by product, in Table 16 shows the principal reason for the popularity of anhydrous ammonia as a fertilizer in the United States. The fob price per ton of nitrogen in the form of ammonia is less than half that for any other nitrogen product. Also, ammonia's relatively high nitrogen content of 82.2% favors low transportation costs, in spite of the need for specialized handling equipment and procedures.

BIBLIOGRAPHY

"Fertilizers" in *ECT* 1st ed., Vol. 6, pp. 376–452, by H. B. Siems, Swift & Co.; in *ECT* 2nd ed., Vol. 9, pp. 25–150, by A. V. Slack, Tennessee Valley Authority; in *ECT* 3rd ed., Vol. 10, pp. 31–125, by E. O. Huffman, Consultant.

1. E. J. Hewitt and T. A. Smith, *Plant Mineral Nutrition*, John Wiley & Sons, Inc., New York, 1975.
2. *Farm Chemicals 1970 Handbook*, Meister Publishing Co., Willoughby, Ohio, p. 134.
3. *Official Methods of Analysis*, Association of Official Analytical Chemists, Inc., Arlington, Va., 1990.
4. *Summary of State Fertilizer Laws*, Product Quality and Technology Committee, The Fertilizer Institute, Washington, D.C., 1991.
5. *Commercial Fertilizers—1990*, Bulletin Y-216, Tennessee Valley Authority, Muscle Shoals, Ala., 1991.
6. *Physical Properties of Fertilizers and Methods for Measuring Them*, bulletin Y-147, Tennessee Valley Authority, Muscle Shoals, Ala., 1979.
7. T. P. Hignett, ed., *Fertilizer Manual*, International Fertilizer Development Center, Muscle Shoals, Ala., 1978.
8. *Storage and Handling of Anhydrous Ammonia*, Bulletin K61.1, American National Standards Institute, Philadelphia, Pa., 1969.
9. *World Fertilizer Consumption Statistics—no. 24*, Bulletin A/92/146. International Fertilizer Industry Association, Paris, 1992.
10. *Fertilizer Grade Ammonium Nitrate: Properties and Recommended Methods for Packaging, Handling, Transportation, Storage and Use*, The Fertilizer Institute, Washington, D.C., p. 24.
11. G. D. Honti, "Urea," in *The Nitrogen Industry*, Akademia Kiado, Budapest, 1976.
12. T. E. Tomlinson, "Urea-Agronomic Implications," *Proceedings of the Fertiliser Society* no. 113, London, 1970.
13. R. W. R. Carter and A. G. Roberts, *Proc. No. 110, The Fertilizer Society (London)*, (Oct. 1969).
14. *Hydrocarbon Process*, **56**(11), 130 (Nov. 1977).
15. R. M. Reed and J. C. Reynolds, "Progress Report on Spherodizer Granulation of Anhydrous Melts," *Proceedings of the 24th Annual Meeting of the Fertilizer Industry Round Table*, Washington, D.C., 1974.

16. O. Skauli, "Pan Granulation of Ammonium Nitrate and Urea," paper presented at the *168th National Meeting of the American Chemical Society*, Atlantic City, N.J., 1974.
17. R. D. Young, and I. W. McCamy, *Can. J. Chem. Eng.*, **45**, 50–56 (1967).
18. G. S. Scott and R. L. Grant, *U.S. Bur. Mines Inf. Circ. 7463*, June 1948.
19. J. J. Burns and co-workers, *U.S. Bur. Mines Rep. Invest. 4994*, Aug. 1953.
20. *Report of the Ammonium Nitrate Working Party*, His Majesty's Stationery Office, London, 1951.
21. G. Perbal, *Proc. No. 124*, The Fertilizer Society, London, 1971.
22. J. D. Stafford, W. E. Samuels, and L. G. Croysdale, "Ammonium Nitrate Plant Safety," *Am. Inst. Chem. Eng.* (Sept. 27–29, 1965).
23. S. M. Lemkowitz, M. G. R. T. deCooker, and P. J. van den Berg, *Proc. No. 131,* The Fertilizer Society, London, Dec. 14, 1972.
24. A. V. Slack and G. M. Blouin, *Urea Technology: A Critical Review, Tennessee Valley Authority Circular Z-4*, TVA, Muscle Shoals, Ala., Dec. 14–16, 1969.
25. A. R. Shirley and co-workers, *Fixed Falling Curtain Granulation of Urea Using Evaporative Cooling, 29th Annual Meeting of the Fertilizer Industry Round Table*, Washington, D.C., Oct.–Nov. 1979.
26. L. M. Nunnelly, and W. C. Brummitt, *TVA's Experience in Producing and Marketing Urea LS, Proceedings of the 37th Fertilizer Industry Roundtable*, New Orleans, La., Nov. 2–4, 1987, pp. 47–53.
27. U.S. Pat. 4,589,904 (May 20, 1986), C. P. Harrison and C. Tittle (to TVA).
28. *Compaction—Alternative Approach for Granular Fertilizers, IFDC Technical Bulletin T-25*, International Fertilizer Development Center, Muscle Shoals, Ala., 1983.
29. M. H. McVicar, *Using Commercial Fertilizers*, 3rd ed., The Interstate Printers and Publishers, Inc., 1970.
30. T. P. Hignett and co-workers, in L. J. Carpentier, ed., *Proc. Tech. Conf. ISMA, Ltd.*, The Hague, Sept. 13–16, 1976, pp. 273–288.
31. J. R. Lehr and co-workers, *Colloque International Sur Les Phosphates Mineraux Solides*, Toulouse, France, Mar. 16–20, 1967.
32. G. H. McClellan and J. R. Lehr, *Am. Mineral.* **54**, 1374 (1969).
33. J. R. Lehr and G. H. McClellan, *A Revised Laboratory Reactivity Scale for Evaluating Phosphate Rock for Direct Application*, Bulletin Y-43, Tennessee Valley Authority, Muscle Shoals, Ala., Apr. 1972.
34. *Superphosphate: Its History, Chemistry, and Manufacture*, U.S. Dept. of Agriculture and Tennessee Valley Authority, Washington, D.C., Dec. 1964.
35. *New Developments in Fertilizer Technology*, 8th demonstration, Bulletin Y-12, Tennessee Valley Authority, Muscle Shoals, Ala., 1970.
36. A. W. Frazier, J. R. Lehr, and E. F. Dillard, *Chemical Behavior of Fluorine in the Production of Wet-Process Phosphoric Acid*, Bulletin Y-113, TVA, Muscle Shoals, Ala., May 1977.
37. *New Process for the Production of Phosphatic Fertilizers Using Hydrochloric Acid*, Fertilizer Industry Series Monograph No. 5, UNIDO, Vienna, 1969.
38. J. R. Lehr, A. W. Frazier, and J. P. Smith. *J. Agric. Food Chem.* **14**, 27 (1966).
39. N. Robinson, in F. E. Khasawneh, E. C. Sample, and E. J. Kamprath, eds., *The Role of Phosphorus in Agriculture*, American Society of Agronomy, Madison, Wis., 1979.
40. R. F. Jameson, in A. V. Slack, ed., *Phosphoric Acid*, Vol. I. Part 1, Marcel Dekker, Inc., New York, 1968, pp. 157–404
41. *Eur. Chem. News* **21**, 526 (Mar. 1972).
42. R. L. Somerville, paper presented to *Div. Fert. Soil Chem., ACS*, Chicago, Aug. 27, 1973.

43. U.S. Pat. 3,642,439 (Oct. 15, 1969), W. P. Moore and co-workers (to Allied Chemical Corp.).
44. *Phosphorus Potassium*, **79**, 25 (Sept.–Oct. 1975).
45. A. Baniel and R. Blumberg, in A. V. Slack, ed., *Phosphoric Acid*, Vol. I, Part II, Marcel Dekker, Inc., New York, 1968, pp. 887–913.
46. *Israel Mining Industries Staff Report*, International Solvent Extraction Conference, The Hague, 1971.
47. R. Blumberg, *Proc. No. 151,* The Fertilizer Society, London, Nov. 13, 1975.
48. *ISMA Technical Conference (Seville)*, Institute for Research and Development, Israel, Nov. 20, 1972.
49. J. F. McCullough, *Proc. 26th Fert. Ind. Round Table*, **49** (Oct. 26–28, 1976).
50. *Eng. Mining J.* **178**(5), 9 (May 1977).
51. F. P. Achorn, B. Wright, Jr., and H. L. Balay, *Fertilizer Solutions Mag.* **20**(3), (June 1976).
52. A. B. Phillips and co-workers, *J. Agr. and Food Chem.* **6**(7), 585 (July 1958).
53. H. L. Thompson and co-workers, *Ind. Eng. Chem.* **42**, 2176–2182 (Oct. 1950). (TVA CD-130)
54. J. G. Getsinger, E. C. Houston, and F. P. Achorn, *J. Agr. Food Chem.* **5**, 433–436 (June 1957).
55. R. D. Young, G. C. Hicks, and C. H. Davis, *J. Agr. Food Chem.* **10**, 442–447 (Nov. 1962).
56. U.S. Pat. 3,153,574 (Oct. 20, 1964), F. P. Achorn, R. D. Young, and G. C. Hicks (to TVA).
57. R. S. Fittel and L. A. Hollingsworth, "Manufacture of Ammonium Phosphates Using a Pipe Reactor Process," *Proceedings of the 27th Meeting of the Fertilizer Industry Round Table*, Washington, D.C., pp. 70–80, 1977.
58. *Phosphorus and Potassium*, **87**, 33–36 (1977).
59. R. J. Danos, *Chem. Eng.* **85**(22), 81–83 (1978).
60. *New Developments in Fertilizer Technology*, Bulletin Y-136, Tennessee Valley Authority, Muscle Shoals, Ala., 1978, pp. 52–55.
61. G. C. Hicks, *Tennessee Valley Authority,* Bulletin Y-119, TVA, Muscle Shoals, Ala., Nov. 1977.
62. R. D. Young and G. C. Hicks, *Comm. Fert.* **114**(2), 26–27 (Feb. 1967).
63. I. S. E. Martin and F. J. Harris, in A. I. More, ed., *Granular Fertilizers and Their Production*, British Sulphur Corp., Ltd., London, 1978, pp. 181–197, 253–270.
64. H. Storen, "The Nitrophosphate Process—an Alternative Route to Phosphate Fertilizers," in proceedings of *Phosphate Fertilizers and the Environment*, International Fertilizer Development Center, Muscle Shoals, Ala., 1992.
65. R. Ewell, paper presented at *UNIDO, 2nd Interregional Fertilizer Symposium*, Kiev, Sept. 1971.
66. A. V. Slack and co-workers, *Farm. Chem.* **130**, (Apr.–July 1967); *Phosphorus Potassium*, Part I (39), 18 (Jan.–Feb. 1969); Part II, 29 (Mar.–Apr. 1969).
67. J. A. Stewart, in R. D. Munson, ed., *Potassium in Agriculture*, ASA-CSSA-SSSA, 1985, p. 83.
68. *Phosphorus Potassium*, (2), 32 (1962).
69. G. R. Hagstrom, *Fert. Sol.* **17**(2), 76, 78, 80, 86, 88 (1973).
70. A. T. Brook, "Developments in Granulation Techniques," *Proceedings of the Fertiliser Society (London)*, no. 47, 1957.
71. L. D. Yates, F. T. Nielson, and G. C. Hicks, *Farm Chemicals*, Part I and Part II (Aug. 1954).
72. U.S. Pat. 2,741,545 (Apr. 10, 1956), F. T. Nielsson (to TVA).

73. R. D. Young, G. C. Hicks, and C. H. Davis, *J. Agr. Food Chem.* **10**, 68–72 (Jan.–Feb. 1962).
74. *New Developments in Fertilizer Technology*, 9th demonstration, Bulletin Y-50, Tennessee Valley Authority, Muscle Shoals, Ala., 1972.
75. *New Developments in Fertilizer Technology*, 11th demonstration, Bulletin Y-107, Tennessee Valley Authority, Muscle Shoals, Ala., 1976.
76. J. Medbery, *Proc. Fert. Ind. Round Table*, 52–55 (1977).
77. K. J. Baggett, and D. J. Brunner, *Proc. Fert. Ind. Round Table*, 64–70 (1977).
78. R. S. Fittell, L. A. Hollingworth, and J. G. Forney, *Proc. Fert. Ind. Round Table*, 70–81 (1977).
79. R. G. Lee, M. M. Norton, and H. G. Graham, *Proc. 24th Fert. Ind. Round Table, 1974*, 79 (Dec. 3–5, 1976).
80. *NPK Fertilizer Production Alternatives*, special report SP-9, International Fertilizer Development Center, Muscle Shoals, Ala., 1988.
81. R. Zisselmar, in Ref. 75.
82. L. Taylor, in Ref. 75.
83. C. Rodriguez, in Ref. 75.
84. N. L. Hargett and R. Pay, "Retail Marketing of Fertilizers in the U.S.," presented at the *Fertilizer Industry Round Table*, Atlanta, Ga., 1980.
85. G. Hoffmeister and G. H. Megar, "Use of Urea in Bulk Blends," *Proceedings of the 25th Annual Meeting of the Fertilizer Industry Round Table*, Washington, D.C., 1975, pp. 212–226.
86. G. Hoffmeister, S. C. Watkins, and J. Silverberg. *J. Agri. Food Chem.* **12**(1), 64–69 (1964).
87. G. Hoffmeister, *Proceedings of the TVA Fertilizer Bulk Blending Conference*, TVA Bulletin Y-62, Tennessee Valley Authority, NFDC, Muscle Shoals, Ala., pp. 59–70.
88. G. Hoffmeister, *Particle Size Requirements for Bulk Blend Materials*, report Z-146 Tennessee Valley Authority, Muscle Shoals, Ala., 1982.
89. *SGN—A System of Materials Identification*, brochure, Canadian Fertilizer Institute, Ottawa, Ontario, Canada, 1982.
90. D. McKnight and M. M. Striplin, *J. Agr. Chem.* **13**, 33–34 (Aug. 1958).
91. J. M. Potts, ed., *Fluid Fertilizers*, Bulletin Y-185, Tennessee Valley Authority, Muscle Shoals, Ala., 1984.
92. A. L. Huhti and P. A. Gartaganis, *Can. J. Chem.* **34**, 785 (1956).
93. R. S. Meline, R. G. Lee, and W. C. Scott, *Fert. Soln.* **16**(2), 32–45 (Mar./Apr. 1972).
94. U.S. Pat. 3,775,534 (Nov. 27, 1973), R. S. Meline (to TVA).
95. E. W. Sawyer, J. A. Polon, and H. A. Smith, *Solutions*, 36–43 (Jan.–Feb. 1960).
96. C. Mann, II, K. E. McGill, and T. M. Jones, *I&EC Prod. Res. Dev.*, 488–495 (Sept. 1982).
97. U.S. Pat. 4,337,079 (June 29, 1982), H. C. Mann and R. S. Meline (to TVA).
98. M. F. Broder, in Ref. 86.
99. J. J. Mortvedt and co-eds., *Micronutrients in Agriculture*, Soil Science Society of America, Madison, Wis., 1972.
100. J. Silverberg, R. D. Young, and G. Hoffmesiter, Jr., in Ref. 99, pp. 431–458.
101. J. R. Lehr, in Ref. 99, pp. 459–503.
102. D. E. Nichols, P. C. Williamson, and D. R. Waggoner, paper presented at *The Steenbock-Kettering International Symposium on Nitrogen Fixation*, Madison, Wis., June 12–16, 1978.
103. Paper presented to *Commission on Natural Resources*, National Academy of Sciences, Washington, D.C., 1975.
104. H. Brown, in J. M. Hollander, ed., *Annual Review of Energy*, Vol. 1, Annual Reviews, Inc., Palo Alto, Calif., 1976, pp. 1–36.

105. J. B. Cathcart, in F. E. Khasawneh, E. C. Sample, and E. J. Kamprath, eds., *The Role of Phosphorus in Agriculture*, American Society of Agronomy, Madison, Wis., 1979, Chapts. 1 and 2.
106. E. A. Imhoff, *U.S. Geological Survey Open File Report No. 76-648*, U.S. Government Printing Office, Washington, D.C., 1976.
107. *World Survey of Potash Resources*, 2nd ed., The British Sulphur Corp. Ltd., London, 1966.
108. "Mineral Facts and Problems," *U.S. Bur. Mines Bull.* **667**, (1975).
109. J. A. Ober, *Sulfur,* annual report, U.S. Bureau of Mines, Washington, D.C., 1991.
110. A. J. Bodenlos, *Sulfur,* U.S. Geological Survey Professional Paper 820, Washington, D.C., 1972.
111. D. W. Bixby, in Ref. 105, Chapt. 1.
112. W. J. Free, B. J. Bond, and J. L. Nevins, *Changing Patterns in Agriculture and Their Effect on Fertilizer Use*, Bulletin Y-106, TVA Fertilizer Conference, July 27–28, 1976, Tennessee Valley Authority, Muscle Shoals, Ala.
113. *The Fertilizer Industry—The Key to World Food Supplies*, International Fertilizer Industry Association, Paris, 1986.
114. *Fertilizer Investment and Production Costs*, report of the World Bank to the 9th Session of the FAO Commission on Fertilizers, Rome, Feb. 1985.

GEORGE HOFFMEISTER
Consultant

FIBERBOARD. See WOOD-BASED COMPOSITES AND LAMINATES.

FIBER OPTICS

Optical communication was long considered as a possibility for high speed data transmission because light at terahertz frequency is capable of enormous bandwidth. The power of lightwave communication could not be tapped, however, until the obstacle of a suitably transparent transmission medium was overcome. For example, in 1880 Alexander Graham Bell patented the photophone, a device that utilized the atmosphere as a transmission medium. Here a narrow beam of sunlight was focused on a thin mirror, soundwaves produced by the human voice caused the mirror to vibrate, and the light was transmitted to a selenium detector. The resistance of the selenium varied proportionally, producing a current variation in the receiver that was akin to speech waves. The transmission distance limit of the system is dependent on the loss encountered along the transmission path, combined with the brightness of the light source and the sensitivity of the detector. In the case of the photophone, disruptions from rain, fog, smoke, clouds, and other atmospheric disturbances severely limited the maximum distance.

The invention of the laser in 1958 prompted the beginning of the story of optical fiber communications. This device was capable of producing a high intensity, coherent beam of light which could be modulated at a high rate (see LASERS). Still, no transmission medium of suitable clarity was available. A number of sophisticated systems for transmitting light signals were produced (1,2) using a closed system of pressurized aluminum pipes, thus solving the problem of opacity caused by atmospheric disturbances. However, the use of lenses to control beam divergence introduced reflection losses that limited the range of such systems. Lens losses were minimized upon the invention of a gas lens (3) created by uniform heating of the gas in the tube, forming a density and therefore a refractive index gradient in the gas to focus the beam. Although this system worked well, systems of buried metal pipes were expensive and could only be economical for very wide band communications networks. The need was evident for a low loss transmission medium which could be produced in long lengths at a reasonable cost.

Transmission of light through thin glass fibers was proposed in the 1960s (4). However, at that time no glass (qv) existed with optical absorption sufficiently low to be practical. In addition to low loss transmission, the light must be guided by a refractive index structure such that a cylindrically symmetric composite, having a core of higher refractive index than the surrounding cladding, confines the light by the process of total internal reflection. It was predicted that losses of less than 20 dB/km could be achieved in glass if transition metals, initially present in the starting materials or added during processing, could be limited to a level of ca 1 ppm. Strides in reducing impurities and fabricating waveguide structures by a variety of methods were made to the point where by the mid-1980s, near-intrinsic optical losses in silica glass were achieved. During this same period the guiding properties in lightguides were improved to realize the high bandwidth capability of the medium. The number of voice channels available has also increased dramatically as higher bit-rate systems are manufactured to use more of the fiber's potential bandwidth. As of 1993, regenerative undersea fiber systems are in operation that have capacities of 80,000 voice channels per fiber pair. These systems are limited more by the electronics driving the laser sources than by the properties of the glass fiber used to transmit the signal. This limitation has precipitated the development of a new generation of fiber-optic systems where the glass fiber, once solely a passive component, is becoming an active amplifying component. Rare-earth-doped fibers are being manufactured for use in optical amplifiers. When pumped by light of an appropriate wavelength, an optical signal triggers emission at the signal wavelength, resulting in amplification.

In early telephones, sound (voice) waves caused a carbon microphone's resistance to vary, thus varying the current flowing in a series external circuit. This d-c current could then be used to regenerate voice waves in a receiver. Two wires were required to carry a single conversation. With time, telecommunications traffic was encoded on a-c carriers, at first using amplitude or frequency modulation, and more recently pulse code modulation.

The growth in the number of circuits necessary to meet the growing demand for capacity prompted the use of ever higher carrier frequencies to increase capacity. This advance was limited by two obstacles. First, the resistance of the wire increases with higher frequencies, as current is carried on a thinner and thinner surface layer of the metallic conductors. Second, noise becomes incorporated onto the signal during analogue transmission and results in both degradation and distortion of the signal. The problems generated by the use of higher carrier frequencies were initially circumvented by the use of single-sideband microwave radio links. These systems first operated from towers spread across the continent and then from Telstar and other satellites which provided up to 1800 voice circuits when operated at 4 GHz (4×10^9 cycles per second). The use of radio was eventually limited by the allocation of suitable frequencies.

The development of digital encoding solved the problem of signal degradation noise. Using this method, the amplitude of an analogue waveform is sampled at frequent intervals and coded into a binary (0s and 1s) sequence of digits. A voice waveform requires the transmission of 64,000 bits/s, which are digitally transmitted by a series of positive and negative electrical pulses. The optical equivalent uses pulses of light transmitted over glass fiber. Unlike the electrical signal, laser light may be modulated at a frequency high enough (10s of gigahertz) for any conceivable communication need.

Figure 1 gives a comparison of analogue and digital transmission schemes. The incoming signal (Fig. 1**a**) can be transformed directly into an intensity variation of the light beam (Fig. 1**b**). A photodetector at the receiver converts this varying intensity into an electrical signal which is then amplified to reproduce the original waveform. Such a signal becomes increasingly degraded and distorted during transmission and amplification. Improved fidelity is provided by digital encoding (Fig. 1**c**). In the digital scheme the signal is encoded by flashes of light at regularly timed intervals. The sampling rate must be twice that of the highest frequency component for accurate representation of the waveform. A voice signal having a maximum frequency of 4000 Hz must be sampled at a rate of 8000/s. The binary coding of 0s and 1s corresponds to the absence and presence of light. Representation of the height of a voice waveform requires eight bits (a bit is a 0 or a 1). Therefore, to sample a voice wave for one second the digital system requires 64,000 bits (8,000 samples $\times$ 8 bits/sample). Although the intensity of the light signal diminishes over distance, as long as it remains above the threshold of the detector the signal can be cleanly regenerated because a pulse is either present or absent. In this manner, noise is eliminated.

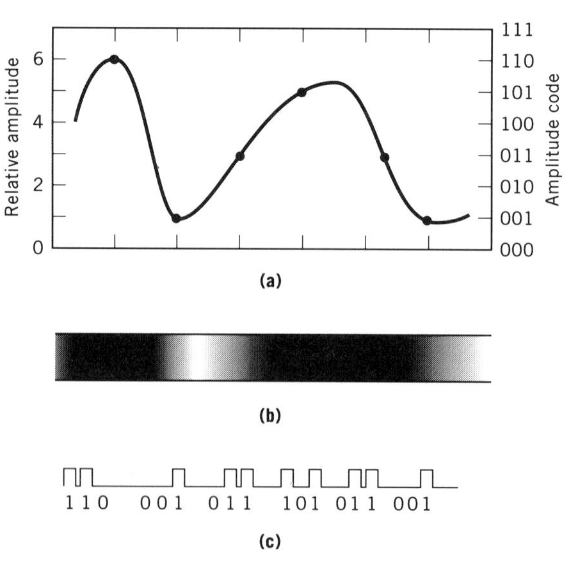

Fig. 1. (**a**) A transmission signal and its (**b**) analogue and (**c**) digital encoding.

Principles of Light Guidance

Light guidance is governed by the structure of the lightguide itself. The refractive index, n, of a material is defined as the ratio of the speed of light in a perfect vacuum to the speed of light through that material. This property is a function of both the composition of the material and the wavelength of the transmitted light. The higher the refractive index of a material, the more light is retarded, or slowed, in passing through it. At the interface of two materials of different refractive indexes, light is refracted, or bent toward the higher index medium by an angle the sine of which is proportional to the relative indexes of the two media. This property is known as Snell's law. Light that travels in a medium of higher refractive index and impinges on the surface of a medium with a lower refractive index at an angle less than the critical angle, θ_c, is totally reflected. As shown in Figure 2 for a variety of waveguide structures, this property of total internal reflection provides a means to transmit light over long distances without radiative losses.

The ability of a waveguide to collect light is determined by the numerical aperture (NA) which defines the maximum angle at which light entering the fiber can be guided.

$$\mathrm{NA} = \sin\theta_c = (n_1^2 - n_2^2)^{1/2} - n_1(2\Delta)^{1/2} \tag{1}$$

where

$$\Delta = (n_1 - n_2)/n_1 \tag{2}$$

and typically $\Delta << 1$; n_1 is the refractive index of the core; and n_2 is the index of the cladding. Lightguide structures are shown in Figure 3. Whereas Figures 3**a** and 3**b** are multimode structures having relatively higher Δs and core diameters on the order of 50 μm, Figure 3**c** is a single-mode fiber of lower refractive index and a core diameter < 10 μm. The number of modes that can propagate in a fiber is governed by Maxwell's equations for electromagnetic fields, and is related to a dimensionless quantity V called the normalized frequency:

$$V = (2\pi a/\lambda)\,\mathrm{NA} \simeq (2\pi a n_1/\lambda)\,(2\Delta)^{1/2} \tag{3}$$

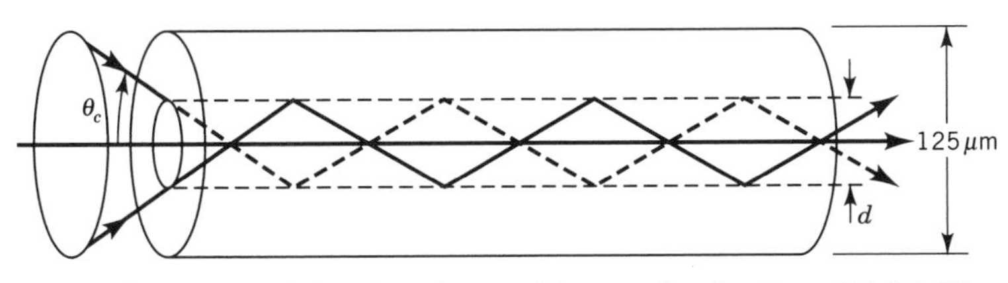

Fig. 2. Waveguide structure showing the total internal reflection of light. The diameter, d, is 50 μm for a standard multimode system, 62.5 μm for a large core multimode, and from 4 to 10 μm for a single mode.

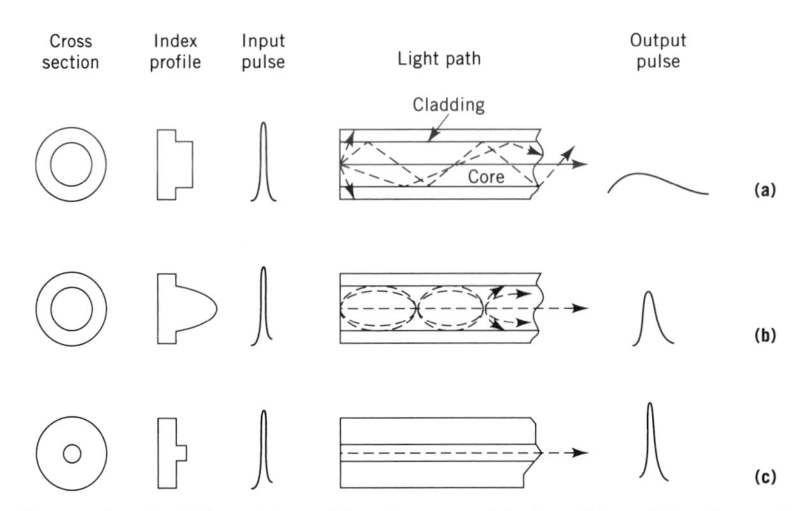

Fig. 3. Types of optical fiber: (**a**) multimode stepped index, (**b**) multimode graded index, and (**c**) single-mode stepped index.

where λ is the wavelength of light in vacuum and a is the radius of the fiber core. For example, as the core becomes smaller, there are fewer paths for the light to undergo total internal reflection. When V is less than 2.405 for a stepped index core profile only a single mode of light, the fundamental mode, can propagate. All other modes are cut off. This governs the design of single-mode fibers as shown in Figure 3**c**.

Attenuation. The exceptional transparency, or low attenuation, of silica-based glass fibers has made them the predominant choice for optical transmission because of the low level of absorption and scattering of light as it traverses the material. Together these comprise optical attenuation, or loss, measured in dB where

$$\text{loss (dB)} = 10 \log (I_0/I) \tag{4}$$

I_0 is the input intensity and I is the output intensity. Values for loss are typically given per kilometer of fiber. The window of transparency for silica-based glass, shown in Figure 4, is bounded at short wavelength because of electronic transitions. At longer wavelengths molecular vibrations cause attenuation. Between these two regions is the transmission window in which the attenuation is limited mainly by Rayleigh scattering. The ultraviolet absorption edge is determined by the electronic band gap of the material. It decays exponentially with increasing wavelength and becomes negligible at infrared wavelengths. Rayleigh scattering results from glass composition and density fluctuations on a scale shorter than the wavelength of light. These losses decrease as the fourth power of the wavelength of light ($\alpha\lambda^{-4}$). Extrinsic loss mechanisms, such as absorption caused by transition-metal or hydroxyl ion, OH^-, contamination, may dominate over Rayleigh scattering in the transmission window of interest. At longer wavelengths absorption arises from oxygen–cation multiphonon vibrational modes in the glass lattice.

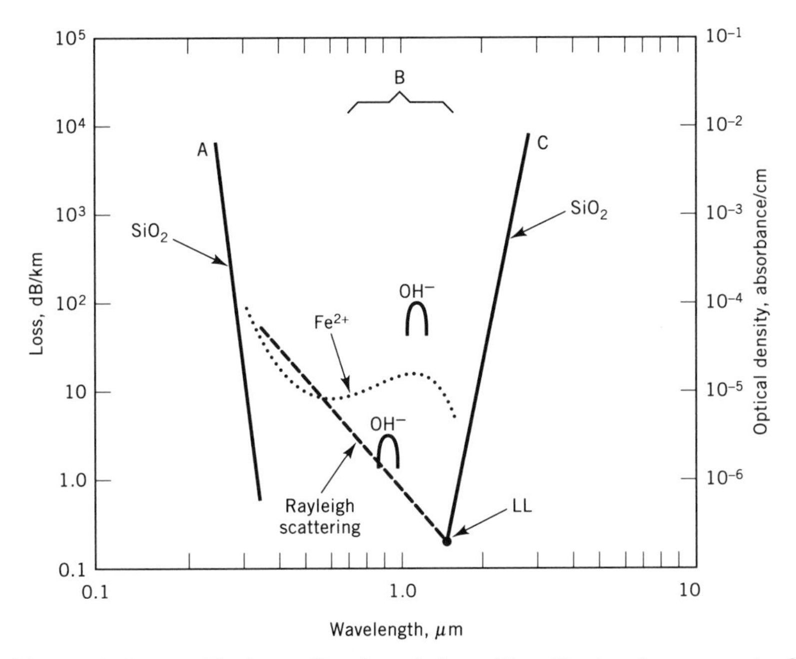

Fig. 4. Transmission profile for a silica-based glass fiber. Region A represents electronic transitions; B, the transmission window; and C, molecular vibrations. Point LL is the lowest loss observed in an optical fiber. Absorption profiles for (——) OH^- and (···) Fe^{2+} are also shown. See text.

Additional optical attenuation may result from large-scale imperfections or defects in the glass structure as well as waveguide imperfections formed during processing of the glass. Fluctuations longer than the wavelength of light, such as diameter variations, may cause Mie scattering. Even in a nearly perfect glass, absorptions from low level cation impurities, eg, on the order of 1 ppm as shown in Figure 4, are detrimental. Similarly, point defects in the anion network can play a role in controlling loss; suboxides of germanium can be formed at high temperature in an oxygen-deficient reaction and result in coloration, especially after exposure to radiation (5). If these extrinsic losses are avoided, as is typical of current production, then fiber attenuation is dominated by Rayleigh scattering and decreases as λ^{-4} until the multiphonon edge is intersected. The lowest loss for SiO_2 glasses is thereby achieved between 1.3 and 1.55 μm, although the OH^- overtone absorption at 1.38 μm often limits loss around that wavelength. Even for low attenuation, system performance was limited by pulse broadening or dispersion effects and designers generally exploited the shorter wavelength side of the curve because chromatic dispersion can be made zero at 1.3 μm. However, by proper fiber design the minimum dispersion can be shifted to 1.55 μm where higher transparency is advantageous for long-distance transmission.

Dispersion. The effects of dispersion on the ultimate system performance are as important as the attenuation. Dispersion arises from the variation of the velocity of light with the wavelength of the light. Two types of dispersion which occur are intermodal, found only in multimode fibers, and chromatic, which is important for single-mode performance. Intermodal dispersion relates to the delay differences experienced by modes traveling in different regions of the waveguide. For example, the fundamental mode travels in a straight path, whereas higher order modes travel a helical route (Fig. 3**b**). To minimize this effect, multimode fibers are designed having refractive index gradients such that the lower order modes are more heavily retarded by higher index glass at the center of the core. A minimum in intermodal dispersion is achieved when the index distribution varies as $n_1[1 - \Delta(r/a)^{\alpha}]$, where n_1 is the refractive index of the core; Δ is the relative index difference; r is the radial position at which the calculation is made (at the outer edge of the core $r = a$; in the center $r = 0$); a is the core radius; and α, the profile coefficient, is approximately 2. The optimum value for α is wavelength- and compositionally dependent.

In single-mode fibers intermodal dispersion is not a factor; however, pulse broadening occurs owing to chromatic dispersion. Optical sources are not spectrally pure and typically emit light over a narrow range of wavelengths. The speed of light, controlled by the refractive index, is dependent on wavelength, and slightly different wavelengths travel at different velocities because of two effects: material dispersion and waveguide dispersion. Fortuitously, as can be seen in Figure 5**a**, these effects are opposite in sign and may be cancelled. Material dispersion, which is essentially independent of the fiber's structure or, for high silica glasses, its glass composition, is a nonlinear property and rises steeply with wavelength. Waveguide dispersion, controlled by the fiber design, can be optimized to cancel material dispersion. For waveguide designs, where the glass consists primarily of silica doped with small amounts of other ions, the zero dispersion crossover occurs near 1.3 μm.

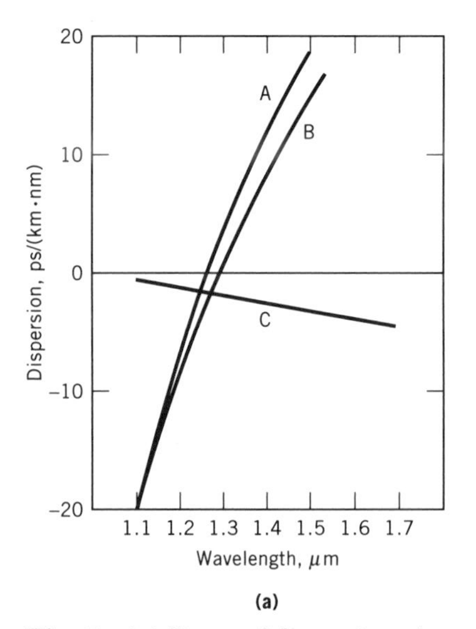

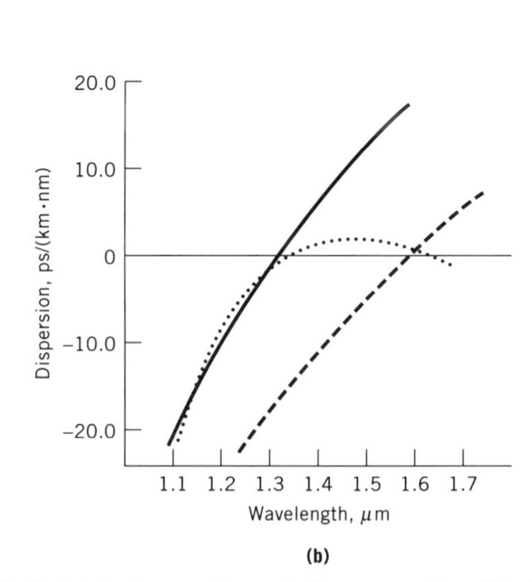

Fig. 5. (**a**) Types of dispersion: A, material; B, total chromatic; and C, waveguide. (**b**) Dispersion in single-mode lightguides where (—) corresponds to a 1.3 μm operation having a profile of (⎍); (- - -) to a dispersion shifted smaller core higher index system having a (⎍⋀) profile; and (····) to an ultrabroad band of more complex structure (profile).

Waveguide dispersion depends on how much of the power travels in the cladding of the fiber relative to the amount traveling in the core. Proper design of the fiber's core diameter and refractive index profile may be fine-tuned to completely cancel the material dispersion at a given wavelength, or to flatten the total dispersion over a range of wavelengths (6). In Figure **5b** the relationship between fiber profile and waveguide dispersion is illustrated. The shift to longer wavelength operation to take advantage of lower optical losses has led to more complex designs, as seen in the dotted curve where the zero dispersion crossover occurs at two wavelengths of low loss in the fiber allowing multichannel operation at both 1.3 μm and 1.55 μm in a single fiber.

Optical Fiber Fabrication

Viable glass fibers for optical communication are made from glass of an extremely high purity as well as a precise refractive index structure. The first fibers produced for this purpose in the 1960s attempted to improve on the quality of traditional optical glasses, which at that time exhibited losses on the order of 1000 dB/km. To achieve optical transmission over sufficient distance to be competitive with existing systems, the optical losses had to be reduced to below 20 dB/km. It was realized that impurities such as transition-metal ion contamination in this glass must be reduced to unprecedented levels (see Fig. 4).

Double Crucible. The earliest attempts at producing high purity glass employed the double crucible technique. Low optical losses were achieved by purifying the starting materials to a high degree and taking care not to contaminate them during the forming process (7,8). Soda lime silica glasses and sodium borosilicate glasses were prepared from materials purified to parts per billion levels of transition metals. The methods used in the purification included solvent extraction, recrystallization, ion-exchange, and electrolysis. After the bulk material was melted and fined, a process by which gas is removed from the melt to minimize bubble formation, it was drawn into cane. This cane was then fed to a continuous casting system of concentric platinum crucibles, shown in Figure 6. The glass for the core was fed into the inner crucible, exiting as a thin glass stream which entered the second crucible where it was coated with glass of a different composition (lower index) to make a guiding structure. By controlling the draw rate, ionic diffusion between the glasses in the lower crucible produced the gradient in the refractive index necessary to minimize intermodal dispersion.

In spite of the elegance of this technique, contamination occurred during processing, leading to impurity levels in the glass on the order of parts per million rather than the ppb levels necessary. Many ingenious methods were devised in an attempt to eliminate these impurities. The most successful of these used control of the oxygen partial pressure of the processing atmosphere during manufacture. Absorption owing to the two principal contaminants, iron and copper, could be reduced by altering the valence state. The iron could be oxidized to the Fe^{3+} and copper reduced to Cu^{+}, diminishing the strong absorptions in the near-ir by Fe^{2+} and Cu^{2+}.

The best fibers, installed in systems operating at 0.9 μm, had losses of 5 dB/km. The lower intrinsic losses in the 1.3 to 1.55-μm window were unattainable

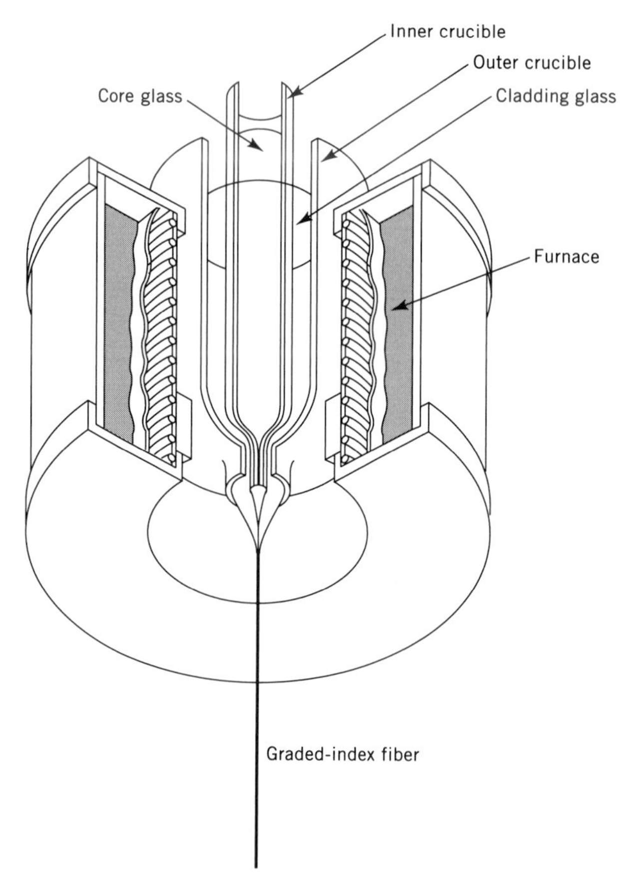

Fig. 6. Double crucible configuration showing core glass (upper crucible) flowing through molten cladding glass in lower crucible to yield graded-index fiber.

by this technique. Fundamental cation–oxygen vibrational modes as well as OH^- contamination were intrinsic to the compositions.

High silica glasses having lower losses at wavelengths from the visible to the infrared became available through technology using vapor-phase techniques where silicon tetrachloride [*10026-04-7*], $SiCl_4$, is the precursor to silicon oxide [*7631-86-9*], SiO_2, of near-intrinsic purity. Other compositions could be produced as well using other chloride vapors such as phosphorus oxychloride [*10025-87-3*], $POCl_3$, and germanium tetrachloride [*10038-98-9*], $GeCl_4$, to dope the silica and provide changes in the refractive index of the glass. Two methods evolved. Inside vapor-phase deposition followed from chemical vapor deposition (CVD) processes used in the electronics industry (see ELECTRONIC MATERIALS; INTEGRATED CIRCUITS). When adapted for glass fabrication, chloride reactants were mixed with oxygen inside a silica tube which was externally heated. The chlorides were oxidized rather than hydrolyzed, producing particles of SiO_2 which adhered to the inner wall of the tube. The outside processes stemmed from the work in the 1930s (9) utilizing flame hydrolysis of SiO_2 to produce silica articles such as mirror blanks. Here $SiCl_4$ together with dopant precursors form a soot by passing the chlorides through a fuel–oxygen flame and depositing submicrometer-sized oxide particles on a mandrel.

Inside Processes. *Modified Chemical Vapor Deposition.* Inside processes such as modified chemical vapor deposition (MCVD) followed CVD techniques. In these the concentration of the reactants was kept low to inhibit homogeneous gas-phase reaction in favor of heterogeneous wall reactions which produce a vitreous (amorphous), particle-free deposit on the tube wall. Deposition was continued until a sufficient thickness of glass was produced; then the tube was collapsed to form a solid rod, or preform. This preform was then drawn to a relatively low loss fiber (10). Deposition rates by this method were impractically low, and attempts to increase them invariably led to the formation of particles and, subsequently, bubbles, resulting in excess scattering loss. The solution was to radically increase the reactant concentration and flow rates to more than 10 times those used for CVD. This led to the formation of SiO_2 and germanium oxide [*1310-53-8*], GeO_2, particles in the gas stream that were then deposited on the tube wall interior and fused by a traversing external torch. Multiple layers were deposited to achieve the desired refractive index profile before the tube was collapsed to a preform. The process was refined (11) to that depicted in Figure 7. First a high quality

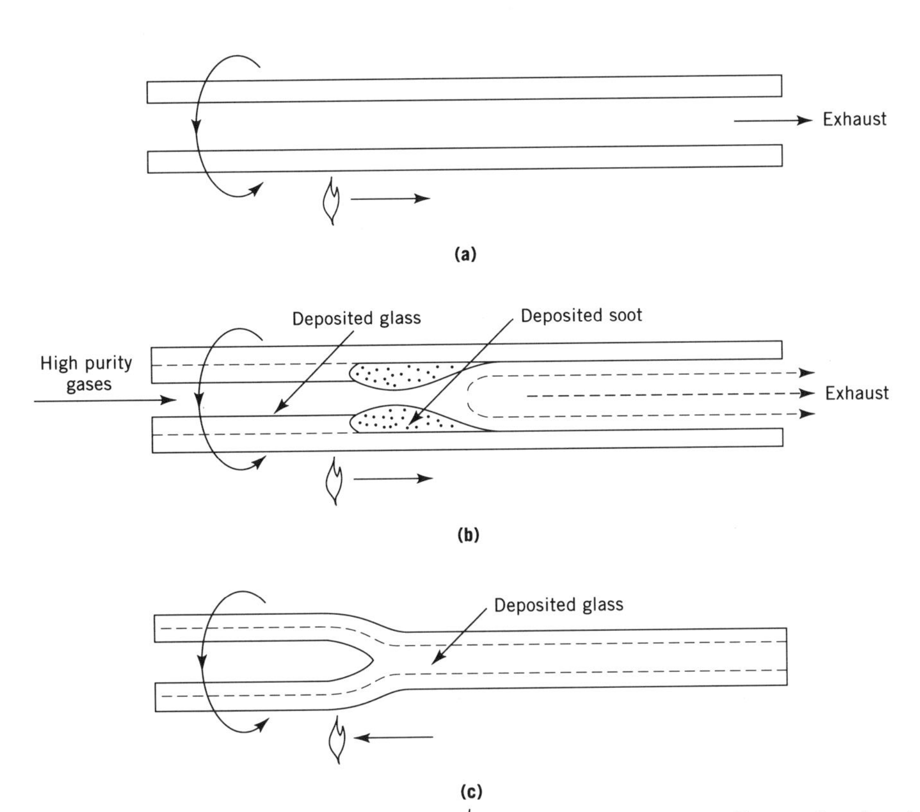

Fig. 7. Fiber formation by MCVD, where (flame symbol) represents the O_2–H_2 flame, showing (**a**), tube setup; (**b**), deposition; and (**c**), collapse.

fused silica tube is mounted on a glass working lathe with a rotating joint for the injection of the reactants. High purity gases are then introduced into the rotating tube, which is traversed by an oxy–hydrogen torch. A homogeneous gas-phase reaction occurs in the hot zone created by the torch. The particles are small enough to be carried in the gas stream and deposited onto the tube walls downstream of the torch. As the torch passes, the particulate layer sinters to form a thin layer of highly pure oxide glass. The torch temperature must be kept high enough to fuse the glass layer without deformation of the substrate tube, as repeated deposition cycles are necessary to build up, layer by layer, the desired refractive index profile. The composition of each layer may be changed in order to manufacture highly complex structures. Typically 25 to 100 layers may be deposited to form either a graded-index multimode or single-mode optical fiber preform.

One of the areas critical to the MCVD process was understanding the chemistry of the oxidation reactions. It was necessary to control the incorporation of GeO_2 while minimizing OH^- formation. Additionally, understanding the mechanism of particle formation and deposition was critical to further scale-up of the process.

Thermophoretic Deposition. A fundamental understanding of how particles form and deposit on the tube walls was necessary for optimization of the deposition process. SiO_2 particles which form homogeneously in the gas phase have particle sizes of 0.02 to 0.1 μm and are carried in the gas stream. Without the imposition of a thermal gradient in the tube the particles remain in the gas stream and pass out of the tube; however, owing to temperature gradients caused by the traversing torch, thermophoresis occurs (12,13). Thermophoresis is a process by which particles in a temperature gradient travel toward the cooler region when bombarded by more energetic particles on the hot side. In this case travel is outward from the center to the tube wall. Within the MCVD substrate tube the wall is only hotter than the gas in the hot zone of the torch where the oxidation reaction is occurring. Downstream the gas is hotter than the tube because of differences in thermal conductivity of the gas and silica, as well as to the flow of the gas. In the MCVD process (Fig. 7) the reactants enter the tube, are reacted in the hot zone of the torch, deposit thermophoretically downstream of the torch, and are subsequently sintered to a clear glass as the torch passes over the deposited particulate layer. Once the desired structure has been deposited, the direction of the torch is reversed and the tube is collapsed to form a solid preform.

Chemical Equilibria. The chemistry of the MCVD process has been studied using a variety of techniques. Infrared (ir) spectroscopy (14) was used to investigate the oxidation reaction of $SiCl_4$ and $GeCl_4$ (see INFRARED AND RAMAN SPECTROSCOPY). The effluent gases from an MCVD reactor were fed into an infrared cell and analyzed to determine the reaction products as a function of temperature and composition. As the hot-zone temperature reached 1300 K, $SiCl_4$ began to react and form silicon oxychloride [*14986-21-1*], Si_2OCl_6. Figure 8 shows that the partial pressure of silicon oxychloride increases to a maximum at 1450 K, above which concentrations of $SiCl_4$, Si_2OCl_6, and $POCl_3$ decrease until by 1750 K these concentrations become negligible in the effluent because all of the species have been converted to oxides. Germanium tetrachloride, $GeCl_4$, behaves some-

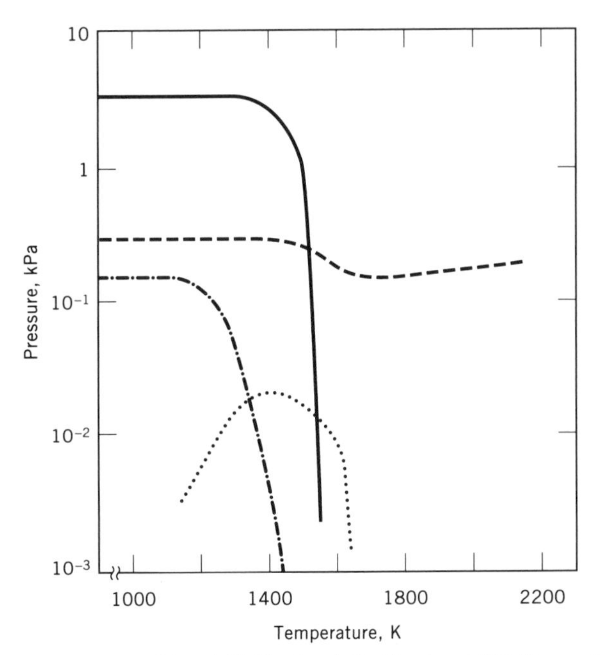

Fig. 8. Partial pressures of (——) $SiCl_4$, (- - -) $GeCl_4$, (–·–) $POCl_3$, and (·····) Si_2OCl_6.

what differently. Up to 1500 K the pressure is constant; it then decreases until the reactor temperature reaches 1700 K, at which point it again stabilizes and grows slightly up to 2200 K. The majority of the initial species remains unreacted and is swept out of the reactor in the effluent. It was concluded from these results that at temperatures lower than 1600 K the degree of reaction of the $SiCl_4$, $GeCl_4$, and $POCl_3$ is controlled by reaction kinetics, whereas at higher temperatures the reaction is dominated by thermodynamic equilibria. Rate studies have shown that the residence times typically experienced in the hot zone are sufficient to reach equilibrium above 1700 K. The oxidation reactions for $SiCl_4$ and $GeCl_4$ are

$$SiCl_4(g) + O_2(g) \rightarrow SiO_2(s) + 2\ Cl_2(g) \tag{5}$$

$$GeCl_4(g) + O_2(g) \rightarrow GeO_2(s) + 2\ Cl_2\ (g) \tag{6}$$

The equilibrium constants for these reactions may be written as

$$K_{SiO_2} = (a_{SiO_2})\ (P_{Cl_2})^2/\ (P_{SiCl_4})\ (P_{O_2}) \tag{7}$$

and

$$K_{GeO_2} = (a_{GeO_2})\ (P_{Cl_2})^2/\ (P_{GeCl_4})\ (P_{O_2}) \tag{8}$$

where P_i are the partial pressures of the gaseous species, and a_i are the activities of the solid species. An approximation for the activities may be made using $\gamma_i \chi_i$, where χ_i are the mole coefficients. An activity coefficient of 1 implies ideal behavior, ie, obeying Raoult's law. The equilibrium constants for these reactions have been determined as a function of temperature and it has been shown that equation 5 strongly favors the formation of SiO_2 at high temperature. The formation of GeO_2 according to equation 6 is repressed by Cl_2 formed from oxidation of $SiCl_4$ (eq. 5). The equilibrium constant for the $GeCl_4$ reaction approaches 1 above 1400 K. As a result only a portion of the $GeCl_4$ is converted to GeO_2. This fraction decreases with increasing temperature or chlorine partial pressure. Decreasing the partial pressure of oxygen shifts the germanium reaction even farther toward more unreacted $GeCl_4$.

Incorporation of OH is another critical aspect of the oxidation chemistry. Reduction to the ppb level is necessary for the manufacture of low loss optical fiber. Hydrogen is incorporated into the glass according to the reaction

$$H_2O + Cl_2 \rightarrow 2\ HCl + \tfrac{1}{2}\,O_2 \tag{9}$$

with the equilibrium constant

$$K_{OH} = \frac{(P_{HCl})^2 (P_{O_2})^{1/2}}{(P_{H_2O})\,(P_{Cl_2})} \tag{10}$$

The amount of OH incorporated into the glass, C_{OH}, is

$$C_{OH} = \frac{(P_{H_2O_{initial}})\,(P_{Cl_2})^{1/2}}{(P_{O_2})^{1/4}} \tag{11}$$

During the deposition phase of MCVD the chlorine level is between 3 and 10% owing to the oxidation of the chloride reactants. This level of chlorine leads to a reduction in the OH incorporation by a factor of about 4000. During the collapse phase of the process the chlorine level is significantly reduced and OH incorporation can be high (Fig. 9) as a result of the presence of hydrogen from either the oxy–hydrogen torch or impurities in the starting gases.

Sintering. The process by which the glass consolidates was found to be viscous sintering (15) controlled by the viscosity of the glass at the consolidation temperature. The driving force is the reduction of surface energy via a decrease in the surface area. Because the soot layer is sintered by a moving torch, large thermal gradients may be present in the deposited layer (~300°C/cm). The time and temperature of consolidation must be controlled to allow a pore-free glass to result.

Plasma Chemical Vapor Deposition. Another process closely related to MCVD is plasma chemical vapor deposition (PCVD) (see PLASMA TECHNOLOGY) (16). This inside process uses the same precursor chemicals as MCVD to form a glass coating inside a tube which is collapsed to form a solid glass preform. The reaction within the tube, however, is quite different. Here the reaction is initiated

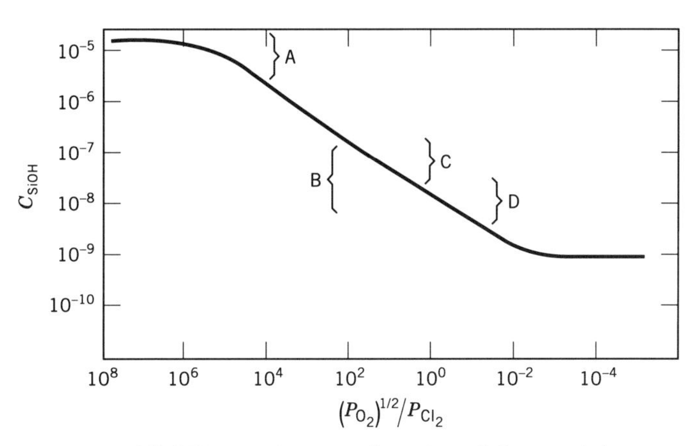

Fig. 9. Concentration of SiOH moieties as a function of the partial pressures of oxygen and chlorine when 10 ppm of H_2O is in the starting gas, where region A represents MCVD collapse conditions; B, MCVD Cl_2 collapse; C, MCVD deposition and consolidation; and D, soot consolidation.

by a nonisothermal microwave plasma which traverses the inside of the tube, as shown in Figure 10. The plasma requires a pressure of a few hundred Pascals and is generated by a microwave cavity which operates at 2.45 GHz. The glass is deposited not as a soot, but as a thin glass layer with efficiencies approaching 100%. In addition to the high efficiency, complex waveguide structures may be formed because many thin layers are produced by rapidly traversing the plasma. This method provides a smoother refractive index variation, which is especially advantageous for multimode fiber preforms.

Outside Processes. *Outside Vapor Deposition.* The outside vapor deposition (OVD) process developed by Corning Glass Works (17), is depicted in Figure 11. Soot is deposited layer by layer on a rotating mandrel at a temperature such that the soot particles are partially sintered. The precursor chemicals are the

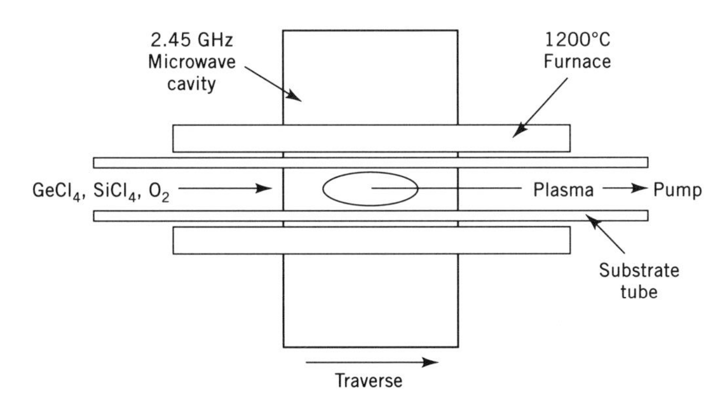

Fig. 10. Schematic diagram of PCVD deposition apparatus.

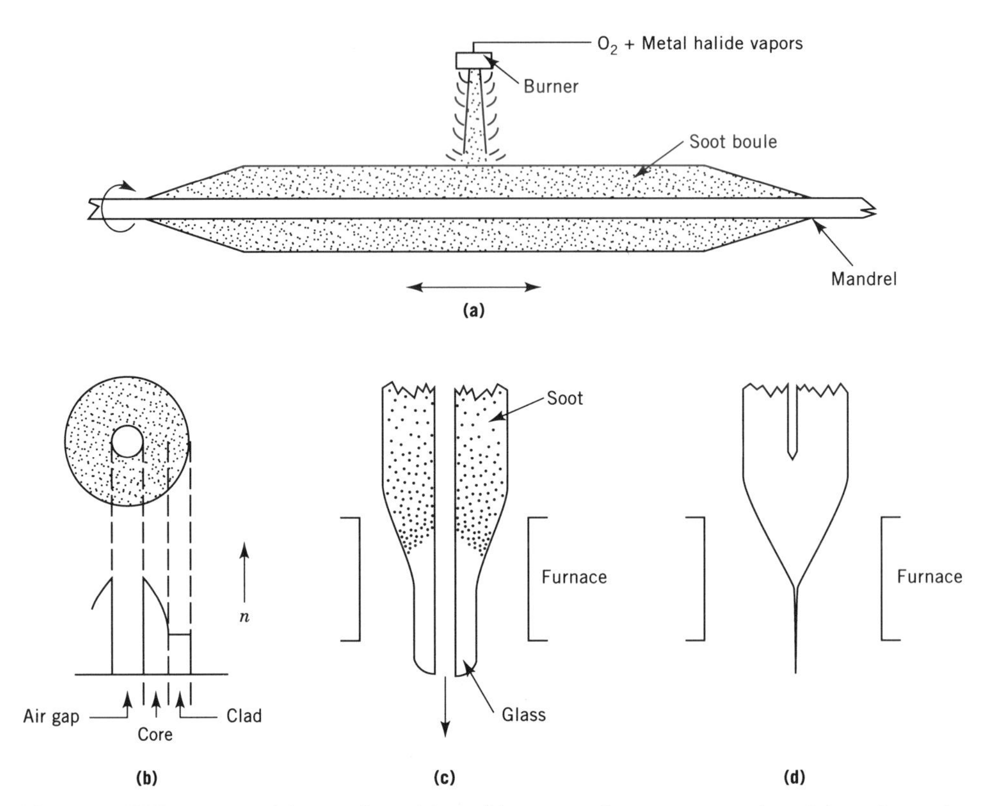

Fig. 11. OVD process: (**a**) soot deposition, (**b**) soot preform cross section, (**c**) preform sintering, and (**d**) fiber drawing.

same as those used in the MCVD process but are oxidized by a gas–oxygen torch by similar chemical reactions. A doped core is deposited first followed by a SiO_2 cladding. The mandrel is removed and the porous preform consolidated at 1500–1600°C in a furnace with a controlled atmosphere containing helium, oxygen, and chlorine. The central hole may be collapsed either during sintering or fiber drawing, eliminating the need for a substrate tube. Additionally, dopants such as titania, TiO_2, may be added to the outer layers of the boule to improve fatigue resistance of the fiber drawn from it.

Vertical Axial Deposition. The vertical axial deposition (VAD) process (18) was developed by a consortium of Japanese cable manufacturers and Nippon Telephone and Telegraph (NTT). This process also forms a cylindrical soot form. However, deposition is achieved end-on without use of a mandrel and subsequent formation of a central hole. Both the core and cladding are deposited simultaneously using more than one torch (Fig. 12). Whereas the OVD, PCVD, and MCVD processes build a refractive index profile layer by layer, the VAD process uses gaseous constituents in the flame to control the shape and temperature distribution across the face of the growing soot boule.

The design and engineering of the VAD torch, consisting of a series of concentric silica tubes, have been critical to its success. Reactants pass through the

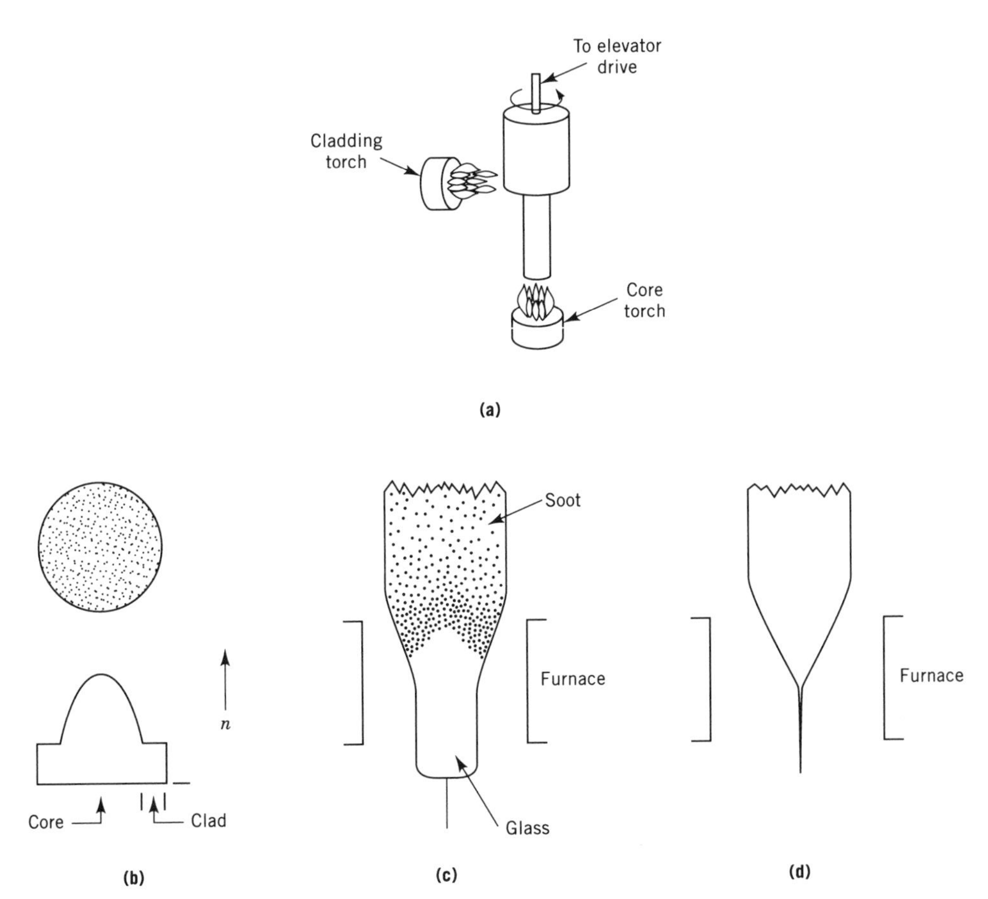

Fig. 12. VAD process: (**a**) schematic of deposition apparatus, (**b**) soot preform cross section, (**c**) preform sintering, and (**d**) fiber drawing.

innermost rings. Shield gases separate these from oxygen and hydrogen in the outer rings, which form the flame. The soot plume and surface temperature of the boule are controlled by gas flows; temperatures and particle distributions in the flame are manipulated to provide the optimum surface temperature profile and tip shape.

Although the control in VAD is more difficult, this process has an advantage over OVD, especially for multimode fiber. The thermal mismatch between core and cladding materials caused by the heavy doping necessary to achieve the desired refractive index profile causes cracking in the OVD preforms at the inner surface as the glass cools through the glass transition, T_g. The VAD preforms withstand the stress because they possess no central hole to result in tensile stress. The primary obstacle for VAD was the creation of an optimized refractive index profile to minimize intermodal dispersion. Then it was discovered that the composition could be graded by control of the boule's surface-temperature distribution. This temperature distribution depends on the shape of the boule's growing face. Through understanding the relationship between the temperature and germania concentration (19), fiber properties equivalent to those formed by OVD and MCVD were achieved.

Fiber Drawing and Strength

Preforms manufactured by MCVD, PCVD, OVD, and VAD all must be drawn into fiber in a similar manner. Standard fibers are drawn to 125 μm in diameter from preforms on the order of 2 to 7.5 cm diameter. Fibers are drawn by holding the preform vertically and lowering it into a furnace. The preform is heated to a temperature at which the glass softens (2200°C) until a gob of glass stretches from the tip of the preform and drops under the force of gravity. A neck-down region is formed at this point, providing the transition between preform and fiber. Fiber is drawn by means of a capstan system, and its diameter is controlled by a diameter monitor that adjusts the draw speed at a fixed furnace temperature. The result is long lengths of uniform fiber.

To preserve the intrinsic strength of the pristine glass surface, a polymer coating must be applied before the fiber is contacted by the capstan. The basic requirements for fiber drawing were known prior to the invention of low loss glass manufacturing techniques. However, stringent loss and strength requirements for optical fiber transmission systems necessitated optimization of the draw process. Figure 13 shows a fiber draw tower. Preforms are lowered by a feed mechanism into the furnace to keep the transition region at a constant temperature. The furnace may be a graphite resistance or a zirconia induction furnace, or any other clean heat source capable of achieving temperatures of 1950–2200°C. Graphite furnaces require the use of an inert gas shield to guard against oxidation of the heating elements. Zirconia furnaces must be maintained at a temperature greater than 1600°C owing to a large-volume change associated with a crystallographic transition in zirconia, which causes stress-induced fracture of the furnace tube. The advantage of the zirconia furnace is that an inert atmosphere is not necessary. Fiber diameter is controlled by a servo-system which controls the fiber draw speed using a monitor placed directly below the furnace. This device provides high (0.1-μm) resolution monitoring of the fiber diameter using a high update rate, eg, 500 Hz, which provides control to 0.29 μm (20). Such control is required for low loss fiber interconnection. Uniformity of the fiber diameter depends on the temperature control, preform feed rate, and pulling tension. Uniform preform diameters are required for low fiber diameter variability over long ($>$ 100 cm) distances whereas variations with a short period are caused by temperature fluctuations in the neck-down region. To control these perturbations the convective currents in the furnace may be controlled by appropriate baffling or controlled gas flow.

The strength of optical fiber is critical to its usefulness. Silica's intrinsic strength is very high (ca 14 GPa (2×10^6 psi)). However, in practice long lengths of fiber fail at considerably lower stress because of flaws which act as stress concentrators. The strength of the fiber is therefore determined by the largest flaw present in the length under test. Flaws may result from surface contamination, corrosion, and abrasion. In all cases cleanliness and care to avoid any physical contact with the unprotected fiber are required.

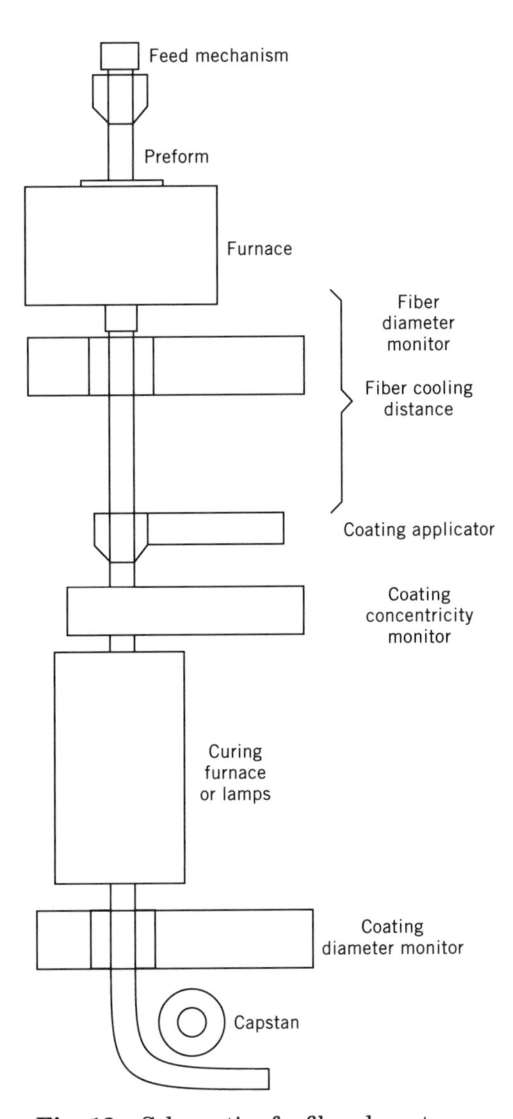

Fig. 13. Schematic of a fiber draw tower.

To protect the fiber surface a coating applicator is placed directly under the furnace as close as possible to the diameter monitor. The distance is controlled by the point at which the fiber temperature has dropped to 80°C (21). The fiber is coated upon passing through the cup containing a liquid polymeric coating and then passes through curing lamps or ovens before being taken up on a drum or spool. Typical coatings are uv-curable urethane acrylates and thermally cured silicone rubbers applied to a diameter of 250-μm (see RUBBER CHEMICALS; URETHANE POLYMERS). Beyond protecting the fiber from abrasion, coatings must be clean and free of particulate matter, concentric to the fiber, and have no voids. Low modulus coatings can improve the microbending performance of sensitive fibers by effectively cushioning the fiber. Dual coatings made up of a low (10^6–10^7

Pa (150–1500 psi)) modulus elastomeric primary coating surrounded by a high modulus secondary coating are used for sensitive applications. Single coatings of high (10^9 Pa (150,000 psi)) modulus are used where strength is of primary concern.

An additional issue in fiber strength is that of fatigue (22), which can produce delayed failure of a fiber. Fatigue is thought to be caused by a surface reaction of fiber and OH causing the growth of subcritical flaws to the point where fracture occurs.

Although exact mechanisms have yet to be established, hermetic coatings (23) are being directly applied to keep contaminants from reaching the fiber surface. Polymeric coatings may be permeated by atmospheric moisture. A number of materials, metals and carbon, may be applied during the fiber drawing process, using vapor deposition. But only an amorphous carbon hermetic coating has been commercialized. These coatings are primarily intended to protect the fiber from corrosion by water, but also prevent hydrogen-induced loss increases. Accelerated testing at high temperature or high H_2 pressure have been used to predict effects on the fiber over its lifetime.

The draw process itself, if not properly controlled, can lead to excess loss in the fiber. High draw tension (a combination of temperature and draw speed) can lead to losses associated with the breaking of Si–O bonds. High draw temperature may cause germanium suboxides to form which have well-defined absorption bands. Similar defects can be caused by the uv-curing lamps; however, short-uv filtering and uv-absorbing coatings can be used to ameliorate that problem (see COATINGS).

Overcladding. Fiber manufacturing and drawing technology have advanced to the point that the optical losses are limited almost entirely by the intrinsic loss of the glass. Initially all of the fiber manufacturing processes (MCVD, PCVD, OVD, and VAD) produced preforms yielding on the order of 10 km of fiber. This situation changed as single-mode fiber usage grew. The proportion of core glass to the total amount of glass in single-mode fiber is much lower than in multimode fiber. Single-mode core diameter is ca 8 μm, multimode core diameter is 50 or 62.5 μm. This led to a desire for a method of manufacturing larger preforms by shrinking a second silica tube over the preform or depositing a thick soot layer on the preform to provide additional cladding. This procedure, known as overcladding, increased the length of fiber drawn from a single VAD preform to more than 100 km, for example, and significantly reduced the cost of producing fiber.

Sol–Gel Processing. Fibers can be designed so that light only travels in the inner 30–40 μm of the fiber (24), which accounts for only about 5% of the fiber mass. Thus, using a core rod, the remaining 95% could be manufactured from less expensive, lower purity materials typically obtained by sol–gel processing (see SOL–GEL TECHNOLOGY). In the sol–gel technique silicon alkoxide [*78-10-4*], $Si_2(OC_2H_5)$, reacts with water in the presence of a catalyst and alcohol to form a sol. The sol can then be cast into a cylindrical mold where polycondensation of the silanol groups produces a siloxane gel network which eventually forms a semirigid gel. The gel body must then be dried and consolidated to form a glass. Alternatively, colloidal powders known as fumed silica can be formed into tubes by compaction (25), casting–gelation (26,27), and centrifugation (28). The mechanical compaction method uses dry silica which is packed into a mold, then consolidated at high temperature to form a glass. Centrifugation uses a cylindrical mold rotating at a speed of up to 35,000 rpm, forming a body which is subsequently dried and sintered. There are two methods for the casting–gelation technique. The first uses colloidal silica dispersed in an alkoxide sol. The sol acts as a binder during drying to prevent cracking. An alternative route uses fumed silica (colloidal) mixed with water and stabilized against agglomeration by the addition of surfactants (qv) (steric stabilization) or by control of the pH (electrostatic stabilization). The sol is then gelled (often by changing the pH), dried, and fired to a glass much as in OVD or VAD. During drying of gel bodies large stresses develop from shrinkage and capillary forces. These often cause the partially dried gel to fracture. Controlled drying is necessary for the formation of large gel bodies without cracking.

There has been a considerable effort to form all gel preforms from alkoxide gels using dopant elements such as germanium. Fibers produced using germanium ethoxide [*14165-55-0*], $Ge(OC_2H_5)_4$ (29), have not shown losses comparable to those made by vapor deposition techniques. A typical fiber produced by this technique contains included bubbles and an only slightly increased index in the core owing to germanium vaporization during firing (30). More success has been achieved in a silica-core–fluorine-doped cladding structure (31). The hydrolysis and polycondensation of monofluorosilicon ethoxide [*358-60-1*], $Si(OC_2H_5)_3F$, incorporates fluorine into the glass and this lowers the refractive index. This material is cast into a cylindrical body, gelled, and dried to form a porous tube having a surface area between 200 and 650 m^2/g. The high surface area makes it possible to consolidate the tube in a fluorine-containing atmosphere at low temperature, resulting in a down-doped tube having $\Delta = -0.62\%$. By collapsing this tube with a stream of oxygen flowing through the center, a core of higher index is formed as the oxygen removes fluorine from the inner surface of the tube. Losses of 0.4 dB/km have been reported.

For the production of large bodies the colloidal approach has yielded more success. These approaches are aimed at overcladding preforms produced by other methods. The starting material is a commercially available fumed silica such as Aerosil OX-50* (Degussa AG, Frankfurt, Germany). The unique nature of this material is the large particle size, with a mean diameter of 40 nm. This product is manufactured by flame hydrolysis and produces soot similar to VAD and OVD. The larger particles reduce capillary forces during drying because these forces are inversely proportional to the pore radii. Additionally, gas permeation is enhanced by the larger pore size and sintering is delayed to temperatures above 1000°C. Use of a chlorine-containing atmosphere in a silica muffle furnace effectively removes impurities such as OH and alkali and transition-metal ions (32,33). By these means, glass having impurity levels less than 100 ppb is produced. Two methods of using gel as an overcladding material are shown in Figure 14. In Figure 14**a** a wet gel is granulated, dried, and dehydrated, then sprayed through an oxygen plasma torch onto a core rod (34). The gel is deposited as glass droplets ~100 μm in diameter. Because of the large size the droplets are impacted onto the surface of the preform rather than thermophoretically deposited as soot. This route yields high deposition rates and efficiencies. In Figure 14**b**, an overcladding tube is prepared from gel and then consolidated over a core rod (33). These tubes are formed by dispersion of colloidal silica, milling, casting, gelation, and drying.

The interface between the core rod and the overclad tube must be clean and free of bubbles to obtain a low loss optical fiber. Although high quality low loss fiber, and large (> 1 kg) overcladding tubes have been demonstrated by sol–gel and powder methods, no commercialization of these processes has occurred.

Defects. The ever-increasing demand for high data rate systems is forcing the search for an even greater understanding of those defects which produce attenuation of only hundredths of a dB/km. Profile control to produce zero dispersion at operating wavelengths is necessary, and environmental effects such as radiation and hydrogen exposure must be minimized. In addition, for reliability concerns, higher strength must be achieved with a narrow distribution, necessitating the understanding of flaw distributions and growth of flaws (fatigue).

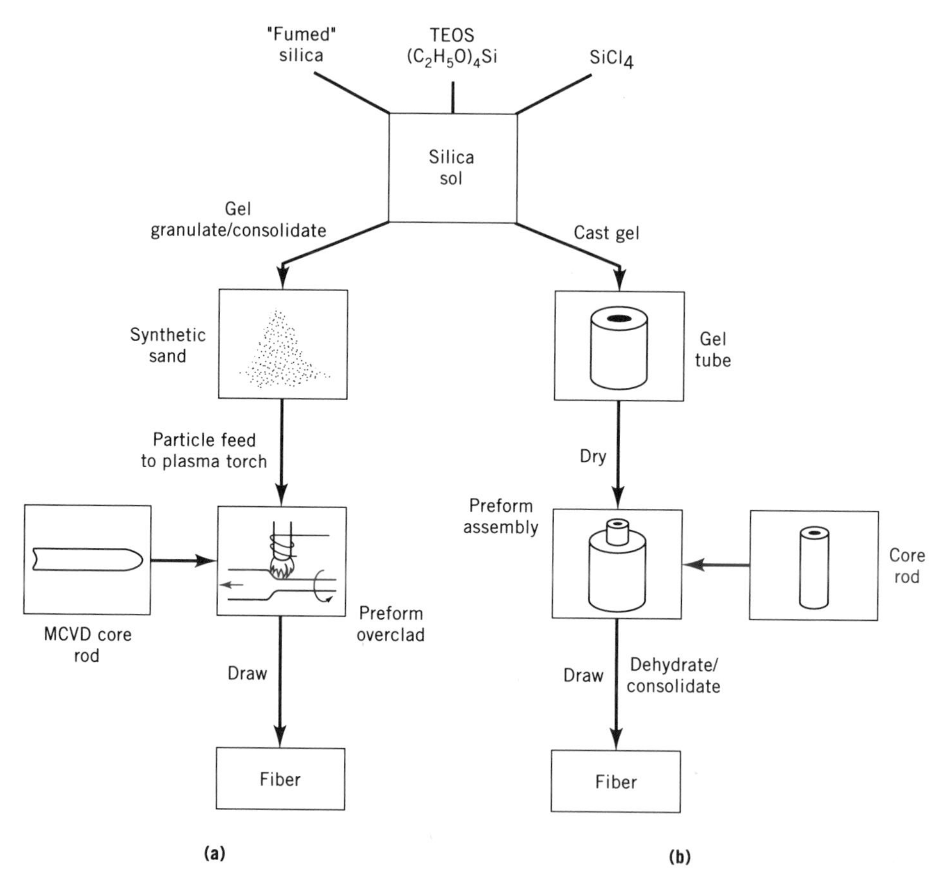

Fig. 14. Sol–gel techniques for overcladding where TEOS = $(C_2H_5O)_4Si$: (**a**) wet gel method and (**b**) rod-in-tube method.

Optical Amplifiers

Throughout the first two decades of their existence optical fibers served a passive role, ie, in the transmission of encoded light signals. In the late 1980s erbium-doped fiber amplifiers (EDFAs) were introduced (35,36), making it possible to amplify a 1.55 μm optical signal without first converting it to an electronic signal. The amplifier consists of a section (tens of meters) of single-mode optical fiber having about 100 ppm of erbium [*7440-52-0*] incorporated into the core. This fiber section becomes an amplifier when a continuous source of pump light, usually 0.98 or 1.48 μm, wavelength is propagating through the fiber. As the optical signal, usually 1.53 to 1.6 μm, travels through the length of fiber containing excited erbium ions, amplification occurs by the stimulated emission of photons from the excited state. Noise in the form of broad-band spontaneous emission accompanies this process; however, the signal-to-noise ratio is kept to an acceptable level even when cascading many of these devices for system applications. The device is shown in Figure 15.

A number of means have been developed to produce erbium-doped optical fibers. The task is complicated by the tendency of rare-earth ions to exsolve from high silica glasses. Rare-earth and alkaline-earth ions tend to compete with silicon for oxygen ion coordination (see LANTHANIDES). Thus high silica glasses tend to phase-separate into two liquids at the high temperatures where fiber is drawn. Even rare-earth concentrations in the 100s of ppm undergo a subtle form of segregation. Association of these ions in the glass network (37) tends to cause interactions which lead to excited-state absorption. The situation is ameliorated by incorporation of homogenizer ions in the glass network. Aluminum is the most common of these. This trivalent ion can be thought to produce a charge deficiency which permits incorporation of erbium in adjacent sites.

The compatibility of aluminum and erbium extends to the vapor phase where complex aluminum–erbium chlorides exist at vapor-pressure orders of magnitude higher than $ErCl_3$. The passage of aluminum chloride, Al_2Cl_6, vapor over a heated erbium oxide, Er_2O_3, or erbium chloride, $ErCl_3$, source permit doping in MCVD reactions. In addition, erbium chelates and other organic precursors can be introduced into VAD or OVD flows to produce doped soot. Finally, doping with solutions containing rare-earth ions or sol–gel doping of soot or MCVD substrate tubes prior to the final collapse to a solid preform also provide the means for controlled introduction of the ions.

EDFAs are being introduced into long-distance, particularly undersea, systems which operate at wavelengths near 1.6 μm. In addition to providing an inexpensive means of amplification EDFA also make it possible to amplify numerous wavelengths near 1.6 μm, whereas semiconductor amplifiers suffer crosstalk when amplifying more than one wavelength (see SEMICONDUCTORS). The first

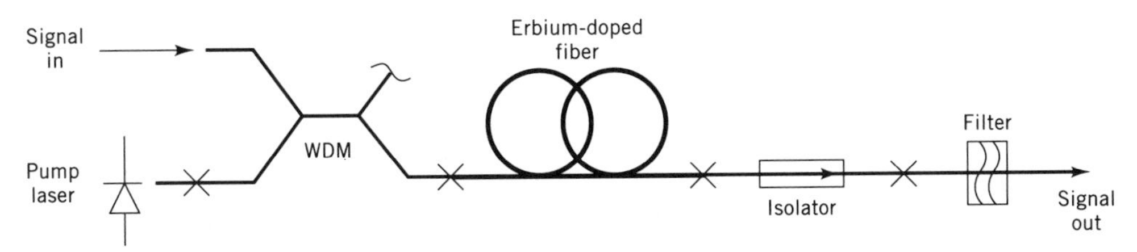

Fig. 15. Erbium-doped fiber amplifier where × represents splices and WDM is a wavelength division multiplexer combining the signal and pump wavelengths.

transatlantic all-optical system is expected to be in operation in 1995. It should have a potential capacity of greater than 700,000 voice channels per fiber pair.

A commercially attractive fiber amplifier to be used in existing terrestrial networks, operated at wavelengths near 1.3 μm, is being sought. Praseodymium and neodymium in fluoride-based glasses have shown some promise, but such amplifier fibers are not in commercial use as of this writing. Future generations of optical systems are expected to make use of additional active fiber components including fiber grating filters, fiber sensors (qv), and fiber lasers. The ever-increasing need for high capacity transmission to satisfy the growth in digital information transfer should lead to implementation of soliton transmission and 1.3-μm optical amplification.

Economic Aspects

The market for optical fiber worldwide in 1992 was $2.8 billion corresponding to 10 million fiber kilometers (Mfk) (38). This can be broken down into the U.S. market (3.7 Mfk), the rest of North America (0.4 Mfk), northern Europe (4.1 Mfk), eastern Europe (2.6 Mfk), the Pacific Rim (2.8 Mfk), and elsewhere (0.3 Mfk). Most of the optical fiber is manufactured by only a few companies, the largest of which are AT & T and Corning. Other producers include Alcatel, Fujikura, Furakawa, Northern Telecom, Pirelli, and Sumitomo. The market for optical fibers is projected to reach $3.5 billion by 1998. In addition, according to ElectroniCast (San Mateo, Ca.), the total market for passive optical components, optical electronics, connectors, and fiber-optic cable is predicted to increase from $1.76 billion (U.S.) in 1992 to over $4 billion in 1997, and $10 billion by 2002.

BIBLIOGRAPHY

"Fiber Optics" in *ECT* 3rd ed., Vol. 10, pp. 125–147, by A. D. Pearson and J. D. MacChesney, Bell Laboratories, Inc., and W. G. French, 3M Co.

1. S. Miller, *Sci. Am.* **214**, 19–27 (1966).
2. S. E. Miller and L. C. Tillotson, *Appl. Opt.* **5**(5), 1538–1548 (1966).
3. P. Kaiser, *Bell Syst. Tech. J.* **49**, 137–153 (1970).
4. E. A. J. Marcatili and R. A. Schmeltzer, *Bell Syst. Tech. J.* **43**, 1783–1809 (1964).
5. H. Hosono, Y. Abe, D. L. Kinser, R. A. Weeks, and H. Kawazoe, *Phys. Rev. B.*, **46**, 1144–11451 (1992).
6. L. G. Cohen "Ultrabroadband Single-Made Fiber", Paper No. MF4 in the *Technical Digest of Fiber Communications Converence, New Orleans, La.*, Optical Society of America, Washington, D.C., 1983.
7. A. D. Pearson and W. G. French *Bell Lbs. Rec.* **50**, 103–106 (1972).
8. K. J. Beals and C. R. Day, *Phys. Chem. Glasses* **21**, 5–19 (1980).
9. U.S. Pat. 2,272,342 (Aug. 27, 1934), J. F. Hyde (to Corning GlassWorks).
10. J. B. MacChesney and co-workers, *Phys. Lett.* **23**, 340–341 (1973).
11. J. B. MacChesney, P. B. O'Connor, F. V. DiMarcello, J. R. Simpson, and P. D. Lazay, "Preparation of Low-Loss Optical Fibers Using Simultaneous Vapor-Phase Deposition and Fusion," in *Proceedings of the Tenth International Congress on Glass, Kyoto, Japan*, Vol. 6 Ceramics Society, Japan, 1974, pp. 50–54.

12. P. G. Simpkins, S. G. Kosinski, and J. B. MacChesney, *J.Appl. Phys.* **50**, 5676–5681 (1979).
13. K. L. Walker, F. T. Geyling, and S. R. Nagel, *J. Am. Ceram. Soc.* **63**, 96–102 (1980).
14. D. L. Wood, K. L. Walker, J. B. MacChesney, J. R. Simpson, and R. Csencits, *J. Lightwave Technol.* **LT-5**, 277–283 (1987).
15. K. L. Walker, J. W. Harvey, F. T. Geyling, and S. R. Nagel, *J. Am. Ceram. Soc.* **63**, 92–96 (1980); S. G. Kosinski, L. Soto, and S. R. Nagel, "Characterization of Germanium Phosphosilicate Films Prepared by MCVD," presentation, at the fall meeting of the *American Ceramic Society, 1981*, Bedford, Pa., 1981.
16. D. Kuppers and H. Lydtin, *Preparation of Optical Waveguides with the Aid of Plasma-Activated Chemical Vapor Deposition at Low Pressures in Topics in Current Chemistry*, Vol. 89, Springer-Verlag, Berlin, 1980, pp. 108–130.
17. U.S. Pat. 3,737,292 (1973), D. B. Keck, P. C. Schultz, and F. Zimar.
18. T. Izawa and N. Inagaki, *Proc. IEEE*, **68**(10), 1184–1187 (1980).
19. T. Edahiro, M. Kawachi, S. Sudo, and S. Tomaru, *Jpn. J. Appl. Phys.* **19**, 2047–2054 (1980).
20. D. H. Smithgall and R. E. Frazee, *Bell Syst. Tech. J.* **60**, 2065 (1981).
21. V. C. Paek and C. M. Schroeder, *Appl. Opt.* **20**(7), 1230–1233 (1981).
22. C. R. Kurkjian, J. T. Krause, and M. J. Matthewson, *J. Lightwave Technol.* **7**(9), 1360–1370 (1989).
23. R. G. Huff, F. V. DiMarcello, and A. G. Hart, Jr., "Amorphous Carbon Hermetic Optical Fiber," paper no. TuG-2 in *Technical Digest of Optical Fiber Communications Conference, New Orleans, La.*, Optical Society of America, Washington, D.C., 1988.
24. J. B. MacChesney, D. W. Johnson, P. J. Lemaire, L. G. Cohen, and E. M. Rabinovich, "Fluorosilicate Substrate Tubes to Eliminate Leaky-Mode Losses in MCVD Single-Mode Fibers with Depressed Index Cladding," paper no. WH2 in *Technical Digest of Optical Fiber Communications Conference, San Diego, Calif.*, Optical Society of America, Washington, D.C., 1985.
25. R. Dorn and co-workers, *Glastech. Ber.* **66**, 29–32 (1987).
26. P. Bachmann, P. Geitner, H. Hydton, G. Ronanowski, and M. Thelen, "Preparation of Quartz Tubes by Centrifugal Deposition of Silica Particles," in *Proceedings of the 14th European Conference on Optical Communications, Brighton, UK*, IEE, London, 1988, pp. 449–453.
27. T. Mori and co-workers, *J. Non-Cryst. Solids* **100**, 523–525 (1988).
28. N. Okazaki, T. Kitagawa, S. Shibata, and T. Kimura, *J. Non-Cryst. Solids* **116**, 87–92 (1990).
29. S. Shibata and T. Kitagawa, *J. Appl. Phys.* **25**, L323–L324 (1986).
30. K. Susa, I. Matsuyama, S. Satoh, and T. Suganuma, *J. Non-Cryst. Solids* **119**, 21–28 (1990).
31. S. Shibata, T. Kitagawa, and M. Horiguchi, "Wholly Synthesized Fluorine-Doped Silica Optical Fiber by the Sol-Gel Method," in *Technical Digest of the 13th European Conference on Optical Communication, Helsinki, Finland*, Association of Electrical Engineers of Finland, Helsinki, 1987, pp. 147–150.
32. J. B. MacChesney, D. W. Johnson, Jr., D. A. Fleming, F. W. Walz, and T. Y. Kometani, *Mater. Res. Bull.* **22**, 1209–1216 (1987).
33. J. B. MacChesney, D. W. Johnson, Jr., D. A. Fleming, and F. W. Walz, *Electron. Lett.* **23**, 1005–1007 (1987).
34. J. W. Fleming, "Sol-Gel Techniques for Lightwave Applications," paper no. MH-1 in *Technical Digest of Optical Fiber Communications Conference, Reno, Nev.*, Optical Society of America, Washington, D.C., 1987.
35. E. Desurvire, J. R. Simpson, and P. C. Becker, *Optics Lett.* **12**, 888–890 (1987).
36. R. J. Mears, L. Reekie, I. M. Jauncey, and D. N. Payne, *Electron. Lett.* **23**(19), 1026–1028 (1987).

37. U.S. Pat. 466,624 (1987), J. B. MacChesney and J. R. Simpson (to AT&T).
38. J. Kessler, KMI Corp., Newport, R.I., personal communication, 1992.

General References

E. J. Friebele and D. L. Griscom, in M. T. Tomazawa and R. H. Doremus, eds., *Treatise on Materials Science and Technology*, Vol. 17, Academic Press, Inc., New York, 1979, pp. 257–351.

E. M. Dianov and co-workers, *Sov. J. Quantum Elec.* **13**(3), 274–289 (1983).

D. L. Griscom, *SPIE* **541** (1985).

P. Urquhart, *IEEE Proc.* **135**(J,6), 385–407 (Dec. 1988).

E. Desurvire, *Sci. Am.*, 114–121 (Jan. 1992).

M. J. F. Digonnet, *Rare Earth Doped Fiber Lasers and Amplifiers*, Marcel Dekker, Inc., New York, 1993.

J. L. Zyskind, C. R. Giles, J. R. Simpson, and D. J. DiGiovanni, *AT&T Tech. J., 53–62 (Jan./Feb. 1992).*

R. J. Mears and S. R. Baker, *Opt. Quant. Elec.* **24**, 517–538 (1992).

J. Auge, J. Chesnoy, P. M. Gabla, and A. Weygang, *Microwave J.*, 62–74 (June 1993).

G. Meltz and W. W. Morey, "Bragg Grating Formation and Germanosilicate Fiber Photosensitivity," *Proceedings of the SPIE Workshop on Photoinduced Self-Organization in Optical Fiber*, May 10–11, 1991, SPIE, Quebec City, Canada, pp. 185–189.

V. Mizrahi, P. J. Lemaire, T. Erdogan, W. A. Reed, D. J. DiGiovanni, and R. M. Atkins, *J. Appl. Phys.* (Sept. 27, 1993) (in press).

V. Mizrahi, D. J. DiGiovanni, R. M. Atkins, Y. K. Park, and J-M. P Delavaux, *J. Lightwave Technol.* (1993) (in pres).

K. O. Hill, B. Malo, F. Bilodeau, D. C. Johnson, and J. Albert, *Appl. Phys. Lett.* **62**, 1035 (1993).

D. Z. Anderson, V. Mizrahi, T. Erdogan, and A. E. White, *Electron. Lett.* **29**, 566 (1993).

SANDRA KOSINSKI
JOHN B. MACCHESNEY
AT&T Bell Laboratories

FIBERS

SURVEY

This overview of fibers and fiber products introduces the underlying concepts that govern the manufacture and properties of these materials. The field of fibers is an evolving one, with new technologies being developed constantly. With the increasing use of fibers in nontraditional textile applications, such as geotextiles (qv), fiber-reinforced composites (qv), specialty absorption media, and as materials of construction, new fiber types and new processing technologies can be anticipated.

Classification

A fiber may be described as a flexible, macroscopically homogeneous body having a high ratio of length to width and being small in cross section. A significant segment of the world's agricultural activity is concerned with the growth and harvesting of natural fibers, and the production of synthetic fibers is an important activity of the worldwide chemical industry. The textile and paper industries are the prime converters of fibers into end products whose properties can be directly related to the unique combination of properties characteristic of fibers. The paper industry uses almost exclusively a natural cellulosic fiber derived from wood. The textile industry, on the other hand, uses a variety of natural and synthetic fibers in the manufacture of its wide range of products. Textile fibers may be classified according to their origin, as follows:

Naturally occurring fibers

vegetable: based on cellulose, eg, cotton, linen, hemp, jute, and ramie
animal: based on proteins, eg, wool, mohair, vicuna, other animal hairs, silk
mineral: eg, asbestos

Synthetic fibers

based on natural organic polymers
rayon: regenerated cellulose
acetate: partially acetylated cellulose derivative

triacetate: fully acetylated cellulose derivative
azlon: regenerated protein

based on synthetic organic polymers
acrylic: polyacrylonitrile (also modacrylic)
aramid: aromatic polyamides
nylon: aliphatic polyamides
olefin: polyolefins
polyester: polyesters of an aromatic dicarboxylic acid and a dihydric alcohol
spandex: segmented polyurethane
vinyon: poly(vinyl chloride)
vinal (or vinylon): poly(vinyl alcohol)
carbon/graphite: derived from polyacrylonitrile, rayon, or pitch
specialty fibers: poly(phenylene sulfide) and polyetheretherketone

based on inorganic substances
glass, metallic, ceramic

The natural fibers are those derived directly from the animal, vegetable, and mineral kingdoms. With the exception of silk, which is extruded by the silkworm as a continuous filament, natural fibers are of finite length and are used directly in textile manufacturing operations after preliminary cleaning. These fibers are known as staple fibers, and an estimate of their average length is referred to as their staple length. Other quality factors that affect the utility of natural fibers for textile purposes are fineness, presence of foreign matter, color, and spinnability. The latter term denotes the ability of a fiber to be spun economically into yarns by conventional textile processing procedures. Chemical properties, such as ability to absorb common dyes, thermal and environmental stability, and resistance to chemical degradation are also important.

The other main category of textile fibers comprises those fibers manufactured from natural organic polymers, synthetic organic polymers, and inorganic substances. Glass fiber is the only inorganic synthetic fiber in common use, although other ceramic and metallic fibers are being developed, particularly for use in high performance fiber-reinforced composites.

Fibers manufactured from natural organic polymers are either regenerated or derivative; historically they are designated man-made fibers and distinguished from fibers based on synthetic organic polymers. A regenerated fiber is one formed when a natural polymer, or its chemical derivative, is dissolved and extruded as a continuous filament, and the chemical nature of the natural polymer is either retained or regenerated after the fiber formation process. A derivative fiber is one formed when a chemical derivative of the natural polymer is prepared, dissolved, and extruded as a continuous filament, and the chemical nature of the derivative is retained after the fiber formation process. These fibers based on natural polymers constitute a most important class of textile fibers; they were the first fabricated fibers to gain industrial and consumer acceptance. They are available in a wide range of properties, making them suitable for many textile products. Since almost all of these natural polymer fibers are derived from cellulose, this group of fibers is frequently referred to as the cellulosics. Cellulose is a nearly ideal polymer for the formation of fibers. In contrast, proteins are not well suited for fiber formation, despite the fact that animal fibers, such as wool and hair, are ubiquitous.

Fibers based on synthetic organic polymers, generally referred to as the synthetic fibers, have revolutionized the textile industry since their initial commercialization by the chemical industry in 1940. The most widely used synthetic fibers are based on polyamides, polyesters, acrylics, polyolefins, and polyurethanes. A relatively small number of polymer types can produce synthetic fibers with wide ranges in fiber properties and characteristics because of the enormous versatility of fiber manufacturing processes. In recent years, numerous specialty high performance polymers have been synthesized, and some have been produced in fiber form. Among these are poly(phenylene sulfide), polyetheretherketone (PEEK), several aromatic polyamides and polyimides, and ultrahigh molecular weight polyethylene. The generic names of synthetic fibers are defined and controlled by the Federal Trade Commission.

Fiber manufacture is based on three methods of fiber formation or spinning. The term spinning should more properly be reserved for that textile manufacturing operation where staple fibers are formed into continuous textile yarns by several consecutive attenuating and twisting steps. A yarn so formed from staple fibers is referred to as a staple or spun yarn. In the context of synthetic fiber manufacture, spinning refers to the overall process of polymer liquefaction (dissolution or melting), extrusion, and fiber formation. The three principal methods are melt spinning, dry spinning, and wet spinning, although there are many variations and combinations of these basic processes.

In melt spinning the polymer is heated above its melting point and the molten polymer is forced through a spinneret. Spinnerets are dies with many small orifices that may be varied in diameter and shape. The jet of molten polymer emerging from each orifice in the spinneret is guided to a cooling zone where the polymer solidifies to complete the fiber formation process. In dry spinning the polymer is dissolved in a suitable solvent, and the resultant solution is extruded under pressure through a spinneret. The jet of polymer solution is guided to a heating zone where the solvent evaporates and the filament solidifies. In wet spinning the polymer is also dissolved in a suitable solvent and the solution is forced through a spinneret which is submerged in a coagulation bath. As the polymer solution emerges from the spinneret orifices in the coagulating bath, the polymer is either precipitated or chemically regenerated. In most instances the filaments formed by melt, dry, or wet spinning are not suitable textile fibers until they have been subjected to one or more successive drawing operations. Drawing is the hot or cold stretching and attenuation of fiber filaments to achieve an irreversible extension to induce a molecular orientation with respect to the fiber axis, and to develop a fiber fine structure. This fine structure is generally characterized by a high degree of crystallinity and by an orientation of both the crystallites and the polymer chain segments in the noncrystalline domains. The fine structure and physical properties of synthetic textile fibers are frequently further modified by a variety of thermomechanical annealing treatments, including processes known as texturing. Texturing provides crimp to the filaments, and bulk and softness to textile products manufactured from them. In the case of melt spinning, the two-step process of spinning (extrusion) and drawing (structure development) is increasingly being combined in a single high speed spinning process.

Bicomponent fibers, where two different polymers are extruded simultaneously in either side-by-side or skin/core configurations, are also an important category of fibers.

In the traditional methods of fiber manufacture the filaments are obtained in continuous form. When several such filaments are combined together and slightly twisted to maintain unity, the product so obtained is called a multifilament yarn. A typical yarn may contain 100 single filaments. Individual filaments, considerably larger in cross section than those used in multifilament yarns, may also be used in certain applications, and these are referred to as monofilaments. Frequently it is desired to obtain fibers in finite lengths for subsequent manufacture into spun yarns by conventional textile spinning operations. In this case thousands of continuous filaments are collected together into a continuous rope of parallelized filaments called a tow. The tow is converted into staple length fiber by simply cutting it into specified lengths. The staple length produced in this conversion process depends on the system of yarn manufacture to be used. The cotton yarn manufacturing system requires lengths of approximately 1.5 in. (3.8 cm), whereas the woolen or worsted systems require lengths between 3 and 5 in. (7.6–12.7 cm). When synthetic fibers are produced in staple form for blending with natural fibers, dimensional properties such as length and fineness are matched to the natural fiber which is to be used in the blend.

Fiber Consumption Trends

For centuries the textile industry relied exclusively on the natural fibers, particularly cotton (qv), wool (qv), and silk (qv). With the commercialization of synthetic fibers, starting with those based on natural polymers in the 1930s, and then with

Table 1. Worldwide Fiber Production by Region, 1990,[a] 10^6 t

	Synthetic fiber[b]		Natural fibers	
Region	Synthetic polymer	Cellulosics	Cotton	Wool
United States	2.87	0.23	3.38	0.04
USSR	0.91	0.56	2.65	0.55
Europe	3.35	0.93	0.30	0.16
other Americas	0.86	0.13	1.73	0.24
Japan	1.42	0.28		
People's Republic of China	1.31	0.20	4.52	0.29
Asia/Oceania	3.93	0.49	5.14	1.40
Middle East/Africa	0.20	0.03	1.25	0.67
Total	*14.9*	*2.85*	*19.0*	*3.35*

[a]Ref. 1.
[b]Not including olefin fiber and glass fiber.

Table 2. Worldwide Synthetic Fiber Production by Fiber,[a] 10^6 t

Fiber	1986	1990
polyester	6.84	8.62
polyamide	3.50	3.76
acrylic	2.44	2.33
olefin	1.93	2.70
glass	1.40	1.84
other	0.14	0.16
Total	*16.3*	*19.4*

[a]Ref. 1.

those based on synthetic polymers in the 1940s, the textile industry has truly undergone a revolution. The 1990 total world production of textile fibers amounted to approximately 40×10^6 t, distributed among the principal types of fibers as shown in Table 1. Nearly all regions of the world are involved with fiber production. Polyester and polyamide fibers alone account for 63% of the synthetic fibers produced worldwide in 1990, as shown by the data in Table 2. The textile industry finds itself with an ever-increasing number of fibers from which to manufacture its products. Not only are there many more fiber types for the textile industry to use, but also it is recognized that many advantages are to be gained by blending the various fibers in an almost infinite number of combinations. Many current textile products are blends of natural and synthetic fibers that incorporate the desirable attributes of both.

There has been a rapid increase in worldwide fiber production to its current level of over 40×10^6 t. This reflects not only an increasing world population but also an increasing per capita consumption of textile fibers. World population has increased during the past several decades at an average annual rate of about 2% to its current level of nearly 5.0 billion. The average per capita fiber consumption increased from 5.6 kg in 1965 to 7.3 kg in 1986. As might be expected, because of variations in climate and socioeconomic conditions, there are large variations in the per capita fiber consumption for different regions of the world, as is shown by the data in Table 3. The per capita fiber consumption of different fibers in the United States for the years 1980 and 1990 is as follows (1):

Fiber	1980	1990
synthetic	15.3	17.9
cotton	6.7	10.9
wool	0.4	0.5
other	1.6	1.0
Total	*24.0*	*30.3*

Table 3. Per Capita Fiber Consumption, 1986,[a] kg

Country	Natural fibers	Synthetic fibers
	Developed nations	
North America	10.1	15.1
Western Europe	7.7	8.1
Eastern Europe	9.1	6.4
Japan	9.6	8.1
Australia	10.2	11.2
	Developing nations	
Africa	0.7	0.5
Latin America	3.0	2.3
South America	2.8	2.0
Near East	4.2	2.8
Far East (Asia)	1.7	0.9
China	3.1	1.1

[a]Ref. 2.

Types of Fibrous Materials

Fibers are used in the manufacture of a wide range of products that can generically be referred to as fibrous materials. The properties of such materials are dependent on the properties of the fibers themselves and on the geometric arrangement of the component fibers in the structure.

Several types of fibrous materials can be distinguished primarily on the basis of fiber organization. At one extreme are isotropic assemblies where the fibers are arranged in a completely random fashion with no preferred orientation in any of the three principal spatial axes. The physical properties of an isotropic fiber assembly are independent of the test direction, and any point in a unit volume space of such an assembly is indistinguishable from another. Isotropic fiber assemblies are quite rare in most end products, and in fact are quite difficult to achieve because of the high length to width ratio of fibers. This high aspect ratio causes fibers to align themselves in a stress field, thereby creating a preferred orientation of the fibers in the processing direction. Thus considerable effort must be exerted to overcome the tendency for preferred orientation when an isotropic fiber assembly is desired. Alternatively, and more commonly, there are various anisotropic fibrous materials with the fibers arranged in well-defined spatial patterns.

Textile Yarns. Textile yarns and several preliminary linear structures, such as roving and sliver, are typical of those structures where there is a high degree of fiber orientation with respect to the principal axis of the material (3). The degree of fiber parallelization and orientation varies from one type of yarn to another, and can be controlled by variation in the fiber type, in the fiber geometric properties, and in the processing conditions used in the yarn production process. The high degree of structural anisotropy of textile yarns is reflected in large differences between axial and transverse physical properties. The structural anisotropy of yarns, and of the individual fibers, imparts a unique combination of high strength and low bending rigidity to textile yarns.

Textile yarns are produced from staple (finite length) fibers by a combination of processing steps referred to collectively as yarn spinning. After preliminary fiber alignment, the fibers are locked together by twisting the structure to form the spun yarn which is continuous in length and remarkably uniform. Depending on the specific processing conditions, the degree of fiber parallelization and surface hairiness can vary over a considerable range which strongly influences yarn physical properties. Sorption properties in particular are strongly affected by the fiber organization in a spun yarn. The staple fibers may be either natural fibers, such as cotton or wool, or any of a number of synthetic fibers.

Textile yarns are also produced from continuous filament synthetic fibers. Such yarns are referred to as multifilament yarns, and are characterized by nearly complete filament alignment and parallelization with respect to the yarn axis. The degree of twist introduced into a multifilament yarn is usually quite low and just adequate to produce some level of interfilament cohesion. Such yarns are quite compact and smooth in appearance. A variety of processes have been developed to introduce bulk and texture in multifilament yarns. These processes are designed to disrupt the high degree of filament alignment and parallelization and to produce yarns with properties generally associated with spun yarns. Schematic representations of typical yarns are given in Figure 1.

Textile Fabrics. Yarns are used principally in the formation of textile fabrics either by weaving or knitting processes (4). Fabrics are a form of planar fibrous assembly where the high degree of structural anisotropy characteristic of yarns is minimized but not totally eliminated. In a woven fabric, two systems of yarns, known as the warp and the filling, are interlaced at right angles to each other in various patterns. The woven fabric can be viewed as a planar sheet-like

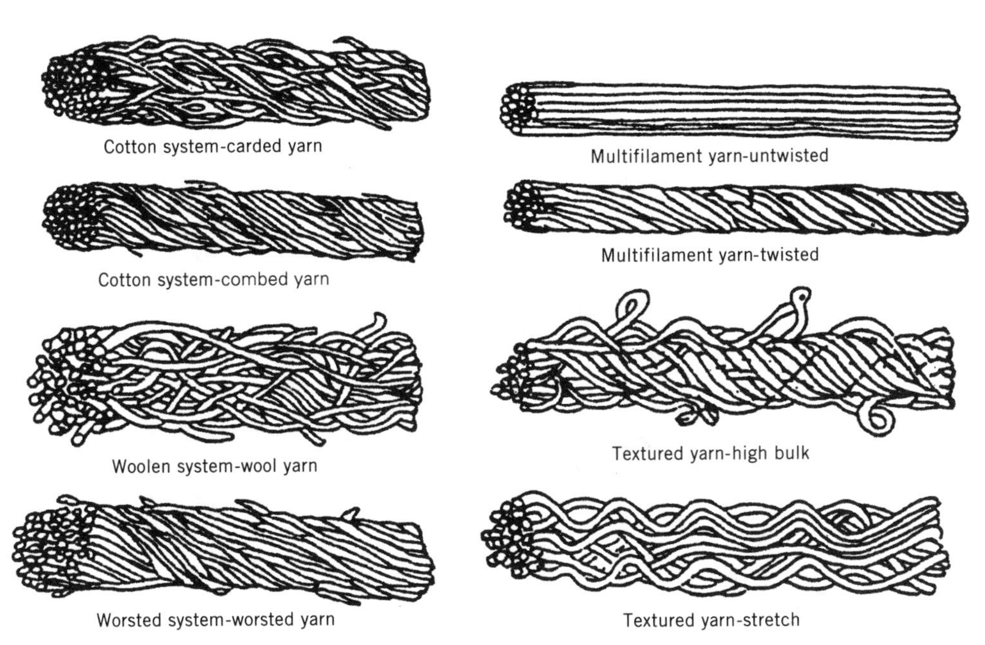

Fig. 1. Schematic description of spun and multifilament yarns (3).

material with pores or holes created by the yarn interlacing pattern. The dimensions of the fabric pores are determined by the yarn structure and dimensions, and by the weaving pattern. The physical properties of woven fabrics are strongly dependent on the test direction.

Knitted fabrics are produced from one set of yarns by looping and interlocking processes to form a planar structure. The pores in knitted fabrics are usually not uniform in size and shape, and again depend largely on yarn dimensions and on the numerous variables of the knitting process. Knitted fabrics are normally quite deformable, and again physical properties are strongly dependent on the test direction.

Nonwovens. The term nonwoven simply suggests a textile material that has been produced by means other than weaving, but these materials really represent a rather unique class of fibrous structure (5,6). In nonwovens the fibers are processed directly into a planar sheet-like fabric structure, bypassing the intermediate one-dimensional yarn state, and then either bonded chemically or interlocked mechanically (or both) to achieve a cohesive fabric such as shown in Figure 2. Typically, staple length fibers are dispersed in a fluid (liquid in the wet-laid process of manufacture or air in the dry-laid process of manufacture) and deposited in sheet-like planar form on a support base prior to bonding or interlocking. The papermaking process is a well-known example of a wet-laid nonwoven process that utilizes short paper (wood) fibers. The spunbond process differs from the dry- and wet-laid processes in that continuous length filaments are extruded, collected in a randomized planar network, and bonded together to form the final product. Spunbonded nonwovens are generally thin, strong, and almost film-like in appearance (see NONWOVEN FABRICS).

Within the plane of a nonwoven material, the fibers may be either completely isotropic or there may be a preferred fiber orientation or alignment usually with

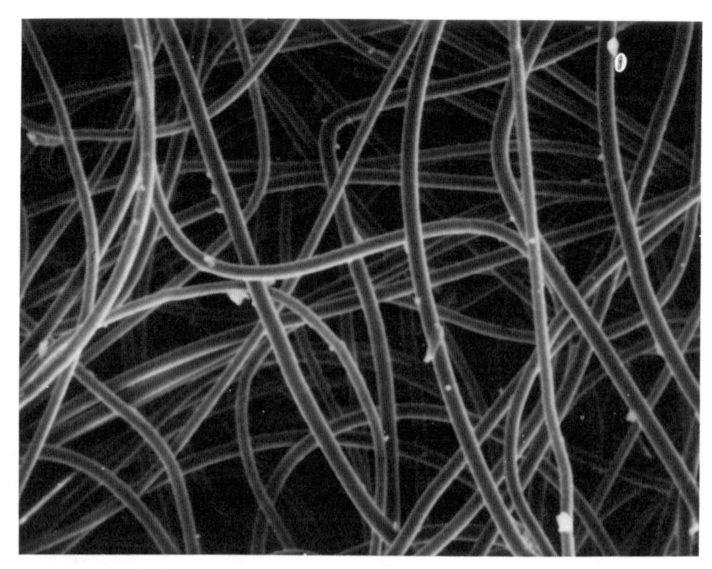

Fig. 2. Scanning electron photomicrograph of a polyester nonwoven fabric. Courtesy of S.B. Reutsch, TRI/Princeton.

respect to a machine or processing direction. In the case of thicker dry-laid nonwovens, fiber orientation may be randomized in the third dimension, ie, that dimension which is perpendicular to the plane of the fabric, by a process known as needle-punching (7). This process serves to bind the fibers in the nonwoven by mechanical interlocking.

The type of nonwoven material that is produced depends largely on the fiber type used and on the method of manufacture. Typically, air-laid nonwovens are less dense and compact and tend to be softer, more deformable, and somewhat weaker. Wet-laid or paper-like nonwovens are more dense, stronger, more brittle, and less permeable to fluids. It is dangerous, however, to generalize. Nonwovens by either process can be produced to achieve a wide variety of products with a broad range of physical properties. The methods used for bonding and for interlocking the constituent fibers to form the nonwoven material may be as important as the method used for producing the initial nonwoven web. The degree of bonding or interlocking also has a critical influence on the final properties. In general, strength increases, extensibility decreases, softness decreases, and sorptive capacity decreases with increasing degree of bonding or interlocking. When thermoplastic fibers are used in the nonwoven, thermal or heat bonding is generally preferred to the use of chemical binding agents.

This brief overview of the types of fibrous materials is intended to indicate the broad range of materials that can be produced from fibers. Since the properties of fibrous materials depend both on the properties of the fibers themselves and on the spatial arrangement of the fibers in the assembly, a given type of fiber may be used in many different end products, and similarly a given end product can be produced from different fiber types.

Fiber Properties

The properties of textile fibers may be conveniently divided into three categories: geometric, physical, and chemical, as shown in Table 4.

Geometric Characteristics. These properties pertain primarily to staple fibers and include various aspects of fiber dimensions and form such as length, crimp, cross-sectional area, and cross-sectional shape. They are of particular importance in processing spun yarns. For example, machine settings in textile spinning operations are based on fiber length, and the efficiency of processing is related to length uniformity. In general, it is desirable that staple fibers not contain an excessive proportion of fibers significantly shorter than the average.

Cross-sectional area or fiber fineness also affects textile processing efficiency and the quality of the end product. The number of fibers in a cross section of yarn of a given size is related to fiber fineness, that is, the smaller the fiber cross section the more fibers will be needed in the yarn. Other factors being equal, yarn strength increases as the number of fibers in the yarn cross section increases. However, fibers with too small a cross section cannot be processed efficiently.

The cross-sectional shape of fibers is an important characteristic that influences many end use properties. The cross-sectional shape of cotton fibers is essentially that of a flat ribbon, although there are cotton fibers that have circular cross sections. Wool fibers are elliptical in shape, although again nearly circular shapes can be found. The earlier synthetic fibers were generally circular in cross-

Table 4. Classification of Fiber Properties

Geometric	Physical	Chemical
Length	*Density*	*Response to*
average value	linear	acids
distribution	bulk	alkalies
		oxidation
Cross section	*Thermal*	reduction
average value	melting point	heat
distribution	transitions	
shape	conductivity	*Sorption*
		moisture
Crimp	*Optical*	dyes
frequency	birefringence	
form	refractive index	*Swelling*
	luster and color	anisotropy
	Electrical	
	resistivity	
	dielectric constant	
	Surface	
	roughness, friction	
	Mechanical	
	tension, compression,	
	torsion, bending, shear	

sectional shape, but it is now possible to produce fibers with modified cross sections. This is accomplished by designing the shape of the spinneret orifice used in melt spinning. Fibers (primarily nylon, polyester, and polypropylene) can be produced with trilobal, multilobal, flat, triangular, and a variety of other shapes. These fibers have special optical effects, and are also useful in modifying mechanical properties of textile products. Hollow fibers can also be produced by special fiber extrusion processes.

Crimp is a form factor that describes the waviness of a fiber or its longitudinal shape. Crimp of a fiber can be quantified either in geometric terms, such as wave height, wave amplitude, and frequency, or in energetic terms, such as the force or energy required to uncrimp a fiber. A certain level of fiber crimp is essential in order for a fiber to be processible on conventional textile manufacturing equipment. All natural fibers are crimped, particularly animal fibers, and synthetic fibers must be crimped artificially to be processible into spun yarns. Crimp is imposed on the filaments by various mechanical means, usually in the tow stage before cutting into staple. In certain fibers it has been possible to develop a crimp that is structurally based. In these cases the crimped configuration is a reflection of molecular structure and may be considered the equilibrium state.

Length, fineness, and crimp of natural fibers cannot be controlled easily, and the economic value of a fiber is highly dependent on the level and uniformity of these properties. In synthetic fibers, these fiber properties can be controlled quite readily during manufacture. The length is simply set to the desired value, and length uniformity is always high. The cross-sectional area or fineness is controlled by the fiber spinning (extrusion) process, and the amount of attenuation in drawing. Similarly, the amount of crimp imposed on a fiber can be specified.

In general, the geometric properties of the natural fibers are highly variable from fiber to fiber, both within a given lot and among lots of the same fiber type. In the synthetic fibers, the geometric properties are extremely uniform in view of the production control possible in a chemical plant but not in an agricultural product.

Physical Properties. Relationships between fiber properties and their textile usefulness are in many cases quite obvious. Since fibers are frequently subjected to elevated temperatures, it is necessary that they have high melting or degradation points. It is also necessary that other fiber properties be relatively constant as a function of temperature over a useful temperature range.

In general, textile fibers should be optically opaque so that their refractive indexes need to be significantly different from those of their most common environments, namely, air and water. Luster and color are two optical properties that relate to a fiber's aesthetic quality and consumer acceptance.

Electrical properties also affect fiber utility. In view of their chemical structure most textile fibers are nonconducting, and their high resistivities readily classify them as insulators. Fibers are subject to static electrification and some means of discharging them needs to be employed in high speed textile manufacturing processes and in certain end uses. Static electrification is a particular problem with hydrophobic fibers that do not absorb water from the atmosphere to form an electrically conducting system whereby the static can be dissipated. A number of antistatic finishes and other special methods have been developed that reduce the static charge generated during textile processing and end use.

Textile fibers are always used in aggregates of a large number of single fibers that are caused to interact with each other through their surfaces. The surface characteristics of fibers, therefore, are of singular importance. Interfiber friction and geometric roughness or rugosity are two surface properties that relate directly to processibility and product performance.

Chemical Properties. Fibers must be resistant to the effects of acids, alkalies, reducing agents, and oxidizing agents, as well as to electromagnetic and particulate irradiation. Resistance to these degradative systems is required for both short-term exposure to large doses or concentrated reagents, and for long-term exposure to weak or dilute systems.

Most textile fibers are hygroscopic at least to some extent, and therefore capable of absorbing moisture from the atmosphere, which is a direct reflection of chemical structure. Textile fibers vary from those that may be considered hydrophilic to those that are essentially hydrophobic (8–10). Moisture sorption follows the Langmuir isotherm, that is, the extent of sorption of water vapor is proportional to the partial pressure of water vapor in the atmosphere (Fig. 3). The moisture sorption properties of textile fibers are generally expressed as the moisture content (wet basis) or the moisture regain (dry basis) under a given set of conditions. Atmospheric conditions of 65% rh and 21°C are used for determining the standard moisture regain of textile fibers, although it is frequently of great

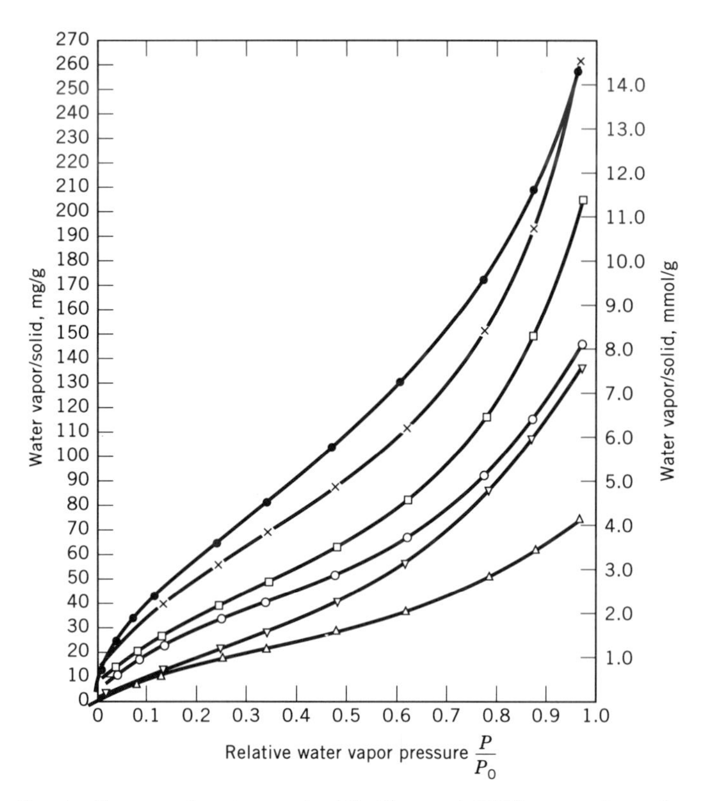

Fig. 3. Sorption isotherms of water on textile fibers at 25°C: ●, wool; × viscose; □, silk; ○, cotton; ▽, acetate; △ nylon.

importance to know the regains at other than these standard conditions. Mechanical and other fiber properties vary as a function of moisture content, and therefore need to be evaluated under controlled and specified atmospheric conditions, usually 65% rh and 21°C. It is of course important to know the properties of fibers under all atmospheric conditions to which they may be subjected.

The sorption of molecular species other than water from the atmosphere or from solution must also be considered. Paramount among general sorption properties of textile fibers are their dyeing characteristics, which describe the rate and extent of dye sorption. In order that a fiber have substantial usefulness for textile purposes, it must be capable of interacting with at least one of the well-known classes of dyes. The fiber–dye complex must be colored to an acceptable depth and must be capable of withstanding exposure to light, actinic rays, washing, dry cleaning, and other degrading influences. Wash- and lightfastness are particularly important; these and other fastness characteristics are properties of the fiber–dye complex and not of the dye alone. Dyeability is dependent on the presence of functional groups as part of the fiber structure which are capable of interacting with the dye. The affinities of various dyes for textile fibers are thermodynamic quantities which depend only on the chemical structures of the dye and the fiber, and should not be confused with the rate of dyeing. The rate is also related to the fiber's supramolecular organization, and the ability of a dye to penetrate and diffuse into the internal regions of the fiber. Rate of dyeing is also strongly influenced by dye concentration in the external solution (or vapor phase), temperature, pH, and degree of dye aggregation in the external solution and in the fiber. The effects of these and other variables on the rate of dyeing, as well as other aspects of the physical chemistry of dyeing, have been summarized (11–13) (see also DYES, APPLICATION AND EVALUATION).

Swelling is a chemical property closely related to a fiber's sorption characteristics. It may be defined as the reversible dimensional changes that occur when fibers undergo an absorption process. Since fibers are structurally anisotropic, they invariably undergo greater transverse (diametral) than longitudinal swelling. The swelling of fibers in aqueous systems or in atmospheres of high relative humidity is of particular importance since fibers are frequently subjected to these environments. The dimensional stability of textile fabrics in laundering is related to fiber swelling phenomena, although other factors such as altered mechanical properties of fibers in the wet state are known to be important in this connection. True thermodynamic swelling of a fiber is reversible. Frequently, however, the absorption of certain molecular species causes a disruption of the internal fiber structure and the dimensional changes accompanying this process are not reversible in the accepted sense. It is important, although frequently quite difficult, to differentiate between reversible thermodynamic swelling and irreversible alterations in fiber structure.

Mechanical Behavior. Because textile fibers are used primarily as elements of construction, a sound engineering approach to their mechanical properties is necessary (14). The mechanical properties of fibers describe the response of a fiber to deforming loads under conditions that induce tension, compression, torsion, or bending. Torsional deformations are normally analyzed in terms of shearing stresses, whereas bending may be considered to produce simultaneous tensile and compressive stresses around a neutral fiber axis. Mechanical properties are usually evaluated under standard conditions of temperature and humidity (65% rh, 21°C), and under closely specified conditions of load application. Since textile fibers are exposed to a variety of chemical environments in the normal course of manufacturing and processing into yarns and fabrics, as well as during their ultimate use by the consumer, it is frequently important to evaluate mechanical properties in relation to a fiber's chemical environment. Not only do such measurements make possible a more probable prediction of fiber behavior under other than standard conditions, but also it may be possible to deduce important information about fiber structure.

The mechanical properties of fibers or of any other material may be described in terms of six factors: strength, elasticity, extensibility, resilience, stiffness, and toughness. Information about these mechanical properties is obtained from stress–strain or load-deformation curves that are graphical records of tensile, compressive, or shearing stresses as a function of deformation. In view of a fiber's geometric shape and dimensions, these curves are usually evaluated under uniaxial tension. The procedure for obtaining a load-deformation curve is to subject the fiber to increasing loads while recording the extension. Alternatively, the fiber can be subjected to controlled extension while recording the force generated by some suitable device. Several commercial constant rate of loading and constant rate of extension tensile testers are available with sufficient sensitivity to permit the accurate evaluation of fiber mechanical properties.

The load-extension curves obtained on these devices must be normalized by the fiber cross-sectional area so that the stress–strain curve may be derived. In fiber science the initial cross-sectional area of a fiber is used for normalization of the load-extension curve. Thus, by not taking into account the fiber attenuation during extension, the true stress–strain curve is not obtained. However, the nominal or engineering stress–strain curve obtained reflects the number of stress-bearing units in the initial fiber cross section, and thus provides significant information. The nominal or engineering stress can be easily converted into true stress by multiplying the nominal stress at any given value of the strain, ϵ, by the factor $(1 + \epsilon)$.

It is difficult to determine the cross-sectional area of a fiber. Direct observation and measurement of a cross section under a microscope is the most accurate method (15). This is a destructive test that does not allow subsequent study of fiber mechanical properties, and is slow and tedious. Also, it does not take into account any variations in the cross-sectional area along the fiber length. Measurement of fiber diameters from microscopic observations of longitudinal views is somewhat easier, but the ellipticity of the cross section in certain fibers can lead to serious errors.

The more usual method of designating fiber size is by its linear density in units of mass per unit length. If the bulk density (mass per unit volume) is known, then the cross-sectional area can be computed by dividing the linear density by the bulk density. In fiber and textile terminology three special linear density units are used: denier, grex, and tex. The tex system has been adopted by the ASTM as the standard unit for designating the linear density of textile fibers and yarns, although the denier unit continues to be widely used. The tex is also acceptable as part of the SI system, although rigorously, the SI unit is kg/m. The tex as a measure of fineness or linear density is defined as 1 g/1000 m (10^{-6} kg/m). Denier is defined as 1 g/9000 m (1.111×10^{-7} kg/m); 1 denier = 0.1111 tex. A grex is 0.1 tex.

Typical textile fibers have linear densities in the range of 0.33–1.66 tex (3 to 15 den). Fibers in the 0.33–0.66 tex (3–6 den) range are generally used in nonwoven materials as well as in woven and knitted fabrics for use in apparel. Coarser fibers are generally used in carpets, upholstery, and certain industrial textiles. A recent development in fiber technology is the category of microfibers, with linear densities <0.11 tex (1 den) and as low as 0.01 tex. These fibers, when properly spun into yarns and subsequently woven into fabrics, can produce textile fabrics that have excellent drape and softness properties as well as improved color clarity (16).

Normalized fiber mechanical properties are expressed in terms of unit linear density. For example, in describing the action of a load on a fiber in a tensile test, units of N/tex or gram force per denier (gpd) are generally used. If this is done, the term tenacity should be used in place of stress. The true units of stress are force per unit cross-sectional area, and the term stress should be reserved for those instances where the proper units are used.

The linear density of fibers is determined by direct weighing of a known length of fiber or by means of a vibroscope. Direct weighing is a feasible method in the case of continuous synthetic filaments where a sufficient length for accurate weighing can be obtained. Direct weighing of staple fibers, either natural or synthetic, is not practicable in view of the uncertainties involved in the length measurement and in the weighing. The vibroscopic method of determining linear density is well suited to fibers of all lengths (17). The vibroscope consists of a system for applying an oscillatory force of known frequency to a fiber while it is held under tension, and also a means of detecting mechanical resonance under the applied oscillatory force. From the values of frequency, tension, and fiber length, the linear density can be computed from the classical vibrating string formula. Corrections taking into account fiber stiffness, cross-sectional shape, and nonuniformity along the length of the fiber have been developed.

A schematic stress–strain curve of an uncrimped, ideal textile fiber is shown in Figure 4. It is from curves such as these that the basic factors that define fiber mechanical properties are obtained.

The strength of a fiber, which measures the ability of the fiber to withstand a load, is expressed in terms of the stress required to produce rupture in units of mass per unit cross section. In more common fiber and textile terminology, strength is expressed in terms of the tenacity at break or ultimate tenacity in units of N/tex or gram force per denier. Extensibility describes the deformation of the fiber produced by a given stress. It is quantitatively defined by the ultimate strain, which is the fractional increase in length of a fiber when subjected to a stress that causes it to rupture. The units for extensibility or strain are length per unit length, which may be expressed in percentage units. Stiffness describes the resistance of the fiber to deformation. This is measured by the elastic stiffness, which is the ratio of the stress to strain at the yield point. The elastic stiffness constitutes the slope of the initial Hookean region of the stress–strain curve where the stress and strain are directly proportional. The elastic stiffness is equivalent to the elastic modulus or Young's modulus of elasticity, and carries the units of stress per unit strain. In textile terminology, where tenacity is used in place of

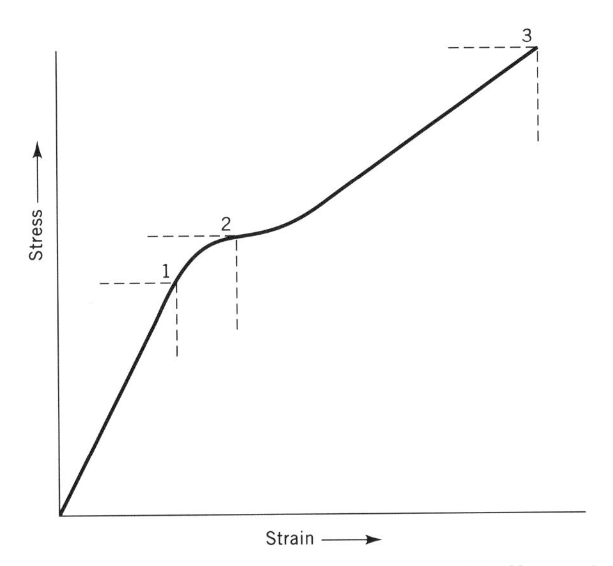

Fig. 4. Idealized stress–strain curves of an uncrimped textile fiber; point 1 is the proportional limit, point 2 is the yield point, and point 3 is the break or rupture point.

stress, the units for stiffness are gram force per unit linear density. On the assumption that the stress–strain curve of a textile fiber is essentially linear, fiber stiffness may be approximated by the average stiffness, which is the ratio of the breaking stress to the breaking strain.

The resilience of a fiber describes its ability to absorb work or mechanical energy elastically, that is, without undergoing permanent deformation. This property is related to the area under the Hookean portion of the stress–strain curve, which can be considered as the energy of elastic deformation. Rigorously, resilience must be defined as the ratio of the energy of recovery to the energy of deformation in a cyclic loading experiment which may be carried to any given value of load or extension. Frequently, resilience may be evaluated in terms of dimensional recovery from a given deformation. Resilience is dimensionless, and is usually expressed in percentage units.

The elasticity of a fiber describes its ability to return to original dimensions upon release of a deforming stress, and is quantitatively described by the stress or tenacity at the yield point. The final fiber quality factor is its toughness, which describes its ability to absorb work. Toughness may be quantitatively designated by the work required to rupture the fiber, which may be evaluated from the area under the total stress–strain curve. The usual textile unit for this property is mass per unit linear density. The toughness index, defined as one-half the product of the stress and strain at break also in units of mass per unit linear density, is frequently used as an approximation of the work required to rupture a fiber. The stress–strain curves of some typical textile fibers are shown in Figure 5.

When an ideal fiber is subjected to a deformation within the Hookean region of the stress–strain curve, the extension is recoverable upon removal of the load. In this region of the stress–strain curve, the ideal fiber behaves as an elastic body. When a fiber is subjected to a load that places it under a stress higher than its yield value, the fiber does not return completely to its original length upon removal of the deforming load. The fiber in this case develops a permanent set. Not

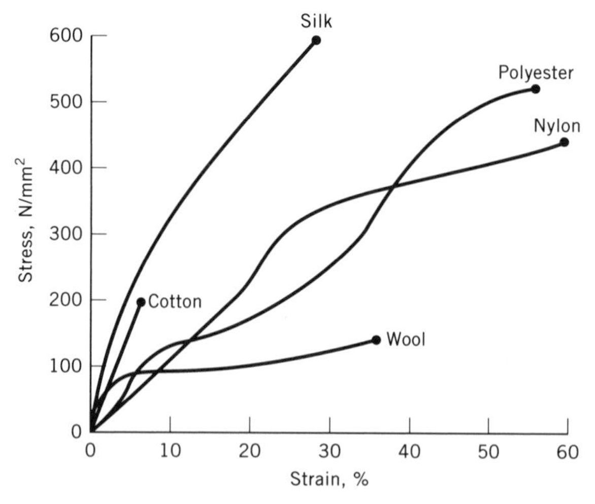

Fig. 5. Stress–strain curves of some textile fibers (17). To convert N/mm^2 to psi, multiply by 145.

all of the original deformation is retained as permanent set; some of it is recoverable. Recovery from deformation may be divided into two components (18). The first takes place immediately upon removal of the deforming load and is called the immediate elastic deflection. The other is the delayed deflection, which can be further subdivided into a time dependent recoverable component and a nonrecoverable component. The latter is also known as permanent set or secondary creep. Permanent nonrecoverable deformation is a manifestation of plasticity or internal stress-induced plastic flow. The tensile recovery behavior of a large number of textile fibers has been studied (19).

Another aspect of plasticity is the time dependent progressive deformation under constant load, known as creep. This process occurs when a fiber is loaded above the yield value and continues over several logarithmic decades of time. The extension under fixed load, or creep, is analogous to the relaxation of stress under fixed extension. Stress relaxation is the process whereby the stress that is generated as a result of a deformation is dissipated as a function of time. Both of these time dependent processes are reflections of plastic flow resulting from various molecular motions in the fiber. As a direct consequence of creep and stress relaxation, the shape of a stress–strain curve is in many cases strongly dependent on the rate of deformation, as is illustrated in Figure 6.

The resistance to plastic flow can be schematically illustrated by dashpots with characteristic viscosities. The resistance to deformations within the elastic regions can be characterized by elastic springs and spring force constants. In real fibers, in contrast to ideal fibers, the mechanical behavior is best characterized by simultaneous elastic and plastic deformations. Materials that undergo simultaneous elastic and plastic effects are said to be viscoelastic. Several models describing viscoelasticity in terms of springs and dashpots in various series and parallel combinations have been proposed. The concepts of elasticity, plasticity, and viscoelasticity have been the subjects of several excellent reviews (21,22).

An important aspect of the mechanical properties of fibers concerns their response to time dependent deformations. Fibers are frequently subjected to con-

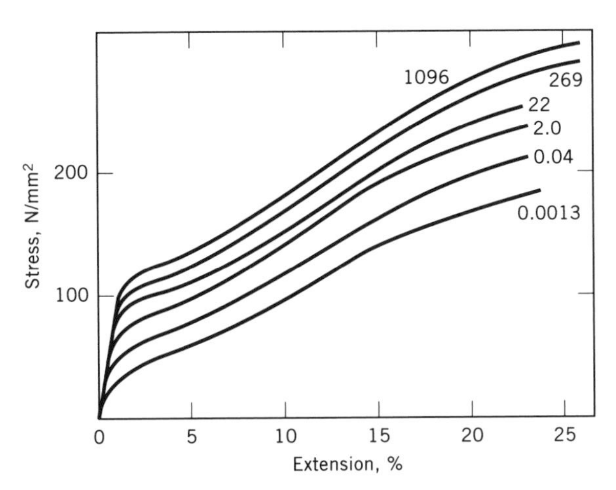

Fig. 6. The effect of rate of extension on the stress–strain curves of rayon fibers at 65% rh and 20°C. The numbers on the curves give the constant rates of extension in percent per second (20).

ditions of loading and unloading at various frequencies and strains, and it is important to know their response to these dynamic conditions. In this connection the fatigue properties of textile fibers are of particular importance, and have been studied extensively in cyclic tension (23). The results have been interpreted in terms of molecular processes. The mechanical and other properties of fibers have been reviewed extensively (20,24–27).

Structure of Fiber-Forming Polymers

With the exception of glass fiber, asbestos (qv), and the specialty metallic and ceramic fibers, textile fibers are a class of solid organic polymers distinguishable from other polymers by their physical properties and characteristic geometric dimensions (see GLASS; REFRACTORY FIBERS). The physical properties of textile fibers, and indeed of all materials, are a reflection of molecular structure and intermolecular organization. The ability of certain polymers to form fibers can be traced to several structural features at different levels of organization rather than to any one particular molecular property.

Fiber structure is described at three levels of molecular organization, each relating to certain aspects of fiber behavior and properties. First is the organochemical structure, which defines the chemical composition and molecular structure of the repeating unit in the base polymer, and also the nature of the polymeric link. This primary level of molecular structure is directly related to chemical properties, dyeability, moisture sorption, and swelling characteristics, and indirectly related to all physical properties. The chemical structure also determines the magnitude of intermolecular forces, which is important in terms of properties and overall fiber structure. The macromolecular level of structure describes the chain length, chain-length distribution, chain stiffness, molecular size, and molecular shape. The supramolecular organization is the arrangement of the polymer chains in three-dimensional space. The physical properties of fibers are strongly influenced by the organization of polymeric chains into crystalline and noncrystalline domains, and the disposition of these domains with respect to each other. In the case of natural fibers, a further level of structural organization related to the natural growth and development of the fiber must be considered. This morphology is quite complex, with fibrils, microfibrils, and similar structural subunits frequently surrounded by a matrix material in a composite configuration.

Polymeric fibers can be differentiated from other polymeric materials, such as films, rubbers, plastics, and powders on the basis of geometric form and physical properties, but it is more instructive to differentiate fibers from other polymeric substances on the basis of structure. Certain underlying structural features are fiber-forming, ie, the ability to assume the special supramolecular organization characteristic of fibers. All polymeric fibers that are useful in textile applications are semicrystalline, irreversibly oriented polymers, ie, the polymeric chains are partially ordered into regions or domains with near perfect registry so that the laws of x-ray diffraction are obeyed. In other regions of the fiber, the molecular chains or segments of chains are not perfectly ordered and may approach random coil configurations. This is the simplest statement of the two-phase crystalline–amorphous model of the structure of semicrystalline polymers that was for a long time used to describe fiber structure and to interpret fiber properties.

The original two-phase model, with sharp boundaries between crystalline and amorphous regions, was replaced by various models postulating gradual transitional regions. One formulation of this idea is the concept of a lateral order distribution, according to which structural regularity varies from the perfectly ordered or crystalline state to the completely disordered or amorphous state in a continuous manner. Other formulations of the gradual transition from crystalline to amorphous regions are the fringed micelle structure and the fringed fibril structure. Another model considers fibers to be fully crystalline, and associates regions of higher reactivity and accessibility (previously the amorphous regions or regions of low lateral order) with imperfections and defects in the crystalline system. These imperfections and defects may be associated with molecular dislocations, chain ends, and chain folds. In many cases such models have been postulated for specific fiber types.

All models of fiber structure picture a fiber as a polymeric substance with a high degree of three-dimensional structural regularity, leading directly to the concept of the degree of crystallinity, ie, the fractional crystalline content of a partially crystalline polymeric material. The degree of crystallinity is difficult to evaluate. X-ray diffraction techniques probably provide the best estimates, but unsolved questions remain because the degree of crystallinity is actually a composite value that reflects not only the fractional quantity of ordered material, but also the size or dimensions and the perfection of the crystallites. Thus a highly ordered polymer may be evaluated by x-ray diffraction as being of low crystallinity if the crystallites are either highly imperfect or so small that well-resolved diffraction patterns cannot be obtained. Other methods of crystallinity evaluation depend on the fact that the polymer chains in the crystalline domains are more densely packed and less accessible to penetrants than the chains in the noncrystalline domains. Measurements of density and moisture sorption are frequently used to estimate degree of crystallinity in polymeric fibers, as are other specialized techniques such as deuterium exchange, chemical reactivity, infrared spectroscopy, and differential thermal analysis.

The requirement of at least partial crystallinity limits the number of polymers suitable for fiber formation. To a large extent, what is meant by the term fiber-forming polymer is actually crystallizable polymer. The structural characteristics of polymers that allow them to crystallize under appropriate conditions have been summarized as follows. *Regularity*. The polymer chains must be uniform in chemical composition and stereochemical form. *Shape and Interaction*. The shape of the polymer chains must allow close contact or fit to permit effective and strong intermolecular interaction. This is generally achieved by linear macromolecules with no bulky side groups or with side groups that are regularly spaced along the backbone chain. *Repeat length*. The ease of crystallization decreases with increasing length of the polymer repeating unit. *Chain directionality*. Since certain polymer chains have a directionality, the mode of chain packing in a crystallite can take either parallel or antiparallel configurations. *Single-chain conformation*. Crystallization is favored if the chain conformation is compatible with its form in the crystallite. *Chain stiffness*. An optimal stiffness of the polymer is necessary.

Attention must also be focused on the noncrystalline domains. Many important properties of fibers can be directly related to these noncrystalline or amorphous regions. For example, absorption of dyes, moisture, and other penetrants occurs in these regions. These penetrants are not expected to diffuse into the crystalline domains, although there may be adsorption on crystallite surfaces. The extensibility and resilience of fibers is also directly associated with the noncrystalline regions.

Noncrystalline domains in fibers are not structureless, but the structural organization of the polymer chains or chain segments is difficult to evaluate, just as it is difficult to evaluate the structure of liquids. No direct methods are available, but various combinations of physicochemical methods such as x-ray diffraction, birefringence, density, mechanical response, and thermal behavior, have been used to deduce physical quantities that can be used to describe the structure of the noncrystalline domains. Among these quantities are the amorphous orientation function and the amorphous density, which can be related to some of the important physical properties of fibers.

Fiber structure is a dual or a balanced structure. Neither a completely amorphous structure nor a perfectly crystalline structure provides the balance of physical properties required in fibers. The formation and processing of fibers is designed to provide an optimal balance in terms of both structure and properties. Excellent discussions of the structure of fiber-forming polymers and general methods of structure characterization are available (28–31).

BIBLIOGRAPHY

"Fibers" in *ECT* 1st ed., Vol. 6, pp. 453–467, by H.F. Mark, Polytechnic Institute of Brooklyn; "Fibers, Man-Made" in *ECT* 2nd ed., Vol. 9, pp. 151–170, by H.F. Mark, Polytechnic Institute of Brooklyn, and S.M. Atlas, Bronx Community College; "Fibers, Chemical" in *ECT* 3rd ed., Vol. 10, pp. 148–166, by W.J. Roberts, Consultant.

1. *Fiber Organon*, Vol. 62, Fiber Economics Bureau, Inc., Roseland, N.J., 1991.
2. *World Apparel Fiber Consumption Survey*, Food and Agriculture Organization of the United Nations, New York, 1989.
3. B. C. Goswami, J. G. Martindale, and F. L. Scardino, *Textile Yarns: Technology, Structure and Applications*, John Wiley & Sons, Inc., New York, 1977.
4. J. W. S. Hearle, P. Grosberg, and S. Backer, *Structural Mechanics of Fibers, Yarns, and Fabrics*, John Wiley & Sons, Inc., New York, 1969.
5. J. Lunenschloss and W. Albrecht, *Nonwoven Bonded Fabrics*, Halstead Press, a division of John Wiley & Sons, Inc., New York, 1985.
6. R. Krcma, *Manual of Nonwovens*, Textile Trade Press, Manchester, UK, and W. R. C. Smith Publishing Co., Atlanta, Ga., 1971.
7. V. Mrstina and F. Fejgl, *Needle Punching Technology*, Elsevier Publishers, Amsterdam, the Netherlands, 1990.
8. R. Jeffries, *J. Text. Inst.* **51**, T339, T399, T441 (1960).
9. J. W. S. Hearle and R. H. Peters, *Moisture in Textiles*, The Textile Institute, Butterworths Scientific Publications, Manchester, UK, 1960.
10. J. F. Fuzek, *Ind. Eng. Chem. Prod. Res. Dev.* **24**, 140–144 (1985).
11. T. Vickerstaff, *The Physical Chemistry of Dyeing*, 2nd ed., Interscience Publishers, Inc., New York, 1954.
12. R. H. Peters, *Textile Chemistry; Volume III*, Elsevier Publishers, New York, 1975.
13. H. Zollinger, *Color Chemistry: Syntheses, Properties and Applications of Organic Dyes and Pigments*, VCH Publishers, New York, 1987.
14. H. D. Smith, *Proc. Am. Soc. Test. Mater.* **44**, 542 (1944).

15. L. C. Sawyer and D. T. Grubb, *Polymer Microscopy*, Chapman and Hall, London and New York, 1987.
16. *Text. World*, 37–48, (Aug. 1992).
17. D. J. Montgomery and W. T. Milloway, *Text. Res. J.* **22**, 729 (1952); J. H. Dillon, *Ind. Eng. Chem.* **44**, 2115 (1952).
18. W. J. Hamburger, *Tex. Res. J.* **18**, 102 (1948).
19. G. Susich and S. Backer, *Text. Res. J.* **21**, 482 (1951).
20. R. Meredith, *Mechanical Properties of Textile Fibers*, Interscience Publishers, New York, 1956.
21. J. D. Ferry, *Viscoelastic Properties of Polymers*, 2nd ed., John Wiley & Sons, Inc., New York, 1970.
22. W. J. MacKnight and J. J. Aklonis, *Introduction to Polymer Viscoelasticity*, 2nd ed., John Wiley & Sons, Inc., New York, 1983.
23. D. C. Prevorsek and W. J. Lyons, *Rubber Chem. Technol.* **44**, 271–293 (1971).
24. W. J. Lyons, *Impact Phenomena in Textiles*, M.I.T. Press, Cambridge, Mass., 1963.
25. W. E. Morton and J. W. S. Hearle, *Physical Properties of Textile Fibers*, 2nd ed., The Textile Institute and Butterworths Scientific Publications, London, 1975.
26. L. R. G. Treloar, *Physics Today*, 23–30 (Dec. 1977).
27. J. W. S. Hearle, *Polymers and Their Properties, Volume 1; Fundamentals of Structure and Mechanics*, Halstead Press, a division of John Wiley & Sons, Inc., New York, 1982.
28. A. V. Tobolsky, *Properties and Structure of Polymers*, John Wiley & Sons, Inc., New York, 1960.
29. R. J. Samuels, *Structured Polymer Properties*, John Wiley & Sons, Inc., New York, 1974.
30. L. Mandelkern, *Crystallization of Polymers*, McGraw-Hill Book Co., Inc., New York, 1964.
31. F. Happey, *Applied Fiber Science*, Vols. 1, 2, and 3, Academic Press, Inc., New York, 1978, 1979.

LUDWIG REBENFELD
TRI/Princeton

ACRYLIC

The first reported synthesis of acrylonitrile [*107-13-1*] (qv) and polyacrylonitrile [*25014-41-9*] (PAN) was in 1894. The polymer received little attention for a number of years, until shortly before World War II, because there were no known solvents and the polymer decomposes before reaching its melting point. The first breakthrough in developing solvents for PAN occurred at I. G. Farbenindustrie where fibers made from the polymer were dissolved in aqueous solutions of quaternary ammonium compounds, such as benzylpyridinium chloride, or of metal salts, such as lithium bromide, sodium thiocyanate, and aluminum perchlorate. Early interest in acrylonitrile polymers (qv), however, was based primarily on its use in synthetic rubber (see ELASTOMERS, SYNTHETIC).

Du Pont introduced the first acrylic fiber under the trade name of Orlon, using dimethylformamide as the spinning solvent, in 1944. Shortly afterward, Chemstrand (now Monsanto Fibers and Intermediates Co.) introduced Acrilan acrylic, Suddeotsche Chemiefaser (Hoechst) introduced Dolan acrylics, and Bayer introduced Dralon. Union Carbide introduced the first flame-resistant modacrylic fiber in 1948 under the trade names Vinyon N and Dynel. Vinyon was a continuous filament yarn; Dynel was the staple form. Both were based on 60% vinyl chloride and 40% acrylonitrile. Later Tennessee Eastman Co. introduced Verel, based on vinylidene chloride, and Monsanto introduced its SEF modacrylic. Monsanto's SEF is the only U.S. produced modacrylic. The distinction between acrylic and modacrylic fibers is that acrylics are composed of at least 85% by weight of acrylonitrile units; modacyclics have $< 85\%$, but at least 35% by weight of AN units.

During the 1950s at least 18 companies introduced acrylic fiber products. By 1960 annual worldwide production had risen to over 100 million kg. In the 1950s and 1960s world production was concentrated in Western Europe and the United States. Once staple processes were developed, acrylic fibers became a significant competitor in markets held primarily by woolen fibers. By 1963 the carpet and sweater markets accounted for almost 50% of the total acrylic production. In the 1970s the growth rate decreased sharply. This was due to the maturing of the wool replacement market in the United States. In addition, nylon became the dominant carpet fiber, and easy care fibers and blends cut into the acrylic share of the apparel market. Acrylics retain a significant share of the sweater, half hose (socks), and handcraft markets. During the 1970s there was rapid growth of acrylic fiber production in Japan, Eastern Europe, and developing countries. By 1981 an estimated overcapacity of approximately 21% had developed. This overcapacity is expected to decrease through the 1990s with continued increases in world production balanced by markets opening in China, Eastern Europe, Russia, and the Americas. Acrylics retain their traditional markets and, in addition, have small market shares in many new areas, such as carbon fiber precursors, asbestos replacement fibers, and conductive/metallized fibers.

Physical Properties

Acrylic and modacrylic fibers are sold mainly as staple and tow products with small amounts of continuous filament fiber sold in Europe and Japan. Staple lengths may vary from 25 to 150 mm, depending on the end use. Fiber deniers may vary from 1.3 to 17 dtex (1.2 to 15 den); 3.2 dtex (3.0 den) is the standard form. The appearance of acrylics under microscopical examination may differ from that of modacrylics in two respects. First, the cross sections (Fig. 1) of acrylics are generally round, bean-shaped, or dogbone-shaped. The modacrylics, on the other hand, vary from irregularly round to ribbon-like. The modacrylics may also contain pigment-like particles of antimony oxide to enhance their flame-retardant properties.

The physical properties of these fibers are compared with those of natural fibers and other synthetic fibers in Table 1. Additional property data may be found in compilations of the properties of natural and synthetic fibers (1). Apart from the polyolefins, acrylics and nylon fibers are the lightest weight fibers on the market. Modacrylics are considerably more dense than acrylics, with a density about the same as wool and polyester.

The elastic properties of these fibers can be characterized as wool-like, with high elongation and elastic recovery. Cotton (qv) is notably deficient in this regard, having an elongation of less than 10% and an elastic recovery of only 74% from a

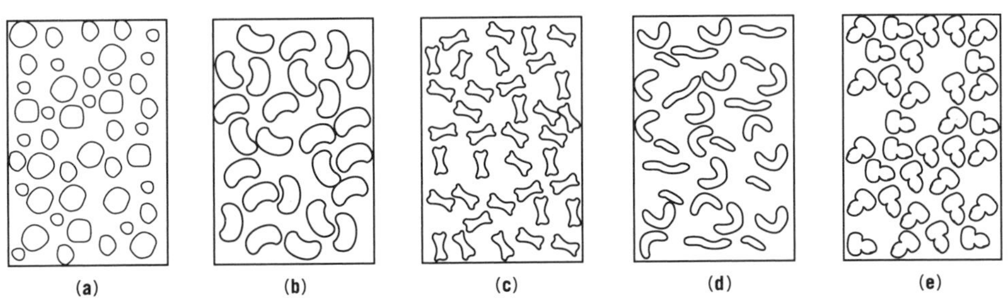

Fig. 1. Principal cross-section types found in acrylic and modacrylic fibers: (**a**), round; (**b**), kidney bean-shaped; (**c**), dogbone-shaped; (**d**), ribbon-like; and (**e**), mushroom-shaped bicomponent.

2% stretch. The tensile strength of acrylics and modacrylics is about the same, both considerably lower than other synthetics but higher than wool and about the same as cotton. This combination of elastic properties ranks the acrylics, modacrylics, and wool as compliant fibers, yielding fabric with a characteristically soft handle. Acrylics with tenacities as high as 0.53 N/tex (6 gf/den) can be produced by stretch orientation. However, these fibers suffer from fibrillation which produces changes in color shade on abrasion. Fibrillation is eliminated by steam annealing and composition changes which also diminish the tensile strength of the fiber.

The mechanical properties of acrylic and modacrylic fibers are retained very well under wet conditions. This makes these fibers well suited to the stresses of textile processing. Shape retention and maintenance of original bulk in home laundering cycles are also good. Typical stress–strain curves for acrylic and modacrylic fibers are compared with wool, cotton, and the other synthetic fibers in Figure 2.

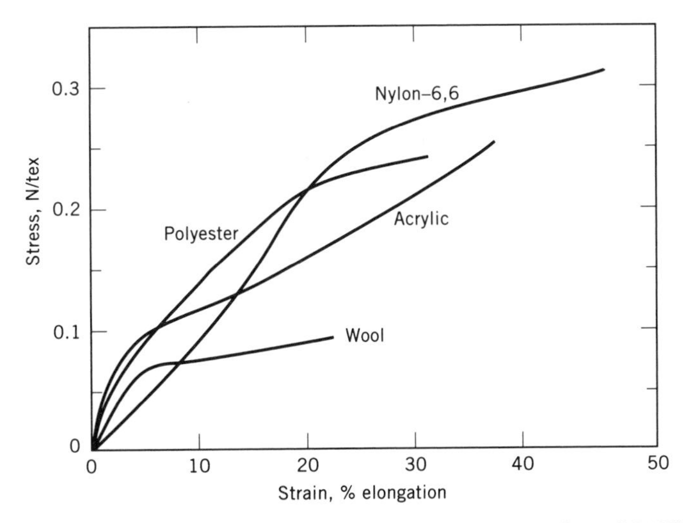

Fig. 2. Typical stress–strain properties of staple fibers at 65% rh and 21°C. Rate of elongation is 50%/min. To convert N/tex to gf/den, multiply by 11.3.

Table 1. Physical Properties of Staple Fibers

Property	Acrylic	Modacrylic	Nylon-6,6	Polyester	Polyolefin	Cotton	Wool
sp gr	1.14–1.19	1.28–1.37	1.14	1.38	0.90–1.0	1.54	1.28–1.32
tenacity, N/tex[a]							
dry	0.09–0.33	0.13–0.25	0.26–0.64	0.31–0.53	0.31–0.40	0.18–0.44	0.09–0.15
wet	0.14–0.24	0.11–0.23	0.22–0.54	0.31–0.53	0.31–0.40	0.21–0.53	0.07–0.14
loop/knot tenacity	0.09–0.3	0.11–0.19	0.33–0.52	0.11–0.50	0.27–0.35		
breaking elongation, %							
dry	35–55	45–60	16–75	18–60	30–150	<10	25–35
wet	40–60	45–65	18–78	18–60	30–150	25–50	
average modulus, N/tex[a]							
dry	0.44–0.62	0.34	0.88–0.40	0.62–2.75	1.8–2.65		
elastic recovery, %							
2% stretch	99	99–100		67–86		74	99
10% stretch		95	99	57–74	96		
20% stretch							65
electrical resistance	high	high	very high	high	high	low	low
static buildup	moderate	moderate	very high	high	high	low	low
flammability	moderate	low	self-extinguishing	moderate	moderate	spontaneous ignition at 360°C	self-extinguishing
limiting oxygen index	0.18	0.27	0.20	0.21		0.18	0.25
char/melt	melts	melts	melts, drips	melts, drips	melts	chars	chars
resistance to sunlight	excellent	excellent	poor; must be stabilized	good	poor; must be stabilized	fair; degrades	fair; degrades
resistance to chemical attack	excellent	excellent	good	good	excellent	attacked by acids	attacked by alkalies, oxidizing, and reducing agents
abrasion resistance	moderate	moderate	very good	very good	excellent	good	moderate
index of birefringence	0.1		0.6	0.16			0.01
moisture regain, std %	1.5–2.5	1.5–3.5	4–5	0.1–0.2	0	7–8	13–15

[a]To convert N/tex to gf/den, multiply by 11.3.

Moisture regain (65% rh, 21°C), a property that has a great effect on wear comfort, is reasonably good though not as high as cotton and wool. This property can be enhanced by adding hydrophilic or hygroscopic comonomers and by incorporating a porous internal structure in the fibers. This property plus their high compliancy make acrylics competitive in the wear comfort markets. However, acrylics cannot compete with nylon and polyester in the easy care markets. This is primarily due to the fact that the wet T_g of acrylonitrile copolymers is lower than the boiling point of water. As a result, in carpets, these fibers lose their bulk and resiliency during the carpet dyeing process and in use under repeated hot and humid conditions. In apparel, acrylics and modacrylics cannot match the wrinkle resistance and crease retention of polyester. In the cotton and wool markets, however, these fibers are outstanding. Acrylics have captured most of the wool-like market. In the cotton replacement market the growth potential of acrylics is limited mainly by consumer acceptance.

Chemical Properties

Among the outstanding properties of acrylic fibers is the very strong resistance to sunlight. In one study, for example, the number of months' exposure required for a 50% loss in yarn strength for Acrilan 3.3 dtex (3 den) acrylic fiber was measured and compared to the other natural and synthetic fiber types (3). It was found that the acrylic resisted degradation eight times longer than olefin fibers, over five times longer than either cotton or wool, and almost four times longer than nylon. This property makes the acrylics particularly useful for outdoor applications, such as in awnings, tents, and sandbags as well as upholstery for autos and outdoor furniture. Pigmented fibers with lightfast colors are particularly useful for outdoor applications.

Acrylic fibers are also resistant to all biological and most chemical agents. Acrylics are affected very little by weak acids, weak alkalies, organic solvents, oxidizing agents, and drycleaning solvents. These fibers are attacked only by strong base and highly polar organic solvents like dimethylacetamide (DMAC), dimethylformamide (DMF), dimethyl sulfoxide (DMSO), and ethylene carbonate. Acrylic fibers tend to be much more susceptible to chemical attack by alkalies than by acids or oxidizing agents. For example, acrylic fibers are stable for up to 24 h at 100°C in 50% sulfuric acid: these same fibers begin degrading with less than 0.5% sodium hydroxide at the same exposure time and temperature (4).

In terms of resistance of acrylic fibers to oxidizing agents, Orlon acrylic has been compared to nylon, cotton, and acetate yarns (4). The acrylic yarn is far superior to the others in retention of strength. The loss in yarn tenacity was measured as a function of bleaching time. After 300 minutes the acrylic had lost 8% of its original tenacity; the nylon had lost 30%. Cotton and acetate yarns were even less resistant. Both of these yarns lost 65–70% of their original tenacity. After 360 minutes of exposure the cotton and acetate yarns had completely deteriorated, whereas the acrylic retained approximately 92% of its original strength.

Acrylic fibers discolor and decompose rather than melting when heated, but they have very good color and heat stability at temperatures less than 120°C. In a study by American Cyanamid (using Federal Test Specification TT-P-141a.

Method 425.2) the yellowness of acrylic fiber was measured as a function of temperature. Compared to a value of 0.0 for a pure white body, the original fiber had a yellowness index of 0.04–0.10. After 30 minutes of exposure at 115°C the yellowness increased only slightly to 0.11–0.17. After 6 h at 130°C, however, the yellowness increased to 0.38–0.41.

The excellent chemical resistance of acrylic fibers may stem from its unique laterally bonded structure. Dipole bonds, formed between nitrile groups of adjacent chains, must be broken before chemical attack, melting, or solvation can occur. In addition the repulsive forces between adjacent nitriles result in a very stiff polymer backbone which yields very little entropy gain when the bonds between adjacent chains are broken in solvation or melting. Therefore, relatively high temperatures are required for solvation and melting.

Flammability

A most important property of acrylic and modacrylic fibers is their flammability and ignition behavior. Fibers used in textiles must not ignite readily when placed in contact with a flame. In this respect acrylic fibers compare favorably to the other natural and synthetic fibers currently on the market. There are, however, significant differences in the burning characteristics of the various fiber types. Cotton and rayon, for example, burn with the formation of a char. Nylon, polyester, olefin, wool, and acrylics, on the other hand, burn and melt simultaneously. More rigorous standards are required for end uses such as carpets, sleepwear, draperies, and bedding. Fibers for these applications must also be self-extinguishing after removal from the ignition source. The modacrylics, eg, SEF, Verel, and Kanekalon, melt and self-extinguish. This is generally achieved in acrylonitrile-based fibers by incorporating halogen comonomers such as vinylidene chloride, vinyl chloride, and vinyl bromide.

Another property, used to compare the flammability of textile fibers, is the limiting oxygen index (LOI). This measured quantity describes the minimum oxygen content (%) in nitrogen necessary to sustain candle-like burning. Values of LOI, considered a measure of the intrinsic flammability of a fiber, are listed in Table 2 in order of decreasing flammability.

Table 2. Limiting Oxygen Index of Textile Fibers

Fiber	LOI	Ignition temperature, °C
cotton	18.0	400
acrylic	18.2	560
rayon	19.7	420
nylon	20.1	530
polyester	21.0	450
wool	25.0	600
modacrylic	27.0	690
verel modacrylic	33.0	self-extinguishing
100% PVC	37.0	self-extinguishing

Polymer Analysis

Many techniques are available for characterizing acrylic and modacrylic materials in order to establish the dye site content, molecular weight, and chemical composition. The dye acceptance of the polymer may be determined by treating a suspension of the polymer with a known aqueous solution of a dye standard. The amount of dye absorbed per gram of polymer can then be calculated from a photometric measurement of the residual dye in solution. The dye sites themselves, in most acrylics and modacrylics, are sulfonate and sulfate end groups derived from the free-radical initiator used in polymerization. Therefore, the dye site content of the polymer can be measured by potentiometric titration of the strong acid groups or by determining the sulfur content of the polymer. The low levels of sulfur normally required for fiber dyeability can be measured accurately by methods based on x-ray fluorescence (see DYES, APPLICATION AND EVALUATION).

The molecular weight distribution of the polymer can be measured using gel permeation chromatography with low angle light scattering detection coupled with special solvents to eliminate ionic effects and computer enhanced analysis of the elution data. Although osmometry and light scattering techniques are used to obtain number and weight average molecular weights, methods based on solution viscosity are still the most popular in commercial practice. The simplest method is to measure the viscosity of a solution of the polymer at a specified concentration and temperature. This may be done using either a Brookfield, capillary, or falling ball viscometer. The viscosity average molecular weight may be obtained from such measurements by extrapolating the specific viscosity to zero solvent concentration. The intrinsic viscosity, thus derived, is then used in the Staudinger equation (5) or Cleland-Stockmayer equation (6) to give the viscosity average molecular weight.

Typical acrylic polymers have number average molecular weights in the 40,000 to 60,000 range or roughly 1000 repeat units. The weight average molecular weight is typically in the range 90,000 to 140,000, with a polydispersity index between 1.5 and 3.0. The solution properties of the polymer and rheological properties of the dope must be precisely defined for compatibility with dope preparation and spinning. The molecular weight of the polymer must be low enough that the polymer is readily soluble in spinning solvents, yet high enough to give a dope of moderately high viscosity.

Fiber dyeability is critically dependent on molecular weight distribution of the polymer because most acrylic fibers derive their dyeability from sulfonate and sulfate initiator fragments at the polymer chain ends. Thus the dye site content of the fiber is inversely related to the number average molecular weight of the polymer and very sensitive to the fraction of low molecular weight polymer. A critical balance must be maintained between the molecular weight distribution required for good rheological properties and the distribution required for good fiber dyeability. Where such a balance cannot be achieved it is usual practice to incorporate one of the sulfonated monomers as a means of establishing the required fiber dyeability. Du Pont's Orlon 42, for example, is believed to contain a small amount of sodium styrene sulfonate as a supplemental dye receptor. The very dense fiber structure, produced by Du Pont's dry spinning process, results in very low dye diffusion rates. The addition of a sulfonated monomer, therefore, compensates by increasing the total dye site content of the fiber.

Analytical methods for determining the chemical composition of acrylic and modacrylic materials are as varied as the possible compositions themselves. The nitrile group has a strong ir absorbance and can also be detected by elemental analysis for nitrogen. Many of the usual comonomers found in acrylics can also be detected by ir. Vinyl acetate, methyl acrylate, and methyl methacrylate, for example, can be detected by ir absorbance of the carbonyl group. Sulfonated monomers, such as sodium styrene sulfonate can be detected by strong uv absorbance due to the phenyl group; these monomers can also be measured by sulfur analysis using x-ray fluorescence. Direct measurement of sulfur is imprecise, however, because the polymer end groups may contribute an indeterminant amount of sulfur, usually 20 to 50 μeq/g. Halogen monomers can be quantitatively measured by pyrolyzing the polymer and analyzing the pyrolysis products by gas chromatography or halide titration. X-ray fluorescence alone may also be used to determine the concentration of specific halogens.

Fiber Identification

Although visual and microscopical examination, together with simple manual tests, are still the primary methods of identification, there are many new sophisticated instrumental methods available based on chemical and physical properties. These methods are able to distinguish between closely related fibers which differ only in chemical composition or morphology.

Visual and Manual Tests. Synthetic fibers are generally mixed with other fibers to achieve a balance of properties. Acrylic staple may be blended with wool, cotton, polyester, rayon, and other synthetic fibers. Therefore, as a preliminary step, the yarn or fabric must be separated into its constituent fibers. This immediately establishes whether the fiber is a continuous filament or staple product. Staple length, brightness, and breaking strength wet and dry are all useful tests that can be done in a cursory examination. A more critical identification can be made by a set of simple manual procedures based on burning, staining, solubility, density determination, and microscopical examination.

Burning. Acrylics form a black bead that is easily crushed in the fingers. There is usually no smell of acetic acid as with cellulose acetate. Many of the modacrylics, such as Dynel, do not form a bead in this test. In general the modacrylics do not support combustion when removed from the source of ignition.

Solubility. Acrylic fibers are insoluble in methanol, acetone, and methylene chloride and soluble in dimethylformamide, dimethylacetamide, and dimethyl sulfoxide at room temperature. Modacrylics, such as Dynel and Verel, are exceptions since they may be soluble in acetone but insoluble in methanol. All of the modacrylics are soluble in butyrolactone at room temperature.

Staining. Two useful staining tests can be carried out to help identify various acrylic and modacrylic fibers by manufacturer and product type: the Meldrum staining test and the Sevron Orange staining test. The Meldrum stain consists of 0.5 g of Lissamine Green SFS (CI Acid Green 5), 0.5 g of Sevron Orange L (CI Basic Orange 24), a surfactant, and 10 mL of 2 N sulfuric acid in a total aqueous volume of 1000 mL. An undyed acrylic fiber test sample is immersed in the stain at 90°C. The sample is removed after 5 minutes, and the color of the fiber is observed after washing and drying the test fiber. The Sevron Orange staining test is carried out in a similar fashion with a second undyed acrylic fiber sample; the colors of the two fiber samples are compared with a standard chart of commercial acrylic fibers. Detailed information on these tests is available (7).

Density. Acrylics have a low specific gravity (1.12–1.19) compared to all of the primary natural fibers and most synthetic fibers. Nylon has a similar specific gravity (1.14) and the polyolefins have lower specific gravities, eg, 0.90 for polypropylene. Again the modacrylics and some acrylics with high levels of comonomer of low molar volume are exceptions. Verel and Dynel, for example, have specific gravities of 1.37 and 1.31, respectively.

Microscopical Examination. All fibers have distinguishing features which either allow outright identification or classification into narrower grouping for specialized analysis. Fiber cross sections are particularly useful for identification.

Instrumental Analysis. It is difficult to distinguish between the various acrylics and modacrylics. Elemental analysis may be the most effective method of identification. Specific compositional data can be gained by determining the percentages of C, N, O, H, S, Br, Cl, Na, and K. In addition the levels of many comonomers can be established using ir and uv spectroscopy. Also, manufacturers like to be able to identify their own products to certify, for example, that a defective fiber is not a competitor's. To facilitate this some manufacturers introduce a trace of an unusual element as a built-in label.

General schemes for the identification of natural and synthetic fibers have been established by the Textile Institute and by the American Association of Textile Chemists and Colorists (8). A comprehensive treatment of burning, solvent, staining, microscopy, and density techniques has been given (9) and a general discussion of procedures for identifying synthetic fibers has been presented (10).

Fiber Characterization

In addition to characterizing the many properties introduced by the choice of monomers and the polymerization process itself, considerable further characterization is required to quantitatively describe the properties imparted by spinning and subsequent downstream processing. These important properties relate to the crystalline order and microstructure of the fibers, and the resultant performance characteristics, such as crimp retention, abrasion resistance, mechanical properties, etc. The physical and mechanical properties of acrylic and modacrylic fibers are measured by standard test methods (11). Many important physical properties are measured by simple gravimetric techniques. Tex or denier, for example, can be measured exactly as it is defined (see FIBERS, SURVEY). Fiber density is also measured gravimetrically, using a pycnometer, by weighing the amount of mercury displaced by a known weight of fiber. Moisture regain, a critical factor affecting wear comfort and static buildup, is measured by first weighing an oven-dry sample of yarn. Then the sample is equilibrated in a chamber at specified conditions, usually 21°C and 65% relative humidity, and weighed again. The difference in weights, expressed as a percentage of the dry weight, represents the equilibrium moisture content or moisture regain of the fiber. Dry heat shrinkage and shrinkage in boiling water are measured by simply determining the difference in the length of a section of fiber after treatment at specified conditions.

The mechanical testing, required to establish properties such as breaking elongation, tenacity, and modulus of elasticity, is carried out using devices such as the Instron. The fiber is gripped at the ends by two jaws which are then pulled apart at a specified rate or load. As the Instron jaws are pulled apart the mechanical stress and strain on the fiber is recorded continuously until the fiber breaks. The result is a stress–strain curve for the fiber from which the breaking elongation, tenacity, and elastic modulus can be derived. Other properties, such as elastic recovery, stress relaxation, creep, cyclic loading effects, and hysteresis can also be evaluated with this equipment.

Acrylonitrile Polymerization

Except for fibers designed for industrial applications where resistance to chemical attack is of prime importance, all acrylic fibers are made from acrylonitrile combined with at least one other monomer. The comonomers most commonly used are neutral comonomers, such as methyl acrylate [*96-33-3*] and vinyl acetate [*108-05-4*] to increase the solubility of the polymer in spinning solvents, modify the fiber morphology, and improve the rate of diffusion of dyes into the fiber. Sulfonated monomers, such as sodium styrenesulfonate [*27457-28-9*] (SSS), sodium methallyl sulfonate [*1561-92-8*] (SMAS), and sodium sulfophenyl methallyl ether [*1208-67-9*] (SPME) are used to provide dye sites or to provide a hydrophilic component in water reversible crimp bicomponent fibers. Halogenated monomers, usually vinylidene chloride [*75-35-4*], vinyl bromide [*593-60-2*], and vinyl chloride [*75-01-4*], impart flame resistance to fibers used in the home furnishings, awning, and sleepwear markets. Modacrylic compositions are used when the end use requires high flame resistance. Almost all of the modacrylics are flame-resistant fibers with high levels of halogen monomers.

POLYMERIZATION METHODS

Acrylonitrile and its comonomers can be polymerized by any of the well-known free-radical methods. Bulk polymerization is the most fundamental of these, but its commercial use is limited by its autocatalytic nature. Aqueous dispersion polymerization is the most common commercial method, whereas solution polymerization is used in cases where the spinning dope can be prepared directly from the polymerization reaction product. Emulsion polymerization is used primarily for modacrylic compositions where a high level of a water-insoluble monomer is used or where the monomer mixture is relatively slow reacting.

Solution Polymerization. Solution polymerization is widely used in the acrylic fiber industry. The reaction is carried out in a homogeneous medium by using a solvent for the polymer. Suitable solvents can be highly polar organic compounds or inorganic aqueous salt solutions. Dimethylformamide [*68-12-2*] (DMF) and dimethyl sulfoxide [*67-68-5*] (DMSO) are the most commonly used commercial organic solvents, although polymerizations in γ-butyrolactone, ethylene carbonate, and dimethylacetamide [*127-19-5*] (DMAC) are reported in the literature. Examples of suitable inorganic salts are aqueous solutions of zinc chloride and aqueous sodium thiocyanate solutions. The homogeneous solution polymerization of acrylonitrile follows the conventional kinetic scheme developed for vinyl monomers (12) (see POLYMERS).

Thermally activated initiators (qv) such as azobisisobutyronitrile (AIBN), ammonium persulfate, or benzoyl peroxide can be used in solution polymerization, but these initiators (qv) are slow acting at temperatures required for textile-grade polymer processes. Half-lives for this type of initiator are in the range of 10–20 h at 50–60°C (13). Therefore, these initiators are used mainly in batch or semibatch processes where the reaction is carried out over an extended period of time.

Chain transfer is an important consideration in solution polymerizations. Chain transfer to solvent may reduce the rate of polymerization as well as the molecular weight of the polymer. Other chain-transfer reactions may introduce dye sites, branching, chromophoric groups, and structural defects which reduce thermal stability. Many of the solvents used for acrylonitrile polymerization are very active in chain transfer. DMAC and DMF have chain-transfer constants of $4.95–5.1 \times 10^{-4}$ and $2.7–2.8 \times 10^{-4}$, respectively, very high when compared to a value of only 0.05×10^{-4} for acrylonitrile itself. DMSO ($0.1–0.8 \times 10^{-4}$) and aqueous zinc chloride (0.006×10^{-4}), in contrast, have relatively low transfer constants; hence, the relative desirability of these two solvents over the former. DMF, however, is used by several acrylic fiber producers as a solvent for solution polymerization.

Considering the two most common comonomers, methyl acrylate is the least active in chain transfer whereas vinyl acetate is as active in chain transfer as DMF and DMAC. Vinyl acetate is also known to participate in the chain transfer-to-polymer reaction (14). This occurs primarily at high conversion, where the concentration of polymer is high and monomer is scarce. Polyacrylonitrile can also participate in branching reactions. In a study (15) of acrylonitrile polymerization in magnesium perchlorate, branch formation occurred by a reaction in which a growing radical chain abstracts a hydrogen atom from the α-carbon, thereby starting a side chain by monomer addition. Branch formation occurred when the ratio of polymer to monomer concentrations was greater than one or at a conversion of 80% or more. At this condition one branch occurs for every 2000 growth steps. Thus, at a molecular weight of 10^5, each molecule shows only one branch on the average. PAN branching by polymerization through the nitrile group has also been suggested (16). Branch formation can also occur as a result of radical termination by disproportionation or other chain-transfer reactions. Any reaction that leaves a terminal double bond can lead to long-chain branching if the double bond subsequently reacts with a growing polymer radical.

The advantage of solution polymerization is that the polymer solution can be converted directly to spinnable dope by removing the unreacted monomer. However, it is more difficult to achieve high molecular weight. The solvents required are often chain-transfer agents and chain termination is more rapid. Incorporation of nonvolatile monomers, such as the sulfonated monomers commonly used as basic dye acceptors, can also be a problem. The sulfonated monomers, in particular, have poor solubility in organic solvents and must be solubilized by converting them to a soluble form such as the amine salt form. Nonvolatile monomers are also difficult to recover from the reaction medium since the usual distillation techniques are unsuitable. Monomer recovery systems based on carbon adsorption have been developed. However, the usual practice is to maximize the single-pass utilization of these monomers.

Bulk Polymerization. The bulk polymerization of acrylonitrile is complex. Even after many investigations into the kinetics of the polymerization, it is still not completely understood. The complexity arises because the polymer precipitates from the reaction mixture barely swollen by its monomer. The heterogeneity has led to kinetics that deviate from the normal and which can be interpreted in several ways.

When initiator is first added the reaction medium remains clear while particles 10 to 20 nm in diameter are formed. As the reaction proceeds the particle size increases, giving the reaction medium a white milky appearance. When a thermal initiator, such as AIBN or benzoyl peroxide, is used the reaction is autocatalytic. This contrasts sharply with normal homogeneous polymerizations in which the rate of polymerization decreases monotonically with time. Studies show that three propagation reactions occur simultaneously to account for the anomalous autoacceleration (17). These are chain growth in the continuous monomer phase; chain growth of radicals that have precipitated from solution onto the particle surface; and chain growth of radicals within the polymer particles (13,18).

Although bulk polymerization of acrylonitrile seems adaptable, it is rarely used commercially because the autocatalytic nature of the reaction makes it difficult to control. This, combined with the fact that the rate of heat generated per unit volume is very high, makes large-scale commercial operations difficult to engineer. Lastly, the viscosity of the medium becomes very high at conversion levels above 40 to 50%. Therefore commercial operation at low conversion requires an extensive monomer recovery operation.

Emulsion Polymerization. The mechanism of emulsion polymerization was first developed qualitatively (19) and later quantitatively (20,21). It was shown that the emulsifier disperses a small portion of the monomer in aggregates of 50 to 100 molecules approximately 5 nm in diameter called micelles. The majority of the monomer stays suspended in droplet form. These droplets are typically 1000 nm in diameter, much larger than the micelles. Since a water-soluble radical initiator is used, polymerization begins in the aqueous phase. The micelle concentration is normally so high that the aqueous radicals are rapidly captured (22). The micelle is essentially a tiny reservoir of monomer, therefore polymerization proceeds rapidly, converting the micelle to a polymer particle nucleus. The ability of emulsion polymerization to segregate radicals from one another is of great importance commercially. The effect is to minimize the rate of radical recombination, allowing high rates of polymerization to be achieved along with very high molecular weight. In practice many commercial processes employ a chain-transfer agent to control molecular weight. Processes of commercial importance are the copolymerization of butadiene with styrene or acrylonitrile to produce synthetic rubber and the polymerization of acrylic esters, vinyl chloride, and vinylidene and vinyl acetate, to produce latices for adhesives and paints. Its use in the textile industry is limited to the manufacture of modacrylic compositions. One notable example of an emulsion process is the old Union Carbide process for Dynel (23,24). Comprehensive reviews of emulsion polymerization technology have been published (25,26), and emulsion polymerization reactor modeling has been reviewed (27).

Aqueous Dispersion Polymerization. By far the most widely used method of polymerization in the acrylic fibers industry is aqueous dispersion. When inorganic compounds such as persulfates, perchlorates, or hydrogen peroxide are used as radical generators, the initiation and primary radical growth steps occur mainly in the aqueous phase. Chain growth is limited in the aqueous phase, however, because the monomer concentration is normally very low and the polymer is insoluble in water. Nucleation occurs when aqueous chains aggregate or collapse after reaching a threshold molecular weight. If many polymer particles are present, as is the case in commercial continuous polymerizations, the aqueous radicals are likely to be captured on the particle surface by a sorption mechanism. The particle surface is swollen with monomer. Therefore, the polymerization continues in the swollen layer and the sorption becomes irreversible as the chain end grows into the particle.

Since polymer swelling is poor and the aqueous solubility of acrylonitrile is relatively high, the tendency for radical capture is limited. Consequently, the rate of particle nucleation is high throughout the course of the polymerization, and particle growth occurs predominantly by a process of agglomeration of primary particles. Unlike emulsion particles of a readily swollen polymer, such as polystyrene, the acrylonitrile aqueous dispersion polymer particles are massive agglomerates of primary particles which are approximately 100 nm in diameter.

The kinetics of aqueous dispersion polymerization differ very little from acrylonitrile bulk or emulsion polymerization. Redox initiation is normally used in commercial production of polymers for acrylic fibers. This type of initiator can generate free radicals in an aqueous medium efficiently at relatively low temperatures. The most common redox system consists of ammonium or potassium persulfate (oxidizer), sodium bisulfite (reducing agent), and ferric or ferrous iron (catalyst). This system gives the added benefit of supplying dye sites for the fiber. This redox system works at pH 2–4 where the bisulfite ion predominates. Two main reactions account for radical production, the oxidation of ferrous iron by persulfate and the reduction of ferric iron by SO_2 in the bisulfite form. The sulfate and sulfonate radicals, thus produced, react with monomer to initiate rapid chain growth. Termination generally occurs by radical recombination, though in most commercial processes chain-transfer agents are used to control molecular weight and impart acid dyesites. Bisulfite ion, the most widely used chain-transfer agent, apparently reacts rapidly since the bisulfite feed has a pronounced effect on polymer molecular weight with virtually no effect on the overall rate of polymerization (28). The ratio of bisulfite to persulfate in the reaction mixture has a strong effect on the dye site content of the polymer (28,29). In the absence of chain-transfer reactions, all dye sites are derived from initiator radicals. Thus if termination occurs exclusively by radical recombination, then each polymer chain contains a dye site at each end. Sulfate and sulfonate radicals are produced at equal rates, so the total dye site content must be an equimolar mixture of these two distributed among the chain ends at random. Chain transfer to bisulfite, however, terminates one chain with a hydrogen atom while starting another with a sulfonate radical. This increases the total dye site content of the polymer by reducing the polymer molecular weight, but at the same time this reaction produces chains with just one dye site. At a given molecular weight the dye site content of the polymer can, in theory, vary from two per chain at low bisulfite levels to one per chain at very high bisulfite levels.

In commercial practice reducing agent to oxidizing agent ratios are used which are equivalent to ratios of bisulfite to potassium persulfate ranging from 8 to 15. These high ratios give narrower molecular weight distributions and the combination of high activator and low oxidizer gives a relatively low conversion reaction. Low conversion is an effective means of minimizing branching and color producing side reactions.

A comprehensive review of aqueous polymerization has been published (30). Reviews of acrylonitrile polymerization are many (31–34).

COPOLYMERIZATION

Homogeneous Copolymerization. Nearly all acrylic fibers are made from acrylonitrile copolymers containing one or more additional monomers that modify the properties of the fiber. Thus copolymerization kinetics is a key technical area in the acrylic fiber industry. When carried out in a homogeneous solution, the copolymerization of acrylonitrile follows the normal kinetic rate laws of copolymerization. Comprehensive treatments of this general subject have been published (35–39). The more specific subject of acrylonitrile copolymerization has been reviewed (40). The general subject of the reactivity of polymer radicals has been treated in depth (41).

The monomer pair, acrylonitrile–methyl acrylate, is close to being an ideal monomer pair. Both monomers are similar in resonance, polarity, and steric characteristics. The acrylonitrile radical shows approximately equal reactivity with both monomers, and the methyl acrylate radical shows only a slight preference for reacting with acrylonitrile monomer. Many acrylonitrile monomer pairs fall into the nonideal category, eg, acrylonitrile–vinyl acetate. This is an example of a nonideality sometimes referred to as kinetic incompatibility. A third type of monomer pair is that which shows an alternating tendency. This tendency is related to the polarization properties of the monomer substituents (42). Monomers that are dissimilar in polarity tend to form alternating monomer sequences in the polymer chain. An example is the monomer pair acrylonitrile–styrene. Styrene, with its pendent phenyl group, has a relatively electronegative double bond whereas acrylonitrile, with its electron-withdrawing nitrile group, tends to be electropositive.

Copolymer composition can be predicted for copolymerizations with two or more components, such as those employing acrylonitrile plus a neutral monomer and an ionic dye receptor. These equations are derived by assuming that the component reactions involve only the terminal monomer unit of the chain radical. The theory of multicomponent polymerization kinetics has been treated (35,36).

Heterogeneous Copolymerization. When copolymer is prepared in a homogeneous solution, kinetic expressions can be used to predict copolymer composition. Bulk and dispersion polymerization are somewhat different since the reaction medium is heterogeneous and polymerization occurs simultaneously in separate loci. In bulk polymerization, for example, the monomer swollen polymer particles support polymerization within the particle core as well as on the particle surface. In aqueous dispersion or emulsion polymerization the monomer is actually dispersed in two or three distinct phases: a continuous aqueous phase, a monomer droplet phase, and a phase consisting of polymer particles swollen at the surface with monomer. This affects the ultimate polymer composition because the monomers are partitioned such that the monomer mixture in the aqueous phase is richer in the more water-soluble monomers than the two organic phases.

Where polymerization occurs predominantly in the organic phases these relatively water-soluble monomers may incorporate into the copolymer at lower levels than expected. For example, in studies of the emulsion copolymerization of acrylonitrile and styrene, the copolymer was richer in styrene than copolymer made by bulk polymerization, using the same initial monomer composition (43–45). Analysis of the reaction mixtures (46) showed that nearly all of the styrene was concentrated in the droplet and swollen particle phases. The acrylonitrile, on the other hand, was distributed between both the aqueous and organic phases. The monomer compositions in the droplet and particle phase were found to be essentially the same. The effect of monomer partitioning on copolymer composition is strongest with the ionic monomers since this type of monomer is usually soluble in water and nearly insoluble in the other monomers. Reviews of emulsion copolymerization kinetics and the effects of reaction heterogeneity on reaction locus have been published (47,48).

COMMERCIAL POLYMERIZATION METHODS

Aqueous media, such as emulsion, suspension, and dispersion polymerization, are by far the most widely used in the acrylic fiber industry. Water acts as a convenient heat-transfer and cooling medium and the polymer is easily recovered by filtration or centrifugation. Fiber producers that use aqueous solutions of thiocyanate or zinc chloride as the solvent for the polymer have an additional benefit. In such cases the reaction medium can be converted directly to dope to save the costs of polymer recovery. Aqueous emulsions are less common. This type of process is used primarily for modacrylic compositions, such as Dynel. Even in such processes the emulsifier is used at very low levels, giving a polymerization medium with characteristics of both a suspension and a true emulsion.

The most common reactor type in the 1990s is the continuous stirred tank. However, in the early years of acrylic fiber production, during the 1950s and early 1960s, the semibatch polymerization process was commonly used for the commercial production of acrylonitrile copolymer. In this type of process the reaction vessel is charged with a portion of the reactants and the reaction is induced by using radical initiators, such as potassium persulfate. Control of copolymer composition is difficult in this type of reactor because the comonomers most frequently used have vapor pressures, solubilities, or reactivities which differ greatly from those of acrylonitrile. Acid-dyeable monomers, such as sodium styrene sulfonate, are nonvolatile solids with high water solubility. Monomers used to impart flame retardancy, such as vinyl chloride, vinyl bromide, and vinylidene chloride, are nearly insoluble in water and have high vapor pressures. As a result, the various monomers employed in a given reaction mixture often react at widely differing rates, and the copolymer formed in the early stages of the reaction has a different composition and molecular weight than that formed at the later stages.

An example of a commercial semibatch polymerization process is the early Union Carbide process for Dynel, one of the first flame-retardant modacrylic fibers (23,24). Dynel, a staple fiber that was wet spun from acetone, was introduced in 1951. The polymer is made up of 40% acrylonitrile and 60% vinyl chloride. The reactivity ratios for this monomer pair are 3.7 and 0.074 for acrylonitrile and vinyl chloride in solution at 60°C. Thus acrylonitrile is much more reactive than vinyl chloride in this copolymerization. In addition, vinyl chloride is a strong chain-transfer agent. To make the Dynel composition of 60% vinyl chloride, the monomer composition must be maintained at 82% vinyl chloride. Since acrylonitrile is consumed much more rapidly than vinyl chloride, if no control is exercised over the monomer composition, the acrylonitrile content of the monomer decreases to approximately 1% after only 25% conversion. The low acrylonitrile content of the monomer required for this process introduces yet another problem. That is, with an acrylonitrile weight fraction of only 0.18 in the unreacted monomer mixture, the low concentration of acrylonitrile becomes a rate-limiting reaction step. Therefore, the overall rate of chain growth is low and under normal conditions, with chain transfer and radical recombination, the molecular weight of the polymer is very low.

The low rate of copolymerization and tendency for low molecular weight polymer are overcome by using emulsion polymerization. The rate of polymerization and polymer molecular weight are then controlled by varying the rate of initiation and the surfactant concentration. The copolymer composition is controlled by adding acrylonitrile monomer to the reactor at a rate that maintains a constant pressure of ~520 kPa (75 to 76 psi) at 40°C. This pressure is produced by a free-monomer phase consisting of 18% acrylonitrile. Thus as long as there is a free-monomer phase the 18% acrylonitrile level can be maintained by holding the reaction pressure constant at 520 kPa (76 psi). The Union Carbide process requires 77 hours and 19 additions of acrylonitrile. The yield of copolymer is 65.8% and the final vinyl chloride content of the polymer is 60.5%.

Processes using a continuous stirred tank reactor have replaced the semibatch process except where low volume specialty products are made. For start-up, the reactor is charged with a certain amount of the reaction medium, usually solvent or pH adjusted water. In more sophisticated processes the start-up period may be minimized by filling the reactor with overflow from a reactor already operating at steady state. The reactor feeds are metered in at a constant rate for the entire course of the production run, which normally continues until equipment maintenance is needed. A steady state is established by taking an overflow stream at the same mass flow rate as the combined feed streams. The main advantage of this process over the semibatch process is that control of molecular weight, dye site level, and polymer composition is greatly improved.

An example of a continuous aqueous dispersion process is shown in Figure 3 (49). A monomer mixture composed of acrylonitrile and up to 10% of a neutral comonomer, such as methyl acrylate or vinyl acetate, is fed continuously. Polymerization is initiated by feeding aqueous solutions of potassium persulfate (oxidizer), sulfur dioxide (reducing agent), ferrous iron (promoter), and sodium bicarbonate (buffering agent). The aqueous and monomer feed streams may be fed at rates which give a reactor dwell time of 40 to 120 minutes and a feed ratio of water to monomer in the range from 3 to 5. The product stream, an aqueous slurry of polymer particles, is mixed with ethylenediaminetetraacetic acid (iron chelate) or oxalic acid to stop the polymerization. This slurry is then fed to the top section of a baffled monomer separation column. The separation of unreacted monomer can be affected by contacting the slurry with a countercurrent flow of steam introduced at the bottom of the column. The monomer is condensed from the overheads stream and separated from the resulting water mixture using a decanter.

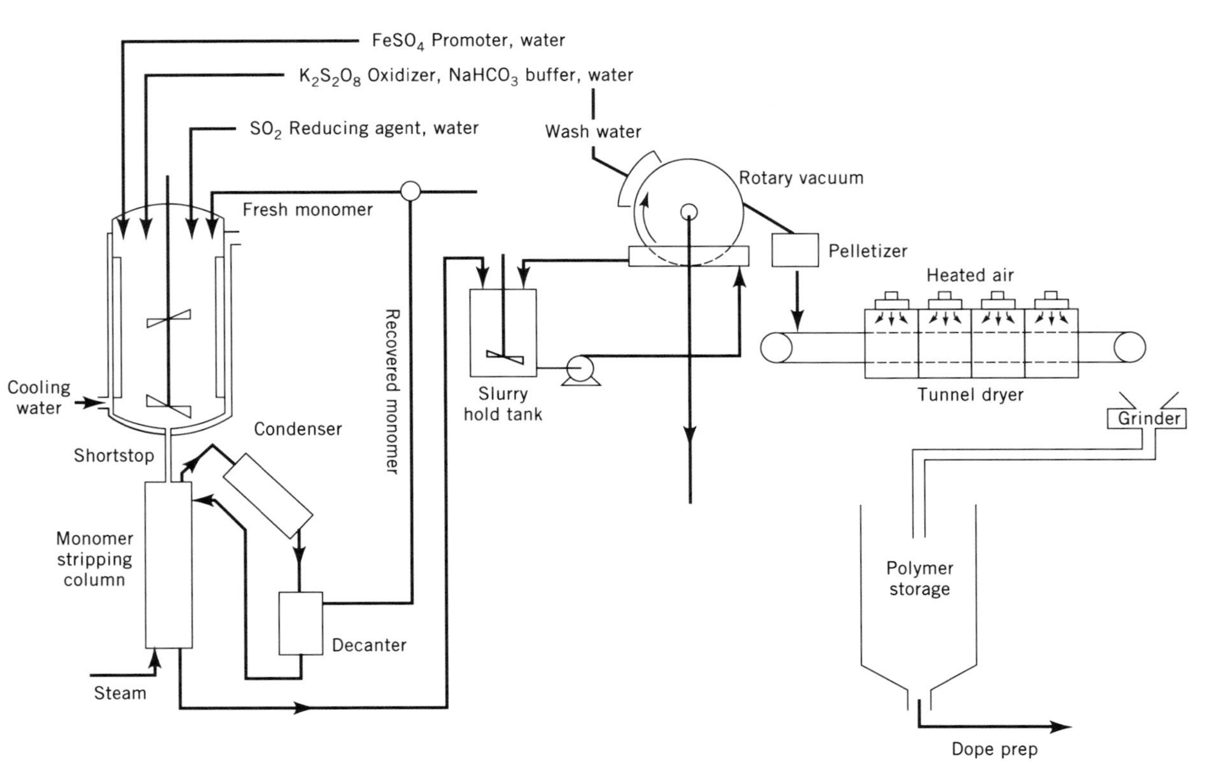

Fig. 3. An aqueous dispersion polymerization process used in the manufacture of acrylic and modacrylic fibers.

The stripped slurry is taken from the column bottoms stream and separated from the water using a continuous vacuum filter. After filtration and washing, the polymer is pelletized, dried, ground, and then stored for later spinning.

The monomer recovery process may vary in commercial practice. A less desirable sequence is to filter or centrifuge the slurry to recover the polymer and then pass the filtrate through a conventional distillation tower to recover the unreacted monomer. The need for monomer recovery may be minimized by using two-stage filtration with filtrate recycle after the first stage. Nonvolatile monomers, such as sodium styrene sulfonate, can be partially recovered in this manner. This often makes process control more difficult because some reaction by-products can affect the rate of polymerization and often the composition may vary. When recycle is used it is often done to control discharges into the environment rather than to reduce monomer losses.

Cost reduction has been a focus of fiber producers since the overall market for acrylic fibers has not grown significantly. A significant cost reduction is realized by operating continuous aqueous dispersion processes at very low water-to-monomer ratios. Mitsubishi Rayon, for example, has reported ratios as low as 1.75. This compares to ratios of 4–5 widely used in the 1970s. The low water-to-monomer ratios produce a change in the nucleation and particle growth mechanisms which yields denser polymer particles. The cost reduction comes in the drying step. While conventional water-to-monomer ratios give wet cake moisture levels of 200% (dry basis) the modified process yields wet cake moisture levels of 100% or less. Thus a savings in drying cost is realized. The low water-to-monomer process has the added advantage of increased reactor productivity (50,51).

The only other commercial polymerization process used for acrylic fibers is solution polymerization. This type of process can be implemented by feeding the monomers to a continuous mixing tank along with a solvent for the polymer. The overflow stream from this tank is then routed to a form of continuous reactor where the polymerization is carried out in a homogeneous solution. Monomer is removed from the product stream and the resulting polymeric solution is used directly for spinning. An obvious advantage of this process is that considerable cost savings can be achieved by eliminating the filtration, drying, and dope making steps required in the aqueous dispersion process. There are two drawbacks associated with solution polymerization. First, it is difficult to produce dopes of high solids, particularly with organic solvents. Second, most of the effective solvents have very high chain-transfer constants, making it difficult to produce polymer of high molecular weight. Solvents suitable for this type of commercial polymerization are DMF and DMSO. Solution polymerizations based on polymer solutions of aqueous zinc chloride and thiocyanate salt have also been reported. DMAC, a popular wet spinning solvent, is not suitable for solution polymerization, however, because of the powerful chain-transfer activity of this solvent.

Because of the highly exothermic nature of acrylonitrile polymerization, bulk processes are not normally used commercially. However, a commercially feasible process for bulk polymerization in a continuous stirred tank reactor has been developed (52). The heat of reaction is controlled by operating at relatively low conversion levels and supplementing the normal jacket cooling with reflux condensation of unreacted monomer. Operational problems with thermal stability are controlled by using a free-radical redox initiator with an extremely high decomposition rate constant. Since the initiator decomposes almost completely in the reactor, the polymerization rate is insensitive to temperature and can be controlled by means of the initiator feed rate. Polymer molecular weight and dye site content are controlled by using mercapto compounds and oxidizable sulfoxy compounds.

The polymer can easily be recovered by simple vacuum filtration or centrifugation of the polymer slurry. This can be followed by direct conversion of the filter cake to dope by slurrying the filter cake in chilled solvent and then passing the slurry through a heat exchanger to form the spinning solution and a thin-film evaporator to remove residual monomer.

The problems of monomer recovery, reaction medium viscosity, and control of reaction heat are effectively dealt with by the process design of Montedison Fibre (53). This process produces polymer of exceptionally high density, so although the polymer is still swollen with monomer, the medium viscosity remains low because the amount of monomer absorbed in the porous areas of the polymer particles is greatly reduced. The process is carried out in a CSTR with a residence time, Q, such that the product $k/d \times Q$ is greater than or equal to 1. k_d is the initiator decomposition rate constant. This condition controls the autocatalytic nature of the reaction because the catalyst and residence time combination assures that the catalyst is almost totally expended in the reactor.

The following conditions are stipulated: the catalyst decomposition rate constant must be one hour or greater; the residence time of the continuous reactor must be sufficient to decompose the catalyst to at least 50% of the feed level; the catalyst concentration must be greater than or equal to $0.002 \times Q$, where the residence time, Q, is expressed in hours. An upper limit on the rate of radical formation was also noted; that is, when the rate of radical formation is greater than the addition rate of the primary radicals to the monomers, initiation efficiency is reduced by the recombination of primary radicals.

Suitable catalysts are *t*-butylphenylmethyl peracetate and phenylacetylperoxide or redox catalyst systems consisting of an organic hydroperoxide and an oxidizable sulfoxy compound. One such redox initiator is cumene–hydroperoxide, sulfur dioxide, and a nucleophilic compound, such as water. Sulfoxy compounds are preferred because they incorporate dyeable end groups in the polymer by a chain-transfer mechanism. Common thermally activated initiators, such as BPO and AIBN, are too slow for use in this process.

Solution Spinning

One of the principal problems in early commercialization of acrylic fibers was the lack of a suitable spinning method. The polymer cannot be melt spun, except possibly at high pressure in the presence of water. Solution spinning was the only feasible commercial route. However, hydrogen bonding between the α-hydrogens and the nitrile groups, dipole bonding between pairs of nitrile groups and the resulting chain stiffness, are so great that only the most powerful polar solvents are effective. Early research led to successful spinning from aqueous solutions containing inorganic salts. However, the first real breakthrough occurred at

Du Pont in 1948 when dimethylformamide (DMF) was used to commercialize Orlon. DMF is still the most important spinning solvent for acrylic fibers, but other effective solvents are in wide use. The key solvents and their spinning concentrations are as follows:

Solvent	Polymer, %
dimethylformamide (DMF)	20–32
dimethylacetamide (DMAC)	20–27
dimethyl sulfoxide (DMSO)	20–30
ethylene carbonate (EC)	15–18
sodium thiocyanate (NaSCN)	
45–55% in water	10–15
zinc chloride ($ZnCl_2$)	
55–65% in water	8–12
nitric acid (HNO_3)	
65–75% in water	8–12

Dope is prepared by chilling the solvent to 5°C or lower to reduce the dissolving power of the solvent. The polymer, finely ground after the normal filtration, washing, and drying steps, is blended into the solvent and mixed to form a uniform dispersion. The mixture is then heated to 80–150°C for several hours until a clear, slightly amber-colored solution is obtained for spinning. Additives, such as pigments, delustrants (TiO_2), antioxidants (qv), flame retardants, and optical brighteners are added at this stage. The resulting dope is degassed and filtered before being fed to the spinneret through special metering gear pumps.

Dry Spinning. This is the process first employed commercially by Du Pont in 1948. The process is similar to early acetate fiber spinning. The dope is pumped through spinnerets with 200–900 holes placed at the top of a solvent drying tower. The solvent is removed by circulating an inert gas through the tower at 150–300°C. However, unlike acetate spinning which employs a low boiling solvent (acetone), acrylic solvents are all high boiling and hard to extract completely in the drying tower. Consequently, the fiber from the bottom of the drying tower contains 10–25% solvent. This is removed in a second step by passing the threadline through a hot water bath and possibly by a series of wash rolls. Vapor from the drying tower and washwater from the secondary solvent extraction are routed to a solvent recovery plant where the solvent and washwater are recovered and purified for reuse. This is essential for economic reasons, but environmental protection also requires recovery rather than discharge.

The filament microstructure in dry spinning is derived from gelation exclusively. No nonsolvent is used so precipitation does not occur during solvent removal. Gelation occurs gradually as the solvent is evaporated. Fiber densities from wet and dry spinning have been compared (54). Gel density is much higher in the dry spun fiber. Moreover, very little of the fibrillar structure, characteristic of wet spun fibers, is observed in the dry spun filaments. During stretching, however, the dry spun fibers develop a fibrillar network similar to that of the wet spun fibers.

Wet Spinning. Wet spinning differs from dry spinning primarily in the way solvent is removed from the extruded filaments. Instead of evaporating the solvent in a drying tower, the fiber is spun into a liquid bath containing a solvent/nonsolvent mixture called the coagulant, as shown in Figure 4. The solvent is almost always the same as the solvent used in the dope and the nonsolvent is usually water. Filament fusion is less of a problem in wet spinning so the number of capillaries in wet spinning spinnerets is much larger than in dry spinning. The spinnerets in commercial processes may have anywhere from 10,000 to 60,000 capillary holes which may range in diameter from 0.05 to 0.4 mm, and it is common to use multiple spinnerets in a single spinbath.

The fiber microstructure is established in the spinbath, therefore the coagulation conditions are generally the result of extensive optimization and study. The critical part of this process is the transition from a liquid to a solid phase within the filaments. Two liquid-to-solid phase transitions are possible in the spinbath. One is precipitation of the polymer to form a solid phase. This is undesirable because it yields fiber of poor mechanical strength. The desirable solid phase is the gel state, characterized by hydrogen and dipole bonding between the polymer and solvent. The gel state is desirable because it is elastic and gives rise to a finer microstructure once the solvent is removed. Thus the conditions in the spinbath should be optimized so that gelation of the polymer precedes precipitation. Studies (55) have shown that gelation occurs more rapidly at high dope solids and lower temperatures.

Spinbath concentration can be adjusted to obtain the desired microstructure. Low spinbath concentration promotes rapid solvent extraction but this also produces a thick skin on each filament which ultimately reduces the rate of solvent extraction and may lead to the formation of macrovoids. High spinbath concentrations give a denser microstructure, but solvent extraction is slow and filament fusion can occur. Other spinbath conditions that affect coagulation and microstructure are dope solids, spinbath temperature, jet stretch, and immersion time.

The fiber emerging from the spinbath is a highly swollen gel, containing both the solvent and nonsolvent from the spinbath. The fibers are essentially unoriented. The microstructure consists of a fibrillar network. The spaces between fibrils are called microvoids. Depending on the conditions of coagulation the filaments may also contain large voids radiating out from the center of the fiber. The best combination of tensile properties, abrasion resistance, and fatigue life is realized when the coagulated fiber has a homogeneous, dense gel structure with small fibrils and no macrovoids (56).

Fiber cross sections are also determined by the coagulation conditions or, in the case of dry spinning, by the solvent evaporation process. The skin that forms early in the solvent removal process may remain intact as the interior of the filament deflates from solvent removal. Wet spun fibers from organic solvents are often bean shaped, while those from inorganic solvent systems are often round. Dry spun fibers, such as Du Pont's Orlon, are dogbone shaped. Examples of these shapes are shown in Figure 1. Special cross sections, however, can be made from nonround capillaries by controlling coagulation conditions. Control of die-swell is of critical importance. To maintain a nonround shape the tension on the filament at the spinneret face must be great enough to eliminate the effects of die-swell.

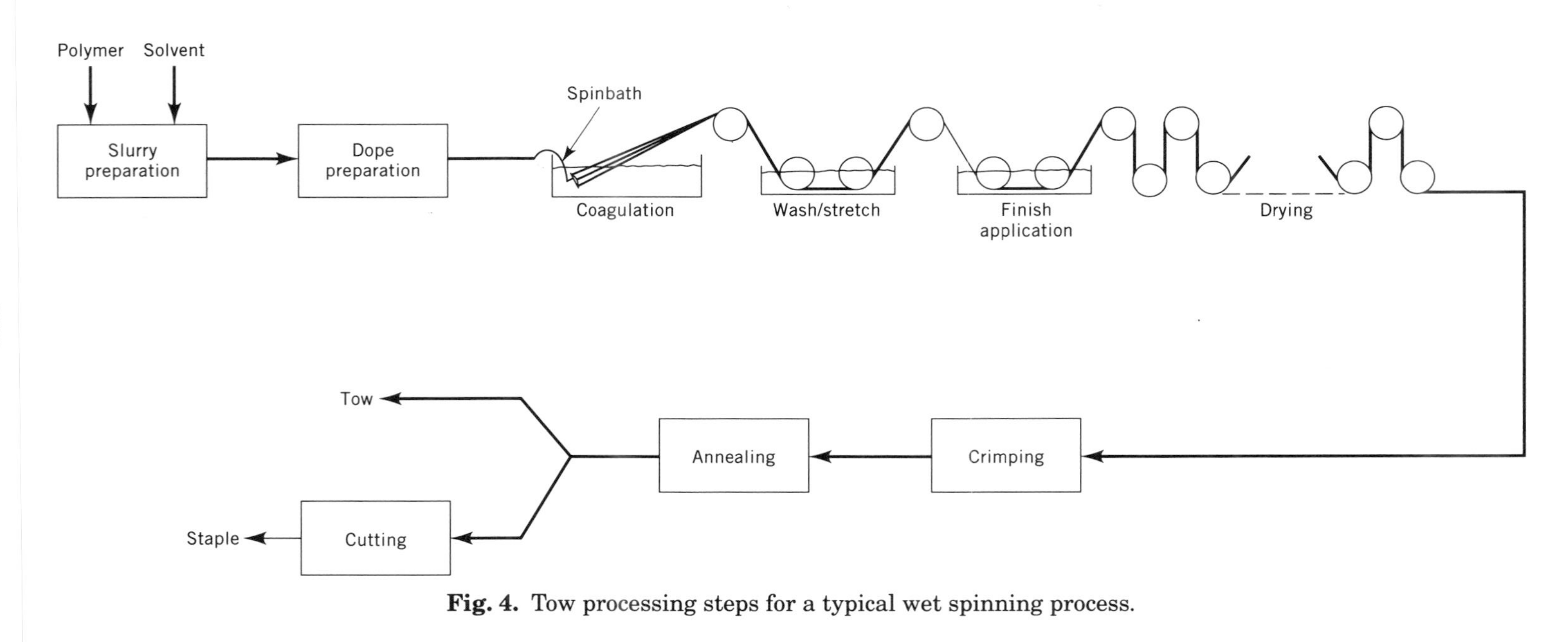

Fig. 4. Tow processing steps for a typical wet spinning process.

Dry and wet spinning each have their advantages. In wet spinning the solvent is removed from the extruded filaments by diffusion into a relatively cool solvent/nonsolvent coagulation bath. Some nonsolvent always diffuses into the filaments creating a skin which reduces the rate of solvent removal. In many cases this skin may become so thick that it fixes the internal volume of the filaments thereby creating voids when all the solvent is eventually removed. Macrovoids, as they are called, can be controlled by adjusted dope solids, solvent:nonsolvent ratio in the coagulation bath, bath temperature, and jet stretch. In dry spinning the solvent is removed by evaporation into hot inert vapor phase. Skin formation is less severe and the process yields denser filaments free of macrovoids. On the other hand, the relatively small number of capillary holes required in dry spinning to avoid filament fusion is an economic disadvantage compared to wet spinning, where as many as 60,000 holes per spinneret can be used.

Tow Processing. After the spinbath step the tow processing is similar for both wet and dry spun yarns. Wet spun yarns, however, may contain 100 to 300% of solvent/nonsolvent mixture per dry pound of fiber, while dry spun tows generally hold only 10 to 30% solvent. Therefore, the initial washing steps differ in their details. Figure 4 shows typical tow processing sequences. The key tow processing steps are washing, stretching, finish application, collapse, drying, crimping, and relaxing. The washing step may consist of several stages with the effluent being recycled to a solvent recovery process. The tow may be taken up on a wash roll with water sprayed directly on the roll. The wash step may also be combined with an initial stretching operation. In one variation of this method the tow is drawn between sets of godets and passed through a series of solvent extraction baths at the same time. A more efficient method employs wash chambers designed to pass water through the tow at high velocity (57). In multistage washing processes fresh water enters the last stage and proceeds countercurrent to the tow.

The washing step is followed by additional stretching and application of finish. A common practice at this point is to stretch the tow in boiling water between two sets of godets. This type of stretch is known as cascade stretching. The density of dry spun fibers actually decreases during this process and becomes similar to the fibrillar structure of wet spun fibers. The porous fibrillar structure of wet spun fibers increases in density with stretching. Therefore, it is common to carry out in-line treatments at this stage, either before or during the stretching process. Dyes, cross-linking agents, heat and light stabilizers, and flame retardants can be applied effectively at this stage.

The drying step comes next. This can be done with the tow held at constant length by contacting the tow on heated rolls or with simultaneous relaxation by passing the tow through an in-line oven using a conveyor belt. As part of the drying step the tow is heated to bring about a stabilization of the oriented fibrillar network. In this process, known as the collapse stage, the network junctures reform and the density of the fiber increases. After collapse, boiling water shrinkage is reduced and higher temperatures are required for subsequent relaxation.

Crimp. The tow is usually relaxed at this point. Relaxation is essential because it greatly reduces the tendency for fibrillation and increases the dimensional stability of the fiber. Relaxation also increases fiber elongation and improves dye diffusion rates. This relaxation can be done in-line on Superba equipment or in batches in an autoclave. Generally saturated steam is used because the moisture reduces the process temperatures required. Fiber shrinkage during relaxation ranges from 10 to 40% depending on the temperature used, the polymer composition used for the fiber, and the amount of prior orientation and relaxation. The amount of relaxation is also tailored to the intended application of the fiber product.

Modifications of Properties. *Handle.* For a given fabric construction the denier, compliance, cross-sectional configuration, degree of crimp, moisture absorption, and surface smoothness of the fibers all influence the softness of the final product. Very fine filament deniers are effective where good draping, anticrease properties, and softness of handle are desired. Softness of handle can also be achieved through modifying the fiber cross section. Flattened cross sections reduce the bending modulus. A flattened cross section with a three-to-one ratio of principal axes has approximately one-third the bending modulus of a round cross-section fiber with equal cross-sectional area. These cross sections can be achieved in wet spinning by modifying the coagulation conditions and/or the shape of the spinneret holes. Sheath-core spinning, conjugate spinning, and drawing into modified spinbath compositions also yield modified handle.

Improved Comfort Properties. Wear comfort generally means cotton-like properties. The ability to absorb moisture from the skin and the softness of cotton fabrics are considered to be the two key properties for comfort. The extremely fine denier of cotton fibers accounts for its softness. Both properties can be achieved in acrylic fibers. Improved moisture retention can be achieved by incorporating hydrophilic comonomers that decrease ultimate fiber density, by modifying the fiber spinning process or by using after-treatments, such as modified finishes. Bayer's process for making porous, moisture absorbent fibers, for example, uses copolymers containing carboxyl groups in the acid form. The polymer is dry spun from a spinning dope containing a high boiling glycol or alcohol derivative. The glycol or alcohol derivative is retained within the fibers in the dry spinning tower where a porous fiber morphology is established and subsequently extracted during the fiber washing step. After spinning, the carboxyl groups are converted to the salt form, thereby enhancing their hydrophilic character. The resulting fibers have a unique porous sheath-core structure that allows water to diffuse into the fiber leaving the surface relatively dry.

Reduced Pilling. Staple fabrics, in general, develop small balls of fiber or pills as a result of abrasive action on the fabric surface. However, the pills build up more on acrylic fabrics than on comparable woolens. Pilling can be reduced by increasing the likelihood that the pills will break or wear off. Thus the most effective approaches include reducing fiber strength, incorporating defects in the fiber, increasing fiber brittleness, and reducing shear strength. The polymer can be modified to give more brittle fibers by decreasing the comonomer content, adding an ionic monomer, or inducing cross-linking during the spinning process. Wet spinning processes can be modified by using low solvent concentration in the spinbath and high spinbath temperature to give brittle fibers with high void content. Low draw ratios result in low tensile strength, drying the fibers under tension increases fiber brittleness, and annealing the fibers at reduced temperatures reduces fiber elongation to break.

Improved Hot–Wet Properties. Acrylic fibers tend to lose modulus under hot–wet conditions. Knits and woven fabrics tend to lose their bulk and shape in dyeing and, to a more limited extent, in washing and drying cycles as well as in high humidity weather. Moisture lowers the glass-transition temperature T_g of acrylonitrile copolymers and, therefore, crimp is lost when the yarn is exposed to conditions required for dyeing and laundering.

A number of polymer and fiber modifications have been devised to overcome this problem, though none has been successful enough to allow acrylics to compete successfully in easy care apparel markets. The carpet market is one area in which acrylics have failed as a result of inadequate hot–wet stability. Although acrylic fibers produce a carpet of superior wool-like appearance and quality, the pile loses its resilience and bulk during dyeing and under high humidity conditions. To compensate, acrylic carpets must be made in expensive, high density constructions. As a result, acrylics have lost popularity as a carpet fiber since the peak year of 1968.

Improved Abrasion Resistance. Abrasion resistance is generally improved by reducing the microvoid size and increasing the fiber density. Abrasion-resistant fibers have been produced by incorporating hydrophilic comonomers or comonomers with small molar volumes. Sulfonated monomers, acrylamide derivatives, and *N*-vinylpyrrolidinone are some of the hydrophilic comonomers that can be used to reduce void content; vinylidene chloride, with its relative small molar volume, is effective in increasing fiber density. The spinning process itself has a significant effect on fiber density and abrasion resistance. Dry spinning, for example, is known to produce a denser fiber structure than conventional wet spinning. Wet spinning techniques used to improve abrasion resistance generally do so by approximating dry spinning conditions in some way. Examples include modified spinbath, high solvent concentration, low spinbath temperature, and additives to the dope or spinbath. Spinbath additives include nondiffusing nonsolvents, such as poly(ethylene glycol) or high molecular weight alcohols, such as *t*-butyl alcohol. Acrilan II, an acrylic fiber produced by Monsanto, gives a 50% improvement in wear life in men's socks. This fiber is produced by a modified spinning technology which optimizes the fiber microstructure at the spinbath and increases the fiber density.

Fiber Whiteness and Thermal Stability. The most effective route to improved whiteness and thermal stability is modification of the polymerization. The polymer must be linear and free of conjugated or unstable chemical structures formed by side reactions during the polymerization process. By maintaining the unreacted monomer concentration in the reactor as high as possible, free radicals in the reaction mixture are more likely to polymerize rather than attack the polymer backbone. Though productivity is reduced, this is best accomplished by operating the polymerization process at low overall conversion of monomer to polymer. Side reactions can also be minimized by using comonomers having reactivity ratios with acrylonitrile close to unity and by avoiding highly reactive comonomers which undergo side reactions themselves. Vinyl acetate, for example, is known to undergo a branching reaction at the carbonyl group. For dispersion polymerization, it is also helpful if all monomers have water solubility similar to acrylonitrile. The sulfonated monomers may be an exception since these appear to be incorporated primarily in the aqueous phase in the early stages of chain growth. Polymerization reaction conditions are also important. High pH is known to cause hydrolysis of the nitrile groups in the polymer and high temperature reaction conditions tend to favor undesirable side reactions. Reaction temperatures above 50°C are normally avoided by using redox initiators, such as the persulfate–bisulfite–iron system. Reducing agents such as organophosphorus compounds and high levels of sodium bisulfite used as the redox reducing agent appear to improve polymer color.

Commercial Products

The majority of acrylic fiber production is 3.3–5.6 dtex (3 to 5 den) staple and tow furnished, undyed, in either bright or semidull luster. The principal markets are in apparel and home furnishings. Within the apparel sector these fibers find extensive use in sweaters and in single jersey, double-knit, and warp-knit fabrics for a variety of knitted outerwear garments such as dresses, suits, and children's wear. Other large markets for acrylics in the knit goods area are hand knitting yarns, deep pile fabrics, circular knit, fleece fabrics, half-hose, coarse-cut knitwear, and deep pile fabrics for blankets. Acrylics also hold a strong position in most broadwoven fabric categories including apparel and home furnishings for area rugs, carpets, curtains, and upholstery.

Much of the growth in acrylics fibers usage has come from the replacement of wool. Acrylics have many of wool's desirable properties and are clearly superior in many areas where wool is deficient. Like wool, acrylic fibers are valued for their warmth, softness of hand, generous bulk and pile qualities, and ability to recover from stretching. At the same time acrylics are less costly, more resistant to abrasion and chemical attack, and more stable toward degradation from light and heat. In addition, acrylics are not attacked by moths and biological agents and show very little of wool's tendency to felt.

Until the mid-1970s, acrylic fiber in 17–22 dtex (15 to 20 den) form was a primary competitor in the carpet market. Strict flammability regulations put in effect during this period provided the impetus for the development of flame-resistant acrylics and modacrylics. Fibers with high levels of vinyl chloride, vinyl bromide, or vinylidene chloride were developed to pass government tests, such as the tunnel test and the methenamine pill test. These acrylic and modacrylic carpet fibers allow exceptional versatility in styling and color patterns. Although acrylic carpets can also be superior to nylon aesthetically, dense and expensive constructions are required to match the pile height and durability of nylon carpets. In less dense constructions acrylic carpets can develop wear patterns and lose resilience and pile height in the dye bath or in service under hot humid conditions. The carpet market is dominated by nylon staple fiber in the 1990s. A small market still exists in the United States and Europe; in Japan acrylic carpets are still relatively popular. Numerous studies have been made to find ways of improving hot–wet and durability properties of acrylic carpet fibers. Fiber density and fibrillar structure can be improved by using modified compositions and spinning processes for high abrasion resistance. Producer dyed fibers and blends of acrylic and nylon have been developed to improve hot–wet performance.

In the general apparel market, filament polyester has a clear advantage because of its easy care qualities and low cost. Although 100% polyester has limited appeal because of its less desirable handle, it is extremely popular in blends with cotton. The lack of a well-defined crystalline structure and low wet T_g limit the attainment of easy care properties in acrylic fibers. Methods have been devised which improve hot–wet and easy care properties of acrylics by cross-linking the oriented fiber molecules and by modifying the polymer composition or spinning process, but these methods increase the cost of the fiber. Also, even though researchers have been able to match the wear-comfort qualities of cotton, consumer demand for genuine cotton remains the dominant market factor. A blend of acrylic bicomponent and monocomponent fibers, Comfort 12 (58) developed by Du Pont, is designed as a cotton replacement for the knitted outerwear market, now dominated by cotton. The fiber is claimed to match the crisp handle and wear comfort properties of cotton.

Acrylics and modacrylics are also useful industrial fibers. Fibers low in comonomer content, such as Dolan 10 and Du Pont's PAN Type A, have exceptional resistance to chemicals and very good dimensional stability under hot–wet conditions. These fibers are useful in industrial filters, battery separators, asbestos fiber replacement, hospital cubical curtains, office room dividers, uniform fabrics, and carbon fiber precursors. The excellent resistance of acrylic fibers to sunlight also makes them highly suitable for outdoor use. Typical applications include modacrylics, awnings, sandbags, tents, tarpaulins, covers for boats and swimming pools, cabanas, and duck for outdoor furniture (59).

Besides the standard staple and tow products, acrylic and modacrylic fibers are offered in many forms for specialized applications. Fibers with enhanced properties are in great demand. Yarn bulk is enhanced by using bicomponent or biconstituent fibers. Pilling is reduced by producing fibers that are more brittle, and yarns with exceptionally soft handle are produced by using fine denier fibers or fibers treated with special friction reducing finishes. There have been many efforts to develop premium products based on proprietary technology. Specialty fibers, such as ultrafine denier fibers, acid-dyeable, producer dyed, and pigmented fibers, ion-exchange fibers, metallized and semiconducting fibers, hollow fibers for apparel, coarse denier fibers for wigs, and simulated animal hair, are available. Other examples of specialty applications are the functional fabrics, eg, antimicrobial fabrics. This type of fabric is made from microporous fibers, such as Bayer's Donova, which are capable of absorbing large amounts of moisture. The void structure is treated with chemicals which retard the growth of bacteria and fungus. These fibers are valuable for apparel subjected to perspiration such as socks, sportswear, underwear, and baby wear. Other functional fabrics are made from moisture repellent fibers, moisture absorbent sheath-core fibers, antistatic fibers, and flame-resistant fibers.

Acrylic Filament Yarns. Continuous filament acrylic yarns face stiff competition from nylon, polyester, and polyolefin fibers. Since they are more costly, acrylics have penetrated only those markets where they have a clear advantage in a critical property. Acrylics, for example, are clearly superior in resistance to sunlight. This makes continuous filament acrylics valuable in outdoor applications such as convertible tops, tents, and awnings. Pigmented acrylic can be used in applications where maximum resistance to fading and loss of strength is needed. Modacrylic fibers may be used in applications such as awnings, which require high resistance to flame. Acrylics with low comonomer content are highly resistant to chemical attack and thermal degradation. These properties make acrylics suitable for industrial filters and battery separators.

In spite of their many desirable properties, acrylics have not penetrated the industrial fibers market to any significant degree. As a filter medium, for example, acrylics lack sufficient abrasion resistance and dimensional stability at elevated temperatures to find universal acceptance. As a fiber for ropes and fish nets, acrylics excel in resistance to weathering but lack sufficient tensile strength. High tenacity continuous filament nylon, polyester, and uv-stabilized polyolefin are generally chosen even though the stability of these fibers toward sunlight and chemical exposure is inferior to acrylic fibers. Geotextiles (qv) is another large volume market. Here again acrylics have excellent properties but are too costly.

In the general apparel markets continuous filament acrylics and modacrylics find use as artificial fur or in fabrics with a fur-like appearance. These fibers are made in coarse deniers with special cross sections and surface modifications, eg, surface inclusions or roughening, to simulate natural animal hairs. Uniform fabrics and silky fabrics are also produced from continuous filament acrylic and modacrylic yarns. In Japan continuous filament yarns in very fine deniers are valued as a silk replacement. In this market continuous filament is a premium product used in high fashion dress fabrics, satins, and poplins or to produce a cloth suitable for surface raising to give a suede or fine velour effect. Fine denier filaments may also be used to produce fabrics with very high fiber density and low air permeability. These fabrics are useful for insulating material, for example, in quilted interliners for parkas and outerwear for skiing.

Fibers with High Bulk and Pile Properties. High bulk acrylic fibers are commonly made by blending high shrinkage and low shrinkage staple fibers. The high shrinkage fiber can be spun using a hot-stretching process in which the fiber is drawn during washing, drying, or after drying. The low shrinkage component is relatively unstretched or relaxed fiber. When the resulting yarn is allowed to relax, the high shrink component causes the unstretched fiber to buckle and add bulk to the yarn. High bulk acrylic yarns are also made by the turbo staple process in which tow is converted to staple by a stretch–break process which produces high shrinkage staple of uneven cut lengths.

Another method of producing high bulk yarns is with bicomponent fibers. Bicomponent fibers have developed from a desire to match the bulkiness and handle of wool. The three-dimensional crimp of animal fibers in general comes from the presence of two components on the fiber surface. Bicomponent fibers achieve this by spinning two polymer compositions concurrently in each capillary of the spinneret. The two components generally have different reactions to heat or moisture. For example, if one component is more moisture absorbent than the other, a crimp develops when the fiber is dried. The copolymers for this type of bicomponent may be acrylonitrile–vinyl acetate and acrylonitrile–vinyl acetate–sodium styrene sulfonate. In this combination the sulfonated copolymer is the more moisture absorbent and therefore shrinks the most on drying. This type of crimp is reversible because it can be renewed by wetting and redrying. Though no longer produced, Du Pont's Wintuk is an example of a commercial water-reversible crimp fiber. Crimp can also be imparted by using polymers that react differently to heat. By using copolymers of different compositions the crimp is imparted permanently when the fiber is heated. The copolymers for this type of bicomponent have a single comonomer, such as vinyl acetate or methyl acrylate, incorporated at two different levels. Du Pont's Orlon 21, a thermally crimped bicomponent, was the first commercial acrylic bicomponent. The most effective bicomponents are those which have two well-defined polymer types incorporated in a continuous track along the threadline. A limitation of this technology is that high productivity spinnerets with large numbers of capillaries cannot be used. Random bicomponents allow the fiber producer to overcome this limitation. With this technology the second component is incorporated through a special spinneret at random along the threadline. This concept is based on the fact that viscous spinning solutions can be merged without complete mixing (60). When passed through a standard spinneret, bicomponent fibers are produced ranging in composition from 100% component A to 100% component B. This type of process gives a bicomponent fiber with good bulk development while yielding a substantial gain in productivity for the fiber producer. Monsanto's Paquel and Remember are commercial examples of water-reversible crimp random bicomponent fibers. With Du Pont's withdrawal from the acrylic market, Monsanto now markets both types of water-reversible crimp fiber as well as bicomponents with thermally activated crimp.

Toray and Asahi have reported new types of bicomponent fibers which employ more than two functional components. Toray, for example, has announced the development of a water-repellant antipilling bicomponent fiber for the sport sweater market (61). This product, which derives its water repellency by the addition of a fluorine-type resin, exhibits an exceptionally soft handle along with long lasting water repellency. Asahi has developed a three-component fiber made by spinning one polymer as a continuous phase and intermittently spinning the remaining two polymers as dispersed phases (62). When polymers of differing dyeability are used, this reportedly gives a fiber with a unique periodically varying shade.

Flame-Resistant Fibers. Acrylics have reasonably good flame resistance compared to cotton and regenerated cellulosic fibers. In addition, acrylics tend to char when burning rather than forming melts as do polyesters and nylons. Additional flame resistance is required for certain end uses, such as children's sleepwear, blankets, carpets, outdoor awnings, and drapery fabrics. This can be achieved by copolymerizing acrylonitrile with halogen-containing monomers such as vinyl chloride, vinyl bromide, and vinylidene chloride. Modacrylics are used where a high resistance to burning is required. In such fibers the level of halogen-containing monomers may be over 50%, as in Dynel, one of the earliest modacrylics. This fiber, no longer produced, was 50/60 acrylonitrile–vinyl chloride copolymer. Tennessee Eastman's Verel, an acrylonitrile–vinylidene chloride copolymer, has also been discontinued.

Monsanto's SEF modacrylic is the only remaining U.S. produced modacrylic flame-resistant fiber. This fiber is a combination of halogen monomers and a dye receptor to offset the poor dye penetration characteristic of acrylonitrile–vinyl halogen copolymers.

Flame-resistant acrylic fibers are also widely produced. These fibers, which by definition contain less than 15% comonomer, are made by copolymerizing or blending with halogen-containing polymers or by using dope additives, such as antimony, halogen, or phosphorus compounds. Panox, an acrylic fiber designed for military and industrial use, is heat-stabilized by an oxidative process so that it does not burn or melt on heating.

There have been reviews of flammability (63–67), methods that can be used to enhance the flame resistance of acrylic and modacrylic fibers (68), and the mechanism of flame-retardant additives (69).

Microfibers. A new trend in the synthetic fiber industry is the increasing popularity of microdenier fibers in Western Europe and the United States (70). These are fibers in which the denier per filament (dpf) is less than 1.0 (0.11 tex). Microdenier fibers have been produced by the Japanese for many years to make fabrics with a silk-like handle. The Japanese, however, produced these fibers using a conjugate spinning technique wherein a single-spun filament breaks into multiple filaments in a special finishing step. For example, a 70 denier 40 filament yarn would be converted to a 70 denier 320 filament yarn, with each filament being 0.2 denier. In the United States and Europe, however, fiber producers are extruding very fine denier filaments. These are drawn down to microdenier size cross sections and then made into fabrics. These fibers are used for their soft handle and dense construction. In addition to the obvious silk-like fabrics, microdenier is useful in stretch fabrics, lingerie, and rainwear. Even industrial products and nonwovens are seeing the use of microdenier fabrics. All of the important synthetic fibers are being used, both in filament and staple form. Acrylic microfibers on the market are all staple products, with American Cyanamid in the United States producing a 0.8 dpf acrylic staple product named Microsupreme. American Cyanamid is also reportedly developing a 0.5 dpf microfiber. In Germany Bayer AG has been marketing Dralon X160, a 0.6 dpf acrylic microfiber. Monsanto's Fi-Lana is another acrylic example, though at 1.2 dpf it does not fit the definition of a microfiber precisely.

High Strength Fibers by Gel Spinning. Gel spinning has been used for some time to produce high strength fibers from ultrahigh molecular weight polyethylene. Spectra 900 and 1000 fibers from Allied and Dyneema SK60 and SK65 fibers from DSM are two commercial examples. Extensive studies have also been carried out on polypropylene, poly(vinyl alcohol), and polyacrylonitrile. Commercial gel spun acrylics are not far off.

The key to gel spinning technology is the ability to spin dilute solutions at reasonably high viscosity levels. At normal dope solids levels the drawability of the fiber is limited by chain entanglements. By using dilute solutions in the range 2–15% the chain entanglement problem is greatly reduced, and by using polymers with molecular weights (M_w) in the range of 1 to 4 million the solution viscosity is increased. With polyethylene, drawability is less of a problem because dipole and hydrogen bonding forces are weak and the polymer chains are highly flexible. Thus high draw ratios are possible with polyethylene, even without gel spinning.

In the Allied process two solvents are used. The first has low volatility and is designed to be an effective solvent for the high molecular weight PAN. The resultant solution or xerogel is extruded with a jet stretch of 10 $\times$ or less through an air gap into a cooling bath where the extrudate is converted into gel fibers. Gelation is important because it restricts chain mobility and prevents chain entanglements from forming as the solvent is extracted. The first solvent is extracted with a volatile solvent, and the solvent is then removed by evaporation. The fibers are drawn in several stages starting with the usual jet stretch at the spinneret face. Subsequent drawing is done after the solvent is removed in stages of increasing temperature from 130 to 230°C. Typical wet spinning solvents are used for solvent 1 and water is used as solvent 2. The Allied and DSM processes are similar, but DSM adds zinc chloride to the spinning solution to prevent phase separation and aid in drawing. The total draw ratio may be anywhere from 8–29 ×, with the highest tenacities achieved at the highest draw ratios. Table 3 compares the properties of gel spun PAN with other high performance fibers (qv) (71,72).

High Strength Fibers by Conventional Solution Spinning. As a reinforcing material for ambient-cured cement building products, acrylics offer three key properties: high elastic modulus, good adhesion, and good alkali resistance (73). The high modulus requires an unusually high stretch orientation. This can be accomplished by stretching the fiber above its glass-transition temperature, T_g. Normally this is done in boiling water or steam to give moduli of 8.8 to 13 N/tex (100 to 150 gf/den) (74,75). Alternatively, the stretch orientation can be achieved by a combination of wet stretch at 100°C and plastic stretch on hot rolls or in a heat-transfer fluid such as glycerol. This technique is reported to give moduli as high as 17.6 N/tex (200 gf/den) (76,77).

Mitsubishi Rayon Co. has developed an acrylic asbestos replacement fiber with a tensile strength of almost 600 MPa (87,000 psi) (78,79). In addition, patents for acrylic asbestos replacement fibers have been obtained by Asahi (80), Wuestefeld (81), REDCO (82), and Hoechst (83). The Hoechst fiber, marketed under the trade name Dolanit (originally Dolan 10), is offered in two forms as shown in Table 4.

Acrylic fibers, such as Dolanit (84) and Courtaulds' Sekril X (85), are blended in ambient-cured cement at a rate of 1 to 3%, compared with 9–15% by weight of

Table 3. Gel Spun PAN Fibers Compared with Other High Performance Fibers

Fiber	Tenacity, N/tex[a]	Elongation, %	Initial modulus, N/tex[a]
PAN[b]			
wet spun	0.3	9–11	14–16
gel spun	2.0	7–10	18–30
polyethylene	2.7	3.5	90
polypropylene	0.6	20	6
nylon-6,6	0.8	20	5
aramid[c]LM[d]	1.9	3.7	40
aramid[c]HM[e]	1.9	1.9	83
carbon HM[e]	1.9	1.4	134
carbon HS[f]	1.2	0.5	210
E-glass	0.8	2.0	28

[a]To convert N/tex to gf/den, multiply by 11.3.
[b]Hoechst Dolanite.
[c]Aromatic polyamide.
[d]Low modulus.
[e]High modulus.
[f]High strength.

Table 4. Properties of Dolanit Asbestos Replacement Fibers

Property	Dolanit 10	Dolanit VF
staple length, mm	6.0	1–24
fineness, dtex[a]	3.0	1.5–25
tenacity, N/tex[b]	0.7–0.8	0.7–0.8
elongation to break, %	9–11	9–11
initial modulus, N/tex[b]	14–16	14–16
wet strength, % of dry	90–98	90–98

[a]To convert dtex to den, multiply by 0.9.
[b]To convert N/tex to gf/den, multiply by 11.3.

asbestos. The flexural strength of cement sheets of acrylic reinforced cement is equivalent to asbestos reinforced cement and nearly double that of untreated cement (86). Two factors limiting the rapid development of acrylic asbestos replacement fibers are a somewhat high manufacturing cost (compared to asbestos) and uncertainty as to the long-term stability of the acrylic fiber. Loss of modulus and chemical degradation may be significant over a period of decades. Accelerated time testing and long-term evaluation are in progress. Other studies of acrylic fibers for concrete reinforcement have been carried out (87).

Another application where acrylics have proven to be effective is in brake and clutch linings. Here thermal stability is the key property. Hoechst has developed an experimental heat-resistant acrylic fiber (VF 1003) which can be used in combination with glass fibers. The French company Valco is developing a friction material for brakes and clutches that is also based on the use of acrylic fibers. Here the acrylic is used as a cross-linkable and fusible additive along with glass fibers, fillers, and binders. The fusible acrylic acts as a lubricant by melting when the friction material is rubbed. This allows the lining to compact, improving its wear resistance without affecting friction properties. Badische Corp. has also reported an asbestos replacement fiber for reinforcement in coatings, sealants, and plastics where outdoor applications require resistance to sunlight and weather.

Carbon and Graphite Fibers. Carbon and graphite fibers (qv) are valued for their unique combination of extremely high modulus and very low specific gravity. Acrylic precursors are made by standard spinning conditions, except that increased stretch orientation is required to produce precursors with higher tenacity and modulus. The first commercially feasible process was developed at the Royal Aircraft Establishment (RAE) in collaboration with the acrylic fiber producer, Courtaulds (88). In the RAE process the acrylic precursor is converted to carbon fiber in a two-step process. The use of PAN as a carbon fiber precursor has been reviewed (89,90).

Other Acrylic Fiber Products. *Antistatic Fibers.* Acrylic fibers present less of a problem with static and cling than other fibers since they are naturally somewhat hydrophilic and tend to dissipate static charge more readily than other fiber types. Antistatic properties have been improved, however, by making the fibers hygroscopic with hydrolyzing and cross-linking agents. Surface treatments are also used. Typically, ionic or metallic finish treatments are used. Thunderon-SSN is one type of antistatic fiber for floor coverings. This fiber, produced by Nippon

Sammo Sensyoku Co., has a high electrical conductivity imparted by treating the fiber with copper compounds in the finish step (91).

Antisoiling Fibers. Low dirt-absorbing fibers have been made by incorporating fluorinated comonomers, and porous fibers with reduced staining tendencies have been made from acrylic copolymers containing sulfonated comonomers. Antisoiling properties can also be achieved by using finishes, either by treating the fiber during spinning, or by applying finishes directly to the fabric. Treatment of acrylic fabrics with sodium hydroxide also gives improved soil release (92). Soil adhesion and soil removal (93) and the effect of fiber properties on soiling resistance (94) have been reviewed. An example of an acrylic stain-resistant fiber process is Stainorain, jointly developed by Pharr Yarn and Du Pont. Using Du Pont's Teflon technology, the fibers are impregnated with a water–oil repellant and stain-resistant chemical. This is done early in the fiber production process to lock in the protective properties. The fiber is designed for use in knitted garments for children and active sportswear such as golf and ski sweaters (95).

Moisture Absorbent Synthetic Paper. Processes for making a water absorbent synthetic paper with dimensional stability have been developed by several companies. In a process developed by Mitsubishi Rayon, acrylic fiber is insolubilized by hydrazine and then hydrolyzed with sodium hydroxide. The paper, formed from 100 parts fiber and 200 parts pulp, has a water absorption 28 times its own weight (96). Processes for making hygroscopic fibers have also been reported in the patent literature. These fibers are used in moisture absorbing nonwovens for sanitary napkins, filters, and diapers.

Electrically Conducting Fibers. Electrically conducting fibers are useful in blends with fibers of other types to achieve antistatic properties in apparel fabrics and carpets. The process developed by Nippon Sanmo Dyeing Co., for example, is reportedly used by Asahi in Cashmilon 2.2 dtex (2 den) staple fibers. Courtaulds claims a flame-resistant electrically conductive fiber produced by reaction with guanadine and treatment with copper sulfide (97).

Metallized textiles are being studied as a means to improve radar screens, making them more sensitive to small objects (98). The ability to be seen by radar is greatly increased by sheet-like materials containing thin layers of metallized fibers. Bayer has developed an acrylic filament yarn of 238 dtex (214 den) which gives radar reflectance of as much as 90%. This fiber is coated with a 0.02–2.5 μm layer of nickel by a currentless wet chemical deposition process (99,100). Monsanto's Flectron is an example of one such commercial product.

Coated Fabrics. Acrylic fibers are also used in coated fabrics (qv) which are produced by impregnating or coating the face or back of a woven, nonwoven, or knitted base fabric. The resulting products are used in apparel, home furnishings, and industrial applications. Acrylics are generally used when the application requires resistance to abrasion, chemical inertness, or resistance to weathering in outdoor exposure. Other fibers are used when high strength is essential. Although it is not the principal market for acrylics, geotextiles may have the greatest potential for high volume. These applications include erosion control, paving, holding ponds, roofing, and waterproofing barriers for below-grade structures. Home furnishing products include bedding, draperies, awnings, and upholstery (101).

Chemical Applications. Courtaulds has developed a series of acrylic-based fibers for controlled release of chemical reagents. The trade name of these fibers is Actipore. The reagents are entrapped within the fiber and slowly released at a rate dependent on the exact porosity of the fiber (102). These fibers may find use in controlled release of drugs, bactericides, and corrosion prevention chemicals (103). Fibers with different active groups have been made for sorption of chemicals. These fibers are designed to replace granular sorbents for air purification, for example, in air filtration masks (104). Acrylonitrile fibers treated with hydroxides have been reported to be useful for adsorption of uranium from seawater (105). Tubular fibers for reverse osmosis gas separations, ion exchange, ultrafiltration, and dialysis are a significant new application of acrylic fibers and other synthetics. Commercial acrylic fibers have already been developed by Nippon Zeon, Asahi, and Rhône-Poulenc.

Economic Aspects

Because of the rapid capital investment in acrylics that occurred in the early 1970s, there is a large excess capacity. In 1981 worldwide demand was 2.1×10^9 kg whereas worldwide capacity was 2.6×10^9 kg. Prices have consequently been soft since 1977. Since that time there has been only minimal investment in plants or equipment and a curtailment in research and development work.

Du Pont, one of the early pioneers in acrylics, has already withdrawn from the business, citing low profitability. Du Pont's Orlon was once the dominant trade name in the acrylic fibers business. Other companies have cut back acrylic production in an effort to adjust to world demand. Three U.S. producers remain. Monsanto, best known for its Acrilan acrylic fiber, is by far the largest and most diversified. With new spinning technology, originally introduced as Acrilan II, Monsanto now produces Duraspun, a fiber with enhanced abrasion resistance for the sock market. Other new products utilizing new spinning technology are Softlon and Ultrette, part of Monsanto's HP apparel line. American Cyanamid continues to do well with Creslan acrylic and other premium products like Micro-Supreme, a microfiber comparable to Monsanto's Fi-Lana. The third U.S. producer is Mann Industries, Inc., having purchased its facilities from BASF. Mann's products include traditional fibers for knitwear, producer dyed fiber for home furnishings and outdoor markets, plus unique products like Biokryl, an antimicrobial fiber designed for medical end uses.

Industrial use of acrylic fiber broadwoven goods has increased in recent years. The advantages are uv stability and its wide range of options for dyeing. The primary products are marine fabrics, awnings, and outdoor furniture. High performance fibers such as Kevlar, Spectra, and Vectran may eventually see competition from acrylic fiber. Gel spinning of ultrahigh molecular weight PAN is capable of producing fibers of comparable strength and modulus. The technology has been proven in the laboratory, but no commercial products have appeared thus far. Acrylics have also captured a share of the market for woven fabrics used in casual furniture. Sunbrella is one such product. The market for outdoor fabrics, such as awnings and marine fabrics, has been a boon to acrylic fiber producers. Brightly colored acrylics and innovative products, such as backlit awnings, have created a new surge of business. The marine fabrics market has been hurt by the effects of recession, but acrylics have been hurt the least. As in other markets, the colorful styles of acrylics have been a key to success. Acrylic fiber is also enjoying a comeback in the carpet market. Monsanto's Acrilan Plus and Traffic Control are examples of current carpet products. Acrilan Plus offers the wool-like hand, rich color, and styling variety possible with acrylics. Traffic Control, while primarily a nylon product, employs a high shrink acrylic blend to enhance appearance retention in high traffic areas. The worldwide picture for acrylics is summarized in Tables 5–7 (106).

Table 5. Synthetic Fiber Production for 1992, 10^3 t

Fiber	Capacity	Production	Utilization, %
acrylic	2,989	2,383	79.7
polyester	11,974	9,765	81.6
nylon	4,894	3,663	74.8
polypropylene	3,459	2,769	80.1

Table 6. Worldwide Synthetic Fiber Capacity and Production, 1991–2001, 10^3 t[a]

Year	Acrylic	Polyester	Nylon	Polypropylene
		Capacity		
1991	2,870	11,200	4,700	3,170
1995	3,360	14,710	5,310	4,000
1998	3,540	15,650	5,410	4,310
2001	3,550	15,980	5,470	4,550
		Production		
1991	2,350	9,170	3,670	2,580
1995	2,620	11,250	4,100	3,230
1998	2,730	13,280	4,390	3,540
2001	2,790	15,050	4,710	3,850

[a]Ref. 106.

Table 7. Regional Production of Acrylic Fibers in 1992, 10^3 t

Area	Production	Utilization, %
North America	220	9.2
Latin America	156	6.5
Western Europe	674	28.3
Eastern Europe	248	10.4
Middle East/Africa	135	5.7
South and Southeast Asia	87	3.7
East Asia	505	21.2
Japan	358	15.0
Total		*100.0*

By 1991 Western Europe was still the largest producer of acrylic fibers, with Bayer of Germany and Enichem of Italy the world's largest acrylic fiber companies. Worldwide, 24% of acrylic production was used for home furnishings and carpet, 75% for apparel, and 1% for industrial. A somewhat higher percentage, 6%, was used for industrial and automotive applications in Western Europe. The regional breakdown of world production should shift in the 1990s, with East Asian production increasing to levels comparable to Western Europe. Production in the People's Republic of China should rise dramatically, making that country self-sufficient in acrylic fiber. Exports from Western Europe and Japan will be hurt the most from this shift in world production. In general, production is forecast to decline in the most developed countries and in regions that depend heavily on acrylic exports. Production appears to be shifting to the next generation of low cost producers, namely South and East Asia, the Middle East, Africa, and Eastern Europe. Countries such as Taiwan and South Korea are losing their competitive advantage to rising wages and prices (107,108).

The principal thrust of Japan's joint research programs is the development of process technologies to reduce conversion costs and development of high value-added products. Energy cost reduction is a prime concern. Opportunities exist in low energy consuming polymerization and spinning technology. High productivity can be achieved by high speed polymerization and spinning, robotization, automation, multi-end spinning, and high speed crimping. Other strategies are withdrawal from unprofitable market sectors and consolidation of production and research and development effort into areas where special cost or property advantages seem apparent. New high volume markets are also being studied. These include acrylics for asbestos replacement, cement reinforcement, and geotextile applications. Products that have limited volume potential but have potential as high value-added premium products are carbon and graphite fiber precursors, high strength fibers, fibers for reverse osmosis, ion-exchange fibers, and functional fibers such as antimicrobial, antistatic, water-repellant, and highly reflectant fibers. Chief among the new products being sought are high strength, high toughness and high modulus fibers, and functional fabrics. Recent developments in man-made fiber technology have been reviewed (109).

BIBLIOGRAPHY

"Textile Fibers" in *ECT* 1st ed., Vol. 13, "Acrilan, Orlon, X-51," pp. 824–830, by P. M. Levin, E. I. du Pont de Nemours & Co., Inc., "Dynel and Vinyon," pp. 831–836, by H. L. Carolan, Union Carbide and Carbon Corp.; "Acrylic and Modacrylic Fibers" in *ECT* 2nd ed., Vol. 1, pp. 313–338, by D. W. Chaney, Chemstrand Research Center, Inc.; in *ECT* 3rd ed., Vol. 1, pp. 355–386, by P. H. Hobson and A. L. McPeters, Monsanto Triangle Park Development Center, Inc.

1. W. E. Morton and J. W. S. Hearle, *Physical Properties of Textile Fibers,* 2nd ed., John Wiley & Sons, Inc., New York, 1975.
2. "Textile World Manmade Fiber Chart 1992," *Text. World* (Aug. 1992).
3. G. H. Fremon, *Fibers from Synthetic Polymers,* Elsevier, New York, 1953, Chapt. 19.
4. J. B. Quig, *Papers Am. Assoc. Text. Tech.* **4**, 61 (1948); *Can. Text. J.* **66**(1), 42,46 (1949); *Rayon Synth. Text.* **30**(2), 79 (1949); **30**(3), 67 (1949); **30**(4), 91 (1949).
5. H. Staudinger and W. Heur, *Ber.* **63**, 222 (1930).

6. R. L. Cleland and W. H. Stockmayer, *J. Polym. Sci.* **18**, 473 (1955).
7. J. E. Ford, G. Pearson, and R. M. Smith, eds., *Identification of Textile Materials,* 6th ed., Textile Institute, Manchester, UK, 1970.
8. *Fibers in Textiles: Identification,* AATCC Test Method 20-1973, Technical Manual 50 50, American Association of Textile Chemists and Colorists, Research Triangle Park, N.C., 1974.
9. W. G. Wolfgang, in J. J. Press, ed. *Man-Made Textile Encyclopedia,* 1959, Chapt. IV, p. 141.
10. R. W. Moncrieff, *Man-Made Fibers,* 6th ed, John Wiley & Sons, Inc., New York, 1975.
11. R. H. Heidner and M. E. Gibson, in F. S. Snell and C. L. Hilton, eds., *Encyclopedia of Industrial Chemical Analysis,* Vol. 4, John Wiley & Sons, Inc., New York, 1967, pp. 219–360; G. C. East, *Text. Prog.* **3**(1), 67 (1971).
12. G. Vidotto, S. Brugnaro, and G. Talamini, *Die Macromol. Chem.* **140**, 263 (1970); J. C. Bevington, *Radical Polymerization,* Academic Press, Inc., New York, 1961; C. H. Bamford, W. C. Barb, A. D. Jenkins, and P. F. Onyon, *The Kinetics of Vinyl Polymerization by Radical Mechanisms,* Academic Press, Inc., New York, 1958.
13. K. E. J. Barret and H. R. Thomas, in K. E. J. Barret, ed., *Dispersion Polymerization in Organic Media,* John Wiley & Sons, Inc., New York, 1975, Chapt. 4.
14. N. Friis, D. Goosney, J. D. Wright, and A. E. Hamielic, *J. Appl. Polym. Sci.* **18**, 1247 (1974).
15. J. Ulbricht, *Z. Phys. Chem. (Leipzig)* **5/6**, 346 (1962).
16. L. H. Peebles, Jr., *J. Am. Chem. Soc.* **80**, 5603 (1958).
17. C. H. Bamford and A. D. Jenkins, *Proc. R. Soc. London Ser.* **A226**, 216 (1953); *Ibid.,* **228**, 220 (1920); C. H. Bamford, A. D. Jenkins, M. C. R. Symons, and M. G. Townsend, *J. Polym. Sci.* **34**, 181 (1959); A. D. Jenkins, in G. Ham, ed., *Vinyl Polymerization,* Part I, Marcel Dekker, Inc., New York, 1967, Chapt. 6.
18. L. H. Peebles, Jr., *Copolymerization,* Wiley-Interscience, New York, 1964, Chapt. IX.
19. W. D. Harkins, *J. Am. Chem. Soc.* **69**, 1428 (1947).
20. W. V. Smith and R. H. Ewart, *J. Chem. Phys.* **16**, 592 (1948); W. V. Smith, *J. Am. Chem. Soc.* **70**, 3695 (1948); *Ibid.,* **71**, 4077 (1949).
21. J. L. Gardon, *J. Polym. Sci., Pt. A-1,* **6**, (1968); *Ibid.,* 643, 665, 687, 2853 and 2859; J. L. Gardon, *Brit. Polym. J.* **2**, 1 (1970).
22. R. M. Fitch and Lih-bin Shih, *Prog. Col. Polym. Sci.* **56**, 1 (1975).
23. E. W. Rugeley, T. A. Field, Jr., and G. H. Fremon, *Ind. Eng. Chem.* **40**, 1724 (1948).
24. U.S. Pat. 2,420,330 (May 13, 1947), L. C. Shriver and G. H. Fremon (to Carbide and Carbon Chemicals Corp.).
25. D. C. Blackley, *Emulsion Polymerization Theory and Practice,* John Wiley & Sons, Inc., New York, 1975; D. R. Bassett and A. E. Hamielec, eds., *Emulsion Polymers and Emulsion Polymerization,* ACS Symposium Series 165, American Chemical Society, Washington, D.C., 1981.
26. I. Piirma, ed., *Emulsion Polymerization,* Academic Press, Inc., New York, 1982.
27. K. W. Min and W. H. Ray, *J. Macromol. Sci.-Revs. Macromol. Chem.* **C11**(2), 177 (1974).
28. L. H. Peebles, Jr., *J. Appl. Polym. Sci.* **17**, 113 (1973).
29. L. H. Peebles, Jr., R. B. Thompson, Jr., J. R. Kirby, and M. Gibson, *J. Appl. Polym. Sci.* **16**, 3341 (1972).
30. S. R. Palit, T. Guha, R. Das, and R. S. Konar, in N. M. Bikales, ed., *Encyclopedia of Polymer Science and Technology,* 1st ed., Vol. 2, John Wiley & Sons, Inc., New York, 1965, p. 229.
31. K. Stueben, in E. C. Leonard, ed., *Vinyl and Diene Monomers,* Part I, High Polymers Series Vol. XXIV, Wiley-Interscience, New York, 1970, Chapt. 1, p. 181.
32. W. M. Thomas, *Adv. Polym. Sci.* **2**, 401 (1961).

33. *The Chemistry of Acrylonitrile,* 2nd ed., American Cyanamid Publisher, New York, 1959.
34. A. D. Jenkins, in G. E. Ham, ed., *Vinyl Polymerization,* Part I, Marcel Dekker, New York, 1967, pp. 369–400.
35. G. E. Ham, in *Copolymerization,* Wiley-Interscience, New York, 1964, Chapt. I, pp. 1–64.
36. A. Valvassori and G. Sartori, *Adv. Polym. Sci.* **5**, 28 (1967).
37. T. Alfrey, Jr., J. J. Bohrer, and H. Mark, eds., *Copolymerization,* Wiley-Interscience, New York, 1952.
38. G. E. Ham, ed., *Copolymerization,* Wiley-Interscience, New York, 1964.
39. G. E. Ham, in J. I. Kroschwitz, ed., *Encyclopedia of Polymer Science and Technology,* 2nd ed., Vol. 4, John Wiley & Sons, Inc., New York, 1966, p. 165.
40. L. H. Peebles, Jr., in G. E. Ham, ed., *Copolymerization,* Wiley-Interscience, New York, 1964.
41. A. D. Jenkins and A. Ledwith, eds., *Reactivity, Mechanism and Structure in Polymer Chemistry,* Wiley-Interscience, New York, 1974.
42. F. R. Mayo and C. Walling, *Chem. Rev.* **46**, 191 (1950).
43. R. G. Fordyce and E. C. Chapin, *J. Am. Chem. Soc.* **69**, 581 (1947).
44. R. G. Fordyce, *J. Am. Chem. Soc.* **69**, 1903 (1947).
45. R. G. Fordyce and G. E. Ham, *J. Polym. Sci.* **3**(6), 891 (1948).
46. W. V. Smith, *J. Am. Chem. Soc.* **70**, 2177 (1948).
47. D. W. Ley and W. F. Fowler, Jr., *J. Polym. Sci.* **A2**, 1863 (1964).
48. V. I. Eliseeva, S. S. Ivanchev, S. I. Kuchanov, and A. V. Lebedev, *Emulsion Polymerization And Its Applications In Industry,* English Translation Consultants Bureau, New York, 1981.
49. U.S. Pat. 3,454,542 (July 8, 1969), D. W. Cheape and W. R. Eberhardt (to Monsanto Co.).
50. S. Ito, Y. Kawai, T. Oshita, *Kobunshi Ronbunshu* **43**(6), 345 (1986).
51. S. Ito, *Sen'i Gakkaishi* **43**(5), 236 (1987); S. Ito and C. Okada, *Sen'i Gakkaishi* **42**(11), T618–T625 (1986).
52. Can. Pat. 911,650 (Oct. 3, 1972), P. Melacini, R. Tedesco, L. Patron, and A. Moretti, (to Montedison Fibre).
53. U.S. Pat. 3,787,365 (Jan. 22, 1974); U.S. Pat. 3,821,178 (June 28, 1974); U.S. Pat. 3,879,360 (Apr. 22, 1975), L. Patron and co-workers (to Montefibre).
54. J. P. Craig, J. P. Knudsen, and V. F. Holland, *Text Res. J.* **33**, 435 (1962).
55. D. R. Paul, *J. Appl. Polym. Sci.* **11**, 439 (1967); *Ibid.,* **12**, 383 (1968).
56. J. P. Knudsen, *Text. Res. J.* **33**, 13 (1963).
57. U.S. Pat. 3,353,381 (Nov. 21, 1967), E. A. Taylor, Jr. (to Monsanto Co.); U.S. Pat. 3,791,788 (Feb. 12, 1974), E. A. Taylor (to Monsanto Co.).
58. *Knitting Times,* **53**(10), 23 (Apr. 1984); *Text. World,* 29 (June 1984).
59. *Ind. Fabr. Prod. Rev.* **59**(10A) (Feb. 1983).
60. W. E. Fitzgerald and J. P. Knudsen, *Text. Res. J.* **37**, 447 (1967).
61. *Text. Daily* (Apr. 8, 1983).
62. Jpn. Pat. 80 137,219 (Oct. 25, 1980) (to Asahi Chemical Co., Ltd.).
63. A. A. Vaidya and S. Chattopadhyay, *Tex. Dyer Printer* **10**(8), 37–41 (1977).
64. M. Lewin, in M. Lewin and S. B. Sello, eds., *Handbook of Fiber Science and Technology,* Vol. II, Part B, Marcel Dekker, Inc., New York, 1985, pp. 1–141.
65. A. A. Kumar, K. V. Vaidya and K. V. Datye, *Man-Made Text. (India)* **24**(1), 23 (1981).
66. M. M. Gauthier, R. D. Deanin, and C. J. Pope, *Polym. Plast. Technol. Eng.* **16**(1), 39 (1981).
67. C. Hsieh, *Hsin Hsien Wei* **20**(11), 12 (1979); M. Hatano and H. Ogawa, *Kagaku Keizoi* **27**(7), 77 (1980).

68. R. C. Nametz, *Ind. Eng. Chem.* **62**, 41 (1970).
69. N. A. Khalturinskii, T. V. Popova, and A. A. Berlin, *Russ. Chem. Rev.* **53**(2), 197 (1984).
70. *Text. World,* 37–48 (Aug. 1992); *Am. Text. Int.* 40–54 (Nov. 1991); *Ibid.,* 50–51 (Jan. 1992).
71. E. Maslowski and A. Urbanska, *Am. Text. Int. (Fiber World Section),* 6–7 (May 1989); *Ibid.,* 2–4 (July 1989); *Ibid.,* 2–4 (Sept. 1989); *Ibid.,* (Nov. 1989).
72. E. Maslowski, *Fiber World,* 15–23 (Mar. 1987).
73. *High Perf. Text.* **8**(11), 7 (1988); *Ibid.,* **3**(10), 3 (1983).
74. Br. Pat. 2,018,188A (Nov. 24, 1978) (to American Cyanamid).
75. U.S. Pat. 3,814,739 (June 4, 1974), H. Takeda (to Toray Ind., Inc.).
76. R. Moreton, *Carbon Fibres: Their Composites and Applications,* Paper No. 12, The Plastics Institute, London, 1971.
77. Eur. Pat. Appl., 44,534 (Jan. 27, 1982) (to Hoechst AG.).
78. *High Perf. Text.* **8**(11), 7 (1988); **4**(12), 3 (1984).
79. *Adv. Comp. Bull.* **9**(1), 8 (1988).
80. Jpn. Pat. 17,966 (Feb. 20, 1981); 60,051 (May 6, 1980); 98,025 (Feb. 13, 1976) (to Asahi Chem. Ind. Co., Ltd.).
81. Ger. Pat. DE 3,012,998 (Oct. 15, 1981), A. Wuestfeld and C. Wuestfeld.
82. S. A. Belg. Pat. 889,260 (Oct. 16, 1981) (to REDCO).
83. Eur. Pat. Appl. 44,534 (Jan. 27, 1982) (to Hoechst AG).
84. *Hoechst High Chem. Mag.* **1**, 66 (1986).
85. *Text. Month* (Oct. 18, 1986).
86. *High Perf. Text.* **3**(10), 3 (1983).
87. J. Wang, S. Backer, and V. C. Li, *J. Mat. Sci.* **22**(12), 4281 (1987).
88. Brit. Pat. 1,110,791 (Apr. 24, 1964), W. Watt, L. N. Phillips, and W. Johnson, and Brit. Pat. 1,148,874 (June 15, 1966) (to National Research Development Corp.); U.S. Pat. 3,412,062 (Nov. 1968) (to National Research Development Corp.).
89. P. Rajalingam and G. Radhakrishnan, *J. Macromol. Sci., Part C. Rev.* **C31**(2) (1991).
90. R. Prescott, *Mod. Plast. Ency.* **66**(11), 232 (1989).
91. G. Aviv, *J. Soc. Dyers Colour.* **105**(11), 406 (1989).
92. K. Sen, P. Bajaj, and J. S. Rameshbapu, *Melliand Textilberichte/Int. Text. Rep.* **72**(12), 1034, E416 (1991).
93. H. J. Jacobasch, *Tenside Deterg.* **17**(3), 113 (1980).
94. P. A. Glubish and I. E. Kucher, *Khim. Teknol. (Kiev)* **2**, 32 (1981).
95. *Am. Text. Int.* K/A 2–4 (Oct. 1991).
96. Jpn. Pat. 144,299 (Nov. 10, 1981) (to Mitsubishi Rayon Co.).
97. *High Perf. Text.* (Feb. 1992).
98. F. Marchini and V. Massa, *Chemiefasern/Textilindustrie* **38/90**(3), T5-10 and E26-27 (Mar. 1988).
99. U.S. Pat. 4,320,403 (to Bayer AG).
100. *High Perf. Text.* (Apr. 1992).
101. *Am. Text. Int.,* 80–84 (June 1992).
102. *J. Coated Fabr.* **18**, 259 (Apr. 1989).
103. *Med. Text.* **4**(4), 8 (1987).
104. Z. M. Petrovic, *Hemijska Vlakna* **31**(3–4), 3 (1991).
105. T. Kago, A. Goto, K. Kusakabe, and S. Morooka, *Ind. Engr. Chem. Res.* **31**(1), 204 (1992).
106. *World Synthetic Fibers 1991–2001,* Tecnon UK, Ltd., Plantation Wharf, London, Aug. 1992.
107. *Int. Fiber J.,* 85–89 (Aug. 1993).

108. *Chemiefasern/Textilindustrie* **41/93**(12), 1350, E177 (1991).
109. V. B. Gupta, *Man-Made Text. (India)* **31**(6), 249 (1988).

General References

S. K. Basu, *Man-Made Text. (India)* **34**(2), 453 (1991).
H. C. Bach and R. S. Knorr, *Polymers—Fibers and Textiles, A Compendium,* John Wiley & Sons, Inc., New York 1990, pp. 1–54.
B. G. Frushour and R. S. Knorr, "Acrylic Fibers," in M. Lewin and Eli M. Pearce, eds. *Handbook of Fiber Science and Technology,* Vol. IV, *Fiber Chemistry,* Marcel Dekker, Inc., New York, 1985.
K. K. Garg, *Synth. Fib.* **14**(2), 29 (1985).
S. Balasubramanian, *Man-Made Textiles (India) Acrylic Fibres-I,* 221–231 (May 1977); *Acrylic Fibres-II,* 277–290 (June 1977).
R. W. Moncrieff, *Man-Made Fibers,* 6th ed, John Wiley & Sons, Inc., New York, 1975.
E. Cernia, "Acrylic Fibers," in H. F. Mark, S. M. Atlas, and E. Cernia, eds., *Man-Made Fibers,* Vol. III, Wiley-Interscience, New York, 1968.
R. K. Kennedy, "Modacrylic Fibers," in H. F. Mark, S. M. Atlas, and E. Cernia, eds., *Man-Made Fibers,* Vol. III, Wiley-Interscience, New York, 1968.

RAYMOND S. KNORR
Monsanto Company

CELLULOSE ESTERS

The predominant cellulose ester fiber is cellulose acetate, a partially acetylated cellulose, also called acetate or secondary acetate. It is widely used in textiles because of its attractive economics, bright color, styling versatility, and other favorable aesthetic properties. However, its largest commercial application is as the fibrous material in cigarette filters, where its smoke removal properties and contribution to taste make it the standard for the cigarette industry. Cellulose triacetate fiber, also known as primary cellulose acetate, is an almost completely acetylated cellulose. Although it has fiber properties that are different, and in many ways better than cellulose acetate, it is of lower commercial significance primarily because of environmental considerations in fiber preparation.

Polymer Characteristics

Cellulose triacetate is obtained by the esterification of cellulose (qv) with acetic anhydride (see CELLULOSE ESTERS). Commercial triacetate is not quite the precise chemical entity depicted as (**1**) because acetylation does not quite reach the maximum 3.0 acetyl groups per glucose unit. Secondary cellulose acetate is obtained by hydrolysis of the triacetate to an average degree of substitution (DS) of 2.4 acetyl groups per glucose unit. There is no satisfactory commercial means to acetylate directly to the 2.4 acetyl level and obtain a secondary acetate that has the desired solubility needed for fiber preparation.

(1)

The degree of acetylation is specified by two separate terms: acetyl value (%) and combined acetic acid (%). The ratio of these two values is always 43:60, reflective of the molecular-weight ratio of the acetyl group to acetic acid. Acetyl and combined acetic acid values over the possible range of acetyl contents are shown in Figure 1. Commercial cellulose triacetate [*9012-09-3*] has a combined acetic acid content of 61.5%, corresponding to 2.92 acetyl groups per glucose unit. Cellulose acetate [*9004-35-7*], with 2.4 acetyl groups per glucose unit, has a combined acetic acid content of approximately 55%.

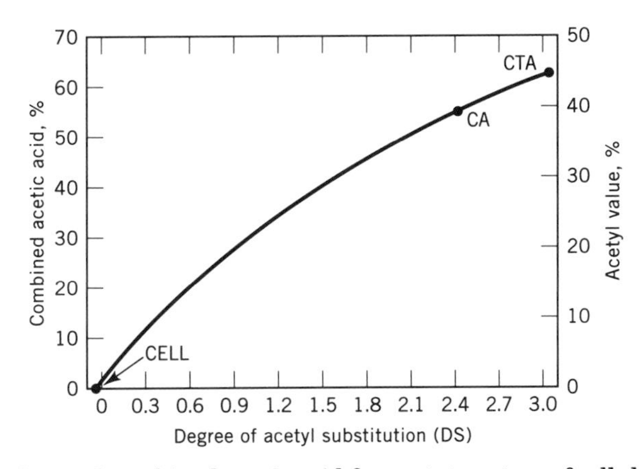

Fig. 1. Acetyl value and combined acetic acid for acetate esters of cellulose. Designations are as follows: CA = acetate (DS = 2.4); CELL = cellulose (DS = 0); CTA = triacetate (DS = 3.0). Combined acetic acid (%) = 6005(DS)/(159.1 + 43.04(DS)); acetyl value (%) = 4304(DS)/(159.1 + 43.04(DS)).

Fiber Properties

The performance of a textile fabric is characterized by terms such as strength, hand, drape, flexibility, moisture transport, and wrinkle resistance. Although the interactions among fibers in a fabric array are complex, its properties reflect in part the inherent properties of the fiber as well as how the fibers are assembled.

Mechanical Properties. The mechanical properties of a fiber are characterized by classical stress–strain and recovery behavior under conditions of tensile, torsional, bending, and shear loading. Typical stress–strain curves indicative of most commercial acetate and triacetate yarns are shown in Figures 2 and 3. It is common to use the ultimate stress or breaking strength to characterize the

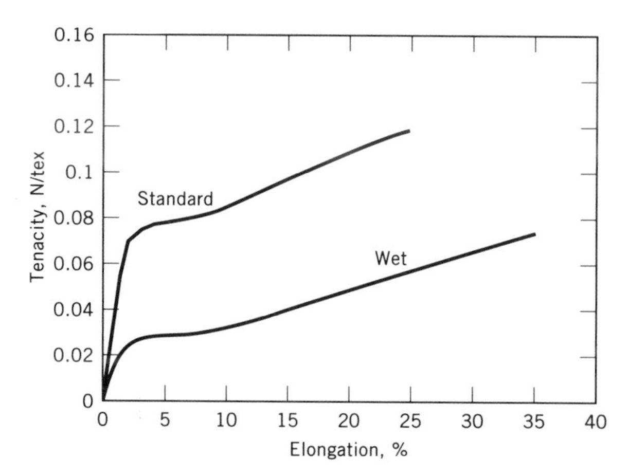

Fig. 2. Cellulose acetate stress–strain properties at standard and wet conditions, tested at 60% min extension rate, 3.9 cm gauge length. Sample conditions: standard, 21°C, 65% rh; wet, 21°C, water wet. To convert N/tex to gf/den, multiply by 11.33.

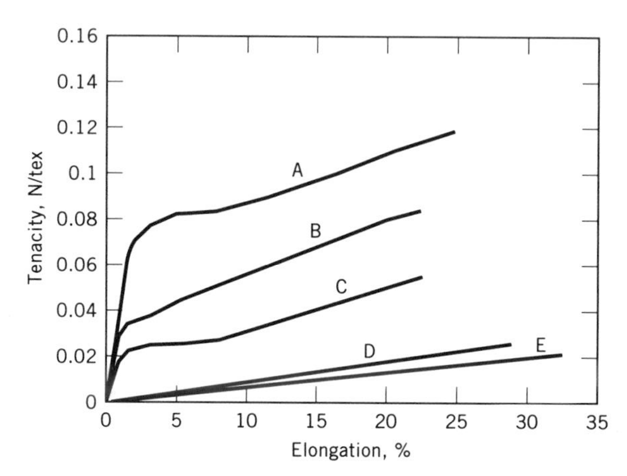

Fig. 3. Cellulose triacetate stress–strain properties at different temperatures (1). A, 21°C, 65% rh; B, 100°C; C, 150°C; D, 175°C; and E, 205°C. Tested at 60% min extension rate, 3.9 cm gauge length. To convert N/tex to gf/den, multiply by 11.33.

fibers' tensile properties. The units of tensile stress, or tenacity, are Newtons/tex (gram force/denier) and the strain is given as a percentage of elongation. Temperature and moisture content of the fiber affect viscous behavior and hence modify the stress–strain relationship. Most stress–strain data are reported under standard conditions of 21°C and 65% rh.

Acetate and triacetate have a tenacity in the range of 0.10–0.12 N/tex (1.1–1.4 gf/den) with a breaking elongation of about 25–30%. Compared to other common textile fibers, acetate and triacetate are relatively weak, eg, 20–25% the tenacity of polyester. This is not necessarily a disadvantage because fabric construction can be used to obtain the desired fabric performance targets. Pilling, the accumulation of fuzz balls on the fabric with wear, is not a problem as it is with the higher tenacity fibers. Figures 4 and 5 also show that moisture or heat can significantly impact the stress–strain behavior.

The ratio of stress to strain in the initial linear portion of the stress–strain curve indicates the ability of a material to resist deformation and return to its original form. This modulus of elasticity, or Young's modulus, is related to many of the mechanical performance characteristics of textile products. The modulus of elasticity can be affected by drawing, ie, elongating the fiber; environment, ie, wet or dry, temperature; or other procedures. Values for commercial acetate and triacetate fibers are generally in the 2.2–4.0 N/tex (25–45 gf/den) range.

The wet modulus of fibers at various temperatures influences the creasing and mussiness caused by laundering. Figure 6 shows the change with tempera-

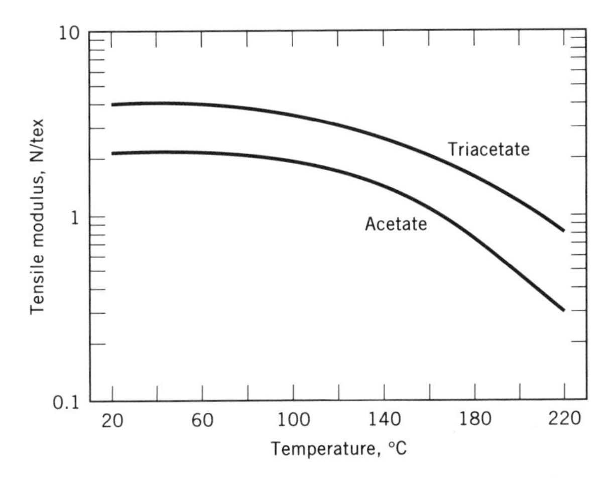

Fig. 4. Tensile modulus as a function of temperature. To convert N/tex to gf/den, multiply by 11.33.

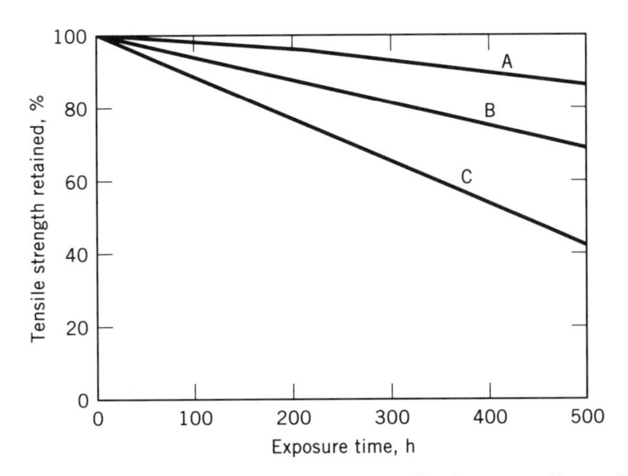

Fig. 5. The effect of dry heat exposure on acetate and triacetate fibers. Tested at 65% rh, 21°C after exposure. A, acetate, 100°C; B, triacetate, 130°C; and C, acetate, 120°C.

ture of the wet modulus of acetate and triacetate, and compares them with a number of other fibers (2). Acetate, triacetate, and rayon behave quite similarly, with a lower sensitivity than acrylic.

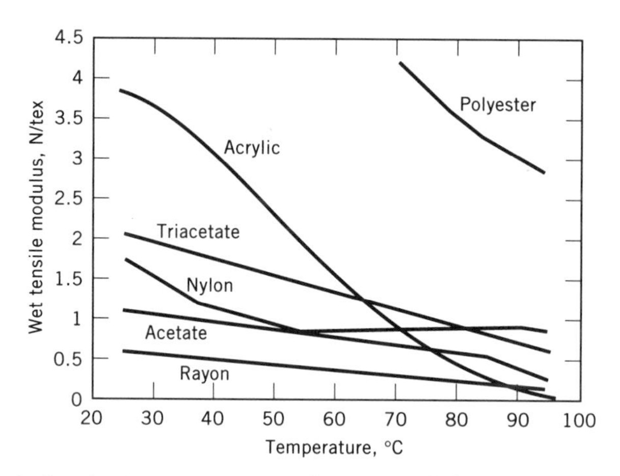

Fig. 6. The effect of water temperature on the wet modulus of fibers. To convert N/tex to gf/den, multiply by 11.33.

The ability of a fiber to absorb energy during straining is measured by the area under the stress–strain curve. Within the proportional limit, ie, the linear region, this property is defined as toughness or work of rupture. For acetate and triacetate the work of rupture is essentially the same at 0.022 N/tex (0.25 gf/den). This is higher than for cotton (0.010 N/tex = 0.113 gf/den), similar to rayon and wool, but less than for nylon (0.076 N/tex = 0.86 gf/den) and silk (0.072 N/tex = 0.81 gf/den) (3).

A fiber that is strained and allowed to recover releases a portion of the work absorbed during straining. The ratio of the work recovered to the total work absorbed, measured by the respective areas under the stress–strain and stress–recovery curves, is designated as resilience.

The elongation of a stretched fiber is best described as a combination of instantaneous extension and a time-dependent extension or creep. This viscoelastic behavior is common to many textile fibers, including acetate. Conversely, recovery of viscoelastic fibers is typically described as a combination of immediate elastic recovery, delayed recovery, and permanent set or secondary creep. The permanent set is the residual extension that is not recoverable. These three components of recovery for acetate are given in Table 1 (4). The elastic recovery of acetate fibers alone and in blends has also been reported (5). In textile processing strains of more than 10% are avoided in order to produce a fabric of acceptable dimensional or shape stability.

The bending properties of a fiber generally depend on the viscoelastic behavior of the material. In most textile applications, the radius of curvature of bending is relatively large, and the imposed strains are of a low order of magnitude. As a first approximation, the bending stiffness or flexural rigidity of a fiber is the product of the bending modulus and the moment of inertia of the cross

Table 1. Elongation Recovery of Acetate Fibers[a]

Fiber	Immediate elastic recovery, %	Delayed recovery, %	Permanent set, %
acetate multifilament			
at 50% of breaking tenacity	74	26	0
at breaking point	14	16	70
acetate staple yarn			
at 50% of breaking tenacity	58	42	0
at breaking point	12	18	70

[a]Ref. 4.

section. For fibers of round cross section and constant modulus, the flexural rigidity varies directly with the square of the tex. Torsional and shear properties of acetate and other fibers are discussed in Reference 6. Table 2 shows some additional mechanical properties characteristic of commercial acetate and triacetate fibers.

Absorption and Swelling Behavior. The absorption of moisture by acetate and triacetate fibers generally depends on the relative humidity and whether equilibrium is approached from the dry or wet side. This hysteresis effect is noted over the entire range of relative humidities, as shown in Figure 7 (7). The percentage of moisture regain of commercial fibers (ASTM D1909-68), taken at 65% relative humidity for the absorption cycle, is 6.5 for acetate fiber and 3.5 for triacetate (8). Heat treatment can lower the moisture regain of triacetate fiber, and values of 2.5–3.2% have been observed (9,10). Sorption isotherms have been analyzed in terms of the BET adsorption isotherms (11).

Percentage of water imbibition is an important property in ease-of-care and quick-drying fabrics. This value is determined by measuring the moisture re-

Table 2. Tenacity and Elongation of Commercial Acetate and Triacetate Fibers

Properties	Value
Tenacity, N/tex[a]	
standard conditions[b]	0.10–0.12
knot	0.09–0.10
loop	0.09–0.10
wet	0.07–0.09
bone dry	0.12–0.14
Elongation at break, %	
standard conditions[b]	25–45
wet	35–50

[a]To convert N/tex to gf/den, multiply by 11.33.
[b]Standard conditions are 65% relative humidity, 21°C.

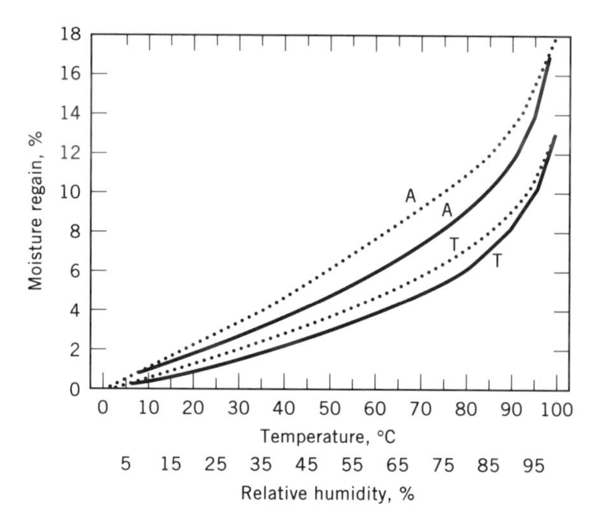

Fig. 7. Moisture regains of cellulose acetate (A) and triacetate (T) fibers. Moisture content on a bone-dry basis, measured at 22°C: (——), absorption cycle; (····), desorption cycle.

maining in a fiber in equilibrium with air at 100% rh while the fiber is being centrifuged at forces up to 1000 g. The average recorded value for acetate is 24%; triacetate not heat-treated, 16%; and heat-treated triacetate, 10%.

Absorption of water by fibers generally causes swelling roughly proportional to the moisture content. The average increase in length of acetate fibers due to water absorption is ca 1%, and the average cross-sectional increase is ca 10%. The corresponding values for triacetate fiber are lower, ie, a 1.5% increase in cross-sectional area for heat-treated triacetate fiber and a 4.0% increase for fiber not heat-treated (12). Increasing the relative humidity from 0 to 100% causes acetate to elongate about 2% as compared to 3.4% for rayon, 2.4% for nylon, and less than 0.3% for acrylics and polyester.

Specific Gravity. Fiber cross sections are often irregular and specific gravity is measured by an immersion technique. The values of 1.32 for acetate and 1.30 for triacetate are accepted for fibers of combined acetic acid contents of 55 and 61.5%, respectively (13–15).

Refractive Index. The refractive index parallel to the fiber axis (ϵ) is 1.478 for acetate and 1.472 for triacetate. The index perpendicular to the axis (ω) is 1.473 for acetate and 1.471 for triacetate. The birefringence, ie, the difference between ϵ and ω, is very low for acetate fiber and practically undetectable for triacetate.

Thermal Behavior. As with most thermoplastic fibers, acetate sticks, softens, and melts. Sticking and softening temperatures depend on geometric factors, eg, yarn diameter and fabric construction, and are not necessarily directly related to the fiber melting point. Acetate softens and sticks in the 190–205°C range, and fuses at ca 260°C. The apparent shining or glazing temperature is usually lower than the sticking temperature. The sole-plate temperature of an iron should not exceed 170–180°C when used on acetate fabrics. The sticking and glazing temperatures of untreated triacetate fiber are in the same range as those of acetate, whereas those exhibited by heat-treated triacetate fibers are considerably higher. Fabrics made of the latter can be ironed at temperatures as high as 240°C. The melting point of triacetate is ca 300°C.

Acetate and triacetate exhibit moderate changes in mechanical properties as a function of temperature. As the temperature is raised, the tensile modulus of acetate and triacetate fibers is reduced, and the fibers extend more readily under stress (see Fig. 4). Acetate and triacetate are weakened by prolonged exposure to elevated temperatures in air (see Fig. 5).

Light Stability. The resistance of textile fibers to sunlight degradation depends on the wavelength of the incident light, relative humidity, and atmospheric fumes. Acetate and triacetate fibers have essentially the same light-absorption characteristics in the visible spectrum; absorption in the uv region is slightly higher. Both fibers, when exposed under glass, behave similarly to cotton and rayon, ie, they are somewhat more resistant than unstabilized pigmented nylon and silk and appreciably less resistant than acrylic and polyester fibers. When acetate and triacetate are exposed to weathering, their resistance is lower than when exposed under glass. Certain pigments, eg, carbon black or the rutile form of titanium dioxide, offer some increased protection from sunlight exposure.

Electrical Behavior. The resistivity of acetate varies significantly with humidity with typical values ranging from 10^{12} ohm·cm at 45% rh to 10^{7} ohm·cm at 95% rh (16). Because of the high resistivity both acetate and triacetate yarns readily develop static charges and an antistatic finish is usually applied to aid in fiber processing. Both yarns have also been used for electrical insulation after lubricants and other finishing agents are removed.

Dyeing Characteristics. Disperse dyes, high melting crystalline compounds with low solubility in the dye bath, are most frequently used for cellulose acetate and triacetate fibers. They are milled to very small particle size, permitting effective dispersion without agglomeration in the dye bath, and diffuse into the fiber to give a uniform color. Dye-bath temperature and fiber composition affect the diffusion-controlled dyeing rate. Triacetate fibers are dyed more slowly than acetate fibers; dye carriers accelerate the rate (1,17–22). Selection of the appropriate azo, anthraquinone, or diphenylamine disperse dye ensures good colorfastness. Fading inhibitors are used to counteract the effects of nitrogen oxides and ozone. Dyed triacetate fabrics, which are subsequently heat-treated to raise the safe ironing temperature, drive the dye further into the fiber to increase fading resistance and improve colorfastness (see DYES, APPLICATION AND EVALUATION).

Colored acetate and triacetate yarns are produced by incorporating colored pigments (inorganic or organic), soluble dyes, or carbon in the polymer solution before extrusion. Solution-dyed acetate and triacetate yarns are extremely colorfast to washing, dry cleaning, sunlight, perspiration, seawater, and crocking, and usually surpass the performance of vat-dyed yarns. In addition, acetate and triacetate dyed by conventional methods are susceptible to gases or fumes and fade; such fading is absent in solution-spun, pigment-dyed yarn.

Chemical Properties. Under slightly acidic or basic conditions at room temperature, acetate and triacetate fibers are resistant to chlorine bleach at the concentrations normally used in laundering.

Triacetate fiber is significantly more resistant than acetate to alkalies encountered in normal textile operations. Temperatures no higher than 85°C and pH no more than 9.5 are recommended for dyeing. Under normal scouring and dyeing conditions, alkalies up to pH 9.5 and temperatures up to 96°C may be used with triacetate with little saponification or delustering. Heat-treated triacetate fiber exhibits even higher alkali resistance (23). Strong alkalies and boiling temperatures saponify triacetate as well as acetate fiber.

Acetate and triacetate are essentially unaffected by dilute solutions of weak acids, but strong mineral acids cause serious degradation. The results of exposure of heat-treated and untreated triacetate taffeta fabrics to various chemical reagents have been reported (9). Acetate and triacetate fibers are not affected by the perchloroethylene dry-cleaning solutions normally used in the United States and Canada. Trichloroethylene, employed to a limited extent in the UK and Europe, softens triacetate.

Resistance to Microorganisms and Insects. Resistance of triacetate to microorganisms, based on soil-burial tests, is high, approaching that of polyester, acrylic, and nylon fibers. Soil-burial test results on acetate, triacetate, and cotton are shown in Figure 8. Neither acetate nor triacetate fiber is readily attacked by moths or carpet beetles.

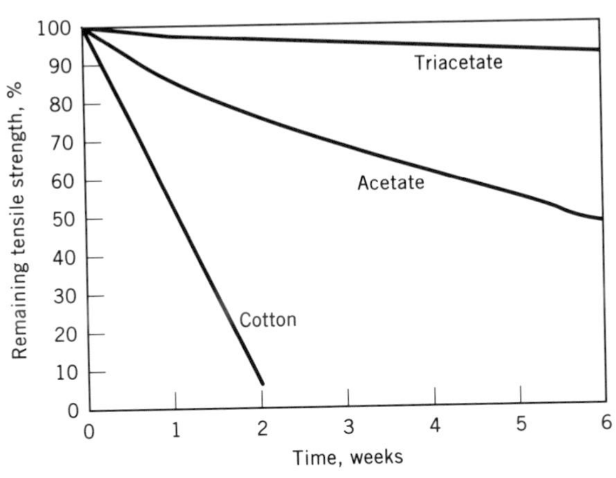

Fig. 8. The resistance of cellulose fibers to biological attack via soil-burial testing.

Manufacture

CELLULOSE ACETATE AND TRIACETATE POLYMER

The production of acetate and triacetate polymer is accomplished by the esterification of high purity chemical cellulose with acetic anhydride (24–26); wood pulp is the principal source of the chemical cellulose, except for special plastic-grade acetates requiring low color and high clarity, where cotton linters are used. A high degree of cellulose purity is needed to assure complete polymer solubility for the preparation of fibers, since the hemicellulose impurities form undesirable gels

(27). Acetylation-grade wood pulp ideally has a 95–98% alpha-cellulose content, but strong economic incentives have led to the commercialization of 92–93% alpha-cellulose wood pulps. The other raw materials, ie, acetic acid, acetic anhydride, and sulfuric acid, are commercially available in high purity grades.

Secondary Acetate Processes. There is no commercial process to directly produce secondary cellulose acetate sufficiently soluble in acetone to produce fiber. Hence, the cellulose is completely acetylated to the triacetate during the dissolution step and then hydrolyzed to the required acetyl value.

Most cellulose acetate is manufactured by a solution process, ie, the cellulose acetate dissolves as it is produced. The cellulose is acetylated with acetic anhydride; acetic acid is the solvent and sulfuric acid the catalyst. The latter can be present at 10–15 wt % based on cellulose (high catalyst process) or at ca 7 wt % (low catalyst process). In the second most common process, the solvent process, methylene chloride replaces the acetic acid as solvent, and perchloric acid is frequently the catalyst. There is also a seldom used heterogeneous process that employs an organic solvent as the medium, and the cellulose acetate produced never dissolves. More detailed information on these processes can be found in Reference 28.

The solution process consists of four steps: preparation of cellulose for acetylation, acetylation, hydrolysis, and recovery of cellulose acetate polymer and solvents. A schematic of the total acetate process is shown in Figure 9.

Preparation for Acetylation. Wood pulp is customarily supplied in rolls weighing up to 300 kg, but can also be obtained in bales of individual sheets. The pulp sheet must be dispersed without damaging individual fibers to generate sufficient surface area for complete acetylation. Some manufacturers use a disk refiner, others use wet methods. In one example, to further increase accessibility the fluffed pulp is agitated with an acetic acid–water mixture for ca 1 h at 25–40°C. An activation stage is generally included in the low catalyst process using an acetic acid–sulfuric acid mixture with a sulfuric acid concentration of ca 1–2% of the pulp weight. The activation may last for 1–2 h, during which the degree of polymerization of the cellulose is reduced. A controlled combination of activation time and temperature achieves the desired degree of polymerization. A separate activation step is usually not included in the high catalyst procedure. The degree of polymerization is further controlled by the conditions selected during acetylation and hydrolysis. The pulp is subsequently charged to the acetylation reactor after pretreatment and activation.

Acetylation. The esterification of cellulose with acetic anhydride liberates 1.03 kJ/g (246 cal/g) of cellulose, and the reaction of acetic anhydride with water from the pretreatment generates 3.30 kJ/g (789 cal/g) of water. Hence, a heat sink is needed for the two exothermic reactions that occur in the acetylation process. In high catalyst acetylation, heat is generated quickly, and a jacketed vessel does not provide the necessary cooling capacity for the viscous mixture. Hence, the acetylation mixture is prechilled in a separate vessel called a crystallizer, and some of the acetic acid freezes and provides a readily accessible heat sink. This is particularly important in the early stages of acetylation, in which a rapid temperature rise would uncontrollably reduce the degree of polymerization. To improve process economics, attempts are made to acetylate at the highest possible temperature (29). In the low catalyst process, heat is generated more slowly, and

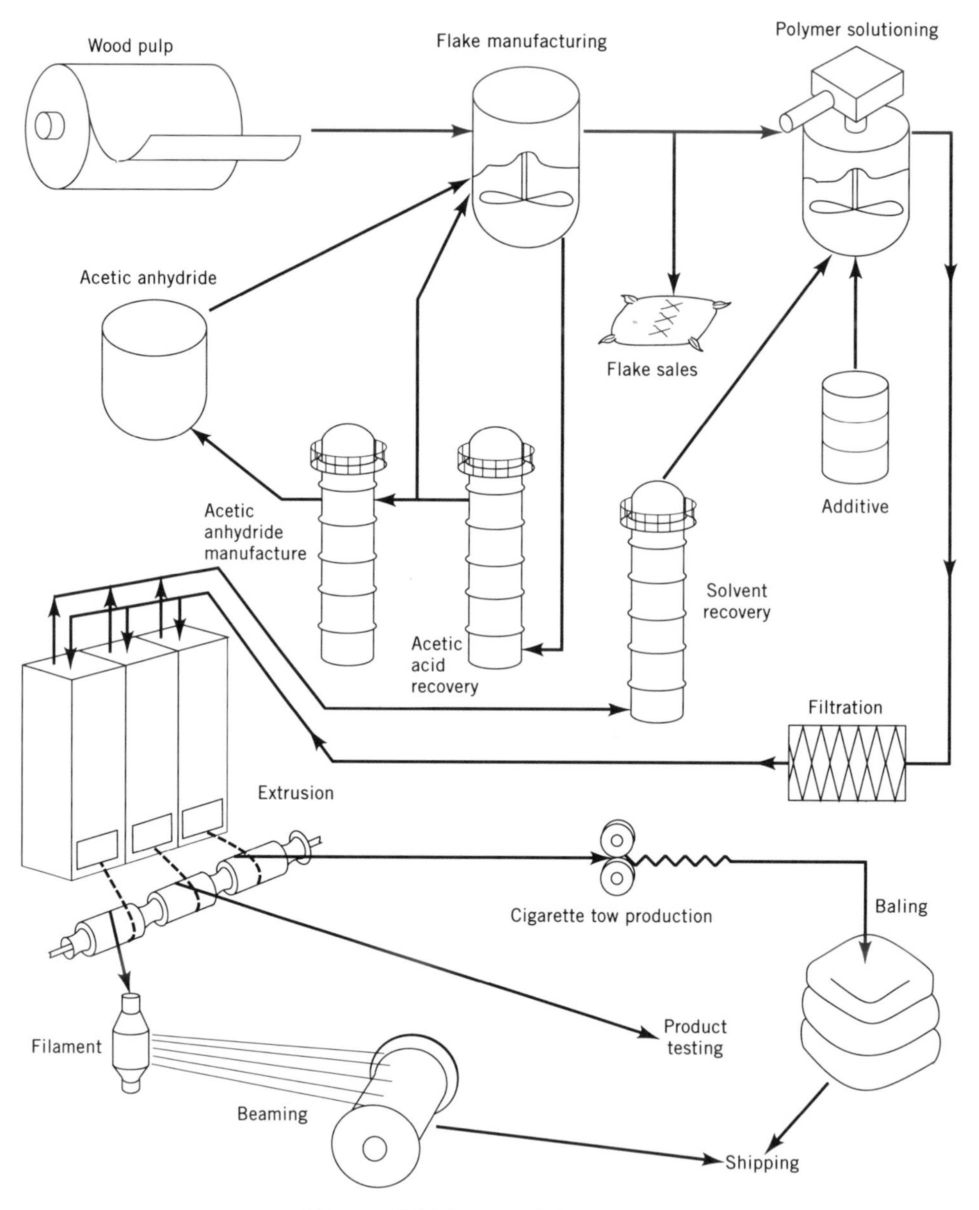

Fig. 9. Cellulose acetate process.

cooling in a jacketed vessel is satisfactory. In the methylene chloride process, the required heat transfer is provided by refluxing the solvent. In either process, a 5–15 wt % excess of acetic anhydride ensures complete reaction. A series of simultaneous, complex reactions occurs during acetylation (24–26).

When the acetylation is completed, microscopic examination of the solution should reveal no undissolved residues. The reaction is terminated by adding water to destroy the excess anhydride and provide a water concentration of 5–10% for hydrolysis. A 10–25% cellulose acetate concentration is typical.

All commercial acetylation processes are essentially heterogeneous, even though the final polymer is soluble in the solvent. Much of the original cellulose crystalline structure is retained in the acetate or triacetate (30). A truly homogeneous process, although not commercial, is attractive from a technical standpoint. Several solvents permit nondegradative dissolution of the cellulose, eg, dimethylformamide–dinitrogen tetroxide, dimethyl sulfoxide–formaldehyde, dimethylacetamide–lithium chloride, and *N*-methylmorpholine oxide (31–38). Cellulose in solution can be esterified uniformly and directly to the desired degree of substitution; hydrolysis to obtain solubility in acetone is not necessary. Homogeneous acetylation with acetyl chloride in dimethylformamide–chloralpyridine reportedly produces a chloral-containing cellulose acetate (39). Dissolution of cellulose in dimethylformamide–dinitrogen tetroxide followed by sulfuric acid treatment produces an acetic acid-soluble cellulose sulfate which is directly acetylated to an acetone-soluble cellulose acetate (40). The homogeneous process offers the advantage of good quality acetate flake from lower purity wood pulps. The disadvantage is the cost of solvent recovery.

Hydrolysis. The number of acetyl groups present on each anhydroglucose unit at the completion of the acetylation is slightly less than 3.0 and must be reduced to ca 2.4 to prepare secondary cellulose acetate soluble in acetone. The number of acetyl groups is reduced and the combined sulfate groups are minimized by acid hydrolysis under controlled time, temperature, and acidity. The sulfate groups, which are hydrolyzed more easily than the acetyl groups, increase the acidity of the reaction. In high catalyst acetylation, for example, a portion of the sulfuric acid is neutralized, eg, by the addition of sodium acetate or magnesium acetate, to prevent excessive depolymerization of the cellulose acetate. The hydrolysis temperature is normally 50–100°C and the reaction time varies from 1 to 24 h. Hydrolysis can also be conducted at other conditions. For example, when conducted at higher temperatures (41) there is polymer degradation and reduced yield.

Polymer Recovery. Precipitation, washing, and drying are the final steps in polymer preparation. Precipitation is initiated by treating the hydrolyzed cellulose acetate solution with a stream of dilute (10–15%) acetic acid to the point of incipient precipitation. More dilute acetic acid is rapidly added and the solution vigorously agitated. To obtain a powder precipitate, the agitated solution is slowly diluted until precipitation occurs. Another process involves extrusion of the hydrolyzed solution through small holes into a precipitating acid bath; this produces fine strands which are cut into pellets.

The precipitated cellulose acetate is filtered from the dilute (25–36%) acetic acid. The acetic acid and salts remaining from the sulfuric acid neutralization are removed by washing. The wet polymer is typically dried to a moisture content of 1–5%. The dilute acetic acid obtained from the washing and precipitation steps cannot be used in other stages of the process. Its efficient recovery and recycle are an economic necessity.

If thermal stability and absence of yellowing upon heating are critical, for example in thermoplastic molding applications, any remaining sulfate group must be removed. This is accomplished by heating the acetate polymer in deionized water at 1.4 MPa (200 psi) for ca 1 h and is called pressure stabilization.

Acetate and triacetate polymers are white amorphous solids produced in granular, flake, powder, or fibrous form. They are used as raw materials in the preparation of fibers, films, and plastics. Polymer density varies and ranges from 100 kg/m^3 for the fibrous form to 500 kg/m^3 for granules. Acetate polymer is shipped by trailer truck, railroad freight car, or multiwall bags.

Acid Recovery. Approximately 4.0–4.5 kg of acetic acid per kg of cellulose acetate is used in the solution process; ca 0.5 kg is consumed by the product and the remaining 3.5–4.0 kg is recovered as an aqueous solution of 25–35% acetic acid. This solution may also contain dissolved salts from sulfuric acid neutralization, and dissolved and suspended low molecular weight cellulose and hemicellulose acetates. Acetic acid is recovered from the clarified weak acid stream by solvent extraction with solvents such as ethyl acetate or methyl ethyl ketone. Benzene, also formerly used, is being phased out because of carcinogenic concerns. The organic extract is sent to a distillation column, and the aqueous raffinate phase, containing most of the inorganic salts, is discarded. The extraction solvent is distilled off, leaving glacial acetic acid. The energy requirements for acid recovery depend on the organic solvent used and may be in the 4.2–10.5 kJ/g (1–2.5 kcal/g) range of acid recovered. A portion of the acetic acid may be subsequently converted to acetic anhydride by catalytic pyrolysis in good yield and at low cost. A new acetic anhydride process uses synthesis gas obtained from coal as feedstock (42,43).

Processes for Triacetate. There are both batch and continuous process for triacetate. Many of the considerations and support facilities for producing acetate apply to triacetate; however, no acetyl hydrolysis is required. In the batch triacetate sulfuric acid process, however, a sulfate hydrolysis step (or desulfonation) is necessary. This is carried out by slow addition of a dilute aqueous acetic acid solution containing sodium or magnesium acetate (44,45) or triethanolamine (46) to neutralize the liberated sulfuric acid. The cellulose triacetate product has a combined acetic acid content of 61.5%.

In the batch methylene chloride process, the sulfuric acid concentration can be as low as 1% and only limited desulfonation is required to reach a combined acetic acid content of 62.0%. With perchloric acid catalyst, the nearly theoretical value of 62.5% combined acetic acid is obtained.

A lesser employed batch heterogeneous process employs a liquid that does not dissolve the triacetate and gives products of nearly theoretically combined acetic acid.

The continuous triacetate process has been described in detail (9); the economic impact of triacetate has significantly diminished recently, particularly in the United States (47–50).

CELLULOSE ACETATE AND TRIACETATE FIBERS

Extrusion Processes. Polymer solutions are converted into fibers by extrusion. The dry-extrusion process, also called dry spinning, is primarily used for acetate and triacetate. In this operation, a solution of polymer in a volatile solvent is forced through a number of parallel orifices (spinneret) into a cabinet of warm air; the fibers are formed by evaporation of the solvent. In wet extrusion, a polymer solution is forced through a spinneret into a liquid that coagulates the filaments and removes the solvent. In melt extrusion, molten polymer is forced through a multihole die (pack) into air, which cools the strands into filaments.

The dry-extrusion process consists of four operations: dissolution of the polymer in a volatile solvent; filtration of the solution to remove insoluble matter; extrusion of the solution to form fibers; and lubrication, yarn formation, and packaging.

Polymer Dissolution. Acetone is the universal solvent for cellulose acetate in dry extrusion. Although methylene chloride is a better solvent (39), it is more expensive and there are more environmental concerns. The optimum concentration for an acetate spinning solution depends on a balance between the highest possible solids concentration and the resulting high solution viscosity. Though high concentrations of solids produce fibers with better properties and reduce the relative amount of solvent to be recovered, practical limits of viscosity are quickly reached. Solutions in aqueous acetone exhibit minimum viscosity at an acetone–water ratio of ca 9:1. A cellulose solvent, *N*-methylmorpholine oxide, reportedly reduces the solution viscosity (51). Typical solvent composition is ca 95% acetone and 5% water, with a typical polymer solids concentration of 20–30%, depending on the polymer molecular weight. The viscosity of the solution at room temperature is ca 100–300 Pa·s (1000–3000 P). The solubility of acetates of different combined acetic acid content in acetone–water and acetic acid–water solution has been well-studied (52).

Cellulose triacetate is insoluble in acetone, and other solvent systems are used for dry extrusion, such as chlorinated hydrocarbons (eg, methylene chloride), methyl acetate, acetic acid, dimethylformamide, and dimethyl sulfoxide. Methylene chloride containing 5–15% methanol or ethanol is most often employed. Concerns with the oral toxicity of methylene chloride have led to the recent termination of the only triacetate fiber preparation facility in the United States, although manufacture still exists elsewhere in the world (49).

Acetate or triacetate polymer is charged to heavy-duty mixers along with the solvent and a filter aid such as wood-pulp fibers. Concentration, temperature, and mixing uniformity are closely controlled. For a delustered or dull fiber, 1–2% of a finely ground titanium dioxide pigment may be added in the mixing process or by injection of a pigment slurry after filtration. The polymer concentration and the composition of the extrusion solvent strongly affect the uniformity and tensile properties of the acetate fiber and must be closely controlled.

Solution Filtration. The polymer solution, free of unacetylated cellulose, rigid particle contaminants, and dirt, must pass through spinnerets with holes of 30–80 μm diameter. Multistage filtration, usually through plate-and-frame filter presses with fabric and paper filter media, removes the extraneous matter before extrusion. Undesirable gelatinous particles, such as the hemicellulose acetates from cellulose impurities, tend to be sheared into smaller particles rather than removed. The solution is also allowed to degas in holding tanks after each state of filtration.

Extrusion. The filtered, preheated polymer solution is delivered to the spinneret for extrusion at constant volume by accurate metering pumps. The spinnerets are of stainless steel or another suitable metal and may contain from thirteen to several hundred precision-made holes to provide a fiber of desired size and shape. Auxiliary filters are inserted in front of the fixture that holds the spinneret and in the spinneret itself to remove any residual particulate matter in the extrusion solution.

Before entering the spinneret, the extrusion solution, also called a dope, is heated to reduce the viscosity and provide some of the heat necessary to flash the solvent from the extruded filament. A thermostatically controlled heat exchanger may be used to heat the dope, or the filter–spinneret assembly may be located inside the heated extrusion cabinet.

The heated polymer solution emerges as filaments from the spinneret into a column of warm air. Instantaneous loss of solvent from the surface of the filament causes a solid skin to form over the still-liquid interior. As the filament is heated by the warm air, more solvent evaporates. More than 80% of the solvent can be removed during a brief residence time of less than 1 s in the hot air column. The air column or cabinet height is 2–8 m, depending on the extent of drying required and the extrusion speed. The air flow may be concurrent or countercurrent to the direction of fiber movement. The fiber properties are contingent on the solvent-removal rate, and precise air flow and temperature control are necessary.

A feed roll applies tension to the bundle of fibers to withdraw them from the extrusion cabinet. The product of one extrusion position is called a continuous-filament yarn, as distinguished from staple. Cellulose acetate yarns are generally produced in a weight range of 5–100 tex (45–900 den). Feed-roll speed, metering pump output, and column conditions are carefully balanced to produce a yarn of specified and uniform tex (denier).

Only a small quantity of triacetate yarn is made by wet extrusion because extrusion speeds are much lower than for dry extrusion and the process is not attractive for producing filament yarns. Melt extrusion is only used for the production of a small quantity of triacetate yarn.

The solvent used to form the dope is evaporated during the extrusion process and must be recovered. This is usually done by adsorption on activated carbon or condensation by refrigeration. For final purification, the solvent is distilled. Approximately 3 kg of acetone, over 99%, is recovered per kg of acetate yarn produced. Recovery of solvent from triacetate extrusion is similar, but ca 4 kg of methylene chloride solvent is needed per kg of triacetate yarn extruded.

Finish. A finish or lubricant gives the extruded yarn the frictional and antistatic properties required for further processing. The finish, applied at concentrations of 1–5% as the yarn leaves the cabinet, depends on the intended use of the yarn; proprietary formulations are generally used. The lubricated yarn, containing only a small amount of residual solvent, can be taken up on a ring twister, which imparts just enough twist to prevent the handling difficulties of untwisted yarn, or with no twist by being wound on a cylindrical tube. In addition, the yarn filaments may be entangled, ie, compacted, by passing the yarn through a device which intermingles the filaments with air jets (53–55).

Anisotropic Solutions

Many cellulosic derivatives form anisotropic, ie, liquid crystalline, solutions, and cellulose acetate and triacetate are no exception. Various cellulose acetate anisotropic solutions have been made using a variety of solvents (56,57). The nature of the polymer–solvent interaction determines the concentration at which liquid crystalline behavior is initiated. The better the interaction, the lower the concentration needed to form the anisotropic, birefringent polymer solution. Strong organic acids, eg, trifluoroacetic acid, are most effective and can produce an anisotropic phase with concentrations as low as 28% (58). Trifluoroacetic acid has been studied with cellulose triacetate alone or in combination with other solvents (59–64); concentrations of 30–42% (wt vol) triacetate were common.

Ternary-phase diagrams for cellulose triacetate, trifluoroacetic acid, and either water, methylene chloride, or formic acid demonstrate the narrow concentration region over which the solutions are anisotropic (65). Cellulose triacetate in a nitric acid–water solution is reported to give very high strength fiber, eg, a tenacity of 0.7–0.9 N/tex (8–10 gf/den) (66). It has been demonstrated that the inclusion of a low concentration of a flexible polymer, eg, poly(ethylene terephthalate), in a trifluoroacetic acid–methylene chloride solution of cellulose triacetate enhances the ordering of the triacetate, reducing the concentration of triacetate needed to form an anisotropic solution. The viscosity of the solution is also lowered (67,68). Subdenier triacetate fibers were prepared from a trifluoroacetic acid solution by attenuating the fibers with a pressurized gas stream (69). Another study characterized the crystal structure of deacetylated triacetate produced from a lyotropic solution (70). Anisotropic solutions were not limited to just acetate esters. Formate and formate–acetate esters from anisotropic solutions have been described (71). Although fiber has been prepared experimentally from almost all these liquid crystalline polymeric solutions, no significant commercial products exist.

Products

Yarns and Fibers. Many different acetate and triacetate continuous filament yarns, staples, and tows are manufactured. The variable properties are tex (wt in g of a 1000-m filament) or denier (wt in g of a 9000-m filament), cross-sectional shape, and number of filaments. Individual filament fineness (tex per filament or denier per filament, dpf) is usually in the range of 0.2–0.4 tex per filament (2–4 dpf). Common continuous filament yarns have 6.1, 6.7, 8.3, and 16.7 tex (55, 60, 75, and 150 den, respectively). However, different fabric properties can be obtained by varying the filament count (tex per filament or dpf) to reach the total tex (denier).

Although the cross-sectional shape of the spinneret hole directly affects the cross-sectional shape of the fiber, the shapes are not identical. Round holes produce filaments with an approximately round cross section, but with crenelated edges; triangular holes produce filaments in the form of a "Y." Different cross sections are responsible for a variety of properties, eg, hand, luster, or cover, in the finished fabric. Some fibers may contain chemical additives to provide lightfastness and impart fire retardancy. These are usually added to the acetate solution before spinning.

A metier is an array of individual extrusion positions on one machine. The yarn is collected as a package, eg, bobbin, tube, pirn, at each metier position and removed from the machine at regular intervals to maintain a constant supply of yarn on each package. The package may contain 0.5–7.0 kg of yarn. Bobbin yarn may contain a slight twist, ie, ca 0.08 turns per cm, whereas yarn taken up on tubes may have zero twist. The yarn is transferred from bobbins to different packages for sale. The product may contain a twist of 0.3–8.0 turns per cm. At present, compacted yarn is more popular than low twist yarn.

Yarn Packages. The principal package types used by the textile industry are tubes, cones, and beams. Tubes are wrapped with 1.0–7.0 kg of yarn. The package is built on winders to provide package integrity and easy removal. Some packages are provided with a magazine wrap at the start of winding so that packages can be changed automatically.

Cones contain 0.6–4.0 kg of yarn. The tip of the cone tube must have a smooth finish to prevent damage to the yarn, which is drawn over the top. Again, a magazine wrap may be provided for automatic package transfer. Both compacted and twisted yarns are packaged on cones.

Beams, ie, large spools, are usually constructed of an aluminum alloy and vary from 50 to 170 cm in length and from 50 to 90 cm in flange diameter. A beam holds 100–700 kg of yarn. The beam most commonly used by the warp knitting trade is 107 cm long and 53–76 cm in diameter. Section beams for weaving are usually 137 cm long and 76 cm in diameter. Both types are parallel wound with a large number (up to 2400) of individual yarn ends. The length of yarn on beams varies with yarn denier, beam capacity, and intended use. Lengths are ordinarily 11,000–78,000 m. When beam winding is complete, the ends are taped in position and the beam is wrapped with a protective cover.

Staple and Tow. The same extrusion technology that produces continuous filament yarn also produces staple and tow. The principal difference is that spinnerets with more holes are used, and instead of winding the output of each spinneret on an individual package, the filaments from a number of spinnerets are gathered together into a ribbon-like strand, or tow. A mechanical device uniformly plaits the tow into a carton from which it can be continuously withdrawn without tangling.

Staple is produced by cutting a crimped tow into short lengths (usually 4–5 cm) resembling short, natural fibers. Acetate and triacetate staple are shipped in 180–366 kg bales, but production is quite limited. Conventional staple-processing technology applied to natural fibers is used to process acetate and triacetate staple into spun yarn.

Fibrillated Fibers. Instead of extruding cellulose acetate into a continuous fiber, discrete, pulp-like agglomerates of fine, individual fibrils, called fibrets or fibrids, can be produced by rapid precipitation with an attenuating coagulation fluid. The individual fibers have diameters of 0.5 to 5.0 μm and lengths of 20 to 200 μm (Fig. 10). The surface area of the fibrillated fibers are about 20 m^2/g, about 60–80 times that of standard textile fibers. These materials are very hydrophilic;

an 85% moisture content has the appearance of a dry solid (72). One application is in a paper structure where their fine fiber size and branched structure allows mechanical entrapment of small particles. The fibers can also be loaded with particles to enhance some desired performance such as enhanced opacity for papers. When filled with metal particles it was suggested they be used as a radar screen in aerial warfare (73).

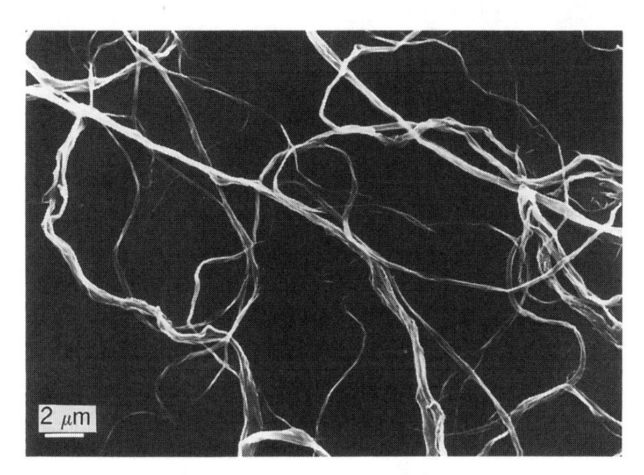

Fig. 10. Cellulose acetate fibrets.

Economic Aspects

Cellulose acetate, the second oldest synthetic fiber, is an important factor in the textile and tobacco industries; 731,000 metric tons were produced worldwide in 1991 (Fig. 11) (74). Acetate belongs to the group of less expensive fibers; triacetate is slightly more expensive. An annual listing of worldwide fiber producers, locations, and fiber types is published by the Fiber Economics Bureau, Inc. (74).

The principal textile applications of both acetate and triacetate fibers are in women's apparel and home-furnishing fabrics. Although the use of acetate fiber for textile applications has generally declined (Fig. 11), the total worldwide production of cellulose acetate increased owing to tow for cigarette filters, which rose from 335,000 to 481,000 t between 1980 and 1991 (Table 3) (74). Much of the increase in cigarette filter tow production came from the conversion of textile acetate capacity. A new cigarette tow facility in the People's Republic of China was commissioned in 1989 (10,000 tons), and there are other announced plans for additional capacity by Hoechst Celanese, Eastman/Rhône-Poulenc, and Daicel/Mitsui (75–78). Because of its superior filtration, effect on cigarette taste, and low cost, acetate is expected to supply over 90% of the filter-cigarette market.

A list of world acetate and triacetate producers is given in Reference 74. The combined annual world acetate production (filament, staple, and tow) peaked in 1980 with 672,000 t, dropped to 574,000 t in 1984, and rose to 731,000 t in 1991. The United States accounted for ca 45% of the world total. Other principal acetate producing countries include the UK, Japan, Canada, Italy, and the former USSR.

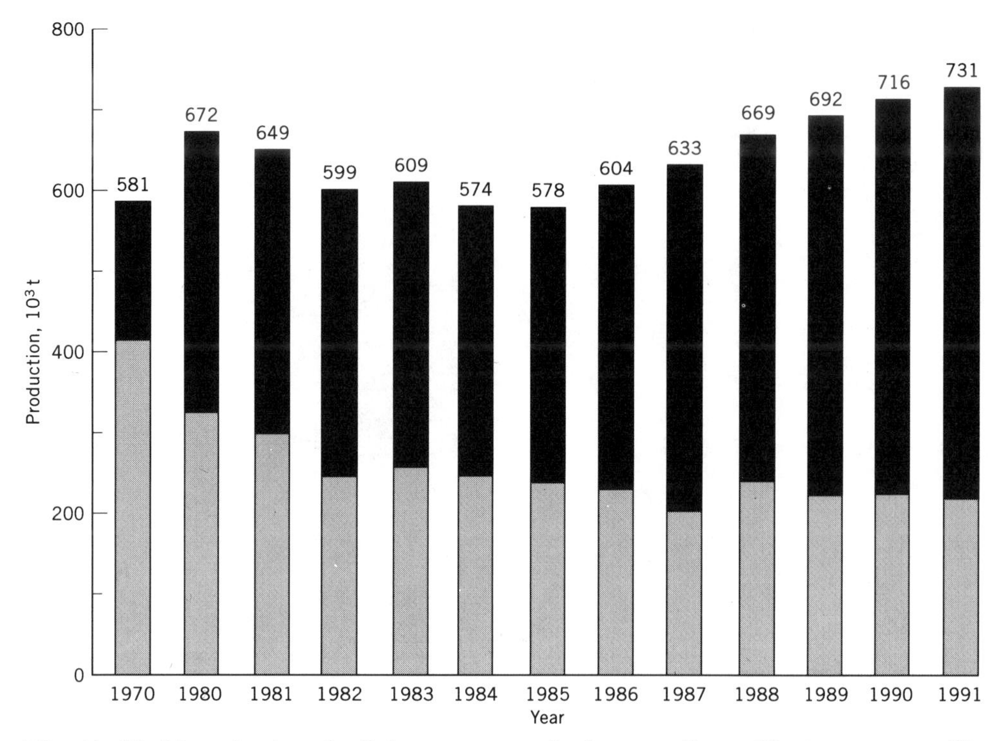

Fig. 11. World production of cellulose acetate and triacetate fibers: ■, cigarette tow; □, textile (74).

Table 3. World Production of Cellulose Acetate Cigarette Filter Tow,[a] 10^3 t

Year	Americas[b]	Europe[c]	Asia[d]	China	Totals
1970	99	35	20		154
1980	220	74	41		335
1981	226	81	42		349
1982	216	81	42		339
1983	211	80	43		334
1984	187	80	47		314
1985	210	83	44		337
1986	216	104	42		362
1987	248	107	46		401
1988	250	114	49		413
1989	265	118	55	5	443
1990	275	120	58	12	465
1991	280	120	69	12	481

[a]Ref. 74.
[b]United States, Brazil, Canada, Colombia, Mexico, and Venezuela.
[c]UK, Belgium, Germany, and the former USSR.
[d]Japan and the Republic of Korea.

Uses

The two principal markets for cellulose acetate are textiles and cigarette filters.

Textiles. A unique combination of desirable qualities and low cost accounts for the demand for acetate in textiles. In the United States, acetate and triacetate fibers are used in tricot-knitting and woven constructions, with each accounting for approximately half the total volume. This distribution changes slightly according to market trends. The main markets are women's apparel, eg, dresses, blouses, lingerie, robes, housecoats, ribbons, and decorative household applications, eg, draperies, bedspreads, and ensembles. Acetate has replaced rayon filament in liner fabrics for men's suits and has been evaluated for nonwoven fabrics (79–81).

Acetate and triacetate fibers have lower strength and abrasion resistance than most other synthetic fibers and are frequently combined with nylon or polyester in yarns or specific fabric constructions. This permits them to be used in applications not suitable for 100% acetate fabrics, eg, men's shirts. Combination yarns can be prepared by twisting or air-entanglement and bulking. Yarns prepared by air-entanglement and bulking have unique characteristics and aesthetic properties that allow their use in casement and upholstery fabric markets. With chemical additives, both acetate and triacetate fibers can pass U.S. government flame-retardant fabric regulations, eg, DOC FF 3-71. The flammability of acetate and triacetate is compared to that of other textiles in Reference 82.

Triacetate offers better ease-of-care properties than secondary acetate in many apparel applications. Of particular importance are surface-finished fabrics, eg, fleece, velour, and suede for robes and dresses. These fabrics offer superb aesthetic qualities at reasonable cost. Triacetate is also desirable for print fabrics, where it produces bright, sharp colors. The recent discontinuance of triacetate fiber in the United States has led to the use of acetate with fibers such as polyester (47–50).

Cigarette Tow. Acetate fiber used in the production of cigarette filters is supplied in the form of tow (83). Tow is a continuous band composed of several thousand filaments held loosely together by crimp, a wave configuration set into the band during manufacture (Fig. 12**a**). A tow is formed by combining the output of a large number of spinnerets and crimping the combined filaments to create an integrated band of continuous fibers. The tow is dried and baled. The wide range of available acetate filter tow products permits control of properties in the finished cigarette-filter rod.

A tow product is characterized by cross section, tex (denier), and crimp. The shape of the cross section is related to the shape of the minute orifices in the spinneret used to form the filament (Fig. 13).

Several tex terms are important in filter-tow processing: tex (denier) per filament, total tex (denier) of the uncrimped tow (the product of the tex per filament multiplied by the number of filaments in the tow band), and crimped total tex (denier), which is somewhat higher than the total tex (denier).

A tow product described as 0.9 tex, 5000 total tex, may therefore be interpreted as an uncrimped tow band that weighs 5000 g for each kilometer of length

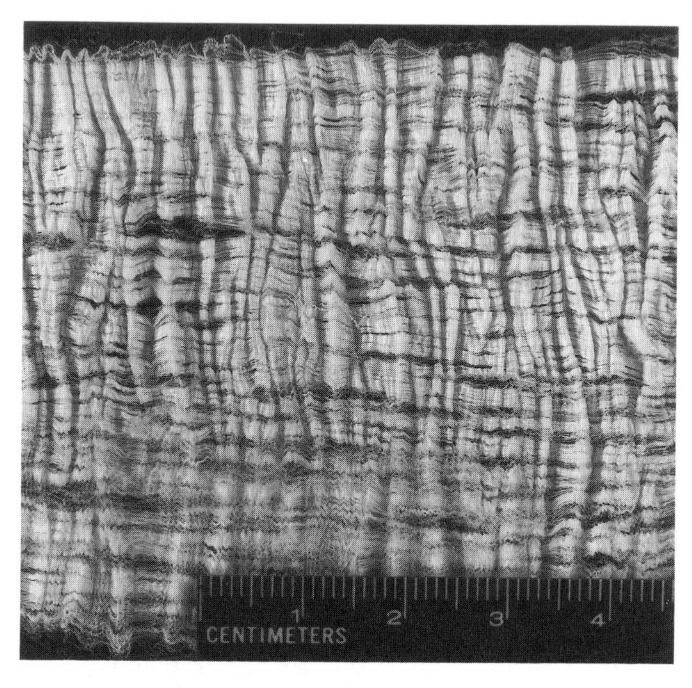

(a)

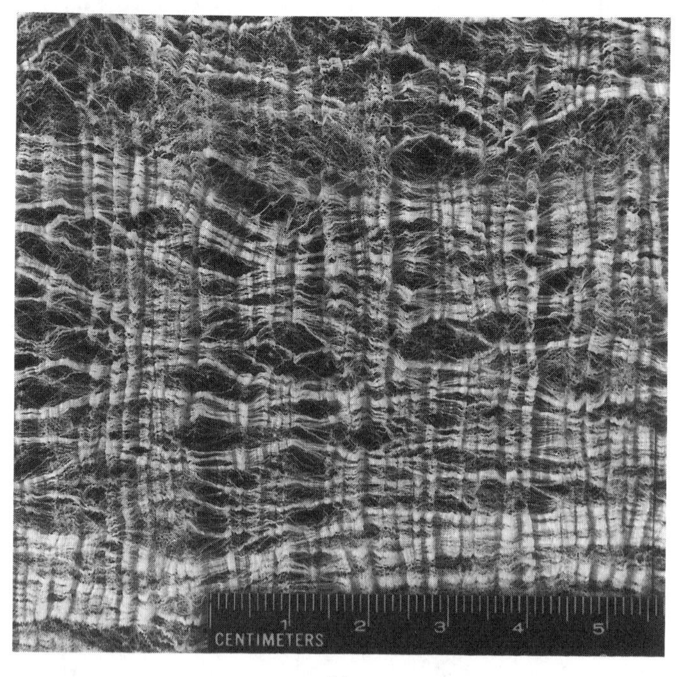

(b)

Fig. 12. Cellulose acetate cigarette filter tow (**a**) as supplied and (**b**) "opened" for cigarette-filter rod preparation.

and that is composed of 5555 individual filaments (5000/0.9) each weighing 0.9 g for each kilometer of length.

The crimp imparted to the tow has a sawtooth or sinusoidal wave shape. Because the filaments are usually crimped as a group, the crimp in parallel fibers is in lateral registry, ie, with the ridges and troughs of the waves aligned, as shown in Figure 14.

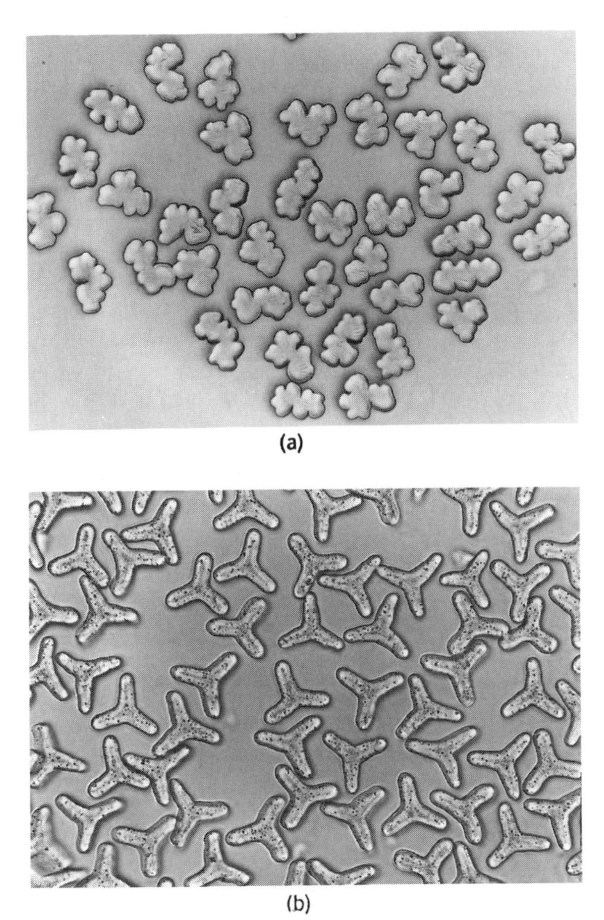

Fig. 13. Cross sections of cellulose acetate fiber from (**a**) circular and (**b**) triangular spinneret holes.

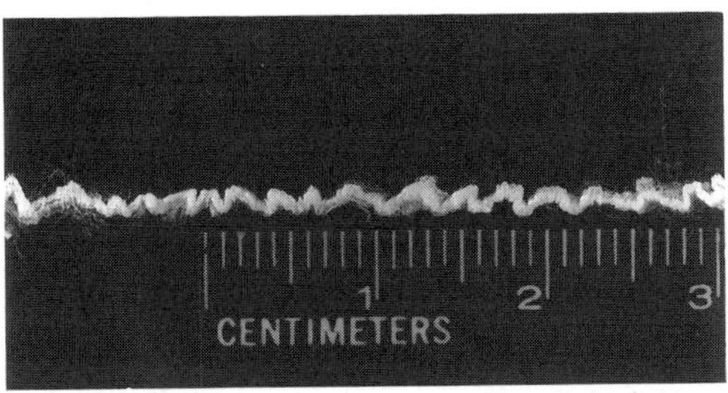

Fig. 14. Section of cellulose acetate cigarette filter tow showing crimp configuration.

The presence of crimp in the tow is necessary to ensure that the tow can be packaged, processed, and handled easily and to impart bulk to the finished filter. To achieve the latter in the production of the filters, it is necessary to open the tow band (Fig. 12**b**) to the desired bulk so that the fibers completely fill the paper wrap without voids and soft spots.

A linear relationship, constant for any given tow item, exists between the weight of tow in a cigarette-filter rod and certain of its properties, such as pressure drop and smoke-removal efficiency. By preparing filter rods over a range of rod weights and testing the rods for pressure drop, a linear curve characteristic of the specific tow item is obtained. This characterization provides a means of designing the required cigarette-filter characteristics. For example, a typical linear plot shows that to achieve a certain target pressure drop, either a 3.0- or 3.9-denier (tex) per filament of the same total denier can be selected. However, the latter will produce a heavier filter rod. At the same denier per filament, increasing the total denier can give the same pressure drop, but this also gives heavier rods and shifts the pressure drop range upward. Other factors such as economics, rod firmness, and rod diameter affect the final selection of the tow product used.

Other Applications. Other large-volume, nontextile, or cigarette-filter tow applications for cellulose acetate are filament fibers for decorative packaging ribbons and tows which are converted into ink dispensers for felt-tip pens. Additional commercial and development applications for acetate and triacetate polymers and fibers continue to be pursued. Cellulose acetate has been a beneficial membrane material in film or hollow fiber form (84–97). Water-soluble acetate polymers and fibers (98–101) can be produced by selecting a combined acetic value in the range of 20–27%. A technique for producing them directly, rather than by hydrolysis of the triacetate, was recently described (102,103). Techniques for improving the antistatic characteristics of acetate fiber have been evaluated (104–110). Other references describe work aimed at improved water and solvent resistance (111–114), abrasion resistance (115–118), grafting (119), selective adsorption (120), timed release of additives (121), microporous fiber (122), reinforcing fiber (123), absorbent (124), blends with polyester (125), photochromic fibers (126), irrigation system (127), odor adsorbing (128), blood filtration (129–131), immobilizing substrate (132–135), hygroscopic additive for polyester fibers (136), and makeup applicators (137) and offer potential for future applications.

BIBLIOGRAPHY

"Acetate Fibers" treated in *ECT* 1st ed., under "Rayon and Acetate Fibers," Vol. 11, pp. 552–569, by G. W. Seymour and B. S. Sprague, Celanese Corp. of America; "Acetate and Triacetate Fibers" in *ECT* 2nd ed., Vol. 1, pp. 109–138, by L. I. Horner and A. F. Tesi, Celanese Fibers Co., Division of Celanese Corp. of America; "Cellulose Acetate and Triacetate Fibers" in *ECT* 3rd ed., Vol. 5, pp. 89–117, by G. A. Serad and J. R. Sanders, Celanese Fibers Co.

1. T. Vickerstaff and E. Water, Jr., *J. Soc. Dyers Colour.* **58,** 116 (1942).
2. J. C. Guthrie, *J. Text. Inst.* **46**(6), T193 (1957).
3. R. Meredith, *J. Text. Inst.* **37,** T107 (1945).
4. G. Susich and S. Backer, *Text. Res. J.* **21,** 482 (1951).
5. W. Zurek, I. Sobieraj, and T. Trzesowska, *Text. Res. J.* **49**(8), 438 (1979).

6. E. R. Kaswell, *Textile Fibers, Yarns, and Fabrics,* Reinhold Publishing Corp., New York, 1953, p. 57.
7. D. K. Beever and L. Valentine, *J. Text. Inst.* **49,** T95 (1958).
8. *ASTM Standards, ASTM D1909-E1*, Vol. 17.01, American Society for Testing and Materials, Philadelphia, Pa., 1992, pp. 500–501.
9. G. A. Serad, in J. I. Kroschwitz, ed., *Encyclopedia of Polymer Science and Engineering,* Vol. 3, 2nd ed. Wiley-Interscience, New York, 1985, pp. 200–226.
10. A. Mellor and H. C. Olpin, *J. Soc. Dyers Colour.* **71,** 817 (1955).
11. K. Kohata, M. Miyagawa, A. Takaoka, and H. Kawai, *Proceedings of the International Symposium on Fiber Science and Technology, Hakone, Japan, August 24–25, 1985,* Society of Fiber Science and Technology, Tokyo, p. 270.
12. A. F. Tesi, *Am. Dyest. Rep.* **45,** 512 (1956).
13. F. Fortess, *Text. Res. J.* **19,** 23 (1949).
14. P. M. Heertjes, W. Colthof, and H. I. Waterman, *Rec. Trav. Chim.* **52,** 305 (1933).
15. C. J. Malm, C. R. Fordyce, and H. A. Tanner, *Ind. Eng. Chem.* **34,** 430 (1942).
16. *Effect of Conditioning Humidity on the Electrical Resistance of Rayon Yarns,* British Cotton Industry Research Association, London, 1945.
17. C. L. Bird and co-workers, *J. Soc. Dyers Colour.* **70,** 68 (1964).
18. R. K. Fourness, *J. Soc. Dyers Colour.* **72,** 513 (1956).
19. F. Fortess and V. S. Salvin, *Text. Res. J.* **28,** 1009 (1958).
20. F. Fortess, *Am. Dyest. Rep.* **44,** 524 (1955).
21. R. J. Mann, *J. Soc. Dyers Colour.* **76,** 665 (1960).
22. J. Boulton, *J. Soc. Dyers Colour.* **71,** 451 (1955).
23. L. G. Ray, Jr., *Text. Res. J.* **22,** 144 (1952).
24. L. Segal, in N. M. Bikales and L. Segal, eds., *Cellulose and Cellulose Derivatives,* High Polymers Series, Vol. V, Wiley-Interscience, New York, 1971, Chapt XVII-A.
25. G. D. Hiatt and W. J. Rebel in Ref. 24, Chapt. VI1-B.
26. C. L. Smart and C. N. Zellner in Ref. 24, Chapt. XIX-C.
27. W. B. Russo and G. A. Serad, in A. L. Turbak, ed., *Solvent Spun Rayon, Modified Cellulose Fibers and Derivatives,* ACS Symposium Series 58, American Chemical Society, Washington, D.C., 1977, p. 96.
28. C. J. Maim and G. D. Hiatt, in E. Ott, H. M. Spurlin, and M. W. Graffin, eds., *Cellulose and Cellulose Derivatives,* High Polymers Series, 2nd ed., Vol. V, Wiley-Interscience, New York, 1954, Pt. II.
29. U.S. Pat. 4,306,060 (Dec. 15, 1981), Y. Ikemoto (to Daicel Chemical Industries, Ltd.).
30. S. E. Doyle and R. A. Pethrick, *J. Appl. Polym. Sci.* **33**(1), 95–106 (1987).
31. B. Philipp, H. Schleicher, and W. Wagenknecht, *Chem. Tech. Leipzig* **7**(11), 79 (1977).
32. A. F. Turbak, R. B. Hammer, R. E. Davis, and H. L. Hergert, *Chem. Tech. Leipzig* **10**(1), 51 (1980).
33. S. M. Hudson and J. A. Cuculo, *J. Macromol. Sci. Chem.* **18**(1), 1 (1980).
34. U.S. Pat. 4,302,252 (Nov. 24, 1981), A. L. Turbak, A. El-Kafrawy, T. W. Snyder, Jr., and A. B. Auerbach (to International Telephone and Telegraph Corp.).
35. U.S. Pat. 4,324,593 (Apr. 13, 1982), J. K. Varga (to Akzona, Inc.).
36. C. L. McCormick and T. C. Chen, preprint, *American Chemical Society National Meeting,* Division of Organic Coatings and Plastics, New York, Aug. 1981.
37. Brit. Pat. Appl. GB 2,055,107 (Feb. 25, 1981), A. L. Turbak, A. El-Kafrawy, T. W. Snyder, Jr., and A. B. Auerbach (to International Telephone and Telegraph Corp.).
38. U.S. Pat. 4,265,675 (May 5, 1981), G. T. Tsno, B. E. Dale, and M. R. Ladisch (to Purdue Research Foundation).
39. A. Isogai, A. Ishizu, and J. Kanako, *Cellul. Chem. Technol.* **17,** 123 (1983).
40. W. B. Russo and G. A. Serad in Ref. 27, p. 115.

41. Brit. Pat. 2,105,725 (Mar. 30, 1983), H. Yabune and M. Uchida (to Daicel Chemical Industries Ltd.).
42. *Chem. Eng. News,* 6 (Jan. 14, 1980).
43. *Chem. Week,* 40 (Jan. 16, 1980).
44. U.S. Pat. 2,259,462 (Oct. 21, 1941), C. L. Fletcher (to Eastman Kodak Co.).
45. Brit. Pat. 566,863 (Feb. 26, 1945), H. Dreyfus (to Celanese Corp.).
46. U.S. Pat. 3,525,734 (Aug. 25, 1970), A. Rajon (to Societé Rhodiaceta).
47. *Daily News Record,* 7 (Jan. 10, 1986).
48. *Women's Wear Daily,* 38 (Jan. 22, 1986).
49. *Text. Marketing,* 5 (June 1986).
50. *Women's Wear Daily,* 6 (July 14, 1986).
51. U.S. Pat. 4,118,350 (Oct. 3, 1978), A. F. Turbak and co-workers (to International Telephone and Telegraph Corp.).
52. C. J. Maim and co-workers, *Ind. Eng. Chem.* **49,** 79 (1957).
53. U.S. Pat. 2,985,995 (May 30, 1961), W. W. Bunting, Jr., and T. L. Nelson (to E. I. du Pont de Nemours & Co., Inc.).
54. U.S. Pat. 3,110,151 (Nov. 12, 1963), W. W. Bunting, Jr., and T. L. Nelson (to E. I. du Pont de Nemours & Co., Inc.).
55. U.S. Pat. 3,364,537 (Jan. 23, 1968), W. W. Bunting, Jr., and T. L. Nelson (to E. I. du Pont de Nemours & Co., Inc.).
56. U.S. Pat. Appl. 656,359 (Feb. 9, 1976), M. Panar and O. Willcox (to E. I. du Pont de Nemours & Co., Inc.).
57. J. Takahashi, A. Kiyose, S. Nomura, and M. Kurijawa, in Ref. 11, pp. 20–24.
58. S. M. Aharoni, *Mol. Cryst. Liq. Cryst. Lett.* **56,** 237 (1980).
59. J. LeMatre, S. Dayan, and P. Sixou, *Mol. Cryst. Liq. Cryst.* **84,** 267 (1982).
60. P. Navard, J. M. Haudin, S. Dayan, and P. Sixou, *J. Polym. Sci. Part B* **19,** 379 (1981).
61. G. H. Meeten and P. Navard, *Polymer* **23,** 177 (1982).
62. S. Dayan, P. Maissa, M. J. Vellutini, and P. Sixou, *J. Polym. Sci. Part B* **20,** 33 (1982).
63. G. H. Meeten and P. Navard, *J. Polym. Sci. Part B* **24**(7), 815 (1983).
64. D. L. Patel and R. D. Gilbert, *J. Polym. Sci. Part A-2* **21,** 1079 (1983).
65. U.S. Pat. 4,501,886 (Feb. 26, 1985) John P. O'Brien, (to E. I. du Pont de Nemours & Co., Inc.)
66. U.S. Pat. 4,725,394 (Feb. 16, 1988) John P. O'Brien, (to E. I. du Pont de Nemours & Co., Inc.)
67. R. D. Gilbert, R. E. Fornes, and Y. K. Hong, *Tappi Nonwoven Conference,* Tappi, Norcross, Ga., 1986, pp. 231–236.
68. Y. K. Hong and co-workers, *Polymer Associated Structures: Microemulsions and Liquid Crystals,* ACS Symposium Series, 384, American Chemical Society, Washington, D.C., 1989, pp. 184–203.
69. U.S. Pat. 4,963,298 (Oct. 16, 1990) S. R. Allen, A. A. Mian, S. L. Samuels (to E.I. du Pont de Nemours & Co., Inc).
70. E. J. Roche and J. P. O'Brien in Ref. 11, p. 281.
71. U.S. Pat. 4,839,113 (June 13, 1989) P. Vallaine, C. Janin (to Michelin Recherche et Technique SA).
72. J. E. Smith, *Nonwovens World* **3**(6), 40–42 (1989).
73. Eur. Pat. 0,390,580 (Mar. 30, 1990) J. C. Hornsby (to E. I. du Pont de Nemours & Co., Inc.).
74. *Data of Fiber Economics Bureau,* as published in *Text. Organon,* 112–113, and 115 (June 1986); *Fiber Organon,* 138–139 and 144 (July 1992); and author information.
75. *Chem. Mark. Rep.,* 52 (July 23, 1990).
76. *Chem. Week* **150,** 13 (Feb. 12, 1992).
77. *Chem. Week* **150,** 30 (March 18, 1992).

78. *Chem. Week* **150,** 6 (April 8, 1992).
79. A. A. Lukoshaitis and co-workers, *Fibre Chem. USSR* **6,** 441 (1974); Y. A. Matskevichene and co-workers, *Fibre Chem. USSR* **6,** 446 (1974).
80. R. R. Rhinehart, *Symposium Papers, March 1975,* Miami Beach Meeting, International Nonwovens and Disposables Association, New York, p. 25.
81. Ger. Pat. 2,502,519 (July 31, 1975), C. K. Arisaka and co-workers (to Daicel, Ltd.).
82. B. Miller, J. R. Martin, and R. Turner, *J. Appl. Polym. Sci.* **28**(1), 45 (1983).
83. *Acetate Tow Production and Characterization,* Filter Products Division, Technical Bulletin FPB-4, Hoechst Celanese Corp., Charlotte, N.C., 1989.
84. G. Rakhmanberdiev and co-workers, *Zh. Prikl. Khim. Leningrad* **46**(2), 416 (1973).
85. U.S. Pat. 3,763,299 (Oct. 2, 1973), W. S. Stephen (to FMC Co.).
86. G. Rakhmanberdiev and co-workers, *Fibre Chem. USSR* **6,** 219 (1974).
87. Brit. Pat. 1,418,115 (Dec. 17, 1975), R. L. Leonard (to Monsanto Co.).
88. G. Tanny, *Appl. Polym. Symp.* **31,** 407 (1977).
89. U.S. Pat. 4,127,625 (Nov. 28, 1978), K. Arisaka, K. Watanabe, and K. Sasazima (to Daicel, Ltd.).
90. Brit. Pat. 2,001,899 (Feb. 14, 1979), M. Mishiro and S. Kasai (to Asahi).
91. Brit. Pat. 2,000,722 (Jan. 17, 1979), M. J. Kell and R. D. Mahoney (to Cordis Dow Corp.).
92. Brit. Pat. 2,002,679 (Feb. 28, 1979), M. Mishiro and K. Nakata (to Asahi).
93. Brit. Pat. 2,035,197 (June 18, 1980), K. Hamada, Z. Takade, and K. Numata (to Toyobo Co., Ltd.).
94. U.S. Pat. 4,219,517 (Aug. 26, 1980), R. E. Kesting (to Puropore, Inc.).
95. Brit. Pat. 2,065,546 (July 1, 1981), D. T. Chen and R. D. Mahoney (to Cordis Dow Corp.).
96. U.S. Pat. 4,276,173 (June 30, 1981), M. J. Kell and R. D. Mahoney (to Cordis Dow Corp.).
97. S. Kim, J. Cha, J. Kim, and U. Kim, *J. Membrane Sci.* **37**(2), 113–129 (1988).
98. U.S. Pat. 3,482,011 (Dec. 2, 1969), T. C. Bohrer (to Celanese Corp.).
99. G. Rakhmanberdiev, G. A. Petrotpsavlovskii, and K. U. Usmanov, *Cellul. Chem. Technol.* **12**(2), 153 (Mar.–Apr. 1978).
100. C. Rakhmanberdiev, K. Dustmukhamedov, M. Mukhamedzhanov, and K. U. Usmanov, *3rd Mezhdunar. Simpoz. PO Khim. Voloknam* **5,** 281 (1981).
101. A. K. Mukherjee, R. K. Agarwal, H. K. Chaturvedi, and B. D. Gupta, *Indian J. Text. Res.* **6,** 120 (1981).
102. C. M. Buchanan, K. J. Edgar, J. A. Hyatt, and A. K. Wilson, *Macromolecules* **24,** 3050–3059 (1991).
103. C. M. Buchanan, K. J. Edgar, and A. K. Wilson, *Macromolecules,* **24,** 3060–3064 (1991).
104. O. G. Pikovskaya and Z. G. Serebryakova, *Fibre Chem. USSR* **2,** 378 (1970).
105. F. A. Ismailov and co-workers, *Fibre Chem. USSR* **4,** 584 (1972).
106. P. A. Chakhoya and co-workers, *Fibre Chem. USSR* **5,** 184 (1973).
107. A. V. Kuchmenko, *Fibre Chem. USSR* **5,** 210 (1973).
108. Brit. Pat. 1,381,334 (May 31, 1973), W. Ueno, H. Kawaguchi, and N. Minagawa (to Fuji Photo Film Co., Ltd.).
109. A. P. Fedotov and co-workers, *Fibre Chem. USSR* **6,** 87 (1974).
110. M. N. Skarnlite and Y. Y. Shlyazhas, *Fibre Chem. USSR* **6,** 557 (1974).
111. *Text. World* **123**(9), 37 (1973).
112. U.S. Pat. 3,816,150 (June 11, 1974), K. Ishii and co-workers (to Daicel, Ltd.).
113. U.S. Pat. 3,839,617 (Oct. 1, 1974), A. F. Turbak and J. R. Thelman (to International Telephone and Telegraph Corp.-Rayonier).

114. U.S. Pat. 3,839,528 (Oct. 1, 1974), A. F. Turbak and J. R. Thelman (to International Telephone and Telegraph Corp.-Rayonier).
115. C. F. Kiseleva and co-workers, *Fibre Chem. USSR* **4,** 257 (1972).
116. M. Papikyan and co-workers, *Fibre Chem. USSR* **4,** 441 (1972).
117. M. Mirzaev and co-workers, *Khim. Volokna* **3**, 21 (1975).
118. Brit. Pat. 1,414,395 (Nov. 19, 1975), N. V. Mikhailov and co-workers (to USSR).
119. M. A. Siahkolah and W. K. Walsh, *Text. Res. J.* **44**(11), 895 (1974).
120. Ger. Pat. 2,507,551 (Feb. 28, 1975), B. V. Chandler and R. L. Johnson (to Commonwealth Scientific Organization).
121. U.S. Pat. 3,846,404 (Nov. 5, 1974), L. D. Nichols (to Moleculon Research Corp.).
122. Brit. Pat. 2,009,667 (June 20, 1979), R. Noguchi, S. Suzuki, Y. Meede, and R. Minami (to Mitsubishi Acetate Co.).
123. Brit. Pat. 2,040,956 (Sept. 3, 1980), D. William and P. Grey (to Wycombe Marsh Paper Mills, Ltd.).
124. U.S. Pat. 4,289,130 (Sept. 15, 1981), A. Usami, T. Uebayashi, K. Sato, and T. Goda (to Daicel Ltd.).
125. U.S. Pat. 4,770,931 (Sep. 13, 1988), M. A. Pollock, W. J. Stowell, J. J. Krutak (to Eastman Kodak Co.).
126. Eur. Pat. 0,328,320 (Feb. 3, 1989), P. Wright (to Courtaulds).
127. U.S. Pat. 4,928,427 (May 29, 1990) J. A. Patterson.
128. Brit. Pat. 2 229 364 A (Sept. 26, 1990) H. Nakel, S. Edagawa (to O.K. Trading Co. Ltd.).
129. U.S. Pat. 4,919,823 (Apr. 24, 1990), L. A. Wisdom (to Miles Inc.).
130. U.S. Pat. 4,767,541 (Aug. 30, 1988), L. A. Wisdom (to Miles Laboratories, Inc.).
131. U.S. Pat. 4,596,657 (Jun. 24, 1986), L. A. Wisdom (to Miles Laboratories, Inc.).
132. A. A. Sedov, *Fibre Chem. USSR* **9**(3), 281 (May–June 1977).
133. A. A. Sedov, *Fibre Chem. USSR* **9**(4), 367 (July–Aug. 1977).
134. A. V. Matveev, and co-workers, *Fibre Chem.* **21**(1), 29–33 (1989).
135. K. G. Raghavan, T. P. A. Devasagayam, and V. Ramakrishnan, *Analyt. Lett.* **19**(1&2), 163–176 (1986).
136. Jpn. Pat. 2,221,413 (Feb. 17, 1989), O. Kimihiro and Y. Hironori (to Teijin Ltd.).
137. Jpn. Pat. 1,119,203 (Nov. 2, 1987), N. Kenji.

GEORGE A. SERAD
Hoechst Celanese Corporation

ELASTOMERIC

Elastomeric fibers can be made from natural or synthetic polymeric materials that provide a product with high elongation, low modulus, and good recovery from stretching. Currently, these fibers are made primarily from polyisoprenes (natural rubber) or segmented polyurethanes and to a lesser extent from segmented polyesters. In the United States the generic designation spandex has been given to a manufactured fiber in which the fiber-forming substance is a long-chain synthetic polymer comprised of at least 85% of a segmented polyurethane (1); in Europe the equivalent term elastane is commonly used.

The experimental production of elastomeric fibers based on segmented polyurethanes was first reported in the early 1950s by Farbenfabriken Bayer, a pioneer in urethane and diisocyanate chemistry (2,3). This was followed by semicommercial-scale production of polyurethane-based fibers by the Du Pont Co., in the late 1950s (4,5). Prior to development of the polyurethanes, most elastomeric fibers were made with natural rubber. Two processes were used: slitting rubber sheets to produce cut rubber threads or extruding rubber latex into an acid coagulation bath followed by washing, drying, and curing. Smaller amounts of cut rubber threads have also been produced from synthetics such as neoprene and nitrile rubbers especially where improved solvent resistance is required. Fiber cross-sections are square or rectangular for cut rubbers and essentially round for extruded latex threads.

Thermoplastic, inelastic fibers, such as nylon and polyester, may be processed to provide spring-like, helical, or zigzag structures. These fibers can exhibit high elongations as the helical or zigzag structure is stretched, but the recovery force is very low. This apparent elasticity results from the geometric form of the filaments as opposed to elastomeric fibers whose elastic properties depend primarily on entropy changes inherent within their polymer structure. Thus processed inelastic fibers must comprise a significant portion of a stretch fabric whereas an elastomeric fiber provides the necessary stretch properties at 5–20% of fabric weight.

Other elastomeric-type fibers include the biconstituents, which usually combine a polyamide or polyester with a segmented polyurethane-based fiber. These two constituents are melt-extruded simultaneously through the same spinneret hole and may be arranged either side by side or in an eccentric sheath–core configuration. As these fibers are drawn, a differential shrinkage of the two components develops to produce a helical fiber configuration with elastic properties. An applied tensile force pulls out the helix and is resisted by the elastomeric component. Kanebo Ltd. has introduced a nylon–spandex sheath–core biconstituent fiber for hosiery with the trade name Sideria (6).

Nonspandex elastomeric fibers based on segmented polyesters and polyethers are currently being developed that can be melt-spun into threads (7). Teijin Ltd. produces an elastomeric fiber of this type with the trade name Rexe.

Mechanical Properties

In both rubber thread and spandex fibers, mechanical properties may be varied over a relatively broad range. In rubber, variations are made in the degree of cross-linking or vulcanization by changing the amount of vulcanizing agent, usually sulfur, and the accelerants used. In spandex fibers, many more possibilities for variation are available. By definition spandex fibers contain urethane linkages with the following repeat structure (1):

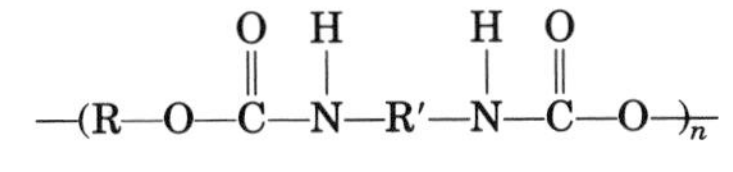

$$-\!\!\left(R-O-\overset{\displaystyle O}{\overset{\|}{C}}-\overset{\displaystyle H}{\overset{|}{N}}-R'-\overset{\displaystyle H}{\overset{|}{N}}-\overset{\displaystyle O}{\overset{\|}{C}}-O\right)_{\!n}$$

The number of polymers in the classification is obviously very large. Most urethane polymers (qv) in current use for the manufacture of spandex fibers are made by the reaction of 1000–4000 molecular weight hydroxy terminated polyethers or polyesters with a diisocyanate at a molar ratio of ca 1:1.4 to 1:2.5, followed by reaction of the resulting isocyanate-terminated prepolymer with one or more diamines to produce a high molecular weight urethane polymer. Small amounts of monofunctional amines may also be included to control polymer molecular weight. Mechanical properties may be affected by changing the particular polyester or polyether glycol, diisocyanate, diamine(s), and monoamine used; they can be further modified by changing the molecular weight of the glycol and by changing the glycol–diisocyanate molar ratio (8,9).

The long-chain urethane polymer molecules in spandex fibers are substantially linear block copolymers comprising relatively long blocks in which molecular interactions are weak, interconnected by shorter blocks in which interactions are strong. The weakly interacting blocks, commonly referred to as soft segments, are from the polyether or polyester glycol component whereas the blocks having strong interactions result from diisocyanate and chain extender reactions and are referred to as hard segments. The hard segments are usually aromatic–aliphatic ureas that connect with the soft segment through urethane linkages. With fiber formation, hard segments from several chains associate into strongly bonded cluster domains. These form islands of a discontinuous phase and convert the polymer to a three-dimensional network (10). The principal interchain forces are hydrogen bonds between NH groups and carbonyls, but crystallizability is also favored by the rigid and planar configuration of the aromatic rings. Interchain bonding must be not only strong enough to prevent molecular slippage, but also concentrated so that the connecting soft segments can comprise a large fraction of the polymer chain. This results in high stretch along with low modulus. Urea hard segments that comprise less than 25% of polymer mass provide this needed concentrated bonding force. In contrast, the network structure in rubber depends on covalent bonds between chain molecules that result from vulcanization with sulfur. In both polyurethanes and rubber, modulus is directly related to tie-point density. Similarly, the relationship for maximum elongation is an inverse function of tie-point density. In rubber fibers, tie-point density is controlled by the amount of vulcanizing agent, accelerant, and reaction conditions. In polyurethanes, tie-point (hard segment) density is controlled by the soft segment molecular weight and the molar ratio used to prepare the glycol–diisocyanate prepolymer.

The physical characteristics of current commercial rubber and spandex fibers are summarized in Table 1. Typical stress–strain curves for elastomeric fibers, hard fibers, and hard fibers with mechanical stretch properties are compared in Figure 1.

Manufacture

Cut Rubber. To produce cut rubber thread, smoked rubber sheet or crepe rubber is milled with vulcanizing agents, stabilizers, and pigments. This milled stock is calendered into sheets 0.3–1.3 mm thickness, depending on the final size of the rubber thread desired. Multiple sheets are layered, heat-treated to vulcanize, then slit into threads for textile uses (Fig. 2). Individual threads have either square or rectangular cross-sections.

Table 1. Physical Properties of Elastomeric Fibers

Property	Spandex	Extruded rubber	Cut rubber
sizes available[a]	1.1–250 tex[b]	16–610 tex[b]	2.5–21 μm dia[c]
tenacity, N/tex[d]	0.05–0.13	0.02–0.03	0.01–0.02
elongation, %	400–800	600–700	600–700
modulus[e], N/tex[d]	0.013–0.045	0.004–0.005	0.002–0.004
stability[f]			
uv light	good	fair	fair
ozone	good	poor	poor
NO_x	fair, yellows	poor	poor
active Cl	fair, yellows	poor	poor
body oils	fair	poor	poor
cosmetics	good	fair	fair
dyeability	dyeable	not dyeable	not dyeable
abrasion resistance	very good	poor	poor

[a]Spandex size is usually expressed in denier which is weight in g/9000 m length. However, the SI unit is tex, the weight in g/1000 m. Rubber size is expressed as gauge, which is the reciprocal of diameter or size in inches.
[b]To convert tex to den, multiply by 9.09.
[c]1,200–10,000 gauge.
[d]To convert N/tex to gf/den, multiply by 11.33.
[e]First cycle stress at 300% elongation.
[f]Both spandex fibers and rubber threads normally contain antioxidants and other stabilizers.

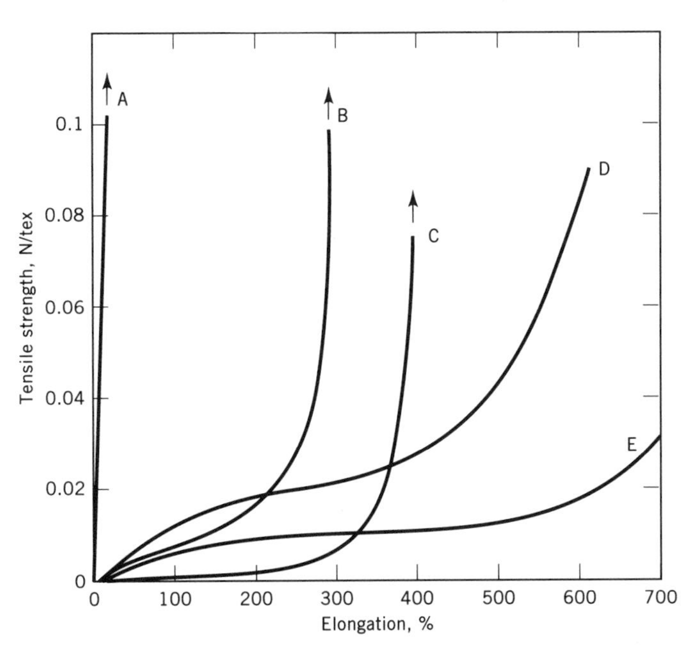

Fig. 1. Stress–strain curves: A, hard fiber, eg, nylon; B, biconstituent nylon–spandex fiber; C, mechanical stretch nylon; D, spandex fiber; E, extruded latex thread. To convert N/tex to gf/den, multiply by 11.33.

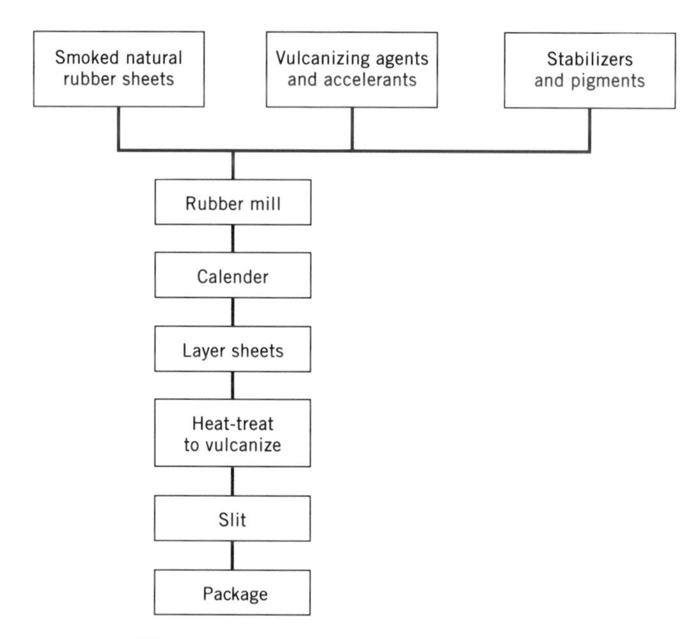

Fig. 2. Cut rubber thread manufacture.

Extruded Latex Thread. In the manufacture of extruded latex thread, a concentrated (up to ca 50% solids) natural rubber latex is blended with aqueous dispersions of vulcanizing agents, stabilizers, and white pigments. This compounded latex is held under controlled temperature conditions until partial vulcanization occurs. This has the effect of increasing wet strength and thus the processibility of the extruded threads. The matured latex is extruded at constant pressure through precision-bore glass capillaries into a 15–55% acetic acid [*64-19-7*] bath where coagulation into thread form occurs. Threads are removed from the coagulation bath by transfer rollers, washed free of excess acid with water, and conducted through a dryer, after which a silicone oil-based finish is applied and the threads are formed into multiend ribbons. The ribbons are then vulcanized by multiple passes on a conveyer belt through an oven that can increase curing temperature in stages up to about 150°C. After vulcanization the multiend ribbons are packed without support in boxes for shipment to the customer. A typical extruded latex thread production line is shown in Figure 3. Latex thread production rates vary with thread size and equipment but, owing to hydrodynamic drag and the weak nature of the coagulating thread, maximum line take-up speeds are about 30 m/min.

Spandex Fibers. Four different processes are currently used to produce spandex fibers commercially: melt extrusion, reaction spinning, solution dry spinning, and solution wet spinning. As shown in Figure 4, these processes involve different practical applications of basically similar chemistry. An isocyanate terminated prepolymer is formed by the reaction of a 1000–4000 molecular weight macroglycol with a diisocyanate at a glycol–diisocyanate ratio that may range from 1:1.4 up to about 1:2.5. The soft segment macroglycol can be either a poly-

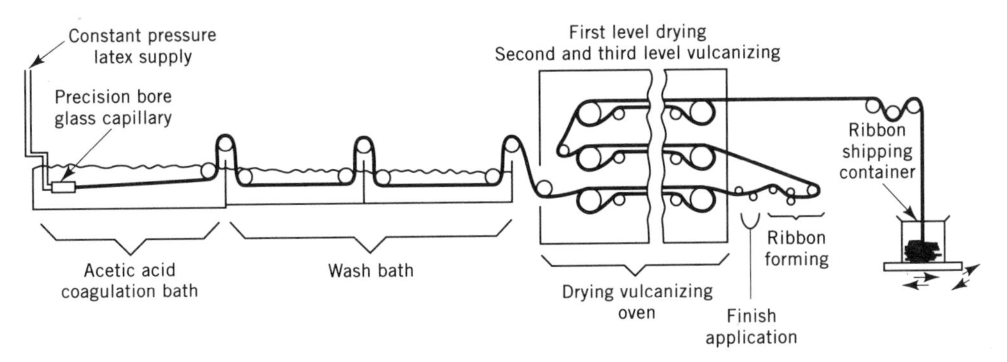

Fig. 3. Extruded latex thread production.

ether, a polyester, a polycarbonate, hydroxyl-terminated polycaprolactone, or a combination of these. The prepolymer subsequently reacts with either a glycol or diamine(s) at near stoichiometry; a small amount of monofunctional amine may be included to control final polymer molecular weight. If the diol or diamine(s) reaction with the prepolymer is carried out in a solvent, the resulting block copolymer solution may be wet or dry spun into fiber. Alternatively, the prepolymer may be reaction spun by extrusion into a bath containing diamine to form a fiber,

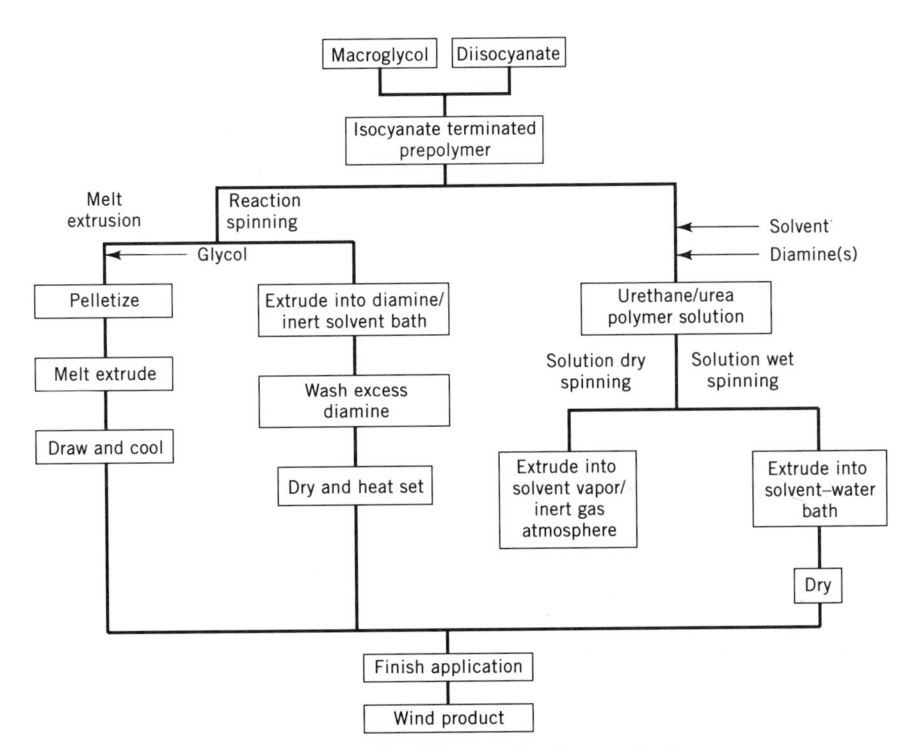

Fig. 4. Spandex fiber production methods.

or the prepolymer may be permitted to react in bulk with a diol and the resulting polymer melt extruded in fiber form.

Melt Spinning. Currently, the principal producer of melt-spun spandex fibers is Nisshin Spinning Co. in Japan. However, melt-spun spandex fibers are also produced by Kanebo and Kururay Co., also in Japan. Because of thermal stability constraints, only polymers that contain all urethane hard segments (glycol extended) can be melt-extruded. The intermolecular association between all urethane hard segments is inherently weaker compared with urea-based hard segments produced from diamine extenders; melt-spun fibers are normally made at higher diisocyanate–glycol ratios which in effect produce a relatively longer hard segment to compensate for the weaker intermolecular bonding forces.

More recently, melt-spun biconstituent sheath–core elastic fibers have been commercialized. They normally consist of a hard fiber sheath (polyamide or polyester) along with a segmented polyurethane core polymer (11,12). Kanebo Ltd. in Japan currently produces a biconstituent fiber for hosiery end uses called Sideria.

Reaction Spinning. Several commercial spandex fibers were produced by reaction spinning in the 1950s. However, only one producer, Globe Manufacturing Co., currently uses reaction spinning techniques to produce spandex fibers Glospan and its unpigmented counterpart, Cleerspan (13). Reaction-spun fibers include only products 7.7 tex (70 den) or higher; finer Glospan spandex fibers are dry spun.

To produce a spandex fiber by reaction spinning, a 1000–3500 molecular weight polyester or polyether glycol reacts with a diisocyanate at a molar ratio of about 1:2. The viscosity of this isocyanate-terminated prepolymer may be adjusted by adding small amounts of an inert solvent, and then extruded into a coagulating bath that contains a diamine so that filament and polymer formation occur simultaneously. Reactions are completed as the filaments are cured and solvent evaporated on a belt dryer. After application of a finish, the fibers are wound on tubes or bobbins and rewound if necessary to reduce interfiber cohesion.

Trifunctional hydroxy compounds, eg, glycerol [*56-81-5*] or 2-ethyl-2-(hydroxymethyl)-1,3-propanediol [*787-99-6*], may be added with the macroglycol to produce covalent cross-links in the reaction-spun spandex fiber. Also, covalent cross-links may result from allophanate and/or biuret formation during curing by reaction of free isocyanate end groups with urethane or urea–NH groups along the polymer chain. A multiplicity of filaments are normally extruded from each spinneret of about 1.1–3.3 tex (10–30 den), then collected in bundles of the desired tex at the exit of the reaction bath. This approach makes the surface area to mass ratio and diamine diffusion into the prepolymer cross-section substantially constant irrespective of the final tex produced, thus minimizing condition changes required in changing tex. Because the individual filaments have reacted incompletely and are in a semiplastic state at the exit of the diamine bath, they interbond quite tightly into a fused multifilament. Production speeds in reaction spinning are limited by filament weakness in the bath along with hydrodynamic drag. Take-up speeds are limited to about 100 m/min.

Stabilizers and pigments are normally slurried with macroglycol and added to the polymeric glycol charge, prior to diisocyanate addition. Therefore, care must be taken to avoid additives that react significantly with diisocyanates or diamines under processing conditions. Also, stabilizers should be chosen that have no adverse catalytic effect on the prepolymer or chain-extension reactions.

Reaction spinning equipment is quite similar to that of solution wet spinning. It differs principally in the use of fewer wash baths and in the use of belt-type dryers instead of heated cans.

Solution Spinning. The initial step to prepare polyurethane polymers for wet or dry solution spinning includes reaction of 1000–3500 molecular weight macroglycol with a diisocyanate at molar ratios of between about 1:1.4 and 1:2.0. Reaction conditions must be carefully selected and controlled to minimize side reactions, eg, allophanate and biuret formation, which can result in trifunctional branched chains and ultimately to insoluble cross-linked polymers. For the prepolymer reaction, poly(tetramethylene ether) glycol [*25190-06-1*] (PTMEG) and bis(4-isocyanatophenyl) methane [*101-68-8*] (MDI) are currently the most commonly used macroglycol and diisocyanate. Several types of polyester-based macroglycols are included in spandex producers product lines, but with the exception of Dorlastan, made by Bayer AG in Germany, the polyester-based products represent only a minor part of their spandex fiber production.

In the polymerization reaction, called chain extension, the prepolymer is dissolved in a solvent and reacts with diamine(s) to form a urethane–urea polymer in solution. In all commercial processes, the solvent used is either *N,N*-dimethylformamide [*68-12-2*] (DMF) or *N,N*-dimethylacetamide [*127-19-5*] (DMAc). Normally, one or two diamines are used as chain extenders. Because the reaction of diamine with diisocyanate is exceedingly rapid, prepolymer is normally diluted with solvent so that mixing in the polymer reactor is optimized. An improved mixing and chain-extension process has been described whereby the solvent-diluted prepolymer is separated into two groups with chain extension being initiated by reaction of diamine solution with one of these groups, after which the prepolymer of the other group is mixed in and chain extension is completed (14). Molecular weights of polymers made in solution can be controlled by adding small amounts of a secondary monoamine to provide dialkylurea end groups. Branching reactions must be minimized in order to obtain a stable polymer solution for spinning. Stoichiometry of polymerization is normally adjusted to provide a urethane polymer solution of 20–40% solids and viscosity of 20–200 Pa·s (200–2000 P). The viscosities and solids of solutions for dry spinning are generally higher than those used for wet spinning.

Stabilizers, pigments, and other additives are milled in spinning solvent, normally along with small amounts of the urethane polymer to improve dispersion stability; this dispersion is then blended to the desired concentration with polymer solution after chain extension. Most producers combine prepolymerization, chain extension, and additive addition and blending into a single integrated continuous production line.

Dry Spinning. On a worldwide basis, about 90% of all spandex fibers are produced by various adaptations of dry spinning (15,16). The solution dry spinning process is illustrated in Figure 5. The polymer spinning solution is metered at a constant temperature by a precision gear pump through a spinneret into a cylindrical spinning cell 3–8 m in length. Heated cell gas, made up of solvent vapor and an inert gas, normally nitrogen, is introduced at the top of the cell and

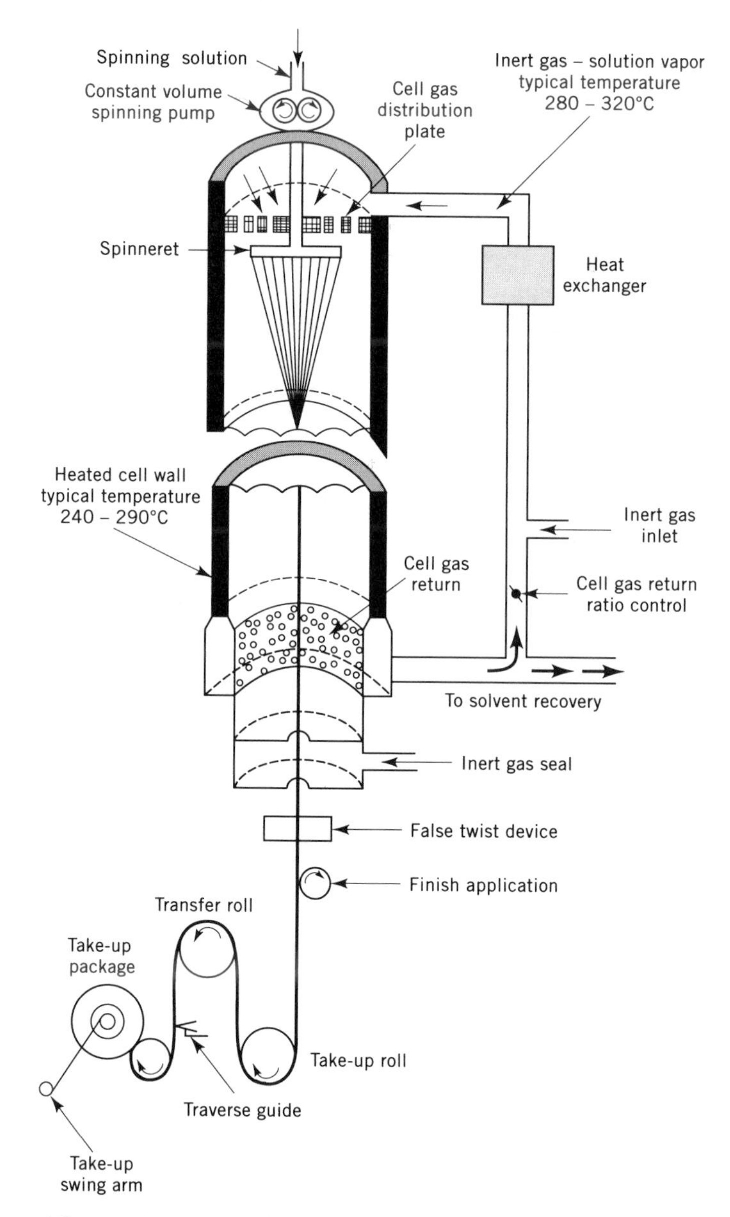

Fig. 5. Spandex production, solution dry spinning process.

passed through a distribution plate behind the spinneret pack. Because both cell gas and cell walls are maintained at high temperatures, solvent evaporates rapidly from the filaments as they travel down the spinning cell. The spinning solvent is then condensed from the cell gases, purified by distillation, and returned for reuse. Individual filament size is normally maintained in the range of 0.6–1.7 tex (5–15 den) to maximize, within operable limits, surface-to-mass ratio and solvent removal rate. Individual filaments are grouped into bundles of the desired final tex at the exit of the spinning cell by a coalescence guide. A commonly used guide employs compressed air to create a minivortex which imparts a false twist and rounded cross section to the filament bundle. Solution dry-spun spandex fibers are normally referred to as continuous multifilaments or coalesced multifilaments. However, the individual filaments do not coalesce into larger structures but remain discrete; they adhere to one another because of natural elastomer tack at their surface.

After coalescence, a finish is applied to the multifilament bundle before it is wound onto a tube. Commonly used finishing agents include poly(dimethylsiloxane) [*9016-00-6*] (17) and magnesium stearate [*557-04-0*] (18) which provide lubrication for textile processing and prevent fibers from sticking together on the package. Windup speeds are in the range of 300–500 m/min depending on tex and producer.

Wet Spinning. Any urethane–urea polymer that can be dry spun may also be wet spun; however, the productivity constraints of wet spun processes have limited their utility. A typical wet spinning process line is shown in Figure 6. Spinning solution is pumped by precision gear pumps through spinnerets into a solvent–water coagulation bath. As with dry spinning, individual filament size is maintained at about 0.6–1.7 tex (5–15 den) in order to optimize solvent removal rates. At the exit of the coagulation bath, filaments are collected in bundles of the desired tex. A false twist may be imposed at the bath exit to give the multifilament bundles a more rounded cross section. After the coagulation bath, the multifilament bundles are countercurrently washed in successive extraction baths to remove residual solvent, then dried and heat-relaxed, generally on heated cans. Finally, as in dry spinning, a finish is applied and the multifilaments wound on individual tubes. A typical spinning line may produce 100–300 multifilaments at side-by-side filament spacings of less than 5 mm.

Water is continuously added to the last extraction bath and flows countercurrently to filament travel from bath to bath. Maximum solvent concentration of 15–30% is reached in the coagulation bath and maintained constant by continuously removing the solvent–water mixture for solvent recovery. Spinning solvent is generally recovered by a two-stage process in which the excess water is initially removed by distillation followed by transfer of crude solvent to a second column where it is distilled and transferred for reuse in polymer manufacture.

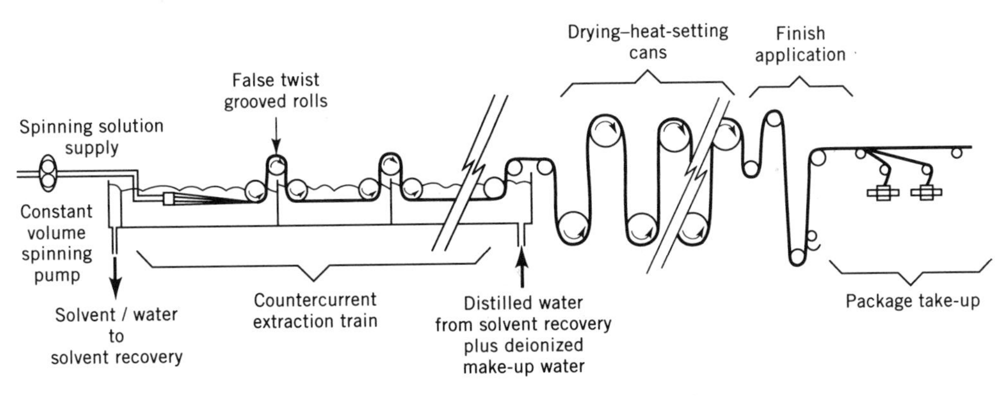

Fig. 6. Spandex production, solution wet spinning process.

In wet spinning processes, spinning speeds are limited to about 100–150 m/min by hydrodynamic drag of the bath medium. It is this limitation that has apparently caused most spandex fiber producers to have chosen dry spinning techniques. However, this limitation has been minimized by subjecting the spandex filament to drawing as much as three to four times after the spinning bath (19,20). Temperatures and residence times are selected so that the filaments are brought to temperatures above their second-order transition points, ie, the hard segment melting points. This allows the molecular chains to move freely to relieve stresses and results in filaments of fine tex but with similar mechanical properties as the heavier tex feed. Thus it is possible to windup fibers from a wet spinning process at speeds in excess of 300 m/min by continuously drawing and heat-relaxing the filaments after drying.

Chemical Properties

Stabilization. Both rubber and spandex fibers are subject to oxidative attack by heat, light, atmospheric contaminants such as NO_x, and active chlorine. Rubber is especially subject to oxidative degradation from exposure to ozone whereas urethane polymers are relatively inert. Both rubber and spandex fibers are likely to contain antioxidants; the spandex fibers may also be stabilized to uv light and to atmospheric contaminants that cause discoloration. Spandex fibers use a variety of monomeric and polymeric hindered phenolic-type antioxidants (qv). Many spandex fibers also include uv screeners based on hydroxybenzotriazoles. Several producers include the more recently developed hindered amine-type light stabilizers (21) that apparently act as radical scavengers and therefore also possess antioxidant activity. Compounds with tertiary amine functionality are commonly added to spandex fibers to inhibit discoloration from atmospheric pollutants such as oxides of nitrogen and chemicals that develop under smog-like conditions. Spandex producers have been designing stabilizers that are highly compatible with the segmented urethane polymer to enhance their effectiveness and durability (22–24). Spandex fibers based on polyester soft segments are susceptible to mildew attack in end uses such as swimwear; this type of degradation is minimized by either using antimildew additives or by soft segment structural modifications (25).

Solvent Resistance. Elastomeric fibers tend to swell in certain organic solvents; rubber fibers swell in hydrocarbon solvents such as hexane. Spandex fibers become highly swollen in chlorinated solvents such as tetrachloroethylene [*127-18-4*] (Perclene). Although the physical properties of spandex fibers return to normal after the solvent evaporates, considerable amounts of its stabilizers may have been extracted. Therefore, the development of stabilizers that are more resistant to solvent extraction has become important as solvent scouring during mill processing replaces aqueous scouring at many mills, especially in Europe (26).

Dyeing. Spandex fibers have an affinity for dispersed or acid dyes; rubber fibers normally cannot be dyed. Perfect dye matches between spandex and hard fibers are usually not necessary because the elastomer is well hidden in the fabric. Clear spandex fibers can be left undyed when plied with dyed hard fiber yarns, thus avoiding loss of stretch properties from conditions of dyeing or bleaching.

Nylon–spandex combinations are often dyed with disperse dyes, or, for better fastness, with acid dyes (27). Retarders may be needed to prevent the nylon from depleting the acid dye from the bath before the spandex fiber is dyed. With polyester–spandex fabrics, disperse dyes have been used with pressure dyeing or carriers to increase the dyeing rate for the hard fiber (see DYE CARRIERS). However, spandex fibers exhibit relatively poor wetfastness to disperse dyes, and their retractive power can be reduced under pressure dyeing conditions needed for full shades on disperse dyed polyesters. For this reason, many polyester spandex fabrics now contain cationic dyeable polyester in combination with clear (transparent) spandex fiber. Since the spandex fibers have low affinity for cationic dyes, fastness is not a problem, and the fabrics can be dyed at lower temperatures to preserve spandex retractive power (see DYES, APPLICATION AND EVALUATION).

Economic Aspects

A worldwide list of spandex fiber and related elastomer producers is shown in Table 2. Most process developments have occurred in the United States, Germany, Japan, and Korea. A large proportion of worldwide capacity is controlled by Du Pont, either directly or through subsidiaries and joint ventures. These include three plants in North America, two in South America, two in Europe, and two in Asia.

Commercially, elastomeric fibers are almost always used in combination with hard fibers such as nylon, polyester, or cotton. Use levels vary from a low of about 3% in some filling stretch cotton fabrics to a high of about 40% in some warp-knit tricot fabrics. Raschel fabrics used in foundation garments normally contain 10–20% spandex fiber.

Prices of spandex fibers are highly dependent on thread size; selling price generally increases as fiber tex decreases. Factors that contribute to the relatively high cost of spandex fibers include (*1*) the relatively high cost of raw materials, (*2*) the small size of the spandex market compared to that of hard fibers which limits scale and thus efficiency of production units, and (*3*) the technical problems associated with stretch fibers that limit productivity rates and conversion efficiencies.

Uses

Cut Rubber and Extruded Latex. The manufacturing technology for cut and extruded rubber thread is much older and more widely known than that for spandex fibers. Because production facilities can be installed with relatively modest capital investment, manufacture of rubber thread is fragmented and more widely distributed with a few major and many minor producers. On a worldwide basis, Fillattice of Italy is the largest rubber thread producer with modern extruded latex plants in Italy, Spain, Malaysia, and the United States. Second in production capacity is the Globe Manufacturing Co., Fall River, Massachusettes with production operations in the United States and the UK. These firms also produce spandex fibers.

Table 2. Producers of Spandex and Related Fibers

Company	Process	Trade name
	North America	
Canada		
Du Pont	dry-spun	Lycra
Mexico		
Du Pont/Nylmex	dry-spun	Likra
United States		
Du Pont	dry-spun	Lycra
Globe Manufacturing	reaction-spun and dry-spun	Glospan, Cleerspan
	South America	
Argentina		
Du Pont	dry-spun	Lycra
Brazil		
Du Pont	dry-spun	Lycra
Venezuela		
Spandhaven	wet-spun	Gomelast
	Europe	
Germany		
Bayer AG	dry-spun	Dorlastan
Italy		
Fillattice	wet-spun	Lyneltex
the Netherlands		
Du Pont	dry-spun	Lycra
Russia		
(licensed from Toyobo)	dry-spun	VCE
United Kingdom		
Du Pont	dry-spun	Lycra
	Asia	
Japan		
Asahi Chemical Industry	dry-spun	Roica
Du Pont/Toray	dry-spun	Opelon
Fuji Spinning	wet-spun	Fujibo
Kenbo	melt-extruded	Loobell
	melt-extruded	Sideria[a]
Kuraray	melt-extruded	Spantel
Nisshinbo	melt-extruded	Mobilon
Teijin	melt-extruded	Rexe[b]
Toyobo	dry-spun	Espa
Korea		
Dong Kook	wet-spun	Texlon
Tae Kwang	dry-spun	Acelan
People's Republic of China		
(licensed from Toyobo)	dry-spun	Yantai
Singapore		
Du Pont	dry-spun	Lycra

[a]Nylon–spandex sheath–core biconstituent fiber.

[b]Segmented polyester elastomeric fiber.

Most extruded latex fibers are double covered with hard yarns in order to overcome deficiencies of the bare threads such as abrasiveness, color, low power, and lack of dyeability. During covering, the elastic thread is wrapped under stretch which prevents its return to original length when the stretch force is removed; thus the fiber operates farther on the stress–strain curve to take advantage of its higher elastic power. Covered rubber fibers are commonly found in narrow fabrics, braids, surgical hosiery, and strip lace.

Spandex Fibers. Spandex fibers are supplied for processing into fabrics in four basic forms as outlined in Table 3. Bare yarns are supplied by the manufacturer on tubes or beams and can be processed on conventional textile equipment with the aid of special feed and tension devices. In covered yarns, the spandex fibers are covered with one or two layers of an inelastic filament or staple yarn; the hard yarn provides strength and rigidity at full extension, which facilitates knitting and weaving.

Table 3. Spandex Fiber Uses

Fiber form	Fabric types	Uses
bare	warp knits, circular knits, narrow fabrics (woven, knits, and braids), and hosiery (knit)	foundation garments, swimwear, control tops for pantyhose, brassieres, elastic gloves, waist and leg bands, sportswear, and upholstery
covered	warp knits, circular knits, hosiery (knit), and narrow fabrics	hosiery, elastic bandages, sportswear, upholstery, and sock tops
core-spun	wovens, circular knits, and men's hosiery (knit)	shirting, slacks, and sportswear
core-plied	wovens	blouses and trousers

With core-spun yarns, the spandex fibers are stretched and combined with a roving of inelastic cotton or cotton–polyester staple fibers; twisting action of the hard sheath fibers around the elastomeric fiber core produces a spun yarn that contracts when tension is relieved. Woven fabrics with core-spun yarns generally contain small amounts of spandex fibers; stretch characteristics are the result of the intrinsic properties of the spandex fibers in their final form and to interactions with hard fibers which take place during weaving and heat setting (28). This interaction provides a permanent weave crimp in the hard fiber which together with the spandex component imparts stretch and recovery to the fabric. Core-plied yarns are formed when stretched spandex fibers are plied with extended textured continuous filament yarns on a twisting machine; these yarns are used in high stretch woven fabrics.

Spandex fibers are available as fine as 1.1 tex (10 den), and the finest extruded latex thread available is about 16 tex (140 den). The availability of spandex fibers in such fine sizes and their unique properties compared to rubber, eg, dyeability, high modulus, abrasion resistance, and whiteness, has allowed extensive penetration into hosiery and sportswear markets.

BIBLIOGRAPHY

"Spandex" in *ECT* 2nd ed., Vol. 18, pp. 614–633, by R. A. Gregg, Uniroyal, Inc.; "Fibers, Elastomeric" in *ECT* 3rd ed., Vol. 10, pp. 166–181, by T. V. Peters, Consultant.

1. *Textile Fibers Products Identification Act, U.S. Public Law 85-897*, U.S. Federal Trade Commission, Washington, D.C., effective Mar. 3, 1960.
2. Ger. Pat. 826,641 (1952), E. Windemuth (to Farbenfabriken Bayer).
3. Ger. Pat. 886,766 (1951), W. Brenschede (to Farbenfabriken Bayer).
4. U.S. Pat. 2,957,852 (Oct. 25, 1960), P. Frankenburg and A. Frazer (to Du Pont Co.).
5. U.S. Pat, 2,929,804 (Mar. 22, 1960), W. Steuber (to Du Pont Co.).
6. Jpn. Appl. 63 189132 (1988), I. Matsuya (to Kanebo Ltd.).
7. G. Richeson and J. Spruiell, *J. Appl. Polym. Sci.* **41**, 845 (1990).
8. D. Allport and A. Mohajer, in D. Allport and W. H. Janes, eds., *Block Copolymers*, John Wiley & Sons, Inc., New York, 1973, pp. 443–492.
9. R. Bonart, *Angew. Makromol. Chem.* **58–59**(1), 259 (1977).
10. H. Lee, Y. Wang, and S. Cooper, *Macromolecules* **20**, 2089 (1987).
11. Jpn. Pat. Appl. 138, 124 (1975), T. Hidaka, K. Ikawa, and S. Mizutani (to Toray Industries, Inc.).
12. Jpn. Pat. Appl. 63-256719 (1988), S. Tanaka, Y. Yamakawa, and K. Hirasa to (Kanebo Ltd.)
13. U.S. Pat. 3,387,071 (June 4, 1968), J. Cahill, J. Powell, and E. Gartner (to Globe Mfg. Co.).
14. Jpn. Appl. 63 231244 (1988), K. Tani, K. Katsuo, and H. Tagata (to Toyobo Co. Ltd.).
15. H. Ishihara and co-workers, *J. Polym. Eng.* **6**, 237 (1986).
16. T. Kotani and co-workers, *J. Macromol. Sci-Phys.* **831**, 65 (1992).
17. U.S. Pat. 3,296,063 (Jan. 3, 1967), C. Chandler (to Du Pont Co.).
18. U.S. Pat, 3,039,895 (June 19, 1962), J. Yuk (to Du Pont Co.).
19. U.S. Pat, 4,002,711 (Jan. 11, 1977), T. Peters.
20. Jpn. Pat. Appl. 76-04,313 (1976), Y. Ikeda, T. Hirukawa, and Y. Ishiki (to Fuji Spinning Co., Ltd.).
21. H. Miller, *Polymer Preprints, ACS* **25**(1), 21 (1984).
22. Jpn. Pat. Appl. 62-86,047 (1987), H. Hanabatake and A. Kitsuri (to Asahi Chemical Industries).
23. U.S. Pat. 5,028,642 (July 2, 1991), C. Goodrich and W. Evans (to Du Pont Co.).
24. U.S. Pat. 4,824,929 (Ap. 25, 1989), G. Arimatsu and co-workers (to Toyobo Co., Ltd.).
25. Ger. Pat. 3,641,703 (1988), M. Kausch and co-workers (to Farbenfabriken Bayer).
26. Jpn. Pat. Appl. 61-218,659 (1986), Y. Fujimoto, S. Gotou and Y. Fujita (to Asahi Chemical Industries).
27. C. Pernetti, *Tinctoria* **80**, 133 (1983).
28. S. Ibrahim, Ph.D. dissertation, University of Leeds, UK, 1969.

General References

A. J. Ultee, in J. I. Kroschwitz ed., *Encyclopedia of Polymer Science and Engineering*, 2nd ed., Vol. 6, John Wiley & Sons, Inc., New York, 1986, pp. 733–755.

M. Couper, in M. Lewin and J. Preston, eds., *High Technology Fibers*, Marcel Dekker, Inc., New York, 1985, pp. 51–85.

M. Joseph, *Essentials of Textiles*, 3rd ed., Holt, Rinehard and Wilson, New York, 1984.

John E. Boliek
Arnold W. Jensen
E. I. du Pont de Nemours & Co., Inc.

OLEFIN

Olefin fibers, also called polyolefin fibers, are defined as manufactured fibers in which the fiber-forming substance is a synthetic polymer of at least 85 wt % ethylene, propylene, or other olefin units (1). Several olefin polymers are capable of forming fibers, but only polypropylene [*9003-07-0*] (PP) and, to a much lesser extent, polyethylene [*9002-88-4*] (PE) are of practical importance. Olefin polymers are hydrophobic and resistant to most solvents. These properties impart resistance to staining, but cause the polymers to be essentially undyeable in an unmodified form.

The first commercial application of olefin fibers was for automobile seat covers in the late 1940s. These fibers, made from low density polyethylene (LDPE) by melt extrusion, were not very successful. They lacked dimensional stability, abrasion resistance, resilience, and light stability. The success of olefin fibers began when high density polyethylene (HDPE) was introduced in the late 1950s. Yarns made from this highly crystalline, linear polyethylene have higher tenacity than yarns made from the less crystalline, branched form (LDPE) (see OLEFIN POLYMERS). Markets were developed for HDPE fiber in marine rope where water resistance and buoyancy are important. However, the fibers also possess a low melting point, lack resilience, and have poor light stability. These traits caused the polyethylene fibers to have limited applications.

Isotactic polypropylene, based on the stereospecific polymerization catalysts discovered by Ziegler and Natta, was introduced commercially in the United States in 1957. Commercial polypropylene fibers followed in 1961. The first market of significance, contract carpet, was based on a three-ply, crimper-textured yarn. It competed favorably against wool and rayon–wool blends because of its lighter weight, longer wear, and lower cost. In the mid-1960s, the discovery of improved light stabilizers led to the development of outdoor carpeting based on polypropylene. In 1967, woven carpet backing based on a film warp and fine-filament fill was produced. In the early 1970s, a bulked-continuous-filament (BCF) yarn was introduced for woven, texturized upholstery. In the mid-1970s, further improvement in light stabilization of polypropylene led to a staple product for automotive interiors and nonwoven velours for floor and wall carpet tiles. In the early 1980s, polypropylene was introduced as a fine-filament staple for thermal bonded nonwovens.

The growth of polyolefin fibers continues. Advances in olefin polymerization provide a wide range of polymer properties to the fiber producer. Inroads into new markets are being made through improvements in stabilization, and new and improved methods of extrusion and production, including multicomponent extrusion and spunbonded and meltblown nonwovens.

Properties

Physical Properties. Table 1 (2) shows that olefin fibers differ from other synthetic fibers in two important respects: (*1*) olefin fibers have very low moisture absorption and thus excellent stain resistance and almost equal wet and dry properties, and (*2*) the low density of olefin fibers allows a much lighter weight product

Table 1. Physical Properties of Commercial Fibers[a]

Polymer	Standard tenacity, GPa[b]	Breaking elongation, %	Modulus, GPa[b]	Density, kg/m^3	Moisture regain[c]
olefin	0.16–0.44	20–200	0.24–3.22	910	0.01
polyester	0.37–0.73	13–40	2.1–3.7	1380	0.4
carbon	3.1	1	227	1730	
nylon	0.23–0.60	25–65	0.5–2.4	1130	4–5
rayon	0.25–0.42	8–30	0.8–5.3	1500	11–13
acetate	0.14–0.16	25–45	0.41–0.64	1320	6
acrylic	0.22–0.27	35–55	0.51–1.02	1160	1.5
glass	4.6	5.3–5.7	89	2490	
aramid	2.8	2.5–4.0	113	1440	4.5–7
fluorocarbon	0.18–0.74	5–140	0.18–1.48	2100	
polybenzimidazole	0.33–0.38	25–30	1.14–1.52	1430	15

[a]Ref. 2.
[b]To convert GPa to psi, multiply by 145,000.
[c]At 21°C and 65% rh.

at a specified size or coverage. Thus one kilogram of polypropylene fiber can produce a fabric, carpet, etc, with much more fiber per unit area than a kilogram of most other fibers.

Tensile Strength. Tensile properties of all polymers are a function of molecular weight, morphology, and testing conditions. The effect of temperature on the tensile properties of a typical polypropylene fiber is shown in Figure 1 (3). Tensile properties are also affected by strain rate, as shown in Figure 2 (3). Lower temperature and higher strain rate result in higher breaking stresses at lower elongations, consistent with the general viscoelastic behavior of polymeric mate-

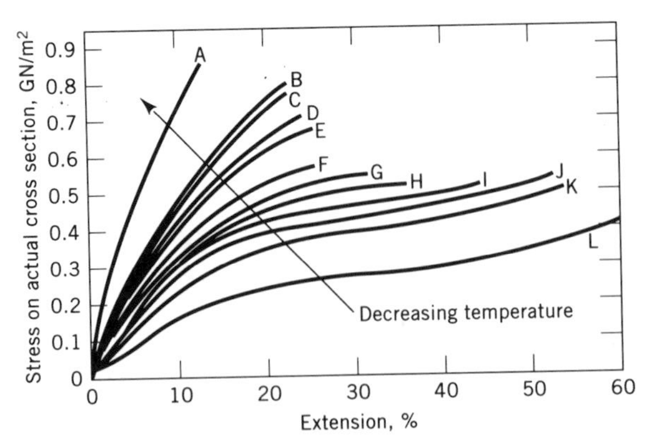

Fig. 1. Effect of temperature on tensile properties of polypropylene (3); strain rate = $6.47 \times 10^{-4}\ s^{-1}$. In degrees Kelvin: A, 90; B, 200; C, 213; D, 227; E, 243; F, 257; G, 266; H, 273; I, 278; J, 283; K, 293; and L, 308 (broken at 74.8% extension, $4.84 \times 10^8\ N/m^2$ stress). To convert GN/m^2 to $dyne/cm^2$, multiply by 10^{10}.

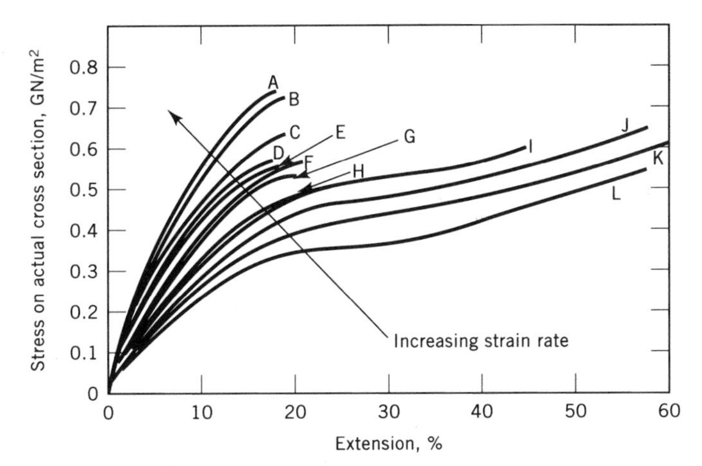

Fig. 2. Effect of strain rate on tensile properties of polypropylene at 20°C (3). In s^{-1}: A, 4.9×10^2; B, 2.48×10^2; C, 1.26×10^2; D, 6.3×10^1; E, 3.2×10^1; F, 2.87×10^1; G, 2.3×10^{-1}; H, 3.3×10^{-2}; I, 1.33×10^{-2}; J, 4.17×10^{-3}; K, 1.67×10^{-3}; and L, 3.3×10^{-4}. To convert GN/m^2 to dyne/cm^2, multiply by 10^{10}.

rials. Similar effects are observed on other fiber tensile properties, such as tenacity or stress at break, energy to rupture, and extension at break (4). Under the same spinning, processing, and testing conditions, higher molecular weight results in higher tensile strength. The effect of molecular weight distribution on tensile properties is complex because of the interaction with spinning conditions (4,5). In general, narrower molecular weight distributions result in higher breaking tenacity and lower elongation (4,6). The variation of tenacity and elongation with draw ratio for a given spun yarn correlates well with amorphous orientation (7,8). However, when different spun yarns are compared, neither average nor amorphous orientation completely explains these variations (9–11). Recent theory suggests that the number of tie molecules, both from molecules traversing the interlamellar region and especially those resulting from entanglements in the interlamellar region, defines the range of tensile properties achievable using draw-induced orientation (12,13). Increased entanglements (more ties) result in higher tenacity and lower elongation.

Creep, Stress Relaxation, Elastic Recovery. Olefin fibers exhibit creep, or time-dependent deformation under load, and undergo stress relaxation, or the spontaneous relief of internal stress. Because of the variety of molecular sizes and morphological states present in semicrystalline polymers, the creep and stress relaxation properties for materials such as polypropylene cannot be represented in one curve by using time–temperature superposition principles (14). However, given a spun yarn and thus a given structural state, curves for creep fracture (time to break under variable load) can be developed for different draw ratios, as shown in Figure 3 (15), indicating the importance of spun-yarn structure in a crystallizable polyolefin fiber. The same superposition can be carried out up to 110°C, where substantial reordering of polymer crystalline structure occurs (16).

High molecular weight and high orientation reduce creep. At a fixed molecular weight, the stress-relaxation modulus is higher for a highly crystalline

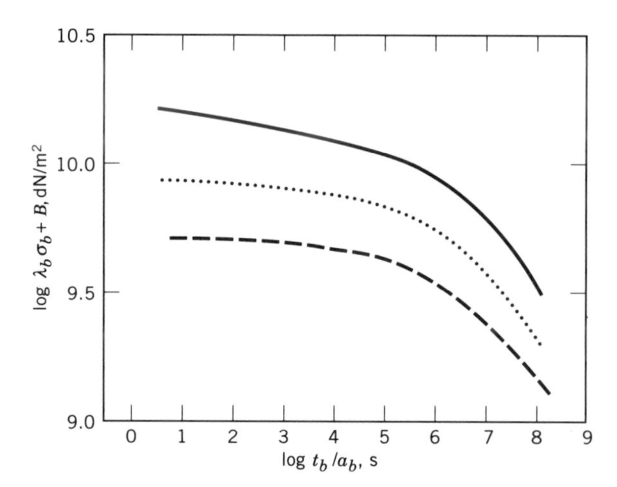

Fig. 3. Composite curve of true stress at break $\lambda_b\sigma_b$ at 40°C vs reduced time to break t_b/a_b for polypropylene fibers of three draw ratios (15): (——), 2.7 × draw, B = 0; (····), 3.5 × draw, B = 0.2; (– – –), 4.5 × draw, B = 0.4. Values of B are arbitrary.

sample prepared by slow cooling than for a smectic sample prepared by rapid quench (14). Annealing the smectic sample raises the relaxation modulus slightly, but not to the degree present in the fiber prepared by slow cooling.

Elastic recovery or resilience is the recovery of length upon release of stress after extension or compression. A fiber, fabric, or carpet must possess this property in order to spring back to its original shape after being crushed or wrinkled. Polyolefin fibers have poorer resilience than nylon; this is thought to be partially related to the creep properties of the polyolefins. Recovery from small strain cyclic loading is a function of temperature, as shown in Figure 4, and found to be a

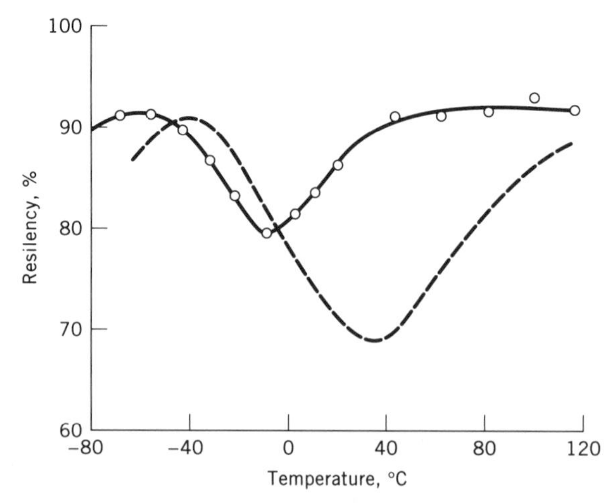

Fig. 4. Resiliency of polypropylene, (—o—), and polyethylene, (– – –), multifilament yarns as a function of temperature (17).

minimum for polypropylene at 10°C, near the glass-transition temperature, T_g (17). The minimum for polyethylene is at 30°C, higher than the amorphous T_g. This minimum is thought to be associated with motions in the crystalline phase of the highly oriented crystalline structure (17).

Chemical Properties. The hydrocarbon nature of olefin fibers, lacking any polarity, imparts high hydrophobicity and consequently resistance to soiling or staining by polar materials, a property important in carpet and upholstery applications. Unlike the condensation polymer fibers, such as polyester and nylon, olefin fibers are resistant to acids and bases. At room temperature, polyolefins are resistant to most organic solvents, except for some swelling in chlorinated hydrocarbon solvents. At higher temperatures, polyolefins dissolve in aromatic or chlorinated aromatic solvents, and show some solubility in high boiling hydrocarbon solvents. At high temperatures, polyolefins are degraded by strong oxidizing acids.

Thermal and Oxidative Stability. The thermal transitions of several polyolefins are compared to other polymers in Table 2. In general, polyolefins undergo thermal transitions at much lower temperatures than condensation polymers, thus the thermal and oxidative stability of polyolefin fibers are comparatively poor (18). They are highly sensitive to oxygen, which must be carefully controlled in all processing. The tertiary hydrogen in polypropylene imparts sensitivity to oxidative degradation by chain scission resulting in molecular weight degradation. Polyolefins are stabilized by hindered phenols or phosphites. Hindered phenol stabilizers provide moderate melt stability and good long-term heat aging, but undergo gas yellowing, which is a chemical reaction of phenolic compounds and nitrous oxide gases producing yellow-colored compounds. Typical sources of nitrous oxides are gas-fired heaters, dryers, and tenters, and propane-fueled lift trucks used in warehouses. Phosphites are good melt stabilizers, do not gas yellow, but have poor long-term heat aging. Preferred stabilizers are highly substituted phenols such as Cyanox 1790 and Irganox 1010, or phosphites such as Ultranox 626 and Irgafos 168 (see ANTIOXIDANTS; HEAT STABILIZERS).

Ultraviolet Degradation. Polyolefins are subject to light-induced degradation (19); polyethylene is more resistant than polypropylene. Although the mechanism of uv degradation is different from thermal degradation, the resulting chain scission and molecular weight degradation is similar. In fiber applications, sta-

Table 2. Thermal Properties of Olefins and Other Fiber-Forming Polymers

Polymer[a]	T_g,°C	T_m,°C	Softening temperature,°C	Thermal degradation temperature,°C
high density polyethylene (HDPE)	−120	130	125	
i-polypropylene (PP)	−20	170	165	290
i-poly(1-butene)	−25	128		
i-poly(3-methyl-1-butene)		315		
i-poly(4-methyl-1-pentene)	18	250	244	
poly(ethylene terephthalate) (PET)	70	265	235	400
nylon-6,6	50	264	248	360

[a]i = isotactic.

bilization against light is necessary to prevent loss of properties. The stabilizer must be compatible, have low volatility, be resistant to light and thermal degradation itself, and must last over the lifetime of the fiber. Chemical and physical interactions with other additives must be avoided. Minimal odor and toxicity, colorlessness, resistance to gas yellowing, and low cost are additional requirements (see UV STABILIZERS).

Stabilizers that act as uv screens or energy quenchers are usually ineffective by themselves. Because polyolefins readily form hydroperoxides, the more effective light stabilizers are radical scavengers. Hindered amine light stabilizers (HALS) are favored, especially high molecular weight and polymeric amines that have lower mobility and less tendency to migrate to the surface of the fiber (20,21). This migration is commonly called bloom. Test results for some typical stabilizers are given in Table 3 (22).

Flammability. Flammability of polymeric materials is measured by many methods, most commonly by the limiting-oxygen-index test (ASTM D2863), which defines the minimum oxygen concentration necessary to support combustion, or the UL 94 vertical-burn test, which measures the burn length of a fabric. Most polyolefins can be made fire retardant using a stabilizer, usually a bromine-containing organic compound, and a synergist such as antimony oxide (23). However, the required loadings are usually too high for fibers to be spun. Fire-retardant polypropylene fibers exhibit reduced light and thermal resistance. Commercial fire-retardant polyolefin fibers have just recently been introduced, but as expected the fibers have limited light stability and poor luster. Where applications require fire retardancy it is usually conferred by fabric finishes or incorporation of fire retardants in a latex, such as in latex-bonded nonwovens and latex-coated wovens.

Dyeing Properties. Because of their nonionic chemical nature, olefin fibers are difficult to dye. Oil-soluble dispersed dyes diffuse into polypropylene but readily bloom and rub off. In the first commercial dyeing of olefin fibers, nickel dyes such as UV-1084, also a light stabilizer, were used. The dyed fibers were colorfast but dull and hazy. A broad variety of polymeric dyesites have been blended with polypropylene; nitrogen-containing copolymers are the most favored (24–26). A commercial acid-dyeable polypropylene fiber is prepared by blending the polypropylene with a basic amino–polyamide terpolymer (27). In apparel applications

Table 3. Stabilization of Polypropylene Fiber by Polymeric HALS[a]

HALS	Manufacturer	Carbon arc T50,h[b]	Florida T50, kJ/m^2 [c,d]
none		70	<105
Chimassorb 944	CIBA-GEIGY Corp.	300	293–418
Cyasorb 3346	American Cyanamid Co.	320	418
Spinuvex A-36	Montedison Corp.	370–400	293

[a]Test specimens were 0.5 tex (4.5 den) filaments containing 0.25% specified HALS (22).
[b]Hours to 50% retention of initial tensile strength under carbon arc exposure.
[c]kJ/m^2 to 50% retention of tensile strength; Florida under glass exposure.
[d]To convert kJ/m^2 to Langley, multiply by 239.

where dyeing is important, dyeable blends are expensive and create problems in spinning fine denier fibers. Hence, olefin fibers are usually colored by pigment blending during manufacture, called solution dying in the trade.

Manufacture and Processing

Olefin fibers are manufactured commercially by melt spinning, similar to the methods employed for polyester and polyamide fibers. The basic process of melt spinning is illustrated in Figure 5. The polymer resin and ingredients, primarily stabilizers, pigments, and rheological modifiers, are fed into a screw extruder, melted, and extruded through fine diameter holes. The plate containing the holes is commonly called a spinneret. A metering pump and a mixing device are usually installed in front of the spinneret to ensure uniform delivery and mixing to facilitate uniform drawdown at high speeds. In the traditional or long spinning process, the fiber is pulled through a long cooling stack-type quench chamber by a take-up device at speeds in the range of 50–2000 m/min and discontinuously routed to downstream finishing operations. In the short spinning process, filaments are cooled within a few centimeters of the spinneret at speeds of 50–150 m/min. Because of the lower speeds, fiber can be continuously routed to downstream finishing operations in a one-step process (28). Finishing operations include drawing the fiber to as much as six times its original length, heat treatment to relieve internal stresses, and texturizing processes, which are combinations of deformational and heat treatments. These treatments were developed to impart specific characteristics to the olefin fiber dependent on its end use. Commercial olefin fibers are produced in a broad range of linear densities, from 0.1 to 12 tex (1.1–110 den), to fit a variety of applications, as shown in Figure 6 (29).

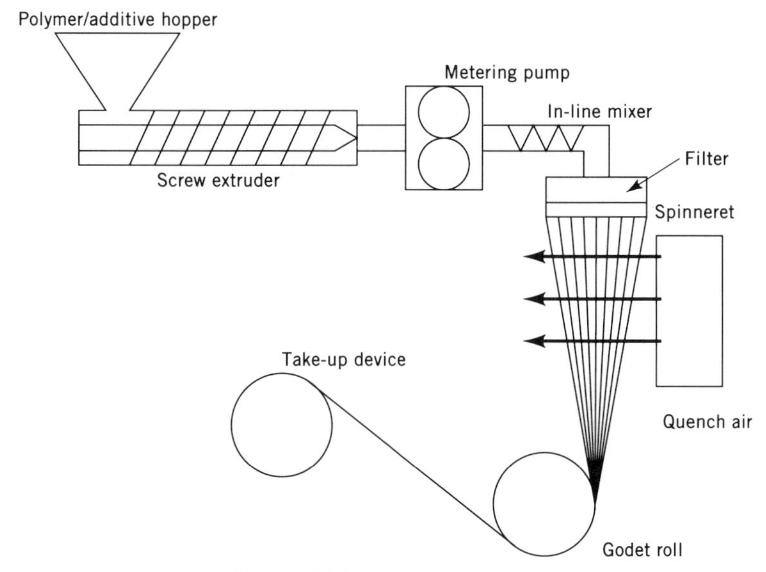

Fig. 5. Melt spinning process.

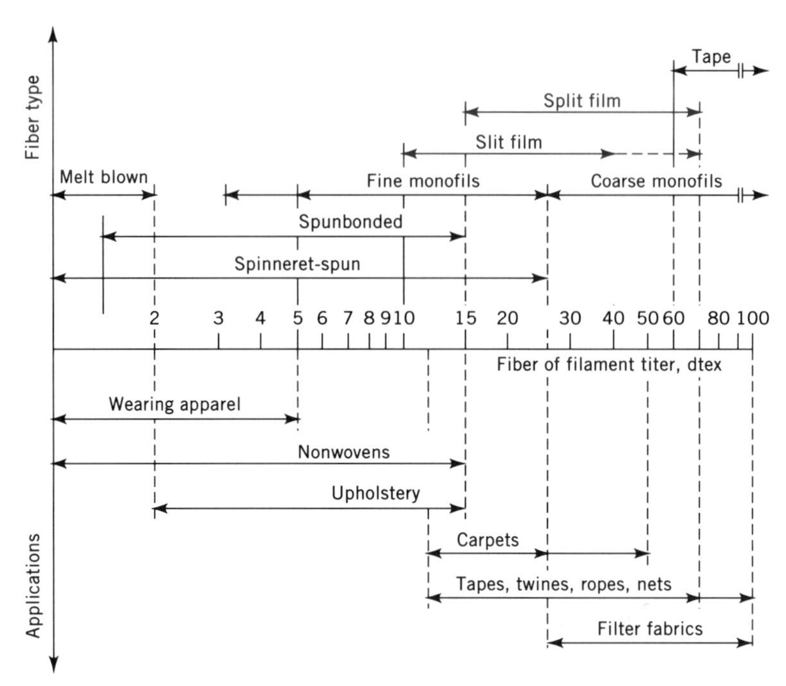

Fig. 6. Linear density of olefin fibers in various applications (29); dtex = 1.1 den.

Extrusion. Polymer resin and additives are melted and pumped through an extruder into a spinning pump. The pump meters the molten polymer through a filter system which removes particles from the molten polymer stream that might clog the capillaries of the spinneret or cause discontinuities in fine diameter fiber. These filters are typically either sand packs or metal screens. The polymer continues to a spinneret where it is extruded through holes under pressure. These holes or capillaries define the size and shape of the extruded fiber. A spinneret may contain up to several thousand capillaries, typically 0.3 to 0.5 mm in diameter. The length of the capillary is sized to the melt viscosity of the polymer. Typical length-to-diameter ratios of spinneret capillaries are 2:1 to 8:1. The spinneret holes can be arranged in a variety of hole spacings and patterns including rectangular, round, and annular. Considerations include throughput rate, heat transfer required to quench the fiber, and fiber diameter.

The extrusion of olefin fibers is largely controlled by the polymer. Polyolefin melts are strongly viscoelastic, and melt extrusion of polyolefin fibers differs from that of polyesters and polyamides. Polyolefins are manufactured in a broad range of molecular weights and ratios of weight-average to number-average molecular weight (M_w/M_n). Unlike the condensation polymers, which typically have molecular weights of 10,000–15,000 and M_w/M_n of approximately 2, polyolefins have weight-average molecular weights ranging from 50,000 to 1,000,000 and, as polymerized, M_w/M_n ranges from 4 to 15. Further control of molecular weight and distribution is obtained by chemical or thermal degradation. The full range of molecular weights used in olefin fiber manufacture is above 20,000, and M_w/M_n varies from 2 to 15. As molecular weight increases and molecular weight distribution broadens, the polymer melt becomes more pseudoplastic as indicated in Table 4 and shown in Figure 7 (30). In the sizing of extrusion equipment for olefin fiber production, the wide range of shear viscosities and thinning effects must be considered because these affect both power requirements and mixing efficiencies.

Table 4. Molecular Weight Characterization Data for Polypropylene Samples[a,b]

Code[c]	Melt flow rate[d]	$M_w \times 10^{-5}$	M_w/M_n	M_z/M_w	$M_v \times 10^{-5}$
	High molecular weight polypropylene				
○ narrow	4.2	2.84	6.4	2.59	2.40
△ regular–broad	5.0	3.03	9.0	3.57	2.42
□ broad–regular	3.7	3.39	7.7	3.54	2.71
	Middle molecular weight polypropylene				
⦶ narrow	11.6	2.32	4.7	2.81	1.92
⚠ regular	12.4	2.79	7.8	4.82	2.13
◫ broad	11.0	2.68	9.0	4.46	2.07
	Low molecular weight polypropylene				
⊕ narrow	25.0	1.79	4.6	2.47	1.52
◑ regular–narrow	23.0	2.02	6.7	3.18	1.66

[a]Ref. 30.
[b]Figures 7–9.
[c]Narrow, regular, and broad refer to molecular weight distribution.
[d]ASTM D1238 (Condition L; 230/2.16).

Fiber spinning is a uniaxial extension process, and the elongational viscosity behavior, which is the stress–strain relationship in uniaxial extension, is more important than the shear viscosity behavior. The narrower molecular weight distributions tend to be less thinning, and as shown in Figure 8 (30), elongational viscosity increases at higher extension rates. This leads to higher melt orientation, which in turn is reflected in higher spun fiber orientation, higher tenacity, and lower extensibility. In contrast, the broad molecular weight distributions tend to be more thinning and hence more prone to necking and fracture at high spinning speeds (30,31), but yield a less oriented, higher elongation spun fiber. The choice of an optimum molecular weight and molecular weight distribution is determined by the desired properties of the fiber and the process continuity on available equipment.

Because of the high melt viscosity of polyolefins, normal spinning melt temperatures are 240–310°C, which is 80–150°C above the crystalline melting point. Because of the high melt temperatures used for polyolefin fiber spinning, thermal stabilizers such as substituted hindered phenols are added. In the presence of pigments, the melt temperature must be carefully controlled to prevent color degradation and to obtain uniform color dispersion.

Polyolefin melts have a high degree of viscoelastic memory or elasticity. First normal stress differences of polyolefins, a rheological measure of melt elasticity,

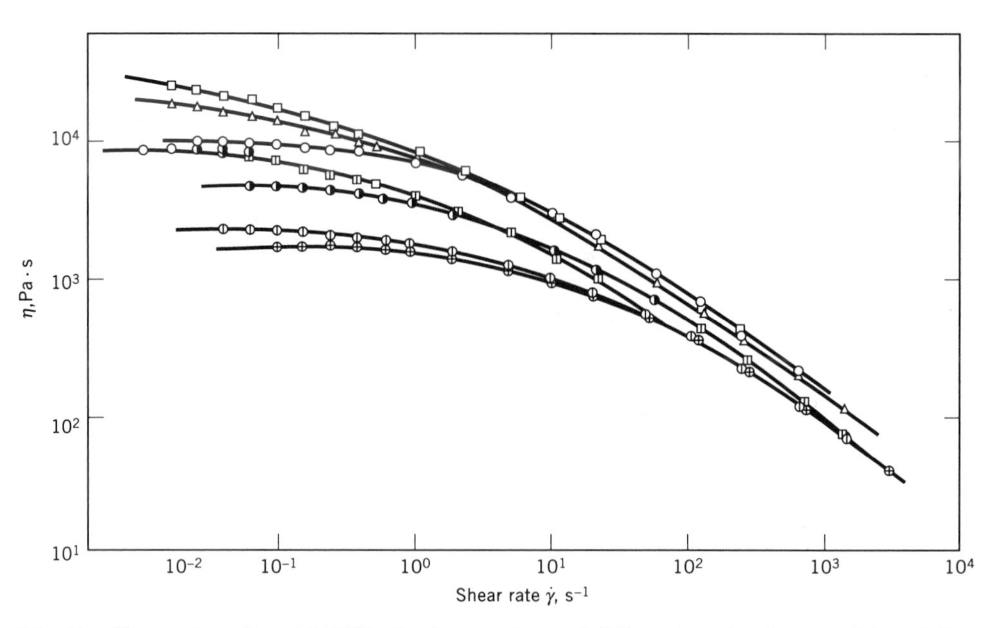

Fig. 7. Shear viscosity at 180°C of polypropylene of different molecular weight and distribution vs shear rate (30); see Table 4 for key. Pa·s = 0.1 P.

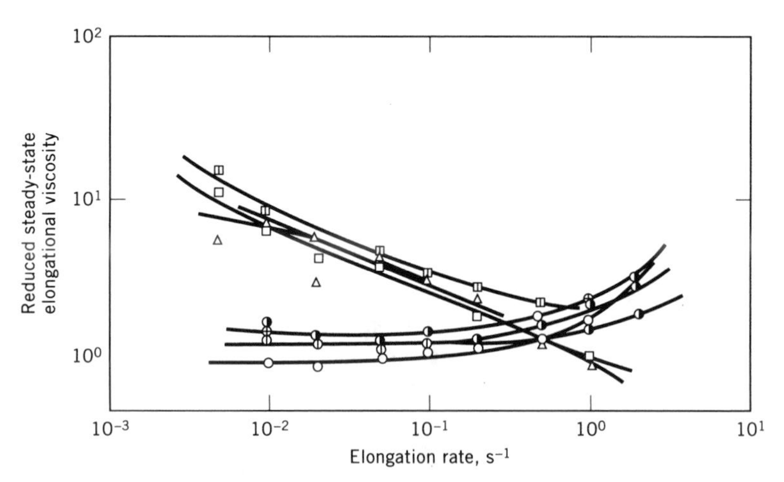

Fig. 8. Elongational viscosity at 180°C of polypropylene of different molecular weight and distribution (30); see Table 4 for key.

are shown in Figure 9 (30). At a fixed molecular weight and shear rate, the first normal stress difference increases as M_w/M_n increases. The high shear rate obtained in fine capillaries, typically on the order of 10^3–10^4 s^{-1}, coupled with the viscoelastic memory, causes the filament to swell (die swell or extrudate swell) upon leaving the capillary. On a molecular scale, the residence time in the region of die swell is sufficient to allow relaxation of any shear induced orientation. How-

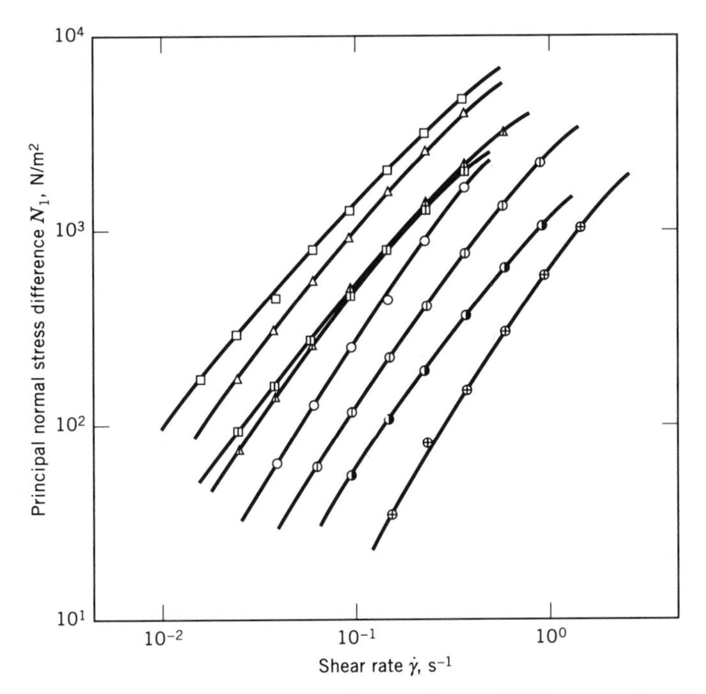

Fig. 9. First normal stress differences of polypropylene of different molecular weight and distribution (30); see Table 4 for key. To convert N/m^2 to dyne/cm^2, multiply by 10.

ever, high die swell significantly affects the drawdown or extension rate, leading to threadline breaks. Die swell can be reduced by lower molecular weight, narrower molecular weight distribution, or higher melt temperature.

Quench. Attempts have been made to model this nonisothermal process (32–35), but the complexity of the actual system makes quench design an art. Arrangements include straight-through, and outside-in and inside-out radial patterns (36). The optimum configuration depends on spinneret size, hole pattern, filament size, quench-chamber dimensions, take-up rate, and desired physical properties. Process continuity and final fiber properties are governed by the temperature profile and extension rate.

Polypropylene and other linear polyolefins crystallize more rapidly than most other crystallizable polymers. Unlike polyester, which is normally amorphous as spun, the fiber morphology of polyolefins is fixed in the spinning process; this limits the range of properties in subsequent drawing and annealing operations. In a low crystallinity state, sometimes called the paracrystalline or smectic form, a large degree of local order still exists. It can be reached by extruding low molecular weight polyolefins, processing at low draw ratio, or by a rapid quench such as by using a cold water bath (37).

Quench is more commonly practiced commercially by a controlled air quench in which the rate of cooling is controlled by the velocity and temperature of the air. During normal cooling, crystallization occurs in the threadline. In-line x-ray scattering studies demonstrate that crystallization is extremely rapid; the full crystalline structure is almost completely developed in fractions of a second (38).

Fiber spinning is an extensional process during which significant molecular orientation occurs. Under rapid crystallization, this orientation is fixed during the spinning process. Small-angle neutron-scattering studies of quiescent polypropylene crystallization show that the chain dimensions in both melt and crystallized forms are comparable (39). Although there may be significant relaxation of the amorphous region after spinning, the primary structure of the fiber is fixed during spinning and controls subsequent drawing and texturizing of the fiber. For fixed extrusion and take-up rates, a more rapid quench reduces the average melt-deformation temperature, increases relaxation times, and gives a more entangled melt when crystallization begins. The rapidly quenched fiber usually gives lower elongation and higher tenacity during subsequent draw (40). Using a very rapid quench, the melt may not be able to relax fast enough to sustain drawdown, resulting in melt fracture. Under conditions of a slow quench, the melt may totally relax, leading to ductile failure of the threadline.

A common measurement useful in predicting threadline behavior is fiber tension, frequently misnamed spinline stress. It is normally measured after the crystallization point in the threadline when the steady state is reached and the threadline is no longer deformed. Fiber tension increases as take-up velocity increases (38) and molecular weight increases. Tension decreases as temperature increases (41). Crystallinity increases slightly as fiber tension is increased (38). At low tension, the birefringence increases as tension is increased, leveling off at a spinline tension of 10 MPa (1450 psi) (38).

Take-Up. Take-up devices attenuate the spinline to the desired linear density and collect the spun yarn in a form suitable for further processing. A godet wheel is typically used to control the take-up velocity which varies from 1–2 m/s for heavy monofilaments to 10–33 m/s for fine yarns. The yarn can be stacked in cans, taken up on bobbins, or directly transferred to drawing and texturizing equipment.

In the spunbond process (Fig. 10), an aspiratory is used to draw the fibers in spinning and directly deposit them as a web of continuous, randomly oriented filaments onto a moving conveyor belt. In the meltblown process (Fig. 11), high velocity air is used to draw the extruded melt into fine-denier fibers that are laid down in a continuous web on a collector drum.

Draw. Polyolefin fibers are usually drawn to increase orientation and further modify the physical properties of the fiber. Linear density, necessary to control the textile properties, is more easily reduced during drawing than in spinning. The draw step can be accomplished in-line with spinning in a continuous spin–draw–texturing process (36,42) or in a second processing step. This second processing step allows simultaneous mixing of colors in a multi-ply continuous filament yarn for textiles. For staple fiber production, large bundles or tows consisting of up to a million or more filaments are stretched, texturized (crimped), and cut.

In secondary drawing operations, the aging properties of the spun yarn must be considered. Because polypropylene fibers have a low T_g, the spun yarn is restructured between spinning and drawing; this is more important as the smectic content is increased (43). The aging process depends on whether the yarn is stored on bobbins under tension or coiled in cans with no tension on the fiber. The aging of quick-quenched (smectic) polypropylene films has been studied (43). Stored at

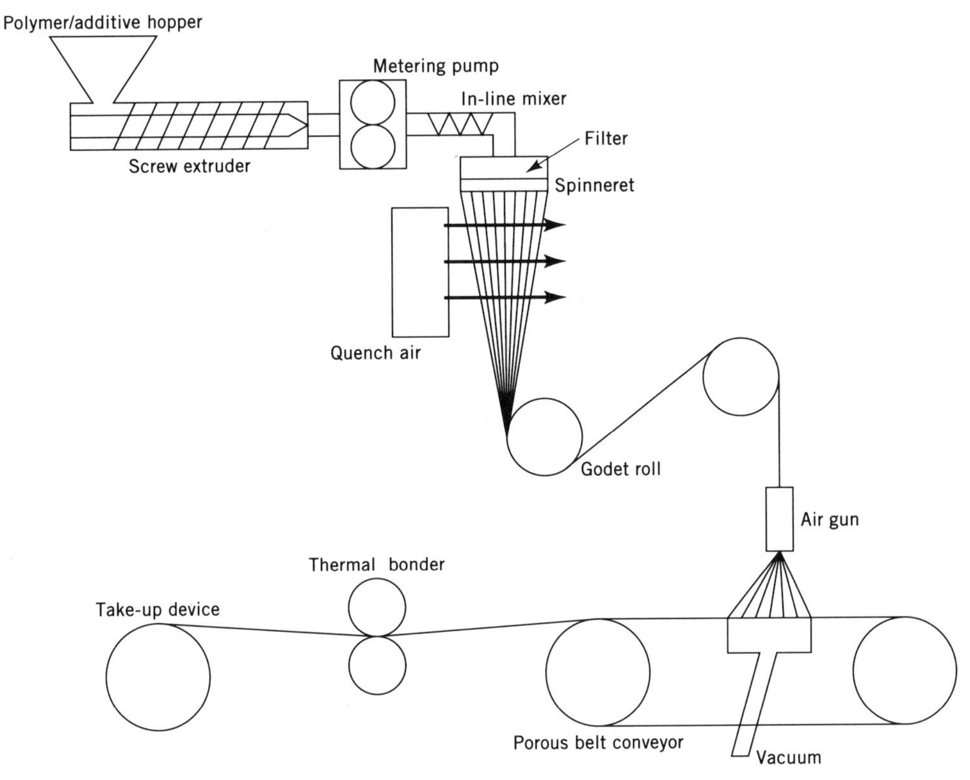

Fig. 10. Flow sheet for typical spunbond fabric manufacture.

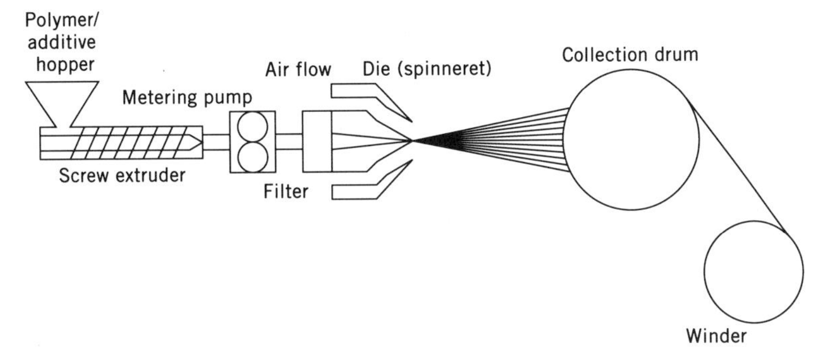

Fig. 11. Flow sheet for typical meltblown fabric manufacture.

room temperature, the increase in yield stress is 5% in 24 hours. Similar data on polypropylene spun fibers have not been published, but aging effects are similar. Drawn fiber properties, such as density, stress relaxation modulus, and heat of fusion, age because of collapse of excess free volume in the noncrystalline fraction (44).

The crystalline structure of the spun yarn affects the draw process. Monoclinic yarns tend to exhibit higher tenacity and lower elongation at low draw ratios than smectic yarns (6). They exhibit lower maximum draw ratios, undergo brittle fracture, and form microvoids (45) at significantly lower draw temperatures, which creates a chalky appearance. Studies of the effect of spun-yarn structure on drawing behavior show that the as-spun orientation and morphology determine fiber properties at a given draw ratio, as shown in Figure 12 (9,45,46). However, final fiber properties can be correlated with birefringence, a measure of the average orientation, as shown in Figure 13 (9,45). Fiber properties and amorphous orientation show good correlation in some studies (Fig. 14) (7,8), but in most studies the range of spun-yarn properties is limited. Such studies suggest that the deformation during draw primarily affects the interlamellar amorphous region at low draw ratio. At higher draw ratio, the crystalline structure is substantially disrupted.

Texturing. The final step in olefin fiber production is texturing; the method depends primarily on the application. For carpet and upholstery, the fiber is usually bulked, a procedure in which fiber is deformed by hot air or steam jet turbulence in a nozzle and deposited on a moving screen to cool. The fiber takes on a three-dimensional crimp that aids in developing bulk and coverage in the final fabric. Stuffer box crimping, a process in which heated tow is overfed into a restricted outlet box, imparts a two-dimensional sawtooth crimp commonly found in olefin staple used in carded nonwovens and upholstery yarns.

Slit-Film Fiber. A substantial volume of olefin fiber is produced by slit-film or film-to-fiber technology (29). For producing filaments with high linear density, above 0.7 tex (6.6 den), the production economics are more favorable than monofilament spinning (29). The fibers are used primarily for carpet backing and rope

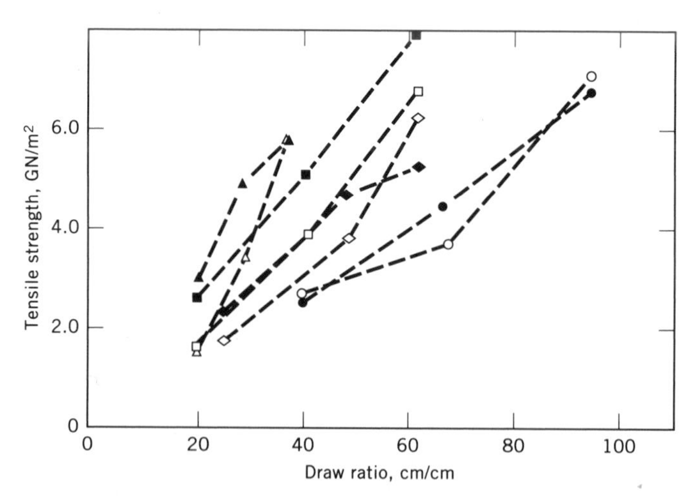

Fig. 12. Tensile strength vs draw ratio (6): 0.42 melt index spun at 50 m/min, ■, □; and 500 m/min, ▲, △; 12.0 melt index spun at 100 m/min, ●, ○; and 500 m/min, ◆, ◇. Open symbols = cold drawn and annealed at 140°C; filled symbols = drawn at 140°C. To convert GN/m^2 to $dyne/cm^2$, multiply by 10^{10}.

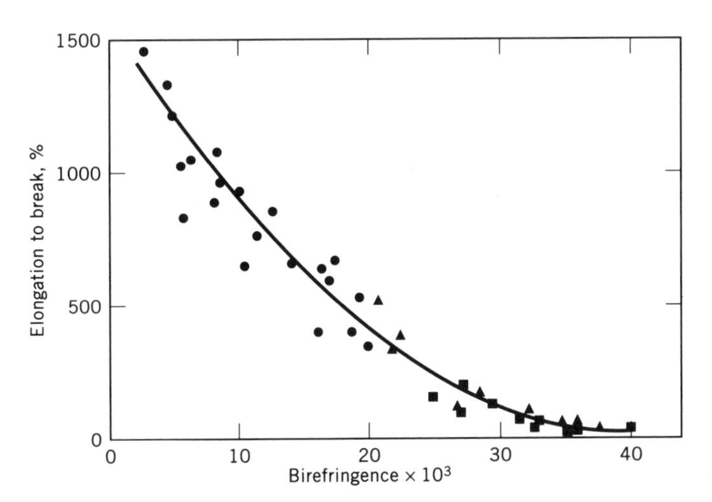

Fig. 13. Elongation to break as a function of birefringence for undrawn, hot-drawn, and cold-drawn annealed fibers (6): ●, undrawn; ▲, cold-drawn, annealed at 140°C; ■, hot-drawn at 140°C

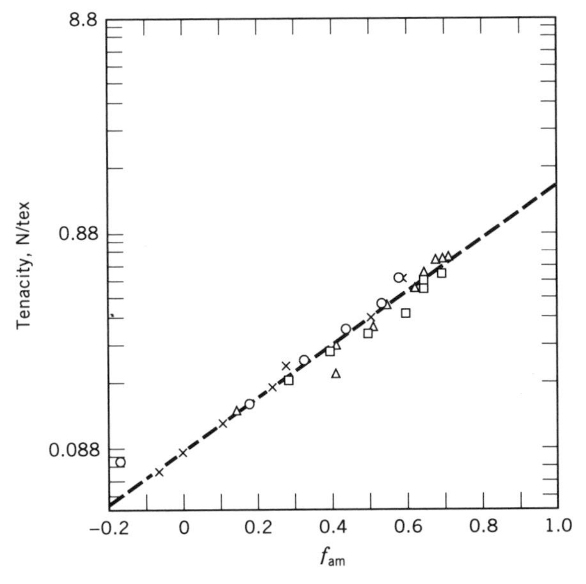

Fig. 14. Tenacity as a function of amorphous orientation (f_{am}) for polypropylene fibers and films (7). Film drawn at: ○, 135°C; ×, 110°C; □, 90°C. △, Heat-set fiber. To convert N/tex to gf/den, multiply by 11.3. Tenacity_{max} = 1.3 N/tex (15 gf/den).

or cordage applications. The processes used to make slit-film fibers are versatile and economical.

The equipment for the slit-film fiber process is shown in Figure 15 (29). An olefin film is cast, and as in melt spinning, the morphology and composition of the film determine the processing characteristics. Fibers may be produced by cutting or slitting the film, or by chemomechanical fibrillation. The film is fibrillated mechanically by rubbing or brushing. Immiscible polymers, such as polyethylene or polystyrene (PS), may be added to polypropylene to promote fibrillation. Many common fiber-texturing techniques such as stuffer-box, false-twist, or knife-edge treatments improve the textile characteristics of slit-film fibers.

Several more recent variations of the film-to-fiber approach result in direct conversion of film to fabric. The film may be embossed in a controlled pattern and subsequently drawn uniaxially or biaxially to produce a variety of nonwoven products (47). Addition of chemical blowing agents to the film causes fibrillation upon extrusion. Nonwovens can be formed directly from blown film using a unique radial die and control of the biaxial draw ratio (48) (see NONWOVEN FABRICS).

Bicomponent Fibers. Polypropylene fibers have made substantial inroads into nonwoven markets because they are easily thermal bonded. Further enhancement in thermal bonding is obtained using bicomponent fibers (49). In these fibers, two incompatible polymers, such as polypropylene and polyethylene, polyester and polyethylene, or polyester and polypropylene, are spun together to give a fiber with a side-by-side or core–sheath arrangement of the two materials. The lower melting polymer can melt and form adhesive bonds to other fibers; the higher melting component causes the fiber to retain some of its textile characteristics.

Bicomponent fibers have also provided a route to self-texturing (self-crimping) fibers. The crimp results from the length differential developed during proc-

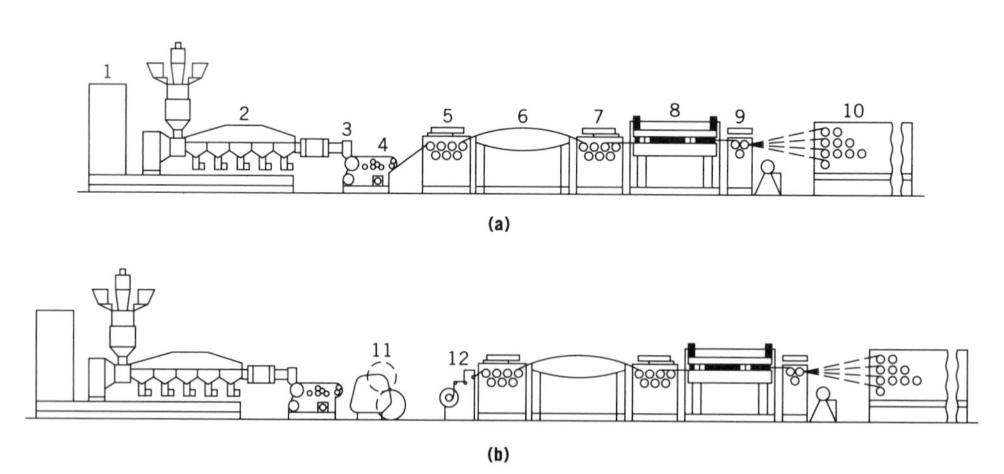

Fig. 15. Production lines for stretched film tape (29): (**a**) continuous production line for film tape; (**b**) discontinuous production lines for film and film tape. 1, Control cabinet; 2, extruder; 3, flat die; 4, chill roll; 5, septet (seven rolls); 6, hot plate; 7, septet (seven rolls) 8, heat-setting oven; 9, trio (three rolls); 10, bobbin winder; 11, film winder; and 12, film-unrolling stand.

essing caused by differential shrinkage in the two polymers in side-by-side or eccentric core–sheath configurations (50).

Conventional spinning technology is limited in the production of very fine denier filaments because of spinning and mass uniformity problems as the melt drawdown is increased. Ultrafine filaments (microfibers) can be produced through bicomponent technology by extruding two or more components together as a single fiber and later separating the components through chemical or mechanical processes. Fibers of 0.1 to 0.001 tex (~ 1–0.01 den) per filament can be produced (50,51).

Meltblown, Spunbond, and Spurted Fibers. A variety of directly formed nonwovens exhibiting excellent filtration characteristics are made by meltblown processes (52), producing very fine, submicrometer filaments. A simple schematic of the die is shown in Figure 11. A stream of high velocity hot air is directed on the molten polymer filaments as they are extruded from a spinneret. This air attenuates, entangles, and transports the fiber to a collection device. Because the fiber cannot be separated and wound for subsequent processing, a nonwoven web is directly formed. Mechanical integrity of the web is usually obtained by thermal bonding or needling, although other methods, such as latex bonding, can be used. Meltblown fabrics are made commercially from polypropylene and polyethylene. The webs are soft, breathable, and drapable (53–55).

In the spunbond process, the fiber is spun similarly to conventional melt spinning, but the fibers are attenuated by air drag applied at a distance from the spinneret. This allows a reasonably high level of filament orientation to be developed. The fibers are directly deposited onto a moving conveyor belt as a web of continuous randomly oriented filaments. As with meltblown webs, the fibers are usually thermal bonded or needled (53).

Pulp-like olefin fibers are produced by a high pressure spurting process developed by Hercules Inc. and Solvay, Inc. Polypropylene or polyethylene is dissolved in volatile solvents at high temperature and pressure. After the solution is released, the solvent is volatilized, and the polymer expands into a highly fluffed, pulp-like product. Additives are included to modify the surface characteristics of the pulp. Uses include felted fabrics, substitution in whole or in part for wood pulp in papermaking, and replacement of asbestos in reinforcing applications (56).

High Strength Fibers. The properties of commercial olefin fibers are far inferior to those theoretically attainable. Theoretical and actual strengths of common commercial fibers are listed in Table 5 (57). A number of methods, including superdrawing (58), high pressure extrusion (59), spinning of liquid crystalline polymers or solutions (60), gel spinning (61–65), and hot drawing (66) produce higher strengths than those given in Table 5 for commercial fibers, but these methods are tedious and uneconomical for olefin fibers. A high modulus commercial polyethylene fiber with properties approaching those of aramid and graphite fibers (Table 6) (67) is prepared by gel spinning (68). Although most of these techniques produce substantial increases in modulus, higher tensile strengths are currently available only from gel spinning or dilute fibrillar crystal growth. Even using these techniques, the maximum strengths observed to date are only a fraction of the theoretical strengths.

Table 5. Theoretical and Actual Strengths of Commercial Fibers[a]

Polymer	Density, kg/m^3	Molecular area, nm^2	Theoretical strength, GPa[b]	Strength of commercial fiber, GPa[b]
polyethylene	960	0.193	31.6	0.76
polypropylene	910	0.348	17.6	0.72
nylon-6	1140	0.192	31.9	0.96
polyoxymethylene	1410	0.185	32.9	
poly(vinyl alcohol)	1280	0.228	26.7	1.08
poly(*p*-benzamide)	1430	0.205	29.7	3.16
poly(ethylene terephthalate)	1370	0.217	28.1	1.15
poly(vinyl chloride)	1390	0.294	20.8	0.49
rayon	1500	0.346	17.7	0.69
poly(methyl methacrylate)	1190	0.667	9.2	

[a]Ref. 56.
[b]To convert GPa to psi, multiply by 145,000.

Table 6. Properties of Commercial High Strength Fibers[a]

Fiber	Density, kg/m^3	Strength, GPa[b]	Modulus, GPa[b]	Elongation to break, %	Filament diameter, mm
polyethylene	970	2.6	117	3.5	0.038
aramid	1440	2.8	113	2.8	0.012
S-glass	2490	4.6	89	5.4	0.009
graphite	1730	3.1	227	1.2	0.006
steel whiskers	7860	2.3	207	1.3	0.250

[a]Ref. 66.
[b]To convert GPa to psi, multiply by 145,000.

Hard-Elastic Fibers. Hard-elastic fibers are prepared by annealing a moderately oriented spun yarn at high temperature under tension. They are prepared from a variety of olefin polymers, acetal copolymers, and polypivalolactone (69,70). Whereas the strengths observed are comparable to those of highly drawn commercial fiber, in the range 0.52–0.61 N/tex (6–7 gf/den), the recovery from elongation is substantially better. Hard-elastic fibers typically exhibit 90% recovery from 50% elongation, whereas highly drawn, high tenacity commercial fibers exhibit only 50–75% recovery from 5% elongation. The mechanism of elastic recovery differs from the entropic models normally used to explain plastic properties. The hard elastic fibers are thought to deform through opening of the lamellae stacked structure, resulting in void formations; recovery is controlled by energy considerations. Although there are potential uses in applications involving substantial deformation, products such as stretch fabrics and hard-elastic fibers are not yet used commercially.

Economic Aspects

In the United States, olefin fiber consumption has risen steadily since its introduction in 1961. The growth of olefin fiber usage is high, as shown in Figure 16 (71,72). U.S. olefin fiber consumption as a percentage of total synthetic fibers consumed is compared to other fiber types in Figure 17 (71,72). Olefin fiber is the only synthetic fiber showing market growth in recent years.

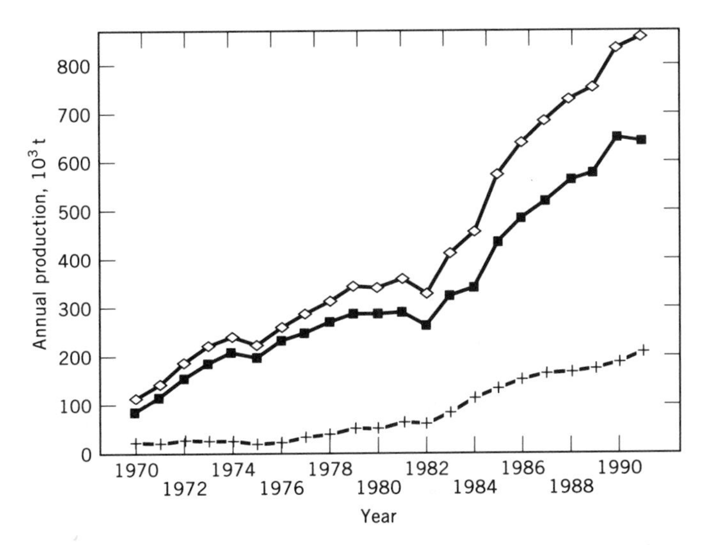

Fig. 16. U.S. olefin fiber annual production: +, staple plus tow; ■, yarn plus monofilament; ◇, total (71,72).

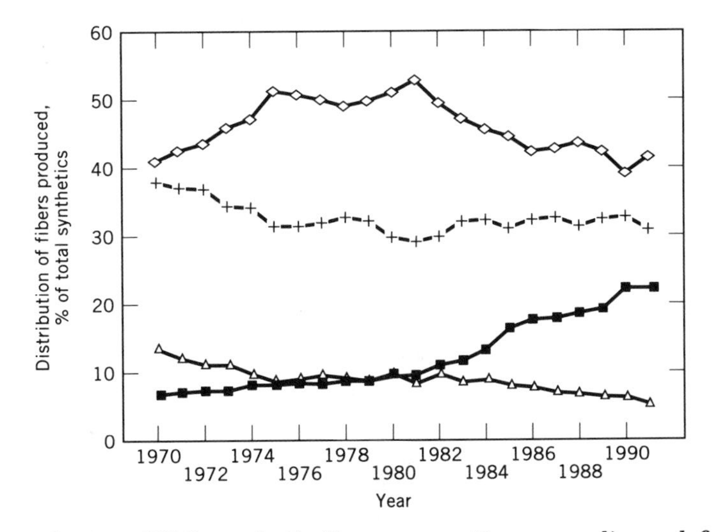

Fig. 17. Distribution of U.S. synthetic fiber consumption: △, acrylic; ■, olefin; +, nylon and aramid; ◇, polyester (71,72).

Applications

Olefin fibers are used for a variety of purposes from home furnishings to industrial applications. These include carpets, upholstery, drapery, rope, geotextiles (qv), and both disposable and nondisposable nonwovens. Fiber mechanical properties, relative chemical inertness, low moisture absorption, and low density contribute to desirable product properties. Table 7 gives a breakdown of olefin fiber consumption by use (73–75). Olefin fiber use in apparel has been restricted by low melting temperatures, which make machine drying and ironing of polyethylene and polypropylene fabrics difficult or impossible. However, this market is increasing as manufacturers take advantage of the wicking properties (moisture transport) as in lightweight sportswear (76).

Polypropylene fibers are used in every aspect of carpet construction from face fiber to primary and secondary backings. Polypropylene's advantages over jute as carpet backing are dimensional stability and minimal moisture absorption. Drawbacks include difficulty in dyeing and higher cost. Bulked-continuous-filament (BCF) carpet yarns provide face fiber with improved crimp and elasticity. BCF carpet yarns are especially important in contract carpets, characterized by low dense loops, where easy cleaning is an advantage.

Olefin fiber is an important material for nonwovens (77). The geotextile market is still small, despite expectations that polypropylene is to be the principal fiber in such applications. Disposable nonwoven applications include hygienic coverstock, sanitary wipes, and medical roll goods. The two competing processes for the coverstock market are thermal-bonded carded staple and spunbond, both of

Table 7. Annual Domestic Shipments of Olefin Fibers,[a] 10^3 t

Fiber	1984	1986	1988	1990	1991[a]
		Filament yarn			
carpet facing	70.1	98.1	121	164	188
carpet backing	104	184	194	203	187
other broad woven	86.5	123	153	189	175
narrow woven[b]	6.8	8.0	8.6	7.9	8.1
rope, cordage	48.6	48.3	54.7	54.9	45.3
other[c]	14.4	15.5	13.5	20	18.5
		Staple plus tow			
broad woven	10.1	8.2	7.8	8.3	17
carpet face	43.9	53.3	62.8	57.9	54.1
nonwovens[d]	54.2	75.6	82.2	106	121
other[e]	1.4	5.1	3.6	4.4	3.1

[a]Ref. 72–74. 1991 annual data estimated from first three quarters.
[b]Includes ribbons, braids, seat belts, and furniture webbing.
[c]Includes sewing thread, sliver knit backing, and other industrial applications.
[d]Includes wet laid, dry laid, and needled batts.
[e]1984–1986 data include vinyon.

which have displaced latex-bonded polyester because of improved strength, softness, and inertness.

A special use for meltblown olefin fiber is in filtration media such as surgical masks and industrial filters (78). The high surface area of these ultrafine filament fibers permits preparation of nonwoven filters with effective pore sizes as small as 0.5 μm.

Other applications, including rope, cordage, outdoor furniture webbing, bags, and synthetic turf, make up the remaining segments of the olefin fiber market. Spunbond polyethylene is used in packaging applications requiring high strength and low weight. Specialty olefin fibers are employed in asphalt and concrete reinforcement (79–82). Hollow fibers have been tested in several filtration applications (83,84). Ultrafine fibers are used in synthetic leather, silk-like fabrics, and special filters (50,51). These fibers are also used in sports outerwear, where the tight weaves produce fabrics that are windproof and waterproof but still are able to pass vapors from perspiration and thus keep the wearer cool and dry (51). If the economics of the high modulus olefin fibers becomes more favorable, substantial markets could be developed in reinforced composites such as boat hulls (67).

BIBLIOGRAPHY

"Polypropylene Fibers" in *ECT* 2nd ed., Suppl., pp. 806–836, by F. C. Cesare, M. Farber, and G. R. Cuthbertson, Uniroyal, Inc.; "Olefin Fibers" in *ECT* 3rd ed., Vol. 16, pp. 357–385, by D. R. Buchanan, North Carolina State University.

1. *The Textile Fiber Products Identification Act, Public Law 85-897*, Washington, D.C., Sept., 1958.
2. *Text. World* **134**, 49 (Nov. 1984).
3. I. M. Hall, *J. Polym. Sci.* **54**, 505 (1961).
4. F. Lu and J. E. Spruiell, *J. Appl. Polym. Sci.* **34**, 1521 (1987).
5. J. E. Flood and S. A. Nulf, *Polym. Eng. Sci.* **30**, 1504 (1990).
6. H. S. Brown, T. L. Nemzek, and C. W. Schroeder, "MWD and Its Effect on Fiber Spinning," paper presented at *The 1983 Fiber Producer Conference*, Greenville, S.C., Apr. 13, 1983, sponsored by *Fiber World*, Brilliam Publishing Co., Atlanta, Ga.
7. R. J. Samuels, *Structural Polymer Properties*, Wiley-Interscience, New York, 1974.
8. F. Geleji and co-workers, *J. Polym. Sci. Polym. Symp.* **58**, 253 (1977).
9. H. P. Nadella, J. E. Spruiell, and J. L. White, *J. Appl. Polym. Sci.* **22**, 3121 (1978).
10. A. J. de Vries, *Pure Appl. Chem.* **53**, 1011 (1981).
11. A. J. de Vries, *Pure Appl. Chem.* **54**, 647 (1982).
12. D. T. Grubb, *J. Polym. Sci. Polym. Phys. Ed.* **21**, 165 (1983).
13. D. Thirion and J. F. Tassin, *J. Polym. Sci. Polym. Phys. Ed.* **21**, 2097 (1983).
14. G. Attalla, I. B. Guanella, and R. E. Cohen, *Polym. Eng. Sci.* **23**, 883 (1983).
15. A. Takaku, *J. Appl. Polym. Sci.* **26**, 3565 (1981).
16. A. Takaku, *J. Appl. Polym. Sci.* **25**, 1861 (1980).
17. G. M. Bryant, *Text. Res. J.* **37**, 552 (1967).
18. L. Reich and S. S. Stivala, *Rev. Macromol. Chem.* **1**, 249 (1966).
19. D. J. Carlsson and D. M. Wiles, *J. Macromol. Sci. Rev. Macromol. Chem.* **14**, 65, (1976).
20. D. J. Carlsson, A. Garton, and D. M. Wiles, in G. Scott, ed., *Developments in Polymer Stabilisation*, Applied Science Publishers, London, 1979, p. 219.
21. F. Gugumaus, in Ref. 20, p. 261.

22. L. M. Landoll and A. C. Schmalz, private communication, Hercules Inc., Oxford, Ga., 1986.
23. J. Green, in M. Lewin, S. M. Atlas, and E. M. Pierce, eds., *Flame-Retardant Polymeric Materials*, Vol. 3, Plenum Press, New York, 1982, Chapt. 1.
24. U.S. Pat. 3,873,646 (Mar. 25, 1975), H. D. Irwin (to Lubrizol Corp.).
25. U.S. Pat. 3,653,803 (Apr. 4, 1972), H. C. Frederick (to E. I. du Pont de Nemours & Co., Inc.).
26. U.S. Pat. 3,639,513 (Feb. 1, 1972), H. Masahiro and co-workers (to Mitsubishi Rayon Co., Ltd.).
27. U.S. Pat. 3,433,853 (Mar. 18, 1969), R. H. Earle, A. C. Schmalz, and C. A. Soucek (to Hercules Inc.).
28. K. Hawn and R. Meriggi, *Nonwovens World*, 44 (Sept. 1987).
29. H. Krassig, *J. Polym. Sci. Macromol. Rev.* **12**, 321 (1977).
30. W. Minoshima, J. L. White, and J. E. Spruiell, *Polym. Eng. Sci.* **20**, 1166 (1980).
31. O. Ishizuka and co-workers, *Sen-i Gakkaishi* **31**, T372 (1975).
32. S. Kase and T. Matsuo, *J. Polym. Sci. Part A* **3**, 2541 (1965).
33. S. Kase and T. Matsuo, *J. Appl. Polym. Sci.* **11**, 251 (1967).
34. C. D. Han and R. R. Lamonte, *Trans. Soc. Rheol.* **16**, 447 (1972).
35. R. R. Lamonte and C. D. Han, *J. Appl. Polym. Sci.* **16**, 3285 (1972).
36. F. Fourne, *IFJ* **3**, 30 (Aug., 1988).
37. C. Prost and co-workers, *Makromol. Chem., Macromol. Symp.* **23**, 173 (1989).
38. H. P. Nadella and co-workers, *J. Appl. Polym. Sci.* **21**, 3003 (1977).
39. D. G. H. Ballard and co-workers, *Polymer* **20**, 399 (1979); **23**, 1875 (1982).
40. W. C. Sheehan and T. B. Cole, *J. Appl. Polym. Sci.* **8**, 2359 (1964).
41. Ref. 4, p. 1541.
42. R. Wiedermann, *Chemiefasern Textilind.* **28/80**, 888 (1978).
43. D. M. Gezovich and P. H. Geil, *Polym. Eng. Sci.* **8**, 210 (1968).
44. C. P. Buckley and M. Habibullah, *J. Appl. Polym. Sci.* **26**, 2613 (1981).
45. H. Bodaghi, J. E. Spruiell, and J. L. White, *Intern. Polym. Processing* **III**, 100 (1988).
46. A. Garten and co-workers, *J. Polym. Sci. Polym. Phys. Ed.* **15**, 2013 (1977).
47. U.S. Pat. 3,137,746 (June 16, 1964), D. E. Seymour and D. J. Ketteridge (to Smith and Nephew, Ltd.).
48. U.S. Pat. 4,085,175 (Apr. 18, 1978), H. W. Keuchel (to PNC Corp.).
49. *Text. Month*, 10 (Aug. 1983).
50. D. O. Taurat, *IFJ* **3**, 24 (May 1988).
51. W. R. Baker, *IFJ* **7**, 7 (Apr. 1992).
52. R. R. Buntin and D. T. Lohkamp, *TAPPI* **56**, 74 (1973).
53. J. Zhou and J. E. Spruiell, in *Nonwovens—An Advanced Tutorial*, TAPPI Press, Atlanta, Ga., 1989.
54. L. C. Wadsworth and A. M. Jones, "Novel Melt Blown Research Findings," paper presented at *INDA/TEC, The International Nonwovens Technological Conference*, Philadelphia, Pa., June 2–6, 1986.
55. A. M. Jones and L. C. Wadsworth, "Advances in Melt Blown Resins," paper presented at *TAPPI 1986 Nonwovens Conference*, Atlanta, Ga., Apr. 21–24, 1986.
56. T. W. Rave, *Chemtech* **15**, 54 (Jan. 1985).
57. T. Ohta, *Polym. Eng. Sci.* **23**, 697 (1983).
58. M. Kamezawa, K. Yamada, and M. Takayanagi, *J. Appl. Polym. Sci.* **24**, 1227 (1979).
59. H. H. Chuah and R. S. Porter, *J. Polym. Sci. Polym. Phys. Ed.* **22**, 1353 (1984).
60. J. L. White and J. F. Fellers, *J. Appl. Polym. Sci. Appl. Polym. Symp.* **33**, 137 (1978).
61. P. Smith and P. J. Lemstra, *Makromol. Chem.* **180**, 2983 (1979).
62. P. Smith and P. J. Lemstra, *J. Mater. Sci.* **15**, 505 (1980).
63. P. Smith and P. J. Lemstra, *Polymer* **21**, 1341 (1980).

64. B. Kalb and A. J. Pennings, *Polymer* **21**, 3 (1980).
65. B. Kalb and A. J. Pennings, *J. Mater. Sci.* **15**, 2584 (1980).
66. A. F. Wills, G. Capaccio, and I. M. Ward, *J. Polym. Sci. Polym. Phys. Ed.* **18**, 493 (1980).
67. R. C. Wincklhofer, "Extended Chain Polyethylene Fiber: New Technology/New Horizons," paper presented at *TAPPI 1985 Nonwovens Symposium*, Myrtle Beach, S.C., Apr. 21–25, 1985.
68. U.S. Pat. 4,413,110 (Nov. 1, 1983), S. Kavesh and D. C. Prevorsek (to Allied-Signal Inc.).
69. R. J. Samuels, *J. Polym. Sci. Polym. Phys. Ed.* **17**, 535 (1979).
70. S. L. Cannon, G. B. McKenna, and W. O. Statton, *J. Polym. Sci. Macromol. Rev.* **11**, 209 (1976).
71. *Text. Organon* **57**, 6 (1986).
72. *Text. Organon* **63**, 4 (1992).
73. *Text. Organon* **62**, 298 (1991).
74. *Text. Organon* **58**, 286 (1987).
75. Ref. 71, p. 294.
76. *Am. Tex.* **13**, 44 (Dec. 1984).
77. R. G. Mansfield, *Nonwovens Ind.* **16**, 26 (Feb. 1985).
78. W. Shoemaker, *Nonwovens Ind.* **15**, 52 (Oct. 1984).
79. D. J. Hannant, *Fiber Cements and Fiber Concretes*, John Wiley & Sons, Inc., New York, 1978.
80. *Nonwovens Rept.* **87**, 1 (July 1978).
81. U.S. Pat. 4,492,781 (Jan. 8, 1985), F. J. Duszak, J. P. Modrak, and D. Deaver (to Hercules Inc.).
82. J. P. Modrak, "Fiber Reinforced Asphalt," paper presented at *19th Paving Conference and Symposium*, Albuquerque, N.M., Jan. 12, 1982.
83. A. G. Bondarenko and co-workers, *Fibre Chem.* **14**, 246 (May–June 1982).
84. *Daily News Record* **11**, 12 (May 4, 1981).

General References

M. Ahmed, *Polypropylene Fibers: Science and Technology*, Elsevier Science Publishing Co., Inc., New York, 1982, pp. 344–346.

V. L. Erlich, "Olefin Fibers," in N. M. Bikales, ed., *Encyclopedia of Polymer Science and Technology*, 1st ed., Vol. 9, John Wiley & Sons, Inc., 1968, pp. 403–440.

L. M. Landoll, "Olefin Fibers," in J. I. Kroschwitz, ed., *Encyclopedia of Polymer Science and Engineering*, 2nd ed., Vol. 10, John Wiley & Sons, Inc., New York, 1987, pp. 373–395.

C. J. WUST, JR.
L. M. LANDOLL
Hercules Incorporated

POLYESTER

Polyesters were initially discovered and evaluated in 1929 by W. H. Carothers, who used linear aliphatic polyester materials to develop the fundamental understanding of condensation polymerization, study the reaction kinetics, and demonstrate that high molecular weight materials were obtainable and could be melt-spun into fibers (1–5).

$$n\ \mathrm{HOOC{-}R{-}COOH} + n\ \mathrm{HO{-}R'{-}OH} \rightarrow \left(\overset{\mathrm{O}}{\overset{\|}{\mathrm{C}}}{-}\mathrm{R}{-}\overset{\mathrm{O}}{\overset{\|}{\mathrm{C}}}{-}\mathrm{O}{-}\mathrm{R'}{-}\mathrm{O} \right)_n$$

However, because of the low melting points and poor hydrolytic stability of polyesters from available intermediates, Carothers shifted his attention to linear aliphatic polyamides and created nylon as the first commercial synthetic fiber. It was nearly 10 years before J. R. Whinfield and J. T. Dickson were to discover the merits of poly(ethylene terephthalate) [*25038-59-9*] (PET) made from aromatic terephthalic acid [*100-21-0*] (TA) and ethylene glycol [*107-21-1*] (2G).

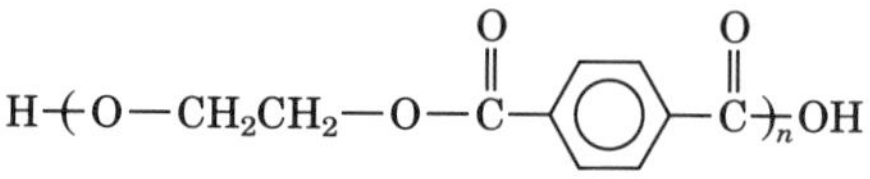

$$\mathrm{H}\left(\mathrm{O{-}CH_2CH_2{-}O{-}}\overset{\mathrm{O}}{\overset{\|}{\mathrm{C}}}{-}\mathrm{C_6H_4}{-}\overset{\mathrm{O}}{\overset{\|}{\mathrm{C}}} \right)_n\mathrm{OH}$$

The Whinfield and Dickson patents (6,7) dominated the art. The U.S. patent rights were assigned to E. I. du Pont de Nemours & Co., Inc., and Imperial Chemical Industries Ltd. (ICI) obtained the rights for the rest of the world. These patents were quickly followed by patents for improved catalysts for exchange or polymerization reactions (8–12) and for improved fiber properties by drawing (13–15). PET is a fiber of great commercial significance, useful in cordage, apparel fabrics, industrial fabrics, conveyor belts, laminated and coated substrates, and numerous other areas.

Properties

The Textile Fiber Product Identification Act (TFPIA) requires that the fiber content of textile articles be labeled (16). The Federal Trade Commission established and periodically refines the generic fiber definitions. The current definition for a polyester fiber is "A manufactured fiber in which the fiber-forming substance is any long-chain synthetic polymer composed of at least 85% by weight of an ester of a substituted aromatic carboxylic acid, including but not restricted to terephthalate units, and para substituted hydroxybenzoate units."

Poly(ethylene terephthalate), the predominant commercial polyester, has been sold under trademark names including Dacron (Du Pont), Terylene (ICI), Fortrel (Wellman), Trevira (Hoechst-Celanese), and others (17). Other commercially produced homopolyester textile fiber compositions include poly(1,4-cyclohexanedimethylene terephthalate) [*24936-69-4*] (Kodel II, Eastman), poly(butylene terephthalate) [*26062-94-2*] (PBT) (Trevira, Hoechst-Celanese), and poly(ethylene

4-oxybenzoate) [*25248-22-0*] (A-Tell, Unitika). Other polyester homopolymer fibers available for specialty uses include polyglycolide [*26124-68-5*], polypivalolactone [*24937-51-7*], and polylactide [*26100-51-6*].

In the late 1980s, new fully aromatic polyester fibers were introduced for use in composites and structural materials (18,19). In general, these materials are thermotropic liquid crystal polymers that are melt-processible to give fibers with tensile properties and temperature resistance considerably higher than conventional polyester textile fibers. Vectran (Hoechst-Celanese and Kuraray) is a thermotropic liquid crystal aromatic copolyester fiber composed of *p*-hydroxybenzoic acid [*99-96-7*] and 6-hydroxy-2-naphthoic acid. Other fully aromatic polyester fiber composites have been introduced under various trade names (19).

Most polyester fiber produced is standard molecular weight (ca 0.6 dL/g intrinsic viscosity), round cross-section PET. However, to engineer specific properties for special uses, many product variants have been developed and commercialized. These variants include using alternative cross sections, controlling polymer molecular weight, modifying polymer composition by using comonomers, and using additives including delusterants, pigments, and optical brighteners.

Changing the cross section of standard PET by the use of specially designed spinneret capillaries can change fabric visual and tactile aesthetics. Fabrics with luster and hand ranging from silk or cotton to fur have been made from nonround cross-section fibers (20,21). Trilobal, pentalobal, octalobal, and scalloped-oval fiber cross sections are currently offered commercially. Fibers containing single or multiple holes are currently used in filling products for improved bulk and thermal management (22–24).

High molecular weight polymer is used for high strength industrial fibers in tires, ropes, and belts. High strength and toughness are achieved by increasing the polymer molecular weight from 20,000 to 30,000 or higher (DP = 150–200) by extended melt polymerization or solid-phase polymerization. Special spinning processes are required to spin the high viscosity polymer to high strength fiber (25). Low molecular weight fibers are weak but have a low propensity to form and retain pills, ie, fuzz balls, which can be formed by abrasion and wear on a fabric surface (26). Most pill-resistant fibers are made by spinning low molecular weight fibers in combination with a melt viscosity booster (27).

Standard polyester fibers contain no reactive dye sites. PET fibers are typically dyed by diffusing dispersed dyestuffs into the amorphous regions in the fibers. Copolyesters from a variety of copolymerizable glycol or diacid comonomers open the fiber structure to achieve deep dyeability (7,28–30). This approach is useful when the attendant effects on the copolyester thermal or physical properties are not of concern (31,32). The addition of anionic sites to polyester using sodium dimethyl 5-sulfoisophthalate [*3965-55-7*] has been practiced to make fibers receptive to cationic dyes (33). Yarns and fabrics made from mixtures of disperse and cationically dyeable PET show a visual range from subtle heather tones to striking contrasts (see DYES, APPLICATION AND EVALUATION).

In addition to dyeability, polyesters with a high percentage of comonomer to reduce the melting point have found use as fusible binder fibers in nonwoven fabrics (32,34,35). Specially designed copolymers have also been evaluated for flame-retardant PET fibers (36,37).

Fibers spun from two different polyesters placed side-by-side or in a sheath–core arrangement have found utility (18,35,38,39). Bicomponent fibers produced from PET and a copolymer can be used as a binder fiber. Bicomponent fibers made from PET and PBT homopolymers are used in apparel applications which take advantage of the dyeability and high recovery of the PBT polyester.

Most textile fibers are delustered with 0.1–3.0 wt % TiO_2 to reduce the glitter and plastic appearance. Many PET fibers also contain optical brighteners (17). Through the use of soluble dyes or pigments, including photochromic pigments (19), a wide variety of producer-colored fibers and effects is available.

Physically or chemically modifying the surface of PET fiber is another route to diversified products. Hydrophilicity, moisture absorption, moisture transport, soil release, color depth, tactile aesthetics, and comfort all can be affected by surface modification. Examples include coating the surface with multiple hydroxyl groups (40), creating surface pores and cavities by adding a gas or gas-forming additive to the polymer melt (41), roughening the surface by plasma treatment of fibers coated with fine particles (42), forming grooves and rough surfaces by combining alkaline treatment with special fiber cross sections (43), and increasing water sorption and improving comfort by alkaline treatment of a freshly extruded fiber surface (44,45).

Fine Structural Properties. The performance and properties of PET fibers are significantly impacted by the relative amounts of amorphous and crystalline structures, the orientation of the structures with respect to the fiber axis, and the size distribution of the crystalline regions. By x-ray diffraction, the unit cell of crystalline PET has been determined to be triclinic ($a = 0.456$ nm; $b = 0.594$ nm; $c = 1.075$ nm (46)) with one monomer unit per crystalline unit cell. In the crystalline regions, the molecular chains are almost fully extended (1.075 nm unit cell length vs 1.090 nm theoretical fully extended chain length) with the aliphatic segments in a trans configuration and the aromatic rings in a planar side-by-side register perpendicular to the fiber axis (47).

Density, mechanical, and thermal properties are significantly affected by the degree of crystallinity. These properties can be used to experimentally estimate the percent crystallinity, although no measure is completely adequate (48). The crystalline density of PET can be calculated theoretically from the crystalline structure to be 1.455 g/cm^3. The density of amorphous PET is estimated to be 1.33 g/cm^3 as determined experimentally using rapidly quenched polymer. Assuming the fiber is composed of only perfect crystals or amorphous material, the percent crystallinity can be estimated and correlated to other properties.

Mechanical Properties. Polyester fibers are formed by melt spinning generally followed by hot drawing and heat setting to the final fiber form. The molecular orientation and crystalline fine structure developed depend on key process parameters in all fiber formation steps and are critical to the end use application of the fibers.

Molecular orientation and crystallinity generally increase with draw ratio, increasing break tenacity and Young's modulus while decreasing fiber break elongation. Typical properties of continuous filament and staple poly(ethylene terephthalate) fibers are shown in Table 1. Fiber dimensional stability and Young's modulus also are dependent on the heat-setting process. Fibers that are relaxed,

Table 1. Mechanical Properties of PET Fibers

	Staple/tow		Continuous filament		
Property	Regular tenacity	High tenacity	POY[a]	Regular tenacity	High tenacity
break tenacity, N/tex[b]	0.3–0.5	0.5–0.6	0.2–0.3	0.4–0.5	0.6–0.9
elongation, %	40–60	20–30	110–250	20–40	10–25
elastic recovery, % at 5% elongation	75–80	90		88–93	90
stiffness, N/tex[b]	1–2	5–6	0.2–0.5	1–3	5–7
toughness, N/tex[b]	0.02–0.15	0.02–0.10	0.10–0.20	0.04–0.10	0.04–0.07

[a]POY = partially oriented yarn.
[b]To convert N/tex to gf/den, multiply by 11.33.

or heat-set under no restraint, show a low shrinkage and low initial modulus. Annealed fibers, which are heat-set under tension at constant length, have a low shrinkage and maintain a high initial modulus. Typical stress–strain curves for poly(ethylene terephthalate) fibers are shown in Figure 1. Other factors including polymer molecular weight or the presence of comonomers can significantly affect the fiber mechanical properties.

Chemical Properties. The hydrolysis of PET is acid- or base-catalyzed and is highly temperature dependent and relatively rapid at polymer melt temperatures. Treatment for several weeks in 70°C water results in no significant fiber strength loss. However, at 100°C, approximately 20% of the PET tenacity is lost in one week and about 60% is lost in three weeks (47). In general, the hydrolysis and chemical resistance of copolyester materials is less than that for PET and depends on both the type and amount of comonomer.

At room temperature, PET is resistant to organic and moderate strength mineral acids. At elevated temperatures, PET strength loss in moderate strength acids can be appreciable. Strong acids such as concentrated sulfuric acid dissolve and depolymerize PET.

Polyester fibers have good resistance to weakly alkaline chemicals and moderate resistance to strongly alkaline chemicals at room temperature. PET fibers are attacked by strongly alkaline substances in one of two ways (47). Surface etching is caused by strongly alkaline chemicals including sodium hydroxide (caustic soda) or sodium silicate. Caustic reduction has been used to produce fine diameter fibers from mono- or multicomponent starting materials (17,47). Other organic bases including ammonia and methylamine penetrate the structure, initiating attack of the polymer molecule in the amorphous regions and resulting in significant loss of strength.

Polyester fibers have excellent resistance to soap, detergent, bleach, and other oxidizing agents. PET fibers are generally insoluble in organic solvents, including cleaning fluids, but are soluble in some phenolic compounds, eg, *o*-chlorophenol.

Thermal Properties. The melting point of poly(ethylene terephthalate) is generally reported to be 258–265°C and is generally considered to be independent of molecular weight (17,47–50). Copolymerization with generated or added co-

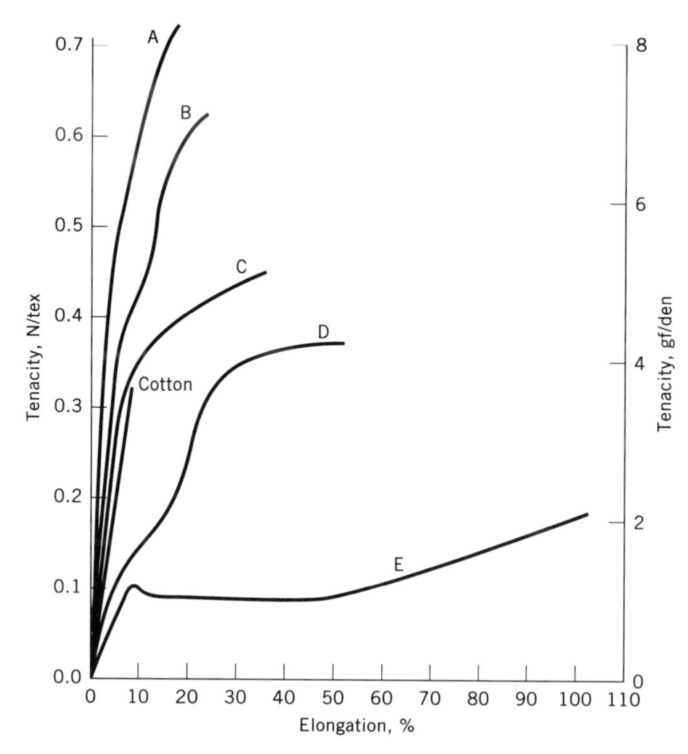

Fig. 1. Typical stress–strain curves for cotton and PET fibers. A, industrial; B, high tenacity, staple; C, regular tenacity, filament; D, regular tenacity, staple; E, partially oriented yarn.

monomers results in a decrease in the melting point and a disruption of the crystalline order (lower crystallinity) dependent on the amount of comonomer present. The melting point of the copolymer can be estimated by:

$$\frac{1}{T_m} - \frac{1}{T_m^0} = -\frac{R}{\Delta H_m} \ln n$$

where T_m = melting point of polyester copolymer; T_m^0 = melting point of pure PET polymer; R = ideal gas law constant; ΔH_m = heat of fusion of PET; and n = mole fraction of PET. The heat of fusion of PET is reported (49) to be approximately 24 kJ/mol (30 cal/g) and is dependent on the crystallinity of the polymer. The melting points of most copolymers are 2.5–3.5°C per mol % lower than PET and are independent of the chemical nature of the comonomer.

The glass-transition temperature, T_g, of dry polyester is approximately 70°C and is slightly reduced in water. The glass-transition temperatures of copolyesters are affected by both the amount and chemical nature of the comonomer (32,47). Other thermal properties, including heat capacity and thermal conductivity, depend on the state of the polymer and are summarized in Table 2.

Other Properties. Polyester fibers have good resistance to uv radiation although prolonged exposure weakens the fibers (47,51). PET is not affected by

Table 2. Thermal Properties of PET

Property	Value
melting point, T_m, °C	255–265
glass-transition temperature, T_g, °C	60–77 67 (amorphous) 81 (crystalline)
stick temperature, °C	230–240
heat capacity, C_p, kJ/(kg·K)[a]	
molten polymer (270–290°C)	$1.357 + 2.364 \times 10^{-3}\,T$
undrawn fiber (−5 to 60°C)	$1.033 + 4.213 \times 10^{-3}\,T$
thermal conductivity, W/(m·s·K)	37.5×10^{-3}
thermal diffusivity, cm²/s	9.29×10^{-4}
heat of fusion, ΔH_f, kJ/mol[a]	24
heat of combustion, ΔH_c, kJ/kg[a]	2.16×10^4

[a]To convert kJ to kcal, divide by 4.184.

insects or microorganisms and can be designed to kill bacteria by the incorporation of antimicrobial agents (19). The oleophilic surface of PET fibers attracts and holds oils. Other PET fiber properties can be found in the literature (47,49).

Manufacturing and Processing

Terephthalic acid (TA) or dimethyl terephthalate [*120-61-6*] (DMT) reacts with ethylene glycol (2G) to form bis(2-hydroxyethyl) terephthalate [*959-26-2*] (BHET) which is condensation polymerized to PET with the elimination of 2G. Molten polymer is extruded through a die (spinneret) forming filaments that are solidified by air cooling. Combinations of stress, strain, and thermal treatments are applied to the filaments to orient and crystallize the molecular chains. These steps develop the fiber properties required for specific uses. The two general physical forms of PET fibers are continuous filament and cut staple.

Raw Materials. For the first decade of PET manufacture, only DMT could be made sufficiently pure to produce high molecular weight PET. DMT is made by the catalytic air oxidation of *p*-xylene to crude TA, esterification with methanol, and purification by crystallization and distillation. After about 1965, processes to purify crude TA by hydrogenation and crystallization became commercial (52) (see PHTHALIC ACID AND OTHER BENZENEPOLYCARBOXYLIC ACIDS). In Japan, oxidation conditions are modified to give a medium purity TA suitable to manufacture PET, provided color toners such as bluing agents or optical brighteners are added during polymerization (53). Compared to DMT, advantages of TA as an ingredient are lower cost, no methanol by-product, lower investment and energy costs, higher unit productivity, and more pure polymer because less catalyst is used (54–56). Ethylene glycol is made by oxidizing ethylene to ethylene oxide (qv) followed by hydrolysis (see GLYCOLS, ETHYLENE GLYCOL AND DERIVATIVES). Catalysts are used in the transesterification reaction of DMT with 2G and in polycondensation. Many compounds have catalytic activity (29). Divalent zinc and manganese are the prevalent transesterification catalysts. Antimony, titanium, and germanium are the predominant polycondensation catalysts. Up to 3% delusterant is added to many PET fiber products to make them more opaque and scatter light; titanium dioxide is the most common delusterant. PET fiber blended with cotton for apparel frequently contains small amounts of fluorescent optical brighteners added during polymerization.

Polymerization. Commercial production of PET polymer is a two-step process carried out through a series of continuous staged reaction vessels. First, monomer is formed by transesterification of DMT or by direct esterification of TA with 2G:

$$\underset{\text{DMT}}{CH_3O\overset{O}{\overset{\|}{C}}-C_6H_4-\overset{O}{\overset{\|}{C}}OCH_3} + \underset{\text{2G}}{2\,HOCH_2CH_2OH} \xrightarrow{-2\,CH_3OH} \underset{\text{BHET}}{HOCH_2CH_2O\overset{O}{\overset{\|}{C}}-C_6H_4-\overset{O}{\overset{\|}{C}}OCH_2CH_2OH}$$

$$\underset{\text{TA}}{HO\overset{O}{\overset{\|}{C}}-C_6H_4-\overset{O}{\overset{\|}{C}}OH} + \underset{\text{2G}}{2\,HOCH_2CH_2OH} \xrightarrow{-2\,H_2O} \underset{\text{BHET}}{HOCH_2CH_2O\overset{O}{\overset{\|}{C}}-C_6H_4-\overset{O}{\overset{\|}{C}}OCH_2CH_2OH}$$

Starting with DMT, methanol is removed from the reaction; starting with TA, water is removed. Catalysts are used to transesterify DMT but not for direct esterification of TA. The second step is the polycondensation reaction which is driven by removing glycol. A polycondensation catalyst is used.

$$\underset{\text{BHET}}{n\,HOCH_2CH_2O\overset{O}{\overset{\|}{C}}-C_6H_4-\overset{O}{\overset{\|}{C}}OCH_2CH_2OH} \xrightarrow{\text{catalyst}}$$

$$\underset{\text{PET}}{H\!\left(OCH_2CH_2O\overset{O}{\overset{\|}{C}}-C_6H_4-\overset{O}{\overset{\|}{C}}\right)_{n}OCH_2CH_2OH} + \underset{\text{2G}}{(n-1)\,HOCH_2CH_2OH}$$

In general, esterification is conducted in one or two vessels forming low molecular weight oligomers with a degree of polymerization of about 1 to 7. The oligomer is pumped to one or two prepolymerization vessels where higher temperatures and lower pressures help remove water and 2G; the degree of polymerization increases to 15 to 20 repeat units. The temperatures are further increased and pressures decreased in the final one or two vessels to form polymer ready to spin into fiber. For most products, the final degree of polymerization is about 70 to 100 repeat units. Number average molecular weight is about 22,000; weight average molecular weight is about 44,000. Typical process conditions are shown in Figure 2.

Esterification and prepolymerization vessels may be agitated. Polycondensation vessels have agitators designed to generate very thin films. As PET poly-

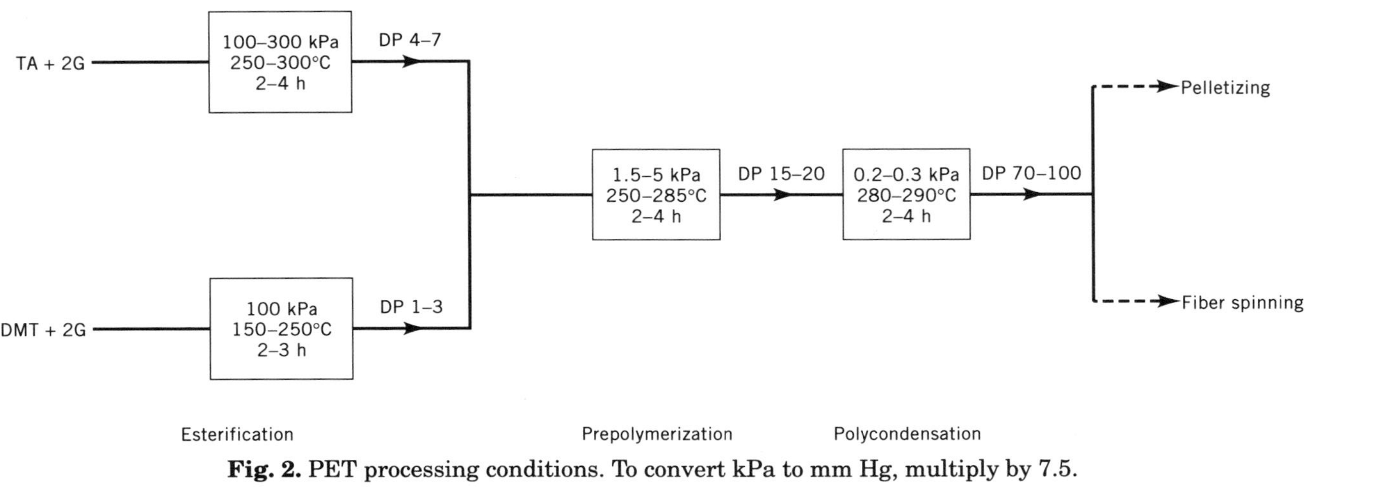

Fig. 2. PET processing conditions. To convert kPa to mm Hg, multiply by 7.5.

merizes, viscosity increases rapidly to more than 250 Pa·s. Generating surfaces increases the rate of removal of 2G from the melt and, consequently, the rate of polymerization. Agitator designs include screws, ribbons, rotating disks, and wiped films (56,57). Temperatures and hold-up times are optimized for each set of vessels to minimize the occurrence of side and degradation reactions. By-products in PET manufacture are diethylene glycol, acetaldehyde, water, carboxyl end groups, vinyl end groups, and anhydride end groups (58–60). Oligomers, mostly cyclic trimer, also are present in the range of 1–3% (61). Reaction kinetics are complex; they involve heat and mass transfer and multiple reactions (62–64). Control and removal of the by-products are required to maintain polymer purity and consistency and to protect the environment. Polymer is either pumped directly to fiber spinning units or is solidified and collected as pellets for later remelting and spinning.

For some uses, higher molecular weight polymer consisting of 150–200 repeat units is required. Such polymer usually is prepared by solid-state polymerization in which pellets are heated under an inert atmosphere to 200–240°C. The 2G is removed continuously. The rate of polymerization depends on particle size, end group composition, and crystallinity (65).

Older polyester plants generally were based on a batch or semicontinuous polymerization process. Polymer was extruded, solidified in water, and cut to chips or pellets. Capacity was 30 to 60 t/day. In the 1970s and 1980s, polymerization unit capacities increased to 150 to 225 t/day which reduced investment and operating costs per weight of polymer (54,55). Coupling polymerization and spinning, which eliminated the chipping step, also significantly reduced costs.

Spinning. PET fibers are made either by directly spinning molten polymer or by melting and spinning polymer chip as shown schematically in Figure 3. A special, precise metering pump forces the molten polymer heated to about 290°C through a spinneret consisting of a number of small capillaries, typically 0.2 to 0.8 mm in diameter and 0.3 to 1.5 mm long, under pressures up to 35 MPa (5000 psi). After exiting the capillary, filaments are uniformly cooled by forced convection heat transfer with laminar-flow air (66). Air flow can be transverse across the bundle, radial from outside-in or from inside-out, or a combination of transverse and radial along the threadline length. Solidification generally occurs from 0.2 to up to several meters from the spinneret (67,68). Following solidification, the threadline is passed over a finish applicator and collected at speeds of 100 to 7000-plus m/min for subsequent processing. Continuous filament products are small bundles of up to 300 individual filaments. Each bundle is collected on an individually wrapped package for further processing or for direct use. Staple products generally are bundles of 200 to 3000 individual filaments. Each bundle is combined with up to 50 or more others and collected in a large container for further processing. Staple is spun at speeds up to about 2000 m/min. A spin finish is applied to reduce friction and eliminate static.

Flow processes inside the spinneret are governed by shear viscosity and shear rate. PET is a non-Newtonian elastic fluid. Spinning filament tension and molecular orientation depend on polymer temperature and viscosity, spinneret capillary diameter and length, spin speed, rate of filament cooling, inertia, and air drag (69,70). These variables combine to attenuate the fiber and orient and sometimes crystallize the molecular chains (71).

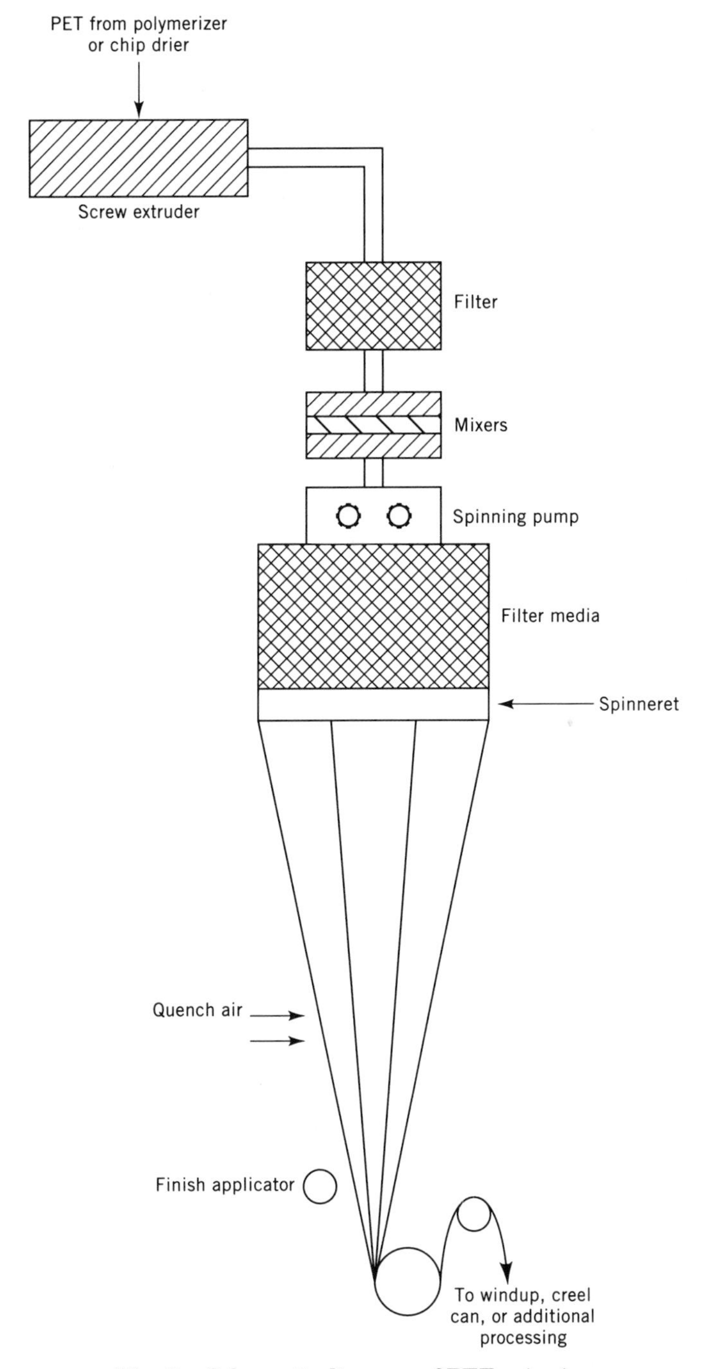

Fig. 3. Schematic diagram of PET spinning.

It is convenient to classify commercial PET spinning processes according to the degree of molecular orientation developed in the spun fiber. Generally, the classification is a function of spinning speed (67,72): low oriented yarn (LOY) is spun at speeds from 500 to 2500 m/min; partially oriented yarn (POY) is spun at 2500 to 4000 m/min; highly oriented yarn (HOY) is spun at 4000 to 6500 m/min; and fully oriented yarn (FOY) is spun at greater than 6500 m/min. Figures 4 and 5 show some trends in fiber physical properties and fine structure response to spinning speed.

LOY is characterized by low spinning tension, mostly rheological effects, little orientation, amorphous structure, low tensile strength, and high elongation. The spun filament must be drawn, usually three to six times its initial length, and heat-treated before it develops useful properties. Nearly all PET staple is spun this way.

At POY spinning speeds, orientation increases rapidly. Crystallinity begins to increase. The combination gives a fiber of moderate strength and dimensional stability. The discovery of these properties, the development of winders capable of collecting yarn at these speeds, and the discovery of sequential or simultaneous draw texturing led to the explosive growth of textured polyester filament markets in the 1970s and 1980s (73,74).

At HOY speeds, the rate of increase in orientation levels off but the rate of crystallization increases dramatically. Air drag and inertial contributions to the threadline stress become large. Under these conditions, crystallization occurs very rapidly over a small filament length and a phenomenon called neck-draw occurs

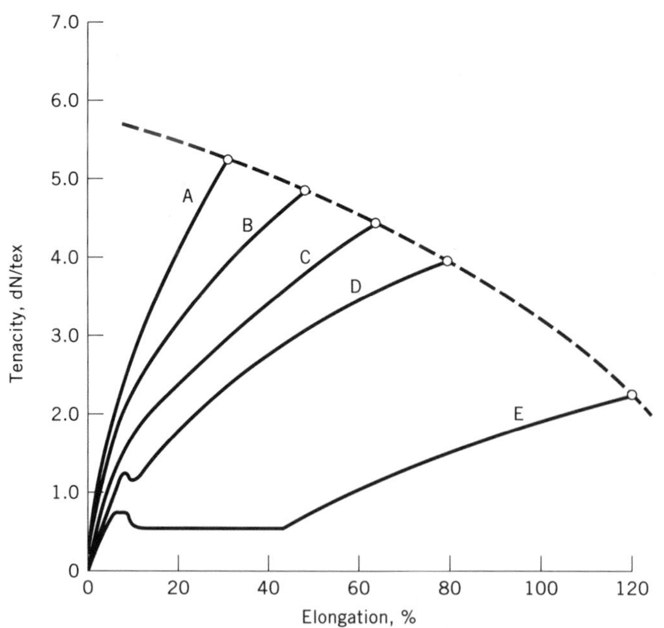

Fig. 4. Representative stress–strain curves of spun and drawn PET: A, low speed spun–mechanically drawn yarn; B, 6405 m/min; C, 5490 m/min; D, 4575 m/min; E, 3202 m/min. To convert dN/tex to gf/den, multiply by 1.13.

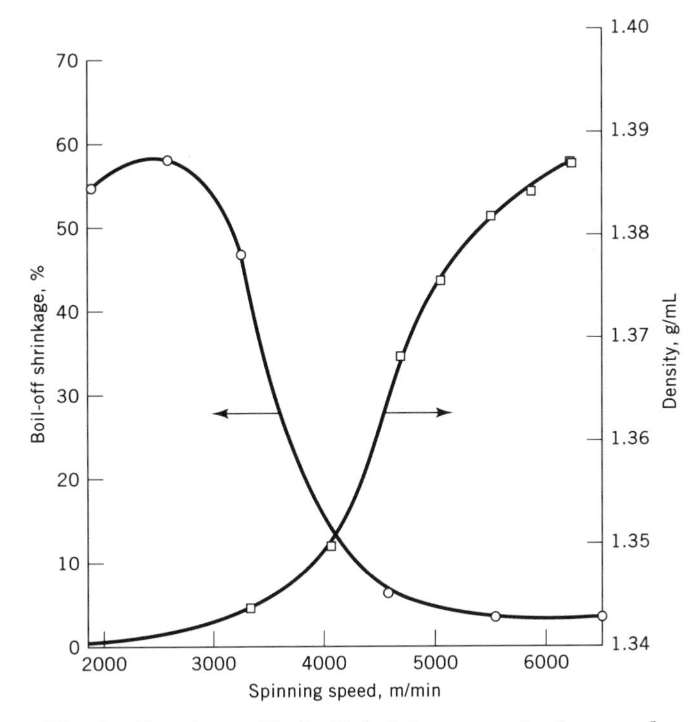

Fig. 5. Density and boil-off shrinkage vs spinning speed.

(68,75,76). The molecular structure is stable, fiber tensile strength is adequate for many uses, thermal shrinkage is low, and dye rates are higher than traditional slow speed spun, drawn, and heat-set products (77).

FOY speeds are the most recent development in PET spinning (78). Properties are similar to HOY and appear to be limited by the differential cooling rate from filament surface to filament core. This leads to radial distribution of viscosity, stress, and, consequently, molecular orientation (75). Fiber tensile strength is limited. Nevertheless, speeds up to 7000 m/min are commercial and forecasts are for speeds up to 9000 m/min by the year 2000 (79). Speeds to 9000 m/min have been studied (68,80,81).

Drawing and Stabilization. Drawing is the stretching of low orientation, amorphous spun yarns (LOY) to several times their initial length. This is done to increase their orientation and tensile strength. The temperature is above T_g, about 80°C, to ensure plastic deformation and maximum elongation after drawing. Molecular chains orient and heat initiates crystallization. Precise control of the temperature profile is necessary to orient the molecular chains before crystallization limits the amount of draw available and the fiber ruptures. Drawing in two or more stages is useful to optimize tensile properties and process continuity (82). Stabilization is heating the fiber to release stress within the molecular chains, melt and reform crystals, and increase the level of crystallinity in order to stabilize the fiber structure. Heating without tension allows the fiber to shrink and reduce orientation. Tensile strength and modulus decrease; thermal dimensional stability and elongation increase. Heating at constant length, with tension, maintains the fiber tensile strength and modulus. Crystallinity increases, but some residual shrinkage remains if the fiber is subsequently heated to above the stabilization temperature. Tensile stress in these structures is borne primarily by the tie molecules bridging the amorphous regions between the crystalline lattices. Consequently, tensile strength is only a small fraction of theoretical bond strengths because tie molecules are a small fraction of the total molecules (69).

Staple Processes. In staple processing, the containers of combined spun ends are further combined to form a tow band of one to 300,000 tex (2.7×10^6 den) and fed to a large draw line as shown schematically in Figure 6. The tow band is spread out into a flat band tracking over multiple feed and draw rolls. The surface speed of the rolls is from 100 to 400 m/min. Natural staple fibers, such as wool or cotton, have a three-dimensional configuration such that groups of them are held together by cohesive forces. Synthetic fibers are not cohesive unless special steps are taken. Crimping is the process by which two-dimensional configuration and cohesive energy is imparted to synthetic fibers so they may be carded and converted to spun yarns. Many proposals have been made to crimp synthetic fibers, but the stuffer box principle dominates commercial operation (69). Two rolls force the tow band into a chamber. Geometrical and frictional restraining forces in the chamber cause the tow band filaments to buckle and form a plug. Crimp frequency, amplitude, and permanence of the bent filament are functions of roll and chamber geometry, frictional restraining forces, and filament temperature. Many staple products are relaxed after crimping under low tension in a continuous-belt convection oven. These products generally are used in applications where high tensile strength and modulus are not critical properties. High strength, high modulus staple fibers are made by passing the flat rope band at high tension over heated multiple rolls (83) or through high pressure saturated steam (84,85) prior to crimping. Because these fibers have been crystallized in a linear configuration, crimping them is more difficult than crimping products which are later stabilized. The tow band is cut to precise lengths using a radial multiblade cutter, normally 30 to 40 mm for blending with cotton, 50 to 100 mm for blending with wool, and up to 150 mm for making carpets. Cut staple is packaged in up to 500 kg bales at densities greater than 0.5 g/cm^3 and shipped to a mill for further conversion. For some uses, uncut tow is boxed and shipped. Finishes specially designed to facilitate the various conversion processes are applied at one or several stages in the drawing, stabilizing, cutting, and baling processes.

In the last few years, small, 10–15 t/day, compact staple spinning/drawing/cutting/baling units have been offered for plants producing specialty, small-volume products (86,87). These units offer high flexibility for rapid production changes and simplicity of operation. These units often are used to reprocess polyester polymer recovered from bottles to make carpet fiber or filling products (88).

Filament Processes. Filament-spun products are generally processed in one of two ways: in the first, spun filaments are drawn and stabilized by passing individual ends over sets of heated rolls to draw, twist or interlace, and stabilize the fiber. Initially, individually packaged ends were unwound, drawn, twisted, and rewound to new individual packages. Later, higher speed processes to draw and stabilize the threadlines during spinning were developed (89). Multiple spun

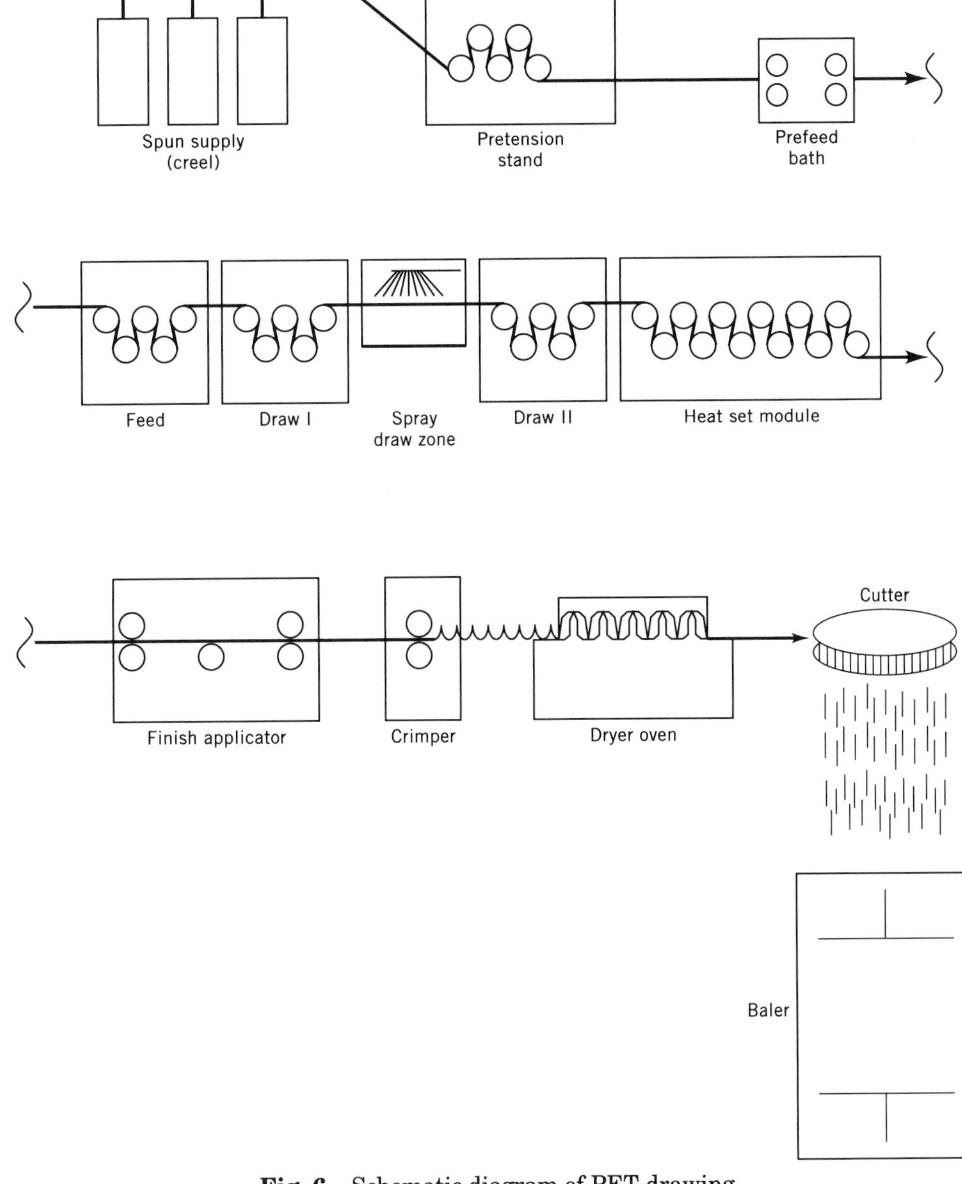

Fig. 6. Schematic diagram of PET drawing.

threadlines have been combined into a warp sheet, drawn, and wound up on beams (90). These products have no crimp. The second process introduces crimp into filament products; the process is called texturing. Its purpose is to increase yarn bulk and elasticity and give the fabric tactile aesthetics more like those from staple. The loops and bends created in the yarn by the texturing process impart fabric tactile aesthetics similar to those created by the free fiber ends in spun staple yarns.

Texturing. This is a process applied to continuous filament yarns to introduce loops and bends in the individual filaments. The first process to bulk yarn was twisting followed by untwisting and was disclosed in the early 1930s (91). Stuffer box, knit–deknit, false twist devices, and air jets also are used to texture yarns. In the 1970s, with the advent of POY, false twist texturing grew explosively and accounts for most of the textured yarn produced in the 1990s.

In false twist texturing (FTT), shown schematically in Figure 7, a device twists yarn upstream of its location as the threadline passes across a heater, and the yarn untwists downstream from the device and is wound up. If the yarn is not heated downstream from the twist device, it has bulk and high elasticity (stretch). If the yarn is heated downstream from the twist device, it has bulk, but much less stretch. FTT machines initially used two steps to sequentially draw and texture. Later machines combined those steps to simultaneously draw and texture (92). Initial machines used pin spindles as the false twist device. Texturing speeds were about 150 m/min and slowly increased to about 300 m/min. As POY quality and structural stability improved, new friction twist devices have appeared that allow higher speed texturing. Pin twisters control twist directly,

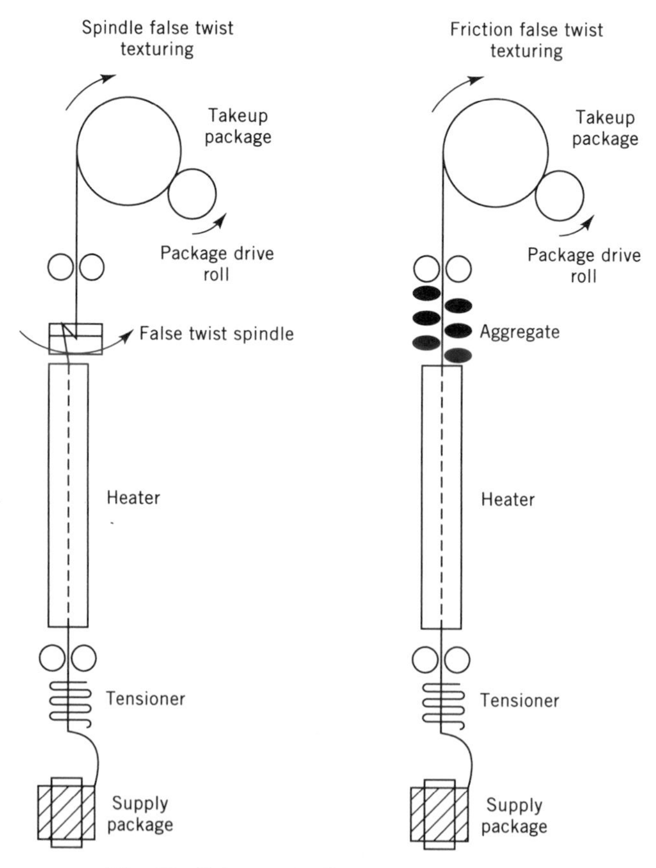

Fig. 7. Schematic diagram of texturing.

but friction twisters control it indirectly by controlling yarn torque. Stacked disks generate rotational friction at the nip points between sets of vertically arranged disks (93). The disks have polyurethane or ceramic surfaces. Crossed belts generate rotational friction between belt surfaces traveling in opposite directions (94). In ring twisters, overlapping disks (rings) rotating in opposite directions generate rotational friction. The yarn path is between the overlapping surfaces of two rings. Normal pressure applied to the ring surface creates frictional twisting forces. The yarn crossing angle, external pressure, and ring-to-yarn speed ratio are key variables (95). Texturing speeds reach up to 1200 m/min. The current barrier to higher speeds is a problem called surging. As speeds and yarn tensions increase, threadline instability caused by intermittent variation in yarn tension develops (96,97). Texturing speed becomes unstable and yarn quality is unacceptable. Shortening the texturing zone and increasing the rate of heat transfer by using condensing steam are potential routes to increase speeds further (98,99).

The air jet textured yarn process is based on overfeeding a yarn into a turbulent air jet so that the excess length forms into loops that are trapped in the yarn structure. The air flow is unheated, turbulent, and asymmetrically impinges the yarn. The process includes a heat stabilization zone. Key process variables include texturing speed, air pressure, percentage overfeed, filament linear density, air flow, spin finish, and fiber modulus (100). The loops create visual and tactile aesthetics similar to false twist textured and staple spun yarns.

Analytical Test Methods

Physical testing applications and methods for fibrous materials are reviewed in the literature (101–103) and are generally applicable to polyester fibers. Microscopic analyses by optical or scanning electron microscopy are useful for evaluating fiber parameters including size, shape, uniformity, and surface characteristics. Computerized image analysis is often used to quantify and evaluate these parameters for quality control.

Polyester composition can be determined by hydrolytic depolymerization followed by gas chromatography (28) to analyze for monomers, comonomers, oligomers, and other components including side-reaction products (ie, DEG, vinyl groups, aldehydes), plasticizers, and finishes. Mass spectroscopy and infrared spectroscopy can provide valuable composition information, including end group analysis (47,101,102). X-ray fluorescence is commonly used to determine metals content of polymers, from sources including catalysts, delusterants, or tracer materials added for fiber identification purposes (28,102,103).

Gel permeation chromatography can be used to determine the molecular weight and molecular weight distribution of polyester polymers. Polymer molecular weight can also be evaluated using wet chemistry techniques. Polyester polymers are dissolved in strong solvents such as phenol, *o*-chlorophenol, dichloroacetic acid, tetrachloroethane–phenol mixtures, and hexafluoro-2-propanol (17,47,49). Relative viscosities, comparing the solution viscosity of polymer solutions at standard concentrations vs the solvent viscosity, are commonly used for quality assurance and control. Intrinsic viscosity, η, is measured from solution and several studies have correlated the number average molecular weight to intrinsic viscosity by a variety of mathematical equations (19,47,49,50).

A variety of analytical techniques have been used to study the structure of polyester fibers. Microscopic refractometry and interferometry can be used to determine the fiber birefringence. Spectroscopic methods including infrared, near infrared, Fourier transform infrared, mass, and Raman spectroscopy have been used to evaluate structural details of polyesters (17,47,49,101–103). Wide angle and small angle x-ray diffraction have been used to obtain information about the crystalline nature of polyesters (17,47,49,102). Thermal mechanical analysis and nuclear magnetic resonance (nmr) have been used to determine second-order transition temperatures of polyester polymers and fibers. Differential scanning calorimetry (dsc) and differential thermal analysis are useful in detecting thermal properties as well as structural or composition information. Advances in computer control and data acquisition have allowed the on-line monitoring and control of various polymer and fiber parameters such as polymer melt viscosity and fiber spinning tension.

Economic Aspects

Since the initial commercial production of polyester fiber at Du Pont's plant in Kinston, North Carolina, in 1953, polyester production expanded to an annual worldwide production of approximately 9.85 million tons in 1992 (104). The growth of worldwide and U.S. polyester fiber production is shown in Figure 8. These data show that although worldwide growth in polyester production has been rapid, polyester production in the United States has been level throughout most of the 1980s. Polyester production has currently leveled in the industrialized countries, with real growth occurring in the developing countries. With several engineering companies designing and installing turnkey operations, polyester production has become a reliable means of creating jobs and supplying feed materials for development of garment-based exports.

Worldwide, the production capacity for polyester fiber is approximately 11 million tons; about 55% of the capacity is staple. Annual production capacity in the United States is approximately 1.2 million tons of staple and 0.4 million tons of filament. Capacity utilization values of about 85% for staple and about 93% for filament show a good balance of domestic production vs capacity (105). However, polyester has become a worldwide market with over half of the production capacity located in the Asia/Pacific region (106). The top ranked PET fiber-producing countries are as follows: Taiwan, 16%; United States, 15%; People's Republic of China, 11%; Korea, 9%; and Japan, 7% (107–109). Worldwide, the top producing companies of PET fibers are shown in Table 3 (107–109).

PET is based on petroleum and the price of polyester fiber fluctuates with the price of *p*-xylene and ethylene raw materials as well as with the energy costs for production. With the ability to interchange with other fibers, especially cotton in cotton blends, the price of polyester is affected by the price and availability of cotton as well as the supply and demand of polyester.

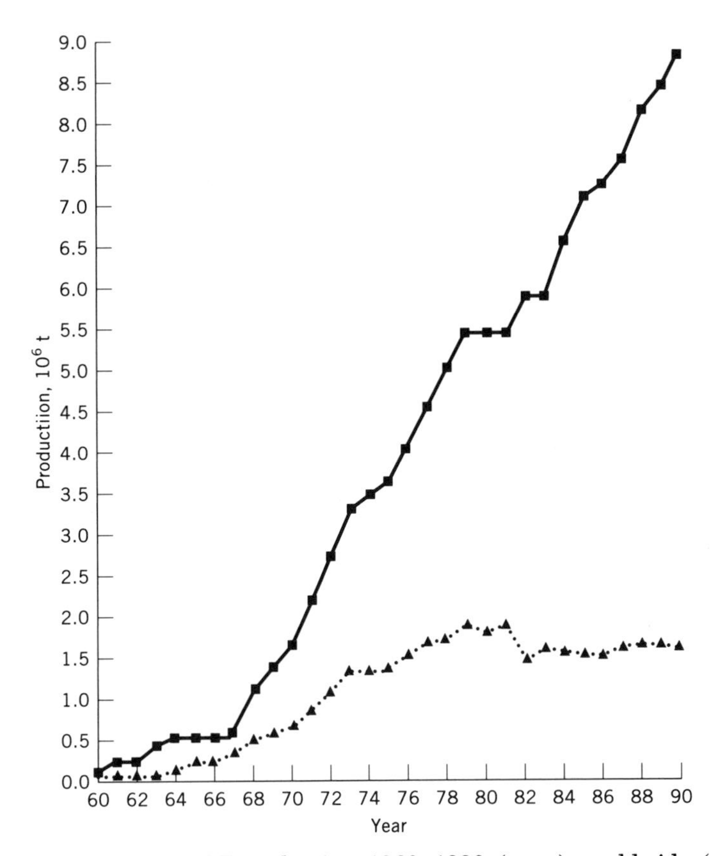

Fig. 8. Worldwide and U.S. PET production, 1960–1990; (—■—), worldwide; (· · ▲ · ·), U.S.

Table 3. 1990 PET Fiber Production Capacity

Manufacturer	Capacity, 10^3 t	Location
Hoechst	700	United States, Europe
Du Pont	650	United States, Europe
Nan Ya	361	Asia/Pacific
Teijin	352	Asia/Pacific
Wellman	337	United States, Europe
Far Eastern	312	Asia/Pacific
Sam Yang	243	Asia/Pacific

Safety and Environmental Factors

Health Safety. PET fibers pose no health risk to humans or animals. Fibers have been used extensively in textiles with no adverse physiological effects from prolonged skin contact. PET has been approved by the U.S. Food and Drug Administration for food packaging and bottles. PET is considered biologically inert and has been widely used in medical inserts such as vascular implants and artificial blood vessels, artificial bone, and eye sutures (19). Other polyester homopolymers including polylactide and polyglycolide are used in resorbable sutures (19,47).

When PET is extracted with water no detectable quantities of ethylene glycol or terephthalic acid can be found, even at elevated extraction temperatures (110). Extractable materials are generally short-chained polyesters and aldehydes (110). Aldehydes occur naturally in foods such as fruits and are produced metabolically in the body. Animal feeding studies with extractable materials show no adverse health effects.

Environmental Factors. PET materials are not dangerous to the environment and cannot contaminate surface or ground water. During polymerization, noncondensible organic by-products are stripped from the process outflow streams and burned. Glycol and water are separated by refining. The water is treated in a standard wastewater facility. The glycol is reused. The methanol from the DMT transesterification is refined and reused. Like all materials, polyesters should be disposed of properly to avoid litter and can be disposed of by landfill or incineration. In sanitary landfills, PET produces no leachate problems and the packing of materials promotes aeration, accelerating breakdown of biodegradable materials present. In incineration, PET has a relatively high fuel value, ca 23 MJ/kg (9900 BTU/lb), promoting efficient combustion and energy recovery.

A key environmental advantage for PET material is the ability to recycle. Polyester materials, especially bottles, can be separated from contaminated materials such as aluminium caps and paper labels and remanufactured by direct remelt extrusion into fibers for filling products or carpets (111) or into layered constructions for food-grade bottles. In alternative recycling processes, PET can react with methanol (methanolysis) or with ethylene glycol or 1,4-cyclohexanedimethanol (glycolysis) to produce low molecular weight monomer or oligomers (112,113) that are recycled back through the polymerization process.

Applications

Staple. PET staple is widely used in 100% polyester or cotton-blend fabrics for apparel. Typical cotton-blend polyester staple fibers have a linear density of about 0.08 to 0.4 tex (0.7 to 3 den) per filament, a tenacity of about 0.4–0.6 N/tex (5–7 gf/den), and a crimp frequency of 3–6 crimps per cm (7–14 crimps per in.). The fibers are coated with about 0.05–0.25 wt % of a finish to reduce friction and control static electricity, cut to about 25–75 mm (1–3 in.), and packaged into 300–500 kg bales. Light, topweight apparel fabrics are commonly 35–65 wt % polyester, and heavier bottomweight fabrics are generally 50/50 blends.

Along with cotton blends, polyester blends with rayon or wool are also important. Wool–polyester blends are widely used in men's suiting materials. For these fabrics, PET staple or tow can be used with a linear density typically about 0.16–0.45 tex per filament (1.5–4 dpf) and a staple length of 50–75 mm (2–3 in.).

In addition to fabrics, PET staple is used in a wide variety of other applications. High tenacity staple fibers are widely used in sewing thread. Staple PET fibers have been engineered for use in rugs, carpets, and filling products including furniture, pillows, mattresses, sleeping bags, and stuffed toys. Polyester staple fibers are commonly used in nonwovens for applications in diaper coverstock, filters, linings and interfaces, and disposable towels and wipes.

Filament. Fully drawn flat yarns and partially oriented (POY) continuous filament yarns are available in yarn sizes ranging from about 3.3–33.0 tex (30–300 den) with individual filament linear densities of about 0.055 to 0.55 tex per filament (0.5–5 dpf). The fully drawn hard yarns are used directly in fabric manufacturing operations, whereas POY yarns are primarily used as feedstock for draw texturing. In the draw texturing process, fibers are drawn and bulked by heat-setting twisted yarn or by entangling filaments with an air jet. Both textured and hard yarns are used in apparel, sleepwear, outerwear, sportswear, draperies and curtains, and automotive upholstery.

High molecular weight polyester is commonly used to make high strength industrial fibers. Typical yarn bundle sizes of 111–222 tex (1000–2000 den) and single filament sizes of 0.55–1.11 tex per filament (5–10 dpf) are available with tenacities on the order of 0.7–1.0 N/tex (8–10 gf/den). These fibers are commonly used in applications requiring high strength and stability, including tire cord, seat belts, industrial belts and hoses, ropes, cords, and sailcloth.

Polyesters are also used in continuous filament spunbonded nonwovens (see NONWOVEN FABRICS). Reemay spunbonded fabric is composed of continuous filament PET with a polyester copolymer binder. These spunbonded fabrics are available in a wide range of thicknesses and basis weights and can be used for electrical insulation, coated fabric substrates, disposable apparel for clean rooms, hospitals, and geotextiles (qv).

BIBLIOGRAPHY

"Polyester" under "Textile Fibers" in *ECT* 1st ed., Vol. 13, pp. 840–847, by S. A. Rossmassler, E. I. du Pont de Nemours & Co., Inc.; "Polyester Fibers" in *ECT* 2nd ed., Vol. 16, pp. 143–159, by G. Farrow and E. S. Hill, Fiber Industries; in *ECT* 3rd ed., Vol. 18, pp. 531–549, by G. W. Davis and E. S. Hill, Fiber Industries.

1. W. H. Carothers and J. A. Arvin, *J. Amer. Chem. Soc.* **51,** 2560 (1929).
2. U.S. Pat. 2,071,250 (Feb. 16, 1937), W. H. Carothers (to E. I. du Pont de Nemours & Co., Inc.).
3. W. H. Carothers, *Chem. Rev.* **8,** 353 (1931).
4. W. H. Carothers and J. W. Hill, *J. Amer. Chem. Soc.* **54,** 1559 (1932).
5. Ref. 4, p. 1579.
6. Brit. Pat. 578,079 (Aug. 29, 1946), J. R. Whinfield and J. T. Dickson (to Imperial Chemical Industries, Ltd.).
7. U.S. Pat. 2,465,319 (Mar. 22, 1949), J. R. Whinfield and J. T. Dickson (to E. I. du Pont de Nemours & Co., Inc.).
8. U.S. Pat. 2,534,028 (Dec. 12, 1950), E. F. Izard (to E. I. du Pont de Nemours & Co., Inc.).
9. U.S. Pat. 2,518,283 (Aug. 8, 1950), E. F. Casassa (to E. I. du Pont de Nemours & Co., Inc.).
10. U.S. Pat. 2,578,660 (Dec. 18, 1951), L. A. Auspos and J. B. Dempster (to E. I. du Pont de Nemours & Co., Inc.).
11. U.S. Pat. 2,647,885 (Aug. 4, 1953), H. R. Billica (to E. I. du Pont de Nemours & Co., Inc.).

12. U.S. Pat. 2,662,093 (Dec. 8, 1953), H. R. Billica (to E. I. du Pont de Nemours & Co., Inc.).
13. U.S. Pat. 2,533,013 (Dec. 5, 1950), H. F. Hume (to E. I. du Pont de Nemours & Co., Inc.).
14. U.S. Pat. 2,556,295 (June 12, 1951), A. Pace (to E. I. du Pont de Nemours & Co., Inc.).
15. U.S. Pat. 2,578,899 (Dec. 18, 1951), A. Pace (to E. I. du Pont de Nemours & Co., Inc.).
16. *Rules and Regulations Under the Textile Fibers Product Identification Act,* as amended July 9, 1986, Federal Trade Commission, Washington, D.C.
17. G. Farrow and E. S. Hill, in N. M. Bikales ed., *Encyclopedia of Polymer Science and Technology,* Vol. 11, Wiley-Interscience, New York, 1969, p. 1.
18. A. Ziabicki, *ATI,* 66 (May 1988).
19. T. Hongu and G. O. Phillips, *New Fibers,* Ellis Horwood Ltd., New York, 1990.
20. U.S. Pat. 3,846,969 (Nov. 11, 1974), J. B. McKay (to E. I. du Pont de Nemours & Co., Inc.).
21. U.S. Pat. 3,914,488 (Oct. 21, 1975), A. A. Goraffa (to E. I. du Pont de Nemours & Co., Inc.).
22. U.S. Pat. 3,772,137 (Nov. 13, 1973), J. W. Tolliver (to E. I. du Pont de Nemours & Co., Inc.).
23. U.S. Pat. 3,924,988 (Dec. 9, 1975), J. D. Hodge (to E. I. du Pont de Nemours & Co., Inc.).
24. U.S. Pat. 4,836,763 (June 6, 1989), C. R. Broaddus (to E. I. du Pont de Nemours & Co., Inc.).
25. U.S. Pat. 3,216,187 (Nov. 9, 1965), W. A. Chantry and A. E. Molini (to E. I. du Pont de Nemours & Co., Inc.).
26. D. Gintis and E. J. Mead, *Textile Res. J.* **29,** 578 (1959).
27. U.S. Pat. 3,335,211 (May 8, 1967), E. J. Mead and C. E. Reese (to E. I. du Pont de Nemours & Co., Inc.).
28. S. Morimoto, in H. F. Mark, S. M. Atlas, and E. Cernia, eds., *Man-Made Fibers, Science & Technology,* Vol. 3, John Wiley & Sons, Inc., New York, 1968, p. 21.
29. R. E. Wilfong, *J. Poly. Sci.* **54,** 385 (1961).
30. Bjorksten Res. Labs., *Polyesters and Their Application,* Reinhold Publishing Corp., New York, 1956.
31. N. C. Pierce, in P. W. Harrison ed., *Textile Progress* **3,** 38 (1971).
32. J. N. Kerawalla and S. M. Hansen, *Proceedings of the Polyester Textile Conference,* Shirley Institute, Blackpool, UK, May 1988.
33. U.S. Pat. 3,381,058 (Jan. 23, 1962), J. M. Griffing and W. R. Remington (to E. I. du Pont de Nemours & Co., Inc.).
34. U.S. Pat. 4,418,116 (Nov. 29, 1983), P. T. Scott (to E. I. du Pont de Nemours & Co., Inc.).
35. *Nonwovens Industry* **18,** 26 (1987).
36. U.S. Pat. 3,719,727 (Mar. 19, 1970), Y. Masai, Y. Kato, and N. Fukui (to Toyo Spinning Co., Ltd.).
37. U.S. Pat. 3,941,752 (Mar. 2, 1976), H. J. Kleiner and co-workers, (to Hoechst Aktiengesellschaft).
38. U.S. Pat. 3,671,379 (June 20, 1972), E. F. Evans and N. C. Pierce (to E. I. du Pont de Nemours & Co., Inc.).
39. R. Jeffries, *Bicomponent Fibers,* Merrow Publishing Co., Ltd., Watford Herts, UK, 1978.
40. U.S. Pat. 4,569,974 (Feb. 11, 1986), G. E. Gillberg-LaForce and R. N. DeMartino (to Celanese).
41. U.S. Pat. 4,380,594 (Apr. 19, 1983), E. Siggel, G. Wich, and E. Kessler (to Akzona).

42. U.S. Pat. 4,522,873 (June 11, 1985), T. Akagi, S. Yamaguchi, and A. Kubotsu (to Kuraray).
43. U.S. Pat. 4,954,398 (Sept. 4, 1990), S. Bagrodia and B. M. Phillips (to Eastman Kodak).
44. U.S. Pat. 5,069,847 (Dec. 3, 1991), T. H. Grindstaff (to E. I. du Pont de Nemours & Co., Inc.).
45. U.S. Pat. 5,069,846 (Dec. 3, 1991), T. H. Grindstaff and C. E. Reese (to E. I. du Pont de Nemours & Co., Inc.).
46. R. L. Miller, in J. Brandup and E. H. Immergut, eds., *Polymer Handbook,* 2nd ed., John Wiley & Sons, Inc., New York, 1975.
47. K. W. Hillier, in Ref. 28.
48. G. Farrow and I. M. Ward, *Polymer* **1,** 330 (1960).
49. Ref. 46, p. V–71.
50. F. W. Billmeyer, *Textbook of Polymer Science,* 3rd ed., Wiley-Interscience, New York, 1984.
51. M. Mohammadian and co-workers, *Textile Res. J.* **61,** 690 (1991).
52. Brit. Pat. 994,769 (June 10, 1965) (to Standard Oil); U.S. Pat. 3,584,039 (June 8, 1971), D. H. Meyer (to Standard Oil).
53. Jpn. Pat. 54-120,699 (Sept. 19, 1979), H. Hashimoto and co-workers (to Toyo Boseki).
54. C. H. Ho, *Man-Made Fiber Year Book,* Chemiefasern Textilindustrie, Frankfurt, 1988, p. 44.
55. H. D. Schumann, *Chemiefasern Textilindustrie,* **40/92,** 1058 (1990); English translation, **41/93,** E1 (1991).
56. K. Ravindrath and R. A. Mashelkar, in A. Whelan and J. L. Craft, eds, *Developments in Plastics Technology-2,* Elsevier Applied Science Publishers, Ltd., Essex, UK, 1985, p. 1.
57. L. Gerking, *Int. Fiber J.* **6**(2), 52 (1991).
58. H. A. Pohl, *J. Amer. Chem. Soc.* **73,** 5660 (1951).
59. L. H. Buxbaum, *Angew. Chem. Internat.* **7**(3), 182 (1968).
60. J.-M Besnoin and K. Y. Choi, *JMS-Rev. Macromol. Chem. Phys.* **C29**(1), 55 (1989).
61. W. S. Ha and Y. K. Choun, *J. Poly. Sci: Poly. Chem. Ed.* **17,** 2103 (1979).
62. K. Ravindrath and R. A. Mashelkar, *Chem. Eng. Sci.* **41**(9), 2197 (1986).
63. T. Yamada, *J. Appl. Poly. Sci.* **41,** 565 (1990).
64. T. Yamada, Y. Imamura, and O. Makimura, *Poly. Eng. Sci.* **25,** 788 (1985).
65. Ref. 54, p. 40.
66. F. Fourne in Ref. 54, p. 52.
67. G. W. Davis, A. E. Everage, and J. R. Talbot, *Fiber Producer,* 22 (Feb. 1984).
68. G. Vassilatos, B. H. Knox, and H. R. E. Frankfort in A. Ziabicki and H. Kawai, eds., *High Speed Fiber Spinning,* John Wiley & Sons, Inc., New York, 1985, p. 383.
69. H. Luckert and co-workers, *Ullman's Encyclopedia of Industrial Chemistry,* Vol. 10, 5th ed., VCH Publishers, New York, 1987, p. 511.
70. A. Ziabicki, in Ref. 68, p. 21; S. Kase, in Ref. 68, p. 67; S. Kubo, in Ref. 68, p. 115; M. Matsui, in Ref. 68, p. 137.
71. G-Y. Chen, J. A. Cuculo, and P. A. Tucker, *J. Appl. Polym. Sci.* **44,** 447 (1992).
72. S. C. Winchester, D. A. Shiffler, and S. M. Hansen, in J. J. McKetta, ed., *Encyclopedia of Chemical Processing and Design,* Vol. P, Marcel Dekker, Inc., New York, to be published.
73. U.S. Pat. 3,771,307 (Nov. 13, 1973), D. G. Petrille (to E. I. du Pont de Nemours & Co., Inc.).
74. U.S. Pat. 3,772,872 (Nov. 20, 1973), M. J. Piazza and C. E. Reese (to E. I. du Pont de Nemours & Co., Inc.).

75. A. Ziabicki, *Proceedings Fiber Producer Conference 1984,* Clemson University, Clemson, S.C., p. 1-1.
76. J. Shimizu, N. Okui, and T. Kikutani, in Ref. 68, p. 429.
77. U.S. Pats. 4,134,882 (Jan. 16, 1979) and 4,195,051 (Mar. 25, 1980), H. R. E. Frankfort and B. H. Knox (to E. I. du Pont de Nemours & Co., Inc.).
78. N. Ohya, *International Man-Made Fibers Congress,* Dornbirn, Austria, Sept. 1990.
79. R. Beyreuther, G. Schauer, and A. Schoene, *Acta Polymerica* **40,** 695 (1989).
80. J. Shimizu, N. Okui, and T. Kikutani, *Society of Fiber Sci. Tech. (Japan) J.* **37**(4), 29 (1981).
81. J. Shimizu, *J. Text. Mach. Soc. Japan* **38**(6), 243 (1985).
82. U.S. Pat. 3,816,486 (June 11, 1974), O. R. Vail (to E. I. du Pont de Nemours & Co., Inc.).
83. U.S. Pat. 3,044,250 (July 17, 1962), H. H. Hebler (to E. I. du Pont de Nemours & Co., Inc.).
84. U.S. Pat. 4,639,347 (Jan. 27, 1987), J. A. Hancock, W. D. Johnson, and A. D. Kennedy (to E. I. du Pont de Nemours & Co., Inc.).
85. U.S. Pat. 4,704,329 (Nov. 3, 1987), J. A. Hancock, W. D. Johnson, and A. D. Kennedy (E. I. du Pont de Nemours & Co., Inc.).
86. H. Luckert and W. Stibal, *Chemiefasern Textilindustrie* **36/88,** 24 (1985).
87. M. Navrone, in Ref. 54, 1990 ed., p. 44.
88. A. Schweitzer, *Chemiefasern Textilindustrie* **40/92,** 1066 (1990).
89. U.S. Pat. 3,452,132 (Apr. 22, 1963), G. Pitzl (to E. I. du Pont de Nemours & Co., Inc.).
90. J. F. Hagewood, *Inter. Fiber J.* **6**(3), 5 (1991).
91. Ger. Pat. 618,050 (Apr. 10, 1932), (to Heberlein Maschinenfabrik AG).
92. D. K. Wilson, in J. C. Ellis ed., *Textile Progress* **10**(3), 1 (1978).
93. K. Greenwood and P. J. Grigg, *J. Textile Inst.* **76,** 244 (1985).
94. T. J. Kang and A. El-Shiekh, *Textile Res. J.* **53,** 1 (1983).
95. T. J. Kang, W. Li, and A. El-Shiekh, *Textile Res. J.* **58,** 653 (1988).
96. J. Lunenschloss and J. Bruske, *Fiber and Textile Tech.,* 14, (Apr. 1989).
97. G. W. Du and J. W. S. Hearle, *J. Textile Inst.* **81,** 36 (1990).
98. J. Bruske, *Chemiefasern Textilindustrie* **37/89,** 102 (1987).
99. S. Li, S. Backer, and P. Griffith, *Textile Res. J.* **60,** 619 (1990).
100. D. K. Wilson and T. Kollu, in P. W. Harrison ed., *Textile Progress* **16**(3), 1 (1987).
101. J. W. S. Hearle and R. Muldith, eds., *Physical Methods of Investigating Textiles,* Interscience Publishers, Inc., New York, 1959.
102. F. Happey, ed., *Applied Fiber Science,* Vol. 1, Academic Press, Inc., New York, 1978.
103. F. Happey, ed., *Applied Fiber Science,* Vol. 2, Academic Press, Inc., New York, 1979.
104. *Fiber Organon* **64**(7), 1 (1993).
105. *Fiber Organon* **63**(1), 1 (1992).
106. *Japan Chemical Annual 1991,* The Chemical Daily Co., Ltd., Tokyo, 1991, p. 52.
107. *1990 Directory of Chemical Producers—United States,* SRI International, Menlo Park, Calif., 1990, p. 627.
108. *1990 Directory of Chemical Producers—Western Europe,* Vol. 2, SRI International, Menlo Park, Calif., 1990, p. 1225.
109. *1990 Directory of Chemical Producers—East Asia,* SRI International, Menlo Park, Calif., 1990, p. 437.
110. B. I. Turtle, *Dev. Soft Drink Tech.* **2,** 49 (1989).
111. K. B. Bryan, *PEP Review 90-3-1,* SRI International, Menlo Park, Calif., 1991.
112. U.S. Pat. 3,907,868 (Sept. 23, 1975), R. M. Currie (to E. I. du Pont de Nemours & Co., Inc.).
113. U.S. Pat. 4,605,762 (Aug. 12, 1986), J. W. Mandoki (to Celanese Mexicana SA).

General References

H. Luckert and co-workers, in *Ullman's Encyclopedia of Industrial Chemistry,* Vol. A10, 5th ed., VCH Publishers, New York, 1987, p. 511.
A. Ziabicki and H. Kawai, eds., *High Speed Fiber Spinning,* John Wiley & Sons, Inc., New York, 1985.
C. J. Heffelfinger and L. K. Knox, in O. J. Sweeting, ed, *Science & Technology of Polymer Films,* Vol. 1, John Wiley & Sons, Inc., New York, 1971.

S. M. HANSEN
P. B. SARGEANT
E. I. du Pont de Nemours & Co., Inc.

POLY(VINYL ALCOHOL)

The principal fiber types that fall under the category of vinyl fibers are fibers that contain at least 85% by weight of vinyl chloride, known generically as vinyon fibers, and those that are composed of at least 50% by weight of vinyl alcohol and are referred to as vinal fibers. The latter are by far larger volume commercial products. Other fibers in this category are based on vinylidene chloride or tetrafluoroethylene (see VINYL POLYMERS, VINYL CHLORIDE AND POLY(VINYL CHLORIDE); VINYLIDENE CHLORIDE AND POLY(VINYLIDENE CHLORIDE); FLUORINE COMPOUNDS, ORGANIC–POLYTETRAFLUOROETHYLENE).

Vinal fibers, or poly(vinyl alcohol) fibers, are not made in the United States, but the fiber is produced commercially in Japan, Korea, and China where the generic name vinylon is used. These materials are the subject of this article (see also VINYL POLYMERS, POLY(VINYL ALCOHOL)).

History of Poly(vinyl alcohol) Fiber

Vinyl alcohol does not exist as a monomer, but Herrmann and Haehnel (1) were able to obtain the desired product poly(vinyl alcohol) [*9002-89-5*] (PVA), by polymerizing vinyl acetate and then hydrolyzing the resultant poly(vinyl acetate). This process is employed for the commercial production of PVA even now. The principal concern of the discoverers was development of a suture for surgical operations; the fiber then obtained was not suited for clothing use (2).

In Germany (3), the UK (4), the United States (5), and France (6) study began to improve the water resistance of the fiber so that it could be used for general-purpose items including clothing. In Japan, Sakurada and his co-researchers succeeded in obtaining a water-insoluble PVA fiber in 1939, by wet spinning an aqueous PVA solution into a coagulating bath of concentrated aqueous sodium sulfate, followed by formalization of the fiber thus spun (7). At about the same time, Kanegafuchi Spinning Co. published a process for producing a PVA fiber by wet spinning an aqueous PVA solution into a specific coagulating bath (8).

In 1942, Kyoto University, with government support, constructed a pilot plant to promote research and development of PVA fiber; during World War II they continued developing techniques for the industrial production of PVA fiber. After the war, Kurashiki Rayon Co., Ltd. (now Kuraray Co., Ltd.) constructed a pilot plant and started commercial production in 1950. Many chemical companies and fiber manufacturers in Japan, such as Dainippon Spinning Co., Ltd. (now Unitika Co., Ltd.) and The Nippon Synthetic Chemical Industry Co., Ltd. subsequently began production of PVA materials including fibers. PVA fiber was given the generic name of vinylon in Japan in 1948, and vinal in the United States.

Commercial production of PVA fiber was thus started in Japan, at as early a period as that for nylon. However, compared with various other synthetic fibers which appeared after that period, the properties of which have continuously been improved, PVA fiber is not very well suited for clothing and interior uses because of its characteristic properties. The fiber, however, is widely used in the world because of unique features such as high affinity for water due to the —OH groups present in PVA, excellent mechanical properties because of high crystallinity, and high resistance to chemicals including alkali and natural conditions.

The People's Republic of China introduced Kuraray technology and started production of PVA fiber by a wet spinning process in 1965. Its annual capacity reached 165,000 tons in 1986 (9). The Democratic People's Republic of Korea produce PVA and reportedly have an annual production capacity of 50,000 tons (9).

Manufacture of Fiber

Raw Material. PVA is synthesized from acetylene [*74-86-2*] or ethylene [*74-85-1*] by reaction with acetic acid (and oxygen in the case of ethylene), in the presence of a catalyst such as zinc acetate, to form vinyl acetate [*108-05-4*] which is then polymerized in methanol. The polymer obtained is subjected to methanolysis with sodium hydroxide, whereby PVA precipitates from the methanol solution.

PVA used for the manufacture of fiber generally has a degree of polymerization of about 1700 and, for general-purpose fiber, a high degree of hydrolysis of vinyl acetate units of at least 99 mol %.

Pure PVA dissolves in water but does not fluidize by melting. Commercial production of PVA fiber is therefore carried out by wet spinning or dry spinning, utilizing aqueous PVA solution. In either case, purified PVA is dissolved in hot water and the solution is extruded through fine holes of a spinneret; the extruded streams are coagulated to form continuous filaments, which are then heat-treated to have adequate mechanical properties.

Since PVA fiber as spun is soluble in water, it is necessary to improve the water resistance of the as-spun fiber (10). Heat treatment followed by acetalization is a classic method to provide high water resistance.

Tows and staples, as well as short cut chips, are being manufactured by wet spinning and filament yarns by both dry and wet spinning.

Wet Spinning Process. Figure 1 is a flow diagram for producing PVA fiber by means of the wet spinning process utilizing a coagulating bath of an aqueous sodium sulfate solution. After PVA has been washed to remove impurities, it is dissolved in hot water to yield an aqueous solution with a prescribed concentration and viscosity. A dissolution-improving agent such as calcium chloride may be

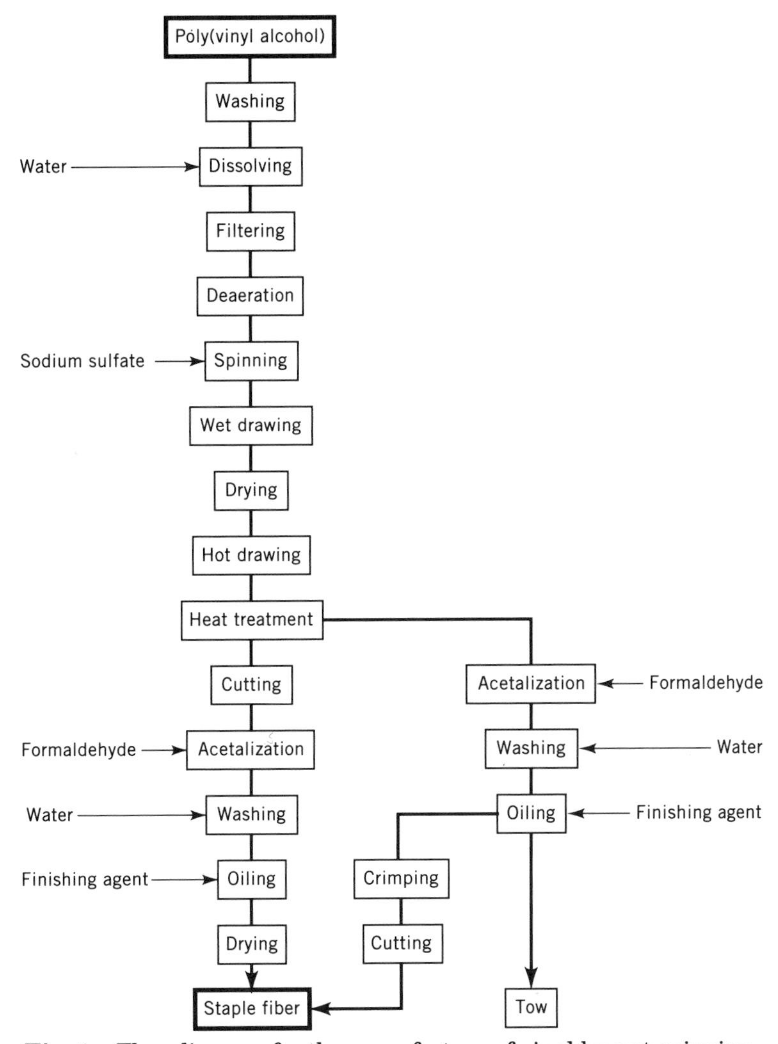

Fig. 1. Flow diagram for the manufacture of vinal by wet spinning.

added (11) or, instead of water, organic solvents capable of dissolving PVA can be used (12,13). The solution is, after filtration and deaeration, sent to a spinning machine. In the spinning machine, the aqueous PVA solution is extruded through spinnerets into a coagulating bath of an aqueous sodium sulfate solution with high concentration and coagulated therein; the continuous filaments formed are withdrawn from the bath.

The spinning machine is, generally, of the vertical type because of its stable operation and compactness (14). The filaments as spun are slightly drawn with guide rolls, then drawn in the coagulating bath in a ratio of 2:3 and dried. The filaments are then heat-drawn and heat-treated at high temperatures to obtain necessary mechanical properties and dimensional stability. The sodium sulfate deposited on the filaments from the coagulating bath does not hinder heat processing. The heat-treated filaments are provided with hot water resistance by acetalization with formaldehyde or the like.

For staple, the heat-treated filaments in the form of tow are cut to prescribed lengths and then acetalized to develop crimp on individual cut fibers. A suitable finish is selected from conventional ones and applied to the acetalized fibers to improve their spinnability and other properties required for the intended use.

Filament yarn is produced basically by the same process except that the spinning machine is so constructed as to prevent individual yarns leaving spinnerets from contacting each other.

Mechanism of Fiber Formation. After extrusion of a spinning dope through the orifices of a spinneret into a coagulating bath, the extruded streams undergo dehydration and coagulation by action of the salt contained in the bath. Network structure forms in each of the streams as a result of secondary bonding between the molecules of PVA. The fiber structure that forms is fixed depending on the stage during the coagulation and to what extent the network structure is formed. For instance, with salt in the bath, dehydration from the outer surface of each of the extruded streams proceeds rapidly just after extrusion and the outer surface of each of the extruded streams proceeds rapidly just after extrusion and the outer surface becomes a dense skin layer. The inner portion of the fiber solidifies only insufficiently because of the skin suppressing outward diffusion of interior water. As a result, the finished fiber has a two-layer (skin/core) structure as shown in the cross section in Figure 2**a.** On the other hand, in what is known as alkali spinning using a bath of sodium hydroxide solution or the like, gelation is considered to precede dehydration and the fiber cross section becomes circular and uniform (Fig. 2**b**).

The cross-sectional shape and uniformity of as-spun fiber influence the drawability in the succeeding heat-drawing process. As-spun fiber from dehydration–coagulation spinning is difficult to draw to a high draw ratio so that the finished fiber is limited in tensile strength and like properties. However, finished fiber from alkali spinning has high strength and low elongation as a result of the circular, uniform cross section.

An aqueous PVA solution containing a small amount of boric acid may be extruded into an aqueous alkaline salt solution to form a gel-like fiber (15,16). In this process, sodium hydroxide penetrates rapidly into the aqueous PVA solution extruded through orifices to make it alkaline, whereby boric acid cross-links PVA molecules with each other. The resulting fiber is provided with sufficient strength to withstand transportation to the next process step and its cross section does not show a distinct skin/core structure.

After the washing, the fiber is dried, and then is heat-drawn in the same manner as in the case of dehydration–coagulation with salt but to a much higher draw ratio. As a result the finished fiber has high strength and modulus and is, without acetalization, sufficiently resistant to boiling water. Figure 3 shows schematic fiber structures (17).

Boric acid/alkali spinning has been commercialized in Kuraray Co. and Unitika Co. in Japan, and is reportedly under research and development also in the People's Republic of China as a process for producing high strength PVA fiber to be used for replacing asbestos (9).

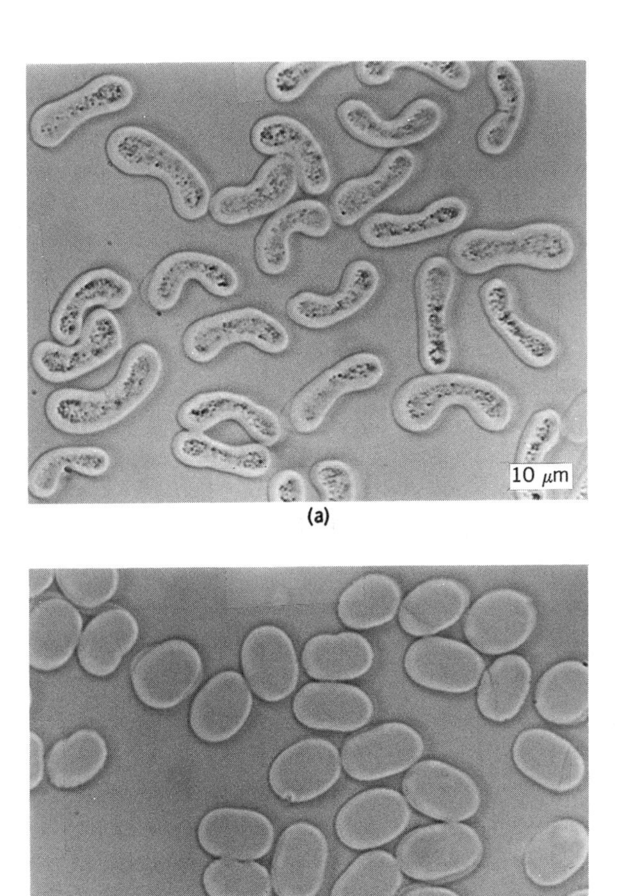

Fig. 2. Photograph of the cross sections of PVA fiber manufactured by wet spinning with a coagulating bath of sodium sulfate (**a**) and sodium hydroxide (**b**).

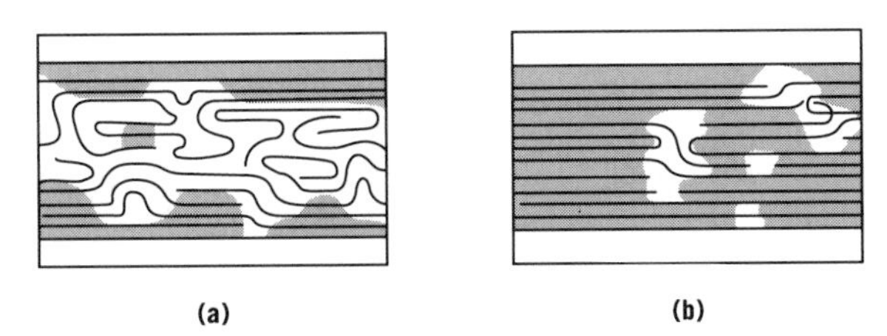

Fig. 3. Schematic comparison of the structures of PVA fibers formed by salt coagulation and alkali coagulation. (**a**) Low orientation, low crystallinity; (**b**) high orientation, high crystallinity.

Fine adjusting and optimization of each step of this process is still underway, and a PVA fiber having a single fiber strength as high as 2 N/tex (21 gf/dtex), which is close to that of aramid fiber, has been reported (18).

Dry Spinning. This process comprises extruding a high concentration aqueous PVA solution into conditioned air and drying the extruded streams to obtain fibers. Details of this process have been given (19,20). In dry spinning, the extruded streams coagulate by cooling, rather than by evaporation of water. Figure 4 shows a flow diagram of the manufacture of filament yarn by dry spinning.

This process starts with preparation of raw material PVA chips. Chip preparation is conducted for the purposes of securing prescribed water content and improving operation efficiency during extrusion dissolution. Ordinary PVA, containing sodium acetate, is washed with cool water to reduce the sodium acetate content. After washing and squeezing off of excess water, the PVA is fed to a kneader, where it is kneaded to give a prescribed water content and then formed into chips having a prescribed size through a granulator.

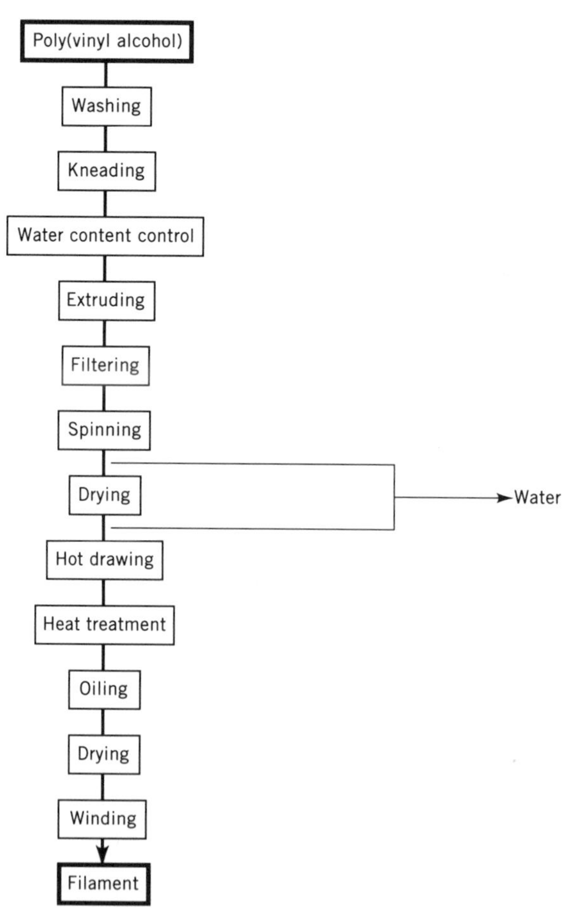

Fig. 4. Flow diagram for the manufacture of vinal by dry spinning.

The chips are fed to an extruder, where they are dissolved and the air contained therein is mostly removed by backflow. The concentrated solution leaving the extruder is filtered and then extruded through orifices into a vertical, several meter long spinning tube, where part of the water evaporates and the extruded filaments obtain the strength to withstand their own weight. The fluidity of the spinning solution is an important factor in fiber formation in spinning tubes, and a number of rheological studies have been conducted (21–24). One determination (25) of the relationship between the apparent viscosity upon passage through an orifice and the shearing force indicates that higher shearing force is favorable insofar as it decreases apparent viscosity, but it increases ballooning (die swell) of the extruded solution, which damages spinnability.

After being dried through a drier, the fiber is heat-drawn at a draw ratio of around 10, to obtain a serviceable strength. Various modifications have been applied to this process, to produce a variety of filament yarns including a monofilament having a diameter of 200 to 700 μm (for seaweed cultivation material and for reinforcing cement) and water-soluble yarns (26). Water-soluble yarns mostly use PVAs with low degree of polymerization, and the dissolution temperatures are adjusted by controlling the degree of hydrolysis of the raw material PVA and/or the heat-treatment conditions of the yarns (27).

Other Spinning Processes. The following examples are of scientific interest but have not been employed on an industrial scale.

Dimethyl Sulfoxide (DMSO) Solvent. The spinning solution is a 13–15% solution in DMSO, and the coagulating bath comprises acetone, methanol, toluene, or mixtures of the foregoing with DMSO (28). The fiber obtained as-spun can be drawn in a drawing ratio of 20 at 200°C, to give a finished fiber having a strength of 0.88 N/tex (9 gf/dtex) and a Young's modulus of 44 N/tex (450 gf/dtex).

Solvent of Ethylene Glycol and Other Polyols. Highly crystalline PVAs are obtained using a solvent of ethylene glycol (EG), diethylene glycol (DEG), triethylene glycol, or glycerol, which are called single-crystal solvents (29,30).

Dry-Jet Wet Spinning (Gel Spinning). This is a new spinning process that comprises dissolving a PVA with a high degree of polymerization, extruding the solution into air, immediately thereafter introducing the extruded streams into an aqueous salt solution or an organic solvent to permit them to gel therein, and drawing the gelled filaments in a high draw ratio (31,32). Fibers having high strength and modulus not hitherto achieved are reportedly obtained.

Fiber Mechanical Properties

The mechanical properties of PVA fiber vary depending on the conditions of fiber manufacture such as spinning process, drawing process, and acetalization conditions, and the manufacture conditions of raw material PVA. Table 1 shows the mechanical properties of PVA fibers commercially available in Japan under the name of vinylon (33). As apparent from the table, PVA fibers are characterized by high strength, low elongation, and high modulus. In addition to general-purpose types, high strength types with strength of at least 1.47 N/tex (15 gf/dtex) are produced by alkali spinning (34); material with a yarn strength of nearly 1.77 N/tex (18 gf/dtex) has become available (35).

Table 1. Mechanical Properties of Vinylon[a,b]

Property	Staple fiber		Filament yarn	
	Regular	High strength	Regular	High strength
tensile strength, N/tex[c]				
dry	0.35–0.58	0.6–0.93	0.3–0.35	0.84–1.20
wet	0.27–0.46	0.47–0.79	0.19–0.28	0.40–0.93
dry/wet strength ratio, %	72–85	78–85	70–80	75–90
loop strength, N/tex[c]	0.28–0.46	0.44–0.51	0.40–0.53	0.62–1.15
knot strength, N/tex[c]	0.22–0.35	0.40–0.46	0.2–0.26	0.24–0–4
elongation, %				
dry	12–26	9–17	17–22	6–22
wet	12–26	9–17	17–25	8–26
elastic recovery,[d] %	70–85	72–85	70–90	70–90
initial modulus, N/tex[c]	2.3–6.2	6.2–22	5.3–7.9	10.6–35.3
apparent Young's modulus, MPa[e]	2,940–7,850	7,850–28,440	6,865–9,320	13,240–45,100
moisture regain, %				
20°C, 65% rh[f]	4.5–5.0	3.5–4.5	3.5–4.5	2.5–4.5
20°C, 20% rh	regular 1.2–1.8; high strength 1.0–1.5			
20°C, 95% rh	regular 10.0–12.0; high strength 8.0–10.0			

[a]Specific gravity = 1.26–1.30; official moisture regain = 5%.
[b]Ref. 33.
[c]To convert N/tex to gf/dtex, multiply by 10.2.
[d]At 3% elongation.
[e]To convert MPa to kgf/mm^2, multiply by 0.102.
[f]Standard conditions.

Physical and Chemical Properties

Moisture Absorbency. PVA fiber is more hygroscopic than any other synthetic fiber. The hygroscopicity varies depending on how the fiber is processed after spinning, ie, in heat-drawing, heat-treatment, acetalization, and the like.

Dimensional Stability. The wet heat resistance of PVA fiber is indicated by the wet softening temperature (WTS) at which the fiber shrinks to a specified ratio. At one time, the WTS was not more than 95°C for nonacetalized PVA fiber, but improvement of WTS has been achieved by improvement in heat-drawing and -treating techniques; other methods proposed include suppression of polymerization temperature of vinyl acetate (36) and employment of alkali spinning (37).

On the other hand, water-soluble PVA fibers are available on the market. They are stable in cool water but shrink in warm water and dissolve at 40 to 90°C. The dissolution temperature is controlled by the degree of polymerization and hydrolysis of PVA, heat-treatment conditions after spinning, etc.

PVA fiber is better in dimensional stability under dry heat than other synthetic fibers.

Thermal Resistance and Flammability. Thermal analysis of PVA filament yarn shows an endothermic curve that starts rising at around 220°C; the endothermic peak (melting point) is 240°C, varying a little depending on manufacture conditions. When exposed to temperatures exceeding 220°C, the fiber properties change irreversibly.

With respect to flammability, PVA fiber has a limiting oxygen index (LOI) of 20, the same as that of polyester and polyamide fibers, and does not drip even when burnt, a feature which is highly valued. A flame-retardant-grade PVA fiber is available; it is obtained by mix-spinning with poly(vinyl chloride). This fiber, having an LOI of 30–34 and high strength, is used for bedding and workclothes.

Weather Resistance. It has been shown for over 40 years that PVA fiber has excellent resistance against exposure to sunlight. The resistance to uv light is also excellent, which has been proven with the tents used in Antarctic expeditions and by Himalayas' climbing parties. Nets for golf practice ranges maintained 90% of their initial strength after actual exposure for seven years (38). A PVA nonwoven fabric used for reinforcing an asphalt roofing maintained, after 10 years of actual use, 100% of the initial fiber strength (39).

Chemical Resistance. Table 2 shows the chemical resistance of PVA fiber (40). The fiber exhibits markedly high resistance to organic solvents, oils, salts, and alkali. In particular, the fiber has unique resistance to alkali, and is hence widely used in the form of a paper principally comprising it and as reinforcing material for cement as a replacement of asbestos.

Table 2. Comparison of Various Fibers' Resistance to Chemicals

	Testing conditions			Retention of strength[a], %			
Chemical	Concentration, %	Temperature, °C	Time, h	Kuralon[b]	Cotton	Nylon	Polyester
acid							
sulfuric	10	20	10	100	51	56	100
hydrochloric	10	20	10	100	70	77	95
alkali							
sodium hydroxide	40	20	10	100	84	82	97
salt							
sodium carbonate	1	100	10	97	75	99	95
sodium chloride	3	100	10	100	90	86	98
organic solvent							
benzene	100	20	1000	100	100	88	98
carbon tetrachloride	100	20	1000	100	90	82	87
miscellaneous							
mineral oil	100	100	10	100	70	100	100
cottonseed oil	100	20	1000	100	100	93	100
lard	100	20	1000	100	100	76	100

[a]Control showed 100% retention.
[b]Kuralon is the registered name of Kuraray's PVA fiber.

Application

In the 1950s and 1960s PVA fiber had been used for clothing and interiors, but after commercialization of polyester fiber and acrylic fiber, PVA fiber gradually lost its market share in these fields because of problems in elastic recovery, dyeability, cost, etc. On the other hand, the fiber has found uses in a variety of industrial fields thanks to its superior mechanical properties of high strength and modulus and low elongation, resistance to chemicals, in particular to alkali, high durability against uv light, etc.

Rubber Industry. This is the oldest industrial area in which PVA fiber has been successfully applied. The fiber is used for reinforcement for belts, hoses, and the like, where its high strength and modulus and low elongation are highly appreciated. The fiber is also easy to treat with resorcinol–formaldehyde latex and has high adhesiveness.

PVA fiber is best fit for reinforcement of oil brake hoses for cars that require high reliability, because of excellent mechanical properties and good chemical resistance to pressure–transmission liquid contained in the hose.

Agricultural Materials. PVA fiber is used in this field mostly in the form of shade cloth. The mesh cloth protects plants and vegetables not only from vermin but also from frost or excessive sunlight. In some cases the mesh cloth, by reflecting far infrared light, is used for warmth keeping purposes. In these applications the high weather resistance of the fiber plays an important role, in addition to high strength and excellent dimensional stability.

Fishing Materials. Since the late 1960s, the fiber has been losing its market share in the field of fishing nets, because of other fiber materials superior in water-separatability and resistance to waves. At present, the fiber is used, because of chemical and physical uniqueness of the fiber surface, for nets for cultivating seaweed, an important food in Japan and Korea. Long lines for fishing tuna also use PVA fiber.

PVA fiber ropes are widely used in fishing and on ships, because of excellent weather resistance, coiling property, ease of handling, twist stability, etc. For this purpose spun yarns obtained directly from tow by the Perlok spinning system are used.

Sewing Thread. Spun yarns and filament yarns of PVA fiber with their characteristics of low elongation and high strength are used as industrial sewing threads for leather materials such as shoes and bags and for similar items. In Japan, the PVA fiber threads are also used for sewing tatami mattress.

Nonwoven Fabric. Crimped PVA staple is being used for the manufacture of dry-laid nonwoven. Also, as an example utilizing the uniqueness of the fiber, a soft sheet is prepared by shrinking and partly dissolving in hot water a nonwoven from water-soluble PVA fiber and then insolubilizing the fabric by acetalization or similar processes. This sheet is used as car wipers, wipers for high grade furniture, and for similar purposes.

Slivers obtained by directly stretch-breaking tow through the Perlok spinning system are formed into a nonwoven by a random web process. The nonwoven is impregnated with asphalt (qv) and this composite is used as roofing material. This usage makes the most of the excellent asphalt-impregnability, as well as the high strength and modulus, of PVA fiber.

Paper. Both warm-water-soluble and water-insoluble PVA fibers are used as raw materials for paper, the former as a fiber-shaped binder and the latter as a main, structural-member fiber singly or in combination with other fibrous materials, to yield unique species of paper. The insoluble type fiber is used, for example, in combination with wood pulp to improve paper strength or in combination with the water-soluble type to give an alkali-resistant paper. Paper containing PVA fiber is excellent in folding characteristics and has markedly high tensile strength. Because its porosity is readily controllable and it has high alkali resistance, the paper is widely used in the world as separator for alkali dry battery.

Reinforcement. PVA fiber in the form of short cut chips having a length of several millimeters to several tens of millimeters is widely used as raw material for paper and for reinforcing plastics, cement, and the like, and has been acquiring more and more significance.

Fiber-Reinforced Plastics. PVA fiber, with its high strength and toughness, has become widely used as reinforcement fiber for FRP. Strength as high as 1.8 N/tex (18 gf/dtex) has been reported (32,41). Penetration in this field will therefore become more active.

Fiber-Reinforced Cementitious Material. Use of asbestos (qv) has been legally restricted in Europe and the United States as being hazardous to health. In asbestos cement, which had consumed 70–80% of total asbestos, PVA fiber has been used in large amount as a replacement for asbestos. PVA fiber has a strength of at least 0.88 N/tex (9 gf/dtex) and can therefore provide the necessary reinforcement for cement; the fiber has excellent adhesiveness to cement (qv) and alkali resistance, and is not a health hazard.

Based on the technology developed for using PVA fiber as a replacement for asbestos in cement products, Kuraray has been developing thick fibers for reinforcing concrete (42). Super-thick fibers with a thickness of 39 tex (350 den) (200 μm in diameter) to 444 tex (4000 den) (660 μm in diameter) are now available; the 39 tex material is used for reinforcing various mortar-based cement products and the 444 tex material for reinforcing concrete in civil engineering works such as tunnels, roads, harbors, and bays.

BIBLIOGRAPHY

1. Ger. Pat. 450,286 (1924), W. Haehnel and W. O. Herrmann.
2. Ger. Pat. 685,048 (1931); U.S. Pat. 2,072,302 (1932); Brit. Pat. 386161 (1932), W. O. Herrmann and W. Haehnel.
3. Ger. Pat. 666,264 (1931), E. Hubert, H. Dabst, and H. Hecht.
4. U.S. Pat. 2,399,970 (May 7, 1946), D. L. Wilson, (to Courtaulds Ltd.).
5. U.S. Pat. 2,083,628 (June 15, 1937), G. D. Zelger (to Eastmann Kodak Co.).
6. Fr. Pat. 875,864 (Oct. 7, 1941), F. Fiorillo (to Secretariat d'Etat a la Production).
7. Jpn. Pat. 147,958 (Feb. 20, 1941), I. Sakurada, S. Lee, and H. Kawakami (to Inst. of Japan Chemical Fiber).
8. Jpn. Pat. 153,812 (Feb. 16, 1942), M. Yazawa and co-workers.
9. Z. Zhou, *Int. Man-Made Fibers Congress,* Dornbirn, Austria, Sept. 1986.
10. S. Kobinata and co-workers, *Gosei Sen-i,* 203 (1964).
11. Jpn. Pat. 43-22,355 (Sept. 25, 1968), T. Ashikaga, H. Kurashige, and T. Endou (to Kurashiki Rayon Co.).
12. U.S. Pat. 3,102,775 (Sept. 3, 1963), N. V. Seeger (to Diamond Alkali Co.).
13. U.S. Pat. 3,063,787 (Nov. 13, 1962), M. J. Rynkiewicz and W. K. Eugene (to Diamond Alkali Co.).

14. U.S. Pat. 2,988,802 (June 20, 1961), T. Tomonari, T. Osugi, and K. Taka (to Kuraray Co.).
15. J. Arakawa, *Sen-i Gakkaishi* **16,** 849 (1960).
16. U.S. Pat. 3,660,556 (May 20, 1972), T. Ashikaga and S. Kosaka (to Kuraray Co.).
17. A. Mizobe, private communication, Sept. 7, 1991.
18. H. Fujiwara and co-workers, *J. Appl. Polymer Sci.* **37,** 1403 (1989).
19. H. Suyama, *Sen-i to Kogyo* **21,** 5162 (1965).
20. S. Nagano, *Kobunshi Kogaku Koza* **4,** 301 (1965).
21. K. Kawai, *Zairyou Shiken* **9,** 278 (1960).
22. B. Bernstein, E. Kearsley, and L. Zapas, *Trans. Soc. Rheology* **7,** 391 (1965).
23. *Ibid.,* **9,** 27.
24. B. Bernstein, *Acta Mechanica* **4,** 329 (1966).
25. S. Nagano, *Kobunshi Kogaku Koza* **4,** 305 (1960).
26. Jpn. Pat. 1,016,001 (June 27, 1979), H. Narukawa, M. Okazaki, and T. Komatsu (to Kuraray Co.).
27. M. Uzumaki, *Sen-i Gakkaishi* **33,** 55 (1977).
28. Jpn. Pat. 535,769 (July 13, 1968), K. Matsubayashi and H. Segawa (to Kuraray Co.).
29. Jpn. Pat. 539,682 (Sept. 25, 1968), T. Ashikaga, T. Endou, and H. Kurashige (to Kuraray Co.).
30. Jpn. Pat. 546,553 (Dec. 19, 1968), T. Ashikaga, T. Endou, and H. Kurashige (to Kuraray Co.).
31. U.S. Pat. 4,440,711 (Apr. 3, 1984), Y. D. Quon, S. Kavesh, and D. S. Drevorsech (to Allied Corp.).
32. U.S. Pat. 4,603,083 (July 29, 1986), H. Tanaka, M. Suzuki, and F. Ueda (to Toray Industries, Inc.).
33. *Jpn. Chem. Fiber Assoc.* (Oct. 1975).
34. Jpn. Pat. 47-8186 (Mar. 8, 1972), K. Kawashima and A. Miyoshi (to Unitika Ltd.).
35. Product brochure, Kuraray Co. Ltd., Okayama City, Japan, 1988.
36. H. Kawakami and co-workers, *Sen-i Gakkaishi* **19,** 192 (1963).
37. Unpublished data, Kuraray Co., Ltd., Okayama City, Japan, May 1968.
38. Unpublished data, Kuraray Co., Ltd., Okayama City, Japan, 1982.
39. J. Hikasa, *Asbestos Replacement Symposium,* UMIST, Apr. 1984.
40. Unpublished data, Kuraray Co., Ltd., Okayama City, Japan, Feb. 1961.
41. *Nikkei New Material,* (Aug. 12, 1991).
42. Technical brochure, Kuraray Co., Ltd., Okayama City, Japan, 1990.

JUN-ICHI HIKASA
Kuraray Co., Ltd.

REGENERATED CELLULOSICS

Fibers manufactured from cellulose are either derivative or regenerated; historically they are designated man-made fibers and distinguished from synthetic fibers based on synthetic organic polymers. A derivative fiber is one formed when a chemical derivative of a natural polymer, eg, cellulose, is prepared, dissolved, and extruded as a continuous filament, and the chemical nature of the derivative is retained after the fiber formation process (see FIBERS, CELLULOSE ESTERS). A regenerated fiber is one formed when a natural polymer, or its chemical derivative, is dissolved and extruded as a continuous filament, and the chemical nature of the natural polymer is either retained or regenerated after the fiber formation process. The difficulties of making solutions of natural cellulose from which fibers can be spun has led to most fabricated cellulosic fibers being regenerated from more readily soluble derivatives of cellulose. It is the technology of making these fibers that is the subject of this article.

Originally, the word rayon was applied to any cellulose-based man-made fiber, and therefore included the cellulose acetate fibers. However, the definition of rayon was clarified in 1951 and includes textiles fibers and filaments composed of regenerated cellulose and excludes acetate. In Europe the fibers are now generally known as viscose; the term viscose rayon is used whenever confusion between the fiber and the cellulose xanthate solution (also called viscose) is possible.

All synthetic fibers are produced as continuous filaments, either as yarns or tows. Yarns are fine enough to be woven or knitted directly, but cannot be intimately blended with other fibers on the principal conversion systems used for cotton or wool. For these processes, staple fibers, made by cutting the much larger tows into short lengths, are needed. Tows can also be stretch broken into slivers or tops, which can then be drawn out and twisted into spun-yarns.

Denier and tex are both weight/unit length measures of fiber fineness: denier is the weight in grams of 9000 m of filament, yarn, or tow; tex is the weight in grams of 1000 m. For filaments, as opposed to tows or yarns, the decitex tends to be used as the measure of fineness, 1 denier being equivalent to 1.11 decitex. Filament strengths are measured in gram force/denier (gf/den) or in centiNewtons/tex (cN/tex): 1 gf/den is equivalent to 8.82 cN/tex. Occasionally, the strength of filaments is seen expressed as N/mm^2 to allow comparison with engineering materials. One N/mm^2 is equivalent to the strength in gf/den $\times$ 9 $\times$ 9.807 $\times$ the fiber specific gravity. For cellulose, the specific gravity = 1.51, so N/mm^2 = gf/den $\times$ 133.277.

History

The first successful attempt to make textile fibers from plant cellulose can be traced to George Audemars (1). In 1855 he dissolved the nitrated form of cellulose in ether and alcohol and discovered that fibers were formed as the dope was drawn into the air. These soft strong nitrocellulose fibers could be woven into fabrics but had a serious drawback: they were explosive, nitrated cellulose being the basis of gun-cotton (see CELLULOSE ESTERS, INORGANIC).

Sir Joseph Swan, as a result of his quest for carbon fiber for lamp filaments (2), learned how to denitrate nitrocellulose using ammonium sulfide. In 1885 he exhibited the first textiles made from this new artificial silk, but with carbon fiber being his main theme he failed to follow up on the textile possibilities. Meanwhile Count Hilaire de Chardonnet (3) was researching the nitrocellulose route and had perfected his first fibers and textiles in time for the Paris Exhibition in 1889. There he got the necessary financial backing for the first Chardonnet silk factory in Besancon in 1890. His process involved treating mulberry leaves with nitric and sulfuric acids to form cellulose nitrate which could be dissolved in ether and alcohol. This collodion solution could be extruded through holes in a spinneret into warm air where solvent evaporation led to the formation of solid cellulose nitrate filaments.

Although this first route was simple in concept, it proved slow in operation, difficult to scale up safely, and relatively uneconomical compared with the other routes. Denitration of the fibers, necessary to allow safe use wherever the fabrics may risk ignition, spoiled their strength and appearance. Nevertheless, Chardonnet earned and truly deserved his reputation as the Father of Rayon. His process was operated commercially until 1949 when the last factory, bought from the Tubize Co. in the United States in 1934 by a Brazilian company, burned down.

The second cellulosic fiber process to be commercialized was invented by L. H. Despeissis (4) in 1890 and involved the direct dissolution of cotton fiber in ammoniacal copper oxide liquor. This solvent had been developed by M. E. Schweizer in 1857 (5). The cuprammonium solution of cellulose was spun into water, with dilute sulfuric acid being used to neutralize the ammonia and precipitate the cellulose fibers. H. Pauly and co-workers (6) improved on the Despeissis patent, and a German company, Vereinigte Glanstoff Fabriken, was formed to exploit the technology. In 1901, Dr. Thiele at J. P. Bemberg developed an improved stretch-spinning system, the descendants of which survive today.

Its early commercial success owed much to the flammability disadvantages of the Chardonnet process, but competition from the viscose process led to its decline for all but the finest filament products. The process is still used, most notably by Asahi in Japan where sales of artificial silk and medical disposable fabrics provide a worthwhile income. However, its relatively high cost, associated with the cotton fiber starting point, prevented it from reaching the large scale of manufacture achieved by the viscose rayon process.

In 1891, Cross, Bevan, and Beadle, working at Kew in the UK, discovered that cotton or wood cellulose could be dissolved as cellulose xanthate following treatment with alkali and carbon disulfide (7). The treacle-like yellow solution could be coagulated in an ammonium sulfate bath and then converted back to pure white cellulose using dilute sulfuric acid. They patented their process in 1892 without considering its fibermaking potential. In 1893 they formed the Viscose Syndicate to grant licences for nonfiber end uses, and the British Viscoid Co. Ltd. to exploit the process as a route to molded materials. These companies were later merged to form the Viscose Development Co. in 1902.

In another laboratory at Kew, C. H. Stearn and C. F. Topham, who had worked for Sir Joseph Swan on lamp filaments, developed the continuous filament spinning process (8) and the machinery needed to wash and collect (9,10) the yarns. A fibermaking method was outlined in 1898, and the Viscose Spinning Syndicate was formed to develop the concept into a commercial proposition.

In 1904, the Kew laboratories were visited by representatives of Samuel Courtauld & Co. Ltd., who were silk weavers looking for new raw materials and new opportunities to grow. The success and profitability of Samuel Courtauld had been built on the nineteenth century fashion for black silk mourning crepe, and the company was planning its stock market flotation. The visitors to Kew knew that Chardonnet's process was creating a lucrative market for artificial silk in France. They believed that Cross and Bevan's viscose route could make a similar fiber much more economically. Nevertheless, it took two attempts, the second to a changed Board of Directors after the flotation, before Courtaulds was persuaded to acquire the viscose process rights. On July 14, 1904, the Viscose Spinning Syndicate agreed to sell the viscose process rights and patents to Courtaulds for the sum of £25,000. A new factory was built and the first commercial production was started in Coventry, UK, in November 1905.

The acquisition of the rights to the viscose process became one of the most profitable investments of all time. Interest in the new fiber was intense, and growth of production capacity was exponential. By 1907, the Courtauld company was selling all the artificial silk it could produce and proceeded to expand into the U.S. market. In 1910 they formed the American Viscose Co. and in 1911 started the first U.S. viscose factory at Marcus Hook. By 1939, Courtaulds had six factories in the United States, seven in the United Kingdom, one in France, one in Canada, and joint ventures in Germany and Italy.

From 1920 to 1931, after the expiration of the viscose patents, world output increased from 14,000 to 225,000 t per year, as more than 100 companies entered the cellulose fiber field. In Europe, Vereinigte Glanstoff Fabriken (VGF, Germany), Enka (Holland), I.G. Farben (Germany), Snia Viscosa (Italy), Comptoir des Textiles Artificiels (CTA, France), Rhodiaceta (France), Tubize (Belgium), and Chatillon (Italy) were among the new starters. In the United States the new entrants included Du Pont (with help from CTA), Tubize, Chatillon, American Enka, The Industrial Fibre Corp. (later The Industrial Rayon Corp.), American Glanzstoff (later North American Rayon), and American Bemberg. By 1941 the production had risen to 1,250,000 tons.

Applications

The original yarns were marketed as silk substitutes for use in apparel, hosiery, lace, home furnishings, ribbons, braids, and in a whole range of fabrics using blends with cotton or wool yarns. As the end uses expanded beyond silk replacement, the harsh metallic luster of the yarn proved disadvantageous and dull "matt" fibers had to be developed. Oil dulling was invented (11) in 1926, and an improved method using titanium dioxide was developed (12) in 1929.

Commencing in the late 1930s, new developments to make very strong yarns allowed the viscose rayon to replace cotton as the fiber of choice for longer life pneumatic tires. The pace of this line of development increased during World War II, and by the 1960s a significant part of the production of viscose yarn was for tires and industrial applications.

From 1910 onward waste filament yarn had been chopped into short lengths suitable for use on the machinery designed to process cotton and wool staples into spun yarns. In the 1930s new plants were built specifically to supply the staple fiber markets. During World War II the production of staple matched that of filament, and by 1950, staple viscose was the most important product. The new spun-yarn outlets spawned a series of viscose developments aimed at matching the characteristics of wool and cotton more closely. Viscose rayon was, after all, silk-like. Compared with wool it lacked bulk, resilience, and abrasion resistance. Compared to cotton, it was weaker, tended to shrink and crease more easily, and had a rather lean, limp hand.

For wool outlets, coarser crimped staple fibers were developed. By the late 1960s these were widely used in the new tufted carpets, which for a period became the single largest application for the cellulosic fiber. For cotton outlets, two distinct lines of development were undertaken, the first being to boost the strength and modulus of rayon so that it could match that of cotton, and the second to alter the shape of the fiber to give bulkier, more cotton-like textures. High wet modulus (HWM) rayons of two types were developed. Staple versions of the tougher tire-yarn fibers (modal HWM fibers) were introduced for use in industrial textiles, and for blending with the rapidly growing synthetics. This class was later subdivided into the low, intermediate, and high strength variants (LWM, IWM, and HWM) as producers tuned the process to market niches. Polynosic HWM fibers with even better wet stability and higher wet modulus were introduced to blend with the better grades of cotton.

In the 1960s and 1970s, viscose producers in Europe, the United States, and Japan made various attempts to make rayon behave more like cotton in both woven fabrics and specialty papers. Avtex (U.S.) and Kurashiki Rayon (Japan) commercialized inflated/collapsed papermaking fibers. Courtaulds (UK) developed a range of hollow fibers of which two were commercialized. The first, having an uncollapsed tubular shape, found worthwhile new business in thermal underwear. The other, having an unevenly collapsed tube shape and a very high absorbency, enjoyed large sales in sanitary protection products. These hollow fibers were replaced in the late 1980s by "I" and "Y" shaped solid fibers that performed as well as the hollow versions in the main applications.

The Viscose Process

Approximately 2.5 million t of viscose process regenerated cellulose fibers were produced in 1990 (Table 1). Measured by production capacity in 1990, the leading producers of filament yarns in 1990 were the Soviet Union state-owned factories (255,000 t capacity) and Akzo Fibres in Europe (100,000 t). The leading producers of staple fiber and tow were Courtaulds with 180,000 t capacity split between the UK and North America; Formosa Chemicals and Fibres Co. with 150,000 t in Taiwan; Lenzing with 125,000 t in Austria, and a 40% stake in South Pacific Viscose's 37,000 t Indonesian plant; and Grasim Industries in India (125,000 t). BASF's U.S. capacity of 50,000 t was acquired by Lenzing in 1992.

The main raw material required for the production of viscose is cellulose (qv), a natural polymer of D-glucose (Fig. 1). The repeating monomer unit is a pair

Table 1. 1990 World Production of Regenerated Cellulose Fibers

Fiber type	Production, 10^3 t
industrial filament yarns	220
textile filament yarns	360
regular staple fiber and tow	1830
HWM staple fiber	90
Total	*2,500,000*

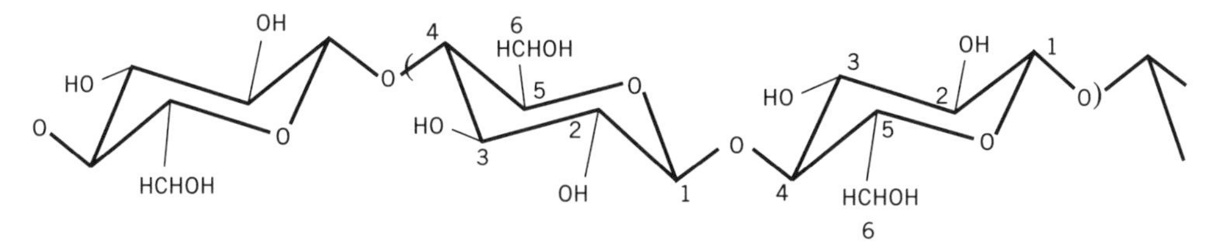

Fig. 1. Anhydro-glucose units with 1–4 beta linkages as in cellulose.

of anhydroglucose units (AGU). Cellulose and starch (qv) are identical but for the way in which the ring oxygen atoms alternate from side to side of the polymer chain (beta linkages) in cellulose, but remain on the same side (alpha linkages) in starch.

Cellulose is the most abundant polymer, an estimated 10^{11} t being produced annually by natural processes. Supplies for the rayon industry can be obtained from many sources, but in practice, the wood-pulping processes used to supply the needs of the paper and board industries have been adapted to make the necessary specially pure grade. Of the 3×10^8 t of wood used by the paper and board industry (13) in 1989, about 6×10^6 t were purified to provide the 2.5×10^6 t of dissolving pulp required by the viscose processes.

The trees used to make dissolving pulp are fast-growing hard or soft woods farmed specifically for their high quality pulping. For example *Eucalyptus grandis* is farmed in relatively poor agricultural areas but grows from seedling to maturity in 7–10 years. It yields about 0.62 kg/m^2 (2500 kg/acre) per year of cellulose for fibermaking, compared with the best cotton fiber yields of 0.74 kg/m^2 (300 kg/acre) per year obtained from prime agricultural land. After sulfite or prehydrolyzed Kraft pulping, the cellulose wood fibers, liberated from the lignin (qv), pass through a sequence of purification stages, eg, chlorination, hot caustic extraction, hypochlorite, peroxide or chlorine dioxide bleaching, and cold caustic extraction.

The final properties of the rayon fiber and the efficiency of its manufacturing process depend crucially on the purity of the pulp used. High tenacity fibers need high purity pulps, and this means pulping to get up to 96% of the most desirable form of cellulose (known as alpha cellulose), and removing most of the unwanted hemicellulose (qv) and lignin. Cellulose molecular weight must be tightly controlled, and levels of foreign matter such as resin, knots, shives, and silica must be very low. Of the two main pulping processes, the sulfite route produces higher yields of lower alpha, more reactive pulp, suitable for regular staple fiber. Prehydrolyzed kraft pulps are preferred for high strength industrial yarn or modal fiber production. The highest purity pulps, up to 99% alpha cellulose, are obtained from cotton fiber. These are no longer used in viscose production but are now the main raw material for cuprammonium rayon.

The flow diagram for the viscose process is given in Figure 2. The sequence of reactions necessary to convert cellulose into its xanthate and dissolve it in soda used to be performed batchwise. Fully continuous processes, or mixtures of batch and continuous process stages, are more appropriate for high volume regular viscose staple production.

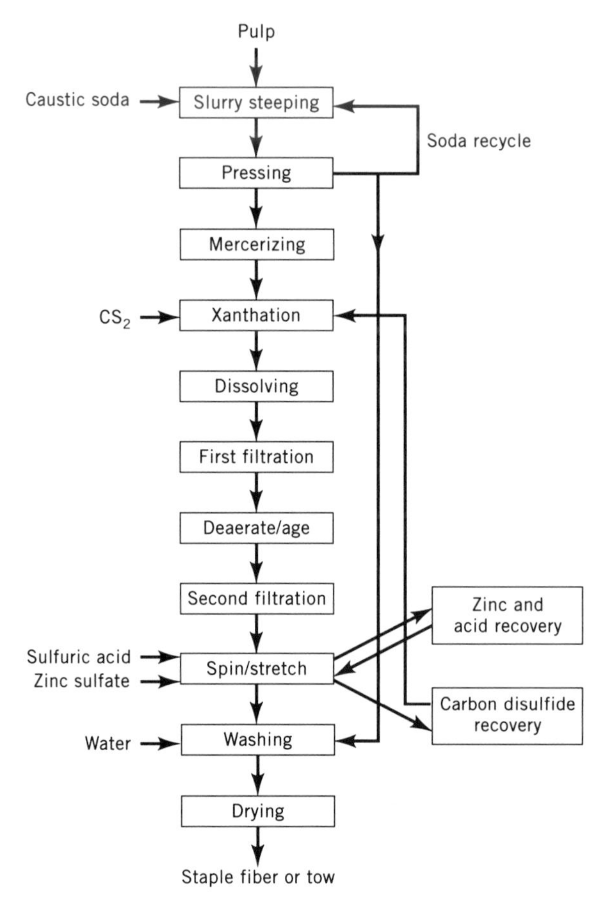

Fig. 2. The viscose process.

Steeping. Sheet, roll, or suitably milled flock pulp is metered into a pulper along with vigorously stirred 18% sodium hydroxide solution at 50°C. The resulting slurry, containing about 5% finely dispersed pulp, passes to a buffer tank from which it is metered to a slurry press that sieves out the swollen fiber and returns the pressings soda for concentration correction and reuse. The cellulose reacts with the soda as a complex alcohol to form the sodium salt or alk-cell.

$$\text{cell–OH} + \text{NaOH} \rightarrow \text{cell–ONa} + H_2O \quad (1)$$

The blanket of alk-cell leaving the slurry press is shredded, and if necessary cooled, before being conveyed to mercerizing (the ripening or aging process).

Hemicelluloses left over from the cellulose pulping process tend to accumulate in the steep soda, and levels of up to 3.0% are tolerable depending on the type of fiber being produced. Equilibrium hemicellulose levels may prove to be too high for the production of stronger fibers, and in this case steps must be taken to purify the soda, eg, by dialysis, prior to reuse. In any case a proportion of the soda from the presses is filtered for later use in dissolving and washing; the steep liquor itself is corrected to the required concentration by the addition of fresh hemicellulose-free strong soda.

Mercerizing. Alk-cell mercerizing (or aging) is ideally carried out on a conveyor belt running through a temperature and humidity controlled tunnel. Towers, silos, cylinders, or even extended treatment in shredders are used as alternatives to belts. During mercerizing, oxidative depolymerization of the cellulose occurs, and this reduces the molecular weight to a level where the final viscose viscosity and cellulose concentration are within acceptable ranges. The optimum conditions for a given pulp depend on the details of the equipment and the cost/quality balance required by the business. The aging temperature is generally around 50°C, and the humidity is adjusted to prevent any drying of the alk-cell. Catalysts such as manganese (a few parts per million in steep liquor) or cobalt have been added to the steep soda to accelerate the aging process (14) or to allow lower temperatures to be used. Catalyzed aging on a conveyor takes 3–4 hours, and at the end of the belt, the alk-cell is blown or conveyed to a fluid-bed cooler.

Whatever system is used, tight control of the aging process is essential for good quality viscose solutions, because the viscosity of the viscose depends on it. Varying viscosity can play havoc with all subsequent operations up to and including spinning. Productivity gains achieved by high temperature or catalyzed aging tend to be accompanied by increased viscosity variability and reduced reactivity in xanthation.

Xanthation. The viscose process is based on the ready solubility of the xanthate derivative of cellulose in dilute sodium hydroxide. The reaction between alkali cellulose and carbon disulfide must therefore be as uniform as possible to avoid problems with incompletely dissolved pulp fibers that will later have to be filtered out of the viscous solution.

In theory, for regular rayon manufacture only one of the hydroxyl groups on each pair of anhydroglucose units needs to be replaced by a xanthate group, ie, the target degree of substitution (DS) is 0.5, which if achievable without waste would need 23% CS_2 on cellulose.

The desired reaction is

$$\text{cell–O}^-\text{Na}^+ + \text{CS}_2 \rightleftarrows \text{cell–OCS}_2\text{Na} \quad (2)$$

but side reactions between the CS_2 and the NaOH also occur, one of which is

$$2\ \text{CS}_2 + 6\ \text{NaOH} \rightarrow \text{Na}_2\text{CO}_3 + \text{Na}_2\text{CS}_3 + \text{Na}_2\text{S} + 3\ \text{H}_2\text{O} \quad (3)$$

It is the sodium trithiocarbonate from this side reaction that gives the viscose dope its characteristic orange color.

In addition to the cellulose xanthate [*9032-37-5*] forming reaction, equation 2 being reversible, cellulose can reform by:

$$\text{cell–OCS}_2\text{Na} + 2\ \text{NaOH} \rightarrow \text{Na}_2\text{CO}_2\text{S} + \text{NaSH} + \text{cell–OH} \quad (4)$$

Clearly, free sodium hydroxide in alkali cellulose leaving the slurry presses causes problems in xanthation, and the presses have to be operated to minimize reactions 3 and 4.

Carbon disulfide [*75-15-0*] is a clear colorless liquid that boils at 46°C, and should ideally be free of hydrogen sulfide and carbonyl sulfide. The reaction with alkali cellulose is carried out either in a few large cylindrical vessels known as wet churns, or in many smaller hexagonal vessels known as dry churns. In the fully continuous viscose process, a Continuous Belt Xanthator, first developed by Du Pont, is used (15).

In a large wet churn operation, a weighed batch of alk-cell (5–10 t) is charged into the churn which is then evacuated. Carbon disulfide (0.5–1 t) is admitted over a 20 min period and vaporizes in the partial vacuum. Churns are rotated slowly to encourage the solid–vapor interaction or they are stirred with a large slow moving paddle. The reaction with alk-cell is slightly exothermic, and churns are generally cooled to maintain the temperature between 25 and 30°C. Under these conditions the xanthation reaction is completed in about 100 minutes and its progress can be monitored by observing the partial vacuum redeveloping as the gas is adsorbed by the solid.

About 25% of the CS_2 is wasted in side reactions (eg, reaction 3), and this means that to achieve the ideal 0.5 DS, the CS_2 charge is about 30% by weight of the cellulose. Higher charges (up to 50%) are needed for the strongest high wet modulus rayon production. Lower charges are possible if steps are taken to increase the reactivity of the alk-cell, or to suppress the undesirable side reactions, or to ensure that viscose filtration is good enough to cope with more undissolved material.

The incomplete accessibility of the cellulose molecules in alk-cell prevents complete xanthation, ie, the formation of the trixanthate, no matter how much CS_2 is used. The trixanthate would imply a DS of 3, sometimes known as a gamma value of 300 CS_2 groups per 100 AGUs. In practice, gamma values over 100 are hard to achieve in the churn, but levels up to 200 can be reached by a combination of churn addition and liquid CS_2 injection into the viscose where the cellulose accessibility is at its greatest. For regular staple production, gamma values tend to be around 50 at the churn and 30–35 at spinning.

At the end of xanthation, any remaining traces of CS_2 are flushed from the wet churn prior to, or in some cases by, admitting a charge of the dissolving or mixer soda in order to commence dissolution. For a dry churn operation, the vessel is opened to allow the golden xanthate crumbs to be discharged into a separate mixer.

Dissolution. Cellulose and its derivatives dissolve more easily in cold alkali than in hot, and the initial contact between cellulose xanthate and the mixer soda should occur at the lowest temperature possible. Two percent NaOH at no more than 10°C is admitted to the wet churn and the paddle speeded up to wet out all the crumb prior to discharge. Some of the required mixing soda charge can be retained to wash out the churn prior to restarting the cycle. Discharge occurs, through cutters that eliminate any large lumps of agglomerated xanthate, into the batch tank where dissolution is completed. Here it is stirred for about two hours to allow the xanthate to dissolve. The mixing conditions are crucial. High shear mixing certainly speeds up the process but can cause local heating which can regenerate cellulose from the xanthate.

The final composition of the solution is controlled by the ratio of xanthate to mixer soda, the concentrations of cellulose and NaOH in the xanthate, and the concentration of NaOH in the mixer soda. For regular staple processes the cellulose:soda ratio is within the range 8.5:6.0 to 10.0:5.0, the latter tending to give lower quality viscose solutions. High quality textile yarns require a 7.5:6.5 ratio, whereas the strongest high wet modulus processes use a 6.0:6.0 ratio. Here lower cellulose concentrations are necessary to allow the use of the higher molecular weights (less mercerizing) which allow higher strength development.

It is possible to add modifiers or delustrants at the dissolving stage. However, modern viscose dope plants feed several spinning machines which are often expected to make different grades of fiber. It is therefore now more common to add the materials needed to make special fibers by injection close to the spinning machines.

Viscose Aging, Filtration, and Deaeration. After the dissolution step, the viscose cannot be spun into fibers because it contains many small air bubbles and particles. Furthermore, the degree of xanthation is too high, with too many of the xanthate groups in positions dictated by their accessibility and not in the ideal positions for uniform dissolution.

Dexanthation, and the redistribution of xanthate groups into the most favorable positions on the cellulose molecules, occurs automatically as the viscose ages. For regular staple fibers the process takes about 18 hours, but for the stronger fibers, shorter times and more difficult spinning conditions have to be tolerated. During this aging process, reaction 2 goes into reverse and the xanthate groups attached to the 2 and 3 positions on the AGU (Fig. 1) hydrolyze 15–20 times as fast as those on C-6. Transxanthation occurs (16); the unoccupied C-6 hydroxyl competing with reaction 2 for the CS_2 which is leaving C-2 and C-3. The reactions favoring more spinnable viscose are in turn favored by aging for longer times at the lowest possible temperature. Cold viscoses are unfortunately much more viscous, and therefore harder to filter and deaerate.

Filtration of viscose is not a straightforward chemical engineering process. The solution of cellulose xanthate contains some easy-to-deal-with undissolved pulp fibers, but also some gel-like material which is retarded rather than removed by the filters. The viscose is unstable and tends to form more gel as it ages. Its flow characteristics make the material close to the walls of any vessel or pipe move more slowly, get older, and gel more than the mainstream viscose. So while filtration can hold back gels arising from incomplete mixing, new gels can form in the pipework after the filters.

The removal of particles that would block the spinneret holes occurs in several stages. The traditional plate-and-frame first-filters dressed with disposable multilayers chosen from woven cotton, cotton wadding, or wood pulp have now been replaced by durable nylon needlefelts that can be cleaned by automatic backwashing (17). Second filtration, usually after deaeration, is also changing to fully automatic systems with sintered steel elements that do not need manual cleaning. Third-stage filtration, close to the spinning machines, is used to provide a final polishing of viscose quality, but is only justifiable for the premium quality fibers. All processes nevertheless use small filters in each spinneret to catch any particulate matter which may have eluded, or been formed after, the main filter systems.

Continuous deaeration occurs when the viscose is warmed and pumped into thin films over cones in a large vacuum tank. The combination of the thinness of the liquid film and the disruption caused by the boiling of volatile components allows the air to get out quickly. Loss of water and CS_2 lower the gamma value and raise the cellulose concentration of the viscose slightly. Older systems use batch deaeration where the air bubbles have to rise through several feet of viscose before they are liberated.

The correct viscose age or ripeness for spinning varies according to the type of fiber being made. Ripeness can be assessed by establishing the salt concentration necessary to just coagulate the viscose dope. The preferred test uses sodium chloride (salt figure) although ammonium chloride is the basis of the alternative method (Hottenroth number).

Spinning. Properly filtered and deaerated viscose of the right salt figure can be coagulated into filament by pumping it through a suitable orifice (jet or spinneret) into a coagulating bath. The forming filaments are pulled through the bath by first godets, which in some systems also serve as an anchor against which stretch can be applied by the second godets or traction units. The assembly of pumps, spin-baths, jets, godets, stretch baths, traction units, and ventilation systems is known as the spinning machine. It can take many forms depending on the type of fiber being made and the priority given to achieving a geometrically similar path for the tows from each jet. Figure 3 illustrates one end of a staple fiber spinning machine which would have hundreds of ends making up the final tow entering the cutter.

Jets for continuous filament textile yarn are typically 1 cm diameter gold–platinum alloy structures with 20–500 holes of 50–200 μm diameter. Tire yarn jets are also 1 cm in diameter but typically use 1000–2000 holes to give the required balance of filament and yarn denier. Staple fiber jets can have as many as 70,000 holes and can be made from a single dome of alloy or from clusters of the smaller textile or tire yarn jets. The precious metal alloy is one of the few materials that can resist the harsh chemical environment of a rayon machine and yet be ductile enough to be perforated with precision. Glass jets have been used for filament production, and tantalum metal is a low cost but less durable alternative to gold–platinum.

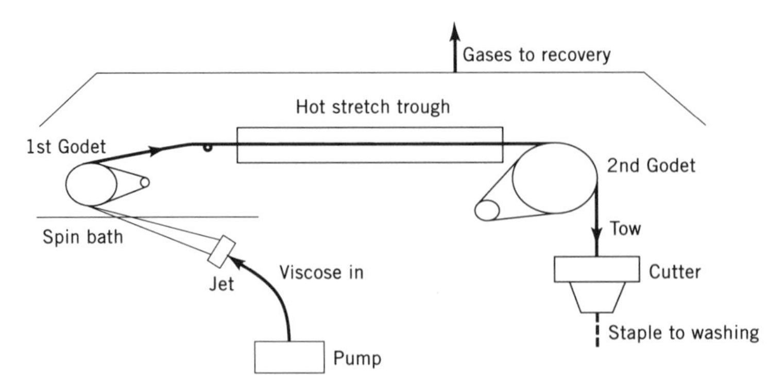

Fig. 3. One "end" of a staple fiber spinning machine.

The basic chemistry of fiber formation is however independent of the hardware. Spin bath liquors are mixtures of sulfuric acid (5–15%), zinc sulfate, (0.05–7%), and sodium sulfate (10–28%) controlled at temperatures ranging from 30 to 60°C. They are circulated past the jets at carefully controlled rates and fully recycled. The liquid filament emerging from the jet coagulates at the interface between acid bath and alkaline viscose to form a cuticle, and later a skin, through which the rest of the coagulation and regeneration is controlled. This regeneration mechanism causes the skin-core nonuniformity which is characteristic of regular rayon fiber cross sections.

$$2\ \text{cell–OCS}_2\text{Na} + \text{H}_2\text{SO}_4 \rightarrow 2\ \text{cell–OCS}_2\text{H} + \text{Na}_2\text{SO}_4 \qquad (5)$$

As coagulation proceeds into the center of the forming fiber, the outside regenerates to cellulose at a rate dependent on the temperature and composition of the bath.

$$\text{cell–OCS}_2\text{H} \rightarrow \text{cell–OH} + \text{CS}_2 \qquad (6)$$

In processes making regular staple fiber, reactions 5 and 6 occur practically simultaneously, but in high wet modulus processes, reaction 5 predominates in the spin bath, with reaction 6 occurring during stretching the filament in a hot water or steam bath. The carbon disulfide liberated in spinning and stretching can be recovered from the extracted air and from the spin-bath liquor.

Zinc salts are added to the spin bath to slow down the regeneration reaction by forming a less easily decomposed zinc cellulose xanthate intermediate. This allows greater stretch levels to be applied and results in fibers with thicker skins. There is still uncertainty as to whether the zinc cellulose xanthate gel acts by hindering acid ingress or water loss. High levels of zinc in the spin-bath allow the production of tough fibers for tire reinforcement and industrial use.

Washing and Drying. In staple fiber production the filaments from all the jets on the spinning machine are combined into a few large tows and cut to length prior to washing. In filament yarn production, the yarns are kept separate and collected either as acid cakes for off-line cake-washing, or wound onto bobbins after on-line washing and drying.

Again, irrespective of the hardware the chemistry is consistent. The partially regenerated fiber from the spinning machine is contaminated with sulfuric acid, zinc sulfate, sodium sulfate, carbon disulfide, and the numerous incompletely decomposed by-products of the xanthation reactions. The washing and drying systems must yield a pure cellulose fiber, suitably lubricated for the end use, and dried to a moisture level of around 10%.

The sequence of treatments is illustrated in Figure 4. Water consumption in washing is minimized by using a countercurrent liquor flow, clean water entering the system at the point where the fiber leaves, and dirty liquor leaving for chemical recovery at the point where the fiber enters.

In the staple fiber system, the tows from spinning are cut to length and sluiced onto a 30–40 meter long wire-mesh conveyor that carries the blanket of fiber through cascades of the various wash liquors. Sluicing is a critical operation because the cut lengths need to be opened just enough to allow good washing and

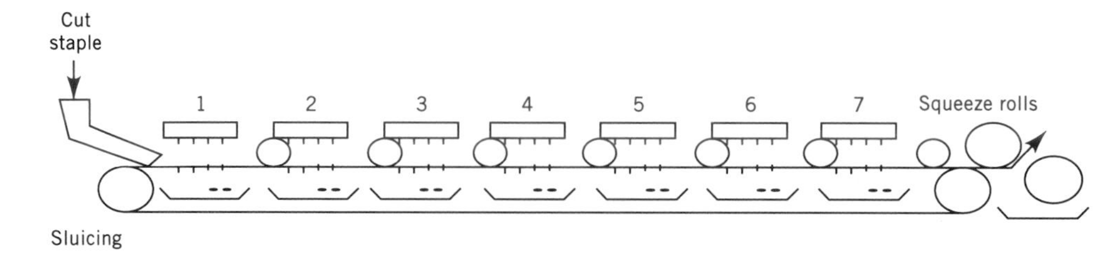

Fig. 4. Staple fiber washing sequence: 1, a *hot acid wash* (2% H_2SO_4 at 90°C) decomposes and washes out most of the insoluble zinc salts. This wash completes the regeneration of xanthate and removes as much sulfur as possible in the form of recoverable CS_2 and H_2S; 2, an alkaline sodium sulfide *desulfurization bath* solubilizes sulfurous by-products and converts them into easily removed sulfides; 3, a *sulfide wash* to remove the sulfides created in bath 2; 4, a *bleach bath* (optional) uses very dilute hypochlorite or peroxide to improve fiber whiteness; 5, a dilute acid or *sour bath* removes any remaining traces of metal ions and guarantees that any residual bleaching chemicals are destroyed; 6, a controlled-pH freshwater *final wash* removes the last traces of acid and salt prior to drying; and 7, a *finish bath* gives the fiber a soft handle for easy drying and subsequent processing.

not so much that the individual fibers entangle. Sluicing must also lay a uniformly permeable blanket of fibers several cm thick and a meter or so wide because uneven laying results in uneven washing. Washing tows before cutting breaks out of the restrictions imposed by sluicing but results in the need to reoptimize processing conditions.

Cakes of textile yarn, collected in Topham boxes immediately after spinning, are processed on cake-wash machines. The cakes are wrapped in a porous paper and mounted in groups on stainless steel tubes through which the wash liquors can be forced. These tubes are mounted on a cake-wash box, which is a trolley on an overhead conveyor carrying the cakes to the different pumping stations on the machine. Each cake-wash box holds 100–200 cakes, and the permeability of each cake is critical to uniform washing. With permeability being determined largely by the degree of swelling of the fiber leaving spinning, very good control of all regeneration parameters is required to make yarn with consistent properties.

Yarn drying procedures are far more critical than staple drying procedures. All cellulosic fibers expand when wetted and contract on drying, the changes of size being greatest the first time the freshly made fiber is dried. If the contraction is hindered in any way, the yarn will possess a latent strain which will recover at a later stage giving inferior weaving and dyeing. Furthermore, freshly formed regenerated cellulose undergoes a permanent change in molecular structure as it is heated and this reduces its absorbency and hence its dye uptake.

Cakes are dried in low temperature, high humidity conditions for a long time in order to minimize strain and absorbency variability. The continuing usage of the cake system almost a century after its invention owes much to the desirable strain-free yarn arising from its washing and drying operations.

Industrial yarn uses a continuous washing and drying process based on self-advancing reels. Whereas the processing of cake yarn can take many days to complete, the continuous process is over in minutes. The process is less labor intensive and more productive than the cake system, but the engineering and maintainance requirements are easily underestimated. It is most used for industrial yarns, still the reinforcement of choice for radial tire carcasses. For these applications the strain built in by fast drying is no disadvantage because the yarns are expected to be at their best when bone dry. Dyeing is not required and an off-white color is tolerable. The main objective of the washing system is to remove acid and sulfur compounds that would affect the strength of the yarn during prolonged operation at high temperature.

Modified Viscose Processes. The need for ever stronger yarns resulted in the first important theme of modified rayon development and culminated, technically if not commercially, in the 0.88 N/tex (10 gf/den) high wet modulus industrial yarn process.

Tire Yarns. A method to increase the strength of viscose yarn from the 0.2 N/tex (2.2 gf/den) standard to levels needed in tires was first patented by Courtaulds in 1935 (18). By raising the zinc concentration in the spin bath to 4% the thread could be stretched more by immersing it in a hot dilute acid bath during extension. Filament strengths increased to about 0.3 N/tex (3.3 gf/den), and the cross section became rounder, with a thicker skin than regular viscose. Pairs of these yarns were capable of being twisted into tire cords which outperformed traditional cotton cords.

The next significant strength improvement followed the 1950 Du Pont (19) discovery of monoamine and quaternary ammonium modifiers, which, when added to the viscose, prolonged the life of the zinc cellulose xanthate gel, and enabled even higher stretch levels to be used. Modifiers have proliferated since they were first patented and the list now includes many poly(alkylene oxide) derivatives (20), polyhydroxypolyamines (21–23), and dithiocarbamates (24).

Fully modified yarns had smooth, all-skin cross sections, a structure made up of numerous small crystallites of cellulose, and filament strengths around 0.4 N/tex (4.5 gf/den). They were generally known as the Super tire yarns. Improved Super yarns (0.44–0.53 N/tex (5–6 gf/den)) were made by mixing modifiers, and one of the best combinations was found to be dimethylamine with poly(oxyethylene) glycol of about 1500 mol wt (25). Ethoxylated fatty acid amines have now largely replaced dimethylamine because they are easier to handle and cost less.

The strongest fibers were made using formaldehyde additions to the spin bath while using a mixed modifier system (26) or using highly xanthated viscoses (50% + CS_2) (27–29). Formaldehyde forms an *S*-methylol derivative with xanthate which decomposes slowly permitting high levels of stretch. It also reacts with the cellulose backbone to form cross-links that render the fiber high in modulus and low in extension. Unfortunately, problems associated with formaldehyde side reactions made the processes more expensive than first thought, and the inevitable brittleness which results whenever regenerated cellulose is highly oriented restricted the fibers to nontextile markets. The commercial operations were closed down in the late 1960s.

The formaldehyde approach is still used by Futamura Chemical (Japan). They make spun-laid viscose nonwovens where the hydroxymethylcellulose xanthate derivative formed from formaldehyde in the spin bath allows the fibers to bond after laying. This process was originally developed by Mitsubishi Rayon (30), who later found that the derivative was thermoplastic, and the web could be calender-bonded (120°C) prior to regeneration (31).

High Tenacity Staple Fibers. When stronger staple fibers became marketable, the tire yarn processes were adapted to suit the high productivity staple fiber processes. Improved staple fibers use a variant of the mixed modifier approach to reach 0.26 N/tex (3 gf/den). The full 0.4 N/tex (4.5 gf/den) potential of the chemistry is unnecessary for the target end uses and difficult to achieve on the regular staple production systems.

The full potential of the mixed modifier tire yarn approach is, however, achieved in the modal or HWM (high wet modulus) staple processes using special viscose making and spinning systems. Courtaulds produced such fibers under the Vincel brand both in the UK and in Canada. Avtex made Avril (Fibre 40) until their closure in 1989. Lenzing still makes HM333, and BASF (whose viscose operation in the United States has been bought by Lenzing) makes Zantrel. The fibers are most popular in ladies' apparel in the United States for their soft handle and easy dyeing to give rich coloration. They are now being made in finenesses down to 1 dtex (0.9 den) for fine yarns and hydroentangled nonwoven production (32).

Polynosic Rayons. The foregoing viscose process variants depend on the presence of zinc in the spin bath and modifiers in the viscose for their efficacy. Another strand of development began in 1952 when Tachikawa (33) patented a method for making strong, high modulus fibers that needed neither zinc nor modifier. The process was in fact related to the Lilienfeld (34) route to high strength, high modulus yarns which had failed in the 1930s because of the need for a spin bath containing 50–85% sulfuric acid. Both processes depend (33,34) on the fact that minimally aged alk-cell can, after xanthation with an excess of carbon disulfide, be dissolved at low cellulose concentration to give very viscous viscoses containing high molecular weight cellulose. These viscoses have sufficient structure to be spun into cold, very dilute spin baths containing no zinc and low levels of sodium sulfate. The resulting gel-filaments can be stretched up to 300% to give strong, round section, highly ordered fibers which can then be regenerated.

Toyobo's Tufcel provides an excellent example of how a modern polynosic fiber process, probably the most difficult viscose process to run efficiently, operates (35). On-line process control allows only four persons per shift to make 10,000 t/yr of a variety of special fibers including Flame Retardant, Deep Dying (two types), Activated Carbon Fiber, and Super Fine 0.55 dtex (0.5 denier). Alk-cell and mixing soda quality are maintained by pressings soda centrifugation, filtration, and dialysis to remove 90% of the hemicellulose. Ion-exchange membranes are used to give 50 times the life and twice the efficiency of the old dialyzer bags used in tire-yarn production. Dissolution of the 500 DP xanthate is augmented by crumb-grinders on the churn outlets and by in-line homomixers, which together reduce the dissolution time from three hours to one. Spinnerets for the finer yarns have 40-μm holes, and these are protected by automatic backflush filters removing gels down to 15 μm diameter.

Bulky Rayons. Unlike the thermoplastic synthetic fibers, viscose rayon cannot be bulked by mechanical crimping processes. Crimpers impart crimp to a regenerated cellulose fiber but it is not a permanent crimp and will not survive wetting out.

Permanent chemical crimp can be obtained by creating an asymmetric arrangement of the skin and the core parts of the fiber cross section. Skin cellulose is more highly ordered than core cellulose and shrinks more on drying. If, during filament formation in the spin bath, the skin can be forced to burst open to expose fresh viscose to the acid, a fiber with differing shrinkage potential from side-to-side is made, and crimp should be obtained (Fig. 5**a**).

Whether or not it is obtained depends on the washing mechanism allowing the shrinkage, and hence the crimp, to develop prior to the completion of regeneration. Crimp development only occurs fully in staple fiber processes where the sluicing operation allows the cut tufts of acid tow to expand freely in ample volumes of hot liquor.

Even when crimp is fully developed it is easy to pull out (low energy) and difficult to translate into noticeably bulkier woven and knitted fabrics. It does however improve the absorbency and the cohesion of the staple (important in spun-yarn and nonwoven making) and gives a subtly different texture to woven fabrics. Coarse crimped rayon was the leading synthetic carpet fiber in Europe in the 1960s, but has since been replaced by the highly durable bulked continuous filament nylon yarns. Crimp is most important in rayon used for hygienic absorbent products.

Process conditions that favor chemical crimp formation are similar to those used for improved tenacity staple (zinc/modifier route). However, spin bath temperature should be as high as possible (ca 60°C) and the spin-bath acid as low as possible (ca 7%). Attempts have been made to overcome some of the leanness of high strength rayons by increasing the crimp levels. ITT Rayonier developed the Prima crimped HWM fiber (36) and made the process available to their customers. Avtex developed Avril III. Neither remain in production.

Cross-sectional modifications of a more extreme nature than skin-bursting, which nevertheless do not form crimp, have grown in importance since the early 1980s. These yield a permanent bulk increase which can be translated into bulky fabrics without the need for special care. The first commercial staple fiber of this type was Courtaulds hollow Viloft, developed in the 1970s using a carbonate inflation technique (37).

Inflation had long been known as an intermittent problem of textile yarn manufacture. It was caused by the skin forming too quickly to allow the escape of gases liberated by the regeneration reactions, or as a result of air in the viscose at the jet. In the 1920s attempts were made to produce hollow naturally dull yarns by inclusion of gas (38) and by the addition of gas generating agents to the viscose (39). It was also possible to cause inflation by capturing the regenerated gases

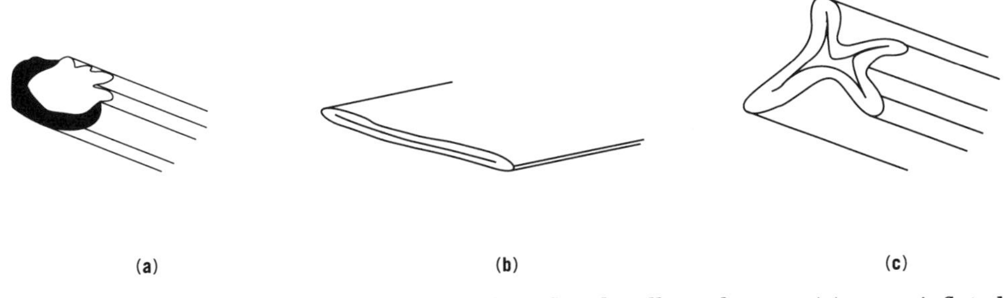

Fig. 5. Bulky rayons: (**a**) crimped rayon; (**b**) inflated–collapsed rayon; (**c**) super inflated rayon.

when spinning a low soda viscose (40). In 1942–1943 Du Pont used air injection to produce Bubblfil yarns as a kapok replacement for lifejackets, as thermal insulation for sleeping bags, and for aviators' uniforms.

All these early inflation processes (41) were difficult to control, and after World War II they were neglected until the 1960s. Companies in Japan, the United States, and Europe then started to develop inflated–collapsed rayons (Fig. **5b**) for speciality papers (42) and wet-laid nonwovens.

Although none of these survived commercially for more than a few years, their development led to an increased understanding of the inflation process and the identification of conditions which could yield a continuously hollow staple fiber in large-scale production (43).

The basis of this process was the injection of sodium carbonate solution into the viscose, although direct injection of carbon dioxide gas that reacts with the viscose soda to form sodium carbonate could also be used (44). The carbonate route yielded a family of inflated fibers culminating in the absorbent multilimbed super inflated (SI) fiber (Fig. 5**c**).

In its original form (45), the fiber was intended for use in toweling where the bulk, firm handle, and high rate of wetting made it one of the few fibers that would blend with cotton without spoiling the wet texture. An even higher absorbency version was developed (46) for use in nonwovens and tampons (47) where it comfortably outperformed cotton by re-expanding like a sponge on wetting.

By the mid-1980s a different approach to the production of fine fibers with novel cross sections became possible. Noncircular spinneret holes, eg, rectangular slots, allowed the large-scale production of flat fibers down to 2.2 dtex (2 den) (Fig. 6**a**). These I-shaped fibers were capable of replacing the inflated hollow fibers in textile applications, providing similar levels of bulk, warmth, and handle while having a much more regular shape. They were followed by the development of solid Y-shaped and X-shaped multilimbed fibers (48) (Fig. 6**b**) which performed like SI fiber but had much lower levels of water imbibition than the inflated version. Their shape and relative stiffness enabled them to absorb more fluid between, as opposed to inside, the fibers. They were therefore as absorbent in use as the inflated versions (49,50) but did not require the extra process chemicals, and were easier to wash and dry in production and use. They are the most important bulky rayons now in production.

Alloy Rayons. It is possible to produce a wide variety of different effects by adding materials to the viscose dope. The resulting fibers become mixtures or

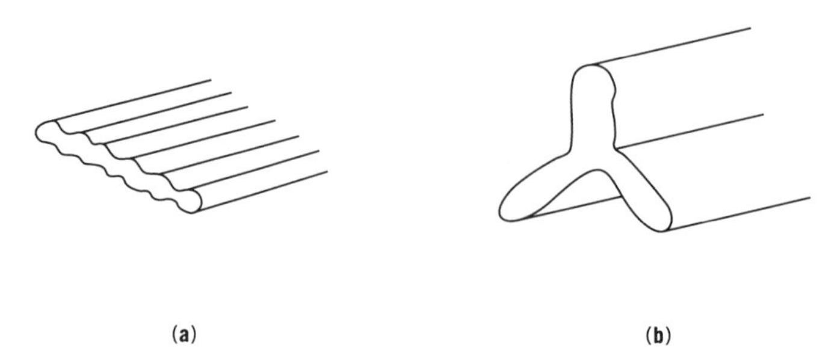

Fig. 6. (**a**) I-section rayon; (**b**) Y-section rayon.

alloys of cellulose and the other material. The two most important types of alloy arise when superabsorbent or flame retardant fibers are made.

American Enka (51,52) and Avtex (53) both produced superabsorbent alloy rayons by adding sodium polyacrylate, or copolymers of acrylic and methacrylic acids, or sodium carboxymethylcellulose to the viscose. The resulting alloys contained about 20% of the water-soluble polymer giving water imbibitions up to double those of the unalloyed rayons. They performed particularly well in tampons where the presence of the slippery polymer at the fiber surface encouraged wet-expansion of the compressed plug. Their use in this, the only real market which developed, declined after the Toxic Shock Syndrome outbreak (54,55) in the early 1980s. Other polymers that have formed the basis of absorbent alloys are starch (56), sodium alginate (57), poly(ethylene oxide) (58), poly(vinyl pyrrolidinone) (59), and sodium poly(acrylamido-2-methyl-2-propane sulfonic acid) (60).

Alloys of cellulose with up to 50% of synthetic polymers (polyethylene, poly(vinyl chloride), polystyrene, polytetrafluoroethylene) have also been made, but have never found commercial applications. In fact, any material that can survive the chemistry of the viscose process and can be obtained in particle sizes of less than 5 μm can be alloyed with viscose.

Flame retardancy can be obtained by adding flame retardant chemicals to make up about 20% of the fiber weight (61). The first commercial products used tris(2,3-dibromo-1-propyl) phosphate, but the P–O bonds made these susceptible to hydrolysis by strong alkali, and the bromine increased the rate of photodegradation. The chemical was found to cause cancer in laboratory tests and in the late 1970s fell from favor (62). Propoxyphosphazine (Ethyl Corp.) retardants were later used in Avtex's PFR fiber, and a bis(5,5-dimethyl-2-thiono-1,3,2-dioxaphosphorinanyl) oxide powder (Sandoz) was the basis of later European FR fiber developments. Alloys with inorganic salts such as silicates or aluminates are possible, the salts being converted to fibrous polyacids when the cellulose is burnt off (63). This latter approach seems to be the basis of the Visil flame retardant fiber introduced by Kemira Oy Saeteri (64).

Cuprammonium Rayon

Asahi Chemical Industries (ACI, Japan) are now the leading producers of cuprammonium rayon. In 1990 they made 28,000 t/yr of filament and spunbond nonwoven from cotton cellulose (65). Their continuing success with a process which has suffered intense competition from the cheaper viscose and synthetic fibers owes much to their developments of high speed spinning technology and of efficient copper recovery systems. Bemberg SpA in Italy, the only other producer of cuprammonium textile fibers, was making about 2000 t of filament yarn in 1990.

The process operated by ACI is outlined in Figure 7. Bales of cotton linter are opened, cooked in dilute caustic soda, and bleached with sodium hypochlorite. The resulting highly purified cellulose is mixed with pre-precipitated basic copper sulfate in the dissolver, and 24–28% ammonium hydroxide cooled to below 20°C is added. The mixture is agitated until dissolution is complete. If necessary, air

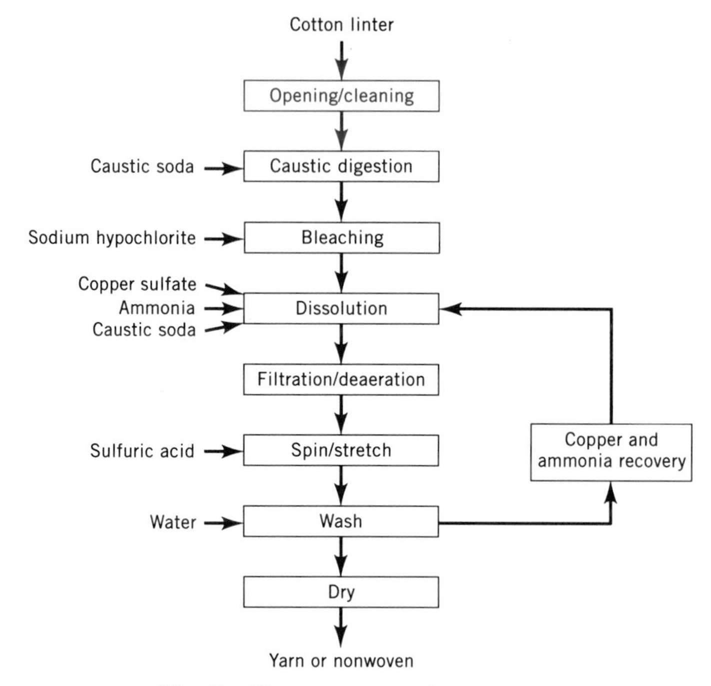

Fig. 7. The cuprammonium process.

is introduced to allow oxidative depolymerization and hence a lowering of the dope viscosity.

The dark blue solution containing 5–10% of cellulose with a DP of 1000–2000 is filtered through a series of plate-and-frame filter presses using fine mesh metal screens to remove any particles that might block the spinneret holes. It is then deaerated under vacuum and stored ready for spinning. Unlike viscose dope, the cuprammonium cellulose [*9050-09-3*] solution is relatively stable.

In the old classical spinning technique developed by Bemberg, the deaerated and filtered dope is extruded into a funnel-shaped glass tube through which slightly alkaline water is forced. The water carries the regenerating cellulose gel while stretching it by about 10,000% as it accelerates to the exit. The yarn then passes through a bath containing about 5% sulfuric acid which completes the regeneration of cellulose and converts the ammonia and copper residues into their sulfates.

In the original batch process, the blue-colored yarn was collected on hanks for separate washing, lubrication, and drying treatments. The continuous process was first patented in the United States in 1934 (66) and 1935 (67), and operated by American Bemberg. This process (68) combined the batch process washing treatments in a series of tanks through which a warp of yarns (from hundreds of spinnerets) were pulled prior to drying and beaming.

The high speed continuous filament process (69) was first used for manufacturing in 1974, and this enabled the yarn production rates to be raised from 150 to 380 m/min. This system uses a pair of net conveyor belts to protect and transport an overfed warp of yarns through washing and drying. After spinning, excess bath is stripped from the yarns by a shaking mechanism that damages the yarn less than the old comb stripper guides. In 1987, Asahi commercialized their "mark 2" high speed spinning system which operated at 1000 m/min, and are now working on systems which will exceed 2000 m/min.

Also in 1974, ACI started the direct spinning of Bemliese nonwoven fabrics using the same conveyor belt washing technique (70). This process uses wide spinnerets and spinning cells to produce a curtain of filaments which fall in the spinning liquor onto the conveyor. The conveyor is oscillated from side to side in order to lay the continuous fibers down in sinusoidal waves. After washing, the webs, which have strength resulting from the self-bonding nature of the fibers as-spun, can be further strengthened and finished by hydroentanglement.

Almost total copper recovery needs to be achieved to meet the economic and environmental criteria for a modern cellulosic fiber process. In the basic process (71), spent spin-bath liquor was passed through ion-exchange resins and the effluent from these resins was reused in the funnels. The copper washed out of the yarn was precipitated with sodium carbonate forming the basic copper sulfate sludge which was reused in cellulose dissolving. Only 95% copper recovery was possible using this system. Recovery efficiencies of up to 98% were claimed for the Russian Kustanai Plant in a document (72) which also records a 99.9% recovery achieved by Japan Organo Co., Ltd. Asahi claims to have perfected copper recovery to a point where it is no longer an issue.

A high percentage of the ammonia can be recovered from the spin-bath effluent and by washing prior to the final acid bath. During acidification, remaining ammonia is converted to the sulfate and recovered when the acid wash liquor is treated with carbonate to recover the copper. Ammonia residuals in the large volumes of washwater can only be removed by distillation. Overall about 75–80% of the ammonia required to dissolve the cellulose can be recovered.

Asahi's innovations have done much to transform the cuprammonium process from an uneconomic competitor for viscose and synthetics into the fastest wet-spinning system in the world. They now claim it to be competitive both economically and environmentally with the viscose filament process.

Direct Dissolution Processes

The routes to regenerated cellulose fibers already described cope with the difficulties of making a good cellulose solution by going through an easy to dissolve derivative, eg, xanthate, or complex, eg, cuprammonium. The ideal process, one that could dissolve the cellulose directly from ground wood, is still some way off, but since the early 1980s significant progress has been made.

There have been reviews of the early work on direct dissolution (73,74). The efforts to dissolve cellulose directly as a *base* using phosphoric, sulfuric, and nitric protonic acids, or zinc chloride, thiocyanates, iodides, and bromides as Lewis acids are recorded (73,74). With regard to cellulose acting as an *acid*, sodium zincate, hydrazine, and sodium hydroxide are listed as inorganic solvents and quaternary ammonium hydroxides, amines, dimethylamine–DMSO mixtures, and amine oxides as organic bases. Cellulose solutions of 16% have been achieved in lithium chloride–dimethylacetamide systems (75), and 14% solutions in dinitrogen tetroxide–dimethylformamide systems (76). Solutions of up to 14% cellulose were achieved in an extensive study of the ammonia–ammonium thiocyanate solvent system (77). The dimethylsulfoxide–paraformaldehyde system (78) is capable of dissolving a wide range of DPs without causing degradation. However, despite early promise, the problems of developing fiber production routes using these systems have, with the exception of the amine oxide route, proved insurmountable.

The cellulose dissolving potential of the amine oxide family was first realized (79) in 1939, but it was not until 1969 that Eastman Kodak described the use of cyclic mono(*N*-methylamine-*N*-oxide) compounds, eg, *N*-methylmorpholine-*N*-oxide [*7529-22-8*] (NMMO), as a solvent size for strengthening paper (80) by partially dissolving the cellulose fibers.

Other patents (81,82) covered the preparation of cellulose solutions using NMMO and speculated about their use as dialysis membranes, food casings (sausage skins), fibers, films, paper coatings, and nonwoven binders. NMMO emerged as the best of the amine oxides, and its commercial potential was demonstrated by American Enka (83,84). Others (85) have studied the cellulose–NMMO system in depth; one paper indicates that further strength increases can be obtained by adding ammonium chloride or calcium chloride to the dope (86).

Both American Enka (87) and Courtaulds set up pilot-plant work in the early 1980s with the objectives of developing fiber spinning and solvent recovery operations. To date, only Courtaulds has proceeded to full commercial scale. The Austrian viscose producer Lenzing studied various systems (88) and commenced pilot operations on an NMMO system at the end of the 1980s, but has yet to announce a commercial plant.

Work on other routes to cellulosic fibers continues, driven by a desire to identify an environmentally benign route to cellulosic fibers that can utilize the large capital investment in the xanthate route and hence cost less than a completely new fiber process.

The Finnish viscose producer Kemira Oy Saeteri collaborated with Neste Oy on the development of a carbamate derivative route. This system is based on work (89) that showed that the reaction between cellulose and urea gives a derivative easily dissolved in dilute sodium hydroxide:

$$\text{cell—OH} + NH_2\text{—}\overset{\overset{\displaystyle O}{\|}}{C}\text{—}NH_2 \rightarrow \text{cell—O—}\overset{\overset{\displaystyle O}{\|}}{C}\text{—}NH_2 + NH_3 \qquad (7)$$

Neste patented an industrial route to a cellulose carbamate pulp (90) which was stable enough to be shipped into rayon plants for dissolution as if it were xanthate. The carbamate solution could be spun into sulfuric acid or sodium carbonate solutions, to give fibers which when completely regenerated had similar properties to viscose rayon. When incompletely regenerated they were sufficiently self-bonding for use in papermaking. The process was said to be cheaper than the viscose route and to have a lower environmental impact (91). It has not been commercialized, so no confirmation of its potential is yet available.

It has been claimed that solutions containing 10–15% cellulose in 55–80% aqueous zinc chloride can be spun into alcohol or acetone baths to give fibers with strengths of 0.13–0.18 N/tex (1.5 to 2 gf/den). However, if these fibers were strain-dried (ie, stretched) and rewetted while under strain, strengths of 0.46 N/tex (5.2 gf/den) were achieved (92).

Asahi has been applying the steam explosion (93) treatment to dissolving-pulp to make it dissolve directly in sodium hydroxide (94), and they claim (95,96) a solution of 5% of steam-exploded cellulose in 9.1% NaOH at 4°C being spun into 20% H_2SO_4 at 5°C. The apparently poor fiber properties (best results being 0.16 N/tex (1.8 gf/den) tenacity dry, with 7.3% extension) probably arise because the fibers were syringe extruded at 8.3 tex/filament (75 den/fil). Asahi feels that this could be the ultimate process for large-scale production of regenerated cellulose fibers.

The Courtaulds Tencel Process. The increasing costs of reducing the environmental impact of the viscose process coupled with the increasing likelihood that the newer cellulose solvents would be capable of yielding a commercially viable fiber process led Courtaulds Research to embark on a systematic search for a new fiber process in the late 1970s.

The project, code named Genesis, did not involve basic research into new solvents for cellulose so much as screening the known solvents against criteria felt to be important for the cellulosic fiber process of the future. The solvent chosen had to be recyclable at a very high level of efficiency, and hence as near to totally containable in the process as possible. It had to be safe to work with and safe in the environment in the event of any losses. It had to be able to dissolve cellulose completely without reacting with it or degrading it, and the resulting process had to be less energy intensive than the viscose route, which was already proven to be less energy intensive than the synthetics in an independent study (97). It was especially important to choose a system which, like the melt-spun synthetics, would not require costly gaseous or liquid effluent treatment systems. Finally, the process would be capable of making good textile fibers to maintain, or even extend, the cellulosics share of the global market against fierce competition from melt-spun synthetics based on cheap but nonrenewable oil reserves.

By 1980, NMMO was shown to be the best solvent, provided well-known difficulties associated with its thermal stability could be avoided by appropriate chemical engineering. Filaments obtained from the first single-hole extrusion experiments had promising properties so Courtaulds committed the resources in 1982 to build the first small pilot plant to test the feasibility of overcoming the solvent handling and recovery problems that had prevented earlier commercial exploitation. This system, capable of making up to 100 kg/week of fiber, met its objectives and allowed the first serious end use development to begin. Scale-up to 1000 kg/week pilot line was possible in 1984, and in 1988 a 25,000 kg/week semi-commercial line was commissioned to allow a thorough test of the engineering and end use development aspects.

Comparisons of Tencel with viscose in both laboratory and test markets proved that the fibers were sufficiently different to deserve separate marketing strategies. Tencel is stronger than any other cellulosic, especially when wet; easy to process into yarns and fabrics alone or in blends; easy to blend (unique fiber presentation); easy to spin to fine count yarns; very stable in washing and drying; thermally stable; easy to dye to deep vibrant colors; capable of taking the latest finishing techniques to give unique drape; and comfortable to wear.

As a 1.7 dtex (1.5 den) fiber, it can be spun into yarns with a better strength conversion factor than other cellulosics, allowing rotor-spun Tencel to outperform ring-spun cotton or modal viscose. Fabrics can be made at high efficiency, and prove to have the anticipated tear and tensile advantages over other cellulosics. Direct, reactive, or vat dyes can be used, and easy care properties can be achieved with less resin finish than normal. Tencel could therefore be positioned as a new premium quality apparel cellulosic and not simply as a long-term replacement for viscose.

The tendency of the strong, highly crystalline fibers to fibrillate, ie, to develop a hairy surface on wet-abrasion has, for the textile applications, been minimized by process changes both in fiber production and fabric manufacture. However, for nonwoven or speciality paper applications, this property can allow potential users to develop cellulosic microfibers during processing.

The Courtaulds semicommercial production system is illustrated in Figure 8. Dissolving-grade woodpulp is mixed into a paste with NMMO and passes through a high temperature dissolving unit to yield a clear viscous solution. This is filtered and spun into dilute NMMO whereupon the cellulose fibers precipitate. These are washed and dried, and finally baled as staple or tow products as required by the market. The spin bath and wash liquors are passed to solvent recovery systems which concentrate the NMMO to the level required for reuse in dissolution.

The new fiber has physical properties (Table 2) sufficiently different from regular rayon to allow an initial market development strategy that does not erode the position of the traditional viscose fiber. The unique strength, texture, and coloration potential of the fiber enable it to command premium prices in up-market mens' and ladies' outerwear. The existing process will also be capable of meeting the needs of the industrial, nonwoven, and specialty paper markets as these end uses are developed. Several studies of its performance in nonwovens show it

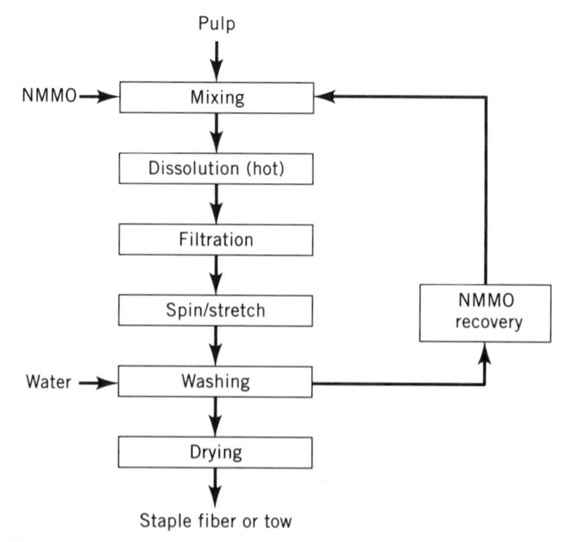

Fig. 8. Direct dissolution: Courtaulds Tencel process.

Table 2. Properties of Selected Commercial Rayon Fibers

Property	Cuprammonium	Regular rayon	Improved rayon	Modal	Polynosic	Y-Shaped rayon[a]	Solvent-spun rayon[b]
fiber cross section							
dry tenacity, cN/tex[c]	15–20	20–24	24–30	34–36	40–65	18–22	40–44
extensibility at break (dry), %	7–23	20–25	20–25	13–15	8–12	17–22	14–16
wet tenacity, cN/tex[c]	9–12	10–15	12–16	19–21	30–40	9–12	34–38
extensibility at break (wet), %	16–43	25–30	25–35	13–15	10–15	23–30	16–18
water imbibition, %	100	90–100	90–100	75–80	55–70	100–110	65–70
cellulose DP[d]	450–550	250–350	250–350	300–500	550–700	250–350	550–600
initial wet modulus[e]	30–50	40–50	40–50	100–120	140–180	35–45	250–270

[a]The Y-shaped rayon data are based on Courtaulds Galaxy fiber.
[b]The solvent-spun rayon data are based on Courtaulds Tencel fiber.
[c]To convert cN/tex to gf/den, divide by 8.82.
[d]DP = degree of polymerization.
[e]The load required to extend the wet fiber by 5% × 20.

capable of yielding fabric strengths 2–3 times higher than hitherto possible with regenerated cellulosics (98,99). It appears particularly suitable for the latest hydroentanglement systems (100) where its basic strength and an ability to develop a microfiber surface under the action of high pressure water jets enables it to make very strong nonwoven fabrics with textile-like properties.

Continuous operation of this realistically sized line provided the necessary confidence in both process and market to justify full commercial operation. An 18,000 t/yr line was therefore designed and built in Mobile, Alabama in 1991–1992. Since its commissioning in May 1992 this plant is approaching its target productivity and quality, and is now supplying fiber to several primary apparel markets. It is set up to be capable of making the range of fiber fineness and lengths required for premium quality textiles and nonwovens.

Fiber Properties

The bulk properties of regenerated cellulose are the properties of Cellulose II which is created from Cellulose I by alkaline expansion of the crystal structure (97,101) (see CELLULOSE). The key textile fiber properties for the most important current varieties of regenerated cellulose are shown in Table 2. Fiber densities vary between 1.53 and 1.50.

A discussion of the fiber properties is complicated by the versatility of cellulose and its conversion routes. Many of the properties can be varied over wide ranges depending on the objectives of the producer. For instance, dry fiber strengths from less than 1 but up to 10 g/den have been made commercially, and at a given dry strength it has proved possible to vary the elongation and wet properties independently. The water imbibition of pure cellulose fibers has been varied from 45 to 450%, and the cross-sectional shape altered from circular all-skin, to multilobal collapsed tubes that reopen on wetting. Alloying adds further scope for adding properties unachievable with pure cellulose. Chemically, cellulose fibers are well-endowed with reactive hydroxyl groups that render them amenable to modification by grafting, which, more so than alloying, allows the properties to be moved in the direction of synthetic polymers if the need arises. The same reactivity makes the fibers more degradable than synthetics, both chemically and biologically, although it must be said that biological degradability is easily halted by inclusion of suitable biocides.

Thermal Properties. Fibers are not thermoplastic and stable to temperatures below 150°C, with the possible exception of slight yellowing. They begin to lose strength gradually above 170°C, and decompose more rapidly above 300°C. They ignite at 420°C and have a heat of combustion of 14,732 J/g (3.5 kcal/g).

Moisture Regain. The fibers are all highly hydrophilic with moisture regains at 65% rh ranging from 11 for the polynosics to 13 for regular rayon. Wetting of cellulose is exothermic, with the regenerated forms giving out 170 J/g (41 cal/g) between 40% rh and 70% rh, compared with 84 J/g (20 cal/g) for cotton. The fiber structure swells as fluid is imbibed, and fiber strength and stiffness falls. The contact angle for unfinished fiber is greater than 150°.

Chemical Properties. The fibers degrade hydrolytically when contacted with hot dilute or cold concentrated mineral acids. Alkalies cause swelling (maximum with 9% NaOH at 25°C) and ultimately disintegration. They are unaffected by most common organic solvents and dry-cleaning agents. They are degraded by strong bleaches such as hypochlorite or peroxide.

Optical Properties. The fibers are birefringent. The refractive index parallel to the fiber length ranges from 1.529 for regular rayon to 1.570 for the polynosics. Refractive indexes perpendicular to the fiber length range from 1.512 (regular) to 1.531 (polynosic). Basic fibers are naturally lustrous, but can be made dull by adding opacifiers or by changing the cross-sectional shape.

Electrical. The electrical properties of the fiber vary with moisture content. The specific resistance of the fibers is around 3×10^6 ohms/cm at 75% rh and 30°C compared with 1×10^{18} ohms/cm for pure dry cellulose. The dielectric constant (100 kHz) is 5.3 at 65% rh and 3.5 at 0% rh. The zeta potential in water is -25 mV.

Environmental Issues

Rayon is unique among the mass produced man-made fibers because it is the only one to use a natural polymer (cellulose) directly. Polyesters, nylons, polyolefins, and acrylics all come indirectly from vegetation; they come from the polymerization of monomers obtained from reserves of fossil fuels, which in turn were formed by the incomplete biodegradation of vegetation that grew millions of years ago. The extraction of these nonrenewable reserves and the resulting return to the atmosphere of the carbon dioxide from which they were made is one of the most important environmental issues of current times. Cellulosic fibers therefore have much to recommend them provided that the processes used to make them have minimal environmental impact.

Liquid Effluents. Recycling of acid, soda, and zinc have long been necessary economically, and the acid–soda reaction product, sodium sulfate, is extracted and sold into other sectors of the chemical industry. Acid recovery usually involves the degassing, filtering, and evaporative concentration of the spent acid leaving the spinning machines. Excess sodium sulfate is removed by crystallization and then dehydrated before sale. Traces of zinc that escape recovery are removable from the main liquid effluent stream to the extent that practically all the zinc can now be retained in the process.

Gaseous Effluents. Twenty percent of the carbon disulfide used in xanthation is converted into hydrogen sulfide (or equivalents) by the regeneration reactions. Ninety to 95% of this hydrogen sulfide is recoverable by scrubbers that yield sodium hydrogen sulfide for the tanning or pulp industries, or for conversion back to sulfur. Up to 60% of the carbon disulfide is recyclable by condensation from rich streams, but costly carbon-bed absorption from lean streams is necessary to recover the remaining 20 + %. The technology is becoming available to deal with this, but there remains the danger that cost increases resulting from the necessary investments will make the fibers unattractively expensive compared to synthetics based on cheap nonrenewable fossil fuels.

Energy Use. Energy consumption in the xanthate process compares favorably with the synthetics. The methodology of assessing the energy usage of products and processes is currently the subject of much debate, and a standardized approach has yet to emerge. Not surprisingly, most of the published work on fibers was carried out during the last energy crises in the 1973–1981 period.

The early attempts to assess the total energy required to make baled staple fiber from naturally occurring raw materials, wood in the case of cellulosics and oil in the case of synthetics, used differing approaches and were insufficiently rigorous to allow hard conclusions to be drawn. In general, the fiber production sequence is broken into monomer making, polymer making, and fiber production, and although a variety of fibers are covered, only viscose rayon and polyester are mentioned in all of them (102–107). Tons of fuel oil equivalent per ton of fiber (TFOE/T) are the most popular units, with rayon requiring from 1.7 to 2.4 TFOE/T and polyester requiring 2.6 to 4.2 TFOE/T.

The studies all concluded that rayon required less energy to make than polyester but there was little agreement on the magnitude of the difference. The wet-spun cellulosic fibers required more energy than melt-spun polyester for the fiber-making step, but they had no monomer energy requirement. The polymerization requirement was therefore minimal. In the case of the 1.7 TFOE/T figure for rayon, full credit was being given for the fact that the pulp mills energy needs were renewable and not dependent on fossil fuels. Pulp could be fed directly into the viscose process without incurring any transport or drying cost, and the pulp mill could be driven entirely by energy obtained from burning the parts of the tree which were not needed in the final product. This free and renewable energy was not counted.

A qualitative consideration of the NMMO route suggests that it will require slightly less energy than the xanthate route, and will emit lower levels of gaseous and liquid effluents.

Fiber Disposal. Cellulosic fibers, like the vegetation from which they arise, can become food for microorganisms and higher life forms, ie, they biodegrade. If necessary they burn with a rather greater yield of energy than natural vegetation. In complete biodegradation or incineration, the final breakdown products are carbon dioxide and water; these disposal methods simply recycle the cellulose to the atmospheric components from which it was made.

It is also possible to liberate and use some of the free solar energy that powered the manufacture of sugars and cellulose during photosynthesis. This can be achieved by burning or by anaerobic digestion. Slow anaerobic biodegradation occurs in all landfill sites dealing with municipal solid waste. This process generates methane from cellulose, which can be burnt to drive gas-turbines. If future landfills are lined and operated with moisture addition and leachate recycling, then energy generation and the return of landfill sites to normal use can be accelerated (108).

Economic Aspects

Since the early 1980s, the viscose-based staple fibers have, like the cuprammonium and viscose filament yarns in the 1970s, ceased to be commodities. They have been repositioned from the low cost textile fibers that were used in a myriad of applications regardless of suitability, to premium priced fashion fibers delivering comfort, texture, and attractive colors in ways hard to achieve with other synthetics. They are still widely used in blends with polyester and cotton to add value, where in the 1980s they would have been added to reduce costs.

Such repositioning inevitably means reduced production volume, and for the first time this century production in the last decade has been below that a decade earlier (Fig. 9). Most capacity reductions have been in North America and especially eastern Europe. This has been offset in part by capacity increases in the Far East. Rayon is no longer a significant component of carpets, and has lost the disposable diaper coverstock business to cheaper and more easily processed poly-

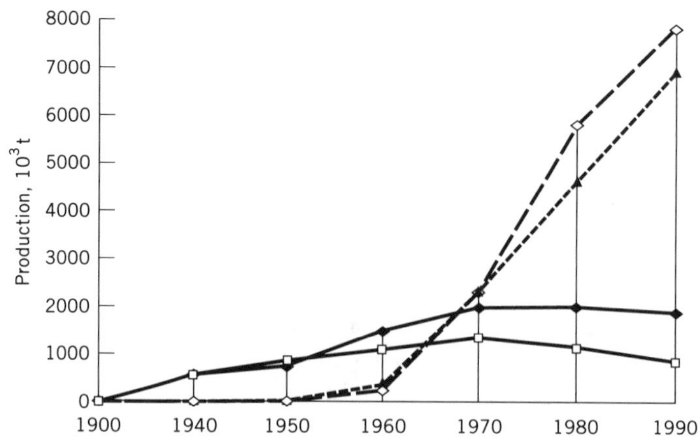

Fig. 9. World synthetic fiber production, 1900–1990: (—□—) cellulosic filament; (—◆—) cellulosic staple; (--▲--) synthetic filament; and (—◇—) synthetic staple.

propylene. It has, however, gained share in health and hygiene products and is now a principal component of tampons worldwide.

Future Possibilities

The cellulose polymer and its conversion routes have already proved to be capable of adaptation to meet a wide range of market demands. The advances being made in getting cellulose into solution with minimal environmental impact augur well for the development of streamlined routes from tree to fiber or fabric.

The progress on understanding the biogenesis of cellulose may yet yield an industrial route from atmospheric carbon dioxide and water, although the simple "plant more forests" option will be more attractive to many. Between these two extremes, in concept and in timing, the production of cellulose from sugar using bacteria, eg, *Acetobacter xylinum* (109), is becoming feasible. Sony Corp. has been using Ajinomoto's bacterial cellulose microfibers for speaker cones (110), and Weyerhauser (111,112) has developed microfiber nonwovens and food thickeners. The costs of cellulose obtained in this way are currently very high compared with growing wood, but cost reduction proposals have been made (113). These involve either identifying, culturing, or genetically engineering strains of bacteria that produce cellulose microfibers while metabolizing a wider range of foods more efficiently. This challenge is at least as tough as the program to develop an industrial-scale direct dissolution process for wood cellulose which was beginning just as the 3rd edition of this *Encyclopedia* went to press in 1978.

BIBLIOGRAPHY

"Rayon and Acetate Fibers" in *ECT* 1st ed., Vol. 11, pp. 522–550, "Rayon," by L. A. Cox, American Viscose Corp., and "High-Tenacity Rayon," by P. M. Levin, E. I. de Pont de Nemours & Co., Inc.; "Rayon" in *ECT* 2nd ed., Vol. 17, pp. 168–209, by R. L. Mitchell and G. C. Daul, ITT Rayonier Inc.; in *ECT* 3rd ed., Vol. 19, pp. 855–880, by J. Lundberg and A. Turbak, Georgia Institute of Technology.

1. Brit. Pat. 283 (Apr. 17, 1855), G. Audemars.
2. Brit. Pat. 5,978 (Dec. 31, 1883) J. W. Swan.
3. Fr. Pat. 165,349 (May 12, 1884), H. de Chardonnet.
4. Fr. Pat. 203,741 (1890), L. H. Despeissis.
5. E. Schweitzer, *J. Prakt. Chem.* **72,** 109 (1857).
6. Ger. Pat. 98,642 (1897), H. Pauly.
7. Brit. Pat. 8,700 (Apr. 8, 1893), C. F. Cross, E. J. Bevan, and C. Beadle.
8. Brit. Pat. 1,020 (Dec. 23, 1898), C. H. Stearn.
9. Brit. Pat. 23,157 (1900), C. F. Topham.
10. Brit. Pat. 23,158 (1900), C. F. Topham.
11. Brit. Pat. 273,386 (June 29, 1926), W. H. Glover and G. S. Heaven (to Courtaulds).
12. U.S. Pat. 1,875,894 (Sept. 6, 1932), J. A. Singmaster.
13. Envirocell, *Eco-profiling of Cellulose-Based Products,* a study prepared for Courtaulds and others, Aug. 1990.
14. U.S. Pat. 2,542,285 (Feb. 20, 1951), R. L. Mitchell (to Rayonier Inc.).
15. H. P. von Bucher, *Tappi* **61**(4), 91 (1978).
16. A. Lyselius and O. Samuelson, *Svensk Papperstidning* **64**(5), 145 (1961).

17. L. Ivnas and L. Svensson, *Tappi* **57**(8), 115 (1974).
18. Brit. Pat. 467,500 (June 14, 1937), J. H. Givens, L. Rose and H. W. Biddulph (to Courtaulds).
19. U.S. Pat. 2,535,045 (Dec. 26, 1950), N. L. Cox (to E. I. du Pont de Nemours & Co., Inc.).
20. U.S. Pat. 2,937,922 (May 24, 1960), R. L. Mitchell, J. W. Berry, and W. H. Wadman (to Rayonier Inc.).
21. U.S. Pat. 2,792,279 (May 14, 1957), M. R. Lytton (to American Viscose Corp.).
22. U.S. Pat. 2,792,280 (May 14, 1957), B. A. Thumm, M. R. Lytton, and J. A. Howsmon (to American Viscose Corp.).
23. U.S. Pat. 2,792,281 (May 14, 1957), C. A. Castellan (to American Viscose Corp.).
24. U.S. Pat. 2,696,423 (Dec. 7, 1954), M. A. Dietrich (to E. I. du Pont de Nemours & Co., Inc.).
25. U.S. Pat. 2,942,931 (June 28, 1960), R. L. Michell, J. W. Berry, and W. H. Wadman (to Rayonier Inc.).
26. U.S. Pat. 3,018,158 (Jan. 23, 1962), R. L. Mitchell, J. W. Berry, and W. H. Wadman (to Rayonier Inc.).
27. U.S. Pat. 3,109,698 (Nov. 5, 1963), E. Klein, H. Wise and W. C. Richardson (to Courtaulds North America Inc.).
28. U.S. Pat. 3,109,699 (Nov. 5, 1963), W. C. Richardson (to Courtaulds North America Inc.).
29. U.S. Pat. 3,109,700 (Nov. 5, 1963), E. Klein, D. S. Nelson and B. E. M. Bingham (to Courtaulds North America Inc.).
30. U.S. Pat. 3,718,537 (Feb. 27, 1973), A. Kawai, T. Katsuyama, M. Suzuki and H. Ohta (to Mitsubishi Rayon Co. Ltd.).
31. U.S. Pat. 3,832,281 (Aug. 27, 1974), A. Kawai, T. Katsuyama, M. Suzuki and H. Ohta (to Mitsubishi Rayon Co. Ltd.).
32. D. Mach, "Experiences with Fine Denier Viscose and Modal Fibres," *Proceedings of the 28th Dornbirn International Man Made Fibres Conference,* Sept. 1989.
33. U.S. Pat. 2,592,355 (Apr. 8, 1952), S. Tachikawa.
34. U.S. Pat. 1,683,199 (Sept. 4, 1928), I. Lilienfeld.
35. T. Nagata, "New Process Developments for Polynosic Fibres," *Proceedings of the 28th Dornbirn International Man Made Fibres Conference,* Austria, Sept. 1989.
36. I. H. Welch, *Am. Text. Rep. Bull. Edn.* AT8, 49 (1978).
37. Lane and McCombes, *Courtaulds Challenge the Cotton Legend,* ACS Symposium Series 58, ACS, Washington, D.C., 1977, Chap. 12.
38. Brit. Pat. 143,253 (Aug. 15, 1920), L. Drut.
39. Brit. Pat. 189,973 (Dec. 14, 1922), J. Rousset.
40. Brit. Pat. 253,954 (July 1, 1926), H. J. Heagan and F. Bayley.
41. C. R. Woodings and A. I. Bartholomew, "The Manufacture, Properties and Uses of Inflated Viscose Rayon Fibres," *23rd Man-Made Fibres Congress,* Dornbirn, Austria, 1984.
42. K. L. Gray and I. A. Mc Nab, "The Behaviour of Rayon Fibre in Papermaking Systems," *TAPPI Plastic Paper Conference,* 1970.
43. Brit. Pat. 1,283,529 (July 26, 1972), C. R. Woodings (to Courtaulds Ltd.).
44. Brit. Pat. 1,310,504 (Mar. 21, 1973), C. R. Woodings (to Courtaulds Ltd.).
45. Brit. Pat. 1,333,047 (Oct. 10, 1973), C. R. Woodings (to Courtaulds Ltd.).
46. Brit. Pat. 1,393,778 (May 14, 1975), C. R. Woodings (to Courtaulds Ltd.).
47. C. R. Woodings, "The Development of Viscose Rayon for Nonwoven Applications," *TAPPI Nonwovens Fibres Seminar 1979,* pp. 15–28.
48. Brit. Pat. 2,208,277B (Nov. 13, 1991), A. G. Wilkes and A. I. Bartholomew (to Courtaulds plc).

49. A. G. Wilkes, "Galaxy-A New Viscose Rayon Fibre for Nonwovens," *Proceedings of the INDA-TEC 89 Conference,* 1989.
50. C. R. Woodings "An Absorbency Overview with Inter-Fibre Comparisons," *Proceedings of the Asia Nonwoven Conference,* 1989.
51. U.S. Pat. 4,066,584 (Jan. 3, 1978), T. C. Allen and D. B. Denning (to Akzona Inc.).
52. U.S. Pat. 4,263,244 (Apr. 21, 1981), T. C. Allen and D. B. Denning (to Akzona Inc.).
53. U.S. Pat. 3,844,287, (Oct. 29, 1974), F. R. Smith (to FMC Corp.).
54. *T.S.S.—Assessment of Current Information and Future Research Needs,* Institute of Medicine, National Academy Press, Washington, D.C., 1982.
55. S. F. Berkley and co-workers, JAMA **258**(7) (Aug. 21, 1987).
56. U.S. Pat. reissue 31,380 (Sept. 13, 1983), F. R. Smith (to Avtex Fibers Inc.).
57. U.S. Pat. 4,063,558 (Dec. 20, 1977), F. R. Smith (to Avtex Fibers Inc.).
58. U.S. Pat. 3,843,378 (Oct. 22, 1974), F. R. Smith (to FMC Corp.).
59. U.S. Pat. 4,179,416 (Dec. 18, 1979), F. R. Smith (to Avtex Fibers Inc.).
60. U.S. Pat. 4,242,242 (Dec. 30, 1980), T. C. Allen (to Akona Inc.).
61. B. V. Hettich, Text. Res. J. **54** (6), 382–390 (June 1984).
62. S. M. Suchecki, Text. Ind. **142** (2), 29 (1978).
63. B. V. Hettich, "Regenerated Cellulose Fibres—Technology for the Future," *50th Annual TRI Meeting,* 1980.
64. Sh. Heidari, *Chemifasern / Textilindustrie* **41/93** (Dec. 1991).
65. Y. Kaneko, *Lenzinger Ber.,* 78–82 (June 1990).
66. U.S. Pat. 1,983,221 (Dec. 4, 1934), W. H. Furness (to Furness Corp.).
67. U.S. Pat. 1,983,795 (Dec. 11, 1934), W. H. Furness (to Furness Corp.).
68. U.S. Pat. 2,587,619 (Mar. 4, 1952), H. Hoffman (to Beaunit Mills Inc.).
69. U.S. Pat. 3,765,818 (Oct. 16, 1973), T. Miyazaki and co-workers (to Asahi Kasei Kogyo Kabushiki Kaisha).
70. U.S. Pat. 3,833,438 (Sept. 3, 1974), T. Kaneki and co-workers (to Asahi Kasel Kogyo Kabushiki Kalsha).
71. "Synthetic Fibre Development in Germany," *Tex. Res. J.* **16**(4) (Apr. 1946).
72. U. V. Grafov and co-workers, *Khim. Volokna,* (3), 63 (May–June 1976).
73. A. F. Turbak and co-workers, *A Critical Review of Cellulose Solvent Systems,* ACS Symposium Series 58, ACS, Washington, D.C., 1977.
74. A. F. Turbak, *Proceedings of the 1983 International Dissolving and Speciality Pulps Conference,* TAPPI, Atlanta, Ga., 1983, p. 105.
75. U.S. Pat. 4,302,252 (Nov. 24, 1981), A. F. Turbak and co-workers (to International Telephone and Telegraph Corp.).
76. U.S. Pat. 4,056,675 (Nov. 1, 1977), A. F. Turbak and R. B. Hammer (to International Telephone and Telegraph Corp.).
77. U.S. Pat. 4,367,191 (Jan. 4, 1983), J. A. Cuculo and S. M. Hudson (to Research Corp.).
78. U.S. Pat. 4,097,666 (June 27, 1978), D. C. Johnson and M. C. Nicholson (to The Institute of Paper Chemistry).
79. U.S. Pat. 2,179,181 (Nov. 7, 1939), C. Graenacher and R. Sallman (to Society of Chemical Indstry in Basle).
80. U.S. Pat. 3,447,956 (June 3, 1969), D. L. Johnson (to Eastman Kodak Co.).
81. U.S. Pat. 3,447,939 (June 3, 1969), D. L. Johnson (to Eastman Kodak Co.).
82. U.S. Pat. 3,508,941 (Apr. 28, 1970), D. L. Johnson (to Eastman Kodak Co.).
83. U.S. Pat. 4,145,532 (Mar. 20, 1979), N. E. Franks and J. K. Varga (to Akzona Inc.).
84. U.S. Pat. 4,196,282 (Apr. 1, 1980), N. E. Franks and J. K. Varga (to Akzona Inc.).
85. H. Chanzy and A. Peguy, *J. Poly. Sci, Poly. Phys. Edn.* **18,** 1137–1144 (1980).
86. H. Chanzy, M. Paillet, and R. Hagege, *Polymer* **31** (Mar. 1990).
87. R. N. Armstrong and co-workers, *Proceedings of the 5th International Dissolving Pulps Conference,* TAPPI, Atlanta, Ga., 1980.
88. L. Lenz, *J. App Poly Sci.* **35,** 1987–2000 (1988).

89. U.S. Pat. 2,134,825 (Nov. 1, 1938), J. W. Hill and R. A. Jacobson (to E. I. du Pont de Nemours & Co., Inc.).
90. Finn. Pat. 61,033 (1982), K. Ekman, O. T. Turenen and J. I. Huttunen.
91. V. Rossi and O. T. Turenen, "Cellulose Carbamate," *PIRA International Conference,* Brighton, UK, Nov. 10–12, 1987.
92. U.S. Pat. 4,999,149 (Mar. 12, 1991), L. Fu Chen (to Purdue Research Foundation).
93. T. Watanabe and co-workers, *Preprint for the 20th Annual Meeting of Polymer Science, Japan,* 427 (1971).
94. K. Kamide and co-workers, *Brit. Polymer J.* **22,** 73–83, 121–128, 201–212 (1990).
95. K. Kamide and T. Yamashiki, Cellulose Sources and Exploitation, Ellis Horwood Ltd., New York, 1990, Chapt. 24.
96. K. Kamide and co-workers, *J. Appl. Poly. Sci.* **44,** 691–698 (1992).
97. A. Grobe, Properties of Cellulose Materials, *Polymer Handbook,* 3rd ed., John Wiley & Sons, Inc., New York, 1989, pp. 117–170.
98. D. A. Smith, S. D. J. Williams, and C. R. Woodings, "Solvent-Spun Cellulosics in Nonwovens," *Proceedings of the EDANA Index 87 Symposium,* Geneva, 1987.
99. D. Cole and C. R. Woodings, "Solvent-Spun Cellulosics in Nonwovens—Update 1987–90," *Proceedings of the EDANA Index 90 Symposium,* Geneva, 1990.
100. C. R. Woodings, "The Hydroentanglement of a Range of Staple Fibres," *Proceedings of the IMPACT89 Nonwovens Conference,* Miller Freeman Publications, Amelia Island, Fla., 1989.
101. A. Sarko, *Wood and Cellulosics,* Ellis Horwood Ltd., New York, 1987, Chapt. 6.
102. Woodhead, *International TNO Conference,* 1976.
103. Lane and McCombes, *Text. Manuf.* (1) (1979).
104. Kogler, *Proceedings of the EDANA AGM,* Munich, Dec. 6, 1980.
105. Armstrong, *Proceedings of the EDANA AGM,* Munich, June 12, 1980.
106. Marini and Six, *Proceedings of the EDANA Nonwovens Symposium,* Milan, June 11, 1985.
107. W. C. Frith, *Energy Balance of Man-Made Fibre Production,* CIRFS, Brussels, 1980.
108. Pohland and Sratakis, *Controlled Landfill Management—Principles and Applications,* Insight 91, Charleston, Oct. 1991.
109. Brit. Pat. 2,169,543 A, (July 16, 1986), E. M. Roberts, L. J. Hardison, and R. M. Brown (to University of Texas).
110. Eur. Pat. 0 200,409 A, (Nov. 5, 1986), M. Iguchi and co-workers, (to Sony Corp. and Ajinomoto Co. Inc.).
111. "Ultrafine Cellulose Fibre from Weyerhauser Co.," *Medical Textiles,* (Nov. 1990).
112. Eur. Pat. 0 380,471 (Aug. 1990), D. Johnson and A. Neogi (to Weyerhauser).
113. R. M. Brown, in *Cellulose-Structural and Functional Aspects,* Ellis Horwood, New York, 1989.

General References

C. M. Deeley, "Viscose Rayon Production," *Notes for the Associateship of the Textile Institute Examination Lectures,* Sept. 14, 1959.

J. Dyer and G. C. Daul, *The Handbook of Fibre Science and Technology,* Vol. 4, *Fibre Chemistry,* Marcel Dekker Inc., New York, 1985, Chap. 11.

W. Albrecht and co-workers, in *Man Made Fibre Yearbook,* CTI, Maryland, 1991, pp. 26–44.

D. C. Coleman, *Courtaulds: An Economic and Social History,* Vols. 2 and 3, Clarendon Press, New York, 1980.

C. R. WOODINGS
Courtaulds

VEGETABLE

Fibers for commercial and domestic use are broadly classified as natural or synthetic. The natural fibers are vegetable, animal, or mineral in origin. Vegetable fibers, as the name implies, are derived from plants. The principal chemical component in plants is cellulose, and therefore they are also referred to as cellulosic fibers. The fibers are usually bound by a natural phenolic polymer, lignin, which also is frequently present in the cell wall of the fiber; thus vegetable fibers are also often referred to as lignocellulosic fibers, except for cotton which does not contain lignin.

Vegetable fibers are classified according to their source in plants as follows: (*1*) the bast or stem fibers, which form the fibrous bundles in the inner bark (phloem or bast) of the plant stems, are often referred to as soft fibers for textile use; (*2*) the leaf fibers, which run lengthwise through the leaves of monocotyledonous plants, are also referred to as hard fibers; and (*3*) the seed-hair fibers, the source of cotton (qv), are the most important vegetable fiber. There are over 250,000 species of higher plants; however, only a very limited number of species have been exploited for commercial uses (less than 0.1%). The commercially important fibers are given in Table 1 (1,2).

The fibers in bast and leaf fiber plants are integral with the plant structure, providing strength and support. In bast fiber plants, the fibers are next to the outer bark in the bast or phloem and serve to strengthen the stems of these reedlike plants. They are strands running the length of the stem or between joints (Fig. 1). To separate the fibers, the natural gum binding them must be removed. This operation is called retting (controlled rotting). For most uses, particularly for textiles, this long composite-type strand fiber is used directly; however, when such fiber strands are pulped by chemical means the strand is broken down into much shorter and finer fibers, the ultimate fibers shown in Figure 1.

The long leaf fibers contribute strength to the leaves of certain nonwoody, monocotyledonous plants. They extend longitudinally the full length of the leaf and are buried in tissues of a parenchymatous nature. The fibers found nearest the leaf surface are the strongest. The fibers are separated from the pulp tissue by scraping because there is little bonding between fiber and pulp; this operation is called decortication. Leaf fiber strands are also multicelled in structure.

Ancient humans used cordage in fishing, trapping, and transport, and in fabrics for clothing. Rope and cord making started in paleolithic times, as seen in cave drawings. Rope, cords, and fabrics were made from reeds and grasses in ancient Egypt (400 BC). Ropes, boats, sails, and mats were made from palm leaf fibers and papyrus stalks and writing surfaces, known as papyrus, from the pith section. Jute, flax, ramie, sedge, rushes, and reeds have long been used for fabrics and baskets. Jute was cultivated in India in ancient times and used for spinning and weaving. The first true paper is believed to have been made in southeastern China in the second century AD from old rags (bast fibers) of hemp and ramie and later from the bast fiber of the mulberry tree.

Scientists at the U.S. Department of Agriculture (USDA) Northern Regional Research Center in Peoria, Ill., carried out an extensive screening of over 500 species of plants as potential sources of papermaking fibers (3–6). The species were rated for agronomic potential, chemical composition, fiber properties, and

Table 1. Vegetable Fibers of Commercial Interest[a]

Commercial name	Source	Botanical name of plant	Growing area
		Bast or soft fibers	
China jute	Abutilon	*Abutilon theophrasti*	China
flax		*Linum usitatissimum*	north and south temperate zones
hemp		*Cannabis sativa*	all temperate zones
jute		*Corchorus capsularis; C. olitorius*	India
kenaf		*Hibiscus cannabinus*	India, Iran, CIS, South America
ramie		*Boehmeria nivea*	China, Japan, United States
roselle		*Hibiscus sabdariffa*	Brazil, Indonesia (Java)
sunn		*Crotalaria juncea*	India
urena	cadillo	*Urena lobata*	Zaire, Brazil
		Leaf or hard fibers	
abaca		*Musa textilis*	Borneo, Philippines, Sumatra
cantala	Manila maguey	*Agave cantala*	Philippines, Indonesia
caroa		*Neoglaziovia variegata*	Brazil
henequen		*Agave fourcroydes*	Australia, Cuba, Mexico
istle		*Agave* (various species)	Mexico
mauritius		*Furcraea gigantea*	Brazil, Mauritius, Venezuela, tropics
phormium		*Phormium tenax*	Argentina, Chile, New Zealand
pineapple	piña	*Ananas comasus*	Hawaii, Philippines, Indonesia, India, West Indies
sansevieria	bowstring hemp	*Sansevieria* (entire genus)	Africa, Asia, South America
sisal		*Agave sisalana*	Haiti, Java, Mexico, South Africa
		Seed-hair fibers	
coir	coconut husk fiber	*Cocos nucifera*	tropics, India, Mexico
cotton		*Gossypium* sp.	United States, Asia, Africa
kapok		*Ceiba pentandra*	tropics
milkweed floss		*Chorisia* sp.	North America
		Other fibers	
broom root	roots	*Muhlenbergia macroura*	Mexico
broom corn	flower head	*Sorghum vulgare technicum*	United States
crin vegetal	palm leaf segments	*Chamaerops humilis*	North Africa
palmyra palm	palm leaf stem	*Brossus flabellifera*	India
pissava	palm leaf base fibers	*Attalea funifera*	Brazil
raffia	palm leaf segments	*Raphia raffia*	East Africa

[a]Refs. 1 and 2.

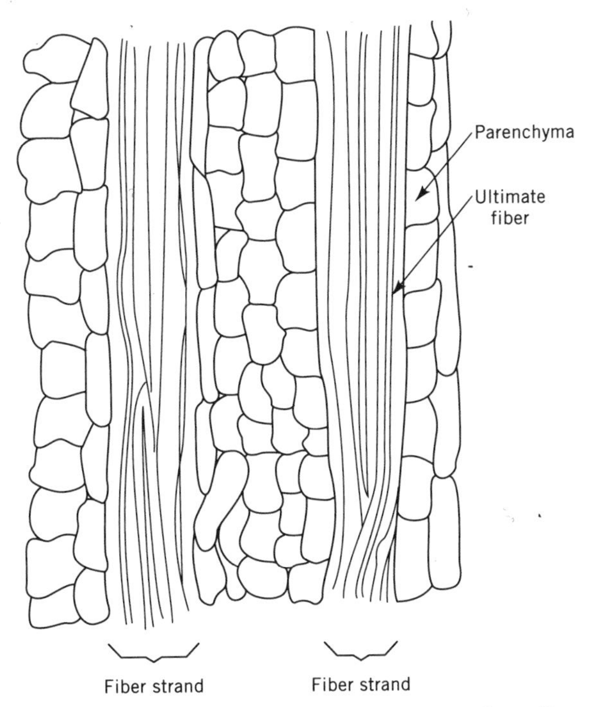

Fig. 1. Typical longitudinal section of bast fiber showing very long fiber strands composed of many ultimate fibers (see Table 3 for dimensions).

physical characteristics. The species with the greatest potential were in the genera *Hibiscus, Crotalaria, Sorghum, Cannabis, Gynerium, Lygerum,* and *Sinarundinaria.*

World markets for vegetable fibers have been steadily declining in recent years, mainly as a result of substitution with synthetic materials. Jute has traditionally been one of the principal bast fibers (tonnage basis) sold on the world market; however, the precipitous decline in jute exports by India (Fig. 2) indicate the decreasing market demand for this fiber that is vitally important to the economies of India (West Bengal), Bangladesh, and Pakistan.

General Properties

Chemical Composition. The chemical composition of the principal commercial vegetable fibers is given in Table 2. Chemically, cotton is the purest, containing over 90% cellulose (qv) with little or no lignin. The other fibers contain 70–75% cellulose, depending on processing. Boiled and bleached flax and degummed ramie may contain over 95% cellulose. Kenaf and jute contain higher contents of lignin, which contributes to their stiffness. Although the cellulose contents shown in Table 2 are fairly uniform, the other components, eg, hemicelluloses (qv), pectins, extractives, and lignin (qv), vary widely without obvious pattern. These differences may characterize specific fibers.

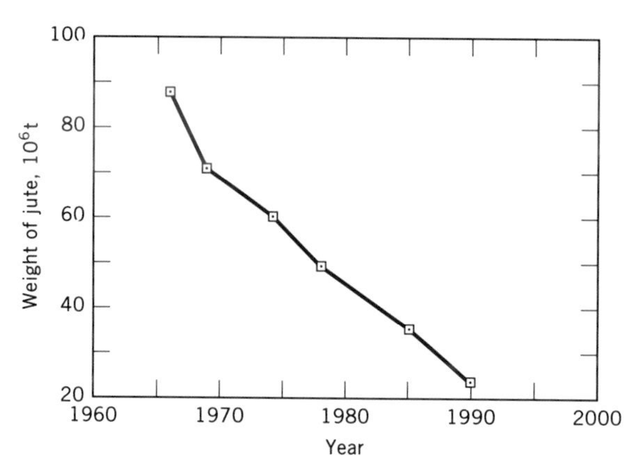

Fig. 2. Export of jute goods from India (1965–1990).

Table 2. Chemical Composition of Vegetable Fibers, wt %[a]

Fiber	Cellulose	Hemicellulose	Pectins	Lignin	Extractives
		Bast fibers			
flax	71.2	18.6	2.0	2.2	6.0
hemp	74.9	17.9	0.9	3.7	3.1
jute	71.5	13.4	0.2	13.1	1.8
kenaf	63.0	18.0		17.0	2.0
ramie	76.2	14.6	2.1	0.7	6.4
		Leaf fibers			
abaca	70.1	21.8	0.6	5.7	1.8
phormium	71.3				
sisal	73.1	13.3	0.9	11.0	1.6
		Seed-hair fibers			
coir	43.0	0.1		45.0	
cotton	92.9	2.6	2.6		1.9
kapok	64.0	23.0	23.0	13.0	

[a]Refs. 1, 7, and 8.

Fiber Dimensions. Except for the seed-hair fibers, the vegetable fibers of bast or leaf origins are multicelled (Fig. 1) and are used as strands (see Table 3). In contrast to the bast fibers, leaf fibers are not readily broken down into their ultimate cells. The ultimate cells are composites of microfibrils, which, in turn, are comprised of groups of parallel cellulose chains.

Physical Properties. Mechanical properties are given in Table 4. Bast and leaf fibers are stronger (higher tensile strength and modulus of elasticity) but lower in elongation (extensibility) than cotton. Vegetable fibers are stiffer but less

Table 3. Dimensions of Ultimate Fibers and Strands[a]

	Ultimate fiber		Cell cross section		Fiber strand	
Fiber	Length, mm	Diameter, μm	Shape	Diameter,[b] μm	Length, cm	Width, mm
			Bast fibers			
Chinese jute	2–6.5	7–33				
flax	4–69	8–31	polygonal	8.8–16.1	25–120	0.04–0.62
hemp	5–55	16	polygonal	13.1–23.6	100–400	0.5–5
jute	0.7–6	15–25	polygonal to oval	12.3–18.6	150–360	
kenaf	2–11	13–33	cylindrical		200–400	
ramie	60–250	16–120	hexagonal to oval	6.2–32.4	10–180	
sunn	2–11	13–64	irregular	13.6–24.6	108–216	
nettle	4–70	20–70		50–50		
			Leaf fibers			
abaca	2–12	6–40	oval to round	14–20	150–360	0.01–0.28
cantala				13.8–16.4		
caroa	2–10	3–13		3.2–8.2		
henequen	1.5–4	8.3–33.2		11.6–22.2		
istle		9.6–16		11.2–13.4	30–75	
mauritius	1.3–6	15–32	cyclindrical		124–210	
phormium	2–11	5.25	round	10.3–12.1	150–240	
sansevieria	1–7	13–40				
sisal	0.8–7.5	8–48	cylindrical	11–16	60–120	0.1–0.5
			Seed-hair fibers			
coir	0.2–1	6.24				1
cotton	10–50	12–25	circular elliptical		1.5–5.6	0.012–0.025
kapok	15–30	10–30	round		1.5–3	0.03–0.036
			Others			
broom root					25–40	

[a]Refs. 1, 8, and 9.
[b]Minor and major diameters, respectively.

tough than synthetic fibers. Kapok and coir are relatively low in strength; kapok is known for its buoyancy.

Among the bast textile fibers, the density is close to 1.5 g/cm^3, or that of cellulose itself, and they are denser than polyester, as shown in Table 5. Moisture regain (absorbency) is highest in jute at 14%, whereas that of polyester is below 1%. The bast fibers are typically low in elongation and recovery from stretch. Ramie fiber has a particularly high fiber length/width ratio.

Table 4. Mechanical Properties of Vegetable Fibers[a]

Fiber	Fineness km/kg	Tensile strength,[b] km	Elongation %	Modulus of elasticity,[c] N/tex[d]	Modulus of rupture, mN/tex[d]
		Bast (soft) fibers			
flax		24–70	2–3	18–20	8–9
hemp	139	38–62	1–6	18–22	6–9
jute	489	25–53	1.5	17–18	2.7–3
kenaf	180	24	2.7		
ramie		32–67	4.0	14–16	11
urena	342	16	1.9		
		Leaf (hard) fibers			
abaca	32	32–69	2–4.5		6
cantala	58	30			
henequen	32	20–42	3.5–5		
istle	34	22–27	4.8		
phormium	38	26			
sansevieria	118	43	4.0		
sisal	40	36–45	2–3	25–26	7–8
		Seed-hair fibers			
coir		18	16	4.3	16
kapok		16–30	1.2	13	10

[a]Refs. 8 and 10.
[b]Based on breaking length, which measures strength per unit area.
[c]Young's modulus.
[d]To convert N/tex to gf/den, multiply by 11.33.

Table 5. Textile-Associated Properties of Bast Fibers Compared to Polyester[a]

Property	Cotton	Flax	Hemp	Jute	Ramie	Polyester
density, g/cm^3	1.54–1.5	1.50	1.48	1.50	1.51	1.22–1.35
moisture regain, %	8–11	12	12	13.7	6	0.4–0.8
fiber tenacity, mN/tex[b]	260–440	230–240	510–600	260–510	450–653	180–790
elongation, dry, %	3–10	2.7–3.3	1.6	1.2–1.9	3.0–7.0	18–67
recovery from 2% elongation, %	75	65		74	52	85–97
length/width ratio	1400	1200			3000	

[a]Ref. 11.
[b]To convert N/tex to gf/den, multiply by 11.33.

The microfibrils in vegetable fibers are spiral and parallel to one another in the cell wall. The spiral angles in flax, hemp, ramie, and other bast fibers are lower than cotton, which accounts for the low extensibility of bast fibers.

Processing and Fiber Characteristics

BAST FIBERS

Bast fibers occur in the phloem or bark of certain plants. The bast fibers are in the form of bundles or strands that act as reinforcing elements and help the plant to remain erect (Fig. 1). The plants are harvested and the strands of bast fibers are released from the rest of the tissue by retting, common for isolation of most bast fibers. The retted material is then further processed by breaking, scutching, and hackling. A general description of the processing of bast fibers is given below with modifications for specific fibers described in the following sections.

Processing. *Harvesting and Pretreatment.* At optimum maturity, the plants are pulled or mowed by hand or machine and, if necessary, threshed to remove seeds. The plants are spread out in a field to dry.

Retting. The removal of the bast fibers from bark and woody stem parts is promoted by a biological treatment called retting (rotting). This is an enzymatic or bacterial action on the pectinous matter of the stem. After retting, the bundles are dried in fields. Retting may be carried out in several ways.

Dew retting involves the action of dew, sun, and fungi on the plants spread thinly on the ground. Dew retting takes 4–6 weeks, but the action is not uniform and it tends to yield a dark-colored fiber. However, it is far less labor intensive and less expensive than water retting. It is commonly used in regions of low water supply and accounts for 85% of the Western European crop, especially in France, and also in the former Soviet Union.

Water retting involves immersion of bundles of plants in stagnant pools, rivers, ditches (dam retting), or in specially designed tanks (tank retting). The biological effect is achieved through bacterial action and takes 2–4 weeks for dam retting. In tanks with warm water (tank retting) the time is reduced to a few days. Water retting gives a more uniform product. In stream retting the plants are immersed in slow moving streams for a longer time and the quality of the product is high.

Chemical retting involves immersion of the dried plants in a tank with a solution of chemicals, such as sodium hydroxide, sodium carbonate, soaps, or mineral acids. The fibers are loosened in a few hours, but close control is required to prevent deterioration. Chemical retting is more expensive and does not produce a superior fiber to that obtained from biological retting.

Breaking and Scutching. The dried retted stalks in bundles are passed through fluted rolls to break or reduce the woody portion into small particles, which are then separated by scutching. The scutching is done by beating with blunt wooden or metal blades by hand or mechanically.

Hackling. The bundles are hackled or combed to separate the short and long fibers. This is done by drawing the fibers through sets of pins, each set finer than the previous one. As a result the fibers are further cleaned and aligned parallel to one another.

Flax. The flax fiber from the annual plant *Linum usitatissimum* (flax family, Linaceae) has been used since ancient times as the fiber for linen. The plant grows in temperate, moderately moist climates, for example, in Belgium, France, Ireland, Italy, and Russia. The plant is also cultivated for its seed, from which linseed oil is produced. A by-product of the seed plant is the tow fiber used in papermaking.

The bast fibers are dew or water retted with dew retting generally yielding a gray fiber. High quality flax fiber is produced by water (stream) retting in the river Lys in Belgium. A cross-sectional view of the flax fiber is shown in Figure 3**a**; the fibers are round to polygonal with a small lumen. The boiled, bleached fiber contains almost 100% cellulose. The flax fiber is the strongest of the vegetable fibers, even stronger than cotton. The fiber is highly absorbent, an important property for clothing, but is particularly inextensible. The most important application is in linen for clothing, fabrics, lace, and sheeting. Flax fiber is also used in canvas, threads and twines, and certain industrial applications such as fire hoses. Chemical pulping of flax provides the raw material for production of high quality currency and writing paper. Flax fiber is also commonly used in cigarette papers. Flax fibers are graded for fineness, softness, stretch, density, color, uniformity, luster, length, handle, and cleanliness.

Hemp. The source of hemp fiber is the plant *Cannabis sativa* (mulberry family, Moraceae) originating in central China. It is grown in central Asia and

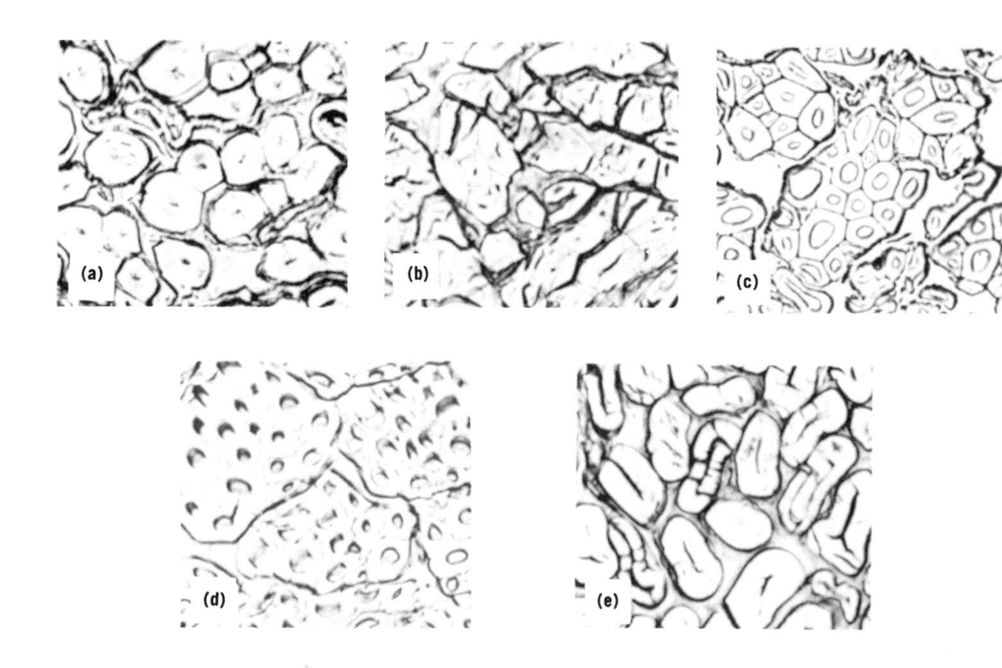

Fig. 3. Cross sections (500×) of bast fibers: (**a**), flax; (**b**), hemp; (**c**), jute; (**d**), kenaf; and (**e**), ramie. Courtesy of the American Association of Textile Chemists and Colorists (12).

eastern Europe. The stem is used for fiber, the seeds for oil, and leaves and flowers for drugs, among them marijuana. The stalks grow 5–7 m tall and 6–16 mm thick. The hollow stems, smooth until the rough foliage at the top, are hand cut and spread on the ground for dew retting for the highest quality product. Water retting is used on sun-dried bundles from which the seeds and leaves have been removed. Strands of hemp fiber can be 2 m in length. The fibers are graded for color, luster, spinning quality, density, cleanliness, and strength.

A cross-sectional view of hemp fiber is shown in Figure 3**b**. It has a Z twist in contrast to the S twist of flax. Hemp is regarded as a substitute for flax in yarn and twine. Its earlier use in ropes has been replaced by leaf and synthetic fibers. Hemp fiber is used in Japan, China, CIS, and Italy to make specialty papers, including cigarette paper, but bleaching is difficult. The fiber is coarser and has less flexibility than flax.

Jute. Jute fiber is obtained from two herbaceous annual plants, *Corchorus capsularis* (linden family, Tiliaceae) originating from Asia, and *C. olitorius* originating from Africa. The former has a round seed pod, and the latter a long pod. Jute is grown mainly in India, Bangladesh, Thailand, Nepal, and Brazil.

The plants are harvested by hand, dried in the field for defoliation, and water (pool) retted for periods up to a month. The depth of the retting pools is dependent on the volume of rainfall during the monsoon season in Southeast Asia. Thus a year with less rainfall results in low water levels in the retting pools and a lower grade jute product due to contamination with sand and silt. The fibers for export are graded for color, length, fineness, strength, cleanliness, luster, softness, and uniformity. The color ranges from cream white to reddish brown, but usually the fiber has a golden luster. A cross-sectional view of the fiber is shown in Figure 3**c**. The fibers are polygonal in cross section with a wide lumen. Jute has traditionally been an important textile fiber, second only to cotton; however, jute is being steadily replaced by synthetics in the traditional high volume uses such as carpet backings and burlap (hessian) fabrics and sacks (Fig. 2). The strands are also used for twine, while kraft pulping of jute gives ultimate fibers for cigarette papers. The Indian Government in cooperation with the United Nations Development Program is involved in a significant jute diversification program to find new uses for jute in finer yarns and textiles, composites and boards, and paper products. A particularly promising new outlet for jute is in interior automobile liners from molded composites with thermoplastic materials.

Kenaf and Roselle. These closely related bast fibers are derived from *Hibiscus cannabinus* and *H. sabdariffa* (mallow family, Malvaceae), respectively. The fibers have other local names. Kenaf is grown for production in the People's Republic of China, Egypt, and regions of the former USSR; roselle is produced in India and Thailand. Plantation-grown kenaf is capable of growing from seedlings to 5 m at maturity in five months. It is reported to yield about 6–10 tons of dry matter per acre, nine times the yield of wood (13).

The plants are hand-cut, mowed, or pulled in developing countries while mechanized harvesting methods are under investigation in the United States. Ribboning machines are sometimes used to separate the fiber-containing bark before retting for recovery of the kenaf strands. For pulping, the kenaf is shredded or hammermilled to 5-cm pieces, washed, and screened.

A cross-sectional view of kenaf is shown in Figure 3**d**. The ultimate cells are nearly cylindrical with thick cell walls. Kenaf fibers are shorter and coarser than those of jute. Both chemical (kraft) and mechanical pulps have been produced from kenaf, and successful demonstration runs of newsprint have been made for the *Dallas Morning News,* the *St. Petersburg Times,* and the *Bakersfield Californian* with a furnish of 82% kenaf chemithermomechanical pulp and 18% softwood kraft pulp. Kenaf fiber is also considered a substitute for jute and used in sacking, rope, twine, bags, and as papermaking pulp in India, Thailand, and the former Yugoslavia. Roselle bleached pulp is marketed in Thailand.

Ramie. The ramie fiber is located in the bark of *Boehmeria nivea,* a member of the nettle family (Urticaceae). The plant is a native of China (hence its name China grass), where it has been used for fabrics and fishing nets for hundreds of years. It is also grown in the Philippines, Japan, Brazil, and Europe. The ramie plant grows 1–2.5 m high with stems 8–16 mm thick. The roots send up shoots on harvesting, and two to four cuttings are possible annually, depending on soil and climate.

The plant is harvested by hand sickle and, after defoliation, is stripped and scraped by hand or machine decorticated. Because of the high gum (xylan and araban) content of up to 35%, retting is not possible. The fibers are separated chemically by boiling in an alkaline solution in open vats or under pressure, then washed, bleached with hypochlorite, neutralized, oiled to facilitate spinning, and dried.

The degummed, bleached fiber contains 96–98% cellulose. A cross-sectional view is shown in Figure 3**e**. The ramie fibers are oval-like in cross section with thick cell walls and a fine lumen. The cell wall constituents in the ramie fiber, like other bast fibers except flax, have a counterclockwise twist. Ramie is the longest of the vegetable fibers and has excellent luster and exceptional strength; however, it tends to be stiff and brittle. Wet strength is high and the fiber dries rapidly, an advantage in fish nets.

Traditional uses for ramie have been for heavy industrial-type fabrics such as canvas, packaging material, and upholstery. Increased production of the fiber in Asia, particularly China, has promoted the use in blended fabrics with silk, linen, and cotton which can now be found on the market.

Sunn. The stems of the herbaceous plant *Crotalaria juncea* (legume family, Fabaceae), called sunn or sunn hemp, provide a bast fiber. The plant is native to India, the chief fiber producer, and it is also grown in Bangladesh, Brazil, and Pakistan. It has a long tap root and grows to a height of up to 5 m. Harvesting is done manually by pulling or cutting. The plant is defoliated in the field, water retted, and processed similarly to jute. The white fiber is graded by color, firmness, length, strength, uniformity, and extraneous matter content. Sunn is used for canvas, paper, fishing nets, twine, and other cordage.

Urena and Abutilon. These are less important vegetable fibers of a jute-like nature. *Urena lobata* (*Cadillo*) of the mallow family (Malvaceae) is a perennial that grows in Zaire and Brazil to a height of 4–5 m with stems 10–18 mm in diameter. Because of a lignified base, the stems are cut 20 cm above the ground. The plants are defoliated in the field and retted similarly to jute and kenaf. The retted material is stripped and washed and, in some cases, rubbed by hand. The soft, near-white fiber is graded for luster, color, uniformity, strength, and cleanliness. It is used for sacking, cordage, and coarse textiles.

Abutilon theophrasti is a herbaceous annual plant producing a jute-like fiber. The plant is native to the People's Republic of China and is commercially grown in China and the former USSR. Because of its association with jute in mixtures and export, it is also called China jute. The plant grows to a height of 3–6 m with a stem diameter of 6–16 mm. After harvesting by hand and defoliation, bundles of the stems are water retted and the fiber is extracted by methods similar to those for jute. The fiber is used for twine and ropes.

LEAF (HARD) FIBERS

Hard or cordage fibers are found in the fibrovascular systems of the leaves of perennial, monocotyledonous plants growing in Central America, East Africa, Indonesia, Mexico, and the Philippines. They are generally of the *Agave* and *Musa* genera. The leaf elements are harvested by cutting at the base with a sickle-like tool, and bundled for processing by hand or by machine decortication. In the latter case, the leaves are crushed, scraped, and washed. The fibers are generally coarser than the bast fibers and are graded for export according to national rules for fineness, luster, cleanliness, color, and strength.

Abaca. The abaca fiber is obtained from the leaves of the banana-like plant (same genus) *Musa textilis* (banana family, Musaceae). The fiber is also called Manila hemp from the port of its first shipment, although it has no relationship with hemp, a bast fiber. The mature plant has 12–20 stalks growing from its rhizome root system; the stalks are 2.6–6.7 m tall and 10–20 cm thick at the base. The stalk has leaf sheaths that expand into leaves 1–2.5 m long, 10–20 cm wide, and 10 mm thick at the center; the fibers are in the outermost layer. The plant produces a crop after five years, and 2–4 stalks can be harvested about every six months.

In the Philippines, the principal supplier of abaca fiber, the fibrous layer in the sheath is separated with a knife between the layers, and the strips of fiber-containing layers, called tuxies, are pulled off and cleaned by hand to remove the pulp. In Indonesia and Central America these operations are performed mechanically. Hand- and spindle-stripped fiber is graded for braids, fine textiles, and cordage; decorticated fiber is another class. A cross-sectional view is shown in Figure 4**a**. The abaca fiber has a large lumen and the presence of silicified plates is not unusual.

Abaca fiber is unique in its resistance to water, especially salt water, and it is used for marine ropes and cables, although it is being replaced by synthetic fibers. Abaca fiber is the strongest of the leaf fibers (Table 4), followed by sisal, phormium, and henequen; it is also the strongest among the papermaking fibers. It is used for sausage casings, selling at \$5.50–\$12 per kg, and it is the preferred fiber for tea bags because of its high wet strength, cleanliness, and structure which permits rapid diffusion of the tea extract.

Cantala, Manila Maguey. *Agave cantala* is a member of the agave family (Agaraceae) which includes sisal. It originated in Mexico and was transported to Indonesia and the Philippines, where it is now produced commercially. The plant

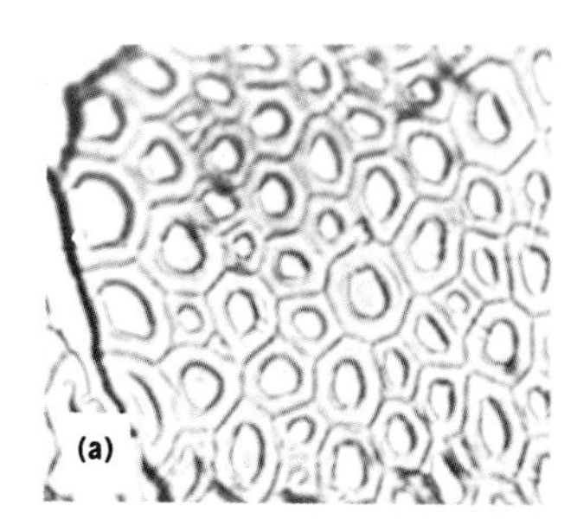

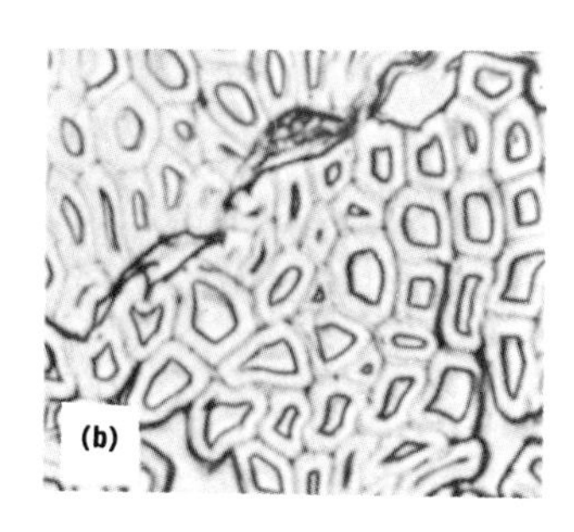

Fig. 4. Cross sections (500×) of leaf fibers: (**a**), abaca and (**b**), sisal. Courtesy of the American Association of Textile Chemists and Colorists (12).

grows in a moist, humid soil. The fiber is extracted in Indonesia mechanically by a decorticator (raspador) and in the Philippines by retting in seawater and cleaning by hand or with a decorticator. The cantala fiber is lighter in color than other agaves, and its strength depends on its preparation.

Caroa. Caroa is a hard leaf fiber, resembling sisal, obtained from *Neoglaziovia variegata,* a plant of the pineapple family (Bromeliaceae) growing wild in eastern and northern Brazil. The sword-shaped leaves are 1–3 m long and 2.5–5 cm wide. The fiber is extracted by hand scraping after beating or retting. The fiber is used for cordage and acoustic material.

Henequen. *Agave fourcroydes* grows in Mexico where it was first cultivated by the Mayans in the Yucatan (Yucatan Sisal). The plant produces for 20–30 years. The lower bottom leaves, which are up to 2 m long and 10–15 cm wide, are cut, machine decorticated, and cleaned. Henequen fibers are white to yellowish red and are inferior to sisal in strength, cleanliness, texture, and length, the other grading criteria. Henequen is grown for local use in Cuba (Cuban Sisal) and El Salvador. Twine, small ropes, coarse rugs, and sacks are made commercially from henequen.

Istle. Istle is the general name for several agave species and related plants producing short, coarse, hard fibers. The commercial name is tampico. The plants from which the fibers are harvested include tula istle (*A. lophantha*), jaumave istle (*A. funkiana*), and palma istle (*Samuela carnerosana*). They are grown in central and northern Mexico. The fibers are recovered by hand scraping and drying; the palma istle is gummy and requires presteaming. The jaumave istle fibers closely resemble animal bristles in brushes.

Mauritius. Mauritius hemp, also called piteira, is obtained from the *Furcraea gigantea,* also a member of the Agavaceae. The plant is mostly grown on the island of Mauritius, but is also harvested in Brazil and other tropical countries. The leaves are longer and heavier than those of the agaves. The fiber is extracted by mechanical decortication. It is whiter, longer, and weaker than sisal fiber. Because of its color it is used in blends.

Phormium. The *Phormium tenax* plant yields a long, light-colored, hard fiber also known as New Zealand hemp or flax, although it has none of the bast fiber characteristics. The plant is a perennial of the Agavaceae with leaves up to 4 m long and 10 cm wide. The fibers are recovered by mechanical decortication.

Sansevieria. This genus of the Agavaceae is a perennial also known as bowstring hemp from its use in bow strings. The plant is native to tropical Africa and Asia but is widely grown, mainly as an ornamental plant. It is of minor importance as a fiber plant, even though the fiber is of high quality. The highest grade of *S. cylindrica* fiber is greenish yellow, very soft and fine, and compares with sisal in strength.

Sisal. The true sisal fiber from *Agave sisalana* is the most important of the leaf fibers in terms of quality and commercial use. Originating in the tropical western hemisphere, sisal has been transplanted to East Africa, Indonesia, and the Philippines. It is named after the port in the Yucatan from which it was first exported.

The sisal plant leaves grow from a central bud and are 0.6–2 m long, ending in a thorn-like tip. The fibers are embedded longitudinally in the leaves, which are crushed, scraped, washed, and dried. The highest grades are further cleaned by a revolving drum. The growth of the sisal plant depends on water availability; it stores water during the wet season and consumes it during periods of drought.

A cross-section of the sisal ultimate fiber is shown in Figure 4**b**. The sisal fiber is coarse and strong, but compared with the abaca fiber it is inflexible, although with a relatively high elongation under stress. It is also resistant to salt water. The principal applications are in binder and baling twine and as a raw material for pulp for products requiring high strength. A large pulp mill is operating in Brazil based on sisal.

SEED- AND FRUIT-HAIR FIBERS

The seeds and fruits of plants are often attached to hairs or fibers or encased in a husk that may be fibrous. These fibers are cellulosic based and of commercial importance, especially cotton (qv), the most important natural textile fiber.

Coir. This fiber, obtained from husks of the fruit of the coconut palm, *Cocos nucifera* (palm family, Arecaceae), is mainly produced in India and Sri Lanka. The fruits are broken by hand or machine, and the fiber extracted from the broken husks from which the coconut has been removed for copra. The husks are retted in rivers, and the fiber separated by hand beating with sticks or by a decortication machine. The fibers are washed, dried, and hackled, and used in upholstery, cordage, fabrics, mats, and brushes.

Kapok. Kapok fiber is obtained from the seed pods of the kapok tree, *Ceiba pentandra,* of the Kapok tree family (Bombacaceae) which is indigenous to Africa and southeast Asia; it is produced mainly in Java. The tree grows to a height of 35 m. The seeds are contained in capsules or pods which are picked and broken open with mallets. The floss is dried and the fiber is separated by hand or mechanically. A nondrying oil is produced from the seeds with properties similar to cottonseed oil. The fiber is exceedingly light, with a circular cross section, thin walls, and a wide lumen. Kapok fibers are moisture resistant, buoyant, resilient, soft, and brittle, but not suitable for spinning. The traditional uses were in life jackets, sleeping bags, insulation, and upholstery; however, synthetics have replaced most of the applications as filling material and now kapok is mainly used for life preservers. Kapok-filled life preservers can support up to three times the weight of the preserver and do not become waterlogged.

PALM AND OTHER FIBERS

Piassava and Crin Vegetal. These palm (palm family, Arecaceae) fibers are obtained from the palm leaf base of *Attalea funifera* growing in Brazil and the palm leaf segments of *Chamaerops humilis* growing in North Africa. The former are used for cordage and brushes, the latter for stuffing.

Broom Root and Broom Corn. Broom root fiber is obtained from the root of the bunch grass, *Muhlenbergia macroura* (Poaceae), found in Mexico, where it grows 1–3 m high. The long fibers are bleached in fumes of burning sulfur before grading according to length. They are used in stiff scrubbing and scraping brushes and whisk brooms.

Broom corn is the fiber obtained from the flower head of another grass, *Sorgum vulgare technicum,* grown in the United States. The fibers are less stiff than those of the broom root and are used in brooms.

Economic Aspects

The principal bast and leaf fibers are produced in yields of 2–5%, with some exceptions such as flax (15%) and kapok (17%), on a green plant basis. The world production of several cordage (leaf) and textile (bast) fibers is given in Table 6. Vegetable fiber production on the world market has dropped 25–35% since 1970 because of periods of economic recession and synthetic fiber replacements. The U.S. imports of vegetable fibers are given in Table 7. Imports of vegetable fibers have dropped 70–90% since 1970 (with the exception of flax wastes). The precipitous decline in jute exports from India since 1965 are shown in Figure 2. These market trends reflect the recessions and substitution with synthetic fibers. Al-

Table 6. World Production of Cordage and Textile Fibers,[a] 10^3 t

Commodity	1970	1975	1981	1991
		Cordage fibers		
abaca	184.2	57.5	86.7	
henequen	353.0	129.0	95.5	
sisal	473.0	477.0	379.8	388
		Textile fibers		
jute	3373	1582	2743[b]	2165[c]
kenaf	1114	473	181[d]	

[a]Refs. 14–17.
[b]1979–1981.
[c]Jute category now includes jute-like fibers such as kenaf, mesta, roselle, and congo jute or paka.
[d]India only.

Table 7. U.S. Vegetable Fiber Imports,[a] t

Commodity	Countries of origin	1975	1980	1984	1991
abaca	Ecuador, Philippines	18,666	26,140	16,032	10,335
coir	Sri Lanka, India		9,216	8,114	7,748
flax (flax waste)	Belgium, Luxembourg	1,921	290	110 (2085)	7,918 (47,128)
hemp	Philippines, Thailand	18	741	43	59
jute and jute waste	Bangladesh, Thailand	21,260	9,476	9,513	5,468
istle or tampico	Mexico	729	2,591	2,446	1,721
kapok	Thailand	6,333	3,057	2,117	479
sisal and henequen	Brazil, Kenya	8,841	4,920	1,552	454
others		25,350	22,398	593	4

[a]Refs. 14–18.

though most vegetable fibers are converted to lower cost commodity products, some of the fibers are converted into the most expensive products in their respective industries, eg, U.S. currency (paper). The total value of U.S. imports and unit prices are given in Table 8. The value of most of the fibers has dropped since 1981.

Table 8. Value and Unit Price of U.S. Vegetable Fibers Imported[a]

	$1000				
Commodity	1981	1982	1983	1991	$/t
abaca	18,866	17,185	11,272	9,715	940
coir	2,311	2,015	1,604	1,310	169
flax (flax wastes)	0 (1,811)	0 (1,655)	0 (1,668)	3,309 (17,788)	418 (377)
hemp	20	240	669	70	
jute and jute waste	976		2,321	2,754	504
kapok	2,055	1,224	1,432	453	946
sisal and henequen	1,120	539	1,101	379	835

[a]Refs. 14–18.

Uses

Vegetable fibers have application in a broad range of fibrous products, including textiles and woven goods, cordage and twines, stuffing and upholstery materials, brushes, and paper (Table 9). The uses for each of the specific fibers has been discussed in the designated sections. The traditional uses for the vegetable fibers have been eroded by substitution with synthetics on the world market. The declining uses include cordage, mats, filling material, brushes, etc. However, the

Table 9. Uses of Vegetable Fibers

	Leaf fibers				Bast fibers					Seed, fruit, and other fibers		
Uses	Abaca	Cantala	Henequen	Sisal	Flax	Hemp	Jute	Kenaf	Ramie	Sunn	Kapok	Coir
Textiles and woven goods												
fine household, clothing					X	X			X			
coarse household					X	X	X	X	X			
bags and sacks			X				X	X				
thread, yarn					X	X	X		X			
canvas, sailcloth							X	X	X	X		
carpet backing, matting[a]	X	X	X	X	X		X	X				X
rugs, hammocks, belting, hose		X	X	X	X		X	X	X			
Cordage and twine												
industrial ropes	X		X			X			X	X		
marine ropes	X					X						
binder, baling twine		X	X	X								
nets									X	X		
twines[a]	X		X		X	X	X	X	X	X		
Upholstery and stuffing												
mattresses, furniture			X	X					X		X	
sleeping bags, life preservers											X	
Brushes												
brooms, brushes												X
pulp for papermaking	X			X	X	X	X	X				

[a]Also phormium leaf fibers.

unique properties of the bast fibers have allowed continued use in such specialty papers as bank notes, some writing papers, and cigarette papers.

Recent work by the USDA and Kenaf International (Texas) has demonstrated the potential of both growing and processing kenaf fibers for newsprint and other paper products in the United States. Another promising potential use for vegetable fibers is in the new lignocellulosic-based composites under development in various parts of the industrialized world (19). The vegetable fibers are mixed with thermoplastic or thermosetting resin matrices and either extrusion or compression molded into a variety of useful shapes. Such products are already utilized in the automotive industry for automobile interior door and head liners and as trunk liners. Although vegetable fibers will continue to provide indigenous populations with traditional uses, new innovative applications need to be developed to maintain international markets for the fibers.

BIBLIOGRAPHY

"Fibers, Vegetable" in *ECT* 1st ed., Vol. 6, pp. 467–476, by D. Himmelfarb, Boston Naval Shipyard; in *ECT* 2nd ed., Vol. 9, pp. 171–185, by D. Himmelfarb, Boston Naval Shipyard; in *ECT* 3rd ed., Vol. 10, pp. 182–197, by J. N. McGovern, University of Wisconsin.

1. M. Harris, ed., *Handbook of Textile Fibers,* Harris Research Laboratories, Inc., Washington, D.C., 1954.
2. *ASTM Annual Book of Standards,* Sect. 7 07 01, D 123, Tables 4–6, American Society for Testing and Materials, Philadelphia, Pa., 1983.
3. H. J. Nieschlag and co-workers, *Tappi* **43**(3), 193 (1960).
4. G. H. Nelson and co-workers, *Tappi* **44**(5), 319 (1961).
5. R. L. Cunningham and co-workers, *Tappi* **53**(9), 1697 (1970).
6. T. F. Clark and co-workers, *Tappi* **56**(3), 107 (1973).
7. L. Rebenfeld, in J. I. Kroschwitz, ed., *Encyclopedia of Polymer Science and Engineering,* Vol. 6, 2nd ed., Wiley-Interscience, New York, 1986, p. 647.
8. T. Zylinski, *Fiber Science,* Office of Technical Services, U.S. Department of Commerce, Washington, D.C., 1964, pp. 27–30.
9. W. Von Bergen and W. Krauss, *Textile Fiber Atlas,* American Wool Handbook Co., New York, 1942.
10. W. E. Morton and J. W. S. Hearle, *Physical Properties of Textile Fibers,* 2nd ed., John Wiley & Sons, Inc., New York, 1975, p. 284.
11. M. L. Joseph, *Textile Science Introductory,* 5th ed., Holt, Rinehart, and Winston, New York, 1986, pp. 63–80.
12. *AATCC Technical Manual,* American Association of Textile Chemists and Colorists, Research Triangle Park, N.C., 1984, pp. 87 and 88.
13. J. Young, *Tappi* **70**(11), 81 (1987).
14. *FAO Quarterly Bulletin of Statistics,* Vol. 5, No. 1, Food and Agricultural Organization of the United Nations, Rome, 1992, pp. 16–17.
15. *Commodity Year Book,* Commodity Research, Inc., Jersey City, N.J., 1984, p. 79 and 1992, p. 169.
16. *USDA Agricultural Statistics,* U.S. Government Printing Office, Washington, D.C., 1990.
17. *Fiber Organon* **63**(4) (Apr. 1992).
18. *Foreign Agricultural Trade of the U.S. (FATUS) Calendar Year and 1991 Supplement,* USDA, U.S. Government Printing Office, Washington, D.C., July 1992 and previous years, p. 307.

19. R. A. Young, in R. Rowell, T. Schultz, and R. Narayan, eds., *Emerging Technologies for Materials and Chemicals from Biomass,* ACS Symposium Series, no. 476, American Chemical Society, Washington, D.C., 1992, p. 115.

General References

E. E. Nelson, in S. P. Parker, ed., *Encyclopedia of Science and Technology,* Vol. 9, McGraw-Hill, Inc., New York, 1982, pp. 8–12.

L. T. Lorimer, ed., *Encyclopedia Americana,* Vol. 11, Grolier Inc., Danbury, Conn. 1992, p. 147.

J. G. Cook, *Handbook of Textile Fibers I. Natural Fibers,* 5th ed., Merrow Publishing, Durham, UK, 1984.

F. Hamilton and B. Leopold, eds., *Pulp and Paper Manufacture,* Vol. 3, *Secondary Fibers and Non-Wood Pulping,* Joint Textbook Committee of the Paper Industry, Tappi, Atlanta, Ga., 1987.

B. P. Corbman, *Textiles, Fiber to Fabric,* 6th ed., McGraw-Hill, Inc., New York, 1983.

B. Hochberg, *Fibre Facts,* Betty Hochberg, Santa Cruz, Calif., 1981.

AATCC Technical Manual, Vol. 59, American Association of Textile Chemists and Colorists, Research Triangle Park, N.C., 1984.

D. M. Colling and J. E. Grayson, *Identification of Vegetable Fibres,* Chapman and Hall Ltd., London, 1982.

J. M. Dempsey, *Fiber Crops,* University Presses of Florida, Gainesville, 1975.

RAYMOND A. YOUNG
University of Wisconsin, Madison

FIBERS, CHEMICAL. See FIBERS, SURVEY.

FIBERS, CARBON. See CARBON AND GRAPHITE FIBERS.

FIBERS, INORGANIC. See ALUMINUM COMPOUNDS, ALUMINUM OXIDE (ALUMINA); ASBESTOS; GLASS; REFRACTORY FIBERS; SILICA.

FIBERS, POLYAMIDE. See POLYAMIDE FIBERS.

FIBERS, SYNTHETIC. See FIBERS, SURVEY.

FIBRINOGEN. See FRACTIONATION, BLOOD.

FILLERS

A filler is a finely divided solid added to a liquid, semisolid, or solid composition, eg, paint, paper, plastics, or elastomers, to modify the composition's properties and reduce its costs. Fillers can constitute either a major or a minor part of a composition. The structure of filler particles ranges from precise geometrical forms, such as spheres, hexagonal plates, or short fibers, to irregular masses. Fillers are generally used for nondecorative purposes, although they may incidently impart color or opacity to a composition. Lower cost materials possessing modest performance-enhancing properties are discussed here, in contrast to functional fillers or engineered materials designed to achieve very specific results. Additives that supply bulk to drugs, cosmetics, and detergents, often referred to as fillers, are more properly referred to as diluents because their purpose is to adjust the dose or concentration of a composition, rather than modify its properties or reduce cost. Fibers and whiskers are not discussed here because they are generally regarded as reinforcements, not fillers. Also, additives which primarily modify or impart electromagnetic properties are not discussed.

Although the use of simple diluents and adulterants almost certainly predates recorded history, the use of fillers to modify the properties of a composition can be traced as far back as early Roman times, when artisans used ground marble in lime plaster, frescoes, and pozzolanic mortar. The use of fillers in paper and paper coatings made its appearance in the mid-nineteenth century. Functional fillers, which introduce new properties into a composition rather than modify preexisting properties, were commercially developed early in the twentieth century when Goodrich added carbon black to rubber and Baekeland formulated phenol–formaldehyde plastics with wood flour.

Fillers can be classified according to their source, function, composition, or morphology. None of these classification schemes is entirely adequate due to overlap and ambiguity of the categories. Morphological distinctions, used here for the discussion of general filler properties, are either crystalline, eg, fibers, platelets, polyhedrons, and irregular masses; or amorphous, eg, fibers, flakes, solid spheres, hollow spheres, and irregular masses. The compositional scheme used for the compilation of data on specific fillers classifies fibers as either inorganic, eg, carbonates, hydroxides, metals, oxides, silicates, sulfates, and sulfides; or organic, eg, carbon, celluloses, lignins, polymers, proteins, and starches.

Properties

The properties of fillers which influence a given end use are many. The overall value of a filler is a complex function of intrinsic material characteristics, eg, true density, melting point, crystal habit, and chemical composition; and of process-dependent factors, eg, particle-size distribution, surface chemistry, purity, and bulk density. Fillers impart performance or economic value to the compositions of which they are part. These values, often called functional properties, vary according to the nature of the application. A quantification of the functional properties per unit cost in many cases provides a valid criterion for filler comparison and selection. The following are summaries of key filler properties and values.

Particle Morphology, Size, and Distribution. Many fillers have morphological and optical characteristics that allow these materials to be identified microscopically with great accuracy, even in a single particle. Photomicrographs, descriptions, and other aids to particle identification can be found (1).

Filler particle size distribution (psd) and shape affect rheology and loading limits of filled compositions and generally are the primary selection criteria. On a theoretical level the influence of particle size is understood by contribution to the total energy of a system (2) which can be expressed on a unit volume basis as:

$$E_{\text{tot}} = e_i + \gamma\,(\text{SA}/V) \qquad (1)$$

where e_i is the internal energy per unit volume (V) and γ is the interfacial energy per unit surface area (SA). When the ratio of surface area to volume becomes large, the second term becomes significant and the particle exhibits colloidal behavior. This can occur when one, two, or three dimensions of the system are or become small, eg, films or platelets, fibers or rods, and fine regular particles. Regular is here defined to mean particles whose dimensions are roughly equal; thus by definition the first two cases above are irregular.

The magnitude of the influence of size can be visualized by contrasting the effect of water in the form of rain and fog on driving visibility. Additionally, to understand rheological phenomena, visualize the mechanical intermingling of particles which might be expected as a filled composition flows. Particles with fibrous, needle-like, or other irregular shapes yield compositions more resistant to flow, as compared to compositions filled with spheres or other fillers with more regular morphologies. Thus fillers with regular shapes can be used in compositions at higher loadings than fillers with irregular shapes, all other properties being equal.

Because of the diversity of filler particle shapes, it is difficult to clearly express particle size values in terms of a particle dimension such as length or diameter. Therefore, the particle size of fillers is usually expressed as a theoretical dimension, the equivalent spherical diameter (esd), ie, the diameter of a sphere having the same volume as the particle. An estimate of regularity may be made by comparing the surface area of the equivalent sphere to the actual measured surface area of the particle. The greater the deviation, the more irregular the particle.

The particle sizes of fillers are usually collected and ordered to yield size distributions which are frequently plotted as cumulative weight percent finer than vs diameter, often given as esd, on a log probability graph. In this manner, most unmodified fillers yield a straight-line relationship or log normal distribution. Inspection of the data presented in this manner can yield valuable information about the filler. The coarseness of a filler is often quantified as the esd at the 99.9% finer-than value. Deviations from linearity at the high and low ends of the plot suggest that either fractionation has occurred to remove coarse or fine particles or the data are suspect in these ranges.

Manufacturers report particle sizes or related information in varying ways: percent passing through or residue on a given mesh-sized screen; median particle size as interpolated from the log probability plot at a % finer-than value of 50;

and Hegman Grind (ASTM D1210), ie, fineness of dispersion value, which is generally close to the particle size at the 99.9% finer-than value. The size distribution of fillers whose particles are larger than 40 μm and which have at least moderate sphericity can be measured conveniently by sieving. The Coulter particle-size technique can be used for fillers whose particles range from 4 to 40 μm. Filler particle-size distributions smaller than 4 μm are measured by sedimentation, permeametry, or light scattering methods (3,4). However, microscopy is usually the most informative technique when characterizing the morphology of irregular particles, and it is advisable to periodically inspect filler lots with direct microscopic observations to validate results from indirect methods.

Surface Area. Surface area is the available area of fillers, be it on the surface or in cracks, crevices, and pores. The values obtained from different methods for measuring the surface area of a filler may vary significantly. These variations are because of the nature of the methods and in many instances yield information related to the heterogeneity of the surface. Understanding the surface area is important because many processing factors are dependent on the surface area, eg, ease of filler dispersion, rheology, and optimum filler loading.

The external surface area of the filler can be estimated from a psd by summing the area of all of the equivalent spheres. This method does not take into account the morphology of the surface. It usually yields low results which provide little information on the actual area of the filler that influences physical and chemical processes in compounded systems. In practice, surface area is usually determined (5) from the measured quantity of nitrogen gas that adsorbs in a monolayer at the particle surface according to the BET theory. From this monolayer capacity value the specific surface area can be determined (6), which is an area per unit mass, usually expressed in m^2/g.

Surface Energy or Wettability. At the phase boundary between a filler particle and a liquid composition, there can be an energy barrier to the liquid's wetting the filler, or the filler's dispersing in the liquid. The simple rule of thumb that like attracts like holds here. Surfaces may be defined, using water affinity as the standard, as hydrophilic or hydrophobic. A hydrophilic surface, like that found on a typical amorphous silica particle, might be expected to disperse readily in water and most liquids miscible in water. This wettability is a function of the surface area and chemical composition of the surface. Contact angle measurements between a drop of liquid and the filler surface are often used as a predictor (7). Many commercial fillers are surface-coated or chemically treated to modify their surface chemistry and wetting properties; eg, a hydrophobized form of the silica discussed above, made by treating it with an organosilane, now resists dispersal in a water-based composition but readily disperses in oil. A filler particle's oil or water absorption values can also be indirect indicators of relative wetting properties.

Acid–Base Behavior. The relative acidity–basicity of the filler, generally determined by measuring the pH value of a slurry of a specific mass of filler in 100 mL of deionized water, can influence the behavior of a filler in some systems. For example, the curing behavior of some elastomers is sensitive to the pH value of carbon black.

Loading and Packing. The amount of filler in a filled composition is termed the loading and is always expressed quantitatively; however, the quantitative indexes vary from industry to industry. Formulators in the plastics industry use parts filler per 100 parts resin or rubber (phr), weight percent filler (wt %), and volume percent filler (vol %). In the paint industry, volume percent pigment (filler) in the dry paint film or the volume ratio of pigment to binder are commonly used. In the paper industry, filler loading is usually expressed as percent of sheet weight; however, it can sometimes be measured simply as percent ash based on a loss-on-ignition method.

The optimal filler loading in a composition is usually determined by balancing the functional physical properties of the overall composition against the cost over a range of loadings. This determination can be aided greatly by the use of predictive models based on the packing geometry of filler particles. Experimental determinations are always necessary because of error introduced by simplifying assumptions in models; at the very least models can be used to establish the maximum values of filler loading.

For large amounts of fillers, the maximum theoretical loading with known filler particle size distributions can be estimated. This method (8) assumes efficient packing, ie, the voids between particles are occupied by smaller particles and the voids between the smaller particles are occupied by still smaller particles. Thus a very wide filler psd results in a minimum void volume or maximum packing. To get from maximum packing to maximum loading, it is only necessary to express the maximum loading in terms of the minimum amount of binder that fills the interstitial voids and becomes adsorbed on the surface of the filler.

For small amounts of filler, eg, in order to alter the rheology of the composition, the same method used to estimate the maximum packing and loading can be used to estimate the minimum packing and loading. For instance, if maximum loading is obtained with a filler having a wide psd and a low surface area, then a narrow psd and a high surface area will yield a filled system at minimum loading.

In modern industrial practice, compositions often contain pigments, reinforcements, rheological modifiers, surfactants, and other materials in addition to fillers. These materials can function synergistically in the system. Hence, more complex models are needed to predict the optimal filler loading. Excellent discussions of filler loading and selection in plastics are given (9,10).

True Density or Specific Gravity. The average mass per unit volume of the individual particles is called the true density or specific gravity. This property is most important when volume or mass of the filled composition is a key performance variable. The true density of fillers composed of relatively large, nonporous, spherical particles is usually determined by a simple liquid displacement method. Finely divided, porous, or irregular fillers should be measured using a gas pycnometer to assure that all pores, cracks, and crevices are penetrated.

Bulk Density. Bulk density, or the apparent density, refers to the total amount of space or volume occupied by a given mass of dry powder. It includes the volume taken up by the filler particles themselves and the void volume between the particles. A functional property of fillers in one sense, bulk density is also a key factor in the economics of shipping and storing fillers.

When determining bulk density, a distinction should be made between loose bulk density and tap density, eg, ASTM B527-81. The latter is a measure of the influence of settling on filler volume at constant mass.

Optical. The optical properties of fillers and the influence that fillers have on the optical properties of filled systems are often misunderstood. The key parameters in understanding the optical properties of fillers themselves are filler psd, color, and index of refraction. These characteristics influence the optical properties of filled composition, such as color, brightness, opacity, hiding power, and gloss.

Color and Brightness. Color and brightness result from the degree of reflectance, absorption, or transmission of light in the visible wavelength region of the electromagnetic spectrum. If light containing the full spectrum of visible light were reflected back fully to the eye by a powder, the filler would appear pure, bright white, MgO being a close approximation. If all light were absorbed, the filler would appear jet black, eg, as does carbon black (see CARBON, CARBON BLACK). Glass beads transmit most visible radiation and are clear or colorless. The actual color and brightness perceived is a result of selective absorption or subtraction of specific wavelengths of light from the reflected light. Color and brightness are frequently determined by measuring the reflection of light of specific wavelengths from a compacted disk of filler. The psd, impurities, and refractive index can influence these reflectance measurements.

Refractive Index. The relative amount of light reflected (L_{rel}) by a surface is dependent on both components of the interface, ie, filler and air. The refractive index is a measure of a substance's tendency to reflect light according to Fresnel's equation:

$$L_{rel} = \frac{[n_D(f) - n_D(m)]^2}{[n_D(f) + n_D(m)]^2} \qquad (2)$$

where $n_D(f)$ is the refractive index of the filler and $n_D(m)$ is the refractive index of the matrix. Air has a refractive index of 1.000 at 25°C. This equation teaches that the closer the n_D values of the components of the interface the less the reflectance. This concept is readily visualized by thinking about a fairly reflective white paper plate made of pulp fiber and clay with many air interfaces. Fiber and clay have similar n_D values, thus most of the reflectance occurs at air–fiber and air–clay interfaces. However, if the air is displaced by vegetable oil, the paper becomes transparent because vegetable oil has a refractive index similar to the fiber and clay. In fact, refractive index values for fillers can be determined microscopically using oils of varying n_D values (11).

Free Moisture. The free moisture of a filler is the water present on the surface of the particles. This weakly bound water can sometimes contribute to interparticle bonding (reinforcing) or filler–matrix interaction, ie, binder adsorption or catalysis. A determination of free moisture is usually made by measuring the percent loss on drying the sample at either 100 or 110°C.

Thermal Stability and Expansion. Fillers for high temperature application must not decompose under conditions of processing or use. Determination of the thermal stability varies from industry to industry. Modern thermogravimetric or scanning calorimetric methods are usually the most practical ways to determine the thermal behavior of fillers. In order to preserve the structural strength of filled compositions over a wide range of temperatures, all components of the composition should have similar coefficients of thermal expansion. This prevents stress-induced damage that occurs when substances expand or contract to different degrees as temperature is increased or decreased. Coefficients of thermal expansion are reported in units of dimension change per degree Celsius for a specified temperature range. Many fillers expand differently in different dimensions.

Hardness. The hardness (qv), or related property abrasiveness, is an important filler property. Hardness is determined by comparison to materials of known hardness on the Mohs' scale. On this nonlinear scale, diamond is rated 10, quartz 7, calcite 3, and talc 1. The abrasiveness of a filler is also dependent on psd and the presence of impurities, eg, kaolin clay (Mohs' hardness of 3) can be quite abrasive because of the presence of quartz impurities.

Filled Polymer Systems

Polymer systems have been classified according to glass-transition temperature (T_g), melting point (T_m), and polymer molecular weight (12) as elastomers, plastics, and fibers. Fillers play an important role as reinforcement for elastomers. They are used extensively in all subclasses of plastics, ie, general-purpose, specialty, and engineering plastics (qv). Fillers are not, however, a significant factor in fibers (qv).

Elastomers. In the rubber industry the terms filler, reinforcement, and pigment have been used for the same material, derived from different sources. Most rubber technologists distinguish inert fillers from reinforcing fillers. Inert fillers, such as clay, improve the workability of the unvulcanized rubber stock but have little effect on the final properties; reinforcing fillers improve the mechanical properties of the vulcanized rubber. In practice, these materials impart abrasion resistance, tear resistance, tensile strength, and stiffness. The reinforcing action of a given material is primarily dependent on its chemical composition and the type of elastomer in which it is compounded. For instance, carbon black in isoprene rubber and pyrogenic silica in silicone rubbers are more active reinforcing agents in these respective elastomers than other fillers with equivalent particle size distributions. In general, as particle size diminishes or surface condition improves, most fillers become more reinforcing. Table 1 lists typical elastomer fillers and their uses.

Since most fabricated elastomer products contain 10–50 vol % of filler, their physical properties and processing characteristics depend to a great extent on the nature and quality of the fillers. Rubber technologists manipulate the formula so as to optimize a large number of properties and keep costs down.

Recovery, Rebound, or Nerve. Uncured latex stock tends to recover its previous shape after being rolled or extruded during processing. This tendency can result in processing difficulties and reduces dimensional accuracy of the finished products. The recovery of latex stock can be reduced by adding large particle-size fillers, or fillers which possess a high degree of particle aggregation, ie, agglomerated or structured fillers such as carbon black or pyrogenic silica, or by increasing the loading of other fillers.

Tack. Tack causes layers to adhere when they are pressed together. This property can be reduced by employing fillers with a finer psd or by dusting the

Table 1. Elastomer Fillers

Filler	Specific gravity	Compatible elastomer[a]	Uses
alumina	2.7	NR, CR, SRs	hose, mats
asbestos	2.4	NR, SRs	mats, tile
barium sulfate	4.3	NR, CR, SRs	O-rings, belts
carbon blacks[b]	1–2.3		
N110		IR, NR	tires, pads
N220		NR, SBR	tire treads
N550		NR, SRs	extruded goods
N660		BR, NR	tubes
N762		NR, SRs	footwear
N774		NR, SRs	tire carcasses
N990		CR, EPDM	extruded goods
calcium carbonate	2.7–2.9	NR, SRs	footwear, mats
kaolin clay	2.6	NR, SBR, EPM, EPDM	flooring, footwear
mica	2.8	NR, SRs	molded goods
resins	1.2	NBR, CR, NR, SBR	footwear, coatings
silicas			
colloidal	1.3	NR, SBR	sponge
diatomaceous	2.2	NR, IIR	carpet backing
novaculite	2.7	NR, SRs	molded goods
wet process	2	IIR, CR, NBR, NR	hygienic goods
pyrogenic	2.2	silicone rubber	electrical goods
surface treated		NR, SRs	specialty goods
talc	2.8	NR, CR, IIR, EPM	molded goods
natural materials[c]	1.1	NR, SBR, CR	tape, extruded goods
wood and shell flour	0.9–1.6	CR, NR, SRs	footwear

[a]NR, natural rubber; CR, chloroprene; SRs, synthetic rubbers; IR, natural isoprene; SBR, styrene–butadiene rubber; BR, butadiene; EPDM, ethylene–propylene–diene; EPM, ethylene–propylene polymer; IIR, isobutylene–isoprene; NBR, nitrile–butadiene.
[b]Carbon black grades identified by four characters (ASTM D1765-67), ie, cure rate of normal (N) or slow (S), digit classifying typical particle size in nm, and two arbitrarily assigned characters.
[c]For example, rice.

stock with a laminar filler such as mica. In a related value, fillers such as mica can inhibit adhesion to the mold during processing.

Retardation (Scorch Resistance). Scorch is the premature vulcanization of an elastomer during processing. The tendency of an elastomer to scorch can be reduced by compounding it with a filler of different acidity or larger particle size. The type of filler also affects the tendency of a latex stock to scorch, ie, highly reinforcing fillers such as carbon black, silica, and zinc oxide promote scorching, whereas most clay fillers retard scorching.

Abrasion Resistance. The tendency of rubber to undergo surface attrition when subjected to frictional force is called abrasion. The abrasion resistance of a given rubber is a function of both filler type and form. If listed in order of their decreasing effectiveness in preventing abrasion, carbon black > silica > zinc oxide > clay > calcium carbonate. Although the abrasion resistance of fillers tends to increase with increasing sphericity and decreasing particle size, fillers should be compared by evaluating their abrasion resistance using samples prepared at equal loading and by a procedure such as ASTM method D1630.

Elongation. The extension produced by a tensile stress applied to an elastomer, ie, elongation, is almost always reduced by fillers. Regardless of what type of filler is used, elongation decreases with increased loading above approximately 5 vol % (13).

Hardness. The resistance of a fabricated rubber article to indentation, ie, hardness, is influenced by the amount and shape of its fillers. High loadings increase hardness. Fillers in the form of platelets or flakes, such as clays or mica, impart greater hardness to elastomers than other particle shapes at equivalent loadings.

Modulus of Elasticity (Stiffness). The stress–strain behavior of a material under tension is an important characteristic. The slope of the stress vs strain curve is called the modulus of elasticity or stiffness. Reinforcing fillers result in higher modulus composites than inert fillers at equal loading. In general, modulus increases with decreasing particle size of a filler. Air voids introduced during processing, possibly as a result of filler agglomeration, can lower modulus.

Permanent Set. When an elastomer is stretched and then allowed to relax, it will not completely recover its original dimensions. This divergence from its original form is called its permanent set. It is principally affected by the affinity of the elastomer for the filler surface and is, therefore, primarily a function of the surface energy or wetting of the filler.

Resilience. The ratio of energy output to energy input in a rapid full recovery of a deformed elastomer specimen is termed resilience. It is a reliable index to the energy lost through internal friction of viscous flow. In general, fillers with the least effect on resilience are those that are least reinforcing. Zinc oxide is an exception because it has good resilience and also gives good reinforcement.

Tear Resistance. The resistance of an elastomer to tearing is affected by the particle size and shape of the filler it contains. Tear resistance generally increases with decreasing particle size and increasing sphericity of fillers.

Tensile Strength. Fillers of small particle size and large surface area increase the tensile strength of a rubber compound. For most fillers, tensile strength increases with loading to an optimum value after which it decreases with increased loading.

Plastics. In the plastics industry, the term filler refers to particulate materials that are added to plastic resins in relatively large, ie, over 5%, volume loadings. Except in certain specialty or engineering plastics applications, plastics compounders tend to formulate with the objective of optimizing properties at minimum cost rather than maximizing properties at optimum cost. Table 2 lists typical plastic fillers and their uses.

Resin Viscosity. The flow properties of uncured compounded plastics is affected by the particle loading, shape, and degree of dispersion. Flow decreases with increased sphericity and degree of dispersion, but increases with increased loading. Fillers with active surfaces can provide thixotropy to filled materials by forming internal network structures which hold the polymers at low stress.

Resin Curing. Fillers with high surface areas can retard the curing of plastics by adsorbing catalysts or promoters. Other fillers can accelerate curing if they

Table 2. Plastic Fillers and Their Primary Functions

Filler	Specific gravity	Compatible resins[a]	Primary function
alumina trihydrate	2.42	varied	flame retardance
carbon blacks	1–2.3	varied	optical, electrical, mechanical
calcium carbonate			
mineral	2.60–2.75	PVC, HDPE	multiple, cost
synthetic	2.7	rigid PVC, PP	impact, weathering
kaolin	2.58–2.63	varied	bloom prevention, asbestos
feldspar	2.61	ABS, EVA, SMC	translucency
organics			
wood flour	0.65	varied	reinforcement
starch	1.5	varied	biodegradability
silicas, synthetic			
fumed silica	2.2	FRP, PVC, epoxy	rheology
silica gel		PVC, LDPE	gloss reduction
precipitated silica	1.9–2.2	PVC, PE, EVA	thixotropy
fused silica	2.18	general	electrical
silicas, natural			
crystalline silicas	2.65	general	mechanical, cost
diatomaceous silica	2.65	LDPE fims	antiblocking
sphericals			
hollow glass	0.15–0.30	PVC, SMC, BMC	weight reduction
fly ash	0.30–1.0	varied	cost reduction
solid glass	2.5	varied	mechanical
talcs	2.7–2.8	PP, HDPE, TPE, PVC	reinforcement

[a]HDPE, high density polyethylene; PP, polypropylene; EVA, ethylene–vinyl alcohol; SMC, sheet-molding compound; FRP, fiber-reinforced plastic; LDPE, low density polyethylene; PE, polyethylene; BMC, bulk molding compound; TPE, thermoplastic elastomer.

contain active sites or trace quantities of catalytic materials. Low density fillers cause high temperatures during curing and subsequent long cool-down periods because of their insulating effects. High loadings of dense solid fillers can appreciably increase the curing time needed to fully cure thermosetting compounds.

Tensile Strength. The tensile strength of filled plastic compositions is affected by filler particle shape and size, psd, surface area, and interfacial bonding. In general, tensile strength increases with decreasing sphericity of fillers. At equivalent volume loadings, small filler particle sizes and a narrow psd give better tensile strengths than large particle sizes and broad distributions. Higher filler surface areas and, in general, stronger filler-to-matrix bonding also result in higher tensile strength compositions. Tensile strength of plastics is measured by ASTM D638-76.

Compressive Strength. The strength of the weakest component of the system, be it filler, matrix, or the bond area between the filler and the matrix, governs the compressive strength of filled compositions. The weak compressive fillers, such as celluloses, reduce the compressive strength of composites, whereas the reverse is true for strong, rigid fillers such as mineral oxides. Compressive strength of plastics is measured by ASTM D621-64.

Fire Resistance. Many fillers, particularly inorganic oxides, are noncombustible and provide a measure of passive fire resistance to filled plastics by reducing the volume of combustible matter in the filled composition. Depending on their density, they may also serve as insulation.

Fillers that contain combined water or carbon dioxide, such as alumina trihydrate, $Mg(OH)_2$, or dawsonite [*12011-76-6*], increase fire resistance by liberating noncombustible gases when they are heated. These gases withdraw heat from the plastic and can also reduce the oxygen concentration of the air surrounding the composition.

Electrical Resistance–Conductivity. Most fillers are composed of nonconducting substances that should, therefore, provide electrical resistance properties comparable to the plastics in which they are used. However, some fillers contain adsorbed water or other conductive species that can greatly reduce their electrical resistance. Standard tests for electrical resistance of filled plastics include dielectric strength, dielectric constant, arc resistance, and d-c resistance.

Other Filler Systems

Paper. Paper is prepared by depositing cellulose pulp fibers on a continuous wire screen from a dilute suspension in water (see PAPER). Fillers, or loading materials, are finely divided solids that are incorporated into the paper sheet structure by adding them to the pulp slurry prior to its deposition on the wire. Finely divided solids dispersed in water containing an adhesive and then coated on the paper after it has been formed are usually termed coating pigments. Here the term filler is used for both loading materials and coating pigments. Table 3 lists typical paper fillers and their uses. Most paper contains 1–40 wt % of fillers (ca 1993). The optical and mechanical properties of filled paper are superior to those of unfilled paper.

Optical Properties. Brightness, or visual whiteness of paper, can be defined as the degree to which light is reflected uniformly over the visible spectrum. Since pulp and typical impurities tend to be yellowish, blue dye is sometimes added in addition to appropriate fillers. The percentage reflectance is usually measured in the blue end of the spectrum at or near 457 nm (14).

The ability of fillers to improve paper brightness increases with their intrinsic brightness, surface area, and refractive index. According to the Mie theory, this ability is maximum at an optimum filler particle size, about 0.25 μm in most cases, where the filler particle size is roughly one-half the wavelength of light used for the observation.

Gloss, or surface luster, is the property of a surface to reflect light specularly. It is associated with such phenomema as shininess, highlight, and reflected images. The gloss of paper is usually quantified with a spectrophotometer which measures light at a variety of angles of incidence and reflection.

Paper gloss is influenced by the size and shape of filler particles at the surface of the paper. A roughness on the order of one-fourth of the wavelength of

Table 3. Paper Fillers

Filler	Specific gravity	Refractive index[a]
alumina trihydrate	2.4	1.57
aluminosilicate, clay, kaolin	2.58	1.55–1.57
calcined	2.63	1.6
calcium carbonate	2.71	1.49 (1.66)
calcium sulfate[b]		
dihydrate	2.32	1.521 (1.530)
anhydrous	2.96	1.569 (1.613)
magnesium silicate, talc	2.6–2.7	1.57
silica, amorphous	2.1	1.5
titanium dioxide		
anatase	3.9	2.55
rutile	4.2	2.7

[a]The refractive index of birefringent crystals is shown in parentheses.
[b]Alternative refractive index of 1.523 for dihydrate and 1.575 for anhydrous calcium sulfate.

light can produce a perceptible reduction of specularly reflected light (15). Since each particle at the surface only protrudes partially above the surface, the average height of surface irregularities does not exceed one-half the particle diameter. Thus gloss reduction does not take place until filler particles exceed roughly 3 μm in diameter (16). Filler particles with laminar plate-like shapes, which tend to orient parallel to the surface of the sheet, thus reducing its roughness more than would be predicted for their particle size distribution, increase gloss more effectively than spherical or other irregular particles.

Opacity, or lack of show through, is a function of the light absorption of the paper and the amount of light that is scattered at the filler–air, filler–fiber, and fiber–air interfaces. Light scattering is the principal cause of opacity in paper containing white fillers, whereas in colored paper opacity depends more strongly on light absorption.

The filler properties that influence opacity are color, particle size, surface area, and refractive index. Dark pigments generally provide better opacity than light pigments because they absorb more light. Light scattering, like brightness, is a reflective phenomenon; the optimal particle size is again defined by the Mie theory. The opacification power of a filler is usually quantified as a specific scattering coefficient which generally increases with surface area and refractive index. Scattering coefficients are determined by measuring the optical properties of papers made from pulps, with known light scattering properties, formed into paper with variable filler loadings.

Mechanical Properties. Fluid Absorbancy. Fluids like ink penetrate into paper during the printing process. The further the ink penetrates, the less glossy the print. The degree of penetration in paper is generally a function of the paper porosity and wettability by the fluid. It can be controlled by the particle size, shape, and chemical nature of the filler or filler surface. In particular, plate-like fillers, such as clays, tend to produce the best fluid holdout because they tend to overlap and reduce the porosity at the paper surface (see INKS).

Retention. After filler is added to the pulp slurry in the course of papermaking, it must remain in the sheet in a uniform manner when the pulp is deposited on the wire. If the filler is not well retained, the paper develops a two-sided character, ie, one side has less and finer particles than the other side. Additionally, there is a greater buildup of filler fines in the water recycle system (white water) which increases processing costs. Retention is reduced by increasing filler density and sphericity. It is increased by increasing the filler particle size, using more favorable shapes, or by using chemical retention aids which flocculate smaller particles. Factors that increase filler retention often reduce opacity.

Sheet Strength. Characteristics of sheet strength of paper are measured as bursting strength, tear resistance, wet and dry tensile strength, stretch, internal bond strength, folding endurance, and abrasion resistance. There are numerous TAPPI and ASTM standards for strength properties. Fillers reduce the paper sheet strength because they dilute the pulp and reduce the number of fiber-to-fiber interactions. However, fillers are not necessarily detrimental. They often increase flexibility and stretchability and thus reduce the number of tears that occur when paper is run on high speed equipment (17). It has also been noted that high filler loadings have a tendency to alleviate pitch problems, ie, tacky rosin and fatty acid deposits (18).

Paint. The liquid phase of paint formulations, usually termed the vehicle, contains volatile and nonvolatile fractions. When paint is applied, the volatile fraction of the vehicle evaporates and the nonvolatile fraction polymerizes to become a film matrix in which prime pigments and fillers are embedded (see PAINT). Prime pigments are coloring and opacifying agents. Fillers in paint are variously referred to as inerts, extender, or supplemental pigments. The properties that fillers impart to paints are similar to the optical and mechanical properties they impart to paper. The primary paint fillers are TiO_2, kaolin clays, calcium carbonate, silica, talc, and zinc oxide. The move toward lower volatile organic carbon (VOC) is changing the nature of the volatile fraction in paints and accelerating the search for new or better fillers (19). In particular, new viscosity, compatibility, gloss, and applications problems are driving innovations in the market.

Optical Properties. Gloss. As in paper technology, gloss is the specularly reflected light typically associated with such phenomena as shininess, highlights, and reflected images. It is usually evaluated by comparing a paint sample to a highly polished, black surface. The degree of gloss of a painted surface depends on the smoothness of the reflecting surface. For this reason, filler size, shape, and surface energy influence the glossiness of filled paints. High gloss paints are formulated with fillers and pigments that have small particle dimensions. An average particle diameter of less than 0.3 μm is necessary to produce high gloss surfaces (20). Low gloss paint films are obtained by using fillers with a large average particle size. As with paper, laminar, plate-like fillers increase the gloss of paints more effectively than spherical or irregularly shaped fillers of equivalent particle size. Pigments that are completely wetted by the vehicle permit higher gloss than pigments that are incompletely wetted (21).

Hiding Power. The hiding power of fillers in paint is determined by the difference between their index of refraction and the index of refraction of the surrounding medium. The index of refraction values for common organic paint vehicles are approximately 1.5, pure water is 1.33, and air has a refractive index of 1.0. Since most fillers used in paint have refractive indexes of 1.4–1.7, they contribute little to the hiding power of a paint when they are completely embedded in the vehicle. However, if the filler particles extend above the vehicle film into the air, they contribute to the hiding power. Hence, hiding power can be expressed as wet hiding power and dry hiding power. Dry hiding and gloss are mutually exclusive.

Mechanical Properties. The stain resistance of paints is directly related to their porosity. Therefore fillers that help to reduce porosity, ie, those with low surface areas, wide size distribution, and laminar shapes, contribute to stain resistance.

The ease with which paint is applied and the uniformity of the paint film depends largely on the paint viscosity. If the viscosity is too high, paint does not level and, therefore, brush, roller, or spray marks result. If a paint is not sufficiently viscous, it sags or runs. High surface area, ie, high vehicle demand, a high degree of filler agglomeration, and low sphericity contribute to paint viscosity.

Weather Resistance. The principal causes of paint weathering are moisture, abrasion, uv radiation, and microorganisms. Fillers that tend to yield porous coatings decrease weathering because of blistering or trapped water. Hard fillers such as silica and clay contribute to abrasion resistance of paints. Since uv radiation oxidizes prime pigments and solvent vehicles in the presence of oxygen and water vapor, fillers that absorb uv radiation and limit paint porosity reduce this type of degradation. Alkaline fillers, such as zinc oxide and calcium carbonate, inhibit the growth of mildew and other microorganisms.

Economic Aspects

Sales of fillers worldwide were estimated to be on the order of 10^6 t and \$9 billion for 1990, with paint (42%), paper (21%), and plastics (16%) the principal consumers on a dollar basis. When viewed on a volume basis, paint and paper are even at 37% each, plastics use 21%, and rubber 6%. Titanium dioxide represented 67% of the dollar volume, with 60% of its use in paint. During 1991 the industry experienced the no-growth effects of the recession; however, 24% growth is expected between 1990 and 1995 (19,22). TiO_2 is expected to grow only 15% on volume as precipitated calcium carbonate makes greater inroads into its end uses, particularly in paper. Other fillers besides TiO_2 represent over 70% of the market based on volume, primarily because of their lower cost. They continue to be upgraded by improvements in precipitation, grinding, beneficiation, and surface-treatment technologies. For example, Pfizer has developed technology to produce calcium carbonate in paper mills using available carbon dioxide and CaO (23).

All markets, except paper, have been adversely affected by the downturn in the automotive and construction industries in the early 1990s. Most large-volume fillers are sufficiently diversified so that their growth trends follow GNP. There are some exceptions. Table 4 gives 1992 price information on specific fillers, including some physical properties and manufacturing processes.

Table 4. Properties of Specific Fillers

Filler	CAS Registry Number	Mohs' hardness	Oil adsorption, g/100 g	Average refractive index	Specific gravity	Price,[a] $/kg	Manufacturing process
alumina trihydrate	[*21645-51-2*]	3.0	20–40	1.57	2.42	0.28	Bayer process
aluminosilicate, clay	[*1327-36-2*]	2.0					
classified		2.0	25–50	1.56	2.58	0.07	mined
calcined		6–8	45–90	1.62	2.63	0.36	mined
barium carbonate	[*513-77-9*]	3.5	15	1.68	4.4	0.59	ppt
barium sulfate							
barite	[*13462-86-7*]	3.3	5	1.6	4.50	0.20–0.30	mined
blanc fixe	[*7727-43-7*]	3.3	15	1.6	4.3	0.44	ppt
calcium carbonate							
mineral	[*1317-65-3*]	3.0	10–15	1.6	2.7–2.9	0.05–0.20	mined
synthetic	[*471-34-1*]	3.0	25–60	1.6	2.7	0.20–0.50	ppt
carbon black	[*1333-86-4*]	2–4	30–400		1–2.3	0.60	furnace
magnesium carbonate	[*546-93-0*]	4.0	80–100	1.7	3–3.5		ppt
magnesium silicate, talc	[*14807-96-6*]	1+	20–45	1.6	2.7–2.8	0.10–0.18	mined
magnesium oxide	[*1309-48-4*]	4–5	55–70	1.74	3.6–3.7	1.65	calcined
metals							
aluminum	[*7429-90-5*]	2.5			2.55	7.00	atomized
zinc	[*7440-86-9*]	2.5			7	1.30	atomized
mica	[*1327-44-2*]	2.7	20–50	1.6	2.8–2.9	0.16–0.36	mined
organics							
wood and shell flour		low	10–30		0.65–1.6	0.15–0.40	ground
starch		low			1.5		natural

resins		low		1.5	1.2	1.00	various
silicas, natural							
crystalline silica	[*14808-60-7*]	7.0	20–30	1.55	2.65	0.10–0.15	ground
diatomaceous silica	[*7631-86-9*]	1.3	ca 100	1.48	2.3	0.20–0.75	crushed
silicas, synthetic	[*7631-86-9*]			1.4			
fumed silica			200+		2.2	6.00–10.00	hydrolysis
silica gel			200	1.35–1.45	1.9–2.2	2.50–4.00	gelation
precipitated silica			150	1.45	1.9–2.2	1.50–3.00	ppt
fused silica	[*60676-86-0*]	7.0	10		2.18	1.00–1.50	melt quartz
sphericals			ca 30	1.5			
hollow glass					0.2–0.6	3.00	remelt–blow
fly ash		5.0			0.30–1.0	0.3–0.5	beneficiation
solid glass		5.5			2.5	0.50–1.00	spheroidization
titanium dioxide	[*1317-70-0*]		20–30				hydrolysis
anatase		5.7		2.55	3.9	1.70	
rutile		6.2		2.7	4.2	2.10	
wollastonite	[*13983-17-0*]	4.7	20–30	1.63	2.9–3.1	0.21–0.59	mined
zinc oxide	[*1314-13-2*]	4.0	10–20	2.01	5.6	1.03–1.12	oxidation

[a]Approximate for 1991.

Health and Safety Factors

The principal hazard involved in the handling and use of many fillers is inhalation of airborne particles (dusts) in the respirable size range, ie, 10 μm and below. Filler dusts may be classified as nuisance particulates, fibrogens, and carcinogens. Nuisance particulates are dusts that have a long history of little adverse effect on the lungs and do not produce significant organic disease or toxic effect when exposures are kept under reasonable control.

The American Conference of Governmental and Industrial Hygienists (ACGIH) establishes TLVs for the airborne concentration of many fillers in workroom air (24). A new manner of occupational and environmental legislation is aimed at the hazards of ultrafine particles, and OSHA regulations effective in 1992 regulate the total workplace, including nonproduction areas (25). In addition, concern for the toxicity of many metals and their compounds is limiting the use of many fillers, eg, Pb, Co, Cr, and Ba compounds, and possibly the use of certain organometallic surface coatings. Suppliers have information on proper usage and handling of their products. The use of NIOSH–OSHA-approved dust masks or respirators is required when dust concentrations exceed permissable exposure limits.

It has been reported that because of concerns by some international agencies over the possibility that crystalline silica might be a carcinogen, a 0.1% max silica (crystalline) specification has been mandated in mineral fillers (26).

BIBLIOGRAPHY

"Fillers" in *ECT* 3rd ed., Vol. 10, pp. 198–215, by J. G. Blumberg, J. S. Falcone, Jr., and L. H. Smiley, PQ Corp., and D. I. Netting, ARCO Chemical Co.

1. W. C. McCrone and co-workers, *Particle Atlas,* Vols. 1–6, Ann Arbor Science, Ann Arbor, Mich. 1973–1978.
2. E. Matijevic, *MRS Bulletin* **14**(12), 18 (1989).
3. T. Allen, *Particle Size Measurement,* Chapman and Hall, London, 1974, p. 93.
4. R. R. Irani and C. F. Callis, *Particle Size: Measurement, Interpretation, and Application,* John Wiley & Sons, Inc., New York, 1963.
5. R. W. Camp and H. D. Stanley, *Am. Lab.* **23**(14), 34 (1991).
6. S. Brunauer, P. H. Emmett, and E. Teller, *J. Am. Chem. Soc.* **60,** 309 (1938).
7. A. W. Adamson, *Physical Chemistry of Surfaces,* 2nd ed., John Wiley & Sons, Inc., New York, 1960.
8. C. C. Furnas, *Ind. Eng. Chem.* **23,** 1052 (1931).
9. T. H. Ferrigno in H. S. Katz and J. V. Milewski, eds., *Handbook of Fillers for Plastics,* Van Nostrand Reinhold, New York, 1987, pp. 8–63.
10. J. V. Milewski in H. S. Katz and J. V. Milewski, eds., *Handbook of Fillers and Reinforcements for Plastics,* Van Nostrand Reinhold, New York, 1978, p. 66.
11. N. H. Hartshorne and A. Stuart, *Crystals and the Polarizing Microscope,* 4th ed., Arnold, London, 1970.
12. F. W. Billmeyer, Jr. *Textbook of Polymer Science,* 4th ed., Wiley–Interscience, New York, 1984, p. 349.
13. T. H. Ferrigno, in Ref. 9, pp. 8–63.
14. B. L. Browning, *Analysis of Paper,* 2nd ed., Marcel Dekker, Inc., New York, 1977, p. 16.
15. R. S. Hunter, *ASTM Bull.* **186,** 48 (1952).

16. P. B. Mitten in T. C. Patten, ed., *Pigment Handbook,* Wiley–Interscience, New York, 1973, p. 289.
17. H. C. Schwalbe, *Paper Web Transactions of the Cambridge Symposium,* Vol. 2, British Paper and Board Makers' Association, London, 1966, p. 692.
18. K. W. Britt, *Handbook of Pulp and Paper Technology,* Reinhold Publishing Corp., New York, 1964, pp. 332–344.
19. *Chem. Week* **149**(14), 40(1991).
20. N. F. Miller, *Off. Dig. Fed. Soc. Paint Technol.* **34,** 465 (1962).
21. E. Singer in R. Myers and J. Long, eds., *Treatise on Coatings,* Vol. 3, Part 1, Marcel Dekker, Inc., New York, 1974, p. 32.
22. *CPI Purchas.* **9**(12), 12 (1991).
23. M. S. Refsch, *C. & E News,* 9 (Apr. 29, 1991).
24. *TLV and Biological Exposure Indices, 1987–1988,* ACGIH, Cincinnati, Ohio, 1988.
25. W. Gregg and J. W. Griffin, *Poll. Eng.,* 80 (Apr. 1991).
26. *Mod. Plast.* (Feb. 1992).

JAMES S. FALCONE, JR.
West Chester University

FILM AND SHEETING MATERIALS

Film and sheet are defined as flat unsupported sections of a plastic resin whose thickness is very thin in relation to its width and length. Films are generally regarded as being 0.25 mm or less, whereas sheet may range from this thickness to several centimeters thick. Film and sheet may be used alone in their unsupported state or may be combined through lamination, coextrusion, or coating. They may also be used in combination with other materials such as paper, foil, or fabrics.

Film or sheet generally function as supports for other materials, as barriers or covers such as packaging, as insulation, or as materials of construction. The uses depend on the unique combination of properties of the specific resins or plastic materials chosen. When multilayer films or sheets are made, the product properties can be varied to meet almost any need. Further modification of properties can be achieved by use of such additives or modifiers as plasticizers (qv), antistatic agents (qv), fire retardants, slip agents, uv and thermal stabilizers, dyes (qv) or pigments (qv), and biodegradable activators.

The first film or sheet materials were manufactured in the early twentieth century from cellulose nitrate, and later from cellulose acetate. These products were essential to the development of the photographic film industry. Colored films were also used from the earliest days as color filters for lighting. These products were followed in 1929 by the universally recognized clear packaging film, cellophane. Following World War II, polyethylene film became the largest volume film product, and later polypropylene, polystyrene, polyester, poly(vinyl chloride), and others joined the list of commercial film products. Sheet markets grew with poly(acrylate), polystyrene, polycarbonate, and polyolefin resins as the predominate materials. By 1990, many other polymers were used in films or sheet applications, such as the specialty resins, polysulfone, poly(acrylate), polyamide, polyimide, and poly(arylketone), each used primarily for high performance market needs.

Properties and Test Methods

Film and sheet materials have an amazing range of properties so that a product may generally be formed or produced to meet the needs of a specific end use. Films as thin as 1.5 μm are produced for capacitor insulation, whereas cast sheet products ranging up to 5.7 cm are used for the construction industry. Films are generally wound on spools or in rolls in widths from 3 mm to several meters. Sheeting materials may be wound in rolls, but more likely are cut and shipped as flat sheet 1–3 meters in width and length. The products may be very stiff and have high tensile strengths, or may be rubbery or flimsy. They may be impermeable to water or gases, or may dissolve or be porous. Some may be crystal clear or opaque, colorless to brilliantly hued. Films may degrade quickly or last indefinitely as the base material for information storage. They may be excellent insulators for use as protective materials, or may be compounded to conduct small currents. Films may be inert to chemical attack or be made easily printable, even degradable. Sheets may be intractable or readily thermoformed into complex shapes. Some films may dissolve in water, whereas others may act as barriers to moisture permeation almost as effectively as metal.

Film and sheeting materials test methods have been standardized by ASTM, DIN, and others. As with all materials, the test specimens must be carefully prepared and conditioned. Thin-film specimens are vulnerable to nicks and tears which mar the results. Moisture and temperature can affect some materials. Common test methods are listed in Table 1.

Tensile properties of importance include the modulus, yields, F^5 (strength at 5% elongation), and ultimate break strength. Since in many uses the essential function of the film may be destroyed if it stretches under use, the yield and F^5 values are more critical than the ultimate strength. This is true, for example, where film is used as the base for magnetic tape or microfilm information storage. In some cases, the tensile properties at temperatures other than standard are critical. Thus if films are to be coated and dried in hot air ovens, the yield at 150°C or higher may be critical.

Tear strength is critical for packaging and other film end uses. Both tear initiation and tear propagation are important. Very high tear initiation may require the use of notching, perforation, or tear strips to facilitate opening a packaging. Low tear strength can be a significant problem in processing ultrathin films such as are used in capacitors.

Impact tests for film and sheet vary. Tensile impact and dart drop are used to measure the force required to rupture films. Izod impact is generally used on

Table 1. **Film and Sheet Test Methods**

Property	Units	ASTM method
general		
thickness and yield	mm	D2103
mechanical		
tensile strength	MPa	D882
elongation	%	D882
stiffness, tensile modulus	MPa	D638
tear strength, propagation (Elmendorf)	g/mm	D1922
tear resistance (initiation)		D1004
burst strength (Mullen)	points	D774
impact (dart drop)	g/mm	D1709
tensile impact	(N·m)/m	D1822
Izod impact	(N·m)/m	D256
folding endurance		D643
coefficient of friction (slip)		D1894
abrasion resistance		
surface (Tabor)		D1044
falling sand		D968, 1003
optical		
refractive index	n_D	D542
haze	%	D1003
luminance transparency	%	D1746
gloss		D523
chemical		
water absorption	%	D570
water-vapor transmission	[a]	E96
gas permeability	[a]	D1434
electrical		
dielectric constant		D150
dissipation (power) factor		D150
dielectric strength	kV/mm	D149
volume resistivity	Ω·cm	D257
permanence		
thermal shrinkage	%	D1204
		D2411
outdoor weatherability		D1435
coefficient of humidity exposure		E104

[a]See BARRIER POLYMERS for a discussion of units of gas permeability and WVTR.

sheet materials. Correlation of test results with end use performance may not be good. Temperature is critical to impact results. Some users of film perform actual package tests to measure performance, where packages of the product are produced and then dropped or thrown under controlled conditions to see if the package will withstand shipping and handling. Actual shipping tests may also be used.

Stiffness of the films and sheeting can be measured as the tensile modulus of elasticity. Droop or drape tests may be used, particularly for multilayer products. The stiffness is strongly influenced by thickness (to the third power) and temperature, and is important to the processing of film in printing, coating, or end use applications where it affects the "hand" of the product.

The slip characteristics of film and sheeting are also critical to processing and use. It may influence the hand, but is particularly critical to the free passage of film over rolls and through equipment. Slip is measured as the static or kinetic coefficient of friction and may be measured as film-to-film, or as film to another surface. Hot slip, or slip over heated platens or rolls, can be an important characteristic of packaging film. Closely related to slip is abrasion resistance, a property difficult to measure. Tabor abrasion is often used but is difficult to compare from one material to another, and correlation with end use performance is only fair. Falling sand abrasion is sometimes used, and this simulates a different abrasion mechanism from the rubbing action of the Tabor. Steel wool abrasion testing has also been used. The effect of abrasion is generally determined by change in haze or gloss, although weight loss can also be used. This property is critical where constant abrasion in use may occur, as with magnetic tape or in glazing, where long-term optical clarity is critical, eg, in solar collectors or vehicle glazing.

Transparency, color, haze, and gloss are all important elements of film and sheet used for optical end uses. These include packaging, reprographic and photographic uses, glazing, solar, etc. Spectral transmission, index of refraction, as well as standard haze and gloss tests are used. Lack of color (yellowness) of film and sheet can be important. Matching specific color or transparency characteristics for end uses as light filters or for a decorative purpose is necessary in some uses. The permanence of these properties after long-term exposure to heat, rain, and sun and uv light lead to the need for such tests as the Weather-O-Meter, Q-UV, and outdoor exposure tests. Some applications call for 10–20-year lifetimes under intense exposure conditions. Predicting performance for such uses accurately is difficult. Although resistance of the film or sheet base material to light or moisture may be key, the selective passage or blockage of light through the material may be an essential requirement. Thus some films or sheets have been developed to prevent uv or ir light transmission for use in glazing applications.

Moisture and gas barrier properties are of prime interest in packaging applications (1). These are measured in standard tests which in recent years have been improved to give more meaningful predictive information (see BARRIER POLYMERS). Sensitive techniques have been developed to measure lower concentrations of permeants and thus get information more quickly. Some films are good barriers to both moisture and gases, whereas others may be barriers to moisture but allow gases to pass (and the product to breathe). High fat food products need to be protected from oxygen or they rapidly become rancid, particularly if uv light also penetrates the packaging. Bland products may be susceptible to pickup of off-taste or flavor if volatile oils or spices permeate the package material, or the reverse can be true if highly seasoned contents are packaged. To achieve the right combination of barriers, combinations of films are often needed. Coating, coextrusion, or lamination with another film, foil, or paper are commonly used.

Other important properties that can be measured in the laboratory include sealability, printability, or coating adhesion. Many of these tests have been developed by the film manufacturer in cooperation with customers and are specifically designed to measure product performance in the end use. Some tests, like sealability, can be standardized to time, pressure, and temperature of sealing with instrument-measured peel values, but other tests are subjective, such as evaluations of printing loss to pulloff by adhesive tape.

Dimensional stability of films or sheet when exposed to temperature or humidity are important. Inherent dimensional change with temperature or humidity (coefficient of thermal or humidity expansion over a selected temperature range) may be measured, but for film in particular this may be less important than shrinkage due to relaxation of stress imparted to the film during manufacture. Often films shrink in one direction and grow in the other direction depending on orientation, heat-setting, and other parameters in their manufacture. These changes may be irreversible in the base film and destroy its usefulness. In other cases, eg, shrink films, the film is designed to shrink with controlled force and amount in order to shrink-wrap a container or packaging without distortion. Creep, or elongation under long-term low or no load, can be measured and controlled for those applications where it is important, such as strapping tape.

Thermoformability is a property required by the many sheet materials used in the thermoforming industry. These properties are unique for the specific forming methods used, and are best determined by actual thermoforming tests on small-scale equipment. The softening or drape temperature of the material, residual stress in the sheet from its manufacture, and its melt strength and viscosity are important parameters relating to this use.

An important property of all film and sheet products is the gauge (or thickness) uniformity. Machine direction uniformity is vital to consistent processibility in web-handling equipment and to controlled economic production. More important, perhaps, is uniformity of transverse gauge. Irregularities in TD gauge, particularly consistent ones, lead to gauge bands, soft spots, honeycomb or chain-like defects, which in turn lead to poor handleability, bad coating or printing, improper tracking, and totally unusable product. Measurements of thickness variation are usually done in-line in order to provide feedback control. Laboratory measurement is also used to calibrate and more precisely determine the film's suitability for use. The average thickness, gauge variation (high and low), and rate of gauge change (ie, slope) across the web are among the characteristics measured.

Performance measures such as slittability (ease of slitting the film web into smaller widths), or cuttability (ease of trimming, stamping, or cutting the sheet by shearing) can be determined usually by actual performance testing on processing machines. Good materials can be so processed with a reasonable force and without generating debris, stringy edges, or chips or tears in the product.

Because of the nature of film and sheet products, ie, large area-to-volume ratio, optical defects in the polymer may be readily evident and unacceptable. Thus the presence of degraded polymer, gels, fish-eyes, contamination, or improperly dispersed additives, pigments, or colorants may result in a product that is aesthetically displeasing or functionally unacceptable. In addition, such contaminants may cause mechanical or electrical failures under stress. Visual inspection, for optical defects on the moving web, or microscopic inspection are used to measure and define these problems. Fortunately, new technology is now available that can determine the presence of relatively small particles within a web moving at speeds of several hundred meters per minute. These devices can alert the production operator to take the necessary action to correct the problem.

Tables 2–5 list some typical properties or ranges of properties for the more common film and sheet products. Although these values are good for comparative purposes, actual performance tests are best to determine suitability for use. Properties of multiple-layer films or sheets in laminar structures cannot always be predicted from values for the individual polymer layers. Use conditions of stress, temperature, humidity, and light exposure all strongly influence performance. Film and sheet manufacturers can recommend product combinations or variations that may provide significant performance advantages to the user.

Materials

Acrylonitrile–Butadiene–Styrene. Available only as sheet, ABS has good toughness and high impact resistance. It is readily thermoformable over a wide range of temperatures and can be deeply drawn. ABS has poor solvent resistance and low continuous-use temperature. It is often used in housings for office equipment (see ACRYLONITRILE POLYMERS).

Acrylic Polymers. A small amount of film is produced, but most acrylic resin is used in sheet and is produced from poly(methyl methacrylate) (PMMA). This material has optical clarity approaching glass and is resistant to uv exposure (see METHACRYLIC POLYMERS). It has good rigidity, impact resistance, and scratch resistance so it finds a large market for signs and safety glazing. It can be stretched to increase its toughness and coated to improve its abrasion resistance. This product is used for airplane canopies and windows. It is, however, susceptible to stress crazing, has poor solvent resistance, and is flammable.

Cellophane. Once the leading clear packaging film, cellophane has a small fraction of the film market in the 1990s. It has largely been replaced by cheaper biaxially oriented polypropylene films and superior polyester films. Cellophane has set the standard for processibility on printing and packaging equipment. Cigarette packaging equipment was specially designed to process it at high speeds. Cellophane is produced from regenerated cellulose and is quite moisture-sensitive. It is necessary to coat it to provide moisture control and sealability. Cellulose nitrate was first used as a coating but this gave way to poly(vinylidene chloride) and to polyethylene. Cellophane has excellent stiffness and strength and its optical properties are outstanding; however, it does tend to yellow with age. Manufacturing cost is high, due in part to the large quantities of hazardous solvents required. It is still used for high quality transparent packaging.

Cellulosics. The four previously common cellulosic films and sheets are less important factors in the markets of the 1990s. These include cellulose acetate, cellulose acetate propionate, cellulose acetate butyrate, and cellulose triacetate. These materials have excellent optical properties, and good scuff and grease resistance. Their barrier properties are midrange. With the exception of cellulose triacetate, they can be readily thermoformed and make fine clear boxes, and blister and skin packages. The butyrate and propionate copolymers are tougher than cellulose acetate, providing better protection. They also weather well and have been used to make outdoor signs. The cellulose triacetate is still widely used as the base for amateur photographic film. Environmental concerns and the rela-

Table 2. General and Thermal Properties of Film and Sheet[a,b]

Material[b]	Abbreviation	CAS Registry Number	Manufacturing method[c]	Thickness range, mm	Max width, m	Max use temperature, °C	Min use temperature, °C	Heat seal range, °C	Dimensional strength at 100°C, % change
acrylonitrile–butadiene–styrene[d]	ABS	[9003-56-9]	EX, CL	0.25–0.75	2.67	97			
cellophane[e]		[9005-81-6]	REG	0.02–0.04	1.19	177	−17	82–177	−0.7 to 3.0
cellulose acetate	CA	[9004-35-7]	CAST, EX	0.02–6.35	1.52	79	−26	177–232	+0.2 to 3.0
cellulose triacetate	CTA	[9012-09-3]	CAST	0.05–0.51	1.19	175			0 to 0.7
fluoroplastics									
ethylene–tetrafluoroethylene copolymer	ETFE	[26770-96-4]	EX	0.01–2.1	1.22			274	
fluorinated ethylene propylene copolymer	FEP	[25067-11-2]	EX	0.01–2.41	1.22	249	−254	282–371	<1
polychlorotrifluoroethylene copolymer	PCTFE	[9002-83-9]	CAST, EX	0.01–0.76	1.37	135	−196	232–260	+2 to −2
polytetrafluoroethylene	PTFE	[9002-84-0]	CAST, EX	0.01–3.18	1.22	260	−253		
poly(vinyl fluoride)	PVF	[24981-14-4]	EX	0.01–0.10	3.51	113	−73	204–218	1
ionomer		[25608-26-8]	EX	0.03–0.25	1.52	66	−73	93–260	
nylon-6		[25038-54-4]	EX, BO	0.01–0.76	2.13	150	−73	193–232	
polycarbonate	PC	[24963-68-3]	EX	0.006–12.7	1.37	132	−101	204–221	
poly(ethylene terephthalate)[f]	PET	[25038-59-9]	EX, BO	0.003–0.36	3.05	149	−79	218–232	<0.5
polyimide[e]	PI	[25036-53-7]		0.008–0.13	0.90	400	−270		0.3
polyethylene[f]	PE								
low density	LDPE	[9002-88-1]	EX	0.0008 up	12.19	88	−57	121–204	−2
linear low density[e]	LLDPE		EX						
medium density	MDPE		EX	0.008 up	6.10	104	−57	121–204	0 to 0.7
high density	HDPE		EX	0.01 up	1.52	121	−45	135–204	−0.7 to 3.0
ultrahigh molecular weight	UHMWPE		EX	0.05 up	1.52	121	−45	135–204	
poly(methyl methacrylate)	PMMA	[9001-14-7]	EX, BO	0.13–0.25	1.09	77			
polypropylene[f]	PP	[25085-53-4]	EX	0.02–0.25	1.52	140	−18	140–200	2
			BO[e]	0.01–0.03	2.03	143	−51	88–150	
polystyrene[f]	PS	[9003-53-6]	BO	0.006–0.51	1.93	88	−63	121–177	
poly(vinyl chloride)	PVC	[9002-86-2]							
rigid			CAL, EX	0.015–1.91	2.13	79		177–216	−7 to 4
plasticized[f]			CAL, EX	0.013–2.54	2.03	79	−46	157–182	−7 to 15

[a]Much of this data was obtained from Refs. 2 and 3. [b]Available as both film and sheet unless otherwise noted. All materials are available in FDA grades except for CTA, PVF, and PI. [c]EX = extrusion; CL = calendering; REG = regeneration; CAST = casting; BO = biaxial orientation. [d]Sheet only. [e]Film only. [f]Heat shrinkable grades available.

Table 3. Physical and Mechanical Properties[a]

Material	Tensile strength, MPa[b]	Elongation, %	Impact strength, (kN·m)/m[c]	Tear strength, N/mm[d]	Burst strength (Mullen)
ABS	50	25			
cellophane	80	30	31–58	0.8–8	30–50
CA	75	15–55	10.8	1.6–3.9	30–60
CTA	86	10–50		1.6–11.8	50–70
fluorocarbons					
ETFE	52	300		235–350	
FEP	20	300		49	10
PCTFE	52	50–150		1–15.7	23–30
PTFE	20	100–400		3.9–39	
PVF	86	115–250		3.9–39	19–70
ionomer	34	250–400		11.8–49	
nylon-6	225	85–120		6.3–11	
PC	67	40–100		8–10	
PET	210	60–165		20–118	55–80
PI	172	70		3	75
PE					
LDPE	15	200–600	27–42	20–118	10–12
LLDPE	38	400–800	31–50		
MDPE	24	200–500	15–23	20–118	
HDPE	38	10–50	4–12	20–118	
UHMWPE	29	300			
PMMA	59	4–12			
PP[e]	50–275	35–500	19–58	1.2–3.9	
PS[f]	70	3–60		0.8–5.9	16–35
PVC					
rigid	58	25–50		3.9–275	30–40
plasticized	52	3–100		2–275	20

[a]See Table 1 for ASTM test methods.
[b]To convert MPa to psi, multiply by 145.
[c]To convert (kN·m)/m to (kgf·cm)/mil, divide by 3.861.
[d]To convert N/mm to gf/25 μm, multiply by 2.549; to convert to ppi, divide by 0.175.
[e]Biaxial orientation.
[f]Oriented.

tively high manufacturing costs have driven most U.S. producers to limit their production (see CELLULOSE ESTERS).

Fluoropolymers. Fluoroplastic film and sheets as a class possess the most chemically inert behavior of the materials generally used. In addition to their good strength, their resistance to thermal and electrical stress give them a unique niche in the market. The resistance to flame is important for use as liners in aircraft interiors. They are very resistant to moisture and provide a good barrier. When combined with fabrics or other materials, they provide a weatherable surface that is easy to clean. On the other hand, the resins tend to be intractable, difficult to handle, and require special bonding techniques. Because of their good slip and inertness, they are often used as gasketing material. Their relatively

Table 4. Optical and Electrical Properties

Material	Refractive index	Transparency, %	Haze, %	Dielectric constant, kHz	Dissipation factor, kHz	Dielectric strength, kV/mm	Volume resistivity, Ωm·cm
ABS	1.53	33	100	2.75		14.3	10^{16}
cellophane			3.5	3.2	0.015	79.99	10^{11}
CA	1.50	88	<1	3.6	0.013	126–197	10^{10-15}
CTA				4.0	0.016	146	10^{13}
fluorocarbons							
ETFE				2.6	0.0008	138	10^{10}
FEP	1.34	<90	4	2.25	<0.002	276	10^{19}
PCTFE	1.43			2.6	0.023	39–146	10^{12}
PTFE	1.35			4.1	0.0002	17	10^{13}
PVF				8.5	1.6	138	3×10^{13}
ionomer				2.4	0.002	39	10^{16}
nylon-6				3.7	0.03	50	
PC	1.59	83–90	0.5–2.0	2.9	0.0015	59	10^{16}
PET		88	1.0–3.0	2.8	0.005	296	10^{18}
PI				3.5	0.003	276	10^{18}
PE							
LDPE	1.51	0–75	4–50	2.2	0.0003	19	10^{16}
LLDPE			5–7				
MDPE	1.52	10–80	4–50	2.2	0.0003	20	10^{16}
HDPE	1.54	0.40	10–50	2.2	0.0005	20	10^{15}
UHMWPE	1.54				2.3×10^{-4}	51	10^{18}
PMMA	1.5	92	1	3.75	0.04	16	10^{15}
PP[a]			1.5–25	2.2	0.0002	276–400	3×10^{16}
PS[b]	1.6	87–92	0.1–30	2.5	0.0005	197	10^{16}
PVC							
rigid	1.53	76–82	8–18	3.0–3.3	0.013	17–50	10^{16}
plasticized				4.0–8.0	0.11	10–40	10^{11-14}

[a]Biaxially oriented.
[b]Oriented.

Table 5. Barrier and Chemical Properties

Water absorption, %	WVTR, $\frac{nmol^a}{m \cdot s}$	Gas permeability, $\frac{nmol^b}{m \cdot s \cdot GPa}$			Resistance[c] to:					
		O_2	CO_2	N_2	Acid	Alkali	Grease	Organic solvent	Water	Sunlight
0.6–1.0		100	300	14–19	G–F	G	G	F–P	G	F
45–115	0.1–32	1	1–100	0.2	P	P	G	P	F	G
3–8.5	2.6–10	180–240	1700–2000	65–95	P	P	G	P	G	G
2–4.5	7.7–10	240	1800	65	F	P	G	F–P	G	G
<0.02	0.4	160	500	65	G	G	G	G	G	G
<0.01	0.1	1200	3500	680	G	G	G	G	G	
nil	0.006	12–24	30–120	5.3	G	G	G	G	G	G
nil	0.013				G	G	G	G	G	G
<0.5	2	4–8	20	0.5	G	G	G	G	E	E
0.4	0.45	2400			G	G	G	G	G	G
9.5	2.7	4–6	20–24	1.9–2.5	P	F	E	G	G–P	F
<0.8	2.8	480	1600	100	G	P	G	G–P	G	F
0.25	0.3	6–8	30–50	1.4–1.9	G	P	G	G	G	F
2.9		40–60	80	10	G	P	G	G	G	G
<0.01	0.35	500–700	2000–4000	200–400	G	G	P	F	G	F
<0.01	0.2	250–600	1000–3000	150–600	G	G	F	F	G	F
nil	0.09	200–400	1200–1400	80–120	G	G	G	G		
nil					G	G	G	G	G	F
0.3–0.4	0.32				G	G	P		G	G
<0.005	3.2	300–500	1000–1600	60–100	G	G		G	G	F
0.04–0.1	1.8	500–800	1400–3000		G	G		G–P	G	F
nil	0.2–1.3	8–30	40–100		G	G		G–P	G	G
nil	1.3–7.7	50–3000	160–5000		G	G		G–F	E	F

[a]To convert nmol/(m·s) to (g·mm)/(m^2·d), multiply by 1.55.
[b]To convert nmol/(m·s·GPa) to (cm^2·mil)/m^2·d·atm), multiply by 0.13.
[c]G = good; F = fair; P = poor.

high cost limits their market penetration (see FLUORINE COMPOUNDS, ORGANIC–POLYTETRAFLUOROETHYLENE).

Ionomer. Ionomer resins are a specialized class of polymers in which ionic bonding units provide cross-linking in the intermolecular structure. They make good film resins because of their strength, flexibility, adhesion, and optical properties. Ionomers (qv) have a relatively low sealability temperature and thus are used in coextrusion to provide a bonding layer between otherwise incompatible layers.

Polyamide. Polyamide or nylon film is primarily made from nylon-6 or nylon-6,6. About one-quarter of the film produced is biaxially oriented. Similar in most properties to polyester, it has somewhat superior resistance to flex-cracking, has good resistance to oils and grease, and is an excellent gas barrier. Nylon films, however, are sensitive to moisture so they are often laminated or extrusion-coated with PVDC, polyethylene, or ionomer. Typical use is for meat and cheese packaging and brown-and-serve cooking bags. It is used industrially for sheet-molding compound carrier and vacuum-bag molding. Nylon is the material used in metallized balloons. Some nylon is extruded into stock sheets used for machining into precision parts (see POLYAMIDES).

Polycarbonate. Polycarbonate sheet is an important material for use in window glazing, aircraft windshields, and outdoor signs because of its excellent impact resistance stability, transparency, and heat resistance. Glazing and sheet uses account for the largest segment of polycarbonate resin. Impetus for its growth has come from the increasing demand for safety and security glazing in banks, government facilities, schools, hospitals, storm doors, and ice rinks. Its resistance to vandalism warrants it premium over glass and acrylics. Polycarbonate's one-third share of the plastic glazing market will grow as new products with abrasion-resistant coatings overcome one of its deficiencies. New, improved uv stabilizers will provide longer useful life in outdoor applications, and new technologies are being introduced to improve smoke and flame properties (see POLYCARBONATES).

Polyester. Poly(ethylene terephthalate) is used in both film and sheet form. Biaxially oriented film is the premier high volume film in terms of performance characteristics, combining such properties as high tensile properties, excellent dimensional stability, good barrier properties, high usage temperature, and excellent optical properties into a range of products for magnetic tape base, reprographic and photographic film base, electrical insulation, capacitors, decorative labeling and laminates, packaging, and many other uses. New films combining in-line coatings and coextrusion to produce multilayers are growing in sales volume based on enhanced printability, sealability, adhesion, or release characteristics.

Polyester sheet products may be produced from amorphous poly(ethylene terephalate) (PET) or partially crystallized PET. Acid-modified (PETA) and glycol modified (PETG) resins are used to make ultraclear sheet for packaging. Poly(butylene terephthalate) (PBT) has also been used in sheet form. Liquid-crystal polyester resins are recent entries into the market for specialty sheet. They exhibit great strength, dimensional stability, and inertness at temperatures above 250°C (see POLYESTERS, THERMOPLASTIC).

Polyimide. Polyimide is a biaxially oriented high performance film that is tough, flexible, and temperature- and combustion-resistant. Its room temperature properties compare to poly(ethylene terephthalate), but it retains these good characteristics at temperatures above 400°C. Its electrical resistance is good and it is dimensionally stable. The principal detriment is fairly high moisture absorbance. The main uses are for electrical insulation, particularly where high temperatures are prevalent or ionizing radiation is a problem. The films may be coated to reduce water absorption and enhance sealing (see POLYIMIDES).

Polyethylene. Polyethylene remains the largest volume film and sheet raw material. It is available in a wide range of types, with variations in copolymers, homopolymers, molecular weight, and other factors contributing to a long list of resins. Resins are designed specifically for end use, and in addition blends of the various types may be used by processors to optimize properties, processibility, and economics. Almost two-thirds of the volume of all polyethylene resins are used in film or sheet applications (see OLEFIN POLYMERS).

Low density polyethylene (LDPE), or as it is more precisely known, high pressure, low density polyethylene (HP-LDPE), the oldest form of polyethylene, still commands a large market share based on its transparency, tear resistance, impact resistance, and moisture resistance. It does have poor resistance to oils and grease, marginal weatherability, relative low resistance to high temperature, and is permeable to odors and gases. It is primarily used in clear packaging of such items as bread, produce, meat, poultry, seafood, and frozen foods, as well as in garment bags. It is also used in industrial applications such as liners, stretch wrap, heavy-duty bags, shrink wrap, and overwrap.

In the late 1980s, linear low density polyethylene (LLDPE) became a significant factor in the polyethylene film business. These resins have somewhat better strength than HP-LDPE and are cheaper to produce. LLDPE has been most successful in nonclear packaging, industrial or trash-bag use where its milky character is not critical and its properties allow downgauging with substantial cost savings. LLDPE is often used in grocery sacks, stretch wrap, liners, and mulch film. There is a growing market for its use in diapers, both infant and adult.

High molecular weight, high density polyethylene (HMW-HDPE) also has a large market share in film and sheet. As molecular weight and density of the polyethylene increase the tensile properties, chemical resistance and barrier properties of the film products increase. Excellent abrasion resistance is also achieved. As film, HMW-HDPE has a large share of the grocery sack business. It is also used for cereal and snack food packaging, where its resistance to moisture penetration is a prime factor.

In the sheeting market, the low density polyethylenes are less important than the high density resins. The high density resins have excellent chemical resistance, stress-crack resistance, durability, and low temperature properties which make them ideal for pond liners, waste treatment facilities, and landfills. In thicker section, HMW-HDPE sheet makes good containers, trays, truck-bed liners, disposable items, and concrete molds. The good durability, abrasion resistance, and light weight are critical elements for its selection.

Polypropylene. Polypropylene (PP) film, the bulk of which is biaxially oriented (BOPP), is characterized by its excellent clarity, strength, and low moisture transmission. It has captured a large share of the packaging market originally held by cellophane and competes with poly(ethylene terephthalate) film, glassine, paper, and other plastics in some markets. The packaging markets of importance are snack foods, baking items, cigarettes, candy, overwrap for boxes or trays, bottle labels, and more. It is used for pressure-sensitive tapes, sheet protectors, and stationery products, overwrap for toys and games, and as label stock. Its excellent electrical characteristics provide for its use as metallized thin-film capacitor dielectric.

Some cast (unoriented) polypropylene film is produced. Its clarity and heat sealability make it ideal for textile packaging and overwrap. The use of copolymers with ethylene improves low temperature impact, which is the primary problem with unoriented PP film. Orientation improves the clarity and stiffness of polypropylene film, and dramatically increases low temperature impact strength. BOPP film, however, is not readily heat-sealed and so is coextruded or coated with resins with lower melting points than the polypropylene shrinkage temperature. These layers may also provide improved barrier properties.

Because of poor thermoformability, there are relatively few applications for polypropylene sheet. New solid-phase pressure forming (SPPF) techniques are under development for forming PP sheet. Polypropylene is used in coextruded sheet to some extent for food packaging containers. Glass-filled, wood-filled, or other modified polypropylene sheet materials are used in limited automotive applications.

Polystyrene. Polystyrene (PS) film and sheet has the third largest production volume, behind only the polyethylenes and poly(vinyl chloride). As biaxially oriented film, it has excellent clarity and a crisp metallic feel, with high gas permeability and moderate moisture vapor permeability. It is used mostly for fresh produce wrapping. Sheet markets for thermoforming predominate, both as crystal clear impact grades and foamed sheet. The high melt strength of clear polystyrene makes it easy to thermoform by pressure or vacuum forming, and it stamps and cuts easily. Natural polystyrene is very clear but brittle, so impact grades are produced using butadiene as a comonomer, or by the addition of impact modifiers. Products produced from sheet range from cups and containers for foods and beverages, trays, furniture components, and lighting fixtures to architectural components.

Foamed polystyrene sheet has excellent strength, thermal resistance, formability, and shock resistance, as well as low density. It is widely known for its use in beverage cups, food containers, building insulation panels, and shock absorbent packaging. Polystyrene products can be recycled if suitable collection methods are established. Foamed polystyrene sheet can also be easily thermoformed (see STYRENE PLASTICS).

Vinyl Films. Vinyl films include a number of polymers, both homo- and copolymers, with a range of properties. They include poly(vinyl chloride) (PVC), vinyl chloride–vinyl acetate copolymers, poly(vinylidene chloride) (PVDC), and poly(vinyl alcohol) (PVOH) (see VINYL POLYMERS). The chemical and physical properties can vary over a wide spectrum. Blending, plasticizers, modifiers, and other additives are used to match properties to end use requirements. The properties are also influenced by the method of manufacture. They are generally made by extrusion, casting or calendering, and in some instances have been biaxially oriented. The products may range from hard, brilliant sheet to soft, pliable films. Poor thermal stability, plasticizer migration, poor solvent resistance, and low yield have acted to restrict applications. Shrinkable films can be made by biaxial orientation, and find use in food and other packaging. Vinyl resins may be blown into film of lesser quality than cast film. Rigid sheet extrusion products are self-extinguishing and their excellent weatherability have led to expanding use in siding for housing.

Calendering is used to produce flexible vinyl sheet and film above 0.8 mm (see POLYMER PROCESSING). These products can be readily thermoformed and have good sealability and are useful in packaging. The calendering of rigid vinyl sheet reached a low point in the mid-1970s, but has regained favor in the 1990s. Crystal clear sheet is used for box formation. Calendered rigid vinyl sheet is used in labels, print stock, credit card stock, and floppy disk jackets. It is easily stamped, cut, embossed, printed, thermoformed, and sealed.

Poly(vinylidene chloride) (PVDC) film has excellent barrier properties, among the best of the common films (see BARRIER POLYMERS). It is formulated and processed into a flexible film with cling and tacky properties that make it a useful wrap for leftovers and other household uses. As a component in coatings or laminates it provides barrier properties to other film structures. The vinylidene chloride is copolymerized with vinyl chloride, alkyl acrylates, and acrylonitrile to get the optimum processibility and end use properties (see VINYLIDENE CHLORIDE AND POLY(VINYLIDENE CHLORIDE)).

Polyurethane. Small quantities of polyurethane film are produced as a tough rubber-like film. Polyurethane is more commonly used to produce foamed sheet, both flexible and rigid. The flexible foam is used as cushioning in furniture and bedding; the rigid foam is widely used for architectural insulation because of its outstanding thermal insulation efficiency (see URETHANE POLYMERS).

Specialty Films. Small quantities of film or sheet are manufactured from high performance resins for a variety of specialty applications. Most of these are for electrical insulation or support structures exposed to temperature, radiation, or environmental extremes. Some are designed to resist burning or flame. Excellent retention of dimensions under environmental changes may also be a prerequisite (see ENGINEERING PLASTICS). Among the polymers of commercial use are polysulfones (PSO), poly(phenyl sulfide) (PPS) (see POLYMERS CONTAINING SULFUR), liquid crystal polyesters (LCP), poly(arylates), polyetherketones, and acetal resins (qv). Another specialty film product is poly(vinyl butyral) film (4). This film is the safety inner layer of automotive safety glass used in windshields and other safety glazing. It is produced under clean room conditions, wound in rolls supported on a carrier web, and laminated to glass. Its excellent flexibility and ability to absorb shock combined with its tenacious bond to glass to prevent dispersal of fragments of broken glass are its special attributes. The use of this film in architectural glass is expected to grow as the emphasis on safety and security continues.

Water-Soluble Films. Water-soluble films can be produced from such polymers as poly(vinyl alcohol) (PVOH), methylcellulose, poly(ethylene oxide), or starch (qv) (see CELLULOSE ETHERS; POLYETHERS; VINYL POLYMERS). Water-soluble films are used for packaging and dispensing portions of detergents, bleaches, and dyes. A principal market is disposable laundry bags for hospital use. Disposal packaging for herbicides and insecticides is an emerging use.

Manufacture

The processes used commercially for the manufacture of film and sheeting materials are generally similar in basic concept, but variations in equipment or process conditions are used to optimize output for each type of film or sheeting material. The nature of the polymer to be used, its formulation with plasticizers (qv), fillers (qv), flow modifiers, stabilizers, and other modifiers, as well as its molecular weight and distribution are all critical to the processibility and final properties of the product. Most polymers are amenable to one or another of the common manufacturing processes, but some are so intractable they can only be made into film by skiving from large casting of the resin.

The basic methods for forming film or sheeting materials may be classified as follows: melt extrusion, calendering, solution casting, and chemical regeneration. Of special note is the use of biaxial orientation as part of the critical manufacturing steps for many film and sheet products.

Melt Extrusion. By far the most important method for producing film and sheeting materials relies on one or another of the various melt extrusion techniques (5). The main variations of melt extrusion are the slot (or flat) die-cast film process, the blown films process, and the flat die sheeting-stack process. These may be combined with one or more steps such as coextrusion wherein multilayer film or sheet is formed, biaxial orientation, and in-line coating (6).

The simplest form of melt extrusion is the use of a slot die to form the molten polymer into a thin flat profile which is then quenched immediately to a solid state (Fig. 1). This is usually done by contacting the hot web very quickly on a chilled roll or drum. A liquid quenching bath may be used in place of or contiguous to the chill roll. Depending on the polymer type or formulation, the quenched web is generally substantially amorphous. In some cases, the web may be drawn down in thickness by overdriving the quenching roll relative to the extrusion velocity.

With liquid bath quenching it is difficult to achieve an optically smooth surface unless all surface ripples are eliminated. In chill roll quenching, it is frequently necessary to pin the hot web to the drum to eliminate air pockets, surface ripples, and other defects. As line speeds increase this can become a controlling

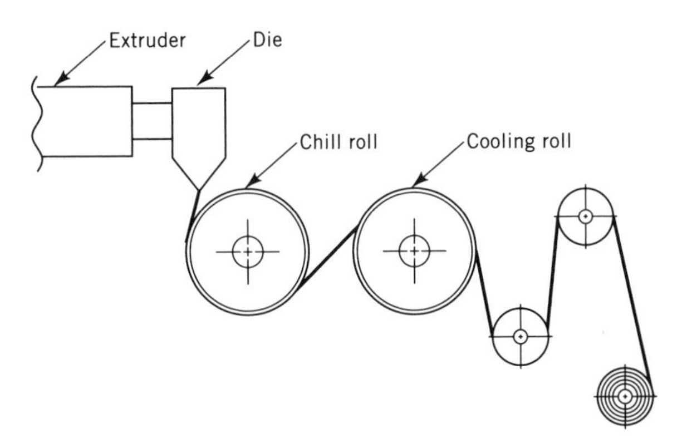

Fig. 1. Slot-die melt extrusion.

factor in quality film production. The most common methods used to pin a web are to use an air knife in close proximity to the emerging melt, or electrostatic pinning, where a high voltage field is established to force the web to the drum. Vacuum assist under the web or hooded quench rolls may be used to control the atmosphere near the die and quench roll(s). Thicker webs or higher line speeds may demand the use of multiple cooling rolls. The quenched film may be further processed (uni- or biaxially drawn, coated, or corona-discharge-treated to enhance adhesion). Generally the film is fed directly in-line to these other processes. If the film is suitable for use, it has the uneven edges trimmed off and is then wound into master rolls for subsequent slitting into narrower rolls, or is wound directly for shipment to customers.

The most critical factor in the slot-film process is the design of the die. The control of transverse gauge is extremely important and requires some form of cross-web gauge adjustment. Mechanical alternation of the die gap across the web may be done by using precisely controlled bolt action on the die lips, or by controlled application of heat in small increments across the die. Interior die geometry must be carefully designed to accommodate the flow and shear characteristics of the particular polymer being used. Care must be taken to prevent melt fracture by adjusting die characteristics to extrusion rate, drawdown, viscosity, and temperature.

When drawdown is high, the film may be uniaxially oriented and the properties of the final film isotropic. In the manufacture of strapping tape this effect is accentuated. If the cast or quenched film is to be used to feed an orientation line, additional attention must be given to the amorphous–crystalline nature of the film in the draw processes so that maximum strength can be achieved and uniform gauge and optical quality maintained. Slot casting is used for the orientation of these resins, polyesters, polyamides, and a variety of others.

Sheet can be produced by melt extrusion, but in this case a three-roll stack of quenching rolls is generally used (Fig. 2). More than three rolls may be used

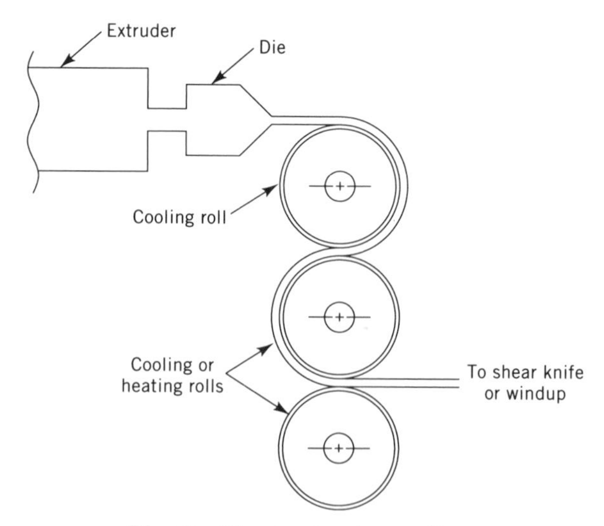

Fig. 2. Sheet extrusion casting.

where necessary. The rolls may be mounted vertically or horizontally. The web is extruded through a slot die in a thickness close to the desired final thickness. The die is in very close proximity to the first chill roll or chill-roll nip. The web may be cast horizontally directly onto the upper chill roll of the stack as shown (Fig. 2), or it may be extruded into the first nip directly. The rolls quench the sheet and provide the surface polish desired. In some applications, matte or embossed rolls may be used to impart special surface characteristics for certain functions. Where the utmost in optical (glazing) quality is desired the trend has been to mount the roll stack horizontally. The hot melt is then extruded vertically down into the first nip. This avoids problems associated with sag of a horizontal hot melt no matter how short the distance between die and quench.

Quenched sheet is pulled horizontally from the stack and is then either wound on rolls or sheared into sheets of the required dimension. Among the polymers made into sheet this way are the polyolefins, poly(vinyl chloride), amorphous polyester, polycarbonate, and polyarylate.

Blown Film. The blown or tubular film process provides a low cost method for production of thin films (Fig. 3). In this process, the hot melt is extruded through an annular circular die either upward or downward and, less frequently, horizontally. The tube is inflated with air to a diameter determined by the desired film properties and by practical handling considerations. This may vary from as small as a centimeter to over a meter in diameter.

As the hot melt emerges from the die, the tube is expanded by air to two or three times its diameter. At the same time, the cooled air chills the web to a solid state. The degree of blowing or stretch determines the balance and level of tensile and impact properties. The point of air impingement and the velocity and temperature of the air must all be controlled to give the optimum physical properties to the film. With some polymers, an internal air cooling ring is used as well, in order to increase throughput rates and optical quality. Rapid cooling is essential to achieve the crystalline structure necessary to give clear, glossy films.

The film tube is collapsed within a V-shaped frame of rollers and is nipped at the end of the frame to trap the air within the bubble. The nip rolls also draw the film away from the die. The draw rate is controlled to balance the physical properties with the transverse properties achieved by the blow draw ratio. The tube may be wound as such or may be slit and wound as a single-film layer onto one or more rolls. The tube may also be directly processed into bags. The blown film method is used principally to produce polyethylene film. It has occasionally been used for polypropylene, poly(ethylene terephthalate), vinyls, nylon, and other polymers.

Downward extrusion of a bubble into a water bath and over an inner water-cooled mandrel is used in a few instances for polypropylene and polyesters. The water is removed prior to slitting and winding.

The double-bubble process may be used to produce biaxially oriented film, primarily polypropylene. In this process the first bubble formation is similar to the conventional blown film, except that the bubble is not collapsed. Rather it is reheated to the orientation temperature and blown and drawn further in a second stage. It is then collapsed, slit, and wound. This process is generally limited to a final film thickness of less than 24 μm.

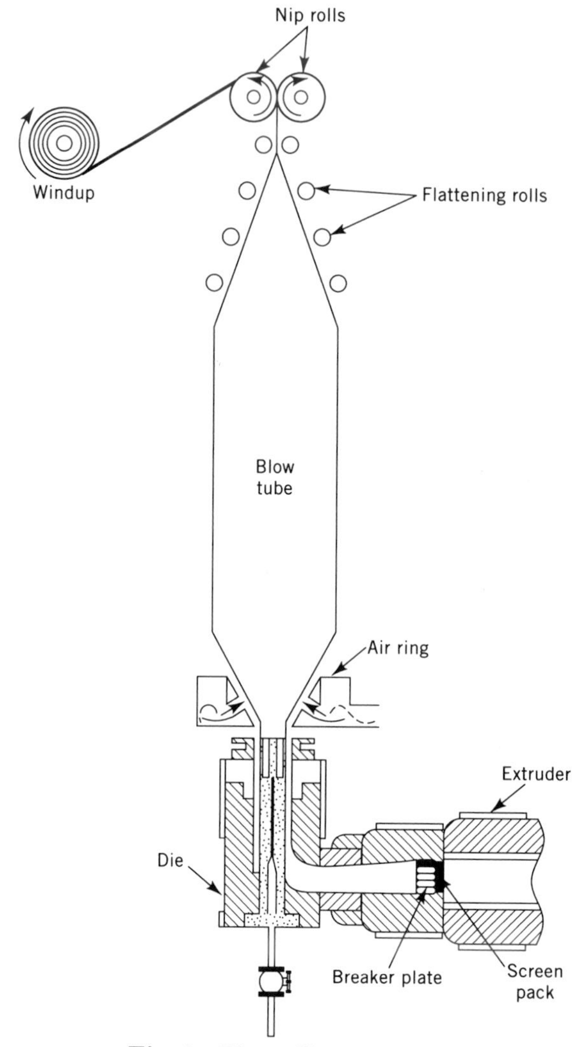

Fig. 3. Blown film extrusion.

Coextrusion. An increasingly popular technique to produce tailored film or sheet products is to coextrude one or more polymer types in two or more layers of melt (6). In this fashion the benefits of specific polymer types or formulations may be combined. Thus high cost barrier resins may be combined with a low cost thicker layer of standard resin to achieve an optimum barrier film at lower cost. Thin slip-control layers may be used on the surface of a bulk layer of optically clear resin to obtain an aesthetic film with good handleability. Lower melting outer layers may be used to provide heat sealing for polymers that seal with difficulty by themselves.

The layers of the different polymers or resins may be combined in one of two ways. One is to use a combining block prior to the slot extrusion die. Parallel openings within the block are fed from two or more extruders, one for each resin. The melts flow in laminar fashion through the die and onto the quench drum. The film is processed conventionally or may then be oriented. Careful control of resin viscosity must be obtained to provide smooth flow, and the resins must be compatible in order to bond together properly. The second method uses a multimanifold die to bring the melt streams together within the die. This allows use of resins with a wider difference in viscosity since fewer changes in flow patterns are necessary. Multimanifold dies may be flat or tubular. The most common types of coextrusion are AB, ABA, or ABC where A is one polymer system, B is another (of the same polymer type or different), and C is a third polymer type. Coextrusions of many, many layers lead to film products with a pearlescent appearance. Where two polymers may not adhere sufficiently, it is possible to extrude a tie or adhesive layer in the coextrusion. Ionomer resins are often used as such tie layers.

The process can be used to recover scrap or low quality resins by using them as the core layer, and using outer layers of virgin resins designed for the specific functional needs of the product such as slip or gloss and appearance. The inner core may be a foamed resin with surface layers of superior finish resins. Coextruded films often eliminate the need for costly lamination processes.

Calendering. Calendering is the process whereby a polymer is heated on hot rolls and squeezed between two or more parallel rolls into a thin web or sheet (Fig. 4) (8). The polymer is blended and masticated in preliminary operations and then fed to a rolling nip between hot, temperature-controlled rolls. The polymer mass is worked further in the nip and flows out to a uniform sheet as it passes through the nip. The web is nipped again and drawn down to a thinner sheet or film, and may undergo a third such step. Draw may be imparted to the web by a slight overdrive between nips. Since the surface of the film tends to take on the nature of the hot nip rolls, special surfaces such as high polish, matte, or embossing may be achieved by changing the rolls in the calender. The film and sheet properties are controlled by the resin composition (polymer–copolymer ratios, use of plasticizers or impact modifiers, stabilizers, molecular weight), by the amount

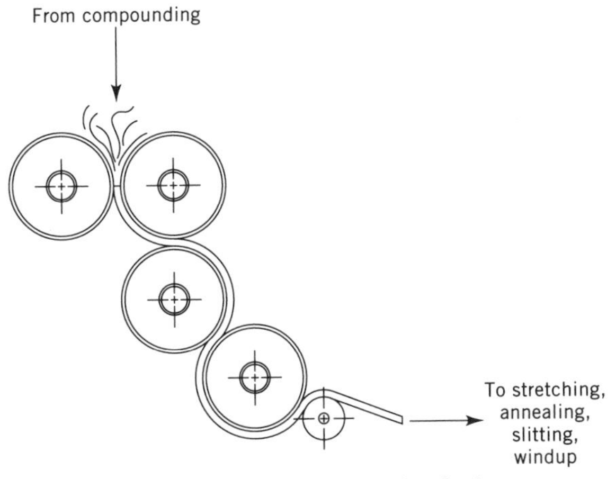

Fig. 4. Melt extrusion, calendering.

of prework, by the temperature of plastic melt, by the amount of squeeze in each nip, and by the degree of draw imparted to the film (9).

Calendering is used principally for poly(vinyl chloride) film and sheet, both flexible and rigid. The use of the process had been in decline since the 1970s, but since 1986 sales have improved so that in 1993 12% of all PVC resin is used in calendering operations. The main products of calendering are flexible sheet for combination with textile backing for use in seating and wall coverings, flooring, and packaging. Rigid vinyl sheet is used for credit cards and packaging.

Biaxial Orientation. Many polymer films require orientation to achieve commercially acceptable performance (10). Orientation may be uniaxial (generally in the machine direction [MD]) or biaxial where the web is stretched or oriented in the two perpendicular planar axes. The biaxial orientation may be balanced or unbalanced depending on use, but most preferably is balanced. Further, this balance of properties may relate particularly to tensile properties, tear properties, optical birefringence, thermal shrinkage, or a combination of properties. A balanced film should be anisotropic, although this is difficult to achieve across the web of a flat oriented film.

There have been many methods and types of equipment developed to orient films, but only two are of real significance. One, the double-bubble process, is mentioned above; the other is the tenter process. In the latter process, a cast amorphous heavy-gauge film is drawn in the machine direction by passing it over heating rolls and then stretching it by speeding up the rolls at a point in the process where the web has reached the proper draw temperature. Immediately upon stretching the web is cooled and then fed into a tenter frame to achieve transverse draw. Here the edges of the film are engaged by a continuous chain of metal chips that tightly grip the film. The film enters an oven where it is reheated to the requisite second draw temperature. At this point the track for the chains of clip diverge, causing the film to stretch in the transverse direction (TD). The drawn web then enters a hotter section of the oven to heat-set or anneal the film. In this step, stresses are relieved and crystallization of the polymer may occur. The web is then cooled and the heavy edges that were within the clips are trimmed off. The film is then wound onto rolls for shipment or reslitting. The amount of stretch in each direction varies by polymer type and with the property balance that is desired in the product. Stretch ratios of from 3-to-1 up to 9-to-1 or more may be used. The degree of heat-set controls the crystallinity of the polymer and the amount and force of shrinkage when the free film is later heated.

In conventional tenter orientation, the sequence of steps is as described above (MD–TD). In some cases it is advantageous to reverse the draw order (TD–MD) or to use multiple draw steps, eg, MD–TD–MD. These other techniques are used to produce "tensilized" films, where the MD tensile properties are enhanced by further stretching. The films are generally unbalanced in properties and in extreme cases may be fibrillated to give fiber-like elements for special textile applications. Tensilized poly(ethylene terephthalate) is a common substrate for audio and video magnetic tape and thermal transfer tape.

Biaxially oriented films have excellent tensile strength properties and good tear and impact properties. They are especially well regarded for their brilliance and clarity. Essentially all poly(ethylene terephthalate) film is biaxially oriented, and more than 80% of polypropylene film is biaxially oriented. Polystyrene film is oriented, and a lesser amount of polyethylene, polyamide, poly(vinyl chloride), and other polymers are so processed. Some of the specialty films, like polyimides (qv), are also oriented.

Solution Casting. The production of unsupported film and sheet by solution casting has generally passed from favor and is used only for special polymers not amenable to melt processes. The use of solvents was generally very hazardous because of their flammability or toxic nature. The cost of recovery and disposal of solvents became prohibitive for many lower price film applications. The nature of the drying operations leads to problems with solvent migration and retention that are not problems with melt-processed polymers.

For solvent casting (Fig. 5), the resin is dissolved in the chosen solvent, along with the required stabilizers, modifiers, plasticizers, dyes, or other additives. Concentration of resin and temperature are critical factors to the film-forming process. The resin solution is pumped through filters to a hopper or die. It emerges onto a belt or drum for continuous processing, or may be passed into a cell or mold for *in situ* sheet production. When continuously cast, the solution is dried under rigorously controlled conditions of temperature and air flow. Care must be taken to avoid forming a skin of dried polymer on the surface which may impede further solvent migration. Also, gradients of solvent retention through the thickness are undesirable. In some processes, once the web is sufficiently strong, it may be removed from the drum or belt and then passed through a festoon dryer where both surfaces are exposed to the drying air flow. When drying is complete, the web is cooled, trimmed, and wound in conventional fashion. Band casting is used to produce cellulose acetate or triacetate film and sheet, and for specialized resins such as polysulfone.

Acrylic resins are often cell-cast (11). In this process no solvent is used, but the fluid monomers are passed into a cell or mold where polymerization takes place to form a solid sheet. Heavy sections can be produced in this fashion. Formation of cast acrylic is moving toward modified continuous-based casting to increase productivity. Drum casting on a commercial scale has almost disappeared.

Regeneration. The regeneration process is the oldest of the film-forming processes and was used exclusively for the manufacture of regenerated cellulose (cellophane) film from viscose. Highly purified cellulose is made into viscose by first converting it to alkali cellulose in strong caustic soda solution, pressing, and drying it into fluff. The fluff is aged, then treated with carbon disulfide to form cellulose xanthate, which is dissolved in dilute caustic to produce viscose solution (12). The cellophane is produced by extruding the viscose solution through a slot

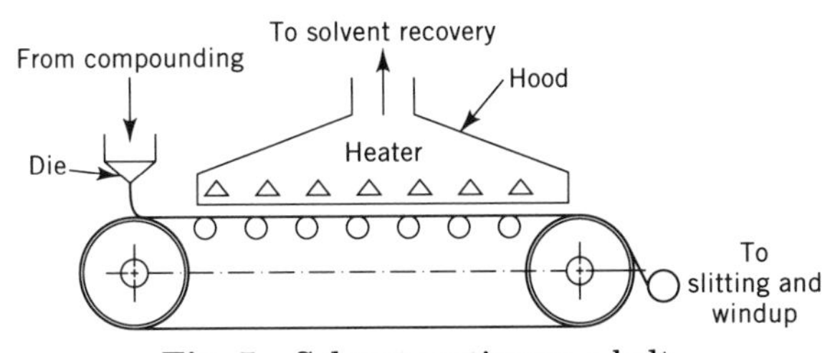

Fig. 5. Solvent casting on a belt.

die of controlled gap size into a regeneration bath consisting of dilute sulfuric acid and sodium sulfate salt, where the viscose coagulates and the cellulose is regenerated. The web then passes through a series of baths where it is washed to remove acid, salts, and other materials. It is desulfurized, washed, bleached, washed again, and then sent to a conditioning bath where glycerol solution is added to plasticize the web. It is then dried to a desired moisture level. Each step of the process must be carefully controlled since the film properties and performance may be influenced by the type and molecular weight of cellulose, the effectiveness of washing out of impurities, the plasticizer added, the rates of regeneration, the degree of stretching and orientation, and the moisture level.

Almost all cellophane is coated with either a moisture barrier or heat-sealable coating. This is generally done in tower coaters after film production.

Casting and Lamination. Film and sheet are often coated or laminated to enhance their functional performance. Coatings may be applied by melt extrusion, solvent-based solution, water-based solution or emulsions, or vacuum deposition, in-line during the film- or sheet-forming process, or off-line as a secondary step (13). In-line coating has gained favor because it offers improved bonding, lower coating weights, and improved economics. Off-line coating may be done by the film or sheet manufacturer, but is more frequently done by their customers. Lamination is the combination of two or more web materials by melt or adhesive bonding.

In extrusion coating a polymer is extruded from a slot die into the nip of two rolls where it is bonded to a substrate under pressure (Fig. 6). A corona discharge may be applied to the substrate just prior to the nip to enhance adhesion. Polyethylene or ionomer are the most common resins used in extrusion coatings. They provide improved moisture barrier (on paper), or sealability (on foil, polypropylene, or polyester). When a second substrate is introduced to the nip, laminated structures may be produced.

Solution or emulsion coatings are used to produce film or sheet materials with special characteristics. The coatings that may be applied include a broad

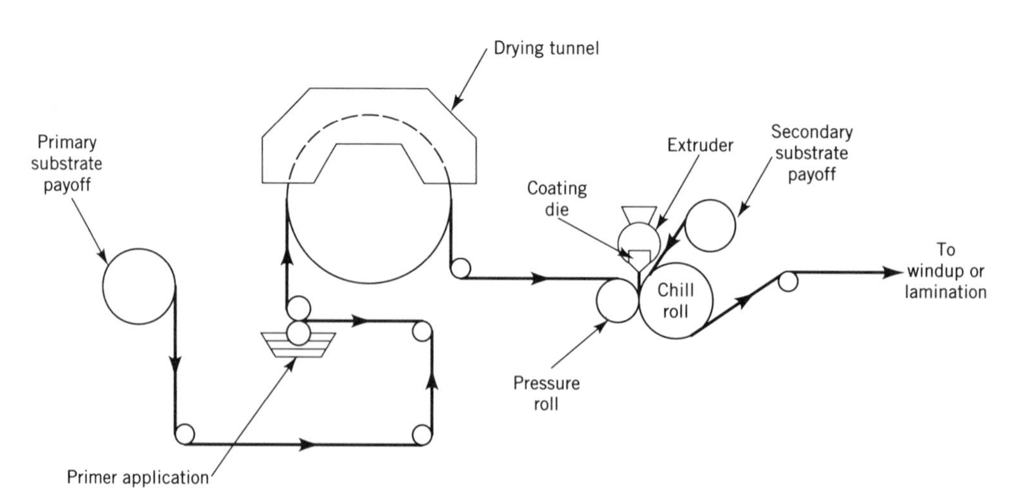

Fig. 6. Extrusion coating on a substrate.

range of polymers, metals, oxides, light-sensitive compounds, pigments, inks, etc. They are designed to improve such properties as adhesion, printability, slip or handleability, sealability, antistatic behavior, water-vapor barrier, gas and odor barrier, opacity, color, conductivity, release characteristics, magnetic susceptibility, light sensitivity, light resistance, and selective absorption or reflection. Such coatings may be in one or several layers, and vary from 1 nm in thickness to a fraction of a mm.

These coatings (qv) are applied to the substrate using gravure roll coaters, reverse roll coaters, wire-wound rod coaters, knife coaters, kiss coaters, or spray coaters, to name a few methods (see COATING PROCESSES). Excess coatings are metered or doctored off the web, and then the solvent is removed by drying in an oven. Radiation drying may also be used, and often the coating is cured or cross-linked to make it permanent. The web may be supported on rollers, a drum, continuous belt, or by air-flotation during drying, depending on the number of sides coated and the nature of the coating system.

The use of vacuum-based systems to apply ultrathin coatings to various film and sheet substrates has grown in importance in recent years, and represents an area where new technology is still appearing. The film is placed in a chamber and a high vacuum is drawn. The film is unwound and passed through one or more zones in which the vaporized coating is applied. Metals may be vaporized by use of resistance heat or ionizing radiation and deposited on the surface to give metallized film products. In other processes, plasmas are used to activate coatings that result in oxides or nitrides being deposited on the film that are useful in selective light-absorption coating. Ion implantation is another technique available. Sputtering of a material may be used to vaporize it for deposition on the substrate surface.

Economic Aspects

Table 6 shows the sales estimates for principal film and sheet products for the year 1990 (14). Low density polyethylene films dominate the market in volume, followed by polystyrene and the vinyls. High density polyethylene, poly(ethylene terephthalate), and polypropylene are close in market share and complete the primary products. A number of specialty resins are used to produce 25,000–100,000 t of film or sheet, and then there are a large number of high priced, high performance materials that serve niche markets. The original clear film product, cellophane, has fallen to about 25,000 t in the United States, with only one domestic producer. Table 7 lists some of the principal film and sheet material manufacturers in the United States.

Packaging (qv) represents the largest market area for film and sheeting materials (15). It is a complex market with so many categories that it is difficult to get an accurate measure of end usage for specific materials (16). The structure of the marketplace which uses both monolayers of film, as well as converted composite structures and laminates, adds to the complexity. The ultimate user or packager may purchase raw film directly from a manufacturer, or use the same film laminated to one or more other films or substrates through a converter. The

Table 6. U.S. Film and Sheet Sales,[a] 1990, 10^3 t

Material	Film		Sheet	Total
	Packaging	Other		
acrylics				
cell-cast			42	42
continuous cast			30	30
extruded		2	93	95
cellulosics		20[b]	7	27
cellophane	24	5		29
HDPE	345	70	150	565
LDP	1700	1140	118	2956
LLDP	560	730	8	1298
polypropylene	180	60	60	300
polyester (PET)	22	265	57	344
nylon	24	6	5	35
polycarbonate		15	54	69
polystyrene				
crystal		20	600	620
foam			335	335
poly(vinyl chloride)				
calendered				475
extruded	140		23	164
poly(vinyl alcohol)	2			2
ionomer	12			12
polyimide		1		1
fluorocarbons		6		6

[a]Based on estimates of resin sales for film and sheet.
[b]Includes estimate of captive use for photographic film and pressure-sensitive tape.

converter may buy film or extrude his own supply. Resin sales to film producers do not always correlate with their film sales, because of scrap and yield losses.

Food packaging, in turn, constitutes a significant segment of the packaging market. Films or sheeting materials are used to package meat and poultry, snack foods, baked goods, dairy products, prepared foods, frozen foods, produce, cereals, and tobacco, to list some of the market areas. Each category has its own requirements for protection and preservation of its food contents, which makes necessary the diverse line of products used for packaging.

Nonfood packaging categories include drugs and pharmaceuticals, textile goods, bags (merchandise, shipping, carry-out, T-shirt, and laundry), shrink wrap, stretch wrap, clear boxes, and many more.

The largest nonpackaging markets for film and sheet include agricultural (mulching) and construction film, trash liners and bags, pressure-sensitive tape base, insulation board, and diaper liners. These are relatively low cost, high volume uses. There are a large number of premium end uses for film and sheet that demand high performance, long life, or high purity. These uses are filled by higher value films such as polyester, BON, BOPP, polycarbonates, acrylics, fluorocarbons, polyurethanes, polysulfone, polyimides, and many new products. These

Table 7. U.S. Manufacturers or Distributors of Film and Sheet

Material	Manufacturer	Trade name
acrylic film and sheet	CYRO Industries	
	Bristech Chemical Corp.	
	Polycast Technology Corp.	
cellophane	E. I. du Pont de Nemours & Co., Inc.	
cellulosics	Eastman Kodak Co.	Kodacel
	Flex-O-Glass, Inc.	
	Technical Plastics Extruders	
	Polymer Extruded Products	
fluoroplastics	Allied Signal Corp.	Aclar
	E. I. du Pont de Nemours & Co., Inc.	Tedlar, Tefrel, Teflon
	Fluoro Plastics Inc.	
nylon	Allied Signal	Capran
	E. I. du Pont de Nemours & Co., Inc.	Dartak
polycarbonate	General Electric Co.	Lexan
	Miles Inc.	Makofol
polyester	E. I. du Pont de Nemours & Co., Inc.	Mylar
		Hostaphan
	Hoechst Diafoil	Melinex
	ICI Americas, Inc.	Lumirror
	Toray	
polyethylene	Exxon Chemical Co., Film Products	
	Bemis Co., Inc., Film Division	
	AEP Industries, Inc.	
	Presto Products Co.	
	James River Corp.	
	Cryovac Division, W.R. Grace & Co.	
	Continental Extrusion Corp.	
	Georgia Pacific	
	Mobil Chemical Co.	
polypropylene	Hercules, Inc.	Bicor
	Mobil Chemical Co.	
	Quantum Performance Films	
	Toray Plastics America, Inc.	
polystyrene	Dow Chemical Co.	Opticite
	Mobil Chemical Co.	
poly(vinyl chloride)	Borden Packaging & Industrial Products	
	Hüls America, Inc.	
	American Mirex Corp.	
	Klockner Pentaplast of America, Inc.	
polyimides	E. I. du Pont de Nemours & Co., Inc.	Kapton
		Apecal
	Allied Signal Corp.	Upilex
	ICI Americas, Inc.	

Table 8. Area Yields and Pricing,[a] 1990

Film material	Density, g/mL	Area yield,[b] m^2/kg[c]	Pricing By weight, \$/kg	By area, ¢/m^2[d]
acrylic	1.26	31.2	13.20	42.3
cellulose acetate	1.30	30.3	7.15	23.6
cellophane	1.44	27.3	4.77	17.5
HDPE	0.95	41.4	2.99	7.2
LDPE	0.915	43.0	1.32–2.20	3.1–5.1
polypropylene				
OPP	0.905	43.5	3.85	8.9
BOPP	0.89	44.2	2.53	5.7
polyester	1.39	28.3	4.84	17.1
nylon	1.13	34.8	7.37	21.2
polycarbonate	1.20	32.8	9.13	27.9
PVC-shrink	1.30	30.3	5.50	18.2
polystyrene	1.05	37.5	2.97	7.9
ionomer	0.95	41.4	3.63	8.8
fluorocarbons	2.20	17.9	22.00	123
polysulfone	1.30	30.3	33.00	109
polyimide	1.42	27.7	128	461

[a]Prices vary by grades, special treatments, or other variables.
[b]Film thickness = 25.4 μm = 1 mil.
[c]To convert m^2/kg to 1000 $in.^2/lb$, multiply by 0.705.
[d]To convert ¢/m^2 to ¢/1000 $in.^2$, multiply by 0.645.

markets include magnetic tape base, photographic film base, microfilm, drafting and layout bases, color separation films, cartoon cells, imaging films, electrical and electronic insulation or support films, capacitor dielectric, permanent labels, solar control window films, laminates for safety glazing, and high impact glazing.

Sheeting materials are utilized in a wide variety of applications (17). About 30% of the sheet market is foamed sheet, mostly all polystyrene, used for egg trays, food and drink containers, and insulation. About 10% of the market for solid sheeting is used for glazing, lighting, and outdoor signs. This market has seen growth with the need for high security, impact-resistant glazing for banks, storefronts, ice rinks, and transportation. Polycarbonate, acrylics, and cellulosics dominate these markets. Thermoformed applications account for about 40% of the use for sheet. Vacuum-formed or pressure-formed products include blister and skin packaging, food and drink containers (cups, tubs, trays, and bowls for single use), toys, auto and appliance parts, and luggage. Polystyrene, polypropylene, HDPE, thermoplastic polyester, ABC, and vinyls are used in these thermoforming markets.

Typical area yields and pricing (1990) are shown in Table 8. Conventional converting practice considers area yield, ie, area per weight of film of specified thickness. In this fashion easy comparison between materials or combinations of material can be made to determine cost of packaging materials. As film thickness diminishes for a given material, price per given weight may increase dramatically to cover production costs and lower production yields. Area pricing is affected to a lesser extent.

BIBLIOGRAPHY

"Film Materials" in *ECT* 2nd ed., Vol. 9, pp. 220–244, by D. T. Surprenant, Union Carbide Corp.; "Film and Sheeting Materials" in *ECT* 3rd ed., Vol. 10, pp. 216–246, by E. L. Crump, Gulf Oil Chemicals Co.

1. R. L. Dimarest, *J. Plast. Film Sheeting* **8**(2), 109 (1992).
2. *Modern Plast.*, 550–554 (mid-Oct. 1991); (Jan. 1991).
3. *Packaging Encyclopedia*, Cahners Publishing Co., Newton, Mass., 1988, pp. 54–57.
4. W. Daniels in J. I. Kroschwitz, ed., *Encyclopedia of Polymer Science and Engineering*, 2nd ed., Vol. 17, John Wiley & Sons, Inc., New York, 1989, pp. 393–425.
5. G. A. Kruder in Ref. 4, Vol. 6, 1986, pp. 571–631.
6. W. J. Schrenk and E. W. Veazey in Ref. 4, Vol. 7, 1987, pp. 106–127.
7. H. C. Park and E. M. Mount III in Ref. 4, Vol. 7, 1987, pp. 88–106.
8. A. W. M. Coaker in Ref. 4., Vol. 2, 1985, pp. 606–622.
9. A. Hannachi and E. Mitrontes, *J. Plast. Film Sheeting* **5**(2), 104 (1989).
10. J. L. White and M. Cakmak in Ref. 4, Vol. 10, 1987, pp. 619–636.
11. W. C. Harbison in Ref. 4, Vol. 2, 1985, pp. 692–707.
12. C. C. Taylor in Ref. 4, Index Vol., 1990, pp. 11–17.
13. J. C. Colbert, ed., *Modern Coating Technology*, Noyes Data Corp., Park Ridge, N.J., 1982.
14. *Facts and Figures of the U.S. Plastics Industry*, Society of Plastics Industry, Washington, D.C., 1988.
15. C. J. Benning, *Plastic Films for Packaging*, Technical Publishing Co., Lancaster, Pa., 1983.
16. R. W. Hall, *J. Plast. Film Sheeting*, **1**, 56 (1985).
17. K. J. Mackenzie in Ref. 4, Vol. 15, 1989, pp. 146–167.

General References

Chemical Economics Handbook, SRI International, Menlo Park, Calif.
Chemical Marketing Reporter
Journal of Plastic Film and Sheeting
Journal of Polymer Science and Engineering
Modern Plastics, McGraw-Hill Book Co., Inc., New York. Particularly the annual market review in the January issues.
Packaging, Cahners Publishing, Newton, Mass. Publishes annual encyclopedia issue.
Paper, Film and Foil Converter, Maclean Newton Publishing Co., Chicago, Ill.
Plastics Technology, Bill Communications, Inc., New York. Publishes annual encyclopedia issue.
Plastics World, Cahners Publishing, Newton, Mass.

K. J. MACKENZIE
Consultant

FILM THEORY OF FLUIDS. See MASS TRANSFER.

FILTERS, OPTICAL. See OPTICAL FILTERS.

FILTRATION

Filtration is the separation of two phases, particulate form, ie, solid particles or liquid droplets, and continuous, ie, liquid or gas, from a mixture by passing the mixture through a porous medium. This article discusses the more predominant separation of solids from liquids. Filtration of solid particles or liquid droplets from gases is dealt with elsewhere (see AIR POLLUTION CONTROL METHODS). The oldest recorded applications of filtration are the purifications of wine and water practiced by the ancient Greeks and Romans. Cake filters, such as the rotary vacuum filter and the filter press, were developed much later from the necessity to filter sewage.

Filtration is often referred to as mechanical separation because the separation is accomplished by physical means. This does not preclude chemical or thermal pretreatment used to enhance filtration. Although some slurries separate well without chemical conditioning, most pulps of a widely varying nature can benefit from pretreatment (see FLOCCULATING AGENTS).

In a filtration system (Fig. 1) the porous filtration medium is housed in a housing, with flow of liquid in and out. A driving force, usually in the form of a static pressure difference, must be applied to achieve flow through the filter medium. It is immaterial from the fundamental point of view how the pressure difference is generated but there are four main types of driving force, ie, gravity, vacuum, pressure, and centrifugal. The two types of filtration used most often in practice are cake, or surface, filtration and deep bed filtration. This division usually is unambiguous but in some cases, such as in cartridge filters, there is no sharp dividing line.

Surface Filters. In surface filters (Fig. 2), the goal is to achieve separation on the upstream side of a relatively thin filter medium. The particles to be separated must be larger than the pores in the medium, ie, in strainers, membrane filters, etc, or the particles must approach the pores in large numbers and bridge over the pores, as in cake filters.

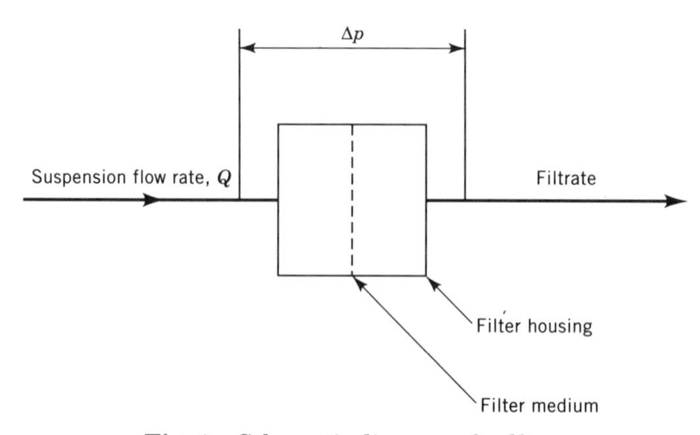

Fig. 1. Schematic diagram of a filter.

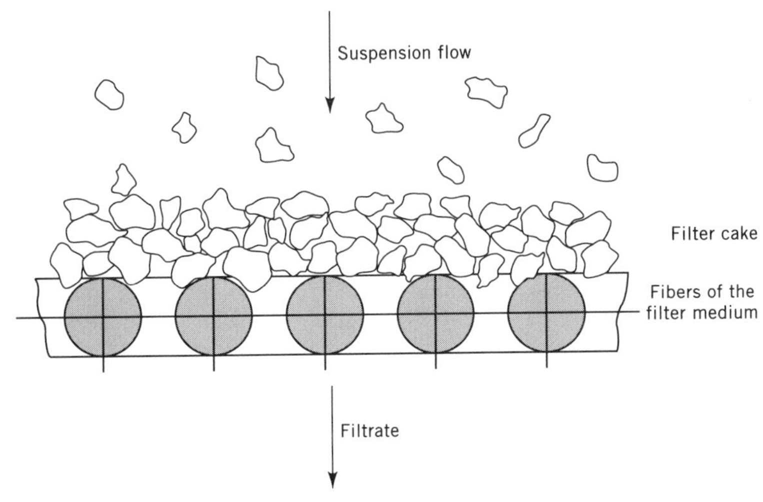

Fig. 2. Schematic diagram of a surface filter, ie, the cake filtration mechanism.

In cake filters, the medium only begins the process. The successive layers of solids deposit on top of the first layer and form a cake; the medium merely supports the cake thereafter. Cake filtration is the most widely used (ca 1993) liquid filtration in the chemical industry. It favors relatively high feed solids concentrations which lead to more porous cakes and better bridging over the pores; hence more open filter media can be used. Cake filters are best used for filtration of suspensions of solids concentrations in excess of 1% by volume to minimize the medium blinding which occurs in the filtration of dilute suspensions. If cake filtration is to be applied to clarification of liquids, which implies low feed concentrations of solids, addition of filter aids is usually necessary.

Cake filters are used in clarification of liquids, recovery of solids, dewatering of solids, thickening of slurries, and washing of solids. They are classified into vacuum, pressure, and centrifugal filters, depending on generation of the required pressure drop across the filter medium. Gravity static head rarely is used in cake filtration, except perhaps for the initial dewatering of highly flocculated sludges in belt presses.

The most important disadvantage of conventional cake filtration, ie, the so-called dead-end filtration where the flow approaches the filter medium at 90°, is the declining rate due to the increasing pressure drop caused by the growth of the cake on the filter medium. The high flow rate of the liquid through the medium can be maintained if no or little cake is allowed to form on the medium; this leads to the thickening of the slurry on the upstream part of the medium. The filters based on this principle are sometimes called filter-thickeners. There are several ways of limiting cake growth but the most widely used method is cross-flow filtration. Here the slurry moves tangentially to the filter medium so that the cake is continuously sheared off (Fig. 3).

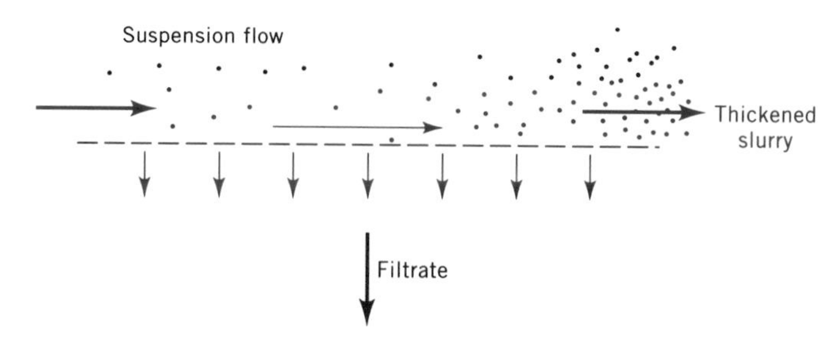

Fig. 3. Schematic diagram of cross-flow filtration.

Deep Bed Filters. Deep bed filtration is fundamentally different from cake filtration both in principle and application. The filter medium (Fig. 4) is a deep bed with pore size much greater than the particles it is meant to remove. No cake should form on the face of the medium. Particles penetrate into the medium where they separate due to gravity settling, diffusion, and inertial forces; attachment to the medium is due to molecular and electrostatic forces. Sand is the most common medium and multimedia filters also use garnet and anthracite. The filtration process is cyclic, ie, when the bed is full of solids and the pressure drop across the bed is excessive, the flow is interrupted and solids are backwashed from the bed, sometimes aided by air scouring or wash jets.

To keep the frequency of backwash and the washwater demand down, and to prevent undesirable cake formation on the filter surface, deep bed filtration is applied to very dilute suspensions of solids concentrations less than 0.1% by volume.

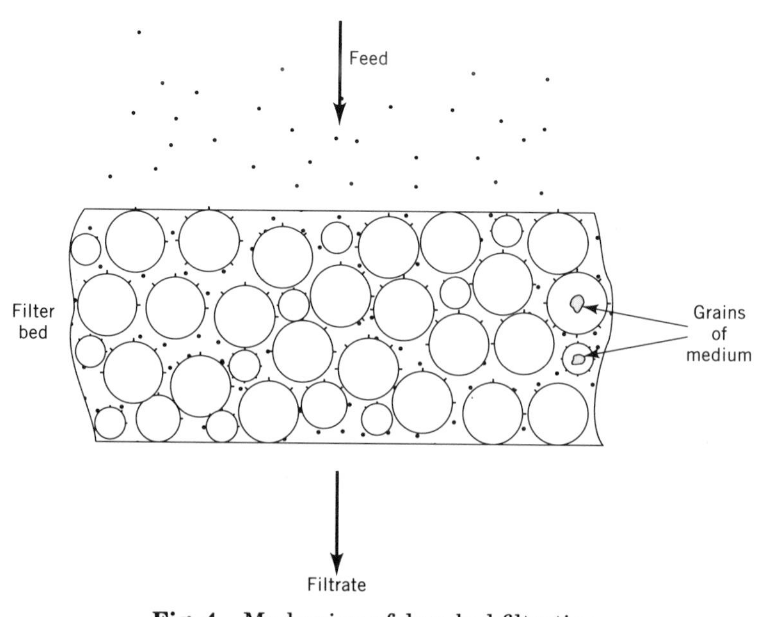

Fig. 4. Mechanism of deep bed filtration.

Deep bed filters were developed for potable water treatment as the final polishing process following chemical pretreatment and sedimentation. They are increasingly applied in industrial wastewater treatment under somewhat harsher operating conditions of higher solids loadings and more difficult backwashing of the media. Most deep bed filters are gravity-fed but some pressure types are used. The most common arrangement is the downflow filter (Fig. 5), with backwash in the upward direction to fluidize the bed.

Stratification of the particles making up the bed, caused by the fluidization (fines on top), is not desirable. The solids holding capacity of the bed is best utilized if the filtration flow encounters progressively finer sand particles. This is achieved in upflow filters where the fluidization due to backwash produces the correct stratification in the bed. Unfortunately, the filtration flow and the backwash take place in the same direction; the disadvantage is that the washwater goes to the clean side of the filter.

The trend in the use of deep bed filters in water treatment is to eliminate conventional flocculators and sedimentation tanks, and to employ the filter as a flocculation reactor for direct filtration of low turbidity waters. The constraints of batch operation can be removed by using one of the available continuous filters which provide continuous backwashing of a portion of the medium. Such systems include moving bed filters, radial flow filters, or traveling backwash filters. Further development of continuous deep bed filters is likely. Besides clarification of liquids, which is the most frequent use, deep bed filters can also be used to concentrate solids into a much smaller volume of backwash, or even to wash the solids by using a different liquid for the backwash. Deep bed filtration has a much more limited use in the chemical industry than cake filtration (see INDUSTRIAL WATER

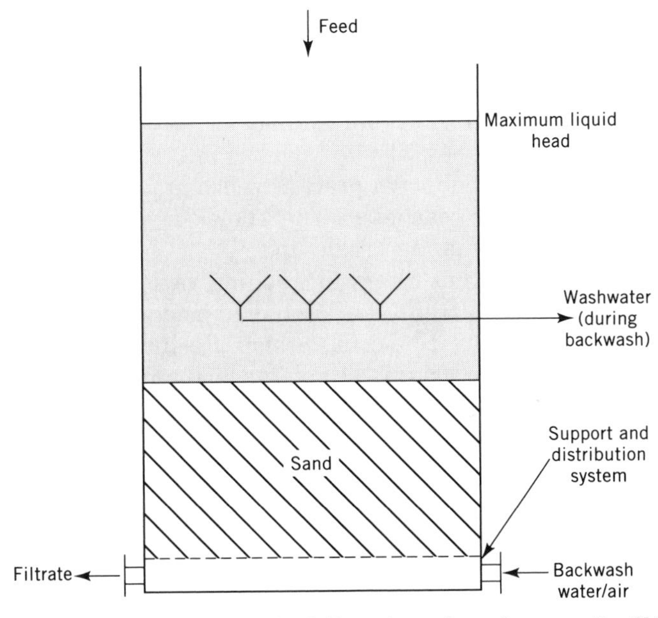

Fig. 5. Example of a deep bed filter, ie, a downflow gravity filter.

TREATMENT; MUNICIPAL WATER TREATMENT; WATER; WATER POLLUTION; and WATER REUSE).

Separation Efficiency. Similarly to other unit operations in chemical engineering, filtration is never complete. Some solids may leave in the liquid stream, and some liquid will be entrained with the separated solids. As emphasis on the separation efficiency of solids or liquid varies with application, the two are usually measured separately. Separation of solids is measured by total or fractional recovery, ie, how much of the incoming solids is collected by the filter. Separation of liquid usually is measured in how much of it has been left in the filtration cake for a surface filter, ie, moisture content, or in the concentrated slurry for a filter-thickener, ie, solids concentration.

Filtration-Related Processes

Several processes are used to enhance the filtration process itself. They may also be related processes in their own right.

Washing of Solids. Washing is a process designed to replace the mother liquor in the solids stream with a wash liquid. The growing importance of this process is due to demands for increased purity of the products combined with the increasingly poorer raw materials available. Washing often may represent a dominant portion of the installation cost because it is usually multistaged and often countercurrent.

The three types of washing are washing of filter cakes by displacement, washing by reslurrying of cakes or sludges, and washing by successive dilution.

Cake washing, used following cake filtration, uses small amounts of wash liquid, usually only a few times more than the volume of the mother liquor. The quality of cake washing is characterized by a washing curve that plots the ratio of the instantaneous to initial mother liquor concentration in the cake against the quantity of washing liquid used. The quantity of wash liquid usually is expressed as the wash ratio, ie, the number of void volumes in the cake. Ideally, the washing curve should be a step function at one void volume, therefore assuming ideal displacement of the mother liquor; in practice it is a gradually declining function that tends to zero but never reaches it. The knowledge of the washing curve is important for equipment design and scale-up.

The suitability of different filters to washing varies widely. Cake washing can be cocurrent or countercurrent, it often has to be done in several stages, and it is advantageous to enhance it by compression. Reslurry washing is a good alternative when the filtration cake is slimy, cracked, or generally not very permeable, and when the solids are not in cake form, eg, deposited in deep bed filters or in the matrix of a high gradient magnetic separator. The solids are simply reslurried from a cake or backwashed from a packed bed into the wash liquid.

Washing by successive dilution is used when the solids are separated into a slurry, such as in filter thickeners. The solids, thickened into a small amount of mother liquor, are diluted into a wash liquid and then separated again, diluted, separated, etc until clean of mother liquor. The consumption of the wash liquid can be reduced in countercurrent washing systems, sometimes referred to as countercurrent decantation. Cocurrent dilution washing, however, can be built into some dynamic filter-thickeners such as the Escher-Wyss filter.

Cake Dewatering. Dewatering (qv), identified as a separate entity in filtration, is used to reduce the moisture content of filter cakes either by mechanical compression or by air displacement under vacuum pressure or drainage in a gravitational or centrifugal system. Dewatering of cakes is enhanced by addition of dewatering aids to the suspensions in the form of surfactants that reduce surface tension.

In dewatering by mechanical compression, the prerequisite compressibility of the cake is usually expressed by an empirical exponential equation which relates cake voidage to applied pressure.

In dewatering by air (or other gas) displacement, the important quantities are threshold pressure, which has to be exceeded in order for air to enter the filter cake; irreducible saturation level, which gives the least moisture content achievable by air displacement; and kinetic dewatering characteristics. In dewatering by gravity or centrifugal action, the irreducible saturation and kinetic dewatering characteristics must be known for effective equipment design and scale-up.

The key to understanding dewatering by air displacement is the capillary pressure diagram. Figure 6 presents an example typical for a fine coal suspension; there is a minimum moisture content, about 12%, called irreducible saturation, which cannot be removed by air displacement at any pressure; and a threshold pressure, about 13 kPa.

The capillary retention forces in the pores of the filter cake are affected by the size and size range of the particles forming the cake, and by the way the

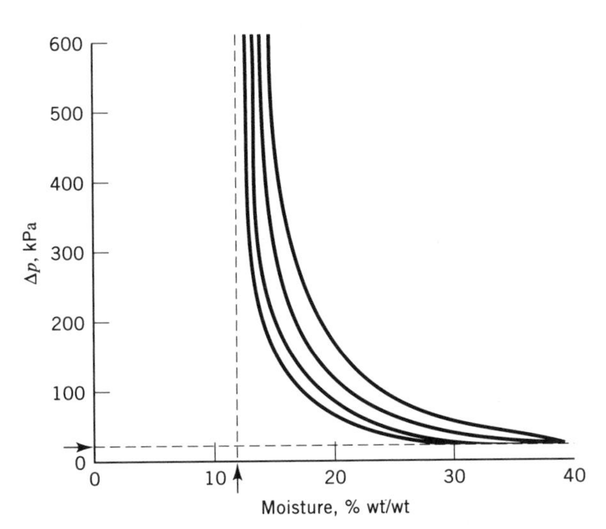

Fig. 6. Typical capillary pressure curves in air displacement cake dewatering of a fine coal suspension at varying dewatering times: from top curve down the time increases 25 s, 50 s, 100 s, ∞. The arrows and dotted lines indicate threshold pressure and irreducible saturation as explained in the text.

particles have been deposited when the cake was formed. There is no fundamental relation to allow the prediction of cake permeability but, for the sake of the order-of-magnitude estimates, the pore size in the cake may be taken loosely as though it were a cylinder which would just pass between three touching, monosized spheres. If d is the diameter of the spherical particles, the cylinder radius would be 0.0825 d. The capillary pressure of 100 kPa (1 bar) corresponds to d of 17.6 μm, given that the surface tension of water at 20°C is 72.75 mN/m (= dyn/cm).

Although most cakes consist of polydisperse, nonspherical particle systems theoretically capable of producing more closely packed deposits, the practical cakes usually have large voids and are more loosely packed due to the lack of sufficient particle relaxation time available at the time of cake deposition; hence the above-derived value of 17.6 μm becomes nearer the 10 μm limit when air pressure dewatering becomes necessary.

The lowest curve in Figure 6 gives the moisture content at different pressure drops which can be achieved given infinite time of dewatering. The dewatering kinetics are such that many such curves can be measured and drawn, depending on the time available to dewatering. Figure 6, or any such diagram, only applies to one cake thickness. Given a suitable filter cloth and sufficiently high pressure differential, the filter speed of a continuous pressure filter can be increased nearly to the speeds giving the minimum practicable cake thickness without increases in moisture content. A capillary pressure diagram obtained for a given cake thickness can be used for any thinner cake also.

The same moisture content of the produced cake can be obtained in shorter dewatering times if higher pressures are used. If a path of constant dewatering time is taken, moisture content is reduced at higher pressures with a parallel increase in cake production capacity. This is an advantage of pressure filtration of reasonably incompressible solids like coal and other minerals.

Pretreatment of Suspensions. Another important aspect of solid–liquid separation is conditioning or pretreatment of the feed suspension to alter some important property of the suspension and improve the performance of a separator that follows. A conditioning effect is obtained using several processes such as coagulation and flocculation, addition of inert filter aids, crystallization, freezing, temperature or pH adjustment, thermal treatment, and aging. The first two operations are considered in more detail due to their importance and wide use.

Coagulation and Flocculation. Both of these processes lead to increased effective particle size with the accompanying benefits of higher settling or flotation rates, higher permeability of filtraton cakes, and better particle retention in deep bed filters. Coagulation brings particles into contact to form agglomerates. The suspension is destabilized by addition of chemicals such as hydrolyzing coagulants like alum or ferric salts, or lime, and the subsequent agglomeration can produce particles up to 1 mm in size. Some of the coagulants simply neutralize the surface charges on the primary particles; others suppress the double layer, ie, indifferent electrolytes such as NaCl or $MgSO_4$; and some combine with the particles through hydrogen bridging or complex formation.

Flocculating agents (qv), usually in the form of natural or synthetic polyelectrolytes of high molecular weight, interconnect and enmesh colloidal particles into giant flocs up to 10 mm in size. Flocculating agents underwent fast development during the 1980s which has led to a remarkable improvement in the use and performance of many types of separation equipment. Flocculating agents are relatively expensive; the correct dosage is critical and has to be carefully optimized. Overdosage is not economical and may inhibit the flocculation process by coating the particles completely, with the subsequent restabilization of the suspension, or may cause operating problems such as blinding of filter media or mudballing and underdrain constriction in sand filters. Overdosage also may increase the volume of sludge for disposal. Optimum dosage has been found to correspond to the point where about one-half of the surface area of the particles is covered with polymer. Surface charges are affected by pH, and the control of pH is essential in pretreatment.

The natural process of bringing particles and polyelectrolytes together by Brownian motion, ie, perikinetic flocculation, often is assisted by orthokinetic flocculation which increases particle collisions through the motion of the fluid and velocity gradients in the flow. This is the idea behind the use of in-line mixers or paddle-type flocculators in front of some separation equipment like gravity clarifiers. The rate of flocculation in clarifiers is also increased by recycling the flocs to increase the rate of particle–particle collisions through the increase in solids concentration.

The type of floc required depends on the separation process which follows, eg, rotary vacuum filtration requires evenly sized, small, strong flocs that capture ultrafines to prevent cloth blinding and cloudy filtrates. The flocs should not be subject to sedimentation in the vat or breakage by the agitator. Such flocs are not likely to cause localized air breakthrough, cake collapse, shrinkage, or cracking in the dewatering stage.

In filtration operations which use initial gravity settling like the belt presses, large and loosely packing flocs are required. The resulting free-draining sediment can then be subjected to a controlled breakdown over a period of time, ultimately leading to a complete collapse of the cake due to mechanical squeezing between the belts.

In gravity thickening, large and relatively fragile flocs are needed to allow high settling rates and fast collapse in the compression zone.

The optimum flocculant type and dosage depend on factors such as solids concentration, particle size distribution, surface chemistry, electrolyte content, and pH value; the effect of these is very complex. Flocculant selection and dosage optimization require extensive experimentation with only general guidance as to the ionic charge or molecular weight (or chain length) required. Anionic flocculants, for example, are known to perform well on coal slurries because the relatively high levels of calcium ions and hydrogen bonding provide salt linkages for the anchorage of the polymer anionic groups to the coal surface. Zeta potential does not seem to be involved because neutral (nonionic) or wrong polyelectrolytes, ie, cationic rather than anionic, or vice versa, can be quite effective.

Addition of Inert Filter Aids. Filter aids are rigid, porous, and highly permeable powders added to feed suspensions to extend the applicability of surface filtration. Very dilute or very fine and slimy suspensions are too difficult to filter by cake filtration due to fast pressure build-up and medium blinding; addition of filter aids can alleviate such problems. Filter aids can be used in either or both of two modes of operation, ie, to form a precoat which then acts as a filter medium on a coarse support material called a septum, or to be mixed with the feed suspension as body feed to increase the permeability of the resulting cake.

In the precoat mode, filter aids allow filtration of very fine or compressible solids from suspensions of 5% or lower solids concentration on a rotary drum precoat filter. This modification of the rotary drum vacuum filter uses an advancing knife continuously to skim off the separated solids and the uppermost layer of the thick precoat until the whole of the precoat is removed and a new layer must be applied. This makes it possible to discharge very thin cakes as well as that part of the precoat which has been penetrated into by the solids and partially blinded.

In the precoat and body feed mode, filter aids allow application of surface filtration to clarification of liquids, ie, filtration of very dilute suspensions of less than 0.1% by volume, such as those normally treated by deep bed filters or centrifugal clarifiers. Filter aids are used in this mode with pressure filters. A precoat is first formed by passing a suspension of the filter aid through the filter. This is followed by filtration of the feed liquid, which may have the filter aid mixed with it as body feed in order to improve the permeability of the resulting cake. The proportion of the filter aid to be added as body feed is of the same order as the amount of contaminant solids in the feed liquid; this limits the application of such systems to low concentrations. Recovery and regeneration of filter aids from the cakes normally is not practiced except in a few very large installations where it might become economical.

Materials suitable as filter aids include diatomaceous earth, expanded perilitic rock, asbestos, cellulose, nonactivated carbon, ashes, ground chalk, or mixtures of those materials. The amount of body feed is subject to optimization, and the criterion for the optimization depends on the purpose of the filtration. Maximum yield of filtrate per unit mass of filter aid is probably most common but longest cycle, fastest flow, or maximum utilization of cake space are other criteria that require a different rate of body feed addition. The tests to be carried out for such optimization normally use laboratory or pilot-scale filters, and must include variation of the filtration parameters such as pressure or cake thickness in the optimization.

Mechanical Squeezing of Cakes. Mechanical squeezing of the cake in the so-called variable chamber filters has been used relatively recently to lower moisture content of the final cake. This is applicable only to cakes that are compressible. Many filters are available in which some form of mechanical expression of the cake is used either to follow a conventional filtration process or to replace it.

There are three basic ways to achieve mechanical expression, ie, with an inflatable diaphragm, with a compression belt, or in a screw conveyor of reducing pitch or diameter.

An inflatable diaphragm or membrane has been used in membrane plate presses closely related to the conventional plate and frame presses. A pressure filtration period is followed by compression with the hydraulically operated membrane or by a hydraulically operated ram if flexible rim seals are fitted. This principle also is used in vertical presses that use either one or two endless cloth belts indexing between plates. Inflatable membrane also may be used on a cylindrical filtration surface with or without a preceding pressure filtration stage.

Compression belts or rollers are sometimes added to conventional vacuum drum or belt filters but the most significant application is in the belt presses. These are usually horizontal belt filters, with one exception of a vertical belt press, which initially use gravity dewatering of the highly flocculated feeds, followed by compression with a second belt which gradually moves closer to the primary belt and squeezes the cake. The cake is also sheared by going through a series of meander rollers; this also aids dewatering.

Screw presses consist of a single or double screw conveyor that has perforated walls. The solids being conveyed are squeezed due to a gradual reduction in the pitch or diameter of the screw. Screw presses are used for dewatering of rough organic materials.

Electro-Kinetic Effects. The application of d-c potential in filtration or sedimentation is known to have a beneficial effect on the separation. Although this has been known and studied since the beginning of the nineteenth century, practical application and development have only accelerated since the late 1980s; commercial application is likely.

Several effects, due to the existence of the double layer on the surface of most particles suspended in liquids, can be used to measure the so-called zeta potential. Table 1 gives a simplified summary of the effects.

Electrophoresis (qv), ie, the migration of small particles suspended in a polar liquid in an electric field toward an electrode, is the best known effect. If a sample of the suspension is placed in a suitably designed cell, with a d-c potential applied across the cell, and the particles are observed through a microscope, they can all be seen to move in one direction, toward one of the two electrodes. All of the particles, regardless of their size, appear to move at the same velocity, as both the electrostatic force and resistance to particle motion depend on particle surface; this velocity can be easily measured.

Electroosmosis often accompanies electrophoresis. It is the transport of liquid past a surface or through a porous solid, which is electrically charged but immovable, toward the electrode with the same charge as that of the surface. Electrophoresis reverts to electroosmotic flow when the charged particles are made immovable; if the electroosmotic flow is forcibly prevented, pressure builds up and is called electroosmotic pressure.

Electrophoresis and electroosmosis can be used to enhance conventional cake filtration. Electrodes of suitable polarity are placed on either side of the filter medium so that the incoming particles move toward the upstream electrode, away

Table 1. Summary of Electro-Kinetic Effects

Effect	Fluid	Particles	Electric field
electrophoresis	still	moving	applied
sedimentation (or migration) potential	still	moving	measured
electroosmosis	moving	still	applied
streaming potential	moving	still	measured
electroosmotic pressure	still	still	applied

from the medium. As most particles carry negative charge, the electrode upstream of the medium is usually positive. The electric field can cause the suspended particles to form a more open cake or, in the extreme, to prevent cake formation altogether by keeping all particles away from the medium.

There is an additional pressure drop across the cake, developed by electroosmosis, which leads to increased flow rates through the cake and further dewatering at the end of the filtration cycle. The filtration theory proposed for electrofiltration assumes the simple superposition of electroosmotic pressure on the hydraulic pressure drop.

Dewatering of high value products and particle systems sensitive to high pressure drops are the most likely candidates for electrofiltration. The Dorr-Oliver Electrofilter is a commercial example of a vacuum filter adapted for electrofiltration.

Magnetic Separation. A magnetic field can be used as a means of removing ferromagnetic particles from a suspension. In mineral processing permanent magnets are used for removing tramp iron and for concentrating magnetic ores.

The application of magnetic separation using permanent magnets can be extended to filtration. Ceramic magnets, metal-alloy magnets, or magnetized steel balls are useful when placed into the flow for the separation of small ferromagnetic particles from hydraulic fluids. Such filters are simple in design and are generally cleaned manually by withdrawing the entire assembly and washing the collecting surfaces. Particles as fine as 1 μm can be removed at low flow rates, and some removal of nonferromagnetic contamination has been observed (1) due to its agglomeration with magnetic particles.

High Gradient Magnetic Separators. Much stronger magnetic fields and gradients are produced by the high gradient magnetic separators (HGMS) which can be used for the separation of very fine and weakly paramagnetic particles on a large scale (2). The design and operation of the HGMS is similar to deep bed filters (Fig. 7). The filtration takes place through a loosely packed bed of fine steel wool, at about 5% packing density, which is placed in a uniform magnetic field generated by an electromagnet. After the filtration stage, the field is switched off and backwash takes place to remove the particles separated on the steel wool. Although primarily batchwise in operation, HGMS can be designed for continuous operation by constructing the matrix into a carousel resembling a rotary table filter, which rotates slowly through a magnet head. As the matrix goes around, it becomes loaded with solids in the magnet head and is cleaned in a separate flush station.

The separation efficiency depends on the magnitude of the magnetic force on particles which is the product of the magnetic field strength, the gradient of the field, the magnetic susceptibility of the particle, and the particle volume. Only the field strength and the gradient are a matter of choice, and both are to be as high as possible. The commercially available HGMS produce fields up to 2 Tesla, using electromagnets with iron in the magnetic circuit. Higher fields up to 6 Tesla have been achieved with superconducting coils. The field gradient is increased by providing poles of convergence in the field, such as steel wires perpendicular to the field.

The field gradient is approximately equal to the saturation magnetization of the wire divided by the diameter of the wire. The gradient can therefore be

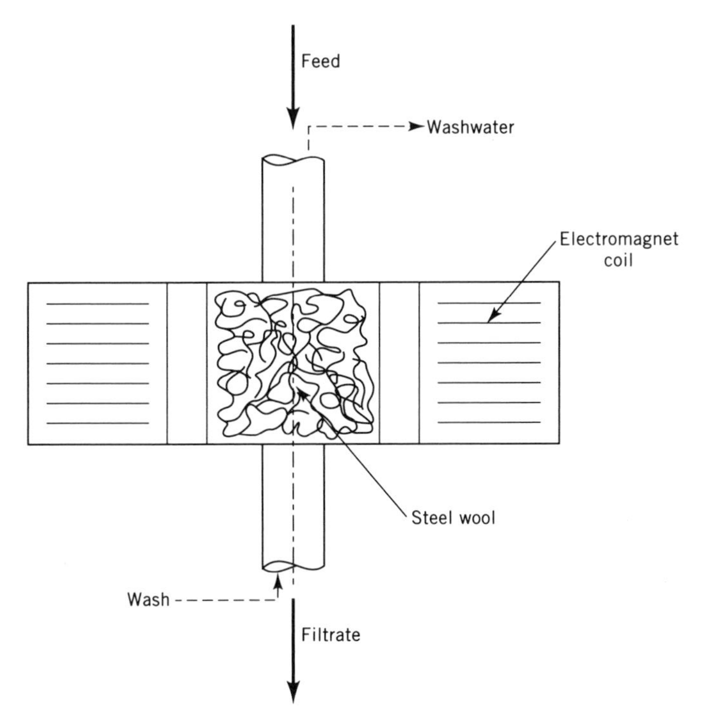

Fig. 7. Principle of high gradient magnetic separation.

increased by reducing the diameter of the steel wool fibers but the latter should not be less than the particle size to be collected since the highest magnetic force has been shown to occur (3) for wire diameters 2.7 times the particle diameter. This provides an argument for matching the size of the wire to the material to be separated. The most frequently used matrices in the clay industry, for example, are felted pads of stainless steel wire from 50 to 70 μm in diameter.

If paramagnetic particles are separated they can be easily washed from the matrix once the field is switched off because the magnetic forces are then small. If ferromagnetic particles are to be separated, there is some advantage in using paramagnetic collectors in order to keep the cleaning characteristics high but at some loss of efficiency of separation.

The main advantage of HGMS is high efficiency of separation even at relatively high flow rates and minimum pressure drops across the filter. The capital cost is very high, and only large installations are attractive economically because capacity increases with the square of the diameter of the canister while the weight of copper conductor increases linearly with diameter.

HGMS applications in industry include, but are not limited to, the brightening of kaolin clay by the removal of feebly magnetic contaminants, the purification of boiler feed water in power generation (both conventional and nuclear), and water treatment in steelmaking (4). Other potential users of HGMS (5) include beneficiation and desulfurization of coal, papermaking, nuclear fuel processing, separation of red blood cells, recovery of asbestos fibers from ducts (3), and potable water treatment. In the treatment of potable water, algae, bacteria, colloidal matter, dissolved colored material, and other essentially nonmagnetic materials are flocculated with, or adsorbed on, finely divided magnetite, ie, waste product of some iron ore processing plants, and the magnetic flocs are then separated in HGMS (6); the magnetite can be recycled after being stripped of the impurities and washed with alkali.

Most of the above applications are in clarification duties. The use of HGMS to dewater relatively concentrated, paramagnetic mineral slurries has been demonstrated on 2 to 12 wt % feed concentration of synthetic malachite (7) concentrated to 40%. The magnetic collection was optimized at flow velocities of 1 mm/s, and product concentrations greater than 40% were not possible unless the collected material could be removed from the matrix with less than the equivalent of one canister of washwater.

Cake Filtration Theory

It does not matter, from the fundamental point of view, how the pressure drop is generated in the filter. In the case of the centrifugal filters there is an additional phenomenon of the mass forces acting on the liquid within the cake. The conventional filtration theory must be amended to include this effect (2).

Carman-Kozeny Equation. Flow through packed beds under laminar conditions can be described by the Carman-Kozeny equation in the form

$$\frac{Q}{A} = \frac{\Delta p}{\mu L} \frac{\epsilon^3}{5\,(1 - \epsilon)^2\, S_0^2} \tag{1}$$

where Q is volumetric flow rate, A is face area of the bed, L is depth of the bed, Δp is applied pressure drop, ϵ is voidage of the bed (porosity), S_0 is volume specific surface of the bed, and μ is liquid viscosity.

The constant given the value 5 in equation 1 depends on particle size, shape, and porosity; it can be assumed to be 5 for low porosities. Although equation 1 has been found to work reasonably well for incompressible cakes over narrow porosity ranges, its importance is limited in cake filtration because it cannot be used for most practical, compressible cakes. It can, however, be used to demonstrate the high sensitivity of the pressure drop to the cake porosity and to the specific surface of the solids.

Darcy's Law and the Basic Filtration Equation. Darcy's law combines the constants in the last term of equation 1 into one factor K known as the permeability of the bed, ie,

$$K = \frac{\epsilon^3}{5\,(1 - \epsilon)^2\, S_0^2} \tag{2}$$

where K is a constant for incompressible solids. For compressible cakes, K depends on applied pressure, approach velocity, and concentration, and therefore presents serious problems in cake filtration testing and scale-up. There are some materials such as highly flocculated beds for which the above linear relationship between the face velocity and pressure drop does not hold; the flow is then called non-Darcian.

Modern filtration theory tends to prefer the Ruth form of Darcy's law, ie,

$$\frac{Q}{A} = \frac{\Delta p}{\mu R} \tag{3}$$

where R is the bed resistance. In cake filtration the bed resistance consists of the medium resistance in series with the resistance of the deposited cake, assuming no penetration of solids into the filtration medium; the general filtration equation is then written as

$$\frac{Q}{A} = \frac{\Delta p}{\mu R + \sigma \mu c\, V/A} \tag{4}$$

where σ is specific cake resistance, μ is liquid viscosity, c is solids concentration in the feed, V is filtrate volume collected since commencement of filtration, and R is medium resistance.

This equation is the basis of cake filtration analysis. Feed liquid flow rate and filtrate volume V are usually assumed to be related as

$$dV/dt = Q \tag{5}$$

The concept of the specific resistance used in equation 4 is based on the assumptions that flow is one-dimensional, growth of cake is unrestricted, only solid and liquid phases are present, the feed is sufficiently dilute such that the solids are freely suspended, the filtrate is free of solids, pressure losses in feed and filtrate piping are negligible, and flow is laminar. Laminar flow is a valid assumption in most cake formation operations of practical interest.

There is hidden assumption in equation 4 that the volume of the solids and liquid retained in the cake is negligible. This is reasonable at low concentrations but can lead to errors at high solids concentrations and moisture contents of cakes. A corrected value of the concentration c can be used in equation 4 to reduce the errors, ie,

$$c \text{ (corrected)} = \frac{1}{(1/c) - (1/\rho_s) - [(m - 1)/\rho]} \tag{6}$$

where m is the mass ratio of wet to dry filter cake, ρ_s is solids density, and ρ is liquid density. This correction is necessary only at feed concentrations of greater than 200 g/L.

The scale-up of conventional cake filtration uses the basic filtration equation (eq. 4). Solutions of this equation exist for any kind of operation, eg, constant pressure, constant rate, variable pressure–variable rate operations (2). The problems encountered with scale-up in cake filtration are in establishing the effective values of the medium resistance and the specific cake resistance.

The medium resistance R, which theoretically should be constant, often varies with time. This behavior results when some of the solids penetrate the medium or when it compresses under applied pressure. For convenience, the resistance of the piping and the feed and outlet ports is sometimes included in R.

The specific cake resistance σ is the most troublesome parameter; ideally constant, its value is needed to calculate the resistance to flow when the amount of cake deposited on the filter is known. In practice, it depends on the approach velocity of the suspension, the degree of flow consolidation that the cake undergoes with time, the feed solids concentration, and, most importantly, the applied pressure drop Δp. This changes due to the compressibility of most cakes in practice. σ often decreases with the velocity and the feed concentration. It may sometimes go through a maximum when it is plotted against solids concentration. The strongest effect on σ is due to pressure, conventionally expressed as:

$$\sigma = \sigma_o \, (\Delta p)^n \tag{7}$$

where σ_o is the cake resistance at unit applied pressure drop, Δp is the pressure drop across the cake, and n is an empirical exponent. Because each layer in a cake is subject to a different pressure drop, the only simple way to deal with the problem is to define an average value, σ_{av}. A rather unusual but widely used definition of the average is as follows:

$$\sigma_{av} = \Delta p \int_0^{\Delta p_c} \frac{d(\Delta p_c)}{\sigma} \tag{8}$$

which, when used with equation 7, gives:

$$\sigma_{av} = (1 - n) \cdot \sigma_o \cdot (\Delta p_c)^n \tag{9}$$

This can be substituted for σ in the basic filtration equation 4, which can then be solved for the filtration operation in question.

Traditionally, the average specific cake and medium resistances have been determined from constant pressure experiments and the solution of the basic filtration equation for constant pressure which relates filtrate volume to time. This relationship is, in theory, parabolic but deviations occur in practice.

The difficulties with approximations used in theoretical derivations and with the experimental techniques have been pointed out (8), and a series of tutorials have been published in which the deviations from parabolic behavior are explained and the correct interpretations of cake filtration data, with many numerical examples, are laid down (8,9).

Conventional filtration theory has been challenged; a two-phase theory has been applied to filtration and used to explain the deviations from parabolic behavior in the initial stages of the filtration process (10). This new theory incorporates the medium as an integral part of the process and shows that the interaction of the cake particles with the medium controls filterability. It defines a cake-septum permeability which then appears in the slope of the conventional plots instead of the cake resistance. This theory, which merely represents a new way of interpreting test data rather than a new method of sizing or scaling filters, is not yet accepted by the engineering community.

Benefits of Prethickening. The feed solids concentration has a profound effect on the performance of any cake filtration equipment. It affects the capacity and the cake resistance, as well as the penetration of the solids into the cloth which influences filtrate clarity and medium resistance. Thicker feeds lead to improved performance of most filters through higher capacity and lower cake resistance.

The effect on solids yield can be easily demonstrated using the following equation (neglecting medium resistance):

$$Y = \left[\frac{2\,\Delta pfc}{\sigma\mu t_c}\right]^{1/2} \tag{10}$$

where Δp is pressure drop, c is feed solids concentration, σ is specific cake resistance, μ is liquid viscosity, Y is the solids yield (dry cake production in kg/m^2s), f is ratio of filtration to cycle time, and t_c is cycle time (11). For the same cycle time, ie, the same speed, if the concentration is increased by a factor of four, production capacity is doubled. In other words, filtration area can be halved for the same capacity.

For given operating conditions and submergence, the dry cake production rate increases with the speed of rotation (eq. 10) and the limiting factor is usually the minimum cake thickness which can still be successfully discharged by the method used in the filter. Equation 11 shows the dependence of the solids yield on cake thickness:

$$Y = \frac{2\,\Delta pfc}{\sigma\mu L\,(1-\epsilon)\,\rho_s} \tag{11}$$

where L is cake thickness, ρ_s is solids density, and ϵ is cake porosity.

As can be seen, for constant cake thickness doubling the feed concentration doubles the yield. So-called high duty vacuum drum filters use a unique cake discharge method to allow very thin cakes to be discharged and can therefore be operated at very high speeds up to 25 revolutions per minute.

For those filters that allow variations in cake formation time within a fixed cycle time, such as the horizontal vacuum belt filter, prethickening shortens cake formation time for thicker feeds, thus giving more time for dewatering, washing, or other cake processing operations. On the other hand, it can be shown that in an optimum cycle time, at constant pressure operation, and where medium resistance is small compared with cake resistance, the cake formation time should be equal to the time the filter is out of service for cake dewatering, washing, or discharge. This result takes no account of any optimum operating conditions for minimum moisture content of the cake produced. In any case, this theoretical result may be outside the operational range governed by the design of a particular continuous filter, eg, design limitations on submergence in vacuum drum filters prevents the application of the above optimum condition. It can, however, be applied to any batch filter.

An additional benefit of prethickening is reduction in cake resistance. If the feed concentration is low, there is a general tendency of particles to pack together more tightly, thus leading to higher specific resistances. If, however, many particles approach the filter medium at the same time, they may bridge over the pores; this reduces penetration into the cloth or the cake underneath and more permeable cakes are thus formed.

This effect of concentration is particularly pronounced with irregularly shaped particles. A possible explanation of the variation in the specific resistance is in terms of the time available for the particles to orient themselves in the growing cake. At higher concentrations, but with the same approach velocities, less time, referred to as particle relaxation time, is available for a stable cake to form and a low resistance results.

An example of the concentration effect on the specific cake resistance is available (12) that reports results of some experiments with a laboratory horizontal vacuum belt filter. In spite of operational difficulties in keeping conditions constant, the effect of feed concentration on specific cake resistance is so strong that it swamps all other effects.

The function, fitted to the measured values for the aluminum hydroxide tested, was in the form:

$$\sigma = 13.47\ 10^{10}\ c^{-0.8031} \tag{12}$$

where σ is in m/kg and c is in g/L (kg/m^3) from 5 to 580 g/L.

The form of equation 12 is in keeping with a similar study (13) using a pressure filter to test a range of inorganic materials at constant pressure, ie, calcium carbonate, silicate, and sulfate, and two grades of magnesium carbonate. A range of values for the numerical constant in equation 12 from 9.14 to 37,800, and for the exponent from 0.08 to 1.92 with feed concentrations ranging from 5 to 190 kg/m^3 (g/L) were obtained.

The benefits of prethickening can be summarized as an increase in dry cake production, reduction in specific cake resistance, clearer filtrate, and less cloth blinding.

Further work is needed to build a physical model that allows prediction of the concentration effect from the primary properties of the slurry or from a limited amount of slurry testing.

Prethickening of filter feeds can be done with a variety of equipment such as gravity thickeners, hydrocyclones, or sedimenting centrifuges. Even cake filters can be designed to limit or completely eliminate cake formation and therefore act as thickening filters and be used in this thickening duty.

Pressure Filtration. High pressure drops have a twofold effect, ie, on capacity and on displacement dewatering which often follows.

The most important feature of the pressure filters which use hydraulic pressure to drive the process is that they can generate a pressure drop across the medium of more than 1×10^5 Pa which is the theoretical limit of vacuum filters. While the use of a high pressure drop is often advantageous, leading to higher outputs, drier cakes, or greater clarity of the overflow, this is not necessarily the case. For compressible cakes, an increase in pressure drop leads to a decrease in permeability of the cake and hence to a lower filtration rate relative to a given pressure drop.

This reduction in permeability due to cake consolidation or collapse may be so large that it may nullify or even overtake the advantage of using high pressures in the first place and there is then no reason for using the generally more expensive pressure filtration hardware. While a simple liquid pump may be cheaper than the vacuum pump needed with vacuum filters, if air displacement dewatering is to follow filtration in pressure filters, an air compressor has to be used and is expensive.

The fundamental case for pressure filters may be made using equation 10 for dry cake production capacity Y (kg/m^{2}s) derived from Darcy's law when the filter medium resistance is neglected. For the same cycle time (same speed), if the pressure drop is increased by a factor of four, production capacity is doubled. In other words, filtration area can be halved for the same capacity but only if σ is constant. If σ increases with pressure drop, and depending how fast it increases, the increased pressure drop may not give much more capacity and may actually cause capacity reductions.

For most industrial inorganic solids such as minerals etc, the increase in σ with Δp is not too great, and thus should the material to be filtered be too fine for vacuum filtration, pressure filtration may be advantageous and give better rates.

Pressure filters can treat feeds with concentrations up to and in excess of 10% solids by weight and having large proportions of difficult-to-handle fine particles. Typically, slurries in which the solid particles contain 10% greater than 10 μm may require pressure filtration, but increasing the proportion greater than 10 μm may make vacuum filtration possible. The range of typical filtration velocities in pressure filters is from 0.025 to 5 m/h and dry solids rates from 25 to 250 kg m^2/h. The use of pressure filters may also in some cases, such as in filtration of coal flotation concentrates, eliminate the need for flocculation.

Optimization of Cycle Times. In batch filters, one of the important decisions is how much time is allocated to the different operations such as filtration, displacement dewatering, cake washing, and cake discharge, which may involve opening of the pressure vessel. All of this has to happen within a cycle time t_c which itself is not fixed, though some of the times involved may be defined, such as the cake discharge time.

If all of the nonfiltration operations are grouped together into a downtime, t_d, assumed to be fixed and known, an optimum filtration time t_{opt} in relation to t_d can be derived by optimizing the average dry cake production obtained from the cycle. For constant pressure filtration and where the medium resistance R and the specific cake resistance σ are constant, the following equation applies:

$$t_{\text{opt}} = t_d \left[1 + \sqrt{\frac{2\mu R}{\sigma c \Delta p t_d}} \right] \tag{13}$$

where Δp and c are the operating pressure drop and the feed solids concentration, respectively (11).

When the medium resistance R is small compared with the specific cake resistance σ, the second term in the above equation becomes negligible and the optimum filtration time t_{opt} becomes equal to downtime t_d. For any other case, t_{opt} is always greater than t_d. It follows, therefore, that the filtration time should be at least equal to the sum of the other nonfiltration periods involved in the cycle.

Vacuum Filters

In vacuum filters, the driving force for filtration results from the application of a suction on the filtrate side of the medium. Although the theoretical pressure drop available for vacuum filtration is 100 kPa, in practice it is often limited to 70 or 80 kPa.

In applications where the fraction of fine particles in the solids of the feed slurry is low, a simple and relatively cheap vacuum filter can yield cakes with moisture contents comparable to those discharged by pressure filters. Vacuum filters include the only truly continuous filters built in large sizes that can provide for washing, drying, and other process requirements.

Vacuum filters are available in a variety of types, and are usually classified as either batch operated or continuous. An important distinguishing feature is the position of the filtration area with respect to gravity, ie, horizontal or nonhorizontal filtering surface.

A number of vacuum filter types use a horizontal filtering surface with the cake forming on top. This arrangement offers a number of advantages: gravity settling can take place before the vacuum is applied, and in many cases may prevent excessive blinding of the cloth due to action of a precoat formed by the coarser particles; if heavy or coarse materials settle out from the feed they do so onto the filter surface, and can be filtered; and fine particle penetration through the medium can be tolerated because the initial filtrate can be recycled back onto the belt. Top-feed filters are ideal for cake washing, cake dewatering, and other process operations such as leaching.

Horizontal filter surfaces also allow a high degree of control over cake formation. Allowances can be made for changed feeds and/or different cake quality requirements. This is particularly true of the horizontal belt vacuum filters. With these units the relative proportions of the belt allocated to filtration, washing, drying, etc, as well as the belt speed and vacuum quality, can be easily altered to suit process changes.

There are, however, two significant drawbacks to horizontal filters, ie, such filters usually require large floor areas, and their capital cost is high. Saving in floor area as well as in installed cost can be made by using a filter with vertical or other nonhorizontal filtration surfaces but at the cost of losing most, if not all, of the advantages of horizontal filters.

Nutsche Filter. The nutsche filter (Fig. 8) is simply an industrial-scale equivalent of the laboratory Buckner funnel. Nutsche filters consist of cylindrical or rectangular tanks divided into two compartments of roughly the same size by a horizontal medium supported by a filter plate. Vacuum is applied to the lower compartment, into which the filtrate is collected. It is customary to use the term nutsche only for filters that have sufficient capacity to hold the filtrate from one complete charge. The cake is removed manually or sometimes by reslurrying.

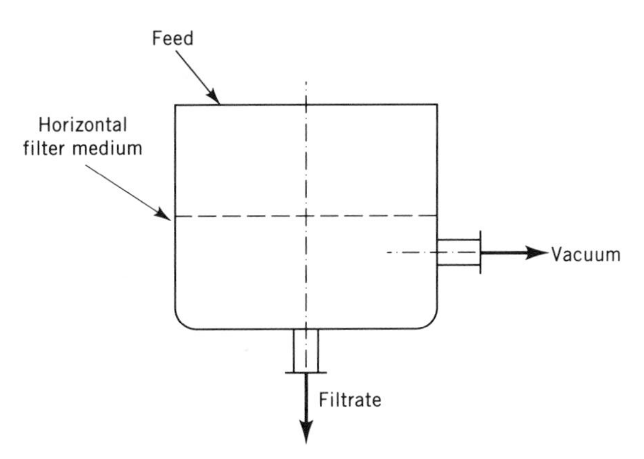

Fig. 8. Schematic diagram of a nutsche filter.

Depending on the chemical nature of the slurry to be filtered, materials of construction include wood, plastics, earthenware, steel, lead-lined steel, and brick-lined cast iron.

These filters are particularly advantageous when it is necessary to keep batches separate, and when extensive washing is required, although separation of wash from the mother liquor is difficult. Washing is carried out either by displacement or by reslurrying, the latter being more conveniently performed in the enclosed agitated nutsches described separately. If cake cracking occurs, the cracks can be closed manually for the subsequent washing.

Nutsche filters are simple in design, but laborious in cake discharge. They are prone to high amounts of wear due to the digging out operation. They are completely open and quite unsuited for dealing with inflammable or toxic materials. For such applications enclosed agitated nutsche filters are used. Throughputs are limited but the range of solids that can be filtered is very wide, ranging from easily filterable precipitates to dyestuffs, beta salt, etc.

Enclosed Agitated Vacuum Filters. These filters, often called mechanized nutsches, are circular vessels provided with a cover through which passes a shaft carrying a stirrer. The stirrer can sweep the whole area of the filter cake and can be lowered or raised vertically as required. During the filtration process the stirrer is in the top position, in motion if necessary to prevent the formation of an unequal layer of cake with fast settling suspensions. The suspension is fed through a connecting piece in the cover. When filtration is completed, any cracks that might occur in the filter cake can be smoothed over by lowering the stirrer on top of the cake and pasting over the cake to cover the cracks. The stirrer is rotated in the reverse direction for this operation and the pressing of the cake by the rotor blades might result in an additional dewatering, particularly with thixotropic cakes.

The filter cake can then be washed either by displacement or by reslurrying. Reslurrying is easily accomplished using the stirring action of the rotor blades when the rotor is lowered into the cake. The cake may also be dried *in situ* by the passage of hot air through it, or may be steam distilled for the recovery of solvent.

For discharge of the cake, a discharge door is provided at the edge and the cake is moved by the rotor toward the door, the stirrer blades gradually being lowered onto the surface so that a small depth of solids is scraped away. Alternatively, the cake may be reslurried and pumped away.

Enclosed agitated filters are useful when volatile solvents are in use or when the solvent gives off toxic vapor or fume. Another significant advantage is that their operation does not require any manual labor. Control can be manual or automatic, usually by timers or by specific measurements of the product. Most filters are made of mild steel, with the exposed surfaces protected by lead, tile, rubber lining, or by coating or spraying with other substances as necessary. Filtration areas up to 10 m^2 are available and the maximum cake thickness is 1 m. Applications are mainly in the chemical industry for the recovery of solvents.

The pressure version of the enclosed agitated filter is known as the Rosenmund filter; it uses a screw conveyor to convey the cake to a central cake discharge hole.

Vacuum Leaf Filter. The vacuum leaf, or Moor, filter consists of a number of rectangular leaves manifolded together in parallel and connected to a vacuum or compressed air supply by means of a flexible hose (Fig. 9). Each leaf is composed of a light pervious metal backing, usually of coarse wire grid or expanded metal set in a light metal frame and covered on each surface with filter cloth or woven wire cloth. The leaves, which are carried by an overhead crane during the filtration sequence, are dipped successively into a feed slurry tank, where the filtration takes place; a holding tank, where washing occurs; and a cake-receiving container, where cake discharge is performed by back-blowing with compressed air. An alternative arrangement is to move the tanks rather than the leaf assembly.

The operating cycle is seldom less than two hours, and several sets of frames can be operated in rotation. The cake thickness should be more than 3 mm, 9 mm being a typical value. Sluicing of the cake with a jet of compressed air has been used to permit thinner cakes and shorter filtration times. The leaves are spaced sufficiently far apart that there is always clearance between the finished cakes.

A static leaf filter is used for cleaning machine tool coolants. These are used on the suction side of a pump circulating system, with the same pump employed for withdrawal of the filtrate as for backflushing the filter elements. Solids in this case are removed from the sump by a scraper conveyor.

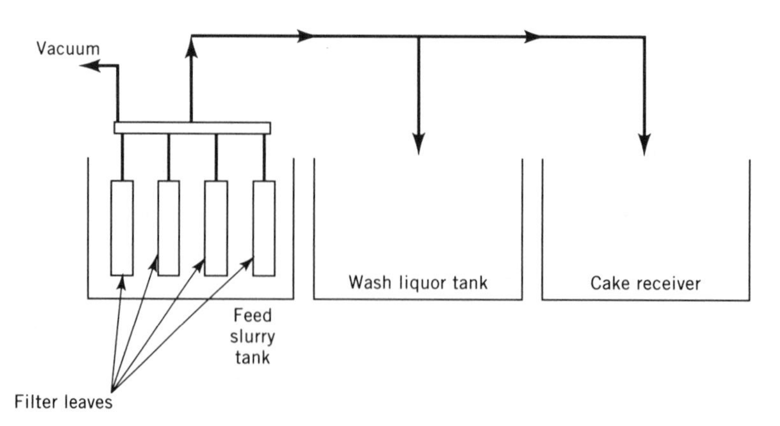

Fig. 9. Schematic diagram of the operation of a vacuum leaf filter.

Simple design, general flexibility, and good separation of the mother liquor and the wash are important advantages of vacuum leaf filters. On the other hand, they are labor-intensive, require substantial floor space, and introduce the danger of the cake falling off during transport. Vacuum leaf filters are particularly useful when washing is important, but washing is difficult with very fine solids, such as occur in titanium dioxide manufacture.

Tipping Pan Filter. This is a nutsche filter with a small filtrate chamber, in the form of a pan built so that it can be tipped upside down to discharge the cake. A separate vessel is used to receive the filtrate; this allows segregation of the mother and wash liquor if necessary.

A variation on this type of filter is the double tipping pan filter, which is a semicontinuous type consisting of two rectangular pans fitted with a filter cloth and pivoted about a horizontal axis. Slurry is first fed onto one pan, which is turned over for cake discharge at the end of the cycle. The second pan is used for filtration while the first is being discharged.

In general, pan filters are selected for freely filtering solids and thick filter cakes. Cake washing can be introduced easily. Most applications are in the mining and metallurgical industries for small-scale batch filtration.

Horizontal Rotating Pan Filters. These filters (Fig. 10) represent a further development of the tipping pan filter for continuous operation. They consist of a circular pan rotating around the central filter valve. The pan is divided into wedge-shaped sections covered with the filter medium. Vacuum is applied from below. Each section is provided with a drainage pipe which connects to a rotary filter valve of the same type as in drum filters. This allows each section, as it rotates, to go through a series of operations such as filtration, dewatering, cake washing, and discharge. Two basic designs exist, depending on the method of solids discharge.

The horizontal pan filter with scroll discharge incorporates a spiral scroll in a radial position just ahead of the feed zone. The scroll scrapes the cake off the medium. The need for a clearance between the scroll and the medium can cause incomplete cake discharge and consolidation of the remaining heel, due to the

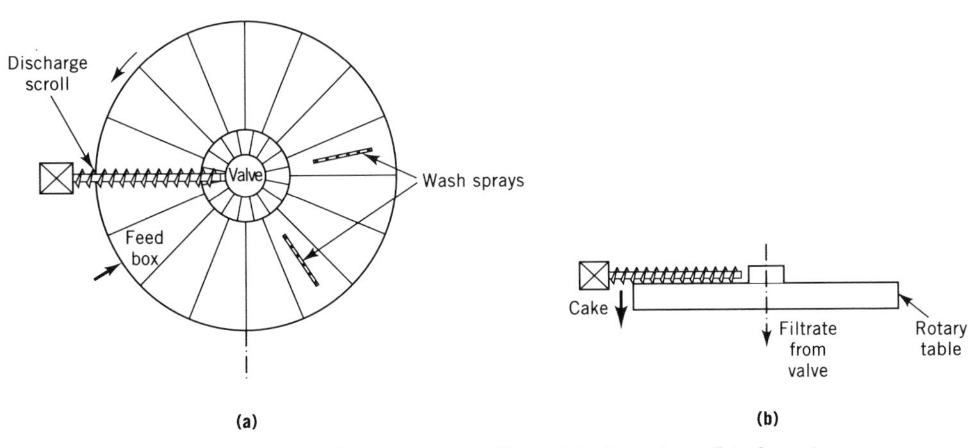

Fig. 10. Horizontal rotating pan filter: (**a**) plan view; (**b**) elevation.

action of the scroll. This problem may be overcome by injection of compressed air in the reverse direction through the cloth under the feed zone or by sucking air through the cloth in the opposite direction, combined with cloth washing. Filtration areas range from 0.5 to 100 m^2.

An important variation of this filter is based on replacing the rigid outer wall necessary for containing the feed and the cake on the rotating table by an endless rubber belt. The belt is held under tension and rotates with the table. It is in contact with the table rim except for the sector where the discharge screw is positioned, and where the belt is deflected away from the table to allow the solids to be pushed off the table. The cloth can also be washed in this section by high pressure water sprays. This filter, recently developed in Belgium, is available in sizes up to 250 m^2, operated at speeds of 2 minutes per revolution, and cake thicknesses up to 200 mm.

In horizontal rotary tilting pan filters, the wedge-like compartments are arranged as independent pans. Each is connected to the center valve by a swivel pipe joint that inverts the pan as it passes the discharge point. Air blowback is often used to assist the cake discharge. The units can also be adapted for cloth washing. Typical applications include filtration of gypsum from phosphoric acid and many mineral processing uses. Areas from 15 to 250 m^2 are available. The tilting pan is more expensive, requires more floor space, and has higher maintenance than the horizontal rotating pan filters. However, its advantages include excellent cake discharge and control of wash liquor, and the availability of larger sizes.

Horizontal Belt Vacuum Filters. This type of filter (Fig. 11) is another development of the pan filter idea. A row of vacuum pans arranged along the path of an endless horizontal belt was the original patented design. This has been superseded by the horizontal belt vacuum filter, which resembles a belt conveyor in appearance. The top strand of the endless belt is used for filtration, cake wash-

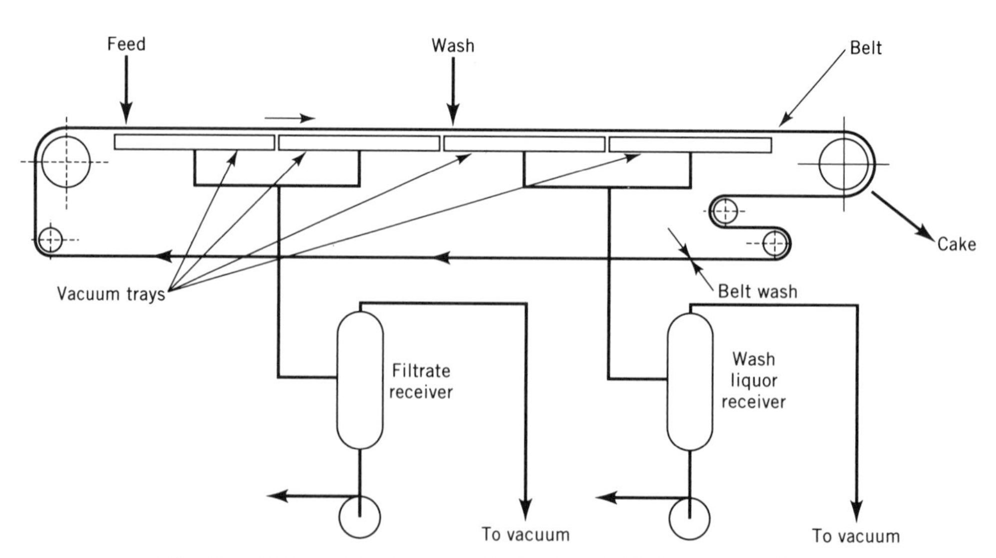

Fig. 11. Schematic diagram of a horizontal belt vacuum filter.

ing, and drying, whereas the bottom return is used for tracking and washing of the cloth. There is appreciable flexibility in the relative areas allocated to filtration, washing, and drying. Hooded enclosures are available wherever necessary. Modular construction of many designs allows field assembly as well as future expansion if process requirements change.

Horizontal belt filters are well suited to either fast or slowly draining solids, especially where washing requirements are critical. Multistage countercurrent washing can be effectively carried out due to the sharp separation of filtrates available. Horizontal belt vacuum filters are classified according to the method employed to support the filter medium.

One common design is typified by a rubber belt mounted in tension. The belt is grooved to provide drainage toward its center. Covered with cloth, the belt has raised edges to contain the feed slurry, and is dragged over stationary vacuum boxes located at the belt center. Wear caused by friction between the belt and the vacuum chamber is reduced by using replacable, secondary wear belts made of a suitable material such as PTFE, terylene, etc.

The rubber belt filters are available in large capacities with filtration areas up to 200 m^2 or more. They can be run at very high belt speeds, up to 30 m/min, when handling fast-filtering materials such as mineral slurries. The main disadvantages of rubber belt filters are the high replacement cost of the belts, the relatively low vacuum levels, and limitations on the type of rubber that can be used in the presence of certain solvents. The vacuum pumps are usually sized to give from 2 to 20 m/min air velocity over the filtration area, at approximately 21 kPa, depending on the filterability of the solids. The solids capacity ranges from 10 to 500 kg of solids per m^2 of filtration area per hour with cake thicknesses from 3 to 150 mm.

Another type of horizontal belt vacuum filter uses reciprocating vacuum trays mounted under a continuously traveling filter cloth. The trays move forward with the cloth as long as the vacuum is applied and return quickly to their original position after the vacuum is released. This overcomes the problem of friction between the belt and the trays because there is no relative movement between them while the vacuum is being applied. The mechanics of this filter are rather complex, and the equipment is expensive and requires intensive maintenance. A range of solvents can be used. Widths up to 2 m and areas up to 75 m^2 are available. The cloth can be washed on both sides.

The indexing cloth machines are a further development. In these, the vacuum trays are stationary and the cloth is indexed by means of a reciprocating discharge roll. During the time the vacuum is applied, the cloth is stationary on the vacuum trays. When the vacuum is cut off and vented, the discharge roll advances rapidly, moving the cloth forward 500 mm. The cycle is then repeated. As with the reciprocating tray types, the cloth can be washed on both sides. The cake discharges by gravity at the end of the belt when it travels over the discharge roll.

The primary advantages of this filter are its simple design and low maintenance costs. The main disadvantage is the difficulty of handling very fast-filtering materials on a large scale. Areas up to 93 m^2 are available.

Some horizontal belt vacuum filter designs incorporate a final compression stage for maximum mechanical dewatering. This is achieved by another compression belt which presses down on the cake formed in the preceding conventional filtration stage.

Rotary Vacuum Drum Filters. This is the most popular vacuum filter (ca 1993). There are many versions available and they all incorporate a drum which rotates slowly, approximately 1 to 10 minutes per revolution, about its horizontal axis and is partially submerged in a slurry reservoir (Fig. 12). The perforated surface of the drum is divided into a number of longitudinal sections of about 20 mm in thickness. Each section is an individual vacuum chamber, connected through piping to a central outlet valve at one end of the drum. The drum surface is covered with a cloth filter medium and the filtration takes place as each section is submerged in the feed slurry. A rake-type slowly moving agitator is used to keep the solids in suspension in the slurry reservoir without disturbing the cake formation. The agitator usually has a variable speed drive. Materials of construction of rotary drum filters include mild steel (sometimes rubber covered), stainless steel, nickel, tantalum, and plastics.

Filtration can be followed by dewatering, washing, and drying. In some applications, compression rolls or belts are used to close possible cracks in the cake before washing or to further dewater the cake by mechanical compression. Final dryness of the cake can also be enhanced by fitting a steam hood. Systems of cake discharge, all of which can be assisted by air blowback, include simple knife discharge, advancing knife discharge with precoat filtration, belt or string discharge, and roller discharge. The type of discharge selected depends on the nature of the material being filtered.

The scraper or knife discharge consists of a blade that removes the cake from the drum by direct contact with the filter cake. It is normally used for granular

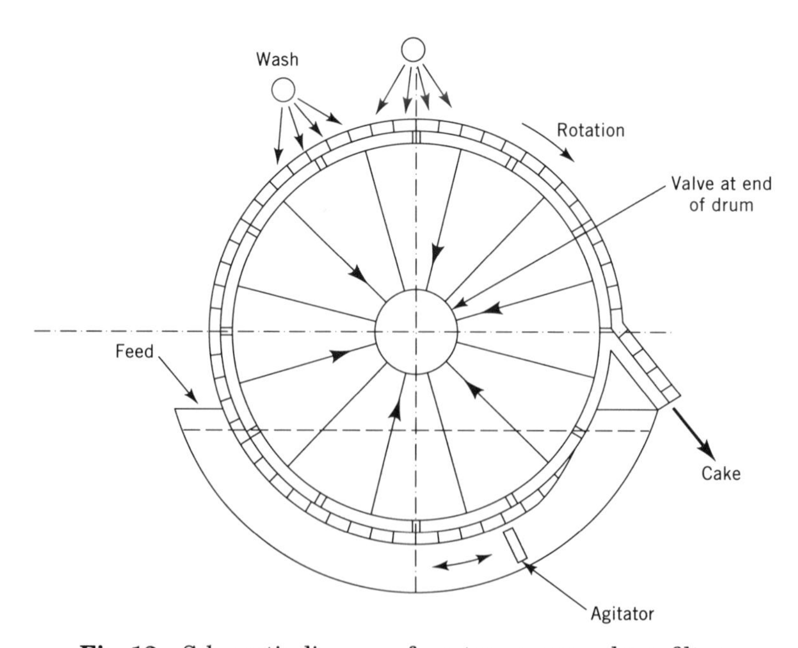

Fig. 12. Schematic diagram of a rotary vacuum drum filter.

materials with cake thickness greater than about 6 mm. In order not to damage the filter cloth, a safety distance of 1 to 3 mm between the blade and the cloth must be observed. If the residual layer is made not of filter aid but of the product, there is danger of its blocking by fine particles and by successive consolidation by the scraper blade.

The string discharge arrangement uses a number of parallel strings tied completely around the filter at a pitch of 1 to 2 cm, passing over the discharge and return rolls. As the strings leave the drum before the discharge point, they lift the filter cake from the medium and discharge it at the discharge roll. This type of discharge is recommended for gelatinous or cohesive cakes. Wire, chains, or coil springs have also been used. The coil spring discharge has been successfully applied to fibrous solids such as waste from a pulp mill or highly flocculated slurries such as in sewage treatment.

Another variation of the string discharge is the use of a thick plastic belt which is perforated by conical openings. The drum is covered with filter cloth and the belt covers the cloth for the filtration and dewatering operations. The solids fill the perforations in the belt and then leave the drum with the belt to be discharged by air blow as pellets. This is an effective way of pelletizing coarse mineral ores.

The cloth belt discharge is based on taking the cloth endless belt off the drum in the same way as with the string discharge. The advantage here is the ease of washing both sides of the cloth before the cloth returns to the drum. The disadvantage is in the need for an additional control device for the guidance of the cloth.

Roll discharge systems use a roll, which rotates at a slightly greater peripheral speed than the cake and is in contact with the cake. The cake is transferred from the drum to the discharge roll by adhesion which, by giving the roll a rough surface or because of the presence of residual cake on the roll, is designed to be greater to the roll than to the drum. The cake is usually removed from the roll by a knife. This type of discharge is designed to discharge tacky cakes that cannot be handled effectively by either of the previously described discharge designs. The cake thickness is small here, from 0.5 to 3 mm.

Multicompartment drum filters range in size from about 1 m^2 to over 100 m^2; they are widely used in mineral and chemical processing, in the pulp and paper industry, and in sewage and waste materials treatment.

Despite their theoretically poor washing performance, due to uneven wash distribution and excessive run-off because the filter surface is not horizontal, many multicompartment drum filters continue to be used as cake washing filters. Effective washing of the filter cloth can be done only with the belt discharge type, where the cloth leaves the drum for a brief period and can thus be washed on both sides.

Because in the most common bottom feed version the feed suspension enters the drum from the bottom, fast settling slurries are not suitable for most rotary drum filters; nor are very fine slurries because of inevitable penetration problems.

In units designed to use a precoat filter aid, the drum can be evacuated over the full 360° and fitted with an advancing knife system that continuously shaves off the deposited solids together with a thin layer of the precoat. The precoat has to be renewed periodically.

No internal piping and no conventional filter valve are needed with single-cell drum filters where the entire drum also operates under vacuum. The cake discharge is effected by air blowback from an internal stationary shoe mounted inside the drum at the point of discharge. There are very close tolerances between the inside surface of the drum and the shoe in order to minimize the leakage. The inside of the drum acts as a receiver for the separation of air and filtrate; conventional multicompartment drum filters require a separate external receiver. This type of filter permits operation of the filter with thin cakes so that high drum speeds, up to 26 rpm, can be used and high capacities can be achieved. Sizes up to 14 m^2 are available.

In most rotary drum filters, the submergence of the drum is usually about one-third of its circumference. Greater submergence is achieved in units equipped with submerged bearings, although this is more costly.

Another option available with rotary vacuum drum filters is full enclosure. This enables operation under nitrogen or other atmospheres, for reasons such as safety, prevention of vapor loss, etc. Enclosure may also be used to prevent contamination of the material being filtered or to confine the spray from washing nozzles. The rotary drum filter also can be enclosed in a pressure vessel and operated under pressure.

Disadvantages of the bottom feed arrangement can be overcome by using top-feed drum filters, which use the nearly horizontal surface on the top of the drum. The area available for filtration is small, however, and such filters are reserved for fast settling solids that dewater readily, and applications that permit precoating. Top-feed arrangements are common in the brewing industry. Drying may also be carried out in top-feed filters in totally enclosed systems.

A variation to the top-feed drum filter is the dual drum filter which uses two drums of the same size in contact with each other and rotating in opposite directions. The feed enters into the V-shaped space formed on top of the two drums and the cake that starts forming initially contains coarser particles due to the settling which takes place in the feed zone. This is beneficial to the clarity of the filtrate because the coarser particles act as a precoat. From the point of view, however, of the final moisture content of the cake the stratification of the solids in the cake may lead to somewhat wetter cakes. Utilization of the area of the drums is poor since there are dead spaces under the two drums. The primary application of the dual drum filters is in dewatering coarse mineral or coal suspensions at feed concentrations greater than 200 kg/m^3.

An interesting but not widely used variation is the internal drum filter which has the filtering surface on the inside of the drum. It has no slurry reservoir, as the feed is supplied to the inside of the drum. The filtrate piping is mounted externally at one end of the drum and terminates in a conventional filter valve. Cake discharge is by a scraper and a chute. Filters of this type are suited to handling fast settling materials and are cheaper in installed cost than conventional drum filters. Cake washing is impractical, however, because it must be done against gravity. There is also danger of the cake falling off and cloth replacement is difficult. Because the disadvantages of this filter tend to outweigh the advantages, this filter has been largely superseded by horizontal vacuum filters.

Rotary Vacuum Disk Filters. An alternative to the drum filter is the disk filter, which uses a number of disks mounted vertically on a horizontal shaft and suspended in a slurry reservoir. This arrangement provides a greater surface area for a given floor space, by as much as a factor of 4, but cake washing is more difficult and cloth washing virtually impossible.

As in the case of drum filters, each disk is divided into a number of separate segmental sectors, normally 12 but sometimes up to 30, that have suitable drainage and cloth support, and the sectors are connected to a filter valve similar to the type used in drum filters. It is necessary to operate disk filters at high submergence because a sector must be completely submerged during cake formation. There is also a top limit to the liquid level to allow the scraper to discharge the cake under gravity. The higher submergence, 40% or even higher (up to 55%), reduces the area available for dewatering.

Total submergence is used in the vacuum disk filter thickener (Fig. 13) in which the cake discharge, by backwashing with filtrate, occurs as each sector passes through the lowest point of the slurry tank.

In conventional disk filters, cake discharge is usually performed by a scraper blade, for cakes thicker than 10 mm, or sometimes by a tapered roll; air blowback is often used to assist the discharge. High pressure sprays also have been used for cake discharge.

As with drum filters, the slurry reservoir has to be agitated to prevent settling. Rather than using one large vat for all the disks, some designs provide individual low volume, closely fitting vats for each disk, thus avoiding the need for an agitator.

Speed of rotation is relatively high, from 20 to as much as 180 rpm. The disk size varies from 0.5 to 5.3 m, with up to 20 disks assembled on one shaft, providing

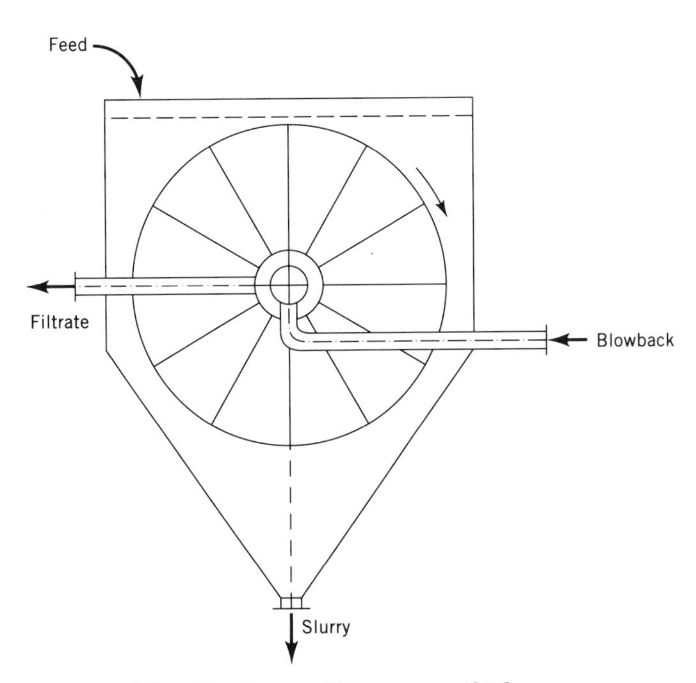

Fig. 13. Rotary disk vacuum thickener.

filtration areas up to 380 m^2. Disk vacuum filters are principally used in mining and metallurgical applications for handling large volumes of free-filtering materials. They have also been successfully operated with cement, starch, sugar, paper and pulp, and flue dusts.

Batch Pressure Filters

Excluding variable chamber presses, which rely on mechanical squeezing of the cake and are discussed in a separate section, pressure filters may be grouped into two categories, ie, plate-and-frame filter presses, and pressure vessels containing filter elements. The latter group also includes cartridge filters; these are discussed separately. All of the above pressure filters are suited to handling different types of cake. Pressure vessel filters (leaf-type) handle incompressible or slightly compressible cakes. Filter presses handle both compressible and incompressible cakes, especially with the flexibility potential of membranes. Variable chamber presses cannot be used on incompressible materials. Cylindrical element filters, ie, candle filters, are used for clarification applications, using precoat and often body-feed, resulting in cakes that are slightly compressible. Cartridge filters are for clarification only, with little if any cake formed.

Plate-and-Frame Filter Presses. In the conventional plate-and-frame press (Fig. 14), a sequence of perforated square, or rectangular, plates alternating with

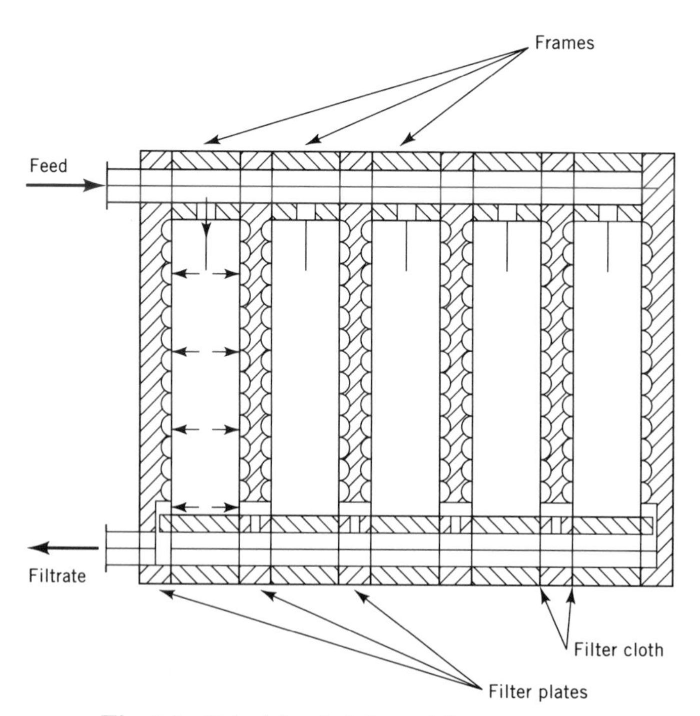

Fig. 14. Principle of plate-and-frame presses.

hollow frames is mounted on suitable supports and pressed together with hydraulic or screw-driven rams. The plates are covered with a filter cloth which also forms the sealing gasket. The slurry is pumped into the frames and the filtrate is drained from the plates.

The drainage surfaces are usually made in the form of raised cylinders, square-shaped pyramids, or parallel grooves in materials such as stainless steel, cast iron, rubber or resin-coated metal, polypropylene, rubber, or wood. Designs are available with every conceivable combination of inlet and outlet location, ie, top feed, center feed, bottom feed, corner feed, and side feed, with a similar profusion of possible positions of discharge points. Each combination has particular advantages, depending on whether washing is required and also on the application and nature of the suspension. The discharge may be through a separate cock on each plate, rather than through a common filtrate port and manifold, to allow observation and sampling of the filtrate from each plate. This enables the operator to spot cloth failure and isolate the plate or segregate the cloudy filtrate. The cocks discharge into open channels or enclosed pipe systems fitted with a sight glass.

Both flush plates and recessed plates can be specified. Recessed plates obviate the need for the frames but are tougher on filter cloths due to the strain around the edges. These presses are more suitable for automation because of the difficulty of the automatic removal of residual cake from the frames in a plate-and-frame press. Recessed plates with no frames limit the chamber width to less than 32 mm to limit the strain on the cloth, whereas plate-and-frame presses allow this to be more than 40 mm if necessary.

Plate sizes range from 150 mm to 2 m^2, giving filtration areas up to 200 m^2. The number of chambers varies up to 100, exceptionally to 200.

Plate-and-frame filters are most versatile since their effective area can be varied simply by blanking off some of the plates. Cake holding capacity can be altered by changing the frame thickness or by grouping several frames together. These filters are available in a variety of semiautomated or fully automated versions that feature mechanical leaf-moving devices, cake removal by vibration, or cake removal by pulling the cloth when the press is open, etc. An operator usually must be present, however, because it is not certain that each and every chamber will discharge its cake unaided every time. Should manual intervention be necessary, the operator must be protected by a suitable safety photoelectric device from injury by the leaf-moving mechanism.

Some attempts have been made to reslurry the filter cake without having to open the filter press. However, a number of problems appear, eg, bending of the plates due to uneven cake deposition or cavitation, uneven dewatering and washing within the frames, and plugging of the inlet ports.

Washing performed in filter presses is either simple or thorough washing. In simple washing, the wash liquid is introduced either through the main feed port or through a separate port into each chamber, and the washing is therefore in the same direction as the filtration process that formed the cake. In thorough washing, the wash liquid enters through a separate port, behind the filter cloth on every other plate, thus passing through the whole thickness of the cake in the chamber. Washing is less efficient with recessed chamber presses than with flush plate frame presses, perhaps because of poorer distribution of the wash liquid. In either case the amount of wash liquid needed is high.

Filter media for plate-and-frame presses include various cloths, mats, and paper. Paper filter media usually must be provided with a backing cloth for support.

The typical operating pressure of filter presses is 600 or 700 kPa, although some manufacturers offer presses for 2000 kPa or higher. As the pressure increases during filtration, it forces the plates apart; this may be offset by a pressure compensation facility offered with some large mechanized presses.

Full mechanization of filter presses started in the late 1950s; this was followed by addition of the mechanical expression, ie, cake squeezing, mechanism. Rubber or plastic membranes are sometimes fitted to compress the cake which is formed by conventional pressure filtration. The membranes normally rest on the plates and have grooves and openings in their surfaces for filtrate collection. They are inflated at the end of the filtration cycle by air or, for pressures higher than 1000 kPa, by water or hydraulic fluid. The membranes should be designed to last up to or in excess of 10,000 pressings. The principal advantages of using mechanical expression of cakes are the additional dewatering usually achieved and the ability to handle thin cakes. The main filtration process can be done at lower pressures so that a relatively cheap, centrifugal pump can be used, and the compression by the membrane then goes to higher pressures.

The automation of filter presses has affected several other advantages and developments. Plate shifting mechanisms have been developed, allowing the cloths to be vibrated; filter cloth washing, on both sides, has been incorporated to counteract clogging from the expression; and downtimes have been reduced with automation, thus increasing capacities.

The vertical recessed plate automatic press is shown schematically in Figure 15. Unlike the conventional filter press with plates hanging down and linked in a horizontal direction, this filter press has the plates in a horizontal plane placed one upon another. This design offers semicontinuous operation, saving in floor space, and easy cleaning of the cloth, but it allows only the lower face of each chamber to be used for filtration.

The filter usually has an endless cloth, traveling intermittently between the plates via rollers, to peel off cakes. Unfortunately, if the cloth is damaged anywhere, the whole cloth must be replaced, which is a difficult process. Each time the filter cloth zigzags through the filter, the filtering direction is reversed; this tends to keep the cloth clean. Most of these filters incorporate membranes for mechanical expression, and cakes sometimes stick to the membranes and remain in the chamber after discharge. Some vertical filters are available with a separate cloth for each frame. The cloths may be disposable and such filters are designed to operate with or without filter aids.

Because the height of the vertical press makes maintenance difficult, the number of chambers is restricted, usually to 20, with a maximum of 40, with filter areas up to 32 m^2.

The application of filter presses spans virtually all areas of the processing industries due to their versatility. Examples of use include clarification of beer and juices; wastewater and activated sludge filtration in breweries, paper mills, and petrochemical plants; dewatering of fine minerals; lime mud separation; and

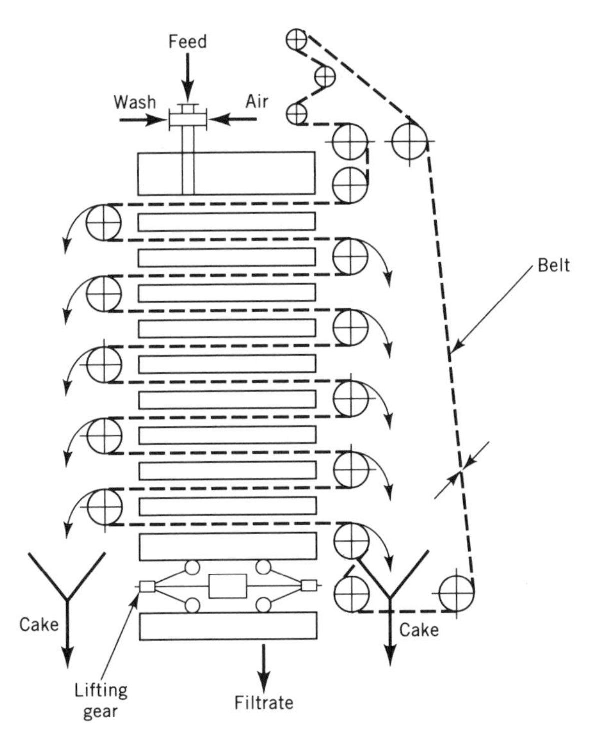

Fig. 15. Vertical automatic filter presses.

washing in the sugar industry. Filtration velocities are usually from 0.025 to 1 m/h. Dry solids handling capacities are less than 1000 kg/m^2 per hour, with the higher values being more usual with the automatic presses due to their shorter down times.

Pressure Vessel Filters. The several designs of pressure vessel filters all consist of pressure vessels housing a multitude of leaves or other elements which form the filtration surface and which are mounted either horizontally or vertically. With horizontal leaves most suitable where thorough washing is required, there is no danger of the cake falling off the cloth; with vertical elements, a pressure drop must be maintained across the element to retain the cake. The disadvantage of horizontal leaf types is that half the filtration area is lost because the underside of the leaf is not used for filtration because of the danger of the cake falling off. Discharge of the cake also may be more difficult in this case.

The elements or leaves normally consist of a coarse stainless steel mesh over which a fine, often metal, gauze or filter cloth is stretched and sealed at the edges. The leaves are in parallel, each connected to a header and, almost without exception, filtration is from the outside in through the gauze. These filters are essentially batch operated, and most require the use of a filter aid for precoating to avoid cloudy filtrates and blinding. A separate filtering element is often installed at the bottom of the vessel as a scavenger filter for the filtration of the heel of unfiltered slurry that is still in the vessel at the end of the filtration period. This residual slurry is particularly troublesome with the vertical leaf filters because compressed air cannot be used to complete the filtration of this heel, as the air would preferentially escape through the tops of the leaves as soon as they emerged from the suspension. During the scavenge filtration the main leaves are usually isolated so that compressed air is not lost through them.

Cylindrical Element Filters. These filters, often referred to as candle filters, have cylindrical elements or sleeves mounted vertically and suspended from a header sheet, which divides the filter vessel into two separate compartments (Fig. 16). The filtration takes place on the outside of the sleeves. The inlet is usually in the bottom section of the vessel and the filtrate outlet in the top section above the header sheet. A less usual design is to locate the filtrate outlet at the bottom of the elements and thus allow the top chamber to be opened for each inspection of the elements during operation.

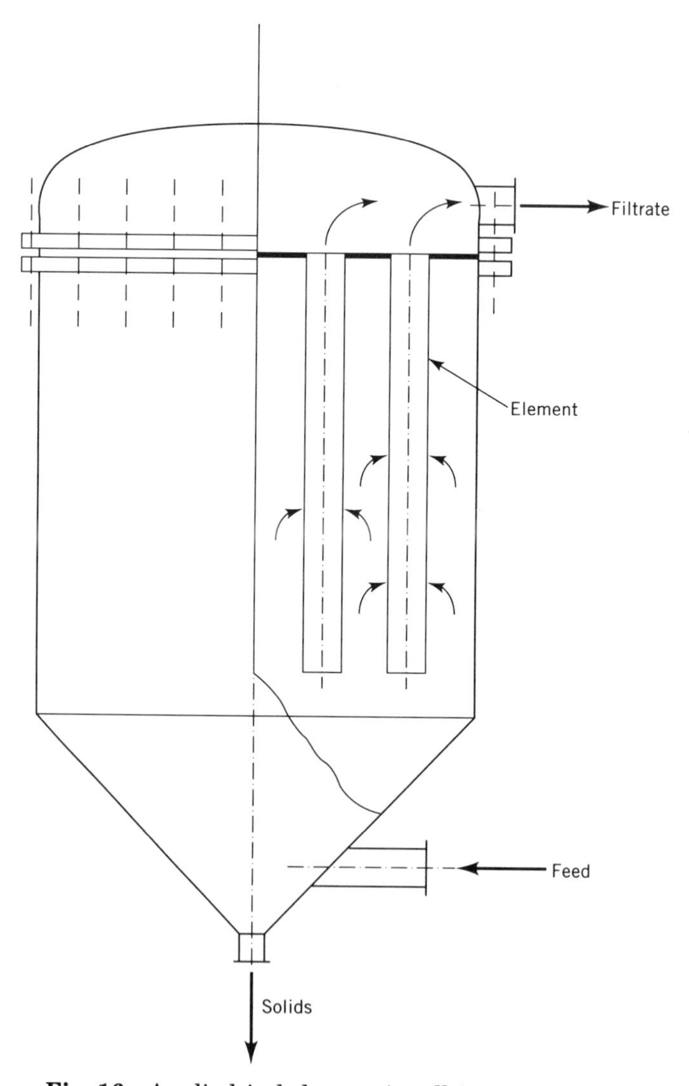

Fig. 16. A cylindrical element (candle) pressure filter.

The tubes generally measure from 25 to 75 mm in diameter, and up to 2 m in length. Made from metal or cloth-covered metal, they provide filtration areas up to 100 m^2. Alternatively, the tubes can be made of stoneware, plastics, sintered metal, or ceramics. The elements may be deliberately made flexible, sometimes filled with loose packing. Tank diameters up to 1.5 m are available. Cake removal is performed by scraping with hydraulically operated scraper rings, by vibration, or by turbulent flow bumping. The mechanical strength of the tubular element makes it ideal for cleaning by the sudden application of reverse pressure. Physical expansion or flexing of the tubular elements on application of the reverse flow aids cake discharge. These filters find wider use where cake washing is not required.

The advantage of candle filters is that as the cake grows on the tubular elements the filtration area increases and the thickness of a given volume of cake is therefore less than it would be on a flat element. This is of importance where a thick cake is being formed; the rate of increase in the pressure drop is less with tubular elements.

The pressure filter with tubular elements has also been used as a thickener, when the cake, backwashed by intermittent reverse flow, is redispersed by an agitator at the bottom of the vessel and discharged continuously as a slurry. In some cases the filter cake builds up to a critical thickness and then falls away without blowback.

Vertical Vessel, Vertical Leaf Filters. These are the cheapest of the pressure leaf filters and have the lowest volume-to-area ratio (Fig. 17). Their filtration areas are limited to less than 80 m^2. Large bottom outlets, fitted with rapid-opening doors, are used for dry cake discharge, and smaller openings are used for slurry discharge. Wet discharge may be promoted by spray pipes, vibrators, reverse flow, bubble rings, scrapers, etc, while dry discharge is usually caused by vibration. As with all vertical leaf filters, these are not suited for cake washing.

In the Scheibler filter, the filter leaves take the form of bags, each suspended in a rectangular pressure vessel from a horizontal tube which acts as the filtrate outlet. The sides of the bag are prevented from meeting by looped chains which are attached at the top of the loop to the horizontal tube and hang downward. With this method of separating the cloth surfaces by chains, the bags can be wide enough to hang in pleats and the filtering area can thus be as much as three times the area of the frames inside the bags. Filtration areas up to 250 m^2 are available with applications being mostly in the chemical industry.

The vertical cylindrical vessel of the pressure version of the Moore filter developed in France for the sugar industry, which houses a set of radially arranged leaves, is twice the height of the leaves. This allows the leaves to be raised, rotated, and lowered into the different compartments in the bottom half of the vessel. Positive air pressure must be maintained throughout the operation to prevent cake fall-off, and the cake is blown off the leaves by air blowback.

Horizontal Vessel, Vertical Leaf Filters. In a cylindrical vessel with a horizontal axis (Fig. 18), the vertical leaves can be arranged either laterally or longitudinally. The latter, less common, arrangement may be designed as the vertical vessel, vertical leaf filters but mounted horizontally. Its design is suitable for smaller duties and the leaves can be withdrawn individually through the opening end of the vessel.

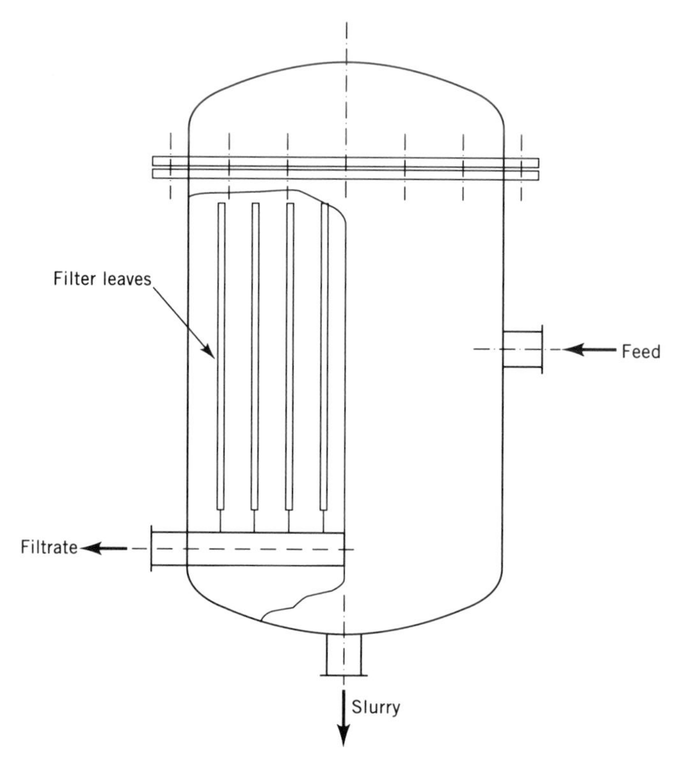

Fig. 17. Vertical vessel, vertical leaf filter.

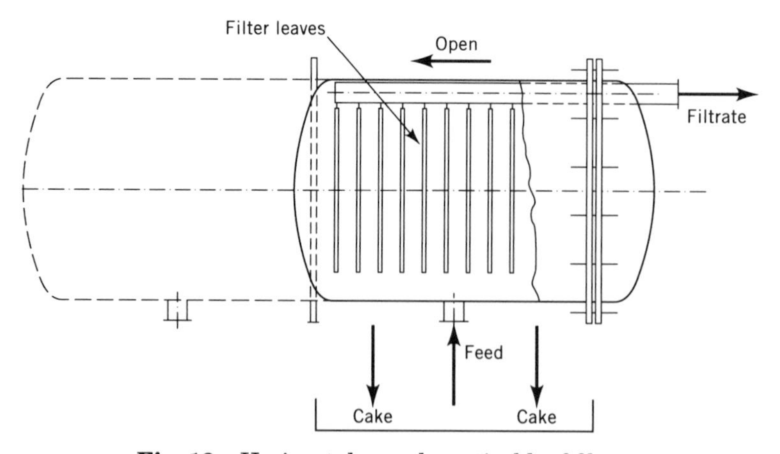

Fig. 18. Horizontal vessel, vertical leaf filter.

Filtration areas up to 120 m^2 are available with the Kelly leaf filter, another longitudinal arrangement and probably the earliest pressure leaf filter. It has been used for the filtration of very viscous liquids such as glycerol and concentrated sugar solutions. The height of the leaves varies according to the space available at their location in the vessel. The leaves are attached to the removable circular front cover, with each leaf having an outlet connection through this cover. The leaves, together with the cover and outlet pipes, are attached to a carriage which can be run into and out of the shell to facilitate cake discharge outside the vessel by air blowback or rapping. This arrangement requires a considerable amount of floor space and to minimize this drawback the presses are often constructed in pairs on a single runway, with their opening ends facing each other. Thus each filter can be opened in turn into the common space between them.

Horizontal vessel filters which have the vertical leaves arranged in a plane perpendicular to the axis of symmetry of the vessel, ie, laterally, have the greatest use because they provide easy access to the leaves. Most of these designs open in a way similar to the Kelly filter. Some designs move the shell rather than the leaf assembly so that the filtrate pipe can remain permanently connected; withdrawal of one leaf or a bundle of leaves at a time is possible. The leaves may be rectangular, circular, or of some other shape, and may be designed to rotate during cake discharge. Sluicing by sprays is used for wet discharge, with or without rotation of the leaves. If the leaves are designed for rotation, they are invariably circular and mounted on a central hollow shaft which serves as the filtrate outlet. Dry cake discharge may be carried out with rotating leaves by application of a scraper blade. If this is to be done without opening the vessel, then the bottom of the vessel must be shaped as a hopper, with a screw conveyor if necessary.

The Vallez filter, originally developed in the United States for the sugar industry, rotates the leaves at about 1 rpm during the filtration operation to keep the solids in suspension and achieve a more uniform cake.

The Sweetland filter, a significant departure from the standard end-opening design, has the cylindrical shell split in a horizontal plane into two parts, where the bottom half can be swung open for cake discharge. The upper half is rigidly supported and both the feed and the filtrate piping are fixed to it. The lower part is hinged to the upper along one side and is counterbalanced for easy opening. Cake discharge is either by sluicing or by dropping, assisted by some scraping. If much scraping is needed, there is not much advantage in using this type of filter. Because the leaves are stationary, the cake deposited on them may be uneven, with greater mass of cake at the bottom of the leaves.

Generally, the horizontal vessel, vertical filters with leaves arranged laterally can be designed up to filtration areas of 300 m^2. Cake washing is possible but must be carried out with caution since there is a danger of the cake falling off.

Horizontal vessel filters with vertical rotating elements have been under rapid development with the aim of making truly continuous pressure filters, particularly for the filtration of fine coal.

Vertical Vessel, Horizontal Leaf Filters. These filters, like all horizontal leaf filters, are advantageous where the flow is intermittent or where thorough cake washing is required. Filtration areas are limited to about 45 m^2.

The pressure versions of the nutsche filter, which falls into this category, are either simple pressurized filter boxes or more sophisticated agitated nutsches, much the same in design as the enclosed agitated vacuum filters described earlier. These are extremely versatile, batch-operated filters, used in many industries, eg, agrochemistry, pharmaceuticals, or dyestuff production.

An obvious method of increasing the filtration area in the vessel is to stack several plates on top of each other; the plates are operated in parallel. One design, known as the plate filter, uses circular plates and a stack that can be removed as one assembly. This allows the stack to be replaced after the filtration period with a clean stack, and the filter can be put back into operation quickly. The filter consists of dimpled plates supporting perforated plates on which filter cloth or paper is placed. The space between the dimpled plates and the cloth is connected to the filtrate outlet, which is either into the hollow shaft or into the vessel, the other being used for the feed. When the feed is into the vessel, a scavenger plate may have to be fitted because the vessel will be full of unfiltered slurry at the end of the filtration period. This type of filter is available with filtration areas up to 25 m^2 and cakes up to 50 mm thick.

Centrifugal discharge filters form another group in this category. As the name suggests, the cake discharge is accomplished by rotating the stack of plates around the hollow shaft. The cake slides off the plates due to the centrifugal action; sometimes it is necessary to supplement this by sluicing with a suitable liquid, in which case the discharge is wet. The filtrate leaves through the hollow shaft. These filters lend themselves to automation and, as opposed to manually operated leaf filters, they can be operated with short cycle times and very short downtimes, which is economical. Many different designs are available, with various ways of driving the shaft and locations of the electric motor as well as other varying constructional details. Sizes vary up to 65 m^2.

Another available design allows discharge of the cake by vibration of the circular plates, which are slightly conical, sloping downward toward the outside of the plates. This design allows higher pressures to be used as there are no rotating seals necessary.

Horizontal Vessel, Horizontal Leaf Filters. These filters consist of a horizontal cylindrical vessel with an opening at one end (Fig. 19). A stack of rectangular horizontal trays is mounted inside the vessel; the trays can usually be withdrawn for cake discharge, either individually or in the whole assembly. The latter case requires a suitable carriage. One alternative design allows the tray assembly to be rotated through 90° so that the cake can fall off into the bottom part, designed in the shape of a hopper and fitted with a screw conveyor.

The trays may be fitted with rims; this is particularly useful for flooding the trays in washing operations. Scavenger leaves are often used. Filtration areas up to 50 m^2 are available. Like all horizontal leaf filters, horizontal vessel, horizontal leaf filters are particularly suitable when thorough washing is needed.

Cartridge Filters. Cartridge filters use easily replaceable, tubular cartridges made of paper, sintered metal, woven cloth, needle felts, activated carbon, or various membranes of pore size down to 0.2 μm. Filtration normally takes place in the direction radially inward, through the outer face of the element, into the hollow core (Fig. 20). Cartridge filtration is limited to liquid polishing or clarification, ie, removing very small amounts of solids, in order to keep the frequency of cartridge replacements down. Typically, suspensions of less than 0.01% volume

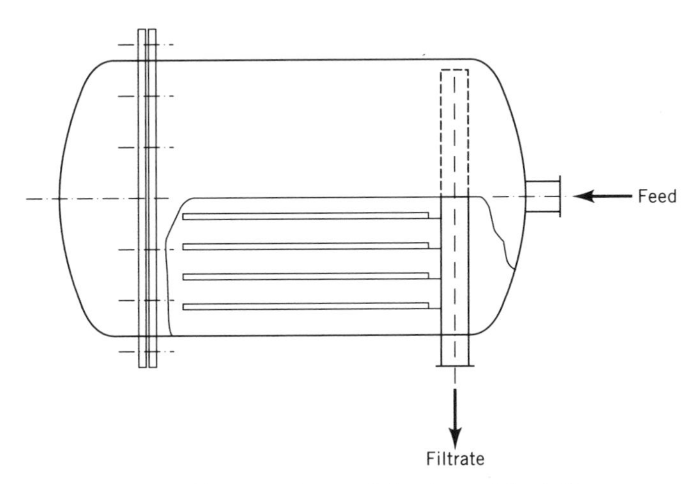

Fig. 19. Horizontal vessel, horizontal leaf filter.

concentration of solids can be treated with cartridge filters, and such filters are favored in small-scale manufacturing applications.

Cartridge filters are either depth type or surface type, according to where most of the solids separate; the precise demarcation line is difficult to draw. The most common depth cartridge is the yarn-wound type which has a yarn wound around a center core in such a way that the openings closest to the core are smaller than those on the outside. The aim is to achieve depth filtration, which increases the solids-holding capacity of the cartridge. The yarn may be made of any fibrous material, ranging from cotton or glass fiber to the many synthetic fibers such as polyester, nylon, or Teflon. The spun staple fibers are brushed to raise the nap and this makes the filter medium. The cores are made from polypropylene, phenolic resin, stainless steel, or other metals or alloys. The nominal μm rating of this type of cartridge varies from 0.5 to 100 μm.

Cartridges for higher viscosity liquids are often made of long, loose fibers, again either natural or synthetic, impregnated with phenolic resin. Such bonded cartridges are usually formed into the shape of a thick tube by a filtration technique and do not require a core because they are self-supporting. The porosity of the medium can be graded during the formation process, again to increase the solids-holding capacity. Bonded cartridges are available in somewhat coarser ratings from 10 to 75 μm.

Depth-type cartridges cannot be cleaned but have high solids-holding capacity, and are cheap and robust. Considerable standardization of the cartridge size throughout industry, approximately 25 cm long, 6.3 to 7.0 cm overall diameter, and 2.5 cm internal diameter, allows testing of different cartridge makes and types in the same holder.

A common surface cartridge is the pleated paper construction type, which allows larger filtration areas to be packed into a small space. Oil filters in the automobile industry are of this type. The paper is impregnated, for strength, with

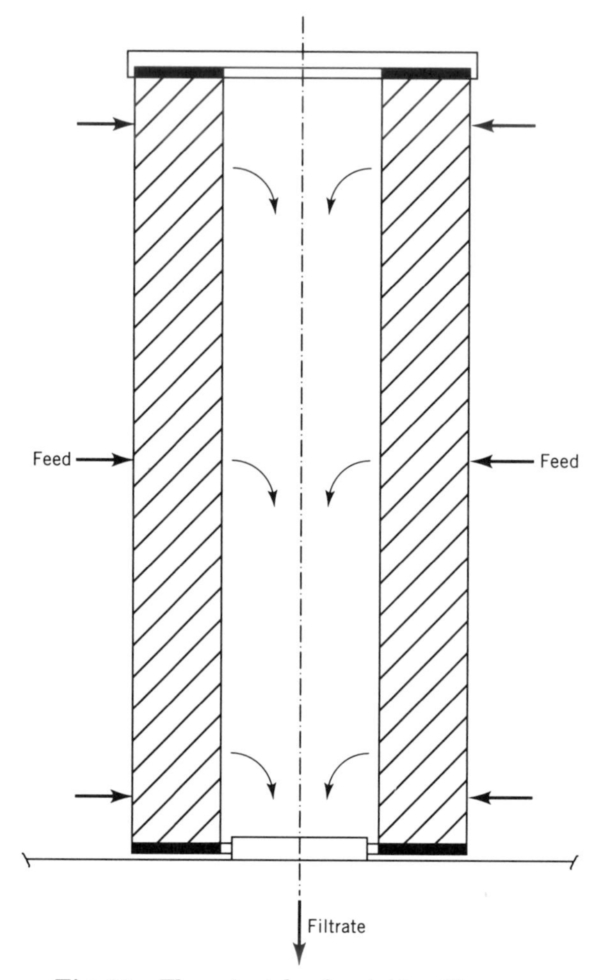

Fig. 20. The principle of cartridge filters.

epoxy or polyurethane resin. Any other medium in sheet form, similar to cellulose paper, such as wool, polypropylene, or glass may be used.

The nominal rating ranges from 0.5 to 50 μm. Pleating in the radial direction is usual, but some cartridges have axial pleating, or, in an effort to pack as much filtration surface into a given space as possible, have hollow disks of lenticular shape; up to 3 m^2 of surface in one cartridge is possible. The solids-holding capacity of pleated paper surface cartridges is low but some applications allow the prolonged buildup of solids on the surface, until the pleats are completely filled up.

Another type of cartridge is the edge filter which contains a number of thin disks mounted on a central core and compressed together. The disks are usually made of metal although paper or plastics are used. Filtration takes place on the surface of the cylinder, with the particles unable to pass between the disks. The principal advantage of the edge-type filter is that it is cleanable. This is done by reverse flow, ultrasonics, or by scraping the outside surface of the cylinder with a mechanical scraper. Edge filters made from paper disks have been known to retain particles as fine as 1 μm but the metal variety retains solids larger than 50 μm or so. Other designs of cartridges use active carbon, Fuller's earth, sintered metal, or other specialized media.

The most important characteristics of cartridge filters are the filter rating, ie, the largest spherical particle (~ 98% retention cut size) which passes through the filter; the relationship between the pressure drop and solids-holding capacity; and the maximum allowable pressure drop beyond which the cartridge fails structurally. Both the retention and the solids-holding capacity depend on filtration velocity and this must be considered when testing cartridges. Thermal or shock stresses can lead to cracking of cartridges, with the subsequent loss of overflow clarity.

The housings of cartridge filters are simple pressure vessels designed for one cartridge, or a number of cartridges in parallel, in multielement filters. Some housings are designed to withstand pressures up to 30,000 kPa. Proper sealing of the elements is a necessary prerequisite of their efficient use. Frequent replacement of cartridges should be facilitated by quick-opening clamping fittings.

Cartridge filters are used to clean power fluids, lubrication oils, wines, fruit juices, or pharmaceutical liquids. They are also used to protect other equipment, eg, in reflux control systems or automatic valves. Low capital and installation costs, low maintenance costs, simplicity, and compactness are the main advantages of cartridge filters. Running costs are high, especially when disposable cartridges are used. It is most important, therefore, that a full economical analysis, based on reliable cartridge replacement frequency, is carried out before adopting a cartridge filtration system; the low cost of the basic hardware may be deceptive.

Mechanical Batch Compression Filters. In conventional cake filtration the liquid is expelled from the slurry by fluid pressure in a fixed-volume filtration chamber; in mechanical compression this is achieved by reduction of the volume of the retaining chamber. This compression of either a slurry or a cake, which might have been formed by conventional filtration, offers advantages to industries handling a variety of different materials. Such materials include highly compressible, sponge-like solids; very fine particles such as clays; fibrous pulps; gelatinous mixtures like starch residues or some pharmaceuticals; and flocculated wastewater sludges.

The compressibility of filter cakes is a nuisance from the point of view of the filtration theory. In practice it means that with increasing pressure cakes become more compact and therefore drier. The resistance to flow increases due to reduced porosities, however, and, with some materials, eg, paper mill effluents, higher pressures do not necessarily give increased flow rates. In cakes undergoing conventional pressure filtration, the bottom layers closest to the medium are subjected to the highest compression forces whereas the top layers are subjected only to light hydraulic forces and are not compacted so tightly. If a mechanical force is applied to the top of the filter cake, the distribution of pressure through the cake is more uniform. Cakes drier than those formed by using high pumping pressures of the feed suspension can thus be achieved.

Since 1980, a number of new filters have appeared on the market, utilizing some form of mechanical compression of the filter cake, either after a conventional pressure filtration process or as a substitute for it. In most designs the compression is achieved by inflating a diaphragm which presses the slurry or the freshly formed filter cake toward the medium, thus squeezing an additional amount of liquid out of the cake.

Other designs squeeze the cake between two permeable belts or between a screw conveyor of diminishing diameter, or pitch, and its permeable enclosure. The available filters which use mechanical compression can be classified into four principal categories, ie, membrane plate presses, tube presses, belt presses, and screw presses.

Membrane plate and tube presses are dealt with here; belt and screw presses are included in the discussion of continuous pressure filters.

The advantages of using mechanical compression with compressible cakes include increased solid content of the cakes, leading to reduced energy requirements if thermal drying has to follow, or to better handling properties; improved washing efficiencies; increased filtration rates; and easier or automatic cake discharge. Invariably, however, the capital cost of such filters is higher than for conventional pressure filters. Whenever a rubber membrane is used for the compression, however, this increases the capital cost and thus variable chamber filters tend to be more expensive than conventional filters in the same category.

Membrane Plate Presses. Membrane presses are closely related to conventional plate and frame presses. They consist of a recessed plate press in which the plates are covered with an inflatable diaphragm which has a drainage pattern molded into its outside surface. The filter cloth is placed over the diaphragm (Fig. 21). During the first stage of filling the filter chambers with the slurry and con-

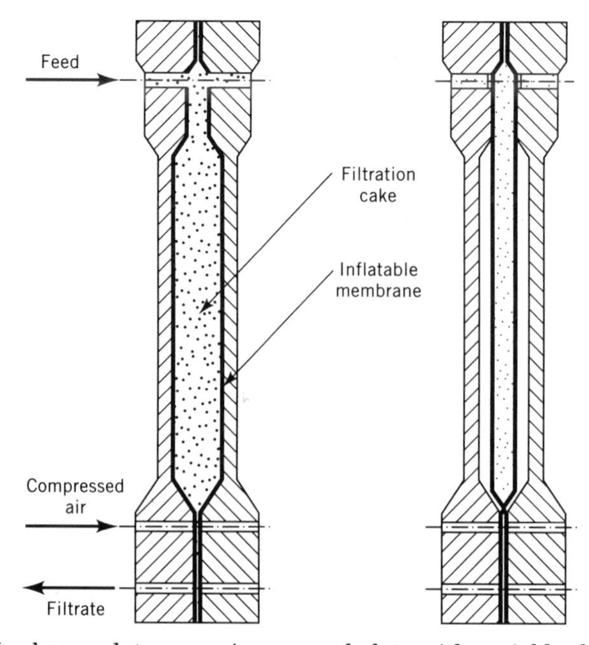

Fig. 21. Membrane plate press, ie, recessed plate with variable chamber press.

ventional pressure filtration, the membrane is pushed against the plate body. When the chambers are full of cake, the feed is terminated and the membranes are inflated by pumping compressed air or hydraulic fluid in between the membranes and the plates; the cake in the chamber is compacted as the two membranes move toward one another. Washing of the cake may follow and can be carried out more effectively than in a normal press because the cake is compacted to a more uniform density by squeezing. The resulting advantage is in the reduction of the washing time and washwater requirements. The squeeze pressures vary from 600 to 1500 kPa. Additional reductions of up to 25% in the moisture content over that obtained with conventional filter presses can be achieved. The membranes are usually made of rubber compounds resistant to solvents.

Another advantage of the membrane plate is its flexibility to cake thickness, ie, thinner cakes can be easily handled without loss of dryness. Cake release characteristics are also improved by deflation of the membrane prior to cake discharge. Alternating arrangements, in which the membrane plates and the normal recessed plates alternate, have been used to reduce cost.

The plate press filter replaces the pneumatically operated membranes with flexible seals and compression by a hydraulically powered ram. The cake is formed in the chambers between the hollow circular frames carrying the filter medium and the flat rectangular discharge plates. The frames are sealed against the discharge plates by rubber rim seals which form the filtration chambers. As the rim seals are compressed by the hydraulic ram, the cake inside is squeezed. The filter is then opened, the cake adheres to the discharge plates and, as the plates are lowered and raised again, the cake is removed by scrapers. The process is fully automated and the filtration areas available go up to 20 m^2. No washing of the cake is possible.

The OMD leaf filter (Stella Meta Filters) is a vertical leaf filter with a rubber diaphragm suspended between the leaves. The cake that forms on the leaves eventually reaches the diaphragm at which point pump pressure is used to inflate the diaphragm and compress the cake. The cake discharge is by vibration.

A variation of the same principle is the DDS-vacuum pressure filter which has a number of small disks mounted on a shaft which rotates discontinuously. The cake is formed on both sides of the disks when they are at the bottom position, dipped into the slurry. When the disks come out of the slurry and reach the top position, hydraulically driven pistons squeeze the cake and the extra liquid then drains from both sides of the cake. The cake is removed by blowback with compressed air.

Cake compression by flexible membranes is also used in the new automated vertical presses that use one or two endless cloth belts, indexing between plates.

Filtration and compression take place with the press closed and the belt stationary; the press is then opened to allow movement of the belt for cake discharge over a discharge roller of a small diameter. This allows washing of the belt on both sides (Fig. 15). Cycle times are short, typically between 10 and 30 minutes, and the operation is fully automated. Sizes up to 32 m^2 are available and the maximum cake thickness is 35 mm.

Washing and dewatering by air displacement of cakes are possible. Applications are in the treatment of minerals, in the sugar industry, and in the treatment of municipal sewage sludge and fillers like talc, clay, whiting, etc.

Cylindrical Presses. Another group of filters that utilize the variable chamber principle are those with a cylindrical filter surface. There are two designs in this category, both of which originate from the United Kingdom.

The VC filter (Fig. 22) consists of two concentric hollow cylinders mounted horizontally on a central shaft. The inner cylinder is perforated and carries the filter cloth, the outer cylinder is lined on the inside with an inflatable diaphragm. The slurry enters into the annulus between the cylinders and conventional pressure filtration takes place, with the cake forming on the outer surface of the inner cylinder. The filtration can be stopped at any cake thickness or resistance, as required by the economics of the process, and hydraulic pressure is then applied to the diaphragm that compresses the cake.

As with other filters of this type, washing can be carried out by deflation of the diaphragm and introduction of washwater into the annulus. Reinflation of the membrane then forces the wash liquor through the cake, thus displacing the mother liquor. At the end of the process, the inner cylinder is withdrawn from the outer shell and the cake is either discharged manually or blown off with compressed air. Sizes available range from small, mobile test units of 0.4 m^2 area to large, fully automated machines with 6.1 m^2 filtration area. Choice of two alternative core sizes is offered, giving annuli of 6 or 2.8 cm available for the cake. The hydraulic pressure for operating the membrane goes up to 1400 kPa. Although originally developed for filtration of dyestuffs, the VC filter has been successfully used for the filtration of gypsum, china clay, cement, industrial effluents, metal oxides, coal washings, nuclear waste, and other slurries.

The ECLP tube press is smaller in diameter and, unlike the VC filter, is operated in a vertical position. It uses compression only, both for the filtration and for squeezing the cake. The space between the cylindrical rubber membrane and the cloth tube is first filled with the slurry and the hydraulically operated membrane is used to drive the liquid through the cloth. It follows, therefore, that this filter is suitable for higher solids concentrations, usually in excess of 10% by weight, in order to obtain the minimum cake thickness necessary for efficient cake discharge of about 4 mm. At the end of the process, the central core is lowered by about 300 mm and the cake is removed by a blast of compressed air from the

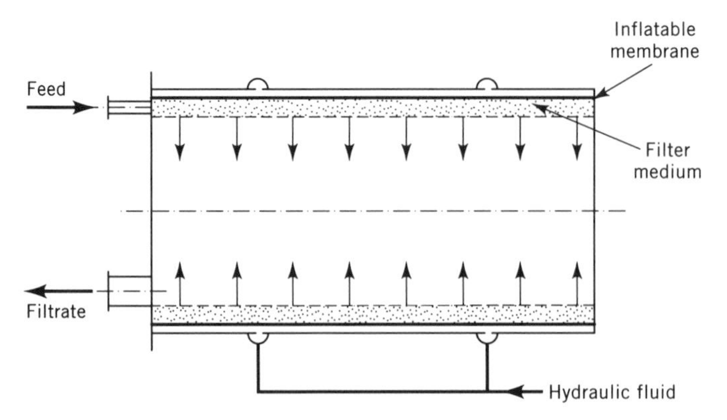

Fig. 22. Principle of the VC filter.

inside. The hydraulic operating pressures are higher than the VC filter at about 10,000 kPa but the single tube area is only about 1.3 m^2. Multiple tube assemblies are used to treat larger flows. Cake washing is possible but with some solids there is a danger of the cake falling off the inner core while the annulus is being filled with water.

The ECLP tube press was originally developed for the filtration of china clay but has been used with many other slurries such as those in mining, TiO_2, cement, sewage sludge, etc. The typical cycle time is about four minutes or more.

Continuous Pressure Filters

A continuous pressure filter may be defined as a filter that operates at pressure drops greater than 100 kPa and does not require interruption of its operation to discharge the cake; the cake discharge itself, however, does not have to be continuous. There is little or no downtime involved, and the dry solids rates can sometimes be as high as 1750 kg/m^2h with continuous pressure filters.

Most continuous pressure filters available (ca 1993) have their roots in vacuum filtration technology. A rotary drum or rotary disk vacuum filter can be adapted to pressure by enclosing it in a pressure cover; however, the disadvantages of this measure are evident. The enclosure is a pressure vessel which is heavy and expensive, the progress of filtration cannot be watched, and the removal of the cake from the vessel is difficult. Other complications of this method are caused by the necessity of arranging for two or more differential pressures between the inside and outside of the filter, which requires a troublesome system of pressure regulating valves.

Despite the disadvantages, the advantages of high throughputs and low moisture contents in the filtration cakes have justified the vigorous development of continuous pressure filters.

Horizontal or vertical vessel filters, especially those with vertical rotating elements, have undergone rapid development with the aim of making truly continuous pressure filters, particularly but not exclusively for the filtration of fine coal. There are basically three categories of continuous pressure filters available, ie, disk filters, drum filters, and belt filters including both hydraulic and compression varieties.

The advantages of continuous pressure filtration are clear and indisputable, particularly with slow-settling slurries and fairly incompressible cakes. Such filters are expensive, both to install and to run, and the most likely applications are either in large-scale processing of products that require thermal drying after the filtration stage, ie, fine coal or cement slurries in the dry process, or in small-scale processing of high value products such as in the pharmaceutical industry.

There are many technical problems to be considered when developing a new commercial and viable filter. However, the filtration hardware in itself is not enough, as the control of a continuous pressure filter is much more difficult than that of its equivalents in vacuum filtration; the necessary development may also include an automatic, computerized control system. This moves pressure filtration from low to medium or even high technology.

Disk Filters. *The McGaskell and Gaudfrin Disk Filters.* One of the earliest machines in this category, the McGaskell rotary pressure filter, is essentially a disk-type filter enclosed in a pressure vessel. The rotating disks are each composed of several wedge-shaped elements connected to a rotary filter valve at the end of the shaft, similar to the vacuum rotary disk filters. Originally designed for the filtration of waxes in the oil industry and equipped for a gradual increase in pressure with cake buildup, the McGaskell rotary pressure filter is said to have produced high filtrate clarity. Pressures up to 700 kPa have been used.

The slurry reservoir is divided into pockets or crenellations that have spring-loaded scrapers. The scrapers press against the disks and direct the cake into the spaces between the pockets around the disks which lead to a chute connected to an inner casing in which is placed a worm gear. The worm conveys and compresses the cake, thus squeezing more liquid from it, through a filter cloth. The compressed cake forms a plug around a spring-loaded, tapered discharge valve, and the plug prevents leakage of gas. The cake can be washed but not very effectively. This filter is reported in filtration literature (14), but is no longer commonly used.

The Gaudfrin disk filter, designed for the sugar industry and available in France since 1959, is also similar in design to a vacuum disk filter but it is enclosed in a pressure vessel with a removable lid. The disks are 2.6 m in diameter, composed of 16 sectors. The cake discharge is by air blowback, assisted by scrapers if necessary, into a chute where it may be either reslurried and pumped out of the vessel or, for pasty materials, pumped away with a monopump without reslurrying.

The Gaudfrin disk filter is designed for only relatively low pressures of 100 kPa on average and it provides for cake washing in two stages, in two separate compartments within the same vessel.

The KDF Filter. The KDF filter (Fig. 23) (Amafilter, Holland) is based on the same principle as disk filters. It was developed for the treatment of mineral raw materials, like coal flotation concentrates or cement slurries, and can produce a filter cake of low moisture content at very high capacities, up to 1750 kg/m^2h. The pressure gradient is produced by pressurized air above the slurry level which provides the necessary driving force for the filtration and also is used for displacement dewatering of the cake.

Assemblies of small disks are rotated in a planetary movement around a central screw conveyor. The disks are mounted on six hollow axles and the axles revolve on overhanging bearings from the gearbox at one end of the vessel where they are driven, via a drive shaft, by an electric motor. The filtrate is collected from the disks via the hollow shafts and a filter valve into a large collecting pipe. The hollow shafts also collect the water and air from the dewatering process, in another part of the rotational cycle. The number of disks mounted on the shafts can be adjusted for different materials, depending on the required capacity and the cake thickness to be used.

As the vessel is only about half filled with slurry, the disks become coated with the cake when immersed, the cake is dewatered when the disks emerge from the slurry, and scraped or blown off, by reverse blow, into the central conveyor which takes the cake to one end of the vessel. The planetary action and the slow movement of the disks through the feed slurry ensure exceptionally good homogeneity of the cake which is critically important for good dewatering character-

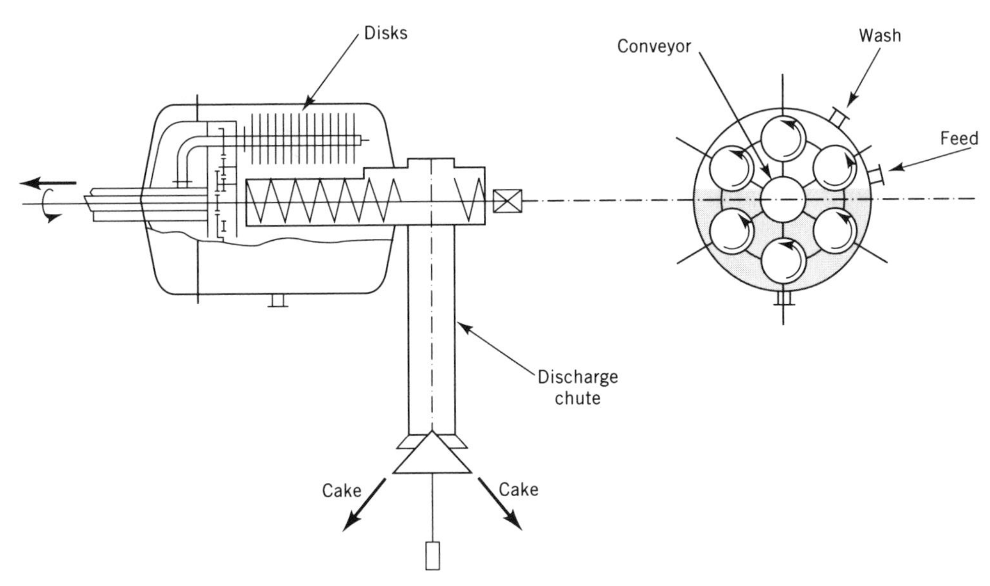

Fig. 23. KDF continuous pressure filter.

istics; the typical speed of rotation of the planetary system of shafts is from 0.8 to 1 rpm.

A screw conveyor was used originally to convey the cake, but this has been replaced with a chain-type conveyor. The first prototype used a tapered screw to form a plug before discharge into atmospheric pressure; this has been replaced by compaction in a vertical pipe.

The cake discharge is initiated and stopped by two level indicators inside the vertical pipe. The cake is actually discharged using the pressure inside the vessel; a specially designed, hydraulically operated discharge valve momentarily opens and the cake shoots out. The air pressure used for driving the slurry through the filter is 600 kPa and filtration areas are available up to 120 m^2. Cake washing is possible but it has not been reported as actually performed.

The KDF filter was first tested in prototype on a coal mine in northern Germany. It was installed in parallel with existing vacuum filters and it produced filter cakes consistently lower in moisture content by 5 to 7% than the vacuum filters. Two production models have been installed and operated on a coal mine in Belgium. The filter is controlled by a specially developed computer system; this consists of two computers, one monitoring the function of the filter and all of the detection devices installed, and the other controlling the filtration process. The system allows optimization of the performance, automatic start-up or shut-down, and can be integrated into the control system of the whole coal washing plant.

The KHD Pressure Filter. Another development of the disk filter has been reported (KHD Humboldt Wedag AG, Germany). A somewhat different system, probably a predecessor, was patented (15).

The patented system (15) has stationary disks mounted inside a pressure vessel (horizontal vessel, vertical disks) which is mounted on rollers and can rotate slowly about its axis. A screw conveyor is mounted in the stationary center of rotation; it conveys the cake, which is blown off the leaves when they pass above the screw, to one end of the vessel where it falls into a vertical chute. The cake discharge system involves two linear slide valves that slide the cake through compartments which gradually depressurize it and move it out of the vessel without any significant loss of pressure. The system relies entirely on the cake falling freely from one compartment to another as the valves move across. This may be an unrealistic assumption, particularly with sticky cakes; when combined with lots of sliding contact surfaces which are prone to abrasion and jamming, the practicality of the system is questionable.

Another significant disadvantage of the patented process is the two large running seals involved in the main body of the filter as the vessel rotates around a stationary central arrangement; this seal is another potential source of trouble. This version has little chance of commercial success and has been shelved in favor of a more conventional system of stationary vessel (16).

The newer version, tested with coal slurries in a pilot-plant facility, and with a 90 m^2 version in production, has the rotating disks and all the driving elements inside a stationary vessel. The disks, according to the manufacturer's literature, range from 1300 to 3000 mm in diameter, with up to 10 in one vessel, giving filtration areas up to 480 m^2. The cake discharge is through a rotary lock discharger which has cylindrical compartments rotating around a vertical axis. Sliding surfaces are involved and the cake is assumed to be nonsticky so that it will fall out when the compartment opens to atmosphere. The filtration area can be varied by changing the size and number of disks; no overall vessel dimensions or further details of the filter construction, other than those quoted above, are given.

The test results reported show the advantages of pressure filtration quite clearly, ie, the dry cake production capacity obtained with the test solids (coal suspensions) was raised 60 or 70% and the final moisture content of the cake reduced by as much as 5 to 7% by increasing the pressure drop from 60 to 200 kPa. Further increases in the operating pressure bring about less and less return in terms of capacity and moisture content.

The operating pressure is kept low to reduce air consumption. To obtain high capacities, the disks spin fairly fast (up to 2 or 3 rpm) and this leads to thin cakes which give little resistance to air flow; the only way to keep this flow within economical limits is to reduce the air pressure in the vessel.

Drum Filters. The rotary drum filter, also borrowed from vacuum filtration, makes relatively poor use of the space available in the pressure vessel, and the filtration areas and capacities of such filters cannot possibly match those of the disk pressure filters. In spite of this disadvantage, however, the pressure drum filter has been extensively developed.

The drum may be mounted in a vertical rather than horizontal vessel, and pressure is created by pumping compressed gas into the vessel. The intake of the compressor may be connected to the filtrate side of the filter and the gas goes around in a closed circuit. The method used to discharge the cake continuously from the high pressure inside the vessel into the atmosphere is the real basis of the design. Variable pitch screws, star valves, alternating or serial decompression chambers, monopumps (for pasty and thixotropic cakes), and other similar devices have been tried with varying degrees of success, depending on the properties of the cake. Another, rather obvious, alternative is the use of two storage vessels into which the cake is alternately discharged at the same pressure as in the filter, the pressure is later released, and the cake discharged from the vessel under atmospheric pressure.

A test unit has been developed of a small drum filter, total filter area of 0.7 m^2 with 30% submergence, housed in a large horizontal pressure vessel. Several interesting concepts have been developed and tested based on this model (17).

The so-called hyperbar vacuum filtration is a combination of vacuum and pressure filtration in a pull–push arrangement, whereby a vacuum pump of a fan generates vacuum downstream of the filter medium, while a compressor maintains higher-than-atmospheric pressure upstream. If, for example, the vacuum produced is 80 kPa, ie, absolute pressure of 20 kPa, and the absolute pressure before the filter is 150 kPa, the total pressure drop of 130 kPa is created across the filter medium. This is a new idea in principle but in practice requires three primary movers: a liquid pump to pump in the suspension, a vacuum pump to produce the vacuum, and a compressor to supply the compressed air. The cost of having to provide, install, and maintain one additional primary mover has deterred the development of hyperbar vacuum filtration; only Andritz in Austria offers a system commercially.

TDF Drum Filter. This is a fairly conventional drum filter housed in a vertical pressure vessel. Test data, obtained with the smallest model of only 0.75 m^2 filtration area, is available (18). Larger models have also been announced, ranging up to the filtration area of 46 m^2 and very large vessels. The operating pressures are moderate, up to 25 or 35 kPa, and the drum speeds fairly conventional from 0.3 to 1.5 rpm. The range of dry cake production quoted is from 250 to 650 kg/m^2h for fine coal.

The cake is scraped off with a conventional knife arrangement, then conveyed in a screw conveyor to one end of the vessel where it enters the discharge system. There are four design alternatives depending on the cake to be processed: tapered rotary valve with a horizontal axis; a pump, presumably a monopump; a rotary valve, vertical axis, with a blow-through, similar to the one used in pneumatic conveying; and a vertical pipe compactor similar in design to the Fuller-Kinyon pump in pneumatic conveying.

A plate-type filter, the PDF filter (18), uses a paddle wheel with radial, longitudinal plates covered with filter cloth and manifolded to the filter valve at one end of the vessel, instead of a drum. This filter uses a horizontal pressure vessel, was built to have only 0.75 or 1.5 m^2 area, and operates at 25 kPa. A central screw conveyor collects the cake blown off the plates and conveys it to the discharge end of the vessel.

The BHS-Fest Filter. A different approach to the use of a drum for pressure filtration is made in the BHS-Fest filter (Fig. 24). This permits a separate treatment of each filter section, in which the pressure may vary from vacuum to a positive pressure; pressure regulation is much less difficult than in the conventional enclosed drum-type pressure filter.

The BHS-Fest pressure filter has a rotating drum divided into sections. The separating strips project above the filter cloth and thus form cells. The drum is almost completely surrounded by an outer shell and the space between the shell and the drum is divided into a number of compartments separated by seals under adjustable pressure. As the drum rotates, each cell on the drum passes progres-

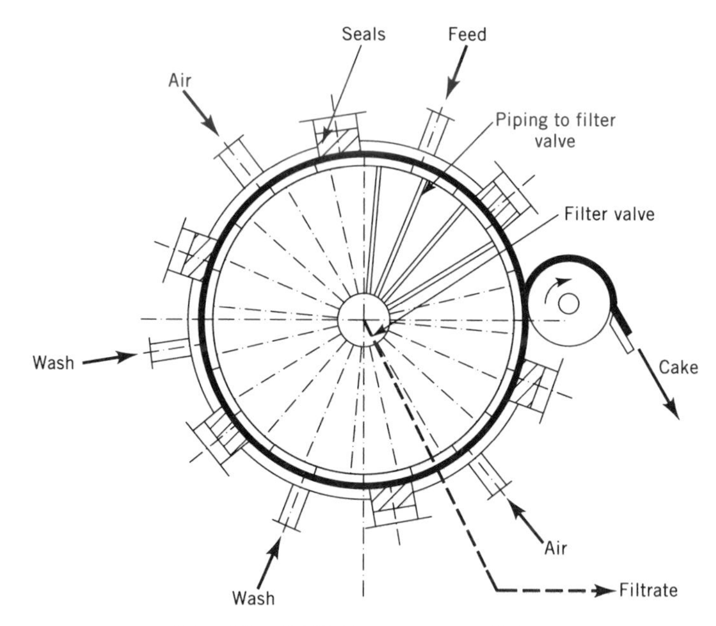

Fig. 24. BHS-Fest pressure filter.

sively through the series of compartments, thereby undergoing different processes such as cake formation, dewatering, cake washing, or cake drying; these can be carried out in several stages under different pressures or even under vacuum.

Cake discharge occurs at atmospheric pressure by the action of a roll or a scraper, assisted by blowback. The cloth may be washed by a spray before the cycle starts again. Filtering areas range up to 8 m^2 and drum diameters up to 2 meters. The necessity for large seals limits the operating pressure to less than 300 kPa, typically. Cake thickness can be from 2 to 150 mm, depending on machine size, and the speed of drum rotation up to 2 rpm, usually from 0.3 to 1 rpm. Applications occur in the manufacture of pharmaceuticals, dyestuffs, edible oils, and various chemicals and minerals.

Horizontal Belt Pressure Filters. Horizontal belt filters have a great advantage in cake washing application due to their horizontal filtration surface. In the context of pressure filtration and the requirements of good dewatering, however, they have a significant disadvantage because the cake is not very homogeneous; gravity settling on the belt and the inevitable problems of distribution of the feed suspension over the belt width cause particle stratification and nonhomogeneous cakes.

A horizontal belt filter has been used in place of the small drum filter in filtration studies (17). The entire filter was placed in a large pressure vessel with no moving parts passing through the filter shell. There is no commercial filter based on this principle; the utilization of the space inside the pressure vessel would be poor and the filtration areas limited.

Another possibility is to enclose only the working, top part of the horizontal belt in a pressure vessel and pass the belt through the sides of the vessel. The operation must be intermittent because the belt cannot be dragged over the support surface with the pressure on, and the entrance and exit ports for the belt must be sealed during operation to prevent excessive losses of air. The movement of the belt is intermittent and is synchronized with decompression in the vessel; therefore, the entire vessel volume must be depressurized in every cycle and this is wasteful. There is also an inevitable downtime. There are no problems with discharging the cake because this is done at atmospheric pressure.

The Flat-bed pressure filter (Hydromation Engineering Co. Ltd.) (19) is based on the above principle. The pressure compartment consists of two halves, top and bottom. The bottom half is stationary while the top half can be raised to allow the belt and the cake to pass out of the compartment, and can be lowered onto the belt during the filtration and dewatering stage. The filter can be considered as a horizontal filter press with an indexing cloth; in comparison with a conventional filter press, however, this filter allows only the lower face of the chamber to be used for filtration.

The same idea has been described except it does not lift the top half of the pressure compartment but opens and closes little gates for the belt to pass through (20). This filter is proposed for dewatering of fine coal in particular.

The vertical recessed plate automatic press, shown schematically in Figure 15 and described previously, is another example of a horizontal belt pressure filter. Cycle times are short, typically between 10 and 30 minutes, and the operation is fully automated. The maximum cake thickness is about 35 mm; washing and dewatering (by air displacement) of cakes is possible. Applications include treatment of mineral slurries, sugar, sewage sludge, and fillers like talc, clay, and whiting.

Continuous Compression Filters. The variable chamber principle applied to batch filtration, as described before, can also be used continuously in belt presses and screw presses.

Belt Presses. Belt filter presses combine gravity drainage with mechanical squeezing of the cake between two running belts. The Manor Tower press (Fig. 25), a Swiss invention also available in the United Kingdom, consists of two acutely angled vertically converging filter belts running together downward. The shallow funnel formed between the belts, and the ends sealed by a special edge-sealing belt, is filled from the top with the slurry to be filtered. The slurry moves with the belts down the vertical narrowing gap where the filtration takes place. The hydrostatic pressure causes the solids to be deposited on the faces of the belts until the two cakes combine to form a continuous ribbon of filter cake which is squeezed in the closing gap at a pressure of 250 kPa, gauge pressure, and discharged at the bottom. To protect the press against overload, the final gap at the bottom is automatically controlled so the above mentioned pressure is not exceeded.

Originally designed for the continuous filtration of conditioned sewage sludges, as were most of the filter belt presses available, the Manor Tower press is increasingly used for the treatment of paper mill sludge, coal, or flocculated clay slurries.

There are probably more than 20 designs of horizontal filter presses available (ca 1993), most developed during the 1980s. They owe their existence to the availability of cationic polyelectrolytes which promote the release of water from

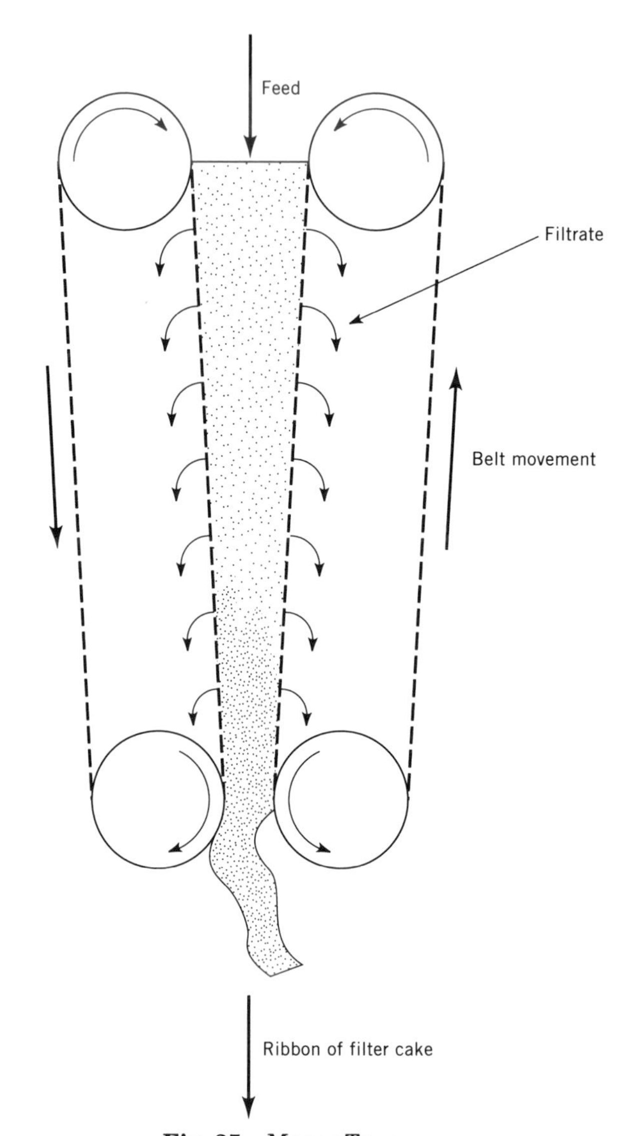

Fig. 25. Manor Tower press.

organic sludges, ie, the area where belt filter presses are most used. Some are therefore called sewage sludge concentrators.

The Unimat belt filter press has most of the features characteristic of many belt presses. The flocculated feed is first introduced onto the horizontal drainage section where the free water is removed by gravity. Sometimes a system of ploughs may be employed to turn the forming cake and allow any free water on the cake surface to drain through the belt mesh. The sludge then is sandwiched between the carrying belt and the cover belt and compression dewatering takes place; liberated water passes through the belts. The third zone, the shear zone, shears the cake by flexing it in opposite directions during passage through a train of rollers in a meander arrangement, to produce drier cake. To prevent the released water from being absorbed back into the cake on the cake release, scrapers or wipers may be installed to remove the water from the outside of the belts.

Belt filter presses are made up to a width of 2.5 m and produce a final solids concentration of the discharge sludge in the range of 35 to 60%.

Apart from the specially designed belt presses described above, the compression principle is utilized in some conventional vacuum drum or belt filters by the addition of compression rollers or belts. The benefit, in terms of further dewatering, is small, if indeed there is any advantage, because the compression time is short and the excluded liquid might be sponged back into the cake before it can be removed. Such devices probably have greater value in fighting cake cracking problems.

Screw Presses. Another way of achieving compression of the cake is by squeezing in a screw press. This is suitable only for the dewatering of rough organic materials, pastes, sludges, or similar materials, because it does not include a filtration stage. The material is conveyed by a screw inside a perforated cage. The volume available continuously diminishes, either by reducing the pitch of the screw in a cylindrical cage or by reducing the diameter of the screw, in which case the cage is conical, as shown in Figure 26. The cage is either perforated or constructed from longitudinal bars in a split casing. The solids discharge is controlled by a suitable throttling device which controls the operating pressure. Washing or dilution liquid can be injected at points along the length of the cage. The power requirements are large.

The Stord twin screw press uses two counter-rotating, intermeshing screws in a perforated cage. The gradual reduction in the space between the screw flights and the strainer plates is achieved by a combination of reduction in the cage, and screw, diameter and of increase in the diameter of the hollow trunk of the screw body. Although more than one type of material can be handled in the same press, the press usually has to be adapted to different raw materials, according to their dewatering characteristics. Typical applications include dewatering of sugar beet pulp, fish meal, distillers and brewers spent grains, starch residues, fruit, potato starch by-products, grass, maize, leaves, and similar materials.

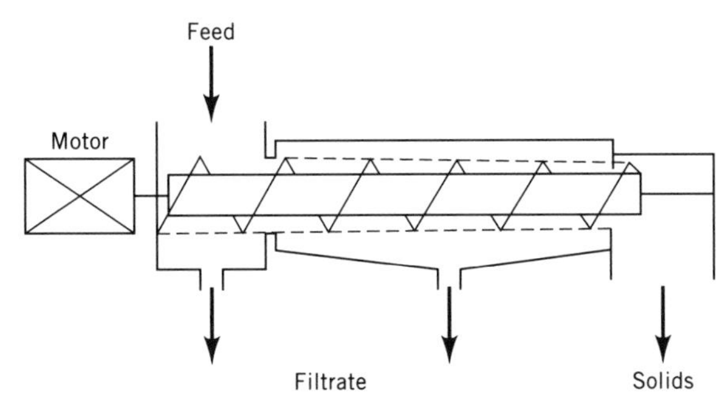

Fig. 26. The principle of a screw press.

Thickening Pressure Filters. The most important disadvantage of conventional cake filtration is the declining rate due to the increased pressure drop caused by the growth of the cake on the filter medium. A high flow rate of liquid through the medium can be maintained if little or no cake is allowed to form on the medium. This leads to thickening of the slurry on the upstream part of the medium; filters based on this principle are sometimes called filter thickeners.

The methods of limiting cake growth are classified into five groups, ie, removal of cake by mass forces (gravity or centrifugal), or by electrophoretic forces tangential to or away from the filter medium; mechanical removal of the cake by brushes, liquid jets, or scrapers; dislodging of the cake by intermittent reverse flow; prevention of cake deposition by vibration; and cross-flow filtration by moving the slurry tangentially to the filter medium so that the cake is continuously sheared off. The extent of the commercial exploitation of these principles in the available equipment varies, but cross-flow filtration is exploited most often.

Removal of Cake by Mass Forces. This method of limiting cake growth employs mass or electrophoretic forces on particles, acting tangentially to or away from the filter medium. Only mass forces are considered here because the electrophoretic effects have been discussed previously.

Because gravity is too weak to be used for removal of cakes in a gravity side filter (2), continuously operated gravity side filters are not practicable but an intermittent flow system is feasible; in this arrangement the cake is first formed in a conventional way and the feed is then stopped to allow gravity removal of the cake. A system of pressure filtration of particles from 2.5 to 5 μm in size, in neutralized acid mine drainage water, has been described (21). The filtration was in vertical permeable hoses, and a pressure shock associated with relaxing the hose pressure was used to aid the cake removal.

Much greater accelerations can be obtained in centrifuges and this makes the idea of a side filter much more attractive. A side filtering centrifuge which combines the conventional centrifugal filtration in a cyclindrical basket with filtration through the sides, as sometimes used in peeler centrifuges, has been described (2), as has the basic theory of the side filtration process, sometimes called by-pass filtration. It has been shown that such filters can reduce the filtration time to less than 35% of the normal value. A planetary filter centrifuge which has two or more baskets rotating about their own respective axes as well as spinning on arms around a parallel central axis in a planetary fashion has been patented (22). This arrangement under certain conditions gives rise to such centrifugal forces which keep a part of the medium free of particles and thus allow some filtrate to by-pass the cake. The side filtration concept seems attractive; it remains to be seen whether it will find wider application in practice.

Mechanical Cake Removal. This method is used in the American version of the dynamic filter described under cross-flow filtration with rotating elements, where turbine-type rotors are used to limit the cake thickness at low speeds. The Exxflow filter, introduced in the United Kingdom, is described in more detail under cross-flow filtration in porous pipes. It uses, among other means, a roller cleaning system which periodically rolls over a curtain of flexible pipes and dislodges any cake on the inside of the pipes. The cake is then flushed out of the curtain by the internal flow.

Dislodging of Cake by Reverse Flow. Intermittent back-flushing of the filter medium can also be used to control cake growth, leading to filtration through thin cakes in short cycles. Conventional vacuum or pressure filters can be modified to counter the effects of the forces during the back-flush (23,24).

A filter based on continuous backwash by spray has been announced (25); the feed is introduced tangentially into a cylindrical vessel, which has a cylindrical screen in the center. The filtrate leaves through the screen and the solids deposit on it. A rotary spray system mounted on the inside of the screen cylinder goes around continuously at 115 rpm and sprays water radially outward, thus removing the solids that fall down and are removed from the filter through the solids outlet. Full backwashing may become necessary in addition to this spray cleaning and provision is made for this. The medium is made of polypropylene or nylon, available down to 20-μm pore size. Flow rates up to 68 m^3 per hour can be treated at feed concentrations up to 0.2%. There is a minimum underflow discharge rate of 1.1 m^3 per hour leading to relatively dilute underflows.

Prevention of Cake Deposition by Vibration. Vibration provides another method for preventing the formation of a dense filter cake. However, commercial exploitation of this principle is uncommon. A notable exception is the vibrating slurry filter (26). Each filter unit consists of three 38-mm dia tubular screens enclosed in a cylindrical housing and made from wedge-shaped wires. The screen openings available go down to 44 μm (325 mesh). The tubular filter elements are continuously vibrated pneumatically at high frequency and low amplitude. This keeps the oversize particles in suspension and reduces premature screen blinding. The solids are removed by intermittent backwash. Two or more filters can be manifolded in parallel to common headers to provide higher flows. In such cases the service can be continuous, as individual filters can be backflushed while all other filters remain on-stream. The filters are designed for pressures up to 100 kPa, the areas available are multiples of 0.3 m^2, and flow rates of 8 m^3 per hour can be handled by a single unit. Applications include clarification of paper coatings, colloidal gels, ceramic slips, or calcium carbonate slurries.

Cross-Flow Filtration. In conventional filtration, the flow of the suspension is perpendicular, ie, dead end, to the surface of the filter medium, with all the liquid passing through it, except for the small amount of moisture retained in the cake. In cross-flow filtration, the slurry flow is tangential to the filtration medium (Fig. 3) at high relative velocities with respect to the medium. The shear forces in the flow close to the medium continuously remove a part or the whole of the cake and mix it with the remaining suspension. The filtration velocity through the medium, due to the pressure drop across it, is small relative to the velocity of the flow parallel with the medium, typically by three orders of magnitude. There is little tendency for any colloidal and particulate matter to accumulate in the boundary layer. In most cases, however, a layer of fines exists within the boundary layer, leading to the so-called dynamic membrane effect which limits the performance.

As more and more of the filtrate is removed, the slurry gradually thickens and may become thixotropic. The solids content of the thickened slurry may be higher than that obtained with conventional pressure filtration, by as much as 10 or 20%. A range of velocity gradients from 70 to 500 L/s has been suggested as necessary to prevent cake formation and to keep the thickening slurry in a fluid state (27).

Cross-Flow Filtration with Rotating Elements. The first patented method of limiting cake growth by this means (28) was a cylindrical, rotating filter element, mounted in a cylindrical pressure vessel. The suspension was then pumped into the annulus between the rotor and the vessel (23). More recently, virtually all possible combinations and arrangements for such dynamic filters have been patented (27). Figure 27 gives an example of one such arrangement. It comprises a pressure vessel with a hollow filter leaf inside, connected to and rotating about a hollow shaft. The slurry is continuously fed at a constant rate into the vessel, and the filtrate passes through the rotating filter medium and out through the hollow shaft. The thickened slurry is discharged via a control valve used to maintain optimum slurry concentration.

As observed from Figure 27, the cake removal by fluid shear is also aided by centrifugal force. Other arrangements include stationary filtration media and rotating disks to create the shear effects, and rotating cylindrical elements; it has also been shown how such filters can be used for cake washing.

Since its conception, the dynamic filter has been widely reported and further developed. Most European designs are comprised of a multistage disk arrangement (Fig. 28) with both the rotating and stationary elements covered with filter cloth, thus utilizing the space inside the pressure vessel. Such filters have been found (29) to be from 5 to 25 times more productive in mass of dry cake per unit area and time than filter presses for the same moisture content of the final slurry. In some cases, the moisture content with the dynamic filter was actually lower than with a filter press. The maximum productivity was achieved with peripheral disk speeds from 2.8 to 4.5 m/s.

The axial filter (Oak Ridge National Laboratory) (30) is remarkably similar to the dynamic filter in that both the rotating filter element and the outer shell

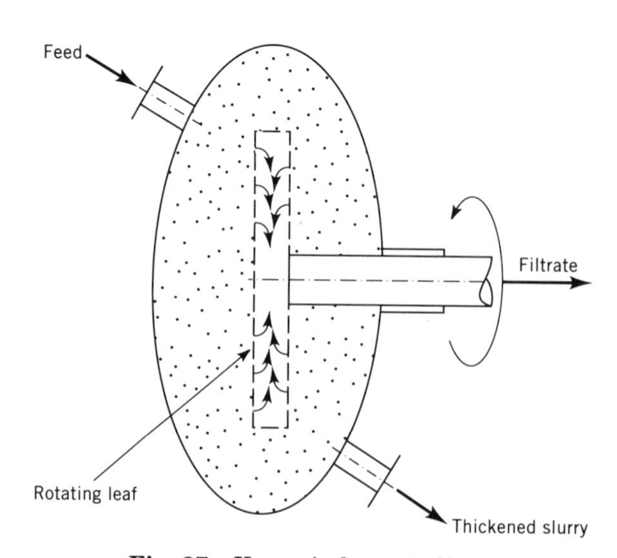

Fig. 27. Kaspar's dynamic filter.

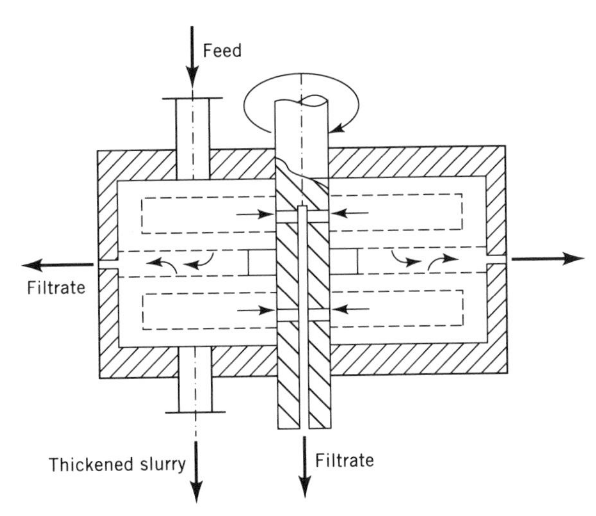

Fig. 28. European dynamic filter.

are also cylindrical. An ultrafiltration module based on the same principle has also been described (31). Unlike the disk-type European dynamic filters described above, the cylindrical element models are not so suitable for scale-up because they utilize the space inside the pressure vessel poorly.

The idea of axial filtration with a cylindrical rotating element is taken a step further in the Escher-Wyss pressure filter (32). In addition to the cylindrical rotating filter surface, it provides a stationary outer filter surface (Fig. 29) so that the suspension, passing through the narrow annulus between the two cylinders, filters through the whole of the wetted surface. Strategically placed washwater entry points along the filter length allow continuous washing, equivalent to reslurrying. This filter is particularly suitable for viscous, plastic, or thixotropic slurries.

All the dynamic filters that use rotating filter elements are subject to several disadvantages. First, the filter medium, usually cloth, is stretched by the centrifugal force, and this imposes special requirements on its fastening and stretch resistance. Second, the filtrate is collected into the element radially inward, ie, against the centrifugal force. This reduces filtration velocities, and the necessity for discharge through a hollow shaft makes the filter more complicated. Finally, any solids that may penetrate through the medium might accumulate inside the rotating elements due to centrifugal settling, with the accompanying danger of partial reduction in the area of the cloth available for filtration.

The three disadvantages described can be avoided by using solid elements, instead of permeable ones, which create the shear to prevent or reduce cake formation. Only the stationary surface inside the filter is then available for filtration and this means a reduction in capacity. This is not a problem because the solid disks can be slimmer and the collection of filtrate does not have to be through a hollow shaft.

The American version of the dynamic filter, known as the Artisan continuous filter (Fig. 30), uses such nonfiltering rotors in the form of turbine-type elements.

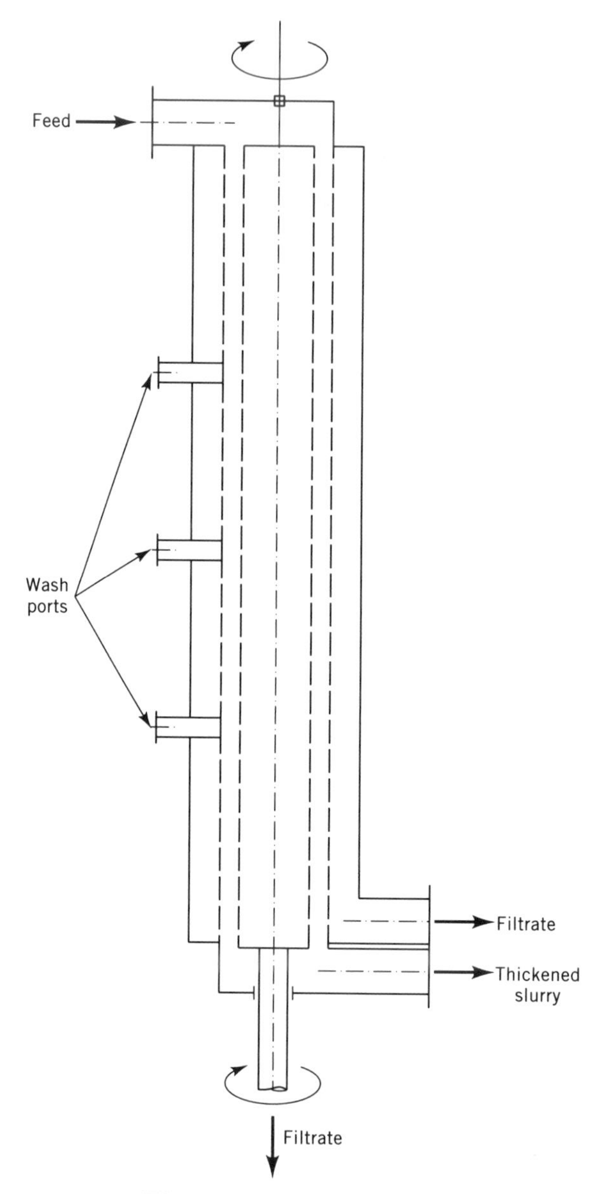

Fig. 29. Escher-Wyss filter.

The cylindrical vessel is divided into a series of disk-type compartments, each housing one rotor, and the stationary surfaces are covered with filter cloth. The feed is pumped in at one end of the vessel, forced to pass through the compartments in series, and discharged as a thick paste at the other end. At low rotor speeds the cake thickness is controlled by the clearance between the scraper and

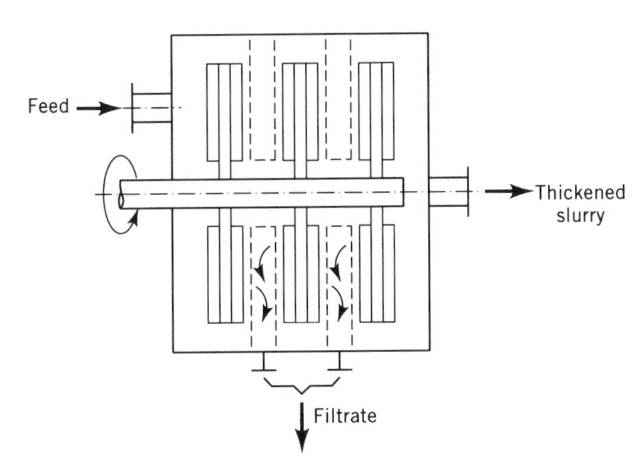

Fig. 30. Artisan continuous filter.

the filter medium on the stationary plate, while at higher speeds part of the cake is swept away and only a thin layer remains and acts as the actual medium.

Results of test work with this filter, producing cakes of 1 mm thickness using a 3 mm clearance, have been published (33,34). The cake formed on the medium was generally stable, giving high filtration rates over long periods of time, and the precoat type cake did not blind with time. There was no evidence of any size selectivity of the process; the only exception was conventional filter aids which were preferentially picked up by the rotating fluid. This was attributed to inertial by-passing of the cake by larger particles and reentrainment of the smaller ones. Good control of the agitator speed and operating pressure is recommended in order to maintain a balance between the overall filtration and slurry flow rates so that the critical mud concentration and excessive torque are avoided.

The Artisan continuous filter operates at pressures from 150 to 200 kPa, it allows washing by injection of wash liquid after the initial filtering stages, and the sizes available range from 0.3 to 19 m^2. Abrasive wear on the cloth is claimed to be eliminated by the thin protective layer of cake on the medium.

Because the energy going into the agitation converts mostly into heat, the extruded cake may be quite hot, from 50 to 80°C in some cases. High torque is needed to drive the agitators and most, if not all, of the energy saved by reducing the cake thickness may go back into the agitation. Little information exists about the actual power requirements but a favorable comparison has been made with a solid bowl centrifuge operated with the same calcium carbonate slurry (35). It is indisputable, however, that higher productivity is obtained per unit area of the filter surface than with conventional pressure filters, ie, the solids dry cake yield varies from 20 to 1000 kg/m^2h for slurries such as pigments, calcium carbonate, magnesium hydroxide, and kaolin. Other materials handled by this filter include dyestuffs, polymer slurries, clays, pharmaceuticals, and metal oxides.

Other tests with the American design of the dynamic filter have been reported (36) in which the cake was eliminated completely by spinning the disks faster. No visible abrasion of the filter medium was observed while the filtration velocities could be maintained high, even at lower pressures below 50 kPa. The work showed clearly the effect of speed and number of vanes on the rotors, on filtration velocities, and specific energy input. The specific energy input in kWh per m^3 of the filtrate is shown to increase with rotor speed for any number of rotor vanes. This increase can be minimized by optimization of the number of vanes. This does not justify the use of higher speeds in terms of running costs but a smaller filter can be used for the same capacity.

Cross-Flow Filtration in Porous Pipes. Another way of limiting cake growth is to pump the slurry through porous pipes at high velocities of the order of thousands of times the filtration velocity through the walls of the pipes. This is in direct analogy with the now well-established process of ultrafiltration which itself borders on reverse osmosis at the molecular level. The three processes are closely related yet different in many respects.

The idea of ultrafiltration has been extended in recent years to the filtration of particles in the micrometer and submicrometer range in porous pipes, using the same cross-flow principle. In order to prevent blocking, thicker flow channels are necessary, almost exclusively in the form of tubes. The process is often called cross-flow microfiltration but the term cross-flow filtration is used here.

The use of various porous tubes for the cross-flow filtration of hydrated alumina and red bauxite was first reported in 1964 (37). The tubes used were ceramic (of 30 to 40% porosity) and titanium, at specific outputs ranging from 0.3 to 8 m/h. Dead-end filtration was compared with cross-flow, the influence of the slurry flow rate and pressure on the filtrate rate was studied, and a multipass model filter resembling a heat exchanger was developed. Reference 37 also describes the first theoretical and empirical formulas for the total length of tubes needed for a specific application, from known average specific output of filtrate, for the filtration velocity from geometric and physical parameters in dimensionless groups, and for the degree of thickening and the variation in the solids concentration along the tube length.

Results of an investigation into the thickening of kaolin clay in woven fiber polyester hoses was published (38). The results achieved production rates from 0.5 to 1.7 m/h and found temperature and slurry velocity to be the primary variables affecting the filtration velocity. The filtration velocity was proportional to the circulation velocity. Sewage filtration in permeable hose pipes (precoated with filter aid), ceramic tubes, carbon tubes, various porous metal, and plastic screens with activated carbon added has been reported (39). A minimum filtration velocity of 0.17 m/h was suggested as a limit of economic viability. The slurry velocities required for particle transport have been studied (40) and the separation of particles using porous stainless steel tubes examined. The relationship between filtration velocity and the slurry velocity was found to be linear, for magnesium carbonate in this case, but not going through the origin. The two constants in the linear equation were found to depend on particle size. Reference 40 also suggests the use of precoating techniques for cross-flow filtration in tubes.

Other investigations of cross-flow filtration include the study of the concentration of bacteria (41), the concentration of fermentation cell debris (42), the concentration of electrocoating paint (43), the chemical effects on cross-flow filtration of primary sewage effluent (44), and the use of tubes of different materials, dimensions, and porosity with several slurries (45).

Centrifugal Filters

The driving force for filtration in centrifugal filters is centrifugal forces acting on the fluid. Such filters essentially consist of a rotating basket equipped with a filter medium. Similar to other filters, centrifugal filtration does not require a density difference between the solids and the suspending liquid. If such density difference exists sedimentation takes place in the liquid head above the cake. This may lead to particle size stratification in the cake, with coarser particles being closer to the filter medium and acting as a precoat for the fines to follow.

In centrifuges, in addition to the pressure due to the centrifugal head due to the layer of the liquid on top of the cake, the liquid flowing through the cake is also subjected to centrifugal forces that tend to pull it out of the cake. This makes filtering centrifuges excellent for dewatering applications. From the fundamental point of view, there are two important consequences of these additional dewatering forces. First, Darcy's law and all of the theory based on it is incomplete because it does not take into account the effect of mass forces. Second, pressures below atmospheric can occur in the cake in the same way as in gravity fed deep bed filters. The conventional filtration theory has been modified to make it applicable to centrifugal filters (7).

Due to good performance and high cost, centrifuges are often referred to as the Rolls-Royces of solid–liquid separation. They have parts rotating at high speeds and require high engineering standards of manufacture, high maintenance costs, and special foundations or suspensions to absorb vibrations. Another feature distinguishing the filtering centrifuges from other cake filters is that the particle size range they are applicable to is generally coarser, from 10 μm to 10 mm. In particular, cake filters that move the cake across the filter medium are restricted to using metal screens, which by their very nature are coarse. No cloth can withstand the abrasion due to the cake forced on the cloth and pushed over its surface. Only the fixed bed, batch-operated centrifuges can use cloth as the filtration medium and be used, therefore, with fine suspensions.

Fixed-Bed Centrifuges. The simplest of the fixed-bed centrifuges is the perforated basket centrifuge (Fig. 31) which has a vertical axis, a closed bottom, and a lip or overflow dam at the top end. Some domestic machines of this type are in use for straining vegetable or fruit juices. In the industrial versions, the basket housing is often supported by a three-point suspension called the three-column centrifuge.

The basket centrifuge is by no means obsolete; it has found a wide range of application in the filtration of slow draining products that require long feed, rinse, and draining times, and for materials sensitive to crystal breakage or that require thorough washing. It can be applied to the finest suspensions of all filtering centrifuges because filter cloths may be of pore size down to 1 μm. The cloth is supported by a mesh.

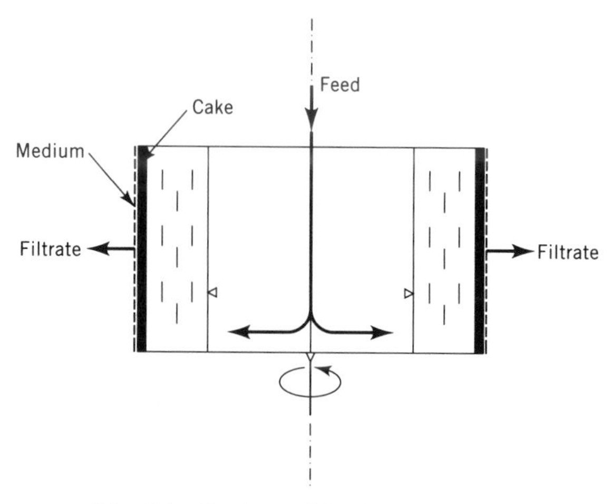

Fig. 31. Perforated basket centrifuge.

The basket centrifuge exists in many different versions. The slurry is fed through a pipe or a rotating feed cone into the basket. The cake is discharged manually by digging it out, or some versions allow the cloth to be pulled inside out by the center causing the cake to discharge; the axis of rotation has to be horizontal in this case. The latter method is particularly suitable for crystalline or thixotropic materials. Alternatively, the cake can be ploughed out by a scraper which moves into the cake after the basket slows down to a few revolutions per minute. The plough directs the solids toward a discharge opening provided at the bottom of the basket, through which the cake simply falls out. The speed of the bowl varies during the cycle, ie, filtration is done at moderate speed, dewatering at high speed, and cake discharge by ploughing at low speed. The frequent changes in speed lead to dead times which limit the capacity.

The plough cannot be allowed to reach too close to the cloth and some residual cake remains. Where this is not acceptable, the cake may be removed by a pneumatic system, by vacuum, or by reslurrying. The cycle can be automated and controlled by timers. The maximum speeds of basket centrifuges vary from 800 to 1500 rpm, and basket diameters are in the range from 10 to 1400 mm. A 1200-mm diameter, 750-mm long basket may handle as much as 200 kg of cake in one charge.

The vertical axis of the basket centrifuge may cause some nonuniformity due to the effects of gravity, with the accompanying problems when cake washing is used. This can be eliminated by making the axis horizontal. This is known as the peeler centrifuge shown in Figure 32. This is designed to operate at constant speed so that the nonproductive periods of acceleration and deceleration are eliminated. The cake discharge is also affected at full speed, by means of a sturdy knife which peels off the cake into a screw conveyor or a chute in the center of the basket. A narrow-bladed reciprocating knife is usually preferred to the faster full-width blade, because it causes less crystal damage and less severe glazing of the heel surface. The presence of the residual heel is not avoidable in either case, and it may be necessary to recondition it by suitable washes in order to restore its drain-

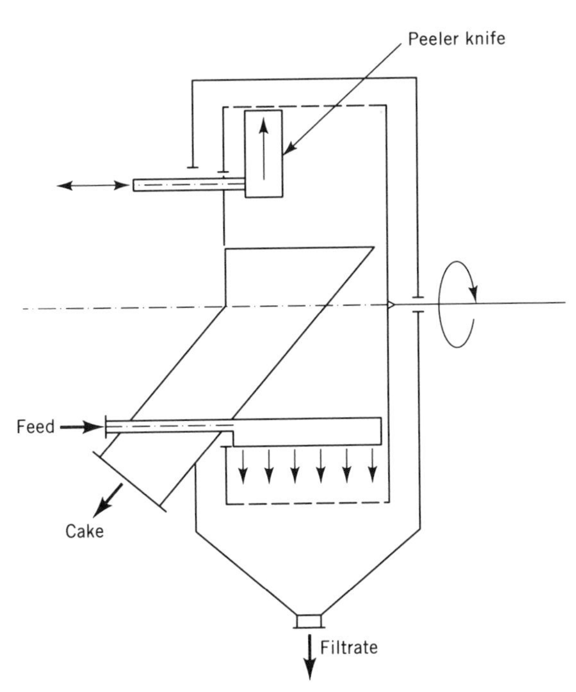

Fig. 32. Peeler centrifuge.

age properties. Some designs incorporate cake feeler/leveler mechanisms which ensure that an even cake is formed.

An interesting improvement of the peeler principle is the addition of a rotating syphon in the form of an additional outer filtrate chamber. This retains the filtrate at a larger diameter than the basket diameter, therefore syphoning higher flow rate through the cloth than would otherwise be achieved. An adjustable syphon pipe controls the total pressure drop across the filter medium. Large increases in throughput are claimed due to the use of the rotating syphon but the additional chamber increases the size of the bowl and similar gains could be obtained by simply making the size of the conventional bowl as large as the one with the syphon. It is also not always desirable to lower the absolute pressure in the cake or the medium because cavitation may result due to the release of gas or liquid vapor.

The peeler centrifuge is attractive where filtration and dewatering times are short. The principal application is for high output duties with nonfragile crystalline materials giving reasonable drainage rates which require good washing and dewatering. Basket diameters range from 250 to 2500 mm and speeds from 750 to 4000 rpm.

Moving Bed Centrifuges. The continously fed, moving bed machines are available with conical or cylindrical screens. As can be seen in Figure 33, the conical screen centrifuges have a conical basket rotating either on a vertical or horizontal axis, with the feed suspension fed into the narrow end of the cone. If the cone angle is sufficiently large for the cake to overcome its friction on the screen, the centrifuge is self-discharging. Such machines cannot handle dilute

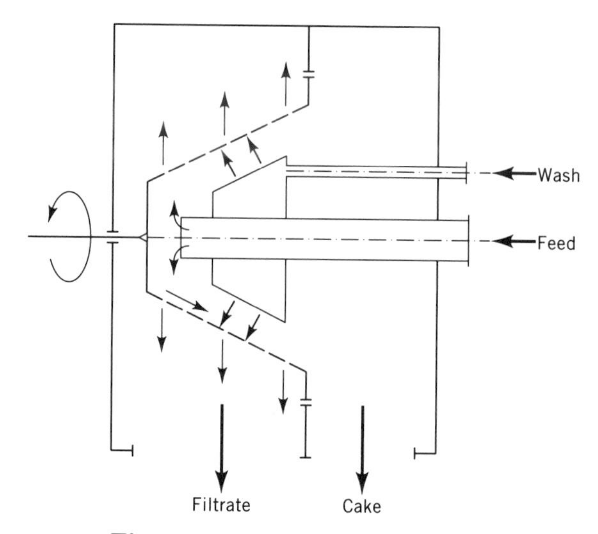

Fig. 33. Conical basket centrifuge.

slurries and require high feed concentrations of the order of 50% by mass, eg, as occurs in coal dewatering. Different products, however, require different cone angles. Unnecessarily large angles shorten the residence time of the solids on the screen surface and thus lead to poor dewatering. The movement of the solids along the screen may be assisted by vibrating the basket either in the axial direction or, in a tumbler centrifuge, in a tumbling action.

A positive control of the solids on the screen is achieved with the scroll-type conical screen centrifuge. Here the scroll moves slowly in relation to the bowl, conveying the solids along the bowl; variable speed differential drive allows good control of the solids movement. Crystal breakage is more severe due to the action of the scroll. Basket speeds vary from 900 to 3000 rpm, while the basket diameters range from 300 to 1000 mm. Washing is possible with all the conical screen centrifuges but it is not very efficient, giving poor separation of the mother liquor and the wash liquid.

The pusher-type centrifuge has a cylindrical basket with its axis horizontal. The feed is introduced through a distribution cone at the closed end of the basket (Fig. 34) and the cake is pushed along the basket by means of a reciprocating piston that rotates with the basket. The screen is made of self-cleaning trapezoidal bars of at least 100 μm in spacing, so that only particles larger than 100 μm can be efficiently separated by this machine. Each stroke of the piston stops just short of the discharge end of the basket, thus allowing a layer of the cake to act as a lip or a dam to effect the axial retention of the feed suspension. If the feedstock is allowed to rise to a height exceeding this remnant cake thickness, the suspension will overflow at the discharge end and rapidly erode a deep furrow in the cake. This is highly undesirable and an overflow limit exists as an upper limit for the feed rate (2).

Pusher centrifuges can be made with multistage screens consisting of several steps of increasing diameter. This is advantageous for liquids of high viscosity, or where the cake is soft, plastic, or of high frictional resistance to sliding. It also

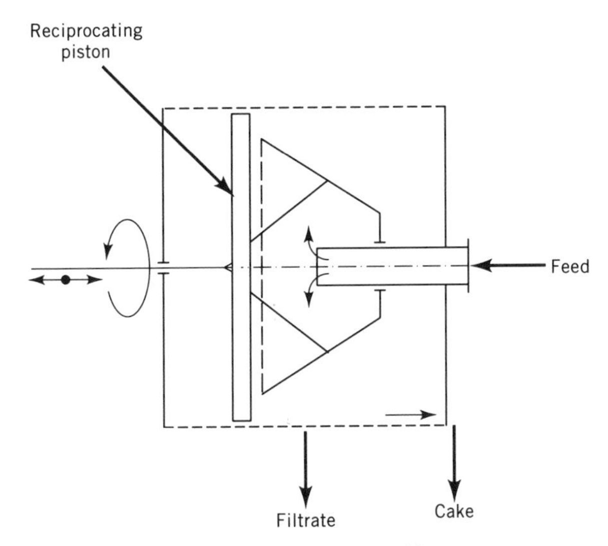

Fig. 34. Pusher centrifuge.

makes washing better, the washing liquid being introduced over the second stage screen. The transit of the solids over the step reorients the particles.

Pusher centrifuges require high feed concentrations to enable the formation of a sufficiently rigid cake to transmit the thrust of the piston. The diameters vary from 150 to 1400 mm, the stroke frequency from 20 to 100 strokes per minute, and the solids handling capacities up to 40 metric tons per hour or more.

The scale-up of filtration centrifuges is usually done on an area basis, based on small-scale tests. Buchner funnel-type tests are not of much value here because the driving force for filtration is not only due to the static head but also due to the centrifugal forces on the liquid in the cake. A test procedure has been described with a specially designed filter beaker to measure the intrinsic permeability of the cake (7). The best test is, of course, with a small-scale model, using the actual suspension. Many manufacturers offer small laboratory models for such tests. The scale-up is most reliable if the basket diameter does not increase by a factor of more than 2.5 from the small scale.

BIBLIOGRAPHY

"Filtration" in *ECT* 1st ed., Vol. 6, pp. 506–530, by L. E. Brownell, University of Michigan; in *ECT* 2nd ed., Vol. 9, pp. 264–286, by L. E. Brownell, University of Michigan; in *ECT* 3rd ed., Vol. 10, pp. 284–337, by R. M. Talcott, C. Willus, and M. P. Freeman, Dorr-Oliver Inc.

1. R. M. Wells, *Filtr. Sep.* **14**(1), 42–43 (Jan./Feb. 1977).
2. L. Svarovsky, ed., *Solid–Liquid Separation*, 3rd ed., Butterworths, London, 1990.
3. J. A. Oberteuffer, IEEE Trans. Magn. **MAG-10**(2) (June 1974).
4. *Ibid.*, **16**(2), 140–141 (Mar./Apr. 1979).
5. J. H. P. Watson, *Filtr. Sep.* **16**(1), 70–75,78 (Jan./Feb. 1979).

6. R. R. Oder and B. I. Horst, *Filtr. Sep.* **3**(4), 363–369,377 (July/Aug. 1976).
7. P. Chakrabarti, S. J. Hill, and D. Melville, *Filtr. Sep.* **19**(2), 105–107 (Mar./Apr. 1982).
8. F. M. Tiller, "Tutorial: Interpretation of Filtration Data, I," *Fluid/Part. Sep. J.* **3**(2) (June 1990).
9. F. M. Tiller, "Tutorial: Interpretation of Filtration Data, II," *Fluid/Part. Sep. J.* **3**(3) (Sept. 1990).
10. M. S. Willis, W. G. Bridges, and R. M. Collins, *Filtech Conference 1981*, Filtration Society, London, 1981, pp. 167–176.
11. L. Svarovsky, *Solid–Liquid Separation Processes and Technology*, Vol. 5, *Handbook of Powder Technology*, Elsevier, Amsterdam, the Netherlands, 1985.
12. L. Svarovsky and A. J. Walker, "The Effect of Feed Thickening on the Performance of a Horizontal Vacuum Belt Filter," *1st World Congress Particle Technology*, Part IV, 331 Event of EFCE, Apr. 16–18, 1986, NMAmbH, Nürnburg, Germany, 1986.
13. A. Rushton, M. Hosseini, and I. Hasan, *Proceedings of the Symposium on Solid–Liquid Separation Practice*, Yorks Branch of IIChemE, Leeds, UK, Mar. 27–29, 1978, pp. 78–91.
14. F. D. Miles, *Report on Filtration*, Report no. 570010 HO/ENS, Imperial Chemical Industries, 1957.
15. Ger. Pat. DE 3316561 A1 (1984), S. Heintges (to KHD AG).
16. D. Blankmeister and Th. Triebert, *Dewatering Ultrafine Coal Slurries by Means of Pressure Filtration*, Aufbereitungstechnik Nr.1/1986, pp. 1–5.
17. R. Bott, H. Anlauf, and W. Stahl, *Continuous Pressure Filtration of Very Fine Coal Concentrates*, Aufbereitungstechnik, Nr.5/Mai 1984, pp. 245–258.
18. K. Maffei, *Information*, B2781, Trommeldruckfilter TDF, Munich, Germany.
19. D. B. Purchas, *Solid/Liquid Separation Technology*, Uplands Press, Croydon, UK, 1981.
20. U.S. Pat. 4,477,358 (1984), S. Heintges (to KHD AG).
21. J. D. Henry, in N. N. Li, ed., *Recent Development in Separation Science*, Vol. 11, Chemical Rubber Co., Cleveland, Ohio, 1972.
22. Ger. Pat. 2357775 (1976), K. Zeitsch (to Krupp AG).
23. M. Muira, *Filtr. Sep.* **4**, 551–554,556 (1967).
24. R. E. Brociner, *Filtr. Sep.* **9**, 562–565 (1972).
25. Q. Coupler, *Memo £8385*, Ronningen-Peter, Portage, Mich., Aug. 27, 1981.
26. *Chem. Eng.* **80**(6) (Mar. 19, 1973).
27. Brit. Pat. 10,577,015 (Feb. 1, 1967), J. Kaspar, J. Soudek, and K. Gutwirth.
28. U.S. Pat. 1,762,560 (June 10, 1930), C. D. Morton.
29. T. A. Malinovskaya and I. A. Kobrinsky, *9th Symposium on Mechanical Liquid Separation*, Magdeburg, Germany, Oct. 1971, pp. 47–51.
30. M. M. Irizarry and D. B. Anthony, *ORNL-MIT-129*, Oak Ridge National Laboratory, Oak Ridge, Tenn., April 28, 1971.
31. B. Halstrom and M. Lopez-Leiva, *Desalination* **24**, 273–279 (1978).
32. W. Tobler, *Filtr. Sep.* **16**(6), 630–632 (Nov./Dec. 1979).
33. F. M. Tiller and K. S. Cheng, *Filtr. Sep.* **14**(1), 13–18 (Jan./Feb. 1977).
34. A. Bagdasarian, F. M. Tiller, and J. Donovan, *Filtr. Sep.* **14**(5), 455–460 (Sept./Oct. 1977).
35. A. Bagdasarian and F. M. Tiller, *Filtr. Sep.* **15**(6), 594–598 (Nov./Dec. 1978).
36. J. Murkes, *The Second World Filtration Congress 1979*, Filtration Society, London, 1979, pp. 15–24.
37. A. I. Zhevnovatyi, *Int. J. Chem. Eng.* **4**(1), 124 (1964).
38. J. A. Dahlheimer, D. G. Thomas, and K. A. Krauss, *IEC Proc. Des. Dev.* **9**(4) (1970).
39. K. A. Krauss, *Eng. Bull.* **145**(2), 1059 (1974).

40. A. Rushton, M. Hosseini, and A. Rushton, *Proceedings of the Symposium on Solid/ Liquid Separation Practice, Supplement*, Yorks Branch of IChemE, Leeds, UK, Mar. 27–29, 1979, pp. 149–158.
41. J. D. Henry and R. C. Allred, *Dev. Ind. Microbiol.* **13**, 78 (1972); M. C. Porter, *A I Ch E Symp. Ser.* **68**, 120 (1972).
42. M. C. Porter, *A I Ch E Symp. Ser.* **68**, 120 (1972).
43. R. L. Goldsmith, R. P. deFilippi, S. Nossain, and R. S. Timmins, in M. Bier, ed., *Membrane Processes in Industry and Biomedicine*, Plenum Press, New York, 1971, p. 267.
44. H. A Mahlman and K. A Krauss, personal communication to J. D. Henry, 1971.
45. A. E. Ostermann and E. Pfleiderer, "Application of the Principle of Cross-Flow in Solid/ Liquid Microfiltration," in the *Proceedings of the Symposium on Economic Optimization Strategy in Solid/Liquid Separation Processes,* Sociétè Belge de Filtration, Louvaine-la-Neuve, Belgium, Nov. 1981, pp. 123–138.

LADISLAV SVAROVSKY
Consultant Engineers and Fine Particle Software

FINE ART AUTHENTICATION AND PRESERVATION.

See FINE ART EXAMINATION AND CONSERVATION.

FINE ART EXAMINATION AND CONSERVATION

SCIENTIFIC EXAMINATION

The application of scientific techniques to the study of art objects is an interdisciplinary undertaking (1–7). The physical scientist is trained to approach a stated problem by analyzing for the identification of measurable variables and devising means to obtain numerical values for these variables. On the other hand, the art historian relies on the trained eye, enabling visual recognition of stylistic characteristics and the more subjective comparison of these with observations about numerous other art works. Communication between these specialists has required mutual efforts. The development of scientific examinations of art objects has had a synergistic relation with the growth of a new profession: that of the art conservator, a specialist having both scientific and artistic training. The conservator consults and collaborates with both scientists and curators, providing appropriate care to objects in the collections to promote long-term preservation.

Early scientific studies were predominantly aimed at objects often referred to as belonging to the fine arts. Subsequently, equal importance and effort has been attached to studies of objects of cultural and historical interest, such as archaeological and ethnographic materials, or manuscripts, documents, photographs, and books in archives and libraries. This article is meant to be inclusive of all such objects as well as of fine arts objects. The term art object when used is an inclusive, generic connotation rather than an exclusive one.

In the twentieth century the number of technical studies in the arts and humanities has grown exponentially. After the great museums were established, bringing together large collections for academic purposes rather than for personal aesthetic gratification of a private owner, the first museum-based laboratories soon followed. The first such laboratory was founded in 1888 at the Royal Museum of Berlin, the second by the British Museum during World War I, and then one at the Boston Museum of Fine Arts in 1929. The explosive development in analytical techniques, especially during the second half of the twentieth century, has also played a role (see ANALYTICAL METHODS). Many instrumental methods, through high sensitivities and small sample requirements, are extremely appropriate for the study of art objects (see TRACE AND RESIDUE ANALYSIS). For example, metal alloy analysis has a long history of application in numismatics, where the evaluation of relative values of contemporary coinages depends on the knowledge of the intrinsic values. Using the traditional wet chemical analysis, evaluation often meant the loss of half a coin and was met by reluctance on the part of a curator. The use of such techniques as energy dispersive x-ray fluorescence or activation analysis allow nondestructive analyses (see NONDESTRUCTIVE TESTING).

The goals of a museum are generally to further scholarly study of the material in the collection, to use the material for educational purposes through the organization of exhibitions, and to preserve these materials for future generations. The responsibility for the objects in a museum's collection belongs to the curator. Because of the great importance of preservation and prevention of damage, it is necessary that any study involving removal of even the tiniest sample of an object be discussed by all parties involved and concerned, ie, the curator, conservator, and scientist. The goals of the proposed project must be determined; the value of the results estimated, especially in furthering the understanding of the object and its historical context; and the extent of the damage that could be inflicted on the object assessed.

There are several possible reasons why a scientific study of an art work may be desirable. An obvious one is in cases where the authenticity of an object is doubted on stylistic grounds, but no unanimous opinion exists. The scientist can identify the materials, analyze the chemical composition, and then investigate whether these correspond to what has been found in comparable objects of unquestioned provenance. If the sources for the materials can be characterized, eg, through trace element composition or structure, it may be possible to determine whether the sources involved in the procurement of the materials for comparable objects with known provenance are the same. Comparative examination of the technological processes involved in the manufacture allows for conclusions as to whether the object was made using techniques actually available to the people who supposedly created it. Additionally, dating techniques may lead to the establishment of the date of manufacture.

Several interesting cases of well-known forgeries are discussed in References 8 and 9. One rather notorious case is that of the forger Hans van Meegeren, who faked paintings by Vermeer and de Hoogh. Several of his works were acquired by reputable Dutch museums; one was sold to Hermann Goering during the German occupation of the Netherlands. When, after World War II, van Meegeren was arrested and charged with collaboration, he claimed his innocence, protesting that this painting, and many others, were not originals, but rather forgeries by his hand. These claims were hotly contested, and van Meegeren had to prove, while in prison, that he could indeed fake Vermeer's style so well as to deceive the stylistic experts. The technical examination which proved these works to be forgeries was the work of the well known Belgian scientific expert Paul Coremans (10). Later lead-210 [*14255-04-0*], ^{210}Pb, dating confirmed the date of manufacture (11).

A famous case of fake ceramic objects is that of the large Etruscan Warriors in the Metropolitan Museum of Art. The proof of nonauthenticity was provided by a spectrochemical analysis of the glaze. Greek and Etruscan artists applied the decoration using a slip of essentially the same clay as the one from which the object was crafted, obtaining the red and black colors through a special cycle of oxidizing and reducing firing conditions. The black color on the Warriors was found to be the result of the use of a manganese black glaze (12).

The Vinland Map, supposed evidence of the Viking discovery of America long before Colombus, was proclaimed a forgery after the detection in the ink of titanium white, a modern pigment (see PIGMENTS, INORGANIC) (13). Subsequently, however, after another analytical study, the interpretation of the earlier results has been questioned, and the matter of authenticity of this unique document still remains an open question (14,15).

The acclaimed bronze Greek horse in the Metropolitan Museum of Art in New York was declared a forgery in 1967, after an examination which led to the erroneous conclusion that the object was cast by means of a sand-casting technique unknown to the ancient Greeks (16). Subsequent examination not only negated all arguments brought against the sculpture, but even confirmed the age of the bronze through thermoluminescence dating of its casting core (17).

Although the cases in which scientific analysis is called upon to assist in the authentication process tend to draw the most attention and have, occasionally, spectacular results, these are by no means the most important applications of scientific examination in the study of art works. More fruitful and satisfying are those interdisciplinary projects in which an object or group of objects, of unquestionable provenance, is studied jointly by the historian and the scientist, in order to better evaluate historical context. Classifications according to material and technical properties may refine stylistic differentiations, and geographic attributions may lead to the evaluation of past cultural cross-fertilizations (18–23). Without the study of such objects of established provenance or unquestionable attribution, the criteria used in authenticity studies could not be established.

Also of value is the study of the history of technology (24,25) that affords insights into the history of the development of civilization. The earliest existing written records, treatises of crafts people and artists on the techniques and materials with which they worked, date back to medieval times (26–29). For prehistoric human activities, the record is in the objects which remain, and only through the study of these can knowledge in this regard be furthered.

From the museum professional's point of view probably the most important need is for scientific support in the preservation of art objects and other cultural materials (30). In order for the conservator to be able to stabilize a deteriorating object, an understanding of the processes involved in its decay must be reached through study of the material properties and identification of the factors that influence the deterioration process. Subsequently, an evaluation of possible treatments with regard to potential effectiveness, both immediate and long term, as well as other possible consequences, is made. Throughout this process, the active involvement of scientists is necessary. Thus, the scientific role extends from the study of an object and possibly samples taken from it for analysis of materials and deterioration products, through evaluation of long-term compatibility and stability of proposed treatment materials and techniques, to basic research into the properties of materials and the formulation of conditions for storage and exhibit that are most appropriate for effective collections preservation.

Methodologies

One of the particularly challenging aspects of the work in a museum laboratory is the enormous variety of problems encountered. Every object examined is unique and for each the questions to be answered differ. Thus the museum laboratory most closely resembles, if anything, the forensic laboratory, and many of the methodologies employed are common (see FORENSIC CHEMISTRY).

Of specific concern in the museum laboratory is the use of nondestructive techniques where the term nondestructive is taken much more literally than in the general analytical field. The term is taken to mean not requiring sample removal at all. The physical integrity of the object is the first and overriding concern in the museum. There is great reluctance to allow sample removal unless absolutely unavoidable, and even then weighty consideration is given as to whether the expected information justifies the damage inflicted by sampling.

Optical Techniques. The most important tool in a museum laboratory is the low power stereomicroscope. This instrument, usually used at magnifications of 3–50 ×, has enough depth of field to be useful for the study of surface phenomena on many types of objects without the need for removal and preparation of a sample. The information thus obtained can relate to toolmarks and manufacturing techniques, wear patterns, the structure of corrosion, artificial patination techniques, the structure of paint layers, or previous restorations. Any art object coming into a museum laboratory is examined by this microscope (see MICROSCOPY; SURFACE AND INTERFACE ANALYSIS).

Higher magnifications are obtainable using the polarizing research microscope. Samples have to be removed, however, and prepared according to the microscopic technique to be used. Reflected light microscopy, for example, may be useful in the study of a polished metal specimen to determine in-depth structure of corrosion or to examine a cross-sectional sample from a painting for the sequence of paint layers (31). Transmitted light microscopy is the preferred tool in the study of thin sections from stone or ceramic objects, eg, for the examination of the crystal structure of a marble for clues to its geographic provenance, or for determining the technology involved in the forming and decoration of a ceramic vessel. The research microscope is also of great importance in the morphological identification of materials used in the making of art objects, such as pigments (qv), fibers (qv), or woods (see WOOD).

The very high powers of magnification afforded by the electron microscope, either scanning electron microscopy (sem) or scanning transmission electron microscopy (stem), are used for identification of items such as wood species, in technological studies of ancient metals or ceramics, and especially in the study of deterioration processes taking place in various types of art objects.

Examinations utilizing uv or ir radiation are frequently used. Ultraviolet light has been in use for a long time in the examination of paintings and other objects, especially for the detection of repairs and restorations (32). Because of the variations in fluorescent behavior of different materials having otherwise similar optical properties, areas of repaint, inpainting, or replacements of losses with color-matched filling materials often can be observed easily. Also, the fluorescence of many materials changes with time as a result of various chemical aging processes, thus providing a means of distinguishing fresh and older surfaces (2,33). Infrared irradiation is often used in the examination of paintings because, owing to the limited absorption by the organic medium, ir light can penetrate deeply into the paint layers. It is reflected or absorbed in varying degrees by different pigments. Study of the reflected ir image may enable the detection of changes in composition, pentimenti, ie, changes made by the artist to already painted areas, restorations, and, especially important, underdrawings, ie, working drawings applied by the artist on the prepared ground surface. Infrared illumination is also used frequently in the examination of paper artifacts, for example, writing or drawing in now faded ink can sometimes be legible under infrared light. Infrared-sensitive photographic films or vidicon cameras are used to transform the reflected ir image into a visible one (34,35). The vidicon camera with its wider spectral sensitivity extends the wavelength range beyond the limits imposed by the sensitivity of photographic emulsions and, when the image is digitized, allows for computer based image manipulation (see INFRARED AND RAMAN SPECTROSCOPY; SPECTROSCOPY).

Structural Analysis. Some of the optical techniques are also used for structural analysis. Microscopic examinations of metallurgical cross sections or of sections through the paint layers of a painting are indeed structural examinations, as is ir reflectography.

The most well known structural examination technique is probably x-radiography (36), one of the earliest scientific techniques used in the examination of art objects (37,38). Typically, the object is placed in a beam of x-rays generated with an x-ray generator, which provides radiation with a continuum spectral distribution up to the energy corresponding to the accelerating voltage applied over the tube (see X-RAY TECHNOLOGY). The optical density for different parts of the objects varies with the specific absorption and thickness of the constituent materials. Exposure of a photographic plate to the x-rays transmitted through the object provides an image that reflects these density variations.

In the case of paintings, the x-rays are primarily absorbed by the heavy metals present in some pigments, ie, mainly by the lead in lead white, which has historically been used as a principal pigment, both by itself and mixed with other colors. Other pigments containing heavy elements, such as lead-tin yellow and vermilion (mercuric sulfide), are also shown in high contrast. A contrast between the pigments containing the heavy elements and the support which is usually organic materials is desired, as is a large gradation in contrast corresponding to the variation in thickness of the various pigment applications. Thus the conditions under which paintings are radiographed differ distinctly from those normally found in medical applications. This difference is reflected in the voltages applied over the x-ray tubes: 15–35 kV for canvas paintings and up to ca 50 kV for panel paintings or icons. X-radiographs of paintings yield information regarding the development of the composition, eg, the blocking out of certain compositional elements, and changes therein, as well as the technique of the artist.

X-radiography is also used extensively in the examination of other types of art objects to obtain information regarding manufacturing techniques or repairs. For example, joining techniques in wooden objects such as furniture or sculpture can be studied; in a ceramic vase, repaired breaks and losses, even if invisible to the naked eye, show clearly. The application of radiography to metal objects is of special interest. Using industrial radiographical equipment with tube voltages of 200–300 kV, or even gamma-ray sources, information relating to shaping techniques, eg, hammering, raising, repousse, lost wax, or piece-mold casting, or joining techniques, eg, soldering, welding, or crimping, can be obtained, thereby revealing important details about the technology involved in the fabrication of the object.

The advantages of three-dimensional imaging and image processing available with computerized tomography (CT) have not gone unnoticed. Whereas CT scan equipment is certainly not one of the standard instruments in the museum laboratory, access to this technology in medical or industrial facilities has resulted in a number of highly informative studies.

In the examination of works of art on paper, the variations in density are often too small to be revealed by conventional x-radiography. Instead, much benefit has been derived from beta-radiography, where the β-radiation is provided by an extended radioisotope source, typically a sheet of plastic in which a certain amount of carbon-14 [*14762-75-5*], ^{14}C, is incorporated homogeneously through labeling on the monomer. The paper is placed on the extended source, with a photographic film in close contact on top of it.

A variation on the conventional technique of x-radiography is xeroradiography, where imaging is obtained by electrostatic rather than photographic means. Because of special merits such as edge enhancement and improved texture rendition, this technique has certain advantages, eg, in the study of ceramics (39).

Neutron activation autoradiography (40) was developed specifically for application in the study of cultural materials, and has been applied very successfully to the examination of paintings. A painting is exposed to a flux of thermal neutrons produced in a nuclear reactor (see NUCLEAR REACTORS). These neutrons, which penetrate the painting completely, interact with a small fraction of the nuclei of the various chemical elements present in the painting. Radioactive isotopes are produced, each having a characteristic half-life. The radiation, especially β- and x-rays, produces a blackening in a photographic emulsion exposed to it. Thus, by placing a sheet of photographic film in close contact with the painting, a so-called autoradiograph is obtained which shows the distribution of the predominant activities within the painting at the time of the exposure of the film. Owing to the different half-lives, different radioisotopes show a dominant activity level with time after activation. Therefore, a series of exposures made consecutively after activation yields distribution patterns within the painting of various radioisotopes, and hence of the various corresponding elements. A fortunate coincidence is that lead, because of its nuclear properties such as low reaction cross sections and unfavorable halflives, does not show on these autoradiographs. Thus, the information obtained relates to a number of pigments other than lead white, and complements rather than duplicates that obtained from x-radiography.

Chemical Analysis. Virtually any technique for chemical analysis, whether qualitative or quantitative, elemental or molecular, organic or inorganic, has been applied in order to solve a wide variety of questions relating to art objects. In authentication studies, the pigments identified in a painting or the alloy composition determined for a metal object may be compared with data obtained from objects of unquestioned provenance, and thus provide evidence with regard to the authenticity of the object under investigation. In addition to providing reference information, similar measurements, when performed on well provenienced objects, also often lead to inferences with regard to the technology involved in the manufacture of the object, and thus the date and place of introduction of these techniques. Trace-element concentrations of many materials indicate the geographic origins of these materials, and trade and exchange relationships. Identification of deterioration products clarifies questions regarding the decay processes acting on an object, enabling the conservator to design an effective treatment and prevent further deterioration.

In the selection of an analytical technique for a specific problem the arguments brought to bear are virtually the same as in any other area of analytical chemistry: information wanted, sensitivity and precision needed, matrix, etc. A very important additional consideration is the required sample size. The damage resulting from the removal of a sample for analytical study has to be weighed carefully against the benefits expected from this study. Clearly, techniques that require smaller samples are preferred. The developments in analytical techniques, with the advent of extremely sensitive instrumentation, have been of great significance. However, the same problem that accompanies all analyses of very small samples, ie, the accurate representation of the whole object by a small sample, is generally more severe in the study of antiquities. Ancient technologies generally produced materials of a much greater compositional heterogeneity than do modern industrial processes. The same holds true for many materials used directly as they occur in nature, such as stones, clays, etc. Therefore, often the sample size is dictated not so much by the requirements of the analytical techniques available as by the nature of the material under study.

Another complication that is quite typical for this type of work lies in the lack of proper reference standards, again because of the age of the materials in question. A typical example is the identification of organic materials such as paint (qv) media or natural dyestuffs (see DYES, NATURAL). In these materials, slow chemical reactions, over a span of time, generate a number of reaction products that do not occur in newly prepared standard materials; thus, the identification of the original material becomes more complicated. In some cases, the chemical behavior of a material may be changed to such a degree that direct comparison with a fresh standard becomes impossible. One partial solution is the artificial, accelerated aging of standard materials. However, because such aging rarely yields the identical mixture of reaction products as do the natural processes, the value of such standards must be assessed cautiously for each application. Occasionally, analysis yields identification of the degradation product rather than the original material. Thus the experience of the analyst is also of importance.

In a number of cases, identifications have been extremely difficult, because the materials were synthetic and knowledge of their existence had actually been lost. For example, several rather commonly encountered synthetic pigments, such as the lead–tin yellow often found in Renaissance and Baroque paintings, were originally misidentified or left unidentifiable until extensive research, including analyses of elemental composition and chemical and physical properties, and replication experiments, led to proper identification of the material and its manufacturing process.

Dating. The usefulness of dating techniques in the study of art objects, either for authentication purposes or in the evaluation of their historical context, is self evident (41,42).

Radiocarbon dating (43) has probably gained the widest general recognition (see RADIOACTIVITY, NATURAL). Developed in the late 1940s, it depends on the formation of the radioactive isotope ^{14}C and its decay, with a half-life of 5730 yr. After ^{14}C forms in the upper stratosphere through nuclear reactions of nitrogen nuclei and neutrons produced by cosmic radiation, the ^{14}C is rapidly mixed into the carbon-exchange reservoir, ie, oceans, atmosphere, and biosphere, yielding a fairly constant concentration of radiocarbon, ca 1 part ^{14}C per 1012 parts ^{12}C in living organisms. No exchange takes place in dead tissue. Thus, when materials from dead plants or animals are used in the manufacture of an artifact, no exchange takes place, and therefore the radioactive decay process results in a continual decrease of the ^{14}C concentration in the material. The time elapsed since the death of the plant or animal can be determined by comparison of the measured radioactive concentration of the artifact with that of the exchange reservoir.

The concentration of ^{14}C is determined by measurement of the specific β-activity. Usually, the carbon from the sample is converted into a gas, eg, carbon dioxide, methane, or acetylene, and introduced into a gas-proportional counter. Alternatively, liquid-scintillation counting is used after a benzene synthesis. The limit of the technique, ca 50,000 yr, is determined largely by the signal to background ratio and counting statistics.

Application of this technique to dating of art objects has been limited severely by the large samples needed. Even in the case of the most ideal materials, such as wood or charcoal, a sample of several grams was required. For ivory and bone, many hundreds of grams might be necessary. However, more recent developments have changed these requirements drastically. The combination of specially designed miniature gas counters and complex electronics that perform elaborate pulse height and shape analysis has resulted in a reduction of the required sample to a few milligrams of carbon (44). Even smaller samples are sufficient when the ^{14}C concentration is determined using a high energy mass spectrometer, which combines a nuclear accelerator and a magnetic mass filter (see MASS SPECTROMETRY) (45).

A fundamental assumption in radiocarbon dating is that the formation rate of ^{14}C is constant. This is only acceptable as a first approximation. In reality, this rate undergoes significant variations over both short and long periods. Hence, for more accurate dating, corrections are in order. These are obtained by dating samples from individual annual growth rings from long-lived trees, such as the bristle-cone pine and the Irish oak. Graphs in which the radiocarbon dates for such samples are plotted against their actual age determined by dendrochronological technique are used to apply what are called the bristle-cone pine corrections to the radiocarbon date obtained for an unknown material. Bristle-cone pines can reach ages of 4000 years. The use of dendrochronological techniques has enabled the determination of the actual age for rings of trees that died in the past. Thus, correction factors have been established for dates up to ca 8000 years ago (46). Because of the variations in radiocarbon production rate, the calibration curve of real age vs radiocarbon age is not smooth, and as a result there are several instances where the radiocarbon date obtained can correspond to two quite different calendar dates. Whereas larger samples, longer counting times, improved signal-to-noise ratios, etc, all can improve the experimental accuracy, this inherent error cannot be overcome.

Dendrochronology depends on the variations in annual growth rates for trees with the climatic conditions during the growing season. Wet summers result in thicker growth rings than dry ones, and because all trees in a climatic region are affected to the same degree, the climatic variations in consecutive summers result in identical patterns of thicker and thinner rings for all trees. The pattern for the outer rings of an older tree are the same as that of the inner rings of a younger tree with overlapping lifespan. Continuous overlaps to successively younger trees allow dating through the counting of the total number of rings.

Dendrochronological sequences have been established for many climatic regions. This is important not only for the corrections thus obtained for radiocarbon dates, but also for the direct dating of the wood used in the manufacture of art objects. For example, this technique is frequently used to date the wood used in paintings on panels (47,48). Measurement of the relative thickness variations in a number of consecutive rings which can be observed on the edge of the panel allows the placing of this pattern in the chronological sequence developed for the area of origin.

Thermoluminescence dating (41,42,49) has, in the relatively short time since its development, had a tremendous impact on the technical study of ceramic materials. Absorption of radiation energy, such as from naturally occurring radioisotopes, in a nonconductor results, via a sequence of primary ionizations and secondary excitations, in the production of electrons and electron-hole pairs having energy levels in the conduction band. Having such energies, the electrons and electron-hole pairs are able to move freely through the crystal and enter crystal imperfections. At such imperfections, metastable energy levels exist between the valency and conduction bands. One such type of metastable level is called a trap. An electron or electron hole partially de-excites until it reaches the trap level. Subsequent de-excitation cannot occur because transitions between the trap level and valency band are forbidden. At the trap level there is restriction both in freedom of movement and in the energy of the trapped electron or hole. Only transfer of extra energy to the system, as through heating of the crystal, can free these particles again via re-excitation to conduction band levels. Although most de-excitations to the valency band are nonradiative in nature, some take place in the luminescence centers, which are crystal imperfections of another type.

In a clay, the radiation dose rate is determined, to a large extent, by the concentration of radioisotopes, eg, ^{40}K, and the members of the decay chains of uranium and thorium contained within the constituent minerals (see CLAYS). This results in fairly constant rates of ionizations, and hence of trap population. Because the process has been taking place over a geological time period, the trap-population density in a clay is very high. However, firing, the last step in the manufacture of a ceramic object, results in a depopulation and resets the clock for thermoluminescence dating. Immediately after cooling, repopulation commences, again at a constant rate determined by the dose rate. Thus, the trap-population density is a function of the time elapsed since the firing.

In thermoluminescence dating, a sample of the material is heated, and the light emitted by the sample as a result of the de-excitations of the electrons or holes that are freed from the traps at luminescence centers is measured providing a measure of the trap population density. This signal is compared with one obtained from the same sample after a laboratory irradiation of known dose. The annual dose rate for the clay is calculated from determined concentrations of radioisotopes in the material and assumed or measured environmental radiation intensities.

Applications of thermoluminescence dating extend well beyond the first human ceramic production. In fact, this technique has been used in geological studies such as the dating of lava flows. The accuracy obtained in the dating of ceramics depends largely on knowledge of the environmental history, eg, burial conditions, and physical behavior of the material. The information on burial conditions is needed for the calculation of certain correction factors on the annual dose rate which otherwise have to be estimated. At best, the accuracy of thermoluminescence dating is ca 5–10%. Though for many historical contexts a far more precise dating is possible on stylistic arguments, thermoluminescence dating is very useful in prehistoric archaeology. Accuracy is of less importance in authenticity questions, and it is here that the technique finds extremely widespread use.

A related technique is based on the same principles, only the trap population is measured by means of esr spectrometry (see MAGNETIC SPIN RESONANCE) (50).

Amino acid racemization dating is applied, in particular, to the dating of shell, bone, or ivory objects. In this technique, the degree to which racemization has progressed since the death of an organism serves as a measure of the time elapsed after that event (51). Because the reaction rates are dependent upon the environmental conditions of the material, eg, average temperature, humidity, and pH of the burial soil, the working range of this technique varies with these locality bound data, whereas the accuracy depends on their variation in time.

Another relative dating technique for fossil materials is known as nitrogen–fluorine dating (52). Under burial conditions, bone or ivory loses nitrogen because of groundwater leaching of the amino acids resulting from hydrolysis of proteins (qv), especially collagen [*9059-25-0*]. At the same time, fluorine is absorbed from the groundwater by the hydroxyapatite [*1306-06-5*] in the bone, forming fluorapatite [*1306-05-4*]. Thus, the ratio of nitrogen to fluorine in the bone decreases with time. Another elemental concentration which changes is that of uranium, which can be absorbed by the hydroxyapatite. The reaction rates are dependent on local burial conditions, and the technique only has value as a relative dating for materials excavated or otherwise recovered from the same site.

A few techniques exist that do not provide for direct dating but rather give information as to whether the object is of modern manufacture. One of these is ^{210}Pb dating (53). In the decay series of uranium, the first long-lived member after ^{226}Ra, with its 1622 yr half-life, is ^{210}Pb which has a 22 yr half-life. In a lead ore, ^{210}Pb is in radioactive equilibrium with ^{226}Ra, but lead refining results in a chemical separation in which the greatest fraction of the radium remains in the slag. Thus, refined lead has a much higher ^{210}Pb than ^{226}Ra activity. This disequilibrium continues until the much faster decay of the excess ^{210}Pb lowers the ^{210}Pb concentration to a level where it is once again in equilibrium with the ^{226}Ra activity. The time needed to reach this state is of the order of 4–5 half-lives, ie, ca 100 yr. If an object contains refined lead, eg, in the pigment lead-white in a painting, measurement of both ^{226}Ra and ^{210}Pb activities can indicate whether equilibrium exists between these two radioisotopes and hence whether the lead was refined within the last 100 years or earlier.

Indirect dating, or dating by inference, is the reaching of certain conclusions regarding the date of manufacture of an object from other information obtained in its study, such as the use of certain materials or working techniques. For example, the presence on a painting of certain pigments of which the date of invention is known provides a *terminus post quem,* ie, the painting could have been executed at any time after the invention. On the other hand, the presence of a material of which the manufacture was discontinued at a certain time can provide a *terminus ante quem.*

Types of Objects and Methods Used

Paintings. Paintings are composed of a wide variety of materials. However, for the purposes herein, paintings may be characterized as a group by the paint layers which produce the image.

Technological History (26,54–61). As a first approach, there are three groups of components: supports, paint media, and pigments. The support is the substrate upon which the paint layers are laid down. This can be a specially prepared area on a wall for a wall painting, a wooden panel as in a panel painting, or a fabric in canvas paintings. Paper is a prevalent support in Oriental painting. Other supports are encountered less frequently, eg, metal panels such as copper sheet.

In the earliest known paintings, the primitive cave paintings, paint was applied directly onto the cave wall, with little or no preparation. As early as the Old Kingdom in ancient Egypt, however, wall surfaces were specially prepared using a coating of plaster. In time, the refinement and complexity of the preparation layers increased until in the Renaissance several layers of different composition and fineness were superimposed. Other preparations used, especially in the Far East, consisted of a clay layer.

The use of wooden panels as supports for painting dates back to ancient Egypt. Although the painting served originally as decoration of a wooden object such as a sarcophagus, later the wooden panel became only a support for the painting, as in the mummy portraits from the Fayum, executed during the Roman period. In Europe's Classical period, wooden panels must have been the supports for portable paintings, and remarks by Pliny indicate so; however, none of these works of art has survived. The tradition was carried through into the Christian era, and wooden panels were virtually the only portable supports used in the Middle Ages and early Renaissance.

The types of wood used in European panel paintings varied with the geographic area. In Italy and southern France, the most popular wood was poplar. In the north, oak was the first choice, followed by pine and linden. Another difference between north and south can be found in the preparation of the panel: whereas the ground layer in the southern paintings consists of real gesso, ie, finely ground burnt gypsum [*13397-24-5*], $CaSO_4{\cdot}2H_2O$, in a glue medium, northern artists prepared their ground layer using chalk.

The earliest remaining example of painting on a fabric support is from the twelfth dynasty in Egypt. In Europe, although the technique was known, these supports were not frequently used until the Renaissance, when the increasing size of paintings resulted in a tremendous rise in the popularity of canvas supports, a popularity which has lasted until the present. The fabric used almost exclusively by western painters was linen; in the Orient silk (qv) became a frequently used support. The nature and composition of the ground or priming layer on European canvases varied significantly.

The medium is the binder which provides for the adhesion of pigments. The most important types are the temper media (glue, egg, and gum), the oils, and wax. In addition, for wall painting there is the true fresco technique, where the pigments are laid down in a fresh, wet plaster preparation layer. Several other media have been used, but much less frequently, eg, casein temper. In modern paints, a number of synthetic resins are used for this purpose. Contemporary artist paints are often based on acrylic polymers (see ACRYLIC ESTER POLYMERS; PAINTS).

The use of the various tempera and of wax has been identified on objects dating back to ancient Egypt. The Fayum mummy portraits are beautiful examples of encaustic painting, ie, using molten wax as medium. A rather special variation was the technique used by the Romans for wall paintings. In these, the medium, referred to by Pliny as Punic wax, probably consisted of partially saponified wax. In Europe, wax ceased to be used by the ninth century.

The polysaccharides in gum arabic formed a medium used for illuminated manuscripts and inks (qv) as well as for painting. Gum is also the binder in watercolors.

The proteinaceous gelatins in the various animal glues were also widely used as paint media, as well as in illuminations. Glues, the traditional media in Oriental painting, remained the prevalent binders for ground layers in European painting long after oils had become virtually the only medium for the color layers.

Of the various tempera, egg was the most important in European painting, both in wall and panel painting. It was little used outside Europe. The main period of its use was in the Middle Ages and early Renaissance. After the sixteenth century, however, it was rarely used, as drying oils (qv) had become the preeminent media.

The earliest written references to the use of oils as paint media date from the twelfth century. The van Eycks, who traditionally have been credited as the inventors of oil painting, improved the technique to such a degree that oil quickly replaced egg tempera as the prevalent medium.

Of the various vegetable drying oils which have been used as paint media, linseed oil is the most well known. However, several others, such as walnut and poppyseed oil, and more recently safflower oil, have also been used extensively (see VEGETABLE OILS). Originally, linseed oil was cold pressed; later oil treatments such as boiling, alkali and acid treatment, and the addition of dryers modified the behavior including drying time, and solvent extraction was introduced to improve yield. For a long time, artists prepared their own paints by grinding the pigments in the oil. Ready made, preground paint in a tube is of rather recent origin. It requires modifications to the preparation of the paint to obtain a satisfactory shelf life for the oil–pigment suspension, such as partial prepolymerization of the oil and/or addition of small amounts of wax.

Pigments can be categorized in two different ways: either as inorganic or organic or as natural or synthetic. Organic pigments generally consist of a dyestuff precipitated on an inorganic substrate. The earlier pigments tend to be inorganic and natural, although important exceptions exist. Thus, a pigment prepared by the precipitation of madder onto a gypsum substrate was an important color in Egypt during the Greco–Roman period, and in the Old Kingdom a synthetic pigment, Egyptian blue, ie, copper calcium silicate [*10279-60-4*], was made in large quantities.

For many pigments, a period of time in which they had their widest use can be indicated (54,62,63). Dates of introduction are known either from documentary sources or from identification on paintings of known dates. For some pigments, an approximate date for the discontinuation of use can be assigned. In some cases, knowledge of the preparation process or even the very existence was lost over an appreciable time span.

A varnish is often applied on top of the paint layers. A varnish serves two purposes: as a protective coating and also for an optical effect that enriches the colors of the painting. A traditional varnish consists of a natural plant resin dissolved or fused in a liquid for application to the surface (see RESINS, NATURAL). There are two types of varnish resins: hard ones, the most important of which is copal, and soft ones, notably dammar and mastic. The hard resins are fossil, and to convert these to a fluid state, they are fused in oil at high temperature. The soft resins dissolve in organic solvents, eg, turpentine. The natural resin varnishes discolor over time and also become less soluble, making removal in case of failure more difficult (see PAINT AND VARNISH REMOVERS). Thus the use of more stable synthetic resins, such as certain methacrylates and cyclic ketone resins, has become quite common, especially in conservation practice.

Examination. A typical technical examination of a painting (64,65), intended to ascertain its period, condition, and the degree of previous restoration would likely entail most of the following steps.

A close inspection under normal illumination reveals many indications of the condition of the painting and previous repairs. Also, because oil paints become more transparent with age, pentimenti, which originally would have been invisible after the overpainting, can be observed. Raking light illumination is very useful to determine the extent of cracking, distortions of the support, delaminations of the paint layers, etc. This stage of the examination is often done in close cooperation with stylistic experts. Thus, obvious problematic areas can be identified before the other tests are started.

Information obtained through examination with the stereomicroscope at low magnification relates to the characteristics of the craquelure, the pigments used (grain size, morphology), buildup of paint layers (visible in damages and along cracks), technique of the artist (brush work, use of pure and mixed colors, use of glazes, etc), and condition (eg, damages and losses, amount of inpainting or overpainting). Examination under uv illumination principally provides information regarding more recent restorations or changes.

The infrared reflectogram, in the form of an ir photograph, or the image from an ir-sensitive vidicon system, gives more evidence of restorations as well as pentimenti. The most important use of this technique is in the revealing of underdrawing. The pigments typically used for underdrawing, such as charcoal or certain earth pigments, absorb ir radiation very efficiently whereas the white pigments in the groundlayer are highly reflective. A variation of this technique is transmitted ir photography, where the light source and the camera are placed on the opposite sides of the painting.

X-radiography reveals evidence of damages and losses in the paint layers and the support. Dimensional changes may also be detected, eg, a cut-down painting on canvas lacks, on one or more sides, the telltale deformations of the canvas caused by the tacking to the stretcher. The radiograph illustrates the artist's compositional approach, eg, blocking out of compositional elements, amount of changes in composition during execution of the painting, as well as the painting technique, eg, preparation of support, brush work, thickness of paint layers, and variations in thickness between different areas in the painting. Through the study of x-radiographs, characteristics of a given artist's works can often be established that can serve as benchmarks in the examination of paintings of uncertain attribution.

Identification of pigments present on paintings of unquestioned attribution provides reference information regarding the use of certain pigments in given periods or schools. In the study of paintings with uncertain provenance, the pigments identified in it may either negate the possibility of the proposed attribution or lend more credibility to it. Most pigment analyses are done in one of three ways: microscopy and microchemical tests, x-ray diffraction-powder analysis, or energy dispersive x-ray fluorescence spectrometry. This last technique does not directly identify the pigment, but rather the chemical elements present, and this information often allows deduction of the identity of the pigment. Moreover, the technique can be used nondestructively, without any need for sample removal. Thus, many areas on a painting can be analyzed, giving better overall information.

Microscopic examination of cross sections through the paint layers gives definite information regarding the paint-layer sequence in the area from which the sample was taken (31, 66). This information illustrates the artist's use of underlayers and glazes, superposition of compositional elements, and changes in composition.

The techniques frequently used for identification of the binding media include relatively simple solubility tests, ir absorption spectrophotometry, especially Fourier transform ir (Ftir) (67), specific staining techniques (68), uv-fluorescence microscopy (69), thin layer chromatography (tlc) (70), high pressure liquid chromatography (hplc), gas chromatography (gc) (71), and mass spectrometry (ms) (see CHROMATOGRAPHY). Significant work, eg, entailing the identification of the particular oils used in a painting, has been made possible through the use of gc/ms (see ANALYTICAL METHODS, HYPHENATED INSTRUMENTS) (72). For the identification of modern, synthetic paint media, pyrolysis-gc is of great utility.

In the case of a panel painting, a small sample of the wooden support can be removed, from which a microscopic specimen can be prepared in order to identify the wood used for the panel.

Metal Objects. *Technological History.* In the history of technology, the developments of metallurgy probably provide the most complicated and important chapter. Although new archaeological evidence continually necessitates changes in accepted hypotheses regarding the developments in the use of metals in various periods and geographic locations, enough is known to allow the sketching of a general, overall picture (73–75). Where the earliest developments in metallurgy (qv) took place, in particular the discovery of bronze-making, is a point of intense discussion. The traditional favorite is Anatolia (Turkey); however, other areas in southeast Asia and Europe are frequently considered to be prime candidates.

The first use of metals, ca 6000 BC, was restricted to those that could be found in the native state, ie, copper (qv) and, subsequently, gold and silver (see GOLD AND GOLD ALLOYS; SILVER AND SILVER ALLOYS). Iron was extremely rare and obtained from meteorites (see EXTRATERRESTRIAL MATERIALS). The shaping technique for metals was beating and hammering, and immediately a difficulty was encountered, especially in the work-hardening of copper. The first real metallurgical breakthrough was the discovery of annealing ca 5000 BC.

The invention of the techniques of melting and casting depended on the development of the technology to produce the requisite high temperatures, which may have been first achieved by potters, in ca 4000–3500 BC. Coinciding with this discovery was the development of the technique of smelting copper from oxidized ores, eg, cuprite [*1308-76-5*], Cu_2O, and malachite [*1319-53-5*] (copper hydroxy carbonate), $Cu_2(OH)_2CO_3$. Smelting from these ores is easier than the refinement of copper from sulfidic ores, eg, covellite or chalcopyrite, which involves a roasting step invented ca 2000 years later.

The first intentional metal alloy, bronze (ca 3000 BC), was probably prepared by the smelting of mixed ores, eg, cuprite or malachite for copper and cassiterite for tin (see TIN AND TIN ALLOYS). However, archeological evidence indicates that sometime in the third millenium BC, tin smelting already occurred in Anatolia (76). The invention of roasting techniques needed for the refinement of metals from sulfidic ores probably occurred in the third millenium BC. This discovery enabled the refinement of lead from galena (lead sulfide), and, more importantly, after the invention of cupellation, the separation of silver, present as a significant impurity in lead prepared from argentiferous galena.

During this period, the first iron (qv) was refined from ores by a smelting technique. Because the temperatures necessary for the melting of iron were not yet attainable, the product of this operation was a spongy mass of reduced metal having high amounts of slag, collected as residue from the smelting furnace. Forming of this new metal was only possible through mechanical working, forging, and welding. In ca 1500 BC, steel (qv) making was discovered and by 1000 BC the iron age was well under way. The development of cast iron did not occur in the West until the Middle Ages; in China, probably because of a natural occurrence of phosphorus in iron ores that result in an appreciable lowering of the melting point, the production of cast iron started ca 1000 years earlier.

Of the early copper alloys (qv), the tin bronzes were the most important and superior in terms of technological characteristics. One other alloy prepared from copper and lead was the arsenical coppers or bronzes, which coexisted with tin bronzes throughout the second millenium. These are alloys in the technical sense, although they may well have come about through the refinement of copper from arsenic-rich copper ores. Brass, from copper and zinc, occurs increasingly during the first millenium BC. The first large-scale use of brass in the West came as coinage metal by the Romans. Refinement of zinc from its ores posed serious problems owing to the metal's immediate oxidation upon exposure to air and the low sublimation point of the oxide (see ZINC AND ZINC ALLOYS). Brass was consequently prepared from copper and zinc ores or, from Roman times, from copper and refined zinc oxide [*1314-13-2*]. The refinement of zinc metal was not achieved until medieval times in India, which exported it to the West, where the technique to refine this metal was an eighteenth century invention.

The first forming technique, hammering of the metal, evolved into many different and highly complex cold-working techniques, such as raising, sinking, spinning, turning, relief application by means of repoussé, and, for the production of coinage, striking. A second class of forming techniques was developed upon the ability to melt metals. Originally, casting was done in open stone molds to produce relatively simple forms such as ax heads. The invention of lost-wax casting was the greatest breakthrough in this area. In its simplest variation, a model of the object is made in wax. Subsequently, an outer layer or investment, made of refractory clay, is applied to the wax model. Heating of the assembly results in the molten wax running out, leaving a hollow space within the investment, into which metal can be poured. This method allows for the casting of solid objects only, requiring large amounts of expensive metal for the fabrication of sizable objects. A significant improvement was the development of casting around a core in which the model of the object is made by forming the approximate shape of the object out of a clay and sand mixture, upon which a thin wax layer is applied. The final modeling is done in the wax. Again, an investment is applied and the wax removed by melting. A narrow space is left between the core and investment, into which metal is poured. The object as cast needs finishing work, to remove traces of the casting process. Then, especially with sculpture, the surface is often embellished through polishing, gilding, or deliberate patination.

Lost-wax casting, as described above, destroys the model, making it impossible to cast more than one piece. Not until the Renaissance were variations on the method developed which allowed the casting of multiple copies of the same model using a lost-wax technique (77). Meanwhile, another casting technique had made its entry in the West. In the Medieval period, piece molding was introduced, probably from China, where its use preceded that of lost-wax casting (78). In this technique, the investment is built up around the model in multiple segments which, after drying, can be removed from the model and subsequently reassembled, resulting in an empty mold. Again, a core can be placed within the mold to cast hollow shapes.

Examination. Examinations of metal objects generally include the characterization of the metal, determination of the techniques involved in the manufacture, and study of aging phenomena. Of the latter, the state of corrosion is especially important, both in the examination of an object with the purpose of determining its authenticity, as well as in making an assessment of its state if conservation is needed (78,79). The layer of corrosion products covering the surface of the metal, the so-called patina, can be studied for its composition and structure (see CORROSION AND CORROSION INHIBITORS). Identification of corrosion products is often performed by means of x-ray diffraction analysis. The structure of the patina is studied using the low power stereomicroscope, where attention is directed to the growth pattern, crystal size, layering, and adherence to the metal. Cross sections prepared from samples can be studied under higher power reflected light microscopes to determine the extent of penetration of corrosion processes into the interior of the metal, or the inter- and intragranular corrosion. The examination with the low power microscope also yields information regarding wear patterns, eg, sharp or rounded edges of incised lines, and manufacturing techniques, tool marks, mold marks, etc. When the structure of the metal is studied through examination of an etched sample under high magnification, conclusions can be drawn regarding metallurgical techniques used in the manufacture such as various types of cold working, annealing, casting with preheated or cold molds, etc (80,81).

An important tool in the analysis of the structure of the object is x-radiography. For metal objects, high energy x-rays are needed, obtained from either an industrial radiography instrument or from radioisotope sources (gamma-radiography). The radiograph may give direct evidence important in a condition analysis, such as of small cracks, casting defects, failing joints, etc. It also provides information relating to technique, such as the type of joining technique and the forming technique. For example, the methods developed by Italian Renaissance sculptors to enable them to produce multiple copies of one model using lost-wax casting were elucidated with the aid of radiographical examination (77).

Chemical analysis of the metal can serve various purposes. For the determination of the metal-alloy composition, a variety of techniques has been used. In the past, wet-chemical analysis was often employed, but the significant size of the sample needed was a primary drawback. Nondestructive, energy-dispersive x-ray fluorescence spectrometry is often used when no high precision is needed. However, this technique only allows a surface analysis, and significant surface phenomena such as preferential enrichments and depletions, which often occur in objects having a burial history, can cause serious errors. For more precise quantitative analyses samples have to be removed from below the surface to be analyzed by means of atomic absorption (82), spectrographic techniques (78,83), etc.

Trace-element analysis of metals can give indications of the geographic provenance of the material. Both emission spectroscopy (84) and activation analysis

(85) have been used for this purpose. Another tool in provenance studies is the measurement of relative abundances of the lead isotopes (86,87). This technique is not restricted to metals, but can be used on any material that contains lead. Finally, for an object cast around a ceramic core, a sample of the core material can be used for thermoluminescence dating.

Ceramics. *Technological History.* Archaeologists often divide the neolithic period, the latter part of what used to be called the Stone Age, into pre- and post ceramic, with reference to when ceramics came into production (88–92). Actually, there are occasions of pre-ceramic pyrotechnology, such as in the case of the fifth millenium BC mideastern plaster production (93). So far the earliest occurrence of ceramics is in the 28th millenium BC in Eastern Europe, although here the technique was not used for the production of vessels but of figurines (94).

Early pottery is rather simple, having little surface decoration. Pyrotechnology was comparatively undeveloped and pottery was fired at relatively low temperatures. As kilns became available and were in turn further perfected, higher firing temperatures became attainable, resulting in harder, stronger wares. The high point, before the introduction of modern electrical and oil- or gas-fired kilns, was the development in China of the dragon kiln, a tunnel kiln built against a hillside, with a draft which resulted in the temperatures needed for the firing of porcelain ware.

A second developmental aspect involved knowledge and choice of raw material. Clays differ in both working behavior and firing properties. Choosing the proper clay, cleaning the clay in order to remove naturally occurring accessory minerals or other impurities that affect its properties, adding other substances to modify its working behavior, eg, tempers or plasticizers, and mixing different clays to obtain a material having the desired properties, are developments of the second kind. Deserving special mention are the developments of various artificial-body materials, such as the Egyptian faience (95,96), a paste made out of sand and natron, the latter acting as binder, and flux, or the white-body material developed in the Islamic Near East in imitation of Chinese porcelain, consisting of a mixture of feldspar, quartz sand, and a little white clay.

Several techniques were developed for the forming of the object out of clay. Pots can be made by working a slab of clay into the desired form, by building up with coils, and by throwing on the wheel. This latter technique was unknown in Precolumbian Central America where artisans nevertheless crafted large, perfectly formed vessels. Sculpture can be hand-formed or, when a production system is devised, pressed in molds, eg, the Greek Tanagra figurines. Clay can also be cast into porous molds, which became the most important manufacturing technique in the porcelain industry (slip casting).

The most spectacular developments took place in the decoration of the ceramic surface using paint, slips, and glazes. In painted ware, the pigments are applied to the object after firing. Slips are very fine-grained fractions of clays, which are applied to the object after it has dried but before firing. Slips of different colors can be used to obtain a color effect. A technical and aesthetic high point of slip decoration was reached with the Greek red- and black-painted pottery, where the painting was done with a slip prepared from the same clay as the one used for the buildup of the vessel, and the two colors were obtained by means of a complex firing procedure involving alternating oxidizing and reducing conditions (97).

Glazes (98) are actually glasses. The first glazes were ash glazes, probably initially made by accident when ash from the fire chamber settled on the pots in the kiln. The flux, needed to lower the melting point for the mixture of minerals including quartz, feldspars, and clay minerals, is, in this case, provided by the alkalies from the ash. However, during the Old Kingdom period, the Egyptians used alkaline glazes, in which the alkalies for the flux were added in the form of natron, to glaze objects carved out of steatite. These glazes were already colored through the deliberate addition of metal oxides. Plant and wood ashes have always served as an important source of alkalies. Lead glazes, which were developed around the Roman period, use lead oxide as their fluxing agent.

The ultimate perfection in glaze technology, obtained in Imperial China, depended on highly developed skills, extensive experimentation with glaze formulations, and precisely controlled firing conditions in high temperature kilns (99). Only one color glaze was not developed and perfected by the Chinese potters themselves. The production of a true pink glaze through the introduction of colloidally suspended gold was a Dutch invention, the knowledge of which was brought to China by Jesuit missionaries.

Because glazes are technically glasses, it is not surprising that the development of glass as a separate material relates to that of glazes (see GLASS) (100). There is a certain amount of debate as to whether the first glass was made in Mesopotamia or Egypt. In Egypt, glass beads were already being made during the fifth dynasty (ca 2500 BC) (96). The greatest development came during the New Kingdom (starting ca 1500 BC), when a multitude of colors was obtained in soda-lime glasses formulated out of quartz sand and natron, the flux which occurs naturally in large quantities in Egypt (101). Calcium, necessary for the chemical stability of the glass, was not added intentionally but happened to be present in large enough quantities in the sand. Indeed, the need for the presence of calcium was not realized until the sixteenth century, as the raw materials had always contained sufficient amounts of this element.

Vessels were formed from this material by application around a sand core. The Egyptian tradition was carried through in the Roman period, when glassmaking technology reached unprecedented heights. During this time, glassblowing was invented and the deservedly famous Venetian glassblowers continued this tradition. The developments in the Near East are closely related to those in the West; indeed, the two are interdependent, though each has its own characteristics (102).

Natron and ash of seaweeds provided the sodium which served as the flux in all glasses till the Medieval period. Wood ashes then came into use, which changed the glass formulation to such a degree that potassium salts became the principal fluxing alkalies.

Lead glass, though used in Roman mosaics and later in enamels, was otherwise never important in the West until the development of flintglass in 1676. In China, however, the first glass was a lead glass, which contained barium as the earth alkali instead of calcium. This was probably a consequence of the natural occurrence of barium in the lead ore (103). During the Han period, contact with the West resulted in the introduction of soda-lime glasses with a Roman formulation. Also in this period the barium–lead glasses were replaced by calcium–lead glasses (104).

Examination. Examinations of ceramic objects involve a variety of techniques, depending on the type of information sought (105). If an assessment of condition and state of repair is made, the most important tools are the low power stereomicroscope, x-radiography, and examination under uv light. For the identification of glass degradation products, a number of chemical analytical techniques can be used, especially x-ray diffraction. X-ray diffraction can also be used to great advantage in the analysis of the mineral composition of a ceramic paste. Not only can the naturally occurring minerals provide clues to the geographic provenance of the clay, but the presence of certain minerals formed at high temperatures can give indications of the firing temperatures.

Elemental chemical analysis provides information regarding the formulation and coloring oxides of glazes and glasses. Energy-dispersive x-ray fluorescence spectrometry is very convenient. However, using this technique the analysis for elements of low atomic numbers is quite difficult, even when vacuum or helium paths are used. The electron-beam microprobe has proven to be an extremely useful tool for this purpose (106). Emission spectroscopy and activation analysis have also been applied successfully in these studies (101).

Trace-element analysis, using emission spectroscopy (107) and, especially, activation analysis (108) has been applied in provenance studies on archaeological ceramics with revolutionary results. The attribution of a certain geographic origin for the clay of an object excavated elsewhere has a direct implication on past trade and exchange relationships.

Microscopic examination of ceramic paste, both at low magnification and at high power with prepared cross sections, can be used for petrographic study of the mineral composition and also for the determination of techniques involved in the manufacturing process. For ware decorated with slips and glazes, microscopic examination of a cross section is irreplaceable as a means of studying the technology. For example, in order to observe the phase separations responsible for the effects of some of the unique glazes on Chinese porcelains, the scanning-electron-beam microscope is the only appropriate tool. Radiographic examination, especially using xeroradiography, has proven to be of outstanding utility in studying forming techniques (109).

In archaeology, the ceramic typologies have always formed the basis for the establishment of chronologies. It has been extremely important that an absolute dating technique for ceramic material became available with thermoluminescence dating. This technique does not yet have enough accuracy to be very useful in areas with well established chronologies, where the typological sequences are supported by historical sources or other evidence. For authentication purposes, however, the error margin is not quite so critical, and it is here that this technique has had an enormous impact.

Stone Objects. *Technological History.* Stone was the first material used by humans to make tools. Flint and obsidian were formed by chipping, called napping, into arrowheads, spearpoints, knives, etc. Obsidian, which, as a volcanic glass, forms and keeps a razorsharp edge, was a prized commodity and traded over great distances. When and where metallurgical developments resulted in superior materials for tool use, especially tin bronze, these supplanted stone for that purpose. In pre-Columbian America, however, metallurgy did not lead to the production of tool metals. Rather obsidian remained for a long time the material for cutting tools. When it ceased to be used for tool use stone remained one of the most important materials in architecture and in sculpture.

Whereas the harder, and hence generally more durable, stones such as granite and basalt have been used extensively in both applications, the relative softness especially of the fine grained limestones and marbles allows a fine detail in carving that has often made these materials the favorites of sculptors (see LIME AND LIMESTONE). Marble was used extensively by ancient Greek sculptors and builders, and through Roman times and the Renaissance remained the most important stone for sculpture in Europe. Reference 110 contains a number of interesting articles on historical techniques for its quarrying, dressing, and carving.

A special class of stones are the precious and semiprecious gemstones, with a long history of use in decorations and jewelry. Especially in the Far East and pre-Columbian America, jade was used extensively for carving ceremonial and luxury items. Jade encompasses two different minerals quite similar in appearance and physical properties: jadeite and nephrite. Only the former was found and used in America; the latter is predominant in the Far East (see GEMSTONE TREATMENT).

Examination. The technical examination of stone objects begins with the use of the low power stereomicroscope. This study yields information regarding toolmarks and, hence, cutting techniques, wear patterns, and wear of toolmark edges. Such information is clearly significant in authenticity studies, but also provides an insight into the skill and the tools of the carver.

When a question exists about whether the carving is ancient, an examination under uv illumination can be very helpful. For example, an aged marble surface exhibits, under a properly filtered mercury lamp, a mellow, brownish fluorescence, whereas a fresh surface, under the same conditions, appears purple. It is not uncommon that old carvings have been sharpened; this is detectable with the microscope and under uv examination.

The first question about a stone object is often which stone was used. Indeed, museums have many labels that carry an erroneous identification. X-ray diffraction provides a relatively easy answer for such questions of identification.

The polarizing microscope is one of the most important tools in the study of stone. Petrographic study of a thin section can be extremely helpful in the identification of a stone, and often yields information that can be used in the assignment of the geographic origin of the materials. Thus, the location of the quarry for many marble sculptures has been deduced. However, this method does not always give conclusive results, eg, the marbles from the Greek islands are notoriously difficult to differentiate by means of microscopic techniques (111). A very interesting application of microscopy is in the structural study of the weathered surface layer of a marble sculpture (112). Again, this provides an indication whether the surface is the undisturbed original or is of recent origin. The use of an attachment to the microscope allows the study of a specimen by cathodoluminescence, thereby improving petrographic study (113).

Trace-element analysis provides another approach to these studies, and activation analysis has been applied successfully in provenance studies of, eg, limestone sculpture (114,115). Marble presents a difficulty because of the large degree of inhomogeneity inherent in this material, which is a consequence of its metamorphic genesis. On the other hand, this same property has been used to advantage in establishing whether fragments of broken objects belong together. Multiple sampling along the break planes provides concentration patterns that indicate matching pieces.

For marble provenance studies, the most successful technique seems to be the measurement, through mass spectrometry, of the abundance ratios of the stable isotopes of carbon and oxygen (116). However, no single technique appears to provide unequivocal results, especially in cases such as the different Mediterranean sources, and a combination is often necessary to arrive at an approximate place of origin (117).

Organic Materials. Museums contain large numbers of objects made out of components from plants or animals, including wood, eg, furniture, carvings; fibers eg, textiles (qv), paper (qv); fruits, skin, eg, leather (qv), parchment; bone; ivory; etc. Several of these materials have properties related to their preservation.

Textiles. *Technological History.* As a class, textiles are characterized as composites of flexible, long and slender fibers. This includes weavings, flat woven fabrics, costumes, embroideries, carpets, tapestries, etc, as well as products from other yarn manipulation processes such as tatting, ie, lace; knitting, crocheting, or matting, ie, felt. Examples in museums range from prehistoric and archaeological to modern fiber art.

Fibers (see FIBERS, SURVEY) used in textile production can have a wide variety of origins: plants, ie, cellulosic fibers (see FIBERS, CELLULOSE ESTERS); animals, ie, protein fibers (see WOOL); and, in the twentieth century, synthetic polymers. Depending on the part of the plant, the cellulosic fibers can be classified as seed fibers, eg, cotton (qv), kapok; bast fibers, eg, linen from flax, hemp, jute; and leaf fibers, eg, agave. Protein fibers include wool and hair fibers from a large variety of mammals, eg, sheep, goats, camels, rabbits, etc, and the cocoon material of insect larvae (silk). Real silk is derived from the cocoon of the silkworm, *Bombyx mori,* and for a long time was only produced in China, from which it was traded widely as a highly valuable material. Tussah silk is derived from other insects.

Before the fibers can be spun into yarns, a certain amount of preparation is necessary for cleaning and removal of undesirable accessory materials such as fat, wax, gum, or pulp. The weighting of silk is a process to counter the weight loss resulting from degumming the fibers using heavy metal salts of tin or bismuth. This process affects the durability and long term preservation.

Textile dyes were, until the nineteenth century invention of aniline dyes, derived from biological sources: plants or animals, eg, insects or, as in the case of the highly prized classical dyestuff Tyrian purple, a shellfish. Some of these natural dyes are so-called vat dyes, eg, indigo and Tyrian purple, in which a chemical modification after binding to the fiber results in the intended color. Some others are direct dyes, eg, walnut shell and safflower, that can be applied directly to the fiber. The majority, however, are mordant dyes: a metal salt precipitated onto the fiber facilitates the binding of the dyestuff. Aluminum, iron, and tin salts are the most common historical mordants. The color of the dyed textile depends on the mordant used; for example, cochineal is crimson when mordanted with aluminum, purple with iron, and scarlet with tin (see DYES AND DYE INTERMEDIATES).

Examination. Specific questions arising in the study of textiles include the identification of the textile fiber. Microscopy is the most important approach. For the identification of historical, ie, nonsynthetic, fibers this would seem simple and straightforward. However, identification can become quite complicated as some different fibers have quite similar morphology and, especially for animal fibers, changes in animal husbandry techniques have changed the appearance significantly, making the use of modern fibers of the same animals as comparative standards not always feasible. Hence, it may only be possible to identify the fiber class or group rather than the specific biological source. For more conclusive identification, a combination of a number of techniques may be necessary: optical microscopy on whole fibers and cross sections, electron microscopy, chemical characterization, and measurements of physical properties.

A second common question relates to the nonfiber components of textiles, which arise from processing and finishing, ie, scouring, weaving, dyeing, glazing. The problem lies in the very low amounts of these materials, even though they have significant effects on properties such as color or hand, ie, the feel of the textile, necessitating the use of highly sensitive techniques. A small sample is removed, and the dyestuff is stripped from the fiber through treatment with acid followed by further extraction in an organic solvent. This solution can then be used to identify the dye by means of spectrophotometric absorption techniques (118) or tlc (119,120). Elemental chemical analysis, such as through emission spectroscopy, is used for identification of the mordant.

Dating of textiles is possible by means of radiocarbon dating. Developments in this technique have greatly improved its utility for that purpose, as exemplified by its application in the dating of the Turin Shroud (121).

Paper. Technological History. The making of paper (qv) (122) is a Chinese invention, dating back to ca 200 BC. The knowledge of papermaking came to the West via the silk route; it took ca 1000 years to reach Europe. Paper is a cellulose fiber product. Until the nineteenth century, Western papers were made primarily of cellulose obtained from rags of cotton and linen textiles and/or cordage, composed of cotton and/or flax and hemp fibers. Oriental papers are primarily made of kozo, mitsumata, and/or gampi bast fibers. Fermented, washed, and stamper-beaten rags provided a pulp of relatively pure, stable alpha-cellulose, in long, easily intertwined fibers. After paper had been hand-cast from the pulp (qv) on wire laid-and-chain or woven paper-molds, it could be surface-sized separately either by brush-coating with starch (qv) or by dipping the paper into a tub or vat of gelatin (qv).

The increasing demand for paper could ultimately not be met by hand-made paper, and the invention of the Fourdrinier paper-making machine signified the start of mass production. The concommitant need for much larger amounts of starting material caused nineteenth century papermakers to seek other sources of cellulose pulp, which they found in wood. However, mechanical or ground wood pulp produced a paper of inferior quality. Other wood components besides cellulose, lignin (qv), and natural wood resins cause yellowing and embrittlement of the paper. The shorter fibers produced in the Hollender beaters result in less strength, particularly after aging, and the breakdown of sizing. Hence, the application of mechanical pulp paper soon became quite restricted, its main use as of this writing being in the newspaper industry which has large volume and no long term stability requirements.

Processing woodpulp using sulfite and sulfate, to separate wood fibers by removing lignin, produced chemical woodpulp paper. However, these processes had the additional effect of diminishing the degree of polymerization of the cellulose and consequently the permanence of the paper. Another source of future trouble introduced during this period occurred with the switch from separate external sizing of paper with gelatin to more efficient internal or engine sizing, in which an alum–rosin mixture was added to the pulp. This increased the intrinsic acidity of the paper, which in turn accelerates deterioration, because the hydrolysis of cellulose is acid catalyzed. Thus the brittle book syndrome can largely be attributed to the use of these acidic sizing materials in the late nineteenth and early twentieth century paper industry.

High alpha-cellulose chemical woodpulp paper, machine-made primarily from fast-growing softwoods, sized using alkaline calcium compounds, and loaded with fillers and other additives, constitutes a presumably more stable material. Different types of paper are used for art, manuscripts, documents, books, etc, each having its own properties of color, texture, feel, etc.

Examination. Microscopic examination (123) can identify the fibers present in the pulp (124). Inks, watercolor pigments and media, etc, are analyzed similarly to the pigments and media for paintings. However, sample removal tends to be far more disfiguring and hence constitutes an even more restrictive factor. Watermarks are studied with the aid of beta-radiography. Examination in infrared illumination can assist in the reading of documents of which the ink has faded.

CONSERVATION

Historical Review

The various chemical and physical processes that play a role in the deterioration of art objects are not restricted to the present, even though the contemporary environment has contributed significantly to the rate of decay. Revered masterpieces have lost splendor throughout the ages. Indeed, from textual evidence, it is known how artists in the Renaissance restored works of art from Classical times. These restorers of past centuries attempted to return the object to its original appearance. The fallacy of that idea lies in the fact that they could not know the exact original appearance of the work, ie, immediately after its creation; therefore, they restored the object according to their subjective opinions.

Also, failure to remove the cause of problems first and deal with symptoms second often resulted in early recurrence of the problems, which were then again treated with the same remedial measures. A spectacular example is DaVinci's *Last Supper*. This wall painting, which started to exhibit serious problems soon after its completion, has undergone virtually continuous restoration; until the latter twentieth century, a significant part of the painting was the work of restorers. The original was largely buried under multiple restorations.

This concept of restoration has been left behind. The modern conservator defines as admissible restoration the compensation of those losses which render the object unreadable, ie, disfigure the object to an extent that the intent of the artist is obscured. The normal visual effects of age are accepted as such and are not *a priori* subject to modification. In the twentieth century, it was realized that a change in approach to restoration was necessary. Expertise in chemistry, physics, and materials-sciences have been brought to bear both to define the roots of the problems that plague a work of art and to devise a means by which to arrest the deterioration processes. A new type of specialist, the conservator, arose who possesses the necessary manual skills and talents, and, moreover, has a good grasp of scientific methodology and a sufficient knowledge of chemistry and physics to be able to understand the basic causes of deterioration mechanisms, to devise effective treatments of objects, and to interact with the scientific experts who are called upon to assist in the analysis of the problems or the testing of proposed treatments.

The conservator is responsible for the physical integrity of the object and the provision of the appropriate care, including such interventions as cleaning, stabilization, consolidation, and restoration, as well as preventive care. This specialist has a thorough training in art history enabling communication on a professional level with the curator, who generally is responsible for the works in the collection and who collaborates with the conservator with regard to considerations of an aesthetic nature.

Ethical Considerations

The practitioners of conservation have formulated a number of ethical professional standards, many of which are basic to the contemporary approach towards the conservation of art objects and other historic materials (125). Central to all of this is a fundamental respect for the integrity of the object *per se*. Careful consideration should be given to the treatment's potential for causing any immediate or future damage to the object. Any proposed treatment should be thoroughly evaluated as to not only immediate but also long-term possible effects besides the intended beneficial ones. Many art objects have existed for periods which are very long in terms of human lifespans, ie, hundreds or even thousands of years. Hence, materials that are to be introduced into an object, eg, as structural consolidants, adhesives, etc, have to be tested rigorously for chemical and physical long-term stability, and for the effects that failure or decomposition might have in terms of the safety of the object. Testing is carried out, often under accelerated aging conditions, by material scientists and chemists, in cooperation with the conservator.

It must be realized that it may eventually be necessary to undo twentieth century repairs, just as past restorations often need to be removed. A common reason is that the materials used in restoration may have aged with unacceptable visual consequences. If the removal of the proposed repairs should ever be necessary for whatever reason, it should be possible without danger to the object. Consequently, extensive testing of the stability of the materials used for the treatment is needed.

The arrest of deterioration and the prevention of its recurrence has higher priority than restoration. Thus, identification of the causes of a problem and the design of measures to stabilize and consolidate the object are primary considerations. Removal of the symptoms and restoration of the visual appearance comes only after the physical integrity has been safeguarded.

Inducing any changes to the original parts in order to minimize the visual impact of repairs and restorations is prohibited. For example, it used to be a common practice for restorers of porcelains to hide repaired breaks and fills of missing fragments by painting these repairs with a matching color. In order to minimize the visibility of this restoration, the area of paint application was extended over parts of the adjacent original surface. Although the inpainting of repairs and fills is certainly quite common and often desirable, overpainting of the original surface is not considered to be an ethical practice.

It is imperative that extensive documentation be kept of any treatment, including photographs during all stages of the work. This documentation should then be stored to be available for consultation in the future, when a renewed conservation assessment of the object may be necessary.

Training and Organization

In the United States, a number of academic graduate conservation programs have been established. Candidates for admission to such schools must satisfy a number of stringent demands, which generally include an undergraduate degree in art history or archaeology, extensive undergraduate science coursework, and demonstrable manual skills and dexterity, as evident from a portfolio. After three to four years of training, successful completion is met by a degree or certificate in conservation. An advanced internship with a practicing conservator for another one or two years is generally deemed necessary for the acquisition of the minimum experience needed to be allowed to work independently. Training programs have been established in many countries. An alternative method of training is that of the apprenticeship, an experienced conservator training and sharing knowledge with future colleagues.

Professional organizations for conservators have been established on both the national and international scales. Most conservators are members of the International Institute for the Conservation of Historic and Artistic Works (IIC). This organization, based in London, organizes biennial international conferences, and also oversees the publication of several professional periodicals, ie, *Studies in Conservation, Art and Archaeology Technical Abstracts*. Several countries or geographic areas have their own subdivisions of IIC. In the United States, an independent but related organization exists, the American Institute for Conservation of Historic and Artistic Works (AIC). The AIC holds annual meetings and publishes its own journal. The International Council of Museums (ICOM), headquartered in Paris, has an international committee on conservation, which sponsors triennial meetings.

Several international organizations have been established that can offer conservation advice or even practical help in areas of the world where such is not readily available. The International Centre for the Study of the Preservation and the Conservation of Cultural Properties (ICCROM), based in Rome, is an intergovernmental organization that serves over 80 member states, among which is the United States. In the United States, the National Institute for Conservation of Cultural Property (NIC) serves as a forum to facilitate information distribution and exchange, coordination and planning between institutions and representatives of the various professions, and to promote public and government awareness of the need for conservation of the cultural patrimony. The Conservation Information Network (CIN) is a collaborative venture of seven institutions in the

United States, Canada, and Europe, providing professionals with worldwide online access to a number of specialized databases, containing bibliographic and materials information.

Deterioration Mechanisms and Conservation

Some of the processes that play a role in the deterioration of particular types of objects are understood (126). The principal techniques employed by conservators to reduce the rates at which these processes take place and, at least partially, undo the damage incurred are mentioned herein.

Metal Objects. *Deterioration.* Apart from physical damage that can result from carelessness, abuse, and vandalism, the main problem with metal objects lies in their vulnerability to corrosion (see CORROSION AND CORROSION CONTROL) (127,128). The degree of corrosion depends on the nature and age of the object. Corrosion can range from a light tarnish, which may be aesthetically disfiguring on a polished silver or brass artifact, to total mineralization, a condition not uncommon for archaeological material.

The corrosion processes for various metals are vastly different, and hence the consequences for the object also differ greatly in nature. For bronze, for example, corrosion starts intergranularly. Cuprite (cuprous oxide) forms as the primary copper corrosion product. Because of the relatively large mobility of copper ions, a crust of cuprite forms on the surface and becomes the basis for continued chemical reactions resulting in the formation of products such as malachite and azurite, which are both basic copper carbonates. This corrosion crust, also known as patina, stabilizes the system to a high degree, slowing the corrosion rate. Because the corrosion progresses slowly, and the copper ions migrate to the surface, the final effect is that the shape of the original surface is preserved, and can be brought to visibility through a careful removal of the corrosion crust. Iron is vastly different. Here, oxides are formed *in situ* and, because of the enormous increase in specific volume of the corrosion penetrating into the interior, this process leads to serious changes in shape and, ultimately, complete disintegration of the object.

In addition to the metal, the other reactive species involved also determine the nature of the corrosion process to an appreciable degree. One primary reactant is water. In the absence of water, corrosion does not take place. Other species influence both the reaction rates and the nature of the corrosion products. For bronze, the simplest process involves water, oxygen, and carbon dioxide. However, when bronze has been buried under conditions that include a sizable chloride content in the groundwater, nantokite [*14708-8517*] (cuprous chloride [*7758-89-7*]), CuCl, forms directly on the residual metal. This compound is highly unstable under conditions of high humidity, leading to the notorious bronze disease (129). This highly aggressive corrosion process takes place when a bronze with a nantokite layer is exposed to humid air. The cuprous chloride reacts with water and oxygen to form copper hydroxy chloride [*16004-08-3*], $Cu_2(OH)_3Cl$, a light-green, powdery material symptomatic of the condition. In the process Cl^- ions are liberated; these continue the corrosive process through formation of cuprous chloride, which in turn hydrolyzes. This series of decomposition reactions stops only with the destruction of the object, unless arrested by desiccation and treatment.

In the outdoor environment, the high concentrations of sulfur and nitrogen oxides from automotive and industrial emissions result in a corrosion having both soluble and insoluble corrosion products and no pacification. The results are clearly visible on outdoor bronze sculpture (see AIR POLLUTION; EXHAUST CONTROL, AUTOMOTIVE; EXHAUST CONTROL, INDUSTRIAL).

Another air pollutant that can have very serious effects is hydrogen sulfide, which is largely responsible for the tarnishing of silver, but also has played a destructive role in the discoloration of the natural patinas on ancient bronzes through the formation of copper sulfide. Moreover, a special vulnerability is created when two metals are in contact. The electromotive force can result in an accelerated corrosion, eg, in bronzes having iron mounting pins.

Conservation. Because the most common conservation problem with metal objects occurs when corrosion processes form a threat to the safety of the object or disfigure its appearance to an unacceptable degree (130,131), many conservation treatments are intended to stabilize the corrosion processes and to remove aesthetically displeasing corrosion crusts. The latter requires a great deal of thought and discussion as to when a corrosion layer ceases to be a desirable patina and becomes unacceptable.

On highly polished surfaces, as on silverware, the slightest tarnish constitutes a disfiguring effect and must be removed and prevented. Often the cleaning is done by careful polishing with a mild abrasive, such as precipitated chalk. On ancient silver, such polishing results in an undesirable shine. Thus to preserve the soft luster of the metal, cleaning is preferably accomplished through chemical or electrochemical means. Chemical removal of the sulfides and chlorides can be done using formic acid, thiourea, thiosulfate, or thiocyanate. Electrolytic treatment involves the reduction of the corrosion product to metal in an electrochemical cell, in which the object forms the cathode.

Removal of disfiguring corrosion crusts can be accomplished in several ways. One approach is mechanical cleaning, which has the advantage of control in the degree and extent of cleaning. A second possibility is the use of chemical means. Finally, electrolytic reduction can be used with great advantage in those cases where the original surface needs to be recovered, eg, in order to discover inscriptions or engraved decorations, or where the object has been dimensionally disfigured. In this last case, the metal can, after reduction, be annealed and subsequently reshaped.

In order to prevent recurrence of the corrosion, a lacquer can be applied. Alternatively, the environment of the object can be strictly controlled with regard to relative humidity and pollutants.

Bronze disease necessitates immediate action to halt the process and remove the cause. For a long time, stabilization was sought by removal of the cuprous chloride by immersing the object in a solution of sodium sesquicarbonate. This process was, however, extremely time-consuming, frequently unsuccessful, and often the cause of unpleasant discolorations of the patina. Objects affected by bronze disease are mostly treated by immersion in, or surface application of, 1*H*-benzotriazole [*95-14-7*], $C_6H_5N_3$, a corrosion inhibitor for copper. A localized treatment is the excavation of cuprous chloride from the affected area until bare metal is obtained, followed by application of moist, freshly precipitated silver oxide which serves to stabilize the chloride by formation of silver chloride. Subsequent storage in very dry conditions is generally recommended to prevent recurrence.

Some special problems are encountered in the treatment of iron, especially that recovered from underwater sites. Corrosion stimulators, especially chlorides, must be removed. One treatment used in cases where large amounts of objects are recovered is hydrogen reduction at high temperatures. Other reduction treatments are electrolytic. Attempts to remove the chlorides by washing and leaching procedures have been unsuccessful. Stabilization of corroded iron has been effected by impregnation with polymers, or by treatment with tannic acid. The latter technique is quite historic and has been used for a long time for the protection of iron tools. Because none of the above treatments is always effective, storage in very dry conditions is essential for these objects.

Stone Objects. *Deterioration.* An important source of damage to stone objects is mechanical in nature. Both breakage and abrasion account for much of the losses on objects made of this relatively fragile material. More difficulties are offered by the processes of a chemical nature which play a role in stone deterioration (132–134).

Types of chemical damage are very much dependent on the nature and composition of the stone. Limestones and marbles, for example, are susceptible to acidic attack. The presence of sulfur oxides in the air and sulfate ions in groundwater has always been high enough to have a marked influence on these stones. The effect of the air pollutants in modern urban environments on these stones is evident. With age, limestone and marble objects often acquire an encrustation consisting of gypsum (calcium sulfate) and redeposited calcite (calcium carbonate), with precipitated silica and carbon (from soot) also often present. The study of the penetration of the sulfation process into the stone through examination of a thin section under the petrographic microscope has proven to be an aid in solving questions regarding the authenticity of marble sculpture. In sandstone, the removal of the binder through mobilization of the clay minerals or, in the case of calcite based sandstones, sulfation, results in a loss of cohesive strength. On the other hand, objects made of several chemically resistant stones, eg, granites and basalts, have survived millennia without much effect.

Besides the chemical composition, porosity is another property of stone which has great influence on its preservation. An increased porosity increases the exposed surface and pores allow movement of materials such as water and its solutes through the stones. If the pores are blocked or reduced in diameter such substances may be trapped within resulting in increased local interior damage. Exposure to the climatic elements is one important source of decay. Freeze–thaw cycles, in particular, result in pressures on the pore walls of the stone's interior from changes in volume during the phase transition of the water, and are extremely harmful.

A similar mechanical effect is caused by the presence of water-soluble salts in the stone. Such salts can be present as a result of the geological history of the stone, as in the case of limestone formed in marine environments, but may also be introduced later into the object or building stone by groundwater. Repeated dissolution and crystallization of these salts, especially those that contain water of crystallization, can result in pressures on the walls of the pores that lead to significant mechanical interior damage. An extra problem arises from the tendency of these salts to percolate toward the exterior, driven by evaporation of the water in which they are dissolved on the surface. If, as is often the case with aged stone objects, the surface has a decreased permeability, the pressures built up below the surface can lead to losses of entire surface segments through exfoliation.

Conservation. The objectives in the treatment of stone objects are primarily cleaning, stabilization, consolidation, repair, and restoration (132–135). Cleaning can vary from a light dusting to the removal of stubborn grime and stains with solvents and detergents. The latter can be applied using a poultice method to increase the efficiency with which the extraneous material is removed from below the surface of the stone.

Stabilization involves the removal of the cause of deterioration which is frequently soluble salts present in the stone. If the structural strength of the stone permits it, this can be done through soaking. The object is placed underwater in a tank, and the water is changed regularly. Another method is by application of poultices on the surface.

Consolidation is the introduction into a stone's interior of a substance which adds extra mechanical strength. Such impregnations have been done using both synthetic resins and inorganic compounds. For objects to be stored and exhibited in a controlled indoor environment, impregnation using an organic material may be deemed acceptable. However, for stone objects displayed outdoors, the inherent instability of the organic materials is a serious drawback to their use. The main practical problems with resin impregnation are the viscosity of the resin solution and the migration of resin to the surface upon solvent evaporation. Many different resins in various solvents have been tried, most notably poly(vinyl acetate), poly(methyl methacrylate), other similar acrylics, and epoxies. Another approach has been to impregnate the stone with epoxy or acrylic monomers and subsequently instigate polymerization *in situ*.

Apart from the instability of the resins, a negative aspect of such impregnations is that all pores of the stone become completely plugged, which makes moisture movement impossible. If, however, any failure in the resin barrier should develop, moisture could penetrate the interior. Subsequent moisture migration, especially when salts are present in the stone, could lead to exfoliation at the depth of resin penetration. For this reason, impregnation using materials which leave the pores open is preferable. Several processes have been devised wherein inorganic compounds are deposited, such as impregnations using barium hydroxide and urea, which result in homogeneous precipitation of barium carbonate, that in turn slowly converts to barium sulfate; barium ethyl sulfate, which results in deposition of barium sulphate, and alkoxysilanes, which result in formation of a silica network after curing of the impregnant, have also been used.

Ceramics. *Deterioration.* Ceramic objects are fragile, and mechanical damages through breakage and abrasions are the most likely source of destruction. Low fired ceramics can suffer through the rehydration of the body material; this process results in a complete loss of mechanical strength. The presence of soluble salts in porous ceramic bodies has the same disastrous results as in stone (136).

The deterioration of glasses, including glazes, involves the devitrification of the glass in which inherently unstable undercooled liquid crystallizes around nucleation centers. Whereas the process is connected to the very nature of the material, it often is worsened through poor formulation of the glass, enhancing its chemical instability. The free oxides that result are, in turn, prone to removal by water. The alkali oxides in particular cause great problems. The weeping of antique glass is a result of the hygroscopic nature of the potassium oxide which results from the devitrification of glasses made with wood-ash flux. The solution of highly concentrated potassium hydroxide and potassium carbonate, which trickle down the objects, also acts as a stimulator for decay because of its potent leaching ability.

On archaeological glass objects, layers of reaction products are formed and the main constituents of these crusts are the less-soluble compounds such as silica and calcium carbonate, which becomes calcium sulfate.

Glazed ceramics can be subject to an additional problem. If the glaze does not fit well, ie, if the thermal expansion coefficient of the glaze does not match that of the ceramic body, changes in temperature can result in loss of adhesion between glaze and body. On archaeological objects, the glaze, through devitrification, can change so much in thermal expansion behavior that it becomes extremely unstable under conditions other than a rigidly controlled constant temperature.

Conservation. The adhesives that have been used historically to mend broken ceramics include stick shellac, a paste of lead white (lead hydroxy carbonate) in oil, and cellulose nitrate. Most are no longer used because of the undesirable properties of these adhesives, notably discoloration upon aging and the large changes in solubility characteristics for some of them. Most conservators use adhesives based on poly(vinyl acetate), acrylic resins, some polyesters, and epoxies. Epoxies and polyesters still have the disadvantage of discoloration, especially noticeable in mended glass. However, there is not much choice in adhesives for glass. Low fired, porous ceramics are subject to loss of structural strength, and become sensitive to moisture. Occasionally, an impregnation with poly(vinyl acetate) or acrylic resins becomes necessary. In archaeological ceramics, salts may have been introduced during burial. These salts must be removed through desalination techniques such as those used for stone (136).

The layers of decay products and hydrated glass formed on the surface of ancient glass often start to spall off. In such cases, an impregnation with an acrylic resin may be helpful. The treatment for weeping glass consists of stabilization through the removal of free alkalies with an acid, followed by thorough rinsing and drying, and subsequent storage under dry conditions.

Organic Materials. Environmental conditions, especially temperature and relative humidity, cause deterioration of objects made from plant or animal materials (137,138). Not only extremely high or low values for these parameters, but also large or sudden changes can cause irreparable harm. The humidity of the object's environment can be especially critical. For example, wooden objects are subject to warping when exposed to high relative humidity. This is a result of absorption of water by the wood fibers and consequent dimensional changes. High humidity can also enhance the growth of molds, which destroy the objects by feeding on the organic materials. On the other hand, low relative humidity is also extremely harmful. The desorption of cell water causes dimensional changes which can result in serious mechanical damage. Wood and ivory crack and split under such conditions. High temperatures, of course, cause increased reaction rates of chemical degradation processes, whereas low temperatures may affect the mechanical properties of materials and render them more brittle.

Thermal expansions and contractions, and dimensional changes resulting from absorption and desorption of water, especially when occurring in temporal cycles, can lead to destruction through loss of mechanical strength and deformations. Organic materials can be conditioned to an environment. If the conditioning happens slowly and is not followed by periodic changes, it may result in little harm. Objects made from various organic substances have survived very well under the extremely dry conditions of a desert climate. It is the sudden changes that are often disastrous.

Materials that have been buried underwater cause a special problem. Waterlogged woods and leathers (139), although quite stable under such burial conditions, are in danger of irreversible damage through drying out upon recovery. Indeed, after excavations from bogs or upon recovery from underwater sites, these items need to be stored underwater until laboratory treatment.

Another source of damage is light (140), especially uv light. Photochemically induced deterioration is a principal cause of damage in textiles, via the fading of dyes and decomposition of fibers. Many of the synthetic fibers are notoriously sensitive in this respect, but light also plays a role in the deterioration of some natural fibers, especially silk. The fading effect of light on dyestuffs also causes problems for several organic pigments used in painting. Moreover, light affects the deterioration of several inorganic mineral pigments, notably red lead. The photoinduced dissociation of the green pigments verdigris (basic copper acetate) and copper resinate (a reaction product of verdigris with oil-resin media) also causes color changes. Besides the pigments, the organic binding media used in painting are also light sensitive, as are textiles and works of art on paper, eg, prints, drawings, photographs, watercolors, and pastels.

Chemical degradation (141), whether thermally or photo-induced, primarily results from depolymerization, oxidations, and hydrolysis. These reactions are especially harmful in objects made from materials that contain cellulose, such as wood, cotton, and paper. The chemistry of these degradation processes is quite complex, and an important role can be played by the reaction products, such as the acidic oxidation products which can catalyze hydrolysis.

A special case is the degradation of cellulose nitrate, which for a long time was used as the base for photographic film. Nitric oxide, one of the reaction products, serves as an accelerator for the degradation and makes this process autocatalytic. This process can render the material subject to spontaneous combustion. Film archives store such films under freezing conditions, and often have extensive programs to copy all old celluloid film onto polyester-based film. Cellulose triacetate, the film base material developed because of the problems with cellulose nitrate, is also quite susceptible to hydrolysis, producing acetic acid which can act as a catalyst for this hydrolysis (the vinegar syndrome).

Cellulose nitrate also has widespread use as an adhesive and coating material. Whereas stabilizers are added to products, eg, sodium carbonate as a neutralizer, many conservators are hesitant to use cellulose nitrate materials because of the inherent instability and the dangers to the object from nitric acid, formed when the nitric oxide combines with moisture.

Modern synthetic polymers are the subject of increasing research by conservation scientists. Not only does their frequent use in conservation treatments require a better understanding of their long term stability, but also many objects, including those in collections of contemporary art and in history and technology museums, are made out of these new materials.

Often when objects are composites of many different materials special problems arise. Conditions that may negatively affect any one component are harmful to the object in its entirety. Then, conditions favorable to the survival of one component may affect the stability of another negatively. A very difficult problem results when one of the components promotes chemical-decay reactions in another of the materials. For example, copper pigments promote the deterioration of proteinaceous binding materials such as glue and supports such as silk fabrics. Similarly, iron has been found to promote degradation of paper, illustrated by the historic iron-gall ink deteriorating the paper on which it was applied.

Paintings. Deterioration. Paintings are composite objects that have high vulnerability. The various materials are adhered to each other, especially in a laminated structure, to form a source of potential trouble. Any dimensional change in one of the components or between the components as a consequence of changes in environmental conditions results in a strain on the adhesion of the various parts. Strains can lead to failure of the adhesion. This is one of the principal causes of losses in panel paintings, where the dimensional changes in the wooden support cause losses in adhesion between the paint layer and the support.

Paintings on canvas dimensionally restrained by the stretcher, are also subject to mechanical stresses and resulting damages, caused by fluctuating climatic conditions. Much research has gone into the characterization and quantification of the processes involved. Careful measurement of the mechanical properties of the individual materials, including tensile strength, maximum elongation, and elastic modulus, as well as the variation of these properties with temperature, relative humidity, and loading rate, has been followed by computer based modeling (using finite element analysis), of the whole structure of the painting (142). It was, for example, found that traditional oil paintings are rather sensitive to cooling, which can result in extensive cracking of the paint layer. Changes in relative humidity are, on the other hand, more indirect. The glue layer, or size, is extremely responsive to changes in humidity and is capable of developing stresses in excess of 580 kPa (4000 psi). At such stress levels, the glue layer has the potential of cracking the paint layer and inducing flaking losses.

Conservation. Conservation problems in paintings can be considered according to the stratum in which these occur, ie, in the varnish, the paint layers, or the support (143–146).

Upon aging, all natural-resin varnishes exhibit problems resulting from chemical deterioration caused mainly by autoxidation and loss of volatile oils. First, the surface may become dirty. The dirt adheres to small blemishes and failures that have developed through the aging of the varnish layer. Second, the varnish film darkens or yellows as a result of oxidation processes. This effect progresses gradually, from a general yellowing which changes the color values of the painting to a darkening which reduces its visibility. The third common problem is called blooming, which is actually a clouding of the varnish caused by small cracks developing in the coating. Treatment can vary from simple dusting to complete removal of the varnish coat. Two frequently used treatment methods are regeneration and replacement (147). Regeneration, which is only applicable to soft resins, involves the exposure of the varnish to solvent vapors in order to partially redissolve the broken resin layer and recover the film as a uniform, unbroken mass. More commonly, however, the old resin coat must be removed. Traditionally, this is done with the aid of suitable organic solvents, with the degree of oxidation and crosslinking of the resin determining the required solvent polarity. Some of the aged natural resins require the use of highly polar solvent systems, use of which involves inherent risks to the safety of the paint layers. Much recent research has been devoted to the development and testing of resin soaps, which would offer the advantage of an aqueous medium and the potential of custom formulations. As of this writing, however, the evaluation of these cleaning agents is ongoing and their use is a matter of controversy. Any method requires extreme care and a generous amount of experience on the part of the conservator.

Once removed, the varnish is replaced, often with a synthetic resin that is not susceptible to yellowing and does not change its solubility properties. The most frequently used polymers are acrylic resins, especially a poly(methyl acrylate-co-ethyl methacrylate), which have undergone a good amount of testing. Regrettably, optical properties are not as good as those of dammar, the most popular natural resin; hence, long term stabilization of dammar through additives has been the subject of much recent research (148). Other work is being done on the development of a synthetic polymer that has satisfactory chemical and optical properties (149).

The most common problem in the paint layers, which can have a wide variety of causes, is loss of adhesion. Upon drying of the medium, the paint layers develop shrinkage cracks. In itself, this is not a particularly worrisome phenomenon, but, if through any cause the adhesion between paint layers and ground or between ground and support is lost, the paint begins to flake. First the flakes curl up, and finally become completely detached and lost.

Flaking paint is treated by infusion of an adhesive in the areas where needed, followed by resetting the flakes on the substrate; the softening of the paint needed to bend it back is effected through solvent action or heat. Losses can only be filled and inpainted. Inpainting may also be necessary when cracks become so wide as to seriously affect the visual appearance of the painting.

Failure of a canvas support occurs when, as a result of aging, the fabric becomes hard and brittle and the fibers break. The fabric loses all strength and is unable to function as a support for the paint layers. The common solution, when this has progressed so far as to endanger the safety of the painting, is the lining or relining. This procedure involves the application of a new fabric to the back of the old one. The front of the painting is provided with a temporary facing, then a new canvas is attached to the back using an adhesive, under gentle, uniform pressure applied by means of a low vacuum system. Traditionally, the adhesive was a paste of animal glue and flour. However, dimensional changes in the various layers frequently induced when this aqueous adhesive was applied and the difficulties in removal of aged glue paste has led to the use of wax-resin adhesives. The wax is applied in the molten state. Wax lining, however, often causes a darkening of the painting, and the adhesive, if it penetrates through the paint layers, is almost impossible to remove. Consequently, much effort has been placed on the development of synthetic adhesives which can be used at lower temperatures and pressures, and use of these materials is commonly accepted as an alternative to wax lining (see ADHESIVES).

Wooden panels have a tendency to warp and eventually crack (150). If the deformation of these supports becomes dangerous to the paint layers, some measures must be taken. In the past, restriction of the movement of the support was attempted through the application of rigid cradles, ie, thick wooden strips running both horizontally and vertically across the back. This was not very successful. Cracks developed between the strips of the cradle and the cleavage between ground and wood often became more aggravated. There is no easy and safe solution to this problem. The best answer lies in prevention through strict climate control of the painting's environment. In cases where the deformation has progressed too far and immediate danger to the painting exists, a transfer can be done in which the painting is again faced on the front, and the original support is tooled down until the groundlayer is exposed. A new support can then be applied to the back. However, these transfers are generally regarded as highly undesirable.

Wall paintings become endangered when the support wall decays. This is often the case with the walls on which the Italian Medieval and Renaissance artists applied their masterpieces. One technique developed for the removal of fresco paintings from the wall, called the strappo technique, involved the attachment of a canvas using a strong glue facing to the front of the painting. When this glue dries and shrinks, the top layer, which includes the upper preparation layer with the pigments and which is penetrated by the glue, tears off the underlayers and the wall. These materials were then adhered to a stable support, after which this assembly could be remounted in the original location, yet free of the wall. Whereas such detachments have undoubtedly saved a number of masterpieces, they do impose significant risks of damage, partial loss, and changes in appearance. Hence, removal of the wall painting from its support is no longer considered to be an acceptable practice except as a measure of last resort, when all efforts to stabilize the work *in situ* have failed (151).

Works on Paper. *Deterioration* Paper is subject to deterioration from several sources (152–158). Hydrolytic degradation of cellulose is acid-catalyzed; hence, the acidity present in aged paper from manufacturing techniques, acidic inks, housing material, and degradation processes, leads to discoloration and embrittlement. Acid attack can be diminished and neutralized somewhat by washing and deacidification, but these treatments pose risks, including altering the paper's original color and texture, or solubilizing media.

A second degradation process is oxidation, often photo-induced especially by exposure to light not filtered for uv. The radicals resulting from this reaction promote depolymerization of the cellulose, as well as yellowing and fading of paper and media. Aging causes paper to become more crystalline and fragile, and this can be exacerbated particularly if the paper is subjected to poor conditions.

Attack by molds in high humidity environments can cause weakening of the support by breakdown of sizing, making the paper susceptible to tearing and blistering, especially in aqueous conservation treatments. Foxing and mold stains result in discolorations that can not be successfully removed, even by chemical bleaching. This latter treatment may revert or diminish strength of paper.

Conservation. In an attempt to save paper, preventive conservation care deserves the highest priority, because it reduces the need for potentially hazardous, complicated, and expensive treatments later (159–162). Problems which have a structural impact on long-term stability of paper should be given a higher priority than problems which are merely cosmetic in nature. For example, infestation by insects, attracted by nutrients in paper, can cause irreparable loss of media and support.

In the paper conservation laboratory, washing is probably the most common treatment. Suction tables are often used to minimize exposure time to water or other solvents used in the treatments. The amount of care bestowed on single works of art on paper in museums would be impractical for individual documents or books in archival and library collections, where the number of items exceeds that in art museum collections by several orders of magnitude. Hence, ongoing development and testing of appropriate technologies for mass deacidification is especially of great importance for libraries and archives. These often contain very large amounts of books and documents on paper produced between the late eighteenth and early nineteenth century from chemically treated wood pulp. Such paper has become extremely brittle as a result of acid hydrolysis, known as the brittle book phenomenon. Proposed mass deacidification processes involve the use of aqueous or nonaqueous solutions of various magnesium and calcium carbonates, or diethylzinc as a gas-phase reagent (163).

Pressure sensitive adhesive tapes, used to mount or mend paper, are particularly dangerous in that these stain paper and cause media to bleed and are hazardous to remove, often requiring solvents that may also adversely affect paper. Brittle, fragile, or torn paper must be reinforced, either by archival quality housing or conservation treatment (such as mending or lining) using appropriate materials and procedures. A great deal of research is underway to investigate the effects of these treatments on the aging of paper.

Textiles. Deterioration. The causes of degradation phenomena in textiles (155–158, 164) are many and include pollution, bleaches, acids, alkalies, and, of course, wear. The single most important effect, however, is that of photodegradation. Both cellulosic and proteinaceous fibers are highly photosensitive. The natural sensitivity of the fibers are enhanced by impurities, remainders of finishing processes, and mordants for dyes. Depolymerization and oxidation lead to decreased fiber strength and to embrittlement.

Light is also the principal cause of damage to both natural and synthetic dyestuffs that fade through photodegradation (165). The dye molecules are more sensitive than the fiber/polymer. Fading is therefore often the initial sign of damage to the textile. Fading is irreversible.

Other causes of worry to the textile conservator are the dangers of mold growth and of insect damage. Whereas environmental control measures can help reduce the risk of mold in temperate or tropical conditions, careful storage and display maintenance are of critical importance to reduce the risk of insect damage. Certain types of textiles are less susceptible to damage than others because of yarn geometry, eg, the semiworsted wool tapestries of Western Europe with their combed yarn. On the other hand, because moths have preferentially attacked carded woolen yarn textiles, the most famous Flemish cloth of the late middle ages, kermes dyed escarleten, no longer exists.

Conservation. Washing is a rather popular treatment for the removal of acidic or otherwise unwanted soluble compounds. For textiles that cannot be exposed to aqueous treatment, solvent based dry cleaning may be the preferred mechanism for cleaning. In either case, thorough testing of the fastness of the dyes in the proposed solvent system is necessary before the treatment. When a textile has lost its mechanical strength, it must be backed with a lining fabric. In the worst cases, consolidation through application of an appropriate synthetic polymer may be necessary, but such treatments are controversial from an ethical point of view. When repairs are made, great care must be taken that the mechanical strength of the repair does not surpass that of the textile, otherwise a condition of stress might cause the original to tear. Another concern is that the repairs are distinguishable from the original. Of course, no repairs are effective unless the original cause of degradation has been addressed (166).

Wooden Objects. Wooden objects (150, 167) cause several special problems, particularly when recovered from underwater sites. Waterlogged wood (139, 168), if allowed to dry out, suffers irreparable damage through warping and cracking. It is, therefore, kept underwater until it arrives in the laboratory, where a few treatments are available. Freeze drying has proven to be extremely effective. Another well established treatment is the immersion of the object in a tank with water, in which poly(ethylene glycol) (PEG) is subsequently introduced in a slowly increasing concentration, until impregnation of the wood with PEG has been achieved. Yet another technique involves impregnation with a solution of rosin in acetone.

Furniture (169,170) is subject to damage by dimensional changes caused by the environment, as well as physical damage through abuse and wear. Insect damage can be severe; because the infestation is in depth, a penetrating treatment must be used. For lesser infestations, an insecticide is sometimes injected into the insect holes; larger infestations, however, require fumigation.

Preventive Conservation

No conservation treatment can completely undo damages to art objects. However, damage can often be prevented (171–173). Many deterioration processes are dependent upon environmental conditions.

Temperature and Humidity. Temperature is probably the easiest environmental factor to control. The main concern is that the temperature remains constant to prevent the thermal expansions and contractions that are particularly dangerous to composite objects. Another factor regarding temperature is the inverse relation to relative humidity under conditions of constant absolute humidity, such as exist in closed areas. High extremes in temperature are especially undesirable, as they increase reaction rates. Areas in which objects are exhibited and stored must be accessible; thus a reasonable temperature setting is generally recommended to be about 21°C.

Although the temperature can be controlled with a well-designed air-conditioning system, the small fluctuations which most cycling systems cause may be very harmful. The temperature–time record should be a continuous, flat graph.

More difficulties are encountered with determining humidity settings. Not only is it important that the relative humidity remain constant, but also the desirable value depends on the materials. Many of the larger institutions have sought to solve the problem through the installation of extensive climate control systems, which serve to keep both temperature and humidity at a constant value by means of heating, refrigeration, humidification, and dehumidification. However, the main difficulties are encountered with humidification in extremely dry conditions such as occur during the heating season along the northeastern seaboard. In refurbished buildings, the installation of a system capable of efficiently regulating humidity becomes extremely complicated and expensive, because of the necessary modifications to the building. Vapor barriers must be installed in order to prevent moisture migration through the walls, which could cause serious damage to the building fabric; multiple glazing is generally needed to prevent condensation, adding to the cost. Especially in historic houses, such modifications may conflict with the historic integrity of the building itself. Moreover, the cost of the energy expended to operate these systems is quite high.

The main advantage of wholesale climate control lies in easy access to the objects, and the absence of differences in conditions between various spaces within the institution, eg, storage areas, conservation laboratories, and exhibition galleries. The actual values set for the rh are a matter of compromise; metals, stone, and ceramics are best served by humidities as low as possible, but organic materials generally require higher values. An accepted compromise is the maintenance of the relative humidity at a strictly controlled level of 50–55%.

An alternative to macroclimate systems is the creation of microclimates. The objects are placed within smaller spaces, such as cases, in which an ideal environment is maintained. One possibility is to install equipment to control the climate in individual cases, or groups of cases with similar materials, by mechanical means.

If the temperature of the space in which an object is placed were truly constant, a sealed case having a constant absolute humidity would also have a constant relative humidity. Because temperature is subject to some variations and totally leakproof cases are not easy to build, a second solution is often sought by placing the objects in reasonably well-sealed cases in which the relative humidity is kept at a constant value by means of a buffering agent. Certain grades of silica gel or selected clay minerals are often used. The buffering material is preconditioned under the selected relative humidity and, after equilibration, installed in the case. This method of microclimate control has proven to be very efficient, not only in exhibition cases and storage spaces, but also in packing crates used for the transportation of sensitive objects.

Rigid monitoring of the climatic conditions is an absolute requirement for a successful preventive conservation program (see TEMPERATURE MEASUREMENT). The desirability of a continuous measurement and a written record favors the use of recording thermohydrographs. Generally, these instruments measure the temperature via the thermal expansion of a metal strip and the relative humidity through dimensional changes induced in humidity-sensitive elements, often a bundle of human hairs. Less expensive elements are used in dial hygrometers, the smaller of which are appropriate for applications within cases. All of these instruments need regular calibration against dry–wet bulb systems, either sling or motor-driven psychrometers. The use of electronic sensors (qv) in sealed cases, has increased.

Biodeterioration. For objects made out of organic materials, mold and insect attack are a principal cause of damage. Microbiological organisms can also be responsible for serious deterioration of outdoor stone. Museums, especially storage areas, are quite conducive to providing the conditions in which infestations can occur. Objects are stored in close proximity, and left untouched for prolonged periods, often with little if any air movement.

In the past, much reliance was placed on early and drastic remedial measures involving the application of residual biocides or insecticides, and fumigation of infested collections. A favorite fumigant was ethylene dioxide, which served both as an insecticide and mold sterilizant (see FUNGICIDES, AGRICULTURAL; INSECT CONTROL TECHNOLOGY). However, the dangers of residual fumigant to museum staff and possibly visitors has led to use restrictions. In the United States usage of this fumigant in museums has all but been eliminated. Moreover, research on the long-term effects on the fumigated materials raised serious concerns about the use of this biocide on museum objects. Similar concerns exist with regard to most commercially available fumigants. Sulfuryl fluoride is used quite often in cases of massive insect infestation, but it too has limitations in its applicability for various collection materials. Much research attention is devoted to the use of carbon dioxide and anaerobic atmospheres (nitrogen, argon) to neutralize insect infestation (174). None of these treatments is effective, however, to arrest mold attack. Sterilization through radiation has been used occasionally for insect control, but the doses required to kill mold are large enough to induce radiation damage in many organic materials.

Most emphasis is placed on a program of rigorous preventive maintenance. Appropriate climate conditions can help to prevent mold attack, which typically only occurs at elevated relative humidity. Storage furniture which provides an effective barrier for insects, regular inspection of the collections, monitoring of all collection areas with insect traps, and access control measures which minimize the chance of insect entry into the collection areas, are some aspects of an effective pest control management program (175).

Light. It is imperative that the exposure of photosensitive materials to uv light is kept to a minimum. In many museums, the use of natural daylight for illumination is preferred on aesthetic grounds, as its spectral distribution results in an ideal color rendition. As the uv component does not play a role in the visualization of the object and is the most harmful, it must be completely removed. This can be done through the use of appropriate filters, available either in the form of rigid sheets of plexiglass doped with a strongly uv-absorbing organic material or as flexible sheets, adhered to the glass windows. A variation of the latter, nowadays often used in museums, is a laminated glass with a uv-absorbing interlayer.

When artificial light is used, those light sources which produce a significant amount of uv radiation, such as fluorescent bulbs, should also be provided with a suitable filter, such as clear transparent sleeves made from a plastic doped with a uv absorber. It is generally accepted that the admissible limit of uv light is its proportion in the uv light emitted by an incandescent tungsten lamp, ca 70 μW/lm. There is also a problem in exposure to visible light. Thus the ideal condition is no light exposure and storage areas should be kept dark. For exhibition galleries, a compromise must be sought between the need for minimum exposure of the object to light and the desirability of good visibility and color perception. Various international organizations involved with the conservation of art objects have adopted a set of two illuminance levels. The higher value of 150 lux is deemed acceptable for the exhibition of oil and tempera paintings, oriental lacquer, ivory, horn, bone, and undyed leather. For the most photosensitive objects, such as textiles of various kinds, watercolors, prints, drawings, manuscripts, dyed materials, etc, a maximum illuminance of 50 lux is recommended. In the latter case these low levels are sufficient to view the objects, but exhibit design should allow visitors to become accustomed to these light levels.

One method of reducing the exposure for sensitive objects is to exhibit them only for limited periods and to maintain a regular rotation schedule. In Japan, for example, some extremely important paintings can only be seen a few days per year.

Pollutants. The problems posed by air pollutants are very serious. Within a museum, measures can be taken to remove harmful substances as efficiently as possible by means of the installation of appropriate filter systems in the ventilation equipment. Proposed specification values for museum climate-control systems require filtering systems having an efficiency for particulate removal in the dioctyl phthalate test of 60–80%. Systems must be able to limit both sulfur dioxide and nitrogen dioxide concentrations to $<10 \mu g/m^3$, and ozone to $<2 \mu g/m^3$.

Hydrogen sulfide has traditionally been a problem in the tarnishing of silver and the discoloration of bronze patinas. This gas can be dealt with in the filters of the climate-control system as well as through the use of proper absorbing agents. For example, a paper treated with activated charcoal is fabricated especially for absorbing H_2S within a microclimate.

Vapors emitted from the materials of closed storage and exhibit cases have been a frequent source of pollution problems. Oak wood, which in the past was often used for the construction of such cases, emits a significant amount of organic acid vapors, including formic and acetic acids, which have caused corrosion of metal objects, as well as shell and mineral specimens in natural history collections. Plywood and particle board, especially those with a urea–formaldehyde adhesive, similarly often emit appreciable amounts of corrosive vapors. Sealing of these materials has proven to be not sufficiently reliable to prevent the problem, and generally their use for these purposes is not considered acceptable practice.

In the preservation of outdoor art objects, the problems caused by air pollution are overwhelming. For metal objects, such as bronze sculpture, a possible solution is the application of a protective barrier layer in the form of a surface coating. Acrylic lacquers, sometimes doped with a corrosion inhibitor such as benzotriazole, and waxes are most frequently used for this purpose. The tremendous cost involved in an effective program of maintenance, necessary in the use of barrier surface layers, is often an insurmountable impediment. For outdoor art objects composed of other materials, eg, stone, there are no efficient protection measures available. Surface coatings are often impossible, because these would also seal the surface against moisture passage. The ravages already inflicted on acid-sensitive stones such as marble and limestone have virtually destroyed innumerable sculptures, and inflicted irreparable damage to architectural monuments.

Physical Safety. Preventive conservation also involves ensuring the physical safety of objects (176). Objects should be guarded against acts of vandalism or damage inflicted by touching them. In many museums, greasy spots on sculpture can be seen, a result of repeated contact with bare hands. An inordinate amount of damage, however, also results from object handling by staff, not necessarily through carelessness but rather as a result of unawareness of the mechanical weaknesses of the various materials. This is especially true for ancient objects, where the material properties have been affected by aging processes. Preventive conservation should therefore include a vigorous training and education program for all who handle art objects.

Severe damage has sometimes been inflicted through faulty design or manufacture of mounting devices. Moreover, objects are especially subject to serious danger of damage during transit, as for example in travelling exhibits. Travel from one climatic environment to another, as well as lack of climate control in the transportation vehicles, may result in large fluctuations of temperature and humidity. During transport, the object also is in danger of shock, eg, from dropping its crate, and vibrations from the transport vehicle. Measurements of mechanical properties and the use of computer modeling have contributed significantly to identification and quantification of these factors, as well as to the design of packing crates that minimize the risk to the objects (177).

Conflicting Interests and Shared Responsibilities. The responsibility of collections holding institutions, museums, archives, libraries, etc, toward their collections is not a simple one. On the one hand, there is the obligation to preserve the collections for future generations. On the other hand, there is the mandate for the use of the collections in promoting and facilitating scholarly studies, education, and public enjoyment. Conditions optimal for preservation often severally limit and inhibit access. The study and exhibit value of objects lies to great part in their continued availability for that same purpose. Thus, curators, collections managers, and conservators share a common goal, ie, to make the collection available for the purposes for which it was brought together, while at the same time taking every measure and precaution possible to preserve that use and value of the objects.

For different types of collections, this balance is differently defined. For example paper conservation treatments commonly undertaken in the museum conservation laboratory would be impractical in a library archive having a far greater collection size. The use of treatments for mass paper quantities would be unacceptable in the art museum. Documents in archives and books in libraries serve a different goal from art objects in a museum. Their use value lies primarily in their information rather than in an intrinsic esthetic value. Whereas optimal preservation of that information value requires preservation of the object itself, a copy or even a completely different format could serve the same purpose.

Clearly, the intended use of a collection item is extremely important to determining the acceptability of a treatment. The degree to which a treatment affects appearance is obviously of the greatest importance for an art object. On the other hand, in natural history collections the collections serve as research resources above all. The effect a preservation or conservation treatment has on these research applications is the main consideration. Collections of art, archaeology, history, science, technology, books, archival materials, etc, all have their own values in terms of balance between preservation needs and collections use, and these values are, moreover, constantly subject to reevaluation and change.

BIBLIOGRAPHY

"Fine Art Examination and Conservation" in *ECT* 3rd ed., Suppl. Vol., pp. 392–442, by L. van Zelst, Museum of Fine Arts, Boston.

1. W. J. Young, ed., *Applications of Science in Examination of Works of Art*, Museum of Fine Arts, Boston, Mass., 1959.
2. *Ibid.*, 1967.
3. *Ibid.*, 1973.
4. P. A. England and L. van Zelst, ed., *Applications of Science in Examination of Works of Art*, Museum of Fine Arts, Boston, Mass., 1985.
5. E. V. Sayre and co-workers, eds., *Materials Issues in Art and Archaeology*, Materials Research Society Symposium Proceedings Vol. 123, Materials Research Society, Pittsburgh, Pa., 1989.
6. P. B. Vandiver, J. Druzik, and G. S. Wheeler, eds., *Materials Issues in Art and Archaeology II*, Materials Research Society Symposium Proceedings Vol. 185, Materials Research Society, Pittsburgh, Pa., 1991.
7. P. B. Vandiver and co-workers, eds., *Materials Issues in Art and Archaeology III*, Materials Research Society Symposium Proceedings Vol. 267, Materials Research Society, Pittsburgh, Pa., 1992.
8. M. Jones, ed., *Fake? The Art of Deception*, British Museum Publications, London, 1989.
9. S. J. Fleming, *Authenticity in Art*, Crane, Russak & Co., New York, 1976.
10. P. Coremans, *Van Meegeren's Faked Vermeers and De Hooghs: A Scientific Examination*, Cassell & Co., Ltd., London, 1949.
11. B. Keisch, *Science* **160**, 413 (1968).
12. D. von Bothmer and J. V. Noble, *An Inquiry into the Forgery of the Etruscan Terracotta Warriors in the Metropolitan Museum of Art*, papers no. 11, The Metropolitan Museum of Art, New York, 1961.
13. W. C. McCrone, *Anal. Chem.* **48**, 676A (1976).
14. T. A. Cahill and co-workers, *Anal. Chem.* **59**, 829 (1987).
15. C. McCrone, *Anal. Chem.* **60**, 1009 (1988).
16. J. V. Noble, *The Forgery of Our Greek Bronze Horse*, The Metropolitan Museum of Art Bulletin, New York, 1968, pp. 253–356.
17. K. C. Lefferts and co-workers, *J. Am. Inst. for Conservation* **21**, 1 (1981).
18. S. G. E. Bowman, ed., *Science and the Past*, British Museum Press, London, 1991.
19. R. H. Brill, ed., *Science and Archaeology*, M.I.T. Press, Cambridge, Mass., 1971.
20. C. W. Beck, ed., *Archaeological Chemistry*, Advances in Chemistry Series No. 138, American Chemical Society, Washington, D.C., 1975.
21. G. F. Carter, ed., *Archaeological Chemistry II*, Advances in Chemistry Series No. 171, American Chemical Society, Washington, D.C., 1978.
22. J. B. Lambert, ed., *Archaeological Chemistry III*, Advances in Chemistry Series No. 205, American Chemical Society, Washington, D.C., 1984.

23. R. O. Allen, ed., *Archaeological Chemistry IV*, Advances in Chemistry Series No. 220, American Chemical Society, Washington, D.C., 1989.
24. H. Hodges, *Artifacts*, J. Baker, London, 1968.
25. H. Hodges, *Technology in the Ancient World*, Penguin Books, Harmondsworth, 1970.
26. C. d'Andrea Cennini, *Il Libro dell'Arte*, D. V. Thompson, Jr., trans., (*The Craftsman's Handbook*), Yale University Press, New Haven, Conn., 1933; republished Dover Publishers, New York, 1960.
27. G. Agricola, *De Re Metallica*, Basel, 1549, H. C. Hoover and L. H. Hoover, trans., London, 1912.
28. V. Biringuccio, *De La Pirotechnia*, Venice, 1540, C. S. Smith and M. T. Gnudi, trans., M.I.T. Press, Cambridge, Mass., 1942.
29. Theophilus, *De Divers Artibus*, J. G. Hawthorne and C. S. Smith, trans., University of Chicago Press, Ill., 1963; republished Dover Publishers, New York, 1979.
30. N. S. Brommelle and G. Thomson, eds., *Science and Technology in the Service of Conservation*, International Institute for Conservation of Historic and Artistic Works, London, 1982.
31. M. H. Butler, *Polarized Light Microscopy in the Conservation of Paintings*, centennial volume, State Microscopical Society of Illinois, Chicago, pp. 1–17, 1970.
32. J. J. Rorimer, *Ultraviolet Rays and their Use in the Examination of Works of Art*, The Metropolitan Museum of Art, New York, 1931.
33. E. R. de la Rie, *Stud. Conserv.* **27**, 1, 65, 102 (1982).
34. J. R. J. van Asperen de Boer, *Appl. Opt.* **7**, 1711 (1968).
35. J. R. J. van Asperen de Boer, Ph.D. Dissertation, University of Amsterdam, Central Research Laboratory for Objects of Arts and Science, Amsterdam, 1970.
36. A. Gilardoni, R. A. Orsini, and S. Taccani, *X-Rays in Art*, Gilardoni SpA, Mandello Lario, Como, 1977.
37. A. Burroughs, *Atlantic Monthly* **137**, 520 (1926).
38. A. Burroughs, *Art Criticism from a Laboratory*, Little Brown & Co., Boston, Mass., 1938.
39. R. E. Alexander and R. H. Johnson, in J. S. Olin and A. D. Franklin, eds., *Archaeological Ceramics*, Smithsonian Institution Press, Washington, D.C., 1982, pp. 145–154.
40. *Art and Autoradiography*, The Metropolitan Museum of Art, New York, 1982.
41. W. J. Young, ed., *Application of Science to Dating of Works of Art*, Museum of Fine Arts, Boston, Mass., 1976.
42. M. J. Aitken, *Science-based Dating in Archaeology*, Longman, London, 1990.
43. S. G. E. Bowman, *Radiocarbon Dating*, British Museum Publications, London, 1990.
44. G. Harbottle, E. V. Sayre, and R. W. Stoenner, *Science* **206**, 683 (1979).
45. R. E. M. Hedges, *Archaeometry* **23**, 3 (1981).
46. E. K. Ralph, in Ref. 41, pp. 77–79.
47. P. Klein, in Ref. 4, pp. 22–28.
48. E. T. Hall, J. M. Fletcher, and M. F. Barbetti, in Ref. 41, pp. 68–73.
49. M. J. Aitken, *Thermoluminescence Dating*, Academic Press, Inc., London, 1985.
50. M. Ikeya, *Archaeometry* **20**, 147 (1978).
51. P. M. Masters and J. L. Bada, in Ref. 21, pp. 117–138.
52. N. S. Baer, T. Jochsberger, and N. Indictor, in Ref. 21, pp. 139–149.
53. B. Keisch, in Ref. 3, pp. 193–198.
54. R. J. Gettens and G. L. Stout, *Paintings Materials: A Short Encyclopedia*, D. van Nostrand, New York, 1942; republished Dover Publishing, New York, 1966.
55. D. V. Thompson, *The Materials and Techniques of Medieval Painting*, Allen & Urwin, London, 1936; republished Dover Publishing, New York, 1956.

56. R. Mayer, *The Painter's Craft*, D. van Nostrand Co., New York, 1948; republished The Viking Press, New York, 1975.
57. R. Mayer, *The Artist's Handbook of Materials and Techniques*, The Viking Press, New York, 1991.
58. C. L. Eastlake, *Methods and Materials of Painting of The Great Schools and Masters* (formerly titled: *Materials for a History of Oil Painting*), Longman, Brown, Green and Longmans, London, 1847; republished Dover Publishing, New York, 1960.
59. D. Bomford and co-workers, *Art in the Making: Italian Painting Before 1400*, National Gallery Publications, London, 1990.
60. D. Bomford, C. Brown, and A. Roy, *Art in the Making: Rembrandt*, National Gallery Publications, London, 1988.
61. D. Bomford and co-workers, *Art in the Making: Impressionism*, National Gallery Publications, London, 1990.
62. R. L. Feller, ed., *Artists' Pigments: A Handbook of Their History and Characteristics*, The National Gallery of Art, Washington, D.C., 1986.
63. R. Harley, *Artist's Pigments c. 1600–1835*, 2nd ed., Butterworths, London, 1982.
64. R. H. Marijnissen, *Paintings, Genuine, Fraud, Fake: Modern Methods of Examining Paintings*, Elsevier, Brussels, 1985.
65. M. Hours, *Conservation and Scientific Analysis of Paintings*, Van Nostrand Reinhold, New York, 1976.
66. J. Plesters, *Stud. Conserv.* **2**, 110 (1956).
67. R. J. Meilunas, J. G. Bentsen, and A. Steinberg, *Stud. Conserv.* **35**, 33 (1990).
68. E. Martin, *Stud. Conserv.* **22**, 63 (1977).
69. J. M. Messinger II, *J. Am. Inst. for Conservation* **31**, 267 (1992).
70. L. Masschelein-Kleiner, J. Heylen, and F. Tricot-Marckx, *Stud. Conserv.* **13**, 105 (1968).
71. J. S. Mills and R. White, in N. S. Brommelle and P. Smith, eds., *Conservation and Restoration of Pictorial Art*, Butterworths, London, 1976, pp. 72–77.
72. J. S. Mills and R. White, *Organic Mass Spectrometry of Art Materials: Work in Progress*, National Gallery Technical Bulletin, London, 1981, pp. 3–19.
73. R. F. Tylecote, *A History of Metallurgy*, The Institute of Metals, London, 1990.
74. R. Maddin, ed., *The Beginning of the Use of Metals and Alloys*, MIT Press, Cambridge, Mass., 1988.
75. P. Craddock, in Ref. 18, pp. 57–73.
76. K. A. Yener and co-workers, *Science* **244**, 200 (1989).
77. R. E. Stone, *Metropolitan Museum Journal* **16**, 37 (1982).
78. R. J. Gettens, *The Freer Chinese Bronzes. Vol. II. Technical Studies*, The Freer Gallery of Art, Smithsonian Institution Press, Washington, D.C., 1969.
79. D. Cushing, in Ref. 2, pp. 53–66.
80. C. S. Smith, in Ref. 2, pp. 20–52.
81. D. A. Scott, *Metallography and Microstructure of Ancient and Historic Metals*, The Getty Conservation Institute, Marina del Rey, Calif., 1991.
82. M. J. Hughes, M. J. Cowell, and P. T. Craddock, *Archaeometry* **18**, 19 (1976).
83. O. Werner, *Spektralanalytische und Metallurgische Untersuchungen an Indischen Bronzen*, E. J. Brill, Leiden, 1972.
84. S. Junghans, E. Sangmeister, and M. Schroeder, *Studien zu den Anfaengen der Metallurgie 2*, Berlin, 1974.
85. P. Meijers, L. van Zelst, and E. V. Sayre, in Ref. 20, pp. 22–33.
86. R. H. Brill, W. R. Shields, and J. M. Wampler, in Ref. 3, pp. 73–83.
87. N. H. Gale, in Y. Maniatis, ed., *Proc. of the 25th Symposium on Archaeometry*, Athens, 1986, pp. 469–502.

88. F. R. Matson, *Ceramics and Man*, Wenner-Grenn Foundation for Anthropological Research, New York, 1965.
89. A. Shepard, *Ceramics for the Archaeologist*, publication no. 609, Carnegie Institution of Washington, Washington, D.C., 1965.
90. J. S. Olin and A. D. Franklin, eds., *Archaeological Ceramics*, Smithsonian Institution Press, Washington, D.C., 1982.
91. F. R. Matson, in D. Brothwell and E. Higgs, eds., *Science in Archaeology*, Thames and Hudson, London, 1963.
92. W. D. Kingery and P. B. Vandiver, *Ceramic Masterpieces: Art, Structure, Technology*, The Free Press, MacMillan Inc., New York, 1986.
93. W. D. Kingery, P. B. Vandiver, and M. Prickett, *J. Field Archaeology* **15**, 219 (1988).
94. O. Soffer and co-workers, *Archaeology* **46**, 36 (1993).
95. P. Vandiver, in Ref. 90, pp. 167–179.
96. A. Lucas and J. R. Harris, *Ancient Egyptian Materials and Industries*, 4th ed., E. Arnold, London, 1962.
97. J. V. Noble, *The Techniques of Painted Attic Pottery*, Watson-Guptill, New York, 1965.
98. P. B. Vandiver, *Sci. Am.* **262**, 105 (1990).
99. M. Medley, *The Chinese Potter: A Practical History of Chinese Ceramics*, Charles Scribners Sons, New York, 1976.
100. S. Frank, *Glass and Archaeology*, Academic Press, Inc., London, 1982.
101. E. V. Sayre, in Ref. 2, pp. 145–154.
102. R. W. Smith, in F. R. Matson and G. E. Rindone, eds., *Advances in Glass Technology, Part 2*, Plenum Press, New York, 1963, pp. 283–290.
103. H. C. Beck and C. G. Seligmann, *Nature* **133** 982 (1934).
104. C. G. Seligmann, P. D. Ritchie, and H. C. Beck, *Nature* **138**, 721 (1936).
105. L. van Zelst, in S. L. Hyatt, ed., *The Greek Vase*, Hudson-Mohawk Association of Colleges and Universities, Latham, N.Y., 1981, pp. 119–134.
106. R. H. Brill and S. Moll, in Ref. 102, pp. 293–302.
107. H. W. Catling and A. Millett, *Archaeometry* **8**, 3 (1965).
108. R. Abascal, G. Harbottle, and E. V. Sayre, in Ref. 20, pp. 81–99.
109. P. Vandiver and co-workers, *Archaeomaterials* **5**, 185 (1991).
110. *Marble. Art Historical and Scientific Perspectives on Ancient Sculpture*, The J. Paul Getty Museum, Malibu, Calif., 1990.
111. C. Renfrew and J. Springer-Peacy, *Annual of the British School at Athens* **83**, 45 (1968).
112. R. Newmann, in Ref. 110, pp. 263–282.
113. V. Barbin and co-workers, in Ref. 6, pp. 299–308.
114. J. M. French, E. V. Sayre, and L. van Zelst, in Ref. 4, pp. 132–141.
115. L. L. Holmes, C. T. Little, and E. V. Sayre, *J. Field Archaeology* **13**, 417 (1986).
116. K. Germann, G. Holzmann, and F. J. Winkler, *Archaeometry* **22**, 99 (1980).
117. N. Herz and M. Waelkens, ed., *Classical Marble: Geochemistry, Technology and Trade*, Vol. 153, NATO Advanced Science Institute, Series E: Applied Sciences, Kluwer, Dordrecht, 1988.
118. M. Saltzman, in Ref. 15, pp. 172–185.
119. H. Schweppe, in H. L. Needles and S. H. Zeronian, eds., *Historic Textile and Paper Materials I*, Advances in Chemistry Series No. 212, American Chemical Society, Washington, D.C., 1986, pp. 153–174.
120. H. Schweppe, in S. H. Zeronian and H. L. Needles, eds., *Historic Textile and Paper Materials II*, Advances in Chemistry Series No. 410, American Chemical Society, Washington, D.C., 1989, pp. 188–219.
121. P. E. Damon and co-workers, *Nature* **337**, 611 (1989).

122. D. Hunter, *Papermaking. The History and Technique of an Ancient Craft*, A. E. Knopf, New York, 1943; republished Dover Publishing, New York, 1978.
123. K. W. Rendell, in E. Berkeley, Jr., ed., *Autographs and Manuscripts: A Collectors Manual*, Charles Scribner, New York, 1978.
124. W. A. Cote, ed., *Papermaking Fibers*, Syracuse University Press, Syracuse, N.Y., 1980.
125. *Code of Ethics and Standards of Practice*, American Institute for Conservation of Historic and Artistic Works, Washington, D.C., 1985.
126. H. J. Plenderleith and A. E. A. Werner, *The Conservation of Antiquities and Works of Art*, 2nd ed., Oxford University Press, London, 1971.
127. B. F. Brown and co-workers, eds., *Corrosion and Metal Artifacts*, NBS Special Publication 479, National Bureau of Standards, Washington, D.C., 1977.
128. W. D. Richey, in Ref. 30, pp. 108–118.
129. D. A. Scott, *J. Am. Inst. for Conservation* **29**, 193 (1990).
130. R. M. Organ, in Ref. 127, pp. 107–142.
131. R. M. Organ, in S. Doeringer, D. G. Mitten, and A. Steinberg, eds., *Art and Technology*, M.I.T. Press, Cambridge, Mass., 1970, pp. 73–84.
132. C. G. Amorosa and V. Fassina, *Stone Decay and Conservation, Cleaning, Consolidation and Protection*, Materials Science Monographs no. 11, Elsevier, Amsterdam, 1983.
133. R. Rossi-Manaresi, ed., *The Conservation of Stone-I*, Centro per la Conservazione della Sculture all'Aperto, Bologna, 1976.
134. R. Rossi-Manaresi, ed., *The Conservation of Stone-II*, Centro per la Conservazione della Sculture all'Aperto, Bologna, 1981.
135. N. S. Brommelle and P. Smith, eds., *Case Studies in the Conservation of Stone and Wall Paintings*, The International Institute for the Conservation of Historic and Artistic Works, London 1986.
136. R. Newton and S. Davidson, *Conservation of Glass*, Butterworths, London, 1989.
137. J. S. Mills and R. White, *The Organic Chemistry of Museum Objects*, Butterworths, London, 1987.
138. M. E. Florian, D. P. Kronkright, and R. E. Norton, *The Conservation of Artifacts Made from Plant Materials*, The Getty Conservation Institute, Marina del Ray, Calif., 1990.
139. B. Muehlethaler, *Conservation of Waterlogged Wood and Wet Leather*, Editions Eyrolles, Paris, 1973.
140. T. B. Brill, *Light: Its Interaction with Art and Antiquities*, Plenum Press, New York, 1980.
141. R. L. Feller, in J. C. Williams, ed., *Preservation of Paper and Textiles of Historic and Artistic Value*, Advances in Chemistry Series no. 164, American Chemical Society, Washington, D.C, 1977, pp. 314–335.
142. M. F. Mecklenburg, in Ref. 6, pp. 105–122.
143. G. L. Stout, *The Care of Pictures*, Columbia University Press, New York, 1948; republished Dover Publishing, New York, 1975.
144. H. Ruehemann, *The Cleaning of Paintings*, Faber & Faber, London, 1968.
145. C. K. Keck, *A Handbook on the Care of Paintings*, American Association for State and Local History, Nashville, Tenn., 1976.
146. J. S. Mills and P. Smith, eds., *Cleaning, Retouching and Coatings, Technology and Practice for Easel Paintings and Polychrome Sculpture*, The International Institute for Conservation of Historic and Artistic Works, London, 1990.
147. R. L. Feller, E. H. Jones, and N. Stolow, *On Picture Varnishes and Their Solvents*, Intermuseum Conservation Association, Oberlin, Ohio, 1959; republished The National Gallery of Art, Washington, D.C., 1985.

148. E. R. de la Rie and C. W. McGlinchev, in Ref. 145, pp. 160–164.
149. *Ibid.*, pp. 168–176.
150. N. S. Brommelle, A. Moncrieff, and P. Smith, eds., *Conservation of Wood in Paintings and the Decorative Arts*, The International Institute for Conservation of Historic and Artistic Works, London, 1978.
151. P. Mora, L. Mora, and P. Philippot, *Conservation of Wall Paintings*, Butterworths, London, 1984.
152. D. van der Reyden, *J. Am. Inst. for Conservation* **31**, 117 (1992).
153. G. Pethersbridge, ed., *Conservation of Library and Archive Materials and the Graphic Arts*, Butterworths, London, 1987.
154. R. Smith and T. Norris, eds., *TAPPI Proceedings: 1988 Paper Preservation Symposium*, Technical Association of the Paper and Pulp Industry, Washington, D.C., 1988.
155. H. L. Needles and S. H. Zeronian, eds., *Historic Textile and Paper Materials I,* Advances in Chemistry Series No. 212, American Chemical Society, Washington, D.C., 1986.
156. S. H. Zeronian and H. L. Needles, eds., *Historic Textile and Paper Materials II,* Advances in Chemistry Series No. 410, American Chemical Society, Washington, D.C., 1989.
157. J. C. Williams, ed., *Preservation of Paper and Textiles of Historic and Artistic Value,* Advances in Chemistry Series No. 164, American Chemical Society, Washington, D.C., 1977.
158. J. C. Williams, ed., *Preservation of Paper and Textiles of Historic and Artistic Value, II*, Advances in Chemistry Series No. 193, American Chemical Society, Washington, D.C., 1981.
159. A. F. Clapp, *Curatorial Care of Works of Art on Paper*, Lyons and Burford, New York, 1987.
160. F. Dolloff and R. Perkinson, *How to Care for Works of Art on Paper*, 3rd ed., Museum of Fine Arts, Boston, Mass., 1979.
161. M. L. Ritzenthaler, *Archives and Manuscripts, Conservation: A Manual on Physical Care and Management*, Society of American Archivists, Chicago, 1983.
162. M. L. Ritzenthaler, G. J. Munoff, and M. S. Long, *Archives and Manuscripts, Administration of Photographic Collections*, Society of American Archivists, Chicago, 1984.
163. D. N. -S. Hon, in Ref. 120, pp. 13–34.
164. J. E. Leene, *Textile Conservation*, Smithsonian Institution Press, Washington, D.C., 1972.
165. T. Padfield and S. Landi, *Stud. Conserv.* **11**, 111 (1966).
166. S. Landi, *The Textile Conservator's Manual*, Butterworths, London, 1992.
167. N. S. Brommelle and A. E. A. Werner, eds., *Deterioration and Treatment of Wood. Problems of Conservation in Museums*, Editions Eyrolles, Paris, 1969.
168. W. A. Oddy, ed., *Problems in the Conservation of Waterlogged Wood*, National Maritime Museum, Greenwich, London, 1975.
169. M. A. Williams, *Keeping It All Together: The Preservation and Care of Historic Furniture*, Ohio Antique Review, Worthington, Ohio, 1988.
170. R. F. McGiffin, Jr., *Furniture Care and Conservation*, American Association for State and Local History, Nashville, Tenn., 1983.
171. G. Thomson, *The Museum Environment*, Butterworths, London, 1978.
172. K. Bachmann, ed., *Conservation Concerns: A Guide for Collections and Curators*, Smithsonian Institution Press, Washington, D.C., 1992.
173. B. Appelbaum, *Guide to Environmental Protection of Collections*, Sound View Press, Madison, Conn., 1991.
174. M. Gilberg, *Stud. Conserv.* **36**, 93 (1991).

175. L. A. Zycherman and J. R. Richard, *A Guide to Museum Pest Control*, Foundation of the American Institute for Conservation of Historic and Artistic Works and the Association of Systematic Collections, Washington, D.C., 1988.
176. F. K. Fall, *Art Objects. Their Care and Preservation*, Laurence McGilvery, La Jolla, Calif., 1973.
177. M. F. Mecklenburg, ed., *Art in Transit: Studies in the Transport of Paintings*, The National Gallery of Art, Washington, D.C., 1991.

General References

References 8, 9, 18, 24, 25, 42, 54, 57, 64, 73, 92, 96, 122, 126, 132, 143, 145, 159, 166, and 170–173.

Lambertus van Zelst
Conservation Analytical Laboratory
Smithsonian Institution

FINE CHEMICALS

PRODUCTION

Until the early 1970s in-house capabilities for production of raw materials and intermediates for products sold were considered a key competitive advantage by the chemical industry. As of this writing, this situation has changed completely. Particularly those chemical companies concentrating on portfolios having high added value specialties consider efficient research and development (qv) (R&D), dynamic marketing, and proper management of human, technical, and financial resources as key success factors rather than production. This change in strategic focus is especially evident in the agrochemical and pharmaceutical industries which together comprise the life science industry. In many instances manufacturing has been regrouped into separate divisions, and in a few cases large life science companies have disinvested their chemical manufacturing activities. In addition to these strategic developments, the requirement for more and more sophisticated organic chemicals has contributed substantially to the emergence of the fine chemicals industry as a distinct entity. Fine chemical manufacturers are backward integrated, production oriented, and service the mega enterprises within the chemical industry. The fine chemicals industry has its own characteristics with regard to R&D, production, marketing, and finance.

In the chemical business products may be described as commodities, fine chemicals, or specialties. Various commodities are also known as petrochemicals, basic chemicals, organic chemicals (large-volume), monomers, commodity fibers, and plastics. Advanced intermediates, building blocks, bulk drugs and bulk pesticides, active ingredients, bulk vitamins, and flavor and fragrance chemicals are all fine chemicals. Adhesives (qv), diagnostics, disinfectants, electronic chemicals, food additives (qv), mining chemicals, pesticides, pharmaceuticals (qv), photographic chemicals, specialty polymers, and water treatment chemicals are all specialties. The added value is highest for specialties.

It is common to both commodities and fine chemicals that these materials are identified according to specifications, according to what they are. These substances are sold within the chemical industry, and customers know better how to use them than suppliers. Specialties are identified according to performance, according to what they can do. Customers are the public, and suppliers have to provide for technical assistance. A particular substance may be both a fine chemical and a speciality. For example, as long as 2-chloro-5-(1-hydroxy-3-oxo-1-isoindolinyl)benzene sulfonamide [*77-36-1*] is sold according to specifications it is a fine chemical. But once it is tableted and marketed as the diuretic chlorthalidone [*77-36-1*], it becomes a specialty (see DIURETICS). The limits between commodities and fine chemicals are not so clearly fixed. Table 1 shows two typical commodities, *o*-xylene and phthalic anhydride, which are relatively low price products manufactured in large quantities by many companies. These materials have many different applications and are usually sold nationally rather than internationally. Two examples of fine chemicals, produced in limited quantities by a limited number of manufacturers, are also shown. These are used exclusively for the preparation of one specific drug. The specialty chlorthalidone is shown for comparison.

In terms of volume the border line between commodities and fine chemicals comes somewhere between about 1,000 and 10,000 t/yr. In terms of unit prices the line typically varies between $2.50/kg, and $10/kg. Establishing more precise demarcations is not practical even though a large number of well-known intermediates fall within these lines, eg, acetoacetanilide, BHT, chloroformates, cyanuric chloride, hydroquinone, malonates, pyridine and the picolines, and sorbic acid. Additionally, for amino acids (qv) and vitamins (qv), two typical groups of fine chemicals, the two largest volume products, L-lysine and methione, and ascorbic acid and niacin, respectively, are sold in quantities exceeding 10,000 t/yr. The prices range beyond the $10/kg level as well.

Research and Development

Product innovation absorbs considerable resources in the fine chemicals industry, in part because of the shorter life cycles of fine chemicals as compared to commodities. Consequently, research and development (R&D) plays an important role. The main task of R&D in fine chemicals is scaling-up lab processes, as described, eg, in the ORAC data bank or as provided by the customers, so that the processes can be transferred to pilot plants (see PILOT PLANTS AND MICROPLANTS) and subsequently to industrial-scale production. Thus the R&D department of a fine chemicals manufacturer typically is divided into a laboratory or

Table 1. Comparison of Chemical Classes

Parameter	Commodities		Fine chemicals		Specialties
example	*o*-xylene	phthalic anhydride	3-amino-2-carboxy-4-chlorobenzophenone	2-chloro-5-(1-hydroxy-3-oxo-1-isoindolinyl) benzenesulfonamide	chlorthalidone[a]
CAS Registry Number	[*95-47-6*]	[*85-44-9*]		[*77-36-1*]	[*77-36-1*]
molecular formula	C_8H_{10}	$C_8H_4O_3$	$C_{14}H_{10}NO_3Cl$	$C_{14}H_{11}ClN_2O_4S$	$C_{14}H_{11}ClN_2O_4S$
applications	>20	>10	1	1	1
price level, $/kg	0.50	1	10	100	1000
production, t/yr	2.5×10^6	$>1 \times 10^6$	100	100	100
producers	100	25	1	1	1
customers	100	50	1	captive	>> consumers
plant type[b]	D, C	D, C	M, B	M, B	F
manufacturing steps	1	2	5	10	1

[a]Also sold under the trade names Hydroton, Regroton, Igrolina, Igroton, and Renon.
[b]B is batch; C, continuous; D, dedicated; M, multipurpose; and F, formulation.

process research section and a development section, the latter absorbing the lion's share of the R&D budget, which typically accounts for 5 to 10% of sales. Support functions include the analytical services, engineering, maintenance, and library.

In the laboratory or process research section a laboratory procedure for a fine chemical is worked out. The resulting process description provides the necessary data for the determination of preliminary product specifications, the manufacture of semicommercial quantities in the pilot plant, the assessment of the ecological impact, an estimation of the manufacturing cost in an industrial-scale plant, and the validation of the process and determination of raw material specifications.

The development section serves as an intermediary between laboratory and industrial scale and operates the pilot plant. A direct transfer from the laboratory to industrial-scale processes is still practiced at some small fine chemicals manufacturers, but is not recommended because of the inherent safety, environmental, and economic risks. Both equipment and plant layout of the pilot plant mirror those of an industrial multipurpose plant, except for the size (typically 100 to 2500 L) of reaction vessels and the degree of process automation.

In development the viability of the process on a semicommercial scale has to be demonstrated, the process viability is tested in terms of quality, semicommercial quantities of the new fine chemical have to be manufactured for market development, etc, the necessary data have to be generated to enable the engineering department to plan the adaptations for construction of the industrial-scale plant and in order to calculate production costs for the expected large-volume requirements, and all questions regarding safety and environment have to be solved. Once a laboratory process has been adapted to the constraints of a pilot plant, has passed the risk analysis, and demonstration batches have been successfully and repeatedly run, the process is ready for transfer to the industrial-scale plant.

Plant Design

The number of products offered by a fine chemicals manufacturer typically exceeds the number of production trains. Yet, for reasons of economy of scale, the production capacity considerably exceeds the yearly requirement for each product. Furthermore, the product portfolio is regenerated at a fast pace. This set of circumstances leads to the multipurpose plant, as opposed to a dedicated plant. A multipurpose plant is capable of handling several types of chemical reactions and performing a series of unit operations. Multiproduct plants are intermediate between multipurpose and dedicated plants. Design of multiproduct plants allows the production of similar products using the same process technology, eg, chlorides or salts of different alkanoic acids or different esters or amides of the same acid. Plant design needs to reflect the required degree of flexibility (see PLANT LAYOUT).

A production train in its simplest form consists of a jacketed reaction vessel made from stainless or glass-lined steel and equipped with an agitator, a manhole, and pipe connections (Fig. 1**a**). Solid raw materials are charged through the manhole, and liquid and solvents feed through a manifold system installed on top of the respective reactor and connected to one of the inlet nozzles of the reactor head.

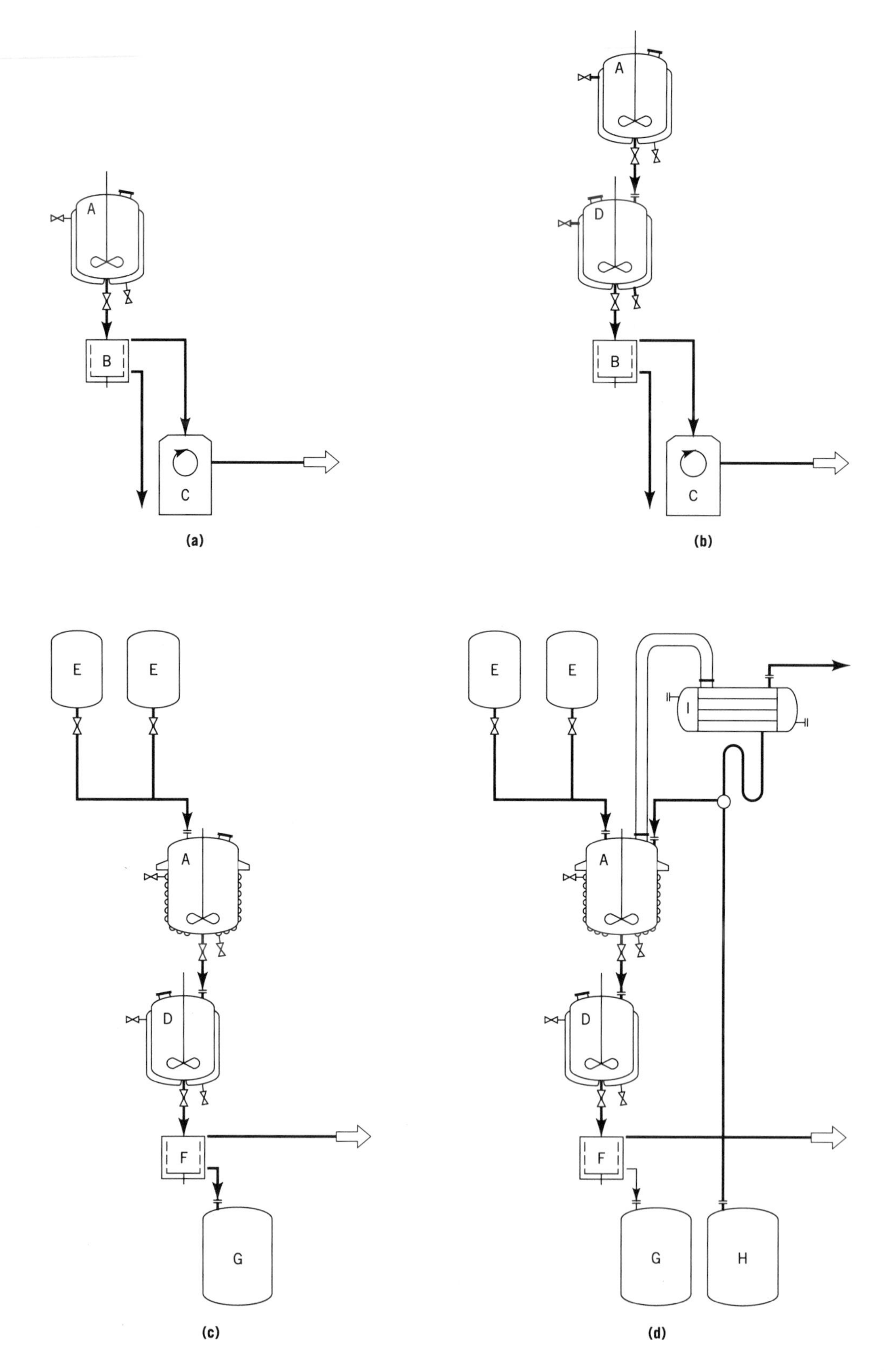

Fig. 1. Fine chemicals plant design showing successive additions of processing equipment, where A represents the reaction vessel with agitator; B, centrifuge; C, dryer; D, crystallization vessel; E, raw material feed tanks; F, centrifuge which may have an automatic discharge; G, mother liquor tank; H, distillate tank; and I, condensor/heat exchanger. (**a**) Module I; (**b**) module II; (**c**) module III; and (**d**) module IV.

After completion the reaction mixture, typically a solid–liquid slurry, is discharged through the bottom valve. The solid–liquid separation is performed in the centrifuge which only holds a part of the reaction mixture. The reaction vessel in this simple configuration also serves as a feeding tank for the centrifuge. A new batch can only be initiated after the centrifugation is completed. This module I-type production train can be used, for example, for the formation and precipitation of salts of organic acids, ie, metal carboxylates, from an aqueous solution.

In module II (Fig. **1b**) a crystallization vessel, jacketed and connected to cooling water, is added. Thus the salt formation step, which may require heating, is separated from the crystallization (qv), which is completed upon cooling. Using module II a substantially increased production capacity can be achieved at only a minor additional capital investment.

In module III (Fig. **1c**), feeding tanks for raw materials and solvents, and a holding tank for the mother liquor, are added. The centrifuge, of the automatic discharge type, is gas tight and allows for inert gas blanketing. All vessels and tanks are connected with a vent system consisting of scrubbers and an off-gas burner. Using this configuration it is possible to use hazardous liquid raw materials and flammable organic solvents.

If the reaction temperature is close or equal to the boiling temperature of the solvent, module IV (Fig. **1d**) is used. It differs from module III by the addition of an overhead condenser connected with a vacuum system, a phase separator, and a distillate receiving tank. Also, the jacket for heating/cooling is substituted for a half coil allowing more rapid heat transfer. For modules III and IV the dryer is located in another part of the plant called the drying section, which also comprises milling/sieving and packaging equipment. In order to comply with regulations, solids handling has to be contained.

Module IV already enables a fair number of chemical reactions, such as an esterification, eg, through the feeding tanks the organic acid, the catalyst (typically an inorganic acid), and the alcohol can be charged to the reaction vessel. The alcohol is used in large excess and serves also as solvent. The reaction mixture is heated to reflux. As a by-product from the reaction, water is formed; therefore, the condensate consists of excess alcohol and water. In the case of higher alcohols, which are not miscible with water, two layers form. The water is discharged and the alcohol returned to the reaction vessel. The reaction is completed when no more water forms. The reaction mixture is cooled down. For the extraction of the catalyst, water is added and the two-phase mixture stirred vigorously. After allowing time for phase separation, the aqueous catalyst solution, which basically is a dilute inorganic acid, is discharged and then neutralized prior to discharge to the wastewater system. In order to obtain a pure ester, the excess alcohol is first distilled followed by distillation of the crude ester under reduced pressure.

In order to make a multipurpose plant even more versatile than module IV, equipment for unit operations such as solid materials handling, high temperature/high pressure reaction, fractional distillation (qv), liquid–liquid extraction (see EXTRACTION, LIQUID–LIQUID), solid–liquid separation, thin-film evaporation (qv), drying (qv), size reduction (qv) of solids, and adsorption (qv) and absorption (qv), may be installed.

Instead of concentrating these functions in one unit, it is also possible to create semispecific production trains, eg, for hydrogenations, phosgenizations,

Friedel-Crafts alkylations (see FRIEDEL-CRAFTS REACTIONS), and Grignard reactions (see GRIGNARD REACTION). The choice between the integrated and the satellite plant design depends primarily on the degree of utilization of the various functions (1).

In the design of a fine chemicals plant equally important to the choice and positioning of the equipment is the selection of its size, especially the volume of the reaction vessels. Volumes of reactors vary quite widely, namely between 1,000 and 10,000 L, or in rare cases 16,000 L. The cost of a production train ready for operation increases as a function of the 0.7 power. The personnel requirement increases at an even lower rate. Thus a large plant using large equipment would be expected to be more economical to run than a small one.

Depending primarily on the differing quantities of the fine chemicals to be produced in the same multipurpose unit, the concentration of substances in the reaction mixture, and the reaction time, there is, however, an upper limit for the size of the reaction vessel and the ancillary equipment. Some factors run countercurrent to the economy of scale and point to small-sized equipment. Six of these are (*1*) length of the production campaign: if the time becomes shorter than about 10 working days, the changeover time for preparing the plant for the production of the next product becomes too long and burdens the production cost too much; (*2*) working capital: if the equipment is oversized with regard to the requirement for any particular fine chemical, the interval between two production campaigns becomes too long and excessive inventory is built up; (*3*) heat transfer: the time required for heating and cooling the reaction mixture and for its transfer among different pieces of equipment becomes too long as compared to the reaction time as such; (*4*) in the case of expensive fine chemicals the value of one batch in one piece of equipment becomes very high, sometimes in excess of $1 million and, therefore, the risk of false manipulations becomes excessive; (*5*) the dimensions of existing buildings, tank farms, and the capacity of utilities often determine an upper limit of the equipment size; and (*6*) consumption of solvents to clean the equipment increases with volume, and this means an additional burden for waste treatment facilities.

Apart from determinating the optimum size of equipment, the degree of flexibility is another key plant design parameter. Flexibility means cost, thus only as much flexibility as required by the processes should be built. Excessive flexibility is counterproductive (2).

Full process control (qv) computerization of multipurpose plants is much more difficult than for single-product continuous units. However, the first computerized fine chemical plants were brought on-line by ICI Organics Division in the early 1970s and there are many others. The additional (ca 10%) cost of computerization has been estimated to give a savings of 30 to 45% in labor costs. However, highly trained process operators and instrument engineers are required (3). The complexity of the plant design, the degree of sophistication and requirements for constant quality of the fine chemicals to be produced, the necessity to process hazardous chemicals, the sensitivity of product specifications to changes of reaction parameters, and the availability of a skilled work force, all determine the degree of automation that is advisable.

The percentage costs associated with a multipurpose plant equipped with 6.3 m^3 reactors are as follows: 27% for building, including the warehouse, where the building has a foundation; 23% for process equipment, ie, reactors, tanks, feeders, hoppers, heat exchangers, condensors, pumps, including vacuum pumps, centrifuges, suction filters (nutsches), filters, and dryers, sieves, mills; 11% for erection, ie, installation of equipment, piping, insulation, and painting; 5.5% for electrical materials and installation; 12% for process control/instrumentation; 16% for engineering, including profit and general overhead; and 5.5% for various other costs.

For this example the cost of the battery limits plant is about four times the purchase cost of the equipment. This number is about two for module I-type plants designed and installed by the fine chemicals company itself, and about six for expanded module IV-type plants designed and built by contractors.

Batchwise operated multipurpose plants are *per definitionem* the vehicle for the production of fine chemicals. There are, however, a few examples of fine chemicals produced in dedicated, continuous plants. These can be advantageous if the raw materials or products are gaseous or liquid rather than solid, if the reaction is strongly exothermic or endothermic or otherwise hazardous, and if the requirement for the product warrants a continued capacity utilization. Some fine chemicals produced by continuous processes are methyl 4-chloroacetoacetate [*32807-28-6*], $C_5H_7ClO_3$, and malononitrile [*109-77-3*], $C_3H_2N_2$, made by Lonza; dimethyl acetonedicarboxylate [*1830-54-2*], made by Ube; and L-2-chloropropionic acid [*107-94-8*], $C_3H_5ClO_2$, produced by Zeneca.

A state-of-the-art fine chemicals plant consisting of four production trains of the expanded module IV type is shown in Figure 2. Each train typically consists of four 10 m^3 reaction and crystallization vessels and is flexibly connected to a tank farm outside the building. The production trains can produce two fine chemicals simultaneously. Furthermore, these trains can be connected with adjacent trains. In the dry section four sets of drying, milling, sieving, and packaging machines are installed in contained departments. All processes are computerized and operated by remote control in a central control room. The total investment amounts to about $\$110 \times 10^6$.

The workforce consists of 92 shift and 8 daily workers. Approximately 20 to 30 different fine chemicals are produced per year which range in volume from 20 to 200 metric tons per train and in campaign length from 20 to 180 days.

The production building is only one part of a full-fledged fine chemicals plant. Apart from the multipurpose plant building there is usually an office and R&D building, the warehouse, the maintenance shop, tank farms, the incinerator, and wastewater treatment facilities.

Plant Operation

Production Planning. Whereas continuous plants run 24 hours per day, there is more freedom in establishing operating schedules for multipurpose plants. Depending on the work load and the flexibility of the workforce, schedules can also be changed during the course of a year. Common schedules include one shift, ie, 8 hours per day for 5 days per week, and two shifts, ie, 16 hours per day for 5 days per week. In this latter case frequently some minimum activity is maintained during the night, such as supervision of reflux reactions, solvent distilla-

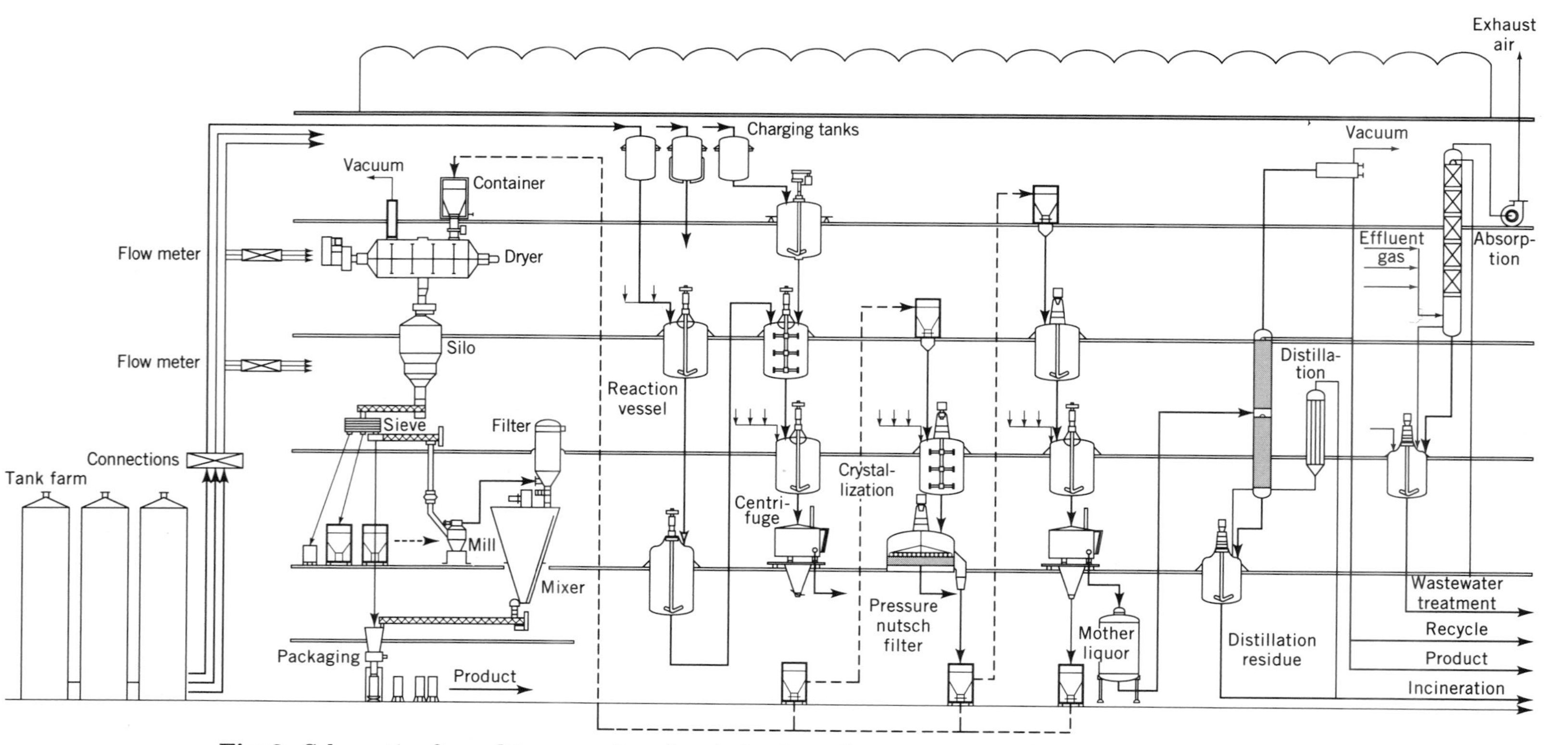

Fig. 2. Schematic of a multipurpose fine chemicals plant. Computer-assisted process control is utilized.

tions, or dryers. Three shifts, ie, 24 hours per day, 7 days per week, are also possible. In terms of production cost this last is the most advantageous scheme. Higher salaries for night work is more than offset by lower fixed costs. Also, only part of the workforce has to adhere to this scheme.

Within the framework of Reference 4, production planning for a fine chemicals company operating multipurpose plants is a demanding task. The goal must be to achieve optimum capacity utilization, important to the profitability of the company. However, conflicting interests of marketing, manufacturing, and controlling have to be aligned.

Particularly critical is the interface between marketing, which determines what quantity of which products can be sold, and manufacturing, which determines how optimum use of existing equipment can be made and what type of plant is needed in the future. There are both short-term and long-term aspects to production planning. A useful tool for the short-term planning is a rolling 18 months sales forecast, which is committing for the first 4 to 6 months and somewhat more flexible for the rest of the period.

Figure 3 shows the capacity utilization resulting from the production program in a multipurpose plant. The annual percentage of occupation is shown on the x-axis reflecting the overall business condition, and the level of equipment utilization is shown on the y-axis, reflecting the degree of sophistication of the fine chemicals to be produced. Several conclusions can be drawn:

(*1*) 100% capacity utilization cannot be achieved in a multipurpose plant. Even in the unlikely event that there is sufficient demand to run the plant for the whole year and that for all products manufactured all available equipment can be used, there is still changeover time that is unproductive. Particularly in the case of frequent product changes, great attention has to be paid to the reduction of changeover time, which may take up to 10 days (2).

(*2*) Product changeover consists of partially overlapping activities, ie, phasing out of the previous product; rinsing the plant; dismantling, adapting, reassembling, repair, and maintenance; cleaning; start-up with new product; and developing analytical procedures to determine the previous product in ppm range in the new product.

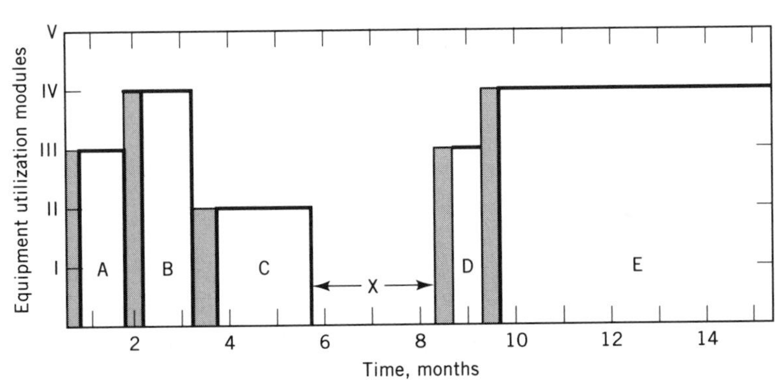

Figure 3. Multipurpose plant capacity utilization where □ represents products A, B, C, D, and E; ■ the changeovers; and X the time the plant was idle.

(*3*) Each new phase of operation may initially perform at below full rate until the plant is operating on a routine basis. Each of the products A to E is produced only once a year resulting in the buildup of substantial inventory. Because the interest on the working capital is lower than the unabsorbed fixed costs during changeover, just-in-time production is not economic in multipurpose plants. It is preferable to have few production campaigns and high inventory than vice versa (5).

Optimum capacity utilization in the two dimensions of time and equipment are crucial to the overall performance, and running a fine chemicals company has been described as gap management. Attempts have been made to develop adequate equations for describing the correlation between processes and plant. However, practical applications in production planning of fine chemical plants have not as of this writing been realized.

Quality Control. Because fine chemicals are sold according to specifications, adherence to constant and strict specifications, at risk because of the batchwise production and the use of the same equipment for different products in multipurpose plants, is a necessity for fine chemical companies. For the majority of the fine chemicals, the degree of attention devoted to quality control (qv) is not at the discretion of the individual company. This is particularly the case for fine chemicals used as active ingredients in drugs and foodstuffs (see FINE CHEMICALS, STANDARDS). Standards for drugs are published in the *United States Pharmacopeia* (USP) in the United States (6) and the *European Pharmacopeia* in Europe (7).

For pharmaceutical (see PHARMACEUTICALS) fine chemicals, typically the following tests must be carried out: aspect, assay, melting point, impurity profile, remaining solvents or loss on drying, sulfated ash, heavy metals, and particle size distribution, if the material is to be used in the solid state. In addition, all impurities present in a concentration above 0.1% have to be identified. Manufacturers of both pharmaceuticals and the corresponding precursors have to adhere to good manufacturing practices (GMP). The standards of GMP are set on an international level, and nearly all countries have comparable national regulations for GMP. A firm producing pharmaceuticals has to be approved by national authorities. For bulk or finished pharmaceuticals sold to the United States, radiopharmaceuticals to Canada, or finished pharmaceuticals to the United Kingdom or Germany, the authorities of these countries also perform international GMP inspections, unless the countries have conventions for the mutual recognition of inspections as, eg, the United States, Sweden, Switzerland, and Canada, or the countries of the EEC.

Standards for food-grade chemicals in the United States are published in the *Food Chemicals Codex* (FCC) (8), for laboratory reagents in *Reagent Chemicals—ACS Specifications,* and for electronic-grade chemicals in the *Book of SEMI Standards* (BOSS) by Semiconductor Equipment and Materials International (SEMI). The latter two product categories, with the exception of reagent chemicals used as diagnostics, are not subject to GMP regulations. In these cases most customers expect the producer to comply with a quality standard such as ISO 9001/9002, ie, only released raw materials can be used, the process must be validated, and documentation of each produced batch is kept.

The analytical laboratory is expected to be certified or at least to comply with one of the quality standards for analytical labs set forth by ISO GLP or EN 45,000 ff (see FINE CHEMICALS, STANDARDS). The laboratories have to fulfill the following criteria: have written procedures for all tests involving instruments, standards, and reagents; perform analytical tests only following the written instructions; use only tested and approved instruments, standards, and reagents; keep documentations of all tests performed; have periodic quality audits and procedures for corrective action; and be absolutely independent of the production. Such a laboratory, able to run routine controls for two to three products simultaneously, needs instruments worth about half a million dollars and not less than six technicians where four of them work in shifts, if the production runs 24 hours per day. Typical instrumentation includes an hplc having a diode array detector, a gc equipped with both a flame ionization and a thermal conductivity detector (fid/tcd), and headspace sampling capability, titrators, ir spectrometer, melting point apparatus, sieve analyzer, etc, besides equipment such as balances, muffel oven, drying oven, refractometer, etc.

During the development of a new process or if unexpected problems arise, the support of a much better equipped laboratory may be necessary. This laboratory must be able to isolate impurities and determine the structures of impurities. Instruments like nmr, ms, and ir, and specialized chemists are a necessity in such a laboratory.

Another quality control problem of multipurpose plants is the clean out for a product change. A test for residual cleaning solvents in the ppm level is a necessity. The best validation of the cleaning process is to develop an analytical method that is able to find the previous product in the new product at a level of not more than 1 ppm. Tests should be run on at least the first three batches.

Apart from the cleanout procedure, the analytical work performed during the production of a bulk pharmaceutical in a three-step synthesis includes 15 different analyses having determination of 22 parameters for raw materials; 15 different analyses having determination of 17 parameters in process controls; and 11 different analyses having determination of 19 parameters for the product.

Cost Calculation. The main elements determining production cost are identical for fine chemicals and commodities (see ECONOMIC EVALUATION). A breakdown of production cost is given in Table 2. In multipurpose plants, where different fine chemicals occupying the equipment to different extents are produced during the year, a fair allocation of costs is a more difficult task. The allocation of the product-related costs, such as raw material and utilities, is relatively easy. It is much more difficult to allocate for capital cost, labor, and maintenance. A simplistic approach is to define a daily rent by dividing the total yearly fixed cost of the plant by the number of production days. But that approach penalizes the simple products using only part of the equipment.

If the daily rent is corrected by an equipment utilization factor, simple products for which only part of the equipment is used can show a good profit margin without providing a good return for the overall investment in the multipurpose plant. For portfolio optimization, not only the profit margin but also the marginal income per day have to be considered. In other words, marketing has to be given the task of finding substitutes for products that have low equipment utilization,

Table 2. Cost Calculations for Fine Chemicals Value-Added Structure

Parameter	Cost, %
Variable costs	
raw materials	26
utilities	3
Manufacturing cost	
direct[a]	14
waste pretreatment[b,c]	6
waste treatment	3
allocated plant costs	4
Capital cost	
plant depreciation	11
plant adaptation[d]	5
inventories	6
Corporate overhead and miscellaneous	
R&D	9
general management and miscellaneous[e]	13
Total	*100*

[a]Labor, maintenance, etc.
[b]Recovery of solvents or reagents, extraction and neutralization of mother liquors, precipitation and separation of salts, and scrubbing of gaseous effluents.
[c]The waste pretreatment frequently (and sometimes dramatically) reduces the output of a process, which is not considered in this cost breakdown.
[d]Specific or new equipment.
[e]Administration, marketing, customer service, packaging, transport, insurance, etc.

such as A, C, and D in Figure 3. Pretreatment and disposal of waste effluents substantially increase the cost of a fine chemical. Up to 50% of the capital cost for a new fine chemical plant has to be earmarked for pollution control equipment.

Only a minority of new products studied in R&D enjoy commercial success, thus allocation of R&D costs is another controversial issue. This problem is usually disguised by not including R&D in the cost calculation of individual products, but by placing R&D in the general overhead.

Industrial Strategies. Outsourcing of the manufacture of fine chemicals by the life science and other specialty chemical companies has been one of the driving forces for the development of the fine chemicals industry. The advantages of making fine chemicals include maintaining control over the whole supply chain as well as expertise in chemical manufacturing; avoiding the risk of dissipating confidential technology to third parties; occupying idle production capacity; avoiding bad experiences with fine chemicals manufacturers; and cost savings. The advantages to buying are allocating financial, technical, and human resources for the core business, eg, drug or pesticide discovery in the life science industry; avoiding the risk of building a new plant at a time in which the ultimate success of a new specialty is not yet known; liberating production capacity for new fine chemicals by transferring production of older ones to fine chemicals manufacturers; past favorable experience with fine chemicals manufacturers; and the ability to pay as earn, ie, expending no capital in anticipation of future sales.

Assuming more or less comparable production costs for the fine chemical manufacturer and the life science industry, and recognizing that the fine chemicals manufacturer needs to make a profit, it appears *a priori* that buying is more expensive than making. This conclusion, however, is only valid for certain cases because outsourcing constitutes a kind of insurance premium in new product development. If everything proceeds according to plan, buying is more expensive than making; if, however, the new product (specialty) launch is delayed or fails, the life science company fares much better with buying.

Make or buy considerations constitute a relatively new element of business planning. As of this writing, some companies, particularly the traditional European ones, tend to consider buying only when there is really no possibility of in-house production. Others, particularly large U.S. pharmaceutical companies, have the opposite view, resorting to in-house production only if there is no possibility of outside sourcing. For all practical purposes, outsourcing is synonymous with single sourcing (9), because of the economy of scale and the managerial difficulty of developing a cooperation with more than one partner.

Economic Aspects

There are about 500 companies worldwide involved in fine chemicals. Most are in Europe, followed by the United States and Japan. For Europe and the United States updated lists of the companies are given (10). Some have developed from forward integration, eg, DSM (the Netherlands) and Ube (Japan) from coal (qv) mining; BASF (Germany) and Lonza (Switzerland) from fertilizers (qv) and medium value commodities; Coalite & Chem. (UK) and Rütgerswerke (Germany) from coal-tar processing; Shell (UK, the Netherlands) from oil exploration, refining, and processing; and Niels Clauson-Kaas (Denmark) and Palmer (UK) from contract research. Others have developed from diversification, eg, Degussa (Germany) from noble metals; Ems Dottikon (Switzerland), Nobel Kemi (Sweden), and SNPE (France) from explosives; or from backward integration, eg, Fermion (Finland), Finorga (France), Upjohn (United States), and Zambon (Italy) from pharmaceuticals.

Fine chemical companies are generally either small and privately held or divisions of larger companies, such as Eastman Fine Chemicals (United States) and Lonza (Switzerland). Examples of large public life science companies, which market fine chemicals as a subsidiary activity to their production for captive use, are Hoffmann-La Roche, Sandoz, and Boehringer Ingelheim, which produce and market bulk vitamins and liquid crystal intermediates, dyestuff intermediates, and bulk active ingredients, respectively. Table 3 lists some representative companies having an important fine chemical business.

Whereas all larger companies are market oriented and offer a list of catalog products, smaller ones are more (single) customer focused and offer primarily toll or custom manufacturing services. In the case of toll manufacturing the customer provides both the manufacturing process and the raw material to be converted, in the custom care the manufacturing process for a given fine chemical is developed from scratch. Over three hundred companies in the United States and Europe offer these services (11).

There are also companies that concentrate on physical operations in connection with fine chemicals manufacturing. Activities include drug and pesticide formulation, distillation, and milling/sieving/drying.

Total sales of the fine chemical industry were estimated to amount to about $12 billion in 1991, by an in-depth analysis of the product portfolio of representative fine chemicals manufacturers (11) and by performing top-down analyses of the life science industry (12). However, some consulting firms gave higher ($\sim\$60 \times 10^9$) figures for the size of the fine chemicals business (13).

Unit sales prices of from 800 to 900 fine chemicals are listed weekly in the *Chemical Marketing Reporter*. This number reflects those fine chemicals produced and sold in industrial quantities. Some market studies on fine chemicals, listing important product families, such as side chains for β-lactam antibiotics (qv), *N*- and *S*-heterocyclic compounds, fluoroaromatics, etc, do exist (14,15).

Products

If fine chemicals are classified according to applications, the most prominent categories in terms of tonnage volumes are the ones used in the production for agrochemicals, followed by pharmaceutical fine chemicals. Within agrochemicals, triazine herbicides (see HERBICIDES), from the key intermediate cyanuric chloride (see CYANURIC AND ISOCYANURIC ACIDS), are produced in quantities exceeding 100,000 metric tons per annum. Chloroacetanilides, from the key intermediates 2,6-diethylaniline and chloroacetylchloride, and phenoxy herbicides, from L-2-chloropropanoic acid rank between 50,000 and 100,000 t/yr. Also phosgene-derived thiocarbamate and urea herbicides, as well as dithiocarbamate fungicides are very large-volume products (see FUNGICIDES, AGRICULTURAL). However, most of the constituents of fungicides are commodities. This is also true for organophosphate insecticides derived from phosphorous oxychloride (see INSECT CONTROL TECHNOLOGY). All of these agrochemicals show zero or even negative growth, because of gradual replacement by more active, lower volume crop protection chemicals. Well-known examples are sulfonyl ureas, many of which contain 2-amino-4,6-disubstituted pyrimidine moieties and imidazolines, the key intermediate of which is pyridine-2,3-dicarboxylic acid.

Within pharmaceuticals, the highest volume categories are vitamins (qv), painkillers (see ANALGESICS, ANTIPYRETICS, AND ANTIINFLAMMATORY AGENTS), and β-lactam antibiotics (see ANTIBIOTICS, β-LACTAMS). Acetylsacetylsalicylic acid (aspirin) and *N*-acetyl-*p*-aminophenol (acetaminophen) have a world production of about 50,000 t/yr; asorbic acid (vitamin C) totals 50,000 t/yr; and niacin

Table 3. Fine Chemicals Industry Data for 1992

Company	Sales, $\$ \times 10^6$	Division	Sales, $\$ \times 10^6$	Business unit	Sales, $\$ \times 10^6$
Alusuisse-Lonza	4,400	Organic Chemicals	1,010	Fine Chemicals	450
Boehringer-Ingelheim	2,960	Chemikalien	220	Feinchemikalien	200[a]
Chemie Linz	704	Industriechemikalien	350	Intermediates	excl. prod. 267
Eastman Kodak	20,183	E.C.C.	3,930	Fine Chemicals	350[a]
Rhone Poulenc	14,800	Spéc. Chimiques	2,430	Organique Fine	790
Shell	109,650	Chemicals	8,250	Fine Chemicals	200[a]
Sandoz	9,610	Biochemie Kundl	560	Fine Chemicals	350[a]
SNPE	750	Chemicals	250	Fine Chemicals	125[a]

[a]Estimate.

(vitamin PP) totals about 25,000 t/yr. There are large production volumes of β-lactam antibiotic precursors such as 6-aminopenicillinic acid [*551-16-6*] (6-APA) and 7-aminocephalosporanic acid [*957-68-6*] (7-ACA), as well as side chains D-phenylglycine [*2935-35-5*] and D-*p*-hydroxyphenylglycine [*22818-40-2*], developed for the first semisynthetic penicillins, manufactured in the multithousand metric tons per year range. 2-Aminothiazolyl alkoximinoacetates, used for third-generation cephalosporins, are made in the several hundred metric ton per annum, as are five-ring heterocycles derived from sodium azide such as 5-mercapto-1-methyltetrazole [*13183-79-4*] and 2-mercapto-5-methyl-1,3,4-thiadiazole [*29490-19-5*].

In terms of types of molecular structures, heterocyclic compounds are the most important fine chemical category, especially fine chemicals having an *N*-heterocyclic structure as found in the vitamins B_2, B_6, H, PP, and folic acid. These and other natural substances have gained a great importance in modern pharmaceuticals and pesticides (16,17). Even modern pigments (qv) such as diphenyl pyrazolopyrazoles, quinacridones, and engineering plastics (qv), such as polybenzimidazoles and triazine resins, exhibit an *N*-heterocyclic structure.

From the point of view of application, pharmaceutical fine chemicals constitute the largest part of all fine chemicals, both in terms of number of products and volume of sales. About 40–50% of the total fine chemicals sales comes from pharmaceutical fine chemicals; about 20 to 25% are agrochemicals, and the rest belong to other categories.

Not many fine chemicals have a production value exceeding $10 million per year. Less than a dozen achieve production volumes above 10,000 metric tons per year and sales of >$100 million per year. Apart from the pharmaceutical and pesticide fine chemicals these comprise the amino acids (qv), L-lysine and D,L-methionine used as feed additives (see FEEDS AND FEED ADDITIVES), and vitamins ascorbic acid and nicotinic acid.

The future development of the fine chemicals industry depends mainly on the development of demand. The growth of the fine chemicals business is mainly fostered by the introduction of new pharmaceuticals, agrochemicals, engineering plastics (qv), and other specialties requiring high value organic intermediates. In the pharmaceutical industry good yardsticks for measuring innovation are the statistics of Investigational New Drug Applications (INDAs) waiting for approval by the Food and Drug Administration in the United States, and New Chemical Entities (NCE) also published by the FDA. The INDA grew from 66 in 1980 to 504 in 1991, and is expected to increase to more than 3000 by 1998 (18). The number of NCEs is expected to stay at least at the present level of about 20 to 25 compounds per year.

The strategy of the life science industry is also important. As strict financial management of all operations is becoming more and more imperative, investing in fine chemical manufacturing becomes less of a good risk/benefit ratio, in part because investment decisions for new plants have to be made years before the anticipated launch of a new drug or pesticide. Therefore, in prioritizing allocations of financial resources between production, marketing, and R&D, the last two are getting preferential treatment. As shown in Figure 4, the R&D expenditure of 12 principal U.S. chemical companies has grown at a faster pace than capital investment. Nevertheless, in the pharmaceutical industry considerable funds are

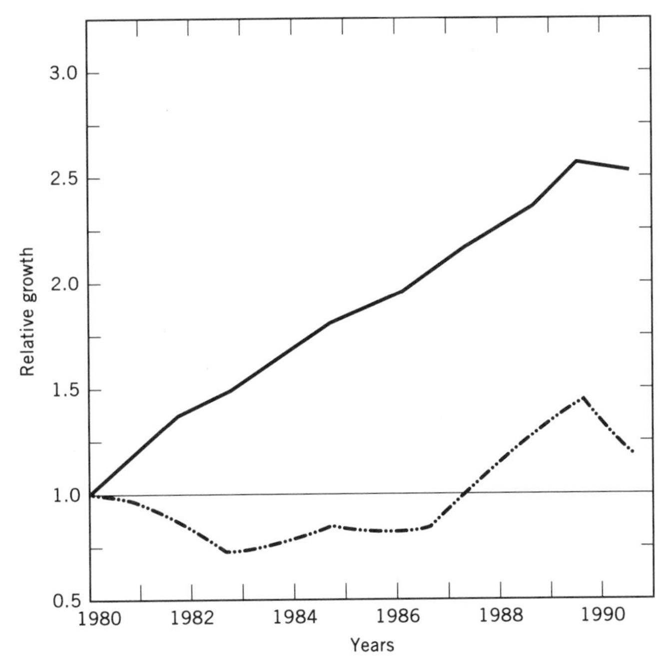

Figure 4. R&D expenditure vs capital spending for 12 U.S. chemical companies where (··—··—) is capital spending and (——) is R&D expenditure.

being invested in a new type of multipurpose plant called launch sites, which are intermediate in size between pilot and industrial-scale plants allowing for a large flexibility both in type and quantity of fine chemicals being produced. These were produced after FDA regulations called for experimental and commercial quantities of new pharmaceuticals to be produced in the same production units.

Future growth of the fine chemicals industry also depends on environmental regulations. In all countries tighter rules are being enforced on the handling of hazardous chemicals and on the quality and quantity of waste allowed to leave a factory. Furthermore, the international industry in general has self-imposed strict standards of quality (the ISO 9000 series) and the U.S. chemical industry in particular is implementing the responsible care program. All fine chemical companies need to allocate an increasing share of both R&D expenditure and capital investment on searching for clean processes and installing the equipment necessary for the reduction of emissions into the air, purification of wastewater, and pretreatment of solid waste (see ENVIRONMENTAL IMPACT). From a customer's perspective there is a trend to rely on a small number of selected, preferred suppliers to whom several tasks, such as inventory control, distribution logistics, and quality control can be delegated.

BIBLIOGRAPHY

"Fine Chemicals" in *ECT* 3rd ed., Vol. 10, pp. 338–347, by S. M. Tuthill and J. A. Caughlin, Mallinckrodt, Inc.

1. G. Schuch and J. König, *Chem.-Ing. Tech.* **64**(7), 587–593 (1992).
2. J. M. Charvat, *Managing Operations in the Chemical Industry by Aggregate Quality,* Peter Lang, Bern, Switzerland, 1990.
3. G. Booth, *The Manufacture of Organic Colorants and Intermediates,* The Eastern Press Ltd., London, 1988.
4. O. W. Wight, *Manufacturing Resource Planning: MRP II,* Oliver Wight Ltd. Publishers, Inc., Essex Junction, Vt., 1984.
5. H. Wiederkehr, *Chimia* **44**(5), 1–8 (1990).
6. *The United States Pharmacopoeia XXI (USP XXII-NF XVII),* The United States Pharmacopeial Convention, Inc., Rockville, Md., 1990.
7. *European Pharmacopoeia,* 2nd ed., Maisonneuve SA, Ste. Ruffine, France 1980.
8. *Food Chemicals Codex,* 3rd ed., National Academy of Sciences, National Research Council, Washington, D.C., 1981.
9. I. P. Morgan, *McKinsey Quart.*, 49–55 (Spring 1987).
10. *Perform. Chem.*, 41–47 (Feb/Mar. 1993).
11. J.-C. Pasquier and J.-M. Weiss, *Informations Chimie* **337,** 138–143 (Apr. 1992).
12. P. Pollak, *Chimica Oggi,* 11–16 (Jan./Feb. 1992); *Eur. Chem. News,* 23–26 (Oct. 5, 1992).
13. A. Boccone, *Chem. Mark. Manag.,* 15 (Spring 1986).
14. *Future Prospects for the Fine Chemicals Business,* ECOPLAN Int., Paris, 1992.
15. *Strategic Developments in the U.S. Fine Chemical Industry,* Kline & Co., Inc., Fairfield, N.J., 1989.
16. H. H. Szmant, *Organic Building Blocks of the Chemical Industry,* John Wiley & Sons, Inc., New York, 1989, pp. 575–637. Desktop reference on production processes, family trees, and economic aspects of commercially important fine chemicals.
17. P. Pollak and G. Romeder, *Perform. Chem.,* 36–38 (Feb. 1988); 44–54 (Apr. 1988).
18. *Chem. Eng. News,* 7 (Aug. 17, 1992).

PETER POLLAK
Lonza Ltd.

STANDARDS

Fine chemicals are generally considered chemicals that are manufactured to high and well-defined standards of purity, as opposed to heavy chemicals made in large amounts to technical levels of purity. Fine chemicals usually are thought of as being produced on a small scale and the production of some fine chemicals is in tens or hundreds of kilograms per year. The production of others, especially fine chemicals used as drugs or food additives (qv), is, however, in thousands of metric tons (see PHARMACEUTICALS). For example, the 1990 U.S. production of aspirin [*50-78-2*] and acetaminophen [*103-90-2*] was on the order of 20,500 t and 15,000 t, respectively.

Fine chemicals are produced by a wide spectrum of manufacturers, largely because the distinction between different kinds of chemicals is not sharp. There are specialty producers of fine chemicals. Many companies that manufacture drugs also manufacture the chemical substances that are used in preparing the dosage forms. A number of companies manufacture drug chemicals and food chemicals. Some fine chemicals are made by manufacturers of heavy chemicals, and either may be simply a segment of their regular production, or some of that production which has been subjected to additional purification steps. Many fine chemicals are imported into the United States from countries such as Japan, Germany, and the Netherlands.

Over the years compendia have been published in many countries to establish the purity of drug, food, and laboratory reagents available within their borders. For instance, Great Britain, France, Germany, and Japan have long published their own pharmacopeias and, more recently, in the interests of harmonization within the European Economic Community, the *European Pharmacopeia* (1) was issued. Compendia describing laboratory reagent chemicals are not so numerous, but two notable examples are the British *AnalaR Standards for Laboratory Chemicals* (2), and the German *Merck Standards* (3).

The fine chemicals standards discussed herein are primarily those originating in the United States. Much discussion has occurred regarding harmonization of the world's standards. It is not yet clear, however, what impact the International Standards Organization Quality Management Standards (ISO 9000) may have on the manufacture and specifications of fine chemicals.

Standards for drugs are established by the United States Pharmacopeial Convention, Inc. (USPC), and have been published in 21 revisions of the *United States Pharmacopeia* (USP). In the past, standards for many drugs that were not in the USP were established by the American Pharmaceutical Association, and were published in the *National Formulary* (NF). The last edition was published in 1975 (4). In that same year, the USPC acquired the NF, and the USP and NF now are published in one volume (5). In this compendium, drug substances and dosage forms of drug substances are designated USP, and pharmaceutic ingredients are designated NF. The latter substances are used to make the active ingredient(s) into suitable dosage forms for use by patients.

Standards for food-grade chemicals in the United States are set by the Committee on Food Chemicals Codex of the National Academy of Sciences (NAS) which publishes them in the *Food Chemicals Codex* (FCC) (6) (see also FOOD ADDITIVES). Standards for laboratory reagents are set by the American Chemical Society (ACS) Committee on Analytical Reagents and are published in *Reagent Chemicals–ACS Specifications* (7). Standards for electronic-grade chemicals, which have extremely low limits for trace ions, are published annually in *The Book of SEMI Standards* (BOSS) by Semiconductor Equipment and Materials International (SEMI) (8) (see also ULTRAPURE MATERIALS).

Standards for sulfuric acid, ranging from technical-grade through drug-, food-, and reagent-grade, to electronic-grade are shown in Table 1. The advances in purity represented by these various grades of chemicals are based on the special uses of the chemicals.

The publications detailing standards (5–8) generally include both specifications and methods of analysis for the substances. The establishment of standards of quality for chemicals of any kind presupposes the ability to set numerical limits on physical properties, allowable impurities, and strength, and to provide the test methods by which conformity to the requirements may be demonstrated. Tests are considered applicable only to the specific requirements for which they

Table 1. Specifications for Various Grades of Sulfuric Acid

Specification	Technical	NF	FCC[a]	ACS[b]	SEMI standard[c]	SEMI guideline[d]
assay, %	97.5–100.0	95.0–98.0		95.0–98.0	95.0–97.0	95.0–97.0
residue after ignition (RAI), ppm		50		5	3	
chloride, ppm		50	50	0.2	0.1	0.1
nitrate, ppm			10	0.5	0.2	0.15
phosphate, ppm					0.5	0.5
substances reducing permanganate, ppm		40	40	2		
arsenic, ppm		1	3	0.01	As + Sb 0.005	As + Sb 0.005
heavy metals,[e] ppm		5	20	1		
lead, ppm			5		0.3	0.01
iron, ppm	40		200	0.2	0.2	0.01

[a]Also can contain up to 20 ppm selenium.
[b]Also can contain up to 2 ppm ammonium and 0.005 ppm mercury.
[c]14 other trace metals at 0.01–0.3 ppm.
[d]31 other trace metals at 0.005–0.01 ppm.
[e]As Pb.

were written. Modification of a requirement, especially if the change is toward a higher level of purity, often necessitates revision of the test to ensure the test's validity.

One of the greatest tasks in providing compendial standards is to obtain, adapt, or develop test methods for determining compliance with the standards. Such methods must be capable of routine use in many laboratories by different personnel and equipment. There is a vast difference between a method that can be used in one laboratory by specialists and one that can be used in many laboratories by generalists to determine whether chemicals pass or fail the established specifications. Additionally, the determinations must be reliable, because the results obtained may determine whether a product is safe or legal.

There are no established specifications for the standard reference samples used in general chemical analysis. Many such substances, however, are analyzed and certified by the National Institute of Standards and Technology (NIST), formerly the National Bureau of Standards (NBS). Specific reference standards are required for many of the analyses included in the USP and NF standards for drugs.

The USP and the NF are recognized in the Food, Drug, and Cosmetic Act (1938) as establishing legal drug standards that the Food and Drug Administration (FDA) is responsible for enforcing. The FCC is recognized by FDA regulations, on an individual substance basis, as defining food grade. The ACS reagent specifications have no special legal status per se, but are used by the USP–NF, the FCC, SEMI, and in government procurement, and are referenced in FDA regulations and in American Society for Testing and Materials (ASTM) methods.

Federal regulation of drugs, food additives, and reagents used in *in vitro* diagnostic tests has increased markedly (see MEDICAL DIAGNOSTIC TESTS). With increasing consumer concern about the safety of fine chemicals, especially those used in foods and drugs, the United States government has become increasingly sensitive to the manner in which standards are set. Freedom of information legislation has confirmed the public's right to know, and this has introduced the objective of due process into the development of standards. The USP–NF and the FCC have mechanisms that make possible public participation in setting the standards with which these agencies are involved. Neither ACS nor SEMI has such a formal mechanism, but individuals from industry, government, and academia serve on the ACS Reagent Committee, and committee meetings of both the ACS and SEMI are open to the interested public upon request.

Drug Chemicals

Standards for drug chemicals are published in USP–NF. Drug substances are chemicals that have therapeutic or diagnostic uses, whereas pharmaceutical ingredients provide preservative action, flavoring, or fulfillment of a function in the formulation of dosage-form drugs. Examples of drug substances are acetaminophen [*103-90-2*], ampicillin [*69-53-4*], aspirin [*50-78-2*], powdered ipecac, riboflavin [*83-88-5*], stannous fluoride [*7783-47-3*], and thyroid. Examples of pharmaceutical ingredients are ethylparaben [*120-47-8*], lactose [*63-42-3*], magnesium stearate [*557-04-0*], sodium hydroxide [*1310-73-2*], starch [*9005-25-8*], and vanillin [*121-33-5*].

Dosage forms include acetaminophen tablets (see ANALGESICS, ANTIPYRETICS, AND ANTIINFLAMMATORY AGENTS), ampicillin capsules (see ANTIBIOTICS), aspirin tablets (see SALICYLIC ACID), ipecac syrup, riboflavin injection (see VITAMINS), and thyroid tablets (see HORMONES). Stannous fluoride (see TIN COMPOUNDS) is used in fluoride-containing toothpastes (see DENTIFRICES). Ethylparaben (see SALICYLIC ACID) is a preservative, lactose (see SUGARS) is a tablet filler, magnesium stearate (see CARBOXYLIC ACIDS) is a tablet lubricant, sodium hydroxide (see ALKALI AND CHLORINE PRODUCTS) is a neutralizing agent, starch (qv) is a tablet binder, and vanillin (qv) is a flavor (see also FLAVORS AND SPICES).

The United States Pharmacopeial Convention. The USP was established in 1820 by the assembly of the first United States Pharmacopeial Convention in Washington, D.C. Among the objectives of the meeting was the inclusion in a pharmacopeia of those drugs most fully established and best understood, and standards of pharmaceutical quality for them. The USPC was incorporated as an independent nonprofit organization in 1900, and is headquartered in Rockville, Maryland. A brief history of the Pharmacopeia of the United States is included in the front of *USP XXII–NF XVII* (5). An excellent discussion of the USPC is given in Reference 9, and additional descriptions are available (10–12).

The early USPC was dominated by physicians who selected the best drugs. This prevented inclusion in the USP of a large number of substances that were widely used, particularly elixirs, a popular dosage form in the late nineteenth century. To fill this gap, in 1888 the American Pharmaceutical Association published the first NF, which provided standards for drugs in wide use but not included in the USP. A history of the National Formulary is also included at the front of the NF section in *USP XXII–NF XVII* (5).

When the Federal Pure Food and Drugs Act was passed in 1906, both the USP and the NF were recognized in the act as setting the legal standards of strength, quality, and purity for the drugs described therein. Since the acquisition of the NF, the USPC has as its objective to provide names and standards of pharmaceutical quality for all United States drugs. This inclusion of such standards for all drugs in the USP is being achieved as rapidly as possible.

Tests described in the early compendia were simple analyses that could be performed by a pharmacist using a minimum of laboratory equipment. Tests in the 1990s employ instrumentation such as ultraviolet and infrared spectrophotometers, various kinds of chromatography including gas chromatography–mass spectrometry (gc–ms), x-ray diffraction, nuclear magnetic resonance spectroscopy (nmr), atomic absorbtion (aa), and thermal analysis (see ANALYTICAL METHODS; CHROMATOGRAPHY; MAGNETIC SPIN RESONANCE; SPECTROSCOPY; THERMAL, GRAVIMETRIC, AND VOLUMETRIC ANALYSES). Such tests are no longer performed by pharmacists but are largely used by manufacturers and the various regulatory agencies, eg, the FDA and comparable state agencies.

The USP Committee of Revision. The work of wholly revising the USP–NF every five years, and making revisions by supplements, is performed by the USP Committee of Revision which now numbers more than 100 experts in the fields of medicine, pharmacy, chemistry, biotechnology, microbiology, etc. The committee members, supported by the Drug Research and Testing Laboratory located in USP headquarters, write the general chapters and product monographs. The *USP XXII–NF XVII* contains well over 3000 monographs.

The USP–NF. In addition to defining the quality of the reagents used in testing, the USP–NF sets the legal standards of strength, quality, purity, and requirements for packaging and labeling for the articles included in USP–NF. Thus much of the text in the USP–NF represents enforceable legal requirements.

The General Notices. These are the basic requirements for the application and interpretation of the tests and specifications that follow in the USP–NF. Many of the terms used in the text are defined, and the majority of procedural questions that may arise within the monograph for each substance are answered.

The Monographs. A separate monograph is provided for each substance included in the USP–NF. Descriptions and solubilities of the substances are given in a table separate from the individual monographs. The body of the monograph includes a statement of assay, followed by a packaging and storage statement, a labeling statement where applicable, identification tests, and then a variety of tests to establish the strength, quality, and purity of the drug. The tests either provide the methods to be followed or refer to methods in the General Tests and Assays section of the book. Whereas the use of alternative methods, as listed in other compendia described herein is permitted, if there is doubt or disagreement the test in USP–NF serves as the referee method, and only that method is authoritative.

The General Tests and Assays. This section of the USP gives methods for tests that are general in nature and apply to a number of the substances. Procedures are included for such tests as heavy metals, melting point, chloride, sulfate, sterility, bacterial endotoxins, and pyrogens. Also included are descriptions of various analytical techniques, such as spectrophotometry, chromatography, and nmr, and descriptions of tests to be used on glass or plastic containers, rubber closures, etc.

Reagents, Indicators, and Solutions. This section includes the specifications and testing methods for reagents to be used in the tests specified in the USP–NF, and directions for making the various indicator, buffer, colorimetric, test, and volumetric solutions used in the testing. Reagents for which ACS specifications exist are referenced to the ACS book (7).

Reference Standards. Many of the identification tests and assays require the use of reference standards. These standards are available for purchase from the USPC.

The Standard-Setting Process. Setting USP–NF standards is a continuing, and by no means unilateral, process. The Committee of Revision not only develops monographs for new substances but also continually reviews the monographs, specifications, and testing methods for existing substances. Results are published in one or more supplements each year. A complete review is done every five years and a revision is published.

When a new substance is admitted to the USP–NF, the first step is to locate producers and to obtain their specifications and testing methods. A new monograph is published by the USP for comment in *Pharmacopeial Forum* (PF), which is published every other month. PF includes proposed changes in existing monographs, the general notices, the general tests, or other sections of the USP–NF, proposed new monographs, and proposals relative to policy or philosophy. The public may comment, protest, or make suggestions, and their views receive due consideration. If significant revisions are necessary after this process, a revised proposal is published. Through *Pharmacopeial Forum*, USP–NF standards are openly developed.

USP Dispensing Information. In the past, USP and NF monographs included information such as the category of use, the usual dose, and package sizes available. This type of information has been greatly expanded, and is now published annually in a four volume book separate from the USP–NF quality standards, entitled *USP Dispensing Information* (USPDI). This information focuses on aspects that enhance the safe and effective use of various medications. It contains information specifically for the health practitioner as well as for the patient, and includes information concerning the dispensing and administration of drugs as well as indications and contraindications related to their use.

Food-Additive Chemicals

The FCC is to food-additive chemicals what the USP–NF is to drugs. In fact, many chemicals that are used in drugs also are food additives (qv) and thus may have monographs in both the USP–NF and in the FCC. Examples of food-additive chemicals are ascorbic acid [*50-81-7*] (see VITAMINS), butylated hydroxytoluene [*128-37-0*] (BHT) (see ANTIOXIDANTS), calcium chloride [*10043-52-4*] (see CALCIUM COMPOUNDS), ethyl vanillin [*121-32-4*] (see VANILLIN), ferrous fumarate [*7705-12-6*] and ferrous sulfate [*7720-78-7*] (see IRON COMPOUNDS), niacin [*59-67-6*], sodium chloride [*7647-14-5*], sodium hydroxide [*1310-73-2*] (see ALKALI

AND CHLORINE PRODUCTS), sodium phosphate dibasic [*7558-79-4*] (see PHOSPHORIC ACIDS AND PHOSPHATES), spearmint oil [*8008-79-5*] (see OILS, ESSENTIAL), tartaric acid [*133-37-9*] (see HYDROXY DICARBOXYLIC ACIDS), tragacanth [*9000-65-1*] (see GUMS), and vitamin A [*11103-57-4*].

Action to compile standards for food-grade chemicals did not take place until after the enactment of the Food Additives Amendment to the Food, Drug, and Cosmetic Act in 1958 (13). This amendment stated that substances added to foods should be of food-grade quality, but it contained no criteria by which such quality could be determined (see also COLORANTS FOR FOOD, DRUGS, COSMETICS, AND MEDICAL DEVICES). The Food Protection Committee of the National Academy of Sciences–National Research Council (NAS–NRC) therefore undertook the project of producing a *Food Chemicals Codex*.

The objective of the FCC is to define food-grade chemicals in terms of the characteristics that establish identity, strength, and quality. It provides specifications in monograph form for some 900 food additives, together with analytical test procedures by which compliance with the specifications can be determined. The third edition was published in 1981; supplements followed in 1983, 1986, 1991, and 1993. The fourth edition is in preparation as of this writing and is to include monographs for almost 1000 food chemicals, including flavors.

Food Chemicals Codex standards are recognized by FDA regulations as defining food grade for many individual chemicals used in foods and food processing. FCC specifications have been adopted by the governments of Australia, Canada, New Zealand, and the United Kingdom. There is extensive international activity in the field of food additives. The FCC is represented at meetings of the Joint Food and Agricultural Organization/World Health Organization (FAO/WHO) Expert Committee on Food Additives (JECFA) and the Food Additive Commission of the International Union of Pure and Applied Chemistry (IUPAC).

The Food Chemicals Codex. The *Food Chemicals Codex* is developed by the Committee on Food Chemicals Codex, which is a part of the Food and Nutrition Board, Institute of Medicine, National Academy of Sciences, under a contract with the U.S. FDA. The Committee has the responsibility for the development and revision of the FCC. To meet this responsibility, the Committee also contacts manufacturers, trade associations, and other knowledgeable parties to obtain comments and criticisms of monographs proposed by the committee. Broader public input is sought by publication, by the FDA in the *Federal Register*, of current committee activity regarding new and revised monographs proposed for inclusion in the FCC.

Reagent Chemicals

Reagent Chemicals—ACS Specifications, in its eighth edition as of 1993 (7), is to reagent chemicals what the USP and the FCC are to drug and food-additive chemicals. The ACS Committee on Analytical Reagents, and its activity relative to specifications for reagents, has a history dating back to 1917, and more tenuously, to 1903 (14). Examples of reagent chemicals are acetone [*67-64-1*] (qv), arsenic trioxide [*1327-53-3*] (see ARSENIC COMPOUNDS), barium chloride [*10361-37-2*] (see BARIUM COMPOUNDS), bromine (qv) [*7726-95-6*], bromthymol blue [*76-59-5*] (see HYDROGEN-ION ACTIVITY), cupferron [*135-20-6*] (see COPPER; IRON), anhydrous ethyl ether [*60-29-7*] (see ETHERS), hexanes (see HYDROCARBONS), hydrochloric acid [*7647-01-0*] (see HYDROGEN CHLORIDE), 70% perchloric acid [*7601-90-3*] (see PERCHLORIC ACID AND PERCHLORATES), silver diethyldithiocarbamate [*38351-46-1*] and silver nitrate [*7783-99-5*] (see SILVER COMPOUNDS), and sodium hydroxide [*1310-73-2*] (see ALKALI AND CHLORINE PRODUCTS).

The ACS Committee on Analytical Reagents is comprised of some 15 members from academia, government, and industry (both manufacturers and users of reagents) and meets twice a year at the ACS headquarters in Washington, D.C. Throughout the year other work is carried out by correspondence. Requirements and details of tests are based on published work, on the experience of committee members in the examination of reagent chemicals, and on studies of test procedures made by committee members.

When a specification for a reagent is first prepared, it generally is based on the highest purity level that is competitively available in the United States. If a higher level of purity subsequently becomes available on a competitive basis, the specification is revised accordingly. An exception to this approach is a need for higher purity in a reagent than is competitively available. In this case the committee sets the standard based on need, presumably stimulating manufacturers to meet that need.

Although the book on reagent chemicals contains many tests for the determination of trace impurities in reagents, it is not intended to be a text on the techniques of trace analysis but rather to provide tests that are reproducible in various laboratories, and which are accurate, economic, and feasible (see TRACE AND RESIDUE ANALYSIS).

The usual reagents suffice for many purposes, but the committee recognizes uses for reagents that require a higher purity than that defined by existing ACS specifications. Sometimes reagents must be further purified. However, when there are reasonably widespread uses the committee defines special grades, such as those suitable for use in high performance liquid chromatography (hplc), pesticide residue analysis, or ultraviolet spectrophotometry. The committee takes into account the practical matters involved in the manufacture of reagents. In general, the higher the purity requirement, the greater the cost. In the combination of high purity requirements with low usage, economics may become an important factor.

The ACS Book. The ACS book, *Reagent Chemicals—ACS Specifications*, establishes a standard of quality for reagents to be used in precise analytical work, for which purpose it contains both specifications and testing methods for some 350 reagent chemicals.

The book is similar to the two compendia already described. It is published approximately every five years and supplements are published between editions. In addition, notices of changes that need to be publicized after adoption at meetings of the committee are published promptly in the ACS journals *Analytical Chemistry* or, for those of a more urgent nature, *Chemical and Engineering News*.

The Standard-Setting Process. The committee has three main lines of endeavor: the improvement of existing limits for impurities, the improvement of present test methods, and the development of specifications and testing methods for additional compounds.

The ACS committee has a general rule: if two producers meet a given specification, the specification is normally so defined. Usually the methods are checked by several laboratories using samples from various sources. Because of the many different matrices in which impurities are determined, even the simplest impurity tests need to be checked for accuracy in the laboratory.

ACS reagents are specified for tests in the USP–NF and the FCC, and in many ASTM methods of analysis. ACS specifications also are used by government agencies for procurement purposes, and are mentioned in FDA regulations pertaining to the labeling of *in vitro* diagnostic reagents. Although these do not have legal status, as is the case for USP–NF and FCC specifications, these are generally recognized, in the United States and elsewhere, as the definitive standards for reagents.

Most of the common reagent chemicals already are covered by ACS specifications. However, as tests become more highly instrumented and complex, it becomes difficult to write instructions that cover them, and to obtain agreement on tests and results among laboratories. For instance, it became necessary to develop tests for specialized uses for existing reagents, eg, tests on selected solvents to ensure that they are satisfactory for spectrophotometric use or for use in the determination of pesticide residues by gas chromatography (gc) (see SOLVENTS, INDUSTRIAL). Large quantities of such solvents are used to extract pesticides from natural products (see INSECT CONTROL TECHNOLOGY). This means that the solvents themselves must show little or no response to the test, if the reagents are not to overwhelm the substance sought. Testing of such solvents involves a manyfold concentration of possible impurities, followed by a gc examination for peaks that would interfere with the analysis for pesticide residues. Similarly, stringent tests control the suitability of solvents for use in hplc.

Electronic Chemicals

Chemicals used in the manufacture of integrated circuits (qv) need to be controlled to even more stringent levels of purity than either USP or ACS grades. In 1973 the Semiconductor Equipment and Materials Institute (SEMI) held its first standards meeting. SEMI standards are voluntary consensus specifications developed by the producers, users, and general interest groups in the semiconductor (qv) industry. Examples of electronic chemicals are glacial acetic acid [*64-19-7*], acetone [*67-64-1*], ammonium fluoride [*12125-01-8*] and ammonium hydroxide [*1336-21-6*] (see AMMONIUM COMPOUNDS), dichloromethane [*75-09-2*] (see CHLOROCARBONS AND CHLOROHYDROCARBONS), hydrofluoric acid [*7664-39-3*] (see FLUORINE COMPOUNDS, INORGANIC), 30% hydrogen peroxide (qv) [*7722-84-1*], methanol (qv) [*67-56-1*], nitric acid (qv) [*7697-37-2*], 2-propanol [*67-63-0*] (see PROPYL ALCOHOLS), sulfuric acid [*7664-93-9*], tetrachloroethane [*127-18-4*], toluene (qv) [*108-88-3*], and xylenes (qv) (see also ELECTRONIC MATERIALS).

ASTM has published a few selected standards for materials used in the electronics industry, such as gold wire for semiconductor lead bonding, but it does not provide a comprehensive set of standards (see ELECTRICAL CONNECTORS).

The Book of SEMI Standards (BOSS). Of the five (5) volumes issued by SEMI, *Chemicals / Reagents* is the one that pertains to fine chemicals. The committee developing these specifications, originally called Reagent Chemicals, in 1991 changed its name to Process Chemicals. In 1988, to reflect the increasingly global importance of SEMI specifications, the industry body changed its name from Semiconductor Equipment and Materials Institute to Semiconductor Equipment and Materials International. The first chemical standards of 1973 borrowed heavily from ACS specifications with the addition of about a dozen trace metals controlled to the ppm (μg/g) level. The 1993 BOSS contains standards for some 38 chemicals such that the most critical materials contain trace metal limits in the sub-ppm range. In addition, SEMI now offers international standards on-line by its international communications network, SEMICOMM. Furthermore, in response to users' needs for ever lower trace metal levels, a novel approach was instituted in 1990. In addition to the 38 standards of base level purity mentioned above which, in accord with USP–NF, FCC, and ACS practices, include referee procedures, BOSS now also includes guidelines for the 20 most critical chemicals in the industry. Guidelines reflect a chemical purity level typically required by semiconductor devices that have geometries of less than 1 micrometer. Standardized test methods are still being developed for some parameters at the purity levels indicated in the guideline. However, until standardized test methods are published, test methodology must be determined by the user and producer. In general these guidelines limit trace level impurities to 5–10 ppb (ng/g) and also limit particles 0.5 micrometers and greater in bottled liquids. To measure trace metals to the levels required in the guidelines involves the use of state-of-the-art instrumentation such as inductively coupled plasma/mass spectrometry (icp/ms).

Chemical and Other Standards Used in Analysis

National Institute of Standards and Technology (NIST). The NIST is the source of many of the standards used in chemical and physical analyses in the United States and throughout the world. The standards prepared and distributed by the NIST are used to calibrate measurement systems and to provide a central basis for uniformity and accuracy of measurement. At present, over 1200 Standard Reference Materials (SRMs) are available and are described by the NIST (15). Included are many steels, nonferrous alloys, high purity metals, primary standards for use in volumetric analysis, microchemical standards, clinical laboratory standards, biological material certified for trace elements, environmental standards, trace element standards, ion-activity standards (for pH and ion-selective electrodes), freezing and melting point standards, colorimetry standards, optical standards, radioactivity standards, particle-size standards, and density standards. Certificates are issued with the standard reference materials showing values for the parameters that have been determined.

Some of the standards are fine chemicals in themselves, and others, such as filters for checking spectrophotometers, are of utility in the testing and control of fine chemicals (see OPTICAL FILTERS).

United States Pharmacopeia. Reference standards are required in many USP and NF tests, and in a few FCC tests. The USPC distributes such standards domestically and has authorized international distribution by a number of organizations or companies. There are well over 1000 USP Reference Standards, including several for melting points, and also specimens of narcotics and other controlled substances. New standards are constantly under development as needed in various USP, NF, and FCC testing methods.

Impact of the Food, Drug, and Cosmetic Act on Fine Chemicals

FDA Quality Standards. Although standards for many drugs and biologicals are included in the USP–NF, and for many food additives in the FCC, the FDA also establishes some specifications of its own. In the drug field, specifications and testing methods for antibiotics and biologicals are set by the FDA. Also, specifications and testing methods are prescribed for colorants. Many food-additive petitions are granted with the requirement that certain specifications are met.

Device Legislation. Regulations covering medical devices define reagents used in *in vitro* diagnostic tests as devices (see MEDICAL DIAGNOSTIC REAGENTS; PROSTHETIC AND BIOMEDICAL DEVICES).

Regulations Concerning Good Manufacturing Practice. Chemicals that are drugs, as defined in the Food, Drug, and Cosmetic Act, are subject to the requirement of the Act that they be made under conditions of Current Good Manufacturing Practice (CGMP). Specific GMP regulations for such chemicals have not been published, but the regulations that have been published for dosage form drugs include many points that should be considered (16).

The primary thrust of GMP is that it is not enough merely to make chemicals to meet USP or other applicable specifications. The chemicals must be made under clean and sanitary conditions, procedures and processes must be validated and documented, and processing and packaging must be carried out under conditions that preclude mixup and mislabeling. Records must be kept of complaints, and the manufacturer must know enough about the storage properties of the products to specify storage conditions and, if necessary, expiration dates on the label.

A manufacturer of drug chemicals is required to register with the FDA, and is subject to FDA inspection at least once every two years. Some manufacturers who make chemicals that incidentally are drugs are impelled to drop the drug designation from their labeling in order to avoid the exposure to inspection that registration entails.

Chemicals used in foods should be made under conditions of GMP similar to those for chemicals that are drugs. Food chemicals specifically are subject to food GMPs (17), which deal mainly with sanitation, but there are no general GMP regulations for food chemicals. For this reason, the FCC includes an informational section, "General Good Manufacturing Practice Guidelines for Food Chemicals," to provide GMP guidance for manufacturers of food chemicals.

Reagent chemicals per se are not subject to any GMP regulations because normally these are not considered to be drugs or devices. However, *in vitro* diagnostic reagents are considered to be devices, and many reagent chemicals can be considered to be *in vitro* diagnostic reagents. Such reagents, to which the *in vitro* diagnostic regulations apply, are subject to the GMPs for diagnostic products (18), and to the labeling requirements for *in vitro* diagnostic products (19). Thus GMP may impinge on certain reagents, unless the manufacturer specifically declares that they are not *in vitro* diagnostic reagents, and does not sell them for this use. Electronic chemicals are not drugs, food additives, or devices, and thus are not subject to any GMP regulations.

BIBLIOGRAPHY

"Fine Chemicals" in *ECT* 3rd ed., Vol. 10, pp. 338–347, by S. M. Tuthill and J. A. Caughlan, Mallinckrodt, Inc.

1. *The European Pharmacopeia*, 2nd ed., Maisonneuve, SA, France, 1990.
2. *AnalaR Standards for Laboratory Chemicals*, 8th ed., BDH Chemicals Ltd., Poole, UK, 1984.
3. *Merck Standards*, E. Merck, Darmstadt, Germany, 1972.
4. *The National Formulary XIV*, The American Pharmaceutical Association, Washington, D.C., 1975.
5. *The United States Pharmacopeia XXII, (USP XXII–NF XVII)*, The United States Pharmacopeial Convention, Inc., Rockville, Md., 1990.
6. *Food Chemicals Codex*, 3rd ed., National Academy of Sciences, National Research Council, Washington, D.C., 1981.
7. *Reagent Chemicals—American Chemical Society Specifications*, 8th ed., American Chemical Society, Washington, D.C., 1993.
8. *The Book of SEMI Standards*, Semiconductor Equipment and Materials International, Mountain View, Calif., 1993.
9. L. C. Miller and co-workers, *Anal. Chem.* **32**, 19A (1960).
10. A. Hecht, *FDA Consumer*, 24 (Oct. 1977).
11. W. M. Heller, *Hosp. Formul.* **12**, 883 (1977).
12. D. Banes, *J. Chem. Inf. Comput. Sci.* **17**, 95 (1977).
13. *Public Law No. 618*, 86th U.S. Congress, Washington, D.C., 1960.
14. S. M. Tuthill, *Anal. Chem.* **42**, 30A (1970).
15. *NIST Special Publication 260*, 1992–1993 ed., National Institute of Standards and Technology, Gaithersburg, Md.
16. *Title 21, Code of Federal Regulations*, Parts 210 and 211, Government Printing Office, Washington, D.C.
17. Ref. 16, Part 110.
18. Ref. 16, Part 820.
19. Ref. 16, Part 809.10, particularly section 809.10(d).

SAMUEL M. TUTHILL
NORMAN C. JAMIESON
Mallinckrodt Specialty Chemicals Company

FIREBRICK. See REFRACTORIES.

FIRE CLAY. See CLAY; REFRACTORIES.

FIRE PREVENTION AND EXTINCTION. See FLAME RETARDANTS; PLANT SAFETY.

FIRE-RESISTANT TEXTILES. See FLAME RETARDANTS FOR TEXTILES.

FIREWORKS. See PYROTECHNICS.

FISCHER-TROPSCH PROCESS. See COAL CONVERSION PROCESSES; FUELS, SYNTHETIC.

FISH FARMING. See AQUACULTURE.

FISH LIVER OILS. See VITAMINS, VITAMIN D.

FISH OILS. See FATS AND FATTY OILS.

FLAME PHOTOMETRY. See ANALYTICAL METHODS; SPECTROSCOPY.

FLAMEPROOFING. See FLAME RETARDANTS.

FLAME RETARDANTS

OVERVIEW

Each year, Americans report over three million fires leading to 29,000 injuries and 4,500 deaths (1). The direct property losses exceed $8 billion (1) and the total annual cost to our society has been estimated at over $100 billion (2). Personal losses occur mostly in residences where furniture, wall coverings, and clothes are frequently the fuel. Large financial losses occur in commercial structures such as office buildings and warehouses. Fires also occur in airplanes, buses, and trains.

Fires occur when an ignition source, a match, cigarette, or stove burner, meets a flammable product such as a chair, wall, or scattered papers. The heat from the source breaks down polymer strands in the material, creating (generally endothermically) chemical fragments that vaporize. At a sufficiently high temperature, these fragments react with the oxygen in the air to release more heat. Some of this heat radiates or convects back to the product, breaking down more polymeric strands, yielding more gas-phase fuel, etc. Life- and property-threatening fires result when the rate of heat feedback to the product exceeds the sum of the heat dispersed from the combustion environment and the marginal enthalpy required to produce a steady stream of vapor-phase pyrolyzate.

Understanding of fires dates to the nineteenth century. The advent of modern fire fighting techniques and equipment has meant less destruction of cities or whole buildings. Additionally, fire-resistant building design usually contains fires to parts of structures. However, a high fuel load in either a residence or a commercial building can overwhelm even the best of building construction.

Terminology

A number of adjectives have been used to describe a product having an apparently low contribution to a fire. Nonquantitative terms such as fireproof, flameproof, self-extinguishing, nonburning, and noncombustible, have been used and have often led to confusion regarding the relative fire safety of different materials. Additionally, a product is sometimes improperly described by a component material rating under a fire test, a V-0 rating in the UL 94 test, or a building code provision, a 25 flame spread limit for wall coverings using the ASTM E84 method.

These ambiguities eventually led the Federal Trade Commission to take action in the case of cellular plastics and to restrict the use of such terminology (3). This action, in addition to the prohibition placed on the use of certain terminology, requires the use of a caveat whenever the results of burning tests are cited. Much of the older literature, however, as well as some of the more recent publications, use this restricted terminology.

Some pertinent definitions include *fire retardant (flame retardant)*, used to describe polymers in which basic flammability has been reduced by some modification as measured by one of the accepted test methods; *fire-retardant chemical,* used to denote a compound or mixture of compounds that when added to or incorporated chemically into a polymer serves to slow or hinder the ignition or growth of fire, the foregoing effect occurring primarily in the vapor phase; *materials,* single substances of which things are constructed that may be composed of single or blended polymers, may be layered or fiber-reinforced, and might contain a variety of additives; and *products,* consumer items made of one or more materials.

Measuring Fire Performance of Products

Laws have been promulgated to improve the fire performance of everyday fuels. Most of the fire test methods in regulations have been developed by consensus standards organizations in response to a particular fire hazard. The two leading entities are the American Society for Testing and Materials (ASTM) and the National Fire Protection Association (NFPA). Methods are then referenced in the model building codes, such as the Standard Building Code (Southern Building Code), Basic Building Code (Building Code Officials Administration International), and the Uniform Building Code (International Conference of Building Officials), as well as NFPA's National Fire Codes, National Electrical Code, and Life Safety Code. Selected portions of these structures are in turn incorporated into laws by a governmental jurisdiction. In addition, there are a number of voluntary practices. For example, Underwriters Laboratories (UL) allows the use of its endorsement on products that meet their test criteria, and the upholstered furniture industry has adopted voluntary cigarette ignition-resistance standards.

Fire test methods attempt to provide correct information on the fire contribution of a product by exposing a small sample to conditions expected in a fire scenario. Methods can be viewed in two ways: the first entails the strategy of the fire test, ignition resistance or low flammability once ignited; the second addresses the test specimen, a sample representative of the product or a sample of a material that might be used in the product. Fire science has progressed markedly since the older test methods were developed and it is known that the basis for many of these tests is doubtful. Results from older tests must be used with great care.

The susceptibility of a product to an ignition source can be measured by flame or heat impingement tests, such as UL 94 (4) or NFPA 260/261 (5), or by ignition delay times in an apparatus such as the Cone Calorimeter (6). In UL 94, a vertical strip of a material is ignited at the bottom and after the burner is removed, one observes whether burning is sustained. This is an example of a material test that results in a simple flammability class assignment. The NFPA cigarette ignition tests are examples of similar tests for a product. There, a cigarette is laid on a reduced-scale mockup of a seat cushion to see whether ignition occurs. The Cone Calorimeter is an apparatus used to measure flammability properties of a product. A cutting representative of the product is exposed to radiant energy typical of a fire of concern. The principal measurement is that of the rate of heat release. Time to ignition can also be determined and used as an indication of ignition susceptibility.

Clearly, fewer ignitions would reduce the number of fires. However, once ignited an ignition-resistant material may burn with a higher intensity than a more easily ignited counterpart (7). Moreover, successful ignition-resistance test performance is not proof of fire prevention. The real world situation may be more severe than the test design, larger ignition sources may occur or thermal radiation from other burning objects could increase the ease of ignition. Thus many elements of fire protection practice presume that ignition can occur. It is then desirable that products burn sufficiently slowly that the fire does not grow rapidly to threatening size, does not ignite adjacent items, and can be readily and simply extinguished. Therefore the controlling characteristic variable is the rate of heat release of the product. Methods have been developed for accurate measurement of rate of heat release (8,6). There is research relating these rates to the performance of products (9,10).

The assessment of the contribution of a product to the fire severity and the resulting hazard to people and property combines appropriate product flammability data, descriptions of the building and occupants, and computer software that includes the dynamics and chemistry of fires. This type of assessment offers benefits not available from stand-alone test methods: quantitative appraisal of the incremental impact on fire safety of changes in a product; appraisal of the use of a given material in a number of products; and appraisal of the differing impacts of a product in different buildings and occupancies. One method, HAZARD I (11), has been used to determine that several commonly used fire-retardant–polymer systems reduced the overall fire hazard compared to similar nonfire retarded formulations (12).

Methods for Improved Performance

The materials of attention in promoting fire safety are generally organic polymers, both natural, such as wood (qv) and wool (qv), and synthetic, nylon (see POLYAMIDES), vinyl, and rubber (qv). Less fire-prone products generally have either inherently more stable polymeric structures or fire-retardant additives. The former are usually higher priced engineering plastics (qv) which achieve increased stability at elevated temperatures by incorporating stronger (often aromatic) chemical bonds in the backbone of the polymer (13). Examples are the polyimides, polybenzimidazoles, and polyetherketones. There are also some advanced polymers, such as the polyphosphazenes and the polysiloxanes, which have strong inorganic backbones. Thermally stable pendent groups are also necessary. Strongly bonded polymers may, however, be brittle or difficult to process.

Fire-retardant additives are most often used to improve fire performance of low-to-moderate cost commodity polymers. These additives may be physically blended with or chemically bonded to the host polymer. They generally effect either lower ignition susceptibility or, once ignited, lower flammability. Ignition resistance can be improved solely from the thermal behavior of the additive in the condensed phase. Retardants such as hydrated alumina add to the heat capacity of the product, thus increasing the enthalpy needed to bring the polymer to a temperature at which fracture of the chemical bonds occurs. The endothermic volatilization of bound water can be a significant component of the effectiveness of this family of retardants. Other additives, such as the organophosphates, change polymer decomposition chemistry. These materials can induce the formation of a cross-linked, more stable solid and can also lead to the formation of a surface char layer. This layer both insulates the product from further thermal degradation and impedes the flow of potentially flammable decomposition products from the interior of the product to the gas phase where combustion would occur.

Flame retardants function in the vapor phase where the enthalpy-generating combustion reactions occur. Halogen-containing species, for instance, can be selected to vaporize at the same temperature as the polymer fragments. Coexisting in the reactive area of the flame, the halogens are effective at decreasing the concentrations of the free radicals that propagate flames, thus reducing the flame intensity, the enthalpy returned to the product, and the burning rate, in that order. For small ignition sources the use of flame retardants can produce self-extinguishment. More intense sources may overwhelm the flame retardant, necessitating either a higher concentration or an alternative choice of additive.

Useful materials incorporating fire-retardant additives are not always straightforward to produce. Loadings of 10% are common, and far higher levels of flame retardants are used in some formulations. These concentrations can have a negative effect on the properties and functions for which the materials were originally intended. Product-specific trade-offs are generally necessary between functionality, processibility, fire resistance, and cost.

Nonetheless a large number of fire-retardant additives are possible. The development of the field of fire-retardant additives has its origins in three efforts: the nineteenth century systematic studies of Gay-Lussac, Perkin's discovery that stannates and tungstates helped make treatment with ammonium salts water-resistant, and the discovery in the 1930s of the effect of mixing antimony oxide with organic halogen compounds (14). Research has led to a diversity of additives and a thriving market. Fire retardants are now the most used plastics additives, exceeding 40% of a $1 billion market in 1991 (15). This market is expected to continue to increase. Table 1 gives the principal groups of chemicals and their relative use. However, there is an ongoing debate over the possible risks of halogenated, especially brominated, fire retardants (15). As can be seen from Table 1, brominated retardants are a significant fraction of the market. Whereas no ban on usage has been issued as of this writing, more data and continued discussions are expected. The issues under debate are (*1*) the burning of halogenated combustibles produces toxic smoke, and international studies show that most fire victims die from smoke inhalation. The smoke from all fires is noxious. It has been shown that if the fire retardant significantly decreases the burning rate of the product, the reductions in smoke and heat yields are more important to survivability than a modest increase in the toxic potency of the smoke (12); (*2*) the burning of halogenated combustibles produces corrosive smoke, which results in additional damage to electronic components, etc. The smoke from nonhalogenated polymers is also corrosive and the fire safety community is in the process of developing methods to characterize this property of smoke; (*3*) the incineration of

Table 1. Flame-Retardant Market Volume[a]

Group	1986, t	1991	
		t	$\$ \times 10^6$
phosphate esters	20	18	50
halogenated phosphates	13	16	46
chlorinated hydrocarbons	15	15	31
brominated hydrocarbons	28	36	160
brominated bisphenol A	16	18	37
antimony trioxide	22	25	85
borates	8	8	10
aluminum trihydrate	140	170	85
magnesium hydroxide	2	3	6
Total	*264*	*301*	*510*

[a]Courtesy of the TPC Business Research Group.

halogenated combustibles may produce significant amounts of dioxin- and furan-like species. Laboratory combustion experiments need to be compared with measurements of the effluent from properly designed and operated incinerators.

Another factor potentially affecting the market for halogenated fire retardants is the waste disposal of plastics (see WASTES, INDUSTRIAL). As landfill availability declines or becomes less popular, two alternatives are incineration and recycling (qv). The nature of the combustion products from halogenated products requires careful construction and maintenance of incinerators (qv) to avoid damage to the incinerator itself and a public health problem from the exhaust. The ease of recycling used products also has a potential effect on fire retardants.

Flame-retardant additives are capable of significant reduction in the hazard from unwanted fires, and techniques are now available to quantify these improvements. Combined with an understanding of fire-retardant mechanisms, polymer–retardant interactions, and reuse technology, formulations optimized for public benefit and manufacturing practicality can be selected.

BIBLIOGRAPHY

"An Overview" under "Flame Retardants" in *ECT* 3rd ed., Vol. 10, pp. 348–354, by J. W. Lyons, National Bureau of Standards.

1. M. J. Karter, Jr., *NFPA J.* **86**(5), 32 (1992).
2. W. P. Meade, *A First Pass at Computing the Cost of Fire in a Modern Society*, The Herndon Group, Inc., Chapel Hill, N.C., 1991.
3. *Fed. Reg.* **40**(12), 255 (1975).
4. *Tests for Flammability of Plastic Materials for Parts in Devices and Appliances*, UL-94, Underwriters Laboratories, Northbrook, Ill., 1991.
5. "Cigarette Ignition Resistance of Components of Upholstered Furniture," NFPA 260, and "Resistance of Mockup Upholstered Furniture Material Assemblies to Ignition by Smoldering Cigarettes," NFPA 261, in *National Fire Codes*, National Fire Protection Association, Quincy, Mass., 1992.
6. *Standard Test Method for Heat and Visible Smoke Release Rates for Materials and Products Using an Oxygen Consumption Calorimeter*, ASTM E1354-90a, American Society for Testing and Materials, Philadelphia, Pa., 1991.
7. J. W. Rowen and J. W. Lyons, *J. Cell. Plast.* **14**(1), 25 (1978).
8. *Standard Test Method for Heat and Visible Smoke Release Rates for Materials and Products*, ASTM E906-83, American Society for Testing and Materials, Philadelphia, Pa., 1991.
9. V. Babrauskas, *SFPE Technology Report 84-10*, Society of Fire Protection Engineers, Boston, Mass., 1984.
10. R. G. Hill, T. I. Eklund, and C. P. Sarkos, *DOT/FAA/CT-85/23*, Federal Aviation Administration Technical Center, Atlantic City, N.J., 1985.
11. R. D. Peacock, W. W. Jones, R. W. Bukowski, and C. L. Forney, *HAZARD I—Fire Hazard Assessment Method*, version 1.1, NIST Handbook 146, National Institute of Standards and Technology, Gaithersburg, Md., 1991.
12. V. Babrauskas and co-workers, *Fire Hazard Comparison of Fire-Retarded and Non-Fire-Retarded Products*, NBS Special Publication 749, National Bureau of Standards, Gaithersburg, Md., 1987.

13. R. G. Gann, R. A. Dipert, and M. J. Drews, in J. I. Kroschwitz, ed., *Encyclopedia of Polymer Science and Engineering*, 2nd ed., John Wiley & Sons, Inc., New York, 1986, pp. 154–210.
14. J. W. Lyons, *The Chemistry and Uses of Fire Retardants*, Wiley-Interscience, New York, 1970, Chapt. 5.
15. S. J. Ainsworth, *Chem. Eng. News*, **70**, 34 (Aug. 31, 1992).

RICHARD G. GANN
National Institute of Standards and Technology

ANTIMONY AND OTHER INORGANIC FLAME RETARDANTS

Flame retardancy can be imparted to plastics by incorporating elements such as bromine, chlorine, antimony, tin, molybdenum, phosphorus, aluminum, and magnesium, either during the manufacture or when the plastics are compounded into some useful product. Phosphorus, bromine, and chlorine are usually incorporated as some organic compound. The other inorganic flame retardants are discussed herein.

Addition of approximately 40% of the halogen flame retardants are needed to obtain a reasonable degree of flame retardancy. This usually adversely affects the properties of the plastic. The efficiency of the halogens is enhanced by the addition of inorganic flame retardants, resulting in the overall reduction of flame-retardant additive package and minimizing the adverse effects of the retardants.

Hydrated metal oxides such as alumina hydrate are usually used alone because these are not synergistic with the halogens. They are useful in applications in which the halogens are excluded or low processing temperatures are used.

Antimony Compounds

Antimony Trioxide. Approximately 20,000 metric tons of antimony trioxide [*1309-64-4*] (commonly referred to as antimony oxide), Sb_2O_3, was used in the United States in 1990 to impart flame retardancy to plastics (see ANTIMONY COMPOUNDS). Although antimony trioxide is found in nature, it is too impure to be used. Flame-retardant grades of antimony oxides are manufactured from either antimony metal or the sulfide ore by oxidation in air at 600–800°C (1). The particle size and chemical reactivity is determined by the processing conditions, enabling the production of several different grades. The physical properties of various grades are listed in Table 1. Antimony trioxide is from 99.0–99.9 wt % Sb_2O_3. The remainder consists of 0.4–0.01 wt % arsenic; 0.4–0.01, lead; 0.1–0.0001, iron; 0.005–0.0001, nickel; and 0.01–0.0001, sulfates. It is insoluble in water and the loss on drying at 110°C is 0.1 wt % max.

Antimony trioxide has been used as a white pigment since ancient times. The pigmentation from antimony oxide in plastics can be controlled and adjusted by the judicious selection of a Sb_2O_3 grade having a specific particle size. The product with the smallest particle size and the narrowest particle-size range im-

Table 1. Physical Properties of Antimony Trioxide

	Grade[a]		
Property	Ultra fine	High tint	Low tint
specific gravity	5.3–5.5	5.3–5.8	5.3–5.8
particle size, μm	0.25–0.45	0.8–1.8	1.9–3.2

[a]All grades are white powders.

parts the whitest color and highest opacity. Translucent plastics can be made by using low tint grades with relatively large particles.

Particle size during manufacture is controlled by adjusting the temperature and rate at which the antimony vapors are precipitated as these vapors exit the furnace. The lower the temperature and the slower the precipitation rate, the larger the particles. Figure 1 shows the particle size distribution of the various commercially available grades of antimony trioxides. Although particle size affects pigmentation, it does not appear to affect flame retardancy efficiency. The cost of antimony trioxide varies from $2.20–3.30/kg, depending on the grade and volume purchased. Suppliers and the grades offered are

Company	Trade name
Amspec	Antimony Oxide KR
	Antimony Oxide KRL
	Twinkling Star
Anzon	Timinox High Tint
	Timinox Low Tint
	Timinox Tru Tint
	Microfine AO3
Asarco	Antimony Oxide High Tint
	Antimony Oxide Low Tint
Elf Atochem NA	Thermoguard UF
	Thermoguard S
	Thermoguard L
	Thermoguard HPM
Laurel Industries	Fireshield H
	Fireshield L
	Ultrafine II
U.S. Antimony Co.	Montana Brand

Antimony Oxide as a Primary Flame Retardant. Antimony oxide behaves as a condensed-phase flame retardant in cellulosic materials (2). It can be applied by impregnating a fabric with a soluble antimony salt followed by a second treatment that precipitates antimony oxide in the fibers. When the treated fabric is exposed to a flame, the oxide reacts with the hydroxyl groups of the cellulose (qv) causing them to decompose endothermically. The decomposition products, water

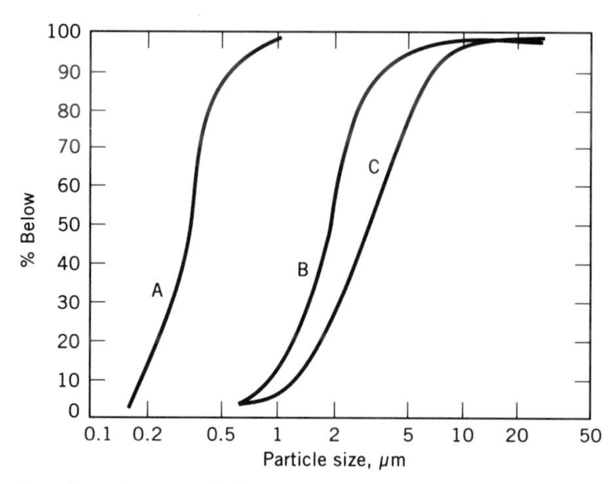

Fig. 1. Particle size distribution of the antimony oxides: A, Thermoguard UF; B, Thermoguard S; C, Thermoguard L.

and char, cool the flame reactions while slowing the production and volatilization of flammable decomposition products (see FLAME RETARDANTS FOR TEXTILES).

Antimony Pentoxide. The second most widely used antimony synergist is antimony pentoxide [*1313-60-9*], Sb_2O_5, produced by the oxidation of the trioxide using either a peroxide or nitric acid (3–8). Antimony pentoxide is available as a nonpigmenting colloidal suspension in either water or organic media or as an agglomerated powder. It is insoluble in water, but soluble in hot concentrated acids. Properties of this unique flame retardant synergist are listed in Table 2.

Submicrometer antimony pentoxide is primarily used to impart flame retardancy to fibers and fabrics. It can be added to the molten or dissolved polymer prior to forming the fiber. The antimony in this form can easily pass through the spinnerets without clogging the openings, whereas standard grades of antimony trioxide would rapidly clog the openings and necessitate frequent shutdowns for cleaning. The submicrometer antimony pentoxide is also more evenly dispersed in the fiber, resulting in better physical properties.

Powdered antimony pentoxide is used primarily in plastics. Stabilizers used to prevent the particles from growing are caustic, and can react with the halogen in the formulation. This can result in color formation and a lower flame-retarding efficiency of the system.

Table 2. Properties of Antimony Pentoxide and Sodium Antimonate

Property	Sb_2O_5	Na_3SbO_4
particle size, μm	0.03	1–2
surface area, m^2/gm	50	
specific gravity	4.0	4.8
surface activity	weakly acidic	basic
refractive index, n_D^{20}	1.7	1.75

Antimony pentoxide is priced about two to three times higher than the trioxide. However, because it is more efficient than the trioxide, the pentoxide is at least cost-equivalent. Antimony pentoxide is manufactured by both Philadelphia Quartz and Laurel Industries under the Nyacol and Fireshield trade names.

Sodium Antimonate. Sodium antimonate [*15593-75-6*], Na_3SbO_4, another antimony synergist of commercial importance, has an antimony content of 61–63 wt % and a bulk density of 39.4–46.4 kg/m^3. Properties are given in Table 2. It is made by oxidizing antimony trioxide using sodium nitrate and caustic. It is a white powder and has a pH of around 9–11 when dissolved in water.

Sodium antimonate contains less antimony than either antimony trioxide or pentoxide and is thus less effective. However, its unique pH and low refractive index makes the antimonate the most desirable synergist for polymers that hydrolyze when processed with acidic additives or in polymers for which deep color tones are specified. Sodium antimonate costs approximately $3.30–4.40/kg and can be obtained from either Elf Atochem NA under the Thermoguard name or from Anzon Inc. as a Timinox product.

Toxicity. Antimony has been found not to be a carcinogen or to present any undue risk to the environment (9). However, because antimony compounds also contain minor amounts of arsenic which is a poison and a carcinogen, warning labels are placed on all packages of antimony trioxide.

Mixed Metal Antimony Synergists. Worldwide scarcities of antimony have prompted manufacturers to develop synergists that contain less antimony. Other metals have been found to work in concert with antimony to form a synergist that is as effective as antimony alone. Thermoguard CPA from Elf Atochem NA, which contains zinc in addition to antimony, can be used instead of antimony oxide in flexible poly(vinyl chloride) (PVC) as well as some polyolefin applications. The Oncor and AZ products which contain silicon, zinc, and phosphorus from Anzon Inc. can be used in a similar manner. The mixed metal synergists are 10 to 20% less expensive than antimony trioxide.

Antimony–Halogen Synergism. Antimony synergists are used almost exclusively with either brominated or chlorinated organic flame retardants. These work in concert with one another and provide a highly effective flame-retardant system. Antimony and the halogens react at flame temperatures to form the corresponding trihalide or oxyhalide. The product formed depends on the mole ratios of the reactants and the structure of the organic halogen compound. The active flame-retarding species, ie, the tribromide and the trichloride (10,11), are formed directly when the mole ratio of halogen to antimony is at least 3-to-1 and the halogen compound is capable of dehydrohalogenating.

$$Sb_2O_3 + 6\ HCl \rightarrow 2\ SbCl_3 + 3\ H_2O \qquad (1)$$

If there are less than three moles of halogen for each mole of antimony present or if the halogen compound cannot dehydrohalogenate, then the oxyhalide forms.

$$Sb_2O_3 + 2\ HCl \rightarrow 2\ SbOCl + H_2O \qquad (2)$$

The formation of antimony trichloride from antimony oxychloride has been described by two different mechanisms. One asserts that formation is via a thermal disproportionation as described in equations 3–6 (12).

$$5\ SbOCl\ (s) \rightarrow Sb_4O_5Cl_2 + SbCl_3 \quad (3)$$

$$4\ Sb_4O_5Cl_2 \rightarrow 5\ Sb_3O_4Cl + SbCl_3 \quad (4)$$

$$3\ Sb_3O_4Cl \rightarrow 4\ Sb_2O_3 + SbCl_3 \quad (5)$$

$$Sb_2O_3\ (s) \rightarrow Sb_2O_3\ (g) \quad (6)$$

The other mechanism proposes that the trichloride is formed by the further chlorination of the oxyhalide as illustrated in equations 7–10 (13).

$$4\ Sb_2O_3 + 2\ HCl \rightarrow Sb_8O_{11}Cl_2 + H_2O \quad (7)$$

$$Sb_8O_{11}Cl_2 + 2\ HCl \rightarrow 2\ Sb_4O_5Cl_2 + H_2O \quad (8)$$

$$Sb_4O_5Cl_2 + 2\ HCl \rightarrow 4\ SbOCl + H_2O \quad (9)$$

$$SbOCl + 2\ HCl \rightarrow SbCl_3 + H_2O \quad (10)$$

The organohalogen compound used in the formulation also influences the formation of the trihalide. If the halogen compound can dehydrohalogenate easily, as for example do hexabromocyclododecane [*25637-99-4*] and poly(vinyl chloride), then the halogen decomposition products react with the antimony directly (14, 15). However if the organohalogen does not form the acid halide easily, as for example decabromobiphenyl oxide, then a third component capable of reacting with the antimony and halogen is needed. This can be the polymer itself. If a third component is not present, such as in the pyrolysis of an antimony–halogen mixture, then the decomposition products of only the halogen and antimony are recovered and no trihalide forms. Two mechanisms have been proposed (16) for this type of reaction. Both require the participation of the polymer in the reaction.

One mechanism has the polymer degrading before the halogen compound does to generate a free radical which replaces the halogen of the flame retardant. The bromine radical can then react with the labile hydrogen of the polymer, forming hydrogen bromide and a new polymeric radical. The hydrogen bromide can react further with antimony oxide and generate antimony tribromide:

$$RHC{-}CHR \rightarrow 2\ RCH^{\cdot} \quad (11)$$

$$RCH^{\cdot} + Br_5O{-}O{-}OBr_5 \rightarrow Br^{\cdot} + Br_4O{-}O{-}OBr_5 + RCH \quad (12)$$

According to the second mechanism the polymer and antimony trioxide form a catalytic complex. The halogen flame retardant then reacts with the complex and expels antimony tribromide.

$$Sb_2O_3 + \text{polymer} \rightarrow [Sb_2O_3{-}\text{polymer}] \quad (13)$$

$$[Sb_2O_3{-}\text{polymer}] + RBr \rightarrow [SbBr_3{-}\text{polymer}] + R^{\cdot} \quad (14)$$

$$[SbBr_3{-}\text{polymer}] \rightarrow SbBr_3 + R \quad (15)$$

Either mechanism can be used to describe how antimony–halogen systems operate in both the condensed and vapor phases. In the condensed phase a char that is formed during the reaction of the polymer, antimony trioxide, and the halogen reduces the rate of decomposition of the polymer; therefore, less fuel is available for the flame (16).

In the vapor phase antimony trihalides inhibit the radical flame reactions. The inhibition reactions of antimony trichloride have been studied by mass spectrophotometry (17). In a flame antimony trichloride was found to decompose in a stepwise manner, enabling the chlorine to remain in the flame zone longer than it would have if it were generated directly from the organic halogen compound. Therefore the halogen becomes more effective. According to this study, chlorine deactivates the very energetic hydrogen, oxygen, and hydroxyl radicals.

$$SbCl_3 \rightarrow SbCl_2^{\cdot} + Cl^{\cdot} \quad (16)$$

$$SbCl_3 + H^{\cdot} \rightarrow SbCl_2^{\cdot} + HCl \quad (17)$$

$$SbCl_3 + CH_3^{\cdot} \rightarrow SbCl_2^{\cdot} + CH_3Cl \quad (18)$$

$$SbCl_2 + H^{\cdot} \rightarrow SbCl^{\cdot} + HCl \quad (19)$$

$$SbCl_2 + CH_3^{\cdot} \rightarrow SbCl^{\cdot} + CH_3Cl \quad (20)$$

$$SbCl^{\cdot} + H^{\cdot} \rightarrow Sb + HCl \quad (21)$$

$$SbCl^{\cdot} + CH_3^{\cdot} \rightarrow Sb + CH_3Cl \quad (22)$$

The chlorine radicals and hydrogen chloride inhibit the reaction in the following manner.

$$Cl^{\cdot} + CH_3^{\cdot} \rightarrow CH_3Cl \quad (23)$$

$$Cl^{\cdot} + H^{\cdot} \rightarrow HCl \quad (24)$$

$$Cl^{\cdot} + HO_2^{\cdot} \rightarrow HCl + O_2 \quad (25)$$

$$HCl + H^{\cdot} \rightarrow H_2 + Cl^{\cdot} \quad (26)$$

$$Cl^{\cdot} + Cl^{\cdot} \rightarrow Cl_2 \quad (27)$$

$$Cl_2 + CH_3^{\cdot} \rightarrow CH_3Cl + Cl^{\cdot} \quad (28)$$

The fine antimony mist formed from the decomposition of the trichloride also participates in the flame-inhibiting process, deactivating oxygen, hydrogen, and hydroxyl radicals.

$$Sb + O^{\cdot} \rightarrow SbO^{\cdot} \quad (29)$$

$$SbO^{\cdot} + 2\,H^{\cdot} \rightarrow SbO^{\cdot} + H_2 \quad (30)$$

$$SbO^{\cdot} + H^{\cdot} \rightarrow SbOH \quad (31)$$

$$SbOH + OH^{\cdot} \rightarrow SbO^{\cdot} + H_2O \quad (32)$$

Boron Compounds

In 1990 approximately 4500 metric tons of boron flame retardants were used in the United States to impart flame retardancy to plastics (see BORON

COMPOUNDS). The most widely used is zinc borate [*1332-07-6*], prepared as an insoluble double salt from water-soluble zinc and boron compounds. Compounds having varying amounts of zinc, boron, and water of hydration are available. The ratio of these components affects the temperature at which the flame-inhibiting powers are activated, as well as the temperature at which they can be processed (18). Zinc borates can either be used alone or in combination with other halogen synergists, such as antimony oxide. In some instances zinc borate is also used with alumina trihydrate to form a glass-like substance that inhibits polymer degradation. Manufacturers of boron flame retardants are listed in Table 3. The cost of the borates varied from $2.00–2.50/kg in the early 1990s.

Barium Metaborate. Barium metaborate is used both as a flame retarder and as an antifungicide for many flexible poly(vinyl chloride) applications (19).

Boric Acid and Sodium Borate. Boric acid [*10043-35-3*] and sodium borate [*1303-96-4*], also known as borax, have been used as flame retardants for cellulose since the 1800s (2). They are quick-fix, nondurable flame retardants. Each is applied by passing the fabric through a solution of the flame retardant. For batts, eg, cotton, a spray applicator is used. Excess solution is removed by passing the fabric through squeeze rollers. Usually an "add-on" of 20% is needed to obtain flame retardancy. However, because the flame retardants are water-soluble, they are removed after several washings or when used in high humidity atmospheres. Boric acid and borax are available from Ashland Chemical, U.S. Borax, and J.H. Henry Chemical Co.

Ammonium Fluoroborate. Ammonium fluoroborate [*13826-83-0*], NH_4BF_4, is unique in that when it is exposed to a flame, it generates both a halogen and a boron flame retardant (20). Antimony oxide is usually recommended as a co-synergist.

$$6\ NH_4BF_4 + Sb_2O_3 \rightarrow 6\ NH_3 + 6\ BF_3 + 2\ SbF_3 + 3\ H_2O \qquad (33)$$

Ammonium fluoroborate is both a condensed and vapor-phase flame retardant. It is available from M & T Harshaw, General Chemical Corp., and Spectrum Chemical Corp.

Boron Mechanism. Boron functions as a flame retardant in both the condensed and vapor phases. Under flaming conditions boron and halogens form the

Table 3. Manufacturers and Trade Names of Boron Flame Retardants

Manufacturer	Trade name	Composition	CAS Registry Number
Climax Performance Materials	ZB 467	$4\ ZnO \cdot 6\ B_2O_3 \cdot 7\ H_2O$	[*12513-27-8*]
	ZB 223	$2\ ZnO \cdot 2\ B_2O_3 \cdot 3\ H_2O$	
	ZB 113	$ZnO \cdot B_2O_3 \cdot 3\ H_2O$	
	ZB 237	$2\ ZnO \cdot 3\ B_2O_3 \cdot 7\ H_2O$	
	ZB 325	$3\ ZnO \cdot 2\ B_2O_3 \cdot 5\ H_2O$	
U.S. Borax	Firebrake ZB	$2\ ZnO \cdot 3\ B_2O_3 \cdot 5\ H_2O$	
		$Zn(BO_2)_2$	[*13701-59-2*]
Buckman Laboratories	Brusan M-11	$Ba(BO_2)_2$	[*27043-84-1*]

corresponding trihalide. Because boron trihalides are effective Lewis acids, they promote cross-linking, minimizing decomposition of the polymer into volatile flammable gases. These trihalides are also volatile; thus they vaporize into the flame and release halogen which then functions as a flame inhibitor (21–24).

Boron also reacts with hydroxyl-containing polymers such as cellulose. When exposed to a flame the boron and hydroxyl groups form a glassy ester that coats the substrate and reduces polymer degradation. A similar type of action has been observed in the boron–alumina trihydrate system.

Alumina Trihydrate

In 1990, approximately 66,000 metric tons of alumina trihydrate [*12252-70-9*], $Al_2O_3{\cdot}3H_2O$, the most widely used flame retardant, was used to inhibit the flammability of plastics processed at low temperatures. Alumina trihydrate is manufactured from either bauxite ore or recovered aluminum by either the Bayer or sinter processes (25). In the Bayer process, the bauxite ore is digested in a caustic solution, then filtered to remove silicate, titanate, and iron impurities. The alumina trihydrate is recovered from the filtered solution by precipitation. In the sinter process the aluminum is leached from the ore using a solution of soda and lime from which pure alumina trihydrate is recovered (see ALUMINUM COMPOUNDS).

Alumina trihydrate is available in a variety of particle size distributions, ranging from less than one up to 100 micrometers. It has a low hiding power and pigmentation because of the large particles and low refractive index. Therefore formulations containing as much as 50% alumina trihydrate are translucent. Physical properties are given in Table 4. Alumina trihydrate is composed of 64.9 wt % Al_2O_3, 0.005 wt % SiO_2, 0.007 wt % Fe_2O_3, and 0.3 wt % Na_2O, of which 0.04 wt % is soluble. Water loss on ignition is 34.6 wt %.

Manufacturers of alumina trihydrate include Solem Industries, Aluchem, Alcoa, Custom Grinding Sales, R. J. Marshall, Georgia Marble, and Hitax. In 1992 the price of alumina trihydrate varied from $0.25–1.35/kg. Alumina trihydrate is the least expensive and least effective of the flame retarders. It is only about one-fourth to one-half as effective as the halogens. Usually about 50–60% of alumina trihydrate is needed to obtain some acceptable degree of flame retardancy. It is also limited to plastics that are not processed higher than 220°C.

Table 4. Physical Properties of Alumina Trihydrate

Property	Value
density, g/mL	2.42
refractive index, n_D^{20}	1.579
average particle size, μm	1–100
Mohs' hardness	2.5–3.5
color	white
water solubility	insoluble

Alumina trihydrate is also used as a secondary flame retardant and smoke suppressant for flexible poly(vinyl chloride) and polyolefin formulations in which antimony and a halogen are used. The addition of minor amounts of either zinc borate or phosphorus results in the formation of glasses which insulate the unburned polymer from the flame (21).

Mechanism. Alumina trihydrate functions as a flame retardant in both the condensed and vapor phases (26). When activated, it decomposes endothermically, eliminating water.

$$2\ Al(OH)_3 \rightarrow Al_2O_3 + 3\ H_2O \tag{34}$$

In the flame phase the water vapor forms an envelope around the flame, which tends to exclude air and dilute the flammable gases. The water vapor reacts endothermically with the flame radicals. The alumina residue becomes a conduit through which heat is conveyed away from the flame area, slowing down polymer decomposition.

Other Inorganic Materials

Magnesium Hydroxide. Magnesium hydroxide [*1309-42-8*] is another metal hydrate that decomposes endothermically, accompanied by the formation of water. It decomposes at 330°C, which is 100°C higher than alumina trihydrate, and can therefore be used in polymers that are processed at higher temperatures.

$$Mg(OH)_2 \rightarrow MgO + H_2O \tag{35}$$

Magnesium hydroxide is white, has an average particle size of 1–10 μm, density of 2.36 g/mL, refractive index of 1.58, and Mohs' hardness of 2.00. Water loss on ignition is 31.8 wt %. Magnesium hydroxide contains 1.0 wt % $Ca(OH)_2$ and is made by Solem Industries and Morton Thiokol (25).

Molybdenum Oxides. Molybdenum was one of the first elements used to retard the flames of cellulosics (2). More recently it has been used to impart flame resistance and smoke suppression to plastics (26). Molybdic oxide, ammonium octamolybdate, and zinc molybdate are the most widely used molybdenum flame retardants. Properties are given in Table 5. These materials are recommended almost exclusively for poly(vinyl chloride), its alloys, and unsaturated polyesters (qv).

Molybdenum trioxide is a condensed-phase flame retardant (26). Its decomposition products are nonvolatile and tend to increase char yields. Two parts of molybdic oxide added to flexible poly(vinyl chloride) that contains 30 parts of plasticizer have been shown to increase the char yield from 9.9 to 23.5%. Ninety percent of the molybdenum was recovered from the char after the sample was burned. A reaction between the flame retardant and the chlorine to form $MoO_2{\cdot}Cl_2{\cdot}H_2O$, a nonvolatile compound, was assumed. This compound was assumed to promote char formation (26,27).

Molybdenum is also a smoke suppressant for poly(vinyl chloride). It promotes the formation of cis- rather than the trans-polymeric decomposition prod-

Table 5. Properties of Molybdenum Flame Retarders

Property	Molybdic oxide[a]	Ammonium octamolybdate[b]	Zinc molybdate
CAS Registry Number	[*1313-27-5*]		[*13767-32-3*]
particle size, μm	25	2	1.88
bulk density, kg/m^3	38	480	
H_2O solubility, g/mL	0.68	4	0.04
specific gravity	4.67	3.18	3.0
oil adsorption per 100 g oil, g	35	20	
loss on ignition, wt %	<0.1	8.29	

[a]The refractive index of MoO_3 is >2.1.
[b]The decomposition temperature is 250°C.

ucts which are the precursors for smoke. The sources for molybdates are Climax Performance Material Corp. and Sherwin Williams.

Tin. Tin has been used as a flame retardant for cellulose since the latter 1800s (2). Only since the 1970s has it been used as a synergist for halogen flame retarders in the same manner as antimony oxide (28,29). Anhydrous and hydrated zinc stannate and stannic oxide are the three most important tin flame retardants. Properties are given in Table 6.

Table 6. Properties of Tin Flame Retarders[a]

Parameter	Zinc hydroxy stannate	Zinc stannate	Stannic oxide
molecular formula	$ZnSn(OH)_6$	$ZnSnO_3$	$SnO_2{\cdot}xH_2O$
CAS Registry Number	[*12027-96-2*]	[*12036-37-2*]	
tin, wt %	47	51	67
zinc, wt %	21	28	
specific gravity	3.4	4.25	
decomposition temperature, °C	180	>570	180

[a]All materials are white powders. Particle sizes range from 1–10 μm.

The mechanism by which tin flame retardants function has not been well defined, but evidence indicates tin functions in both the condensed and vapor phases. In formulations in which there is at least a 4-to-1 mole ratio of halogen to tin, reactions similar to those of antimony and halogen are assumed to occur. Volatile stannic tetrahalide may form and enter the flame to function much in the same manner as does antimony trihalide.

If the tin source is anhydrous, very little volatile tin tetrahalide is formed. If the mole ratio of halogen to tin is less than 4-to-1 or if there is no halogen present, yet flame retardancy is observed, condensed-phase activity is assumed. The only supplier of tin flame retardants is Alcan Inc. The price in 1992 was $7.70/kg.

Applications

Poly(vinyl chloride). PVC is a hard, brittle polymer that is self-extinguishing. In order to make PVC useful and more pliable, plasticizers (qv) are added. More often than not the plasticizers are flammable and make the formulation less flame resistant. Flammability increases as the plasticizer is increased and the relative amount of chlorine decreased, as shown in Table 7. The flame resistance of the poly(vinyl chloride) can be increased by the addition of an inorganic flame-retardant synergist.

Table 7. Effect of Plasticizer[a] on the Oxygen Index of PVC

Plasticizer, wt %	Cl, wt %	Oxygen index (OI)
0	57.0	41–49
16.6	48.5	36.4
28.5	41.0	31.5
37.5	36.4	22.2
44.4	32.0	21.2
47.5	30.0	19.2

[a]Dioctyl phthalate.

Antimony Oxide. The effect of antimony trioxide on the oxygen index of flexible poly(vinyl chloride) containing from 20 to 50 parts of plasticizer is shown in Figure 2. The flame resistance as measured by the oxygen index increases with the addition of antimony oxide until the oxygen index appears to reach a maximum at about 8 parts of Sb_2O_3. Further addition of antimony oxide does not have any increased beneficial effect.

Mixed Metal Antimony Synergists. A similar increase in the oxygen index can be achieved using the mixed metal synergist Thermoguard CPA as illustrated in Table 8.

Zinc Borate. Zinc borate is also effective in enhancing the flame-inhibiting powers of chlorine. Although zinc borate increases flame resistance, it is not as effective as antimony oxide, as is illustrated in Figure 3. However, zinc borate can be used in combination with antimony oxide to obtain equivalent and in some instances enhanced effects over what can be obtained using either of the two synergists alone (Table 9).

Molybdenum Oxide. Molybdenum compounds incorporated into flexible PVC not only increase flame resistance, but also decrease smoke evolution. In Table 10 the effect of molybdenum oxide on the oxygen index of a flexible PVC containing 50 parts of a plasticizer is compared with antimony oxide. Antimony oxide is the superior synergist for flame retardancy but has little or no effect on smoke evolution. However, combinations of molybdenum oxide and antimony oxide may be used to reduce the total inorganic flame-retardant additive package, and obtain improved flame resistance and reduced smoke.

Zinc Stannates. The zinc stannates are also effective synergists for flexible PVC; however, as shown in Figure 4**a**, antimony oxide is more effective. If more chlorine such as in a chlorinated paraffin such as Cereclor is added, then the stannates become more effective and eventually outperform antimony oxide (Fig. 4**b**).

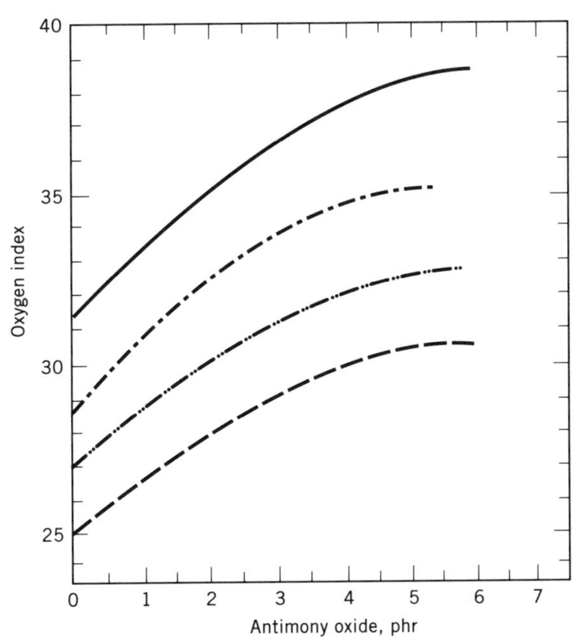

Fig. 2. The effect of antimony oxide on the oxygen index of poly(vinyl chloride) plasticized with dioctyl phthalate (DOP), (——), 20 phr, (—·—), 30 phr; (—··—), 40 phr; (———), 50 phr, where phr = parts per hundred resin.

Table 8. Flammability of Flame-Retardant-Treated Flexible PVC[a]

	Flame retardant	
Parameter	Thermoguard CPA	Sb_2O_3
UL classification	V-0	V-0
flame-out time,[b] s	2.3	2.4
oxygen index	28.9	28.6

[a]Formulation is 100.0 parts by wt PVC; 54.0, plasticizer; 7.0, stabilizer; 10.0, filler; 0.5, lubricant; and 3.0, flame retardant.
[b]UL 94 Test.

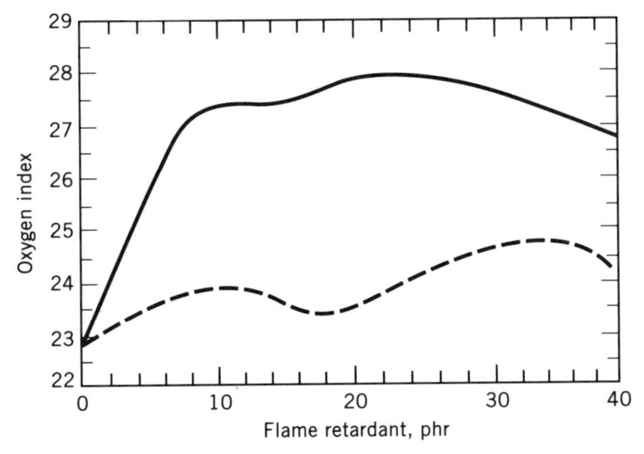

Fig. 3. A comparison of the effect of antimony oxide (——) and Firebrake ZB (———) on the oxygen index of PVC containing 50 phr DOP.

Table 9. Oxygen Index of Flexible PVC Containing Antimony and Boron

Formulation, phr		
Zinc borate	Antimony oxide	Oxygen index
0	3	28.7
0	1.5	27.5
0.5	2.5	28.6
1.0	2.0	28.4
1.5	1.5	28.2
3.0	0	24.1
1.5	0	22.7
0	0	19.8

Table 10. Effect of Flame Retardancy of Molybdenum Oxide and Antimony Oxide[a]

Additive, parts by weight			
Sb_2O_3	MbO_3	Oxygen index	Smoke[b]
		24.5	10.7
2.0		27.5	9.2
3.0		28.3	10.9
5.0		29.0	9.4
	2.0	26.0	6.8
	3.0	27.0	4.4
	5.0	27.5	4.1
2.0	1.0	28.0	7.6
1.5	1.5	27.5	5.0
1.0	2.0	27.0	7.0
2.0	2.0	29.5	6.0

[a]Flexible PVC containing 50 phr dioctyl phthalate.
[b]Results from Araphoe Smoke Test.

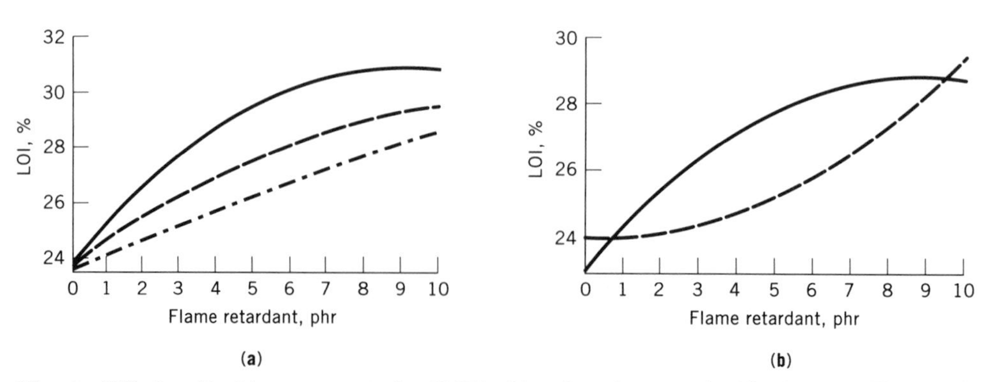

Fig. 4. Effect on limiting oxygen index (LOI) of (——) antimony trioxide, (– – – –) Flamtard S, and (—·—) Flamtard H added to (**a**) the basic unfilled PVC formulation, and (**b**) PVC containing Cereclor and 100-phr chalk (29).

Alumina Trihydrate. Alumina trihydrate is usually used as a secondary flame retardant in flexible PVC because of the high concentration needed to be effective. As a general rule the oxygen index of flexible poly(vinyl chloride) increases 1% for every 10% of alumina trihydrate added. The effect of alumina trihydrate on a flexible poly(vinyl chloride) formulation containing antimony oxide is shown in Figure 5.

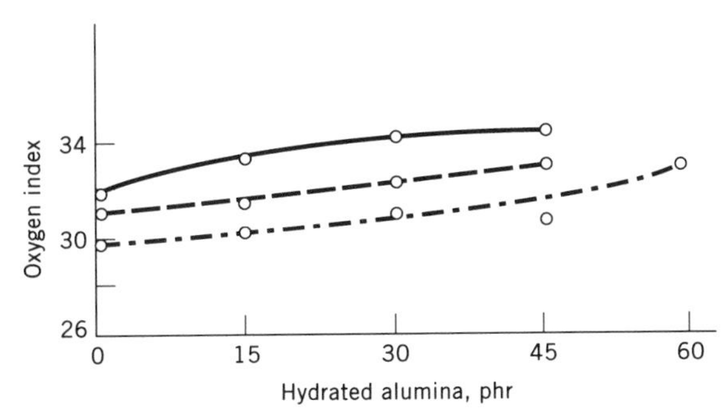

Fig. 5. Effect of alumina trihydrate on the oxygen index of flexible PVC (30) having 5 phr Tribase XL, 3 phr Sb_2O_3, 0.05 phr petroleum wax, and (——) 35, (– – –) 40, and (—·—) 45 phr 711 phthalate.

Unsaturated Polyesters. There are two approaches used to provide flame retardancy to unsaturated polyesters. These materials can be made flame resistant by incorporating halogen when made, or by adding some organic halogen compound when cured. In either case a synergist is needed. The second approach involves the addition of a hydrated filler. At least an equal amount of filler is used.

Synergists. The effect of antimony oxide on the flammability of unsaturated polyesters that contain chlorine is shown in Table 11. A similar effect on the flammability of unsaturated polyester containing zinc stannates and bromine instead of chlorine is given in Table 12.

Table 11. Effect of Chlorine and Antimony Oxide on the Oxygen Index of Unsaturated Polyesters

Chlorine, wt %	Antimony oxide, wt %	Oxygen index
0	0	18.2
12.5	2.5	30.0
20.0	1.0	30.2
25.0	3.0	38.7
25.0	1.0	33.4
25.0	0	30.2

Antimony oxide and zinc borate are also used as synergists for unsaturated polyesters. Their combined effect on the flame spread index (25) is shown in Table 13. The combination of molybdenum oxide and antimony oxide is used to reduce both flammability and smoke evolution as illustrated in Table 14.

Table 12. Oxygen Index of Unsaturated Polyesters Containing Bromine

	Bromine, wt %				
Additive	0	5	10.0	20.0	28.0
no additive	19.1	23.1	26.7	37.2	41.2
$ZnSnO_3$					
1 wt %	19.6	24.1	28.0	42.1	52.2
2 wt %	19.6	25.3	32.7	42.4	54.2
$ZnSn(OH)_6$, 2 wt %	19.3	23.6	28.9	40.9	53.4
Sb_2O_3, 2 wt %	19.1	23.7	29.2	40.8	51.2

Table 13. Fire Test Results of Laminates Containing Zinc Borate and Antimony Oxide[a]

Additive, phr			
Sb_2O_3	Zinc borate	Flame spread[b]	Oxygen index
0	0	15.0	26.0
2	0	9.1	37.3
2	6	3.7	40.2
1	6	12.5	36.8
0.5	6.5	15.4	34.1

[a]26% fiber glass; contains 28% chlorine.
[b]Ref. 25.

Table 14. Oxygen Index and Smoke Reduction of Unsaturated Polyester Containing Molybdenum Oxide and Antimony Oxide

Additive, wt %			
Sb_2O_3	MoO_3	Oxygen index	Smoke reduction
0	0	33.0	
1		34.5	+67
3		42.0	+96
5		43.0	−26
	5	36.0	−36
	7	39.0	−49
1	1	39.5	−54
2	1	41.0	−20
1	2	40.0	−15

Alumina Trihydrate. Alumina trihydrate must be used at high concentrations in order for it to be effective. The flammability and smoke of a general-purpose polyester (75 wt %) containing 20 wt % glass mat gave values of 780 and 1000 on flame spread and smoke in ASTM E84 testing. The same polyester (40 wt %) containing 40 wt % alumina trihydrate and 20 wt % glass mat gave values of 70 and 300 for the flame spread and smoke tests (18).

Olefin Polymers. The flame resistance of polyethylene can be increased by the addition of either a halogen synergist system or hydrated fillers. Similar flame-retarder packages are used for polypropylene (see OLEFIN POLYMERS). Typical formulations of the halogen synergist type are shown in Table 15; the filler-type formulations are in Table 16.

Table 15. Inorganic Synergist–Halogen Formulations

Parameter	Polyethylene				Polypropylene			
	Formulation							
polymer, wt %	82	82	60	60	56	94	60	54
antimony oxide, wt %	6	6	13	13	6	2	10	12
decabromobiphenyl oxide, wt %	12				22			
bis(tetrabromophthalimide) ethane, wt %		12						
Dechlorane Plus			27				30	
ethylene bis(dibromonorborane) dicarbanimide						4		
70% chlorinated paraffin				27				24
talc					14			10
	Flame test results							
UL 94 Class[a]	V-0	V-0	V-1	V-0	V-0	V-2	V-1	V-0
oxygen index					25.0		27.0	26.0

[a]3.175 mm.

Table 16. Hydrated Filler Systems for Polyethylene and Polypropylene[a]

Parameter	Polyethylene		Polypropylene	
polymer, wt %	35	45	35	36
alumina trihydrate, wt %	65		65	
magnesium hydroxide, wt %		55		64

[a]Flame test results for all formulations is V-0 using the UL 94 Class (⅛ in. = 3.175 mm) designations.

Halogen and inorganic synergists are used to impart flame retardancy to polystyrene rather than fillers. ABS is slightly more difficult to flame retard and requires higher concentrations of the inorganic synergist–halogen system. Fillers are rarely if ever used. Table 17 contains typical formulations.

Table 17. Inorganic Synergists–Halogen System for Polystyrene and ABS

Parameter	High impact polystyrene		ABS	
	Formulation			
polymer, wt %	84	84	78.2	79.0
antimony oxide, wt %	4	4	3.1	5.0
brominated biphenyl oxide, wt %	12[a]	0	18.8[b]	0
brominated polystyrene, wt %	0	12	0	16.0
	Flame test results			
UL 94 Class[c]	V-0	V-0	V-0	V-0
oxygen index	26.0	24.5	32.0	

[a]Decabromobiphenyl oxide.
[b]Octabromobiphenyl oxide.
[c]3.175 mm = ⅛ in.

Table 18. Flame-Retarded Nylon-6,6

Parameter	Value			
	Formulation			
nylon-6,6, wt %	48.0	48.0	50.0	55.0
glass fibers, wt %	30.0	30.0	30.0	30.0
antimony oxide, wt %	6.0	4.4		
ferric oxide, wt %			3.0	
zinc oxide, wt %				4.0
brominated polystyrene, wt %	18.0			
Dechlorane Plus, wt %		17.0	16.0	12.0
	Flame test results			
UL 94 Class[a]	V-0	V-0	V-0	V-0
oxygen index	34.8	31.2	36.0	34.2

[a]3.175 mm = ⅛ in.

Nylons, because of the high processing temperatures needed, are generally flame retarded using an inorganic halogen system (Table 18). Linear polyesters are susceptible to acid hydrolysis during processing. Thus sodium antimonate, the synergist having a basic pH, is favored as indicated in Table 19.

Table 19. **Flame-Retarded Linear Polyesters**

Parameter	Value		
Formulation			
polybutylene terephthalate, wt %	57.7	57.7	55.0
glass fibers, wt %	25.0	25.0	25.0
sodium antimonate, wt %	3.4	3.4	3.4
brominated polystyrene, wt %	14.0		
bis(tetrabromophthalimide) ethane, wt %		14.0	
tetrabromobisphenol A carbonate, wt %			18.0
Flame test results			
UL 94 Class[a]	V-0	V-0	V-0
oxygen index	32.7	37.1	35.8

[a]3.175 mm = ¾ in.

BIBLIOGRAPHY

"Antimony and Other Inorganic Compounds" under "Flame Retardants" in *ECT* 3rd ed., Vol. 10, pp. 355–372, by J. M. Avento and I. Touval, M & T Chemicals, Inc.

1. I. Touval and H. H. Waddell, *Handbook of Fillers and Reinforcements for Plastics*, Van Nostrand Reinhold, Co., Inc., New York 1978, p. 219.
2. J. E. Ramsbottom, *Fire Proofing of Fabrics*, His Majesty's Stationary Office, London, 1947.
3. U.S. Pat. 3,873,451 (Mar. 25, 1975), C. C. Combo and Y. C. Yates (to E. I. du Pont de Nemours & Co., Inc.).
4. U.S. Pat. 3,994,825 (Nov. 30, 1976), C. E. Crompton (to Chemetron Chemical Corp.).
5. D. K. Kintz in *2nd International Conference on Flame Retardants 85*, BPF, Luton, Ltd., London, 1985, p. 9/1.
6. U.S. Pat. 4,017,418 (Apr. 12, 1977), C. E. Crompton and A. M. Z. Kazi (to Chemetron Chemical Corp.).
7. U.S. Pat. 4,022,710 (May 10, 1977), T. Kopbashi and H. Shiota (to Japan Exlan Corp.).
8. E. A. Myszak, Jr., *Use of Submicron Inorganic Flame Retardants in Polymeric Systems*, PQ Corp., Valley Forge, Pa., 1991, p. 19,482.
9. A. Sheldon, *FRCA Flammability Conference,* Ponte Vedra Beach, Fla., Oct. 21, 1990, p. 1.
10. R. W. Little, *Flame Proofing Textile Fibers*, American Chemical Society Monograph Series 40, American Chemical Society, Reinhold, N.Y., 1947.
11. J. A. Rhys, *Chem. Ind. London*, 187 (1969).
12. J. J. Pitts, P. H. Scott, and J. Powell, *Cell. Plast.,* 635 (1970).
13. L. Costa, G. Paganetto, G. Bertelli, and G. Camino, *J. Therm. Anal.*, 36 (1990).
14. P. J. Gale, *Int. J. Mass Spectrom. Ion Phys.* **100**, 313–322 (1990).
15. E. R. Wagner and B. I. Joensten, *J. Appl. Polym. Sci.* **20**, 2143 (1976).

16. M. J. Drew, C. W. Jarves, and G. C. Lickfield in G. L. Nelson, ed., *Fire and Polymers*, ACS Symposium Series 425, American Chemical Society, Washington, D.C.
17. J. W. Hastie, *High Temperature Vapors*, Academic Press, Inc., New York, 1975, p. 353.
18. R. W. Sprague, *Systematic Study of Firebrake ZB as a Flame Retarder for PVC, Parts III, IV, and V*, U. S. Borax.
19. Technical data sheet, Buckman Labs. Inc., 1992.
20. J. A. Finn, in *6th International Symposium on Flammability and Flame Retardants*, Nashville, Tenn., 1979.
21. K. K. Shen and T. S. Griffen in G. L. Nelson, ed., *Fire and Polymers*, ACS Symposium Series 245, American Chemical Society, Washington, D.C., 1990, p. 157.
22. C. J. Hilado, *Flammability Handbook for Plastics*, 2nd ed., Technomic Publications, Lancaster, Pa., 1974.
23. G. A. Byrne, D. Gardiner, and F. H. Holmes, *J. Appl. Chem.* **16**, 81 (1966).
24. P. H. Kemp, *Chemistry of Borates, Part 1*, Borax Consolated, Ltd., London, 1956.
25. F. Molesky in *Recent Advances in Flame Retardancy of Polymeric Materials*, Stamford, Conn., 1990; F. Molesky, "The Use of Magnesium Hydroxide for Flame Retarded Low Smoke Polypropylene," *Polyolefins IV International Conference*, Feb. 24, 1991, Houston, Tex.
26. F. W. Moor, *Molybdenum Compounds as Smoke Suppressants for Polyvinyl Chloride*, Society of Plastics Engineers (SPE ANTEC), Montreal, Canada, 1977.
27. *Smoke Suppression of Rigid Vinyl Using Moylbdenum Compounds*, Technical Bulletin No. 2 Climax Performance Materials Corp.
28. I. Touval, *J. Fire Flam.* **3**, 130 (1972).
29. P. A. Cusack and J. Killmeyer, in Ref. 21, p. 189.
30. S. Kaufman and C. A. Landrith, *Development of Improved Flame Resistant Interior Wiring Cables*, 24th ICCS, Nov. 18, 1975.

IRVING TOUVAL
Touval Associates

HALOGENATED FLAME RETARDANTS

The use of synthetic polymers has grown dramatically in the latter part of the twentieth century. Along with that growth has come a growth in the use of flame retardants of all types. Approximately 10% of the volume of any given polymer type is sold as some sort of flame-retarded grade. Halogenated flame retardants typically have high levels of either bromine or chlorine. The high level of activity allows these additives to be used at relatively low loadings, translating to a greater retention of the physical properties of the base polymer.

Halogenated flame retardants fall into two general classes, additive and reactive. Additives are mixed into the polymer in common polymer processing equipment. Other ingredients such as stabilizers, pigments (qv), and processing aids are often incorporated at the same time. Reactive flame retardants literally become part of the polymer by either reacting into the polymer backbone or grafting onto it.

The use of flame retardants came about because of concern over the flammability of synthetic polymers (plastics). A simple method of assessing the potential contribution of polymers to a fire is to examine the heats of combustion, which for common polymers vary by only about a factor of two (1). Heats of combustion correlate with the chemical nature of a polymer whether the polymer is synthetic or natural. Concern over flammability should arise via a proper risk assessment which takes into account not only the flammability of the material, but also the environment in which it is used.

Fundamentals of Flammability

In order for a solid to burn it must be volatilized, because combustion is almost exclusively a gas-phase phenomenon. In the case of a polymer, this means that decomposition must occur. The decomposition begins in the solid phase and may continue in the liquid (melt) and gas phases. Decomposition produces low molecular weight chemical compounds that eventually enter the gas phase. Heat from combustion causes further decomposition and volatilization and, therefore, further combustion. Thus the burning of a solid is like a chain reaction. For a compound to function as a flame retardant it must interrupt this cycle in some way. There are several mechanistic descriptions by which flame retardants modify flammability. Each flame retardant actually functions by a combination of mechanisms. For example, metal hydroxides such as $Al(OH)_3$ decompose endothermically (thermal quenching) to give water (inert gas dilution). In addition, in cases where up to 60 wt % of $Al(OH)_3$ may be used, such as in polyolefins, the physical dilution effect cannot be ignored.

Inert Gas Dilution. Inert gas dilution involves the use of additives that produce large volumes of noncombustible gases when the polymer is decomposed. These gases dilute the oxygen supply to the flame or dilute the fuel concentration below the flammability limit. Metal hydroxides, metal carbonates, and some nitrogen-producing compounds function in this way as flame retardants (see FLAME RETARDANTS, ANTIMONY AND OTHER INORGANIC COMPOUNDS).

Thermal Quenching. Endothermic degradation of the flame retardant results in thermal quenching. The polymer surface temperature is lowered and the rate of pyrolysis is decreased. Metal hydroxides and carbonates act in this way.

Protective Coatings. Some flame retardants function by forming a protective liquid or char barrier. These minimize transpiration of polymer degradation products to the flame front and/or act as an insulating layer to reduce the heat transfer from the flame to the polymer. Phosphorus compounds that decompose to give phosphoric acid and intumescent systems are examples of this category (see FLAME RETARDANTS, PHOSPHORUS FLAME RETARDANTS).

Physical Dilution. The flame retardant can also act as a thermal sink, increasing the heat capacity of the polymer or reducing the fuel content to a level below the lower limit of flammability. Inert fillers such as glass fibers and microspheres and minerals such as talc act by this mechanism.

Chemical Interaction. Halogens and some phosphorus flame retardants act by chemical interaction. The flame retardant dissociates into radical species that compete with chain propagating and branching steps in the combustion process.

Flammability Testing

One problem associated with discussing flame retardants is the lack of a clear, uniform definition of flammability. Hence, no clear, uniform definition of decreased flammability exists. The latest American Society for Testing and Materials (ASTM) compilation of fire tests lists over one hundred methods for assessing the flammability of materials (2). These range in severity from small-scale measures of the ignitability of a material to actual testing in a full-scale fire. Several of the most common tests used on plastics are summarized in Table 1.

Material Tests. Material tests measure some property of the polymer or plastic as opposed to measuring the flammability of the final product which contains the plastic.

Limiting Oxygen Index. The minimum concentration of oxygen in an O_2/N_2 mixture that supports combustion of a vertically mounted test specimen is called the limiting oxygen index (3,4). Test specimens are 0.65×0.3 cm $\times$ 12.5 cm. The principal advantage of this test is its reproducibility which makes it useful for quality control. The main disadvantage is that the results rarely correlate with the results of other fire tests.

Table 1. Flammability Tests

Designation[a,b]	Description or application[c]	Characteristic measured
ASTM E162-87	radiant panel	flame spread
ASTM E119-88	building materials	fire endurance
MVSS 302	materials for automotive interiors	burning rate
ASTM D2863-87	limiting oxygen index	ease of extinction
ASTM E662-83	NBS smoke chamber	smoke
ASTM E84, UL 723, UL 910	steiner tunnel	flame spread and smoke
UL 94	vertical burn	ignition resistance
UL 790, ASTM E108-90	roof burn	flame spread
UL 1715	room burn	flashover potential and smoke
CAL 133	furniture	flashover potential and smoke
CAL 117, ASTM E1353-90, ASTM E1354-90	cigarette ignition for furniture	ignitability
ASTM E1354-90	Cone calorimeter	heat release and smoke
ASTM E906	OSU heat release rate calorimeter	heat release and smoke
factory mutual flammability apparatus	FMRC flammability apparatus	heat, smoke, toxic and corrosive products release, ignition and flame spread

[a]Designations listed together are not meant to imply equivalency.
[b]CAL = California; MVSS = Motor Vehicle Safety Standard; UL = Underwriter's Laboratory.
[c]NBS = National Bureau of Standards; OSU = Ohio State University; FMRC = Factory Mutual Research Corp.

Specific Tests. Federal (United States) Motor Vehicle Safety Standard (MVSS) 302 is used to measure the burning behavior of materials used in automobile interiors. A specimen is mounted horizontally and ignited for 15 seconds. The burning rate should be below 10 cm/min. The test specimen is 35.5 × 10.1 cm by the actual thickness (up to 1.3 cm). Automakers typically impose more severe criteria than the 10 cm/min in the standard.

The Underwriters Laboratory UL 94 Standard for Safety (5) measures the ignitability of plastics by a small flame. The test specimens are mounted vertically and ignited using a Bunsen burner held at a 30° angle. A layer of cotton (qv) is placed under the specimen to test for flaming drips. The specimen dimensions are 12.7 × 1.27 cm by the thickness in the intended application. Typical thicknesses tested are 3.2, 1.6, and 0.8 mm. The flame is applied for 10 seconds, plus a subsequent 10-s application if the specimen self-extinguishes after the first ignition. The flammability is classified as: (*1*) V-0 if no specimen burns longer than 10 seconds after each flame application. The sum of the after-flame times for five specimens (two flame applications per specimen, 10 total flame applications) should not exceed 50 seconds. The cotton cannot be ignited. (*2*) V-1 if no specimen should burn longer than 30 seconds. The sum of the after-flame times for five specimens (two flame applications per specimen, 10 flame applications) should not exceed 250 seconds. The cotton cannot be ignited. (*3*) V-2 if the requirements are the same as for V-1 except that the cotton can be ignited. Most commercial plastics are flame retarded to meet either the V-0 or the V-2 classification.

Heat Release Calorimeters. There are three principal types of heat release calorimeters. The Cone calorimeter measures the rate of heat release of a burning specimen (6–8). Test specimens 10 × 10 cm by up to 2.5 cm thick are exposed to radiant heat of up to 100 kW/m^2. Many parameters in addition to heat release rate may be measured. These include total heat released, mass loss, time to ignition, critical heat flux, and smoke production. The heat released is calculated based on consumption of oxygen, thus this is also sometimes called oxygen consumption calorimetry. The advantage of this test is that materials can be subjected to heat fluxes similar to those encountered in real fires. The main disadvantage is that the results must correlate with large-scale testing before they are useful. Correlations in the area of wire and cable (9,10) and wall coverings (11) have been developed.

The Ohio State University (OSU) calorimeter (12) differs from the Cone calorimeter in that it is a true adiabatic instrument which measures heat released during burning of polymers by measurement of the temperature of the exhaust gases. This test has been adopted by the Federal Aeronautics Administration (FAA) to test total and peak heat release of materials used in the interiors of commercial aircraft. The other principal heat release test in use is the Factory Mutual flammability apparatus (13,14). Unlike the Cone or OSU calorimeters this test allows the measurement of flame spread as well as heat release and smoke. A unique feature is that it uses oxygen concentrations higher than ambient to simulate back radiation from the flames of a large-scale fire.

Product Tests. Tests in which the finished article is subjected to a more or less realistic fire are called product tests. A few examples follow.

Tunnel Test. The tunnel test is widely used to test the flame spread potential of building products such as electrical cable (15) and wall coverings (16). The test apparatus consists of a tunnel 7.62 × 0.445 m × 0.305 m in cross section, one end of which contains two gas burners. The total heat supplied by the burners is 5.3 MJ/min. The test specimen (7.62 m × 50.8 cm), attached to the ceiling, is exposed to the gas flames for 10 minutes while the maximum flame spread, temperature, and smoke evolved are measured. The use of this and other flame spread test methods has been reviewed (17).

Factory Mutual Corner Test. This is a large-scale corner test used to test building products (18–20). The test rig consists of three sides of a cube. The two walls are 15.24 and 11.58 m by 7.62 m tall. The ceiling is 9.14 × 15.24 m. The product to be tested is mounted on the walls and ceilings in a manner consistent with the intended use. The fire source is a 340 kg stack of wood pallets located in the corner. In order to pass the test, no flame can propagate to any extremity of the walls or ceiling. The Factory Mutual flammability apparatus is proposed to replace this test for certain applications (21).

CAL 133. California Technical Bulletin 133 is a test of the fire hazard associated with upholstered furniture (22). The test is carried out by igniting a standard fire source directly on the piece of furniture being tested. In the most recent version of the test, the fire source is a gas flame. Smoke, heat, and toxic gas emissions are measured during the test. A related test, BS 5852, uses various wooden cribs as the fire source (23).

Flame Retardants

Compounds of chlorine and bromine are the halogen compounds having commercial significance as flame-retardant chemicals. Fluorine compounds are expensive and, except in special cases, ineffective. Iodine compounds, although effective, are expensive and too unstable to be used. Halogenated flame retardants can be broken down into three classes: brominated aliphatic, chlorinated aliphatic, and brominated aromatic. As a general rule, the thermal stability increases as brominated aliphatic < chlorinated aliphatic < brominated aromatic. The thermal stability of the aliphatic compounds is such that with few exceptions, thermal stabilizers such as a tin compound must be used (see TIN COMPOUNDS). Brominated aromatic compounds are much more stable and may be used in thermoplastics at fairly high temperatures without the use of stabilizers and at very high temperatures with stabilizers. It is commonly thought that it is desirable for the flame retardant to decompose with the liberation of halogen at a somewhat lower temperature than the decomposition temperature of the polymer. This view is overly simplistic. In fact, in some systems, it is degradation of the polymer that promotes degradation of the flame retardant and not vice versa (24).

Suggested formulations for various polymers using hexabromocyclododecane (HBCD), a brominated aliphatic; a chlorinated paraffin, ie, a chlorinated aliphatic; and decabromodiphenyl oxide, a brominated aromatic, are shown in Tables 2–4. These suggested formulations may not be strictly comparable because of differences in the nature of the base resins. However, the suggestions are specific to a given UL-94 rating.

In selecting a flame retardant for a given application, the cost contribution of the flame retardant to the final polymer compound must be taken into account.

Table 2. **Formulations for Chlorinated Paraffin**[a,b]

Polymer	UL-94 rating at 3.2 mm	Flame retardant, wt %	Antimony oxide, wt %
polyethylene	V-0	24	10
polypropylene	V-0	22	11
polystyrene, high impact	V-0	16	5
polyester, unsaturated	V-0	15	5

[a]Ref. 25.
[b]Based on 70% chlorine content. For chlorinated paraffins having lower chlorine content, the use level must be raised accordingly.

Table 3. **Formulations for Hexabromocyclododecane**[a,b]

Polymer	UL-94 rating at 1.6 mm	Flame retardant, wt %	Antimony oxide, wt %
polypropylene, copolymer	V-2	4	1
polystyrene			
crystalline	V-2	5	0
high impact	V-2	7	0
	V-2	4	1

[a]Ref. 26.
[b]Acid scavenging stabilizers are typically recommended for use with HBCD in thermoplastic applications. Stabilized grades are available from the manufacturers.

Table 4. **Formulations for Decabromodiphenyl Oxide**[a,b]

Polymer	UL-94 rating at 1.6 mm	Flame retardant, wt %	Antimony oxide, wt %
polyethylene			
low density	V-2	6	2
high density	V-2	8	3
cross-linked	V-0	20	10
polypropylene with 14% talc	V-0	22	6
polystyrene, high impact	V-0	12	4
acrylonitrile butadiene styrene	V-0	15	5
polybutylene terephthalate[c]	V-0	10	5
polyamide[c]	V-0	14	5
thermoset epoxy resin	V-0	6	3

[a]Ref. 27.
[b]Decabromodiphenyl oxide contains approximately 83% bromine. For flame retardants having a lower bromine content, the use level must be raised accordingly.
[c]Glass-reinforced grades may require lower levels of flame retardants.

Assessment of cost should be done on a cost per volume basis rather than a simple cost per weight basis.

Antimony–Halogen Synergism. Antimony oxide is commonly employed as a fire-retardant supplement for halogen-containing polymer systems as a means of reducing the halogen levels required to obtain a given degree of flame retardancy. This reduction is desirable because the required halogen content may be so high that it affects the physical properties of the final polymer. In many cases, the antimony oxide is used simply to give a more cost-effective system.

Antimony–halogen systems have been widely studied. The various mechanistic studies of the interaction of brominated flame retardants, antimony oxide, and polymers have been reviewed (28). No completely satisfactory theory to explain the synergistic effects obtained with this combination of elements is available, but it is generally agreed that the active agents, antimony trihalides or antimony oxyhalides, act principally in the gas phase (29–31). The antimony halides are postulated to act as radical traps. Although the antimony halides appear to act principally in the gas phase, some effect on the condensed-phase chemistry cannot be ruled out. Antimony–halogen flame-retardant compositions usually produce a carbonaceous residue, even in polymers such as polypropylene, which produces none in the absence of fire retardants. The production of the carbonaceous residue probably results from the antimony trihalides, strong Lewis acid catalysts, which are capable of promoting the dehydrohalogenation of organic halides as well as coupling and rearrangement reactions in organic systems.

Brominated Additive Flame Retardants. Additive flame retardants are those that do not react in the application designated. There are a few compounds that can be used as an additive in one application and as a reactive in another. Tetrabromobisphenol A [*79-94-7*] (TBBPA) is the most notable example. Tables 5 and 6 list the properties of most commercially available bromine-containing additive flame retardants.

Brominated Diphenyl Oxides. Brominated diphenyl oxides are prepared by the bromination of diphenyl oxide. They are often referred to as diphenyl ethers. Taken together, the class constitutes the largest volume of brominated flame retardants. They range in properties from high melting solids to liquids. They are used, as additives, in virtually every polymer system.

Decabromodiphenyl Oxide. Decabromodiphenyl oxide [*1163-19-5*] (decabrom) is the largest volume brominated flame retardant used solely as an additive. It is prepared by the bromination of diphenyl oxide in neat bromine using $AlCl_3$ as a catalyst (32). The bromination may also be carried out in an inert solvent such as methylene dibromide [*74-95-3*] (33). The commercially available grades are >98% decabromodiphenyl oxide with the remainder being the nonabromo species.

Historically, impurities such as iron and aluminum (catalyst residues) limited use. The levels of these impurities have since been reduced. Decabrom is thermally stable and can be employed with polymer systems that require high processing temperatures. It is almost always used in conjunction with antimony oxide. The largest volume use of decabrom is in high impact polystyrene that is used to manufacture television cabinets. Decabrom is used in virtually every class of polymer, including ABS (see ACRYONITRILE POLYMERS, ABS RESINS); engineering thermoplastics such as polyamides (qv) and polyesters (qv) (see ENGI-

Table 5. Brominated Additive Flame Retardants

Common name	CAS Registry Number	Molecular formula	Bromine, %	Specific gravity	Mp, °C	Sources[a]
ethylenebisdibromonorbornanedicarboximide	[*52907-07-0*] [*41291-34-3*]	$C_{20}H_{20}Br_4N_2O_4$	45	2.07	294	EC
bis(2-ethylhexyl)tetrabromophthalate	[*26040-51-7*]	$C_{24}H_{34}Br_4O_4$	45	1.54		EA, GL
tetrabromobisphenol A	[*79-94-7*] [*6386-73-8*]	$C_{15}H_{12}O_2Br_4$	58.4	2.17	180	ASC, DS, EC, GL, T, TS, CECA
bis(methyl)tetrabromophthalate	[*55481-60-2*]	$C_{10}H_6Br_4O_4$	63			EA
tris-dibromopropylisocyanurate	[*52434-90-9*]	$C_{12}H_{15}Br_6N_3O_3$	65.8		106–108	AC, ASC, T
tetrabromobisphenol A, bis(2,3-dibromopropylether)	[*21850-44-2*]	$C_{23}H_{20}Br_8O_2$	67.7	2.17	95	GL
ethylenebistetrabromophthalimide	[*32588-76-4*]	$C_{18}H_4N_2O_4Br_8$	68	2.66	445	EC
bis(tribromophenoxy)ethane	[*37853-59-1*]	$C_{14}H_8O_2Br_6$	68	2.58	224	GL
tetrabromobisphenol *S*-bis(2,3-dibromopropylether)	[*42757-55-1*]	$C_{18}H_{14}Br_8O_4S$	70.8		52–55	M
pentabromodiphenyl oxide	[*32534-81-9*]	$C_{12}H_9Br_5O$	71	2.27	liquid	AC, DS, GL, ISC, WI
tetrabromocyclooctane	[*3194-57-8*] [*31454-48-5*]	$C_8H_{12}Br_4$	74.7	2.27	73	EC
hexabromocyclododecane	[*3194-55-6*]	$C_{12}H_{18}Br_6$	74.7	2.06	180[b]	DS, EC, GL, ISC
dibromoethyldibromocyclohexane	[*3322-93-8*]	$C_8H_{12}Br_4$	74.7	2.38	70–76	EC
octabromodiphenyl oxide	[*32536-52-0*]	$C_{12}H_4Br_8O$	79	2.75	85	DS, EC, GL, ISC, MI
tetradecabromodiphenoxybenzene	[*58965-66-5*]	$C_{18}Br_{14}O_2$	82	3.25	370	EC
pentabromotoluene	[*87-83-2*]	$C_7H_3Br_5$	82			DS
decabromodiphenyl oxide	[*1163-19-5*]	$C_{12}Br_{10}O$	83	3.00	305	ASC, DS, EC, GL, ISC, MI, MT, NC, TS, WI
decabromobiphenyl	[*13654-09-6*]	$C_{12}Br_{10}$	84.5			EA

[a]Company and country are as follows: Akzo Chemicals BV, Netherlands (AC); Asahi Chemical, Japan (ASC); Dead Sea Bromine, Israel (DS); Elf Atochem, France (EA); Ethyl Corp., United States (EC); Great Lakes, United States (GL); ISC Chemicals, Ltd. United Kingdom (ISC); Manac Inc., Japan (MI); Marubishi, Japan (M); Mitsui Toatsu Fine Chemicals, Inc., Japan (MT); Nippon Chemicals Corp., Ltd., Japan (NC); Teijin, Japan (T); Tosoh, Japan (TS); Warwick International, United Kingdom (WI); CECA, SA, France (CECA).
[b]Commercial products are mixtures; the mp given is for the low end of the range.

Table 6. Polymeric and Oligomeric Brominated Additive Flame Retardants

Common name	CAS Registry Number	Molecular formula[a]	Bromine, %	Specific gravity	Mp, °C	Commercial[b] sources
tetrabromobisphenol A carbonate oligomer, phenoxy end capped	[94334-64-2] [28906-13-0]	$(C_{16}H_{12}O_3Br_4)_n$	52	2.2	210–230	GL, MG, T
tetrabromobisphenol A carbonate oligomer, tribromophenoxy end capped	[71342-77-3]	$(C_{16}H_{12}O_3Br_4)_n$	58	2.2	230–260	GL
epoxy oligomers of tetrabromobisphenol A	[68928-70-1] [400039-93-8]	$(C_{18}H_{16}O_3Br_4)_n$	52–54			DS, DI, DC, H, MI, SY, TK
tetrabromobisphenol A epoxy oligomers, tribromophenoxy end capped		$(C_{18}H_{16}O_3Br_4)_n$	55–58			DI, H, TK
poly(dibromostyrene)	[62354-98-7]	$(C_8H_6Br_2)_n$	59	1.9	155–165[c]	GL
poly(dibromophenylene oxide)	[26023-27-8] [69882-11-7]	$(C_6H_2OBr_2)_n$	62	2.25	225	GL
brominated polystyrene, low molecular weight	[88497-56-7]	$(C_8H_{5.3}Br_{2.7})_n$	66	2.1	130–140	KC
brominated polystyrene	[88497-56-7]	$(C_8H_{5.3}Br_{2.7})_n$	66	2.10	195[d]	KC
	[88497-56-7]		58	1.85	165	KC
poly(pentabromobenzylacrylate)	[59447-55-1]	$(C_{10}H_6Br_5O_2)_n$	70	2.05	210	DS

[a]Formulas for polymeric compounds are for the repeat unit only and ignore the end groups.
[b]Company and country are as follows: Dead Sea Bromine, Israel (DS); Dainippon Ink and Chemical, Japan (DI); Dow Chemical USA, United States (DC); Great Lakes, United States (GL); Hitachi, Japan (H); Keil Chemical Div., Ferro Corp., United States (KC); Manac Inc., Japan (MI); Mitsubishi Gas Chemical Co., Japan (MG); Sakamoto Yukuhin Kogyo Co., Ltd., Japan (SY); Teijin, Japan (T); Tohto Kasei Co., Ltd., Japan (TK).
[c]A higher molecular weight version has a mp of 210–230°C.
[d]Commercial products are mixtures; the mp given is for the low end of the range.

NEERING PLASTICS); polyolefins; thermosets, ie, epoxies and unsaturated polyesters; PVC; and elastomers (qv). It is also widely used in textile applications as the flame retardant in latex-based back coatings.

Decabrom has poor uv stability in styrenic resins and causes significant discoloration. The use of uv stabilizers (qv) can minimize, but not eliminate, this effect. For styrenic applications that require uv stability, several other brominated flame retardants are more suitable. In polyolefins, the uv stability of decabrom is more easily improved by the use of stabilizers.

Octabromodiphenyl Oxide. Octabromodiphenyl oxide [*32536-52-0*] (OBDPO) is prepared by bromination of diphenyl oxide. The degree of bromination is controlled either through stoichiometry (34) or through control of the reaction kinetics (35). The melting point and the composition of the commercial products vary somewhat. OBDPO is used primarily in ABS resins where it offers a good balance of physical properties. Poor uv stability is the primary drawback and use in ABS is being supplanted by other brominated flame retardants, primarily TBBPA.

Pentabromodiphenyl Oxide. Pentabromodiphenyl oxide [*32534-81-9*] (PBDPO) is prepared from diphenyl oxide by bromination (36). It is primarily used as a flame retardant for flexible polyurethane foams. For this application PBDPO is sold as a blend with a triaryl phosphate. Its primary benefit in flexible polyurethanes is superior thermal stability, ie, scorch resistance, compared to chloroalkyl phosphates (see PHOSPHATE ESTERS).

Tetrabromobisphenol A. Tetrabromobisphenol A [*79-94-7*] (TBBPA) is the largest volume brominated flame retardant. TBBPA is prepared by bromination of bisphenol A under a variety of conditions. When the bromination is carried out in methanol, methyl bromide [*74-80-9*] is produced as a coproduct (37). If hydrogen peroxide is used to oxidize the hydrogen bromide [*10035-10-6*], HBr, produced back to bromine, methyl bromide is not coproduced (38). TBBPA is used both as an additive and as a reactive flame retardant. It is used as an additive primarily in ABS systems. In ABS, TBBPA is probably the largest volume flame retardant used, and because of its relatively low cost is the most cost-effective flame retardant. In ABS it provides high flow and good impact properties. These benefits come at the expense of distortion temperature under load (DTUL) (39). DTUL is a measure of the use temperature of a polymer. TBBPA is more uv stable than decabrom and uv stable ABS resins based on TBBPA are produced commercially.

Hexabromocyclododecane. There are three flame retardants obtained from the bromination of cyclic oligomers of butadiene (see BUTADIENE). These are hexabromocyclododecane [*25637-99-4*] (HBCD), tetrabromocyclooctane [*31454-48-5*], and dibromoethyldibromocyclohexane [*3322-93-8*]. Each has a theoretical bromine content of 74.7%. They differ primarily in melting point and solubility. Hexabromocyclododecane is the only one used in large volume. The primary use of HBCD is in polystyrene foam. It is prepared by the bromination of cyclododecatriene (40). The product of the reaction is a mixture of three principal isomers. The gamma isomer is the predominant product from the reaction. Because it has the highest melting point (205°C), purity correlates with melting point. Neither the purity nor the melting point correlate to the stability of the compound during processing (41). Other factors such as levels of minor impurities and particle size are at least as important. For this reason a myriad of grades are commercially available. Compared to brominated aromatic flame retardants, the thermal stability of HBCD is considerably lower, limiting use to situations where the processing temperature is low. Thermal stabilizers are often needed. Commercial grades containing thermal stabilizers are available.

1,2-Bis(2,4,6-tribromophenoxy)ethane. The additive 1,2-bis(2,4,6-tribromophenoxy)ethane [*37853-59-1*] (BTBPE) is a white crystalline powder having the good thermal stability expected of a brominated aromatic. It is prepared by the reaction of tribromophenol [*75-80-9*] and ethylene dibromide [*106-93-4*] in the presence of a base. The principal market for BTBPE is in ABS resins. It is the second most prevalent flame retardant used in ABS. It imparts good impact and reasonable flow properties at the expense of DTUL. Compared to tetrabromobisphenol A, it has somewhat better uv stability.

Ethylenebis(tetrabromophthalimide). The additive ethylenebis(tetrabromophthalimide) [*41291-34-3*] is prepared from ethylenediamine and tetrabromophthalic anhydride [*632-79-1*]. It is a specialty product used in a variety of applications. It is used in engineering thermoplastics and polyolefins because of its thermal stability and resistance to bloom (42). It is used in styrenic resins because of its uv stability (43). This flame retardant has been shown to be more effective on a contained bromine basis than other brominated flame retardants in polyolefins (10).

Chlorinated Additive Flame Retardants. Table 7 is a general listing of chlorinated compounds used as additive flame retardants.

Chlorinated Paraffins. The term chlorinated paraffins covers a variety of compositions. The prime variables are molecular weight of the starting paraffin and the chlorine content of the final product. Typical products contain from 12–24 carbons and from 40–70 wt % chlorine. Liquid chlorinated paraffins are used as plasticizers (qv) and flame retardants in paint (qv) and PVC formulations. The solid materials are used as additive flame retardants in a variety of thermoplastics. In this use, they are combined with antimony oxide which acts as a synergist. Thermal stabilizers, such as those used in PVC (see VINYL POLYMERS), must be used to overcome the inherent thermal instability.

Bis(hexachlorocyclopentadieno)cyclooctane. The di-Diels-Alder adduct of hexachlorocyclopentadiene [*77-47-4*] and cyclooctadiene (44) is a flame retardant having unusually good thermal stability for a chlorinated aliphatic. In fact, this compound is comparable in thermal stability to brominated aromatics in some applications. Bis(hexachlorocyclopentadieno)cyclooctane is used in several polymers, especially polyamides (45) and polyolefins (46) for wire and cable applications. Its principal drawback is the relatively high use levels required compared to some brominated flame retardants.

Oligomeric Flame Retardants. There are several oligomeric flame retardants. The principal advantage claimed for these materials is their resistance to bloom and plate-out. In some cases they are used at levels high enough that the resulting flame-retarded resin should properly be viewed as a polymer blend or alloy. All of the available oligomeric flame retardants are brominated (Table 6).

Brominated Polystyrene. Brominated polystyrene (BrPS) is prepared by the bromination of polystyrene. A variety of processes have been proposed including the use of bromine chloride [*13863-41-7*] as the bromination agent. There are three versions of BrPS available. The original product [*57137-10-7*] has a minimum of 66% bromine, corresponding to 2.8 bromines per aromatic ring. This ma-

Table 7. Chlorinated and Mixed Halogen Additive Flame Retardants

Common name	CAS Registry Number	Molecular formula	Halogen,[a] %	Specific gravity	Mp, °C	Sources[b]
bis(hexachlorocyclopentadieno)-cyclooctane	[*13560-89-9*]	$C_{18}H_{12}Cl_{12}$	65	1.9	350	OCC
chlorinated paraffin	[*63449-39-8*] [*61788-76-9*][d]	[c]	39–70	1.12–1.46		AC, AD, CECA, DC, HAG, H, ICI, KC, TC
pentabromochlorocyclohexane	[*25495-99-2*]	$C_6H_5ClBr_5$	78[e] 7	2.90	170	NCI
bromo/chloro alpha olefin and paraffins	[*82600-56-4*] [*68955-41-9*][d]	[c]	24–35[e] 19–35	1.37–1.66	liquid	DC, KC

[a]The halogen is chlorine unless otherwise noted.
[b]Company and country are as follows: Ajinomoto Co., Inc., Japan (AC); Asahi Denka Kogyo KK, Japan (AD); CECA, SA, France (CECA); Dover Chemical, United States (DC); Hoechst AG, Germany (HAG); Hüls, Germany (H); ICI, United Kingdom (ICI); Keil Chemical Div., Ferro Corp., United States (KC), Nissei Chemical Industries Co., Japan (NCI); Occidental Chemical Corp., United States (OCC); Tosoh Corp., Japan (TC).
[c]No unique formula.
[d]No unique CAS Registry Number.
[e]The halogen is bromine.

terial has a degree of polymerization of about 2000. It is widely used in glass-filled engineering thermoplastics, eg, polyamides and thermoplastic polyesters. Its use in unfilled poly(butylene terephthalate) (PBT) is limited by its low thermodynamic compatibility (47). A version with a bromine content of approximately 60% and a lower molecular weight version [*57137-10-7*] of this material are also available. The lower molecular weight is claimed to be an advantage in styrenic resins, because it improves compatibility with the resin (48,49). These materials are polymeric and thus resistant to bloom.

Poly(dibromostyrene). Poly(dibromostyrene) [*62354-98-7*] (PDBS) is prepared by the polymerization of dibromostyrene [*31780-26-4*] (50). Two versions are available. One has a molecular weight of about 10,000, the other a molecular weight of about 80,000. Information comparing the performance of these materials to the performance of the brominated polystyrenes is not available.

Brominated Epoxy Oligomers. Brominated epoxy oligomers are prepared in several ways. One method involves the reaction of tetrabromobisphenol A and its diglycidyl ether. This can give materials having molecular weights from as low as 1600 to as high as 60,000. The bromine contents range from 52–54%. In some cases, the terminal epoxy groups react with phenol or tribromophenol to form end caps. End capping with phenol lowers the bromine content slightly whereas end capping with tribromophenol raises it slightly. This change in bromine content becomes more significant at lower molecular weight. These products are used primarily in styrenic resins, eg, ABS, although they find some use in engineering thermoplastics, eg, PBT. In ABS, they offer good flow and uv resistance. Their effect on physical properties is difficult to characterize because it is dependent on the molecular weight.

Brominated Carbonate Oligomers. There are two commercial brominated carbonate oligomer (BrCO) products. Both are prepared from tetrabromobisphenol A and phosgene. One has phenoxy end caps [*28906-13-0*] and the other tribromophenoxy [*71342-77-3*] end caps. These are used primarily in PBT and polycarbonate/acrylonitrile–butadiene–styrene (PC/ABS) blends.

Reactive Flame Retardants. Reactive flame retardants become a part of the polymer by either becoming a part of the backbone or by grafting onto the backbone. Choice of reactive flame retardant is more complex than choice of an additive type. The reactive flame retardant can exert an enormous effect on the final properties of the polymer. There are also reactive halogenated compounds used as intermediates to other flame retardants. Tables 8 and 9 list the commercially available reactive flame retardants and intermediates.

Tetrabromobisphenol A. TBBPA is the largest volume reactive flame retardant. Its primary use is in epoxy resins (see EPOXY RESINS) where it is reacted with the bis-glycidyl ether of bisphenol A to produce an epoxy resin having 20–25% bromine. This brominated resin is typically sold as a 80% solution in a solvent. TBBPA is also used in the production of epoxy oligomers which are used as additive flame retardants.

Tetrabromophthalic Anhydride. Tetrabromophthalic anhydride [*632-79-1*] (TBPA) is widely used as a reactive flame retardant in unsaturated polyesters as well as the precursor to a number of other fire retardants. Polyesters prepared from this compound have relatively poor photochemical stability and tend to dis-

Table 8. Brominated Reactive Flame Retardants and Intermediates

Compound	CAS Registry Number	Molecular formula	Bromine, %	Specific gravity	Mp, °C	Sources[a]
diester/ether diol of tetrabromophthalic anhydride	*[77098-07-8]*	$C_{15}H_{16}O_7Br_4$	46	1.80	liquid	EC, GL
tetrabromobisphenol A-bis(allyl ether)	*[25327-89-3]*	$C_{23}H_{20}O_2Br_4$	51.2	1.83	119	DS, GL
tetrabromobisphenol A-bis(2-hydroxyethyl ether)	*[4162-45-2]*	$C_{19}H_{20}O_4Br_4$	51.6	1.8	116	DS, GL
tribromophenylmaleimide	*[59789-51-4]*	$C_{10}H_8NBr_3O_2$	57.9			DS
tetrabromobisphenol A	*[79-94-7]*	$C_{15}H_{12}O_2Br_4$	58.4	2.17	180	CECA, DS, EC, GL, T, TS, ASC
dibromostyrene	*[31780-26-4]*	$C_8H_6Br_2$	60		liquid	GL
disodium salt of tetrabromophthalate	*[25357-79-3]*	$C_8O_4Br_4Na_2$	61	2.8	>500	GL
dibromoneopentylglycol	*[3296-90-0]*	$C_5H_{10}O_2Br_2$	61	2.20	109	DS, EC
tetrabromodipentaerythritol	*[109678-33-3]*	$C_9H_{20}O_2Br_4$	63.2	1.98	81.5–82.5	DS
2,4-dibromophenol	*[615-58-7]*	$C_6H_4OBr_2$	63.4		36	DS
tribromophenyl allyl ether	*[26762-91-4]*	$C_9H_7Br_3O$	64.6	2.20	74	DS
tetrabromophthalic anhydride	*[632-79-1]*	$C_8O_3Br_4$	68	2.87	270	CECA, EC, GL
tribromostyrene	*[61368-34-1]*	$C_8H_5Br_3$	68		65–67	DS
pentabromobenzylacrylate	*[59447-55-1]*	$C_{10}H_5O_2Br_5$	71.8			DS
2,4,6-tribromophenol	*[75-80-9]* *[118-79-6]*	$C_6H_3OBr_3$	72.5	2.22–2.55	95.5	DS, GL, MI
dibromopropanol	*[96-13-9]*	$C_3H_6Br_2O$	73.1	2.14	liquid	GL
tribromoneopentyl alcohol	*[1522-92-5]*	$C_5H_9OBr_3$	73.6	2.28	62–67	DS
vinyl bromide	*[593-60-2]*	C_2H_3Br	74.5	1.52	liquid	EC
pentabromobenzyl bromide	*[38521-51-6]*	$C_7H_2Br_6$	84.8			DS

[a]Company and country are as follows: CECA SA, France (CECA): Dead Sea Bromine, Israel (DS); Ethyl Corp., United States (EC); Great Lakes Chemical, United States (GL); Manac Inc., Japan (MI); Tosoh, Japan (T); Tosoh, Japan (TS); Asahi Chemical, Japan (ASC).

Table 9. Chlorinated Reactive Flame Retardants and Intermediates

Compound	CAS Registry Number	Molecular formula	Chlorine, %	Specific gravity	Mp, °C	Sources[a]
hexachlorocyclopentadiene	[*77-47-4*]	C_5Cl_6	78.0	1.710	11[b]	V
chlorendic anhydride	[*115-27-5*]	$C_9O_3Cl_6$	57.7		240	V
chlorendic acid	[*115-28-6*]	$C_9H_2O_4Cl_6$	55.0		[c]	OC
tetrachlorophthalic anhydride	[*117-08-8*]	$C_8O_3Cl_4$	49.6		255-257	MO, NS

[a]Company and country are as follows: Monsanto, United States (MO); Nippon Soda Co., Inc., Japan (NS); Occidental Chemical, United States (OC); Velsicol, United States (V).
[b]239°C bp.
[c]Decomposes to the anhydride.

color upon exposure to light. This tendency to discolor can be reduced, but not eliminated, by the use of uv stabilizers.

TBPA is prepared in high yield by the bromination of phthalic anhydride in 60% oleum (51). The use of oleum as the bromination solvent results in some sulfonation of the aromatic ring (52). Sulfonated material is removed by hydrolyzing the anhydride with dilute NaOH, filtering and acidifying with dilute HCl. The precipitated acid is washed several times with hot water and reconverted to the anhydride by heating at 150°C for several hours.

Diester/Ether Diol of Tetrabromophthalic Anhydride. This material [*77098-07-8*] is prepared from TBPA in a two-step reaction. First TBPA reacts with diethylene glycol to produce an acid ester. The acid ester and propylene oxide then react to give a diester. The final product, a triol having two primary and one secondary hydroxyl group, is used exclusively as a flame retardant for rigid polyurethane foam (53,54).

Dibromoneopentyl Glycol. Dibromoneopentyl glycol [*3296-90-0*] (DBNPG) is prepared by the stepwise hydrobromination of pentaerythritol [*115-77-5*] to yield a product in which about half of the hydroxyl function has been replaced (see ALCOHOLS, POLYHYDRIC; GLYCOLS). The commercial form contains 80–82 wt % DBNPG, 5–7% monobromoneopentyl triol [*19184-65-7*], and 13–15 wt % tribromoneopentyl alcohol [*1522-92-5*]. Pure DBNPG is also available commercially.

The principal use of DBNPG is in rigid urethane foam systems (see FOAMS). It can be used by itself to produce Class II foams or with phosphorus coagents to produce Class I foams as evaluated by the ASTM E84 test (55). DBNPG is also used in unsaturated polyester resins where it can be used to replace part of the regular glycol to yield a resin having the desired bromine content. A preferred method of using the compound is to prepare a resin using DBNPG as the sole glycol. This technique gives a resin solid containing 45 wt % bromine which may be blended with either a nonhalogenated or halogenated resin to give a product having the desired degree of flame-retardant properties (56).

Brominated Styrene. Dibromostyrene [*31780-26-4*] is used commercially as a flame retardant in ABS (57). Tribromostyrene [*61368-34-1*] (TBS) has been proposed as a reactive flame retardant for incorporation either during polymerization or during compounding. In the latter case, the TBS could graft onto the host polymer or homopolymerize to form poly(tribromostyrene) *in situ* (58).

Vinyl Bromide. Vinyl bromide [*593-60-2*] is prepared by the base-promoted dehydrobromination of ethylene dibromide [*106-93-4*]. It is used as a comonomer in the production of acrylic fibers.

Brominated Phenols. Tribromophenol [*75-80-9*] and dibromophenol [*615-58-7*] are both prepared through bromination of phenol. These are not actually used as reactive flame retardants, but rather as starting materials for other flame retardants such as BTBPE [*37853-59-1*] and epoxy oligomers.

Bromine as a Reactive Flame Retardant. Bromine and chlorine are the starting materials for all of the commercial compounds described. Bromine is also used in a somewhat different way to impart flame retardancy. That is, it is used to brominate the resin in interest directly. This is practiced commercially in the case of unsaturated polyesters (59).

Tetrachlorphthalic Anhydride and Tetrachlorphthalic Acid. Tetrachlorphthalic anhydride [*117-08-8*] (TCPA) is manufactured by the ferric chloride catalyzed chlorination of phthalic anhydride. The relatively low chlorine content and the lower flame retardant efficiency of the aromatic chlorides limit use to unsaturated polyester resin formulations that do not require a high degree of flame retardancy.

Chlorendic Acid. Chlorendic acid [*115-28-6*] and its anhydride [*115-27-5*] are widely used flame retardants. Chlorendic acid is synthesized by a Diels-Alder reaction of maleic anhydride and hexachlorocyclopentadiene (see CYCLOPENTADIENE AND DICYCLOPENTADIENE) in toluene followed by hydrolysis of the anhydride using aqueous base (60). The anhydride can be isolated directly from the reaction mixture or can be prepared in a very pure form by dehydration of the acid. The principal use of chlorendic anhydride and chlorendic acid has been in the manufacture of unsaturated polyester resins. Because the esterification rate of chlorendic anhydride is similar to that of phthalic anhydride, it can be used in place of phthalic anhydride in commercial polyester formulations. The double bond in chlorendic acid is not reactive as a cross-linking site; hence, reactive monomers such as maleic anhydride must be included in the polyester backbone to achieve cross-linking.

Table 10. Consumption of Halogenated Flame Retardants[a]

		Volume	
Region	Flame Retardant	1989	1994[b]
Western Europe	brominated	28.0	32.5
	chlorinated	19.6	22.7
Japan	brominated	28.7	38.5
	chlorinated	4.85	5.25
United States	brominated	50.0	65
	chlorinated	16	17
Total	*brominated*	*106.7*	*136*
	chlorinated	*40.45*	*44.95*

[a]Ref. 62.
[b]Estimated.

Table 11. Toxicity Data for Brominated Additive Flame Retardants

Common name	Toxicity			
	Test type	Animal	LD_{50}, mg/kg	Company[a]
ethylenebisdibromonorbornane-dicarboximide	oral	rat	>10,000	EC
bis(2-ethylhexyl)tetrabromo-	oral	rat	>5,000	GL
phthalate	dermal	rabbit	>2,000	GL
tetrabromobisphenol A	oral	rat	>50,000	GL
	dermal	rabbit	>2,000	EC
	inhalation, 2-h	rat	>2,550[b]	EC
tetrabromobisphenol A bis(2,3-	oral	mouse	>20,000	GL
dibromopropyl ether)	dermal	mouse	>20,000	GL
ethylenebistetrabromophthalimide	oral	rat	>5,000	EC
	dermal	rabbit	>2,000	EC
bis(tribromophenoxy)ethane	oral	rat	>10,000	GL
	dermal	rabbit	>2,000	GL
pentabromodiphenyl oxide	oral	rat	6,200	GL
	dermal	rabbit	>2,000	GL
tetrabromodipentaerythritol	oral	rat	>5,000	DS
tetrabromocyclooctane	oral	rat	7,500	EC
hexabromocyclododecane	oral	rat	>10,000	EC
	dermal	rabbit	>10,000	GL
	inhalation, 1-h	rat	>200[c]	EC
dibromoethyldibromocyclohexane	oral	rat	3,220	EC
	dermal	rabbit	>5,000	EC
octabromodiphenyl oxide	oral	rat	>5,000	EC
	inhalation, 1-h	rat	>52.8[c]	EC
	dermal	rabbit	>2,000	GL
tetradecabromodiphenoxybenzene	oral	rat	>5,000	EC
	dermal	rabbit	>5,000	EC
pentabromotoluene	oral	rat	>5,000	DS
decabromodiphenyl oxide	oral	rat	>5,000	GL
	dermal	rabbit	>2,000	GL
	inhalation	rat	>50[c]	GL

[a]Material safety data sheets from Ethyl Corp. (EC) and Great Lakes (GL). Product data sheet from Dead Sea Bromine.
[b]Value is in units of mg/m^3.
[c]Value is in units of mg/L.

Economic Aspects

There are a relatively small number of producers of halogenated flame retardants, especially for brominated flame retardants, where three producers account for greater than 80% of world production. Table 10 gives estimates of the volumes of brominated and chlorinated flame retardants used worldwide. Volumes of flame retardants consumed in Japan have been summarized (61). Prices of halogenated flame retardants vary from less than \$2.00/kg to as high as \$13.00/kg. Cost to the user depends on the level of use of the specific flame retardant and other factors such as the use of stabilizers.

Health and Safety

In general, the acute toxicity of halogenated flame retardants is quite low. Tables 11–14 contain acute toxicity information from various manufacturers' material

Table 12. Toxicity Data for Polymeric and Oligomeric Brominated Additive Flame Retardants

	Toxicology			
Common name	Test type	Animal	LD_{50}, mg/kg	Company[a]
tetrabromobisphenol A carbonate oligomer, phenoxy end capped	oral	rat	>5,000	GL
	dermal	rabbit	>2,000	GL
epoxy oligomers of tetrabromobisphenol A	oral	rat	>2,000	DC
tetrabromobisphenol A carbonate oligomer, tribromophenoxy end capped	oral	rat	>5,000	GL
	dermal	rabbit	>2,000	GL
poly(dibromostyrene)	oral	rat	>5,000	GL
poly(dibromophenylene oxide)	oral	rat	>21,500	GL
	dermal	rabbit	>3,000	GL
poly(pentabromobenzylacrylate)	oral	rat	>5,000	DS

[a]Material safety data sheets from Great Lakes (GL) and Dow Chemical (DC). Product data sheet from Dead Sea Bromine (DS).

Table 13. Toxicity Data for Chlorinated Flame Retardants and Intermediates

	Toxicology			
Common name	Test type	Animal	LD_{50}, mg/kg	Company[a]
chlorendic anhydride	oral	rat	2,336	V
	dermal	rabbit	>10,000 and <20,000	V
	inhalation, 1-h	rat	>203[b]	V
bis(hexachlorocyclopentadieno)-cyclooctane	oral	rat	>25,000	OC
	dermal	rabbit	>8,000	OC
	inhalation, 4-h	rat	>2.25[b]	OC
liquid chlorinated paraffin, % Cl				
51	oral	rat	4,000	D
70	oral	rat	>4,000	D
resinous chlorinated paraffin, 70% Cl	oral	rat	>4,000	D

[a]Material safety data sheets from Dover Chemical (D), Occidental Chemical Corp. (OC), and Velsicol Chemical Co. (V).
[b]Value is in units of mg/L.

Table 14. Toxicity Data for Brominated Reactive Flame Retardants and Intermediates

	Toxicology			
Common name	Test type	Animal	LD_{50}, mg/kg	Company[a]
diester/ether diol of tetrabromophthalic anhydride	oral	rat	>10,000	EC
	dermal	rabbit	>20,000	EC
tetrabromobisphenol A bis(allyl ether)	oral	rat	>5,000	GL
	dermal	rabbit	>2,000	GL
tetrabromobisphenol A bis(2-hydroxyethyl ether)	oral	rat	>5,000	GL
	dermal	rabbit	>2,000	GL
	inhalation	rat	>12.5[b]	GL
tetrabromobisphenol A	[c]			
disodium salt of tetrabromophthalic acid	oral	rat	=2,874	GL
	dermal	rabbit	>2,000	GL
dibromoneopentylglycol	oral	rat	=3,450	EC
	inhalation, 7-h	rat	>2.49[b]	EC
2,4-dibromophenol	oral	mouse	=2,780	DS
tribromophenyl allyl ether	oral	rat	>5,000	DS
	dermal	rabbit	>2,000	DS
tetrabromophthalic anhydride	oral	rat	>10,000	EC
	dermal	rabbit	>10,000	EC
	inhalation	rat	=10.9[b]	EC
2,4,6-tribromophenol	oral	rat	>5,000	DS
tribromoneopentyl alcohol	oral	rat	=2,823	DS
vinyl bromide[d]				EC

[a]Material safety data sheets from Ethyl Corp. (EC) and Great Lakes (GL). Product data sheet from Dead Sea Bromine (DS).
[b]Value is in units of mg/L.
[c]See Table 11.
[d]American Conference of Government Industrial Hygienists (ACGIH) threshold limit value = 5 ppm on an 8-h time-weighted average.

safety data sheets (MSDS) for some of the flame retardants and intermediates listed in the previous tables. The latest MSDS should always be requested from the supplier in order to be assured of having up-to-date information about the toxicity of the products as well as recommendations regarding safe handling.

Continual use of decabromidiphenyl oxide has been placed in question based on the discovery that under certain laboratory conditions brominated dibenzo-*p*-dioxins are generated (63). The condition most often employed in such studies is pyrolysis of milligram-scale samples at 600°C. This temperature is higher than polymer processing conditions and lower than fire temperatures, ie, the conditions are not representative of actual conditions to which flame-retardant polymers are exposed.

The Brominated Flame Retardants Industry Panel (BFRIP) was formed in 1985 within the Flame Retardant Chemicals Association (FRCA) to address such concerns about the use of decabromodiphenyl oxide. Since 1990 the BFRIP has operated as a Chemical Self-Funded Technical Advocacy and Research (CHEMSTAR) panel within the Chemical Manufacturers Association (CMA) (64). As of

1993, members of BFRIP are Akzo, Amerihaas (Dead Sea Bromine Group), Ethyl Corp., and Great Lakes Chemical. Since its formation, BFRIP has presented updates to industry on a regular basis (65,66), and has published a summary of the available toxicity information on four of the largest volume brominated flame retardants (67,68): tetrabromo bisphenol A, pentabromodiphenyl oxide, octabromodiphenyl oxide, and decabromodiphenyl oxide. This information supplements that summarized in Table 11.

In 1987 the U.S. Environmental Protection Agency (EPA) published a test rule under Section 4 of TSCA, 40 CFR Parts 707 and 766. This test rule required industry to develop analytical protocols to determine ultratrace quantities of brominated dibenzo-*p*-dioxins and dibenzofurans in eight brominated flame retardants. The development of the analytical protocols were nontrivial (69–71). The protocols were approved by the EPA in November of 1991. The results of analyses of composite samples have been summarized (72).

Research sponsored by BFRIP regarding the use of brominated flame retardants shows that there is no evidence that the use of decabromodiphenyl oxide leads to any unusual risk. In addition, a study by the National Bureau of Standards (now National Institute of Science and Technology) showed that the use of flame retardants significantly decreased the hazards associated with burning of common materials under realistic fire conditions (73). Work in Japan confirms this finding (74).

BIBLIOGRAPHY

"Halogenated Flame Retardants" in *ECT* 2nd ed., Suppl., pp. 467–488, by V. A. Pattison and R. R. Hindersinn, Hooker Technical Co.; "Halogenated" under "Flame Retardants" in *ECT* 3rd ed., Vol. 10 pp. 373–395, by E. R. Larson, Dow Chemical.

1. J. Brandrup and E. H. Immergut, eds., *Polymer Handbook*, 2nd ed., John Wiley & Sons, Inc., New York, 1975.
2. *Fire Test Standards*, American Society for Testing and Materials (ASTM), Philadelphia, Pa., 1990.
3. *Standard Test Method for Measuring the Minimum Oxygen Concentration to Support Candle-like Combustion of Plastics*, ASTM D2863-87, ASTM, Philadelphia, Pa., 1987.
4. C. P. Fenimore and F. J. Martin, *Modern Plast.* **44**, 144 (1966).
5. *UL 94 Standard for Safety, Tests for Flammability of Plastic Materials for Parts in Devices and Appliances*, Underwriters Laboratories, Inc., Northbrook, Ill., 1991.
6. *Test Method for Heat and Visible Smoke Release Rates for Materials and Products Using an Oxygen Consumption Calorimeter*, ASTM E1354-90, ASTM, Philadelphia, Pa., 1990.
7. V. Babrauskas, *Development of the Cone Calorimeter. A Bench-Scale RHR Apparatus Based on Oxygen Consumption*, NSBIR 82-2611, U.S. Dept. of Commerce, Gaithersburg, Md., 1982.
8. V. Babrauskas and R. D. Peacock, *Proceedings of the Fire Retardant Chemicals Association* (FRCA) Fall Meeting, Lancaster Pa., 1990, pp. 67–80.
9. M. M. Hirschler, *Proceedings of the FRCA* Fall Meeting, Lancaster, Pa., 1991, pp. 167–195.
10. D. M. Indyke and F. A. Pettigrew, in Ref. 9, pp. 109–117.
11. *Proceedings of EUREFIC Seminar 1991*, Interscience Communications Ltd., London, 1991.

12. *Test Method for Heat and Visible Smoke Release Rates for Materials and Products*, ASTM E906-83, ASTM, Philadelphia, Pa., 1983 (updated periodically).
13. A. Tewarson and S. D. Ogden, *Combustion and Flame* **89**, 237–259 (1992).
14. A. Tewarson, "Heat Release and Surface Flame Spread," *IEC TC 89-WG8 Meeting*, British Standards Institution, London, Oct. 1992.
15. *UL 910 Standard for Safety, Test Method for Fire and Smoke Characteristics of Electrical and Optical Fiber Cables used in Air Handling Spaces*, Underwriters Laboratories, Inc. Northbrook, Ill., 1985.
16. *Test Method for Surface Burning Characteristics of Building Materials*, ASTM E84-89a, ASTM, Philadelphia, Pa., 1989.
17. S. K. Bhatnagar, B. S. Varshney, and B. Mohanty, *Fire Mater.* **16**, 141–151 (1992).
18. *Factory Mutual Building Corner Fire Test Procedure*, Factory Mutual Research, Norwood, Mass. (updated periodically).
19. J. S. Newman, *Analysis of FMRC Building Corner Fire Test*, technical report J.I.005E5.RC, Factory Mutual Research Corp., Norwood, Mass., 1989.
20. J. S. Newman, "Smoke Characterization of Rigid Polyurethane/Isocyanurate Foams," *34th SPI Annual Polyurethane Technical/Marketing Conference*, New Orleans, La., Oct. 21–24, 1992, pp. 307–311.
21. A. Tewarson, Factory Mutual Research Corp., Norwood, Mass., personal communication, 1993.
22. S. Nurbakhsh, G. H. Damant, and J. F. Mikami, *Proceedings of the FRCA Fall Meeting*, Lancaster, Pa., 1989, pp. 91–106.
23. *Method of Test for the Ignitability of Upholstered Composites for Seating*, British Standard 5852: part 2, British Standards Institute, London, 1983.
24. M. J. Drews, C. W. Jarvis, and G. C. Lickfield, *Ternary Reactions Among Polymer Substrate–Organohalogen–Antimony Oxides Under Pyrolytic, Oxidative and Flaming Conditions*, NIST-GCR-89-558, U.S. Department of Commerce, Gaithersburg, Md., 1989.
25. Product literature, Dover Chemical, Dover, Ohio, 1992.
26. S. E. Calewarts, G. A. Bonner, and F. A. Pettigrew, *Proceedings of the FRCA Spring 1990 Meeting*, Lancaster, Pa., pp. 227–236.
27. *SAYTEX 102E Flame Retardant Product Bulletin*, Ethyl Corp., Baton Rouge, La., 1992.
28. M. J. Drews, *Proceedings of the Spring 1992 FRCA Meeting*, Lancaster, Pa., 1992, pp. 55–57 and 249–258.
29. S. K. Brauman and A. S. Brolly, *J. Fire Retardant Chem.* **3**, 66 (1977); S. K. Brauman, *J. Fire Retardant Chem.* **3**, 117,138 (1976).
30. J. W. Hastie and C. L. McBee, in R. G. Gann, ed., *Halogenated Fire Suppressants*, ACS Symposium Series 16, American Chemical Society, Washington, D.C., 1975, p. 118.
31. R. V. Petrella, in M. Lewin, S. M. Atlas, and E. M. Pearce, eds., *Flame Retardant Polymeric Materials*, Plenum Press, New York, 1975.
32. U.S. Pat. 4,717,776 (Jan. 5, 1988), B. G. McKinnie and D. R. Brackenridge (to Ethyl Corp.).
33. U.S. Pat. 4,778,933 (Oct. 18, 1988), B. G. McKinnie and M. S. Ao (to Ethyl Corp.).
34. U.S. Pat. 5,041,687 (Aug. 20, 1992), B. G. McKinnie and D. R. Brackenridge (to Ethyl Corp.).
35. U.S. Pat. 4,740,629 (Apr. 26, 1992), D. R. Brackenridge and B. G. McKinnie (to Ethyl Corp.).
36. Brit. Pat. 1,436,657 (May 19, 1976), L. J. Belf (to I. S. C. Chemicals Ltd.).
37. U.S. Pat. 5,0177,728 (May 21, 1991), B. G. McKinnie and D. A. Wood (to Ethyl Corp.).
38. Ger. Pat. 3,935,224 A1 (Apr. 25, 1991), E. Walter (to Degussa AG).

39. *Standard Test Method for Deflection Temperature of Plastics Under Flexural Load,* ASTM D648-82, ASTM, Philadelphia, Pa., 1988.
40. U.S. Pat. 5,043,492 (Aug. 27, 1991), G. H. Ransford (to Ethyl Corp.).
41. B. M. Valange and co-workers, *Proceedings of Flame Retardants '90*, Elsevier, London, 1990, pp. 67–77.
42. F. A. Pettigrew and J. S. Reed, *Proceedings of the BCC Conference on Flame Retardancy*, Business Communications Company, Inc., Norwalk, Conn., 1992.
43. F. A. Pettigrew, S. D. Landry, and J. S. Reed, *Proceedings of Flame Retardants '92*, Elsevier, London, 1992, pp. 156–167.
44. U.S. Pat. 4,053,528 (Oct. 11, 1977), D. H. Thorpe (to Hooker Chemicals and Plastics Corp.).
45. C. S. Ilardo and R. L. Markezich, *Proceedings of the 15th International Conference on Fire Safety*, Product Safety Corp., Sunnyvale, Calif., 1990.
46. R. L. Markezich, C. S. Ilardo, and R. F. Mundhenke, in Ref. 41, pp. 88–101.
47. J. C. Gill, *Plastics Compounding*, 77–81 (Sept./Oct. 1989).
48. U.S. Pat. 5,112,896 (May 12, 1992), J. L. Dever and J. C. Gill (to Ferro Corp.).
49. J. Gill, *Proceedings of the FRCA Spring 1989 Conference*, Lancaster, Pa., 1989, pp. 17–32.
50. U.S. Pat. 3,474,067 (Oct. 21, 1969), H. E. Praetzel, B. Frankenforst, and H. Jenkner (to Chemische Fabrik Kalk).
51. U.S. Pat. 3,382,254 (May 7, 1968), H. Jenkner, O. Rabe, and R. Strang (to Chemische Fabrik Kalk).
52. R. C. Nametz and R. S. Nulph, *Proceedings of the 20th Anniversary Technical Conference of the Society of Plastic. Industrial, Reinforcement, Plastics Division*, Section 11-C, Chicago, Ill., 1965.
53. E. F. Feske, *Proceedings of UTECH '92*, Crain Communications Ltd., London, 1992.
54. J. G. Uhlman, *Proceedings of the SPI 32nd Annual Technical Marketing Conference, San Francisco, Calif.,* Society of the Plastics Industry, New York, 1989, pp. 352–358.
55. E. F. Feske and co-workers, *Proceedings of the SPI 33rd Annual Technical Marketing Conference,* Society of the Plastics Industry, New York, 1990, pp. 107–113.
56. U.S. Pat. 3,507,933 (Apr. 21, 1970), E. R. Larsen, B. R. Andrejewski, and D. L. Nelson (to The Dow Chemical Company).
57. World Pat. 8,603,508 A1 (June 19, 1986), D. F. Aycock, W. R. Haaf, and G. F. Lee, Jr. (to General Electric Co.).
58. M. Peled and co-workers, *Proceedings of the BCC Flammability Conference*, Business Communications Corp., Inc., Norwalk, Conn., 1992.
59. U.S. Pat. 3,536,782 (Oct. 27, 1970), U. Toggweiler and F. F. Roselli (to Diamond Shamrock).
60. E. Prill, *J. Am. Chem. Soc.* **69**, 62 (1947).
61. H. Baba and T. Taniguchi, *Proceedings of the FRCA Spring 1993 Meeting*, Lancaster, Pa., 1993.
62. *Flame Retardants, Specialty Chemicals Update Program*, SRI International, Menlo Park, Calif., 1990.
63. J. Troitzsch, *Proceedings of the FRCA 1987 International Conference,* Lancaster, Pa., 1987, pp. 141–150, and references cited therein.
64. *Brominated Flame Retardants Industry Panel*, Chemical Manufacturers Association, Washington, D.C.
65. D. L. McAllister, in Ref. 22, pp. 173–182; D. R. Lynam, in Ref. 9, pp. 23–32.
66. D. L. McAllister, *Proceedings of the FRCA Fall Meeting*, Lancaster, Pa., 1992, pp. 29–40.

67. C. J. Mazac, *Proceedings of the Second Annual BCC Flammability Conference*, Business Communications Co., Inc., Norwalk, Conn., 1991.
68. M. L. Hardy, in Ref. 58.
69. Y. Tondeur and co-workers, *Chemosphere* **20**, 1269–1270 (1990).
70. D. L. McAllister and co-workers, *Chemosphere* **20**, 1537–1541 (1990).
71. Y. Tondeur and co-workers, *Chemospere*, in press.
72. C. J. Mazac, in Ref. 58.
73. *Reduction of Fire Hazard Using Fire Retardant Chemicals*, Fire Retardant Chemicals Association, Lancaster, Pa., 1989.
74. T. Morikawa, in Ref. 61, pp. 69–87.

General References

J. Troitzsch, *International Plastics Flammability Handbook*, Hanser Publishers, Munich, Germany, 1990.

M. Lewin, S. M. Atlas, and E. M. Pearce, eds., *Flame Retardant Polymeric Materials*, Plenum Press, New York, 1975.

D. Price, B. Iddon, and B. J. Wakefield, eds., *Bromine Compounds Chemistry and Applications*, Elsevier, Amsterdam, the Netherlands, 1988.

J. A. Barnard and J. N. Bradley, *Flame and Combustion*, Chapman and Hall, London, 1985.

Handbook of Fire Retardant Coatings and Fire Testing Services, Technomic, Lancaster, Pa., 1990.

Directory of Testing Laboratories, ASTM, Philadelphia, Pa., 1992.

R. Gachter and H. Muller, eds. *Plastics Additives Handbook*, Hanser Publishers, Munich, Germany, 1987; H. Jenkner, "Flame Retardants for Thermoplastics," Chapt. 11 in Gachter and Muller, pp. 535–564.

G. L. Nelson, *Fire and Polymers Hazards Identification and Prevention*, ACS Symposium Series 425, American Chemical Society, Washington, D.C., 1990.

C. J. Hilado, *Flammability Handbook for Plastics*, Technomic, Lancaster, Pa., 1990.

ALEX PETTIGREW
Ethyl Technical Center

PHOSPHORUS FLAME RETARDANTS

One of the principal classes of flame retardants used in plastics and textiles is that of phosphorus, phosphorus–nitrogen, and phosphorus–halogen compounds (see also FLAME RETARDANTS FOR TEXTILES). Detailed reviews of phosphorus flame retardants have been published (1–6) (see also PHOSPHORUS COMPOUNDS).

Mechanisms of Action

Condensed-Phase Mechanisms. The mode of action of phosphorus-based flame retardants in cellulosic systems is probably best understood. Cellulose (qv) decomposes by a noncatalyzed route to tarry depolymerization products, notably levoglucosan, which then decomposes to volatile combustible fragments such as alcohols, aldehydes (qv), ketones (qv), and hydrocarbons (qv) (7–9). However, when catalyzed by acids, the decomposition of cellulose proceeds primarily as an endothermic dehydration of the carbohydrate to water vapor and char. Phosphoric acid is particularly efficaceous in this catalytic role because of its low volatility (see PHOSPHORIC ACID AND THE PHOSPHATES). Also, when strongly heated, phosphoric acid yields polyphosphoric acid which is even more effective in catalyzing the cellulose dehydration reaction. The flame-retardant action is believed to proceed by way of initial phosphorylation of the cellulose. Certain nitrogen compounds such as melamines, guanidines, and ureas appear to catalyze the cellulose phosphorylation step and are found to enhance or synergize the flame-retardant action of phosphorus on cellulose (10–13). The nonvolatile phosphorus acids are also able to coat the char, rendering the char less permeable and protecting it from further oxidation. The retention of phosphorus in the char may be aided by the nitrogen synergists.

In poly(ethylene terephthalate) (14–16) and poly(methyl methacrylate) (17–19), the mechanism of action of phosphorus flame retardants is at least partly attributable to a decrease in the amount of combustible volatiles and a corresponding increase in nonvolatile residue (char). In poly(methyl methacrylate), the phosphorus flame retardant appears to cause an initial cross-linking through anhydride linkages (19).

The amount and physical character of the char from rigid urethane foams is found to be affected by the retardant (20–23) (see FOAMS; URETHANE POLYMERS). The presence of a phosphorus-containing flame retardant causes a rigid urethane foam to form a more coherent char, possibly serving as a physical barrier to the combustion process. There is evidence that a substantial fraction of the phosphorus may be retained in the char. Chars from phenolic resins (qv) were shown to be much better barriers to pyrolysate vapors and air when ammonium phosphate was present in the original resin (24). This barrier action may at least partly explain the inhibition of glowing combustion of char by phosphorus compounds.

In polymers such as polystyrene that do not readily undergo charring, phosphorus-based flame retardants tend to be less effective, and such polymers are often flame retarded by antimony–halogen combinations (see STYRENE POLYMERS). However, even in such noncharring polymers, phosphorus additives exhibit some activity that suggests at least one other mode of action. Phosphorus compounds may produce a barrier layer of polyphosphoric acid on the burning polymer (4,5). Phosphorus-based flame retardants are more effective in styrenic polymers blended with a char-forming polymer such as a polyphenylene oxide or polycarbonate.

Phosphorus-containing additives can act in some cases by catalyzing thermal breakdown of the polymer melt, reducing viscosity and favoring the flow or drip of molten polymer from the combustion zone (25). On the other hand, red phosphorus [*7723-14-0*] has been shown to retard the nonoxidative pyrolysis of polyethylene (a radical scission). For that reason, the scavenging of radicals in the condensed phase has been proposed as one of several modes of action of red phosphorus (26).

Several commercial polyester fabrics are flame retarded using low levels of phosphorus additives that cause them to melt and drip more readily than fabrics without the flame retardant. This mechanism can be completely defeated by the presence of nonthermoplastic component such as infusible fibers, pigments, or by silicone oils which can form pyrolysis products capable of impeding melt flow (27,28).

Alkyl diphenyl phosphate plasticizers can exert flame-retardant action in vinyl plastics by a condensed-phase mechanism, which is probably some sort of phosphorus acid coating on the char. Triaryl phosphates appear to have a vapor-phase action (29).

Vapor-Phase Mechanisms. Phosphorus flame retardants can also exert vapor-phase flame-retardant action. Trimethyl phosphate [*512-56-1*], $C_3H_9O_4P$, retards the velocity of a methane–oxygen flame with about the same molar efficiency as antimony trichloride (30,31). Both physical and chemical vapor-phase mechanisms have been proposed for the flame-retardant action of certain phosphorus compounds. Physical (endothermic) modes of action have been shown to be of dominant importance in the flame-retardant action of a wide range of non-phosphorus-containing volatile compounds (32).

Triphenylphosphine oxide [*791-28-6*], $C_{18}H_{15}OP$, and triphenyl phosphate [*115-86-6*], $C_{18}H_{15}O_4P$, as model phosphorus flame retardants were shown by mass spectroscopy to break down in a flame to give small molecular species such as PO, HPO_2, and P_2 (33–35). The rate-controlling hydrogen atom concentration in the flame was shown spectroscopically to be reduced when these phosphorus species were present, indicating the existence of a vapor-phase mechanism.

Physical or chemical vapor-phase mechanisms may be reasonably hypothesized in cases where a phosphorus flame retardant is found to be effective in a noncharring polymer, and especially where the flame retardant or phosphorus-containing breakdown products are capable of being vaporized at the temperature of the pyrolyzing surface. In the engineering of thermoplastic Noryl (General Electric), which consists of a blend of a charrable poly(phenylene oxide) and a poorly charrable polystyrene, experimental evidence indicates that effective flame retardants such as triphenyl phosphate act in the vapor phase to suppress the flammability of the polystyrene pyrolysis products (36).

The question as to whether a flame retardant operates mainly by a condensed-phase mechanism or mainly by a vapor-phase mechanism is especially complicated in the case of the haloalkyl phosphorus esters. A number of these compounds can volatilize undecomposed or undergo some thermal degradation to release volatile halogenated hydrocarbons (37). The intact compounds or these halogenated hydrocarbons are plausible flame inhibitors. At the same time, their phosphorus content may remain at least in part as relatively nonvolatile phosphorus acids which are plausible condensed-phase flame retardants (38). There is no evidence for the occasionally postulated formation of phosphorus halides. Some evidence has been presented that the endothermic vaporization and heat capacity of the intact chloroalkyl phosphates may be a main part of their action (39,40).

Interaction with Other Flame Retardants. Some claims have been made for a phosphorus–halogen synergism, ie, activity greater than that predicted by some additivity model. Unlike the firmly established antimony–halogen synergism, however, phosphorus–halogen interactions are often merely additive and in some cases slightly less than additive (12,13). Some reports of phosphorus–halogen synergism can be shown to be artifactual results of nonlinear response–concentration relationships. Nevertheless, combinations of phosphorus and halogens are often quite useful, and there are data supporting synergism with specific compounds (41).

Antagonism between antimony oxide and phosphorus flame retardants has been reported in several polymer systems, and has been explained on the basis of phosphorus interfering with the formation or volatilization of antimony halides, perhaps by forming antimony phosphate (12,13). This phenomenon is also not universal, and depends on the relative amounts of antimony and phosphorus. Some useful commercial poly(vinyl chloride) (PVC) formulations have been described for antimony oxide and triaryl phosphates (42). Combinations of antimony oxide, halogen compounds, and phosphates have also been found useful in commercial flexible urethane foams (43).

Commercial Phosphorus-Based Flame Retardants

Many thousands of phosphorus compounds have been described as having flame-retardant utility (44). The compounds demonstrating commercial utility are much more limited in number.

Inorganic Phosphorus Compounds. *Red Phosphorus.* This allotropic form of phosphorus is relatively nontoxic and, unlike white phosphorus, is not spontaneously flammable. Red phosphorus is, however, easily ignited. It is a polymeric form of phosphorus having thermal stability up to ca 450°C. In finely divided form it has been found to be a powerful flame-retardant additive (26,45–47). In Europe, it has found commercial use in molded nylon electrical parts in a coated and stabilized form. Handling hazards and color have deterred broad usage. The development of a series of masterbatches by Albright & Wilson should facilitate further use.

Ammonium Phosphates. These salts were first recommended for flame retardancy of theater curtains by Gay-Lussac in 1821. Monoammonium phosphate [*7722-76-1*], $NH_4H_2PO_4$, and diammonium phosphate [*7783-28-0*], $(NH_4)_2HPO_4$, or mixtures of the two, which are more water-soluble and nearly neutral, are used in large amounts for nondurable flame-retarding of paper (qv), textiles (qv), disposable nonwoven cellulosic fabrics, and wood (qv) products (7–9) (see FLAME RETARDANTS FOR TEXTILES). The advantage is high efficiency and low cost. Ammonium phosphate finishes are resistant to dry-cleaning solvents but not to laundering or even to leaching by water. One general advantage of ammonium phosphates and phosphorus compounds as flame retardants, especially in comparison to borax which is also used for nondurable cellulosic flame retardancy, is the effectiveness in preventing afterglow.

Formulations of ammonium phosphates and ammonium bromide are sold for use on cellulosic–synthetic fiber blends. Other ammonium phosphate formulations contain wetting and softening agents. A large-volume, ca 9000 t/yr in 1991, use in the United States (48) for ammonium phosphate is in forest fire control, usually by aerial application (see also AMMONIUM COMPOUNDS).

Insoluble Ammonium Polyphosphate. When ammonium phosphates are heated in the presence of urea (qv), or by themselves under ammonia pressure, relatively water-insoluble ammonium polyphosphate [*68333-79-9*] is produced (49). There are several crystal forms and the commercial products, available from

Monsanto, Albright & Wilson, or Hoechst-Celanese, differ in molecular weight, particle size, solubility, and surface coating. Insoluble ammonium polyphosphate consists of long chains of repeating $OP(O)(ONH_4)$ units.

These finely divided solids are principal ingredients of intumescent paint (qv) and mastics (50). In such formulations, ammonium polyphosphate is considered to function as a catalyst. Thus when the intumescent coating is exposed to a high temperature, the ammonium polyphosphate yields a phosphorus acid which then interacts with an organic component such as dipentaerythritol to form a carbonaceous char. The chemistry has been described in detail (51). A blowing, ie, gas-generating agent such as melamine or chlorowax is also present to impart a foamed character to the char, thus forming a fire-resistant insulating barrier to protect the substrate. In addition, the intumescent formulations typically contain resinous binders, pigments, and other fillers. Mastics are related but generally more viscous formulations, intended to be applied in thick layers to girders, trusses, and decking; these generally contain mineral fibers to increase coherence.

A series of compounded flame retardants, based on finely divided insoluble ammonium polyphosphate together with char-forming nitrogenous resins, has been developed for thermoplastics (52–58). These compounds are particularly useful as intumescent flame-retardant additives for polyolefins, ethylene–vinyl acetate, and urethane elastomers (qv). The char-forming resin can be, for example, an ethyleneurea–formaldehyde condensation polymer, a hydroxyethyl isocyanurate, or a piperazine–triazine resin.

Phosphoric Acid-Based Systems for Cellulosics. Semidurable flame-retardant treatments for cotton (qv) or wood (qv) can be attained by phosphorylation of cellulose, preferably in the presence of a nitrogenous compound. Commercial leach-resistant flame-retardant treatments for wood have been developed based on a reaction product of phosphoric acid with urea–formaldehyde and dicyandiamide resins (59,60).

Additive Organic Phosphorus Flame Retardants. *Melamine and Other Amine Phosphates.* Three melamine phosphate commercial products have been reported (60) including: melamine orthophosphate [*20208-95-1*], $C_3H_6N_6{\cdot}H_3O_4P$, [*41583-09-9*], $C_3H_6N_6{\cdot}xH_3O_4P$; dimelamine orthophosphate; and melamine pyrophosphate [*15541-60-3*], $2C_3H_6N_6{\cdot}H_4O_7P_2$ (60). The pyrophosphate is reported to be only soluble to the extent of 0.09 g/100 mL water, whereas melamine orthophosphate is soluble to 0.35 g/mL. The pyrophosphate is the most thermally stable. Melamine orthophosphate is converted to the pyrophosphate with loss of water on heating. All three are available as finely divided solids. All are used commercially in flame-retardant coatings (qv) and from patents also appear to have utility in a wide variety of thermoplastics and thermosets. A detailed study of the thermal decomposition of the these compounds has been published (61).

A newer self-intumescent phosphoric acid salt has been introduced by Albright & Wilson as Amgard EDAP, mainly as an additive for polyolefins. It is a finely divided solid, mp 250°C, having a reported phosphorus content of 63 wt % as H_3PO_4. It appears to be the ethylenediamine salt of phosphoric acid (1:1). Unlike ammonium polyphosphate, it does not require a char-forming synergist (62).

Trialkyl Phosphates. Triethyl phosphate [*78-40-0*], $C_6H_{15}O_4P$, is a colorless liquid boiling at 209–218°C containing 17 wt % phosphorus. It may be manufactured from diethyl ether and phosphorus pentoxide via a metaphosphate intermediate (63,64). Triethyl phosphate has been used commercially as an additive for polyester laminates and in cellulosics. In polyester resins, it functions as a viscosity depressant as well as a flame retardant. The viscosity depressant effect of triethyl phosphate in polyester resins permits high loadings of alumina trihydrate, a fire-retardant smoke-suppressant filler (65,66).

Trioctyl Phosphate. Trioctyl phosphate [*1806-54-8*], $C_{24}H_{51}O_4P$, has been employed as a specialty flame-retardant plasticizer for vinyl compositions where low temperature flexibility is critical, eg, in military tarpaulins. It can be included in blends along with general-purpose plasticizers (qv) such as phthalate esters to improve low temperature flexibility.

Dimethyl Methylphosphonate. Dimethyl methylphosphonate [*756-79-6*] (DMMP), $C_3H_9O_3P$, a water-soluble liquid, bp 185°C, is made by the Arbuzov rearrangement of trimethyl phosphite. DMMP contains 25% phosphorus, which is near the maximum possible for a phosphorus ester, and therefore on a weight basis is highly efficient as a flame retardant. Applications include use as a viscosity depressant and flame retardant in alumina trihydrate (ATH) filled polyester resins (67) such as used for bathtubs and shower stalls. It also contributes a sizeable increase in flame retardancy in halogenated polyesters (68). Some applications have been found in rigid polyurethane foams and as an intermediate for making other flame retardants such as Fyrol 76, Fyrol 51, and Antiblaze 19. Blends of DMMP with triaryl phosphates are sold as highly flame-retardant plasticizers for synthetic rubber and cellulosics.

Diethyl Ethylphosphonate. A liquid compound introduced for applications similar to those of DMMP is diethyl ethylphosphonate [*78-38-6*], $C_6H_{15}O_3P$. This material is claimed to be less susceptible to undesirable interactions with haloaliphatic components, such as blowing agents, or with amine catalysts.

Halogenated Alkyl Phosphates and Phosphonates. In this important class of additives, the halogen contributes somewhat to flame retardancy although this contribution is offset by the lower phosphorus content. The halogens reduce vapor pressure and water solubility, thus aiding retention of these additives. Efficient manufacturing processes lead to favorable economics.

2-Chloroethanol Phosphate (3:1). Tris(2-chloroethyl) phosphate [*115-96-8*], $C_6H_{12}Cl_3O_4P$ (2-chloroethanol phosphate (3:1)), is a low viscosity liquid product that has found widespread usage because of low cost, low odor, high percent phosphorus, and compatibility with essentially all polymers containing polar groups. Akzo's Fyrol CEF contains 10.8% phosphorus and 36.7% chlorine, and is made from a three-to-one mole ratio of ethylene oxide (qv) and phosphorus oxychloride (69). This phosphate is used in rigid polyurethane and polyisocyanurate foams, carpet backing, flame-laminated and rebonded flexible foam, flame-retardant coatings, most classes of thermosets, adhesives (qv), cast acrylic sheet, and wood–resin composites such as particle board. It is used with melamine in flexible urethane foam cushions and institutional mattresses.

1-Chloro-2-Propanol Phosphate (3:1). Tris(1-chloro-2-propyl) phosphate [*13674-84-5*], $C_9H_{18}Cl_3O_4P$, is a liquid containing 33% chlorine and 9.5% phosphorus. It is produced by reaction of propylene oxide (qv) and phosphorus oxychloride, and most of the alkyl groups are the secondary (isopropyl) isomer. Because of the branchy structure, this phosphate is much lower in reactivity to water and bases than the 2-chloroethyl homologue (70). It is sold as Akzo's Fyrol

PCF or Albright & Wilson's Antiblaze 80 and is a preferred additive for rigid urethane foams where good storage stability in the isocyanate or the polyol–catalyst mixture is required. It is used in isocyanurate foam to reduce friability and brittleness, and is also used in flexible urethane foams in combination with melamine.

1,3-Dichloro-2-Propanol Phosphate (3:1). Tris(1,3-dichloro-2-propyl) phosphate [*13674-87-8*], $C_9H_{15}Cl_6O_4P$, sold by Akzo as Fyrol FR-2 and by Albright & Wilson as Antiblaze 195, contains 49% chlorine and 7.2% phosphorus. It is made from epichlorohydrin and phosphorus oxychloride. The principal structure has 1,3-dichloro-2-propyl groups.

$$O{=}P{-}\!\left(OCH(CH_2Cl)_2\right)_3$$

There is also a small quantity of the compound having a single 2,3-dichloropropyl group and two 1,3-dichloro-2-propyl groups. There are many erroneous literature references where the product is named as if it were the tris(2,3-dichloropropyl) isomer, which has been made by addition of chlorine to triallyl phosphate, an impractical route. Compared to the foregoing chloroalkyl phosphates, this product has a greatly reduced volatility, much lower water solubility, and high stability toward the amine catalysts used in foam manufacture. It is a leading additive for flexible urethane foams (71) and can be added to the isocyanate or the polyol–catalyst mixture. This phosphate shows little tendency to scorch, ie, to cause discoloration and degradation, even in high exotherm flexible foam formulations. Flexible foam formulations containing this and other haloalkyl phosphates can be further stabilized against scorch by appropriate antioxidants (qv) and acid acceptors (72,73). Combinations with melamine are used in cushioning formulations (74). This halogenated phosphate is also useful as a flame retardant in styrene–butadiene and acrylic latices for textile backcoating and binding of nonwovens.

Bis(2-Chloroethyl) 2-Chloroethylphosphonate. The commercial product, Albright & Wilson's Antiblaze 78, is a mixture having various related higher boiling diphosphonates. This product is made by the Arbuzov rearrangement of tris(2-chloroethyl) phosphite [*140-08-9*], $C_6H_{12}Cl_{13}O_3P$, itself made by the reaction of ethylene oxide and phosphorus trichloride. Although bis(2-chloroethyl)-2-chloroethyl phosphonate [*6294-34-4*], $C_6H_{12}Cl_3O_3P$, is not as stable as the corresponding phosphate, it is useful as a flame-retardant additive in rigid urethane foams (75), rebonded foams, adhesives, and coatings. It is also an intermediate which upon dehydrochlorination by the action of bases produces bis(2-chloroethyl) vinylphosphonate, a useful flame-retardant monomer (76).

Diphosphates. Three 2-chloroethyl diphosphates have been sold commercially. These have low volatility and good-to-fair thermal stability, and are thus useful in those open cell (flexible) foams which have requirements for improved resistance to dry and humid aging.

The simplest was Olin's Thermolin 101, tetrakis(2-chloroethyl) ethylenediphosphate [*33125-86-9*], $C_{10}H_{20}Cl_4O_8P_2$ (77), used extensively in flexible foams

(78,79). This compound has been discontinued, reportedly because of by-product problems.

The commercially available tetrakis(2-chloroethyl) ethyleneoxyethylenediphosphate [*53461-82-8*], $C_{12}H_{24}Cl_4O_9P_2$, has the following structure:

$$(ClCH_2CH_2O)_2P(=O){-}OCH_2CH_2OCH_2CH_2O{-}P(=O)(OCH_2CH_2Cl)_2$$

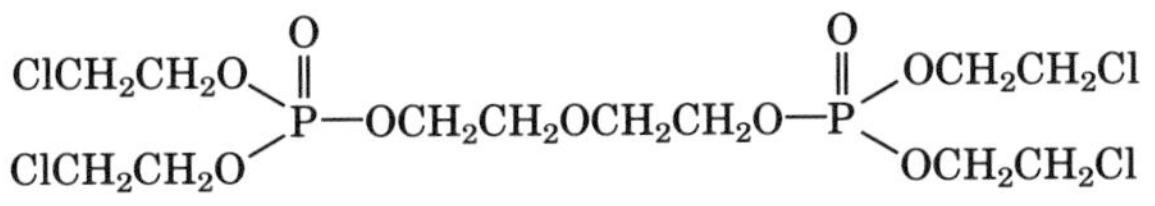

This liquid contains 27% chlorine and 12% phosphorus. It is made from ethylene oxide, diethylene glycol, and phosphorus oxychloride (80). It is available in the United States and Japan from Daihachi.

A third member of this family was introduced originally as Monsanto's Phosgard 2XC20 [*38051-10-4*], $C_{13}H_{24}Cl_6O_8P_2$, but is now available as Albright & Wilson's Antiblaze 100. The synthesis (81) involves reaction of pentaerythritol and phosphorus trichloride to produce a spirobis(chlorophosphite), then chlorination with ring opening, followed by treatment of the resultant bis(phosphorodichloridate) with ethylene oxide. The compound has the following structure:

$$(ClCH_2CH_2O)_2P(=O){-}OCH_2C(CH_2Cl)_2CH_2O{-}P(=O)(OCH_2CH_2Cl)_2$$

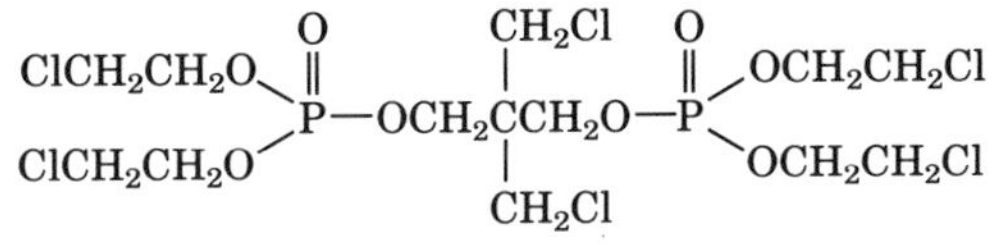

Because of the bulky *neo* structure in the middle of the molecule, this compound has enhanced hydrolytic stability in addition to low volatility. It is useful in many types of flexible foam, as well as in adhesives and epoxy- or phenolic-based laminates.

Oligomeric 2-Chloroethyl Phosphate. Akzo's Fyrol 99 [*109640-81-5*], is produced either by self-condensation of tris(2-chloroethyl) phosphate (82) or by insertion of phosphorus pentoxide into this phosphate (83) followed by ethoxylation. It is low in volatility and useful in resin-impregnated air filters, in flexible urethane foam, rebonded foam, and structural foam.

2-Chloroethyl 2-Bromoethyl 3-Bromoneopentyl Phosphate. This liquid product is made by Great Lakes Chemical Co. as Firemaster 836 [*98923-48-9*], $C_9H_{18}Br_2ClO_4P$, probably by the following reactions (84):

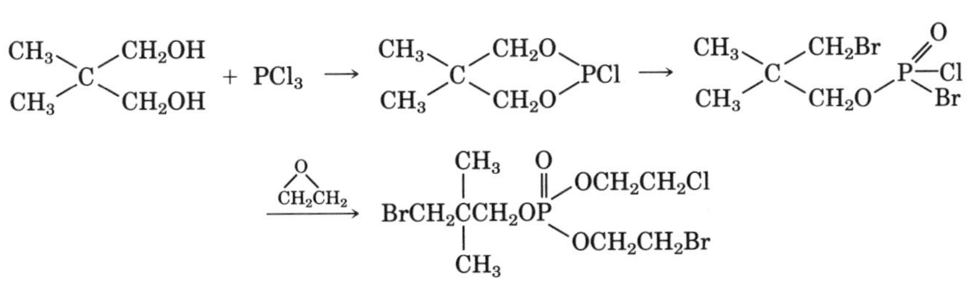

$$(CH_3)_2C(CH_2OH)_2 + PCl_3 \longrightarrow (CH_3)_2C(CH_2O)_2PCl \longrightarrow (CH_3)_2C(CH_2Br)CH_2O{-}P(=O)(Cl)Br$$

$$\xrightarrow{\text{ethylene oxide } (CH_2CH_2O)} BrCH_2C(CH_3)_2CH_2OP(=O)(OCH_2CH_2Cl)(OCH_2CH_2Br)$$

The product is a liquid recommended for flame retarding flexible urethane foams in furniture or automotive seating. It also appears to be useful in polystyrene foam, textile backcoating, and polyester resins.

Oligomeric Cyclic Phosphonates. Albright & Wilson's Antiblaze 19 and 1045 are mixtures of the material shown where $x = 1$ has the CAS Registry Number [*41203-81-0*], $C_9H_{20}O_6P_2$, and $x = 0$, [*42595-49-9*], $C_{15}H_{31}O_9P_3$. The chemistry of manufacture appears to be (85,86):

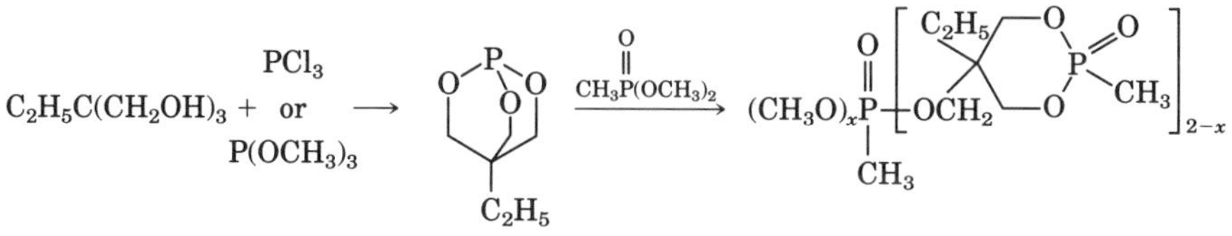

$$C_2H_5C(CH_2OH)_3 + \begin{matrix} PCl_3 \\ \text{or} \\ P(OCH_3)_3 \end{matrix} \longrightarrow \text{(bicyclic phosphite, } C_2H_5\text{)} \xrightarrow{CH_3P(O)(OCH_3)_2} (CH_3O)_xP(O)(CH_3)\left[OCH_2C(C_2H_5)(CH_2O)_2P(O)CH_3\right]_{2-x}$$

Antiblaze 1045 contains a larger amount of [*42595-49-9*]. Both materials are water-soluble thermally stable low volatility liquids having about 20% phosphorus content and no halogen.

The bicyclic phosphite intermediate is highly neurotoxic, but, after the ring-opening step, the product is low in toxicity. Antiblaze 19 and 1045 are used as flame-retardant finishes for polyester fabric. After the phosphonate is applied from an aqueous solution, the fabric is heated to swell and soften the fibers, allowing the phosphonate to be absorbed and strongly held. The lower molecular weight Antiblaze 19 is absorbed better but produces more smoke, ie, vapor, during this thermal process than Antiblaze 1045. These products are also useful flame retardants in polyester resins, polyurethanes, polycarbonates, nylon-6, and textile backcoating.

Pentaerythritol Phosphates. These products take advantage of the excellent char-forming ability of the pentaerythritol structure. The bis-melaminium salt [*70776-17-9*] of the bis-acid phosphate of pentaerythritol was developed as a flame retardant at Borg Warner (87) and has been marketed by Great Lakes Chemical as a component of Charguard 329. This is a high melting solid which acts as an intumescent flame-retardant additive for polyolefins. Synergistic combinations with ammonium polyphosphate have been developed by BF Goodrich for urethane elastomers (88).

A bicyclic pentaerythritol phosphate, CN-1197, has more recently been introduced by Great Lakes Chemical for use in thermosets, preferably in combination with melamine or ammonium polyphosphate (89). It is a high melting solid believed to have the following structure [*5301-78-0*] (87):

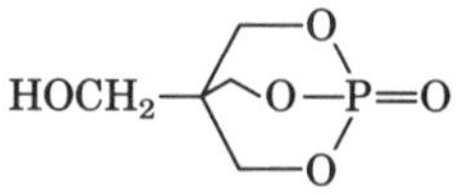

$$HOCH_2-C(CH_2O)_3P{=}O$$

Related esters of this alcohol are disclosed by Akzo as useful flame retardants for polypropylene, particularly in combination with ammonium polyphosphate (90).

Cyclic Neopentyl Thiophosphoric Anhydride. This solid additive, Sandoflam 5060, has been commercialized in Europe by Sandoz for use in viscose rayon (91,92). It has the following structure [*4090-51-1*]:

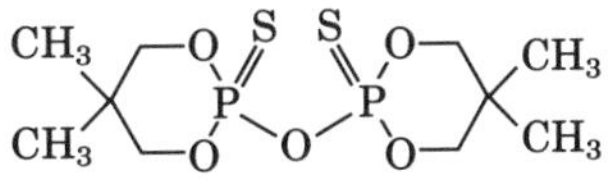

Despite the anhydride structure, it is remarkably stable, surviving addition to the highly alkaline viscose, the acidic coagulating bath, and also resisting multiple laundering of the rayon fabric. The unusual stability may be attributed to the sulfur atoms, which enhance hydrophobicity, and to the sterically hindering neopentyl groups that retard hydrolysis.

Aryl Phosphates. Aryl phosphates were introduced into commercial use early in the twentieth century for flammable plastics such as cellulose nitrate and later for cellulose acetate. Cellulosics are a significant area of use but are exceeded now by plasticized vinyls (93–95). Principal applications are in wire and cable insulation, connectors, automotive interiors, vinyl moisture barriers, plastic greenhouses (Japan), furniture upholstery, conveyer belts (especially in mining), and vinyl foams.

Triaryl phosphates are also used on a large scale as flame-retardant hydraulic fluids (qv), lubricants, and lubricant additives (see LUBRICATION AND LUBRICANTS). Smaller amounts are used as nonflammable dispersing media for peroxide catalysts.

In vinyls, the aryl phosphates are frequently used in combinations with phthalate plasticizers. The proportion of the more expensive phosphate is usually chosen so as to permit the product to reliably pass the flammability specifications. In plasticized vinyls used in automotive interiors, these phosphates are used to pass the Federal Motor Vehicle Safety Standard 302.

Triaryl phosphates are produced from the corresponding phenols (usually mixtures) by reaction with phosphorus oxychloride, usually in the presence of a catalyst (94–96). They are subsequently distilled and usually washed with aqueous bases to the desired level of purity. Tricresyl phosphate was originally made from petroleum-derived or coal-tar-derived cresylic acids, ie, cresols, variously admixed with phenol and xylenols. Discovery of the toxicity of the ortho-cresyl isomers led manufacturers to select cresols having very little ortho-isomer.

An alternative process (97) for direct esterification of cresols using phosphoric acid, a slow reaction, was developed in Israel, where phosphorus oxychloride is not locally available.

In the 1960–1980 period, the use of more economical synthetic isopropyl- and *t*-butylphenols as alternatives to cresols was developed (98,99). Commercial triaryl phosphates such as FMC's Kronitex 100 and Akzo's Phosflex 31P and 41B are based on partially isopropylated or *t*-butylated phenol. The relative volatilities and oxidative stabilities of these phosphates have been compared; the *t*-butylphenyl phosphates are the most oxidatively stable of the alkylphenyl phosphates (100).

Blends of triaryl phosphates and pentabromodiphenyl oxide are leading flame-retardant additives for flexible urethane foams. A principal advantage is their freedom from scorch.

Triphenyl phosphate [*115-86-6*], $C_{18}H_{15}O_4P$, is a colorless solid, mp 48–49°C, usually produced in the form of flakes or shipped in heated vessels as a liquid. An early application was as a flame retardant for cellulose acetate safety film. It is also used in cellulose nitrate, various coatings, triacetate film and sheet, and rigid urethane foam. It has been used as a flame-retardant additive for engineering thermoplastics such as polyphenylene oxide–high impact polystyrene and ABS–polycarbonate blends.

Cresyl diphenyl phosphate [*26444-49-5*], $C_{19}H_{17}O_4P$, is the most efficient plasticizer of the liquid phosphates, but it is relatively volatile. It is used, especially in Europe, in vinyls and in ABS–polycarbonate blends.

The commercial tricresyl phosphate product is essentially a *m,p*-isomer mixture. Typical products of this class are Akzo's Lindol [*1330-78-5*], $C_{21}H_{21}O_4P$, or FMC's Kronitex TCP [*68952-35-2*], nearly colorless liquids, bp about 260–275°C at 1.3 kPa (10 mm Hg). Tricresyl phosphate is used in flexible PVC, cellulose nitrate, ethylcellulose coatings, and various rubbers. Typical applications are vinyl tarpaulins, mine conveyer belts, air ducts, cable insulation, and vinyl films. Trixylenyl phosphate is a related product of lower volatility and less extractability, having advantages for wire and cable insulation.

The plasticizer performance of isopropylphenyl diphenyl phosphate [*28108-99-8*], [*68937-41-7*], [*68782-95-6*], $C_{21}H_{21}O_4P$, is close to that of tricresyl phosphate. It is made from the product of isopropylation of phenol by propylene. The phosphate is a mixture of mainly *o*- and *p*-isomers and contains a distribution of different levels of alkylation (101,102).

tert-Butylphenyl diphenyl phosphate [*56803-37-3*], [*68937-40-6*] is a slightly less efficient plasticizer. It has been used as a flame retardant in engineering thermoplastics and as a fire-retardant hydraulic fluid. The product from different manufacturers (Akzo, FMC) is likely to be somewhat different in isomer distribution and distribution of alkylation levels.

Alkyl diphenyl phosphates are products originally developed to provide improved low temperature flexibility, a fault of triaryl phosphate plasticizers in PVC (103). These phosphates generally provide slightly less flame-retardant efficacy but are generally superior to the triaryl phosphates in regard to smoke when the vinyl formulation is burned. Two commercial products of this family are 2-ethylhexyl diphenyl phosphate [*1241-94-7*], $C_{20}H_{27}O_4P$, and isodecyl diphenyl phosphate [*29761-21-5*], $C_{22}H_{31}O_4P$. The 2-ethylhexyl compound has FDA approval for certain food packaging (qv) applications. A newer member of this group has recently been introduced by Monsanto; it is believed to be a slightly higher alkyl chain length and has improved low temperature properties with reduced volatility (104).

Tetraphenyl resorcinol diphosphate [*57583-54-7*], $C_{30}H_{24}O_8P_2$, is the main component of a relatively new oligomeric phosphate flame retardant, Akzo's Fyrolflex RDP, designed for use in engineering thermoplastics such as polyphenylene oxide blends (105), thermoplastic polyesters, polyamides, vinyls, and polycarbonates. It is a colorless to light yellow liquid, viscosity 400–800 mPa·s(=cP) at 25°C, and a pour point of −12°C. It is less volatile than the triaryl phosphates and has a higher (11%) percentage of phosphorus than triphenyl phosphate. Some closely related diphosphates are also available from Daihachi.

Tris(2,4-dibromophenyl) phosphate [*2788-11-6*], $C_{18}H_9Br_6O_4P$, mp 110°C, was introduced by FMC as Kronitex PB-460 (4% P, 60% Br) for flame-retardant use in thermoplastic polyesters, polycarbonates, ABS, and thermoplastic blends

(106,107). It has outstanding thermal stability and does not discolor at 300°C in air. The use of Kronitex PB-460 without antimony oxide is recommended.

Phosphine Oxides. Development of cyanoethylphosphine oxide flame retardants has been discontinued. Triphenylphosphine oxide [*791-28-6*], $C_{18}H_{15}OP$, is disclosed in many patents as a flame retardant, and may find some limited usage as such, in the role of a vapor-phase flame inhibitor.

A diphosphine of the following structure [*124788-09-6*] has been offered by American Cyanamid.

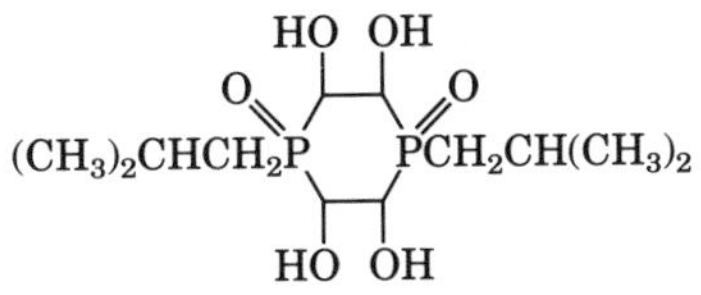

HO OH
O O
$(CH_3)_2CHCH_2P$ $PCH_2CH(CH_3)_2$
HO OH

This compound, designated Cyagard RF1204, has been recommended for use in polypropylene. Despite its high hydroxyl content, it is proposed not as a polyol but as a stable, high melting additive for polypropylene (108).

Reactive Organic Phosphorus Compounds. *Organophosphorus Monomers.* Many vinyl monomers containing phosphorus have been described in the literature (76), but few have gone beyond the laboratory. Bis(2-chloroethyl) vinylphosphonate [*115-98-0*], $C_6H_{11}Cl_2O_3P$, is a commercially available monomer (Akzo's Fyrol Bis-Beta) made from bis(2-chloroethyl) 2-chloroethylphosphonate.

Several applications have been found for bis(2-chloroethyl) vinylphosphonate as a comonomer imparting flame retardancy for textiles and specialty wood and paper applications. Its copolymerization characteristics have been reviewed (76,109). This monomer can be hydrolyzed by concentrated hydrochloric acid to vinylphosphonic acid, polymers of which have photolithographic plate coating utility (see LITHOGRAPHY). It is also an intermediate for the preparation of an oligomeric vinylphosphonate textile finish, Akzo's Fyrol 76 [*41222-33-7*] (110).

Phosphorus-Containing Diols and Polyols. The commercial development of several phosphorus-containing diols occurred in response to the need to flame retard rigid urethane foam insulation used in transportation and construction. There are a large number of references to phosphorus polyols (111) but only a few of these have been used commercially.

Albright & Wilson's Vircol 82 is a diol mixture obtained by the reaction of propylene oxide and dibutyl acid pyrophosphate (112). The neutral liquid has an OH number of 205 mg KOH/g and contains 11.3% phosphorus corresponding to the formula shown where $x + y = 3.4$. The product is a mixture of isomers.

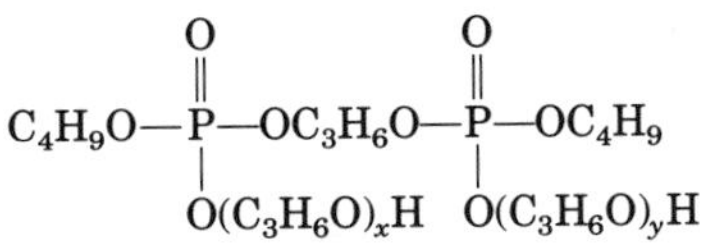

$$C_4H_9O-\overset{\overset{O}{\|}}{\underset{\underset{O(C_3H_6O)_xH}{|}}{P}}-OC_3H_6O-\overset{\overset{O}{\|}}{\underset{\underset{O(C_3H_6O)_yH}{|}}{P}}-OC_4H_9$$

Significant commercial usage has been made of a phosphonate diol containing nitrogen. The nitrogen may exert a synergistic effect on the flame-retardant effect of the phosphorus. This diol, diethyl *N,N*-bis(2-hydroxyethyl)aminomethylphosphonate [*2781-11-5*] (Akzo's Fyrol 6), $C_9H_{22}NO_5P$, is synthesized by the following reaction (113):

$$(HOCH_2CH_2)_2NH + HCHO + HP(=O)(OC_2H_5)_2 \rightarrow (HOCH_2CH_2)_2NCH_2P(=O)(OC_2H_5)_2 + H_2O$$

The product contains 12.6% phosphorus and has an OH number in the 450 mg KOH/g range. Fyrol 6 is used to impart a permanent Class II E-84 flame spread rating to rigid foam for insulating walls and roofs. Particular advantages are low viscosity, stability in polyol–catalyst mixtures, and outstanding humid aging resistance. Fyrol 6 is used in both spray foam, froth, pour-in-place, and slab stock.

A number of commercial phosphorus-containing polyols have been made by the reaction of propylene oxide and phosphoric or polyphosphoric acid. Some have seen commercial use but tend to have hydrolytic stability limitations and are relatively low in phosphorus content. BASF's Pluracol 684 is a high functionality polyol containing 4.5% P, sold for Class II rigid foam use.

Nonreactive additive flame retardants dominate the flexible urethane foam field. However, auto seating applications exist, particularly in Europe, for a reactive polyol for flexible foams, Hoechst-Celanese Exolit 413, a polyol mixture containing 13% P and 19.5% Cl. The patent believed to describe it (114) shows a reaction of ethylene oxide and a prereacted product of tris(2-chloroethyl) phosphate and polyphosphoric acid. An advantage of the reactive flame retardant is avoidance of windshield fogging, which can be caused by vapors from the more volatile additive flame retardants.

Oligomeric Phosphate–Phosphonate. A commercially used reactive oligomeric alcohol, Akzo's Fyrol 51 [*70715-06-9*], has a structure approximately represented by (110):

$$-\!\!\left(\!\left(OCH_2CH_2OP(=O)(OCH_3)\right)_{2x}\left(OCH_2CH_2OP(=O)(CH_3)\right)\!\right)_{x}\!\!-$$

Fyrol 51 is a water-soluble liquid containing about 21% phosphorus. It is made by a multistep process from dimethyl methylphosphonate, phosphorus pentoxide, and ethylene oxide. The end groups are principally primary hydroxyl and the compound can thus be incorporated chemically into aminoplasts, phenolic resins, and polyurethanes. Fyrol 51, or 58 if diluted with a small amount of isopropanol, is used along with amino resins to produce a flame-retardant resin finish on paper used for automotive air filters, or for backcoating of upholstery fabric to pass the British or California flammability standards.

This phosphorus-rich oligomer can also be incorporated into polyurethanes. Combinations with Fyrol 6 permit the OH number to be adjusted to typical values used in flexible foam, urethane coating, or reaction injection molding (RIM) applications (115,116).

Reactive Organophosphorus Compounds in Textile Finishing. Although synthetic fibers can be flame retarded using additives or comonomers, the flame retarding of cotton (qv) requires the application of a textile finish. Markets for such finishes have been in military goods, industrial protective clothing, curtains, hospital goods, and children's sleepwear. An extensive review is available (117).

Tetrakis(hydroxymethyl)phosphonium Salts. The reaction of formaldehyde (qv) and phosphine in aqueous hydrochloric or sulfuric acid yields tetrakis-(hydroxymethyl)phosphonium chloride [*124-62-1*], Albright & Wilson's Retardol C, or the sulfate [*55566-30-8*] (Retardol S), $(C_4H_{12}O_4P)_2SO_4$.

$$PH_3 + 4\ HCHO + HCl \rightarrow (HOCH_2)_4P^+Cl^-$$

These are water-soluble crystalline compounds sold as concentrated aqueous solutions. The methylol groups are highly reactive (118–122) and capable of being cured on the fabric by reaction with ammonia or amino compounds to form durable cross-linked finishes, probably having phosphine oxide structures after post-oxidizing. This finishing process, as developed by Albright & Wilson, is known as the Proban process.

Finishes based on these reagents are durable to numerous launderings. By prereacting the phosphonium salt and urea (123), neutralizing with inorganic base, applying to cotton, curing with ammonia in a special chamber, and post-oxidizing with hydrogen peroxide or perborate, an excellent durable finish with good hand and fabric strength properties can be obtained. The U.S. Navy uses this finish on seamen's cotton uniforms (124).

A competitive cotton finish, CIBA-GEIGY's Pyrovatex CP, was introduced in the 1970s especially for children's sleepwear and other uses. It is based on the following chemistry:

$$(CH_3O)_2\overset{\overset{\displaystyle O}{\|}}{P}H + CH_2{=}CHCONH_2 \rightarrow (CH_3O)_2\overset{\overset{\displaystyle O}{\|}}{P}CH_2CH_2CONH_2 \xrightarrow{HCHO} (CH_3O)_2\overset{\overset{\displaystyle O}{\|}}{P}CH_2CH_2CONHCH_2OH$$

Pyrovatex CP coreacts on cellulose with an amino resin in the presence of a latent acid catalyst, to produce finishes durable to laundering (125,126). A higher assay version, Pyrovatex CP New, has also been introduced.

Oligomeric Vinylphosphonate. A water-soluble oligomer, Fyrol 76 [*41222-33-7*], is produced by reaction of bis(2-chloroethyl) vinylphosphonate and dimethyl methylphosphonate with elimination of all the chlorine as methyl chloride (127,128). This liquid, containing 22.5% P, is curable by free-radical initiation, on cotton or other fabrics. Nitrogen components, such as *N*-methylolacrylamide or methylolmelamines, are usually included in the finish, which can be durable to multiple launderings (129,130).

Phosphorus-Containing Polymers. A large number of addition and condensation polymers having phosphorus built in have been described, but few have been commercialized (131,132). No general statement seems warranted regarding the efficacy of built-in vs additive phosphorus (133). However, in textile fibers, there is greater assurance of permanency.

Polyester Fibers Containing Phosphorus. Numerous patents describe poly(ethylene terephthalate) (PET) flame-retarded with phosphorus-containing difunctional reactants. At least two of these appear to be commercial.

Hoechst-Celanese's Trevira 271 appears to be based on the following chemistry (134,135):

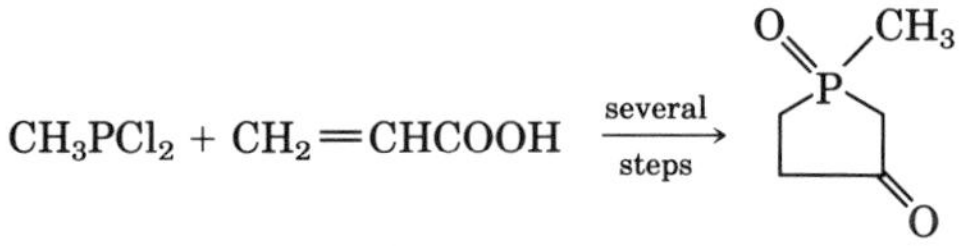

$$CH_3PCl_2 + CH_2{=}CHCOOH \xrightarrow[\text{steps}]{\text{several}}$$

This phosphinic anhydride [*15171-48-9*], $C_4H_7O_3P$, is then reacted with glycol and other precursors of poly(ethylene terephthalate), to produce a flame-retardant polyester [*82690-14-0*], having phosphinate units of the structure —OP(O)-$(CH_3)CH_2CH_2COO$—. Trevira 271 is useful for children's sleepwear, work clothing, and home furnishings. A phosphorus content as low as 0.6% is reported to be sufficient for draperies and upholstery tests if melt-drip is not retarded by print pigments or the presence of nonthermoplastic fibers (28).

Alternative technology for modifying a poly(alkylene terephthalate) by incorporation of a phosphinate structure has been developed by Enichem. Phosphinate units of the structure —$P(C_6H_5)(O)CH_2O$— are introduced into a polyester such as PET or PBT by transesterification with an oligomer comprised of the aforementioned units (136).

Toyobo's HEIM II (former GH) is apparently based on the following intermediate [*63562-31-2*] (137):

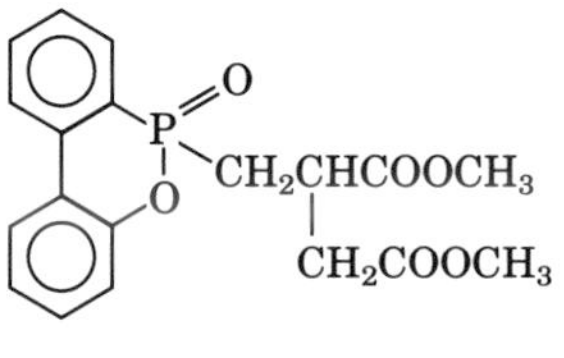

This dicarboxylic ester is then copolycondensed with the other reactants in PET manufacture to produce a flame-retardant polyester [*63745-01-7*]. The advantage of this rather unusual phosphinate structure is its high thermal and hydrolytic stability. The fabric is probably used mainly for furnishings in public buildings in Japan.

Health, Safety, and Environmental Factors

Toxicology. Two factors should be considered when discussing the toxicity of flame-retardant materials: the toxicity of the compounds themselves and the effect of the flame retardants on combustion product toxicity.

Product Toxicology. The structure–toxicity relationships of organophosphorus compounds have been extensively researched and are relatively well understood (138–140). The phosphorus-based flame retardants as a class exhibit only moderate-to-low toxicity. NIOSH or EPA compilations and manufacturers' safety data sheets show the following LD_{50} values for rats, for representative commercial phosphorus flame retardants:

Compound	Oral LD_{50}, mg/kg
mono-/diammonium phosphate	3160–4500
methyl methylphosphonate	>5000
triethyl phosphate	1311
tris(2-chloroethyl) phosphate	390 (female)
tris(2-chloro-1-propyl) phosphate	2800 (female)
tris(1,3-dichloro-2-propyl) phosphate	2830
triphenyl phosphate	>4640
tricresyl phosphate (commercial mixture)	>4640
isopropylphenyl diphenyl phosphate	>10,000
tert-butylphenyl diphenyl phosphate	>4640

A critical review of the toxicity of the haloalkyl phosphates and the potential metabolic products is available (141). The toxicity of flame retardants used in textiles has also been reviewed (142).

A particular mode of neurotoxicity was discovered for tricresyl phosphate that correlated with the presence of the *o*-cresyl isomer (or certain other specific alkylphenyl isomers) in the triaryl phosphates. Many details of the chemistry and biochemistry of the toxic process have been elucidated (139,140,143–146). The use of low ortho-content cresols has become the accepted practice in industrial production of tricresyl phosphate. Standard *in vivo* tests, usually conducted with chickens sensitive to this mode of toxicity, have been developed for premarket testing of new or modified triaryl phosphates. As of 1992, the EPA called for extensive new toxicity and environmental data on this group of products (147). The *Federal Register* document calling for this data has an extensive bibliography on aryl phosphate toxicology as well as a discussion of human exposure.

Mutagenic and later carcinogenic properties were found for tris(2,3-dibromopropyl) phosphate (148–150), a flame retardant used on polyester fabric in the 1970s. This product is no longer on the market. The chemically somewhat-related tris(dichloroisopropyl) phosphate has been intensively studied and found not to display mutagenic activity (148,149,151). Tris(2-chloroethyl) phosphate appears to be a weak tumor-inducer in a susceptible rodent strain (150).

Effects on Combustion Toxicology. There appears to be no documented case of any type of fire retardant contributing to human fire casualties. A survey of data from small-scale combustion or pyrolysis experiments revealed no consistent pattern of decrease or increase in the yields of toxic gases (CO, HCN) when phosphorus flame retardants were present (152,153).

Laboratory experiments using rodents, or the use of gas analysis, tend to be confused by the dominant variable of fuel–air ratio as well as important effects of burning configuration, heat input, equipment design, and toxicity criteria used, ie, death vs incapacitation, time to death, lethal concentration, etc (154,155). Some comparisons of polyurethane foam combustion toxicity with and without phosphorus flame retardants show no consistent positive or negative effect. Moreover, data from small-scale tests have doubtful relevance to real fire hazards.

One noteworthy neurotoxic response was demonstrated in laboratory pyrolysis studies using various types of phosphorus flame retardants in rigid urethane foam, but the response was traced to a highly specific interaction of trimethylolpropane polyols, producing a toxic bicyclic trimethylolpropane phosphate [*1005-93-2*] (152). Formulations with the same phosphorus flame retardants but other polyols avoided this neurotoxic effect completely.

Effects on Visible Smoke. Smoke is a main impediment to egress from a burning building. Although some examples are known where specific phosphorus flame retardants increased smoke in small-scale tests, other instances are reported where the presence of the retardant reduced smoke. The effect appears to be a complex function of burning conditions and of other ingredients in the formulation (153,156,157). In a careful Japanese study, ammonium phosphate raised or lowered the smoke from wood depending on pyrolysis temperature (158). Where the phosphorus flame retardant functions by char enhancement, lower smoke levels are likely to be observed.

Environmental Considerations. The phosphate flame retardants, plasticizers, and functional fluids have come under intense environmental scrutiny. Results published to date on acute toxicity to aquatic algae, invertebrates, and fish indicate substantial differences between the various aryl phosphates (159–162). The EPA has summarized this data as well as the apparent need for additional testing (147).

Tests in pure water, river water, and activated sludge showed that commercial triaryl phosphates and alkyl diphenyl phosphates undergo reasonably facile degradation by hydrolysis and biodegradation (163–165). The phosphonates can undergo biodegradation of the carbon-to-phosphorus bond by certain microorganisms (166,167).

Economic Aspects

Usage of phosphorus-based flame retardants for 1994 in the United States has been projected to be $150 million (168). The largest volume use may be in plasticized vinyl. Other use areas for phosphorus flame retardants are flexible urethane foams, polyester resins and other thermoset resins, adhesives, textiles, polycarbonate–ABS blends, and some other thermoplastics. Development efforts are well advanced to find applications for phosphorus flame retardants, especially ammonium polyphosphate combinations, in polyolefins, and red phosphorus in nylons. Interest is strong in finding phosphorus-based alternatives to those halogen-containing systems which have encountered environmental opposition, especially in Europe.

Trends in the research and development of phosphorus flame retardants have been in the direction of less volatile, less toxic, more stable compounds, and where feasible, in the direction of built-in phosphorus structures. At the same time, there have been an increasing number of regulatory delays in new compounds, and the existent materials are finding increased exploitation in the form of mixtures. Some interest is also noted in encapsulation.

BIBLIOGRAPHY

"Phosphorus Compounds" under "Flame Retardants" in *ECT* 3rd ed., Vol. 10, pp. 396–419, by E. D. Weil, Stauffer Chemical Co.

1. E. D. Weil, in R. E. Engel, ed., *Handbook of Organophosphorus Chemistry*, Marcel Dekker, Inc., New York, 1992, pp. 683–738.
2. R. M. Aseeva and G. E. Zaikov, *Combustion of Polymeric Materials*, Hanser Publishers, Munich and New York (dist. in the United States by Macmillan) 1985, pp. 229–241.
3. A. Granzow, *Accounts Chem. Res.* **11**(5), 177–183 (1978).
4. S. K. Brauman and N. Fishman, *J. Fire Ret. Chem.* **4**, 93–111 (1977).
5. S. K. Brauman, *J. Fire Ret. Chem.* **4**, 18–37 (1977).
6. *Ibid.*, 38–58.
7. M. Lewin, in M. Lewin and S. Sello, eds., *Handbook of Fiber Science and Technology: Chemical Processing of Fibers and Fabrics*, Vol. 2, Part B, Marcel Dekker, Inc., New York, 1984, pp. 1–141.
8. R. H. Barker and J. E. Hendrix, in W. C. Kuryla and A. J. Papa, eds., *Flame Retardancy of Polymeric Materials*, Vol. 5, Marcel Dekker, Inc., New York, 1979, pp. 1–65.
9. S. LeVan, in R. Rowell, ed., *The Chemistry of Solid Wood*, American Chemical Society, Adv. Chem. Ser. 207, Washington, D.C., 1984, pp. 531–574.
10. J. J. Willard and R. E. Wondra, *Textile Res. J.* **40**, 203–210 (1970).
11. D. Bakos and co-workers, *Fire & Materials* **6**(1), 10–12 (1982).
12. E. D. Weil, in W. C. Kuryla and A. J. Papa, eds., *Flame Retardancy of Polymeric Materials*, Vol. 3, Marcel Dekker, Inc., New York, 1975, pp. 186–243.
13. E. D. Weil, "Additivity, Synergism and Antagonism in Flame Retardancy - Recent Developments," paper presented at *3rd Annual BCC Conference on Recent Advances in Flame Retardancy of Polymeric Materials*, Stamford, Conn., May 19–21, 1992.
14. G. Avondo, C. Vovelle, and R. Delbourgo, *Combust. Flame* **31**, 7–16 (1978).
15. T. Suebsaeng and co-workers, *J. Polym. Sci., Polym. Chem. Ed.* **22**, 945 (1984).
16. S. K. Brauman, *J. Fire Ret. Chem.* **7**, 61–68 (1980).
17. I. J. Gruntfest and E. M. Young, *ACS Div. Org. Coatings & Plastics Preprints* **2**(2), 113–124 (1962).
18. G. Camino, N. Grassie, and I. C. McNeill, *J. Polym. Sci., Polym. Chem. Ed.* **16**, 95–106 (1978).
19. C. E. Brown and co-workers, *J. Polym. Sci., Part A, Polym. Chem.* **24**, 1297–1311 (1986).
20. A. J. Papa, in Ref. 12, pp. 1–133.
21. J. J. Anderson, *Ind. Eng. Chem. Prod. Res. Dev.* **2**, 260–263 (1963).
22. H. Piechota, *J. Cell. Plast.* **1**, 186–199 (1965).
23. J. E. Kresta and K. C. Frisch, *J. Cell. Plast.* **11**, 68–75 (1975).
24. K. M. Gibov, L. N. Shapovalova, and B. A. Zhubanov, *Fire & Materials* **10**, 133–135 (1986).
25. E. V. Gouinlock, J. F. Porter, and R. R. Hindersinn, *J. Fire Flamm.* **2**, 206–218 (1971).
26. E. N. Peters, *J. Appl. Polym. Sci.* **24**, 1457–1464 (1979).
27. T. J. Swihart and P. E. Campbell, *Text. Chem. Color.* **6**(5), 109–112 (1974).
28. H.-R. Mach, *Melliand Textilberichte (English)* **1**, E29–E33 (1990).
29. D. Paul, "A New Phosphate Plasticizer for Low Smoke Wire and Cable Applications," paper presented at *3rd Annual BCC Conference on Recent Advances in Flame Retardancy of Polymeric Materials*, Stamford, Conn., May 19–21, 1992.
30. W. A. Rosser, Jr., S. H. Inami, and H. Wise, *Combust. Flame* **10**, 287ff (1966).
31. E. T. McHale, *Fire Research Abstracts & Reviews* **11**(2), 90 (1969).

32. C. T. Ewing, J. T. Hughes, and H. W. Carhart, *Fire and Materials* **8**(3), 148–155 (1984).
33. J. W. Hastie and D. W. Bonnell, *Molecular Chemistry of Inhibited Combustion Systems, Report NBSIR 80-2169*, Natl. Bureau of Standards, Washington, D.C., 1980.
34. J. W. Hastie, *J. Research NBS -A. Physics & Chem.* **77A**(6), 733–754 (1973).
35. J. W. Hastie and C. L. McBee, *Mechanistic Studies of Triphenylphosphine Oxide-Poly(ethylene terephthalate) and Related Flame Retardant Systems, Report NBSIR 75-741*, Natl. Bureau of Standards Washington, D.C., 1975.
36. J. Carnahan and co-workers, "Investigations into the Mechanism for Phosphorus Flame Retardancy in Engineering Plastics," *Proc. 4th Intl. Conf. Fire Safety*, Product Safety Corp., San Francisco, Calif., 1979.
37. K. J. L. Paciorek and co-workers, *Amer. Indust. Hygiene Assn. J.* **39**(8), 633–639 (1978).
38. A. W. Benbow and C. F. Cullis, *Combustion & Flame* **24**, 217–230 (1975).
39. E. D. Weil, "Flame Retardant Chemicals for Polyurethane Applications," *Polyurethane Technol. Conf. (Preprints)*, Clemson Univ., Clemson, S.C., 1987.
40. A. M. Batt and P. Appleyard, *J. Fire Sciences* **7**(5), 338–363 (1989).
41. C.-P. Yang and T.-M. Lee, *J. Polym. Sci., Polym. Chem. Ed.* **27**, 2239–2251 (1989).
42. A. W. Morgan, in V. M. Bhatnagar, ed., *Advances in Fire Retardants*, Vol. 1, Technomic Publishing Co., Westport, Conn., 1972.
43. U. S. Pat. 4,546,117 (Oct. 8, 1985), J. F. Szabat (to Mobay Chemical Corp.).
44. J. W. Lyons, *The Chemistry and Use of Flame Retardants*, Wiley-Interscience, New York, 1970, pp. 29–74.
45. D. Taylor, in *Fire Retardant Engineering Polymers & Alloys*, Fire Retardant Chemicals Association, Lancaster, Pa., 1989, pp. 133–142.
46. H. Staendeke, "Red Phosphorus - Recent Development for Safe and Efficient Flame Retardant Applications," paper presented at *Fire Retardant Chemicals Association National Meeting*, Mar. 1988.
47. E. N. Peters, in M. Lewin, S. M. Atlas and E. M. Pearce, eds., *Flame-Retardant Polymeric Materials*, Vol. 5, Plenum Press, New York, 1978, pp. 113–176.
48. *Chem. Week*, 26–27 (Apr. 15, 1992).
49. C. Y. Shen, N. E. Stahlheber, and D. R. Dyroff, *J. Am. Chem. Soc.* **91**, 62–67 (1969).
50. H. J. Vandersall, *J. Fire Flammability* **2**, 97–140 (1971).
51. G. Camino and L. Costa, *Reviews in Inorganic Chemistry* **8**(1–2), 69–100 (1986).
52. D. Scharf, in *Fire Retardant Engineering Polymers & Alloys*, Fire Retardant Chemicals Association, Lancaster, Pa., 1989, pp. 183–202.
53. R. Mount, in *International Conference on Fire Safety, Papers Presented at Fire Retardant Chemicals Association*, New Orleans, Mar. 25–28, 1990, pp. 253–267.
54. Ger. Pat. Appl. 3,720,094 (Mar. 3, 1988), H. Staendeke and D. Scharf (to Hoechst Celanese).
55. U.S. Pat. 4,579,894 (Apr. 1, 1986), G. Bertelli and R. Locatelli, R. (to Montedison SpA).
56. U.S. Pat. 4,336,182 (June 22, 1982), G. Landoni, S. Fontani, and O. Cicchetti (to Montedison).
57. G. Montaudo, E. Scamporrino, and D. Vitalini, *J. Polym. Sci.* **21**, 3321–3331, 3361–3371 (1983).
58. U.S. Pat. 4,504,610 (Mar. 12, 1985), R. Fontanelli, G. Landoni, and G. Legnani, (to Montedison S.p.A.).
59. U.S. Pat. 3,887,511 (June 3, 1975), S. C. Juneja (to Canadian Patents & Development Ltd.).
60. S. C. Juneja and L. R. Richardson, *Forest Prod. J.* **24**(5), 19 (1974).

61. L. Costa, G. Camino, and M. P. Luda di Cortemiglia, in G. Nelson, ed., *Fire & Polymers*, Am. Chem. Soc. Symposium Ser. 425, Washington, D.C., 1990, pp. 211–238.
62. C. L. Goin and M. T. Huggard, in M. Lewin and G. S. Kirshenbaum, eds., *Recent Advances in Flame Retardancy of Polymeric Materials*, Vol. II, Business Communication Co., Inc., Norwalk, Conn., 1992, pp. 94–101.
63. U.S. Pat. 2,430,569 (Nov. 11, 1947) D. C. Hull and A. H. Agett (to Eastman Kodak Co.).
64. U.S. Pat. 2,407,279 (Sept. 10, 1946), D. C. Hull and J. R. Snodgrass (to Eastman Kodak Co.).
65. R. Nametz, *Ind. Eng. Chem.* **59**(5), 99–112 (1967).
66. U.S. Pat. 3,909,484 (Sept. 30, 1975), A. N. Beavon (to Universal-Rundle Corp.).
67. P. V. Bonsignore and J. H. Manhart, *Proc. 29th Ann. Conf. Reinf. Plast. Compos. Inst. SPE* **23C**, 1–8 (1974).
68. E. R. Larsen and E. L. Ecker, *J. Fire Retardant Chem.* **6**, 182–192 (1979).
69. U.S. Pat. 3,100,220 (Aug. 6, 1963), A. L. Smith (to Celanese Corp.).
70. J. W. Crook and G. A. Haggis, *J. Cellular Plast.*, Mar./Apr. 1969, 119–122.
71. U.S. Pat. 3,041,293 (June 26, 1962), R. Polacek (to Celanese Corp.).
72. U.S. Pat. 4,477,600 (Oct. 16, 1984), G. Fesman (to Stauffer Chemical Co.).
73. U.S. Pat. 4,794,126 (Dec. 27, 1988), G. Fesman, B. Jacobs, and B. Williams (to Akzo America Inc.).
74. O. M. Grace and co-workers, "Improvement in Cigarette Smoldering Resistance of Upholstered Furniture," *Proc. Polyurethanes World Congress*, Sept. 29–Oct. 2, 1987, Aachen, Germany.
75. U.S. Pat. 4,144,387 (Mar. 13, 1979), J. J. Anderson and co-workers, (to Mobil Oil Corp.).
76. E. D. Weil, *J. Fire & Flamm./Fire Retardant Chem.* **1**, 125–141 (1974).
77. U.S. Pat. 3,707,586 (Dec. 26, 1972), R. J. Turley (to Olin Corp.).
78. J. S. Babiec, Jr. and co-workers, *Plastics Technol.*, 47–50 (June 1975).
79. J. E. Puig, F. S. Natoli, and B. W. Peterson, *Proc. Intl. Symposium on Flame Retardants*, Intl. Academic Publ. (Pergamon), Beijing, China, 1989, pp. 139–147.
80. T. M. Moshkina and A. N. Pudovik, *Zh. Obschei Khim.* **32**, 1761ff (1962).
81. U.S. Pat. 3,192,242 (June 29, 1975), G. H. Birum (to Monsanto).
82. U.S. Pat. 3,896,187 (July 22, 1975), E. D. Weil (to Stauffer Chemical Co.).
83. U.S. Pats. 4,382,042 (May 3, 1983) and 4,458,035 (July 3, 1984), T. A. Hardy and F. Jaffe (to Stauffer Chemical Co.).
84. U.S. Pat. 4,696,963 (Sept. 29, 1987), J. A. Albright and T. C. Wilkinson (to Great Lakes Chemical Co.).
85. U.S. Pat. 3,789,091 (Jan. 29, 1974), J. J. Anderson, J. G. Camacho, and R. E. Kinney (to Mobil Oil Corp.).
86. U.S. Pat. 3,849,368 (Nov. 19, 1974), J. J. Anderson, J. G. Camacho, and R. E. Kinney (to Mobil Oil Corp.).
87. Y. Halpern, M. Mott, and R. H. Niswander, *Ind. Eng. Chem. Prod. Res. & Dev.* **23**, 233–238 (1984).
88. D. R. Hall, M. M. Hirschler, and C. M. Yavornitzky, in C. E. Grant and P. J. Pagni, eds., *Fire Safety Science - Proceedings of the First International Symposium*, Hemisphere Publishing, Washington, D.C., 1986, pp. 421–430.
89. E. Termine and K. G. Taylor, "A New Intumescent Flame Retardant Additive for Thermoplastics and Thermosets," in *Additive Approaches to Polymer Modification*, SPE RETEC Conference Papers, Toronto, Ontario, Canada, Sept. 1989.
90. U.S. Pat. 4,801,625 (Jan. 31, 1989), W. J. Parr, A. G. Mack, and P. Y. Moy (to Akzo America Inc.).
91. U.S. Pat. 4,220,472 (Sept. 2, 1980), C. Mauric and R. Wolf (to Sandoz Ltd.).

92. R. Wolf, *Ind. Eng. Chem., Prod. Res. Dev.* **20**(3), 413–420 (1981).
93. D. L. Buszard, *Chem. Ind. (London)*, (16) 610–615 (1978).
94. L. G. Krauskopf, in L. I. Nass and C. A. Heiberger, eds., *Encyclopedia of PVC*, 2nd ed., Vol. 2, Marcel Dekker, Inc., New York, 1988, pp. 143–261.
95. T. W. Lapp, *The Manufacture and Use of Selected Aryl and Alkyl Aryl Phosphate Esters*, Report NTIS PB-251678, Midwest Research Institute, Feb. 1976.
96. F. A. Lowenheim and M. K. Moran, *Faith, Keyes and Clark's Industrial Chemicals*, 4th ed., John Wiley & Sons, Inc., New York, 1975, pp. 849–853.
97. Brit. Pat. 2,215,722 (Sept. 27, 1989), J. Segal and L. M. Shorr (to Bromine Compounds Ltd.).
98. J. M. Heaps, *Plastics (London)* **33**(366), 410–413 (1969).
99. U.S. Pat. 3,919,158 (Nov. 11, 1975), D. R. Randell and W. Pickles (to CIBA-GEIGY AG).
100. S. G. Shankwalkar and D. G. Placek, *Ind. Eng. Chem. Res.* **31**, 1810–1813 (1992).
101. A. J. Duke, *Chimia* **32**(12), 457–463 (1978).
102. E. R. Nobile, S. W. Page, and P. Lombardo, *Bull. Envir. Contam. Toxicol.* **25**, 755–761 (1980).
103. J. R. Darby and J. K. Sears, in N. M. Bikales, ed., *Encyclopedia of Polymer Science and Technology*, Wiley-Interscience, New York, 1975, pp. 229–306.
104. D. H. Paul, "A New Phosphate Plasticizer for Low Smoke Wire and Cable Applications," paper presented at *Fire Retardant Chemicals Association*, Coronado, Calif., Oct. 20–23, 1991, pp. 79–108.
105. Eur. Pat. Appl. 356,633 (Mar. 7, 1990), C. A. A. Claesen and J. H. G. Lohmeijer (to General Electric Co.).
106. J. Green, in Ref. 61, pp. 253–265.
107. J. Green and J. Chung, *J. Fire Sci.* **8**, 254–265 (1990).
108. U.S. Pat. 4,929,393 (May 29, 1990), A. J. Robertson and J. B. Gallivan (to Cyanamid Canada Inc.).
109. R. Gallagher and J. C. H. Hwa, *J. Polym. Sci., Polym. Symp.* **64**, 329–337 (1978).
110. E. D. Weil, R. B. Fearing, and F. Jaffe, *J. Fire Retard. Chem.* **9**(1), 39–49 (1982).
111. A. J. Papa, *Ind. Eng. Chem. Prod. Res. Dev.* **9**(4), 478–496 (1970).
112. Brit. Pat. 954,792 (Apr. 8, 1964), (to Virginia-Carolina Chem. Co.).
113. U.S. Pat. 3,235,517 (Feb. 16, 1966), T. M. Beck and E. N. Walsh (to Stauffer Chemical Co.).
114. U.S. Pat. 3,850,859 (Nov. 26, 1974), J. Wortmann, F. Dany, and J. Kandler (to Hoechst AG).
115. Eur. Pat. Publ. 138,204 (Apr. 24, 1985), G. Fesman, R. Y. Lin, and R. A. Raeder (to Stauffer Chemical Co.).
116. Eur. Pat. Appl. 255,381 (Feb. 3, 1988), D. M. Konkus and P. Kraft (to Stauffer Chemical Co.).
117. A. R. Horrocks, *Rev. Prog. Coloration* **16**, 62–101 (1986).
118. S. L. Vail, D. J. Daigle, and A. W. Frank, *Text. Res. J.* **52**(11), 671–677 (1982).
119. *Ibid.*, Part II, 678–692.
120. A. W. Frank, D. J. Daigle, and S. L. Vail, *Text. Res. J.* **52**(12), 738–750 (1982).
121. D. J. Daigle and A. W. Frank, *Text. Res. J.* **52**(12), 751–755 (1982).
122. W. A. Reeves and co-workers, *Text. Chem. Color.* **2**(16), 283–285 (1969).
123. Ger. Pat. Appl. 2,340,437 (Mar. 7, 1974), R. Cole (to Albright and Wilson Ltd.).
124. E. D. Weil, *Amer. Dyestuff Reporter*, 41–44 (Jan. 1987); 38–39, 49 (Feb. 1987).
125. R. Aenishanslin and co-workers, *Text. Res. J.* **39**(4), 375–381 (1969).
126. R. Aenishanslin and N. Bigler, *Textilveredlung* **3**(9), 467–474 (1968).
127. U.S. Pat. 3,855,359 (Dec. 17, 1974), E. D. Weil (to Stauffer Chemical Co.).
128. U.S. Pat. 4,017,257 (Apr. 12, 1977), E. D. Weil (to Stauffer Chemical Co.).

129. U.S. Pat. 4,067,927 (Jan. 10, 1978), E. D. Weil (to Stauffer Chemical Co.).
130. W. A. Reeves and Y. B. Marquette, *Text. Res. J.* **49**(3), 163–169 (1979).
131. E. D. Weil, in J. I. Kroschwitz, ed., *Encyclopedia of Polymer Science and Engineering*, Vol. 11, John Wiley & Sons, Inc., New York, 1988, pp. 96–126.
132. E. D. Weil, *Proc. Int. Conf. Fire Safety* **12**, 210–218 (1987).
133. R. Stackman, in C. H. Carraher and M. Tsuda, eds., *Modification of Polymers*, ACS Symposium Series 121, Washington, D.C., 1980, pp. 425–434.
134. U.S. Pat. 3,941,752 (Mar. 2, 1976), H. J. Kleiner and co-workers (to Hoechst AG).
135. U.S. Pat. 4,033,936 (July 5, 1977), U. Bollert, E. Lohmar, and A. Ohorodnik (to Hoechst AG).
136. U.S. Pat. 4,981,945 (Jan. 1, 1991), G. Landoni and C. Neri (to Enichem Synthesis, SpA).
137. U.S. Pat. 4,127,590 (Nov. 28, 1978), S. Endo and co-workers (to Toyo Boseki KK).
138. D. C. G. Muir, in O. Hutzinger, ed., *Anthropogenic Compounds, Handbook of Environmental Chemistry*, Vol. 3, Part C, Springer-Verlag, Berlin, 1984, pp. 41–66.
139. M. B. Abou-Donia, in K. Blum and L. Manzo, eds., *Neurotoxicology*, Marcel Dekker, Inc., New York, 1985, pp. 423–444.
140. M. B. Abou-Donia, *Ann. Rev. Pharmacol. Toxicol.* **21**, 511–548 (1981).
141. J. W. Holleman, *Health Effects of Haloalkyl Phosphate Flame Retardants and Potential Metabolic Products*, Oak Ridge National Laboratory, Tenn., 1984, DOE/NBM-4006848 (DE84 006848),
142. A. G. Ulsamer, R. E. Osterberg, and J. McLaughlin, Jr., *Clin. Toxicol.* **17**(1), 103–131 (1980).
143. M. K. Johnson, *CRC Crit. Rev. Toxicol.* 289–316 (June 1975).
144. M. B. Abou-Donia and D. M. Lapadula, *Ann. Rev. Pharmacol. Toxicol.* **30**, 405–440 (1990).
145. G. L. Sprague, T. R. Castles, and A. A. Bickford, *J. Toxicol. Environ. Health* **14**(5–6), 773–788 (1984).
146. M. K. Johnson, *Toxicol. Appl. Pharmacol.* **102**(3), 385–399 (1990).
147. Environmental Protection Agency, *Fed. Regist.* **57**(12), 2138–2159 (Jan. 17, 1992).
148. D. Brusick and co-workers, *J. Envir. Pathol. Toxicol.* **3**, 207–226 (1980).
149. E. J. Soderlund and co-workers, *Acta Pharmacol. et Toxicol.* **56**, 20–29 (1985).
150. M. Sala, G. Moens, and I. Chouroulinkov, *Eur. J. Cancer Clin. Oncol.* **18**(12), 1337–1344 (1982).
151. R. K. Lynn and co-workers, *Drug Metab. Disposition* **9**(5), 434–441 (1981).
152. E. D. Weil and A. M. Aaronson, "Phosphorus Flame Retardants—Meeting New Requirements," in *Proceedings of the 1st European Conference on Flammability and Fire Retardants, Brussels, Belgium, July 1977*, Technomic Publishing Co., Westport, Conn., 1978.
153. E. D. Weil and A. M. Aaronson, "Phosphorus Flame Retardants—Some Effects on Smoke and Combustion Products," lecture at *University of Detroit Polymer Conference on Recent Advances in Combustion and Smoke Retardance of Polymers*, Mich., May 1976.
154. C. Herpol, *Fire and Materials* **1**(1), 29–35 (1976).
155. H. K. Hasegawa, *Characterization and Toxicity of Smoke*, Publication STP 1082, American Association for Testing and Materials, Philadelphia, Pa., 1990.
156. T. C. Mathis and J. D. Hinchen, *31st Ann. Tech. Conf. SPE, Technical Papers* **19**, 343–348 (1973).
157. U.S. Pat. 3,869,420 (Mar. 4, 1975), T. C. Mathis and A. W. Morgan (to Monsanto Co.).
158. Y. Uehara and Y. Ogawa, in V. M. Bhatnagar, ed., *Advances in Fire Retardant Textiles*, Prog. Fire Ret. Series, Vol. 5, Technomic Publishing Co., Westport Conn., 1975, pp. 507–599.

159. V. D. Ahrens and co-workers, *Bull. Env. Contam. Toxicol.* **21**, 409–412 (1979).
160. M. J. Nevins and W. W. Johnson, *Bull. Environ. Contam. Toxicol.* **7**, 250 (1978).
161. L. Cleveland and co-workers, *Envir. Tox. Chem.* **5**, 273–282 (1986).
162. B.-E. Bengtsson and co-workers, *Envir. Tox. Chem.* **5**, 853–861 (1986).
163. W. Mabey and T. Mill, *J. Phys. Chem. Ref. Data* **7**(2), 383–415 (phosphorus compounds on p. 414) (1978).
164. V. W. Saeger and co-workers, *Envir. Sci. Tech.* **13**(7), 840–844 (1979).
165. R. S. Boethling and J. C. Cooper, *Residue Reviews* **94**, 49–95 (1985).
166. O. Ghisalba and co-workers, *Chimia* **41**(6), 206–215 (1987).
167. J. W. Frost, S. Loo, and D. Li, *J. Am. Chem. Soc.* **109**, 2166–2171 (1987).
168. "Frost & Sullivan Market Report," *Chem. Week*, 44 (Nov. 28, 1990).

EDWARD D. WEIL
Polytechnic University

FLAME RETARDANTS FOR TEXTILES

As early as the first century BC, Virgil wrote of the treatment of wood with vinegar to impart fire resistance (1). In the early 1700s a patent was granted for flame retarding cellulose using a mixture containing alum [*7784-24-9*] (aluminum potassium sulfate dodecahydrate), iron(II) sulfate [*7782-63-0*], and borax [*1303-96-4*] (sodium tetraborate decahydrate) (2). In 1821 Gay-Lussac carried out a systematic study of flame proofing (3) using water solutions of ammonium chloride [*12125-02-9*] and ammonium phosphate, or of boric acid and ammonium chloride, to treat cellulosic material such as linen or jute. He concluded that the most effective salts either had low melting points and gave off nonflammable vapors, or they covered the fabric surface with a glassy layer of the salt.

The first known fire-retardant process found durable to laundering was developed in 1912 (4). A modification of an earlier process (5), this finish was based on the formation of a tin(IV) oxide [*18282-10-5*] deposit. Although the fabric resulting from treatment was flame resistant, afterglow was reputed to be a serious problem, resulting in the complete combustion of the treated material through smoldering.

A significant advance in flame retardancy was the introduction of binary systems based on the use of halogenated organics and metal salts (6,7). In particular, a 1942 patent (7) described a finish for utilizing chlorinated paraffins and antimony(III) oxide [*1309-64-4*]. This type of finish was invaluable in World War II, and saw considerable use on outdoor cotton fabrics in both uniforms and tents.

Work began in the 1930s on the development of flame-retardant cottons based on chemical systems that either reacted directly with the cellulosic substrate, or polymerized on or in the cotton fiber. A serious effort in this direction, mounted from the 1950s through the 1970s, resulted in most of the state-of-the-art flame-retardant finishes for cotton available.

Terminology. Flame resistance and fire resistance are often used in the same context as the terms flameproof and fireproof. A textile that is flame or fire resistant does not continue to burn or glow once the source of ignition has been removed, although there is some change in the physical and chemical characteristics. Although the terms resistant and retardant have similar meanings, flame resistant is normally used when referring to that property of a material which prevents it from burning when an external source of flame is removed; flame retardant is used when referring to the chemicals or chemical treatment applied to a material to impart flame resistance. Flameproof or fireproof, on the other hand, refer to materials totally resistant to flame or fire. No appreciable change in the physical or chemical properties is noted. Asbestos (qv) is an example of a fireproof material.

Most organic fibers undergo a glowing action after the flame has been extinguished. From a practical standpoint, flame-resistant fabrics should also be glow resistant because the afterglow may cause as much damage as the flaming itself, in that it can completely consume the fabric.

Another related term is smolder resistance. Smolder resistance implies resistance to ignition by a smoldering source, such as a lit cigarette, placed on the surface of a fabric or in the crevice formed between two butting fabrics. Smolder resistance does not necessarily imply flame resistance, although the material in question may well be flame resistant. A fabric can be smolder resistant and not flame resistant, or vice versa.

Flame Resistance

Factors Affecting Performance. The flame resistance of a textile fiber is affected by the chemical nature of the fiber, its ease of combustion, the fabric weight and construction, the efficiency of the flame retardant, the environment, and laundering conditions. Fibers are classified into natural fibers, eg, cotton, flax, silk, or wool; regenerated fibers, eg, rayon; synthetic fibers, eg, nylons, vinyls, polyesters, and acrylics; and inorganic fibers, eg, glass or asbestos. Combustibility depends on chemical makeup and whether the fiber is inorganic, organic, or a mixture of both (see FIBERS, SURVEY).

The weight and construction of the fabric affect its burning rate and ease of ignition. Lightweight, loose-weave fabrics burn much faster than heavier weight fabrics; therefore, a higher weight add-on of fire retardant is needed to impart adequate flame resistance.

Mechanism of Flame Retardants. The burning process of cellulose depends on both a source of ignition and the presence of oxygen. A low temperature degradation of cellulose proceeds by the formation of levoglucosan, which in turn undergoes dehydration and polymerization, leading to tars, flammable gases, liquids, and other solids (8,9). The flammable gases thus produced ignite, causing the liquids and tars to volatize to some extent. This produces additional volatile fractions which ignite and produce a carbonized residue that does not burn readily. The process continues until only carbonaceous material remains. After the flame has subsided, the carbonized residue slowly oxidizes and glowing continues until the carbonaceous char is consumed.

In general, cotton treated with an effective flame retardant provides the same decomposition products upon burning as does untreated cotton; however, the amount of tar is greatly reduced, with a corresponding increase in the solid char. Consequently, as decomposition takes place, smaller amounts of flammable gases are available from the tar, and greater amounts of nonflammable gases from the decomposition of the char fraction. Char is essentially carbon. Its oxidation causes afterglow. Phosphorus-containing compounds, in some cases polymers, are particularly effective in inhibiting char oxidation. Numerous studies have been made on burning of untreated and flame-retardant-treated cellulose (8,10–14).

Several theories have been postulated to explain the various types of flame retardants for cotton. These theories include coating, gas, thermal, and dehydration or chemical.

Coating Theory. This theory includes fire retardants which form an impervious skin on the fiber surface. This coating may be formed during normal chemical finishing, or subsequently when the fire retardant and substrate are heated. It excludes the air necessary for flame propagation and traps any tarry volatiles produced during pyrolysis of the substrate. Examples of this type of agent include the easily fusible salts such as carbonates or borates.

Gas Theory. The gas theory utilizes two approaches in explaining flame retardancy. In one approach, the flame retardant decomposes to give gases which do not burn. These gases diffuse and mix with the oxygen present around the combustible cellulosic, thereby protecting it from further combustion. Examples of these gases include water vapor, carbon dioxide [*124-38-9*], sulfur dioxide [*7446-09-5*], and hydrogen chloride [*7647-01-0*]. In the second approach, flame retardants function in the gas phase by producing free-radical terminators when the retardant is heated to flaming temperatures. The gas-phase oxidation of cellulose derivatives is a free-radical process utilizing such free radicals as ·H, ·OH, or ·OOH. The function of the flame retardant is to form other free radicals, which trap the active radicals, leading to chain cessation or substitution with less active radicals. The end result is termination of the free-radical chain propagation and hence, flame retardation.

Thermal Theory. The thermal approach to flame retardancy can function in two ways. First, the heat input from a source may be dissipated by an endothermic change in the retardant such as by fusion or sublimation. Alternatively, the heat supplied from the source may be conducted away from the fibers so rapidly that the fabric never reaches combustion temperature.

Dehydration or Chemical Theory. In the dehydration or chemical theory, catalytic dehydration of cellulose occurs. The decomposition path of cellulose is altered so that flammable tars and gases are reduced and the amount of char is increased; ie, upon combustion, cellulose produces mainly carbon and water, rather than carbon dioxide and water. Because of catalytic dehydration, most fire-resistant cottons decompose at lower temperatures than do untreated cottons, eg, flame-resistant cottons decompose at 275–325°C compared with about 375°C for untreated cotton. Phosphoric acid and sulfuric acid [*8014-95-7*] are good examples of dehydrating agents that can act as efficient flame retardants (15–17).

Another aspect of the dehydration theory is that the dehydration process proceeds through either an acid (Lewis acid) or basic mechanism, altering the path of cellulose decomposition to eliminate levoglucosan formation and thus leading to more char and less tar (11). However, it should be pointed out that whereas relatively low levels of applied flame retardant, ie, 2.0%, do inhibit levoglucosan formation, much higher levels, ie, 10.0% or more, are normally required to produce flame resistance. These higher add-ons of flame retardant appear necessary to maximize char formation, thereby imparting flame resistance. In the base-catalyzed dehydration it is postulated that dehydration proceeds with the formation of dehydrocellulose. This is followed by formation of char (15,16).

Durability of Retardant Finishes

Fire resistance of a treated cellulosic fabric is reduced when the retardant contains acid groups and the treated fabric is soaked or laundered in water containing calcium, magnesium, or alkali metal ions. Phosphate- and carbonate-based detergents affect durability of fire retardants (18). Soap-based detergents can result in a substantial loss of fire resistance because of the deposit of fatty acid salts (19). Phosphorus-based flame retardants are adversely affected by water hardness and laundry bleach, sodium hypochlorite [*7681-52-9*]. Exposure to sunlight and weathering can lead to sufficient loss of flame retardant so that the fabric is no longer flame resistant (20). Similarly, a combination of sunlight followed by laundering or autoclaving can also lead to loss of flame resistance in a cellulosic fabric (8,21,22).

Nondurable Finishes. Flame-retardant finishes that are not durable to laundering and bleaching are, in general, relatively inexpensive and efficient (23). In some cases, a mixture of two or more salts is more effective than either of the components alone. For example, an add-on of 60% borax (sodium tetraborate) is required to prevent fabric from burning, and boric acid is ineffective as a flame retardant even at levels equal to the weight of the fabric. However, a mixture of seven parts borax and three parts boric acid imparts flame resistance to a fabric with as little as 6.5% add-on.

The water-soluble flame retardants are most easily applied by impregnating the fabric with a water solution of the retardant, followed by drying. Adjustment of the concentration and regulation of the fabric wet pickup controls the amount of retardant deposited in the fabric. Fabric can be processed on a finishing range which consists of any convenient means of wetting the fabric with the solution, such as a padder or dip tank. This is followed by drying on steam cans, in an oven, on a tenter frame, or merely by tumbling in a mechanical dryer. Water-soluble flame retardants also may be applied by spraying, brushing, or dipping fabrics, as well as a final rinse in a commercial or home laundry (24). The water-soluble flame retardants used most widely for textiles are listed in Table 1. Less commonly used retardants include sulfamates of urea or other amides and amines; aliphatic amine phosphates, such as triethanolamine phosphate [*10017-56-8*], phosphamic acid [*2817-45-0*] (amido phosphoric acid, $H_2PO_3NH_2$), and its salts; and alkylamine bromides, phosphates, and borates. For a blend of polyester and

Table 1. Water-Soluble Flame-Retardant Formulations,[a] % Composition

Formulation	Borax	Boric acid	Diammonium phosphate	Sodium phosphate dodecahydrate	Other
	[*1303-96-4*] $Na_2B_4O_7{\cdot}10H_2O$	[*10043-35-3*] H_3BO_3	[*7783-28-0*] $(NH_4)_2HPO_4$	[*10101-89-0*] $Na_3PO_4{\cdot}12H_2O$	
1	70	30			
2	47	20	33		
3		50		50	
4		50	50		
5	50	35		15	
6			25		75[b]
7	15	47			38[c]

[a]100% ammonium bromide [*12124-97-7*], NH_4Br, is also used.
[b]Ammonium sulfamate [*7773-06-0*], $NH_4OSO_2NH_2$.
[c]18% sodium phosphate [*7601-54-9*], Na_3PO_4, and 20% sodium tungstate dihydrate [*10213-10-2*], $Na_2WO_4{\cdot}2H_2O$.

cellulosic, it is advisable to include a bromine-containing salt such as ammonium bromide in the finish.

Semidurable Finishes. Semidurable fire retardants resist removal from 1 to approximately 15 launderings. Such retardants are adequate for applications such as drapes, upholstery, and mattress ticking. If they are sufficiently resistant to sunlight or can be easily protected from actinic degradation, they can also be applied to outdoor textile products. The principal disadvantage of water-soluble flame retardants is their lack of durability. This undesirable property can be overcome by precipitating their inorganic oxides on the fabric, eg, $WO_3{\cdot}xH_2O$ and $SnO_2{\cdot}yH_2O$:

$$2\ Na_2WO_4 + SnCl_4 + (2x + y)\ H_2O \rightarrow 4\ NaCl + 2\ WO_3{\cdot}xH_2O + SnO_2{\cdot}yH_2O$$

These codeposits add flame- and glow-resistance properties to textile fabrics. However, some insoluble deposits may also degrade the fabrics. Codeposits frequently improve glow resistance, but are usually more soluble than the deposit responsible for flame resistance and more easily removed during the laundering process.

There are several methods for introducing the insoluble deposits into the fabric structure. The multiple bath method, in which the fabric is first impregnated with a water-soluble salt or salts in one bath and is then passed into a second bath which contains the precipitant, is used most often. Most semidurable retardants used on cotton are based on a combination of phosphorus and nitrogen compounds (25).

A research area that has seen considerable activity in nondurable and semidurable smolder-resistant applications is the development of finishes for cotton batting products. Materials such as monoammonium phosphate [*7722-76-1*], sodium borate [*1330-43-4*], urea phosphate [*4861-19-2*], and borated amido polyphosphate have been investigated (26). Subsequent work has focused on the means of depositing boric acid (27). In some cases this was accomplished via vapor-phase deposition. The unstable ester, methyl borate [*121-43-7*], has seen considerable use in batting applications. It is first sprayed onto the batting, and the treated batting is then passed through a steam chamber. The water quickly reacts with the methyl borate, producing boric acid in and on the cotton fibers of the batting (28,29).

Early Durable Finishes. Early studies to produce durable flame retardants for cellulose were based on treatment with inorganic compounds containing antimony and titanium (30–35). Numerous patents have been issued based on these types of treatments, eg, the Erifon process (Du Pont) (36–38) and the Titanox flame-retardant process (Titanium Pigment Corp.) (39). In the Erifon process, titanium and antimony oxychlorides are applied from acid solution (pH 4) to fabric, which is then neutralized by passing through a solution of sodium carbonate [*497-19-8*], followed by rinsing and drying. Fabrics thus finished exhibit good flame resistance but also considerable afterglow. A large amount of tent fabric has been treated by this type of process for the military service. The basic chemicals used in the Titanox flame-retardant process are titanium acetate chloride and antimony oxychloride. In both the Titanox and the Erifon processes, it is difficult to process the fabric without dulling its appearance.

Outdoor Finishes. Excellent fire-resistant fabric has been obtained by treating fabric with a suspension or emulsion of insoluble fire-retardant salts or oxides, eg, antimony(III) oxide [*1309-64-4*], along with a chlorinated organic vehicle such as chlorinated paraffin (7). Antimony(III) oxide alone is a poor flame retardant. However, when used in conjunction with a halogenated compound a very good flame retardant is produced. The active agent is the antimony halide formed *in situ* from Sb_2O_3 and the acid halide under flaming conditions.

In the 1990s, two types of flame retardants are preferred for outdoor fabrics, ie, a system based on phosphorus and nitrogen such as the precondensate–NH_3 finish and an antimony–bromine system based on decabromodiphenyl oxide [*1163-19-5*] and antimony(III) oxide (20,40–42).

FWWMR Finish. The abbreviation for fire, water, weather, and mildew resistance, FWWMR, has been used to describe treatment with a chlorinated organic metal oxide. Plasticizers, coloring pigments, fillers, stabilizers, or fungicides usually are added. However, hand, drape, flexibility, and color of the fabric are more affected by this type of finish than by other flame retardants. Add-ons of up to 60% are required in many cases to obtain adequate flame resistance. Durability of this finish is good and fabric processed properly retains its flame resistance after four to five years of outdoor exposure. This type of finish is suited for very heavy fabrics, eg, tents, tarpaulins, or awnings, but not for clothing or interior decorating fabrics. The metal oxides can be fixed to cotton by use of resins, eg, vinyl acetate–vinyl chloride copolymers (vinylite VYHH) or poly(vinyl chloride) [*9002-86-2*] (PVC) (43–45). A flame retardant has been developed based on an oil–water emulsion containing a plasticizer (PVC latex) and antimony(III) oxide (46,47). High add-ons are necessary to impart adequate flame resistance but the strength of the fabric is little affected.

Test Methods

Numerous tests covering flame retardancy and related matters are available. The requirements most often specified for fire resistance of a textile materials are that it must pass either Federal Specification Method 5903 or NFPA 701.

The most extensive body of tests are provided under the auspices of ASTM Standard methods. Specific ASTM test designations and descriptions are available (48). The other compendium of fire-retardant tests are contained in Federal Test Method Standards 191A (49).

The Fire Tests for Flame Resistant Textiles and Films, issued by the National Fire Protection Association (NFPA) in 1989, is the method most used by industrial fire-retardant finishers (ca 1993) (50). It has been approved by the American National Standards Institute.

Oxygen Index Test. The oxygen index test (OI) is based on the minimum oxygen concentration that supports combustion of various textile materials (51–53). A fabric which would not burn in pure oxygen would have an OI of 1.0. One which would just barely burn in an atmosphere of 20% oxygen and 80% nitrogen would have an OI of 0.20. The higher the OI of a given fabric, the less flammable it is and vice versa. The OI values of various untreated fabrics are listed in Table 2.

A modified method of estimating the OI for fibrous or finely divided cellulosic materials has been developed (53). The fibers or powdered materials are first pressed into disks, then the OI is measured in the usual way.

Table 2. Oxygen Indexes of Fabrics

Fabric	Oxygen index[a]
acrylic	0.182
acetate	0.186
polypropylene	0.186
rayon	0.197
cotton	0.201
nylon	0.201
polyester	0.206
wool	0.252
flame-resistant cotton	0.270
aramid	0.282

[a]A higher OI indicates a less flammable material.

Types of Retardants.

Overview, Fire Retardants for Cellulosics. Phosphorus-containing materials are by far the most important class of compounds used to impart durable flame resistance to cellulose (see FLAME RETARDANTS, PHOSPHORUS FLAME

RETARDANTS). Flame-retardant finishes containing phosphorus compounds usually also contain nitrogen or bromine or sometimes both. A combination of urea [*57-13-6*] and phosphoric acid [*7664-38-2*] imparts flame resistance to cotton fabrics at a lower add-on than when the acid or urea is used alone (8). Other nitrogenous compounds, such as guanidine [*133-00-8*] or guanylurea, could be used instead of urea. Amide and amine nitrogen generally increase flame resistance, whereas nitrile nitrogen can detract from the flame resistance contributed by phosphorus. The most efficient flame-retardant systems contain two retardants, one acting in the solid and the other in the vapor phase. Nitrogen, when used in conjunction with phosphorus compounds, has a synergistic effect on fire retardancy (54–56). In some cases, the phosphorus content can be reduced without changing the efficiency of the flame retardant (57). Bromine in flame-resistant fabric escapes from the tar to the vapor phase during pyrolysis of the textile in air. It appears to have little or no effect on the amount of phosphorus remaining in the char. Bromine contributes flame resistance almost completely in the vapor phase. Another system for making cellulosic fabrics flame-resistant is based on the use of halogens in conjunction with nitrogen or antimony. In the case of systems employing halogen moieties, bromine is more effective than chlorine.

Mesylated and Tosylated Celluloses. It has been established that the flame resistance of cellulose (qv) is improved by oxidation of $—CH_2OH$ groups to —COOH (58–60). To correct some of the shortcomings of this treatment, mesyl or tosyl cellulose was prepared and then the mesyl (CH_3SO_2) or tosyl ($CH_3C_6H_4SO_2$) group was replaced with bromine or iodine (58–60):

$$\text{cell—OSO}_2\text{CH}_3 + \text{NaBr} \rightarrow \text{cell—Br} + \text{CH}_3\text{SO}_3\text{Na}$$

This treatment produced a fabric with durable flame resistance and good strength retention, but an undesirable afterglow; this was eliminated by phosphorylation with diethyl chlorophosphate [*814-49-3*].

Urea–Phosphate Type. Phosphoric acid imparts flame resistance to cellulose (16,17), but acid degradation accompanies this process. This degradation can be minimized by incorporation of urea [*57-13-6*]. Phosphorylating agents for cellulose include ammonium phosphate [*7783-28-0*], urea–phosphoric acid, phosphorus trichloride [*7719-12- 2*] and oxychloride [*10025-87-3*], monophenyl phosphate [*701-64-4*], phosphorus pentoxide [*1314-56-3*], and the chlorides of partially esterified phosphoric acids (see CELLULOSE ESTERS, INORGANIC).

Cellulose phosphate esters are also produced by treatment with sodium hexametaphosphate [*14550-21-1*] by the pad-dry-cure technique. These treated fabrics have high retention of breaking and tearing strength (61). The reaction products contain more than 1.6% phosphorus and are insoluble in cupriethylenediamine [*15243-01-3*], indicating that some cellulose cross-linking occurs. However, since durable-press (DP) levels and wrinkle recovery values are low, it seems reasonable that only limited cross-linking takes place.

Phosphorylated cottons are flame resistant in the form of the free acid or the ammonium salt. Since these fabrics have ion-exchange properties, conversion to the sodium salt takes place readily during laundering if basic tap water is used. However, flame resistance can be restored if the fabric is treated with either acetic acid [*1563-80-8*] or ammonium hydroxide [*1336-21-6*] after washing.

Phosphonomethylated Ethers. A phosphorus-containing ether of cellulose can be prepared by the reaction of cotton cellulose with chloromethylphosphonic acid [*2565-58-4*] in the presence of sodium hydroxide [*1310-73-2*] by the pad-dry-cure technique (62). Phosphorus contents of between 0.2 and 4.0% are obtained. This finish is durable but has high ion-exchange properties and is flame resistant only as the ammonium salt. Durability on medium weight fabrics is obtained with chloromethylphosphonic diamide. This finish has never penetrated the flame retardant market (63).

A durable flame-retardant cellulosic fabric with good hand is obtained by treating phosphorylated or phosphonomethylated cotton with titanium(IV) sulfate [*13825-74-6*] (64):

$$\text{cell—OH} + PO_4^{3-} \xrightarrow{\text{urea}} \text{cell—O}\overset{\overset{\displaystyle O}{\|}}{P}O_2^{2-} \xrightarrow{+Ti^{4+}} \text{cell—O}\overset{\overset{\displaystyle O}{\|}}{P}\begin{matrix}\diagup O \diagdown \\ \diagdown O \diagup\end{matrix}Ti(OH)_2$$

phosphorylated cotton

$$\text{cell—OH} + ClCH_2\overset{\overset{\displaystyle O}{\|}}{P}O_2^{2-} \underset{-HCl}{\xrightarrow{OH^-}} \text{cell—OCH}_2\overset{\overset{\displaystyle O}{\|}}{P}O_2^{2-} \xrightarrow{+Ti^{4+}} \text{cell—OCH}_2\overset{\overset{\displaystyle O}{\|}}{P}\begin{matrix}\diagup O \diagdown \\ \diagdown O \diagup\end{matrix}Ti(OH)_2$$

phosphonomethylated cotton

Amide-Based Systems, Cyanamide. Pyroset CP, originally the trade name for a 50% solution of cyanamide, produces a semidurable fire-retardant finish on cotton fabric when used in conjunction with phosphoric acid (65). Treated fabrics have an excellent hand as well as an increase in wrinkle recovery. Drying and curing is accomplished in a single step, preferably at 150°C for 2–5 minutes. The finish is subject to ion exchange and therefore the hardness of the washwater has a decisive effect on durability of the flame-retardant finish. Fabric flammability fails after seven washings at a water hardness of 70 ppm, whereas in soft water (0 ppm hardness) flame retardancy can withstand 50 wash cycles. Tensile strength losses of Pyroset CP-treated fabric are 40–45%. The treated fabrics are dimensionally stable and have improved rot resistance.

Methylphosphoric Acid–Cyanamide System. In another system (65), based on methylphosphoric acid [*993-13-5*] (MPA) and cyanamide [*420-04-2*], one or more of the hydroxyls in MPA or in its dimer react with cellulose and the water is taken up by the cyanamide, forming urea:

$$CH_3\underset{\underset{\displaystyle OH}{|}}{\overset{\overset{\displaystyle O}{\|}}{P}}OH + \text{cell—OH} \rightarrow CH_3\underset{\underset{\displaystyle OH}{|}}{\overset{\overset{\displaystyle O}{\|}}{P}}O\text{—cell} + H_2O$$

$$H_2NCN + H_2O \rightarrow H_2N\overset{\overset{\displaystyle O}{\|}}{C}NH_2$$

Fabrics are treated by a pad-dry-cure technique; however, smoke is evolved in the curing step. At an add-on of 10%, this flame-retardant finish is durable to 40–50 laundry cycles, has dry wrinkle recovery angles of 200–220°, and wet wrinkle recovery angles of 220–270°. The wrinkle recovery angle reported is the sum of the recoveries in the warp and fill directions (W + F); thus the maximum possible wrinkle recovery is 360°. The same finish gives tensile strength retention of 60–80%, tear strength retention of about 50%, and raises the moisture regain of the fabric about 3%. The system shows a high tolerance for calcium.

Dialkyl Phosphite and Related Retardants. Pyrovatex CP is based on the reaction product of a dialkyl phosphite and acrylamide [*79-06-1*]; the adduct is methylolated with formaldehyde [*50-00-0*] (66–68):

$$(CH_3O)_2\overset{O}{\overset{\|}{P}}H + CH_2{=}CH\overset{O}{\overset{\|}{C}}NH_2 \rightarrow (CH_3O)_2\overset{O}{\overset{\|}{P}}CH_2CH_2\overset{O}{\overset{\|}{C}}NH_2 \xrightarrow{HCHO} (CH_3O)_2\overset{O}{\overset{\|}{P}}CH_2CH_2\overset{O}{\overset{\|}{C}}NHCH_2OH$$

Excess formaldehyde reacts with cotton in the presence of an acid catalyst. A cross-linking agent that induces polymerization and contributes to flame resistance, such as trimethylolmelamine [*1017-56-7*] (TMM), improves the efficacy and durability of this finish. TMM does not greatly affect the hand of the treated fabric. The finish is durable to dry cleaning but not necessarily to washing. Tensile strength losses are between 20 and 30% and tear strength losses are about 30%. Cotton textiles require add-ons of about 20–35%. Because of TMM's questionable safety when used in applications involving contact with the skin, most flame-retardant applications involving it have been discontinued.

Pyrovatex CP "New". A modification of Pyrovatex CP has replaced the original formulation. It is believed that this new product has had TMM removed from the formulation and replaced by a more ecologically friendly cross-linking agent. This finish is much like that achieved with Pyrovatex CP, except that it is more subject to hydrolysis over an extended period of time. Of the heat-cured fire-retardant finishes available (ca 1993), Pyrovatex CP "New" is the most commercially significant (69). It should be noted, however, that NH_3-cured finishes command a much larger share of the domestic market in the 1990s, and that heat-cured finishes are used more in Europe than in the United States (70).

Dialkylphosphonopropionamides. Cellulosic derivatives that closely resemble those based on the dialkylphosphonopropionamides have been prepared (71). The fabric was treated with *N*-hydroxymethylhaloacetamides (chloro, bromo, or iodo) in DMF solution by a pad-dry-cure technique with a zinc nitrate [*10196-18-6*] catalyst. It was then allowed to react in solution with trimethyl phosphite [*121-45-9*] at about 140–150°C; the reaction rates decreased in the order iodo > bromo > chloro. With phosphorus contents above 1.5%, good flame resistance, durable to laundering, was obtained without noticeable loss in fabric strength.

Triazines. When the dialkoxyphosphinyl group is attached to the triazine ring rather than to an alkyl group, 2,4-diamino-6-diethoxyphosphinyl-1,3,5-triazine [*4230-55-1*] (DAPT) is obtained:

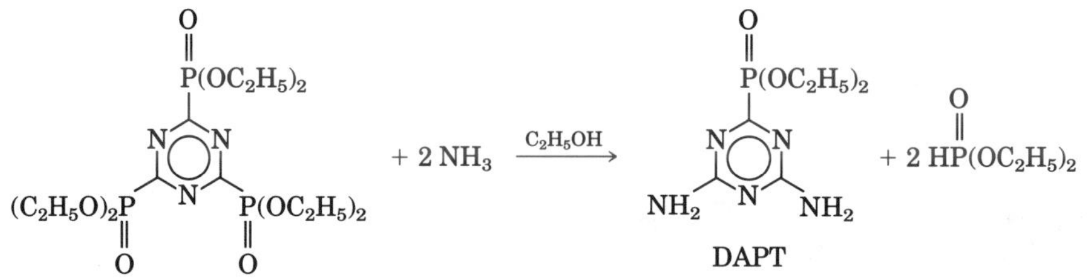

$$\text{2,4,6-tris(diethoxyphosphinyl)-1,3,5-triazine} + 2\,NH_3 \xrightarrow{C_2H_5OH} \underset{\text{DAPT}}{\text{2,4-diamino-6-[}P(O)(OC_2H_5)_2\text{]-1,3,5-triazine}} + 2\,HP(O)(OC_2H_5)_2$$

Formaldehyde and DAPT form a derivative that readily gives an insoluble cross-linked polymer. The fabric is padded with a DAPT solution having an optimum pH of 6.6 (no catalyst), dried, and cured. Fabrics containing from 17.5–20% resin add-on passed the standard vertical flame tests after 35 laundry cycles. Tear strength losses are about 35% and tensile strength losses about 18%. Although the treated fabrics do not yellow on chlorine bleaching and scorching, some strength loss results, indicating chlorine retention (72–74).

An effective, but not very practical, flame retardant for cotton based on 2,4-diamino-6-(3,3,3-tribromopropyl)-1,3,5-triazine [*62160-38-7*] (DABT) was prepared from ethyl γ-tribromobutyrate and biguanide [*56-03-1*]:

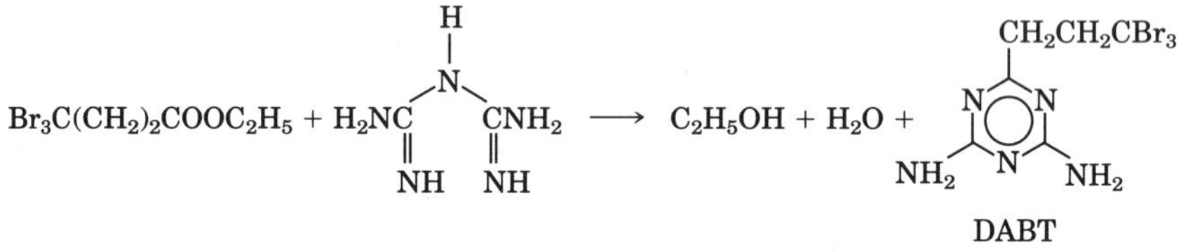

$$Br_3C(CH_2)_2COOC_2H_5 + H_2NC(=NH)NHC(=NH)NH_2 \longrightarrow C_2H_5OH + H_2O + \underset{\text{DABT}}{\text{2,4-diamino-6-(}CH_2CH_2CBr_3\text{)-1,3,5-triazine}}$$

The tetramethylol derivative of DABT, prepared by reaction of DABT with alkaline aqueous formaldehyde, polymerized readily on cotton. It imparted excellent flame retardancy, very durable to laundering with carbonate- or phosphate-based detergents as well as to hypochlorite bleach. This was accomplished at low add-on without use of phosphorus compounds or antimony(III) oxide (75–77).

THPC-Based Retardants. One flame-retardant finish for cotton (78) is based on tetrakis(hydroxymethyl)phosphonium chloride [*124-64-1*] (THPC), $(HOCH_2)_4P^+Cl^-$. It is water-soluble and is prepared by the reaction of phosphine with formaldehyde and HCl. THPC reacts with aminized cotton to impart flame resistance; this has led to further studies with materials such as melamine [*108-78-1*] and urea in an effort to eliminate the costly aminization step. THPC reacts and polymerizes with methylolmelamine [*937-35-9*], as well as with many other materials containing active hydrogens, eg, amines, phenols, and polybasic acids (79), to form insoluble polymers. When these polymers are deposited in the cotton fiber they produce good flame- and glow-resistant finishes (see also AMINO RESINS AND PLASTICS). The methylol groups in THPC react with amines or amides in the following manner:

$$(HOCH_2)_4PCl + RNH_2 \rightarrow RNHCH_2PCl(CH_2OH)_3 + H_2O$$

THPC–Amide Process. The THPC–amide process is the first practical process based on THPC. It consists of a combination of THPC, TMM, and urea. In this process, there is the potential of polymer formation by THPC, melamine, and urea. There may also be some limited cross-linking between cellulose and the TMM system. The formulation also includes triethanolamine [*102-71-6*], an acid scavenger, which slows polymerization at room temperature. Urea and triethanolamine react with the hydrochloric acid produced in the polymerization reaction, thus preventing acid damage to the fabric. This finish with suitable add-on passes the standard vertical flame test after repeated laundering (80).

An improved version of the THPC–amide process, developed in 1972, is based on a finish containing THPC, cyanamide, and disodium phosphate [*13708-85-5*], Na_2HPO_4. It has the advantage of removing the mutagenically suspect TMM from the finish while retaining many of its attributes (81).

THPC–Urea–Disodium Phosphate. A further improved variation uses a combination of THPC, urea, and disodium phosphate (82,83). In this case, the components are mixed, the pH adjusted to about 6.0 with sodium hydroxide, and the solution applied to the textile via a pad-dry-cure treatment. The combination of urea and formaldehyde given off from the THPC further strengthens the polymer and causes a limited amount of cross-linking to the fabric. The Na_2HPO_4 not only acts as a catalyst, but also as an additional buffer for the system. Other weak bases also have been found to be effective. The presence of urea in any flame-retardant finish tends to reduce the amount of formaldehyde released during finishing.

Another modification of this process was reported in 1988 (84). In this process, a precondensate of THPC and urea, plus excess urea, are neutralized to a pH of about 5.7, and the buffer salt is added. The fabric is then given a standard pad-dry-cure process followed by oxidation and laundering. The principal advantage of this modification is a reduction in both formaldehyde vapors and phosphine-like odors released during processing (84).

Finally, a modification has been carried out in which a polyacrylate emulsion is added to a normal tetrakis(hydroxymethyl)phosphonium sulfate [*55566-30-8*] (THPS), urea, and TMM fire-retardant treatment in an attempt to completely alleviate the strength loss during the finishing. Indeed, better retention of tensile properties is achieved with no loss in fire resistance (85).

THPOH–Amide Process. In the THPOH–amide process, the THPC is neutralized to pH 7.2–7.5 with aqueous sodium hydroxide. The specific active species present in solution at this point is not precisely known, but is thought to be a mixture of tris(hydroxymethyl)phosphine [*2767-80-8*] (THP) and formaldehyde, or the hemiacetal adduct of these components (86–89). A disadvantage to this system is that an inactive by-product, tris(hydroxymethyl)phosphine oxide [*1067-12-5*] (THPO), is frequently formed. A further by-product of this reaction is hydrogen gas. In order to avoid formation of THPO, neutralization of THPC is frequently discontinued at $\simeq$ pH 7.2. Fabrics given this treatment (THPOH, TMM, and urea) show less stiffness and better tearing strength than are observed for the same fabrics treated with the THPC–amide process. Good retention of breaking strength and less tendency to yellow when exposed to hypochlorite bleach are also observed for this finish (90–92).

THP–Amide Process. THP has also been made directly from phosphine [*7803-5-27*] and formaldehyde. The THP so generated contains one less mole of formaldehyde than either THPC or THPOH. It can be used in a THP–amide flame-retardant finish. The pad formulation contains THP, TMM, methylol urea, and a mixed acid catalyst (93–95).

Ammonia–Gas-Cured Flame Retardants. The first flame-retardant process based on curing with ammonia gas, ie, THPC–amide–NH_3, consisted of padding cotton with a solution containing THPC, TMM, and urea. The fabric was dried and then cured with either gaseous ammonia or ammonium hydroxide (96). There was little or no reaction with cellulose. A very stable polymer was deposited *in situ* in the cellulose matrix. Because the fire-retardant finish did not actually react with the cellulose matrix, there was generally little loss in fabric strength. However, the finish was very effective and quite durable to laundering.

Two ammoniation processes, ie, the THPOH–NH_3 and the precondensate–NH_3 processes, have seen considerable commercial use.

THPOH–NH_3 Process. In the THPOH–NH_3 process, the fabric is padded with a formulation containing 25–40% THPOH, plus wetting agent and auxiliaries. It is dried to approximately 12–15% moisture, exposed to ammonia gas, and passed through a bath containing H_2O_2 (91). The hydrogen peroxide treatment stabilizes the phosphorus components of the THPOH in the higher oxidation or phosphorous(V) oxide [*1314-56-3*] form. The process can be modified by the addition of copper salts to produce a system stable in the presence of ammonium hydroxide (97). Variations of the ammonia cure have also been reported based on the use of ammonium hydroxide applied by low wet add-on techniques (98,99).

The THPOH–NH_3 process was used extensively for children's sleepwear in the early 1970s. However, the advent of the Tris problem on polyester led to a sharp decline in commercial production of chemically finished children's flame-resistant cotton sleepwear.

Precondensate–NH_3 Process. From a historical point of view, the precondensate–ammonia process is a simplification of the THPC–amide–NH_3 process. In this case, the chemical manufacturer forms a precondensate of THPC and urea by refluxing these components for about 30 minutes. This mixture is padded on the fabric, sometimes with a sodium acetate buffer, dried, and then ammoniated (100–102). Polymer formation is very rapid in the ammoniation step. Subsequent to ammoniation, the polymer is stabilized by oxidation, usually with hydrogen peroxide [*7722-84-1*]. This product has been marketed under the Proban trademark. Somewhat analogous products are available from other producers using the same or similar chemistry. These producers have made further improvements in this type of finish, but published data describing them are proprietary and not available.

In 1992, the precondensate–NH_3 finish was the dominant flame retardant for cotton. Although THPOH–NH_3 is reputed to give a softer hand to finished fabric than precondensate–NH_3, commercial finishers have achieved an excellent hand on fabrics and garments using softening techniques, such as Sanforization or chemical softeners. Apart from hand, two other advantages are possessed by the precondensate–NH_3 finish. First, the odors of phosphorus chemicals and formaldehyde released during padding and drying fabric are greatly reduced when compared to similar treatments with THPOH. Second, the partial replacement of the ammonia moiety with a urea linkage increases the nitrogen content of the P–N polymer. This results in a somewhat more economical finish because cheaper nitrogen is partially used to replace more expensive phosphorus. Westex Corp. (Chicago, Ill.) claims that its version of the THP–precondensate–NH_3 process,

Indura, is comparable if not superior to Nomex, the standard for synthetic inherently fire-resistant textiles (103).

A series of articles takes a comprehensive look at the chemistry of hydroxymethyl phosphorus compounds (86–89).

APO. Research on ethylenimine [*151-56-4*] led to the development of tris(aziridinyl) phosphine oxide [*545-55-1*] (APO) as a flame retardant for cotton. The aziridinyl groups are very reactive, particularly under acidic conditions (see IMINES, CYCLIC). Polymer formation can occur through self-polymerization, but occurs readily with compounds containing active hydrogen atoms such as acids, amines, phenols, and glycols. In addition to conferring flame retardance, these systems confer durable-press performance as well. Although a considerable amount of research and accompanying literature were produced from 1960 to 1990 (104–107), the health problems linked with the aziridinyl system preclude its use in commercial finishing. Similar comments apply to the flame-retardant system based on the reaction of THPC and APO (108,109).

Miscellaneous Consumer Finishes. *Fyrol 76.* This flame retardant from Stauffer is a reactive vinyl phosphorus ester containing 22.5% phosphorus. It can be used alone or with methylolacrylamide (NMA) and a free-radical catalyst, eg, potassium persulfate [*7727-21-1*]. Fabric is treated with the conventional pad-dry-cure technique, rapid steam curing, or radiation curing. Sleepwear fabric requires approximately 25–30% add-on to pass flammability requirements. The fabric has an excellent hand and some permanent press properties. Approximately 80–100% tensile strength and about 65% tearing strength are retained. Chlorine bleach and some carbonate detergents reduce the durability of this finish, whereas perborate bleaches do not affect it. As with the THPC-type finishes, an oxidation step generally is used. Stauffer has withdrawn Fyrol 76 from the market.

Pentamethylphosphorotriamide. Of the phosphoramide derivatives, pentamethylphosphorotriamide [*10159-46-3*] is the most effective finish when applied to fabric in conjunction with dimethylolmelamine and an amine hydrochloride catalyst. The finished fabric passes the FF3-71 flammability test. Its main application is for use on heavyweight clothes since the finish imparts a harsh hand to lightweight fabrics (99).

Application Techniques. *Radiation.* Use of radiation to affect fixation of some flame retardants is being investigated (110). Electron-beam fixation requires the selection of compounds that can be insolubilized inside or outside of the fiber with high yield in a short time. Polyunsaturated compounds, eg, Fyrol 76, have shown promise (see RADIATION CURING).

Incorporation of Flame Retardants in Fiber. Flame retardants suitable for cotton are also suitable for rayon. A much better product is obtained by incorporating flame retardants in the viscose dope before fiber formation. The principal classes of flame retardants used in viscose dope are tabulated annually (111).

Textile-Specific Uses of Flame Retardants

Smolder-Resistant Upholstery Fabric. Chemical finishing to improve the smolder resistance of cotton fabric received some attention during the late 1970s. It was thought that regulatory activity would impact the market position of cotton upholstery fabric; research on semidurable and durable finishes capable of withstanding occasional scrubbing was initiated for cotton upholstery fabric. Two chemical treatments are of particular interest. In one system, a formulation comprised of borax, an acidic compound, and TMM was applied by conventional padding as well as by low wet add-on techniques (112,113). Another successful approach consisted of the application of various polymers as backcoatings to upholstery fabric (114). The most effective polymers for this purpose were copolymers in which one of the monomers contained halogen. One particularly effective copolymer contained butadiene, styrene, and vinylidene chloride. Although the use of synthetic barrier fabrics have largely supplanted the need for this type of finishing, these finishes did provide an effective means for making cotton smolder resistant.

Combination Flame Retardant–Durable Press Performance. Systems using THPC, urea, and TMM can be formulated to give fabrics which combine both flame-retardant performance and increased wrinkle recovery values (80). Another system employs dimethylol cyanoguanidine with THPC under acidic conditions (115). Both of these systems lead to substantial losses in fabric tensile and tearing strength.

The combination of THPOH, urea, and TMM in a ratio of 2:4:1 gives fabrics with flame-retardant performance and high wrinkle recovery values (90). Similarly, trimethylol methylglycoluril [*5001-82-1*] has been used to replace TMM (116). Although this overcomes problems associated with yellowing and bath instability, as well as removing the suspect mutagenic agent, TMM, wrinkle recovery values are not as high as observed when TMM is included.

As previously noted, the APO system leads to fabrics which combine flame resistance and durable press properties; however, the toxicity of the aziridinyl system precludes its use in modern textile finishing.

There has been some effort since 1980 to achieve combined durable press–flame-retardant properties in finished fabrics involving the precondensate–NH_3 system (117–119). In general, adequate durable press performance together with flame resistance can be achieved if one applies the flame-retardant finish (precondensate–NH_3) first, followed by application of a cross-linking agent such as dimethyloldihydroxyethyleneurea [*1854-26-8*]. However, adjustments in finish concentration, acid catalyst, acidity of fabric substrate, and curing conditions are necessary to achieve the desired level of durable press performance. Because a certain amount of the cross-linking agent is diverted to the polymer substrate, losses in strength are much less than they would be if the cross-linking treatment were applied to unmodified cotton.

Another approach to producing durable press–flame-resistant fabric is to utilize an antimony–halogen system. Such a system seems particularly suited for polyester–cotton blends. The Caliban P-44 (White) system, developed in 1972, uses a combination of decabromodiphenyl oxide for the halogen component and antimony(III) oxide for the antimony component (40–42). These materials are usually bound to the fabric by means of a polymeric latex, such as a polyacrylate. If a cross-linking system is included in the finish, it is possible to achieve both durable press and flame-retardant performance on blend fabrics. It should be noted, however, that the basic finish can be applied to a variety of fabrics, especially tenting material and other bottom-weight goods. Care must be taken with polymer binders because pad–roll contamination and whitening of dark fabrics are potential hazards. Caliban P-44 is used extensively on 100% cotton fabric of 10.0 oz. and above as well as many polyester–cotton blends. It is perhaps the best finish available for the flame-retardant treatment of blends (70). Several companies produce commercial finishes similar to Caliban P-44.

Another approach to durable press–flame retardancy uses a combination of a cross-linking system, antimony(III) oxide, and a bromine-containing reactive additive, namely dibromoneopentyl glycol, to achieve dual properties (120,121).

Flame-Retardant Treatments For Wool. Although wool is regarded as a naturally flame-resistant fiber, for certain applications, such as use in aircraft, it is necessary to meet more stringent requirements. The Zirpro process, developed for this purpose (122,123), is based on the exhaustion of negatively charged zirconium and titanium complexes on wool fiber under acidic conditions. Specific agents used for this purpose are potassium hexafluoro zirconate [*16923-95-8*], K_2ZrF_6, and potassium hexafluoro titanate [*16919-27-0*], K_2TiF_6. Various modifications of this process have been made to improve durability and compatibility with wool shrinkproof finishes (124–126).

Thermoplastic Fibers. The thermoplastic fibers, eg, polyester and nylon, are considered less flammable than natural fibers. They possess a relatively low melting point; furthermore, the melt drips rather than remaining to propagate the flame when the source of ignition is removed. Most common synthetic fibers have low melting points. Reported values for polyester and nylon are 255–290°C and 210–260°C, respectively.

Tris. Flame retardancy of the synthetic fibers is obtained by either mechanically building the retardant with the polymer before it is drawn into a fiber, or chemically modifying the polymer itself. Incorporation of chemicals in the dope before spinning the fiber has not been very successful. The most widely used technique for flame-retarding polyester acetate and triacetate is the application of Tris [*126-72-7*] (tris(2,3-dibromopropyl)phosphate, TDBP), by a Thermosol diffusion or an exhaustion technique (127–130). Polyester fiber can only retain about 4–5% TDBP calculated on the weight of the fiber. The use of Tris was banned in 1977 by the Consumer Product Safety Commission (CPSC) as a potential carcinogen (see also FLAME RETARDANTS, HALOGENATED FLAME RETARDANTS; FLAME RETARDANTS, PHOSPHORUS FLAME RETARDANTS). Tris is not effective on cotton fabrics and has only been used as a flame retardant for polyester–cotton blends.

Decabromodiphenyl Oxide–Polyacrylate Finishes. An alternative to the diffusion technique is the application of decabromodiphenyl oxide on the surface of fabrics in conjunction with binders (131). Experimental finishes using graft polymerization, *in situ* polymerization of phosphorus-containing vinyl monomers, or surface halogenation of the fibers also have been reported (129,130,132,133).

Antiblaze 19. Antiblaze 19 (Mobil), a flame retardant for polyester fibers (134), is a nontoxic mixture of cyclic phosphonate esters. Antiblaze 19 is 100% active, whereas Antiblaze 19T is a 93% active, low viscosity formulation for textile use. Both are miscible with water and are compatible with wetting agents, thickeners, buffers, and most disperse dye formulations. Antiblaze 19 or 19T can be diffused into 100% polyester fabrics by the Thermosol process for disperse dyeing and printing. This requires heating at 170–220°C for 30–60 s.

Nylon Finishes. Halogens are less effective flame retardants on nylon than on polyester. Most of the flame retardants effective on cellulosics or on polyester substrates are not effective on nylon. Thiourea [*62-56-6*] and thiourea formaldehyde appear to be the most effective treatments for imparting flame resistance to nylon (135,136).

Polyester–Cotton Fiber Blends. The burning characteristics of synthetic and natural fibers are very different. Synthetic fibers tend to melt, drip, and shrink away when in contact with a flame; cotton does not distort when subjected to a heat source. When synthetic and natural fibers are combined, the cotton serves as a grid for the polyester, preventing it from dripping, and the polyester, which has a higher heat of combustion, accelerates the pyrolysis of cotton. Thus when natural and synthetic fibers are combined, many of the resulting fabrics are more hazardous than fabrics made from the individual components. Table 2 lists the oxygen indexes of a number of commercially available inherently flame-resistant fibers. The flammability behavior of thermoplastic and nonmelting polyester–cellulosic fiber blends cannot be predicted from the flammability of the single-fiber structures (137–140). The treatment of one of the fiber components with a flame-retarding agent specifically effective for that component does not necessarily render the two-component blend flame resistant. The exception to this statement is the situation in which the treated component comprises at least 85% of the blend (137,139,141). In general, an effective flame-retardant treatment must reduce the flammability of each component. Some exceptions do exist, however (142), eg, situations in which the fire-retardant finish is effective on both blend components, but is only present on one.

Considerable effort is being made (ca 1993) to develop satisfactory flame retardants for blended fabrics. It has been feasible for a number of years to produce flame-resistant blended fabrics provided that they contain about 65% or more cellulosic fibers. It appears probable that blends of even greater synthetic fiber content can be effectively made flame resistant. An alternative approach may be to first produce flame-resistant thermoplastic fibers by altering the chemical structure of the polymers. These flame-resistant fibers could then be blended with cotton or rayon and the blend treated with an appropriate flame retardant for the cellulose, thereby producing a flame-resistant fabric. Several noteworthy finishes have been reported since the early 1970s.

THPOH–Ammonia–Tris Finish. By far the most effective finish for polyester–cotton textiles was a system based on the THPOH–NH_3 treatment of the cotton component either followed or preceded by the application of Tris finish to the polyester component. This combined treatment appeared to be effective on almost any polyester–cotton blend. A large amount of fabric treated in this way was sold throughout the United States and much of the rest of the world. Shortly after the introduction of Tris finishing, Tris was found to be a carcinogen. Most of the Tris treated production was in children's sleepwear, and this created a situation in which almost all chemical fire-retardant-treated textiles were unfairly condemned as dangerous. Manufacturers rushed to replace chemically treated textiles with products produced from inherently flame-resistant fibers. Nowhere was the impact more severe than in the children's sleepwear market. New, safer materials have been introduced to replace Tris. Thus far none has been as completely effective.

Decabromodiphenyl Oxide–Polyacrylate Finish. This finish, effective on both polyester and nylon fabrics, is one of the most effective finishes available (ca 1993) for cotton–polyester blends (131). Relatively high cost and difficulty in application may have prevented more widespread use.

THPC–Amide–Poly(vinyl bromide) Finish. A flame retardant based on THPC–amide plus poly(vinyl bromide) [*25951-54-6*] (143) has been reported suitable for use on 35/65, and perhaps on 50/50, polyester–cotton blends. It is applied by the pad-dry-cure process, with curing at 150°C for about 3 min. A typical formulation contains 20% THPC, 3% disodium hydrogen phosphate, 6% urea, 3% trimethylolglycouril [*496-46-8*], and 12% poly(vinyl bromide) solids. Approximately 20% add-on is required to impart flame retardancy to a 168 g/m^2 35/65 polyester–cotton fabric. Treated fabrics passed the FF 3-71 test. However, as far as can be determined, poly(vinyl bromide) is no longer commercially available.

THPOH–NH_3 and Fyrol 76. The THPOH–NH_3 finish and the Fyrol 76 finish also impart flame retardancy to certain polyester–cotton blends if the blends contain at least 65% cotton.

LRC-100 Finish. The use of LRC-100 flame retardant for 50/50 polyester cotton blends has been reported (144). It is a condensation product of tetrakis(hydroxymethyl)-phosphonium salt (THP salt) and *N,N′,N″*-trimethylphosphoramide [*6326-72-3*] (TMPA). The precondensate is prepared by heating the THP salt and TMPA in a 2.3-to-1.0-mole ratio for one hour at 60–65°C. It is applied in conjunction with urea and trimethylolmelamine in a pad-dry-cure oxidation wash procedure. Phosphorus contents of 3.5–4.0% are needed to enable blends to pass the FF 3-71 Test.

Phosphonium Salt–Urea Precondensate. A combination approach for producing flame-retardant cotton–synthetic blends has been developed based on the use of a phosphonium salt–urea precondensate (145). The precondensate is applied to the blend fabric from aqueous solution. The fabric is dried, cured with ammonia gas, and then oxidized. This forms a flame-resistant polymer on and in the cotton fibers of the component. The synthetic component is then treated with either a cyclic phosphonate ester such as Antiblaze 19/19T, or hexabromocyclododecane. The result is a blended textile with good flame resistance. Another patent has appeared in which various modifications of the original process have been claimed (146). Although a few finishers have begun to use this process on blended textiles, it is too early to judge its impact on the industry.

Cotton–Wool Blends. Although they command only a very small fraction of the cotton blend market, cotton–wool blends are easier to make fire resistant than cotton itself. As might be expected, twill fabrics containing both cotton and wool had decreased burning rates and increased OI values both before and after fire-retardant treatment (147).

Core-Yarn Fabric. Core yarns are structures consisting of two component substructures, one of which forms the central axis or core of the yarn, and the other the covering. Generally, the core is a continuous monofilament yarn, while the outer covering is composed of staple fibers, usually cotton. Core yarns were popularized because their strong synthetic core made them stronger than conventional blend yarns and because their 100% cotton staple covering provided enhanced comfort in the fabric form. Although good coverage of the core has always been a problem, new techniques have greatly reduced core grin-through. It has been reported that fabrics made from such core yarns, and constructed to limit their polyester core content to no more than 40% of the total yarn weight, can be made flame resistant. It is only necessary to treat the cotton component with an appropriate flame retardant. No treatment is necessary for the polyester core. It appears that higher level treatments than normally required for pure cotton fabric are necessary for good flame resistance (148–150). Similar results have been reported using glass fiber cores (151) and nylon cores (152).

Economic Aspects

The identification of Tris as a potential carcinogen dealt a resounding blow to the flame-retardant finishing industry. From 1977 to 1984, several principal suppliers of flame-retardant chemicals either reduced the size of their operations or abandoned the market completely. However, Albright and Wilson Corp. (UK) continues to produce THPC–urea precondensate and market it worldwide, and Westex Corp. (Chicago) continues to apply precondensate–NH_3 finish to millions of yards of goods for various end uses. American Cyanamid reentered the market with a precondensate-type flame retardant based on THPS.

American Cyanamid and Albright and Wilson were the primary manufacturers of fire-retardant precondensates in the world in 1993. These agents are distributed in the United States by Albright and Wilson and by Freedom Textile Chemicals (Charlotte, North Carolina). Thor Chemicals, Ltd. markets both precondensate-type and heat-cure-type flame retardants in Europe. The heat-cure finish, marketed as Aflammit KWB, is an acrylamide–dialkyl phosphite agent and has been sold to finishers in the United States. The largest commission finishers of fire-resistant textiles in the United States (ca 1993) are Westex and MF&H Textiles, Inc. (Butler, Georgia). Specialized flame-retardant applications to cotton-wrapped polyester, Kevlar, nylon, and glass core yarns are beginning to attract the interest of the industry for special-purpose fabrics (70).

Health and Safety

Because Tris polyester flame-retardant chemical has been demonstrated to be a potential carcinogen (153–155), workers in this field have tested a number of commonly used chemicals for potential mutagenicity. Neither the THPOH–NH_3 finish nor its extracts caused a significant systematic increase in mutations when tested by the Ames mutagenicity test (156–161). The Hooker Chemical Co. has reported results of tests conducted by an independent laboratory which indicate no significant mutagenic potential from any of the company's proprietary textile flame retardants. Although Fyrol 76 was reported to be nontoxic, results from its mutagenic screening are not known. Stauffer's substitute for Tris, Fyrol FR2, was accused of mutagenic activity by the Environmental Defense Fund, and has been withdrawn from the market by the company. A study has been made by the National Toxicology Program Study on the carcinogenicity of THPC and THPS which concluded that there is no evidence of carcinogenic activity for either compound in rats or mice (162).

Regulatory Legislation. In February 1978, the Consumer Products Safety Commission approved changes in the FF-3 and FF-5 standards for children's sleepwear. It eliminated the melt–drip time limit and coverage for sizes below 1 and revised the method of testing the trim. This permits the use of untreated 100% nylon and 100% polyester for children's sleepwear (157–162).

BIBLIOGRAPHY

"Fire-Resistant Textiles" in *ECT* 1st ed., Vol. 6, pp. 544–558, by G. S. Buck, Jr., National Cotton Council of America; "Fire-Resistant Textiles" in *ECT* 2nd ed., Vol. 9, pp. 300–315, by G. L. Drake, Jr., U.S. Department of Agriculture; "Flame Retardants for Textiles" in *ECT* 3rd ed., Vol. 10, pp. 420–444, by G. L. Drake, Jr., U.S. Department of Agriculture.

1. M. M. Sandholtzer, *Natl. Bur. Std. (U.S.), Circ.* **C455**, 20 (1946).
2. Brit. Pat. 551 (1735), O. Wyld.
3. J. L. Gay-Lussac, *Ann. Chim.* **18**(2), 211 (1821).
4. W. H. Perkin, *J. Ind. Eng. Chem.* **5**, 57 (1913).
5. Brit. Pat. 2077 (1859), F. Versmann and A. Oppenheim.
6. U.S. Pat. 1,885,870 (Nov. 1, 1932), C. Snyder.
7. U.S. Pat. 2,229,612 (Oct. 20, 1942), E. C. Clayton and L. L. Heffner (to William E. Hooper and Sons Co.).
8. R. W. Little, *Flameproofing of Textile Fabrics*, Reinhold Publishing Corp., New York, 1947.
9. J. D. Reid and L. W. Mazzeno, Jr., *Ind. Eng. Chem.* **41**, 2828 (1949).
10. J. M. Church, *Fundamental Studies of Chemical Reactants for Fire-Resistant Treatment of Textiles in U.S.Q.N.C.* Textile Series, Report 38, U.S. Quartermaster Corps, Natick, Mass., 1952.
11. H. A. Schuyten, J. W. Weaver, and J. D. Reid, *Adv. Chem. Ser.* **9**, 7 (June 1954).
12. *A Fundamental Study of the Pyrolysis of Cotton Cellulose to Provide Information Needed for Improvement of Flame Resistant Treatments for Cotton*, Final Report 1959–1964 on PL 480 Project No. URE-29-(20)-9; Grant No. FG-UK-108-59, The Cotton, Silk and Man-Made Fibers Research Association, Didsbury, UK, 1964.
13. J. B. Berkowitz-Mattuck and T. Naguchi, *J. Appl. Polym. Sci.* **7**, 709 (1963).
14. E. Heuser, *The Chemistry of Cellulose*, John Wiley & Sons, Inc., New York, 1944, p. 546.
15. F. J. Kilzer and A. Broido, *Pyrodynamics* **2**, 151 (1965).
16. C. H. Mack and D. J. Donaldson, *Text. Res. J.* **37**, 1063 (1967).
17. M. Leatherman, *U.S. Department of Agriculture Circular 466*, Washington, D.C., 1938.
18. R. J. Brysson, and co-workers, *Proceedings of the 5th Annual Meeting, New York, Dec. 9, 1971*, Information Council on Fabric Flammability, New York, 1971, p. 138.
19. R. M. Perkins, G. L. Drake, and W. A. Reeves, *J. Am. Oil Chem. Soc.* **48**(7), 303 (1971).
20. D. A. Yeadon and R. J. Harper, Jr., *J. Coated Fabr.* **8**(1), 234 (1979).
21. L. W. Mazzeno, Jr., H. M. Robinson, E. R. McCall, N. M. Morris, and B. J. Trask, *Text. Chem. Color* **5**(3), 55 (1973).
22. U.S. Pat. 2,482,755 (Sept. 27, 1949), F. M. Ford and W. P. Hall (to Joseph Bancroft and Sons Co.).
23. W. A. Reeves, G. L. Drake, Jr., and R. M. Perkins, *Fire Resistant Textiles Handbook*, Technomic Publishing Co., Inc., Westport, Conn., 1974, p. 45.
24. J. D. Reid, W. A. Reeves, and J. G. Frick, *U.S. Department of Agriculture Leaflet No. 454*, Washington, D.C., 1959.

25. J. R. W. Perfect, *J. Soc. Dyers Colour.* **74**, 829 (1958).
26. P. A. Koenig and N. B. Knoepfler, *Am. Dyest. Rep.* **58**(17), 30 (1969).
27. J. P. Madacsi, J. P. Neumeyer, and N. B. Knoepfler, *Proceedings of the 14th Textile Chemistry and Processing Conference, New Orleans, April 29, 1974,* Publ. No. ARS-S-60, USDA, Washington, D.C., 1975, p. 146.
28. J. P. Madacsi, J. P. Neumeyer, and N. B. Knoepfler, *J. Fire Ret. Chem.* **4**, 73 (1977).
29. J. P. Madacsi and N. B. Knoepfler, *Therm. Insul.* **3**, 207 (1980).
30. U.S. Pat. 2,570,566 (Oct. 9, 1951), F. W. Lane and W. L. Dills (to E. I. du Pont de Nemours & Co., Inc.).
31. U.S. Pat. 2,607,729 (Aug. 19, 1952), W. L. Dills (to E. I. du Pont de Nemours & Co., Inc.).
32. U.S. Pat. 2,785,041 (Mar. 12, 1957), W. W. Riches (to E. I. du Pont de Nemours & Co., Inc.).
33. U.S. Pat. 2,658,000 (Nov. 3, 1953), W. F. Sullivan and I. M. Panik (to National Lead Corp.).
34. U.S. Pat. 2,691,594 (Oct. 12, 1954), J. P. Wadington (to National Lead Corp.).
35. U.S. Pat. 2,728,680 (Dec. 27, 1955), D. Duane (to National Lead Corp.).
36. U.S. Pat. 2,570,566 (Oct. 9, 1951), F. W. Lane and W. L. Dills (to E. I. du Pont de Nemours & Co., Inc.).
37. H. C. Gulledge and G. R. Seidel, *Ind. Eng. Chem.* **42**, 440 (1950).
38. *J. Text. Inst.* **41**(7), 357 (July, 1950).
39. U.S. Pat. 2,658,000 (Nov. 3, 1953), W. F. Sullivan and I. M. Panik (to National Lead Corp.).
40. V. Mischutin, *Am. Dyest. Rep.* **66**(11), 51 (1977).
41. V. Mischutin and R. Keys, *Text. Chem. Color.* **7**(3), 21 (1975).
42. U.S. Pat. 3,877,974 (Apr. 15, 1975), V. Mischutin (to White Chemical Corp.).
43. R. Van Tuyle, *Am. Dyest. Rep.* **32**, 297 (1943).
44. R. Van Tuyle, *J. Text. Inst.* **34**(10), 587 (Oct. 1943).
45. N. J. Read and E. C. Heighway-bury, *J. Soc. Dyers Colour.* **74**, 823 (1958).
46. U.S. Pat. 2,518,241 (Aug. 8, 1950), J. F. McCarthy (to Treesdale Laboratories).
47. U.S. Pat. 2,591,368 (Apr. 1, 1952), S. H. McAllister (to Shell Development Co.).
48. *1991 Annual Book of ASTM Standards,* Vol. 07.01, 07.02 (Textiles), American Society for Testing and Materials, Philadelphia.
49. *Federal Test Method Standards 191A*, General Services Administration, Washington, D.C., July 20, 1978.
50. *National Fire Protection Association Inc. Standard Method 701*, Quincy, Mass., Rev. Aug. 7, 1989.
51. C. P. Fenimore and F. J. Martin, *Mod. Plast.* **43**, 141, 146, 148, 192 (1966).
52. J. S. Isaacs, *J. Fire Flam.* **1**, 36 (1970).
53. B. J. Trask, J. V. Beninate, and G. L. Drake, Jr., *J. Fire Sci.* **2**, 248 (1984).
54. G. C. Tesoro, *Textilveredlung* **2**, 435 (1967).
55. G. C. Tesoro, S. B. Sello, and J. Willard, Jr., *Text. Res. J.* **38**, 245 (1968).
56. *Ibid.* **39**, 180 (1969).
57. J. E. Hendrix, J. E. Bostic, E. S. Olson, and R. H. Barker, *J. Appl. Polym. Sci.* **14**(7), 1701 (1970).
58. E. Pacsu and R. F. Schwenker, *Text. Res. J.* **27**, 173 (1957).
59. R. F. Schwenker and E. Pacsu, *Ind. Eng. Chem.* **50**, 91 (1958).
60. U.S. Pat. 2,990,232–2,990,233 (June 27, 1961), E. Pacsu and R. F. Schwenker, Jr. (to Textile Research Institute).
61. D. M. Gallagher, *Am. Dyest. Rep.* **53**, 23 (1964).
62. G. L. Drake, Jr., W. A. Reeves, and J. D. Guthrie, *Text. Res. J.* **29**, 270 (1959).

63. M. F. Margavio, A. B. Pepperman, L. A. Constant, E. J. Gonzales, and S. L. Vail, *J. Fire Retard. Chem.* **4**, 192 (1977).
64. R. B. LeBlanc and D. A. LeBlanc, *Proceedings of the Symposium on Textile Flammability*, LeBlanc Research Corp., East Greenwich, R.I., 1974, p. 1.
65. S. J. O'Brien, *Text. Res. J.* **38**, 256 (1968); *Pyroset DO Fire Retardant*, Textile Fin. Bull. No. 130, 2nd ed., American Cyanamid Co., Bound Brook, N.J., 1959.
66. W. A. Sanderson, W. A. Muller, and R. Swidler, *Text. Res. J.* **40**(3), 217 (1970); Fr. Pats. 1,395,178 (Apr. 9, 1965) and 1,466,744 (Jan. 20, 1967) (to Ciba Ltd.).
67. U.S. Pat. 3,374,292 (Mar. 19, 1968), A. C. Zahir (to Ciba Ltd.).
68. R. Aenishanslin, C. Guth, P. Hofmann, A. Maeder, and H. Nachbur, *Text. Res. J.* **39**(4), 375 (1969).
69. D. Allen, CIBA-GEIGY Corp., private communication, June 1992.
70. W. Wyatt, MF&H Textiles, Inc., private communication, June 1992.
71. G. C. Tesoro, S. B. Sello, and J. J. Willard, *Text. Res. J.* **38**(3), 245 (1968).
72. J. P. Moreau and L. H. Chance, *Am. Dyest. Rep.* **59**(5), 37, 64 (1970).
73. L. H. Chance and J. P. Moreau, *Am. Dyest. Rep.* **60**(2), 34 (1971).
74. U.S. Pat. 3,654,274 (Apr. 4, 1972), L. H. Chance and J. P. Moreau (to U.S. Secretary of Agriculture).
75. L. H. Chance and J. D. Timpa, *J. Chem. Eng. Data* **22**(1), 116 (1977).
76. L. H. Chance and J. D. Timpa, *Text. Res. J.* **47**(6), 418 (1977).
77. J. D. Timpa and L. H. Chance, *J. Fire Retard. Chem.* **5**, 93 (1978).
78. W. A. Reeves and J. D. Guthrie, *Text. World* **104**(2), 176, 178, 180, 182 (Feb. 1954).
79. W. A. Reeves and J. D. Guthrie, *Ind. Eng. Chem.* **48**, 64 (1956).
80. J. D. Guthrie, G. L. Drake, Jr., and W. A. Reeves, *Am. Dyest. Rep.* **44**(10), 328 (1955).
81. D. J. Donaldson, F. L. Normand, G. L. Drake, Jr., and W. A. Reeves, *Text. Res. J.* **42**, 331 (1972).
82. D. J. Donaldson, F. L. Normand, G. L. Drake, Jr., and W. A. Reeves, *Text. Ind.* **137**(11), 104, 106, 107 (1973).
83. D. J. Donaldson, F. L. Normand, G. L. Drake, Jr., and W. A. Reeves, *J. Coated Fabr.* **3**, 250 (1974)
84. R. J. Harper, Jr., and J. V. Beninate, *Text. Chem. Color.* **20**(5), 29 (1988).
85. B. J. Trask, J. V. Beninate, and G. L. Drake, Jr., *J. Fire Sci.* **2**, 236 (1984).
86. S. L. Vail, D. J. Daigle, and A. W. Frank, *Text. Res. J.* **52**, 671 (1982).
87. *Ibid.*, 678 (1982).
88. *Ibid.*, 738 (1982).
89. *Ibid.*, 751 (1982).
90. J. V. Beninate, E. K. Boylston, G. L. Drake, and W. A. Reeves, *Text. Res. J.* **38**(3), 267 (1968).
91. J. V. Beninate, E. K. Boylston, G. L. Drake, and W. A. Reeves, *Am. Dyest. Rep.* **57**(25), 981 (1968).
92. J. V. Beninate, R. M. Perkins, G. L. Drake, and W. A. Reeves, *Text. Res. J.* **39**(4), 368 (1969).
93. D. J. Daigle, A. B. Pepperman, G. L. Drake, and W. A. Reeves, *Text. Res. J.* **42**, 347 (1972).
94. D. J. Daigle, A. B. Pepperman, S. L. Vail, and W. A. Reeves, *Am. Dyest. Rep.* **62**(6), 27 (1973).
95. U.S. Pat. 3,796,596 (March 12, 1974), D. J. Daigle, A. B. Pepperman, W. A. Reeves, and G. L. Drake (to U.S. Secretary of Agriculture).
96. U.S. Pat. 2,722,188 (Nov. 27, 1956), W. A. Reeves and J. D. Guthrie (to U.S. Secretary of Agriculture).
97. D. J. Donaldson and D. J. Daigle. *Text. Res. J.* **39**(4), 363 (1969).

98. T. A. Calamari, Jr., R. J. Harper, Jr., and S. P. Schreiber, *Am. Dyest. Rep.* **65**(4), 26 (1976).
99. U.S. Pat. 4,194,032 (Mar. 18, 1980), T. A. Calamari, Jr., R. J. Harper, Jr., and S. P. Schreiber (to U.S. Secretary of Agriculture).
100. U.S. Pat. 2,983,623 (May 9, 1961), H. Coates (to Albright and Wilson Ltd.).
101. Br. Pat. 906,314 (Sept. 19, 1962), H. Coates (to Albright and Wilson Ltd.).
102. Br. Pat. 938,989–938,990 (Oct. 9, 1963), H. Coates and B. Chalkley (to Albright and Wilson Ltd.).
103. *A Performance Comparison—Indura Proban Cotton vs. Nomex*, Westex Inc., Chicago, Ill., 1991.
104. U.S. Pat. 2,915,480 (Dec. 1, 1959), W. A. Reeves, G. L. Drake, Jr., and J. D. Guthrie (to U.S. Secretary of Agriculture).
105. U.S. Pat. 2,870,042 (Jan. 20, 1959), L. H. Chance, G. L. Drake, Jr., and W. A. Reeves (to U.S. Secretary of Agriculture).
106. G. L. Drake, Jr., L. H. Chance, J. V. Beninate, and J. D. Guthrie, *Am. Dyest. Rep.* **51**(8), 40 (1962).
107. G. L. Drake, Jr. and J. D. Guthrie, *Text. Res. J.* **29**(2), 155 (1959).
108. W. A. Reeves, G. L. Drake, Jr., L. H. Chance, and J. D. Guthrie, *Text. Res. J.* **27**, 260 (1957).
109. G. L. Drake, Jr., J. V. Beninate, and J. D. Guthrie, *Am. Dyest. Rep.* **50**(4), 129 (1961).
110. B. Court and co-workers, *Am. Dyest. Rep.* **67**(1), 32 (1978).
111. *Am. Dyest. Rep.* **81**(1), 17 (1992).
112. D. J. Donaldson and R. J. Harper, Jr., *J. Consum. Prod. Flam.* **7**(1), 40 (1980).
113. D. J. Donaldson and R. J. Harper, Jr., *Text. Res. J.* **50**, 205 (1980).
114. D. J. Donaldson, H. H. St. Mard, and R. J. Harper, Jr., *Text. Res. J.* **49**, 185 (1979).
115. J. P. Moreau, L. H. Chance, and G. L. Drake, Jr., *Am. Dyest. Rep.* **62**(1), 31 (1973).
116. D. J. Donaldson, F. L. Normand, G. L. Drake, Jr., and W. A. Reeves, *J. Fire Flam.-Consum. Prod. Flam.* **2**, 189 (1975).
117. R. J. Harper, Jr., and M. F. Demorais, "Durable-Press Flame Retardant Fabrics Based on the THPOH–NH_3 Precondinsate System," *National Technical Conference, AATCC, Montreal, Canada, Oct. 6–9, 1985,* American Association of Textile Chemists and Colorists, Research Triangle Park, N.C., 1986, p. 164.
118. R. J. Harper, Jr. and J. V. Beninate, *Proceedings—74th Annual Industrial Fabrics Association Convention, Boston, Mass., Oct. 19–22,* Industrial Fabrics Association International, St. Paul, Minn., 1987, p. 12.
119. J. V. Beninate and T. J. Harper, Jr., *J. Fire Sci.* **5**(1), 57 (1987).
120. U.S. Pat. 4,536,422 (Aug. 20, 1985), R. J. Harper, Jr. (to U.S. Secretary of Agriculture).
121. U.S. Pat. 4,618,512 (Oct. 21, 1986), R. J. Harper, Jr. (to U.S. Secretary of Agriculture).
122. L. Benisek, *J. Textile Inst.* **65**, 102 (1974).
123. *Ibid.*, 140 (1974).
124. L. Benisek and P. C. Craven, *Text. Res. J.* **50**, 705 (1980).
125. L. Benisek, *Text. Res. J.* **52**, 731 (1982).
126. P. G. Gordon, R. I. Logan, and M. A. White, *Text. Res. J.* **54**, 559 (1984).
127. E. Baer, in E. Greenwich, ed., *Proceedings of the Symposium on Textile Flammability*, LeBlanc Research Corp., East Greenwich, R.I., 1973, p. 117.
128. T. J. McGreehan and J. T. Maddock, *A Study of Flame Retardants for Textiles*, Report No. PB 251-441/AS, National Information Service, Springfield, Va., 1976.
129. H. E. Stepniczka, *Textilveredlung* **10**, 188 (1975).
130. M. Lewin, S. M. Atlas, and E. M. Pearce, *Flame-Retardant Polymeric Materials*, Plenum Publishing Corp., New York, 1975.

131. V. Mischutin, in R. B. Leblanc, ed., *Proceedings of the Symposium on Textile Flammability*, LeBlanc Research Corp., East Greenwich, R.I., 1975, p. 211.
132. R. Liepins and co-workers, *J. Appl. Polym. Sci.* **21**, 2529 (1977).
133. *Ibid.*, 2403 (1977).
134. B. E. Johnston, A. T. Jurewicz, and T. Ellison, *Proceedings of the Symposium on Textile Flammability*, LeBlanc Research Corp., East Greenwich, R.I., 1977, p. 209.
135. D. Douglas, *J. Soc. Dyers Colour.* **73**, 258 (1957).
136. H. Stepniczka, *Ind. Eng. Chem. Prod. Res. Devel.* **12**(1), 29 (1973).
137. G. C. Tesoro and C. Meiser, Jr., *Text. Res. J.* **40**, 430 (1970).
138. W. Kruse and K. Filipp, *Melliand Textilber.* **49**, 203 (1968).
139. W. Kruse, *Melliand Textilber.* **50**, 460 (1969).
140. G. C. Tesoro and J. Rivlin, *Text. Chem. Col.* **3**, 156 (1971).
141. W. A. Reeves and J. D. Guthrie, *U.S. Department of Agriculture Industrial Chemistry Bulletin* 364, Washington, D.C., 1953.
142. P. Linden, L. G. Roldan, S. B. Sello, and H. S. Skovronek, *Textilveredlung* **6**, 651 (1971).
143. D. J. Donaldson, F. L. Normand, G. L. Drake, and W. A. Reeves, *Am. Dyest. Rep.* **64**(9), 30 (1975).
144. R. B. LeBlanc, J. P. Dicarlo, and D. A. LeBlanc, *Text. Res. J.* **40**, 177 (1970).
145. U.S. Pat. 4,732,789 (Mar. 22, 1988), P. J. Hauser, B. L. Triplett, and C. Sujarit (to Burlington Industries, Inc.).
146. U.S. Pat. 4,748,705 (June 7, 1988), J. R. Johnson and C. Sujarit (to Burlington Industries, Inc.).
147. J. V. Beninate, B. V. Trask, Timothy A, Calamari, Jr., and G. L. Drake, Jr., *J. Fire Sci.* **1**, 145 (1983).
148. R. J. Harper, Jr., G. F. Ruppenicker, Jr., and D. J. Donaldson, *Text. Res. J.* **56**(2), 80 (1986).
149. R. J. Harper, Jr., and G F. Ruppenicker, Jr., *Text. Res. J.* **57**(3), 147 (1987).
150. G. F. Ruppenicker, Jr., R. J. Harper, Jr., A. P. S. Sawhney, and K. Q. Robert, *Book of Papers, National Technical Conference of the AATCC, Philadelphia, Oct. 3–6, 1989,* American Association of Textile Chemists and Colorists, Research Triangle Park, N.C., 1990, p. 113.
151. G. F. Ruppenicker, Jr., R. J. Harper, Jr., and C. L. Shepard, *America's Text.* **16**(5), 4 (1987).
152. G. F. Ruppenicker, Jr., R. J. Harper, Jr., A. P. S. Sawhney, and K. Q. Robert, *Text. Technol. Forum* **90**, 71 1990.
153. R. W. Morrow, C. S. Hornberger, A. M. Kligman, and H. I. Maibach, *Am. Ind. Hyg. Assoc. J.* **37**, 192 (1976).
154. F. A. Daniher, *Tenth Annual Meeting, Information Council on Fabric Flammability, New York, Dec. 10, 1976,* Information Council on Fabric Flammability, Galveston, Tex., 1977, p. 319.
155. N. K. Hooper and B. N. Ames in *Regulation of Cancer-Causing Flame-Retardant Chemicals and Governmental Coordination of Testing of Toxic Chemicals, Serial No. 95-33*, U.S. Government Printing Office, Washington, D.C., 1977, p. 42.
156. L. W. Mazzeno, Jr., and N. Greuner, *Text. Chem. Color.* **9**(8), 38 (1977).
157. B. N. Ames, J. McCann, and E. Yamasaki, *Mutat. Res.* **31**(6), 347 (1975).
158. A. F. Kerst, *J. Fire Flam./Fire Retard. Chem.* **1**, 205 (1974).
159. J. McCann and B. N. Ames, *Proc. Nat. Acad. Sci. U.S.A.* **72**(3), 979 (1975).
160. J. McCann and B. N. Ames, *Proc. Nat. Acad. Sci.* **73**, 950 (1976).
161. B. N. Ames, J. McCann, and E. Yamasaki, *Mutat. Res.* **31**, 347 (1975).

162. *Toxicology and Carcinogenesis Studies of THPS and THPC in F344/N Rats and B6C3F$_1$ Mice*, Technical Report Series No. 296, National Toxicology Program, U.S. Dept. of Health and Human Services, Research Triangle Park, N.C., Dec. 1986.

General References

W. C. Kuryla and A. J. Papa, eds. (1973–1975), *Flame Retardancy of Polymeric Materials*, Vol. 5, Marcel Dekker, Inc., New York, 1979.

J. W. Lyons, *The Chemistry and Uses of Fire Retardants*, Wiley-Interscience, New York, 1970.

W. A. Reeves, G. L. Drake, Jr., and R. M. Perkins, *Fire-Resistant Textiles Handbook*, Technomic Publishing Co., Inc., Westport, Conn., 1974.

Textile Flammability, A Handbook of Regulations, Standards and Test Methods, American Association of Textile Chemists and Colorists, Research Triangle Park, N.C., 1975.

TIMOTHY A. CALAMARI, JR.
ROBERT J. HARPER, JR.
United States Department of Agriculture

FLAMETHROWERS. See CHEMICALS IN WAR.

FLARES. See PYROTECHNICS.

FLARE SYSTEMS. See INCINERATORS.

FLAVIANIC ACID. See NAPHTHALENE DERIVATIVES.